P9-DVD-088

CONTENTS

FOREWORD

Why Grow Your Own Food?

TODAY, MORE AMERICANS than ever before want to grow their own fruits and vegetables. The objective is more wholesome, fresh, chemical-free, luscious-tasting foods.

People are tired of mealy, weeks-old, warehouse-ripened produce with a tainting of insecticide or weed killer. They want quality foods for themselves and their family. Yet about the only way to insure this is to *grow them yourself*. This keen interest in growing fruits and vegetables at home is the reason for the publication of this book.

Whenever any space is available, no matter how small, we believe that every family should have a vegetable garden. It is the best investment in recreation, wholesome food and health that a family can make. Many believe that man is by nature a gardener. Working in a garden is healing to both body and spirit. There are both challenges and rewards in the proper preparation of the soil, designing the garden, selecting the seeds and seedlings and establishing them in the garden soil, maintaining conditions favorable to the growth of the plants, dealing intelligently with insects and diseases, harvesting and using the vegetables in ways that provide the body with vitamins and minerals essential to health.

During the years we have been editing *Organic Gardening and Farming* magazine, we have personally interviewed hundreds of organic growers and received reports from a great many others explaining why they grow their own foods. In this introductory section of *How to Grow Vegetables and Fruits by the Organic Method,* we present the personal accounts which we believe best answer the question: "Why do you raise your own fruits and vegetables?"

Why Do I Garden?

Why does a man, living in town, want to raise vegetables when his wife can purchase them at the grocer's much more cheaply? Why does he plan and dig and plant and pull weeds and get blisters and an occasional sore back or split fingernail, when bought vegetables are almost as tasty? He certainly is not compelled to do so, and I would like to try to answer the question.

First: There is the fascination of poring over seed catalogs with their colored pictures of ideal vegetables and glowing descriptions of sizes and colors and tastes, and then the wracking job of deciding which kinds to order, and in what amounts; then the pleasure of planning how to prepare the ground, and where and when to plant each kind of seed, and the making of a diagram to accommodate them.

Second: There is the matter of physical exercise. Most of us would rather watch skillful athletes perform strenuously than weary our own muscles, or else we would rather overexercise one afternoon a week at our favorite sports. But we know that moderate daily exercise is far more beneficial than either, in developing the sparkling eye, the feeling of vigor, the sound sleep and the trim waistline. In our own back yard gardens we have that opportunity for constructive, satisfying and moderate exercise out in the fresh air.

Third: Whatever the manner in which we earn our daily bread, we become mentally exhausted at the shop or store or office, from the many problems we have to try to solve and the tensions we have to endure. What a rest and relaxation, on arrival home, to change into our old, patched clothes and step out into a quiet, happily growing garden. Almost immediately there is an easing of tension within us with the opportunity, as we putter around, to mull over our problems, to see them in their proper perspective and, very often, to reach happy solutions, so that by suppertime we are comfortably tired in body, and at peace, inwardly.

Fourth: And this is the most difficult to get into words. Although we might accurately say that a musician can create a melody or an artist a painting, nevertheless no human being can create an atom or molecule of actual physical matter. Our Creator is the only one who can do that. But we can work with Him and use the things He has given us and, by thought and care and labor, aid in developing them into things of beauty and benefit. Gardening can help us satisfy our creative urge and thrill us, as we see our efforts result in vigorous growth and development.

Fifth: Sinking one's teeth into an ear of juicy sweet corn, picked half an hour previously, convinces one that it really is fresher and sweeter than the bought kind picked a day or two before. It is a pleasure to have vegetables fresher and more tasty and nutritious; and to enjoy the fruits of one's labors furnishes a very gratifying feeling of accomplishment.

Sixth: Most gardeners raise more than they can use, and delight in sharing with their neighbors the delicacies they have produced. This experience is a pure bonus of enjoyment to the donor.

—*C. L. Woodridge*

Health, Fun and Delight

"Ten years ago, when we bought our one and two-thirds acre property near Santa Monica, my wife and I decided to really go in for organic gardening . . . Today I sincerely believe we did the job right."

That's how Dr. John Duge, a California obstetrician, views the result of his gardening efforts. He now produces over $150 worth of quality food each month for his family of 6.

Sheet composting was a key practice in converting the compacted adobe into rich soil. For example, when the corn is a foot high, loads of rotted manure are spread between rows and around plants. Then a ground mulch (mostly leaves) is applied about 3 inches thick.

There's almost a supermarket-like variety growing in the one acre garden. In fruit trees, you'll find peach, cherimoya, sapote, guavas, apples, limequats, kumquats, lemon, lime, orange, tangelo, grapefruit, avocados, plums, papayas, figs, apricots, nectarines, persimmon, tangerine and jujubes. There are also pecans, Chinese chestnuts and even some olive trees. With careful planning, he rotates his corn, beans, squash, tomatoes and other vegetables.

The entire family, wife, John (16), Mildred (14), Karen (9), and Bill (5) also help keep up with the productive garden. "A labor of love" is the way

California obstetrician John Duge (facing the camera) believes that the good health of his four children who are "practically never ill, rarely a cold" is mostly the result of their eating vitamin-rich fruits and vegetables from their organic garden.

Mrs. Duge describes the activity that goes into planting, cultivating, harvesting, as well as her own freezing, drying and canning.

She has a quick freezing unit to process foodstuffs for the ten-by-ten-foot walk-in freezer. Main vegetables she freezes: cut-corn in containers and corn-on-the-cob, tomatoes—whole and pureed, green beans, asparagus, green limas, broccoli. Fruits include peaches, apple sauce, strawberries, plums; of juices, there are strawberry, cherry, boysenberry, grape and tomato.

"Today we are very content to stand in our garden paths and look out over the tall, flourishing plantings of corn, tassels whispering in the breeze, thousands of bees whirring over everything. We look at the long rows of limas and pole beans, the big area of asparagus, the smaller beds of 20-pound cabbages, of giant Sweet Spanish onions, rows of soy beans, squash and tomatoes, beets and broccoli, the trees laden with fruit. No sprays or DDT dusts to worry about. My 4 children, I think, are just about the healthiest and smartest in Southern California. Practically never ill, rarely a cold and," concludes the busy doctor, "mostly the result of eating fruits and vegetables from our organic garden."

—Gordon L'Allemand

Retiring to a More Useful, Wonderful and Relaxing Life

It isn't often that you meet a man who has become so completely dedicated to a purpose in life as Paul Wiegman. Now that he has retired from the sheet metal business, at which he has worked 52 of his 68 years, his sole purpose in life is to grow and eat only those fruits and vegetables grown organically on his own plot in Springdale, Connecticut.

Paul Wiegman of Springdale, Connecticut exhibits his giant-sized peppers. "I couldn't possibly have retired to a more useful, wonderful, and relaxing life," Mr. Wiegman observes. "Organic gardening is just the ticket for an old gent like me."

The plot, which includes a small lawn he calls his "park," measures 60 by 150 feet. In this relatively small space he has growing two apple trees, 4 filbert bushes, 9 blueberry bushes, and one cherry tree. The garden, which extends beyond the small lawn, produces in abundance such winter-eating vegetables as rutabagas, parsnips, carrots, squash (blue hubbard and butternut) and sweet potatoes.

Mr. Wiegman prefers to grow winter vegetables, and has a unique way of conserving them. Instead of pulling and storing them as most of us do, he just allows them to remain right where they are growing. When winter begins to set in, about the last of November, he covers the vegetable rows with at least 12 to 18 inches of leaves, firming them down well. Once a week throughout winter he goes out, scrapes the snow aside, reaches under the leaves and removes a week's supply of each vegetable. Because of the thick layer of leaves which keeps the ground warm, he never has to chop through frozen earth to reach them. His vegetables are always fresh, firm and wholesome, much more so than if they had been pulled and stored. Mr. Wiegman removed his last carrots during the first week in April, and found them just as good and crisp as those he pulled last November.

"Was good eating last winter," he says, "and sure saved on the family budget."

Last season, he grew snap beans over seven inches long. These were from his own organically grown seed—on which he insists. Because there was more than he and his family could eat, he took the surplus to a corner vegetable stand in New Canaan, Connecticut. The owner bought them immediately, greatly surprised by their excellent quality.

Mr. Wiegman now has an assurance from the stand that any vegetables he cares to bring in will be bought immediately. Now that there is a demand for his products, he has since sold the stand many of the vegetables grown on his plot. The townspeople are noting the difference in flavor between those grown by commercial farmers with chemical fertilizers and those grown by Mr. Wiegman on his organically enriched soil.

Because Paul cannot meet the demand, the friend at the vegetable stand saves these vegetables only for his very best customers—who will remain his very best customers as long as they can get vegetables grown by Mr. Wiegman.

Now that he is retired, Paul Wiegman spends all of his time puttering around the garden. His friends have been told time and again, "When you come calling, come to the back door. You'll find me in the garden."

And there you, too, will find him if you chance to pay him a visit. Growing vegetables for the stand now takes up all of his time. This work not only keeps him in the best of health and spirits, but adds to the family income.

"I couldn't possibly have retired to a more useful, wonderful, and relaxing life," Mr. Wiegman says. "Organic gardening is just the ticket for an old gent like me." *—Betty Brinhart*

Back-yard Gardening Pays Off

How would you like to make $65 to $75 a month around the year just on vegetables and fruits raised in your back yard? How would you like—over a 16 year period—to take cement-like soil, develop it to a fine organic tilth, rear 5 children to adulthood—and during all these years spend not over $40 for doctor and dentist bills? And as if anything additional were needed—have a beautiful lawn, flowers of outstanding beauty, shady bench spots and cool patios, and the companionship of dozens of bird families through all those years.

That is just what the E. A. Turlingtons of Venice, California, have done. They say you too can do it. It's health and fun and profit for the joyous taking. Here is the story of the Turlington city "farm."

The Turlington place, 1335 Appleton Way, Venice, is 65 by 175 feet, a charming cream stucco home set amidst green fruit trees, beds of roses and fine dahlias, and a velvety lawn for the front yard, all mulched and thriving to a plant propagator's delight—and that is just what energetic E. A. Turlington is—a plant propagator and seed and bulb specialist with the famous Paul J. Howard Nurseries of West Los Angeles.

Through the years the soil of this garden has been built up by the constant application of humus from two 4 by 4 by 10-foot bins that receive all the kitchen waste, every bit of grass, leaves, old plants, husks, shells, peelings on the place. Nothing is wasted. Kelp is hauled in from nearby beaches. Goat, sheep, chicken, rabbit, steer manures are added to the bins as needed.

"From 4 to 8 tons of compost yearly go into and onto the ground of this garden," explained Mr. Turlington. "We have an iron-clad rule here that every tree and plant must produce, or out it goes."

Walkways of thick, spongy kikuyu grass wind in and out among the many beds of vegetables in the garden. Fruit trees are spotted along the fences and edging the center walkways. The place is indeed a park to walk in and enjoy living.

Here are rows of bantam sweet corn ready to eat in mid-August—the variety—two kinds: Golden Cross Bantam and Aunt Mary's mixed to give larger ears and more rows of grains to the ear. Five kinds of tomatoes—Garden King, Pearson, Stone, Earliana, Ponderosa are planted to give a bearing season lasting from June to the following March.

A large planting of Kentucky Wonder pole beans is just ending a heavy bearing season. Mrs. Turlington says her family can eat fresh and canned green beans five times weekly and like them.

Mr. Turlington says that he and his wife use a rotating planting season like the Chinese farmers. They keep this garden working and producing the year around. Soil enrichment is a steady job, too.

About 25 vegetables are grown: sweet corn, bush and pole green beans, pole and bush limas, four kinds of lettuces, peas, several kinds of summer and winter squashes, green peppers for high vitamin C, cucumbers, pumpkins, banana squash, several kinds of radishes planted frequently, beets, an asparagus bed, artichokes, etc.

"We never sell a thing from this place," says Mrs. Turlington. "We eat all we can and what we can't we can—in glass jars for my pantry." The long colorful rows of jars of beans, chili and barbecue sauces, peaches, beets, jams, jellies are there for good eating, and all are free of condemning sprays and chemical fertilizer residues.

The soil of this garden is a revelation to inspect. It is spongy, moist to touch, sweet smelling. No weeds are allowed to sap the rich food meant for the vegetables.

The peach trees of several varieties were loaded with crops, according to whether early or late varieties. Other fruit trees included plums, oranges, nectarines, apricots, tangerines, lemons, grapefruit, jujubes, pomegranates, limes. Some peach trees were espaliered along the fences for efficient production.

Onions for green and dry use are grown. A new asparagus bed is just coming into production. Nearby are several artichoke bushes that have been doing well.

"The kind of garden you see here," said Mr. Turlington, "is now the exception in America. It should be the rule. People are unfortunately too uninformed, too misinformed, and too indolent to do what we are doing. It is too easy to walk around the corner to the supermarket."

Mrs. Turlington, who says she often serves a half dozen home grown vegetables, cooked or raw, on her dining table at once, also has a fine bed of golden banana squashes, climbing cucumber vines and their fruit, and nearby new rows of bush beans coming into bearing.

The Turlington family has proven over a period of 16 years that the organic way of life is best and that it pays golden profits in health, happiness, and money saved.

"We would like to see American homeowners go back to having a family garden to each place," said Mr. Turlington. "That way lies health, longer life, happiness."

—Gordon L'Allemand

Discovering the Fountain of Youth

Growing his own food supply organically in Florida has helped 87-year-old Charles Weeks establish a "Fountain of Youth." Mr. Weeks has been raising more than 50 different vegetables and fruits on his two-acre plot outside West Palm Beach.

If you're ever near West Palm Beach, Florida, be sure to visit the organic vegetable gardens of Charles Weeks. I don't think I have ever seen such immaculate, well-kept rows of plants, so healthy-looking and so non-insect-bitten. And I don't think I have ever seen such happy enthusiasm. Mr. Weeks is 87 years old and full of zeal and ardor, I might say having an exaltation of soul about the organic method. He is intensely organic.

He came to Florida many years ago from Indiana seeking the fountain of youth and he has found it. Ponce de Leon had nothing on Mr. Weeks. He works stripped to the waist and is outdoors most of the time doing a great part of the work around the place. He grows 50 kinds of vegetables, papayas, avocados, mangoes and bananas, all on two acres located within the city limits of the famous resort town of West Palm Beach.

His vegetable beds are 8 feet wide with two-foot paths between, and the rows of vegetables are planted crosswise. The seeds are planted in shallow trenches filled with earthworm humus from the piles of compost. Around the papayas, avocados, mangoes and bananas is a bed of vegetable waste filled with African night crawlers. His vegetable beds grow vegetables every month in the year. As fast as a bed is past its prime, the old plants go to the compost pile and new seeds are planted of a different variety in order to follow a plan of rotation of crops.

Mr. Weeks has absolutely no lawn. In his opinion many homeowners have too much lawn. He says there is more beauty in rows of crisp lettuce, Brussels sprouts and beets. I can vouch for the fact that the place is most beautiful and does not suffer from the lack of a lawn.

There is no shortage of organic matter for compost-making in this section of Florida. The state keeps harvesting the water hyacinth from the canals and placing it in heaps on the banks for anyone who wishes to haul it away. Seaweed can also be obtained. There are tremendous amounts of palm tree fronds and residues available, but special equipment would be needed to ready them for the compost heap. In the case of the heavy vegetation from the banyan tree, which he sometimes gets from the park system, he puts earthworms on top of them. They eat up the heavy leaves in about a year or more.

Mr. Weeks' little place is a marvelous "pilot plant" that others can very well use as a standard. It is as clean as a whistle, and everything has been worked out for ease of operation. Mr. Weeks does his work, the earthworms do their share, and the plants grow like all get-out. Mr. Weeks is smilingly happy about it all, and his health shines out all over him. He is living long because he is eating much of his vegetables fresh and raw out of his garden. He eats some fish, eggs and milk, but meat rarely. He does not drink or smoke. He likes to quote Elbert Hubbard who said, "Motion must equal emotion." His finest crops, he maintains, are the friends he makes. So if you are ever around Palm Beach way, drop in on Charley Weeks and his delightful wife. They will love to have you visit them.

—*J. I. Rodale*

We Prefer Vegetables to a Lawn

Take a tip from the Parker Richards of Twin Falls, Idaho. Three years ago they decided it would be much nicer to have their own vegetables fresh from the garden. Also the idea of raising something besides lawn and flowers carried a lot of appeal for them. So they picked one corner of the back yard and went to work.

The plot had for many years been a flower garden, later they had extended the lawn to include that space also. Their first job then was to spade the grass and work up the soil. What fertilizer to use was no quandary for them for both had pioneer parents who raised fine vegetables before the day of commercial fertilizers. They decided to follow in their parents' footsteps.

"We picked good aged cow manure," Mr. Richards said, "because weeds seem more prevalent when sheep droppings are used."

Mrs. Richards said they also had used some chicken droppings to augment the other fertilizer occasionally.

Did the venture pan out? Most definitely, yes! Anyone seeing the luxuriant growth and the fine vegetables they raise would be ready to don garden togs, take up their shovel and hoe and start a project of their own. A few lessons from the Richards would be well in line, because besides leading off with the proper fertilizer, it takes a lot of work to till the soil and get it in proper shape for planting. Knowledge of the best varieties for any certain locality, correct time of planting, proper cultivation, all go into the ultimate success of a small vegetable garden, as well as in gardening or farming on a larger scale.

Starting now on their fourth year of "farming" a 20 x 30 foot corner of their yard, they have not only enjoyed all the fresh vegetables (vitamin-rich) that they could use, but have also supplied much produce to their married children and families, and very often treat friends with some extra special tomatoes, cucumbers, beans, etc.

The garden has paid its way, no question about that. Perhaps some people, figuring in dollars and cents against the time, work, seed, fertilizer cost and hauling, weighed beside the cost of the vegetables purchased in a market, would feel it not worth the effort. However, summed up in pleasure, outdoor exercise, satisfaction of picking one's own super-fresh vitamin-charged vegetables with a minimum of time from garden to table, plus the joy of sharing with others, has been to the Richards a very worthwhile investment.

—Olive May Cook

Our Garden Saves Us Ten Dollars a Week

"I've saved at least 10 dollars a week all winter with the food I froze or canned from the garden." That's the way Dorothy Perkins, mother of four from Versailles, Kentucky, describes her gardening results. With food prices going up like impatient satellites, it's more important than ever to get more out of your gardening than pleasure. And you can, too—without sacrificing any of the fun!

"We knew we'd save money with a garden," points out Tom Perkins, "but there are other things, too. The children have been healthier—fewer colds, better health in general—that we feel is due in part to the greater amounts of home-grown vegetables they've been eating."

The Perkins' plot is a pleasant, cooperative effort. Everyone helps, everyone benefits. Because Tom works all day as a garage mechanic and Dorothy is busy with household chores, they garden in the late afternoons and on week ends. Sixteen-year-old Lewis is doing more of the planting each spring, and Carol 13, helps with the weeding. To make preparation of the garden faster and easier, they bought a $2\frac{1}{2}$ horsepower tractor.

"That's paying for itself," Tom declared. "Our neighbor plowed our ground last fall in exchange for the use of the hand tractor to plow his own. How could you invest in anything that pays such high dividends as the money we put into the garden? Altogether we have only a little over an acre, but look what we get out of it!" The Perkins' plot is proof that no one's is too small for fruit trees as well as a plentiful variety of budget-balancing vegetables. The trees are planted at the sides of the garden. Tom mixed sheep manure into the soil around these, and also around the rhubarb.

Mother of four from Versailles, Kentucky, Dorothy Perkins stands in front of her freezer containing some of her garden's harvest—a supply that "saves her at least $10.00 a week in food bills," and contributes much to her family's good health.

Strung across the space among the vegetable rows are strands of wire to serve as supports for the raspberry, blackberry and boysenberry bushes.

As desirable today as the economic advantage of a productive garden is, far more than that stands to be gained by any family.

As Tom Perkins put it, "There's more than a dollar-and-cents satisfaction from a garden—even more than better health and keeping the kids off the street and knowing where they are all the time."

Dorothy nodded. "So many parents and children go their separate ways, seldom sharing interests. We do everything together. The garden is a playground for Joyce and Shirley. For Lewis it's a kind of laboratory and workshop. Carol has her friends in for a barbecue roast on the grill. There's room for the croquet set and we all play. Is it only such a short time since we worried about wholesome recreation for the children?"

Tom sniffed at the fragrant blossoms. "I knew only that I wanted a better life for all of us. I didn't realize how much the garden would . . ." he groped for words.

"Enrich our lives," Dorothy supplied. —*Bertha Newhoff*

The Market Basket and How It Grew

Roger Chamberlain of South Hadley, Massachusetts, is a typical New Englander, conservative in both speech and manner. He is most conservative, however, in his spacious truck garden where he saves hours of hard labor by heavy mulching. It shows up, too in his small country store in which he sells everything from western saddles to organically grown fruits and vegetables.

The "Market Basket," a combination of store and vegetable stand, is neatly situated between the sandy shores of the sparkling, blue Connecticut, and the low-rolling hills of the Holyoke Range. To the left of the store, and on up the sloping hillside toward the purple-gray mountains, were rows upon rows of squash, strawberries, blackberries, sweet corn, etc. And all were carefully mulched with a thick layer of dry, decaying hay.

As Mr. Chamberlain proudly took us on a lengthy tour of his garden, we stopped beside his mammoth plants of yellow, summer squash. The plants looked neat and trim collared by the brown mulch. I asked about his source of hay. He immediately pointed to a large field further up the slope.

"As soon as the grass is tall enough in late spring," he replied, "I cut it, and bring it down to the garden here." He kicked the top few inches of mulch aside. "I put it on at least a foot thick. By fall most of it is gone—rots and goes right into the ground." He bent and quickly cleared a large hole in the mulch. Beneath was the blackest soil I had ever seen up here in New England. (Eastern soil is naturally yellow in color, and very sandy.) Mr. Chamberlain's fingers dug easily into the soft, rich earth. He brought up a double handful of soil that smelled clean and sweet as it fell loosely through his strong fingers. There was no need to ask if earthworms were present. The evidence was all there.

When I inquired about his method of cultivation in spring, he flashed a quick smile.

"Never plow," he replied. "I just move the mulch aside, loosen the soil a bit and plant the seed. Then I push the mulch back in place. When the plants sprout, they break right through. No need to worry about weeds that way."

And he is right. No weeds could possibly grow beneath that heavy, thick covering. Mr. Chamberlain did not say, but, from the appearance of his well-mulched garden, he is a firm believer in the wonders of mulching. And,

New Englander Roger Chamberlain shows the reason why his "Market Basket" has become such a popular place for persons who are interested in better quality vegetables. Organic practices have made his soil fertile and capable of producing nutritious foods.

might I add, the results certainly are rewarding. I have never seen such robust, heavy producing plants before.

I couldn't resist stooping and running my fingers through that rich earth. Good soil has always meant a great deal to me. It felt refreshingly cool and moist even though the outside temperatures hovered above 98 degrees.

"It must take a good-sized compost pile every year to keep this soil producing as it does," I said, looking up. But Mr. Chamberlain shook his head.

"Years ago, I owned a herd of cattle. All the manure, then, did go into the compost bin. After it was broken down, I added it to the garden here. But, since I sold the cows, I use only hay mulch to keep things going."

"How about organic fertilizers?" my husband asked. "Use much of it?"

"Only in early spring to give the new plants a good start. After they catch on, they live off the soil alone."

"And you get squash like this," I said, holding the big leaves aside to expose the large, sunny-bright summer squash beneath. As Mr. Chamberlain was about to answer, a customer drove up to the store.

"You go ahead and wait on her," my husband said. "We'll just look around while you are gone." But Mr. Chamberlain just flashed another quick smile.

"No need to go," he said. "My customers help themselves to what they want and leave the money in the cigar box on the counter."

"And it works out well that way?" I asked in utter astonishment.

"Sure. Works out fine for all concerned." I could hardly believe my eyes as the woman actually selected some of Mr. Chamberlain's organically grown vegetables, weighed them on the scale, then tucked them under her arm, and dropped the money in the waiting box. I have never known a business to operate under such conditions and still survive. But, here was one, and it was doing very well! Mr. Chamberlain is not afraid to trust his many customers. And, put on their honor, they never disappoint him.

Watching the lady-shopper drive away, I asked:

"Do you sell very many of your vegetables?"

"All I can grow."

"Do people buy them only because they want fresh vegetables, or is it because they want organically grown ones?"

"Most buy because they want organically grown vegetables," he replied. "Once a person buys from me, he is sure to come back. Customers claim my vegetables have better flavor than those in the stores. This spring, while peas were in season, I picked thirty pounds a day, and sold them as fast as I could get them on the counter. The customers said they never tasted better peas."

We gradually moved on up past the strawberry patch that had produced very well in spite of the late frost, past the blackberry patch just coming into production, and on up to the rows and rows of sweet corn. Each stalk, I noticed, bore a huge ear. Mr. Chamberlain fondly handled one of them as he talked. He had good reason to be proud. He had grown prize-winning sweet corn without the aid of commercial fertilizers or sprays.

My next question concerned vegetable and flower plants.

"Yes, I grow plenty of both in spring," Mr. Chamberlain replied. "They are my best source of income. People like my plants. Customers say they almost all catch on, and produce better-tasting vegetables. The flower plants, too, seem to bloom better with lots of color. The same people come back year after year to buy my plants." Here, then, is positive proof that plants grown organically have a larger percentage of survival, grow up stronger, and produce better-flavored fruits, and prettier flowers. If only more men like Mr. Chamberlain were scattered across the country so all of us could buy organically grown plants. If it were so, we would, indeed, be able to grow vegetables worth eating.

—Betty Brinhart

Attaining Garden Success

Live out west? Have you given up gardening because you found it: (1) takes too much fertilizer and (2) takes too much water? Well, you can start in again. Experience has shown that in many western areas—especially in the southern sectors of California, Nevada, Arizona and New Mexico—an abundance of vegetables for the householder's own use can be grown on an ordinary city lot, and on nearly a year-around basis.

How can this be done, economically? Simply by putting back into the ground organic materials which come from your own yard or home.

For example, there is the garden plot at the home of Mr. Tony Fido, in Burbank, California. This garden plot is about 25 feet long and from 10 to 15 feet wide. But from this small space Mr. Fido gets just about all of the vegetables that he and his wife can use. Indeed, he does better than that, because he told me, "We have so much that I put some into our freezer and give some to our children for their families!"

A large part of the soil in the nation's western or southwestern area is either sandy or a hard-packed, rocky, clay soil. Both types require considerable humus added to them. Besides its other important roles, this humus acts something like a sponge in absorbing water quickly and then retaining it so that the moisture oozes out slowly for plant roots to use.

What supplies this humus? Natural organic material that is found in any yard. When the Fidos moved into their home the back yard was nothing but fine, barren sand. Now it is a rich, dark loam. In his own case he made a start by digging in some rabbit manure. However, for many years he has put in nothing but compost. Compost piles are the answer for every backyard gardener in the western area. Since material for compost piles is available most of the year—or even all of the year—to western gardeners, that *does* make them lucky!

Tony Fido likes to put his organic materials in compost holes right along the edge of his garden, where it is handy. He covers the material with a layer of dirt and keeps it well watered to hasten the rotting process. Rotting or decomposing is necessary, of course, in order to break down the leaves or grass or whatever the material is to where it can be used again by living plants.

If you want to do even better with your compost piles, simply dig a hole about 2 or 3 feet wide by 4 or 5 feet long and 2 feet deep. Put in a layer of leaves, grass clippings, garbage, chopped up weeds, waste leaves or peelings from vegetables, or whatever organic material might be at hand. Over this layer of perhaps 3 inches in depth sprinkle some fish fertilizer. Cover this with an inch or so of dirt, then put in another layer of organic materials, sprinkle that with fish fertilizer and so on. Be sure to keep the pile well watered. The fish fertilizer also helps to break down the organic materials by encouraging bacterial action.

The point that Tony Fido likes to make is obvious and yet seemingly not understood too well by many would-be back-yard gardeners—that crops take natural elements out of the soil and so those natural elements must be returned to the soil. "I get sort of mad when I see neighbors put out leaves or grass to be hauled away," he said, "and then hear them complain because they can't grow much in their gardens!" Then he grinned and continued, "That's why we don't have a garbage disposal unit. For garbage is organic material, too, you know, and I put it back into my ground.""

His garden is evidence of how well his fertilizing methods work. His vegetable crops are a healthy, dark green and so luxuriant that they are almost jungle-like in thickness. Before he plants each crop he spades some of his composted matter into the ground of his garden plot. "I spade the ground from 8 to 10 inches deep and I space the rows about that same distance apart," he said. "I can put the rows close together because there is plenty of plant food in the ground."

He grows an amazing variety of crops, many of them at the same time. In one year he will have carrots, celery, Italian leaf lettuce, turnips, spinach, onions, radishes, peas, beets, chard, cabbage, and other vegetables.

Early tomato plants are grown by Tony in a different way. He sets the plants out in very early spring, in holes that are from 6 to 8 inches deep, so that the sides of the holes retain moisture and also protect the tender plants against the chilly weather that sometimes hits even in this generally mild climate. He puts a shovelful of well-rotted compost under each plant, with a layer of dirt between the compost and the roots. As the plant grows, he puts more dirt into the holes to provide deeper rooting and greater strength. The plants are trained on wooden supports which give better growth and allow him to grow more of them in a smaller space.

Oh, yes—when the tomato crop has been picked, the vines are chopped up and put into the compost piles! This compost process comes as near to perpetual motion as it is possible to get. The things that grow are put back into the ground to help grow still more things.

Thus, you can see that there is no reason to complain about gardening in the west costing too much for fertilizer and water. You grow your own fertilizer and the use of this compost material cuts down on the amount of water you need. Now watch those gardens grow!

—Ivan F. Hall

Giant Crops from a Midget Homestead

"The bulldozer stole 10 feet of our soil, but we built new soil—better than before."

So says Frank Fiederlein, organic gardener of New Britain, Conn., who has what is virtually a miniature homestead—on a lot 50 by 140!

"People just don't dream of the vast variety of vegetables and fruit they can grow organically on a small plot," says Frank. "They think they need loads of topsoil, bags of chemical fertilizers and all kinds of deadly sprays to grow anything at all.

"But we didn't need any of these things. In fact, we did not even have any topsoil to speak of on the part of our land that now raises our vegetables. But today we have a constant supply of fresh vegetables all through the growing season—plus all the fruit we can eat, a really lush lawn and a huge variety of flowers. And we never spent a cent on chemicals."

When Frank and Cecile Fiederlein moved into their new Cape Cod home 4 years ago, the grounds presented a real problem. The group of houses of which theirs is one is situated on a street cut along the side of a hill. In order to make the street and the rows of homes on each side of it level, the builders bulldozed a cut across the hillside.

This meant that a huge amount of soil was removed from the Fiederlein's place—to a depth of 10 feet in the rear.

Frank Fiederlein of New Britain, Connecticut points out the excellent quality of his soil which had once been only a poor mixture of sand and gravel. After several years of organic gardening, he now grows more produce than his family can eat.

And what was left? "The worst collection of stones, sand and gravel you ever saw," says Frank.

But he had the answer.

"First we took out all the rocks, I used them to make a retaining wall on the bank at the back of our property. Stones and pebbles we removed to a low spot in a nearby woods, or saved for rock mulching.

"Then that autumn we started to build soil, using the trench method. We dug two-feet-wide trenches, about 12 to 14 inches deep. Having no manure, we mixed in half-decayed leaves and pine needles, with a little agricultural lime and cottonseed meal added."

The following spring, Frank planted some tomato plants (mulched with leaves and pine needles), string beans, cucumbers and carrots. The results were surprisingly good. "The tomatoes were the envy of the neighborhood. Besides having enough for our family and friends, my wife put up 55 quarts for the winter."

In the fall of that year (1955), Frank again dug in semi-decayed oak, elm and maple leaves from the woods. He also occasionally buried small amounts of clean garbage—potato and carrot peels, lettuce and cabbage leaves and the like. Then during the winter he was able to get some fairly well-rotted turkey manure, which he spread after sieving. In the spring, small quantities of lime and potash and phosphate rock powders were incorporated into the soil.

"Without realizing it, I had prepared a welcome home for untold numbers of earthworms. Today I can't turn a spadeful of soil without finding as many as two dozen worms in it.

"And the crops—well, I was amazed. I had 20 tomato plants, grown from small packets of seed, that produced tremendous crops. Cecile put up 75 quarts from that 1956 garden. The peas were plentiful, my lettuce, radishes and spinach came in strong. Our supply of Tendergreen beans was constant. My Detroit Dark Red beets grew rapidly and were delicious. I had some wonderfully tasty long slim carrots, grown from seed requested from the Massachusetts Experiment Station. And the harvest of cucumbers from 4 hybrid vines was too much for the family.

"I knew my soil was really coming alive. It was starting to show a wonderful texture, open, well aerated and able to hold plenty of moisture. I could see it was becoming vital and healthy, not like the sad-looking dead stuff you so often see in gardens where chemicals are used."

That fall he put turkey manure on the vegetable plot, digging it in last spring. He also buries the turkey feathers—"they decay very quickly." Mulching materials are constantly applied throughout the growing season. Frank uses grass clippings, garden refuse, leaves and pine needles. A little lime is sprinkled on the pine needles to counteract their acidity. Virtually no weeding has to be done. The mulch is dug in in the fall.

He says there is still some slight unevenness in the texture of his soil, but this is gradually disappearing as more organic materials are incorporated. It is fast becoming a rich, dark loam, far from the dusty, light-colored sand it was originally. Soil tests this spring showed an abundance of nutrients.

The peas produced a bumper crop last year. Four kinds of lettuce, plus radishes, chives and beet greens filled out the vegetable supply through June. The wax beans also came in in June, followed by green beans and then pole beans.

Between the tomato plants he grows red and white cabbage for cole slaw to replace the lettuce, which he will have again in the fall. This year the Fiederleins have three varieties of tomatoes: Peron, a large beefsteak from Argentina; San Marzano, a paste tomato; and Golden Sphere, a large yellow. The first two they combine for tomato sauce.

Frank's real pride and joy is his miniature orchard. In front of the house, as an integral part of the foundation planting, are growing two Hansen bush cherries and two bush plums. Spotted around the rear are 4 dwarf apples—McIntosh, Cortland, Delcon and Red Delicious; two dwarf pears—Bartlett and Clapp's Favorite; two Montmorency cherries; two Korean cherries; two Oka cherries (a cherry-plum hybrid); 3 red currants—two Red Lake and one Wilder; 4 dwarf plums—Abundance, Burbank, Stanley and a Japanese-American hybrid. The last-named plum is growing, along with an Elberta peach, on the bank behind the retaining wall (Frank has also extended the size of his vegetable garden by planting his peas and cucumbers on the bank.)

All of these are doing very well. "The currants bore well last year, and now the dwarf apples and pears and the cherries are starting to come in. Very soon now we will have a plentiful supply of tasty, all-organic fruit in enormous variety.

"Here's a funny thing: the nurseryman told me my Korean cherries would not grow more than 5 to 6 feet high. Already they're over 9 feet tall!"

Another heavy producer is his blueberry bed. A little over 50 feet long, it contains 13 plants of 6 different varieties (to insure a long harvest).

"To make the blueberry bed, I dug out a trench 4 feet wide and about 18 inches deep. I filled it with the top 2 or 3 inches of soil from under some pine trees, plus pine needles, sawdust, semi-decayed leaves and sand. Around the roots I used a mixture of sand, loam and peat moss. Each year I add a two-inch layer of pine needles. The yields are getting better all the time."

The rich lushness of his lawn, despite drought and a municipal ban on watering, is another accomplishment of which Frank is justly proud. He spreads compost on it in early fall, then lightly applies sieved turkey manure on top of the snow or just before a spring rain. The excess is later gathered. By mid-April he had cut the velvet carpet 4 times, which amazed his neighbors no end.

Frank and Cecile grow some 22 annuals in their flower beds, and have a bed of low-growing mums fronting their property—a real car-stopper during blooming season. They are also starting perennials in a bed behind the retaining wall, and have interplanted tulips under the bush plums and cherries. Along the east wall of the house they have also planted 3 filbert nut trees.

Both their adopted daughters, Nancy and Laurie, have their own tiny gardens, and are learning to grow vegetables and flowers the organic way. Every facet of the garden seems to fascinate them—and there's no danger from poisonous chemicals anywhere on the place!

"Everything grows like magic with organics," Frank says. "Three years ago I bought some tiny cactus plants from the five-and-dime. Now one of them is over 4 feet tall."

He stopped to think a moment. "You know, there are untold numbers of homeowners who could and would put their back yards to good use hobby-wise and health-wise, if only they had the know-how. With natural methods, they would—as I did—find it easy to produce huge yields of a vast variety of crops from a small space.

"I treat the soil with enlightened respect and take good care of its marvelous inhabitants. In return, my garden rewards me a hundredfold, not only in produce, but also in satisfaction and untold delight."

—Betty Sudek

We Retired "Down South" in Florida

Five years ago, we came down here to southwest Florida to live beyond the reach of snow, sleet, blizzards and big fuel bills. It is nice to be able to go fishing comfortably almost any day in the year, never to wear an overcoat, never put antifreeze in the radiator.

Best of all, it is nice to be able to bring in something from our organic garden and orchard for any meal in the year. We have ripe citrus fresh off the trees eight months a year, bananas and papayas a few steps from the kitchen ten months of the year, and vitamin-rich fresh vegetables, too. We raise our own beef, chickens and eggs.

We can go dig a peck of Allgold sweet potatoes any time we want them—they keep growing all year long down here—and as I sit here typing we have three more months of late oranges on the trees, two more months of grapefruit, more tomatoes than we can eat or give away, sweet corn, watermelons, turnip greens (the wife is an Alabama gal), lettuce, cucumbers, squash, figs, three kinds of bananas—two dessert kinds and the baking variety—papayas, Surinam cherries, Irish spuds, pole beans, chayotes, flavoring herbs, all of these things organically grown, of course.

Any time I feel like mowing off the tall, bushy asparagus tops and dressing the bed with some compost, I can trim a peck or so of nice, tender shoots before letting the tops grow out tall again. The bees make honey nearly all year long, too. The family is healthy, full of pep, and able to enjoy life outdoors four times as much as folks who live 'way up near the North Pole in such places as Michigan or northern California.

I am a Lieutenant Colonel on the Retired List of the Regular Air Force, having been retired for combat disability in 1947. During 20 years of active duty, I have traveled all over the world, always following my hobby of observing growing things and culture methods under various conditions of soil, climate, folklore, and the loss or maintenance of fertility.

Down here in Florida I have seen so many tragic results of basic ignorance of soil ecology and plant growth. For instance, here in Fort Myers the fine botanical garden planted on 14 acres of Thomas Edison's winter home loses many fine specimens each year because they try to grow everything from exotic tropicals to giant banyan trees with chemical fertilizers and have one employee whose duty is to clean up every leaf that falls, all shrubbery trimmings and grass cuttings, and either burn them or haul them away.

To a lesser extent, the same stupid formula is found in the famous Fairchild Gardens in South Miami; however, they do mulch a little, not for plant nutrition, but as they explained, it keeps down the weeds! Up in the central and northern part of the state, older citrus groves are being bulldozed out and burned at taxpayer's expense due to heavy infestations of the nematode. This pest is a logical result of heavy applications of corrosive fertilizers that have destroyed the soil's ecological balance and the earthworms so that there is nothing to hinder the spread of the nematodes.

Folks are moving down here by the thousands, but there is so much space left that there will be room for many more in the years to come. Unfortunately, there are sharp operators who take unkind advantage of the natural desire of older folks to retire to live inexpensively on modest incomes. These people advertise up north in newspapers, magazines, radio and TV, loudly promoting their "planned communities" and "subdivisions," many of which have only recently been bulldozed out of palmetto scrub wilderness 15 to 30 miles from town.

The wiser folks will take their time in coming here and looking around a good long while before buying a homesite. Organic gardeners want suitable soil for trees and gardens, not swampy lowland that will be under standing water many months during the late summer and fall rainy season. They will get acquainted with people who live here all year around, and stay away from the high-pressure promoters who want to sell them poor land and badly built houses for much more than they could ever be worth.

In my own case, I found that by looking around a few weeks before buying, I was able to get 23 acres of an old orange grove right on a big river, for no more than two acres would have cost me at promoter's prices. I am closer to town than several of these "planned communities," have richer soil than any of them, better drainage in wet weather, and *no* restrictions. In some highly advertised developments, most folks cannot afford the land area needed for a small family-size citrus grove and garden, and most of them will not allow compost piles, bee hives, chickens, milk cows or goats or much of anything any proper organic gardener wants for comfortable and healthful living.

Many older couples now come in a modest house trailer pulled by a pickup truck. They find a nice homesite, buy it, get electric service, have an artesian well driven, then start their retirement home with no hurry and nobody to disturb their careful and leisurely planning, tree planting, and building. Many of these fine folks build their own homes, working slowly as energy and finances permit, enjoying every minute of it, and saving nearly two-thirds of the cost of hiring the work done.

You may say at this point: "I never built anything; I wouldn't know how." Well, pardner, there are lots of folks here with less talent and strength than you, and they proudly live in some extra nice homes they built with their own hands. If you are smart enough to be an organic gardener, there is no limit to what else you can do.

One of my best friends up the river is a young fellow of 72 who has almost completed the nicest and most modern house in his neighborhood during the past year. I built my own home from footing to fireplace chimney top, including plumbing and electric wiring, cypress paneling, sliding doors, aluminum awning windows, concrete block walls, cedar framing and sheathing, sound-absorbing ceiling tile, big garage, two bathrooms plus asbestos shingle fireproof roof, built-in cabinets and ceiling fan. I laid every block and drove every nail and learned how to do the whole job while doing it. Of course you can do it, and save yourself several thousand dollars while having the time of your life.

Growing most anything down here in the subtropics is a little different from the way you did the job up north. It is warmer, more rain falls, vegetation grows faster, decays quicker, and there are more insects, soil fungi and nematodes. Simple organic methods work the same, however. Compost, mulch and earthworms solve any plant problem as long as drainage, shade and soil *p*H are under control. Heavy rainfall leaches out soluble plant foods from the soil; hot sunshine causes the nitrogen in organic matter to evaporate as ammonia unless protected by mulch.

All these lost plant foods may be recovered by adding floating water hyacinths or cattail grasses to the compost pile, and keeping the pile covered to prevent soaking away during the rainy season. Mulch your fruit trees with shavings, old boards, burlap, flat rocks or most anything else to prevent heavy grass and weed growth from robbing the surface roots. Lift up the mulch twice a year to add an inch or so of compost, and your neighbors will be amazed at the size and flavor of your fruit.

This treatment works equally well on bananas, papayas, or any of the dozens of tropical fruits from loquats to cherimoyas. Do these names sound strange to you? After a few months here they will be as familiar to you as apples and pears up north. You will get a real thrill out of showing your "Yankee" guests what you can grow with compost and earthworms: Watermelon vines that bear 8 or more fine melons each; tomato vines that cover the area of an army blanket and keep bearing grape-like bunches of sweet tomatoes for half a year; wild fruit in abundance such as guavas and wild grapes for excellent jams and jellies; or the tropical winged yam with tubers weighing 50 pounds or more. Flower lovers may have blossoms outdoors the year around, and foliage plants grow wonderfully out of doors in this climate.

Wild orchids bloom every summer hanging from oaks and hickory, and in the wet places wild iris in blue and yellow, compete with the soft lavender of floating water hyacinth blossom clusters. Out in the country where stupid spraying programs are avoided, swarms of hungry dragon flies sweep the mosquitoes from the air as fast as they hatch. In town and in the "planned communities" one must retreat behind screens each evening to avoid swarms of little and big mosquitoes since the spraying poisons both the dragon flies and mosquitoes hatching from the same water, and the mosquitoes simply hatch faster.

To sum it all up: You organic folks who want to retire in Florida, come down here and take your time looking around. Stay away from the "fast buck" promoters. Choose the kind of land, neighborhood and neighbors that add value to your choice of a homesite as the years go by. Bring along the old hammer, saw, level and pipe wrench, and build your own dream home, because *you* can do all or most of it with time and patience. As to growing anything down here, organic folks are the only ones who are completely successful year after year because they work *with* Mother Nature, not against her.

—Oliver R. Franklin

"I Like to Work with Plain Dirt"

A white-haired little woman with a country girl's love of the good earth and the things it can yield is amazing her neighbors by harvesting a year's supply of vegetables from the dust of a city's streets.

She is Mrs. H. R. Leversee of Kalamazoo, Michigan, and for nearly 20

(Above) Mrs. H. R. Leversee of Kalamazoo, Michigan dumps a wheelbarrow-load of "dirt" onto her back yard garden. (Below) Mrs. Leversee stands amid the multitude of vegetables and flowers which she grows in her garden with the help of that fertility-laden "dirt" which she collects daily from the neighborhood.

years she has turned a shovel, a wheelbarrow and the courage of her convictions into as beautiful and productive a back-yard garden as there is in the city.

Mrs. Leversee's source of dirt for her garden is unorthodox to say the least. Every morning for the past 17 years, with the exception of the bitter cold days of winter, this amazing little woman has trundled her wheelbarrow into the streets of her neighborhood and shoveled up the dirt which always lines the gutters.

She fills up her wheelbarrow and rolls it back to her home where she spreads it over her back yard or lawn or wherever she needs it most. If it's a good day and she feels spry enough, she will make 2 or 3 trips up and down the streets in search of dirt. And sometimes she's even out in the evening.

She's never kept a record of her work, never thought it would interest anybody. But it's a safe bet that Mrs. Leversee, at an age when most women have slowed down to knitting a fast pair of mittens for a grandchild, has hauled at least 100 tons of dirt from gutter to garden in the last 17 years. It amuses her that anyone should care about her project.

"I know that a lot of my neighbors and others who see me out with my shovel and wheelbarrow think I'm crazy," she says with a smile.

"But I've reached the age where I don't care what others think. If I ever do have any doubts, all I have to do is look at my beautiful garden, my flowers, my cellar full of canned vegetables, and I know I'm right.

"You see, I was born in Arkansas. Down there everyone knows the value of earth. If you are lucky enough to own any, you guard it. My father never let a day pass that he didn't tell us that the earth took care of us and we had to take care of the earth.

"Up here it's different. Folks take their earth for granted. They buy expensive topsoil; it rains, the topsoil drains off into the gutters and they start over again buying more soil.

"If you just stop to think about it, any dirt in the city streets must be top dirt. It may look like dust and be full of grit and cinders, but it's rich and strong. It's a crime to let it flow into the sewers and be lost."

Mrs. Leversee, who attended college in Arkadelphia, Arkansas, and then taught school for years, came to Kalamazoo in 1926. In that time, using her unusual soil-salvaging system, she has raised the level of her lawn at least 6 inches above her neighbors' and turned her barren back yard into a living supermarket.

"I've worn out three wheelbarrows in the process," says Mrs. Leversee, who has been a widow since 1947. "But look what I have.

"I got over 23 dozen ears from 4 30 foot rows of corn this year," she states proudly. "I'll bet a lot of the people who stopped to buy a few ears from me are the same who let their own top dirt run off into the gutters."

The tireless little widow, who can't weigh much more than a bushel of potatoes, cans a good share of her vegetables and then saves the balance of the gutter-born crop in a freezing unit which is as old as time.

"Each fall I dig a shallow hole in the garden. Then I put all extra vegetables in the hole in separate piles. The hole is lined with leaves and the vegetables are covered with leaves. Then I cover it all up with dirt.

"As the winter passes by I can go out to my natural 'deep freeze' and dig up fresh vegetables any time I want them."

This amazing little woman, a combination of pioneer, soil conservation expert, farmer and manual laborer, scoffs at the work involved in her dirt collecting and gardening.

"Working is like walking," she says. "If you enjoy it you never notice it. I like to work with just plain dirt. I like to see things grow in it. Every shovelful I take from along the curbs will produce food some day. I'll never get tired of watching that happen."

—Dan Ryan

Happiness in Gardening

When Paul Trucksis of Tipp City, Ohio, finally retired from his job at Delco Products Division of General Motors, he and his wife looked ahead for a worthwhile and healthgiving hobby. It was then that they became interested in gardening—organically.

In the past two years, he has proved with an abundance of flowers and vegetables that the organic method brings health, fun and profit. In fact, the family has so many flowers they're wondering where to put the next bed for transplanting.

He believes the following steps to be important to the organic gardener:

1. Have a "hidden" compost heap in the back yard where you make your humus. (Compost isn't enough—should be used with earthworms.)
2. Spade garden and cover with organic matter.
3. Plant seeds.
4. After plants have had a chance to grow a little, spread your own mulch around them. The earthworms work it into the soil.

"Once you provide the right soil for your plants (manure and organic matter homogenized by earthworms), you can just sit back and watch your garden grow," he says. "This organic type soil also resists weeds and it grows richer each year by nature's own work."

"I'm just a beginner," he smiles while leading the way through his gardens, "but I CAN tell you, that you, too, can have huge, tasty, Big Boy, acid-free tomatoes for EARLY market if you plant them organically." For vegetables, flowers or any planting, Trucksis mulches ground 4 to 6 inches all winter. Worms carry humus underneath, so no plowing is necessary. If soil is damp in spring there is no need to worry, things will grow.

He made over $100 on 96 tomato plants on the early Dayton market when prices were best. Besides this, the family had plenty of the delicious, smooth fruit for canning, preserving, for gifts to friends, and for table use. "If you're tomato-gardening for profit, you want early market prices. Grow them organically and they'll be ready for early market."

"This is the happiest year of my life," Paul smiled as his wife, Mabel, nodded her approval. "If I can help others with my findings, I'm truly happy." And no one would guess this sincere organic gardener to be past retirement age for he appears 10 years younger. "We've never enjoyed life so much until we began gardening, truly organically," he says.

—Clarissa Schweikert

A Help for Training Our Youth

The Grailville School, of Loveland, Ohio, believes that women have an essential role to play in the renewal of strength of modern society by contact with the earth. The purpose of this forward-looking institute for women is to give a vision of Christian womanhood and to apply this principle to modern young women in all spheres of life.

Agriculture is included in the specialized apprenticeship offered to students —either short courses of a few weeks, or the more extensive, lasting a few years. From the training received here, the girls return to their former work

Lettuce and onions are shown being harvested in slope-planted fields by Grailville (Ohio) School students who not only learn to provide themselves with better food, but will also pass this knowledge along to people in other lands where they travel.

or devote a part or all of their lives to the missionary fields for which they have been trained.

Of special interest to gardeners is the training given the girls in agriculture. Here one can observe the wonderful results of putting into practice the teaching of organic gardening. An abundance of vegetables being harvested by the students leaves no doubt as to the superiority of their harvest and why they believe so profoundly in Nature's way of gardening to produce the best results.

Regardless of the course selected by the student, each one gets some practical training and understanding of gardening. In fact, each group has its own compost pile and sees to it that it is turned at the proper time.

The gardener pointed out one garden which was on a very slight slope.

Here the vegetables were planted on the contour. Even on so small a slope, she remarked of the absence of washing which she had observed with the conventional straight rows. Compost has been applied to all their gardens and mulching is practiced on most vegetables. The gardens are rotated each year and they supply most of the food for the 120 students enrolled during the year. One can well understand the reason for the 1,000 staked tomato plants, the many rows of beans, the abundance of herbs and the well-diversified gardens. Their mulched orchard provides much of their fruit. Any tree destroyed during the winter is replaced the following spring.

The 10 to 15 girls enrolled in agriculture usually attend for at least one year, so that they will have a practical understanding of gardening and the various phases of farming on the 380 acres of land. This group plants, tills and harvests the produce on the farm.

Watching these girls weeding the rows, picking the vegetables and fruits or turning the compost pile, one feels gratified to know that the principles of good gardening will be carried to near and far places; that with their missionary work in Christianity they will also be spreading the organic methods of good gardening and bringing back an awakening to the many peoples with whom they work that God's land cannot be neglected; that in the hearts and minds of the people must be a realization of the priceless heritage to build and maintain the fertility of our earth.

—Marjean Headapohl

Pride in a Job Well Done

It was a very hot summer in McKeesport, Pennsylvania. Most plants were drying up and withering, and so Art Ryden kept an extra close watch on his Marglobe tomato seedling. And it turned out to be a truly fabulous one. Planted in the back yard compost pile, it grew 400 big, juicy tomatoes which totaled 100 pounds. The vine itself attained near record dimensions of 5 feet in height, 7 feet in length and 5 feet in width.

Pleasure in a Bountiful Garden

It may sound a little bizarre at first when you hear that Leroy T. Stratton encourages grass in his family garden. Or, looking at it another way—that he grows vegetables in his lawn.

Why should anyone alternate strips of lawn with strips of vegetables?

The most obvious reason, when you drop in on his place outside Rutland, Vermont, is the charming attractiveness of this setup. Here, the lawn doesn't end abruptly, but becomes lawn with neat beds of succulent edibles growing in it.

Yet Stratton is an intensely practical man (he used to be a banker) and he's not sacrificing food for prettiness. Gardening the organic way, he wants the most food out of the least work. Besides quantity, there's quality—like his winning 6 out of 7 blue ribbons at a recent Rutland Fair.

So he's a master gardener, but a modern one. If he were back in the horse-team era, he'd need more elbow room for mowing machine and plow. But his power mower for grass, and rotary tiller-type plow for vegetable beds make

Art Ryden of McKeesport, Pennsylvania proudly extends his arms to show the huge dimensions of a tomato plant which he started in a backyard compost pile. The vine grew 5 feet high and produced 400 tomatoes totaling more than 100 pounds.

maneuvering easy—and his method adds up to less work instead of more. His third favorite "tool" is the compost pile.

Since his climate is rugged and his soil wasn't much to begin with, here's a gardener well worth visiting for how-to tips.

Nine years ago, the site was just the corner of a rundown meadow near the house. Here, Stratton began with a few beds of vegetables. Stones were a problem, but he liked health-building exercise. Smaller ones were wheelbarrowed away. To handle a large boulder, he first built a brush fire on top of it, then dashed buckets of cold water upon the hot stone, thereby cracking it into pieces he could handle.

It was two-way traffic. Removing stones, he brought back anything he could get in the way of fertile, humus-building material—grass, old hay, any manure that was available, plus enrichening compost. Result is the fine loam he has today, allowing close planting.

Meanwhile, when mowing lawn around the house, he kept on mowing—down those strips between vegetable beds. This kept down weeds, and as clippings decomposed, the grass grew greener. He found he was creating fine grassy walks.

This struck him as lots easier than cultivating between wide rows, or hand-pulling weeds between them. Mrs. Stratton was happy, too. Here was a garden that could be harvested without need to walk in soil that would track mud into the house—all the plots are easily reached from the grass carpet. Another advantage was space-saving. This part of the lawn was doing double duty.

Today, each grass strip is 40 inches wide, right size for cutting with one pass of the power mower's cutter blade. And there are 18 of the vegetable strips, each 54 inches wide. As a result, Stratton's gardening is divided into easy-to-handle bits.

Each of the 150-foot long strips holds, two, sometimes 3 rows of closely planted vegetables. This close planting compensates for space taken by the grass because Stratton can keep the soil highly fertile, using compost that would otherwise be wasted between wide rows.

Another reason he can plant closely is that the grass strips "ventilate" the garden—let more sun and air through to its plants. So fungus diseases don't bother him. At the same time, those grassy walks check erosion nicely. This is a garden without little brooks running through it during a heavy rain. The rain is held there.

As soon as one of the vegetable beds is harvested, the old plants are pulled up to prevent their needlessly draining soil fertility, and are turned into compost. The bed is then prepared with the rotary tiller and sown to winter rye. This provides a soil-enriching humus material when it's turned under next spring. In the meantime, while growing, it keeps the bed green and attractive.

Sometimes, he gets a double "crop" of humus this way. For example, rye sown after peas, around August first, would grow too high and be a nuisance to turn under. It's turned under before it's past a foot high. It starts rotting. Meanwhile, enough of its roots sprout again to provide a winter cover crop which is turned under next spring.

Space between vegetable rows varies according to the plant. For example, members of the cabbage family—cauliflower, broccoli, etc.—are staggered, with plants 18 inches apart in two rows. The same 54-inch-wide strip holds 3 rows of sweet corn. The corn is planted successively, as are other crops, to give successive ripening. These different-age plantings of corn are in different beds, so as to disperse their shading of other vegetables. Along the same idea of giving plants maximum sunlight, the strips were laid out with a compass to run exactly north and south.

Stratton files a tidy planting chart away each year, so there is no lengthy session with seed catalogs. He knows what varieties he has used, which he likes, which were planted where—so that he can rotate vegetables through the different beds.

Here's a gardener who has a place for everything, everything in its place, from the sun dial at the front to compost heaps in the rear. Vines greedy for space, like cucumber and squash, are kept off at one side.

At the other side of the garden are perennials—one strip is occupied by two varieties of strawberries, and the last strip of all is reserved for a bed of asparagus. Beyond this, out by themselves, are the young fruit trees, each properly mulched with a mound of sawdust.

In front of the garden are two attractive beds which serve as sort of an introduction, separating the main lawn from the double-purpose lawn. One bed has rhubarb. The other contains annual flowers grown for cutting.

In the bountiful garden, there is scant waste. Surplus is handled by Mrs.

Stratton in several tasteful ways. Among her "tools" are two faithful recipes —one for canned vegetable soup, the other for 5-vegetable drinking juice— carrot, parsley, tomato, onion, celery.

—*William Gilman*

A Wonderful Way of Living

In 1949 we bought 20 acres of rolling land along a highway two miles from our town of Decatur, Indiana. The location was ideal for us as my husband, Speck, works at the General Electric plant in Decatur and our two eldest children were soon to start at the St. Joseph school there.

The first things I planted were a row of 4 apple trees, two Bartlett pears and a couple of cherry trees. We had a strip plowed up for a garden 75 by 275 feet. The rest of the farm is laid out in 5 small fields, where we rotate corn, oats and hay. This generally provides enough feed for the livestock we raise.

In my garden I grow asparagus, strawberries and rhubarb. Vegetables are sweet corn, tomatoes, green beans, lima beans, red kidney beans, peas, carrots, beets, turnips, potatoes, cabbage, peppers, celery, okra, zucchini, acorn and butternut squash, pumpkins, cucumbers, watermelons, muskmelons, lettuce, radishes and onions. This year I am adding eggplant, cauliflower and Brussels sprouts to my list as I feel that by now my soil is getting good enough to try these harder-to-grow things.

I forgot to mention popcorn. The children love to spend winter evenings eating popcorn while they watch TV. I also grow my own herbs. They are so much better than the ones you buy. I dry basil, marjoram, parsley, thyme, chives, mint and rosemary.

We have some grapevines starting to produce good crops now. I make jelly and can all the juice I can. We get a little fruit from our apple, pear and peach trees.

I've never kept exact figures on how much I spend on my garden, but here are some facts I can remember for this year. My seed orders totaled $12.50; I bought $4.50 worth of colloidal phosphate and potash, plus another $6 or $7 for straw. There won't be any money spent on expensive chemical fertilizers or sprays, you can bet.

Perhaps you are wondering if growing much of our own food saves on the grocery bill. Well, I should say it does! I allow $10 to $15 a week for groceries. In a pinch, I have spent even less. I know a few families the size of ours of moderate means who buy everything and spend at least $30 to $35 a week for food. I'm sure they don't get nearly as good food as we have.

I freeze corn, peas, strawberries, applesauce, lima beans, asparagus and rhubarb. I can green beans, kidney beans, beets, pears, peaches, tomatoes, grape juice, kraut and pickles. For as long as they will keep, we store potatoes, squash, carrots, beets, turnips, celery and cabbage in the basement.

I think this way of living is wonderful. I know we could never afford to be paying for our new house with our present income if we did not save money on our food this way and if we had medical and dental bills to pay all the time. And, as you can see, it really is a wonderful way to raise a family!

—*Lois Hebble*

A Few Facts About Food Costs

The following are statistics issued in a report by the Virginia Polytechnic Institute Extension Service:

1. The average urban family spends about one-fourth of their total income for food.
2. Families with less than $2,000 a year income spend one-half of it for food.
3. Even so, many people are poorly fed; they get enough food to satisfy hunger but not the right kind to meet body needs.
4. The average homemaker in the United States feeds 4 people each day.
5. In 1957, 36 million homemakers spent 37 billion dollars in food stores. That is about $21 a week or $1,021 a year (over 5 million of these are farmers, who produce some of their food).
6. The amount you should spend to get an adequate diet depends on the following:
   a) The time available for preparing meals.
   b) The tastes and food habits of the family.
   c) Facilities at home for preparing and storing food.
   d) What food you produce at home.
7. The United States Department of Agriculture estimates of food costs for a low-cost adequate diet in December, 1954, were as follows:
   a) For a family of two .... $13 a week
   b) For a family of 4 with pre-school children .... $18 a week
   c) For a family of 4 with school children .... $21 a week

This is the lowest amount that could be figured to get an adequate diet. For more variety, the "moderate" cost diet would cost $16, $22, and $26 for the groups above. According to VPI, a 100 by 220-foot plot of good land can produce enough vegetables for a family of 5. That's around $325 worth at the market. Figure you'll put in about 12 days of hand labor after the land's ready for planting.

The *Statistical Abstract of the United States* for 1959, published by the United States Department of Commerce, lists these figures (in pounds) of the per capita consumption of major food commodities:

Meats—152.0 pounds per capita
Fish (edible weight): 10.0 pounds
Poultry products: eggs (number), 348
chicken, 28.5
turkey, 5.6
Fruits: fresh, total, 92.3
processed: canned fruit, 22.4
canned juices (excl. frozen), 12.1
frozen (incl. juices), 8.5
Vegetables and melons: fresh, 130.7
canned, 44.2
frozen, 7.6

To get some idea of the amount of cash income spent by the American consumer on food, look at the following statistics compiled by the editors of *Fortune* magazine. These figures are in terms of 1953 dollars, with income computed after taxes:

Income Group	*Average Income*	*1953*
$0-1,000	$ 510	89.9%
$1,000-$2,000	1,520	52.2
$2,000-$3,000	2,525	33.1
$3,000-$4,000	3,495	30.2
$4,000-$5,000	4,450	23.7
$5,000-$7,500	6,125	25.9
$7,500-$10,000	8,410	22.9
$10,000 and over	18,275	14.7

Section 1.

1—Planning the Vegetable Garden

"Stretch the Harvest" by Planning

THE SUCCESSFUL organic vegetable garden will give you and your family a ready and inexpensive supply of fresh, nutritious vegetables throughout the year. With careful planning, you can get a continuous supply of vegetables from early spring to late fall. Robert Stevens, extension horticulturist at the University of Delaware, calls this "stretching the harvest season."

You'll also be able to double the harvest by interplanting and succession planting. Careful planning will decrease the need for canning, freezing and storing great quantities of food. Vegetables picked fresh for 7 or 8 months of the year not only decrease the amount needed to be preserved, but also provide a fresh source of vegetables for the family table.

Another advantage of having a definite plan for the garden is that the kind and amount of seed can be fairly accurately determined in advance, and proper amounts purchased. If you have no plan you are likely to get too much of

Planning the vegetable garden in early spring is the best way to get the most yield from your garden. This rough outline can save many hours of wasted effort, and help you to a continuous supply of fresh vegetables through the garden season.

some seeds and not enough of others. These 10 points should be considered while drawing the plan, according to Stevens:

1. Perennial crops such as asparagus, strawberries and rhubarb should be located at one side of the garden.

2. Tall-growing crops such as corn must be kept away from small crops like beets and carrots to avoid shading.

3. Provide for succession crops, a fall garden, small fruits and overwintered crops to mature early in the spring. In this way space for spring crops which will be harvested early, may be used again for later crops. Examples: tomatoes after radishes; cucumbers after spinach.

4. Early planted, fast-growing, quick-maturing crops should be grouped together. Examples: radishes, lettuce, early cabbage, scallions, etc.

5. Provide plenty of vegetables for canning, freezing and storing.

6. Do not overplant new varieties, vegetables which the family does not like, or too much of any one vegetable at one time.

7. Rows should follow across the slope (on the contour) in hilly areas.

8. Make sure the plan provides the best spacing between rows for the method of cultivation that you intend to use (hand, tractor, horse).

9. Run rows north and south if possible to prevent plants from shading one another.

10. Long rows save time in care and cultivation. Several crops may be planted in the same row if the distance between rows is the same.

Getting the Most Out of Your Garden

Actually, the entire purpose of careful planning is to get the most out of your garden. Managed correctly, a small garden will yield more, certainly satisfy you much more than a poorly-run large plot. Consider the average suburban home with the yard obviously not large enough to produce everything wished for. Is there anything that can be done about it? Definitely yes! Here's how one Pennsylvania organic gardener, Dr. Lewis Theiss, suggests solving the problem:

"Production in a restricted garden area can be very largely increased by making a two-story garden. In a way, such a garden is like an old-fashioned house with two stories. That type contains twice as many rooms on one plot. Similarly, a two-story garden greatly increases the productive area.

"Some years ago, I had to limit my plantings to an enclosed garden that was just about 50 feet wide. The two-story garden enabled me to produce a very generous supply of vegetables in this restricted area.

"To begin with, I narrowed the space between plant rows. Instead of cultivating, I mulched. That not only did away with a lot of work, but it kept the ground moist in a way that had never happened before. And this generous supply of moisture certainly helped the vegetables to secure more plant food.

"But spacing my rows closer did not entirely meet my needs. So I resorted to the two-story garden. Suppose your garden is 50 by 50 feet. You have 200 feet of fence—that ordinarily goes to waste. Yet that fence will hold plants as well as bean poles, trellises, or other supports. Those 200 feet, of course, are the equivalent of 4 50-foot rows of vegetables. So the thing to do is to make use of the space along your fence.

"What do gardeners usually grow that needs support? There are climbing beans—several varieties of them. There are cucumbers, melons, squashes, and so on. All have tendrils for climbing. So all the gardener has to do is to plant his seeds, or set out young plants, in a long row at the bottom of his fence wire, mulch the ground well, and see that the young plants get hold of the fence wire. The plants will do the climbing.

Vermont garden which has been built "on the contour" is an indication of how a vegetable garden can be planned for a sloping area. By having the rows follow the soil contour, erosion will be avoided and fertility can be maintained.

"You needn't be skeptical when it is suggested that squashes can be grown on a fence. If you have ever seen a farmer's corn field, you must have seen some of his pumpkin vines climbing the cornstalks. And if your farmer is of the thrifty sort, and planted some beans along the corn rows, you will see lots of bean vines going up the cornstalks—and out in a great field at that, where there is ample room for all sorts of plantings. The farmer is simply trying to save himself some work.

"I have suggested climbing beans as an illustration. Some crops that are ordinarily grown on the level will do better if grown vertically. Take cucumbers, for instance. If you get a wet spell, and especially if your ground is heavy, your cucumbers will in all likelihood begin to rot. If you raise

China cucumbers—and every gardener who desires a very superior product *should* raise them—you will be surprised how much can be grown on a fence.

"If you want something more attractive in your garden, try scarlet runner beans on the fence. You will not only get some good beans, but the showy sprays of brilliant scarlet flowers will make your fence a thing of real beauty. Even beans with inconspicuous flowers add living charm to a fence. The long and sightly pods, hanging in heavy clusters, are also a thing of beauty.

"And, when it comes to picking the beans, you never had it so good as you will have it if you put your beans on a fence. There will be no more

A "two-story garden" is an excellent way to double the yield from a small back yard garden. The planting of tomatoes shown above, which adapts itself to a picket fence, is one example of how this two-story idea can be put into practice.

back-breaking stooping to gather the beans. Harvesting them will be as easy as taking a can off a shelf. I know. I have done it.

"You can also tie your tomatoes to the fence. Staked tomato vines always have to be trimmed some. You can trim off the shoots that want to grow through the fence. Fruits on the opposite side of the fence are of little use to you. If you grow some of the little red-fruited tomatoes, you will be amazed at the beautiful picture they will make on your fence. The small red cherry or red pear tomatoes, or the yellow pear or plum tomatoes, are indeed colorful and decorative—as well as tasty. The fruits grow in brilliant clusters. But they need to be hung up to be seen well.

"Climbing peas, such as telephone peas, were really made for fences. They absolutely *have* to have high support. They will cover the woven wire with their beautiful foliage, and the hanging pods are like striking figures worked into a lovely green fabric.

"But you don't need to limit your fence-row growths to climbing vines. A row of corn can be grown hard against the fence—with beans to climb the stalks, at that. And so can any other tall, upright growths successfully occupy the little swath of ground along the foot of your fence. All such growths can be looped back flat against the wire.

"In earlier days, folks grew many products on the sides of their houses, espalier fashion. They thus grew apples, pears, apricots, and other tree growths. Grapes have long been a favorite for the sides of houses. They can be for you, or you can grow them on your garden fence. I have several grape vines on my garden fence.

"This is not intended to be a complete list of things you can grow in your two-story garden. It is meant merely to suggest that, if your garden space is limited, you try two-story gardening—an excellent plan for you to get the most from any garden."

Motto for March: Be Ready!

Here's some advice from Connecticut gardener-mulcher, Ruth Stout, author of *How to Have a Green Thumb Without an Aching Back,* on planning your garden:

"By about March, your seeds should all be arranged in 3 boxes, marked Early, Middle, Late. They should be in alphabetical order, so it will be easier when planting time comes to pick out the ones you are ready to put into the ground.

"But whether you have bought your seeds or not, you can get out the catalogue from which you intend to order them and make out your list so that you will know exactly what space you can allow for each and what you are going to put where. This done, find a big white sheet of good thick paper, a ruler and soft black lead pencil. Unless you are one of those skillful people who never makes a mistake, you will be better off with an eraser at the end of the pencil.

"After outlining the measurements of your vegetable garden, you can settle down to putting your crops on paper. It's a little unbelievable how much time and space a person can save if he has a complete plan to follow when he is ready to plant.

"I rotate my vegetables because the experts tell us to, and hardly anything makes me feel so virtuous as following the advice of the experts on those rare occasions when I feel it isn't hazardous to do so. Every other year I put corn and tomatoes in the upper half of the garden, the following year in the lower half. Why not?

"Maybe you know how many feet of corn you want to plant this year. If you don't, figure how many ears you can use, how many you can get from a foot of corn, and that's it. With my year-round mulch, no plowing, no weeds, I average almost two ears to a stalk and I can also plant very closely because my 14 years of rotting mulch has given me such superb soil.

"Let's say you're going to plant three varieties, two early and one main crop, so that you can eat corn for at least two months. Draw lines on the paper for the rows, and put planting dates and name of the varieties on each line. It's wonderful when planting time comes just to glance at the diagram and waste no time in figuring where you want to put what.

"Next, choose a spot for tomatoes and mark that. Do you plant too many? I do.

"I put peas, including bush edible pod, between the rows of corn. This saves a lot of space and peas being an early crop, corn a late one, the two don't interfere with each other. I find Lincoln peas infinitely more satisfactory than any kind I've ever tried and so do many other people to whom I have recommended them. Keep them well propped up with hay so they won't steal sunshine from the young corn.

"I plant about the same amount of spinach, onions and peppers each year, I always plant these in two rows, the length of the garden, next to the row of strawberries, rotating by putting the onions next to the berries one year, and the spinach there the following year. I put pepper plants in the same row with spinach, since one is so early and the other so late. Spinach is eaten or in the freezer by the time the peppers are old enough to feel the need of a little space and privacy.

"If you are like me you are always dashing out for a handful of parsley, so you may want to do as I do: put it at the near end of the garden. Have you ever tried making a border of it around a flower bed?

"Most of the other vegetables are a matter of ruling your paper, 1, 2, 3 feet apart as each vegetable requires, and writing the names on the lines. When you choose the rows for cabbage, broccoli and cauliflower you can, if you need the space, put kohlrabi seeds between the plants. I drop a few seeds of the larger plants every 18 inches and put the kohlrabi along the rest of the row. You can also crowd the lettuce and bush beans with any of the later crops; they will mature and be out of the way fairly early. I put lettuce and beets extremely close to each other; neither seems to mind.

"Dill seeds itself copiously but if you mulch, it is more practical to plant it each year. We're so fond of it when it's young and fresh that I plant it several times a season. But ignore this when you plant your garden because you can sow the seeds in the same rows with cabbage and so on, unless you have kohlrabi there. Radishes take no space; you can drop their seeds right on top of carrot, parsley, parsnip seeds.

"I've come across very few people who grow edible soy beans, but it seems a shame not to. This year when the frost took all our limas on June 19th, the soy beans, which we think are as delicious as limas, came through bravely. But you have to give them plenty of space. They get huge and sometimes flop around, showing no consideration whatever for any less demanding crop which may be unfortunate enough to be trying to live close to them. If you can plan your garden so that the soy beans are next to the early corn it may be that the corn will get such an early start that it won't mind the soy

PLAN OF RUTH STOUT'S 63′ x 40′ VEGETABLE GARDEN

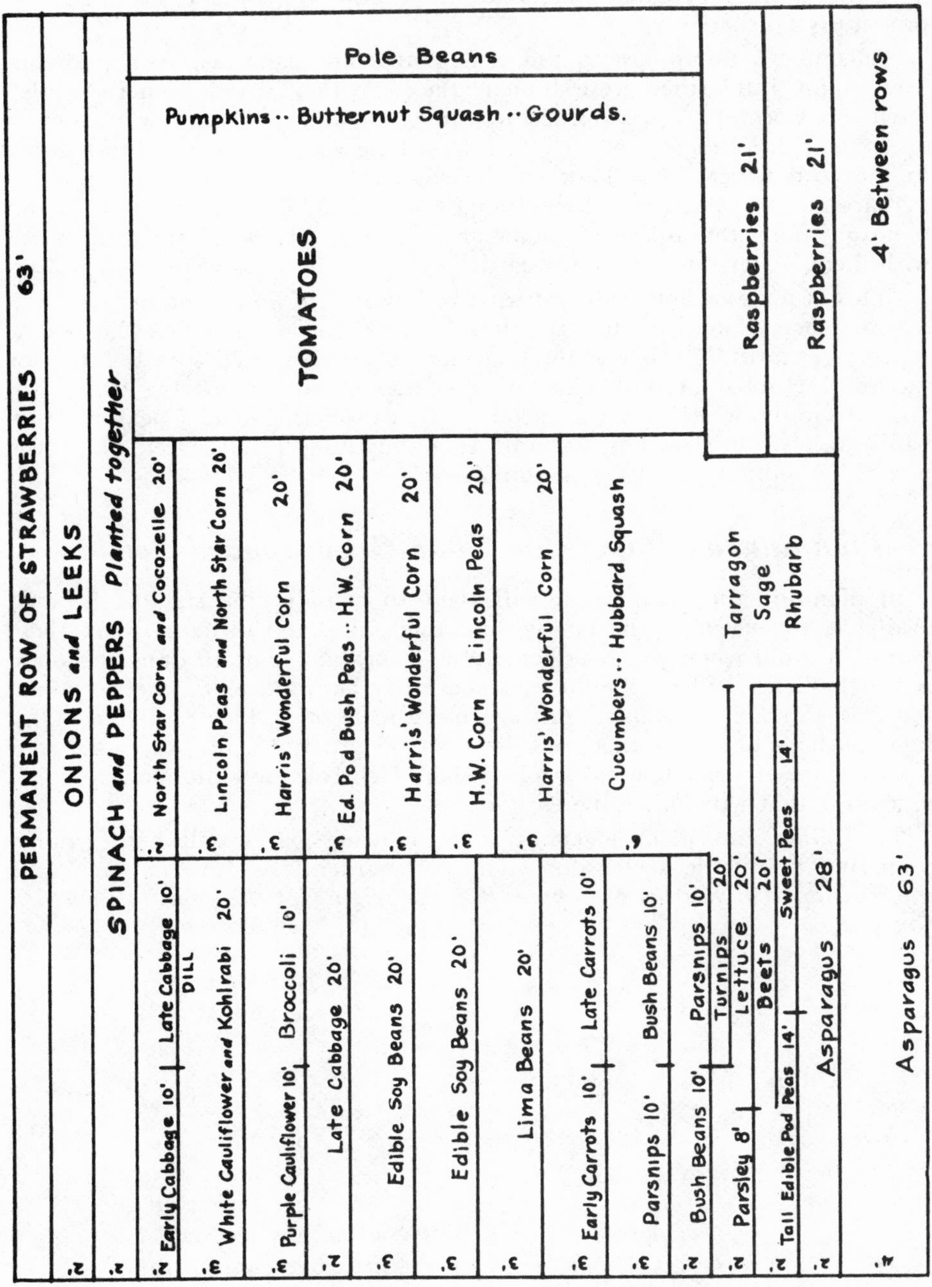

PERMANENT ROW OF STRAWBERRIES 63'
2' ONIONS and LEEKS
2' SPINACH and PEPPERS Planted together
2' Early Cabbage 10' | Late Cabbage 10' — 2' North Star Corn and Cocozelle 20'
DILL
3' White Cauliflower and Kohlrabi 20' — 3' Lincoln Peas and North Star Corn 20'
3' Purple Cauliflower 10' | Broccoli 10' — 3' Harris' Wonderful Corn 20'
2' Late Cabbage 20' — 3' Ed. Pod Bush Peas :: H.W. Corn 20'
3' Edible Soy Beans 20' — 3' Harris' Wonderful Corn 20'
3' Edible Soy Beans 20' — 3' H.W. Corn Lincoln Peas 20'
3' Lima Beans 20' — 3' Harris' Wonderful Corn 20'
3' Early Carrots 10' | Late Carrots 10' — 6' Cucumbers .. Hubbard Squash
3' Parsnips 10' | Bush Beans 10'
2' Bush Beans 10' | Parsnips 10'
2' Parsley 8' | Turnips 20' / Lettuce 20'
2' Tall Edible Pod Peas 14' | Beets 20' / Sweet Peas 14'
2' Asparagus 28' — Tarragon / Sage / Rhubarb
4' Asparagus 63'

Pole Beans
Pumpkins .. Butternut Squash .. Gourds.
TOMATOES
Raspberries 21'
Raspberries 21'
4' Between rows

beans. I somehow have not found it convenient to do that, but I see no reason why it wouldn't work.

"I put poles for beans along one end of the garden and they take up practically no space.

"Cucumbers, pumpkins, squash and gourds are the worst space-grabbers. You might think they were human the way they invade other people's territory. Cocozelle squash (zuccini) can go right in a corn row because it doesn't spread. Besides, very few hills go a long way. But if you grow things in the corn which creep between the rows they become a great nuisance. Perhaps you can put these things along one end or one side, or both. I put them all along the asparagus one year; last year I crowded the pole beans with them. They ran outside the garden over the grass and were most prolific.

"This is no good for Hubbard squash, however, if you want to root it in several places along the vine, as I have learned to do, to outwit the borers. It works beautifully. So I scrabble around to find a special spot for it. One successful Hubbard squash vine provides enough squash for the winter for a small family, which is a very good thing, considering that a healthy Hubbard squash vine has less restraint than anything I have seen, with the possible exception of a Russian sunflower."

Planning on Paper for Your Geographical Section

In planning the garden you will want to consider the size of the area available, the needs of the family, and their likes and dislikes. Keep these points in mind when you take pencil and paper and start to draw the plan. A rough sketch will do, but it must be fairly accurate to be useful. Make the plan to scale if possible. You can use a scale of 1/8 inch to 1 foot. Outline the shape of your garden, put down the length and width, space between rows, names of vegetables to be planted in each row, and the names of late vegetables that will follow the early ones.

The tables and plans given on the following pages will help you in preparing a suitable sketch for your own garden. *Be sure to check the information that has been prepared for the section of the country in which you live.*

NORTHEAST

TABLE 1: Suggested Vegetables for a Garden 50′ x 100′ or Larger

Planned for Power Cultivation

Distance between rows in feet	← 50 feet →		
3	Early Potatoes		
3	Late Potatoes		
3	Late Potatoes		
3	Peas, early as possible	Follow with plantings of any of the following: beets, carrots, snap beans, endive, fall lettuce, turnips, whatever the family likes	
3	Peas, 10 days after 1st		
3	Peas, 20 days after 1st		
3	Spinach (late cabbage Plants)		Parsley
3	Lettuce (follow by late cauliflower)	Radish	Radish (10 days later)
3	Early Cabbage (rutabagas for fall storage)		
3	Broccoli (late beets)	Early cauliflower (late carrots)	
3	Onion Sets	(Fall spinach)	
3	Onion Sets	(Fall lettuce)	
3	Early Carrots (Snap Beans)	Early Beets (Snap Beans)	
3	Early Snap Beans (Lettuce)	Kale or Swiss Chard*	
3	Shell Beans		
3	Shell Beans		
3	Lima Beans (bush)		
3	Parsnips		
3	Peppers	Eggplant	
3	Tomatoes (early)		
3	Tomatoes (midseason)		
3	Tomatoes (late)		
3	Sweet Corn, early	midseason	late
3	Sweet Corn, early	midseason	late
3	Plant early	midseason	late varieties at same time
3	Plant early	midseason	
3	Pole Beans	Pumpkins	
3	Summer Squash		
6	Winter Squash		
6	Cucumbers		
3			

100 feet

* Cut outside leaves only from inside out so as not to cut growing heart.
Vegetables listed in ( ) are second crops.

Prepared by: E. C. Minnum, Extension Vegetable Specialist, Connecticut Agricultural Extension Service

NORTHEAST | TABLE 2: Planting Table

CROP	VARIETY	Seed or Plants per 100 feet of row	Depth to Sow Seed (Inches)	Distance Between rows* (Feet)	Distance Between Plants in row	Date of Outdoor Planting (Average)	When Vegetables Can be Used	Average Days from Planting to Harvest	Average Yield per 100 Feet of Row
CROPS THAT STAND SHARP FROST. PLANT AS SOON AS SOIL CAN BE WORKED.									
Root Crops									
Beets	Crosby's Egyptian Detroit Dark Red	1 oz.	¾	2½	3 in.	Apr. 1 to July 15	1" diam. & up§	55	1½ bu.
Carrots	Nantes—Table Oxheart—heavy soils Chantenay—storage	1 oz.	½	2½	2-3 in.	Apr. 1 to July 15	½" diam. & up	65 by Nov. 1	1½ bu.
Onions (sets)	Yellow	2 lbs.	1	2½	3 in.	Apr. 1	½" stem diam.	50-120	1½ bu.
Parsnips	Improved Hollow Crown	½ oz.	½	2½	3 in.	June 1	Mature	110	1½ bu.
Peas	World's Record, Little Marvel	1 lb.	1	2½	1 in.	Apr. 1	3/16" & up	65	1½ bu. (pods)
Radishes	Scarlet Globe	½ oz.	½	6"	½ in.	Apr. 1	½" & up	30	100 bunches
Rutabaga	Long Island Improved	2 pkts.	½	2½	4 in.	July 1	Mature	by Nov. 1	2 bu.
Turnip	Purple Top White Globe	2 pkts.	½	2½	3 in.	Apr. 1 to July 15	1½" & up	60	1½ bu.
Greens and Salad Crops									
Broccoli	Calabrese	50 plants	½	2½	18-24 in.	Apr. 1 to July 6	While buds are tight	100	30 bunches
Cabbage	Early—Golden Acre Late—Danish Ballhead	50 plants 50 plants	......	2½	2 ft.	Apr. 1 July 5-10	As heads become firm	80 120	35 heads
Endive	Green Curled or Broad-Leaved Batavian	¼ oz.	½	2½	12 in.	Apr. 1 & Aug. 15	Whole plant§ 6-8" spread	90	80 plants
Kale	Scotch Curled Siberian (for wintering)	1 pkt.	½	2½	16 in.	Apr. 1 & Aug. 15 Aug. 15	Leaves 6-8" long	70	60 bunches
Lettuce (leaf)	Grand Rapids	1 pkt.	½	2½	12 in.	Apr. 1 to Aug. 1	Outside leaves§ 3-4" long	60	85 plants
Spinach	Bloomsdale Savoy	½ oz.	½	2½	2-3 in.	Apr. 1 & Aug. 15	4" spread of leaves	50	2 bu.
Swiss Chard	Lucullus	1 oz.	¾	2½	10-14 in.	Apr. 1	Outside leaves 4-5" long	65	60 plants
CROPS THAT STAND LIGHT FROST. PLANT WHEN DANGER OF SOIL FREEZING IS PAST.									
Green, snap beans	Bountiful, Stringless Greenpod	1 lb.	1	2½	3 in.	May 1 to July 15	Pods 3-4" long	60	1½ bu.
Yellow, snap beans	Pencil Pod, Golden Wax	1 lb.	1	2½	3 in.	May 1 to July 15	Pods 3-4" long	60	1½ bu.
Pole, snap beans	Kentucky Wonder	½ lb.	1	2½	Poles—3 ft.†	May 1 to July 15	Pods 3-4" long	80	2 bu.
Sweet Corn	Golden Cross Bantam	¼ lb.	1	2½	15-18 in.	May 1 to June 25	Kernels filled with milk	80-90	60 ears
Squash	Straight Neck (Summer Bush) Butter Nut Green Delicious (Winter)	1 oz.	1	5	4 ft. 8 ft. 8 ft.	May 1 June 1 June 1	Fruits 4-5" long Mature Mature	60 Before frost Before frost	150 fruits 75 fruits 50 fruits
CROPS THAT CANNOT STAND FROST. PLANT WHEN FROST NO LONGER EXPECTED.									
Bush lima beans	Fordhook, Improved Bush	1 lb.	1	2½	8-10 in.	May 10 to June 10	When beans show in pod	85	1½ bu.
Pole lima beans	King of the Garden, Challenger	1 lb.	......	2½	Poles—3 ft.†		When beans show in pod		2 bu.
Peppers	World Beater, Yolo Wonder	50 plants	......	2½	2 ft.	May 15	Green or red	July 25	3 bu.
Tomatoes	Rutgers, Marglobe, Queens	50 plants		2½‡	2 ft.	May 15	Green or red ripe	July 25	3 bu.

*Some vegetables such as peas, spinach, carrots, beets, onions and lettuce may be sown in rows closer than distances shown. See Garden Plan or follow directions on seed packets or in seed catalogs. Time of outdoor planting is approximate. May vary two to three weeks from southern to northern New Jersey.
†6 beans around pole. Thin to four strong plants after second pair of leaves develops.
‡If not grown on stakes, 25 plants 4′ x 5′.
§Thinnings of beets can be used for greens, with small roots attached. Thinnings of lettuce, endive and other salad plants sown in rows can be used as soon as leaves are 3 to 4 inches long.

—*New Jersey Agricultural Extension Service*

NORTHEAST | TABLE 3: Plan of a Farm Garden for a Family of Five Persons

Group	Distance	Crops
1	4*	
		Asparagus—rhubarb—chives—horseradish—herbs—winter onions
	4	
2		Onion sets to mature—thinnings used for green onions (April)
	2½	
		Onion seed to mature—thinnings used for green onions (April)
	2½	
		Early spinach—lettuce—turnips—cress—kohlrabi (April). Followed by snap beans (June 15-July 1)
	2½	
		Early peas (April). Followed by beets and late carrots (June 15-July 1)
	2½	
		Second early peas (April). Followed by late celery—cauliflower—broccoli (June 15-July 1)
	2½	
		Late peas (April). Followed by late endive—Chinese cabbage—lettuce (July 1-15) or late turnips (July 15 to August 1)
	2½	
		Early beets—early carrots (April). Followed by late spinach or snap beans (July 20)
	2½	
		Early cabbage—broccoli—cauliflower (April). Followed by spinach or kale (July 15-August 1)
	2½	
3		Parsnips—salsify—Swiss chard—New Zealand spinach—parsley (April) (Seeded with marker of early and second early radishes.)
	2½	
		Cabbage—Brussels sprouts—early celery (May)
	2½	
		Second plantings of beets—carrots—kohlrabi—lettuce (May)
	4	
		Cucumbers—muskmelons—summer pumpkins—winter squash (May 15)
	4	
		Sweet corn
	2½	
		Sweet corn (May 1) (May 20) (June 15) (July 1)
	2½	
		Sweet corn
	2½	
		Snap beans (May 1-15)
	2½	
		Green shell beans—dry shell beans (May 20)
	2½	
		Lima beans (May 20)
	2½	
		Lima beans (May 20)
	4	
		Tomatoes—sweet and sharp peppers—eggplant (May 20)
	4	
		Late cabbage (June 15)
	2½	
		Late cabbage (June 15)
	2½	

Dates are for Central Pennsylvania.
Grouping shows:—1. Perennial crops. 2. Early maturing crops followed by succession crops. 3. Crops occupying the ground all season.

* Distance between rows may depend upon type of cultivator to be used.

—Pennsylvania Agricultural Extension Service

NORTHEAST | TABLE 4: Plan for a 25 x 50-foot Garden

Practical row arrangements with companion and succession plantings and approximate dates for sowing seed or transplanting plants. Adjust size of plantings to fit needs and preferences of your family.

Feet	ROWS 25 FEET LONG (PLAN NOT DRAWN TO SCALE)
00	
1	Peas, double row, 6 in. apart — April 1. Follow with 1 row snap beans — June 15.
3½	Peas, double row, 6 in. apart* — April 1. Follow with 1 row snap beans — June 15.
6	Spinach, double row, 6 in. apart* — April 1. Interplant with 12 staked tomato plants — May 15.
8½	Spinach, double row, 6 in. apart* — April 1. Interplant with 12 staked tomato plants — May 15.
9¾	Carrots — April 1.
11	Carrots — April 1. Follow with rutabagas — July 1.
12¼	Beets — April 1. Follow with carrots — July 15.
13½	Beets — April 1. Follow with carrots — July 15.
16	Swiss chard, 15 ft. of row — April 1. Peppers, 5 plants 2 ft. apart — May 15.
18½	Cabbage, 12 plants — April 1. Interplant with lettuce, radish seed or onion sets.
21	Broccoli, 12 plants — April 1. Interplant with lettuce, radish seed or onion sets.
23½	Snap beans — May 1. Snap beans — July 10.
26	Corn, ½ row May 1; other half May 10. Follow with 12 kale plants — Aug. 15.
28½	Corn, ½ row May 1; other half May 10. Follow with late lettuce, radish or endive — Aug. 15.
31	Lima beans — May 10.
33½	Snap beans — May 10. Follow with cabbage, 12 plants — July 15.
36	Snap beans — May 10. Follow with snap beans — July 15.
38½	Snap beans — May 20. Follow with beets, 2 rows, 15 in. apart* — July 15.
41	Lima beans — May 20.
43½	Lima beans — May 30.
46	Corn — ½ row May 20; other half May 30. Spinach — Aug. 20.
48½	Corn — ½ row May 20; other half May 30. Spinach — Aug. 20.
50	

*Allow 2½ feet from center of double row.

—New Jersey Agricultural Experiment Station

NORTHEAST

TABLE 5: Facts for the New England Home Gardener

	Seed or Plants Required Per 100' of Row	Dates Seed or Plants May be Sown or Planted in Garden (*)	Planting Distance		Amounts to Plant Per Person Per Planting (1)	Additional Amts. to Plant to Can or Freeze Or Store, Per Person (1)	Best Time to Plant For Canning or Freezing
			Between Rows	Between Plants			
TOMATOES	25-66 (1)	May 15-June 15	36-48 (3)	18-48 (3)	4 plants	6-12 plants	May 30
SNAP BEANS	1 lb.	May 1–July 30	30	2-4	10 ft. row	30 ft.	May 20-30
CARROTS	½ oz.	April 15-July 15	12-15	—	10 ft.	30 ft.	April 10-20
CABBAGE	50-66 (1)	April 1-July 15	24-36	18-24	3 plants	9 plants	—
LETTUCE	¼ oz.	April 15-Aug. 1	12-18	9-15	6 ft.	—	—
SPINACH	½ oz.	April 1-Aug. 15	12	—	8 ft.	10-15 ft.	April 1-15
SWISS CHARD	1 oz.	April 15-July 1	30	6-8	3-5 ft.	3-5 ft.	April 20
BROCCOLI	¼ oz.	April 15-July 15	30	18	3 plants	4-6 plants	April 20-July 1
BEETS	1 oz.	April 15-Aug. 1	12-15	—	10 ft.	30 ft.	April 10-20
SWEET CORN	¼ lb.	May 1-July 1	30-36	24	10 ft.	30 ft.	May 30
PEPPER	66-100 (1)	May 15-June 15	30	12-18	4 plants	6 plants	May 30
RADISH	1 oz.	April 15-Aug. 15	6-12	—	5 ft.	—	—
PEAS	1 lb.	April 1-May 15	24-36	1-2	20 ft.	30 ft.	April 1-15
SQUASH	½ oz.	May 15-June 15	48-120	48-84	2 hills	3 hills	May 30
CELERY	200-300	April 15-July 15	24-30	6-8	4 plants	6 plants	—
ONIONS	1 oz. 1 qt. (2)	April 1-May 15	12-18	2-3	10 ft.	30 ft.	—
CUCUMBERS	½ oz.	May 15-June 15	48	48	3 hills	3 hills	May 25-30
PARSNIPS	½ oz.	April 15-June 1	24	2-3	4 ft.	—	—
TURNIP	¼ oz.	April 15-July 15	24	2-3	5 ft.	—	—

(1) Number of plants, seed planted in greenhouse or hotbed. (2) Onion sets. (3) Shortest distance for trellis culture, longer for flat culture.

(*) Based on statewide conditions and may need to be adjusted for local areas.

—*Massachusetts Experiment Station*

SOUTHEAST

TABLE 6: Virginia Garden Planting Chart

Vegetable	Seed or Plants for 100 Feet of Row	Planting Time	Depth of Planting *(inches)*	Minimum Distance Between Rows *(inches)*	Distance Between Plants in Row *(inches)*	Yield Per 100 Feet of Row
Asparagus	66 roots	March	4-8	36	18	50 lb.
Beet	1 oz.	March to August 15	½	18	3-4	70 lb. (topped)
Broccoli	66 plants	March and July	......	30	18	50 lb.
Brussels sprouts	50 plants	March and July	......	30	24	55 lb.
Cabbage	50 to 66 plants	March, April, July	......	30-36	18-24	135 lb.
Carrot	½ oz.	March to July 15	½	12	3	65 lb. (topped)
Cauliflower	60 plants	March and July	......	30	20	120 lb.
Celery	150 plants	June and July	......	24	8	290 lb.
Chard (Swiss chard)	1 oz.	March to August	1	18	6-8	30 lb.
Chinese cabbage	½ oz.	July	½	18	12-15	130 lb.
Collards	½ oz.	March and July	½	30	18	75 lb.
Corn, Sweet	4 oz.	April to July 15	2	30	10-12	85 lb.
Cucumber*	½ oz.	May and August	½	48	48*	180 lb.
Eggplant	40 plants	May	......	30	30	90 lb.
Endive	1 oz.	March and July	½	18	12	40 lb.
Green pea	1 lb.	February 20 to March 20	2	24	2	20 lb. (in pods)
Kale	½ oz.	March and August	½	30	10-12	75 lb.
Kohlrabi	½ oz.	March and August	½	18	6	65 lb.
Leek	½ oz.	March and April	½	12	3-4	12 lb.
Lettuce, Butterhead	⅓ oz.	March and August	½	18	8	75 heads
Lettuce, Head and Leaf	¼ oz.	March and July	½	18	12	50 heads
Lima Bean, Bush	6 to 10 oz.	May and June	2	30	8-12	25 lb. (in pods)
Lima Bean, Pole*	8 to 10 oz.	May and June	2	36-48	36-48*	35 lb. (in pods)
Muskmelon*	½ oz.	May	1	60	60*	150 lb.
Mustard	½ oz.	March and August	¼	12	5	45 lb.
Okra	2 oz.	May to August	1	36	18	65 lb.
Onion sets (fall)	2 to 3 lb.	September and October	12	......	3-4	75 lb.

SOUTHEAST

TABLE 6: Virginia Garden Planting Chart (Continued)

Vegetable	Seed or Plants for 100 Feet of Row	Planting Time	Depth of Planting *(inches)*	Minimum Distance Between Rows *(inches)*	Distance Between Plants in Row *(inches)*	Yield Per 100 Feet of Row
Onion sets (spring)	2 to 3 lb.	March and April	......	12	3-4	75 lb.
Parsley	½ oz.	March and April	¼	15	6	50 bunches
Parsnip	½ oz.	March and April	½	18	5	65 lb.
Pepper	66 plants	May	......	30	18	70 lb.
Potato	8 lb.	March	4	36	12	150 lb.
Pumpkin*	¾ oz.	May	1	84	84*	300 lb.
Radish	1 oz.	March to October	½	12	1-2	45 bunches
Rhubarb	33 roots	March and April	3½	36	36	25 lb.
Rutabaga	½ oz.	May and June	½	18	6	65 lb.
Salsify	1 oz.	April and May	½	18	3-4	60 lb.
Snap Bean, Bush	10 oz.	April to August	2	24	4	30 lb.
Snap Bean, Pole*	8 oz.	April and May	2	36-48	36-48*	40 lb.
Southern pea	1 lb.	May and June	2	30	2-3	15 lb. (shelled)
Spinach (fall)	1½ oz.	August and September	¾	15	5	40 lb.
Spinach (spring)	1½ oz.	March and April	¾	15	5	40 lb.
Squash, Summer bush*	¾ oz.	May and July	1	48	48*	160 lb.
Squash, Winter vine*	½ oz.	May and June	1	96	96*	360 lb.
Sweet Potato	100 plants	May	......	42	12	100 lb.
Tomato (staked)	50 plants	May and June	......	36	24	210 lb.
Tomato (unstaked)	33 plants	May and June	......	60	36	320 lb.
Turnip	½ oz.	March and August	¼	18	6	65 lb. (topped)
Turnip salad	1 oz.	March and August	¼	12	1-2	45 lb.
Upland Cress	½ oz.	August	¼	12	5	30 lb.
Watermelon*	½ oz.	May	1	72-96	72-96*	200 lb.

* Plant in groups of 6 seeds 2 inches apart, and later thin to 3 plants per group. Space the groups as indicated above. For pole lima and pole snap beans place a pole for each group.

SOUTHEAST

TABLE 7: Planting Guide for Florida Vegetable Garden

Crop	Varieties [1]	Seeds Plants 100′ of Row	Spacing in Inches		Seed Depth Inches	Planting Dates in Florida (Inclusive)			Plant Hardiness [3]	Pounds Yield 100′	Days to Harvest
			Rows	Plants		North	Central	South			
Beans, Snap	Seminole,[2] Tendergreen,[2] Contender,[2] Topcrop,[2] Wade,[2] Cherokee (wax)	1 lb.	18-30	2-3	1½-2	Mar.-Apr. Aug.-Sept.	Feb.-Mar. Sept.	Sept.-Apr.	T	45	50-60
Beans, Pole	Florigreen, US No. 4 (191),[2] McCaslan, Alabama No. 1	1 lb.	40-48	15-18	1½-2	Mar.-June	Feb.-Apr.	Jan.-Feb.	T	80	60-65
Beans, Lima	Fordhook 242,[2] Concentrated,[2] Henderson,[2] Challenger (Pole)	1 lb.	26-48	12-15	1½-2	Mar.-June	Feb.-Apr.	Sept.-Apr.	T	50	65-75
Beets	Early Wonder, Detroit Dark Red, Crosby Egyptian	1 oz.	14-24	3-5	½-1	Sept.-Mar.	Oct.-Mar.	Oct.-Feb.	H	75	60-70
Broccoli	Early Green Sprouting,[2] Freezers, Waltham No. 29, Texas No. 107	60 plts. (¼ oz.)	30-36	16-22	½-1	Aug.-Feb.	Aug.-Jan.	Sept.-Jan.	H	50	60-70
Cabbage	Copenhagen Market, Resistant Detroit, Badger Market, Glory of Enkhuizen, Red Acre, Savoy Chieftain	65 plts. (¼ oz.)	24-36	14-24	½	Sept.-Feb.	Sept.-Jan.	Sept.-Jan.	H	125	70-90
Carrots	Imperator,[2] Touchon,[2] Red Cored Chantenay,[2] Gold Spike	½ oz.	16-24	1-3	½	Sept.-Mar.	Oct.-Mar.	Oct.-Feb.	H	100	70-75
Cauliflower	Snowball Strains	55 plts. (¼ oz.)	24-30	20-24	½	Jan.-Feb. Aug.-Oct.	Oct.-Jan.	Oct.-Jan.	H	80	55-60
Celery	Florida Pascal, Emerson Pascal, Golden Plume, Supreme Golden	150 plts. (¼ oz.)	24-36	6-10	¼-½	Jan.-Mar.	Aug.-Feb.	Oct.-Jan.	H	150	115-125
Chinese Cabbage	Michihli	125 plts. (¼ oz.)	24-36	8-12	¼-½	Oct.-Jan.	Oct.-Jan.	Nov.-Jan.	H	100	75-85
Collards	Vates, Georgia, Florida Savoy, Louisiana Sweet, Georgia 912	75 plts. (¼ oz.)	24-30	14-18	½	Feb.-Mar. Sept.-Nov.	Jan.-Apr. Aug.-Nov.	Sept.-Jan.	H	150	50-55

SOUTHEAST

TABLE 7: Planting Guide for Florida Vegetable Garden (Continued)

Crop	Varieties [1]	Seeds/ Plants 100' of Row	Spacing in Inches		Seed Depth Inches	Planting Dates in Florida (Inclusive)			Plant Hardiness [3]	Pounds Yield 100'	Days to Harvest
			Rows	Plants		North	Central	South			
Corn, Sweet	Ioana,[2] Golden Cross Bantam,[2] Golden Security, Seneca Chief,[2] many others	¼ lb.	34-42	12-18	½	Mar.-Apr.	Feb.-Mar.	Jan.-Feb.	T	15	80-85
Cucumbers	Marketer, Palomar, Santee, Ashley, Stono	1 oz.	48-60	15-24	½-¾	Feb.-Apr.	Feb.-Mar. Sept.	Jan.-Feb.	T	100	50-55
Eggplant	Fort Myers Market, Florida Market, Florida Beauty	30 plts. (¼ oz.)	36-42	36-48	½	Feb.-Mar.	Jan.-Feb. July	Dec.-Feb. Aug.-Sept.	T	200	80-85
Endive Escarole	Deep Heart Fringed, Green Curled, Full Heart Batavian	1 oz.	18-24	8-12	¾	Feb.-Mar. Sept.	Jan.-Feb. Sept.	Sept.-Jan.	H	75	90-95
Kohlrabi	Early White Vienna	¼ oz.	24-30	3-5	½	Mar.-Apr. Oct.-Nov.	Feb.-Mar. Oct.-Nov.	Nov.-Feb.	H	100	50-55
Lettuce (Crisp) (Butterhead) (Leaf)	Premier, Great Lakes, Imperial 44, Bibb, White Boston Black Seeded Simpson, Salad Bowl	½ oz.	12-18	12-18	¾	Feb.-Mar. Sept.	Jan.-Feb. Sept.	Sept.-Jan.	H	75	50-80
Muskmelons Cantaloupes	Smith's Perfect, Hale's Best No. 36 Georgia 47, Rio Gold, PMR 45, Edisto	1 oz.	70-80	48-60	¾	Mar.-Apr.	Feb.-Apr.	Feb.-Mar.	T	150	75-90
Mustard	Southern Giant Curled, Florida Broad Leaf[2]	1 oz.	14-24	4-6	½	Jan.-Mar. Sept.-May	Jan.-Mar. Sept.-Nov.	Sept.-Mar.	H	100	40-45
Okra	Clemson Spineless,[2] White Velvet, Perkins Long Green	2 oz.	24-40	18-24	1-2	Mar.-May Aug.	Mar.-May Aug.	Feb.-Mar. Aug.-Sept.	T	70	50-55
Onions (Bulbing)	Excel, Texas Grano, Granex Creole (hot)	400 plts or sets 1 oz. seed	12-24	3-4	¾	Jan.-Mar. Aug.-Nov.	Jan.-Mar. Aug.-Nov.	Jan.-Mar. Sept.-Nov.	H	100	100-130
(Green)	Excel, Texas Grano, Granex	800 plts. or sets	12-24	1½-2	¾	Aug.-Mar.	Aug.-Mar.	Sept.-Mar.	H	100	50-75
	Shallots (Multipliers)	1½ lb. sets	18-24	6-8	¾	Aug.-Jan.	Aug.-Jan.	Sept.-Dec.	H	100	75-105

SOUTHEAST

TABLE 7: Planting Guide for Florida Vegetable Garden (Continued)

Crop	Varieties [1]	Seeds/Plants 100′ of Row	Spacing in Inches		Seed Depth Inches	Planting Dates in Florida (Inclusive)			Plant Hardi-ness [3]	Pounds Yield 100′	Days to Harvest
			Rows	Plants		North	Central	South			
Parsley	Moss Curled	1 oz.	12-20	8-12	¾	Feb.-Mar.	Dec.-Jan.	Sept.-Jan.	H	40	90-95
Peas	Little Marvel,[2] Dark Skinned Perfection, Laxton's Progress, Emerald[2]	1½ lbs.	24-36	2-3	1-2	Jan.-Feb.	Sept.-Mar.	Sept.-Feb.	H	40	50-55
Peas, Southern	Blackeye,[2] Brown Crowder, Bush Conch,[2] Dixie Lee	1½ lbs.	30-36	2-3	1-2	Mar.-May	Mar.-May	Feb.-Apr.	T	80	70-80
Pepper (Sweet) (Hot)	California Wonder, World Beater Yolo Wonder, Wonder Giant Hungarian Wax, Anaheim Chili	60 plts. (¼ oz.)	20-36	18-24	½	Feb.-Apr.	Jan.-Mar.	Jan.-Feb. Aug.-Oct.	T	50	70-80
Potatoes	Sebago, Kennebec, Katahdin, Pontiac, Red Pontiac	15 lbs.	36-42	12-15	4-8	Jan.-Feb.	Jan.	Sept.-Jan.	SH	150	80-95
Potatoes, Sweet	Gold Rush, Unit No. 1 Porto Rico, Heart-o-Gold, Georgia Red	80 plts.	48-54	18-24	. . .	Mar.-June	Feb.-June	Feb.-June	T	75	120-140
Radish	Early Scarlet Globe, Scarlet Turnip White Tipped, Cherry Belle	1 oz.	12-18	1-2	¾	Oct.-Mar.	Oct.-Mar.	Oct.-Mar.	H	40	20-25
Spinach	Virginia Savoy,[2] Bloomsdale Long Standing, Dark Green Savoy	2 oz.	14-18	3-5	¾	Oct.-Nov. Jan.-Feb.	Oct.-Nov. Jan.	Oct.-Jan.	H	40	40-45
Spinach, Summer	New Zealand	2 oz.	30-36	18-24	¾	Mar.-Apr.	Mar.-Apr.	Jan.-Apr.	T	40	55-65
Squash, Summer	Early Prolific Straightneck, Cocozelle, Zucchini, Table Queen, Patty Pan	2 oz.	42-48	42-48	½	Mar.-Apr. Aug.	Feb.-Mar. Aug.	Jan.-Mar. Sept.-Oct.	T	150	45-60
Squash, Winter	Alagold	2 oz.	90-120	48-72	2	Mar.	Feb.-Mar.	Jan.-Feb.	T	300	95-105
Strawberry	Missionary,[2] Florida 90[2]	100 plts.	36-40	10-14	. . .	Sept.-Oct.	Sept.-Oct.	Oct.-Nov.	H	50	90-110

SOUTHEAST

TABLE 7: Planting Guide for Florida Vegetable Garden (Continued)

Crop		Varieties [1]	Seeds/ Plants 100' of Row	Spacing in Inches		Seed Depth Inches	Planting Dates in Florida (Inclusive)			Plant Hardiness [3]	Pounds Yield 100'	Days to Harvest
				Rows	Plants		North	Central	South			
Tomatoes		Manalucie,[4] Homestead-24, (unstaked)	35 plts. (¼ oz.)	40-60	36-40	½	Feb.-Apr. Aug.	Feb.-Mar. Sept.	Aug.-Mar.	T	125	75-85
		Jefferson,[4] Wilt Resistant Grothen's Globe,[4] Rutgers (staked)	70 plts. (½ oz.	36-48	18-24	½	Feb.-Apr. Aug.	Feb.-Mar. Sept.	Aug.-Mar.	T	200	75-85
Turnips		Japanese Foliage (Shogoin)[2] Purple Top White Globe	1 oz.	12-20	4-6	½-¾	Jan.-Apr. Aug.-Oct.	Jan.-Mar. Sept.-Nov.	Oct.-Feb.	H	150	40-50
Water-melon	(Large)	Congo, Blacklee, Charleston Gray, Fairfax	2 oz.	90-120	60-84	2	Mar.-Apr.	Jan.-Apr.	Feb.-Mar.	T	400	80-100
	(Small)	New Hampshire Midget, Sugar Baby										

[1] Under appropriate crops the varieties included are generally suited to home canning. Certain varieties have resistance to diseases: *Snap Beans*—Contender, Wade, Seminole (mosaic, powdery mildew, several rusts), Topcrop (mosaic). *Pole Beans*—Florigreen (common and Southern bean mosaic, rust), US No. 4 (191) (certain rusts). *Cabbage*—Resistant Detroit (yellows). *Cantaloupe*—Smith's Perfect, Georgia 47 (downy mildew), Rio Gold (downy and powdery mildew). *Celery*—Emerson Pascal (early blight). *Cucumber*—Palomar, Ashley, Stono, Santee (downy mildew). *Eggplant*—Florida Market, Florida Beauty (tip-over). *Pepper*—World Beater (certain strains, leaf spot), Yolo Wonder (tobacco mosaic). *Spinach*—Virginia Savoy (mosaic). *Irish Potato*—Kennebec (late blight). *Tomato*—Manalucie (wilt, early blight, gray leaf spot, leaf mold), Homestead, Jefferson, Wilt Resistant Grothens Globe (wilt). *Watermelon*—Congo (anthracnose), Blacklee (wilt), Charleston Gray, Fairfax (wilt, anthracnose).

[2] Outstanding in freezing trials of Florida Agricultural Experiment Stations, Gainesville. Others may meet home freezing requirements.

[3] H—Hardy, can stand frost and usually some freezing (32° Fahrenheit) without injury. SH—Slightly hard, will not be injured by light frost. T—Tender, will be injured by light frost.

[4] Tomato varieties best adapted to staking.

—Prepared by the University of Florida Agricultural Extension Service

MIDWEST | Illinois

TABLE 8: Small Kitchen Garden — Intensive Culture: 30 by 25 Feet

Planting	Row No. and width	← 30 feet →		
1st	1-12″	Early peas (Snap beans late)		
	2-12″	Second early peas (Lettuce and kohlrabi late)		
	3-12″	Spinach (Spinach late)		
	4-12″	Leaf lettuce (Spinach late)	Turnips (Spinach late)	Kohlrabi (Spinach late)
	5-12″	Onion sets (Radish late)		
	6-12″	Onion seed planted with radish (Turnips late)		
2d	7-24″	Early cabbage plants		
	8-24″	Carrots planted with radish		
	9-18″	N. Z. Spinach	Beets planted with radish	
	10-30″	Tomato seed		
	11-24″	Snap beans		
3d	12-24″	Tomato plants		
	13-24″	Snap beans		
4th	14-18″	Lima beans		
	15-24″	Summer squash or peppers	Cucumbers or eggplant	
	18″	(Border strip)		

25 feet

Crops in parentheses can be planted in the indicated rows after the early crops are harvested.

—*University of Illinois Agricultural Extension Service*

MIDWEST

TABLE 9: Large Farm Garden — Field Culture: 200 by 120 Feet

Planting	Row No. and width	← 200 feet →					
1st	1-4′	Asparagus			Rhubarb		Perennial onions
	2-4′	Early potatoes					
	3-3′	Early potatoes					
	4-3′	Early potatoes					
	5-3′	Early potatoes					
	6-3′	Onion seed planted with radish					
	7-3′	Onion sets			Spinach		
	8-3′	Leaf lettuce		Early turnips		Kohlrabi	
	9-3′	Peas					
	10-3′	Peas					
	11-3′	Early cabbage seed					N. Z. Spinach
2d	12-3′	Early cabbage plants					Head lettuce plants
	13-3′	Early beets		Early carrots			Parsley
	14-3′	Parsnips planted with radish				Swiss chard	
	15-3′	Tomato seed					
3d	16-3′	Early sweet corn		Intermediate sweet corn		Late sweet corn	
	17-3′	Early sweet corn		Intermediate sweet corn		Late sweet corn	
	18-3′	Early sweet corn		Intermediate sweet corn		Late sweet corn	
	19-3′	Snap beans					
4th	20-3′	Snap beans					
	21-3′	Carrots				Beets	
	22-3′	Peppers		Bush lima beans		Bush or pole lima beans	
	23-4′	Tomato plants					
	24-6′	Tomato plants		Muskmelon			
	25-6′	Summer squash	Cucumbers				
	26-9′	Watermelon					
	27-9′	Winter squash					
	28-6′	Sweet potatoes					
Special	29-5′	Late cabbage seed					
	4′	(Border strip)					

(↕ 120 feet)

The special planting of late cabbage is for late foll, sauerkraut, or winter storage. Sow June 1 in northern Illinois, June 15 in central Illinois, and July 1 in southern Illinois.

—*University of Illinois Agricultural Extension Service*

MIDWEST

TABLE 10: Vegetable Planting Guide for Iowa

The planting dates given in the chart are for central Iowa. In southern Iowa, gardeners should plant a week earlier in the spring, and in northern Iowa, gardeners should plant a week later. Reverse this procedure for plantings made after July 1.

Kinds	Feet of row or number of plants needed per person	Plant and row spacing: Inches between plants	Plant and row spacing: Inches between rows	Depth to plant (inches)	Planting date	Seed required to plant 100 feet of row	Planting to eating stage (days)
Asparagus	8 plants	18-24	36-48	4-6	April 1	Plants	3 years
Beans, Bush	40 ft.	2-3	24	1-1½	May 1 to July 15	1 lb.	45-65
Beans, Lima	25 ft.	4-6	24	1-1½	May 15	1 lb.	50-70
Beans, Pole	20 ft.	4-6	24	1-1½	May 1 to 15	1 lb.	45-65
Beets	10 ft. (early)	2-3	12-18	1-1½	April 1 to 15	2 oz.	60-110
	20 ft. (late)	2-3	12-18	1-1½	July 10	2 oz.	60-110
Broccoli	5 plants	18-24	30-36	¼-½	April 1 or set plants April 15	¼ oz.	60-80**
Cabbage	3 plants (early)	18-24	20-28	¼-½	April 1 to 15	¼ oz.	65-70**
	8 plants (late)	18-24	20-28	¼-½	July 1	¼ oz.	80-90**
Carrots	10 ft. (early)	2-3	12-18	¼-½	April 1 to 15	½ oz.	60-100
	20 ft. (late)	2-3	12-18	¼-½	June 15 to July 1	½ oz.	60-100
Cauliflower	6 plants	18-24	24	¼-½	April 1	¼ oz.	60-65**
Celery	10 plants	6	20-24	⅛	Early, April 15	½ oz.	120-150**
					Late, July 15		100-110**
Chinese Cabbage	6 ft.	12-18	20-24	¼-½	July 15	⅓ oz.	80-100
Corn, Sweet	100 ft.	12	30-36	1-2	May 15 to July 7 (Early variety for last planting)	3 oz.	65-110
Cucumbers	4 hills	18-24	4-5 ft.	1	May 15, again in late June	½ oz.	90-130
Eggplant	2 plants	18	24-30	¼-½	May 15	¼ oz.	75-85**
Endive	3 ft.	6	12	½	April 1	½ oz.	65-85
Kale	3 ft.	4	12-18	½	April 1, again July 1	¼ oz.	60-70
Kohlrabi	7 ft.	4-8	15-24	½	April 1, again July 15	¼ oz.	60-75
Lettuce	8 ft.	4-8	12-15	¼-½	April 1 until May 15, again August 15	½ oz.	40-60
Muskmelon	3 hills	18-24	4-5 ft.	1	May 15	½ oz.	90-120

MIDWEST

TABLE 10: Vegetable Planting Guide for Iowa (Continued)

Kinds	Feet of row or number of plants needed per person	Plant and row spacing		Depth to plant (inches)	Planting date	Seed required to plant 100 feet of row	Planting to eating stage (days)
		Inches between plants	Inches between rows				
Mustard	8 ft.	4	12-18	½	April 1 to May 15, again August 15	¼ oz.	40-60
Okra	4 ft.	12	18-24	½-1	May 1	1 oz.	70-90
Onion seed	30 ft.	2-3	12-15	½	April 1 to May 1	1 oz.	100-140
Onion sets	10 ft.	2-3	12-15	1	April 15	2 lb.	45-75
Parsley	2 ft.	4	12-18	⅛	April 1 to May 1	¼ oz.	80-100
Parsnips	15 ft.	3	18-24	½-1	April 1 to 15	½ oz.	140-160
Peas	75 ft.	1-2	6* and 24	1-2	April 1 to 15	1-1½ lb.	65-90
Peppers	4 plants	18-24	24	¼	Plants May 15	¼ oz.	70-75**
Potatoes, Irish	100 ft.	12-15	28-36	3-4	April 1	5-8 lb.	120-140
Potatoes, Sweet	15 ft.	18	3-4 ft.	3-4	May 15	Plants	140-150
Pumpkin	2 hills	7-9 ft.	8-12 ft.	1	May 15	½ oz.	90-120
Pumpkin (summer squash type)	2 hills	7-9 ft.	8-12 ft.	1	May 15	½ oz.	90-120
Radish	10 ft.	1-1½	12	½	April 1 to May 15, again August 15	1 oz.	30-60
Rhubarb	3 plants	36-60	3-5 ft.	2-3	April 1	Plants	1 year
Salsify	12 ft.	2-4	12-18	½	April 1 to 15	1 oz.	140-160
Spinach	20 ft.	3	12-18	½	April 1	1 oz.	50-70
Squash	3 hills	36	3-4 ft.	1	May 15	½ oz.	60-120
Swiss Chard	6 ft.	6-8	15-18	½	April 1 to 15	1 oz.	60-75
Tomatoes	12 plants	36-48 24 if staked	3-4 ft.	¼	Seed April 20 Plants May 15	¼ oz.	150-170 70-80**
Turnips	20 ft.	4	18-24	¼-½	August 1	½ oz.	50-70
Watermelon	3 hills	7-9 ft.	8-12 ft.	1	May 15	1 oz.	90-130

* Plant double rows of peas 6 inches apart with 24 inches between these double rows.

** If you grow your own plants, here is the approximate amount of time required for the plants to grow from seed to transplanting size: cabbage, cauliflower and broccoli, 50 days; tomato, 45 days; pepper and eggplant, 75 days; celery 60 to 70 days; sweet potato, 40 to 42 days.

WEST

TABLE 11: Estimating Your Annual Vegetable Requirements in Wyoming

If you buy from the market, figure that approximately 1 lb. will serve the following number of people:		Average production from 1 foot of row	Length of row for 1 person	Amount 1 person should have in 1 year	Seed required for 1 person
Beans (snap)	4	1/3 lb.	80′-100′	35 lb.	½ lb.
Beets (diced)	4	1 1/3 lb.	35′-40′	46 lb.	1/3 ounce
Broccoli	5	¼ lb.	8′-12′	3.2 lb.	¼ packet
Cabbage (raw, shredded)	7-8	1 lb.	20′-30′	25 lb.	¼ packet
(cooked)	4-5				
Carrots (raw, shredded)	7-8	1 lb.	30′	30 lb.	¼ ounce
(cooked)	5				
Cauliflower	3	⅔ lb.	8′-10′	5.5 lb.	¼ packet
Celery (cooked)	3-4	½ lb.	5′-10′	3.5 lb.	¼ packet
Onions	4	⅔ lb.	50′	30 lb.	¼ lb. sets or ¼ oz. seeds
Parsnips	4	1 lb.	10′-15′	15 lb.	⅛ ounce
Peas (in pod)	2	1/3 lb.	60′	20 lb.	¾ pint
Potatoes	4-5	1 lb.	170′	170 lb.	10 lb.
Spinach	3-4	½ lb.	20′-25′	15 lb.	¼ ounce
Squash	2-3	1 lb.	10′	10 lb.	¼ ounce
Sweet potatoes	3-4				
Tomatoes	3	2 lb.	35′	70 lb.	½ ounce
Turnips	4	1 lb.	10′-15′	13 lb.	⅛ ounce

—University of Wyoming Agricultural Extension Service

WEST

TABLE 12: Schedule for Annual Vegetables in South Dakota

Amount of seeds or plants listed is for 100 feet of row and for row spacing. If shorter rows are desired, reduce the amounts of seed or plants proportionately. Row spacing is based on the assumption of hand cultivation; if machine cultivated, allow an extra foot between rows of most crops. Plant spacing is the distance between plants in the row after thinning or when transplants are set in the field.

CROP	Temperature Classification	Amount or No. required: Seed	Amount or No. required: tubers, plants, or sets	Depth of Planting seed, tubers or sets (Inches)	Plant Spacing	Row Spacing	Suggested Planting dates	First harvest dates
Greens group:								
Chard	Cool	2 oz.		1	6 in.	1½-2 ft.	Apr. 15	June 15
Collards	Cool	1 pkt.		½	1½ ft.	1½-2 ft.	Apr. 15 June 15	June 15 July 20
Kale	Cool	1 pkt.		½	1 ft.	1½-2 ft.	Apr. 15 July 1	June 20 Aug. 20
Mustard	Cool	1 pkt.		½	1 ft.	15-18 in.	Apr. 15 Aug. 1	June 1 Aug. 30
Spinach	Cool	1 oz.		½	3-4 in.	15-18 in.	Apr. 15 Aug. 1	May 20 Sept. 10
New Zealand spinach	Warm	1 oz.		1-1½	1½ ft.	3 ft.	May 20	July 1
Salad group:								
Celery	Cool	1 pkt.	150-200	⅛*	6-9 in.	1½-2 ft.	May 15[a]	Oct. 1
Lettuce, leaf	Cool	1 pkt.		½	4-6 in.	15-18 in.	Apr. 15 May 1 Aug. 1	June 10 June 15 Sept. 10
Lettuce, head	Cool	1 pkt.		½	1 ft.	15-18 in.	Apr. 15[a]	June 20
Parsley	Cool	1 pkt.		⅛	6 in.	15-18 in.	Apr. 15	Aug. 1

WEST

TABLE 12: Schedule for Annual Vegetables in South Dakota (Continued)

CROP	Temperature Classification	Amount or No. required: Seed	Amount or No. required: tubers, plants, or sets	Depth of Planting seed, tubers or sets (Inches)	Plant Spacing	Row Spacing	Suggested Planting dates	First harvest dates
Endive or escarole	Cool	1 pkt.		½	1 ft.	1½-2 ft.	Apr. 15 July 1	July 15 Oct. 1
Root and Tuber Group:								
Beet	Cool	2 oz.		1	2-3 in.	15-18 in.	Apr. 15 June 1 July 1	June 20 July 15 Aug. 15
Carrot	Cool	1 pkt.		½	2-3 in.	15-18 in.	Apr. 15 June 1 July 1	July 15 Aug. 10 Sept. 20
Parsnip	Cool	1 pkt.		½	2-3 in.	1½-2 ft.	May 1	Oct. 15
Radish	Cool	1 oz.		½	1 in.	15-18 in.	Apr. 15 May 15 Aug. 1	May 20 June 15 Sept. 10
Turnip	Cool	½ oz.		¼-½	2-3 in.	15-18 in.	Apr. 15 July 1	June 20 Sept. 30
Rutabaga	Cool	½ oz.		¼-½	2-3 in.	15-18 in.	Apr. 15 May 1	July 15 Oct. 10
Sweet potato	Warm		5 lbs.	**	1 ft.	3 ft.	May 25[a]	Before Frost
Irish potato	Cool		5-6 lbs.	3-4***	9-18 in.	2-2½ ft.	May 1 (early var.) May 15 (late var.)	Aug. 20 Sept. 15
Vine group:								
Cucumber	Warm	1 pkt.		½	3 ft.	6 ft.	May 15	Aug. 1

WEST

TABLE 12: Schedule for Annual Vegetables in South Dakota (Continued)

CROP	Temperature Classification	Amount or No. required: Seed	Amount or No. required: tubers, plants, or sets	Depth of Planting seed, tubers or sets (Inches)	Plant Spacing	Row Spacing	Suggested Planting dates	First harvest dates
Muskmelon	Warm	1 pkt.		1	3 ft.	6 ft.	May 15	Aug. 15
Watermelon	Warm	1 oz.		1-2	3 ft.	6 ft.	May 15	Aug. 20-30
Squash, summer bush	Warm	½ oz.		1-2	15-18 in.	4 ft.	May 15	July 20
Squash, winter vine	Warm	1 oz.		1-2	2-3 ft.	9 ft.	May 15	Sept. 1
Pumpkin	Warm	1 oz.		1-2	3-4 ft.	5-8 ft.	May 15	Sept. 10
Legume group:								
Bean, snap, bush	Warm	½ lb.		1-2	3-4 in.	2 ft.	May 15 June 15 July 1	July 15 Aug. 1 Aug. 20
Bean, snap, pole	Warm	¼ lb.		1-2	3 ft.	2 ft.	May 15	Aug. 1
Bean, lima	Warm	½ lb.		1-2	3-4 in.	2 ft.	May 15	Aug. 15
Pea	Cool	½ lb.		2	1 in.	1½-3 ft.	Apr. 15 July 1	June 20 Aug. 15
Soybean (edible)	Warm	½-1 lb.		1-2	3-4 in.	2-2½ ft.	June 1	Aug. 15
Cabbage group:								
Cabbage	Cool	1 pkt.	75-100	½*	1-1½ ft.	2-2½ ft.	Apr. 15[a] May 15[a] June 1[a]	July 1 July 15 Aug. 20
Broccoli	Cool	1 pkt.	50-75	½*	1½-2 ft.	2-2½ ft.	Apr. 15[a]	June 20

WEST
TABLE 12: Schedule for Annual Vegetables in South Dakota (Continued)

CROP	Temperature Classification	Amount or No. required: Seed	Amount or No. required: tubers, plants, or sets	Depth of Planting seed, tubers or sets (Inches)	Plant Spacing	Row Spacing	Suggested Planting dates	First harvest dates
Brussels Sprouts	Cool	1 pkt.	50	½*	1½-2 ft.	2-2½ ft.	May 15[a]	Sept. 15
Kohlrabi	Cool	1 pkt.		½*	5-6 in.	15-18 in.	Apr. 15 May 15 July 1	July 10 July 15 Sept. 10
Cauliflower	Cool	1 pkt.	50-75	½*	1½-2 ft.	2-2½ ft.	Apr. 15[a] May 15[a]	June 20 July 15
Bulb group:	Cool or warm							
Onion sets			1 lb.	1-2	2-3 in.	15-18 in.	Apr. 15 May 1	May 20 June 1
Onion seeds		1 pkt.	400	½-1*	2-3 in.	15-18 in.	Apr. 15[a]	Sept. 1
Garlic	"	1 pkt.	1 lb.	1-2	2-3 in.	15-18 in.	Apr. 15	Aug. 15
Chives	"	1 pkt.		½	Clusters	15-18 in.	Apr. 15	July 1
Leek	"	1 pkt.		½-1	2-3 in.	15-18 in.	Apr. 15	Sept. 1
Fleshy fruited group:								
Tomato	Warm	1 pkt.	30-50	½*	1½-3 ft.	2-3 ft.	May 15[a]	Aug. 1
Eggplant	Warm	1 pkt.	30-50	½*	2-3 ft.	2-2½ ft.	May 15[a]	Aug. 10
Pepper	Warm	1 pkt.	50-70	½*	1½-2 ft.	2-2½ ft.	May 15[a]	Aug. 10
Okra	Warm	2 oz.	50	1-1½*	2 ft.	3 ft.	June 1[a]	July 20
Sweet corn	Warm	2 oz.		2	1 ft.	2-3 ft.	May 15; every 10-15 days till July 1	Aug. 1 to frost

* Seed generally planted indoors or under glass and the young seedlings later transplanted to the field.
** Plants are obtained by sprouting roots indoors and later transplanting rooted sprouts or "slips" to the field.
*** Pieces of tuber containing one or more "eyes" or buds are used for planting.
[a] Transplanted to field.

NORTHWEST

TABLE 13: Suggested Planting Plan for a Family of Five in Oregon

Dates after crops approximate time of seeding or plant setting in western Oregon.

Row No.	Planting (25′ · 50′ · 75′ · 100′)	Approximate distance between rows (Inches)
1	**Asparagus** (3/15-4/15) ...... (or Early **Potatoes**) ...... **Asparagus**	48
2	**Asparagus** ...... (or Early **Potatoes**) ...... **Rhubarb** (3/15-4/1*)	48
3	**Spinach** (3/10-4/15 followed by late **Beets** (6/15-7/1)	36
4	**Lettuce** (4/1-10*) followed by late **Carrots** ...... **Lettuce** (4/1) followed by late **Carrots** (5/26-6/10)	24-30
5	**Peas** (3/10) followed by **Celery** (6/20*) ...... **Peas** (3/25) followed by **Green Broccoli** (6/25-*)	30-36
6	**Peas** (4/8) followed by late **Cabbage** (7/1-15*) ...... **Peas** (4/20) followed by late **Cabbage** (7/10*)	30-36
7	**Early Cabbage** (3/25*) followed by **Mustard** (9/1) ...... **Early Cabbage** followed by **Lettuce** (8/15-)	30-36
8	**Onions** (sets 3/15) followed by **Turnips** (8/10) ...... **Onions** (4/10*) followed by **Spinach** (8/1-)	30
9	followed by **Mustard** (9/1) ...... **Turnips** (3/20-) ...... followed by fall **Spinach** (9/1)	24-30
10	**Early Beets** (4/10) followed by **Kale** (6/25-*) ...... **Early Carrots** (4/10) followed by **Brussels Sprouts** (7/10-*)	30
11	**Lettuce** in succession (4/10-) followed by **Rutabagas** (8/15)	30
12	**Swiss Chard** (4/10) ...... **Early Cauliflower** and **Broccoli** (4/25-5/1*) followed by **Kohl-rabi** (8/15-)	30
13	**Onions** (seed 4/10-15)	30
14	**Parsnips** (4/25*) ...... **Salsify** (4/25)	30
15	**Bush Beans** (5/1) ...... **Bush Beans** (5/15)	30
16	**Bush Beans** (6/1) ...... **Bush Beans** (7/1)	30
17	**Pole Beans** (5/10) ...... **Pole Beans** (6/1-)	36
18	**Dry Beans** (5/10)	36
19	**Tomatoes** (5/10-25*)	60
20	**Tomatoes** (5/10-25*)	60
21	**Summer Squash** (5/10) ...... **Cucumbers** (5/10-25)	60-72
22	**Winter Squash** and **Pumpkin** (5/10-20)	96
23	**Winter Squash** and **Pumpkin** (5/10-20)	96
24	**Pepper** (5/20*) ...... **Eggplant** (5/20*) ...... **Muskmelon** (5/10)	96
25	**Sweet Corn** (4/25-) ...... **Sweet Corn** (5/10) ...... **Sweet Corn** (5/25) ...... **Sweet Corn** (6/15)	36-42
26	**Sweet Corn** (4/25-) ...... **Sweet Corn** (5/10) ...... **Sweet Corn** (5/25) ...... **Sweet Corn** (6/15)	36-42
27	**Sweet Corn** (4/25-) ...... **Sweet Corn** (5/10) ...... **Sweet Corn** (5/25) ...... **Sweet Corn** (6/15)	36-42
28	**Sweet Corn** (4/25-) ...... **Sweet Corn** (5/10) ...... **Sweet Corn** (5/25) ...... **Sweet Corn** (6/15)	36-42
29	**Sweet Corn** (4/25-) ...... **Sweet Corn** (5/10) ...... **Sweet Corn** (5/25) ...... **Sweet Corn** (6/15)	36-42
30-34	5 rows **Potatoes**	36-42

* Date for setting out plants. Double cropping is suggested only for irrigated gardens.

—*Oregon State College*

NORTHWEST

TABLE 14: Planting Guide for a Family of Five in Washington

Crop	Variety[1]	Planting Dates[2] & Growing Period[2] in Days						Feet of Row[3]	Amount of Seed or Plants Required for Space Stated	Depth of Plantings (in.)	Distance Rows Apart (in.[4])	Distance Between Plants (in.[4])
		Western Washington		Irrigated Central Washington		Eastern Washington						
		Planting Dates	Growing Period	Planting Dates	Growing Period	Planting Dates	Growing Period					
Asparagus	Mary Washington	March	2 yrs.	March	2 yrs.	April	2 yrs.	60	40 plts.	8-10	30-36	18
Beans, pole	Blue Lake FM1	May	80-95	May	65-80			100	¾ lb.	1-1½	36	6-8
	Blue Lake 228	May	85-100	May	70-85							
	Kentucky Wonder	May	85-90	May	65-90							
	Columbia (CTR)			May	65-90							
	Oregon Giant	May	85-90									
Beans, bush* (green)	Stringless Green Pod			May, July	50-60	June	60-65	150	1½ lb.	1-1½	24	2-4
	Idaho Bountiful (CTR)			May	50-60							
	Tendergreen	May-June	75	May, July	50-60	June	60-65					
	Processor	May-June	75									
	Top Crop	May-June	75			June	75					
Beans, bush* (wax)	Golden Gem (CTR)			May, July	50-65			150	1½ lb.	1-1½	24	2-4
	Pure Gold Wax	May-June	75			June	75					
	Pencil Pod Blk Wax	May-June		May, July	50-65	June	60-65					
Beans, lima (bush)	Fordhook			June	100-120			100	½ lb.	1-1½	24	4-6
	Henderson			June	90-100							
Beets*	Early Wonder	March	60-70	Mar.-July	60-70			100	1 oz.	1	24	2-4
	Detroit Dark Red	Mar.-Aug.	70-80	Mar.-July	65-70	April	65-70					
Broccoli (Green sprouting)	Medium Strains	June	65				85-90	50	25 plts.		36	18-24
(Cauliflower-type)	Early Purple Head			July	115							
Cabbage	Golden Acre	January	180	March	105-135	March	65-70	150	30 plts.	½	30	18-24
	Danish Ballhead	March	240	July	150-170	June	105-120		50 plts.			

NORTHWEST

TABLE 14: Planting Guide for a Family of Five in Washington (Continued)

Crop	Variety[1]	Planting Dates[2] & Growing Period[2] in Days						Feet of Row[3]	Amount of Seed or Plants Required for Space Stated	Depth of Plantings (in.)	Distance Rows Apart (in.[4])	Distance Between Plants (in.[4])
		Western Washington		Irrigated Central Washington		Eastern Washington						
		Planting Dates	Growing Period	Planting Dates	Growing Period	Planting Dates	Growing Period					
Carrots*	Nantes	Apr.-July	75-80					100	½ oz.	½	18	2-4
	Imperator	Apr.-July	75-80	Mar.-Aug.	75	Mar.-Aug.	70-75					
	Chantenay Red-Cored	Apr.-July	75-80	Mar.-Aug.	70-90	Apr.-Aug.	70-75					
	Royal Chantenay	Apr.-July	75-80	Mar.-Aug.	70-90	Apr.-Aug.	70-75					
Cauliflower	Early Snowball	February	110-115	July	115	May	105-115	50	35 plts.	½	24	18
Celery	Utah (Green)	February	135-165					25	40 plts.	¼	24	6-8
Chard*	Lucullus	March	60-76	Apr.-Aug.	60-70	April	65-70	25	½ oz.	½	24	6
Cucumbers*	MR-17 Pickling	May	60	May	60-75	May	60-75	25	¼ oz.	½-1	36	12
(Pickling)	National Pickling	May	60	May	60-75	May	60-75					
(Slicing)	Straight Eight	May	75	May	75	May	60	25	¼ oz.	½-1	36	12
Kohlrabi	White Vienna	Mar.-Aug.	50-60	Mar.-July	55-75	Apr.-July	55-70		¼ oz.	½	18	4
Lettuce,	Progress	Feb.-July	70-90					50	¼ oz.	½	18	10-15
head*	Great Lakes	Feb.-July	80	Mar.-Aug.	85	March	75					
Lettuce,	Grand Rapids	March	50-55	Mar.-July	70	Mar.-July	70	50	⅛ oz.	½	18	3-6
leaf*	Salad Bowl	March	50-55	Mar.-July	70	Mar.-July	70					
Muskmelon	Cranshaw			April	100-110	Mid-May	100-110	50	½ oz.	1	36	18-24
	Hale's Best 36			April	85-110	Mid-May	90					
	Heat of Gold			April	100-115	Mid-May	120					
Onion (dry)*	Sweet Spanish	March	165-180	March	165-180			100	1 oz.	1	14	2-4
	Yellow Glb. Danvers	March	130-140	March	130-140	March	130-140					
	Southport Yel. Glb.					March	140-150					
	French (Walla Walla)			Mar. trnsp.	120							

NORTHWEST

TABLE 14: Planting Guide for a Family of Five in Washington (Continued)

Crop	Variety[1]	Planting Dates[2] & Growing Period[2] in Days						Feet of Row[3]	Amount of Seed or Plants Required for Space Stated	Depth of Plantings (in.)	Distance Rows Apart (in.[4])	Distance Between Plants (in.[4])
		Western Washington		Irrigated Central Washington		Eastern Washington						
		Planting Dates	Growing Period	Planting Dates	Growing Period	Planting Dates	Growing Period					
Parsnips	Hollow Crown	May	120	April	100-130	April	130-150	50	¼ oz.	½	24	3
Peas, tall	Thomas Laxton	March	100-120	March	75-90	April	80	150	1½ lbs.	1-1½	30	2
	Tall Alderman	March	100-120									
Peas, dwarf*	Blue Bantam			March	100-120	April	80	200	2 lbs.	1-1½	30	2
	Little Marvel			March	70-80	April	70					
Peppers, sweet	California Wonder	May	180-200	May	175-190	May	170-190	25	15 plts.	½	24	18
	Oakview Wonder	May	180-200	May	160-175	May	170-190					
Potatoes*	Irish Cobbler (early)	Mar.-Apr.	90-110	April	90-110	April	90-120	150	¼ bu.	4	30	10-14
	Bliss Triumph (early)	Mar.-Apr.	90-110	April	90-110	April	90-120					
	White Rose (early)	Mar.-Apr.	90-110	April	90-110	April	90-120					
	Red Warba (early)	Mar.-Apr.	90-110	April	90-110	April	90-120					
	Netted Gem (late)	June	95-125	June	120-150	May	120-160	600	1 bu.	4	30	
Pumpkin	Small Sugar	May	135	May	110-155	May	120	50	½ oz.	1	60	48
	Kentucky Field (CTR)			May	110-155							
Radishes	Early Scarlet Globe	Mar-Sep.	25-35	Mar.-Sep	5-35	April	30	50	½ oz.	½	12	1
	Cherry Belle	Mar-Sep.	25-35	Mar.-Sep	5-35	April	30					
	Comet	Mar-Sep.	25-35	Mar.-Sep.	5-35	April	30					
	White Icicle	Mar-Sep.	25-35	Mar.-Sep.	25-35	April	35-40					
Rhubarb	Victoria, Wine	March	1 yr.	March	1 yr.	April	1 yr.		5 crns.		36-40	24-36
Spinach*	Giant Thick Leaved (Nobel)	March	50-55	Mar.-Aug	5			100	1 oz.	½	18-24	3-4
	Improved Thick Leaved (Viroflay)			Mar.-Aug.	5-60	April	55-60					
Squash (summer)	Cocozelle, Zuccini	May	60-70	April	0	May	60	25	⅛ oz.	1	36	30
	Straightneck	May	65-75	April	0-65	May	75					

TABLE 14: Planting Guide for a Family of Five in Washington (Continued)

Crop	Variety[1]	Planting Dates[2] & Growing Period[2] in Days						Feet of Row[3]	Amount of Seed or Plants Required for Space Stated	Depth of Plantings (in.)	Distance Rows Apart (in.[4])	Distance Between Plants (in.[4])
		Western Washington		Irrigated Central Washington		Eastern Washington						
		Planting Dates	Growing Period	Planting Dates	Growing Period	Planting Dates	Growing Period					
Squash* (winter)	Hubbard	May	135			May	120-130	50	½ oz.	1-1½	60	48
	Sweetmeat	May	135			May	120-130					
	Table Queen, Acorn	May	135			May	120-130					
	Marblehead (CTR)			April	120-150	May	130					
	Butternut (CTR)	May	135	April	120-150	May	130					
Sweet Corn* (Hybrids)	Improved Spancross	May	105-115	May	70-80	May	75-80	350	½ lb.	1	30	12-24
	Golden Beauty	May	110-120	May	70-80	May	75-80					
	Golden Earlipack	May	115-125	May	75-85	May	80-85					
	Carmelcross	May	120-130	May	85-95	May	90-100					
	FM Cross	May	125-135	May	90-100	May	85-105					
	Golden Cross	May	130-140	May	95-105	May	100-110					
Tomatoes*	Chatham	March	115-120					150	50 plts.	½	36	24-36
	Bounty	March	125-130	April	115	May	125-135					
	Bonny Best			April	125	May	140-150					
	Stokesdale			April	125	May	140-150					
	Sioux			April	125							
Turnip	Purple Top	Mar.-Aug.	50-60	Mar.-July	55-75	Apr.-July	55-70	50	¼ oz.	½	18	3
Watermelon	Early Kansas					Mid-May	90-100	50	½ oz.	½	60	48
	Striped Klondike			April	90-125	Mid-May	100-110					
	New Hampshire Midget			April	80-100	Mid-May	90-100					

* Recommended for dry land areas. (CTR) Curly Top (disease) Resistant

[1] Gardeners may have other choices as to varieties, based on their own experiences.

[2] Planting dates and growing season will vary with altitude, etc. Check with your county agent if you are not sure.

[3] Feet of row for a family of five, including canning, freezing, storage. Adjust to family's likes and dislikes.

[4] Plant spacing should be adjusted to soil moisture and fertility. As much as double these spacings may be needed in driest areas. Adjust row spacing and use field cultivators where possible.

—University of Washington

TABLE 15: Planning Guide for California

Climate may vary even in small sections of the state. Since the areas shown here are large, planting dates are only approximate.

Vegetable	PLANTING DATES FOR SECTIONS OF CALIFORNIA: N. Coast: Monterey Co. north	S. Coast: San Luis Obispo Co., south	Interior valleys: Sacramento, San Joaquin, and similar valleys	Imperial and Coachella valleys	W = warm-season crop C = cool-season crop	Moderate planting for family of four	Distance apart in row	Distance between rows without beds	Recommended storage temperatures, degrees F.	Storage period (No. of weeks)
Artichoke[3]	Jan.-May	Jan.-May	......	......	C	3-4 plants	48″	60″	32	1-2
Asparagus[3]	Jan.-Mar.	Jan.-Feb.	Jan.-Feb.	Feb.-Apr.	C	30-40 plants	12″	60″ green 72″ white	32	3-4
Beans (lima)[1]	May-June	Apr.-May	May-June	......	W	15-25′ row	6″ bush; 24″ pole	30″	40	1-3
Beans (snap)[1, 2]	July May-June	Mar.-Aug.	Apr.-May, July	Jan.-Mar. Aug.	W	15-25′ row	3″ bush; 24″ pole	30″[4]	40	1-2
Beets[1]	Feb.-Aug.	Feb.-Aug.	Feb.-Aug.	Sept.-Jan.	C	10-15′ row	2″	24″[4]	32	3-10
Broccoli[1, 3]	June-July	June-July	July	Sept.	C	15-20′ row	24″	36″	32	1-2
Brussels Sprouts[3]	June	June-July	......	......	C	15-20′ row	24″	36″	32	3-4
Cabbage[1, 3]	Jan.-Apr. July-Sept.	Oct.-Feb.	July, Jan.-Feb.	Sept.-Nov.	C	10-15 plants	24″	36″	32	12-16
Cabbage (Chinese)[1]	July-Aug.	Aug.-Sept.	Aug.	Aug.-Nov.	C	10-15′ row	6″	30″[4]	32	2-3
Cantaloupes and similar melons	May	Apr.-May	Apr.-May	Jan.-Apr.	W	5-10 hills	48″	72″	40	2-4
Carrots[1, 2]	Jan.-Aug.	Jan.-Aug.	July-Aug., Feb.	Sept.-Dec.	C	20-30′ row	2″	24″[4]	32	16-20
Cauliflower[3]	June, Jan.	July-Nov.	July-Aug.	Sept.-Oct.	C	10-15 plants	24″	36″	32	2-3
Celeriac	Mar.-June	Mar.-Aug.	June-Aug.	......	C	10-15′ row	4″	24″[4]	32	8-16
Celery[1, 2]	Mar.-June	Mar.-Aug.	June-Aug.	......	C	20-30′ row	5″	24″[4]	32	8-16
Chard[1]	Feb.-May	Nov.-May	Feb.-May	Sept.-Oct.	C	3-4 plants	12″	30″	32	1-2
Chayote	......	Apr.-May	May-June	......	W	1-2 plants	72″	grow along fence	....	....

TABLE 15: Planning Guide for California (Continued)

Climate may vary even in small sections of the state. Since the areas shown here are large, planting dates are only approximate.

Vegetable	PLANTING DATES FOR SECTIONS OF CALIFORNIA: N. Coast: Monterey Co. north	S. Coast: San Luis Obispo Co., south	Interior valleys: Sacramento, San Joaquin, and similar valleys	Imperial and Coachella valleys	W = warm-season crop C = cool-season crop	Moderate planting for family of four	Distance apart in row	Distance between rows without beds	Recom-mended storage temper-atures, degrees F.	Storage period (No. of weeks)
Chives[1]	April	Jan.-Feb.	Feb.-Mar.	Nov.-Jan.	C	1 clump	needs 4 sq. ft.	....	....	....
Corn (sweet)[2]	Apr.-July	Feb.-July	Mar.-July	Jan.-Mar., Aug.	W	20-30′ in 4 rows	15″	36″	32	½-1
Cucumbers	Apr.-June	Apr.-June	Apr.-June	Feb.-May, Aug.	W	6 plants	24″	48″	50	1-2
Eggplant[3]	May	April	Apr.-May	Feb.-Aug.	W	4-6 plants	24″	36″	50	1-2
Endive[1]	Mar.-July	Mar.-Aug.	Aug.	Sept.-Nov.	C	10-15′ row	10″	24″[4]	32	2-3
Florence Fennel	Mar.-July	Mar.-Aug.	Aug.	Sept.-Nov.	C	10-15′ row	4″	30″[4]	32	2-3
Garlic	Nov.-Dec.	Nov.-Jan.	Nov.-Jan.	......	C	10-20′ row	3″	18″[4]	32	24-32
Kohlrabi	July-Aug.	Jan., Aug.	Aug.	Nov.	C	10-15′ row	3″	24″	32	2-4
Leek	Feb.-Apr.	Jan.-Apr.	Jan.-Apr.	......	C	10′ row	2″	24″	32	4-12
Lettuce[1]	Feb.-Aug.	Dec.-Aug.	Aug., Nov.-Feb.	Sept.-Dec.	C	10-15′ row	head 12″; leaf 6″	24″	32	2-3
Mustard	July-Aug.	Aug.-Feb.	Aug.	Nov.	C	10′ row	8″	24″[4]	32	1-2
Okra	May	April	May	Mar.	W	10-20′ row	18″	36″	....	....
Onions	Jan.-Mar.	Nov.-Feb.	Nov.-Feb.	Nov.-Jan.	C	30-40′ row	3″	24″[4]	32	12-32
Parsley[1]	Dec.-May	Dec.-May	Dec.-May	Sept.-Oct.	C	1 or 2 plants	8″	24″	....	....
Parsnips	May-June	June-July	June-July	Oct.	C	10-15′ row	3″	24″[4]	32	8-16
Peas[1]	Jan.-Aug.	Aug., Dec.-Mar.	Nov.-Jan.	Aug.-Nov.	C	30-40′ row	2″	36″ bush 48″ vine	32	1-2
Peppers[1, 3]	May	Apr.-May	May	Mar.	W	5-10 plants	24″	36″	45	4-6
Potatoes (sweet)[3]	May	Apr.-May	May	Mar.-May	W	50-100′ row	12″	36″	55	8-24

TABLE 15: Planning Guide for California (Continued)

Climate may vary even in small sections of the state. Since the areas shown here are large, planting dates are only approximate.

Vegetable	PLANTING DATES FOR SECTIONS OF CALIFORNIA: N. Coast: Monterey Co. north	S. Coast: San Luis Obispo Co., south	Interior valleys: Sacramento, San Joaquin, and similar valleys	Imperial and Coachella valleys	W = warm-season crop C = cool-season crop	Moderate planting for family of four	Distance apart in row	Distance between rows without beds	Recommended storage temperatures, degrees F.	Storage period (No. of weeks)
Potatoes (white)	early: Feb. late: Apr.-May	early: June-Feb. late: Mar.-Aug.	early: Feb.-Mar. late: Aug.	......	C	50-100′ row	12″	30″	40-50	12-20
Pumpkins	May	April	Apr.-June	Mar.	W	1-3 plants	48″	72″	55	8-24
Radish[1,2]	all year	all year	Sept.-Mar.	Oct.-Feb.	C	4′ row	1″	18″[4]	32	....
Rhubarb	Dec.-Jan.	Dec.-Jan.	Jan.-Feb.	......	C	2-3 plants	36″	48″	32	2-3
Rutabagas	July	July, Mar.	July, Aug.	Oct.-Jan.	C	10-15′ row	3″	24″[4]	32	8-16
Spinach[1]	Aug.-Feb.	Sept.-Jan.	Sept.-Jan.	Sept.-Nov.	C	10-20′ row	3″	18″[4]	32	1-2
Squash (summer)	May	Apr.-June	Apr.-June	Feb.-Mar., Aug.	W	2-4 plants	24″	48″	50	2-3
Squash (winter)	May	Apr.-June	Apr.-June	......	W	2-4 plants	48″	72″	55	8-24
Tomatoes[1,3]	May	Apr.-Aug. 15	Apr.-May	Feb.-Mar., Aug.	W	10-20 plants	....	....	50	1-2
Turnips[1]	Jan., Aug.	Aug., Apr.	Aug., Feb.	Oct.-Feb.	C	10-15′ row	2″	24″[4]	32	8-12
Watermelons	May-June	Apr.-May	Apr.-May	Feb.-Mar.	W	6 plants	60″	72″	40	2-3

1. Crops suggested for a small garden.
2. Crops which should be planted more than once in a suitable climate.
3. Transplants used for field planting.
4. If grown on beds, plant two rows per bed, with beds about 36-40 inches apart, and tops of bed 10-16 inches wide.

—*University of California*

SOUTHWEST

TABLE 16: Varieties and Planting Guide for Texas

Name of vegetable	Variety to plant	Seeds or plants per 100 feet of row	Rows apart (inches)	Plants apart in rows	Depth of planting	No. days ready for use	Hardiness
Asparagus	Mary Washington	60-80	24-36	15-20	8-10	2-3 years	Very hardy
Beans, snap, bush	Top Crop Tendergreen Pearl Green Seminole (fall)	1 pint	24-36	2-4	1-2	48-55	Tender
Beans, snap, pole	Bluelake Kentucky Wonder	½ pint	24-36	4-6	1-2	55-65	Tender
Beans, lima, bush	Fordhook 242	½ pint	24-36	3-4	1-2	75	Tender
Beans, lima, pole	Florida Speckled	½ pint	24-36	4-6	1-2	80	Tender
Beets	Detroit Dark Red	1 oz.	24-36	2-3	1	60-65	Hardy
Broccoli	Early Green Sprouting Texas 107	66	30-36	18	2-4	60-80	Hardy
Cabbage	Marion Market Early Round Dutch Golden Acre (early)	75-100	30-36	12-18	2-3	70-75	Hardy
Carrots	Red Core Chantenay Danver's Half-Long	1 oz.	24-36	2-3	½-¾	70-75	Hardy
Chard, swiss	Lucullus	1 oz.	24-36	6-8	½-¾	50	Hardy
Collard	Louisiana Sweet Georgia	¼ oz.	30-36	12-18	½-¾	75	Hardy

SOUTHWEST

TABLE 16: Varieties and Planting Guide for Texas (Continued)

Name of vegetable	Variety to plant		Seeds or plants per 100 feet of row	Rows apart (inches)	Plants apart in rows	Depth of planting	No. days ready for use	Hardiness
Corn, sweet	Aristogold Bantam, Evergreen Calumet 57 Ioana		¼ pint	30-36	18-24	1½-2	80-90	Tender
Cucumbers, slicing	Stono (disease resistant) A & C		1 oz.	48	48-72	1	65-70	Tender
Cucumbers, pickling	Ohio MR 17 Earliest of All		1 oz.	48	48-72	1	55-60	Tender
Eggplant	Black Beauty Florida High Bush		⅛ oz.	30-36	18-24	½	80-85	Tender
Kale	Scotch Curled Siberian		¼ oz.	24-36	6-8	½-¾	55	Hardy
Lettuce, head	Great Lakes		½ oz.	24-36	4-6	½-¾	80	Hardy
Lettuce, leaf	Grand Rapids Salad Bowl	1,2,3 1,2,3	½ oz.	24-36	4-6	½-¾	40-45	Hardy
Muskmelon (cantaloupe)	Smith's Perfect Rio Gold		1 oz.	60-96	24-36	1-2	85-95	Tender
Mustard	Tendergreen (early) Florida Broadleaf Giant Southern Curled	1,2,3 1,2,3 1,2,3	2 oz.	24-36	1	½-¾	30-40	Hardy
Okra	Louisiana Green Velvet Clemson Spineless	1,2,3 1,2,3	1 oz.	30-36	24-36	1	55-60	Tender

SOUTHWEST

TABLE 16: Varieties and Planting Guide for Texas (Continued)

Name of vegetable	Variety to plant	Seeds or plants per 100 feet of row	Rows apart (inches)	Plants apart in rows	Depth of planting	No. days ready for use	Hardiness
Onions	Excel Crystal White Granex	400	24-36	3-4	1½-2½	80-95	Hardy
Parsley	Moss Curled	⅛ oz.	24-36	2-3	½-¾	70	Hardy
Peas, Southern	Extra Early Blackeye Purple Hull 49 Cream 40 Cream 12	1 lb.	24-36	6-8	2-3	60-70	Tender
Peas, English	Laxton's Progress Little Marvel	2 lb.	24-36	3-4	3-4	60-65	Hardy
Pepper	Yolo Wonder Early Calwonder	50-75	36	12-24	2-3	70-80	Tender
Potato, Irish	Bliss Triumph Red Pontiac La Soda Sebago	5-8 lb.	30-36	12-15	3-5	75-100	Half hardy
Potato, sweet	Allgold Texas Porto Rico Red Velvet	50-75	36-42	12-18	3-5	120-150	Half hardy
Radish	Scarlet Globe White Icicle	1 oz.	24-36	½-1	½-¾	20-30	Hardy

SOUTHWEST

TABLE 16: Varieties and Planting Guide for Texas (Continued)

Name of vegetable	Variety to plant	Seeds or plants per 100 feet of row	Rows apart (inches)	Plants apart in rows	Depth of planting	No. days ready for use	Hardiness
Spinach	Bloomsdale Savoy	1 oz.	24-36	3-5	½-¾	40	Hardy
Squash, summer	Early Prolific Straightneck White Bush Scallop	½ oz.	36-48	36	1-2	50	Tender
Squash, winter	Acorn	½ oz.	60-72	36	1-2	85-90	Tender
Tomatoes	Texto 2 Pritchard Firesteel Porter (summer) Summer Prolific (summer)	35-50	48	30-36	2-4	60-85	Very tender
Turnips	Purple Top Shogoin (Greens)	½ oz.	24-36	3-5	½-¾	35-55	Hardy
Watermelon	Charleston Gray Pride O'Texas Tendersweet Sugar Baby	½ oz.	84-96	36-48	1-2	85-95	Tender

—Texas Experiment Station

TABLE 17: Planting Guide for Alaska

PLANTING DISTANCE AND EXPECTED YIELD FOR A 100-FOOT ROW

Vegetable	Seed needed	Row space	Plant space	Seed depth	Days†	Usual yield
		(Inches)				
Beans, bush	1 lb.	24	18	½	85	75 lbs.
Beets	1 oz.	12	2	½	60	2 bus.
*Broccoli	1 pkt.	24	18	½	65	75 lbs.
*Brussells sprouts	1 pkt.	18	6	½	80	100 qts.
*Cabbage	1 pkt.	18	6	½	95	65 hds.
Carrots	½ oz.	18	6	1	125	150 lbs.
*Cauliflower	1 pkt.	18	4	½	75	2 bus.
*Celery	1 pkt.	24	18	½	65	65 hds.
Chard	½ oz.	24	18	½	70	200 lbs.
Collard	¼ oz.	18	14	½	110	200 lbs.
**Cucumber	½ oz.	24	24	¾	75	200 lbs.
**Eggplant	1 oz.	24	24	¾	80	100 lbs.
Kale	1 pkt.	18	10	½	48	90 hds.
Endive	1 pkt.	18	15	¼	55	25 bus.
Kohlrabi	½ oz.	16	6	¼	58	5 bus.
*Lettuce, head	½ oz.	18	12	¼	65	80 hds.
Lettuce leaf	1 pkt.	16	6	¼	45	10 bus.
Mustard greens	1 pkt.	12	6	¼	45	10 bus.
Onion seed	½ oz.	12	1	½	65	100 lbs.
Onion sets	2 lbs.	12	3	2	90	100 lbs.
Parsley	½ oz.	16	4	¼	75	3 bus.
Parsnips	½ oz.	16	3	½	100	100 bus.
Peas	1 lb.	24	1	1	70	50 lbs.
**Peppers	1 oz.	24	24	½	65	100 lbs.
Potatoes	15 lbs.	36	12	3	100	150 lbs.
Radish	1 oz.	12	1	½	30	70 lbs.
Rutabagas	½ oz.	18	8	½	88	400 lbs.
Spinach	1 oz.	24	18	½	60	80 lbs.
Squash	½ oz.	48	48	1	55	100 lbs.
**Tomatoes	1 oz.	24	24	½	65	300 lbs.
Turnips	¼ oz.	16	4	½	50	300 lbs.

* These may be started indoors for later transplanting, or they may be seeded directly into the garden.

** These are usually grown in greenhouses.

† Average number of days for maturity in favorable sites.

—*Alaska Experiment Station*

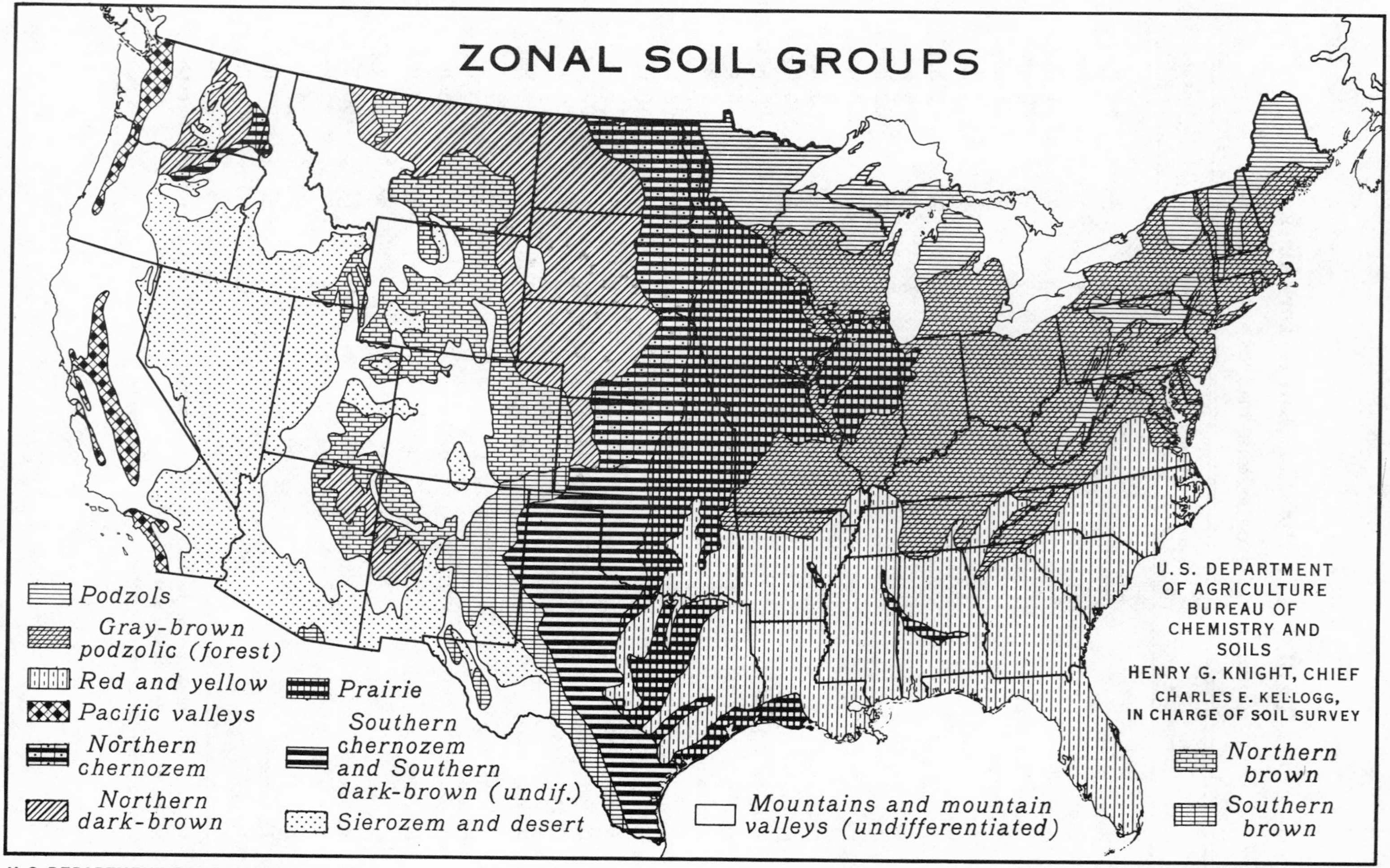

ZONAL SOIL GROUPS
U. S. DEPARTMENT
OF AGRICULTURE
BUREAU OF
CHEMISTRY AND
SOILS
HENRY G. KNIGHT, CHIEF
CHARLES E. KELLOGG,
IN CHARGE OF SOIL SURVEY
Podzols
Gray-brown podzolic (forest)
Red and yellow
Pacific valleys
Northern chernozem
Northern dark-brown
Prairie
Southern chernozem and Southern dark-brown (undif.)
Sierozem and desert
Mountains and mountain valleys (undifferentiated)
Northern brown
Southern brown
U. S. DEPARTMENT OF AGRICULTURE
NEG. 31395
BUREAU OF AGRICULTURAL ECONOMICS

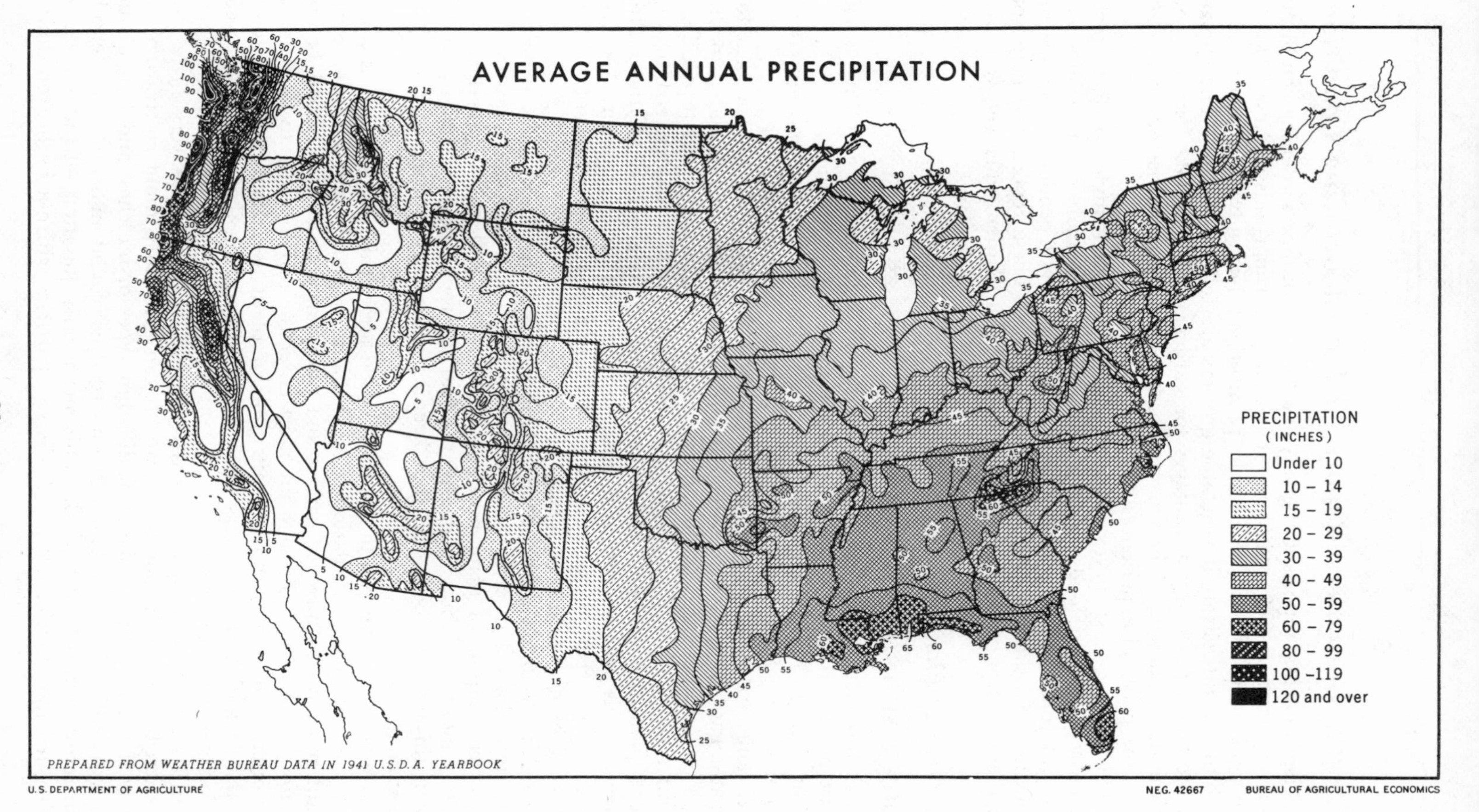
AVERAGE ANNUAL PRECIPITATION
PRECIPITATION
(INCHES)
Under 10
10 – 14
15 – 19
20 – 29
30 – 39
40 – 49
50 – 59
60 – 79
80 – 99
100 –119
120 and over
PREPARED FROM WEATHER BUREAU DATA IN 1941 U.S.D.A. YEARBOOK
U.S. DEPARTMENT OF AGRICULTURE
NEG. 42667
BUREAU OF AGRICULTURAL ECONOMICS

Organic Gardeners Report on Varied Growing Conditions

The following reports will give you some idea of how gardening conditions can vary across the United States. For example, in portions of New England, the sandy loam soil can be a big help; in others the sandy soil structure must be built up intensively if vegetable growers are to have success. The climate in the South has many advantages for gardeners (if you take care of some of the disadvantages). The barren Northwest need not be barren, as evident from the personal experiences of Helen Hoover.

This coast-to-coast survey shows that (1) there is no single "perfect" place to grow your own foods—you can be successful no matter where you live, and (2) your organic gardening methods can vary dependent upon specific growing conditions in your area of the country. These reports can help you in the planning and growing of your vegetable garden wherever you live.

In New England

"DON'T SHUN SANDY SOIL"

Far too many gardeners shun a sandy soil because they feel that a soil must be very dark and somewhat heavy to be highly productive.

But, here in Massachusetts, we have grown fruits, vegetables and flowers on our sandy soil for many years with better-than-average results.

A sandy base, such as is found in the New England soil, can't be beat for organic gardening. When a steady supply of organic matter is added to this type of soil, it releases a constant supply of nitrogen that leads to excellent plant growth. The decaying organic material acts as a sponge, soaking up great quantities of water, and holding it over a long period of time. This safeguards us against drought damage in case rains fail in summer. We also have a good clay subsoil that helps retain moisture, and prevents serious leaching by heavy rains in spring.

Healthy, vigorous plants need a long root system. Roots must penetrate deep into the soil in order to provide the plant with sufficient moisture and a constant supply of plant nutrients. Because the soil always remains moderately moist, the roots of our plants find it easy to penetrate many inches down through the loose soil, thus keeping the plants growing at their best all summer long.

In more soils than you might think, good air circulation is a serious problem and adequate ventilation is important to good plant growth. Such plant nutrients as nitrogen, sulfur and phosphorus must be oxidized before plants can use them. Enough air must enter the soil in order to decompose the organic substance, and free nitrogen for plant use. Sandy soils are loose by nature and fresh air enters them readily, improving the soil tilth as well as increasing fertility.

Still another big advantage is that our sandy loam needs very little cultivation because of its porous nature. It can even do without cultivation entirely if extensive use is made of different types of mulches, and it is about the only soil I know of that can undergo year-round mulching with excellent results and very little work. This means only one thing—if the soil is not cultivated too often or not at all, the soil particles will not be broken down.

And this, in turn, means excellent soil tilth at all times which leads to soil fertility and abundant plant growth.

Because our soil drains so quickly and warms up so fast in spring, we can get a head start on our garden early in April. With a somewhat short growing season, this is a bonus in time and a most welcome one when the itch to garden overtakes us in early spring.

—Betty Brinhart

TRAINING A CAPE COD SAND PATCH

There are some spots here on Cape Cod, Massachusetts, that are as good land as one could find. But most of the Cape is sand and not sandy loam; it is plain sand.

Where my garden is now, there were patches of such pure sand that not a weed grew, and it was such good sand that we used it in mixing our cement. Where there was any growth at all, it was scattered clumps of coarse beach grass, with roots like wire, and a few wild roses and some poison ivy.

The first thing that I did was to grub out all this coarse stuff and pile it in a 10 x 12-foot pile with a liberal amount of hen-house litter scattered all through the pile. This litter is about half dry sugar cane and half poultry droppings. The pile was covered with a few inches of what was nearest to soil of anything that I could get.

In the spring I put the garden rotary tiller over and through the pile until it was well mixed with the sand under it, and set 4 kinds of strawberry plants on it. They grew.

As for the remainder of the garden, I gave it a good coating of the litter and rotary-tilled it in, then sowed rye grass. Some of the grass came up, but there were still many bare spots. More litter was spread in the spring and what grass was there was tilled in and buckwheat sown. As soon as this was one foot tall, it was tilled in, with more litter, and a second crop of buckwheat sown. That began to show some growth and when turned under in fall some of it was two feet tall. More litter was added and winter rye sown.

This spring I had to mow the rye 3 times before it was time to turn it under. Today everything in that garden is jumping, and what was pure sand begins to seem quite like sandy loam and is the easiest land to work that I ever had.

I believe that anyone who can get organic materials as compost or chicken-house litter can tame any kind of ground.

—Eben Wood

In the South

The middle south as here used refers to the two Carolinas, Tennessee, and the northern part of the states of Georgia, Alabama and Mississippi, particularly the piedmont sections of the Carolinas and Georgia and the Tennessee River Valley. It refers in lesser degree to the coastal area of the Carolinas and the mountainous areas of North Carolina and Tennessee.

The climate in this area is mostly controlled by the moisture from the Gulf of Mexico, the Bermuda high pressure area in summer and the recurrent waves of cold Canadian air in winter. By the time the waves of cold air reach this area, however, the extreme conditions are somewhat spent and the result is a period of below freezing weather of only several days' duration.

Occasionally, the temperature may fall as low at 10 degrees, but this is a rare occurrence during a normal winter. The winter average temperature is in the low forties for most of the area, and there are many days in midwinter when the temperature is in the sixties.

Almost all of the water falling on this area falls as rain, much of it during January, February and March, with peaks in early June and August. During the summer months frequent thunder showers generally maintain a high level of moisture for growing crops. Due to the erratic movement of thunderstorms, however, some areas may experience long periods of drought. From September to November, rainfall is much less plentiful, and this period can probably be called the dry season.

For most of this area, gardening begins in February for those crops which will stand a light freeze and ends in October or early November when frost appears. With such a long season, two crops can be harvested in one season if the crops are properly selected and carefully scheduled. Many crops like turnips and collards continue to grow all winter if the temperature does not fall below 25 degrees. The ground seldom freezes deeper than 3 inches and then only for a few days.

In such a wide area as described, there are many types of soil. In the coastal area and for several hundred miles inland, there is a sandy loam, the sand content in general decreasing as one gets away from the coast. In the rest of the area, east of the mountains, the soil is often referred to as red clay. It actually has a low clay content and a better description for it is a red loam. West of the mountains, there is a red loam area and in the Tennessee River Valley, the soil is a brown to black loam.

Nearly all of this area has been under cultivation for many years, without benefit of good soil management such as contour cultivation, crop rotation, and cover crops during the rainy season. As a result, much of the organic material in the soil in the hilly areas has washed away and in the sandy areas has leached out of the upper layers of the soil.

High-powered artificial fertilizer has been used very extensively on cotton and tobacco crops and to a lesser extent on other crops such as corn and small grain. Since about 1930 increasing acreage has been sown to grasses for grazing, but in such a short time, little has been accomplished in replacing the eroded topsoil. In time, however, this practice should restore the soil to normal fertility since the generated organic material is retained by the grass.

A further problem quite general in this area and equally prevalent in other parts of the country is the presence of a "hardpan" layer in the soils which contain little or no sand. This hardpan is almost always found where the soil has been frequently plowed with a moldboard or turn-plow.

To meet the hardpan problem, a change in cultivating methods is indicated. It first needs to be broken up. This can be accomplished by using a subsoiler. After loosening the soil, water will again reach the subsoil and be stored, provided runoff is prevented by keeping the upper surface of the soil porous. This can be done by using a rotary plow or by heavy mulching with organic material.

Leaves are the most plentiful organic material in most of the Midsouth; oak, maple and pine trees are the most prevalent kinds.

Because there are relatively few farms or horse stables in the Middle South, little manure is available. Poultry farms are found here in fairly large numbers and there is almost as heavy competition for the manure as there is for the eggs. The going rate for chicken manure in some of the cities is 10 dollars for a one-half-ton pickup truck load.

In recent years, the professional tree surgeons, largely employed by the electric public utilities, have used power-driven choppers to cut up small limbs and leaves for more easy disposition. Such material can usually be had for the asking. Spoiled hay or straw is hard to find in this area as these products are not harvested so much here since cattle can be grazed all winter.

Along the coast, seaweed is available after storms, but salt hay has not been observed. In the bayous of South Carolina and Georgia, water plants are found but it would be costly to harvest them.

From the foregoing, it can be seen that most organic gardeners in the Middle South are forced to rely upon leaves, pine needles and other wood products such as sawdust and chopped tree limbs for their organic materials.

When the first crop of the season, whatever it may be, has reached a height of 3 to 4 inches, the area between rows should be covered with the available organic material to a depth of two inches. By the time the crop matures, the depth will have gone down to an inch or less if leaves, pine needles or grass clippings are used. The second crop may be planted through the layer of mulch. When the second crop is 3 to 4 inches high, a second layer of organic material two inches deep should be added.

No cultivation during the growing season is needed. Weeds which grow from roots will come through the mulch and should be removed with a hoe. Weeds which come up from seeds are usually smothered by the method proposed. The same may be said of crab grass, but Bermuda and Johnson grass which grow from roots will come through even the 3 inch layer of mulch. These grasses are hard to control but winter cultivation is most helpful. If the roots are brought to the surface by the use of a rotary tiller during a warm spell in winter, an ensuing period of freezing weather will kill those roots left on the surface. Even a heavy infestation of these grasses can be brought under control by cultivating 3 to 4 times during successive winters.

Winter cultivation has further advantages. By keeping 4 to 6 inches of soil loose, the winter rain water is retained and soaks into the subsoil, softening the hardpan. Cultivation mixes the partially decomposed organic material with soil, thus enabling the soil bacteria to complete decomposition.

When spring arrives, the soil is in excellent condition for planting the first crop. As it has been kept porous during the winter, it will be dry enough for planting several weeks earlier than it would be if not cultivated. Grubs and other forms of insect life which spend the winter in the upper layers of the soil will be destroyed during freezing periods following cultivation.

In summary, use the available organic material for mulching during summer, thus eliminating cultivation when it is hot, and turn in the mulch in winter when it is cool and comfortable to work. Be a lazy gardener when the sun is pouring down and work off the fat which accumulates in winter by doing the heavy garden work. That's the way it can be done successfully in the Midsouth.

—*Roy C. Corderman*

In the Northwest

We live on the extreme northern border of Minnesota, with Canada just across Lake Superior and a dense evergreen forest surrounding us. Our little clearing is shaped like a stemmed glass with a path that leads downhill from the road forming the stem. The cleared area which is our yard forms the bowl. Our cabin stands at the rim of the glass.

The topsoil in the region is a thin layer, produced by the decomposition of evergreen needles and wood, but even that is gone in our yard, which some years ago was bulldozed down several feet to level it. The subsurface soil thus exposed was formed by the breakdown of igneous rock and is a mixture of red clay and golden sand, which turns to "ooze" at the least drizzle and bakes into a sort of cracked pottery in the sun. Even earthworms and ants seem to avoid it. Its natural crop is a sparse and spindly growth of dandelions, plantain and grass. In some sections, it is so shadowed by the giant trees that nothing even tries to sprout.

This is one of the coldest spots in the United States. We can expect killing frost as early as September 10th and as late as May 30th, with an occasional freezing wave from the Arctic in June. Although we have warm summer days, the nights tend to be chilly even in July and August.

The gardening prospect, 3 years ago, looked very dim, but fresh vegetables are almost unobtainable here because we are 45 miles from the village and much farther from the large markets. When we do get them, they are often sadly wilted from their long trip, so we couldn't give up the idea of a garden without trying.

Quick-maturing vegetables were a possibility. Hardy ones would be best, but others might be managed if we listened to all weather reports and were prepared to dash out and cover the delicate plants against out-of-season frost.

But did we have enough sunlight? By observing the movements of the shadows, we located 4 small areas that had several hours of bright sun on every clear day. These spots were at the sides of the clearing.

Briskly we dug into the bed sites. Oversized roots on undersized plants told their own story of the poorness of the soil. We piled the plants aside to start a future compost heap and concentrated on removing boulders. At the end of a week we had enough rocks to cobble a terrace—and 4 beds of light-tan dirt.

As we are many miles from farming country, having manure hauled in was impossible because of the prohibitive cost. The woods would have to supply the means of improving our soil.

The snow water in a nearby swamp was too high for us to take out black dirt without a dredge, but along the banks of a little rocky stream were patches of rich soil. In a clearing where deer had yarded during many winters, the ground was covered with a sort of natural compost, made from manure and the fallen leaves of scrub maple and birch. Sphagnum moss grew thickly at the edges of the swamp. We dug and hauled, then worked these finds into our prospective garden spots. The little beds took on a brown, crumbly look.

We were ready to plant!

The weeds came up so thickly that we could not find our infant plants and feared that they would be crowded out. Only the sprouting onions gave us hope. When true leaves formed, sturdy vegetable seedlings peered up from the miniature jungle—and some of the wild sprouts were unmistakably flowers.

Back we went to the stream banks and the deer yard and the swamp for material to make borders along our flagstone paths and little beds in pleasant corners. We weeded with painstaking care and transplanted the wild flowers which had sprung up between the rows of carrots and cabbage and chard.

As our vegetables matured, gray jays dropped from the trees to gobble cutworms, and brown toads sat patiently beside the lettuce rows, feasting on garden slugs. Bees hummed around the opening flowers. Earthworms began to wiggle out of the ground on damp evenings. Nature was busy completing the job we had begun.

Now we can depend on our garden for crisp summer greens and sturdy winter vegetables.

—Helen Hoover

In Northern California

All around us, land lovers from southern California are buying land and building homes here in the foothills of the Sierra Nevadas in northern California.

Gardeners there must be among them, perhaps such as you, and green thumbs will be agitated in the first warm, spring-like days that come. But don't let the weather fool you. Don't make the same mistakes I made.

We bought our property in November—twenty acres of timber and meadow, located in Hidden Valley, 3 miles from the city of Grass Valley. I had expected it to be pretty cold up here by that time, and I was amazed to find the days warm and sunny with very little frost at night.

A generous sprinkling of black oak covered our building site and garden spot. Oak leaf mold had been accumulating through the years and lay several inches deep on the ground around the trees we cleared away. Rain had recently fallen and the moisture in the soil accentuated its rich, black color. I could see seeds sprouting, springing through it, and growing by leaps and bounds. I could hardly wait for the first of the year to arrive. In the San Joaquin Valley where I had lived, we started thinking about gardening in January.

We arrived in Grass Valley around the middle of January, rented a house in town and started making daily trips to the ranch. While the building was going on, I began to poke around in the soil. Several people told me that planting time up here was the first of May, but I pooh-poohed the notion. I argued that it was warmer up here than it was in the valley, that there was more sun and less frost.

Thus reasoning, I allowed my green thumb to overrule what I had been told by old-timers. I would show them I said and, like an eager beaver, I went to work. I put in potatoes, garden peas and onions in February. The latter part of March I planted carrots and various other small seeds. Every seed came up and I was exultant. Every day on arriving I hurried to look at my early garden and gloat. But, after a while, it didn't grow. Though I watered it and applied more and more leaf mold, it still did not grow enough to talk about.

The first of April arrived and I was just about ready to take a sample of the soil in to the agricultural department. I reasoned that it must be lacking in some mineral, because my plants just didn't grow. Then one day a man, who had lived here and gardened for many years, came out to our ranch.

I showed him my garden, and asked, "What is wrong with this soil?"

He looked at my plants sitting there like wizened old people, doing nothing. "There ain't nothin' wrong with the soil. You planted the seed too early."

"But the days are so warm—and we've had scarcely any frost."

He kicked a clod of dirt with his toe. "No frost you could see, but it was there in the ground. The nights are long and cold, and the days are short. There ain't enough time for the ground to really warm up." He lifted his eyes from the plants to me. "Plant some more seed the first of next month, and you'll see how fast things grow, once the ground has warmed up."

For several decades oak leaf mold and other humus had been piling up on the ground we had cleared for a garden spot. The soil was loamy-soft to a considerable depth. There were so many other things to do, and I just couldn't wait for the plow. I decided to work the soil up with a hoe, make rows and plant.

Up here moisture rises at night, and every seed I planted came up pronto. I was thrilled. I thought I had found the ideal garden spot. Here was ground that did not even have to be plowed.

When the seeds were all well up I started to irrigate. While the water was running the soil looked black and rich enough to grow things overnight, but I was soon to make a discovery. I found that no matter how long I let the water run I could go back in a few hours and find it as dry as if I had never irrigated.

This went on for a couple of weeks. Then one day, puzzled and very disappointed, I picked up a hoe and began to dig between the rows. My hoe struck the claylike soil beneath the bed of loamy humus, and I was no longer puzzled. It was so impacted the water could not penetrate it. The water was simply evaporating from the light, airy soil on the surface, and the plants were receiving very little benefit from it.

If crops are to grow in it, virgin soil has to be plowed deeply, turning under the humus, and for best results harrowed until all lumps are pulverized. Around here it is allowed to lie fallow all winter. It soaks up all the rain that falls, and by planting time it has turned into a rich clay loam that holds the moisture for long periods.

Of course, I am talking about virgin soil and the first crop grown on it. After the first year, the minerals that have been taken out by growing plants will have to be put back in the form of composted fertilizers the same as any other soil.

If you have moved, or if you are going to move to northern California this summer or fall, and plan on planting a garden next spring, control that urge until around the first of May. Then you will be amazed and thrilled with the fast growth your plants make. There are one or two exceptions. Potatoes and garden peas can and should be planted around the first of April.

—Regina Hughes Jones

Five-Row-Plan for Vegetables

The Russians may have Five-Year Plans for other things, but Vermonter Roy VanSkoik prefers his Five-Row-Plan when it comes to raising a well-behaved garden. It's his method of solving cucurbit troubles in May rather than fighting an unholy mess in August.

From August on is when you hear those yells in the unplanned garden for elbow room. That's when the cucurbits—standard pumpkins, winter squashes, cucumbers and melons—are aiming their vines all over the lot. No matter how much you love a melon or squash you don't want it infiltrating everything in sight—blanketing beans and onions, strangling cauliflower and tomato, making a helicopter necessary when you want to pick sweet corn. So, if you're ambitous, you plant the ramblers off by themselves, meaning extra garden space, or else you simply do without as many of them as you'd like.

VanSkoik heads the Soil Conservation District at Poultney, Vermont, and is also a practical home gardener on the side. Instead of trouble from his garden, he wants food—as much as he can get. And here's the tip he slips a farm wife after talking fields and pastures to her husband.

When your garden's ready in spring, mark off a strip into 5 3-foot-wide rows. Into the first two, sow seed for early growth, to suit your taste. Van Skoik plants radishes, lettuce and early beets in row one, and early peas in row two. Leave the 3rd row empty for the time being. In row 4, he plants another pea row. Spinach and scallions go into row 5.

While they're starting, he's working plenty of composted manure into skipped row 3. When frost danger is ended, toward the end of May, he sows this row with hills of cucurbits. They're slow starters and he can cultivate them with the rest of the garden.

Toward the end of July, their vines start getting ambitious. But no harm's done because by now those early-crop rows on both sides are being harvested. From now on, instead of the row where they started, the cucurbits have a band over 18 feet wide along which they can sprawl to their heart's content.

—William Gilman

Choosing the Garden Site

As stated before, a location close to the house is more likely to receive the attention it needs to be successful; the crops are near at hand and can be harvested as needed. Even more important in choosing a location is a good plot of ground. A well-protected area of loam sloping slightly to the south is ideal for a vegetable garden. Plenty of sunshine is vital, and the area should be a reasonable distance away from trees and shrubs, which compete vigorously for both plant food and moisture far beyond their shade line.

Vacant lots or blocks in the city that are open to the sunshine and where good soil has not been covered too deep with excavated materials from other areas can be utilized for gardens. But it is usually hopeless to attempt to grow vegetables where tall buildings shut out the sunlight for a large part of the day, or where there are deep fills of excavated material.

In certain sections of the country, because of extreme weather conditions, the garden area should be protected. For example, in South Dakota it is important that the garden be well sheltered with some kind of windbreak, preferably on the sides of the prevailing summer winds.

The best kind of protective windbreak is that provided by trees; however, a few rows of sweet corn, corn cribbings, or shrubs are quite effective in

The ideal location for the home vegetable garden is one that is convenient to the house, can receive plenty of sunshine and is far enough away from trees, which would compete with the growing crops for needed moisture and plant nutrients.

reducing wind velocity. The garden should be located at least 50 feet from a tree windbreak to prevent competition between the vegetable and tree roots for moisture and plant food. In certain cases, it may even be desirable to have the garden partially protected by fencing. Sandy loam soils are considered ideal because they drain well and are easy to work. Very coarse sands are very subject to drought damage and it is difficult to maintain a high state of fertility. Very heavy soils are difficult to work and remain wet until late in the spring. Many town and city gardens do not have the ideal soil. However, it is possible to grow a good garden on practically any soil type, providing it is well-drained and properly managed.

When the garden is located in the best possible place it should be maintained in that location for an indefinite period. It is then possible to develop the soil to a high state of productivity by the addition of organic matter and natural mineral fertilizers. If enough land is available, the garden crops may be alternated between two plots. Soil improving crops can be grown in the alternate years for plowing under.

Soil fertility cannot be maintained where erosion is severe. Since gardens are cultivated intensively every year, there is little opportunity for protection against soil losses. If possible, avoid slopes where erosion will be a problem.

Good air drainage is a protection against frost and sometimes helps to reduce diseases.

If the soil varies in different parts of your garden, locate your vegetables accordingly. Low, moist soil is best for salad plants, such as lettuce and endive. Tomatoes, celery, and beans likewise do well in moist soil. However, the soil must not be wet. Potatoes, sweet corn, and cabbage do well on comparatively dry soils.

Size and Arrangement of Garden

The size of the garden will depend on the size of the family and the fertility of the soil. The total length of row necessary to grow enough vegetables for the average person for a year is about 1,200 running feet. Since there are approximately 15,000 running feet of row (3 feet apart) in an acre, a family of 5 could raise enough vegetables on a scant half-acre garden to supply their needs for a year. The person with a garden many times smaller than a half-acre can plan for an adequate supply of the particular vegetables he is interested in growing.

A century ago, vegetables were regarded as of minor importance in the American diet, reports the University of New Hampshire. With the exception of potatoes, many vegetables grown in the home garden were used for animal feed only. Tomatoes were regarded as "poisonous love apples," interesting for color alone. Carrots were considered fit only for rabbit bait. In 1909 the per capita consumption of leafy green and yellow vegetables was 77 lbs. per year, while in 1945 the per capita consumption had reached 133 lbs. At the present time, due to the many improved facilities for vegetable storage and preservation, the per capita consumption has climbed higher.

The arrangement of the garden depends upon several factors:

1. Whether it is to be cultivated by hand or a tractor.
2. The soil and the slope of the land.
3. The kinds of vegetables to be grown.

Tractor-cultivated gardens should preferably be long and narrow rather than square so as to reduce the amount of turning during cultivation. The distance between the rows is determined by the form of cultivation. If a tractor is used, plant your rows far enough apart so that your equipment can pass through between the rows even when the crops are well developed. Consider the width of your cultivator or garden tractor and the eventual size of the plants. For hand cultivation, you may usually leave less space between rows provided the particular crop allows it.

What Makes Your Vegetables Grow?

A clear understanding of the many factors that make up the "vegetable environment" is necessary if your garden plan is to be successful. These factors decide whether your vegetables will be tender and sweet with better flavor and nourishment than those on sale at the market, or whether they will be flat and tasteless. High quality should be the constant aim whenever vegetables are grown. In some respects, particularly as relates to flavor, the matter of quality is more or less a question of individual taste. However, other factors of quality such as tenderness, succulence, sweetness and the like are generally recognized by all of us. Tough, bitter lettuce does not invite a second bite. Snap beans which are tough and stringy and no longer "snap" will not fool anyone who knows the taste of well-grown, tender beans.

Of course, the particular taste of vegetables differs, depending on the variety grown, but the environmental conditions under which these varieties are grown play an important part in what the harvest product is going to taste like. High quality varieties of some vegetables do not always stand up well on the market. Unfortunately, many of the vegetables you purchase at the market are good shippers rather than high quality vegetables, but this is no excuse for poor quality vegetables from the garden. The gardener has the opportunity of having for his own consumption a much better product than does the person who must buy his supply from the market. In either case, there are two groups of factors which operate to affect the ultimate condition of the vegetables when prepared for the table.

The first is concerned with the conditions affecting the crop while it is growing. The second includes how the vegetables are kept during the interval between harvest and use. The shorter the time from garden to pot, the less chance there will be for changes to take place. All that is thus gained may, of course, be lost by poor cooking. To the gardener, the first group of factors, principally the growing conditions, is of importance. To the housewife it is the second group of factors that she must concentrate on, investigate, and improve.

In order to grow vegetables of highest quality, an understanding of the principles of plant growth is essential. Regardless of whether a plant is growing in a greenhouse, garden, cultivated field, in the sun or in the shade, it is continuously acted upon by a large group of environmental forces which influence the way it grows and what it will be like at harvest time. The environment of living plants is so complex as to defy any complete study of all the factors. However, there are certain physical forces of the environment which ordinarily exert a more or less *direct* effect upon the growth and development of plants that can be rather easily recognized. These factors are: temperature of the soil and air; sunlight; humidity; soil water; soil aeration; concentration of minerals in the soil solution; amount of organic matter present in the soil; and concentration of gases in the atmosphere.

The growth and composition of garden plants will be greatly influenced by the factors mentioned above. An analysis of plant tissue indicates two very important points, namely, the high percentage of water, especially in the leafy vegetables like celery or lettuce, and the relatively small amount of minerals. The cabbage, for example, contains about 91.5 per cent water, 7.5 per cent organic matter, and 1 per cent mineral salts. If the plant is burned, the ash contains most of the minerals which the plant has drawn from the soil. The individual amounts and kinds of minerals taken in will greatly influence the flavor of the cabbage. Besides this, the plant has absorbed water through its roots, which in turn affects the turgidity and crispness of the leaves. The organic material represents combinations of the carbon dioxide obtained from the air in conjunction with water and other materials absorbed by the roots which are combined into the production of sugars through the process of photosynthesis. This process takes place in the leaves of the plant, and any practice which reduces the leaf area, such as defoliation through disease or pruning the top growth, will consequently reduce production. Therefore, all factors affecting plant growth, center around their influence directly or indirectly on the photosynthetic process. Continual growth is dependent upon new supplies of organic compounds built up in the plant.

The organic or vegetable matter that has decomposed in the soil plays an important part in crop production. The incorporation of organic matter in the soil maintains the life of a productive soil and good healthy plants result. Tasty vegetables are never produced by weak, deficient plants, so soils high in organic content are a good assurance of palatable crops. When the land is cultivated frequently, the content of organic matter may be depleted rapidly. This is due to the stimulation of bacterial activity through better aeration. The various oxidation processes are constantly reducing the supplies of organic material in the soil, but they are being renewed by the decomposition of the bodies of soil organisms, old root systems or any top growth and added material which may be plowed into the soil.

Since the oxidation processes work continually to reduce the soil organic content, the return of organic materials to the soil should be equal to or in excess of that lost. The processes of breakdown are most rapid during the summer, resulting in a reduced organic content in the soil in the fall.

The fall season is an excellent time to follow the new organic method of sheet composting in the garden. At this time, organic materials are readily available in the form of garden plant residues, leaves, and many other materials so prevalent around the countryside in the fall. If your crops this year did not come up to your fullest expectations, remember that this fall is the best time to prepare your garden for next spring.

Photosynthesis, or the manufacture of sugars by the plant, is dependent upon the catalytic properties of the green matter (chlorophyll) in the leaves and stems in combining carbon dioxide and water in the presence of sunlight. Only a small percentage of the sunlight striking the leaf is used in the process, but the heating effect of the remainder plays an important part in speeding up the chemical reactions in the plant which are concerned with growth and maturity. Light is necessary for the formation of the green coloring matter of the leaves, as are also sugar, magnesium and nitrogen. Iron is necessary, but does not enter into the composition of the chlorophyll molecule.

Light has still another effect on the type of growth. The number of hours of daylight may determine whether the crop will develop as the gardener desires, or whether it will "bolt to seed" and pass out of an edible and salable condition. This effect of the length of sunlight each day on the growth of plants is called *photoperiodism* and is of considerable importance to the vegetable gardener. In regard to this influence of light, plants in general fall into 3 groups:

Long-Day Species—Species in this group go to seed more or less readily in a range of long-day lengths. Many such species flower and fruit in continuous illumination. At short-day lengths, these plants produce only vegetative organs.

Short-Day Species—Species in this group go to seed when the days are short in duration. Under longer photoperiods, they develop only vegetatively.

Indeterminate Species—The species falling into this category exhibit no critical photoperiod effect, most of them developing both vegetatively and reproductively over a wide range of day-lengths.

The way your vegetables grow, depending on whether they are short-day or long-day plants and whether you hope to pick the fruit or use them for the vegetative growth, can be controlled to a considerable extent by the proper placement of each vegetable in the garden in regard to sunlight or shade. Long-day salad plants should be well shaded while short-day plants of this type do better in full sunlight. Similarly, fruiting vegetables that are of the long-day species should have full sunlight, and those of the short-day species, plenty of shade.

The greater part of the plant tissue in its fresh state is composed of water. If it is possible to say that any factor is the most essential one for growth, then water is that one. Not only is water used in the process of manufacturing the carbohydrate materials of the plant, but it might be said to form the backbone of the whole structure. An adequate water supply for good plant-growth is especially important during this time of season. Ample water retention by the soil particles and a minimum loss to the atmosphere are two factors which can be largely controlled by gardeners under most conditions. To provide good moisture-retention the soil should be high in organic content and should be loose and friable. Compact soils greatly speed up water-loss through capillary action. When soils are packed, the spaces between the soil particles become very small and water rises much more rapidly to the surface and is lost through evaporation than when the soil is loose and the pore spaces in the soil are larger.

Organic matter has a very high affinity for water and holds large amounts of it available for plant use. Thus soils high in organic matter are far more moist than light, sandy soils.

Mulching the garden soil prevents water run-off during rains, so that all of the precious midsummer rain is saved for the garden vegetables. If the mulch is rather heavy, weeds will not grow in between the rows and near the garden plants. Weeds take large quantities of water from the soils which would otherwise be available for the vegetables.

Vegetables differ in their reaction to shortage of moisture. When the supply is abundant a more rapid and succulent growth will result. In general, the so-called salad vegetables will be of higher quality, earlier maturity, and better size if they are constantly supplied throughout their growth with sufficient moisture. Toughness of leaf (lettuce), petiole (celery), or swollen root (beet), will be the result of retarded moisture.

Minerals are essential for the growth of vegetables. Some minerals are needed in relatively large quantities while only a trace of others is sufficient. The supply of these minerals in the soil is dependent upon: (a) The amount added in the form of ground rock products such as raw ground limestone, phosphate rock, potash rock or glauconite, granite, marl or other rock materials. (b) The liberation of those naturally present in the soil by the action of organic acids formed by the plant roots and during the decomposition of organic matter. (c) The action of nitrifying bacteria in making the nitrogen of the air available for absorption by the plant. A considerable growth of weeds among vegetables will provide too much competition because of the use of water and nutrients, especially nitrates, by the weeds.

The chemical fertilizer people think primarily in terms of nitrogen, phosphorus, potassium and calcium, but little consideration is given the many other elements which are needed for healthy plant-growth and the resulting flavorable food products. Excesses of these primary elements result in deficiencies in other elements.

A reduced moisture supply will often have an effect similar to that of a deficient nutrient supply, especially nitrogen. It is evident that any diminution of the nitrogen will result in a reduction in the quality of salad vegetables. It is not a case of the roots searching out for nutrients or water when these are short, but they can grow better because they receive more carbohydrates. Even in a leafy vegetable, an oversupply of nitrogen will result in a very tender, succulent growth which wilts quickly when cut. Heavily nitrated celery will not keep well in storage. Thus it is evident that there must be a balance between soil nutrient supply and the manufacture of sugar by the leaves.

The acidity or alkalinity of the soil has an important bearing on plant growth. Most vegetables make their best growth on soils that are slightly acid. Beets, celery and asparagus respond to lime on very acid soils. Watermelons, sweet potatoes, radishes and peanuts may be injured by liming.

Efforts may be made to adjust the reaction of the soil as a means of preventing the growth of some specific disease organism, like that causing the club root of cabbage or the potato scab organism, or as a means of rendering certain toxic substances ineffective. When this occurs, there is danger of bringing the reaction to a point where manganese or iron is rendered unavailable. A chlorotic condition results, and with chlorophyll deficient, growth is poor.

—*William Ackerman*

Consult State Experiment Station

Each state has an experiment station which does research and publishes reports of particular interest to the gardeners and farmers in the state. For example, you can find out which plant varieties are best suited for your particular climate and soil by writing to the experiment station. Experiment stations also make soil tests either free or at a nominal charge. Here are the addresses of state colleges and experiment stations:

Alabama Agricultural Experiment Station, Auburn, Alabama.
Alaska Agricultural Experiment Station, College, Alaska
Arizona Agricultural Experiment Station, Tucson, Arizona
Arkansas Agricultural Experiment Station, Fayetteville, Arkansas
California Agricultural Experiment Station, Berkeley 4, California
California Agricultural Experiment Station, Davis, California
Colorado Agricultural Experiment Station, Fort Collins, Colorado
Connecticut Agricultural Experiment Station, New Haven 4, Connecticut
Connecticut Agricultural Experiment Station, Storrs, Connecticut
Delaware Agricultural Experiment Station, Newark, Delaware
Florida Agricultural Experiment Station, Gainesville, Florida
Georgia Agricultural Experiment Station, Experiment, Georgia
Hawaii Agricultural Experiment Station, Honolulu 14, Hawaii
Idaho Agricultural Experiment Station, Moscow, Idaho
Illinois Agricultural Experiment Station, Urbana, Illinois
Indiana Agricultural Experiment Station, Lafayette, Indiana
Iowa Agricultural Experiment Station, Ames, Iowa
Kansas Agricultural Experiment Station, Manhattan, Kansas
Kentucky Agricultural Experiment Station, Lexington 29, Kentucky
Louisiana Agricultural Experiment Station, University Station, Baton Rouge 3, Louisiana
Maine Agricultural Experiment Station, Orono, Maine
Maryland Agricultural Experiment Station, College Park, Maryland
Massachusetts Agricultural Experiment Station, Amherst, Massachusetts
Michigan Agricultural Experiment Station, East Lansing, Michigan
Minnesota Agricultural Experiment Station, University Farm, St Paul 1, Minnesota

Mississippi Agricultural Experiment Station, State College, Mississippi
Missouri Agricultural Experiment Station, Columbia, Missouri
Montana Agricultural Experiment Station, Bozeman, Montana
Nebraska Agricultural Experiment Station, Lincoln 3, Nebraska
Nevada Agricultural Experiment Station, Reno, Nevada
New Hampshire Agricultural Experiment Station, Durham, New Hampshire
New Jersey Agricultural Experiment Station, New Brunswick, New Jersey
New Mexico Agricultural Experiment Station, State College, New Mexico
New York (Cornell) Agricultural Experiment Station, Ithaca, New York
New York State Agricultural Experiment Station, Geneva, New York
North Carolina Agricultural Experiment Station, State College Station, Raleigh, North Carolina
North Dakota Agricultural Experiment Station, State College Station, Fargo, North Dakota
Ohio Agricultural Experiment Station, Wooster, Ohio
Ohio Agricultural Experiment Station, Columbus, Ohio
Oklahoma Agricultural Experiment Station, Stillwater, Oklahoma
Oregon Agricultural Experiment Station, Corvallis, Oregon
Pennsylvania Agricultural Experiment Station, University Park, Pennsylvania
Puerto Rico Agricultural Experiment Station, Rio Piedras, Puerto Rico
Rhode Island Agricultural Experiment Station, Kingston, Rhode Island
South Carolina Agricultural Experiment Station, Clemson, South Carolina
South Dakota Agricultural Experiment Station, State College Station, South Dakota
Tennessee Agricultural Experiment Station, Knoxville 16, Tennessee
Texas Agricultural Experiment Station, College Station, Texas
Utah Agricultural Experiment Station, Logan, Utah
Vermont Agricultural Experiment Station, Burlington, Vermont
Virginia Agricultural Experiment Station, Blacksburg, Virginia
Virginia Truck Experiment Station, Norfolk 1, Virginia
Washington Agricultural Experiment Station, Pullman, Washington
Washington Agricultural Experiment Station, Puyallup, Washington
West Virginia Agricultural Experiment Station, Morgantown, West Virginia
Wisconsin Agricultural Experiment Station, Madison 6, Wisconsin
Wyoming Agricultural Experiment Station, Laramie, Wyoming

2—Starting with the Soil

YOUR SOIL IS THE FIRST thing to think about when growing plants in your garden. A knowledge of the science of the soil, how to enrich and maintain its fertility, is the foundation of good gardening practices. In essence, good soil means good results.

How Soil Was Formed

In order to understand how to fertilize a soil it might be a good thing to study how that soil was formed and what it consists of. If we know

A well-drained, humus-rich soil has all of the needed microörganisms and plant nutrients needed to grow large amounts of quality vegetables. The earthworms shown above help a great deal in building up soil and maintaining adequate aeration.

what it is made up of, we can better understand what would be the effect of introducing certain substances into it. The soil consists of organic and inorganic matter. Let us see how they originated.

Organic matter is an extremely important substance and is the basis of all life. The farmer and gardener cannot do without it, for through its use he will obtain healthier plants besides a greater yield of plant matter.

Organic matter originates from matter that was once alive, although that was not the case at the beginning of the formation of the earth.

Originally the earth was one mass of rock, there being no plants of any kind or soil. The rock was the parent, the precursor of the soil as we know it today. Through the action of certain agencies part of this rock was transformed into soil. Over millions of years this rock has been "weathering," that is decomposing by the action of heat and cold air, winds, rains, fogs, movement of glaciers, and climatic changes. Professor Albrecht of the University of Missouri said some years ago: "What is soil after all? It is a temporary rest stop while the rock is on its way to the sea." That is, the rock gradually forms into soil which is washed by erosion into the bottom of the seas where over a period of millions of years, it will harden again into rock.

Rocks are porous, more so than can be noticed with the naked eye. They therefore absorb water, which, upon the alternate action of heat and cold with expansion and contraction, causes a crumbling and a breaking. Running water and the action of glaciers moving over the rocks soften and grind them. These continuing actions keep grinding, breaking down and disintegrating the rocks into finer and finer powdery masses. The action of the carbonic acid gas which entered the water from the air helped with its acidic solvent action to soften the rock.

Practically none of these rocks contained nitrogen—in fact practically no rocks today contain nitrogen, which is in the air and in the soil. But, in order for plants to grow they must have nitrogen. The rocks contained all the *minerals* needed for plant growth, but nitrogen had to come from the air at the beginning of the process of formation of soil. The air, containing 78 per cent nitrogen, had more than enough for that purpose.

It is important in this study to attempt to understand how the first organic matter might have been created. Science does not know for certain and there are many theories extant about it. It is all tied up with carbon, for carbon is the principal constituent of organic matter. It is certain however, that even before the simplest lichens and mosses could have grown there must have been bacteria or enzymes to aid them, and since the latter consist of organic matter they represent an early existence of life.

But evidently organic matter was in existence before life, for bacteria and enzymes were no doubt created out of organic matter, although it was an inferior dead type. It lacked the living qualities given to it by enzymes, bacteria and fungi. The organic matter no doubt came into existence first, obtaining its carbon from the atmosphere by means of electric and lightning discharges. Eventually enzymes appeared, a crude sort which could not regenerate themselves. Later, there were evolved self-regenerating enzymes, and then bacteria. Millions of years must have passed between each stage from enzyme to self-regenerating enzyme to bacteria.

The evolution of the development of plant life was an elaborate, long-time process, taking millions of years of time. The first types of plant that were able to grow were extremely simple ones, such as the lichens and the mosses that grew on the face of the rocks. The requirements for sustenance were not too great. In helping them to grow, certain bacteria which had already come into existence took a prominent part, feeding upon the minerals that were in the rock and nitrogen from the air. It is certain that bacteria came before the lichens and the mosses. As these lichens and mosses died, their remains were the first source of organic matter, outside, of course, of the tissues of dead bacteria. The remains of such dead lichens and mosses mixed with the minuscule rock fragments, becoming soil. Thus the soil was a mixture of rock particles and organic matter.

Process of Decay

During the process of decay of the lichens and the mosses, certain acid substances were given off such as carbonic acid and other humus acids which worked upon the rock to make more mineral food available for the future plants. The action of air and carbonic acid on the rock particles and on the organic matter turned the substances dark, which is characteristic in the formation of humus. Soon there was sufficient soil so that plants higher up in the scale of plant life could live and by that inexorable process of slow evolution in living matter, a desire upon nature to adjust and to improve, ferns came into existence and could grow. Over millions of years of slow evolution, still higher plants began to evolve such as grasses and shrubs. Finally trees began to grow.

The soil consisted primarily of weathered rock fragments, water, organic matter and dusts which fell upon it from the air. The lightning charged nitrogen into it. The rains washed nitrogen and other elements into it. Soil bacteria extracted nitrogen from the air. But basically you can visualize the soil as being made up mainly of weathered rock particles and organic matter, closely associated and mixed together. Therefore in our farming and gardening we should supply the soil with those same two ingredients, namely finely ground up rock and organic matter.

Soil requires the major elements—nitrogen, phosphorus and potash—besides the minor and the trace mineral elements. They can all be obtained through ground up rocks, except nitrogen, which is furnished through organic matter and by bacteria which extract it from the air. In the organic method, therefore, we have a means of adequately feeding the soil. In the organic method the consideration and the study of rocks assumes great importance because in their use is a safe means of furnishing to the soils minerals in a form which the soil is used to.

Today, however, agricultural science advises that artificial fertilizers, which contain other principles beside the above two mentioned classes of substances, be added to the soil.

In the rocks the various elements are safely diffused by nature, so that there is no dangerous concentration in one place of substances which may assume the quality of a poison when concentrated in one spot. Acids are sometimes employed in these artificial fertilizers. Wherein, in natural soil, there may be formed the same acids, they are usually present in extremely small quantities. In the case of the trace mineral elements which are required in such small amounts as only 3 or 4 parts per million, it is dangerously easy to oversupply them by indiscriminate application in chemical compound form and damage can result.

In an artificial fertilizer, after the plant takes up what it needs, unneeded residues, such as chlorides, sodas and sulfur, pile up in the soil. They would not be dangerous if they were diffused properly as nature does in rocks. There are some rocks which may contain certain elements in too large a quantity, but then they would not be recommended for use as a ground rock fertilizer.

This "diffusion" principle, which is a valuable device used by nature, can be described by citing how a fish swims in ocean water, which contains all these minerals. It can exist safely in such waters. But should something happen to make some of the elements concentrated in one place, it would become poisonous and the fish would die. This diffusion principle is an important one in the general study of the organic method. Both in organic matter and in rocks, all the elements and compounds are safely diffused

Rocks are the parents of all our soil. The above photograph shows four rocks which are frequently used by organic gardeners; below each rock is the powdered form ready for use by gardeners. (Left to right) Phosphate rock, Dolomite, Limestone and Granite. (Rock specimens: Courtesy of the Department of Geology, Lehigh University.)

throughout the mass. This is something that man cannot do in the laboratory or the factory, however much he mixes or homogenizes matter manually. If he lets nature enter into the process, like in the making of yeast, a true diffusion can take place.

Classifying Soils

Soils are classified into groups, series, and types. The groups are based largely on climatic factors and associated vegetation, the series on parent material, and the soil types on the texture of the soil. The following information is based on government agricultural reports:

TEXTURE: By texture is meant the relative amounts of the various sizes of particles making up the soil. These particles range in size from stones and gravel, through sand and silt to clay, the particles of which may be too small to be seen under the strongest microscope.

STRUCTURE: Refers to the grouping of individual particles into larger pieces, or granules. Good granulation or crumb structure of the heavier soils is essential to good results. Sandy soils show little if any granulation, due to the coarseness of their component particles. With soils containing a substantial percentage of clay, working them when wet results in destruction of the granular structure. Excessive tramping by livestock under like conditions is likely to have a similar effect.

Alternate freezing and thawing, or wetting and drying, and penetration of the soil mass by plant roots, are natural forces which favor the formation of soil granules, or aggregates. Such aggregation is most highly developed in soils near neutrality in their reaction: both strongly acid and strongly alkaline soils tend to "run together" and lose their structural character. Tillage also tends to break down the structure of many soils.

POROSITY: Associated with both texture and structure is pore space, or porosity. These spaces may be large, in the case of coarse, sandy soils or those with well-developed granulation. In heavy soils, containing mostly finer clay particles, the pore spaces may be too small for plant roots or soil water to penetrate readily. Good soils have 40 to 60 per cent of their bulk occupied with pore space, which may be filled with either water or air, neither of which can truly be said to be more important than the other.

Here, as in all other soil relationships, a satisfactory balance is important for productivity. Too much water slows the release of soil nitrogen, depletes mineral nutrients, and otherwise hinders proper plant growth. Too much air speeds nitrogen release beyond the capacity of plants to utilize it, and much of it is lost. The stored water in an overly aerated soil evaporates into the atmosphere and is lost to plants.

WATER: Soil water occurs in 3 forms, designated as hygroscopic, capillary and gravitational. The hygroscopic soil water is chemically bound in the soil constituents and is unavailable to plants. Gravitational water is that which normally drains out of the pore spaces of the soil after a rain. If drainage is poor, it is this water which causes the soil to be soggy and unproductive. Excessive drainage hastens the time when capillary water runs short and plants suffer from drought.

It is the capillary water upon which plants depend very largely for their supply of moisture. Hence the capacity of a soil to hold water against the pull of gravity is of great importance in ordinary agriculture. Organic matter and good structure add to this supply of water in soils.

But plants cannot extract the last drop of capillary water from the soil, since the attraction of soil materials for it is greater than the pull exerted by the plant roots. The term "wilting coefficient" is used to express the percentage of water in a soil at the time the loss from transpiration exceeds the renewal of the water by capillary means. Medium-textured loams and silt loams, because of their faster rate of movement of moisture from lower depths to the root zone and the fact that they can bring up moisture from greater depths than either sands or clays, provide the best conditions of available but not excessive soil moisture for best plant growth.

A well-drained soil generally is brown or yellowish brown to a depth of 2 or 3 feet. A poorly drained soil is gray or pale brown to that depth, according to Ohio State University Extension agronomists.

EROSION: Generally erosion works this way: first, the main loss is by sheet erosion; each time it rains, the runoff water removes a thin layer of surface soil. Then, as the topsoil becomes thinner, miniature gullies appear. After most of the surface soil is gone, gullies become the main problem.

Usually there's a clear difference between the topsoil and subsoil. The subsoil is finer-textured, more plastic, and lighter in color than the topsoil. Here's how erosion is classified:

No apparent erosion. All or nearly all the surface soil is present. Depth to subsoil is 14 inches or more. The surface may have received some recent deposits as the result of erosion from higher ground.

Slight. Depth to subsoil varies from 7 to 14 inches. Plowing at usual depths will not expose the subsoil.

Moderate. Depth to subsoil varies from 3 to 7 inches. Some subsoil is mixed with the surface soil in plowing.

Severe. Depth to subsoil is less than 3 inches. Surface soil is mixed with subsoil when the land is plowed. Gullies are beginning to be a problem.

Very severe. Subsoil is exposed. Gullies are frequent.

Very severe gullies. Deep gullies or blowouts have ruined the land for agricultural purposes.

There is a direct relationship between erosion and a soil's ability for intake of air and water. For example, when the soil surface becomes compacted, the danger of erosion increases, while the intake of water and air decreases.

SOIL SERIES: This refers to a subdivision of soil groups. A series is often given the name of a town, river or other geographical feature near which the soil was first identified. Since many soils in this country are young, the original geological characteristics of the soil materials is still evident. Thus members of the same soil series, or subdivisions, signify soils which have developed from the same kind of parent material and by the same processes.

A soil phase is a subdivision on the basis of some important deviation such as erosion, slope or stoniness. It indicates the departure or difference from the overall soil description.

SOIL GROUPS: All soils are composed of particles varying greatly in size and shape. In order to classify them by texture as well as physical properties, 4 fundamental soil groups are recognized: gravels, sands, loams and clays. (The last 3 make up most of the world's arable lands.)

The sand group includes all soils of which the silt and clay make up less than 20 per cent by weight. Its mineral particles are visible to the naked eye and are irregular in shape. Because of this, their water-holding capacity is low, but they possess good drainage and aeration, and are usually in a loose, friable condition.

In contrast, particles in a clay soil are very fine (invisible under ordinary microscope), become sticky and cement-like.

Texture of the loam class cannot be as clearly defined, since its mechanical composition is about midway between sand and clay. Professors T. Lyon and Harry Buckman in their book, *The Nature and Properties of Soil* (Macmillan, New York), describe loams "as such a mixture of sand, silt and clay particles as to exhibit light and heavy properties in about equal proportions. . . . Because of this intermixture of coarse, medium and fine particles, usually they possess the desirable qualities both of sand and clay without exhibiting those undesirable properties, as extreme looseness and low water capacity on the one hand and stickiness, compactness, and very slow air and water movement on the other."

Fortunately for the gardeners and farmers in the United States, most soils are in the loam classification. The majority of soils are mixtures; the more common class names appear below. (Combinations are given when one size of particles is evident enough to affect the texture of the loam. For example, a loam in which sand is dominant will be classified as a sandy loam of some kind.)

Sandy Soils

Gravelly sands
Coarse sands
Medium sands
Fine sands
Loamy sands

Loamy Soils

Coarse sandy loams
Medium sandy loams
Fine sandy loams
Silty loams and stony silt loams
Clay loams

Clayey Soils

Stony clays
Gravelly clays
Sandy clays
Silty clays
Clays

You can get a good idea of your soil's texture and class by rubbing it between the thumb and the fingers or in the palm of the hand. Sand particles are gritty; silt has a floury or talcum-powder feel when dry and is only moderately plastic when moist, while the clayey material is harsh when dry and very plastic and sticky when wet.

Observe Professors Lyon and Buckman: "This method is used in all field operations, especially in soil survey, land classification and the like. Accuracy . . . can be acquired by the careful study of known samples." If you're interested in developing an ability to classify soils, contact the local county agent for soil samples that are correctly classified.

The ideal structure is granular, where the rounded aggregates (or clusters) of soil lie loosely and readily shake apart. When the granules are especially porous, the term crumb is applied.

Air in the Soil

The average soil includes about 25 per cent air in its make up—an obviously significant proportion while fertility and optimum root functioning depend directly on the extent of soil ventilation.

TABLE 18: Physical Characteristics of Soil

Kind of structure	Description of aggregates (clusters)		Horizon
Crumb	Aggregates are small, porous, and weakly held together	Nearly spherical, with many irregular surfaces	Usually found in surface soil or A horizon
Granular	Aggregates are larger, harder, and strongly held together		
Platy	Aggregates are flat or plate-like, with horizontal dimensions greater than the vertical. Plates overlap, usually causing slow permeability		Usually found in subsurface or A_2 horizon of timber and claypan soil
Blocky or cube-like	Aggregates have sides at nearly right angles, tend to overlap	Nearly block-like, with 6 or more sides. All 3 dimensions about the same	Usually found in subsoil or B horizon
Subangular blocky or nut-like	Aggregates have sides forming obtuse angles, corners are rounded. More permeable than blocky type		
Prismatic	Without rounded caps	Prism-like with the vertical axis greater than the horizontal	
Columnar	With rounded caps		
Structure lacking Single grain	Soil particles exist as individuals such as sand and do not form aggregates		Usually found in parent material or C horizon
Massive	Soil material clings together in large uniform masses, as in loess		

—Prepared by E. D. Walker and W. F. Purnell
University of Illinois

Air is needed in the soil for the proper workings of bacteria and fungi. It aids in the breakdown of organic matter. This is important in considering the decomposition of the roots of the previous crop. With sufficient air, these roots will turn to humus in time to feed the next crop. Air aids in the oxidation of mineral matter. In an air-poor soil not much of the minerals would be available for plant sustenance. The presence of sufficient air acts as a regulator of the supply of carbon dioxide, too much of which is detrimental to plants. In the process of soil respiration, oxygen is fed to the roots. Better aeration provides a bigger root system and higher yields. In the process of plant growth, the leaves absorb carbon dioxide from the atmosphere and give off oxygen. The reverse takes place in the roots, which take in oxygen and give off carbon dioxide. In the decay of organic matter, carbon dioxide is given off.

The composition of the soil air differs somewhat from that of the atmosphere above ground. In the soil much of the air is dissolved in the soil water, but as such it is available to the needs of plants. The humidity is greater in the soil, which is a condition necessary for the optimum well-being of the soil organisms. The carbon dioxide content is much higher in the soil air and therefore the percentage of oxygen and nitrogen is less. In the soil there may be hundreds of times more carbon dioxide than in the air above it. While too much of this gas is detrimental, enough is needed to provide the needs for biochemical activities, the processes similar to human digestion, which in plant functioning begin to take place before the nutrients enter the roots.

DEVICES FOR BETTER AERATION: Among the methods used for increasing the air supply in soils are the addition of organic matter, the application of rock powders, soil drainage, subsoiling, cultivation, mixed cropping, etc. By far, the most important of all is to see that the soil is supplied with sufficient organic matter. It is axiomatic in agricultural literature that the more humus present in the soil, the better the aeration, the more pore spaces it will contain. There seems to be a direct relation between the amount of humus and the volume of pore space. The more of the former—the more of the latter.

SOIL NUTRIENTS: The presence of sufficient air in the soil is necessary for the transformation of minerals to forms usable by plants. In an experiment it was found that forced aeration increased the amount of potassium taken in by plants. Nitrate formation in soils can take place only in the presence of a liberal supply of oxygen, and many of the processes in the soil are oxidatious—sulfur transformed to sulfur dioxide, carbon to carbon dioxide, ammonia to nitrate. Oxygen is essential to these processes.

THE EARTHWORM: Here we find one of the greatest aids to soil aeration. The earthworm will burrow down 6 feet and more, leaving his passageways as means for the entry of air. Russell, in his book *Soil Condition and Plant Growth,* suggests planting earthworm eggs in farm soils to increase aeration where it is poor. But applications of organic matter would be better because humus itself brings about better aeration, while the presence of organic matter automatically multiplies the earthworm population.

The hardpan that forms from lack of organic matter impedes the aeration of the soil below it, but the earthworm, if present in sufficient numbers can destroy it. In well-run organic farms and gardens there should be millions of earthworms per acre.

In helping to increase the soil's aeration, the earthworm serves a valuable purpose in one respect. There are many disease-producing bacteria that can thrive only under anaerobic conditions, that is, where there is a lack of oxygen. With a teeming earthworm population the conditions are kept aerobic.

Burrowing insects also help in maintaining soil aeration. For example, termites make tunnels which roots have been known to use.

A legume crop is a good natural subsoiler. Alfalfa roots have been known to go down 15 feet or more. The decaying roots of this crop, after the plants are harvested, provide air passageways to very low depths. Tremendous increases in crop yields have been obtained by subsoiling.

CULTIVATION AND TILLING: Plowing is an excellent means of improving soil aeration and reducing the carbon dioxide content of the soil. The increased air produces more nitrate nitrogen. Thus when one plows for corn and then cultivates 3 times, as the crop grows, much air is mixed with the soil, and crop yields increase. In the recent "no plowing" vogue many failures occurred and one of the reasons was the lack of aeration in heavy clay soils. Drainage by means of soil tiles is an effective means of restoring aeration.

SOIL ORGANISMS: Air is an urgent need of many beneficial soil organisms that aid in transforming soil nutrients for plant use and that take part in the various oxidation processes, including the oxidation of humus. Oxygen gives the bacteria part of their energy. If it were not for the aerobic bacteria, no organic matter would decompose in the soil. These valuable organisms manufacture protein from these organic residues.

Aeration aids the formation of mycorrhiza on the roots of many plants. The mycorrhiza is a fungus organism that acts in partnership with the roots of plants to feed it valuable nutrients.

CARBON DIOXIDE: If the soil air contains too much carbon dioxide, given off by the roots, there will be a drop in crop yields. There are various methods of ridding the soil of an oversupply of this gas such as changes in soil temperature, and rain water, the latter bringing oxygen into the soil. Soil that is more porous because of its organic matter content will absorb more rain water.

In the decay of manure much carbon dioxide is produced, but the effect of the application of manure is to increase the pore spaces in the soil which increases the rate of diffusion of this gas into the atmosphere. Sir E. John Russell has shown that green manuring, particularly if the crop is fairly succulent, will also raise the carbon dioxide content of the soil air. If seeds are sown too soon afterwards, the carbon dioxide may inhibit germination or harm the very young root system of the seedling.

How Air Fertilizes the Soil

Air is a complete fertilizer. It contains all of the nutrient elements needed by plants for proper growth. It costs nothing, and is easy to use. It will make soil more fertile and will boost crop yields.

The University of Illinois, Urbana, has found that adding air to soil increased yields of corn from 94 to 144 bushels per acre!

A gain of 50 bushels of corn per acre just from air is a lot! Tests with soybeans showed that yields rose from 41 bushels per acre to 44 just by adding air. The University added air by pumping it into the soil through drainage ditches.

Tests run by G. Ingham, Pretoria, South Africa, show that air contains large quantities of the nutrients needed by plants. In fact, Ingham says that air contains *all* of the nutrients needed by plants for proper growth under natural conditions.

Ingham has measured the nutrients in rain. He has measured the nutrients under trees and bushes, and the nutrients in cleared spaces. He has made a special study of the top two inches of soil. He has tried adsorbing nutrients from air on cellulose. Ingham has found that the air is just chock-full of nutrients!

Air is made up of about 78 per cent nitrogen, 21 per cent oxygen, and one per cent other gases. These other gases include carbon dioxide. Carbon dioxide and water are the raw materials from which all sugars, starches and fats are made. Add nitrogen and we have protein. Sugars, starches, fats and proteins make up all plant and animal matter. But for life something else is needed. . . .

This other something is minerals. Only a very small amount of mineral matter is needed. And air, or atmosphere, contains sufficient of this mineral

1. Hard subsoil. Note the root scantiness past the 10-inch mark. Use of heavy equipment is held responsible for this unhealthy condition. 2. Subsoil loosened. The 11-week-old oats on this plot sent roots down to the two-foot mark when compacted layer was loosened.

matter according to Ingham. His tests show that as much as 30 pounds of nitrogen, 338 pounds of calcium, and 22 pounds of phosphorus are delivered free by the atmosphere to each acre of soil every year!

A part of the mineral matter plus oxides of nitrogen come to the soil with rain. A part is collected by plants on their surfaces. Some of this is absorbed by the plants with the aid of dew and helps nourish them. Some of it washes off the plants onto the soil beneath when it rains.

Ingham has found that the soil beneath plants is especially rich in nutrients which have come from air. This has been verified by experiments conducted at the University of Michigan, East Lansing. Here, Dr. Tukey and several associates ran tests on apple trees and berry bushes.

Dr. Tukey found that nutrients were leached from leaves of plants by rain.

These nutrients fell to the soil below the plants. Here they stayed—providing there was absorbent material. This is one of the big secrets of how Nature uses air for fertilizer – absorbent material to catch the nutrients that air supplies.

Other scientists around the world have verified the fact that the atmosphere furnishes the nutrients needed by plants. Swedish investigator Egner says that tests he has made show that one to five gamma of nitrogen, phosphorus, potassium, calcium, sodium, magnesium, chlorine and sulfur are present in each cubic meter of air.

Agriculturist Robert Stewart, University of Illinois, says that the atmosphere supplies all of the carbon and sulfur needed by plants for growth. Sulfur does not need to be added to soil! He found that on an average 41 pounds of sulfur was supplied to each acre of land per year. A 100-bushel crop of corn removed 9 pounds of sulfur, 50 bushels of wheat—five pounds. The atmosphere supplies more than enough sulfur to the soil!

A special study made by the Ministry of Local Industries, Budapest, Hungary, proves that air has another important function other than adding nutrients. Scientists at the Ministry tried excluding air from soil. They found that the soil became alkaline. Nitrogen and organic matter decreased. Humus became deactivated, and the soil became less fertile. They had proven that air is necessary for a healthy, fertile soil!

Farmers and gardeners have been making use of air for fertilizer, even though many may not have realized it.

We have all noticed how bushes and plants get dusty. The dust that covers these plants is pulverized nutrient material. Some of the nutrients are dissolved by the plant with the aid of dew, or of moisture on a humid day. This helps nourish the plant. Thus, the plant absorbs the nutrient material. *If, then, we use the plant in compost we make use of this nutrient material that has come from air.*

This is free fertilizer that has been supplied by the atmosphere. It is over and above any nutrient material that has been taken from the soil. Though this dust may appear slight, it will add all of the nutrients that the soil requires under normal conditions. As in the case of sulfur and corn, the atmosphere adds 4 times the amount of sulfur removed by the edible portion of a large crop.

Compost that has been added to soil, or organic matter that has been worked into soil, performs another function. Scientist Ingham found that air works into soil to be adsorbed directly on organic and inorganic colloids (small, fine particles) in the soil. Organic colloids come from decomposing organic materials.

Thus, not only the dust and nutrients absorbed by the plants from the dust are added to soil, but the very organic material itself in the soil works with air to form more fertilizer!

Another function of adding organic material to soil is to open up the soil and form paths whereby air can work down into the soil. Organic matter forms colloids and humic acids as it decomposes. These colloids adsorb nutrients from air that enters the soil. Organic acids dissolve nutrients from rocks in the soil. This makes minerals available for the nourishment of plants. The whole lot of humus, undecomposed organic material, soil nutrients and air fertilizer is eaten by microörganisms. These tiny, microscopic bits of life build up the fertility of soil.

Microörganisms aid in the direct conversion of nitrogen from air into compounds that can be used by plants for growing. They produce stimulants that speed up the growth of plants. They produce materials that help protect plants from insects and diseases. They even help to aggregate soil so that more air and rain water can enter!

Organic matter is a mechanical aid, also. Before it becomes decomposed it is in coarse chunks. Thus, when it is added to soil, it breaks up clods. It helps make soil porous and open. This allows air and water to enter and make the soil moist and fertile.

Plenty of rough organic material worked into soil and used as mulch is very important where flash or heavy rains occur. The organic material breaks up the drops of rain and allows the rain to enter the soil. It helps absorb the extra water until it is needed by plants on hot, dry days. It holds particles of soil so that soil erosion does not occur. It prevents leaching out of nutrients from soil and helps hold the fertilizer being added from air.

Everyone has noticed that a little extra puttering around pet plants helps them to thrive. A part of this puttering is a stirring of soil around the roots. This stirring adds air to the soil. The University of Illinois says that even under the best of conditions on very fertile soil, plants may be starving for oxygen. This is why the University went so far as to rig up a pump to force air through a water drainage ditch up into the soil. They ran the pump intermittently so that the soil had a complete change of air every 3 hours, or 8 times a day. Did it pay off? It did. *Air gave 50 bushels of corn more per acre!*

We don't have to go so far as to pump air into the soil. But things that we can do to make use of the natural fertilizer that Mother Nature is dumping on our farms and gardens every minute of every day of the year is to use good organic methods. Organic matter used as compost and added to soil adds fertilizer that has come from the air, helps build new fertilizer in the soil, and helps the soil retain that fertilizer.

Rain carries oxygen and brings it into the soil, soaking down to subsoil levels. In regions where there is plenty of rainfall there is much better soil aeration than in arid regions where the soil becomes silt, lacking porosity and forming crusts. Some soils that are low in organic matter may lose their permeability during heavy rains, and become waterlogged. Percolation stops, asphyxiation sets in, and the root growth of crops is seriously affected.

—Charles Coleman

Knowing Your Soil's Temperature

It may not be too long until you'll be reading in your local newspaper—Forecast for today: sandy soil—90 degrees; clay soil—80 degrees.

Soil temperatures vary just as much as air temperatures. The difference, however, is that soils don't change in the same degree as air; one part of your garden is probably hotter (or colder) than another depending upon its location, chemical and physical make-up.

There's a fascinating story about soil temperatures, and an understanding of it will make you a better gardener. So join me in taking a careful look underground.

First, let's find out just how soil temperatures influence your gardening results. Here are some examples:

1. Germination of seeds depends upon warmth of soil below as well as upon air above.
2. Planting your first crop as soon as soil has warmed up enough in spring can mean that you'll have time to have a late planting in the same spot.
3. A mulch or cover crop regulates your soil's temperature to your advantage.
4. Small animals, such as rabbits, pick certain sites in your garden to dig their burrows because they know that surface conditions above will protect them from winter's cold.
5. You'll learn how to save plants from frost damage.
6. You can aid the work of helpful soil bacteria, if you know at what soil temperatures they work best.

Perhaps the most dramatic example of how an understanding of soil temperatures can improve crops took place some 50 years ago in Wisconsin. Briefly, *two inches of sand* transformed the entire cranberry growing industry in that state. Here's how it was done:

Frost in the Wisconsin marshes was undoubtedly the worst enemy of the cranberry. In early August, 1904, a 26 degree frost ruined 40 per cent of that year's crop. Something had to be done, and it was decided that the Weather Bureau should make an investigation on the possibility of frost protection.

Cranberries were grown in marshes, and in all these marshes, weeds and natural vegetation were generally dense and effectively screened the soil from the sun. Thus it was heated very little by day and had only a small supply of heat stored up at night to counterbalance the outgoing radiation. As a result, the temperature of the air above the bog fell to a low level on every calm clear night.

Many of the factors involved were natural phenomena outside the control of man—the draining of the cold air from the hills to the marsh, the moist air, dew and fogs—but man could at least do something to check evaporation and eliminate the weeds.

After much experimenting, it was found that soil temperatures did not vary as much—and that protection against frost was greatest—when the cranberry plots were well weeded, thinly vined and covered with a two-inch layer of coarse sand.

Here's why the sand helped: its lack of capillarity reduced the moisture brought to the surface and evaporated; its low specific heat made it warm up quickly in the sun and pass this heat down to the layers of peat below; its absorption of the sun's radiation was good, but it gave back its heat at night slowly and so modified the night temperature by warming the air above it during most of the night.

Now don't go out and throw several tons of sand all over your garden. It won't work—unless your soil is mostly a peat bog that can keep the soil always moist. Otherwise your sand will be a parched covering after the first dry spell.

Organic gardeners are well aware that their soil is alive—providing a home for millions of beneficial bacteria. These bacteria require special conditions of warmth, moisture and free aeration of soil to do their best job. These conditions are found only in the upper cultivated layers of the soil, and are more easily obtained in sandy loams than in clays where the moisture content is too high and the supply of oxygen is lacking.

According to T. Bedford Franklin, author of *Climates in Miniature,* "Fields of corn turn yellow, especially on the clay lands, when cold and drying east winds cool the soil in the spring, because the bacteria produce too little nitrates in a cold soil to feed the crop; it is only when warmer conditions arrive and the bacteria resume activity that the bright green color returns to the crop." He points out that during the summer months a light, well-warmed soil, kept well-cultivated, may produce 8 times as much nitrates as the same soil left to itself and never stirred with a hoe.

Soil bacteria like heat and, Mr. Franklin believes, work best at about 100 degrees F.; they stop work below 41 degrees and above 130 degrees. What can you do to keep your soil in that 41 to 130 degrees range? You don't have any control over the air temperature, but you do have an effective method for your soil.

That method is a mulch! In effect, a two-inch mulch can give you a 7-degree boost. Let's suppose that you want to get your peas planted as early in March as you can. Your outdoor thermometer has been registering a fairly consistent 43 degrees for a week during the daytime. Strong March winds with little humidity have scared you out of doing any planting.

If you haven't been mulching, you're properly afraid and should wait a while longer. If you have been mulching, put your coat on and get out and start planting your seeds. Evaporation caused by low-humidity winds cools the soil off by as much as 7 degrees—too cold for planting in this instance. But a mulch will prevent excessive evaporation—with a drop in soil temperature—and will maintain the 43 degree temperature.

In early spring, within those depths underground where the roots of crops are to grow, a growing temperature of above 42 degrees for several hours in the day (even if it falls below that during the night) means that you can begin planting your early crops.

There's no reason why you can't become an amateur soil researcher, using your home garden as a testing ground just as the author of *Climates in Miniature* did. I'm sure you will find it interesting and worthwhile.

It's important to get the right kind of thermometer. (It will be simpler for you if it's scaled in Fahrenheit degrees rather than Centigrade.) One type that would serve you well is an Armored Thermometer, 12 inches long, with a range from 30 degrees below zero to 120 degrees Fahrenheit. (If not available locally, you can obtain one from the Arthur H. Thomas Co., 230 South 7th Street, Philadelphia, Pa. The Taylor Instrument Co. in Rochester, N. Y., is another source.) Less expensive ones, mounted in a wooden or metal frame, can also be used.

A Geologist Looks at the Soil

The United States is blessed by having within its borders large areas which produce food in great abundance. Several of these stand out because of the high quality and high nutrient content of the food which is grown therein. These latter areas include the corn belt across Illinois and Iowa, the high-protein wheat belt in Kansas, the grain and fruit areas in Oregon and Washington, the fertile "Great Valley" of California, the major "river bottoms" or the flood plains of the large streams which flow across our country, and the stock grazing areas of our western states. Other areas might also be included but they would serve no additional purpose in illustrating the point to be made in this article.

What do these areas have in common which may be responsible for the nutrients in their soils? Can it be climate? The corn belt is in a region of moderate to heavy rainfall, with hot summers and cold winters. The surface of the land is flat to rolling. The elevation is low, being only a few hundred feet above sea level.

The Oregon grain land, however, is in a radically different climatic setting than the corn belt. In the dry farm country, a crop is recovered only every other year. Each field of grain this year is matched by a counterpart which lies fallow and picks up its scanty measure of moisture which is held for the next year. The growing season is entirely different from that in Illinois.

The Washington fruit country is still different, and the "Valley" of California is unlike all the rest. Irrigation is necessary here. Even in California alone the temperature ranges through a cool wintry season in the north to a subtropical climate in the Imperial Valley to the south.

Clearly then, temperature and rainfall are not the common denominators in the development of fertile soil.

If climate is not the common factor in the fertility of these different regions, perhaps their soils have a common heritage in abundant black humus as occurs in the dark, corn belt soils. Or, are all these soils at the same stage of maturity, or are they texturally favorable so that they will "raise anything?" The answer is definitely "No" to all of these questions.

Whereas the corn belt soil is typically rich with dark humus, the Oregon soil is grayer, and some irrigated, desert California soil may be so light colored as to appear as if it contained no organic matter. The soils range from a mature stage of development to others where a profile is hardly apparent. They vary in texture from heavy clay to almost gravel. Their common denominator still has not been found.

The solid rock beneath the Oregon-Washington rich soils is commonly basaltic lava, but the deeply buried, solid country rock in the corn belt region is sedimentary rock, such as limestone, sandstone or shale. In the grazing areas, the country rock may be sedimentary or it may be igneous granite. Under the alluvial stream bottoms it may be any kind. But here we come to a clue. The fertile soil is not a product of the solid, underlying rock; it may develop from a stream-washed or other secondary deposit above the solid rock. Let us look into that possibility.

The parent material underlying the soil of the corn belt is a glacial deposit which is more or less covered by a wind-blown mantle. The well-known, rich Palouse soil of the Oregon-Washington region is derived from wind-blown silt, clay and sand, which overlies the solid rock. In the "Valleys" of California, thick alluvial deposits built of rock waste which was carried in from the surrounding high mountains, rise high in the intermontane troughs, and these are the parental sources of the soil. The alluvial flood plain deposits in the river bottoms give rise to their fabulously rich soils. The thinner, but nutrient rich, soils in the western grazing areas may lie on rocky, mountainous slopes, without benefit of depositional processes which accounted for the other thick deposits.

As far as superficial classification goes, these parental sources are different: glacial, windblown, alluvial, and slope washed. Is there anything in common? The answer, if we look closer at the constituents of the fertile soils, is "Yes, all of them contain *finely pulverized, little weathered, native rock and mineral* particles.

All of the before-mentioned areas of fertile soils are developed upon parental geological deposits which contain abundant physically weathered (little chemical weathering), finely pulverized, well-mixed native rocks and minerals that contain a reserve of nutrient elements in an adequately available condition. We will review them, one by one.

The corn belt soils are underlaid by a thick glacial deposit which was brought in from the north and from nearby rocks, by a thick, grinding, churning glacial ice sheet that later melted away and left its crushed and pulverized rock load, fabulously rich in slightly weathered rock and mineral particles. Let us contrast the situation here with that of an old leached and worn soil. In the latter soil, usually, only barren quartz (silica), sand and silt, coated with a thin film of reddish to brownish iron oxide, is mixed with some exhausted, degraded clay and leached gravel fragments composed of chert or flint. Nothing of nutrient value to plants remains.

Not so with the glacial deposit of the corn belt. The ice sheet which moved southward from Canada, across the Great Lakes Region, picked up, shoved, carried and mixed mechanically the igneous (heat-formed) rocks of the north with the sedimentary rocks to the south, and sweetened them with various concentrations of copper, zinc, cobalt, manganese, iron, and other ores (trace elements) over which it occasionally passed.

The resistant potash feldspars of the igneous and metamorphic rocks were mixed with potassium-containing sedimentary shale which releases potassium easily. Calcium and sodium igneous feldspars were mixed with the calcium-rich, easily soluble calcite of sedimentary limestone. Magnesium, calcium and iron in crisp, hard, hornblende and pyroxene minerals were blended with magnesium, calcium and iron in sedimentary, easily soluble dolomite. Slowly soluble phosphate of calcium in the igneous mineral apatite was churned alongside of quickly soluble, phosphatic, sedimentary limestone and shale.

All of these rocks and minerals were crushed by the grinding and abrasive work of the ponderous, slow-moving, but relentless heavy ice sheet. They were pulverized to particles so fine that the melt waters from the glacial ice were milky white with rock flour, yet these microscopically fine particles were unchanged, unleached, and as rich as the original rock in mineral nutrients! This point is most important; the unchanged, native rock and mineral particles were pulverized under icy, cold, non-reactive chemical conditions and were thoroughly mixed and blended with all other kinds of rocks and minerals, so that the final product constituted a rich potential feast of *a wide variety of inorganic nutrients to be chosen as desired by plant rootlets,* in a wide range of availabilities. The fine pulverization of the particles made them easily susceptible to attack by plant rootlets, and the particles had not lost any of their nutrient constituent elements because they had been pulverized mechanically in the cold, without undergoing chemical weathering or loss.

It is no wonder that the corn belt soils have high reserve fertility. Think of the fertility reserve, silt-size minerals which can be drawn upon year after year and crop after crop. The pulverized glacial product is a naturally perfect, fully blended agstone, widely spread for development into soil. The modern gardener and farmer who wishes to add a balanced agstone to his depleted soil need only duplicate in his crushing plant the formula that Mother Nature worked out and produced from her glacial mill! There are no patents on this fertilizer.

The quick development of a fertile soil on the glacial deposits included a favorable climate for organic growth, of weathering to a high-exchange type of clay mineral, and flat topography which did not lose by erosion the fertility that developed.

The Palouse soil parent of the northwest states was redeposited by the wind after a glacial ice sheet pulverized the surface of the widespread basalt rock flows which underlay the region. Again, the native, unweathered, nutrient-rich basalt was mechanically pulverized by the ice, then picked up, mixed, and scattered far and wide by the wind, and deposited to give rise to a fertile soil today. Finely pulverized, native rocks and minerals are the crux of the fertility.

The Great Valley and the lesser intermontane valleys of California owe the inorganic fertility of their soils to mechanically pulverized and weathered rock fragments that have been washed in by water, blown in by wind or carried part way by glacial ice to their present resting place. The arid climate has preserved without chemical weathering or leaching the nutrients in the fine rock particles. Upon being irrigated, the moisture and the weathering effect of plant rootlets now releases and pulls out the stored-up, adequately available potassium, calcium and other nutrient elements. To repeat, the native rocks were reduced in size, the arid climate protected and preserved the pulverized particles, and man is now capturing the inorganic fertility prize which long awaited development.

The soils of eastern Colorado and western Kansas, notable for high protein food production, have developed upon deposits rich in unweathered native rocks and minerals which were derived from the Rocky Mountains, and were washed out eastward on to vast, wide-spread, coalescing alluvial fans. These alluvial deposits thin eastward, downslope across western Kansas. Tremendous quantities of rock fragments were washed away from high, primitive Rocky Mountains. The unweathered to partly weathered rock fragments scattered across the plains bear mute testimony to the process of deposition just described.

Fertile soils in river bottoms, and the wind blown loess which occurs on adjacent valley walls are replete in silt and clay size particles of unweathered to partly weathered native rock and mineral fragments. These fragments were washed down by the rivers, or were pulverized by the abrading stream, and then spread out over their flood plains (bottoms) in time of flood, or on their sand bars during high water. Wind blowing across the dried deposits picks up the silt and clay, carrying them upward to the land alongside the river. The theme here again emphasizes unweathered to partly weathered native rock and mineral particles, with clay and organic matter, carried by a geologic agent and deposited where soil can develop upon it.

Thin, but nutrient-rich soils in the stock grazing ranges of our western states develop upon the primitive underlying rock, or they accumulate from slope wash and gravity creep of the mechanically weathered rock in the higher elevations. Here, as before, the nutrient elements of the native rocks and minerals are the sources of the inorganic fertility of the soil.

Numerous examples from Nature show how soil fertility has been evolved over the past. Man may duplicate artificially, on any size scale which is consistent with economics, the same process of developing fertility which has been so manifestly successful before.

—Dr. W. D. Keller, Dept. of Geology, University of Missouri

Glossary of Geologic Terms

Alluvial—Relating to mud, clay, or other material left by running water.

Alluvial Fan—The alluvial deposit of a stream where it issues from a gorge upon an open plain.

Abrade—To scrape, wear, or rub away or to remove by friction.

Calcite—A calcium carbonate occurring in many crystalline forms, such as chalk or marble.

Dolomite—A calcium magnesium carbonate of varying proportions found in limestone and marble.

Feldspar—A closely related group of minerals, all silicates of aluminum, with either potassium, sodium or calcium.

Hornblende—A common dark mineral, generally black, composed of a light to dark green silicate of iron, magnesium and calcium with aluminum.

Igneous—"Fire formed" rock, the result of the cooling of molten rock below the earth's crust.

Iron Oxide—The combining of iron with oxygen—rust.

Loess—A fine grained, erosional sediment, deposited by wind, covering vast areas in Asia, Europe and North and South America.

Pyroxene—Any of a common and important group of iron, magnesium, and calcium silicates found in many igneous rocks and molten lava.

Sedimentary—Designating rocks formed by simple precipitation from solution, as rock salt and rocks formed of organic materials, as limestones, coal and peat.

Weathered—Rocks whose appearance is changed by exposure to the atmosphere.

Brief Description of Soils

Clay Soils—A typical "clay" soil may be composed of approximately 60 per cent actual clay, 20 per cent silt, and 20 per cent sand. The particles in a clay soil are so fine that it tends to compact, which makes cultivation difficult and interferes with the oxygen supply for plant roots. Water can do little to enter the impervious clay soil, and runoff is very common during rainfalls.

Sandy Soils—A typical "light sandy" soil may be composed of approximately 70 per cent sand, 20 per cent silt, and 10 per cent clay. The particles in a sandy soil are comparatively large, permitting water to enter the soil and to pass through it so quickly that it dries out very rapidly, and often carries nutrients with it. Sandy soils are generally very difficult to correct.

Loam Soils—A typical "loam" soil may be composed of approximately 40 per cent sand, 40 per cent silt, and 20 per cent clay. Loam is the ideal garden soil, combining the best percentages of sand, silt, and clay. It is easily worked, and retains water and nutrients, which are slowly absorbed by plant roots. Every gardener can strive for a soil close in structure to that of loam.

Adobe Soils—In hot, dry areas of the nation, a heavy clay soil, not being able to absorb what little rainfall there may be, will become very hard, dry, and cracked. Adobe is perhaps the most difficult of all soil for the gardener to cope with. It has all the disadvantages of a heavy clay soil, plus the fact that it will be much drier.

3—Improving Soil Structure

HAVE YOU EVER NOTICED tomatoes and other plants suffering from lack of moisture in midsummer even though there was sufficient rain . . . or soils puddling, washing away instead of storing the needed moisture for plant growth? These are a few symptoms of poor soil structure. Your job is to change your soil so that it will furnish your plants with enough air and water. Here's how A. M. O'Neal and A. A. Klingebiel, soil scientists of the Soil Conservation Service, sum it up:

Soils need a certain amount of air if plants are to grow. If soils are saturated with water or have "run together" at the surface, air cannot get in. As a result, poisonous gases may accumulate, less plant food may be available, and plant growth and yields are reduced. Plants, like people, require air, water, and food. Reducing the amounts of any one of these will cut growth and yield of crops.

If you can make more water enter the topsoil, a number of problems will be solved. In the first place, the more water that enters the soil the less there is to run off. Less runoff means less erosion of soil and less loss of plant food by erosion. Secondly, if most of the rain that falls in July and August soaks into the soil, there will be more water available for plant growth during this critical period. Thirdly, since air and water movement in soils are closely related to each other, any condition that causes more water to move into the topsoil will also insure better air circulation.

Tilth Affects Air and Water Movement

The physical condition of the soil that has most to do with air and water movement in the soil is called "soil tilth." A soil in good tilth breaks up easily into crumbs, or granules, about the size of wheat grains or soybeans. These crumbs are porous. They are made up of tiny bits of soil linked together something like popcorn in a popcorn ball. They hold this structure even when soaked. Because of the pores in the crumbs themselves, and because the size of the crumbs keeps them from fitting together as tightly as smaller bits of soil, there is space for air and water. In other words, to have good soil tilth is to have proper air-water relationships in the soil.

Cultivation Has Changed Soil Tilth

Soil tilth has changed a lot during the years that the land has been farmed. You can appreciate this change better if you compare virgin soils with soils found under today's farming conditions.

There are few areas of virgin soil left—that is, soil that has never been disturbed. The soils that come nearest to being like the soils the pioneers found lie along old undisturbed fence rows, in road banks that have never been plowed, or in woods that have not been severely grazed or burned.

Dig up some undisturbed soil. Then, dig some soil from a cultivated field adjoining the spot where you dug up the undisturbed soil. Take this sample from the plow layer of the cultivated field. Be sure to get it from the same kind of soil as that in the fence row or woods, but take it a few rods away.

Compare these two samples of soil for color, organic matter, tilth, weight, and compaction. The soil from the fence row, for example, is porous and

Many successful gardeners start off their garden year by opening a trench, raking into it leaves and other materials that have overwintered on top of the ground and finally spading the earth back into the trench. In this way, much organic matter is turned into the soil.

full of grass roots. It is high in organic matter (or humus), very crumbly, and easily broken apart. By comparison, the soil taken from the cultivated field is harder, heavier, and lighter in color. It is not uncommon for cultivated soils to have lost from 20 to 50 per cent of their original organic matter and to be 10 to 30 per cent heavier than they were originally. After making this comparison, examine soils that have been farmed under good soil management.

Adding Humus to Your Soil

The best method to improve soil structure is to increase its organic matter content. There are four general ways to do this:

(1) Mulch regularly with organic materials, working them into the soil after each crop;

(2) Turn under cover crops;

(3) Work surface plant residues, additional materials, with fertilizer and manures into the soil where it will decompose. This method is known as sheet composting;

(4) Make a compost heap and work finished compost into the soil later.

Mulching

Mulching around plants is a slow but sure way to increase the humus content of your soil. It can be done gradually at your own convenience and whenever

Mulching around plants is a slow but sure way to increase the humus content of your soil. Above, a mulch of straw is being spread evenly between plants and rows about 5 inches deep and will also serve to keep the soil cool and hold moisture.

mulching materials are available. This way organic materials serve two purposes: as a mulch and then as a soil improver.

Here is a partial list of materials suitable for mulch: grass clippings, weeds (expose to air before applying to prevent rooting) leaves, peat moss, compost, hay, straw, corn stalks, corn cobs, plant residues, wood chips, sludge, spent hops (available from many breweries), sawdust, pine needles, bark, and many industrial food wastes.

When judging how good a material is for mulching, some of the points to consider are (1) does it have enough weight not to blow away? (2) is it free of any toxic chemicals which might cause plant injury? (3) is it fairly easy and inexpensive to obtain and apply? and (4) would it remain loose enough to allow enough air circulation down to the soil surface?

Cover Cropping

Cover cropping (growing plants and turning them under) is one of the best ways to increase soil organic matter as well as prevent erosion. Since most growing seasons are short, you can grow cover crops whenever the land is free before or after the main crop.

In this way you can grow both the plants you want as well as a soil-building

Making a compost heap and working the decomposed material into the soil later is an excellent way to improve the soil's physical structure. Well-made compost can be a concentrated fertilizer as well as a soil conditioner; it can be spread over the entire vegetable garden prior to planting, or side-dressed around growing plants.

and soil-conserving crop *in one season.* Summer is an excellent time to practice cover cropping, since the crop can be turned under in late fall or early the next spring.

Choices are many, but a good cover crop should make as large an amount of growth as possible in the shortest time. According to extension horticulturist Edward Hume of Vermont, for most gardens it's better to grow grasses for cover crops instead of legumes, since grasses will produce much more organic matter in the same time than would legumes. He believes that the extra nitrogen produced by the legume crop can be added more easily than the extra organic matter produced by the grasses.

For early spring or fall sowing, he suggests rye and rye grass as outstanding producers. Since these plants grow for only a short time, they should be seeded thickly—about 25 to 30 pounds per 1,000 square feet.

Other cover crops include crimson clover, oats, wheat, buckwheat, cow-peas, soybeans, millet and Sudan grass. Planting of these can be done at vegetable harvest time or somewhat earlier by planting between the vegetable rows.

When perennial crops are grown, cover cropping usually must wait until it's time to clear out at least a bed. Then this should be cover cropped for at least a full season to build up the organic matter which has to last during the years that section will be growing the next perennial crop. "Some folks sow a cover crop between the perennial plants in the fall and cultivate it very early the next spring," points out Mr. Hume. "This is a good idea if it does not interfere with the growth of your crop plants."

Rotary tillers, garden tractors and the like, make cover cropping a lot more practical—not to mention easier on the back—since the plant material can be readily worked into the soil.

Sheet Composting

Here again, tillers and tractors make sheet composting in the garden more feasible. By this method, leaves, weeds, manure, and other waste organic materials are spread over the garden and worked into the soil to decompose. It's a good idea to make several passes with a rotary mower over this material as cut-up material breaks down faster. Also, add any lime, nitrogen fertilizers, bone meal, tankage, dried blood, phosphate or potash rock at this time.

Incidentally, you can go one step further by growing and turning under a cover crop in the same garden area where you are sheet composting. If power equipment is unavailable, the organic matter can be spaded under.

Compost Heap

Materials to use: vegetable matter, such as leaves, grass clippings, garbage and many of the items mentioned previously in the mulch section; animal manures; mineral fertilizers; and soil.

Nitrogen is essential to fast composting, so use enough manure, cottonseed meal, dried blood, bone meal, etc.

Mix together in your heap two or more types of raw material. Leaves alone mat down and decay very slowly but, when mixed with grass clippings or other material, will compost well.

Shred your material, using a rotary mower if a compost shredder is unavailable.

Sufficient moisture in the heap is needed, but don't keep it soggy.

Turn it with a fork as often as you care to.

Add limestone, if the soil is acid. No matter what kind of soil you have, mix in rock phosphate and potash to increase mineral content.

Observes Vermonter Hume: "Your finished compost is a concentrated fertilizer as well as a soil conditioner. You can spread it evenly one-half inch deep over gardens, lawns or around trees and shrubs."

All vegetable material, such as grass clippings shown being used as a mulch, and leaves spread in a trench, will help convert a heavy, light-colored soil into a dark, loose fertile condition. Decaying organic matter in the soil helps make nutrients more available to growing plants.

List of Organic Materials

Occasionally, gardeners live close to processing plants which have enormous amounts of waste materials, free for the hauling. The wise gardener will avail himself of these various forms of organic matter and use them in the compost heap. Many gardeners get the names of the best sources of free organic materials, and make periodic trips, hauling several different kinds of materials in each trip.

The following is an extensive list of organic materials, along with the nitrogen, phosphorus and potash content of each material.

TABLE 19: Percentage Composition of Various Materials

Material	Nitrogen	Phosphoric Acid	Potash
Alfalfa hay	2.45	0.50	2.10
Apple, fruit	.05	.02	.10
Apple, leaves	1.00	.15	.35
Apple pomace	.20	.02	.15
Apple skins (ash)	...	3.08	11.74
Banana skins (ash)	...	3.25	41.76
Banana stalk (ash)	...	2.34	49.40
Barley (grain)	1.75	.75	.50
Bat guano	1–12	2.5–16	...
Beet roots	.25	.10	.50
Bone, ground, burned	...	34.70	...
Brewer's grains (wet)	.90	.50	.05
Brigham tea (ash)	...	...	5.94
By-product from silk mills	8.37	1.14	.12
Cantaloupe rinds (ash)	...	9.77	12.21
Castor-bean pomace	5– 6	2–2.5	1.0–1.25
Cattail reed and stems of water lily	2.02	.81	3.43
Cattail seed	.98	.39	1.71
Coal ash (anthracite)	...	.1– .15	.1– .15
Coal ash (bituminous)	...	.4– .5	.4– .5
Cocoa-shell dust	1.04	1.49	2.71
Coffee grounds	2.08	.32	.28
Coffee grounds (dried)	1.99	.36	.67
Corncobs (ground, charred)	...	...	2.01
Corncob (ash)	...	...	50.00
Common crab	1.95	3.60	.20
Corn (grain)	1.65	.65	.40
Corn (green forage)	.30	.13	.33
Cottonseed	3.15	1.25	1.15
Cottonseed-hull ashes	...	7–10	15–30
Cottonseed hull (ash)	...	8.70	23.93
Cotton waste from factory	1.32	.45	.36
Cowpeas, green forage	.45	.12	.45
Cowpeas, seed	3.10	1.00	1.20
Crab grass (green)	.66	.19	.71
Cucumber skins (ash)	...	11.28	27.20
Dog manure	1.97	9.95	.30
Dried jellyfish	4.60	...	...
Dried mussel mud	.72	.35	...
Duck manure (fresh)	1.12	1.44	.49
Eggs	2.25	.40	.15
Eggshells (burned)	...	.43	.29
Eggshells	1.19	.38	.14
Feathers	15.30	...	...
Field bean (seed)	4.00	1.20	1.30

TABLE 19: Percentage Composition of Various Materials (Continued)

Material	Nitrogen	Phosphoric Acid	Potash
Field bean (shells)	1.70	.30	.35
Fire-pit ashes from smokehouses	...	...	4.96
Fish scrap (red snapper and grouper)	7.76	13.00	.38
Fish scrap (fresh)	2- 7.5	1.5- 6	...
Fresh-water mud	1.37	.26	.22
Garbage rubbish (New York City)	3.4–3.7	.1–1.47	2.25–4.25
Garbage tankage	1 –2	.5–1	.5 –1
Greasewood ashes	...	...	12.61
Garden beans, beans and pods	.25	.08	.30
Gluten feed	4 –5	...	...
Greensand	...	1 –2	5.00
Grapes, fruit	.15	.07	.30
Grapefruit skins (ash)	...	3.58	30.60
Hair	12 –16	...	...
Harbor mud	.99	.77	.05
Hoof meal and horn dust	10 –15	1.5–2	...
Incinerator ash	.24	5.15	2.33
Kentucky bluegrass (green)	.66	.19	.71
Kentucky bluegrass (hay)	1.20	.40	1.55
King crab (dried and ground)	10.00	.26	.06
King crab (fresh)	2 –2.5	...	...
Leather (acidulated)	7 –8	...	...
Leather (ground)	10 –12	...	...
Leather, scrap (ash)	...	2.16	.35
Lemon culls, California	.15	.06	.26
Lemon skins (ash)	...	6.30	31.00
Lobster refuse	4.50	3.50	...
Lobster shells	4.60	3.52	...
Milk	.50	.30	.18
Mussels	.90	.12	.13
Molasses residue in manufacturing of alcohol	.70	...	5.32
Oak leaves	.80	.35	.15
Oats, grain	2.00	.80	.60
Olive pomace	1.15	.78	1.26
Olive refuse	1.22	0.18	0.32
Orange culls	.20	.13	.21
Orange skins (ash)	...	2.90	27.00
Pea pods (ash)	...	1.79	9.00
Peanuts, seeds or kernels	3.60	.70	.45
Peanut shells	.80	.15	.50
Peanut shells (ash)	...	1.23	6.45
Pigeon manure (fresh)	4.19	2.24	1.41
Pigweed, rough	.60	.16	...
Pine needles	.46	.12	.03
Potatoes, tubers	.35	.15	.50
Potatoes, leaves and stalks	.60	.15	.45
Potato skins, raw (ash)	...	5.18	27.50
Poudrette	1.46	3.68	.48
Powder-works waste	2–3	...	16–18
Prune refuse	.18	.07	.31
Pumpkin, flesh	.16	.07	.26
Pumpkin seeds	.87	.50	.45
Rabbit-brush ashes	...	...	13.04
Ragweed, great	.76	.26	...
Red clover, hay	2.10	.50	2.00
Redtop hay	1.20	.35	1.00
Residuum from raw sugar	1.14	8.33	...
Rockweed	1.90	.25	3.68
Roses, flower	.30	.10	.40

TABLE 19: Percentage Composition of Various Materials (Continued)

Material	Nitrogen	Phosphoric Acid	Potash
Rhubarb, stems	.10	.04	.35
Rock and mussel deposits from sea	.22	.09	1.78
Salt-marsh hay	1.10	.25	.75
Salt mud	.40	...	...
Sardine scrap	7.97	7.11	...
Seaweed (Atlantic City, N. J.)	1.68	.75	4.93
Sewage sludge from filter beds	.74	.33	.24
Shoddy and felt	4–12	...	...
Shrimp heads (dried)	7.82	4.20	...
Shrimp waste	2.87	9.95	...
Siftings from oyster-shell mound	.36	10.38	.09
Silkworm cocoons	9.42	1.82	1.08
Soot from chimney flues	.5–11	1.05	.35
Spanish moss	.60	.10	.55
Starfish	1.80	.20	.25
String bean strings and stems (ash)	...	4.99	18.03
Sunflower seed	2.25	1.25	.79
Sweet potato skins, boiled (ash)	...	3.29	13.89
Sweet potatoes	.25	.10	.50
Tanbark ash	...	.24	.38
Tanbark ash (spent)	...	1.5–2	1.5–2.5
Tea grounds	4.15	.62	.40
Tea-leaf ash	...	1.60	.44
Timothy hay	1.25	.55	1.00
Tobacco leaves	4.00	.50	6.00
Tobacco stalks	3.70	.65	4.50
Tobacco stems	2.50	.90	7.00
Tomatoes, fruit	.20	.07	.35
Tomatoes, leaves	.35	.10	.40
Tomatoes, stalks	.35	.10	.50
Waste from hares and rabbits	7.00	1.7–3.1	.60
Waste from felt-hat factory	13.80	...	.98
Waste product from paint manufacture	.028	39.50	...
Waste silt	8–11	...	...
Wheat, bran	2.65	2.90	1.60
Wheat, grain	2.00	.85	.50
Wheat, straw	.50	.15	.60
White clover (green)	.50	.20	.30
White sage (ashes)	...	...	13.77
Wood ashes (leached)	...	1–1.5	1–3
Wood ashes (unleached)	...	1–2	4–10
Wool waste	5–6	2–4	1–3

—Compiled by the U. S. Dept. of Agriculture

Growing Green Manure Plants

Green manure plants are one of the best soil conditioners ever discovered. They cost little, take little time to use and provide the answer to good soil tilth.

Gardeners can make use of green manure plants to build soil fertility. Thanks to power equipment—especially the rotary tiller—every gardener, who may not have time to make enough compost for the whole garden area, can get the fertility build-up, biological activation, disease and insect resistance, by planting green manures.

Set aside a separate section each year for green manure treatment, or interplant, when practical, between row crops. In warm weather green manure plowed into the ground decomposes rapidly. If you start early in spring, you may even be able to use two successive green manure plantings this summer.

Here are some tips for efficient green manuring:

1. Legumes make good green manures, because they increase soil nitrogen supplies.

2. It is good to plant a vegetable crop within a short time after working in green manure—to take advantage of the nutrients given off by the decaying plant matter.

A rotary tiller is extremely valuable for sheet composting—working surface plant residues and additional materials into the soil where it will decompose and eventually loosen a heavy soil. Green manuring can easily be done with a tiller.

3. With sandy or porous soils, green manures can be left at or near the surface to retard leaching. In clay soils, it's best to incorporate the green manure into the soil where it will break down faster, since leaching in such soils takes place slowly.

Study the following list to see the plants best suited to your needs:

Alfalfa—deep-rooted perennial legume; grown throughout the United States. Does well in all but very sandy, very clayey, acid, or poorly drained soils. Inoculate when growing it for the first time, apply lime if *p*H is 6 or below, and add phosphate rock. Sow in spring in the North and East, late summer elsewhere, 18 to 20 pounds of seed per acre on a well-prepared seedbed.

Alsike Clover—biennial legume; grown mostly in the northern states. Prefers fairly heavy, fertile loams, but does better on wet, sour soil than most clovers. Sow 6 to 10 pounds per acre in spring. May be sown in early fall in the South.

Allyce Clover—annual legume; lower South. Prefers sandy or clay loams with good drainage. Sow in late spring, 15 to 20 pounds of scarified seed per acre.

Austrian Winter Pea—winter legume in the South, also grown in early spring in the Northwest. Winter-hardy north to Washington, D.C. For culture, see *field pea.*

Barley—annual nonlegume, grown in the North. Loams, not good on acid or sandy soils. In colder climates, sow winter varieties, elsewhere spring varieties, 2 to 2½ bushels per acre.

Beggarweed—annual legume; South, but growth fairly well north to the Great Lakes. Thrives on rich sandy soil, but is not exacting; will grow on moderately acid soils. Inoculate when not grown before. Sow 15 pounds of hulled and scarified seed or 30 pounds of unhulled seed when all danger of frost is past. Volunteers in the South if seed is allowed to mature.

Berseem (Egyptian clover)—legume for dry and alkali regions of the Southwest. Usually grown under irrigation. Will not stand severe cold.

Black Medic—legume; throughout the United States. A vigorous grower on reasonably fertile soils. Sow 7 to 15 pounds of scarified, inoculated seed in the spring in the North, fall in the South. Needs ample lime.

Buckwheat—nonlegume; grown mostly in the Northeast. Tops for rebuilding poor or acid soils; has an enormous, vigorous root system, and is a fine bee plant. Sow about two bushels to the acre, any time after frost. Can grow 3 crops, 40 tons of green matter per acre, in a season. Uses rock fertilizers very efficiently.

Bur Clover—a fine winter legume as far north as Washington, D. C., and on the Pacific Coast. Prefers heavy loams, but will grow on soils too poor for red or crimson clover, if phosphate is supplied. Sow in September, 15 pounds of hulled seed or 3 to 6 bushels of unhulled seed per acre. Volunteers if allowed to set seed.

Cowhorn Turnip—nonlegume; widely adapted. Its value lies in its enormous long roots that die in cold weather and add much organic matter in the spring. Plant in late summer, two pounds per acre.

Cowpea—very fast-growing annual legume. Thrives practically anywhere in the United States on a wide range of soils. A fine soil builder, its powerful roots crack hardpans. Inoculate when planting it the first time. Sow any time after the soil is well warmed, broadcasting 80 to 100 pounds or sowing 20 pounds in 3-foot rows.

Crimson Clover—winter annual legume; from New Jersey southward. Does well on almost any fairly good soil; on poor soil, grow cowpeas first for a preliminary build-up. Sow 30 to 40 pounds of unhulled seed or 15 to 20 of hulled, about 60 days before the first killing frost. Inoculate if not previously grown. Dixie hard-seeded strain volunteers from year to year in the South.

Crotalaria—annual legume; for very poor soil in the South and as far north as Maryland. Sow scarified seed in the spring, 10 to 30 pounds, depending on the variety. Makes sandy soil like loam.

Dalea (Wood's clover)—legume; northern half of the United States. Still being tested, but shows promise for strongly acid, sandy soils. Volunteers for many years.

Domestic Rye Grass and Italian Rye Grass—nonlegume; many areas. Wide range of soils. Sow 20 to 25 pounds in the spring in the North, fall in the South.

Fenugreek—winter legume; Southwest. Loam soils. Sow 35 to 40 pounds in the fall.

Field Brome Grass—nonlegume; northern half of the United States. Widely adapted as to soils. Good winter cover, hardier than rye. Sow in early spring or late summer, 10 to 15 pounds per acre.

Field Peas—annual legume; wide climatic range. Well-drained sandy to heavy loams. Sow 1½ to 3 bushels, depending on the variety, in early spring in the North, late fall in the South. Inoculate first time grown.

Hairy Indigo—summer legume; deep South. Moderately poor sandy soil. Makes very tall, thick stand. Sow in early spring, 6 to 10 pounds broadcast, 3 to 5 drilled.

Kudzu—perennial legume; South to Central states. All but the poorest soils. Commonly allowed to grow for several years before plowing under. Seedlings planted in early spring.

Lespedeza—legume; South and as far north as Michigan (Korean and sericea varieties in the North). All types of soil, but sericea is particularly good for poor, sour soils—for these, it's one of the best fertility builders available. Sow in spring, 30 to 40 pounds. Benefits from phosphate rock. Inoculate first time grown. Will volunteer if seed is allowed to set.

Lupine—legume; Southeast to North. Sour, sandy soils. Blue lupine is a fine winter legume in the South; white and yellow are most often grown in the North. Sow in spring in the North, late fall in the South, 50 to 150 pounds, depending on the variety. Always inoculate.

Oats—nonlegume; widely grown. Many soils. Winter oats suitable for mild winters only. Sow two bushels in the spring.

Pearl Millet—nonlegume; as far north as Maryland. Fair to rich soils. Commonly planted in 4-foot rows, 4 pounds per acre.

Persian Clover—winter annual legume; South and Pacific states. Heavy, moist soils. Sow in the fall, 5 to 8 pounds. Inoculate. Volunteers well.

Quaker Comfrey—a new crop currently being tested. Prefers clays, loams and sandy loams. Its huge leaves are generally chopped up for green manure. Rootstocks planted in spring or fall.

Rape—biennial nonlegume; many areas. A rapid grower in cool, moist weather. Sow 5 to 6 pounds per acre.

Red Clover—biennial legume; practically all areas, but does not like high temperatures, so is most useful in the North. Any well-drained, fair to rich soil; needs phosphorus. Its decay is of exceptional benefit to following crops.

A two-inch layer of leaf mold or compost can be spread over the surface of the vegetable plot. This material can then be raked over and worked into the topsoil to provide abundant humus for feeding plant roots which are located close to the surface.

Sow early in the spring to allow time for two stands, 15 pounds of seed per acre. Inoculate the first time grown.

Roughpea (caley pea, singletary pea)—winter annual legume; southern half of the United States and the Northwest. Many soils, but best on fertile loams. Sow 30 pounds of inoculated, scarified seed in the fall. Needs phosphorus. Will volunteer.

Rye—nonlegume; grown mostly in the Northeast and South. Many soil types. Sow 80 pounds in the fall. Tetra Petkus is an excellent new giant variety.

Sesbania—legume; as far north as Washington, D. C. Prefers rich loam, but will grow on wet or droughty land, very poor or saline soils. Very rapid grower in hot weather. Broadcast or drill 25 pounds in the spring.

Sour Clover—winter legume; South and West. Many soils. Sow in early fall, 15 to 20 pounds of scarified, inoculated seed.

Soybeans—summer legume; deep South to Canada. Nearly all kinds of soil, including sour soils where other legumes fail. Will stand considerable drought. Use late-maturing varieties for best green manure results. Sow 60 to 100 pounds, spring to midsummer. Inoculate first time grown.

Sudan Grass—nonlegume; all parts of the United States. Any except wet soils. Very rapid grower, so good for quick organic matter production. Use Tift Sudan in Central and Southeastern states to prevent foliage disease damage. Sow 20 to 25 pounds broadcast, 4 to 5 drilled, in late spring.

Sweet Clover—biennial legume; all parts of the United States. Just about any soil, if reasonably well-supplied with lime. Will pierce tough subsoils. Especially adept at utilizing rock fertilizers, and a fine bee plant. Sow 175 pounds of scarified or 25 pounds of unscarified seed, fall to early spring. Fast-growing Hubam, annual white sweet clover, can be turned under in the fall; other varieties have their biggest roots in the spring of the second year, so turn them under then.

Velvet Beans—annual legume; South. One of the best crops for sandy, poor soils. Produces roots 30 feet long, vines up to 50 feet long. Sow when the soil is well warmed, 100 pounds, or 25 to 30 pounds in wide rows.

Vetches—annual and biennial legumes; varieties for all areas. Any reasonably fertile soil with ample moisture. Hairy vetch does well on sandy or sour soils and is the most winter-hardy variety. Hungarian is good for wet soils in areas having mild winters. Sown in the North in spring, elsewhere in the fall, 30 to 60 pounds, depending on the variety.

Weeds—whenever weeds will not be stealing needed plant food and moisture, they can be used as green manures. Some produce creditable amounts of humus, as well as helping make minerals available and conserving nitrogen.

Plant a Winter Cover Crop

Here is a program for improving home vegetable garden soils suggested by E. C. Wittmeyer, extension horticulturist, Ohio State University:

Rye is the standard crop. Seed, between the rows, at the rate of 3 to 4 pounds per 1,000 square feet as soon as possible after August. Sow elsewhere in the garden as soon as crops are removed.

Rye grass is equally as good and may be used in place of rye, if the seeding can be made before September 15. Seed at the rate of 1 to 2 pounds per 1,000 square feet.

In either case, the green crop should be turned under the following spring, preferably before the "grass" is knee-high. Prepare the seedbed for the cover crop by cultivating it lightly; then sow the seed. Raking will cover the seed sufficiently.

If the rye or rye grass makes a very heavy growth and turns light green in color in the spring, apply an organic fertilizer high in nitrogen.

Grow Improvement Crop in Alternate Years

If the garden is a large one, it may be divided into two equal parts. Use one part for the vegetables and the other half for growing a soil improvement crop. Reverse the two sections the following year and continue this procedure until the soil is in good physical condition.

A. Mixtures of alfalfa, clover and grass may be seeded in March and plowed down the following spring. These materials may be seeded in the rye, which has been used for a winter cover crop; however, the rye should be cut in May or early June and left on the ground. Other legumes can also be used.

B. Corn may be planted thickly in early summer and turned under before rye is seeded in the fall. Buckwheat (1½ to 2 pounds per 1,000 square feet) and soybeans (3 to 4 pounds per 1,000 square feet) have also been used as summer cover crops.

The Best Conditioners

Recent experiments reveal that, after the addition of green manures and other crop residues, the soil bacteria produce materials called polysaccharides. These are the glue-like materials which stick the soil particles into aggregates so essential for a good structure.

The amount of these valuable polysaccharides, produced by decomposing green manures, can be tremendous. For example, agronomists at the University of Delaware report that decaying alfalfa and oat straw produced as much as 5,500 and 4,000 pounds per acre, respectively, of these glue-like materials only one week after they had been added to the soil!

4—Fertilizing Your Soil

A HIGHLY FERTILE SOIL is important to the growth of vegetables. And here again—just as with structure—organic matter plays a vital role.

Wesley Chaffin and Robert Woodward, agronomists at Oklahoma A. and M. College, have written: "Nearly all of the nitrogen and sulfur, and more than one-third of the phosphorus that become available for plant use are supplied by the organic matter.

"Smaller quantities of the other plant nutrients also come from this source: consequently, an increase in the rate of organic matter decomposition likewise increases the quantities of nitrogen, phosphorus, potassium, calcium, magnesium, and other plant nutrients in the soil solution."

All soils differ widely in their content of plant nutrients. This difference is caused not only by the fact that soils differ greatly in the original content of nutrients which they had but also because soils lose these nutrients through erosion, leaching and the harvesting of crops. Some of these losses are made

According to agronomists at Oklahoma A. and M. College, "Nearly all of the nitrogen and sulfur and more than one-third of the phosphorus that become available for plant use are supplied by organic material." Compost, especially that fortified with rock fertilizers, as shown above is the best way to supply that organic material to your plants.

up by the weathering of minerals, rainfall, action of earthworms and bacteria in the soil, and other natural soil phenomena. But serious deficiencies must be corrected if the soil is to produce adequate and healthy crops. This is the reason why organic fertilizers and mineral nutrients should be added—to increase crop yields as well as to produce crops with the proper nutrients.

Some gardeners and farmers have been led to believe that water-soluble fertilizers are most effective in feeding plants. Thus, since organic fertilizers are not water-soluble, these people consider them less useful. But whether or not a mixed fertilizer dissolves in water has little if any, effect on crops, notes Alva Preston, University of Missouri extension soils specialist.

To test the value of water-soluble plant foods, different University experimental plots were treated with mixed fertilizers of varying ability to dissolve in water and seeded to barley. All of the plots of barley appeared to be equal in growth whether they were treated with a mixed fertilizer that was 5 or 95 per cent water-soluble.

Despite considerable discussion on the subject of water-soluble fertilizers, the important thing is that plants get the needed nutrients whether or not they dissolve in water, he says.

Testing Soil

A good gardener wants to know as much about his soil as possible. He wants to know what general type of soil it is, what plants will grow best on it, and how to fertilize it. The simple truth is that most gardeners know practically nothing about the makeup of soil. If something doesn't grow right they merely guess at the proper remedy. It is easy to see that many benefits can be gained by learning more about the composition and capabilities of your soil.

There are two basic philosophies of soil management. Both have advantages and weaknesses, in varying degrees.

The Pure Chemical Philosophy: The pure chemical soil manager considers his soil to be basically a support for plant roots. He tries to find out exactly what each type of plant needs, and attempts to supply these needs with chemicals. Aside from the harm done by a heavy chemical diet, he fails because modern soil methods do not reveal all that goes on in good soil, and possibly never will.

The Pure Humus Philosophy: Some organic gardeners feel that since they give their soil continuous and heavy applications of humus, they do not have to study the individual nutrient needs of their soil. Most pure humus gardeners have a great deal of success, but they miss some of the fun of soil study, and could probably achieve even better results if they took the time to analyze their soil.

Do you know whether your soil is acid or alkaline?

Do you know what its humus content is?

What trace elements are lacking in your soil?

Does your soil have a sufficient amount of the major nutrients—nitrogen, phosphorus, potassium and calcium?

Does your soil have a hardpan below the surface and do you know where it's located?

If you can answer those questions, you are really interested in the welfare of your soil. If you don't have a perfect score, it will take only a small amount of time and effort to find the answers. There are two ways to find out what nutrients your soil is hungry for. You can send a sample of your soil to a laboratory or to your state college, or you can buy a testing kit and make many of the necessary tests yourself. It doesn't hurt to use both methods, because you will be able to double check your results. A home testing kit

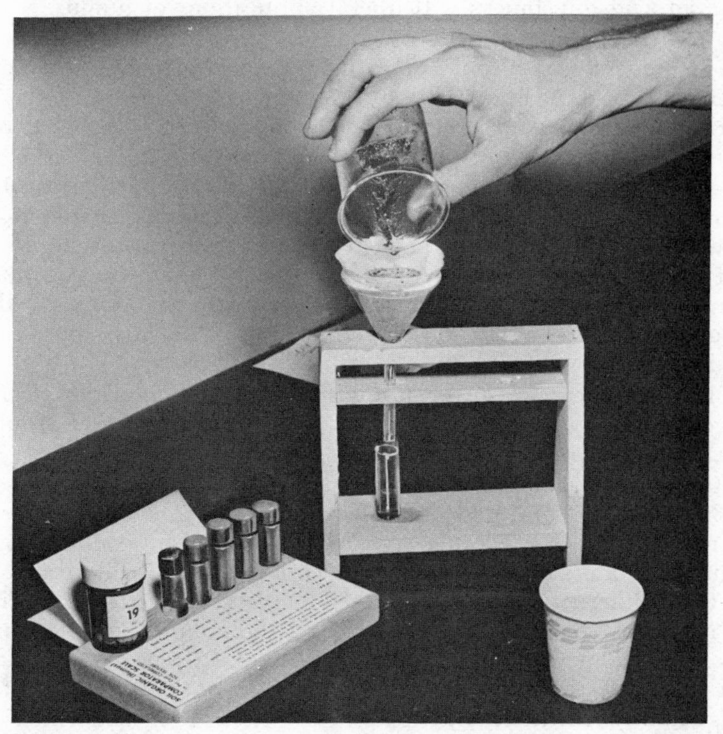

Soil testing will answer many questions about the make-up of your soil. You can learn if your soil is acid or alkaline, and if it has a sufficient amount of the major nutrients. Many gardeners prefer a home testing kit since it permits more frequent tests.

is valuable because it enables you to make frequent periodic tests of your soil. Most people don't realize that the nutrient supply in the soil varies greatly from one season to another.

A soil test will tell you what nutrients are "available" or soluble in your soil. There are also "unavailable" or insoluble nutrients in your soil which plants can feed on, but the supply of soluble and insoluble nutrients is usually quite similar. If your soil is low in soluble phosphorus, it will probably be low in insoluble phosphorus, etc.

The best place to start your soil analysis is with its *p*H. Is it alkaline or acid or neutral? This is important, because correcting the *p*H will often release supplies of major and minor plant foods. A soil that is too acid will not release "unavailable" nutrients properly, and the same is true of a soil that is too alkaline.

Many Western soils are very alkaline and, if owners of those alkali soils could correct their *p*H problems, most of their problems would be over. Western soils are mineral rich, but high (alkaline) *p*H locks up the nutrients that should be feeding plants. When considering your soil's *p*H, remember that humus or organic matter tends to neutralize soil. Organic matter is the best (and sometimes the only) remedy for alkaline soils. Of course, lime will neutralize an acid soil, but so will liberal applications of humus.

A test for the humus content of your soil is also important, because it indicates its general condition and will show you how successful have been your attempts to add organic matter to your soil. American soils today average 1.5 per cent humus, but 4 per cent is what we started with when our land was virgin.

A good way to gain an understanding of the major nutrient problems of your own soil is to use a portable testing kit. They are simple to operate and require no knowledge of chemistry or laboratory procedure. One of the chief advantages of a home test kit is the fact that you can make tests right on the spot—when and where you want them. Many of the state colleges give very slow service on soil tests during the busy months, so you can avoid the wait by making your own tests.

Portable kits will test for nitrogen, phosphorus, potash and *p*H. Deficiencies of nitrogen, phosphorus and potash can cause a great deal of trouble, and are in fact indicative of the basic value of your soil.

Collecting the soil sample is the first step in making a test. In fact, collecting the sample could be considered the most important and critical part of the testing procedure. There is often a great variation in soil condition nutrients in various parts of a garden or field, so it is important to collect a number of samples from different locations. These samples can be tested individually and the results averaged, or you can mix all the samples together and use a portion of this "homogenized" sample for your test. Always be careful to make sure that your collecting shovel or container is not contaminated by a fertilizer. That might throw off the results.

The test is made by putting a small portion of the sample in a test tube and then introducing one or two reagents. A reagent is a chemical which reacts with the nutrient being tested and shows the quantity of the nutrient available by changing color. Color charts are supplied with the test kits and the final analysis is made by checking the color of the solution in the tube with the test chart for the nutrient being tested.

Many test kits indicate the quantity of fertilizers to be applied to counteract the indicated deficiencies. Some of these recommendations are made in terms of chemical fertilizers, but you can easily transpose these recommendations into terms of natural or organic fertilizers. A general application of compost will correct an overall deficiency of nutrients, but it helps to act on individual deficiencies with natural rock fertilizers or nitrogen supplements.

Making Best Use of Organic Fertilizers

Fertilizing garden soil with organic materials is a simple job—except, perhaps, for the person doing it for the first time. For example, take the case of Mr. J. D., an organic gardener from Louisiana. He's just had his soil tested by the State University and was advised to use 800 pounds of 8-8-8 per acre. He writes: "Since I'm gardening the organic way, and don't want to use commercial fertilizers, I thought of the following substitutes that I can easily get: cottonseed meal (7 per cent nitrogen, 2 to 3 per cent phosphorus, 1.5 per cent potash), phosphate rock (30 per cent phosphorus) and wood

Natural rock fertilizers, such as rock phosphate, colloidal phosphate, greensand and granite dust, will supply plants with both major plant nutrients as well as many trace elements. They can be applied with a regular fertilizer spreader and mixed in with the first inch of the surface soil, or else worked deeply into the soil.

ashes (1.5 per cent phosphorus, 7 per cent potash). How many pounds of each should I apply to equal the fertilizer amount in 8-8-8?"

Like so many other organic growers, Mr. D. has a good, clear idea of what materials to use and where to get them. His main trouble, as we see it though, it that he's suffering from a "number" complex. Because some expert told him to use 8-8-8, he's going to struggle to match it—only with organic fertilizers instead of chemical. Aside from the needless mental activity (perhaps even anguish) caused by this complex, there are other disadvantages:

First, it's often difficult to equate the organic ratio with the chemical one. The result may be that the "new" organic gardener may say that the organic method is difficult or confusing.

Secondly, and perhaps the most common trouble, a lot of gardeners and farmers make the big mistake of not using organic fertilizers heavy enough on their first applications. We've found this to be true time and again. Advertisements of chemical companies who have just come out with an expensive fancy mix or super blend may advise applying at the rate of 200 pounds per acre, or a pound or two for the whole vegetable garden. Then when the organic grower wants to convert, he still thinks of such applications.

That's what harm the "number" complex can do in terms of worry, confusion and inefficiency. Our advice to Mr. D. of Louisiana (and every organic gardener and farmer worrying about chemical ratios) is to forget about the numbers, and concentrate on a long-range fertilizing program. Once you do this, you'll find yourself growing better plants . . . and sleeping better too.

Soil tests are a valuable guide to all gardeners and farmers, as they will indicate when and how to increase fertilizer applications. At first, these tests will show what elements are needed most; future tests will tell how well your fertilizing program is working out. But when these results are accompanied by suggestions for chemical fertilizers, don't feel you *must* come up with the *exact* mixture in organic fertilizers.

Nitrogen is a major element in plant nutrition. It is responsible for producing leaf growth and greener leaves; lengthening the growth period tends to increase set of fruits. Deficiency causes yellow leaves and stunted growths; excess delays flowering, causes too much elongation of stems, reduces quality of fruits, causes lodging of wheat, renders crops less resistant to disease.

If you believe your soil is deficient in nitrogen, you can correct it by adding compost, manure or other nitrogen-rich organic fertilizers such as dried blood, tankage, cottonseed meal, cocoa bean and peanut shells, bone meal or sewage sludge. Returning weeds, grass clippings and other garden wastes to the soil will add to its humus content and improve its nitrogen content at the same time.

ORGANIC VERSUS CHEMICAL NITROGEN: Organic forms of nitrogen are more stable in the soil and become available for plant growth more gradually than nitrogen from chemical fertilizers. When concentrated chemical nitrogen is applied to the soil, it produces a shot-in-the-arm effect to plant growth. The plants are subjected to too much nitrogen at one time. Then, if a sudden heavy rain storm drenches the field, the chemical form of nitrogen is to a large extent washed out, and the plants can become starved for lack of the element.

Soil Care and Nitrogen Content

Following is a report from New Mexico State University: A chemical test for organic matter present in the soil is routinely made when a soil sample is sent to the state soil test laboratory at New Mexico State University for analysis. The presence of crop residues and soil organic matter (crop residue which has been decomposed until its original identity is undetectable) have an important influence on the nitrogen status of a soil.

During the decomposition process of organic residues, microörganisms responsible for the decomposition "eat at the first table." Essentially, the soil microörganisms have the same kind of mineral requirements as do the crops you grow, although not necessarily in the same proportion. Every organism eats, grows, and reproduces as fast and as long as the environment and food supply permit. Thus, any nitrogen present in a residue is used first to satisfy the microörganism's requirements. This microbiological tie-up and release of nitrogen during the decomposition process exerts a large influence on the availability to growing crops of both the soil nitrogen and the fertilizer nitrogen.

Crop residues have wide ranges in nitrogen content. Wood wastes have low nitrogen content (1 to 3 pounds per ton of residue), whereas young alfalfa hay may have up to 60 pounds per ton of residue; and soil organic matter may contain up to 100 pounds of nitrogen per ton of residue.

When crop residues are added to the soil, a rapid build-up of the microbiological population takes place and a concurrent breakdown of the residue occurs. Within the first 8 weeks, under favorable conditions, the most rapid rate of decay occurs. This rapid decay occurs whether the residue is easily decomposed, such as green manures with a high carbohydrate and protein content, a good energy source for microörganisms, or the residue is mainly sawdust, a poor energy source. It is during this period that the soil is either enriched with a nitrogen release such as from succulent legume residues or depleted due to nitrogen tie-up during the decomposition of such residues as corn stalks or sawdust.

Following this initial rate of decay, there is a long period of slow rate of decay with very little change in the rate and a slow rate of release of nitrogen regardless of the kind of residue. Only a small fraction of the nitrogen content of soil organic matter becomes available to a crop each year. This is usually estimated to be 5 per cent, but it probably is more nearly 2 to 4 per cent. Thus, the year following the addition of a residue, such as corn stalks, one should expect to receive little or no benefit to a crop from nitrogen released from the residue. In fact some nitrogen from the soil may become unavailable to the crop because the nitrogen requirement of the microörganisms during decomposition must be satisfied first.

Some immediate benefit to a crop in the spring may be obtained from succulent green manures turned under during the previous fall. It is often assumed that if a fresh residue has more than 1 per cent nitrogen some inorganic nitrogen release will take place during the first 6 weeks; if the residue contains less than 1 per cent nitrogen, then additional soil nitrogen would be tied up in unavailable organic form. A rule of thumb is that 1 per cent nitrogen in the residue is needed to satisfy the microörganisms' appetites during residue decomposition.

What happens to the unavailable nitrogen that is tied up in an organic form? It eventually becomes available to crops grown on that soil but only very slowly. For instance, let us take a hypothetical, but reasonable situation: In the corn belt often 2,000 pounds per acre of corn stalks with .7 per cent nitrogen are plowed under. After 8 weeks, about 1,000 pounds of residue with 2.4 per cent nitrogen are left; after 18 months say, 500 pounds of residue with a 4 to 5 per cent equilibrium nitrogen content are left. What is our net nitrogen situation?

2,000 pounds at .7 per cent = 14 pounds of nitrogen in the fresh residue.

1,000 pounds at 2.4 per cent = 24 pounds of nitrogen tied up in organic form. This will require that 10 pounds of soil nitrogen be removed from availability and tied up in organic form.

500 pounds at 4 to 5 per cent = 20-25 pounds nitrogen. Either no additional loss of nitrogen to tie up in organic form or a net release of 4 pounds to the soil.

After 18 months, the decomposition rate is very slow. However, there is a net release of about one-half to one pound of nitrogen per acre per year. Legumes and meadow grasses will have a much higher net release of nitrogen during the first 8 weeks. However, after 18 months or after they become soil organic matter and unidentifiable, they too contain about 4 to 5 per cent organic nitrogen and release about one-half to one pound of nitrogen per year per acre for every 500 pounds of residue remaining.

A soil with 2 per cent organic matter has the potential of releasing 40 or 50 pounds of nitrogen to the soil during a growing season. However, all of this will not be available to the crop. Some of the nitrogen will be lost during the off season, and some will be lost through leaching, volatilization and fixation.

Fertilization to increase crop yields and the return of the increased residue is an excellent way to maintain a high level of soil organic matter and thus obtain a potential nitrogen supply.

Good soil management and an understanding of soil organic matter-nitrogen relationships can materially increase the efficiency with which this potential nitrogen supply of the soil is used.

—Vernal H. Gledhill, Extension Soil Specialist, New Mexico State University.

Phosphorus

All growing plants need phosphorus. It is important for a strong root system, for brighter, more beautiful flowers and for good growth. If plants are unusually small and thin, with purplish foliage, it may be an indication of a phosphorus deficiency in the soil. Phosphorus is also said to hasten maturity, increase seed yield, increase fruit development, increase resistance to winter kill and diseases, increase vitamin content of plants, while a deficiency causes stunted growth and sterile seed.

SOURCES OF PHOSPHORUS: You can best add phosphorus to your soil with rock phosphate, which is a natural rock product containing from 30 to 50 per cent phosphorus. When the rock is finally ground, the phosphate is available to the plant as it needs it. Rock phosphate is especially effective in soils which have organic matter.

Besides rock phosphate, other phosphorus sources are basic slag, bone meal, dried blood, cottonseed meal and activated sludge.

HOW MUCH TO APPLY: Vegetable gardens—apply liberally around plants and carefully mix with first inch of the top soil. In most areas, barring any great deficiencies or excesses, one pound for 10 square feet is a good amount to use and would be sufficient for 3 to 5 years.

One thing a farmer or gardener must remember: to get best results, you must have your phosphate level built up along with your organic matter. The normal supply of phosphate in our virgin soils averaged about 1,200

Broadcasting the rock powders and other natural fertilizers while preparing the seedbed, raises the general fertility level of the soil. The fertilizer can be applied by hand or with a spreader and should then be mixed in with the soil.

pounds per acre. In many areas, this level is now down to 125 pounds per acre.

The trouble of "too light applications" often pops up with rock phosphate. After a soil test shows a phosphorus deficiency, the new organic grower may continue to apply less than 500 pounds per acre, only substituting rock phosphate for superphosphate. It is impossible to build up the phosphate level with this application.

When we recommend phosphate to a farmer, we want him to use 1,500 to 2,000 pounds per acre on his first application. Say the phosphate cost him $34 per ton. If he is on a 4-year rotation, that figures out at $8.50 per year for phosphate. Then follow this up on each rotation with 1,000 pounds plus crop residue and you're increasing the mineral content as well as the fertility of your soil. Many farmers go by the rule: "Feed your phosphate to your clover, feed your clover to your corn and you won't go wrong."

Tests with different crop rotations at the University of Illinois showed that corn, beans and clover did just as well on applications of rock phosphate as superphosphate. The work also revealed that a 4-year rotation of corn, soybeans, wheat and legume hay uses fertilizers more efficiently than a two-year rotation of corn and soybeans.

Green manuring, with deep-rooted crops like sweet clover and alfalfa, helps to restore potash to depleted land. One to two thousand pounds per acre of greensand and granite dust will definitely help correct the problem. Many farmers feed phosphate to small grains and greensand to their corn. Many we know use 500 to 1,000 pounds per acre through the corn drill when they plant and are harvesting 100 bushels per acre.

The suggested amounts for applying the rock powders—colloidal or rock phosphate, greensand and granite dust—in the garden are the same: spread about 10 to 15 pounds per 100 square feet. In fall or winter, you can spread it right on top of the ground. In spring, dust a mixture of phosphate and potash (equal amounts) in the row; mix a good handful in the soil around plants such as tomatoes, cabbage, etc. The following year, use in the row only. In this way, you're returning more than what you drew out the previous year.

Other recent experiments at Illinois showed that a single application of potash per acre—applied ahead of either corn or wheat in a 4-year rotation of corn, soybeans, wheat and mixed hay—is just as effective as split applications on both corn and wheat. Therefore, growers can save time and money by applying the full amount of rock potash once instead of two or three times in a rotation.

Potash

THE VALUE OF POTASSIUM: Potassium is the third major nutrient and is very important to the strength of the plant. It carries carbohydrates through the plant system, helps form strong stems, and helps to fight diseases which may attack. If plants are slow-growing and stunted, with browning leaves and stunted fruits, there is probably a potassium deficiency. Potassium is said to accomplish the following: improves keeping quality of fruit; aids in the production of starches, sugar and oils; decreases water requirement of plants; makes plants more resistant to disease; reduces winter kill; promotes color of fruit; is essential for cell division and growth; aids plants to utilize nitrogen; balances effect of excess nitrogen or calcium; reduces boron requirements.

Deficiencies can cause firing of the edges of leaves which later turn brown and die. This affects lower leaves first, causing shriveled, sterile seeds. Corn ears also can fail to fill out when there is a potassium deficiency.

Potash Sources: Natural mineral fertilizers supply insoluble potash, which the plant can only take up as it needs it. That's why rock potash and other rock powders rich in potash are recommended. It has been found that the more soluble potash there is in the soil, the more plants will take up. And this "potash feast" will actually prevent the plant from taking up other elements it needs. Since the natural mineral fertilizers are insoluble, there is no worry about this occurring.

There are 3 main sources of potassium used by organic gardeners and farmers:

1. Plant residues.
2. Manures and compost.
3. Natural mineral sources, like granite dust, greensand and basalt rock.

Included in these categories are wood ashes (6 to 10 per cent), hay (1.2 to 2.3 per cent) and leaves (0.4 to 0.7 per cent). The best plan is to use both organic and mineral potash fertilizer—organic for short term potash release and mineral for the longer period.

Other plant foods needed by crops include calcium, magnesium, sulfur, iron, zinc, molybdenum, tin and iodine. These are called trace or minor elements.

These trace elements are very important to proper growth of plants, even though they are only needed in small amounts. Some in fact have been found to serve as partial substitutes for other nutrients and also have been found to increase plant resistance to diseases.

Place Fertilizers Deep: An important idea recently developed is the concept of the deep placement of fertilizer. Particularly in regions where the soil is poor or there is scarcely enough rainfall during the growing season for proper growth, deep placement is most effective.

The whole concept of deep placement is based on the placing of fertilizers deep down at the very bottom of the root system of plants. If this is done, the roots of the plants work themselves deep down in the soil to reach the nutrients. This means that the plants will become especially well-rooted. When there is a shortage of rainfall, the plants thus well-rooted will be able to stand up under the drought conditions. Plants that are inadequately rooted in the shallow topsoil will dry up and die when the topsoil dries out.

Another important reason why deep placement of organic fertilizer is better than chemical fertilizer is water retention. A good compost has billions of pores and absorptive plant cells per cubic yard. Chemical fertilizers have few and will retain no water except what little is held by the surface wetting of the chemicals. Good compost is like a sponge taking up water. It soaks it up when there is plenty of rainfall and releases it to the growing plants when there is insufficient rainfall.

List of Fertilizers and Soil Builders

Following is a description of 20 organic soil builders. There are many more than 20, but this list gives an indication of what they are, where they are available, and how to use them. Unless otherwise noted, these fertilizers can either be worked into the soil in spring or fall, top-dressed around growing plants, added to the compost heap or used as a mulch.

Basic Slag

An industrial by-product, resulting when iron ore is smelted to form pig iron. Rich in calcium, slag also includes valuable trace elements as boron, sodium, molybdenum, copper, zinc, magnesium, manganese and iron. Its efficiency varies with its fineness. Alkaline in action, slag is most effective on moist clays and loams, and on peaty soils deficient in lime. For light soils, tests show that it's better to mix with greensand or granite dust. Apply in fall or winter; especially effective on legumes as beans, peas, clovers and alfalfa. Available commercially.

Bloodmeal and Dried Blood

Blood collected in slaughterhouses, later dried and ground. Bloodmeal analyzes 15 per cent nitrogen, 1.3 per cent phosphorus, 0.7 per cent potash; dried blood contains 12 per cent nitrogen, 3 per cent phosphorus.

These materials can be used directly in the ground or composted. Because of its high nitrogen content, use very sparingly—a sprinkling is enough to stimulate bacterial growth. Some gardeners add dried blood to water before applying to plants or sprinkle plants thoroughly before adding. Excellent for hastening plant breakdown in compost heaps, dried blood is available from mail order firms, garden store and fertilizer suppliers, feed supply houses and slaughterhouses.

Bone Meal

Years ago, great amounts of buffalo bones were collected on western plains for use as fertilizer; nowadays, the main source comes from the slaughterhouse. Consisting mostly of calcium phosphate, the phosphorus and nitrogen content depends mostly on the kind and age of the bone. Raw bone meal has between 2 and 4 per cent nitrogen, 22 to 25 per cent phosphoric acid. The fatty materials in raw bone meal somewhat delay its breakdown in the soil.

Steamed bone meal, the type generally used, is made from green bones that have been boiled or steamed at high pressure to remove fats. Since bones are steamed, they are ground more easily and therefore considered in better condition for the soil. Steamed bone meal contains one to two per cent nitrogen, up to 30 per cent phosphorus.

There is also bone black or charred bone that has a nutrient content similar to steamed bone meal.

Bone meal is more effective when used with other organic materials. Generally, it acts more quickly when applied to well-aerated soils. It also tends to reduce soil acidity. Commercially available.

Compost

Compost can include any or all of the other materials listed here. Like humus in the soil, compost is the storehouse for plant nutrients. Here are some general instructions about using compost: if notably fibrous, it can be applied in fall to soil, where it will decompose sufficiently by planting time; most recommended time for applying is about a month prior to planting; if storing in open, keep heap covered to retain nutrients. For best results, apply compost liberally, from 1 to 3 inches per year. The analysis of compost, of course, varies with the materials of which it is composed. For example, adding dried blood or cottonseed meal, tankage, activated sludge or bone meal will increase nitrogen value. Rock phosphate, basic slag, bone meal and sludge will add to phosphorus content, while potash will be increased through additions of greensand, granite dust, manure and wood ashes. Generally the trace element content is good.

Certain materials and techniques act as compost activators and will speed up rate of decomposition. As previously mentioned, the high nitrogen and protein materials such as dried blood, bone meal and manure do an effective job. It's also worthwhile to shred materials before adding to the heap, using either a rotary mower or a compost shredder.

Cottonseed Meal

Made from the cottonseed which has been freed from lints and hulls, then deprived of its oils. (Cottonseed cake is one of the richest protein foods for animal feeding.) Its low *pH* makes it especially valuable for acid-loving crops. Cottonseed meal analyzes 7 per cent nitrogen, 2 to 3 per cent phosphorus, 1.5 per cent potash. A truly excellent fertilizer; available commercially.

Other vegetable meals which are also rich in nitrogen are soybean, linseed and peanut.

TABLE 20: Feed Analyses

FEEDS	Crude Protein	Total Digestible Nutrients	Mineral & Fertilizing Constituents		
			N	P	K
Vegetable Protein Concentrates:	%	%	%	%	%
Cottonseed meal, 43%	43.3	74.2	6.92	1.04	1.39
Cottonseed meal, 41%	41.2	69.8	6.60	1.22	1.48
Cottonseed meal, 36%	37.1	69.2	5.94	1.10	1.83
Cottonseed, whole-pressed	28.2	59.8	4.51	0.64	1.25
Coconut oil meal	21.3	77.7	3.41	0.64	1.95
Corn gluten meal	43.1	80.2	6.90	0.38	0.02
Distillers' dried corn grains	28.3	82.4	4.53	0.47	0.24
Linseed meal (all analyses)	35.4	77.2	5.66	0.87	1.24
Peanut meal (all analyses)	43.5	82.4	6.96	0.54	1.15
Soybean meal (all analyses)	44.3	78.4	7.09	0.66	1.77

—Taken from Feeds and Feeding, 21st Edition, by F. B. Morrison

TABLE 21: How to Use Meal as a Garden Fertilizer

WHEN	QUANTITY	HOW
LAWNS — Before seeding new lawns; or if already seeded, wait until grass is 2 inches high. For old lawns, early in spring preferred; but any time during growing season will give excellent results. One application is usually sufficient in all cases.	5 lbs. (4 qts.) per 100 sq. ft. or 100 lbs. per each 2000 sq. ft. of lawn (buildings excluded). Allowing for normal house areas: 40 ft. x 125 ft. needs 150#; 50 ft. x 125 ft. needs 200#; 75 ft. x 150 ft. needs 350#; 100 ft. x 150 ft. needs 500#; 150 ft. x 150 ft. needs 850#.	In all cases, apply by hand or with a fertilizer spreader. *New seed bed:* work thoroughly into topsoil with rake or harrow. *New and old grass:* apply when grass is dry; brush grass with broom or back of rake and soak thoroughly.
SHRUBS AND TREES — Early spring preferred but may be done later during growing season.	*Shrubs:* 1 lb. (.8 qts.) applied to soil for each 20 sq. ft. of perpendicular top area of shrubs; or 2 level tablespoonfuls per each sq. ft.	*Shrubs:* Apply to soil around shrub distributing most of it in a circle directly below the outer extremities of the top growth. Work it down into the subsoil as much as possible to encourage roots to grow downward and outward.
	Trees: 1 lb. to each 1 inch diameter of tree trunk.	*Trees:* Attempt to incorporate it as deeply as possible into the ground to encourage roots to penetrate deeply. A good method is to drive holes (1″-2″ in diameter and 18″-24″ deep) into the ground at the outer extremities of the top growth directly below the main top branches. Distribute the allotment for entire tree equally between the various holes, pack in bottom of hole and fill remainder of hole with water. Then refill hole with soil after water has settled.
FLOWERS AND VEGETABLES — Spring, before planting or any time during growing season.	*Spring:* 5 lbs. (4 qts.) per 100 sq. ft or 2 level tablespoonfuls per each sq. ft. For row or trench, 1 lb. per 25 running feet.	Before planting, broadcast evenly by hand or with fertilizer spreader, and work well into topsoil. For growing plants, apply around plants and work lightly into soil.

—*Courtesy of Archer-Daniels-Midland Co.*

Grass Clippings

Fairly rich in nitrogen, grass clippings are useful as a green manure to be worked into the soil, for adding to compost heaps or for mulching. Clippings from most lawns contain over one pound of nitrogen and two pounds of potash for every 100 pounds of clippings in the dry state.

Greensand and Granite Dust

Highly recommended sources of potash. Materials can be applied as a top dressing or worked directly into the soil. General recommendation is 10 pounds to 100 square feet. Tests have shown that the granite dust or stone meal supply adequate amounts of potash to growing crops. Granite and other potash-bearing rocks occur all over the country, and it is hoped that use of these materials for agricultural purposes will increase.

Glauconite greensand or greensand marl is an iron-potassium-silicate that gives a green color to the minerals in which it occurs. It contains about 6 or 7 per cent potash.

Hulls and Shells

Hulls and shells of cocoa beans, buckwheat, oats, rice and cottonseed are commonly used as a fertilizer and mulch. They decay readily and may be spaded into the ground. The coarse shells are excellent as mulches, while the finer ones (sometimes almost in dust form) can be applied to lawns and elsewhere with a spreader. Cocoa shell dust analyzes 1 per cent nitrogen, 1.5 per cent phosphorus and 2.5 per cent potash. In general, the hulls are richest in potash, although peanut shells analyze 3.6 per cent nitrogen, 0.7 per cent phosphorus and 0.45 per cent potash. Hulls and shells make an exceptionally attractive mulch, and are most effective when about one inch thick. Available commercially from nurseries, stores or by mail.

Leaf Mold

As fresh leaves are placed in a container (snow-fencing or one of wood or stone), shred them, if possible, and keep them damp. Apply ground limestone to offset the acidity, unless you plan to use the leaf mold around acid-tolerant plants only. Leaf mold from deciduous trees has been found to be somewhat richer in potash and phosphorus than that made from conifers. Nitrogen content varies, sometimes as high as 5 per cent.

Leaves

An abundant source of humus and mineral material, including calcium, magnesium, as well as nitrogen, phosphorus and potassium. They are especially valuable for use around acid-loving plants such as azaleas, rhododendrons, and hollies. (Some leaves, as sugar maple, are alkaline.) They may be applied directly in the soil, as a mulch, for leaf mold and for composting.

Manure, Fresh

A basic fertilizer used for centuries. Some manures, such as horse, hen, sheep and rabbit, are considered *hot* manures because of their relatively high nitrogen content. Rabbit manure, for example, analyzes 2.4 per cent nitrogen, 1.4 per cent phosphorus and 0.6 per cent potash. It's best to allow these manures to compost before applying directly to plants. Cow and hog manure, relatively wet and correspondingly low in nitrogen, are called *cold* manures and ferment slowly. All manures are excellent and should be included in an organic fertilizing program, when available.

Sheep manure is seldom obtained in a fresh, unleached condition, but is available in dried and ground form commercially. It analyzes about 1.25 per cent nitrogen, 1 per cent phosphoric acid and 1 per cent potash.

Commonly available at just about every fertilizer store, dried manure is always useful in the garden. Pulverized sheep, goat and cattle manure have about 1 to 2 per cent nitrogen, 1 to 2 per cent phosphorus, 2 to 3 per

(Top) Sowing the fertilizer in a narrow row where seeds will be planted puts the nutrients close to the roots and makes them readily accessible to the growing plant. (Bottom) Side-dressing, placing the fertilizer along the row after the plants have made some growth, can be done once or twice during the season as the plants need it.

cent potash, while poultry manure analyzes 5 per cent nitrogen, 2 to 3 per cent phosphorus, 1 to 2 per cent potash.

You can make broadcast applications of manure and work it into the soil by plowing, disking or spading for all garden crops.

TABLE 22: Rate of Application

	Per Acre	Per 1000 sq. ft.	Per 100 sq. ft.
Barnyard manure	15 to 20 tons	750 to 1000 pounds	75 to 100 pounds
or			
Poultry litter	7 to 10 tons	350 to 500 pounds	35 to 50 pounds
or			
Dried, pulverized manure	3 to 4 tons	150 to 200 pounds	15 to 20 pounds

Oregon State College horticulturist, A. G. Bouquet, who specializes in vegetable crops, has this to say about using manure in the garden:

In the earlier days of gardening, manures were a readily available form of fertilizer. In recent years, however, scarcity of this material has greatly reduced the available supply especially for city and urban gardeners. It is necessary that manures be stored under cover to prevent leaching of valuable plant food. Nitrogen, always a valuable element in gardening, is easily lost in the manure by leaching and hot fermentation. (To reduce the loss of nitrogen by fermentation in the form of ammonia, about 100 pounds of rock phosphate, 18 per cent, should be mixed with each ton of manure as it is being composted or piled.)

When manures are applied to the soil in a fresh condition, there is the least loss of plant food from leaching and fermentation. Manures in general should be applied early enough, possibly about 8 weeks before the planting season, to be rotted in the soil.

The rate of applying manure depends upon the kind and supply available, the kind of crops grown and the condition in which the manure enters the soil. At the rate of 8 to 12 tons per acre, the application on the home garden area will be about 100 to 150 pounds or 3 to 5 wheelbarrow loads per square rod (272 square feet). A small amount of manure on garden ground is better than none, for it acts as an inoculant and stimulant to the soil microörganisms as well as providing plant food and organic matter.

Under most conditions broadcasting is the best method of applying manure. On the farm a manure spreader can be used to advantage. Most manures are plowed or spaded under the soil so that the material is rotted in the soil before planting operations begin. Manure seldom is so finely rotted that it can be satisfactorily worked into the soil by disking. For some crops such as cucumbers, melons, and tomatoes, rotted manure is often applied in hills and thoroughly mixed with the soil of the hill rather than put in the bottom of the hole and covered with soil.

TABLE 23: Approximate Average Composition of Fresh Manures

		Contents per ton		
Source	Water	Nitrogen	Phosphoric Acid P_2O_5	Potash K_2O
	Per Cent	*Pounds*	*Pounds*	*Pounds*
Poultry	55	20.0	16.0	8.0
Sheep	68	19.0	7.0	20.0
Horse	78	14.0	5.0	11.0
Swine	87	10.0	7.0	8.0
Cow	86	12.0	3.0	9.0
Mixed	80	10.0	5.0	10.0

Poultry manure is the most concentrated of farm manures. One hundred hens in a year will produce 4 tons of droppings, exclusive of the bedding material, which may be straw, peat moss, sawdust or shavings. (It is advisable when composting poultry manure to add two pounds of phosphate to droppings of 100 hens per day. This prevents loss of nitrogen in the form of ammonia, and adds phosphoric acid.)

Because of the concentration of chicken manure more than usual care is necessary in applying it to garden plants. The application should be held to approximately 25 to 30 pounds to a square rod, or 2 to 3 tons per acre. As with other manures, it is advisable to apply poultry manure and bedding early enough for it to be decomposed before planting seed or setting out plants.

Peat Moss

Partially decomposed remains of plants accumulated over centuries under relatively airless conditions. Though it doesn't contain any nutrients, peat moss serves to aerate the soil, to improve drainage, ultimately to help plants absorb nutrients from other materials. Established lawns can be top-dressed with a one-half inch layer of peat moss twice a year, and an inch or more can be spread and worked into vegetable gardens and flowers beds. Extremely useful as a mulch. Commercially available.

Phosphate Rock and Colloidal Phosphate

An excellent source of phosphorus, phosphate rock also contains many valuable trace elements, including calcium, iron, sodium, magnesium, boron and iodine.

Sawdust

Since sawdust is low in nitrogen, containing about 1/10 of 1 per cent nitrogen, one of the objections against its use is that it may cause a nitrogen deficiency. However, many gardeners report fine results applying sawdust as a mulch to the *soil surface* without adding any supplementary nitrogen fertilizer. If your soil is of low fertility, watch plants carefully during the growing season. If they become light green or yellowish in color, side-dress with an organic nitrogen fertilizer as cottonseed meal, bloodmeal, compost, manure or tankage.

Some people are afraid that the continued application of sawdust will sour their soil, that is, make it too acid. A very comprehensive study of sawdust and wood chips made from 1949 to 1954 by the Connecticut Experiment Station, reported no instance of sawdust making the soil more acid. It is possible, though, that sawdust used on the highly alkaline soils of the western United States would help to make the soil *neutral.* That would be a very welcome effect.

Plentiful quantities of wood chips are becoming available in many sections of the country and are being widely used by gardeners and farmers. In some ways wood chips are superior to sawdust. They contain a much greater percentage of bark, and have a higher nutrient content.

The general verdict on sawdust and wood chips is that both materials are safe and effective soil improvers. They do a fine job of aerating the soil and increasing its moisture-holding capacity.

Sawdust is valuable as a mulch to the home gardener provided it is used intelligently. Raw sawdust is not a manure or compost and cannot be used as a substitute, that is, it should not be deeply worked into the soil.

Plant vegetable seeds in the usual way after the soil has been fertilized according to its needs. Then spread a band of sawdust about 4 inches wide and ¼ inch thick on top of the planted row. The mulch will help reduce crusting of the soil and allow the young seedlings to push through easily.

When the plants are about two inches high, apply a one-inch layer of sawdust over the entire area between the rows. Weeds more than one inch tall should be killed by cultivating or hoeing before the mulch is added. If weeds are less than an inch tall, the sawdust will smother them. Some weeds will continue to grow up through the mulch. These can easily be pulled out by hand when the ground is moist after a rain. Do not cultivate as this will mix the sawdust with the soil and destroy its value as a mulch.

Seaweed and Kelp

High in potash (about 5 per cent) and trace elements. Many seaweed users apply it fresh from the sea; others prefer washing first to remove salt. It can be used as a mulch, worked directly into the soil or placed in the compost heap. Dehydrated forms are available commercially.

Sludge

Activated sludge, produced when sewage is agitated by air rapidly bubbling through it, contains about 5 per cent nitrogen, 3 to 6 per cent phosphorus; that is similar to cottonseed meal. Digested sludge, formed when sewage is allowed to settle over filter beds, has about the same fertilizer value as barnyard manure—2 per cent of both nitrogen and phosphorus. Sludge is usually on the acid side. It can be worked into the soil in fall or in early spring at time of initial cultivation.

Tankage

Large quantities of slaughterhouse refuse, including rejected animals. These materials are placed in steel tanks and steam pressured for several hours.

Meat meal carries about 10 per cent nitrogen, up to 3 per cent phosphorus. Bone tankage contains 3 to 10 per cent nitrogen, about 10 per cent phosphorus.

Wood Ashes

Containing 1.5 per cent phosphorus, 7 per cent or more potash, wood ashes should never be allowed to stand in the rain, as the potash would leach away. They can be mixed with other fertilizing materials, side-dressed around growing plants or used as a mulch. Apply about 5 to 10 pounds per 100 square feet. Avoid contact between freshly spread ashes and germinating seeds or new plant roots by spreading ashes a few inches from plants. Wood ashes are alkaline.

Wood Chips

Like sawdust and other wood wastes, wood chips are useful in the garden. They have a higher nutrient content than sawdust and do a fine job of aerating the soil and increasing its moisture-holding ability.

Other Soil Building Materials

Depending upon where you live, there will still be many other soil-building materials available to you which are not listed above. Soybean meal, castor pomace, tobacco stems, hay and straw, hoof and horn meal, dry fish scrap, brewery and cannery wastes, feathers, industrial wastes—all are excellent and can fit into your organic fertilizer program.

TABLE 24: Quantities of Plant Food in the Composition of Average Yields of Vegetable Crops

			Plant food content			
Crop	Part of plant	Yield per acre	Water	Nitrogen	Phosphoric Acid	Potash
			Per cent	*Pounds*	*Pounds*	*Pounds*
Asparagus	Young shoots	3 tons	85	21	6	15
Rhubarb	Leaves and stems	10 tons	85	80	24	170
Snap beans	Leaves, pods	10 tons	85	150	35	160
Beets	Roots	10 tons	85	51	20	105
Cabbage	Heads	10 tons	90	60	20	80
Cauliflower	Heads	300 crates	90	57	25	86
Carrots	Roots	20 tons	85	92	52	212
Celery	Leaves and stalks	500 half-crates	84	70	50	187
Corn, sweet	Cobs and kernels	6 tons	75	90	38	72
Cucumbers	Fruit and vines	5 tons	95	30	14	80
Lettuce	Heads	200 crates	94	30	12	45
Onion	Bulbs	300 sacks	86	79	31	75
Parsnip	Roots	10 tons	80	44	40	130
Peas	Green peas	3 tons	85	69	18	27
Pumpkin	Flesh, vines, seeds	10 tons	90	58	17	83
Muskmelons	Fruit	200 crates	90	35	8	60
Spinach	Leaves, stems	6 tons	92	60	18	30
Tomato	Fruit, leaves, stalks	10 tons	92	170	53	250

—*Oregon State College*

A Review of Rock Fertilizers

There are 4 important natural rock fertilizers. Two of them, rock phosphate and colloidal phosphate, respectively contain 30 to 50 and 18 to 30 per cent phosphoric acid. Granite dust and greensand, which originally was an ocean deposit, yield 3 to 5 and from 6 to 7 per cent potassium respectively.

Rock phosphate is a natural mineral, ground to a meal or dust. In recent years it has been reduced to a fine powder to facilitate release of its nutrients in the soil. It does particularly well on acid soils and we will see that, combined with raw animal or green vegetable manures, it is a most effective fertilizer.

It has been determined that, due to the reciprocal action of the manures and phosphate powder, the fertilizing availability of each agent is increased when they are mixed together. The manures ripen more quickly and the nitrogen in the manure is absorbed more readily by the plant. The manure acids correspondingly work on the phosphate rock, making it more assimilable.

A series of experiments conducted in the Soviet Union reveal that soil fertilized with phosphate rock mixed with fresh manure yields 60 per cent more potatoes; 158 bushels per acre, compared to 97. When the manure and phosphate rock were not mixed, but applied separately, 145 bushels were raised. It has been estimated that combining the phosphate and manure increases the availability of the phosphorus to the plant by 150 to 200 per cent.

The next step is to mix potash rock and manure. Reports on the value of this extra procedure are lacking, but it is almost self-evident that the same benefits which derive from mixing phosphate and manure may be obtained. In any case, the experiment of mixing the two rock powders, phosphate and granite, with animal or vegetable manures should be made.

When you spread natural mineral fertilizer, mix it in under the ground. A rotary tiller is the ideal power tool for this vital chore, while the plow or disk can be used in the larger operations.

The principle of mixing fertilizer with soil must also be followed with flowers, shrubs and vegetables. The rock mixture should be spread under trees out to the drip line and then turned under at least one inch of soil.

One final point: when in doubt—apply more rock phosphate. It's cheap and slow-working; you won't burn tender roots and you are putting in a supply of fertilizer that will work for you from 5 to 10 years.

Colloidal phosphate is also a natural mineral product, found chiefly in Florida in sedimentary deposits of soft phosphate with colloidal clay. It is also obtained from ponds which occur where phosphate rock is mined hydraulically. It contains 18 to 25 per cent phosphoric acid and you should apply 50 per cent more than you would with rock phosphate.

Granite dust is an excellent source of organic, slow-working potash. Its potash content varies from 3 to 5 per cent, but granite dust has been obtained from a Massachusetts quarry with a potash content of 11 per cent.

The potash-bearing minerals in granite dust are the potash feldspars and micas, the latter containing the most readily released potash. Small amounts of trace elements are also present. Two tons of granite dust should be applied to the acre. However, as noted under the rock phosphate section, we believe granite dust should previously be mixed with the manure-phosphate combination, spread and then turned under. This should be done in spring, before planting, when you would normally start working the land. Smaller applications call for 20 pounds to 100 square feet and 200 pounds to 1,000 square feet of this complete fertilizer.

Greensand or greensand marl contains more potash than granite dust; 6 to 7 per cent compared with 3 to 5. Being an undersea deposit, it contains most if not all of the elements which are found in the ocean. It is an excellent soil builder.

Superior deposits of greensand contain 50 per cent silica, 18 to 23 per cent iron oxide, 3 to 7½ per cent magnesia and small amounts of lime and phosphoric acid.

Greensand has the ability to absorb large amounts of water and provides an abundant source of plant-available potash. Its minerals or trace elements are also essential to plant growth. Because it is versatile, it may be applied directly to the plant roots, left on the surface as a combined mulch-compost or used in compost heaps to stimulate bacterial action and to enrich the heap. Once again, combining it with the manure-phosphate rock mixture should be seriously considered.

Don't be afraid to use relatively large amounts of natural minerals. They are slow-working and cannot burn young plants. If the soil is alkaline, increase the amount of manure either in the mixture or put it directly into the ground before applying the mixture.

Rotating crops, growing legumes, will also reduce an alkaline condition. But if you apply rock phosphate, remember to turn it under thoroughly so the nitrogen-fixing bacteria growing on the legume roots can go to work on it.

Applying Fertilizers

Here's how horticulturists at the University of Washington describe the 5 common methods of applying fertilizer:

1. Broadcasting the fertilizers during seedbed preparation before plantings.
2. Sowing the fertilizer along the seed row at the time of planting.
3. Placing the fertilizer in or around the hill before or at planting time.
4. Applying the fertilizer along the plant row at some time during the growing season.
5. Applying in solutions.

Broadcasting fertilizers before or during seedbed preparation raises the general fertility level of the soil. Spread the fertilizer by hand or with a fertilizer spreader, but mix it well with the soil by disking or raking. You can make a broadcast application annually and, depending upon need, may or may not supplement it by applying fertilizer along the row after the plants have started to grow. With bulky organic materials as partially-rotted manure, you can spread one-half to one inch over the garden.

Sowing the fertilizer in a narrow furrow along the seed row at planting time puts the plant nutrients close to the roots and makes them readily accessible to the growing plant. The fertilizer should be placed slightly deeper than the seed, but never in contact with the seed. Placement about ½ to one inch to the side of the seed row and one inch deeper than the seed usually results in effective use of the fertilizer.

Placing the fertilizers around the hill when the seed is planted is an efficient way to use fertilizers for widely spaced plants. Place the fertilizer in a circular band around the hill 2 or 3 inches away from the seed and one inch deeper than the seed.

Fertilizers applied along the row after the plants have made some growth should be placed in furrows about 3 inches deep. If the plants are small make the furrows about two inches to the side of the plants. If the plants are larger make the furrow as close to the roots as possible without cutting or breaking them. After the fertilizer is placed in the furrow fill it with soil and moisten the soil by sprinkling. This method is useful when heavy fertilization is practiced. Garden crops may be "side-dressed" in this way once, twice or 3 times during the season if plants show need for it.

Fertilizers applied in solutions. Fertilizers which dissolve readily in water can be applied in solutions. Avoid strong concentrations on the foilage or too close to the roots. Beware of extravagant claims as to the value of fertilizer products, whether liquid or dry.

Converting Fertilizer Recommendations

Generally speaking, bagged fertilizers weigh about a pound to the pint. Some are slightly heavier; others that are bulky weigh less.

One pint is equal to two cups, 32 level tablespoons, or 96 teaspoons.

One acre contains 43,560 square feet, but for most purposes, the figure 40,000 is more convenient and yet is accurate enough in determining amounts of fertilizer for small areas.

Using these figures, you can figure fertilizer applications for small areas from given rates per acre. For each 100 pounds per acre, equal rates for small areas would be as follows:

for 1,000 square feet —— 2½ pounds or 2½ pints
100 square feet —— ¼ pound or ½ cup
1 square yard —— ½ ounce or 2½ teaspoonfuls

For each 2,000 pounds or one ton per acre, equal rates for small areas would be:

for 1,000 square feet —— 50 pounds
100 square feet —— 5 pounds or 5 pints
1 square yard —— ½ pound or 1 cup

TABLE 25: Amount of Fertilizer Per Row for Various Rates of Application

	Approximate amounts per 100 feet, for rows different distances apart					
Rate per acre	12 in.	15 in.	18 in.	24 in.	30 in.	36 in.
250 pounds	9 oz.	12 oz.	14 oz.	1 lb.	1¼ lbs.	1½ lbs.
500 pounds	1 lb.	1¼ lbs.	1½ lbs.	2 lbs.	2½ lbs.	3½ lbs.
750 pounds	1½ lbs.	2 lbs.	2½ lbs.	3 lbs.	3¾ lbs.	4½ lbs.
1000 pounds	2¼ lbs.	2½ lbs.	3 lbs.	4½ lbs.	5¾ lbs.	7 lbs.
1500 pounds	3½ lbs.	4 lbs.	5 lbs.	6½ lbs.	8½ lbs.	10½ lbs.
2000 pounds	4½ lbs.	5 lbs.	6½ lbs.	9 lbs.	11 lbs.	13½ lbs.

The amounts of fertilizer required on small areas or for row feet may be calculated from this table.

—Pennsylvania Agricultural Extension Service

Applying Lime to Your Soil

More than 300,000,000 acres of crop-producing land in the United States *need lime*. Despite that, leading soil experts say we're using only one-fourth as much as we should and point out that liming has declined steadily since 1947. This is foolish economy, they add, since lime not only has a yard-long list of advantages, but actually pays about $6 for every dollar invested.

Except for the highly alkaline areas of the Southwest and West Coast, most farm and garden soils require lime at one time or another. The penalty for

Ground limestone can be spread during any season. The way to determine whether your soil needs lime is by an accurate soil test, since too much can be as bad for your plants as a deficiency. Limestone can be spread by hand or with a fertilizer spreader.

neglecting or delaying its application when it is needed may be a gradual one, but it's certain and progressive. Lower yields, erosion, soil depletion—these are the costly results.

Why? Well, let's take a closer look at lime and how it functions. In agriculture, the chief source of this is limestone, rock and shell deposits which consist of calcium carbonate. Lime supplies calcium, an essential element for plant growth. Furthermore, it unlocks soil fertility, makes other nutrients available in acid soils. Where the *p*H (acid-alkaline balance) is too low, phosphorus and other elements are tied up, unavailable to crops. By acting on the aluminum and iron particles which cause this, lime releases the blocked plant foods. And while doing all this, lime helps to decompose organic matter, release nitrogen and stimulate the work of microbes and root bacteria.

Of course, the more widely recognized benefit from liming is that of neutralizing acid soil, raising the *p*H. Some call this making sour (acid) land sweet (more alkaline), which is literally what it does. Hydrogen in soil makes it acid. The calcium supplied in lime displaces some of this, reducing the acid effect. But at the same time, it improves soil structure by helping to grow bigger crops, increasing organic matter and providing a mellowing effect. Thus, we can see that lime is important not merely to control acidity but much more to bring vital calcium and its many related services to soil and plant life.

What Is pH?

*p*H is the chemists' shorthand method of expressing the amount of acidity or alkalinity. Just like the secretary's shorthand, it cuts out a lot of unnecessary writing. It is very confusing to have someone tell you that their soil is sour or bitter. It doesn't tell you how much sour or how much bitter. But with numbers standing for amounts of acidity or alkalinity there is no such confusion. Hence, we use the chemists' *p*H scale just like we would a yardstick. The difference is that with the yardstick we measure inches or feet, whereas with the *p*H scale we measure amounts of acidity or alkalinity.

The *p*H scale runs from 0 to 14. The 0 end of the scale is the acid end, while the 14 end of the scale is the alkaline end. Logically then, halfway between 0 and 14 should be the exact neutral point where there is just as much acid as alkali. This is true and 7.0 is the neutral point. A soil testing a *p*H of 7.0 will be exactly neutral. A soil testing any number greater than 7.0, for example, 8.5, will be an alkaline soil. Likewise, a soil testing less than 7.0, for example, 6.0, will be an acid soil.

How is the need for lime determined? Most crops do best on soil within a 6.5 to 6.8 *p*H range—that is, just slightly acid. If it is lower than that—even only a little—the difference in yield can be tremendous. Corn, for example, that will yield 100 bushels to the acre at a *p*H of 6.8, will return but 83 bushels when this drops to 5.7. Food value in what is produced is affected, too. Crops supplied the right amount of calcium are richer in minerals and in protein.

The way to find out whether or not your soil needs lime—and how much—is by an accurate soil test. Adding it by guesswork is definitely a mistake. (Too much can be as detrimental as a deficiency.) Thorough soil tests are available from your County Agent, from the nearest USDA state experimental station, from a number of reliable soil-testing laboratories and through use of several available test kits and *p*H gauges. Since different sections of a farm or even a garden can easily vary, it's best to get an analysis of each before liming or planting.

Once the need for lime is discovered, there are several important factors governing how much.

East of the "lime line" soils are generally acid and lime is often needed. X—soils rich in lime; A—fairly rich in lime, but some surfaces may be acid; B—acid soils but lime is available to deep-rooting plants; C—acid, poor in lime; AB—soils mixed high and medium lime; BC—mixed medium and low lime; ABC—mixed high, medium and low lime. *(Prepared by the United States Department of Agriculture.)*

pH Preferences of Common Plants

QUITE ACID (*p*H from 4.0 to 6.0)

Azalea
Bayberry
Blackberry
Blueberry
Chrysanthemum
Cranberry
Fescue
Flax
Heath
Heather
Huckleberry
Lupine
Lily
Lily of the Valley
Marigold
Mountain laurel
Oak
Peanut
Pecan
Potato, Irish
Radish
Raspberry
Rhododendron
Spruce
Sweet potato
Watermelon
Yew

SLIGHTLY ACID (pH 6.0 to 6.5)

Apple	*Gooseberry*	*Rape*
Barley	*Grape*	*Rice*
Beans, lima	*Kale*	*Rye*
Bent grass	*Lespedeza*	*Salsify*
Bluegrass	*Millet*	*Snap bean*
Buckwheat	*Mustard*	*Soybean*
Cherry	*Oats*	*Squash*
Collards	*Pansy*	*Strawberry*
Corn	*Parsley*	*Sudan grass*
Cotton	*Parsnip*	*Timothy*
Cowpeas	*Pea*	*Tomato*
Eggplant	*Peach*	*Turnip*
Endive	*Pear*	*Vetch*
Gardenia	*Pepper*	*Wheat*
Gloxinia	*Pumpkin*	

NEUTRAL TO ALKALINE (pH 7.0 to 7.5)

Alfalfa	*Cantaloupe*	*Leeks*
Alyssum	*Carrot*	*Lettuce*
Asparagus	*Cauliflower*	*Okra*
Beet	*Celery*	*Onion*
Broccoli	*Clover*	*Quince*
Brussels sprouts	*Cucumber*	*Spinach*
Cabbage	*Iris*	*Swiss chard*
Carnation		

Other Factors

1. *Soil Type*—Sandy and light-colored soils need less lime than silt or clay loams of the same *p*H. Soils containing a high percentage of organic matter need heavier liming than those deficient in this.

2. *Weather Pattern*—Areas receiving comparatively low amounts of rain require less lime than others; where the annual rainfall is 20 inches or better, more is needed.

3. *Crops*—What's to be grown has an appreciable influence on lime requirements. Legumes (which depend on calcium to form their nitrogen-supplying nodules) and some grasses need liberal amounts. A few vegetables and ornamental plants prefer an alkaline (*p*H above 7) soil. And some, like potatoes, blueberries and rhododendrons, demand a quite acid soil to thrive.

4. *Kind of Lime*—As noted, the primary source is ground limestone, which is usually 90 per cent pure. Dolomite lime, also a natural ground rock, has some magnesium along with its calcium. Either of these is suitable for liming and will not leave harmful residues in the soil. Hydrated lime, burned lime or quicklime act faster and are caustic; these will burn seeds and seedlings, destroy soil bacteria—and should not be used. In some localities, marl, marble, oyster shells, etc. (all of which contain some calcium) are used for liming. Limestone, in most areas, is easiest to get, costs the least.

5. *Fineness of Grind*—Effectiveness of limestone varies with how fine it is ground or crushed. Finer grades act sooner; coarser ones remain longer in the soil and are effective over a greater number of years. Limestone which passes through a 10-mesh screen and is pulverized fine enough so that 40 to 50 per cent will pass a 100-mesh sieve will give good results within 3 years.

6. *Time of Application*—Lime may be applied during any season that the soil needing it can be reached. From custom, farmers limed in the spring; many now do this in the fall or winter when there's more time and fields are more accessible. The more time that lime has to start its beneficial action the better, especially for acid-sensitive crops or in calcium-deficient soil.

7. *Method of Application*—Several ways to apply limestone are in common practice. For sizeable areas, the easiest means is by using a lime spreader made for this purpose. Other methods include adapting a manure spreader, using a shovel or distributing by hand in the garden. Best results come from having the lime thoroughly mixed with the soil by disking it in or using a subsoil chisel to get it plowed down deep. In plantng legumes, some farmers drill limestone along with the seed. Lime is also added in making compost, particularly for fertilizing acid soils.

Each of these factors should be considered carefully in determining how much lime is needed. And while doing this, *don't* forget any of the other fertilizer essentials, compost, mulch, rock fertilizers, etc. Lime may be urgently important, but it's not a cure-all. There's an old saying that "liming makes the farmer rich and the son poor" (by making crops grow larger and thereby taking more minerals out of the soil). By all means, add lime when it's needed, but don't substitute lime alone for good, all-round soil care and fertility.

Generally, if a soil is acid enough to require liming, it will pay to apply at least one ton of ground limestone to the acre. Some soils may require as much as 3 to 4 tons or more per acre depending on their acidity test and soil type. On a garden scale, one-half to one pound of limestone to each 10 square feet is a usual rate.

In liming, it is best not to add all of the required limestone at one time. Just like adding salt to the soup, it is easier to add a little more later, than to try and take out any excess. Limestone of about 60 mesh is the grade generally considered best. This lasts for about 3 years on average soil before it has been completely used up. About one ton per acre is required to raise a clay loam soil containing a medium amount of organic matter up one unit on the *p*H scale, say from a *p*H of 5.5 to a *p*H of 6.5.

Overliming is easiest on a light sandy soil, whereas soils with plenty of organic matter tend to resist injurious overliming. Liming should be done only when it is required to correct too acid a soil. For gardens limestone is best applied in compost or just before spading so that it can be worked into the soil. Limestone may be applied to shrubs and trees and worked into the soil as needed.

One little considered aspect resulting in a tendency towards excess soil alkalinity is that introduced by watering during the hot, dry months. Test the water to see whether it gives an alkaline reaction. If it does, some extra

TABLE 26: Weights of various fertilizing materials per acre, per 1,000 square feet, and per 100 square feet and the approximate equivalent-volume measures for 100 square feet, grouped according to weight in comparison with that of water.

Materials	Weights Specified per			Volume Measure for 100 Sq. Ft.
	Acre	1,000 Sq. Ft.	100 Sq. Ft.	
	Pounds	*Pounds*	*Pounds*	*Pints*
Weight about 8/10 that of water	1,740	40	4	5
Example: Bone meal.	650	15	1½	2
			Ounces	*Cups*
	175	4	6½	1
				Tbs.
	44	1	1½	4
			Pounds	*Pints*
Weight about 7/10 that of water	1,740	40	4	6
Example: Activated sewage sludge.	1,525	35	3½	5
	650	15	1½	2
			Ounces	
	300	7	11	1
				Cup
	150	3½	5½	1
				Tbs.
	44	1	1½	4
		Ounces		
	11	5	½	1
		Pounds	*Pounds*	*Pints*
Weight about 6/10 that of water	1,300	30	3	5
Examples: Cottonseed meal, fish scrap.	545	12½	1¼	2
			Ounces	
	260	6	10	1
				Cup
	130	3	5	1
Manure (moist):	*Tons*		*Pounds*	*Bushels*
Loose	13	600	60	2
Packed	13	600	60	1
Dry straw or leaves packed tightly with hands	5	250	25	2

—Prepared by Soil and Water Conservation Research Division, Agricultural Research Service

organic matter should be added to the soil to correct for this extra addition of alkali, that is, provided the soil is sufficiently alkaline already.

Soils which are too alkaline may be brought back to a favorable *p*H range by the addition of organic matter. Organic matter contains natural acid-forming material and produces acids directly on decomposition. These acids combine with any excess alkali, thus neutralizing it.

The nice thing about using organic matter is that a lot of it can be used. And it doesn't hurt the soil. It acts in a manner to control either excess alkalinity or excess acidity. Whichever way the soil is bad, either too acid or too alkaline, organic matter will tend to correct it. This is one of the reasons why gardeners get such good results by following the organic method. Regardless of what is wrong, plenty of organic matter will tend to correct it.

The incorporation of abundant amounts of good compost helps correct an adverse soil *p*H to the range wherein most common plants thrive. Even though the *p*H of the soil is not the optimum for the plants being raised, many plants will do well in a soil having plenty of good compost thoroughly incorporated into it.

Deficiencies in Plants

If a plant is not getting a balanced diet of all the elements it needs, it expresses its hunger in signs that can be seen on its leaves, fruit, stems and roots. If you know what these *hunger signs* mean, you can tell just how healthy your soil itself is and what additional elements it needs. And it is important not to wait until a whole field begins to complain of hunger by turning different shades of green, yellow or red or by dying prematurely. One or two unusual plants may hold the key to a sickness that is affecting a whole field and will cut its yield drastically.

If you do notice that your corn is turning yellow in the summer, what can you do about it? Not much at this time. It is hardly practical or logical to make a last minute side-dressing of organic fetilizer. The hunger signs your crops express *now* should dictate your *future* fertilizing, green manuring and cropping policy. A careful survey of all the abnormal plants on your garden or farm at this time will prevent your making expensive mistakes in future years.

But the true organic grower has one important consolation. A soil that is rich in organic matter and that has been fertilized with natural minerals (rock phosphate, granite dust and greensand) achieves a balance and a freedom from hungry crops completely foreign to the chemically fertilized soil. The organic gardener or farmer does not add a pinch of this and a pinch of that and hope that he has the right formula, but he gives to the soil the natural ingredients from which it was originally formed. Balance is the keyword of the organic method.

Soil Nutrients and Plant Composition

If a growing plant is short of one of certain essential mineral nutrients, it may not metabolize some amino acids or form the same amount of protein as plants supplied with all essential nutrients. But the proteins have the same amino-acid composition as that of normal plants of the same variety.

This was shown in studies at the USDA Plant, Soil and Nutrition Laboratory at Ithaca, N. Y., where researchers J. F. Thompson, C. J. Morris, and Rose K. Gering grew turnip plants in solutions containing all—or all but one—of the essential nutrients. Nutrients tested were nitrogen, phosphorus, sulfur, potassium, calcium, and magnesium. Plants short in any one nutrient were stunted and showed deficiency symptoms.

Visual Signs of Mineral Deficiencies

The visual symptoms of nutrient deficiencies may be expressed in all organs of the plant, including the roots. It will be seen that deficiency effects for nitrogen, phosphorus and magnesium always begin at the bottom leaves and develop in an upward direction towards the tips of the shoots, whereas those for calcium and boron begin at growing tips and progress towards the older parts of the plant.

Nitrogen Deficiency

The most common symptoms of nitrogen deficiency are restricted growth of both tops and roots; spindly plant; small leaves which are yellowish green when young but orange, red or purple when mature with the discolorations proceeding from the older to the younger leaves; premature falling of the leaves, beginning with the older leaves; reduction in the number of flowers with a corresponding reduction in yield of grain and fruits; and delay in such processes in spring as the opening of buds and the development of leaves and flowers.

Phosphorus Deficiency

Phosphorus deficiency produces all the modifications in plants as indicated above for nitrogen, but the leaf color usually is a dull, bluish green with tints of purple rather than yellow or red. Or the leaves may assume a dull bronzing with purple or brown spots. The leaf margins also often show brown scorching, especially in potatoes. In some plants, as the tomato for instance, it is difficult to distinguish between nitrogen and phosphorus deficiencies by means of leaf color. Fruit colors are helpful. They have a green ground color and may be highly flushed; the flesh is soft and the fruit is acid in flavor and has poor keeping qualities. In cereal crops, phosphorus deficiency may be recognized by the dark, bluish green color of the stems and leaves, which later become strongly tinted with purple. The purple pigment may also develop in the heads.

Potato leaves lose their luster, tend to curl forward and develop marginal scorching. There is a reduction in the number of tubers which are developed when phosphorus is deficient.

Calcium Deficiency

Calcium deficiency is expressed in the growing tips, in the young leaves particularly, and in the roots. The young leaves are severely distorted and may be rolled forward along the margins or curved backward toward the under surface. The edges of the leaves are irregular in form so that often they appear ragged, and may have thin, yellowish marginal bands or show brown scorching. The roots are poorly developed. In potatoes, the leaflets near the stem tips are very small and blanched, and are rolled toward the upper surface parallel with the midrib. Tuber formation is seriously affected and they may fail to develop altogether.

Potassium Deficiency

The effects produced by potassium deficiency vary somewhat with the degree of the deficiency. In grains, a mild deficiency causes the shoots to be thin, whereas a marked deficiency causes the shoots to die back and to develop into a stunted growth with numerous tillers which produce few or no flowers. In trees a slight deficiency is expressed in restricted shoot growth, whereas a greater deficiency inhibits shoot growth altogether and may even cause die-back.

The leaf symptoms are rather striking and variable, but characteristic for the different kinds of plants. As a rule the leaves assume a dull bluish green color and often show some yellowing in the areas between the veins. This is followed by browning of the leaf tips, scorching of the leaf margins and the development of brown spots in the leaves, especially near the margins.

There is a considerable stunting of the plants with poor root growth, short internodes and poor development of flowers, fruits and seeds. Whenever broad-leaved crop plants show scorched, rolled leaves, potassium deficiency should be suspected. This causes potatoes to assume a squat rather than upright habit with the leaf discolorations already mentioned. Tomato plants grow slowly and become woody, the leaves become greyish green and eventually become scorched and curl forward. The fruits do not ripen evenly, having green patches especially near the attachment to the stem. In clover there is a yellowish spotting and marginal browning of the leaves.

In all fruits, stunted growth and die-back of shoots and branches are common while the leaves have the characteristic symptoms. Flower buds may be plentiful in the early stages of the deficiency, but fruit yields are low and the quality is poor. Apples are sub-acid and have a woody taste. Other fruits are similarly deficient in quality.

Magnesium Deficiency

Magnesium is the only earth element which is used in the building up of chlorophyll, the green pigment which enables plants to make sugars and starches. When magnesium is deficient, the leaves tend to turn yellow and assume brilliant tints and drop from the trees prematurely. These symptoms first appear in the older leaves and then in the younger leaves. Leaf fall in apple may occur in August or September to such an extent that a heavy cropping tree may be left without leaves except those on the stem tips, and the fruits do not ripen properly.

Magnesium deficiency is expressed in cereals by dwarfed growth and yellow stripes between the veins in the leaves; in potatoes by a yellowing of the leaves at the base of the plant, followed by a dying and browning of some of the leaf blade; in tomatoes by a yellowing of the leaves which roll toward the upper surface and develop patches of dead tissue, the leaves finally withering and hanging vertically on the stems. In apples the yellowing is most marked in the intervenal areas around the midrib, leaving this part of the leaf blade more or less completely brown and dead. The fruits are small, poorly developed, and lack flavor. In other fruits, magnesium deficiency causes similar yellowing and dying of the leaves and lack of development in the fruits which are distinct and easily recognized.

Boron Deficiency

The young leaves of the terminal bud become light green (paler at the base than at the tip), cease to grow, and soon show signs of breakdown at the base. The upper leaves tend to roll in a half-circle downward from the tip toward the base. When boron shortage does not become serious until the flowering stage, the flower buds are shed and no fruits or seeds are produced.

Boron deficiency diseases are known as "heart rot" of sugar beets, "canker" of table beets, "brown rot" of swedes and turnips, "hollow stem" of celery, "yellows" in alfalfa, "internal and external cork" in apples, "top sickness" of tobacco, and "hard fruit" of citrus. An exceedingly small amount of boron is needed by the plant.

Manganese Deficiency

All plants respond to a deficiency of manganese by a yellowing of the leaves accompanied by other signs which are varied in different crop plants. In oats the base portions of the leaves, yellow at first, assume grey specks or streaks which tend to elongate and combine to form areas of brown tissue. The leaves may break down and hang vertically and finally wither and turn brown. This is sometimes referred to as the *grey speck* disease and may cause a complete crop failure.

In wheat and barley the leaves become somewhat pale green and may assume faint streaks and yellowing. Affected leaves on the lower part of the stem finally turn brown and drop off. In potatoes, the young leaves become pale and tend to roll toward the upper surface. Small brown spots may form along the veins. In fruits the leaves develop blanching between the veins which begins near the margins and extends towards the midrib, and finally only the veins remain green.

Sulfur Deficiency

There may be a reduction of growth and often there is a characteristic crimping downward of the leaves at the tips. The plants do not lose their lower leaves by firing, as they do in case of nitrogen shortage. This fact serves to distinguish the two deficiencies.

Iron Deficiency

Iron deficiency is revealed much more commonly in fruit trees than in vegetables. It can be recognized by a severe yellowing of the leaves, especially in the new growth. Since iron deficiency is often caused by overliming, it is called lime sickness.

Zinc Deficiency

When zinc deficiency develops in a peach tree for the first time, it is expressed as a yellowish mottling in the leaves late in the summer. It is usually manifested in trees that are two years old or older. The blanched leaves have, between the veins, yellow-green areas which later become light yellow or purple-red. The leaves drop prematurely so that many of the branches are entirely bare of leaves. The leaves assume a characteristic crinkling, and in severe cases become abnormally rigid and form red spots near the stem tips.

Other Deficiencies

Deficiencies of trace elements such as copper, cobalt and others do show up in some plants. However, these hunger signs are not common or important enough to be discussed here.

It must be realized that hunger signs are difficult to interpret accurately. Color reproductions of deficient plants are helpful. However, with a little practice and careful observation, any gardener or farmer can learn to know for what elements his plants are hungry.

—Dr. William Eyster

TABLE 27: Nutrient Chart

		Pounds per acre per year needed				
Crop	**Yield**	**Nitrogen**	**Phosphorus**	**Potassium**	**Calcium**	**Magnesium**
Corn	50 bu.	78.4	27.6	55.2	22.3	17.0
Oats	50 bu.	48.0	18.0	40.8	10.7	9.7
Wheat	35 bu.	59.5	23.3	29.4	11.8	10.3
Onions	300 bu.	39.3	15.4	37.6	23.1	5.8
Potatoes (white)	300 bu.	63.0	27.0	90.0	5.0	12.0
Potatoes (sweet)	300 bu.	41.2	16.5	82.5	5.5	17.1
Tobacco	1000 lbs.	59.0	7.7	78.0	64.0	11.3
Beans	25 bu.	88.0	24.0	57.5	45.2	11.9
Tomatoes	10 T.	48.0	16.8	84.0	...	6.0
Celery	5 T.	25.0	20.0	75.0	...	...
Cabbage	10 T	60.0	20.0	80.0	81.2	4.2
Beets	15½ T.	83.5	31.0	158.0	24.5	16.0
Turnips	12½ T.	70.0	26.0	117.5	35.5	7.6
Alfalfa	4 T.	190.4	43.2	178.4	126.6	53.4
Clover	2 T.	82.0	15.6	65.2	69.2	19.5
Timothy	1½ T.	37.5	16.5	30.0	8.1	5.6
Apples	35 trees	51.5	14.0	55.0	57.0	23.0
Peaches	120 trees	74.5	18.0	72.0	114.0	35.0

5—How to Make Compost

AT THE VERY BASE of the organic method lies compost. In its many forms and variations compost is the beautiful substance which gives fertility to soil, productivity to plants and health to man. It is the combination soil conditioner-fertilizer of the organic gardener, and the hub of all his gardening activities. If you are a successful compost maker, chances are 100 to 1 that you are a successful organic gardener.

In the past two decades there has been a great amount of research in composting methods, resulting in the 14-day method, sheet composting, anaerobic methods and many more variations of these. Behind them all, how-

The large compost heap in the foreground indicates the reason for the success of this Pennsylvania vegetable garden shown at the height of its growing season in August. The combined soil conditioning-fertilizing ability of compost makes it the key to good results.

ever, lies the original Indore method, invented by the father of organic gardening, Sir Albert Howard. The Indore method is still the most widely used, and is still practical and productive.

Sir Albert Howard found that by layering different organic materials, decomposition took place more quickly and more completely. He first placed down a 5 or 6-inch layer of green matter, then a two-inch layer of manure (bloodmeal, bone meal or sewage sludge may be substituted), and a layer of rich earth, ground limestone and phosphate rock. This simple formula produced a rich, crumbly compost, rich in nutrient value and valuable as a soil structure builder. In further research, Howard found that a heap 5 to 10 feet wide, and 5 feet high, was ideal (the length is optional). He also found that decomposition was facilitated by aeration, and so he placed pipes or thick stakes through the pile as it was being built, then pulled them out when the heap was 5 feet high. He then lightly pressed the entire outside surface to prevent blowing, formed a shallow basin on top to catch rainwater, covered the entire surface with a thin layer of earth, and left it to decay.

Organic gardeners have taken Howard's core of compost research and produced beautiful compost and beautiful gardens. Take the example of O. A. Severance of Watertown, New York, who transformed a completely unproductive piece of land into a lush garden spot, all through the use of compost. Mr. Severance makes compost in a pit, surrounded by a wall of

The first step in compost making is choosing a suitable location that is convenient to the garden. Into your composting heap can go a tremendous variety of materials, such as grass clippings, weeds, garbage, manure, leaves and many similar waste materials.

loose field stone 7 feet square on the inside and two feet high. The wall is laid on top of the ground and the soil inside is dug out a foot deep.

Into this pit go hen and stable manure, leaves, weeds, garbage, lawn clippings, sunflower stalks, some sod and ground limestone. This pit, layered according to Howard's Indore process, is level with the top of the stones when it is completed. Severance turns the pile in 3 weeks when he estimates the temperature has reached 150 degrees. Four weeks later he turns the pile again, in order to be sure all material has a chance to get into the center of the heap where decomposition is proceeding most rapidly. In a total of 3 months he takes out well over two tons of finished compost. In this way he can make two piles each season.

From time to time we get reports from readers indicating the tremendous power of compost. A sunflower seed dropped into a heap will grow to gigantic heights, or some other seed accidentally dropped into the heap will amaze the gardener. Art Ryden of McKeesport, Pa., planted a tomato seedling in a compost heap. At the end of the season the single plant had grown 5 feet high, 5 feet wide, and produced 400 tomatoes weighing 100 pounds. Glen Ruchty of Olympia, Washington, similarly grew a 129-pound squash in a compost pit, and Dorothy Hewett grew a 63-foot squash vine in an earthworm-compost bed. Compost alone is too rich for most plants, but on occasion will produce tremendous results.

From experimenting, gardeners often find ways to improve the Indore method, at least in their own gardens. Lois Hebble of Decatur, Indiana, uses a strip composting method. In the middle of the growing season, she lays heaps of organic materials on top of vegetable rows from which she has just harvested early crops. The material is partially composted by the next spring, but is not broken down enough for small seeds. Into these rows she plants melons, squash or cucumbers.

"For each hill," says Mrs. Hebble, "I scoop out a small hole and fill this with a shovelful or two of garden soil, then plant the seeds in this. Later in the summer, just before the vines start spreading out too much, I cover the strip with a good weed-smothering layer of old hay. By the following spring the soil under this strip has become mellow and homogenized enough to plant the smaller seeds. This method also keeps the garden crops in constant rotation."

Other gardeners use variations of the earthworm bed, sheet composting, mulching, pits, bins, plastic, shredding and numerous devices in trying to find the best method for them. You, too, can experiment with different methods to find *your* way of composting. But remember that the key to success is the Indore method. Learn it well and anything is possible.

Composting Under Plastic

Several years ago, a gardener came upon the idea of sealing his compost heap in plastic. This simple invention solved the perennial problem of compost odor, provided a simple way to discourage insects and rodents, and gave the home gardener a wonderfully simple and inexpensive way of composting anaerobically (without air). Now the suburban gardener can freely accept manure for his heap without fear of sanitation officials knocking on his door. He simply covers the compost heap with a large sheet of black polyethylene plastic (preferable to the transparent type), seals the bottom with a layer of soil, and forgets about it. No odors, no flies, no rats. And the anaerobic compost is finished in a shorter time than exposed heaps.

Composting under plastic is not only faster than regular methods, but it is easier, too. There is no turning, no watering, no extra work of any kind—except perhaps keeping the cover in place. First the ground is tilled, to loosen the soil and expose bacteria. The heap is then built upon this soil in the usual manner. It is watered down once, then covered with a sheet of polyethylene.

Users report wonderful results. Adrienne Bond, one of the early experimenters of the method, reported, "Just two and one-half months after we had sealed off the manure, we raised the plastic and found a pile of the blackest, crumbliest, sweetest-smelling compost you could hope for. The pile is almost exactly the same size it was before the composting began and although we have no way of being sure, we believe that the percentage of nutrients lost is very low indeed."

One word of caution—many children have been accidentally killed by smothering in polyethylene bags—the kind used by cleaners and for supermarket vegetables. If you use polyethylene of any kind, be sure not to leave any scraps around where children may play with them.

The 14-Day Method

You can make *18 tons* of crumbly compost in 6 months, in a space *8 x 4 feet!* This is enough compost to put a two-inch layer—an ample annual dosage—over 6,000 square feet of garden area, or a space 75 by 80 feet!

How is it done? Easy—using a shredder or rotary mower with the 14-day composting method. You begin in early spring, making a 4-foot high heap in a space 8 x 4 feet. After 3 weeks (we have allowed an extra week for possible slow heating), remove the finished compost—110 cubic feet, weighing about two tons—and begin the next pile. By mid-fall—six months later—your 9th heap should be finished, giving you a 6-month total of 18 tons.

The 14-day method has passed many trials with flying colors. Initially devised by an agricultural team at the University of California, the method was tried first at the Organic Experimental Farm in 1954, and since that time many readers have reported success with it. Here is the basic formula and procedure. This is not absolute, of course, and it may be altered to suit your individual needs and supply of materials:

First Day: Your basic material can be one of a number: leaves, spoiled hay, weeds or grass clippings. To this should be added one of the nitrogenous materials (manure is best) and any other material available. In the experiments, equal parts of leaves, grass clippings and manure, with a liberal sprinkling of natural rock powders, were found to work very well. Remember, too, that leaves tend to mat down, slowing the composting process; they should be mixed with other material. Shred everything—including the manure—with a compost shredder or a rotary mower; mix materials together and place in a mounded heap. If your materials are low in nitrogen, be sure to add a sprinkling of dried blood, cottonseed meal or other nitrogen supplement on each series of layers.

Second and Third Days: By now, the heap should have begun to heat up. If not, add more nitrogen. Bury a thermometer in the heap to check temperature. Keep the heap moist, but not soggy.

Fourth Day: Turn the heap, check temperature and keep moist.

Seventh Day: Turn it again, check temperature and keep moist.

Tenth Day: Turn it once more. The heap should now begin to cool off, indicating that it is nearly finished.

Fourteenth Day: The compost is ready for use. It will not look like fine humus, but the straw, clippings and other materials will have been broken down into a rich, dark crumbly substance. You may want to allow the heap to decay further, but at this stage it is perfectly good for garden use.

Earthworms Help Compost Process

Want an ever-ready supply of rich humus, and built-in humus beds? Want to stop giving away kitchen garbage to the city truck? If so (and who doesn't), then you should try this simple, but wonderfully efficient, composting method devised by J. J. Bartlett. He combines kitchen garbage, manure, green matter and earthworms into a neat, workable system, especially suited to the small-space residential gardener.

Many types of enclosures can be used to contain a compost heap. Some gardeners prefer a "fence" of chicken wire, since it can easily be removed when the compost is ready to be applied to the garden. Cement blocks are also often used for this purpose.

The first step is to build or find a box (no bottom or top). It can be of any size, but Mr. Bartlett has found that a long narrow one is more suited to inconspicuous placement along fences, in front of hedges, in borders and other small spaces. His boxes are 4 feet long, one foot high and one or two feet wide.

Next, pick a spot and dig a rectangular hole about 18 inches deep and just slightly smaller in dimensions than the box, so that the box will rest firmly on the ground above the hole. After this is done, you are ready to begin composting. The hole is filled in layer style—kitchen garbage, manure and green matter, in that order. Mr. Bartlett has found kitchen garbage to compost faster when run through a meat grinder, but this is not essential. He keeps a bag of pulverized manure and a pile of shredded green matter at the side of the box, and each time garbage is introduced he follows immediately with the other two layers. After each addition, he covers the pit with a burlap bag and wets it down, then places a board on top of the box. In this way there is absolutely no odor, nor any pest problem.

In about 3 weeks when the bottom layers have decomposed to a great extent, introduce about 500 earthworms. You may have to buy these, but they will be the only ones you'll ever have to buy; you'll soon have thousands, to use all around the garden. These little composters will work through successive layers which you add to the pit, mixing and breaking down the heap for you. Be careful, though, not to introduce the earthworms during the terrific heat of the initial breakdown. Heat of successive layers won't bother them, because they'll remain below until the above layer cools.

Continue this layering process until the pit is filled all the way to the top of the box. Then allow it to decompose for 5 or 6 more weeks, keeping it moist. In the meantime you can start another box.

After the first pit is fully composted, remove the compost down to soil level, place it in several cone-shaped piles on a large board or tarpaulin, and leave it exposed to the sun for one or two hours. The worms will have balled up at the bottom of each pile, and can be easily removed and introduced to the second pit. The finished compost piles may be sifted and bagged for future use. Meanwhile, you have just created the richest 18-inch-deep soil in town, which you can use as is, or tempered down or any way you like.

Now you may begin pit number 3, using the box you have just emptied. Using this system, you'll have a new load of compost each 6th week, you'll be able to make rich soil from the worst clay or sand; you'll be raising thousands of earthworms for general garden use; and you'll have done it with little trouble or expense in a small space, with no odor or pest problem.

Sheet Composting for Large Scale Gardens

The conception of sheet composting grew out of the needs of organic farmers and large-scale gardeners. These people found it too difficult and time-consuming to construct the compost heaps necessary to maintain the fertility of large acreages, and so necessity again became the mother of invention. Instead of composting in heaps, organic farmers began to spread compost materials directly on the soil, in raw form, then turn them under to compost right where they were to be used. Here are the basic principles of sheet composting, still the most efficient way to compost large areas:

1. Always have the soil protected by a crop of some kind, or by weeds if no crop is growing. Vegetable matter must be constantly produced by the soil. Only by such continual growth will the full energy of the sun be harnessed for the benefit of the soil.

2. Whenever possible, hasten the soil's production of humus by mixing what is growing into the surface of the soil, by any tillage method you find convenient or practical.

3. Help the soil produce humus yourself by replacing that organic matter which is taken off by cropping. Only by doing so can the permanent fertility and productivity of the soil be maintained.

Green manuring—the growing of cover crops to be turned under—is the most practical way to add substantial amounts of organic matter to a large land area. The green manure crop—usually one of the legumes—is generally planted after a food crop has been harvested, and is turned under about 6 weeks before the next crop is to be planted.

Sheet composting is the most efficient way to add organic material to large-scale gardens and farms. Instead of being composted in heaps, the materials are spread directly on the soil, and then turned into it with a machine such as a rotary tiller.

The most important thing in selecting a green manure crop is to select the one which will produce the greatest amount of organic matter in the time allowed. This may mean growing several stands in one season. Many gardeners with a large area rotate food crops with green manure crops, thus building up a healthy soil with little trouble or expense. If you have the room, green manuring should certainly be included as the mainstay in your soil-building program. The crops most widely used for green manuring are buckwheat, common sesbania, cowpeas, hairy indigo, red clover, soybeans, sudan grass and sweet clover. Even weeds are better than bare soil, and can add large amounts of organic material when turned under the soil.

Besides cover crops, the gardener or farmer must add any other materials readily available: manure, spoiled hay, corncobs, cannery wastes, seaweed, spent hops, sludge, wood chips or any other organic waste product. The addition of these materials is essential to increase the organic and nutrient content of the soil. Green manure and crop residues will help, of course, but they cannot do the job alone. Supplementary materials should be broken up by passing over them several times with a rotary mower. Then materials are worked into the soil with a rotary tiller.

Left over winter, the material will be greatly decayed by spring. It is a good idea to add limestone, phosphate rock, granite dust or other natural

The mineral content of compost can be increased by spreading bone meal or other natural fertilizers over each layer of materials as the heap is being built. Such applications also help to prevent loss of nutrients during decomposition process.

mineral fertilizers along with the other material, because the decay of the organic matter will facilitate the release of the nutrients locked up in these relatively insoluble fertilizers.

Because of the variances in soil needs, crops and available materials, no set program may be given. It is up to the individual farmer or gardener to determine his soil needs by soil testing, and to plan a fertility program to fill these needs. The farmer who lives in a lumbering region, for instance, is fortunate to be able to load up his manure spreader with sawdust several times a week.

The sheet composting idea has been used successfully for many years, in many variations. It is still the most efficient, most practical way to improve the tilth and productivity of large acreage.

Compost Questions and Answers

HEATING

Q. My heap does not seem to heat up at all. What is the reason for this?

A. *The most common cause of compost heap "failures" is a lack of nitrogen. A heap which doesn't heat up or decay quickly is usually made from material which is low in nitrogen, since nitrogen is essential as a source of energy for the bacteria and fungi which do the composting work. Thus it is advisable to add nitrogen supplements such as bone meal, cottonseed meal, tankage, manure or bloodmeal.*

Here compost is placed in trenches before plants are set in ground. This procedure is especially recommended for crops that prefer heavy applications of fertilizer. Drainage along the row is greatly improved by adding humus before planting.

ACTIVATORS

Q. How do activators help compost?

A. *So-called bacterial activators—"canned" cultures of bacteria—have been tested by several universities, including the University of California and*

Michigan State University. Their general conclusion is that these products do little or nothing to speed up the composting process. This conclusion was confirmed by experiments at the Organic Experimental Farm. The necessary bacteria will be found existing naturally in the composting materials, and none need be added. Since, however, the bacteria derive their energy from nitrogenous materials, it is essential that this food be present in sufficient amounts.

ACID LEAVES

Q. Will large amounts of leaves make my compost too acid for general garden use?

A. *All leaves are acid, some more than others. It is possible that large amounts of them may make compost too acid for most garden uses. This condition may be counteracted easily with additions of ground limestone.*

TURNING THE HEAP

Q. I would like to get away from turning the compost heap. How can this be done?

A. *The compost heap need not be turned at all if a sufficient number of earthworms are introduced just after the heat of fermentation has subsided. Sufficient moisture, nitrogen and aeration will also expedite the decay process without turning.*

WEEDS

Q. I find many weeds growing out of my composted garden areas. How may this be prevented?

A. *The high heat of the well-made compost heap should be sufficient to destroy all weed seeds which could otherwise sprout up in the garden after compost is applied. Be sure that the compost is fully decomposed before applying it. This may mean turning it several times, to make sure all material is exposed to the inner heat, or earthworms may be introduced to facilitate decay.*

VALUE OF COMPOST

Q. Are all composts of equal value as organic fertilizers?

A. *All composts add to the soil the much needed organic matter, but differ in plant nutrients according to the plant and animal residues used in making the compost heap. To make compost having the highest fertilizer value it is well to collect a wide variety of plant materials from different sources, including leaves of deep-rooted trees and natural rock powders.*

IN A GARBAGE CAN

Q. Could I make a compost heap in a new galvanized garbage can? I have only a small back yard and not much room.

A. *A garbage can can be made into an ideal composter. Punch holes in the bottom and in the sides as well as the lid. Place about 4 inches of fertile soil in the bottom of the can. Then layer your materials in the same way as for a large heap in the garden. Such a composter is not objectionable even in the best residential parts of the city. If desired, plants can be grown and trained over the can so that it becomes a mound of green leaves.*

WHEN TO APPLY

Q. Should compost be applied before or after planting?

A. *If possible, compost should be added before planting, so that it can be worked into the surface layer of the soil without injuring the plant rootlets. It is a good practice also to mulch the soil lightly with compost after the seedlings are an inch or two high. This mulch will help hold the soil against erosion and will discourage the growth of weeds. Half-finished compost should be applied in October or November, so that by the following spring it will have completed its decomposition in the soil and be ready*

Placing compost around young plants such as cabbage seedlings is most efficient, especially when there is a limited supply. Compost can be worked into the surface layer of the soil without "burning" the rootlets, since its nutrients are not water soluble.

to supply growth nutrients to the earliest plantings made. Otherwise, for general soil enrichment, the ideal time of application is a month or so before planting.

APPLYING COMPOST

Q. What is the best way to apply compost?

A. *For general application, the soil should be stirred or turned thoroughly. Then the compost is added to the top 4 inches of soil. For flower and vegetable gardening, it is best to pan the compost through a half-inch sieve. Coarse material remaining may then be put into another compost heap. Where compost is desired to aid a growing crop, it should be mixed with soil and applied as a mulch. In this way roots will not be disturbed and the top-dressing will gradually work itself down to plants.*

HOW MUCH COMPOST?

Q. For general garden fertility, how much compost should I use each year?

A. *For best results, compost should be applied liberally, perhaps from 1 to 3 inches in thickness per year. Of course, you can get by with as little as half an inch of compost, but in gardening with small plots, it should be applied heavily. There is no danger of burning because of overuse, such as is always the case with chemical fertilizers. You can apply compost either once or twice a year. The amount would depend, of course, on the fertility of your soil and how it is being used.*

6—Preparing the Soil for Planting

TWO FACTORS DETERMINE the best way for you to prepare your garden soil for growing plants—the type of soil and the size of your vegetable garden. The physical condition of soils depends upon their sand and clay content. If your soil is on the sandy side, you'll be able to cultivate it rather easily and relatively early in the season. If it's mostly clay, it's going to be heavy and therefore more difficult to work.

During late winter or in early spring before the garden plot is spaded and plowed, all coarse plant remains that would interfere with plowing or spading should be removed. Any large stones and debris should be hauled away or so placed that it will not detract from the overall attractiveness of the garden.

For small areas, most gardeners use a spading fork, since it's lighter to use than a spade and breaks up clods with less effort. For loose, sandy soils, regular garden spades will do a satisfactory job.

An important thing to remember when cultivating with either spade or fork is that your object is never to pulverize the soil. You do want to loosen and aerate it, but this does not mean that, when you're through, the soil has to pass through a sieve.

If there are many weeds, it's best not to dig them all under. It's helpful to chop up some of them with the sharp end of the spade and work them into the soil, but the majority should be pulled out, roots and all, before planting. You can either add the weeds to your spring compost pile or lay them in a pile for use later as a mulch around the plants.

Use a rake to make the top 2 or 3 inches fine and smooth enough for planting and sowing.

When loosening the topsoil, you may be tempted to go all the way, and put the subsoil on top. This is a dangerous practice, especially since—as in the case of so many new homeowners—the topsoil layer is no more than

Preparing soil for planting can actually begin the previous fall by turning under organic matter such as leaves into the surface soil and permitting them to overwinter. Most of this material will decompose in time for spring planting of vegetables.

several inches deep. Remember, your object is to loosen, not invert. This again is another reason why a fork is a better tool than a spade, since there is less tendency to invert the soil when using a fork.

In using a spade or spading fork, the beginner is generally inclined to dig up too big a "bite" of soil at one stroke, with the result that it cannot be turned over and broken to pieces as well as it should be. Effective spading is an art that requires some practice and judgment for properly working over different kinds of soils. Upon lifting up a spadeful of soil, one should overturn and shatter or slice it to pieces, leaving a loose crumbled mass.

Not more than an inch or so of the hard, unweathered subsoil should be turned up to the surface, even though the usual recommended depth for loosening the soil is 8 to 10 inches. If spading or forking to 8 inches does bring up too much subsoil to the surface, double spading may be necessary. This is a laborious practice, more widely used in Europe than in this country, designed to loosen up the subsoil beneath a shallow surface soil. Here's how double spading is done, as described by United States Department of Agriculture agronomists Victor R. Boswell and Robert Wester:

Starting at the end, spade out a strip a foot or so wide across the plot, removing only the surface soil. On removing this soil, throw it out at the end of the plot, out of the way, but where it can be picked up and later returned to the plot. This will leave an open trench just as deep as the surface layer. The second step is to dig up the subsoil at the bottom of this trench, leaving the subsoil in place after breaking it up. If well-rotted manure or other suitable organic matter is available it should be worked into this subsoil layer at this point. The third step is to spade out the next strip of surface soil, of course continuing on from the strip first spaded out. This strip of surface soil is thrown on top of the recently worked subsoil described in the second step. Naturally, moving the succeeding strip of surface soil exposes a fresh strip of subsoil, which is to be spaded up like the first. This sequence is continued until the plot is all spaded, thus working the surface soil and some of the subsoil but without mixing the two.

Some apparently good surface soils are underlain with a very heavy, tight claypan or hardpan, a layer that is very poor in nutrients and that the roots and moisture can hardly penetrate. When such a layer is very near the surface of the soil, few vegetables can be grown successfully. It is usually necessary to break it up and work into it some surface soil and manure, compost, ground limestone and possibly sand to "lighten it up." Claypans or hardpans are often so very hard that a pick must be used instead of a fork or spade in digging into them.

A true hardpan is formed by the cementing together of the soil grains into a hard stone-like mass which is impervious to water. A more common condition is an impervious layer in the subsoil caused by the pore spaces becoming filled with fine clay particles. Such "tight clay" subsoils, called claypans, are generally associated with an extremely acid condition, so that from both the physical and chemical standpoint they are objectionable.

When hard or claypans exist, the surface soil is cut off from the subsoil; no new minerals are added to the lower part of the soil; plant roots often are unable to penetrate these layers. Plant roots usually grow down to this hard layer and then extend horizontally over the top of it. This results in shallow-rooted plants which may suffer from lack of nutrient elements otherwise available in the subsoil and from water during the dry summer months. Often such shallow-rooted plants die out completely from lack of water during dry periods, while plants nearby where there is no hardpan flourish and grow vigorously.

If land is characterized by poor drainage and large water run-off after a rain or if water stands in ponds after a rain, there is good reason to suspect the presence of hardpans in your soil. Hardpans are horizontal layers sometimes one-sixteenth to one-half inch in thickness. These structures are usually detected by passing a blunt-pointed steel rod, attached to a handle, down through the soil profile. If resistance to the downward descent of the rod is observed at a certain depth, the depth of the obstruction is then observed. The procedure is repeated many times in the general area of the first test. If resistance to the downward thrust of the rod is observed at about the same depth on all tests, a hardpan is undoubtedly present.

The best and most universally used method of breaking up hardpans is by subsoiling. The basic idea of subsoil tillage or chisel plowing is to get more water and organic matter down deeper into the soil and to disturb the land's surface as little as possible. Generally, the word "subsoiling" has come to mean working or cutting up the land with a subsoil chisel to a depth of from 16 to 30 inches and with heavy single chisel implements to over 5 feet. Power requirements for these heavy deep-working tools are high. Sometimes the combined power of 3 of the biggest tracklayer tractors is needed. But for the lighter tools, with a single chisel, penetrating up to 30 inches deep, the power of the ordinary wheeled 2 to 3 plow farm tractor is usually sufficient.

When preparing a seedbed for the first time, spade or rotary-till the subsoil so that it is sufficiently loosened. Well-rotted compost or manure should be added to the soil during this preparatory stage to improve tilth.

Subsoiling can be carried out at any time when the land is not frozen or covered with a growing crop. It generally is advisable to subsoil at the driest time of the year, during the late summer or early fall months, so that the cuts may be open to receive the water of the fall rains and winter snows.

Since subsoiling leaves residues on top of the soil, it serves to reduce run-off. Tests at the University of Nebraska Experiment Station have shown that the run-off from plowed land is from 2 to 3 times as great as where subsoiling was carried on.

Research at the Idaho Experiment Station has revealed that 4 per cent more of the moisture received was retained under subtillage methods compared to one-way disking and moldboard plowing.

Besides controlling erosion, the subsoiling practice most times boosts yields. In southern Idaho, where the annual rainfall is about 15 inches, experiments have clearly shown yield advantages of 3 to 4 bushels of wheat per acre gained from subtillage. In areas of greater rainfall, subsoiling has not made appreciable differences in the yield figures, but here it is well to remember that the practice saves the soil and helps insure continued high yields.

The idea of subsoil improvement by means of plant growth can well be regarded as an accompaniment to other methods of subsoil treatment. Growing deep-rooting legumes supplements the mechanical steps and can make an effective combination to prevent water run-off and erosion. As the roots of these plants decompose, they put organic matter down deep into the subsoil and leave channels to facilitate water intake. Alfalfa and sweet clover have long been the standard plants for this soil conditioning service and other legumes are also being tested.

When to Prepare Soil

The first preparing of soil in spring should never be done when the soil is wet, unless your soil is extremely well drained. It's not only a lot more work to cultivate a wet soil, but it's also a harmful practice. Wet soils tend to bake, puddle and, after drying, form brick-like clods that defeat the entire purpose of cultivation.

Many gardeners, in their desire to get an early start, spade or plow the garden while it is still too wet. Because of the puddling and packing that takes place as a result, the gardener will find that he has to wait for weeks, or even months, until the soil's hardness and cloddiness break down.

Your soil is ready for digging when a handful, firmly compressed into a ball, can be broken apart by the fingers when the pressure is released. If it remains in a wet sticky mass, with moisture on the surface, it's still too wet to work. Sandy soils dry out more quickly than silty and clay soils; also sandy soils are damaged less by working when somewhat wet than are heavy silty or claylike soils.

Using Power Equipment

Rotary tilling has become a widely used and proven method of seed-bed preparation, particularly fine for working cover crops into the soil and building up the humus content. If the cover crop is heavy, cut it first with a rotary mower, as the shorter pieces are more easily worked into the ground.

Whenever possible, use tilling as an opportunity to work organic matter into the ground. If you cleared your garden the previous fall of all plant remains and there are only a few weeds growing, spread sawdust, compost or other organic materials before tilling.

Starting with Sod

Areas in sod or that have not been in cultivation for many years should be tilled, spaded or plowed in the fall, and such manure or other organic matter as is feasible should be turned under. A cover crop adapted to the locality should be sown if there is time for it to make enough growth to

1. For small areas, a spade or spading fork is most commonly used.

2. A spading fork is recommended to break up clods in heavy soil.

3. The object is to loosen and aerate soil so that plant rootlets can easily develop.

4. Overturn the forkful of soil; try to avoid placing subsoil on the surface.

5. Slice any clods of soil so that it is reduced to a loose, crumbled texture.

6. Rake the top two or three inches fine enough for planting seeds and seedlings.

7. Some little roughness on the surface improves water penetration into the soil.

8. Boards across garden provide a convenient platform when soil is damp.

protect the soil before cold weather. If the soil is spaded too late for a cover crop, the surface should be left rough, and generous amounts of straw, strawy manure, leaves, or other coarse organic matter should be chopped into the surface. This may appear a little unsightly, but it will increase absorption of rain and help prevent blowing and washing of the soil. The coarse material that might interfere with working the ground can be raked off in the spring and put on the compost pile.

No-Digging Method

There is a school of gardening which adheres to the belief that no-digging or soil spading is necessary to achieve best garden results. In 1946, F. C. King, an Englishman, published a book entitled *Is Digging Necessary?* which set forth a no-digging method dependent on optimum soil life activity and heavy composting. In 1949 another Englishman, A. Guest, published *Gardening Without Digging,* setting forth essentially the same principles as Mr. King.

Here are the basic principles:

1. To imitate nature closely by not inverting the soil.

2. To economize on compost and other organic materials by using them as a surface mulch, where nature keeps her fertility promoting materials.

3. To reduce weed growth by not bringing more and more seeds to the surface.

4. By all these methods, to maintain a balance of air, moisture, biological life and plant foods.

Not everyone who has experimented with the no-digging method has had success to rave about, so try it out with some caution. For example, one grower found that after a few years, soil fertility seemed to decline; he also was unable to get good yields of root crops (especially radishes) on his undug plots.

The System of Ruth Stout

The foremost exponent of the "no-digging" or "year-round mulch" system of gardening in this country is Ruth Stout of West Redding, Connecticut. Here is Miss Stout's description of how she got started in mulch gardening:

"One maddening thing about gardening is the plowing or spading, whichever one you do. If you hire someone to plow and harrow, it is not possible to have it done piecemeal. This means that the whole plot should be plowed early in April in order to get the peas, spinach, lettuce and parsley started on time. But if you do that, the part where you will put your late crops is lying idle through April, May and part of June. No, not idle, that's the trouble; it is growing a fine crop of weeds if there is rainy weather. And if it is dry the sun is baking the soil.

"If you spade your garden, of course you can dig up just as much as you are ready to use. However, that means that the rest of it is producing weeds, some of them perennial and mean to handle."

Each year—for 14 years—Miss Stout had her large (240 feet by 100 feet) garden plot plowed by a neighboring farmer. Then in April of 1944, the inevitable happened: the farmer was ready to plow, but the tractor wouldn't go.

"So now on this perfect morning I stood there in the garden, longing to put in some seeds. I wandered over to the asparagus bed and said to it affectionately: 'Bless your heart, you don't have to wait for anyone to plow you. You merely . . .'

"I stopped short as a thought struck me like a blow. One never plows asparagus and it gets along fine. Except for new sod, why plow anything, ever? Why turn the soil upside down? Why not just plant?

A furrow for large seeds such as beans can be made with the edge of a hoe as illustrated above. This soil, which has been fertilized and conditioned with organic soil conditioners, is now ready to receive seeds and adequately nourish plant growth.

". . . It was my good fortune that, in spite of all warnings against it, I had formed the habit of leaving all the vegetable waste, such as corn stalks, right there in the garden and had spread leaves all over it in the fall and vegetable garbage all winter long. Now, when I raked this mass of stuff aside to make a row for the spinach I found the ground so soft and moist that I made a tiny drill with my finger.

". . . If it really worked, in May and June the ground would surely be soft enough to put in corn, beans and the other late things. With all these leaves no weeds would come through. Some did, however. The mulch wasn't thick enough.

". . . Even that first year I began to visualize the Utopia I had thought up for everyone who wanted to grow his own vegetables. Besides the expense, it is not easy to find someone to plow and harrow for you and often impossible to get it done just when you want it. The alternative, spading, is quite a job. Eliminate these things and eliminate also hoeing, weeding, cultivating–it sounded like science fiction and yet I believed in it.

" . . . The first 3 years were a struggle and a mess . . . but I must reassure all prospective mulchers; those first years needn't have been difficult. Once you get it into your head that you have to put on enough mulch, 6 to 8 inches deep, you do it, relax and enjoy your leisure.

"It took me a long time to realize how much was necessary. It was hard to find enough; then I remembered that I had heard farmers talk about spoiled hay and I asked a neighboring farmer if he had any. He did and said if I could get someone to bring it down to me I was welcome to it. I got a great load of it, and that autumn I covered the whole garden thickly with leaves and hay to lie there over the winter. . . ."

Whenever Miss Stout wants to put in some seeds, she rakes the mulch back and plants; later when the seeds have sprouted, the mulch is pulled close around the plants, thereby keeping the ground around them moist and outwitting the weeds.

Now, after some 15 years of gardening according to her "overall year-round mulch" system, Miss Stout says emphatically: "I know it works. I know that it saves at least nine-tenths of the labor of gardening. I know the results are splendid and the other advantages tremendous."

Every organic gardener is well aware of the virtues of mulching, and is, undoubtedly, using mulches to some degree. However, Miss Stout's method calls for a thick mulch to be used for every flower and vegetable, shrub and tree. It is never turned under, never disturbed; it is, in effect, "a constantly rotting compost pile" spread over all the places where you want rich earth to abound.

"Some people hesitate to adopt overall mulching because they prefer the looks of a neatly cultivated garden. I used to," Miss Stout points out, "but now a garden with the earth exposed to the burning, baking sun looks helpless and pathetic to me. It looks fine if someone has just cultivated it after a good rain, but how often is that the case? At all other times, an unmulched garden looks to me like some naked thing which, for one reason or another, would be better off with a few clothes on."

After reflecting upon the droughty summers we've been having in recent years, I can't see how anyone could find fault with the above analysis.

Actually, there is no reason why a mulched garden should appear ugly. For example, Miss Stout has a row of mulched rosebushes which look exactly like any other well-kept row of roses. As she observes, "a flourishing garden with clean hay spread neatly between the rows looks attractive to me, and comfortable."

In England, the "no-digging" technique has been advocated for a number of years by several persons; F. C. King has been the foremost spokesman. He is "firmly convinced that if two gardeners of equal skill were to cultivate two plots of equal size and comparable fertility, one digging and trenching and the other practicing the principles of surface cultivation . . . it would be possible for the no-digger to secure the best results at less cost in time and cash." Mr. King also cites his own experiences as head gardener at Levens Hall, South Westmoreland to prove that crops grown by the no-digging method have greater resistance to insects and disease.

The primary difference between Miss Stout's overall mulch program and Mr. King's no digging method is the use of compost. While Miss Stout believes that any organic material such as hay will suffice to cover the soil and provide fertility, Mr. King writes: "One can hardly think of no-digging without at the same time associating in one's mind the necessity for composting; indeed, success with the no-digging technique is quite impossible unless and until the soil has previously received generous applications of properly made compost—say a pailful to the square yard as an initial top dressing—on any kind of soil, sandy or clay, heavy or light, shallow or more or less exhausted. The first step, then, to be taken to ensure an adequate supply of compost for use during September is to make a thorough collection of all organic wastes during the spring and summer months."

Using a Tiller

Have you ever wanted to garden as easily as you mow a lawn with a power mower? To be able to keep your garden free from weeds by safe, easy cultivation? To produce the best possible seed bed in your neighborhood? All this can be yours with a combination rotary tiller and garden tractor.

My greatest interest has been raising strawberries and raspberries, along with all kinds of vegetables. Using my garden tractor-tiller, I now prepare my beds by plowing to a good depth the fall before planting. In the spring I harrow with a cutter harrow attachment and then with a smoothing harrow; next I apply compost and old stable manure.

Then I make a pass over the plot with the tiller, which makes a nice, mellow bed for any annual crop. On what is to be a strawberry bed the following year, I plant either beans or potatoes or some other crop that needs constant cultivation and can be grown in rows. This helps to discourage the growth of weeds, and rids the soil of many of the undesirable insect pests such as grubs, mites and root lice.

After the annual crop has been harvested, I sow to winter rye. The following spring, after the fast-growing rye has reached a height of 6 inches, I rotary-till this into the soil. I also apply a good layer of compost and manure and till again.

Next the plants have to be set. During the summer I give them constant cultivation with the tiller. Later in summer I apply another thin coat of compost and stable manure and start my mulch base with 4 inches of sawdust. Mid-November I apply a 6- to 8-inch mulch of straw. The strawberry bed needs no further attention from me until after it has borne its first crop, after which I mow off the tops of the old plants.

Next I apply another thin coat of manure and till through the center of the rows to separate the old plants. I give them plenty of water, and after new growth starts I use the same method of preparing them for winter as the previous year.

After the plants have borne a second season, I till the whole bed into the soil and plant to another row crop (never to tomatoes as they simply don't like soil where berries have been raised previously). I like to use the green manure crops such as winter rye or clover on most of my soil to be tilled in the following spring.

In my opinion the most important implement that comes with the attachments for the garden tractor is the tiller. The other attachments are work-savers too, but the tiller is the backbone of my whole gardening setup. I have owned and used such a machine for 8 years and wouldn't think of trying to garden without it.

For larger-sized gardens, a rotary tiller is often the most practical way to prepare the soil for planting. In spring, a layer of compost and any surface vegetation can be easily worked into the surface soil in a short amount of time with a tiller.

Once familiar with the maintenance and operation of the rotary tiller, gardening is fun! There are many advantages. Work is done with little effort; soil can be prepared to the best possible tilth in a short time. Cultivation can be done closely to any depth desirable and practical for the crop being raised.

Using the basic principles of good cultivation plus a little common sense will avoid root damage or moisture loss. The green crop manures can be chopped and worked into the soil in one operation, and we all know how valuable they are. Compost and other manures can be worked into the soil easily at any season of the growing months. For the highest possible productivity from your soil, these steps are essential.

There is a widespread belief among gardeners that the rotary tiller is a complicated machine to use or maintain. This simply isn't true. I had never seen a tiller in use until I bought my own. I started from scratch to learn its operation and how it could be used to my best advantage.

After a few trial runs and getting the feel of the machine, I found that I had a wonderful helpmate in my tiller. Cost of operating the machine is small, I found, and repairs slight. Aside from wearing out the tilling "teeth" in a season, my cost of maintenance hasn't been worth mentioning. My first advice to a tyro is to get the feel of the machine in ground that has been worked, otherwise he may have a few rough minutes taming it and be disappointed.

Once man and machine work as a team, which is surprisingly fast, the rotary tiller is no more difficult to operate than a power mower. Rougher tilling can be done after one is familiar with the machine. I have used my tiller to till more than an acre in my garden when it was "green" sod with grass 6 to 8 inches high, but this is not advisable for this size garden as it takes a lot of work to get the soil in the best condition for planting.

If one wishes to till green sod, first till north to south, then east to west, and lastly corner to corner. Repeat this operation until you have a mellow bed. After a seedbed has been sown to a green manure crop, 2 or 3 passes are enough to put the soil in fine condition.

—Warren C. Goodwin

7—Starting Plants from Seed

What Seeds Are

Seeds are embryo plants with enough food stored around them to last until they can make their own food. If you soak a bean seed in water for a day or two, then carefully open it along the two halves, you can see the young plant at one end of the been seed. It will be very small and delicate. You'll easily be able to see the first few leaves, as well as the small round, pointed root.

All seeds are alike in that they have a small plant in them. The rest of the seed contains stored food for the young plant. This little plant needs certain conditions in order for it to grow; these are air, warmth, light, plant food and moisture. Given these conditions, the plant should become well established.

As soon as the plant has a root and *green* leaves, it can start to make its own plant food. This usually takes from 10 to 30 days. The root takes up water and minerals from the soil. These *raw materials* are carried in the stream of water or sap up to the *green leaves,* where they are made

1. Place coarse soil in the bottom of flat to permit free drainage and to form a base for the finer soil in which seed is sown.

2. Fill the flat with sifted planting media, comprising one-third sand, garden loam and compost. Sift through quarter-inch screen.

3. Scrape off the excess soil, leaving the surface flush with the edges of the flat. Be careful not to tamp soil down tightly.

4. Use a half-inch board to make rows for seeds. Leave two inches between the rows and keep them shallow rather than too deep.

5. Sprinkle the seeds along the trenches directly from the seed container. It is best to be liberal; you can thin later.

6. Cover the seeds lightly with extra sifted soil. Depth of soil should not be more than 3 times the width of the seeds.

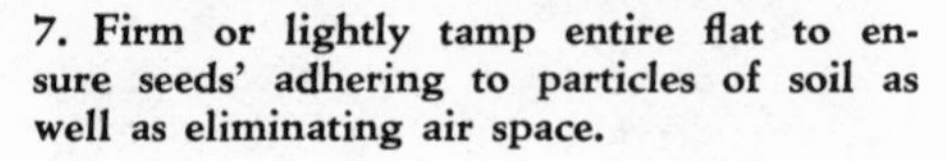

7. Firm or lightly tamp entire flat to ensure seeds' adhering to particles of soil as well as eliminating air space.

8. Cover flat with burlap and water it gently. Burlap may remain as cover until seedlings show through. Water daily.

into plant food. The water and minerals in the soil, carbon dioxide in the air, and light must all come together in a green leaf before they can be used by the plant as food.

The green color in the leaf is caused by chlorophyll, which has the ability to transform raw materials into starch in the presence of light. Without light, true plants cannot continue to grow because plant foods cannot be manufactured. Both chlorophyll and light are necessary.

Leaves turn green in the light. Seeds germinated and kept dark will have white sprouts which will turn green when brought into the light for a few days.

The leaves and other green parts of the plant are a sort of kitchen or manufacturing plant for preparing the plant food. After the green parts have prepared the food, it is sent back to the roots and other parts of the plant and is used for growth. The leaves or "plant kitchen" must have sunlight in order to work.

On the following pages, you'll learn how to grow your own plants from seeds and how to provide seedlings with the best growing conditions.

Growing Your Own Plants

Plant growing was once an art and a must with gardeners throughout the land. They knew their work, did it well and, as a result, reaped a bountiful harvest.

Today, it is surprising how few gardeners and farmers do grow their own plants. Even those who stand to profit greatly by doing so, turn their backs on the cold frame. They have come to depend upon outside sources for their needs, in turn, gambling their entire crop upon another's method of growing the all-important item.

In the opinion of G. J. Raleigh of Cornell University's Department of Horticulture:

"Plants of similar quality usually can be grown cheaper than they can be purchased. Moreover, the grower who produces his own plants is much less likely to be troubled by such diseases as club root and yellows of cabbage which are commonly introduced by purchasing plants from infested soils. Southern-grown and inferior locally grown plants may sell at low prices, but too often results with such plants are disappointing.

"When early maturity is of importance, as in most market-garden sections, the kind of plants used often determines whether or not the crop will be profitable. Large, sturdy plants with good root systems commence growth quickly after careful transplanting and produce crops earlier than do poorly grown plants."

Another great advantage in growing your own plants in hotbeds and cold frames is that you can send off for seeds from any part of the world, and thus raise and experiment with varieties and with new plants unobtainable from local seedmen. Nurseries usually limit themselves to a few good selling varieties of plants. If you want to experiment with some new hybrids, or with new varieties developed abroad or elsewhere in the United States, you should by all means learn to grow your own plants.

Starting Out

Vermiculite gives renewed hope to all home gardeners who want to grow their own vegetable and flower plants, but who have neither the time, place nor funds for a cold frame. Vermiculite can also be a solution to the damping-off problem for those who do maintain a cold frame, but have trouble with this and other diseases caused by contaminated soil. A ½-inch layer of vermiculite is simply spread evenly over the entire surface in the cold frame, just as in the flat, and the seeds then planted.

If just a small number of plants is desired, the vermiculite seed flat is your best bet. Plants growing in such a flat have less chance of being stunted by insufficient water, improper temperatures, or lack of good air circulation.

Seeds can be germinated well just in damp vermiculite—especially the tiny, hard-to-sprout ones such as double petunias. If only vermiculite is used, the seedlings must be transplanted to good soil as soon as they are large enough to handle. If transplanting is inconvenient at the time, the plants must be fed a liquid fertilizer high in nitrogen, such as one made by dissolving 3 tablespoons of dehydrated cow manure in one quart of water. When two inches tall, the plants should be transplanted to soil to prevent crowding. When feeding, saturate the flats well with liquid.

A combination of vermiculite on top of a mixture of organically enriched soil is best, however, for all-around seed germination. Even before sprouting seeds develop top growth, they send a long taproot down through the vermiculite into the rich soil below. This gets them off to a good start and develops strong plants for transplanting.

Vermiculite is heat-expanded mica, which is a form of rock. While in

1. If building your own seed flat, leave large cracks in the bottom for drainage and air. Before putting in soil and mixture, cover cracks with sphagnum or similar material so soil will not wash out when water is added.

2. After adding mixture of soil, lime, humus and sand, sprinkle well with water to settle soil. After excess water drains off, cover with vermiculite to a ½ inch depth.

3. Use a ½ inch strip of wood to make ¼ inch depressions about two inches apart in the damp vermiculite. These are the "drills" into which the seeds are later sown.

4. Sow the seeds thinly in the depressions. Row planting is better than scattering the seed over the entire flat as it's simpler to remove and replant seedlings later.

5. Cover seeds with a layer of dry vermiculite 3 times their diameter in depth. This helps control the possibility of damping off disease, a common trouble.

6. After all the seeds are covered with dry vermiculite, sprinkle the entire flat with water, then place it in a warm room or a semi-shaded area in a greenhouse.

the process of being treated, it becomes sterile, thus rendering itself to good seed germination. Other reasons, such as fine water-holding capacity, good aeration, and lightness in weight, also make it an excellent medium with which to work.

Some organic gardeners have a mistaken impression that vermiculite is not acceptable in the natural method. Since it is formed from a rock-based, naturally occurring mineral without the application or addition of chemicals, there is nothing objectionable about its use for starting seeds or for improving the air- and water-holding capacity of soil. (It will hold several times its own weight of water, and even when thoroughly soaked will permit ample air circulation around plant roots, helping to avoid damping-off.)

Vermiculite does not supply plant-growth nutrients, and should not be counted on to do this. It's an excellent seed-germinating material, and can also be used to store bulbs and winter vegetables or provide a base for flower arrangements. But it cannot substitute for humus or for any of the food needs of growing plants. Neither is it as good for mulching as organic matter, which lightens, aerates and helps hold moisture in soils—along with feeding plants and aiding topsoil.

Another wonderful thing about vermiculite is its cost. The small amount needed to grow enough plants for the average garden, costs no more than a dozen tomato plants at your garden shop. You can purchase it in small as well as in large amounts, depending upon your needs.

Damping-Off

If you have been troubled with damping-off, your soil is most likely contaminated. If you like, you can sterilize your soil at this time by pouring boiling water slowly through the soil. One gallon of boiling water is sufficient for a standard-sized flat. If you would rather, you can add one cup of white (5 per cent) vinegar to one quart of water and pour this slowly through the prepared soil. Although it is not as effective as the boiling water in stamping out damping-off, it does help to control it. Allow the soil to rest for 24 hours, then place a ½-inch layer of vermiculite over it. Apply loosely without tamping with your hand.

Building a Cold Frame and Hotbed

What is the difference between a hotbed and a cold frame? If you want to grow peppers, tomatoes, eggplants, or any of the other heat-loving plants, a hotbed is best to grow them in. A cold frame has the same construction as a hotbed, except that there is no heat used inside it. In a cold frame you can propagate such cold-loving plants as cabbage, the broccoli family, cauliflower. Or you can use your cold frames to taper off and harden plants that have been moved into them from the hotbeds, to get them hardened between the hotbed and setting out into open garden or field.

There are two types of hotbeds. One is heated by a great deal of fermenting straw or fresh manures (preferably horse or chicken) which has been placed in a pit two and one-half feet deep. The manure is packed down to a depth of 18 inches, well watered to soak. Then you shovel into the pit 5 to 6 inches of composted soil or good rich top soil. This soil—which will make the seedbed—must be sieved fine.

Manure Hotbeds

The making of a manure hotbed is described by New Mexico extension horticulturists as follows:

The first essential in preparing a manure hotbed is to have fresh horse manure, preferably from grain fed animals. The manure should contain one-third straw or other similar litter. Sometimes there is insufficient straw in the manure for proper heating. If it does not have sufficient straw in the manure it may not ferment or, if fermentation does take place, the heat may be evolved rapidly and be of only short duration. About 10 to 12 days before the manure is to be put in the pit it should be placed in a flat pile 4 to 5 feet high. If it is dry it should be dampened with water, but not made soggy. The manure should begin to heat in 3 or 4 days after which it should be turned placing the inside of the pile on the outside of the new one. In 3 or 4 more days the manure should be ready to be placed in the pit. The manure is filled into the pit in successive layers of 4 to 6 inches and tramped firmly to secure uniform heating and prevent excessive settling. It is also desirable to place the soil on top of the manure at the same time, since higher temperatures that develop when the bed is first made up tend to kill some of the weed seeds that may be present in the soil. Since a high temperature is likely to develop the first few days after the bed is made up, the planting should be delayed until the temperature drops to about 85 degrees or slightly lower.

If the seeds are planted when the bed is first made up, the high temperature is likely to kill the seeds or at least injure them. Manure heated hotbeds are usually economical to operate after the initial construction costs are paid. Temperature cannot be properly controlled in the manure heated hotbed because the rate of fermentation, and hence the rate of heat formation, is more or less the same on warm days as on the colder days. Thus the bed may be too hot or too cold depending upon the weather. The only means of controlling temperature is by ventilation, but unless much time is spent regulating the ventilation, the temperature is likely to vary considerably from the optimum for the growth of the seedlings.

Electrically-Heated Hotbed

This kind of growing bed is made by arranging electric heating cables, 5 inches below the surface of the topsoil seedbed. No manure is needed for heating this type bed. The coils produce a steady heat day and night, while the manure is effective for a few weeks.

Here is the story of how one Pennsylvania organic gardener, Maurice Franz, built a hotbed and his reasons for building it:

We built a hotbed because we wanted to extend our growing season and garden activity and also because we wanted an assured source of organic, home-grown vitamins when we need them most—during the winter.

The hotbed will make this possible. Our garden activities will no longer be determined by the frost dates, October 15 to May 1. And equally important to us, we will have a steady supply of organically-grown vitamins A and C through the short, dark, severe winter days.

We finished the hotbed in two week ends which included some unexpected last-minute scurrying for window frames. Except for excavating and leveling the ground and perhaps hauling topsoil 100 yards-plus from the garden, the work could fairly be described as light, pleasant and instructive.

Our hotbed cost us $45; it is electrically heated. But we know now that we could have done it for $36—about 20 per cent less—with better planning. Against this, chalk up a big saving of $5-$6 a month for 5 months for salad greens.

The hotbed is approximately 6 feet wide and 6 feet long and is covered by two window frames. It is two feet high in the back and one foot in front with a taper of 12 inches. It is set 4 to 6 inches in the ground and faces due south.

It should be stressed that this is a rugged, durable hotbed which should last for 10 years without any additional cost. This means that our extra vitamins will cost us $4.50 a year for 10 years and practically nothing after that. But you should be able to cut this down to $3.50 a year, as you will see.

Because we wanted a hotbed that would function right through the winter, on bad days and good, we decided on electric heat. We also voted against halfway measures like using electric lights because we wanted a steady source of heat coming from below, automatically regulated by the varying temperature needs of the growing soil.

The General Electric ground heating cable set plus the automatic thermostatic control unit fills the bill. It is rated capable of maintaining soil temperatures from 30 to 100 degrees Fahrenheit in a hotbed measuring 6 feet by 6 feet—36 square feet total.

This heating system will add $1.50 a month to our light bill—about $7.50 a year—the local power company advises. But we think it well worth our while because we would spend 4 times that amount—$30—over the winter months on salad greens. And we are convinced that the home-grown produce will not only be better tasting, but because it is picked just before eating, it will be packed with vitamins A and C just when we need them most.

We further wanted a frame for this heating unit that would stand handling without breaking up, that would resist soil rot sturdily and would also retain heat thriftily. On the advice of the local lumber dealer we chose yellow or western fir, ordering planks two inches thick for the extra insulation and strength. We reinforced this rugged frame at the corners with two-by-twos and then set this massive box 4 to 6 inches into the ground.

For bottom insulation, beneath the heating cables, we spread a two-inch carpet of vermiculite, covering the cables in an inch of sand (you can use ordinary soil). Over this we placed a protective wire netting and finally added 4 inches of rich topsoil as a growing medium.

Building the Hotbed

Putting the frame together is actually the easiest part of the operation. Order the wood you need, specifying that it is pre-cut to size by a power saw. This makes for accuracy and square, even ends. You will also find that hand-sawing 2- by 12-inch boards is no cinch, especially if you order weathered, used wood. Also drill nail or screw holes before assembling to speed the job and eliminate splitting.

You must also weather- and ground-proof your wood with a protective covering. We used linseed oil because it will not give off any chemical reaction which can hurt soil life or the healthy organic growth of our vegetables. It was also recommended by the lumber dealer as a highly satisfactory protective agent. Painting it on with a large brush should hold ground rot off and double the life of our already sturdy frame. Do not neglect painting your frame with linseed oil or a similar protective medium; 20 minutes' painting can add years to the life of your hotbed.

Digging the excavation for the frame is harder and comes first. After picking your site, study the soil before you start digging. If it is clayey and packed, try to loosen it with a rotary tiller—you'll save time and energy for other tasks. The amount of soil you have to remove will amount to 18-20 cubic feet.

Be sure to make the excavation at least 3 inches wider and longer than the size of your frame. This oversize pit will make it easy to fit your frame into the ground, about 4 inches below the top of the soil. If your hotbed will be 70 inches wide by 67 long, make your excavation 73 inches wide by 70 long.

Setting your frame level on a level, secure ground footing is not hard if you plan ahead. Do not put your frame together until your excavation is ready for it. Instead, take the 4 baseboards of your frame and *fit them into place just where they will go*. Take the back board, upend and place it in the back of your excavation, from corner to corner. Check it with your spirit level and make the necessary adjustments until it is level and rests firmly in place.

Now take the board for the left side. Upend it and place it where it will go. Check the level and make your adjustments. Repeat with the front board and the other side. If you keep them level as you go from corner to corner, your frame will have a secure, satisfactory footing. Then, and only then, assemble your frame, give it its protective coat and put it into place.

Installing the Heating Unit

Start the actual growing bed off with a two-inch layer of vermiculite. It is an excellent insulator and will hold downward loss of heat to a minimum and also permits ready drainage. It is a matter of moments to spread it and then level it gently using a board. Then, working slowly and carefully, lay your heating cable on it. This is a single continuous wire ending in a conventional wall plug which connects with the thermostat. Attach the thermostat securely to the frame, but only after you are satisfied with the layout of your heating cables. The General Electric people include a full set of instructions and diagrams which are extremely easy to follow.

Next, cover the cables with one inch of sand or soil. You can save $2 by using soil instead of sand, and there is no reason why you should not. Over the soil place a protective wire screening with about one-half inch mesh. This will keep your trowel out of the heating system.

Finally, top off your hotbed with 4 inches of rich, fertile topsoil. We took ours from the garden and screened it for evenness, then mixed the following growing medium: topsoil—3 parts; wood mold and sand—one part each. We then enriched this with 3 pounds of cottonseed meal, bone meal and granite dust mixed on a 2½-6-6 formula. This is twice as rich as seems necessary, but the roots will not be able to reach down as deep in our rather shallow bed as they would in the garden.

And now that we have spent our $45 and stand committed to spend $1.50 more on electricity each month—what have we gained?

To begin with, we have extended our season 4 months. We'll get an early start in February with our greens, lettuce, mustard and spring onions. By March 15 we will move our early cabbage and celery from the window flats to the hotbed until they can go into the garden—about a month later.

When the cabbage and celery are hardened off we will replace them with peppers, tomatoes and eggplant from the inside window flats. Indoors we'll start cucumbers and melons in pots, moving them into the hotbed about May 15 when we transplant the peppers, tomatoes and eggplant outdoors.

When the fall comes, we'll get ready for a winter supply of vitamins by starting endive, escarole, romaine, kale and lettuce in the garden. That should be before Labor Day, and we figure that we'll transplant them to the bed by mid-October—just about the time of the first hard frosts hereabouts.

Construction Materials

Concrete and cinder blocks are often used for the construction of the walls of hotbeds, cold frames, and small plant houses, on account of the ease with which they can be laid. In some localities concrete blocks 4 inches thick can be obtained, and walls constructed of these occupy less space than those made of the 6- or 8-inch blocks. Where concrete blocks are used for the front and back walls of the beds, it is necessary to put up forms and pour the top of the ends in order to obtain the correct slope. The 8- by 8- by 16-inch concrete and cinder blocks are especially adapted for constructing the walls of double beds with a ridge through the center and covered by means of two lines of hotbed sash or cloth. The top layer of blocks should be crowned with a layer of cement mortar about an inch in thickness to prevent water from entering the cavities of the blocks. Cinder blocks are better insulators against the passage of heat than concrete and, in addition, have the advantage that nails can be driven into them where wood is being fastened to the block walls.

Location

Hotbeds and cold frames should always be located on well-drained land that is free from depressions or danger of flooding during heavy rains. A location near the house where the beds can be given frequent attention is desirable. The beds should be protected on the north by a group of buildings, a grove of trees, a tight board fence, or an evergreen hedge. In many cases windbreaks, consisting of pine boughs or bundles of corn fodder set against supports, are employed for protecting the beds from cold winds. A location with a southern exposure and adequate wind protection on the north and west is ideal. In all cases, protection from cold winds, the securing of direct sunlight, and convenience in the matter of tending should be the main determining factors in the selection of a location for any type of plant bed.

Plant beds and small greenhouses built well below the level of the ground require less heat than those that are entirely above ground and more exposed. This principle applies especially to the construction of sash-covered houses that are heated by flues, stoves, or electricity. Special care must be taken, however, in the selection of a location for beds and houses that are partially below ground, to provide good drainage in order to avoid flooding the beds during rainy seasons.

Testing Seeds for Germination

Some seed will deteriorate with age, and though it is carefully prepared, may refuse to germinate. Before planting homegrown seed and depending upon it in spring to produce a crop, it is wise to make a germination test. This is done by duplicating as far as possible the conditions of warmth and moisture that the plants need when growing in the garden.

Twenty or more seeds are counted out for the test, and the quantity noted. They are then placed on a layer of absorbent material in a saucer. Absorbent cotton, blotting paper, or heavy cloth may be used. Sterile cotton is recommended, because it will not mold as quickly. A second layer of the material is placed over the seeds, it is sprinkled with water and excess water is poured off. The seed may start to germinate in 24 hours, or it may take 2 or 3 weeks, but meanwhile the covering material should not be allowed to dry out. The top layer should be lifted from time to time to allow air to reach the seed. If mold appears near the seed, the covering layer may have to be removed entirely. Seed is germinated, for testing purposes, when it has swelled and put forth a sprout.

The germination tests will show whether the seed should be used at all, and how thickly it should be sown. Even if only a small number of seeds germinate, it may be wise to use the seed, because unusually strong plants are said to result from planting old seed.

The following is a list of vegetables with the number of years indicated during which their seeds can be expected to be viable. If the seed is older, it should be tested before using: corn, onion, parsnip, soybean, salsify, two years; bean, leek, parsley, peas, 3 years; carrot, mustard, pepper, tomatoes, 4 years; broccoli, cabbage, cauliflower, kohlrabi, lettuce, okra, pumpkin, radish, spinach, turnip, 5 years; beets, eggplant, melons, squash, 6 years; celery, 8 years; cucumber, endive, 10 years. Whether the seed is purchased or homegrown, tests should be made if it exceeds the above ages.

Few seeds will germinate as soon as they ripen. Most of them require a rest period of at least a month, and in some cases as much as a year. There are exceptions to this. Some grains, for instance, will sprout in the seed head if the weather is wet when they ripen. A few plants such as beans, sunflowers, lilies and mustard, will start to sprout in a few days. Others, like carrots and parsnips, need a month or more before they are ready to grow. And some trees and shrubs will not germinate in less than a year.

Planting Seeds

In the germinating hotbeds or cold frames you plant seeds plentifully, cover them lightly, and keep the rows one inch apart. You may later thin out to suit. In planting lots of seeds remember this: while most seeds will produce sturdy plantlets, others will be thin or weaklings. With plentiful planting you may pluck out all but the finest, sturdiest plants.

A *must* to remember: The grower must constantly watch moisture and heat with an eagle eye. You must not allow the heat to rise too high, or above 75-80 degrees while the plants are small. If the growing beds get too hot—and there is no ventilation—the hundreds of plants may easily damp-off and die. To ventilate, simply raise the lids a bit. Leave lids closed at night.

Starting in Flats

In a flat, the bottom quarter to half of the flat should be sphagnum moss. Soil which is added to cover these layers should be moist, not wet or dry. Seed should be thinly sown in rows or in circles in the pot, and in rows in the flat. If more than one kind of seed is planted in a container, the seeds should be chosen to germinate in about the same length of time, and to grow at the same rate so that all will be ready for transplanting together. After the seed has been placed in its rows, sand or fine compost is sifted over it to the correct depth. The soil may then be firmed and watered, either with a very fine misty spray from above, or by plunging the container into water almost as deep as the soil. When wet patches begin to appear on top of the soil, the container should be removed from the water and drained.

Seed flats may be covered to preserve surface moisture until germination starts. Temperature for germination of seed may usually be somewhat higher than the plants will bear after growth has started. (Exceptions are the seeds which need a period of frost before they will germinate.) Soil should be kept moist, but not wet, during this period. If the top of the container is covered with glass or paper, the cover should be lifted occasionally to permit air to circulate. At the first sign of fungus growth, the cover should be removed.

As soon as the first green begins to appear, covering should be removed from the seed pot and it should be placed in a southern window. Gradually, as the seedlings sprout and the roots stretch down into the pot, watering may be lighter and less frequent, but the container should never be permitted to become dry. If seedlings are too thick they must be thinned immediately. Occasionally when fine seed is planted it will come up unevenly, with thick patches in places in the pot. These patches should be thinned by means of tweezers, because crowding at this stage will almost inevitably result in damping-off.

Watering

Water the beds daily after planting. Use a fine can sprinkler and *tepid* water. Don't muddy the seedbed; just water it enough to be nicely damp. One gallon sprinkling can to a 6-foot section is generally adequate. For added moisture, beds can be opened to a warm, quiet rain.

Once the plants have sprouted and are several weeks old—and there is the delightful feeling of their crowding one another—lift the lids more and more. As the growing season progresses, and the bedding plants grow faster, there will be nice days when you should take the lids off and get the full benefit of the sun.

First transplanting: When your plants have grown to a size large enough to be handled, they are ready to transplant over into the cold frame beds. There they will grow rapidly and harden off so that the shock of final planting into the open garden won't hurt them.

About the only thing that can harm growing plants in the hotbeds or cold frames is overheating, drying out for lack of water, or being attacked by the fungus disease known as *damping-off*. This damping-off or "black root" or "wire stem" in the seedbed is caused by about a half dozen or more fungus parasites. They usually grow near the surface, and enter the tiny plants at the point where they emerge from the ground. All of these fungus parasites require a high moisture content of soil and air for quick growth.

To prevent trouble from *damping-off* the best defense is to keep the air

The cold frame should have 4 inches of rich, fertile topsoil which has been screened and mixed with sand and humus. Prior to starting plants in the spring, the soil should be turned lightly and receive an application of organic fertilizer.

and surface of the seedbed as dry as is consistent with good growth of the plants. Getting the beds heated without proper ventilation, and not allowing the moisture to escape, is what causes damping-off.

Proper Soil for Plants

Make a mixture of two parts good garden loam, one part fine, but sharp sand, and one part leaf mold, peat, or old decayed compost or manure. Mix well and put 8 inches of it into the bed. It's a good idea to let it age for several months before seeding. Sift the soil mixture through a one-fourth inch mesh screen to get it in condition for planting. When screening the mixture, place coarse screenings in the bottom of the flats to provide better drainage.

There is enough plant food in this mixture without adding manure or organic fertilizers high in nitrogen. Used too soon, these often cause the young plants to grow too rapidly, unbalancing their natural growth.

When to Plant: Two separate compartments are helpful in the cold frame. All vegetables do not require the same temperature for germination. Where some will sprout in the heat, others will rot. To get the maximum number of plants from the seed, the plants must be divided into two groups. Celery,

Water the cold frame daily after planting, using a fine can sprinkler and tepid water. One gallon sprinkling can to a 6-ft. section is generally adequate, and beds can be opened to a warm, light rain for added moisture when necessary.

cabbage, lettuce, cauliflower and broccoli require low temperatures for germination. Plant these together in one section of the cold frame, dropping the seeds evenly in rows two inches apart. Plant about the first week in April or 8 weeks before transplanting time. An inch of fine mulch may be spread evenly over the seeded area to keep the ground moist and to discourage early weeds. The growing plants will push their way through as they grow. Water lightly, then close the frame.

About the middle of April, when the sun is warmer, plant tomatoes, peppers, eggplants, muskmelons, summer squash and cucumbers. Plant these in the other half of the bed. The plants in the first half of the bed should have sprouted by now and need ventilation. Close the cover on the tomato bed until the plants sprout, then ventilate along with the other half.

Cucumbers, muskmelons and summer squash are grown in parts of milk cartons, and set out in the open after danger of frost. This is not absolutely necessary. It merely helps produce an earlier yield.

Ventilation: No matter what the germination temperatures had been, all growing plants do best around 70 degrees. Regulate temperature each day by slightly lifting the two sashes. On cold days, allow the beds to remain closed. Never lift covers into the wind. A cold wind can damage the tender shoots. Open away from the draft, and secure the windows so they will not be broken.

Nevada's experiment station points out that flats, cold frames and hotbeds will bring many plants into production a month to 6 weeks earlier than when they are sown outdoors.

The minimum maturity dates given here indicate the length of time required for maturing after the plants are set in Nevada's high altitude gardens. At elevations below 5,000 feet, and in favorable seasons, some fruit may ripen in a shorter period.

Kind	*Height (Inches)*	*Maturity Dates*	*Depth to Cover the Seeds*
Broccoli*	10-16	65 to 100 days	1/4 to 1/2 inch
Cabbage*	10-15	65 to 125 days	1/4 to 1/2 inch
Cauliflower*	10-15	90 to 150 days	1/4 to 1/2 inch
Celery*	8-12	115 to 150 days	1/4 to 1/2 inch
Eggplant	8-12	90 to 125 days	1/4 to 1/2 inch
Peppers	8-12	70 to 90 days	1/4 to 1/2 inch
Tomatoes	12-30	100 to 150 days	1/4 to 1/2 inch

* Frost resistant and can be set out while weather is still cool.

A cold frame is also useful for hardening off seedlings prior to planting them outside. This step can be started about one week before transplanting time, placing the plants in the cold frame for a longer time each successive day.

TABLE 28: Time, Spacing and Temperature Recommended for Growing Plants in the Home or the Hotbed

Vegetable	Weeks from Seeding to Setting in Garden	Inches to Space After First Leaves Are Formed	Degrees of Temperature	
			Day	Night
Broccoli Brussels sprouts Cabbage Cauliflower Head Lettuce	5 to 7	2 x 2	65	55
Tomato	5 to 7	3 x 3	75	65
Eggplant Pepper	6 to 8	3 x 3	75	65

—Virginia Polytechnic Institute

TABLE 29: Approximate Dates for Sowing Vegetable Seeds Under Glass, and Ranges of Day Temperatures

Vegetable	Long Island and southeastern New York	Southern-tier counties	Remainder of New York other than mountain regions	Approximate temperatures (day)
Beets	Feb. 15-28	March 15-31	March 1-15	60-65
Broccoli	Feb. 10-20	March 1-15	Feb. 20-28	60-65
Cabbage, early	Feb. 10-20	March 1-15	Feb. 20-28	60-65
Cauliflower	Feb. 10-20	March 1-15	Feb. 20-28	60-65
Celery	Feb. 10-20	March 1-15	Feb. 20-28	60-65
Eggplant*	March 10-20	March 25-April 5	March 15-25	70-75
Endive	Feb. 10-20	March 10-20	Feb. 20-28	60-65
Kohlrabi	Feb. 10-20	March 1-15	Feb. 20-28	60-65
Leeks	Feb. 10-20	March 1-15	Feb. 20-28	60-65
Lettuce	Feb. 10-20	March 1-15	Feb. 20-28	60-65
Melons	April 10-15	April 25-May 5	April 15-25	70-75
Onions, Sweet Spanish	Jan. 20-31	Feb. 10-20	Feb. 1-10	60-65
Peppers*	March 10-20	March 25-April 5	March 15-25	70-75
Squash	April 10-15	April 25-May 5	April 15-25	65-70
Tomatoes*	March 10-20	March 25-April 5	March 15-25	65-70

* If to be transplanted twice or if transplanted once but grown in bands or pots, the seeds should be started two weeks earlier.

—Prepared by G. J. Raleigh, Cornell University

Using the Cold Frame for Winter Vegetables

A cold frame is really a protected seedbed and can be used in the first frosty days of autumn just as advantageously as in early spring. Tender lettuce and crisp endive can be enjoyed until after Thanksgiving by properly utilizing the cold frame.

Plant seeds in early autumn. For lettuce, endive and parsley, plant seeds in one compartment in rows 3 inches apart and about one-quarter inch deep, covering the seed with vermiculite to prevent damping-off. Then water the rows with a fine spray, and adjust the sash and place a covering of light boards over it. As soon as the seeds sprout, remove this cover and raise the sash several inches to allow good circulation of air.

About a week before starting the "hardening off" procedure, young plants should be blocked out with a large kitchen knife. This is done by slicing the soil between plants in both directions, as shown in the photograph above.

Seedlings will grow rapidly, and when they begin to crowd each other, they are ready to transplant. Set the little plants in another compartment, about 3 inches apart each way. Mulch again with vermiculite and keep them shaded for a few days. As plants grow, they can be thinned and the thinnings used in salads. The remaining plants will then begin to form heads.

When nights grow frosty, close the frame tightly before sundown to hold the day's heat within, opening it again each morning. If nights are very cold, cover the frame with a blanket and bank leaves around the sides. If the day remains cold, remove the blanket to allow light to enter, but keep the frame closed.

TABLE 30: Time to Plant Seeds

(For Earliness)

Name of Vegetable	Average number of seeds per ounce	In greenhouses or hotbeds	In cold frames	In open ground
Beans, dwarf	100		Last of April	May-July
Beans, pole	100		Last of April	May-July
Brussels sprouts	6500	Mid-March	April	May-June
Beet, garden	1750	First of March	Early April	April 5-15
Cabbage (early)	5000	Last of Feb.	March 15-30	April 5-10
Cauliflower early)	14000	Last of March	April 20-25	May 1-5
Celery (early)	100000	First of March	April 25-30	May 5-10
Cucumber	1000		Last of April	May 20-25
Eggplant	5000	April 1-10	May 1-15	June 1-5
Endive	13500	April 1-10	May 1-5	May 5-10
Kohlrabi	7000	April 1-10	Last of April	May 5-10
Leek	8000	April 1-10	May 1-5	May 5-10
Lettuce	16000	March 1-10	April 1-10	April 5-15
Melon, musk	1200		April 25-30	May 25-30
Melon, water	225		April 25-30	May 25-30
Onion	12500	March 10-15	April 15-20	May 1-5
Parsley	17500	March 10-15	April 20-25	May 1-5
Pepper	4000	March 10-15	April 25-30	May 20-30
Squash (early)	300		Last of April	May 10-20
Tomato	7500	March 10-15	April 25-30	May 15-25

—*University of Connecticut*

You'll be amazed at how much cold the plants could stand under glass. As the season grows late, plants may be frozen occasionally, but by sprinkling with cold water and keeping in the dark for a while, they will revive and come back in good condition. It is possible to have fresh salads until winter.

You can also use the cold frame in winter for storing vegetables. Before the ground freezes, remove about 18 inches of soil and place in vegetables like turnips, rutabagas, beets, carrots and celery on a layer of straw. Over the vegetables cover another layer of straw with sash boards. Vegetables should keep crisp and unfrozen all winter.

Saving Seeds

If you grow your own seeds, you can often develop highly productive and disease-resistant strains in your own back yard. Here are a few simple rules that will help.

1. At the beginning of the season prepare some stakes or strings to label the plants you select to keep for seeds. Make sure everyone who works in the garden knows the meaning of these markers.

2. Watch your plants as they develop and select a few, only a few, of the outstandingly healthy ones to preserve for seed. Make your selection on the basis of the *whole plant.* Seed development is an all-season job for you as well as the plant, not just a hurried trip through the frost-bitten garden to salvage the largest seed pods that happen to be left. Make a habit of observing your plants with seed selection in mind. You'll form new ideals of what the well-balanced plant really can be.

3. Remember there are different characteristics to encourage in each kind of plant. In spinach or Chinese cabbage, for example, do not mark the first few nice ones which go to seed and consider the job done. Those are the very ones you do *not* want. The desirable plants are those which continue producing leafy growth longest and send up seed stalks *latest* in the season. On the other hand, with broccoli or cauliflower, we like best the plants that produce flower heads promptly. Likewise with radishes, save the plant you want most to eat, the first fat root. A healthy top growth should balance the root, but not a premature flower stalk. The same with carrots, beets, turnips and the like, except that these are biennials and normal seed production takes place only the second year, adding the problem of how to keep your best plants safely through the winter. However, it's worth the trouble as over the years you have seed you can rely on and gradually you may find that you are developing new and valuable plant characteristics.

4. Hybrid plants are the result of crossbreeding of two pure-strain varieties. Seeds from these should not be saved as they will revert to the parent varieties and often be useless for your purposes, or may not even be fertile at all. (Hybrid corn, for instance, is not fertile.) Crossbreeding, or crosspollination, means that the pollen of a flower has come in contact with the flower of a plant of another variety or species. The pollen can be blown by the wind, or carried by bees. The seed which results will not be true to type of either variety.

You will have to keep in mind, then, that many vegetables will crossbreed. Do not, for instance, plant seed from sweet corn that has been planted near field corn, or beets near chard or sugar beets.

Corn, cucumber, melon, squash, pumpkin, mustard, Brussels sprouts, collards, kale, kohlrabi, onion, radish, beet, turnips . . . all are vegetables which crosspollinate easily.

Less easily crossbred are tomatoes, eggplant, pepper, celery and carrot.

Beans, okra, peas and lettuce are generally self-pollinated.

It is necessary then, that you plant only one variety or strain of the vegetable you wish to seed. If you do plant more, be sure the plants are at least 100 feet apart, though even at this distance bees or wind may effect cross-pollination.

5. Let the seeds ripen as long as possible in the garden, but not so long as to risk shattering out or feeding birds. Bring in the whole plant and hang it or lay it in a dry place until the pods are brittle and the seed comes out easily.

6. The easiest homemade way to clean seed is to keep a collection of pieces of wire screening of various sizes of mesh. With one size or another you can sift out either the seed or the chaff sufficiently for storage purposes.

7. Keeping seed safely over the winter requires consideration, as mold or insects can spoil it. Store in a dry, cool place, covered but *not* in airtight containers. To prevent insects from taking hold in wheat or bean seed, stir the seed frequently. Also look it over occasionally, removing any spoiling material.

8. To make sure your seed is fertile, test it long enough before planting time so that you can buy more if necessary. Take about 30 seeds and place in a shallow dish on a piece of blotting paper or raw cotton that is kept moist. Do not let the seed actually stand in water, but on the other hand if it once dries out, the test will have to be started again with fresh seed. Best cover the container to shut out light and prevent evaporation, but not so as to keep out air, or mold may develop. *Keep in a warm place.* Record the date and see how long germination takes, also what proportion of seeds grow. Naturally some will be slower, like parsley; some will need more heat, like tomatoes, peppers and eggplant. Any batch from which the test seeds mold without sending out shoots is, of course, not fit to plant.

There are different things to watch for and do in connection with the culture of each type of vegetable for seed.

The Cabbage Family

Remember that all types of cabbage cross with one another. If it is not possible to grow different varieties at great distance from one another then one kind only is grown for seed. The firmest heads are selected and stored very carefully over winter. These are set out early in the following April. They must be set quite deeply and support provided for the seed stalks. The seed must be protected from the birds. The mature seed stalks are hung indoors and allowed to dry until it is possible to thresh them or remove the seed by hand.

Carrots

Carefully keep the best developed roots with medium tops and small cores over the winter. Set these out in March, 12 to 20 inches apart, according to the variety. The cultural directions are the same as those given under summer care. (Regular weeding and shallow cultivating. A mulch of composted oak leaves and bark rich in tannin is helpful if the carrots are attacked by root lice.) Seed stock must not be forced! Cut the ripe umbels, dry them in the shade, rub the chaff out by hand. Seed propagation only makes sense if the area is not surrounded with meadow land, since wild carrot (Queen Anne's Lace) crosses with garden carrot.

Cucumbers

Let the earliest maturing well-formed fruit from healthy plants hang until ripe. The seeds are then scraped out of the fully ripened cucumbers and put into a warm place in a little water to ferment. The next step is to sieve them and wash them, drying gently with a cloth and spread out on a blotting paper to dry.

Beets

The perfectly formed beets, showing no white markings inside, are kept over winter and set out early in April about 20 inches apart. Cut the ripe stalks and hang in the shade to dry. Later, seeds can be rubbed out by hand.

Head Lettuce

Select two or three of the firmest heads from amongst those that are slowest to bolt to seed, and leave them in the garden to blossom. The large flowering stalks should be well staked. When most of the seeds are already developed pull the plants out and hang them in the shade to dry. The removing and cleaning of the seed can be done in winter.

Leeks

Seed bearers are hilled very high to protect them from freezing in the winter. The plant should stand at least a foot and a half apart in all directions. The blossom stem will need supporting. The seeds mature fully only in favorable summers and in a warm situation. Keep in mind that leeks cross easily with pearl onions.

Onions

Select solid, well-ripened onions which were grown from sets the year before. These should be set out as early as possible in a sunny, protected location. It's a good idea to apply a compost mulch. Support the blossom stems. Seeds ripen slowly and must be dried under cover since they fall out easily. Ripe seeds are coal black.

Peas and Beans

Mark a few of the finest plants with a bit of cloth at the beginning of the harvesting season. Allow entire crop of these plants to ripen. It is best to choose plants most alike in variety and earliness.

Potatoes

For a seed stock choose only perfectly scab-free potatoes from healthy plants surrounded by other healthy plants. The chosen ones should be marked beforehand and the potatoes dug before harvesting the others. Store these carefully over the winter in a frost-free dry cellar or pit. They require good ventilation too. Burying in dry sand is one of the best ways to store them.

Radishes

Seed culture occasionally succeeds with early radishes. The plants which bolt to see *quite late* are left standing. The blossoming plants are very brittle and must be carefully staked. Harvested seed stalks ripen completely when hung in shade, but protect them from finches!

Sweet Corn

The earliest, best developed, full-grained ears are allowed to ripen on the stalk. (Hybrid varieties do not produce fertile seed.) When the husks are bleached and straw-like, pick the ears, pull the husks back, tie two ears together and hang up to dry.

Tomatoes

Choose healthy well-formed fruit. Allow tomatoes to ripen beyond the edible stage but do not let them become rotted. Pick off and scrape out the seeds. Set these in a little water and allow them to ferment several days. Then place them in a sieve and wash under a faucet. Spread on blotting paper to dry. The fermentation process is effective in preventing the seed distribution of the bacterial canker disease. Tomatoes intercross only to a small extent, so more than one variety may be planted within a few yards of each other.

Early Turnips and Rutabagas

The chosen plants should be lightly covered and left to remain in the ground over the winter. In the spring uncover early. Further treatment is the same as for beets. Turnips should be harvested early in the morning to avoid shattering by noon heat.

—Evelyn Speiden

8—Seeding, Planting and Transplanting Outside

When to Plant Outside

GETTING PLANTS OFF to an early start outside in the garden is an objective of every gardener. All of us want to make the first planting of each vegetable as soon as it can be safely done—as soon as there's little chance of its being damaged by cold. There's good reason for this early planting—in many cases, it will mean that a second crop can be planted later in the season in the same spot. And—just as important—the latent gardening energy must find its release.

Many vegetables are so hardy to cold that they can be planted a month or more before the average date of the last freeze or about 6 weeks before the frost-free date. (According to the USDA's Victor Boswell, the frost-free date in spring is usually 2 to 3 weeks later than the average date of the last freeze in a locality and is approximately the date that oak trees leaf out.)

It should also be remembered that most, if not all, of the cold-tolerant crops actually thrive better in cool weather than in hot weather and should not be planted late in the spring in the southern two-thirds of the country where hot summers occur. Therefore, the gardener should time his planting not only to escape cold but, with certain crops, also to escape heat. Some vegetables that will not thrive when planted in late spring in areas having rather hot summers may be sown in late summer, however, so that they will make most of their growth in cooler weather.

The specific planting dates for each section of the United States are given under the individual descriptions of vegetables. The maps, drawn from United States Weather Bureau originals, show the average dates of the last killing frosts in spring and the average dates of the first killing frosts in fall. You can determine safe planting dates for different crops by using these maps.

TABLE 31: Some Common Vegetables Grouped According to the Approximate Times They Can Be Planted and Their Relative Requirements for Cool and Warm Weather

Cold-hardy plants for early-spring planting		Cold-tender or heat-hardy plants for late-spring or early-summer planting			Hardy plants for late-summer or fall planting except in the North (plant 6 to 8 weeks before first fall freeze)
Very hardy (plant 4 to 6 weeks before frost-free date)	Hardy (plant 2 to 4 weeks before frost-free date)	Not cold-hardy (plant on frost-free date)	Requiring hot weather (plant 1 week or more after frost-free date)	Medium heat tolerant (good for summer planting)	
Broccoli	Beets	Beans, snap	Beans, lima	Beans, all	Beets
Cabbage	Carrots	Cucumbers	Eggplant	Chard	Collards
Lettuce	Chard	Okra	Peppers	Soybeans	Kale
Onions	Mustard	New Zealand spinach	Sweet potatoes	New Zealand spinach	Lettuce
Peas	Parsnips	Soybeans		Squash	Mustard
Potatoes	Radishes	Squash		Sweet corn	Spinach
Spinach		Sweet corn			Turnips
Turnips		Tomatoes			

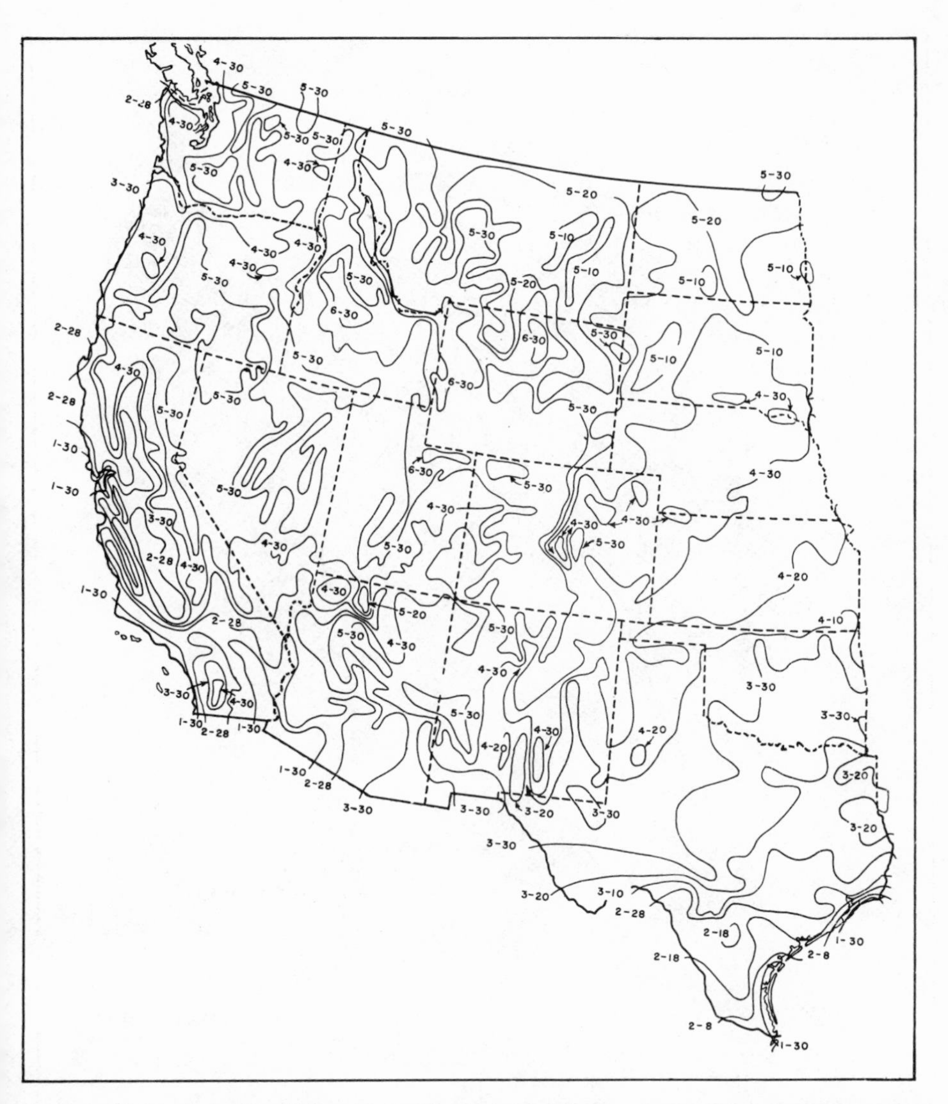

Average dates of the last killing spring frosts in the western United States, 1899 to 1938. Gardeners should locate the line nearest to the locality in which they live, note the date on that line (the first figure indicates the month, the second the day; thus 3-20 is March 20) and then refer to tables 32 and 33. (Redrawn from U. S. Weather Bureau original.)

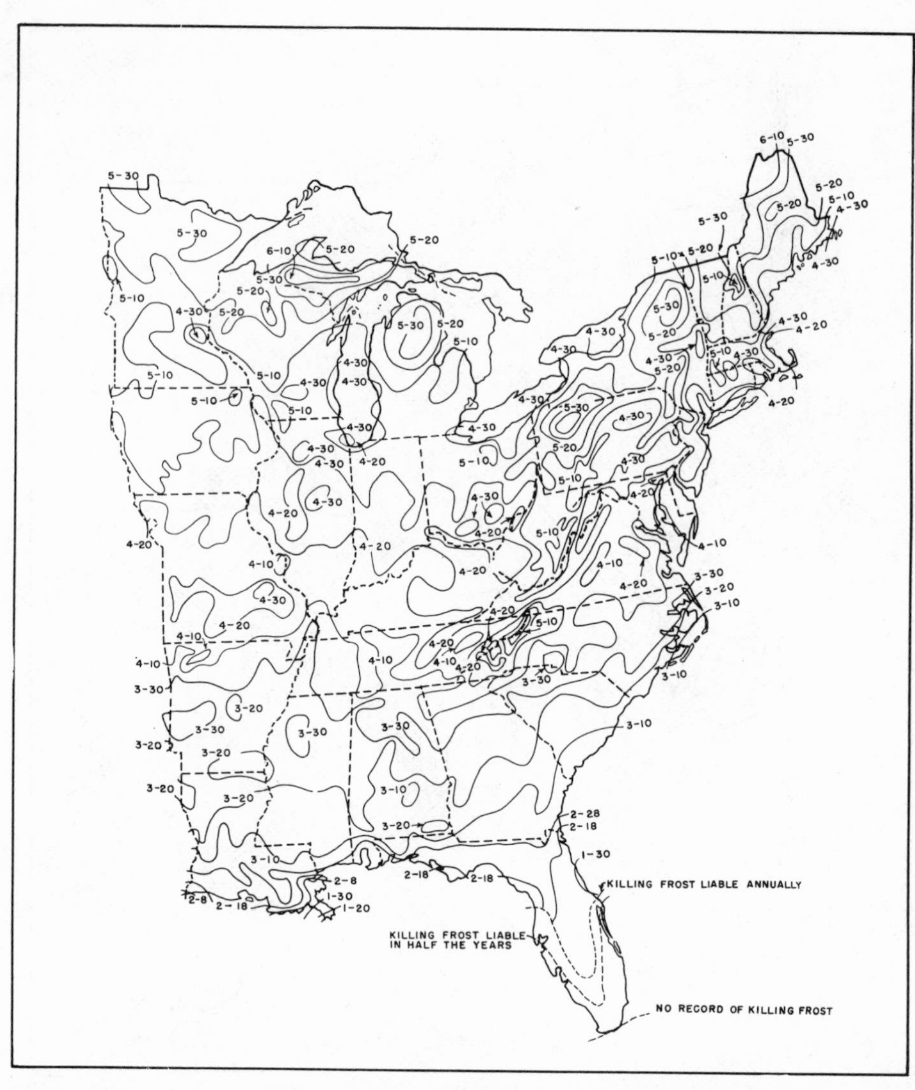

Average dates of the last killing spring frosts in the eastern United States, 1899 to 1938. Gardeners should locate the line nearest to the locality in which they live, note the date on that line (the first figure indicates the month, the second the day; thus 3-10 is March 10) and then refer to tables 32 and 33. (Redrawn from U. S. Weather Bureau original.)

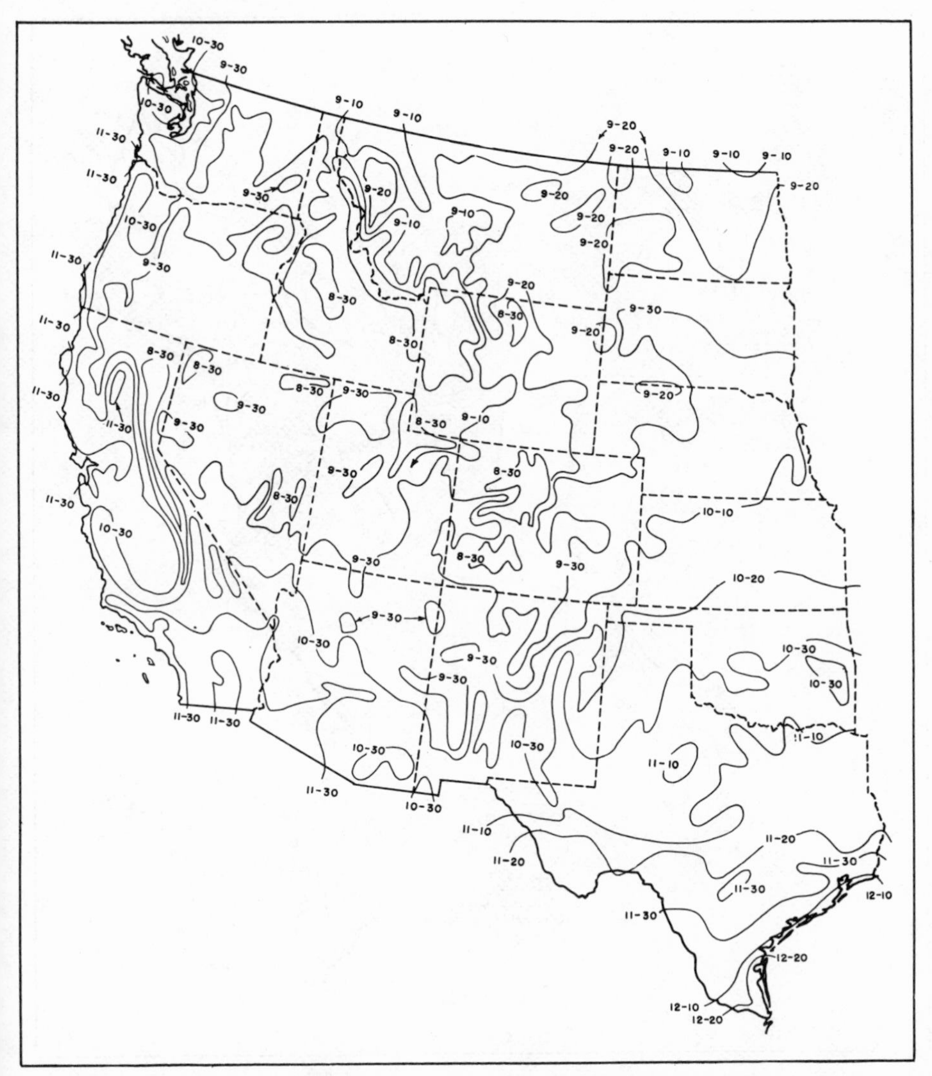

Average dates of the first killing fall frosts in the western United States, 1899 to 1938. Gardeners should locate the line nearest to the locality in which they live, note the date on that line (the first figure indicates the month, the second the day; thus 11-30 is November 30) and then refer to tables 34 and 35. (Redrawn from U. S. Weather Bureau original.)

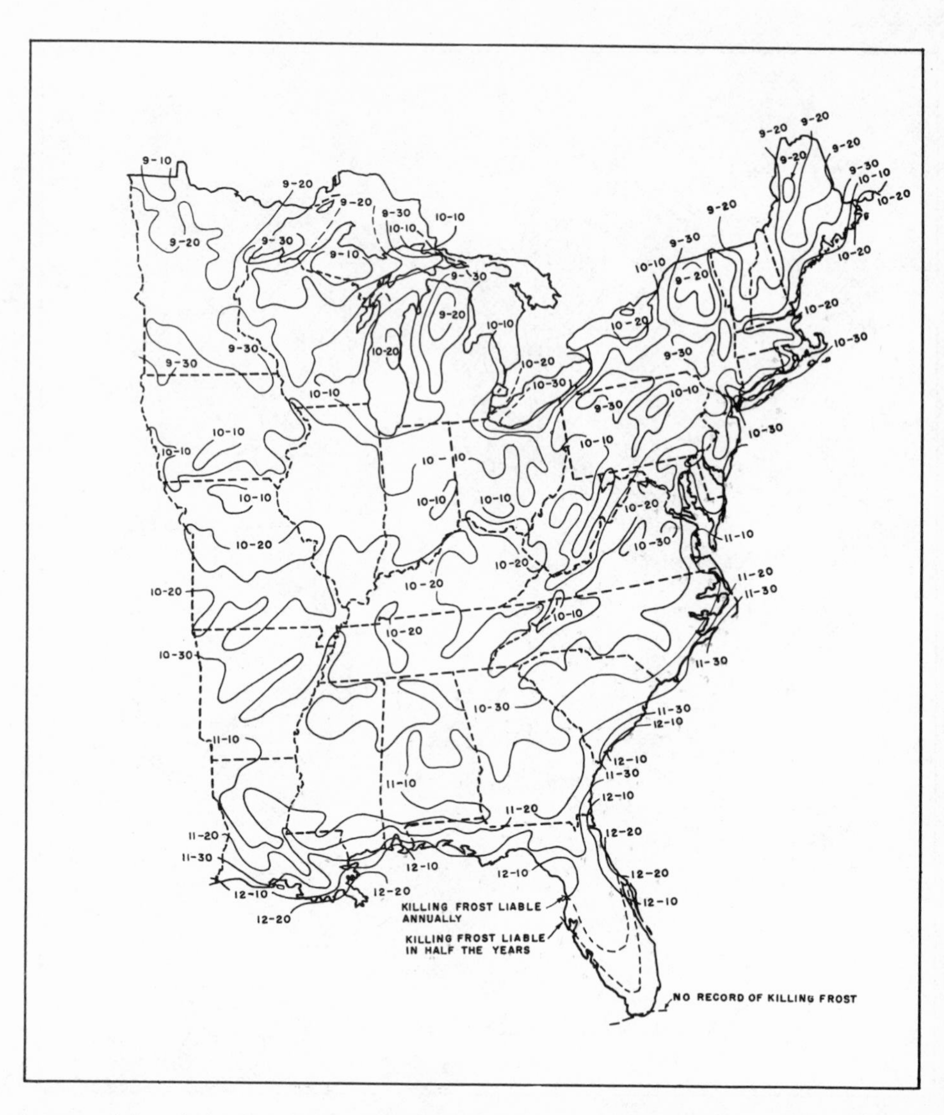

Average dates of the first killing fall frosts in the eastern United States, 1899 to 1938. Gardeners should locate the line nearest to the locality in which they live, note the date on that line (the first figure indicates the month, the second the day; thus 11-10 is November 10) and then refer to tables 34 and 35. (Redrawn from U. S. Weather Bureau original.)

TABLE 32: Earliest Safe Planting Dates and Range of Spring-Planting Dates for Vegetables in the Open

Crop	Planting dates for localities with average last freeze on—						
	Jan. 30	Feb. 8	Feb. 18	Feb. 28	Mar. 10	Mar. 20	Mar. 30
Asparagus [1]					Jan. 1–Mar. 1	Feb. 1–Mar. 10	Feb. 15–Mar. 20.
Beans, lima	Feb. 1–Apr. 15	Feb. 10–May 1	Mar. 1–May 1	Mar. 15–June 1	Mar. 20–June 1	Apr. 1–June 15	Apr. 15–June 20.
Beans, snap	Feb. 1–Apr. 1	Feb. 1–May 1	do	Mar. 10–May 15	Mar. 15–May 15	Mar. 15–May 25	Apr. 1–June 1.
Beets	Jan. 1–Mar. 15	Jan. 10–Mar. 15	Jan. 20–Apr. 1	Feb. 1–Apr. 15	Feb. 15–June 1	Feb. 15–May 15	Mar. 1–June 1.
Broccoli, sprouting [1]	Jan. 1–30	Jan. 1–30	Jan. 15–Feb. 15	Feb. 1–Mar. 1	Feb. 15–Mar. 15	Feb. 15–Mar. 15	Mar. 1–20.
Brussels sprouts [1]	do	do	do	do	do	do	Do.
Cabbage [1]	Jan. 1–15	Jan. 1–Feb. 10	Jan. 1–Feb. 25	Jan. 15–Feb. 25	Jan. 25–Mar. 1	Feb. 1–Mar. 1	Feb. 15–Mar. 10.
Cabbage, Chinese	([2])	([2])	([2])	([2])	([2])	([2])	([2])
Carrots	Jan. 1–Mar. 1	Jan. 1–Mar. 1	Jan. 15–Mar. 1	Feb. 1–Mar. 1	Feb. 10–Mar. 15	Feb. 15–Mar. 20	Mar. 1–Apr. 10.
Cauliflower [1]	Jan. 1–Feb. 1	Jan. 1–Feb. 1	Jan. 10–Feb. 10	Jan. 20–Feb. 20	Feb. 1–Mar. 1	Feb. 10–Mar. 10	Feb. 20–Mar. 20.
Celery and celeriac	do	Jan. 10–Feb. 10	Jan. 20–Feb. 20	Feb. 1–Mar. 1	Feb. 20–Mar. 20	Mar. 1–Apr. 1	Mar. 15–Apr. 15.
Chard	Jan. 1–Apr. 1	Jan. 10–Apr. 1	Jan. 20–Apr. 15	Feb. 1–May 1	Feb. 15–May 15	Feb. 20–May 15	Mar. 1–May 25.
Chervil and chives	Jan. 1–Feb. 1	Jan. 1–Feb. 1	Jan. 1–Feb. 1	Jan. 15–Feb. 15	Feb. 1–Mar. 1	Feb. 10–Mar. 10	Feb. 15–Mar. 15.
Chicory, witloof					June 1–July 1	June 1–July 1	June 1–July 1.
Collards [1]	Jan. 1–Feb. 15	Jan. 1–Feb. 15	Jan. 1–Mar. 15	Jan. 15–Mar. 15	Feb. 1–Apr. 1	Feb. 15–May 1	Mar. 1–June 1.
Corn salad	do	do	do	Jan. 1–Mar. 1	Jan. 1–Mar. 15	Jan. 1–Mar. 15	Jan. 15–Mar. 15.
Corn, sweet	Feb. 1–Mar. 15	Feb. 10–Apr. 1	Feb. 20–Apr. 15	Mar. 1–Apr. 15	Mar. 10–Apr. 15	Mar. 15–May 1	Mar. 25–May 15.
Cress, upland	Jan. 1–Feb. 1	Jan. 1–Feb. 15	Jan. 15–Feb. 15	Feb. 1–Mar. 1	Feb. 10–Mar. 15	Feb. 20–Mar. 15	Mar. 1–Apr. 1.
Cucumbers	Feb. 15–Mar. 15	Feb. 15–Apr. 1	Feb. 15–Apr. 15	Mar. 1–Apr. 15	Mar. 15–Apr. 15	Apr. 1–May 1	Apr. 10–May 15.
Dandelion	Jan. 1–Feb. 1	Jan. 1–Feb. 1	Jan. 15–Feb. 15	Jan. 15–Mar. 1	Feb. 1–Mar. 1	Feb. 10–Mar. 10	Feb. 20–Mar. 20.
Eggplant [1]	Feb. 1–Mar. 1	Feb. 10–Mar. 15	Feb. 20–Apr. 1	Mar. 10–Apr. 15	Mar. 15–Apr. 15	Apr. 1–May 1	Apr. 15–May 15.
Endive	Jan. 1–Mar. 1	Jan. 1–Mar. 1	Jan. 15–Mar. 1	Feb. 1–Mar. 1	Feb. 15–Mar. 15	Mar. 1–Apr. 1	Mar. 10–Apr. 10.
Florence fennel	do	do	do	do	do	do	Do.
Garlic	([2])	([2])	([2])	([2])	([2])	Feb. 1–Mar. 1	Feb. 10–Mar. 10.
Horseradish [1]							Mar. 1–Apr. 1.
Kale	Jan. 1–Feb. 1	Jan. 10–Feb. 1	Jan. 20–Feb. 10	Feb. 1–20	Feb. 10–Mar. 1	Feb. 20–Mar. 10	Mar. 1–20.
Kohlrabi	do	do	do	do	do	do	Mar. 1–Apr. 1.
Leeks	do	Jan. 1–Feb. 1	Jan. 1–Feb. 15	Jan. 15–Feb. 15	Jan. 25–Mar. 1	Feb. 1–Mar. 1	Feb. 15–Mar. 15.
Lettuce, head [1]	do	do	Jan. 1–Feb. 1	do	Feb. 1–20	Feb. 15–Mar. 10	Mar. 1–20.
Lettuce, leaf	do	do	Jan. 1–Mar. 15	Jan. 1–Mar. 15	Jan. 15–Apr. 1	Feb. 1–Apr. 1	Feb. 15–Apr. 15.
Mustard	Jan. 1–Mar. 1	Jan. 1–Mar. 1	do	Feb. 1–Mar. 1	Feb. 10–Mar. 15	Feb. 20–Apr. 1	Mar. 1–Apr. 15.
Okra	Feb. 15–Apr. 1	Feb. 15–Apr. 15	Mar. 1–June 1	Mar. 10–June 1	Mar. 20–June 1	Apr. 1–June 15	Apr. 10–June 15.
Onions [1]	Jan. 1–15	Jan. 1–15	Jan. 1–15	Jan. 1–Feb. 1	Jan. 15–Feb. 15	Feb. 10–Mar. 10	Feb. 15–Mar. 15.
Onions, seed	do	do	do	Jan. 1–Feb. 15	Feb. 1–Mar. 1	do	Feb. 20–Mar. 15.
Onions, sets	do	do	do	Jan. 1–Mar. 1	Jan. 15–Mar. 10	Feb. 1–Mar. 20	Feb. 15–Mar. 20.
Parsley	Jan. 1–30	Jan. 1–30	Jan. 1–30	Jan. 15–Mar. 1	Feb. 1–Mar. 10	Feb. 15–Mar. 15	Mar. 1–Apr. 1.
Parsnips			Jan. 1–Feb. 1	Jan. 15–Feb. 15	Jan. 15–Mar. 1	do	Do.
Peas, garden	Jan. 1–Feb. 15	Jan. 1–Feb. 15	Jan. 1–Mar. 1	Jan. 15–Mar. 1	Jan. 15–Mar. 15	Feb. 1–Mar. 15	Feb. 10–Mar. 20.
Peas, black-eye	Feb. 15–May 1	Feb. 15–May 15	Mar. 1–June 15	Mar. 10–June 20	Mar. 15–July 1	Apr. 1–July 1	Apr. 15–July 1.
Peppers [1]	Feb. 1–Apr. 1	Feb. 15–Apr. 15	Mar. 1–May 1	Mar. 15–May 1	Apr. 1–June 1	Apr. 10–June 1	Apr. 15–June 1.
Potatoes	Jan. 1–Feb. 15	Jan. 1–Feb. 15	Jan. 15–Mar. 1	Jan. 15–Mar. 1	Feb. 1–Mar. 1	Feb. 10–Mar. 15	Feb. 20–Mar. 20.
Radishes	Jan. 1–Apr. 1	Jan. 1–Apr. 1	Jan. 1–Apr. 1	Jan. 1–Apr. 1	Jan. 1–Apr. 15	Jan. 20–May 1	Feb. 15–May 1.
Rhubarb [1]							
Rutabagas				Jan. 1–Feb. 1	Jan. 15–Feb. 15	Jan. 15–Mar. 1	Feb. 1–Mar. 1.
Salsify	Jan. 1–Feb. 1	Jan. 10–Feb. 10	Jan. 15–Feb. 20	Jan. 15–Mar. 1	Feb. 1–Mar. 1	Feb. 15–Mar. 1	Mar. 1–15.
Shallots	do	Jan. 1–Feb. 10	Jan. 1–Feb. 20	Jan. 1–Mar. 1	Jan. 15–Mar. 1	Feb. 1–Mar. 10	Feb. 15–Mar. 15.
Sorrel	Jan. 1–Mar. 1	Jan. 1–Mar. 1	Jan. 15–Mar. 1	Feb. 1–Mar. 10	Feb. 10–Mar. 15	Feb. 10–Mar. 20	Feb. 20–Apr. 1.
Soybeans	Mar. 1–June 30	Mar. 1–June 30	Mar. 10–June 30	Mar. 20–June 30	Apr. 10–June 30	Apr. 10–June 30	Apr. 20–June 30.
Spinach	Jan. 1–Feb. 15	do	Jan. 1–Mar. 1	Jan. 1–Mar. 1	Jan. 15–Mar. 10	Jan. 15–Mar. 15	Feb. 1–Mar. 20.
Spinach, New Zealand	Feb. 1–Apr. 15	Feb. 15–Apr. 15	Mar. 1–Apr. 15	Mar. 15–May 15	Mar. 20–May 15	Apr. 1–May 15	Apr. 10–June 1.
Squash, summer	do	do	do	do	Mar. 15–May 1	do	Do.
Sweetpotatoes [1]	Feb. 15–May 15	Mar. 1–May 15	Mar. 20–June 1	Mar. 20–June 1	Apr. 1–June 1	Apr. 10–June 1	Apr. 20–June 1.
Tomatoes [1]	Feb. 1–Apr. 1	Feb. 20–Apr. 10	Mar. 1–Apr. 20	Mar. 10–May 1	Mar. 20–May 10	Apr. 1–May 20	Apr. 10–June 1.
Turnips	Jan. 1–Mar. 1	Jan. 1–Mar. 1	Jan. 10–Mar. 1	Jan. 20–Mar. 1	Feb. 1–Mar. 1	Feb. 10–Mar. 10	Feb. 20–Mar. 20.

TABLE 33: Earliest Safe Planting Dates and Range of Spring-Planting Dates for Vegetables in the Open (Continued)

Earliest safe planting dates and range of spring-planting dates for vegetables in the open—Continued

Crop	Planting dates for localities with average last freeze on—						
	Apr. 10	Apr. 20	Apr. 30	May 10	May 20	May 30	June 10
Asparagus [1]	Mar. 10–Apr. 10	Mar. 15–Apr. 15	Mar. 20–Apr. 15	Apr. 10–Apr. 30	Apr. 20–May 15	May 1–June 1	May 15–June 1.
Beans, lima	Apr. 15–June 30	May 1–June 20	May 15–June 15	May 25–June 15			
Beans, snap	Apr. 10–June 30	Apr. 25–June 30	May 10–June 30	May 10–June 30	May 15–June 30	May 25–June 15	
Beets	Mar. 10–June 1	Mar. 20–June 1	Apr. 1–June 15	Apr. 15–June 15	Apr. 25–June 15	May 1–June 15	May 15–June 15.
Broccoli, sprouting [1]	Mar. 15–Apr. 15	Mar. 25–Apr. 20	Apr. 1–May 1	Apr. 15–June 1	May 1–June 15	May 10–June 10	May 20–June 10.
Brussels sprouts [1]	do	do	do	do	do	do	Do.
Cabbage [1]	Mar. 1–Apr. 1	Mar. 10–Apr. 1	Mar. 15–Apr. 10	Apr. 1–May 15	do	May 10–June 15	May 20–June 1.
Cabbage, Chinese	(2)	(2)	(2)	do	do	do	Do.
Carrots	Mar. 10–Apr. 20	Apr. 1–May 15	Apr. 10–June 1	Apr. 20–June 15	May 1–June 1	May 10–June 1	Do.
Cauliflower [1]	Mar. 1–Mar. 20	Mar. 15–Apr. 20	Apr. 10–May 10	Apr. 15–May 15	May 10–June 15	May 20–June 1	June 1–June 15.
Celery and celeriac	Apr. 1–Apr. 20	Apr. 10–May 1	Apr. 15–May 1	Apr. 20–June 15	do	do	Do.
Chard	Mar. 15–June 15	Apr. 1–June 15	Apr. 15–June 15	do	do	do	Do.
Chervil and chives	Mar. 1–Apr. 1	Mar. 10–Apr. 10	Mar. 20–Apr. 20	Apr. 1–May 1	Apr. 15–May 15	May 1–June 1	May 15–June 1.
Chicory, witloof	June 10–July 1	June 15–July 1	June 15–July 1	June 1–20	June 1–15	June 1–15	June 1–15.
Collards [1]	Mar. 1–June 1	Mar. 10–June 1	Apr. 1–June 1	Apr. 15–June 1	May 1–June 1	May 10–June 1	May 20–June 1.
Corn salad	Feb. 1–Apr. 1	Feb. 15–Apr. 15	Mar. 1–May 1	Apr. 1–June 1	Apr. 15–June 1	May 1–June 15	May 15–June 15.
Corn, sweet	Apr. 10–June 1	Apr. 25–June 15	May 10–June 15	May 10–June 1	May 15–June 1	May 20–June 1	
Cress, upland	Mar. 10–Apr. 15	Mar. 20–May 1	Apr. 10–May 10	Apr. 20–May 20	May 1–June 1	May 15–June 1	May 15–June 15.
Cucumbers	Apr. 20–June 1	May 1–June 15	May 15–June 15	May 20–June 15	June 1–15		
Dandelion	Mar. 1–Apr. 1	Mar. 10–Apr. 10	Mar. 20–Apr. 20	Apr. 1–May 1	Apr. 15–May 15	May 1–30	May 1–30.
Eggplant [1]	May 1–June 1	May 10–June 1	May 15–June 10	May 20–June 15	June 1–15		
Endive	Mar. 15–Apr. 15	Mar. 25–Apr. 15	Apr. 1–May 1	Apr. 15–May 15	May 1–30	May 1–30	May 15–June 1.
Florence fennel	do	do	do	do	do	do	Do.
Garlic	Feb. 20–Mar. 20	Mar. 10–Apr. 1	Mar. 15–Apr. 15	Apr. 1–May 1	Apr. 15–May 15	do	Do.
Horseradish [1]	Mar. 10–Apr. 10	Mar. 20–Apr. 20	Apr. 1–30	Apr. 15–May 15	Apr. 20–May 20	do	Do.
Kale	Mar. 10–Apr. 1	Mar. 20–Apr. 10	Apr. 1–20	Apr. 10–May 1	Apr. 20–May 10	do	Do.
Kohlrabi	Mar. 10–Apr. 10	Mar. 20–May 1	Apr. 1–May 10	Apr. 10–May 15	Apr. 20–May 20	do	Do.
Leeks	Mar. 1–Apr. 1	Mar. 15–Apr. 15	Apr. 1–May 1	Apr. 15–May 15	May 1–May 20	May 1–15	May 1–15.
Lettuce, head [1]	Mar. 10–Apr. 1	Mar. 20–Apr. 15	do	do	May 1–June 30	May 10–June 30	May 20–June 30.
Lettuce, leaf	Mar. 15–May 15	Mar. 20–May 15	Apr. 1–June 1	Apr. 15–June 15	do	do	Do.
Mustard	Mar. 10–Apr. 20	Mar. 20–May 1	Apr. 1–May 10	Apr. 15–June 1	do	do	Do.
Okra	Apr. 20–June 15	May 1–June 1	May 10–June 1	May 20–June 10	June 1–20		
Onions [1]	Mar. 1–Apr. 1	Mar. 15–Apr. 10	Apr. 1–May 1	Apr. 10–May 1	Apr. 20–May 15	May 1–30	May 10–June 10.
Onions, seed	do	Mar. 15–Apr. 1	Mar. 15–Apr. 15	Apr. 1–May 1	do	do	Do.
Onions, sets	do	Mar. 10–Apr. 1	Mar. 10–Apr. 10	Apr. 10–May 1	do	do	Do.
Parsley	Mar. 10–Apr. 10	Mar. 20–Apr. 20	Apr. 1–May 1	Apr. 15–May 15	May 1–20	May 10–June 1	May 20–June 10.
Parsnips	do	do	do	do	do	do	Do.
Peas, garden	Feb. 20–Mar. 20	Mar. 10–Apr. 10	Mar. 20–May 1	Apr. 1–May 15	Apr. 15–June 1	May 1–June 15	May 10–June 15.
Peas, black-eye	May 1–July 1	May 10–June 15	May 15–June 1				
Peppers [1]	May 1–June 1	May 10–June 1	May 15–June 10	May 20–June 10	May 25–June 15	June 1–15	
Potatoes	Mar. 10–Apr. 1	Mar. 15–Apr. 10	Mar. 20–May 10	Apr. 1–June 1	Apr. 15–June 15	May 1–June 15	May 15–June 1.
Radishes	Mar. 1–May 1	Mar. 10–May 10	do	do	do	do	Do.
Rhubarb [1]	Mar. 1–Apr. 1	Mar. 10–Apr. 10	Mar. 20–Apr. 15	Apr. 1–May 1	Apr. 15–May 10	May 1–20	Do.
Rutabagas			May 1–June 1	May 1–June 1	May 1–20	May 10–20	May 20–June 1.
Salsify	Mar. 10–Apr. 15	Mar. 20–May 1	Apr. 1–May 15	Apr. 15–June 1	May 1–June 1	May 10–June 1	Do.
Shallots	Mar. 1–Apr. 1	Mar. 15–Apr. 15	Apr. 1–May 1	Apr. 10–May 1	Apr. 20–May 10	May 1–June 1	May 10–June 1.
Sorrel	Mar. 1–Apr. 15	Mar. 15–May 1	Apr. 1–May 15	Apr. 15–June 1	May 1–June 1	May 10–June 10	May 20–June 10.
Soybeans	May 1–June 30	May 10–June 20	May 15–June 15	May 25–June 10			
Spinach	Feb. 15–Apr. 1	Mar. 1–Apr. 15	Mar. 20–Apr. 20	Apr. 1–June 15	Apr. 10–June 15	Apr. 20–June 15	May 1–June 15.
Spinach, New Zealand	Apr. 20–June 1	May 1–June 15	May 1–June 15	May 10–June 15	May 20–June 15	June 1–15	
Squash, summer	do	do	May 1–30	May 10–June 10	do	June 1–20	June 10–20.
Sweetpotatoes [1]	May 1–June 1	May 10–June 10	May 20–June 10				
Tomatoes [1]	Apr. 20–June 1	May 5–June 10	May 10–June 15	May 15–June 10	May 25–June 15	June 5–20	June 15–30.
Turnips	Mar. 1–Apr. 1	Mar. 10–Apr. 1	Mar. 20–May 1	Apr. 1–June 1	Apr. 15–June 1	May 1–June 15	May 15–June 15.

[1] Plants.

[2] Planted in fall only.

TABLE 34: Latest Safe Planting Dates and Late Range of Planting Dates for Vegetables in the Open

Latest safe planting dates and late range of planting dates for vegetables in the open

Crop	Planting dates for localities with average first freeze on—					
	Aug. 30	Sept. 10	Sept. 20	Sept. 30	Oct. 10	Oct. 20
Asparagus [1]					Oct. 20–Nov. 15	Nov. 1–Dec. 15.
Beans, lima				June 1–15	June 1–15	June 15–30.
Beans, snap		May 15–June 15	June 1–July 1	June 1–July 10	June 15–July 20	July 1–Aug. 1.
Beets	May 15–June 15	do.	do.	do.	June 15–July 25	July 1–Aug. 5.
Broccoli, sprouting	May 1–June 1	May 1–June 1	May 1–June 15	June 1–30	June 15–July 15	July 1–Aug. 1.
Brussels sprouts	do.	do.	do.	do.	do.	Do.
Cabbage [1]	do.	do.	do.	June 1–July 10	June 1–July 15	July 1–20.
Cabbage, Chinese	May 15–June 15	May 15–June 15	June 1–July 1	June 1–July 15	June 15–Aug. 1	July 15–Aug. 15.
Carrots	do.	do.	do.	June 1–July 10	June 1–July 20	June 15–Aug. 1.
Cauliflower [1]	May 1–June 1	May 1–July 1	May 1–July 1	May 10–July 15	June 1–July 25	July 1–Aug. 5.
Celery [1] and celeriac	do.	May 15–June 15	May 15–July 1	June 1–July 5	June 1–July 15	June 1–Aug. 1.
Chard	May 15–June 15	May 15–July 1	June 1–July 1	do.	June 1–July 20	Do.
Chervil and chives	May 10–June 10	May 1–June 15	May 15–June 15	(2)	(2)	(2)
Chicory, witloof	May 15–June 15	May 15–June 15	do.	June 1–July 1	June 1–July 1	June 15–July 15.
Collards [1]	do.	do.	do.	June 15–July 15	July 1–Aug. 1	July 15–Aug. 15.
Corn salad	do.	May 15–July 1	June 15–Aug. 1	July 15–Sept. 1	Aug. 15–Sept. 15	Sept. 1–Oct. 15.
Corn, sweet			June 1–July 1	June 1–July 1	June 1–July 10	June 1–July 20.
Cress, upland	May 15–June 15	May 15–July 1	June 15–Aug. 1	July 15–Sept. 1	Aug. 15–Sept. 15	Sept. 1–Oct. 15.
Cucumbers			June 1–15	June 1–July 1	June 1–July 1	June 1–July 15.
Dandelion	June 1–15	June 1–July 1	June 1–July 1	June 1–Aug. 1	July 15–Sept. 1	Aug. 1–Sept. 15.
Eggplant [1]				May 20–June 10	May 15–June 15	June 1–July 1.
Endive	June 1–July 1	June 1–July 1	June 15–July 15	June 15–Aug. 1	July 1–Aug. 15	July 15–Sept. 1.
Florence fennel	May 15–June 15	May 15–July 15	June 1–July 1	June 1–July 1	June 15–July 15	June 15–Aug. 1.
Garlic	(2)	(2)	(2)	(2)	(2)	(2)
Horseradish [1]	(2)	(2)	(2)	(2)	(2)	(2)
Kele	May 15–June 15	May 15–June 15	June 1–July 1	June 15–July 15	July 1–Aug. 1	July 15–Aug. 15.
Kohlrabi	do.	June 1–July 1	June 1–July 15	do.	do.	Do.
Leeks	May 1–June 1	May 1–June 1	(2)	(2)	(2)	(2)
Lettuce, head [1]	May 15–July 1	May 15–July 1	June 1–July 15	June 15–Aug. 1	July 15–Aug. 15	Aug. 1–30.
Lettuce, leaf	May 15–July 15	May 15–July 15	June 1–Aug. 1	June 1–Aug. 1	July 15–Sept. 1	July 15–Sept. 1.
Mustard	do.	do.	do.	June 15–Aug. 1	July 15–Aug. 15	Aug. 1–Sept. 1.
Okra			June 1–20	June 1–July 1	June 1–July 15	June 1–Aug. 1.
Onions [1]	May 10–June 10	May 1–June 1	(2)	(2)	(2)	(2)
Onions, seed	May 1–June 1	do.	(2)	(2)	(2)	(2)
Onions, sets	do.	do.	(2)	(2)	(2)	(2)
Parsley	May 15–June 15	May 1–June 15	June 1–July 1	June 1–July 15	June 15–Aug. 1	July 15–Aug. 15.
Parsnips	May 15–June 1	do.	May 15–June 15	June 1–July 1	June 1–July 10	(4)
Peas, garden	May 10–June 15	May 1–July 1	June 1–July 15	June 1–Aug. 1	(2)	(2)
Peas, black-eye					June 1–July 1	June 1–July 1.
Peppers [1]			June 1–June 20	June 1–July 1	do.	June 1–July 10.
Potatoes	May 15–June 1	May 1–June 15	May 1–June 15	May 1–June 15	May 15–June 15	June 15–July 15.
Radishes	May 1–July 15	May 1–Aug. 1	June 1–Aug. 15	July 1–Sept. 1	July 15–Sept. 15	Aug. 1–Oct. 1.
Rhubarb [1]	Sept. 1–Oct. 1	Sept. 15–Oct. 15	Sept. 15–Nov. 1	Oct. 1–Nov. 1	Oct. 15–Nov. 15	Oct. 15–Dec. 1.
Rutabagas	May 15–June 15	May 1–June 15	June 1–July 1	June 1–July 1	June 15–July 15	July 10–20.
Salsify	May 15–June 1	May 10–June 10	May 20–June 20	June 1–20	June 1–July 1	June 1–July 1.
Shallots	(2)	(2)	(2)	(2)	(2)	(2)
Sorrel	May 15–June 15	May 1–June 15	June 1–July 1	June 1–July 15	July 1–Aug. 1	July 15–Aug. 15.
Soybeans				May 25–June 10	June 1–25	June 1–July 5.
Spinach	May 15–July 1	June 1–July 15	June 1–Aug. 1	July 1–Aug. 15	Aug. 1–Sept. 1	Aug. 20–Sept. 10.
Spinach, New Zealand				May 15–July 1	June 1–July 15	June 1–Aug. 1.
Squash, summer	June 10–20	June 1–20	May 15–July 1	June 1–July 1	do.	June 1–July 20.
Squash, winter			May 20–June 10	June 1–15	June 1–July 1	June 1–July 1.
Sweetpotatoes [1]					May 20–June 10	June 1–15.
Tomatoes [1]	June 20–30	June 10–20	June 1–20	June 1–20	June 1–20	June 1–July 1.
Turnips	May 15–June 15	June 1–July 1	June 1–July 15	June 1–Aug. 1	July 1–Aug. 1	July 15–Aug. 15.

[1] Plants.

[2] Generally spring-planted only.

TABLE 35: Latest Safe Planting Dates and Late Range of Planting Dates for Vegetables in the Open (Continued)

Crop	Planting dates for localities with average first freeze on—					
	Oct. 30	Nov. 10	Nov. 20	Nov. 30	Dec. 10	Dec. 20
Asparagus [1]	Nov. 15–Jan. 1	Dec. 1–Jan. 1				
Beans, lima	July 1–Aug. 1	July 1–Aug. 15	July 15–Sept. 1	Aug. 1–Sept. 15	Sept. 1–30	Sept. 1–Oct. 1.
Beans, snap	July 1–Aug. 15	July 1–Sept. 1	July 1–Sept. 10	Aug. 15–Sept. 20	do	Sept. 1–Nov. 1.
Beets	Aug. 1–Sept. 1	Aug. 1–Oct. 1	Sept. 1–Dec. 1	Sept. 1–Dec. 15	Sept. 1–Dec. 31	Sept. 1–Dec. 31.
Broccoli, sprouting	July 1–Aug. 15	Aug. 1–Sept. 1	Aug. 1–Sept. 15	Aug. 1–Oct. 1	Aug. 1–Nov. 1	Do.
Brussels sprouts	do	do	do	do	do	Do.
Cabbage [1]	Aug. 1–Sept. 1	Sept. 1–15	Sept. 1–Dec. 1	Sept. 1–Dec. 31	Sept. 1–Dec. 31	Do.
Cabbage, Chinese	Aug. 1–Sept. 15	Aug. 15–Oct. 1	Sept. 1–Oct. 15	Sept. 1–Nov. 1	Sept. 1–Nov. 15	Sept. 1–Dec. 1.
Carrots	July 1–Aug. 15	Aug. 1–Sept. 1	Sept. 1–Nov. 1	Sept. 15–Dec. 1	Sept. 15–Dec. 1	Sept. 15–Dec. 1.
Cauliflower [1]	July 15–Aug. 15	do	Aug. 1–Sept. 15	Aug. 15–Oct. 10	Sept. 1–Oct. 20	Sept. 15–Nov. 1.
Celery and celeriac	June 15–Aug. 15	July 1–Aug. 15	July 15–Sept. 1	Aug. 1–Dec. 1	Sept. 1–Dec. 31	Oct. 1–Dec. 31.
Chard	June 1–Sept. 10	June 1–Sept. 15	June 1–Oct. 1	June 1–Nov. 1	June 1–Dec. 1	June 1–Dec. 31.
Chervil and chives	(2)	(2)	Nov. 1–Dec. 31	Nov. 1–Dec. 31	Nov. 1–Dec. 31	Nov. 1–Dec. 31.
Chicory, witloof	July 1–Aug. 10	July 10–Aug. 20	July 20–Sept. 1	Aug. 15–Sept. 30	Aug. 15–Oct. 15	Aug. 15–Oct. 15.
Collards [1]	Aug. 1–Sept. 15	Aug. 15–Oct. 1	Aug. 25–Nov. 1	Sept. 1–Dec. 1	Sept. 1–Dec. 31	Sept. 1–Dec. 31.
Corn salad	Sept. 15–Nov. 1	Oct. 1–Dec. 1	Oct. 1–Dec. 1	Oct. 1–Dec. 31	Oct. 1–Dec. 31	Oct. 1–Dec. 31.
Corn, sweet	June 1–Aug. 1	June 1–Aug. 15	June 1–Sept. 1			
Cress, upland	Sept. 15–Nov. 1	Oct. 1–Dec. 1	Oct. 1–Dec. 1	Oct. 1–Dec. 31	Oct. 1–Dec. 31	Oct. 1–Dec. 31.
Cucumbers	June 1–Aug. 1	June 1–Aug. 15	June 1–Aug. 15	July 15–Sept. 15	Aug. 15–Oct. 1	Aug. 15–Oct. 1.
Dandelion	Aug. 15–Oct. 1	Sept. 1–Oct. 15	Sept. 1–Nov. 1	Sept. 15–Dec. 15	Oct. 1–Dec. 31	Oct. 1–Dec. 31.
Eggplant [1]	June 1–July 1	June 1–July 15	June 1–Aug. 1	July 1–Sept. 1	Aug. 1–Sept. 30	Aug. 1–Sept. 30.
Endive	July 15–Aug. 15	Aug. 1–Sept. 1	Sept. 1–Oct. 1	Sept. 1–Nov. 15	Sept. 1–Dec. 31	Sept. 1–Dec. 31.
Florence fennel	July 1–Aug. 1	July 15–Aug. 15	Aug. 15–Sept. 15	do	Sept. 1–Dec. 1	Sept. 1–Dec. 1.
Garlic	(2)	Aug. 1–Oct. 1	Aug. 15–Oct. 1	Sept. 1–Nov. 15	Sept. 15–Nov. 15	Sept. 15–Nov. 15.
Horseradish [1]	(2)					
Kale	July 15–Sept. 1	Aug. 1–Sept. 15	Aug. 15–Oct. 15	Sept. 1–Dec. 1	Sept. 1–Dec. 31	Sept. 1–Dec. 31.
Kohlrabi	Aug. 1–Sept. 1	Aug. 15–Sept. 15	Sept. 1–Oct. 15	do	Sept. 15–Dec. 31	Do.
Leeks	(2)	(2)	Sept. 1–Nov. 1	Sept. 1–Nov. 1	Sept. 1–Nov. 1	Sept. 15–Nov. 1.
Lettuce, head [1]	Aug. 1–Sept. 15	Aug. 15–Oct. 15	do	Sept. 1–Dec. 1	Sept. 15–Dec. 31	Sept. 15–Dec. 31.
Lettuce, leaf	Aug. 15–Oct. 1	Aug. 25–Oct. 1	do	do	do	Do.
Mustard	Aug. 15–Oct. 15	Aug. 15–Nov. 1	Sept. 1–Dec. 1	do	Sept. 1–Dec. 1	Sept. 15–Dec. 1.
Okra	June 1–Aug. 10	June 1–Aug. 20	June 1–Sept. 10	June 1–Sept. 20	Aug. 1–Oct. 1	Aug. 1–Oct. 1.
Onions [1]	(2)	Sept. 1–Oct. 15	Oct. 1–Dec. 31	Oct. 1–Dec. 31	Oct. 1–Dec. 31	Oct. 1–Dec. 31.
Onions, seed	(2)	(2)	Sept. 1–Nov. 1	Sept. 1–Nov. 1	Sept. 1–Nov. 1	Sept. 15–Nov. 1.
Onions, sets	(2)	Oct. 1–Dec. 1	Nov. 1–Dec. 31	Nov. 1–Dec. 31	Nov. 1–Dec. 31	Nov. 1–Dec. 31.
Parsley	Aug. 1–Sept. 15	Sept. 1–Nov. 15	Sept. 1–Dec. 31	Sept. 1–Dec. 31	Sept. 15–Dec. 31	Sept. 1–Dec. 31.
Parsnips	(2)	(2)	Aug. 1–Sept. 1	Sept. 1–Nov. 15	Sept. 1–Dec. 1	Sept. 1–Dec. 1.
Peas, garden	Aug. 1–Sept. 15	Sept. 1–Nov. 1	Oct. 1–Dec. 1	Oct. 1–Dec. 31	Oct. 1–Dec. 31	Oct. 1–Dec. 31.
Peas, black-eye	June 1–Aug. 1	June 15–Aug. 15	July 1–Sept. 1	July 1–Sept. 10	July 1–Sept. 20	July 1–Sept. 20.
Peppers [1]	June 1–July 20	June 1–Aug. 1	June 1–Aug. 15	June 15–Sept. 1	Aug. 15–Oct. 1	Aug. 15–Oct. 1.
Potatoes	July 20–Aug. 10	July 25–Aug. 20	Aug. 10–Sept. 15	Aug. 1–Sept. 15	Aug. 1–Sept. 15	Aug. 1–Sept. 15.
Radishes	Aug. 15–Oct. 15	Sept. 1–Nov. 15	Sept. 1–Dec. 1	Sept. 1–Dec. 31	do	Oct. 1–Dec. 31.
Rhubarb [1]	Nov. 1–Dec. 1					
Rutabagas	July 15–Aug. 1	July 15–Aug. 15	Aug. 1–Sept. 1	Sept. 1–Nov. 15	Oct. 1–Nov. 15	Oct. 15–Nov. 15.
Salsify	June 1–July 10	June 15–July 20	July 15–Aug. 15	Aug. 15–Sept. 30	Aug. 15–Oct. 15	Sept. 1–Oct. 31.
Shallots	(2)	Aug. 1–Oct. 1	Aug. 15–Oct. 1	Aug. 15–Oct. 15	Sept. 15–Nov. 1	Sept. 15–Nov. 1.
Sorrel	Aug. 1–Sept. 15	Aug. 15–Oct. 1	Aug. 15–Oct. 15	Sept. 1–Nov. 15	Sept. 1–Dec. 15	Sept. 1–Dec. 31.
Soybeans	June 1–July 15	June 1–July 25	June 1–July 30	June 1–July 30	June 1–July 30	June 1–July 30.
Spinach	Sept. 1–Oct. 1	Sept. 15–Nov. 1	Oct. 1–Dec. 1	Oct. 1–Dec. 31	Oct. 1–Dec. 31	Oct. 1–Dec. 31.
Spinach, New Zealand	June 1–Aug. 1	June 1–Aug. 15	June 1–Aug. 15			
Squash, summer	do	June 1–Aug. 10	June 1–Aug. 20	June 1–Sept. 1	June 1–Sept. 15	June 1–Oct. 1.
Squash, winter	June 10–July 10	June 20–July 20	July 1–Aug. 1	July 15–Aug. 15	Aug. 1–Sept. 1	Aug. 1–Sept. 1.
Sweetpotatoes [1]	June 1–15	June 1–July 1	June 1–July 1	June 1–July 1	June 1–July 1	June 1–July 1.
Tomatoes [1]	June 1–July 1	June 1–July 15	June 1–Aug. 1	Aug. 1–Sept. 1	Aug. 15–Oct. 1	Sept. 1–Nov. 1.
Turnips	Aug. 1–Sept. 15	Sept. 1–Oct. 15	Sept. 1–Nov. 15	Sept. 1–Nov. 15	Oct. 1–Dec. 1	Oct. 1–Dec. 31.

Spring Planting Dates

To determine the best time to plant any particular vegetable in your locality in spring, proceed as follows:

1. Find your location on the spring frost map.

2. Find the solid line (called an *isotherm*) on this map that comes nearest your locality.

3. The date shown on this line is the average date of the last killing frost (3-10, for instance, means 3rd month and 10th day or March 10; 4-15 means April 15, etc.).

4. On the table which follows the map, find the column having this date as a heading. Mark this column as it is the only one that you will use in your locality for determining spring planting dates.

5. Find the dates in this column opposite the crop you wish to plant. These dates show the period during which this crop can be planted for best results, but the earlier in this period the planting is made the better.

To determine the *latest* safe planting dates, follow the same procedure with the fall frost map and table showing the latest safe planting dates.

Seeding

Mark the corners of the garden with permanent stakes or pipes. Carefully set up a line for the first row and mark it out with your rake handle or the corner of the hoe. Keep this line handy and always use it for laying out rows for sowing seed or setting plants. The handle may open up a furrow sufficiently deep for small seeds. A deeper furrow can be made with a hoe or regular furrower for planting larger seed.

Straight rows in the garden may be obtained by stretching a string between stakes set up at the ends of the row. Stretch the string tight enough to lift it above the surface of the soil or it will not be straight.

Most vegetable seeds should be planted very shallow. The smaller seeds, such as lettuce and radishes, should not be planted more than one-quarter inch deep. For such small seeds, the hoe or rake handle is an excellent tool for opening the shallow seed furrow. For the larger seeds, such as peas and beans, the proper depth is from one-half to one inch. Late plantings may be somewhat deeper than early plantings.

As a general rule, plant seeds somewhat less than the recommended depth if the soil is heavy or the temperature low, and a little deeper in light soil or during warm weather. Bear in mind that more than one vegetable can be planted in a row, if necessary, for frequent small sowings to insure a continuous supply of tender vegetables.

It seldom pays to use a seed planter in the back yard garden, as it will not handle small quantities of seed efficiently. Sow your seeds by hand—it is one of the most pleasurable experiences of gardening. Sow the seed thick enough to get an even stand. Too much seed is wasteful and means twice the work of thinning.

Sow the seed thinly by shaking it from a cut corner of the package or individually dropping the seeds from the fingers. Many gardeners waste seed

Steps in Transplanting

1. Carefully remove plants from flat with ball of earth clinging to roots, trimming off any broken roots. Watering the flat makes removal easier.

2. Soil in planting bed should be fine and well-fertilized. Make small holes with a trowel, large enough to accommodate roots.

3. After planting, firm the earth around stem with both hands. This will prevent air pockets which tend to dry the fine feeder roots.

Steps in Transplanting

4. Mulch should be placed around seedlings to keep down weeds and retain soil moisture necessary to encourage plant growth.

5. Insect damage can be prevented on cabbage and other plants by putting a cardboard collar around the stem and into the soil.

6. Protect seedlings from both sun and frost with paper hats, boxes, pots and slats, in addition to applying layer of mulch.

and make extra work by sowing seed too thick. After seeding, the soil is drawn over the seed, using the corner of the hoe. In case the soil is somewhat cloddy, try to draw in the finer soil to cover the seed.

Sandy soils should be firmed above the seed in order to hasten germination. On heavy soils, where crusting is likely, packing of the soil above the seed increases the danger of crusting. On such soils a shower or irrigation may be necessary to facilitate germination and crusts will be less severe if the soil is not packed.

Late planted seeds sown during the summer may not germinate unless provisions are made to retain moisture. When sowing late seeded crops, water along the rows and cover with a board or a light mulch. To water such seeds without protection against evaporation will usually result in failure, as the soil will dry rapidly and the young seedlings die before emerging. Uncover the rows when the first plants appear.

The question arises frequently on how long does seed remain viable. This will depend greatly upon the particular vegetable involved as well as conditions under which the seed is stored. Seed will retain viability much longer when stored under cool, dry conditions as compared with storage that is warm and moist.

Vegetable seeds may be divided into the following general groups:

1. Comparatively short-lived, usually not good after one to two years: sweet corn, leek, onion, parsley, parsnip and salsify.
2. Moderately long-lived, often good for 3 to 5 years under favorable conditions: asparagus, bean, Brussels sprouts, cabbage, carrot, cauliflower, celery, chicory, cress, endive, kale, kohlrabi, lettuce, okra, peas, pepper, radish, spinach, turnip and watermelon.
3. Comparatively long-lived under favorable conditions, may be good for more than 5 years: beets, cucumber, muskmelon, mustard and tomato.

You can plant seeds in drills, in hills or by broadcasting. Following is a description of these three methods, prepared by members of Nevada's Horticultural Department:

The drill method usually is the best. Using the corner of the hoe, make a straight drill or trench down each row. It should be deep enough to reach into moist ground. The seeds then are dropped into the drill row by hand. Directions usually are found on seed packages indicating the right depth. It is not necessary to fill the drill with soil when covering small seeds. Simply cover the seeds to the right depth and leave the rest of the drill unfilled. Large seeds, such as peas or beans, should be planted from 2 to 4 inches apart. Very small seeds should be mixed with about 3 times their bulk of dry sand so they will not fall too thickly. This is helpful especially with such small seeds as those of turnips and carrots, which are lost easily out of the hand between the fingers. Seeds should be covered with fine, moist soil, which should be firmly pressed down over the seeds with the back of the hoe. This prevents drying out around the seeds.

Hills frequently are used for plants that spread on the ground. A hole about a foot across and a foot deep is made and a 4- to 6-inch layer of *well-rotted* manure is put in the bottom of the hole. Don't use fresh manure, since

Planting Seeds

1. Straight rows can be obtained by stretching a string tightly between stakes set up at the end of each row.

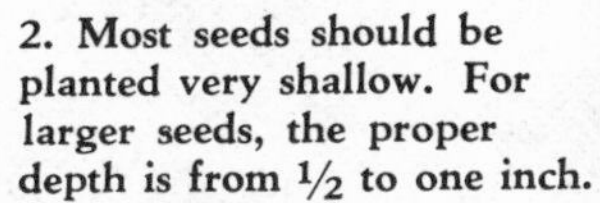

2. Most seeds should be planted very shallow. For larger seeds, the proper depth is from ½ to one inch.

3. Small seeds should be covered very lightly with fine soil. A hoe is an adequate tool for this purpose.

Planting Seeds

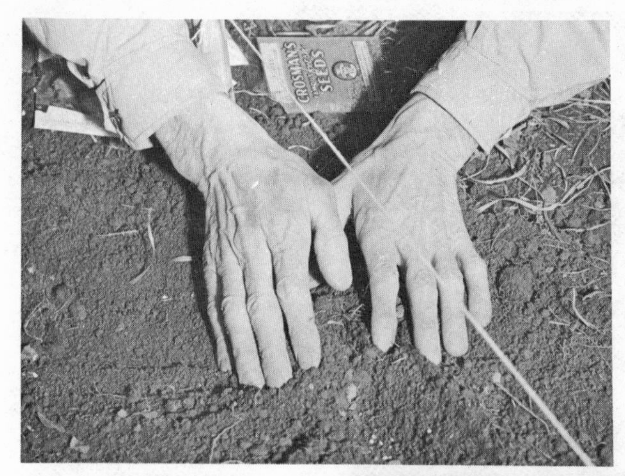

4. About a quarter inch of soil is firmed over radish seeds in their shallow trench, right after seeding.

5. The back of a rake can be used to push soil over a seeded furrow. If ground is cloddy, cover with finer soil.

6. Applying organic matter on top of the row right after seeding prevents crust formation and keeps soil moist.

it injures the plants. The manure is covered with 4 to 6 inches of fine soil, packed down, and the seeds are planted in it. Hills are used because they afford a convenient way to give the plants food material, and because one may know where the roots are after the vines have covered the ground. The distance apart to make the hills will depend upon the size of the vines. Hills for vine squash or pumpkins may be 8 to 12 feet apart, while cucumber hills may be only 3 to 6 feet apart. Unless extra fertilizer is used under the hills, there is little advantage to the hill method. Corn usually is planted in hills because of convenience in planting and because it saves the labor of making a drill.

When the broadcasting method is used, the seeds are scattered thinly over the top of the soil and are then covered by light raking and packing with the back of a hoe. The seedbed should be well prepared first. This method is

Many gardeners have found peat pots a useful means of starting seedlings since transplanting out of flats is unnecessary. Seeds can be planted indoors in the peat pot containing a mixture of soil, sand and peat and later set in the vegetable garden.

used with many small seeds, such as those of turnips, lettuce, etc. It is well adapted to use in small vegetable gardens where the sprinkling method of irrigation is to be used. It eliminates the need for irrigation furrows. It makes hand-weeding necessary because the plants are not in rows and cannot be cultivated with power equipment or tools.

Seeding the Garden with Less Work

Richard V. Clemence, an inventive gardener from Massachusetts, has been testing the hay-mulch method for a number of years. Recently, he has successfully tried a series of experiments on surface-planting with the year-round mulch system. Here is his report:

For the past few years, I've been trying to find a way of starting a hay-mulched garden in the spring without stirring the ground. To the best of my knowledge, this has been impossible. A heavy hay mulch laid on nearly any area in the summer or fall will produce an excellent seedbed for the following spring. But unless this foresighted action is taken, the garden must be prepared in the conventional way, and the plants well started before mulch can be successfully applied.

Now my tests have convinced me that a no-work, permanent-mulch vegetable garden can be started "from scratch," in spring! There's no need to wait until your garden has been planted to spread mulch; *you can seed and plant right on top of a heavy mulch.*

My experiments were made on a small plot, 12 by 20 feet, where grass, weeds and small bushes were allowed to grow until about the first of June. By that time, the growth was waist high and gave a depressing spectacle to anyone meaning to plant the area without doing a lot of hard work. The tests were as follows:

1. Potatoes—the small, whole, Green Mountain variety—were placed on the ground, in rows roughly 14 inches apart, where they were completely hidden by the tall vegetation. This high growth was then flattened, and some 6 inches of hay spread over, leaving the seed potatoes at the bottom. After two weeks, the potato sprouts appeared, and grew into exceptionally large vines, yielding an excellent crop of fine potatoes.

Green Mountain potatoes are notoriously subject to insect and disease attacks, but these were troubled hardly at all. Potatoes, as is well known, thrive on potash, while requiring comparatively little nitrogen. The chemical "potato fertilizers" used by many commercial growers typically have formulas containing twice as much potash as nitrogen.

Now, rotting vegetation tends to lock up nitrogen, and a heavy hay mulch over weeds and grass would mean a nitrogen shortage for many crops. For potatoes, however, this is not a problem. Moreover, the same potash that can be recovered in wood ashes is released in the decay of any vegetation, so this scheme provides potatoes with plenty of that. (This is such a perfect way to grow potatoes that it is a shame I cannot claim credit for having originated it.)

2. Starting again with the same rugged growth of weeds, grass, and so on, I flattened it to the ground as before with hay mulch over it, but this time with nothing under it. I then dug through the cover with a shovel in a half-dozen places, and planted "hills" of squash, cucumbers, pumpkins and melons. All these crops did very well and were notably free from attacks by insects and disease.

The vines developed their fruit on the surface of the hay, where it remained clean and dried quickly after rains. Hardly a weed grew anywhere in the whole area. The weed seeds near the surface of the ground had germinated and grown tall before they were smothered by the hay. Most of these were therefore gone for good. Other weed seeds still in the soil were not brought up by spading or plowing, so they did not germinate at all. The only foreign plants that showed up were 4 or 5 dandelions that pushed through the hay in midseason.

3. The third project was hardly an experiment, since it was an obvious procedure and certain to work. All it amounted to was setting tomato plants through the fresh mulch in the same way as through a mulch that had been

longer in place. It involved moving the hay a little and cutting through the vegetation under it to make holes for the plants. I put dried cow manure and bone meal in the holes, and ran the sprinkler for a few hours as soon as all the plants were set.

As you would expect, the results were all that could be desired, giving evidence that tomatoes may be readily grown in a new garden without turning the soil. This may not be astounding news, but I wonder how many have ever done it. It should work equally well with peppers and other plants, too.

Perhaps I should add that I have also tried setting the tomato plants in grass and weeds *before* applying the hay. This turned out to be considerably more work, particularly where the existing growth was heavy. With all the vegetation flattened under the mulch, nothing impedes the gardener's efforts.

4. The fourth experiment—growing other vegetables such as peas, beans, lettuce, etc.—began like the others with hay mulch laid over tall grass, weeds and similar vegetation. In one test, I included cornstalks I had left standing for the purpose the year before. In another, I included strong raspberry canes spreading several feet from the main row.

In earlier trials, I had applied the hay very liberally. For the potatoes, this was desirable and for the other crops it made no special difference. This time, however, I used only enough hay to cover the existing growth thoroughly, holding it flat and excluding the light. Even so, the mulch was some 3 or 4 inches deep and I used 3 bales of hay on the 12 by 20 foot area.

Before describing the rest of the experiment, I should like to insert a few remarks about the problems (and solutions) involved in planting seed right on top of a heavy mulch.

Seeds, particularly small ones, would vanish at once if placed on the hay, and would never be heard from again. The answer is to sprinkle peat moss on top of the hay along rows marked with a string, and drop the seeds on the moss. For the smallest seeds, a little loam could be mixed with the peat moss to close the spaces between coarse particles.

Seeds would tend to germinate slowly, if at all, on account of their distance above the moisture in the soil. This was an easy one and running a sprinkler on the peat moss was the obvious answer.

Now for the procedures and results. In general, the planting procedure consisted of sprinkling peat moss, dried manure and loam (for the smallest seeds) along a wide row marked with stakes and string, scattering the seed, and pressing it into the prepared surface by walking slowly along the row. Rows were planted about 18 inches apart.

Peas. Two 20-foot rows of Lincoln peas were planted side by side in late June. One row was planted as I have indicated; the other was planted in the same way, but with the manure omitted. The row with manure grew strongly from the beginning, blossomed during a heat wave in mid-August and bore a fine crop of peas. Not so heavy a crop, of course, as peas planted early in the spring, but fully up to the late crops I get by the usual methods. The row without manure started just as well as the other, but the plants soon turned light yellow and later became brown and dry. Only two plants ever blossomed and each of these bore one pod containing one pea.

Beans. Bush snap beans were comparable to peas as grown by this method.

With manure, they seemed to do just as well as if they had been planted directly in good soil.

Beets and Radishes. As I suspected in advance, the method is not satisfactory with either beets or radishes and I assume that it would not work any better with turnips or rutabagas. Germination and top growth are promising, but both radishes and beets develop long, thin roots that contrast most unfavorably with the globe-shaped specimens ordinarily grown.

Lettuce. Black-Seeded Simpson is the only variety of lettuce I have grown by this method. With manure it did very well, but it is a strong and rapid-growing variety that is hard to discourage anyway.

Carrots. Since the main objective of the technique is to save time and labor, I think it should be regarded as successful with any crop that yields well, even if not quite so well as with conventional methods. On this criterion, the one test I have made with long-rooted carrots gave spectacular results.

I chose Tendersweet carrots for the experiment, and mixed a little loam with the peat moss and manure when planting them. They germinated very strongly and grew so rapidly that thinning was necessary in about three weeks. The roots were unusually long and straight, smooth and well-shaped. On the basis of a single test it would be unsafe to attempt any generalizations, but it does seem possible that the method has special virtues in producing long-rooted carrots.

For everyone interested in this program, two points should be given special emphasis:

1. Adequate moisture is essential to satisfactory results. Unless nature is unusually cooperative, a plastic sprinkler or some good substitute must be made available. Ordinary soil tests for moisture are out of the question, but the hay mulch should be damp from just below the surface on down. This reproduces conditions in any good soil, encouraging the roots to continue their development. At the same time, it speeds the decay of the lower levels of vegetation, insuring that the roots will find their long journey worth the effort.

Keeping the mulch damp requires much less water than you might think. A first thorough soaking for quick germination, and a few more if rains are insufficient, will be enough. The hay cover dries from the top downward, so there is always more moisture in the lower layers than near the surface. If plants show no sign of wilting, irrigation is unnecessary.

Adequate moisture early in the game saves water in the end, for once the roots of your plants get well developed, they will find moisture enough in the lower levels during even a severe drought.

2. A liberal supply of a dried manure, compost or some other complete organic fertilizer, is needed to feed the seedlings until their roots reach the soil. The commercial dried and shredded manure has, for once, an advantage over the real thing. Being virtually a powder, it readily penetrates down through the hay, and provides the best of nourishment for the plants as their roots develop.

Instructions for Surface Planting

1. Mark a row with stakes and string over the exposed soil, and scatter your seeds on the surface.

(Left) Scatter seeds on the surface of the soil in the area indicated by the string

(Below) Spreading a thin layer of fine peat moss is recommended for surface plantings.

(Right) The seeds are then pressed into the moss with a roller or by walking over row.

2. Fill a large basket with peat moss, and walk down your row of seeds, shaking the moss onto them as you proceed and, employing your feet (preferably large) to step on the mixture of seeds and moss, pressing them firmly together. (A board or roller will also do the job.) Having reached the end of the row, you return by the same route, stepping on the spots you missed before, and leaving a perfectly planted crop behind you.

3. The function of the sprinkler is merely to guarantee results. The peat moss needs a good wetting to keep the seeds damp until they sprout and rain may well be forthcoming at the proper moments. The sprinkler, however, removes all doubts.

As you can see, this system of surface planting eliminates rotting, blowing, and washing of seeds, and gets nearly every "viable" specimen started at once. No rain short of a cloudburst can move the seeds, for the peat moss absorbs water like a sponge. At the same time, the smallest sprouts reach the air and light immediately and decay can hardly occur. Finally, sun and wind cannot dry or move your seeds if you have a sprinkler ready for use. It might be added that the time and labor of planting are also reduced to a minimum.

In surface planting no furrows are made, of course, and the question thus arises as to how you get seeds like lettuce, carrots, and beets laid in single straight rows. The answer is that you don't. You don't even try to do so. On the contrary, you aim at spreading them over a space about a foot wide, using your marking string merely as a guide for one edge of the row. You thereby get many more plants into a given area than you can by other methods, and when you thin your crops, you save time and effort again. You can thin wide rows nearly as fast as narrow ones and space your plants in both directions in one operation. After thinning carrots, for instance, you will have rows 4 or 5 carrots in width rather than one thin line of plants; and similarly with other crops.

I find that many gardeners shrink from the thought of rows more than one plant wide. They seem uncertain what is wrong about it, but they feel sure that something is. Perhaps the difficulty is that wide rows are not in accord with traditional practice. But, there is certainly no good reason why a plant should require more space from north to south than from east to west, or vice versa. It is hard to believe that many plants can tell one direction from the other. Narrow rows with wide aisles between them merely waste space in the home garden, without fulfilling any useful purpose. They have been copied from commercial practice which requires space for the operation of machinery. Broad alleyways in the kitchen garden are about as logical as 6 lanes of cement in the driveway. Such arrangements were designed for other situations.

Seeds like cucumber, melon and squash may be surface-planted in "hills." As the word may suggest to some, though not to me, a "hill" is a shallow depression in the soil, an improvement on the genuine mound once used. The depression is usually made by filling a small excavation with compost or manure, spreading a little soil on top, and stepping on the area to press it down. In surface planting, you can prepare a hill in this way or, if your soil is rich enough, you can simply scatter some seeds in any convenient place. Then shake on peat moss, and press the moss and seeds well together against the soil. Again, the sprinkler provides insurance of quick results.

You can economize on your time and strength by scattering your seeds rather carelessly and thinning hills to preserve only the most promising plants later on. If you conduct the thinning by clipping off unwanted plants instead of pulling them out by the roots, you will not retard the growth of any of the specimens you choose to save.

Surface planting, as I have described it, has real advantages over more common methods. It can be used successfully with nearly all sorts of vegetable crops, and I know of only one important exception. In planting such crops as corn and pole beans, it saves trouble in the end to take pains enough to place the seeds in exactly the spots they should occupy, and to make sure that they stay there. The easiest way of doing this is to push each seed slightly into the soil with your fingers. Since they are not lying loose on the surface, peat moss is not needed to anchor them, and it is no better than any other good mulch for holding the moisture.

Thinning

Most gardeners sow too much seed. Consequently plants are too close if seed has good germination. The natural tendency is to seed crops like beets, carrots, radishes, lettuce, turnips and endive too thickly. This is because it is difficult to sow small seeds thinly enough to eliminate hand thinning, says E. C. Wittmeyer, Ohio State University extension horticulturist. Beet seeds, for example, are actually fruits containing several seeds. If root crops are not thinned, poorly shaped roots will result.

The easiest time to do the thinning job is when the plants are small and the soil is moist. Turnips should be thinned before their tap roots become fleshy. They need 3 inches between plants for best development. Radishes, on the other hand, can be left in the ground until those that are to be thinned are large enough to eat unless they have been seeded very thickly. If this is the situation, thinning should be done early.

Surplus beet plants can be pulled when they are 4 to 5 inches tall and used for greens. Beets should have 3 inches between plants. Carrots should be thinned early and allowed to stand one inch apart. Later alternate carrots can be pulled and used.

Lettuce, Swiss chard, endive, kohlrabi and similar crops also may need thinning. With lettuce, however, thinning at harvest will help produce high quality lettuce for a long time.

Here are some general rules regarding thinning:

1. Thinning should be done while plants are small and when the soil is moist, so they can be pulled out easily without injuring those that are left.

2. Root crops should be thinned before their taproots become fleshy. Onions from seeds, and radishes, can be left in the ground until those that are thinned out are large enough to eat.

3. Carrots should be thinned first when they are 2 to 3 inches tall, so as to stand about one inch apart. They can then be left to develop until large enough to be eaten, when alternate plants can be pulled and used, leaving more room for those that are left.

4. Swiss chard can be thinned at first to 3 inches, then, as plants develop, harvest alternate ones. Cucumbers and melons should be thinned to 2 or 3 plants per hill or to stand 12 to 15 inches apart in the row.

5. Thinning is rarely practiced with beans, peas, corn and some other large-seeded plants which produce vigorous seedlings, for a good stand usually is obtained by planting only a few more seeds than the number of plants required.

There are many ways to protect plants from early frost. A glass jar (a miniature greenhouse) placed over tender seedlings is a popular method to protect against damage from late spring frost.

Transplanting Vegetables

The few minutes spent transplanting are the most important minutes in the life of the plant.

Some plants are more resistant to shock than others, but all react to some degree. Among the easiest plants to transplant are broccoli, cabbage, cauliflower, lettuce and onions. Some loss may be expected with beans, eggplant, peppers, tomatoes and celery, and among the very difficult to transplant successfully are carrots, corn, cucumber, melons, squash and all other plants with taproots.

The idea in transplanting, of course, is to lessen the shock as much as possible. The plant is an amazingly well-integrated mechanism in which each function of each cell is interdependent on the actions of other cells. Water and soil materials are moving constantly through the roots, stems and leaves, feeding the cells of the leaves which manufacture the plant food. When transplanted, some of the vital roots and tiny root hairs are almost always torn away, the entire water chain system may be disrupted and changed light conditions and temperature may disrupt other normal plant functions. It's up to you to help the seedling make as smooth an adjustment as possible.

If possible, try to transplant on a cloudy, moist day, when light and heat shock will be small. Before beginning to transplant, loosen soil in the new location, and about an hour or two before removing seedlings, thoroughly water soil in the flats to make loosening roots easier.

A heavy paper collar is useful for protecting young tomato plants from cutworm damage. The bottom of the collar is about an inch below the soil surface, and a clip holds the collar together as protection against those insects.

When taking up seedlings, take plenty of soil along. Some root damage is almost inevitable, but it can be kept to a minimum by special care in handling. Root damage can be virtually eliminated by use of special little containers made of dried manure. The seedling is usually placed in the container—or the seed is germinated therein—and when ready, the entire container is set into the soil. This method not only lessens transplanting shock, but also preserves roots intact and provides an initial stimulus for developing roots.

When making the final transplant, make sure roots are in good firm contact with the soil. A great deal depends on this, as roots depend on firm contact for absorption of essential soil ingredients. After the transplant, water well and shade seedlings for a few days—if the weather is hot and sunny, keep them shaded for as long as a week.

If tender seedlings were to be transplanted from indoor flats directly into the open ground, the inevitable result would be serious injury, often fatal. Therefore, to lessen the shock of temperature, wind and sun, a hardening off process is used, in which seedlings are gradually exposed to the natural elements.

If plant protectors such as the plastic ones shown above are used, earlier plantings can be made. To support the plastic cover, gardeners use a variety of material as branches, metal coat hangers or "stands" which can be bought.

Usually, the hardening off process takes about two weeks, and should be timed so that the young seedlings end the two weeks in a period warm enough for them to be planted outside safely. The best way to harden off seedlings (or cuttings, for that matter) is to move them, flats and all, from their indoor location to the cold frame. (The cold frame should be an essential piece of equipment for any gardener. It is inexpensive, not difficult to construct and provides a perfect medium for the hardening off process.)

The frame should be closed for the first two days of the period. On the third day, the top should be opened just a crack, and over the remainder of the period, the air space should be widened more each day until, on the last day, the top is removed.

If you haven't gotten around to constructing a cold frame, you can use your imagination to simulate the same conditions it provides. On the first day of the period, place flats outside for only a few minutes during the warmer part of the afternoon. Cover them with a burlap screen or preferably something more closely knit and place flats in a sheltered place. Then, gradually expose them more each day until the process is completed.

There have been scores of gadgets, most of which work pretty well, marketed to protect seedlings during the hardening off process. In England most popular are cloches—little glass tents—which are set over seedlings planted in open ground. In America, bell jars and various other similar coverings are used for the same purpose. The cloche idea allows you to place seedlings in permanent location one step sooner, cutting out the last transplanting which is the most dangerous to growing roots. And, if a late unexpected frost is announced during an afternoon, you can save your seedlings by setting up the protectors again.

If plant protectors made of plastic, waxed paper, etc., are used to cover plants when first set out, hardening off is not as important. Also you'll be able to plant earlier. Tomato and strawberry baskets, as well as glass cloches are effective in shielding plants from sun or windburn. You can place them to cover the plants entirely or tilt them to allow air to circulate underneath.

To support the plastic or other cover material, gardeners often use such a variety of materials as branches (bent back with both ends in the ground), metal coat hangers or a "stand" that can be purchased along with a plastic cover.

Tennessee University experimenters have found hill covers made of paper cloth are good for melons, cucumbers, squashes and even tomatoes. They were able to plant seeds two weeks earlier than would be ordinarily safe in open ground; the covers also kept insects away for a longer time.

Cloth covers can be made of squares of cheese cloth, supported on 2 or 3 arches made from wire or pieces of veneer wood out of old baskets. The ends of these are stuck in the ground two inches deep. Cover cloth edges with soil. Since the edges decay in 3 or 4 weeks, the cloth as well as the paper covers cannot be used again.

About a week before it is safe to have plants in open air uncovered, cut or tear open the paper cover tops for ventilation. The usual practice is to cut a slit on the south side first, then later enlarge the hole. If the plants grow through the holes, it's all right to leave the cover in place. Here are a list of points to remember when using plant covers:

1. Use early varieties of vegetables for planting under covers since your objective is to harvest as soon as possible.

2. Cabbage, onions and other hardy crops will not be helped very much by covers.

3. A garden that has rather sandy or "early" soil, that is protected from strong winds and perhaps slopes to the southeast, would be the very best place to use hill covers. Clay soil in a naturally cold spot, would seldom produce profitable gains from use of hill covers.

Homemade "Greenhouses" for Early Plants

Following is a report on how one New England gardener provided protection for his crops:

We live in a part of northern Vermont where the growing season is especially short, and we have tried to think of ways to extend the number of days between killing frosts in our garden. Last year we experimented with homemade plastic "tents" which we call miniature greenhouses.

A friend of ours had given us several lengths of heavy, clear plastic about 10 millimeters in thickness. We used some of it to cover old window frames from which the glass was broken. Each of these was placed over a box the same size (open on top and bottom) and with about 8-inch walls. We banked the sides with dirt or hay and placed the portable frames in the proper place in the garden with the window frame on top. They worked very well in starting melons.

Our other method was simpler. We cut lengths of brush about 3 feet long and one-half inch in diameter and stripped off the leaves. By placing one end in the ground and slowly bending the brush and forcing the other end into the ground about 18 inches distant, a bow or arch was made that was 18 inches wide and 10 inches high. In heavy soil the ends do not have to be forced into the ground very far, but in lighter soil they would be safer if pushed in about 6 inches.

We placed these arches every foot or so along the row for whatever length we needed. The soil must be prepared and ready to plant before inserting the arches. Early in May we planted such seed as corn, melons and squash in rows under the middle of the arches and covered them with the plastic. Dirt was used to hold down the sides and ends of the plastic.

We snipped a small slot in each end of the "tent" to allow some air circulation. The corn and squash grew extremely well in these tents. In fact, the plants were touching the top of the plastic before we felt it was entirely safe to remove the protection. On very hot, sunny days, we opened up one end of the little greenhouses by simply removing the earth, as the plastic was heavy enough to hold its shape when an opening was made at one end. If it becomes very dry, the plastic can be opened up or quickly removed for watering, although we found that the plants seemed to be getting enough moisture without help.

The tents are portable and we moved them to various parts of the garden as needed. They can be used very early in the season for lettuce, radishes, beets and carrots and then moved to a new section of the garden to start early corn and melons. Larger ones can be made for tomato, eggplant or pepper transplants.

Coat hangers can be cut and straightened out to make the arches described or any stiff wire can be used if you do not have a ready supply of willowy brush. The plastic should last several seasons and since most plastics seem to admit ultraviolet light, the plants do not appear to be endangered by sunburn. In England they use similar "tents" made of glass (which are called "cloches") in large quantities both for the home gardener and commercially, although plastic is also becoming popular as breakage is always a problem using glass.

We think that other gardeners will find these miniature greenhouses, which can be made at home, very useful in starting early plants.

—Robert F. Stowell

Setting Out Seedlings

When you are ready to transplant the seedlings to their final places in the garden, again wet down beds or flats well.

When plants have been grown in flats, cutting the soil between the plants with a knife or blocking hoe about a week or 10 days before transplanting increases the number of small roots within the block of soil. This tends to bind the soil together and protects the roots when the plant is finally moved from the flat or transplanting bed to the field.

When plants have been grown in starter boxes or plant bands, there's no need to cut them apart. But it is advisable to run a knife between the bands since they sometimes tend to stick together and the roots grow through the bands of one into an adjoining one.

You may have some difficulty removing the first 2 or 3 seedlings from a flat without injuring their roots. One solution to the problem is to make flats with one easily removable side. By using a single short nail at each end, the side can be removed. Then it's a simple matter to slip a trowel underneath the roots and lift up the plants without damaging any roots.

Another suggestion for removing plants is to tilt the flat at an angle of about 45 degrees and jar the lowest end on the ground. This causes some compression of soil and roots, leaving a space of about one-half inch between the soil and edge of the flat at the upper end, making it easy to get your fingers beneath the roots and remove individual plants.

"Box Bed" Gardening

Many California gardeners have found gardening in boxes to have many advantages, such as helping to:

1. Grow more food crops and flowers in less space—even in a lawn.
2. Control weeds, insects and diseases in "limited-area" gardens.
3. Mulch, irrigate and fertilize for specific, separate plants.
4. Start seeds and seedlings with less trouble.
5. Extend the growing season by shading, covering or heating.

This report by Warner G. Tilsher describes how you also can make use of the box bed method.

Basically, a box bed is a self-contained garden about 3 feet wide by about 6 feet long, closed in on all sides. These sides stick out of the ground about 6 to 10 inches and are set dead level all around.

Out here in Southern California we use one- or two-inch rough redwood; back East, cypress would be perfect. Or you can make it up out of concrete slabs. Or pour it on site. At any rate, the material should be able to take rough treatment and all kinds of weather. I would say you could make and set a box bed in an hour and a half at most—without strain or sweat.

If this sounds like more bother than it's worth for such a little piece of land and you are asking why not garden it as is, bear a bit further with us.

First, of all, these box bed gardens have a nice neat look about them. You can set one or even a battery of them, into the back end of your lawn without hurting its looks too much.

There is hardly a back yard so small that a body can't find room for a couple of these beds. Weeds are easy to control and they don't keep crowding over the edges as in an ordinary garden.

Add an extension and a glass or plastic-covered sash and you have a cold frame. Add a soil cable and you have a hotbed—you're in business all year 'round.

It's easy to protect seeds or baby plants in a box bed. Simple screens keep out slugs and worms until the plants are rugged enough to take care of themselves.

Most herbs reach their peak of savor when they are kept a bit on the dry side. Since you may not grow enough of them to fill a whole bed, you can partition off an end. In this "sub-miniature" box bed you can control your watering exactly to the herbs' taste.

If you live in a hot climate, you can keep right on growing lettuce and cool-weather crops by shading them with canopies of nursery cloths. These will also keep your seedlings from burning up.

A nice feature about box beds is that you can segregate crops according to their needs. Leafy crops take to overhead shower baths. Beans, peppers, etc., may do better with flooding. If you like to stand around in the cool of the evening with a hose, box beds are for you; the water stays where you put it. Or you can lay in a hose with something to break the force of the water and the bed soaks up beautifully. Or loop a "soil-soaker" (porous canvas hose) around the rows and go about your other jobs with no worries. In about 20 to 30 minutes drop it into the next bed.

Speaking of water—although we have never experienced a time during even the heaviest rain when a box bed flooded over—you might drill a hole about an inch in diameter in one end about an inch or so above the soil line and plug it with a cork. "Just in case" insurance.

But where box beds really shine is in production. In a regular garden we would have to space rows about two or more feet apart to allow for an irrigation furrow. In a box bed we space carrots, beets, even lettuce only 8 inches apart. We figure we can get two or more times as much out of a box bed as we would from a regular garden plot. Think of the savings in water alone.

To use your box bed to full advantage, you should start vegetables like lettuce, chard and beets in flats. With a bit of planning you can time things so your box beds are working pretty much all of the time during your growing season.

After a bed has been first set, we level off the inside. Then we add manure, compost, organic fertilizer such as sewage sludge and ground rock and spade it over lightly. We give it a good soaking, cover it over (let some air get at it) and let it set a week or so before planting.

Usual method is to use a plug seeder which automatically spaces the seed in the row, drop in three seeds and cover them with coarse sand. We roll the row down tight. Whether it rains or you sprinkle irrigate, this system is a sure way of preventing the soil from caking up over your seeds. It's a fool-proof way of getting 100 per cent stands, no mean trick out here in 90-degree-plus weather.

Once the plants are 3 or 4 inches tall, we mulch the whole bed with leaves, alfalfa hay or straw. About halfway to harvest we give the crop our usual booster shot of fish emulsion.

Once the crop is out we just clean off the top of the bed, lay on a layer of manure and fertilizer, rake it up a bit, soak it and soon it's on the way again. We usually rotate among leafy, root and fruiting crops. Incidentally, a box bed makes an efficient starting bed for items like coles, onions, beets, chard, annual flowers, etc.

Box beds tend to grow on you. Each year you find yourself adding another one or more. This is especially true of city folks who find themselves with steadily shrinking *Lebensraum*.

When we first started with them, we raised only small root crops. Soon we were growing beans, cabbage, peppers, strawberries. Now we're to the point where the only things we grow on the flat are tomatoes, corn, squash and potatoes.

Once every couple of years we grow a crop of rye and clover in the box beds. When this is a foot or more high, we spade it under and keep it moist. Don't plant until all the turned-under material is digested. After a vacation like that, the bed will really go to work.

Starter Solutions

Oregon's Ralph Clark, extension horticultural specialist, has found that starter solutions have a very definite use in the culture of vegetable crops. The home gardener as well as the commercial grower will benefit from the rapid and unchecked growth of his plants if such a solution is used just before they are moved to the field from the greenhouse, hotbed or cold frame. Greenhouse operators have long made use of this help in bringing their crops to a rapid and profitable production. Earlier yields have also been reported from this treatment in the field.

The main benefit received from these solutions is that of providing the plant with immediately available food. This stimulates leaf and root growth, giving the plant a quick pickup after transplanting. These solutions are used especially on young lettuce, tomatoes, celery, peppers, melons, eggplant, cabbage, cauliflower and all kinds of transplanted plants.

Here is a recommended starter solution for organic gardeners:

Fill barrel or other container ¼ full of barnyard manure. (Use ½ as much if poultry manure is used.) Continue to fill container with water, stirring several times during next 24 to 48 hours.

In using, dilute liquid to a light amber color with water. Pour one pint around each plant when setting out or later as necessary to force growth. Liquid manure can be used at 10-day to 2-week intervals especially when soils are not high in fertility.

Some vegetables that are ordinarily started early (cabbages and related plants, tomatoes and peppers) benefit from at least one transplanting before being set out in the garden. Turnips, kohlrabi and Chinese cabbage often grow faster when they have been transplanted than when left in the original rows. (If transplanting is to be done successfully, however, the roots must not be allowed to dry out.) Others will do well without any preliminary transplanting. Many gardeners like using plant bands or starter boxes for individual plants such as melons which do not transplant readily in the seedling stage. Here is a list of plants and their transplanting preferences:

Beans (pole)—sow 6 to 8 seeds; thin to 3; plant two pots to each pole.

Beets—transplant to flats or directly from seedbeds.

Chinese cabbage—transplant to flats.

Corn—sow 4 or 5; thin to 3; plant two feet apart.

Cucumber, melons—sow 6 to 8 seeds in 4-inch (or larger) pots or bands; thin to 3 or 4.

Leeks—transplant into deep trench; fill in gradually.

Okra—sow 3 or 4 seeds per pot; thin to one; set 1½ feet apart in garden.

Onions—transplant directly to garden; trim back about one-third.

Spinach, Swiss chard—transplant to individual pots or bands.

Squash—sow 4 to 6 seeds in each pot or band; thin to 2 or 3.

Inoculate Garden Legumes

If you are planning to grow any beans, peas or peanuts in your garden this spring, why not take time out to inoculate the seed with nitrogen-gathering bacteria? It will certainly be worth your while.

Inoculating legume seed, to both increase the yield and to fortify the soil with added nitrogen, is nothing new. Farmers have been doing it for centuries. The value of using root-nodule inoculants is, of course, well-known today to all who raise legume crops for livestock feeding, for hay or for soil-improving green manures.

Early farmers did not have cultured bacteria to aid them in their work as we do today. In spring they took a handful of soil from a field or garden plot that had produced a good stand of beans the year before, and mixed it with their bean seed. Or, if they wanted to plant peas and wished a good stand, they took soil from a plot that produced good peas the year before and mixed it with the seed. They were inoculating that particular legume with the right strain of nitrogen-gathering bacteria.

Now, however, the benefits of inoculating leguminous vegetables can easily be reaped by any gardener. Bacteria inoculants in cultures prepared and packaged for garden-scale use are available from several firms. They are very inexpensive and simple to apply to seed. Used with reasonable care, *they will improve plant growth, increase yields of beans or peas, etc., from 25 to 30 per cent—and help the soil at the same time.*

Research reported by the United States Department of Agriculture recently shows that *"an inoculated stand will yield more without using up the nitrogen reserve in the soil."* Stands will be more vigorous, continues the report, and protein content may be raised 1 or 2 per cent. Seed inoculation may not be spectacular, but it has a real dollar-and-cents value.

Last spring we inoculated our green snap beans to see if they would pro-

1. Some gardeners use table syrup to coat seeds before applying nitrogen-fixing bacteria. Seeds should be stirred well to make sure all are coated with syrup.

2. Sprinkle the desired amount of bacteria over the seeds after applying syrup. Each seed should be entirely covered with inoculating material.

3. Inoculated peas are planted in prepared furrows, one seed to every two inches. Plenty of growing room should be allowed plants for best development.

duce a noticeable increase in yield. Until then we had been getting a moderate yield with average-sized pods. Last summer our yield was so heavy from the treated seed that the many pods actually pulled the plants over. From a small garden plot, which measured 8 by 10 feet, we harvested a full bushel and one-half at the first picking. Not only were these beans the best we had ever seen, but also the tastiest.

Any legume, including those grown as forage, greatly benefits from inoculation. Even flowers of the pea family, such as the sweet pea, can be inoculated to produce larger and more blooms. The one important thing to keep in

Southern California gardeners have found that gardening in boxes has many advantages. A box bed is a self-contained garden about 3 feet wide by 6 feet long, closed in on all sides, and is 6 to 10 inches above ground level.

mind here is that there is a different strain of bacteria for every type of legume. Those that work best on clover, for instance, will not benefit the snap bean. When buying the inoculant, specify what legume you intend to treat with it or read the material carefully on the packages of bacteria if buying at a self-service counter. One inoculant will not do for all legumes, although some companies have started marketing a "garden mix" culture usable on seeds of peas and beans for garden planting.

Nitrogen-gathering bacteria of all kinds come packed in plastic bags of moist humus to keep them alive. Sometimes the culture arrives in bottles of nutrient agar. Either one gives excellent results. Purchase the needed amount of inoculant as near planting time as possible to insure active bacteria. Store unopened at room temperature until ready to use.

If inoculating a large amount of seed, place them in a tub, or directly upon the clean floor. Sprinkle a little water over them, then stir until all seeds become slightly moist. Your next step will be to sprinkle the inoculant over the pile and then stir it well with a wooden paddle or turn it several times with a grain shovel so that all seeds become coated with bacteria.

We have found a method of moistening the seed that causes far more bacteria to cling to them. This is especially good for the home gardener, or for smaller plantings by hand. We place the seed in a small container, pour a little corn syrup over them, then stir them until all of the seeds are covered with a sticky jacket. Because of the sticky coating, each seed becomes completely covered with bacteria which remain in place until planted. This syrup also provides the necessary moisture so vital to the well-being of the bacteria and aids in prolonging their life in dry soil.

Chemical fertilizers, especially those placed directly in the furrow with the coated seed, can quickly undo the benefits of inoculation. Bacteria will die if they come in direct contact with the strong chemicals. Organic fertilizer should be applied in furrows on either side of the planted row or 2 to 3 inches below the planted seed and covered with fresh soil.

If your soil is highly acid, nitrogen-gathering bacteria will have great difficulty in producing nitrogen. They work best in a *p*H of around 6.5, which is an almost neutral soil. If your soil tests too acid, apply natural ground limestone. Not only will the bacteria benefit from the added calcium, but so will the legumes which require a great deal of calcium in order to carry on their growth processes.

One important thing to remember when planting inoculated seed is that these bacteria are much like ourselves and are actually tiny animals who feed chiefly on organic matter in the soil. If your soil is low in humus, turn in cow manure, compost, leaf mold, a green cover crop or any other organic material you have on hand well in advance of planting time.

Just how these microscopic bacteria extract nitrogen from the air, and put it to use to benefit themselves and the legumes they rest upon, is not actually known. But we do know this much: When the right strain of bacteria for a certain legume comes in contact with the root hairs of that plant, they enter them and form small houses, or nodules, which become factories for extracting nitrogen from the air. They also become storehouses for the obtained nitrogen.

When a plant takes in food through its root system, it also absorbs some of this stored nitrogen. Any plant that has been inoculated contains a greater percentage of protein because of the readily available nitrogen in its root system. This, of course, makes it a more valuable food product for both man and animal.

Now, it is a known fact that all types of nitrogen-gathering bacteria are found in any ordinary field and garden soil. Then why inoculate? The answer—there are not enough of the desired strain present to noticeably increase the yield.

When inoculating, you concentrate the right strain of nitrogen-gathering bacteria in the immediate root zone of your legumes. The more bacteria you have working for you, the greater will be the amount of nitrogen produced. And the greater amount of nitrogen stored in the root nodules, the greater will be the yield. Crops inoculated with these little workers usually come up with a nice increase of as much as 25 to 30 per cent over uninoculated seed. This, in itself, is worth the time and effort it takes to inoculate before planting.

Here is another good reason for treating legume seed with commercial inoculants: The bacteria they contain are improved strains that work far better than those found normally in the soil. These inoculants are prepared in agricultural laboratories. Laboratory technicians obtain these bacteria directly from the root nodules of healthy legumes. They are then studied for efficiency and vigor, then the best strains are reproduced in cultures of nutrient agar under favorable conditions.

Considering the great amount of good they do in increasing yield and protein content, inoculants are very cheap. A small 25-cent package will treat anywhere from 5 to 10 pounds of seed depending upon size.

Legume inoculation is a natural process. When inoculating your legumes you are merely helping Nature put more of the right kind of nitrogen-gathering bacteria into your garden plot.

—Betty Brinhart

9—How to Use Mulches

Mulch is a layer of material, preferably organic material, that is placed on the soil surface to conserve moisture, hold down weeds and ultimately improve soil structure and fertility. As with composting, mulching is a basic practice in the organic method; it is a practice which nature employs constantly, that of always covering a bare soil. In addition, mulching also protects plants during winter, reducing the dangers of freezing and heaving.

Experiments have shown these advantages:

1. We know that a mulched plant is not subjected to the extremes of temperatures of an exposed plant. Unmulched roots are damaged by the heaving of soil brought on by sudden thaws and sudden frosts. The mulch acts as an insulating blanket, keeping the soil warmer in winter and cooler in summer.

2. Certain materials used for a mulch contain rich minerals and, gradually, through the action of rain and time, these work into the soil to feed the roots of the plants. Some of the minerals soak into the ground during the first heavy rain. Therefore mulch fertilizes the soil while it is on the soil surface as well as after it decays.

3. For the busy gardener mulching is a boon indeed. Many back-breaking hours of weeding and hoeing are practically eliminated. Weeds do not have a chance to get a foothold, and the few that might manage to

come up through the mulch can be hoed out in a jiffy. And since the mulch keeps the soil loose, there is no need to cultivate.

4. The mulch prevents the hot, drying sun and wind from penetrating to the soil, so its moisture does not evaporate quickly. A few good soakings during the growing season will tide plants over a long dry spell. It also prevents erosion from wind and hard rains. Soil underneath a mulch is damp and cool to the touch. Often, mulched plants endure a long, dry season with practically no watering at all.

5. At harvest time vegetables which sprawl on the ground, such as cucumbers, squash, strawberries, unstaked tomatoes, etc., often become mil-

Clarence Holden of Guilford, Vermont solves his weed problem in his potato field by smothering them with a hay mulch. Loading his truck with the hay, Holden drives between the newly planted rows, throwing mulch in a layer around 5 inches deep.

dewed, moldy, or even develop rot. A mulch prevents this damage by keeping the vegetables clean and dry. This is the season when most gardens begin to look unkempt. But the mulched garden always looks neat and trim, no matter what the season. In addition, mud is less of a problem when walking on mulched rows and low-growing flowers are not splashed with mud.

Here are some potential disadvantages of mulching:

1. Seedlings planted in very moist soil should not be mulched immediately. The addition of any organic matter which keeps the soil at a high humidity

encourages damping-off of young plants. Damping-off is a disease caused by a fungus inhabiting moist, poorly ventilated soil and can be 90 per cent fatal. Allow seedlings to become established then, before mulching.

2. It is wise, too, to consider the danger of crown-rot in perennials. This disease is also caused by a fungus. If there has been especially heavy rains, postpone mulching until the soil is no longer water-logged. Do not allow mulches composed of peat moss, manure, compost or ground corn cobs to touch the base of these plants. Leave a circle several inches in diameter. The idea here is to permit the soil to remain dry and open to the air around the immediate area of the plant.

3. Do not mulch a wet, low-lying soil or, at most, use only a dry, light type of material, such as salt hay or buckwheat hulls. Leaves are definitely to be avoided as they may mat down and add to the sogginess.

The heavy mulching method stands a better chance of success if the soil contains some humus (well-decayed organic matter) and is fairly high in nitrogen content.

Where the soil is poor and mostly clay in composition, it is well to test the soil and apply the needed elements, as nitrogen, phosphate and potash, according to test results. Then spread the mulch in thin layers without packing, so as to permit air and moisture to start breaking down the raw material. When the first layer of mulch shows signs of decay, sprinkle some cottonseed meal, blood meal or other nitrogen-rich material and apply another thin layer of mulch. By this method, any danger of the heavy mulch taking too much nitrogen from the soil is avoided.

With the instructions given above, it is simple enough to know when and where not to mulch. Except for these instances, the gardener really can't do without mulching as a wonderful labor-saving helpmate.

—Virginia Brundage

What Mulches to Use

When you set out to mulch a home garden there are 3 factors to be considered:

1. How the material will affect the plants most intimately concerned.
2. How the completed mulch will look.
3. How easily and inexpensively the mulch may be obtained.

Following is a list of commonly used mulch materials that have been found beneficial by many organic gardeners:

Grass Clippings
Leaves
Leaf Mold
Hulls and Shells
Peat Moss
Sawdust
Seaweed and Kelp
Wood Chips
Corn Stalks
Straw
Corn Cobs
Pine Needles
Alfalfa Hay
Oak Tow
Rotted Pine Wood
Packing Materials
Weeds and Native Grasses
Salt Hay
Dust Mulch
Stones
Aluminum Foil
Glass Wool

Descriptions of many of these materials appear in Chapter 4, "Fertilizing Your Soil."

Sawdust as Mulch

Following is a Virginia Polytechnic Institute report on the use of sawdust as mulch:

A sawdust mulch may help you have a fine garden. It conserves moisture, keeps the soil cooler in the summer, and promotes healthy vigorous growth of your vegetables. It does not increase soil acidity or harm plants. A

Wood shavings and sawdust make an excellent mulch for many plants such as strawberries. Prior to spreading wood wastes around their plants, many gardeners side-dress with a nitrogen fertilizer of cottonseed meal, bloodmeal, tankage or manure.

deficiency of nitrogen may develop if a large amount of organic material, such as sawdust, is mixed with soil. A nitrogen deficiency is usually indicated by the leaves becoming a paler green than is normal. This trouble can be readily corrected though, by the addition of an organic fertilizer containing nitrogen.

Plant your garden seeds and cover them with soil in the usual way. Then apply a quarter-inch layer of sawdust in a band about 4 inches wide on top of the row.

Additional sawdust should be applied when the vegetables are about two inches high. Destroy all weeds by cultivating or hoeing and apply a one-inch layer of sawdust over the entire planted area. A garden 50 by 50 feet will require nearly 8 cubic yards of sawdust.

After the sawdust has been used as a mulch, plow it into the ground during the fall or winter to improve the workability of the soil. Do not wait until spring to plow because the sawdust will keep the ground from drying out.

Suggestions for Mulching

Here are some general suggestions for mulching offered by Norman Oebeker, University of Illinois vegetable crops specialist:

Fairly rich in nitrogen, dried lawn clippings make an ideal mulch for plants in the vegetable garden. Pictured are lawn clippings being spread around squash vines to keep the weeds down and prevent soil moisture from evaporating.

Spread the material evenly over the soil surface between the rows and around the plants. Mulch thickness will depend on the material and its cost. Such mulches as leaves or straw are usually 4 to 8 inches deep. A mulch of peat moss, shavings or wood chips is seldom more than two inches thick.

Newspapers 8 or 10 sheets thick also make a fairly good mulch. Use small stones or a band of soil along the outer edge to hold the papers in place. Aluminum foil is sometimes used in the same way. Black polyethylene plastic is a new idea in mulches that appears promising for the home garden. Sheets of the plastic are laid over the garden soil, and vegetables are planted through holes in the material.

Plastic tends to warm the soil rather than cool it. This is an advantage early in the season, but a disadvantage later. Plastic works best in midsummer on such crops as tomatoes that will shade the mulch surface. This shading will prevent the soil from becoming too warm.

Mulches sometimes cause a nitrogen deficiency to develop. This deficiency develops because the organism causing organic matter (mulching material) to decompose, uses nitrogen. That's why it's a good idea to apply extra nitrogen before mulching. When plants tend to have light green or yellowish leaves, apply several pounds of cottonseed meal per 100 square feet.

Growing Mulch in the Garden

This experience of California's Warner Tilsher indicates the benefits of growing mulch in the garden.

For some years now we have been trying out a sort of "do-it-yourself" or rather "grow-it-on-site" mulching program. It goes like this:

We plant the garden area with a mixture of rye and clover; the first for its heavy top growth, the clover for its roots, which not only gather and hold nitrogen but dig deep for fresh minerals and loosen up the subsoil.

Material for mulch can be grown right in your own garden. California organic gardener Warner G. Tilsher shows how he cuts a mixture of rye and clover when it is about a foot high. This same area is planted later to different vegetables.

We treat this "garden lawn" like a regular lawn, keeping it mowed down to about 2 or 3 inches with an ordinary reel-type hand or power mower. We leave the clippings as they fall; pretty soon the whole garden is covered with a mat of mulch gradually rotting down at the bottom to a rich, black, living humus.

Some two weeks before planting row crops, we rotary-till a slot about 12 to 18 inches wide. For vegetables like squash, melons, tomatoes, etc., we spade up little plots about 18 inches in diameter. The rows and plots are then fertilzed with minerals, compost and fish emulsion.

We make our planting as usual, mulching the seeds with sawdust or a like material to keep them moist. For irrigating we use a porous canvas hose (soil-soaker) which not only takes care of off-level ground, but unforeseen gopher holes as well.

Meanwhile, all around the crop, the rye and clover keep on growing; when they are about a foot high we cut them down with a hand sickle and scatter the cuttings around the vegetables. So here we have a ready-grown, first-class mulch right where and when it is needed.

With this system you may need a bit more water but I feel the gain in organic matter far outweighs this, especially in warm climates where we are constantly fighting a battle to replenish the humus which disappears so fast.

We use the same system in our orchard—for us the results have been fabulous. It completely does away with tillage, the making and keeping up of irrigation basins or furrows. And we have been able to stretch the time between irrigations from 3 weeks to 4 to 6 weeks. Just before an irrigation we hand sickle the cover crop down around the tree to the drip line, keeping the area near the trunk clear of all mulch for a foot or so. In between the trees we use a power rotary mower.

Besides producing unblemished, healthy fruit that is out of this world, there is another point. Ripe fruit falling on the soft mulch doesn't split or smash—you can never know how really good a plum or an apricot can taste until you try one that has just dropped.

A Report on Year-Round Mulching

When I first read of the year-round mulch system, I was inclined to embrace it with wild enthusiasm. It seemed to be the answer to so many of my garden problems that I decided right then and there to try it. It really sounded too good to be true and in some ways it was too good to be true.

Most of my garden is heavy clay mixed with subsoil, and in some spots it's all subsoil. In my second year of mulching when I pulled the mulch aside in early spring, far from finding a mellow, friable soil underneath, I was greeted with the depressing sight of cold, wet clay.

I do have a few spots of sandy soil and I noticed that these spots responded at once to year-round mulching. Also here and there were spots of good, rich loam and these also did wonderfully. I was forced to come to the conclusion that mulching will not act as a cure-all for poor and compacted soil.

The first year I mulched, I had the garden plowed first and then I put the permanent mulch cover on as I planted it. The second year, I tried to plant without working the ground as Ruth Stout said she did, blindly ignoring the fact that she has sandy soil and I am stuck with clay. I used a hoe to make furrows in the hard ground and planted. Well, a lot of the seed didn't come up at all and what did come up didn't yield the way it should have. For the later crops I pulled the mulch aside and spaded a seedbed. The results were much better. It was obvious that until my soil's humus content was greatly increased, I was going to have to spade. If you have a rotary tiller, of course, this is no problem.

Since the ground proved to be so wet in the spring, I could see that I couldn't very well spade then for the early crops, so I decided to do it in the fall. I pulled the mulch aside whenever I had a little time during October and November and spaded compost and manure into 5 rows, 40 feet long for the first peas, radishes, lettuce, cabbage and onions. These rows, left in the rough over winter, dried out rapidly in the spring and I was able to get them planted just as soon as we had a warm spell in the spring.

If you are accustomed to having your garden plowed each year, all this spading may sound discouraging. It's true that spading is hard work, but actually not a great deal is necessary. For plants such as tomato, pepper and cabbage, only a hole need be dug for each plant; thus only a small percentage of the ground taken up by these plants is spaded. For corn you need only to turn over a shovelful of earth for each corn plant. It isn't necessary to spade between the plants. Soil preparation for potatoes is simple. I spade an area about a foot square, give it a quick working over with a hoe, tuck my tuber shallowly into the loose soil and cover with a heavy mulch. The potatoes won't grow deep and can be picked by just lifting up the mulch.

In my garden, these take up most of the space and just leave a few crops such as peas, beans, carrots, lettuce, okra, onions and radishes for which long rows or sometimes beds must be spaded.

The nice thing about a mulched garden is that you can spade what must be done a little at a time. I am now able to plant from April to September. Before, it all had to be done in June. Then I would have nothing from the garden until the middle of the summer when I would be swamped for a few weeks—and then there would be nothing again.

And how wonderful it is not to be swamped with weeds all summer long! I never could keep up with them before. Now I am able to care for a garden 4 times as large as I did before. Since I also use the same space for 2 or 3 crops in many instances, you can see that I have been able to cut my grocery bill by a very substantial amount.

Much of my garden is on a slope and suffered badly from erosion. Before I heard of year-round mulching, I read many books on different methods of stopping erosion, trying to find some way that I could handle myself. Most of the things suggested were too difficult for a woman to tackle. Building terraces, for instance, was beyond my strength, and although I tried to plant on the contour, I wasn't very successful. We usually had a hard rainstorm right after the garden was planted and the topsoil would wash before any plants had a chance to come up to hold the soil. I have often had little new seedlings at the lower edge of the garden completely buried with topsoil after a storm. Mulching stops erosion completely and you don't need a degree in engineering to figure it out, either.

Mulching can make you feel very superior to your nonmulching neighbors at the end of a hot, dry summer. When driving along the countryside, pay special attention to the vegetable gardens. Some of them are lovely, but all too often they'll be choked with weeds and badly dried up. Last summer much of the late sweet corn looked stunted and beyond the help of any amount of rain that might come later. It felt very good to go home and look at my garden, free of weeds and growing vigorously.

Year-round mulching has brought me some problems I didn't have before, but the benefits far outweigh the disadvantages and I expect to be mulching all the rest of my life.

—Lucille Shade

Questions on Mulching

People are always asking Ruth Stout a great many questions about the year-round-mulch method: *what kind of mulch to use; how much; when to apply it.* Here are some of her answers:

Kind: Hay, leaves, straw, seaweed, pine needles, sawdust, weeds, garbage—any vegetable matter which rots. Some people write me and complain that the bales of "hay" they bought were all coarse stalks, weeds, etc. That is all good mulch.

All kinds of leaves? Yes. *Aren't oak leaves too sour? People say so.* Put

Frederick S. Eaton of Wallingford, Connecticut applies a layer of hay over his home vegetable garden. General advice is to spread material evenly over soil surface; thickness depends on material used, with hay varying from 4 to 8 inches deep.

on a little more lime if your soil is acid. *Don't some leaves decay too slowly?* Then they remain mulch just that much longer. *Don't leaves mat down?* Somewhat but that makes no difference, since they are *between* the rows of growing things, not on top of them. *Can you use leaves without hay?* Yes, but I prefer a combination of the two, if you can get it.

How much mulch do you need? The answer to that is: more than you would think. You should start with a good 8 inches of it. Then I'm asked: *How can tiny plants survive between 8-inch walls?* And the answer to that is: the mulch is trampled on, rained on, and packed down by the time you are ready to plant; it doesn't stay 8 inches high.

How often do you put it on? Whenever you see a spot that needs it. If weeds begin to peep through, don't bother to pull them; just toss an armful of hay on them.

What time of year do you start to mulch? NOW, whatever the date may be. *Would it be better to wait until the crop is up?* No, for by that time the sun will have had a chance to bake the soil, and the weeds will be making progress.

Now the million-dollar question: *Where do you get your mulch?* That is difficult to answer but I can say this: if enough people in any community demand it, I believe that someone will be eager to supply it. At least that is what happened in this part of Connecticut. Anyone within a hundred miles of us can get all the spoiled hay he wants, delivered, for 65 cents a bale. (Spoiled hay is simply hay which has gotten wet or very old and isn't fit to feed to livestock; it is perfect for mulching.) If you belong to a garden club, why can't you all get together and create a demand for spoiled hay? If you don't, surely you know a number of people who garden who would be pleased to make a success of it with only one-tenth of the labor which is involved in the old-fashioned way of growing things.

Also, use all your leaves. Clip your cornstalks into foot-length pieces and use them. Utilize your garbage, tops of perennials, everything. In many localities the utility companies chop up the branches which they cut when they clear their wires; you can probably get these for nothing if you can haul them away.

Now for the drawbacks. People have complained to me that mulching does not kill everything. I just got a letter from someone saying that it won't kill cockleburs, morning glories, Johnson grass, nut grass. She left out witch grass. I know it won't kill *that* and neither will it pick your peas or plant your seeds. I am just saying (in a friendly, sarcastic way) that just because it does 100 things for you, should it be expected to do 101?

Here are questions I get over and over: *Doesn't mulch draw mice, moles, slugs?* I even got one letter asking me if mosquitoes didn't "hide under the hay, ready to jump out at you when you walk into your garden?" My answer to all this is: No. And I've gardened this way for 13 years. I see a slug or two now and then on lettuce or cabbage, but I've had no special trouble with mice or mosquitoes.

Moles deserve a separate paragraph. If I had time to go through my files, I would count the mole letters I've received; there are a lot. The interesting thing about them is that half of their writers reproach me because they have mulched and now have moles, and the other half are grateful to me because since they began to mulch, the moles have disappeared from their gardens. So I suppose we will have to assume that mulching neither attracts nor chases away moles. However, I will tell you this for what it is worth: two summers ago moles built a subway in our lawn, but completely avoided all of the mulched flower beds right there in the same area.

Many people ask: *shall I use manure and plow it under and then mulch?* Yes, if your soil isn't very rich. Otherwise, just mulch.

Rain-spoiled hay is a popularly used mulch material and is often available for free. Growers find that mulching permits an earlier start for planting in spring; young tomato plants above, set out in a mulched field, are protected by baskets.

When shall I put on lime, how much, should it be put on top of the hay, or under it? It doesn't matter whether it's on top or under the mulch. As to when, and how much, that depends entirely on your soil; you can have it tested. Your county agricultural agent, whose office is usually in the post office, will tell you exactly how much lime you need. Nor can I tell you how much cottonseed meal to use; I can only say that I sprinkle it along rows of lettuce, spinach, beets and corn each time I plant them.

Questions about planting are the most numerous. *How can seeds come up through the mulch if weeds can't?* Of course you have to pull back the mulch to plant your seeds, which you plant exactly as you would if you weren't mulching. Don't put the mulch back on the row of seeds; wait until the little plants appear, then tuck the mulch *around* them.

I put a narrow board or strip of cardboard on the row as soon as I plant it. (This is not essential and has nothing to do with mulching.) It keeps the ground soft and moist. I watch carefully and the moment the seeds sprout, I remove the board, and pull the mulch up closely around the tiny plants, thus choking out any weeds. For a short time keep an eye on these newly-sprouted plants; the wind, or some animal, might pull mulch on top of them.

Onion sets are planted in the usual way; then you can cover them immediately with hay and they will come through it. Potatoes can be thickly covered with hay; they needn't have any dirt on them at all.

How far apart are the rows? Exactly the same as if you weren't mulching. However, after you have used the mulching system for a few years your soil will become so rich from rotting vegetable matter that you can plant more closely than in the old-fashioned way of gardening.

What kind of soil does mulch improve? Every kind, that I know of. My soil is sandy, but I have neighbors who have improved their clayey soils tremendously. And many letters saying the same.

A few people have asked: *don't you have to clear a large space in transplanting things such as lettuce and cabbage, so the plant leaves can spread?* No. Pull the mulch back slightly, put in the plant, and tuck the mulch cozily right up close to it, to prevent weeds and hold in moisture. I wish you could see the size of my heads of Great Lakes lettuce with this treatment.

Doesn't hay full of weed seeds cause trouble? Not if you put the hay on thickly enough; the weeds won't sprout. *Don't you have to hand-weed?* If you don't put the mulch up close around your plants, you will of course get weeds. If you do get some, don't bother to pull them, smother them with mulch. Right in the row, among the plants, you may have to hand-weed a little, but very little.

Doesn't mulching look awful? Almost everyone (400 of them) who has seen my vegetable garden liked the appearance of it—neatly covered with clean hay. One woman started to change her shoes before going into the garden, but I said: "You don't need to do that. Even if you had evening slippers on, they wouldn't get soiled in my garden."

When any new thing is presented in any field, there are those who accept it instantly, those who keep an open mind and give it a try and those who call it bunk. Recently a man stood and gazed at my fine row of carrots, then said that you just can't grow carrots without plowing. I don't think he thought I was lying when I said my garden hadn't been plowed for 13 years and I even think he had a good mind, but I am afraid it was simply closed and locked tight and therefore no new idea had a chance of getting in.

Make up your minds to one thing: if you are the only person in your neighborhood who is using this no-plow, no-spade, no-cultivating method, your friends and neighbors will say you are crazy. Ignore them. They will change their tune. Fifteen years ago everyone called *me* crazy, but just the other day, a neighbor said to me:

"Doesn't it make you feel good, when you drive around, to see the great piles of hay in so many yards?"

It does. It makes me feel *very* good.

10—Summer Garden Care

Why Plants Wilt

HAVE YOU EVER LOOKED at your neighbor's garden, and perhaps your own, when the summer sun was shining down hotly on it and wondered why, exactly, the plants looked so tired and droopy? Of course, you say, they wilt. But why do they wilt? And most important of all, what, if anything, can be done to prevent wilting?

A fancy word that the plant physiologists use for wilting is loss of *turgidity*. When plant cells are filled with water, they are swelled up like so many tiny balloons. This makes celery and lettuce crisp, and corn stand tall and straight. It makes leaves appear wide and lush, and flowers lift up their heads. Loss of turgidity means droopy plants, leaves that are curled up tight or folded and twisted or wrinkled stalks.

One reason why it's so important to avoid wilting in vegetable plants is to prevent vitamin loss. Studies with fresh vegetables conducted by the Department of Agriculture have shown that vitamin C is lost rapidly when vegetables wilt. Their research reveals that vegetables that lose moisture readily and wilt appreciably tend to be affected more by humidity and to lose vitamin C more rapidly than resistant vegetables.

Even those that wilt most easily are affected much less by humidity than by temperature. Cabbage loses less vitamin C and moisture than either spinach, collards, turnip greens or rape. Also low temperature seems to affect loss of the vitamin in snap beans more than wilting conditions. (Chilling injury may cause this as well.)

But why is it that plants do or don't have turgidity? The answer, of course, depends on the amount of water that happens to be in the plant cells. But behind this answer is the technique that makes the difference between the good and the poor gardener. And adding more water to the droopy plants may or may not do some good.

Plant Structure

In general, plants consist of roots, stems and leaves. Other than their appearance as in the case of flowers, we are interested in plants primarily as food which we get from the roots as in radishes and potatoes, the stems which we eat as in celery and rhubarb, the seeds which we eat as in beans and peas, the whole fruit which we eat as in tomatoes and strawberries, the leaves which we eat as in cabbage and spinach or the flowers which we eat as in the case of cauliflower and broccoli.

The roots of the plant serve to provide the plant with food and water. Root hairs absorb water by osmosis. Also mineral elements diffuse through the root hairs. These two functions depend upon the rate by which oxygen is taken up by the roots and by the rate at which they respire. And right here, right now, we can give one of the secrets of why organic gardening pays off. The amount of oxygen that is around the roots of the plants depends on how porous the soil is. As we all know, there is nothing like plenty of good organic matter for breaking up packed earth and making lots of open spaces so that air can get down to plant roots.

It is not wise to flood and keep flooded the soil around plants. If the soil is kept flooded then air has no chance to get down through the water to the roots of the plants. We see examples of this when corn plants are flooded in fields that have low spots in them. After heavy rains, the water collects in the low spots and stands in pools for a day or so at a time. Air can't get down to the roots of the plants. In a few days there is a bare spot where the corn plants have died. And then the farmer has to reseed the bare spot.

Drainage systems and plenty of coarse organic matter worked deep down are ways of taking care of low spots that have pools of water standing on them for a long time after heavy rains. A few hours in water once in a while doesn't hurt plants much, but standing in water for a day or so may drown them. The nice thing about organic matter is that it soaks up lots of excess water during periods of flooding. It is like a sponge in this respect. Then when dry periods come, it releases the water to thirsty plants.

There is an optimum amount of water in plant cells and the water content of plants is usually fairly constant. This is desirable from the standpoint of growth because there should be neither an excess nor deficit of water. The organic matter sponge reservoir that is in the soil is one of the best guarantees of neither too much nor too little water.

Temporary Wilting

Unless gardens have plenty of organic matter in the soil that is capable of taking up excess water and then releasing it to thirsty plants, and unless tender plants are shielded from the direct rays of the hot summer sun and hot, dry winds, we may expect some wilting. This is called temporary wilting. It is a result of loss of water in the plant cells. As soon as the conditions are removed that caused the wilting, as with the coming of night, the plant cells immediately regain their turgidity, and are bright and fresh and crisp looking again.

Temporary wilting does not necessarily mean a loss of vital activity. If water is restored to the plant soon enough, turgidity is reestablished and the plant resumes its vital activity. However, temporary wilting does slow down growth because photosynthetic activity is retarded. Also yields may be somewhat reduced.

Permanent Wilting

If plants don't get water soon enough to recover from temporary wilting, plant physiologists say that the plants are permanently wilted. Permanent wilting is very injurious to plants. New root hairs must grow to replace the ones that died from lack of water. If wilting has progressed to the stage where water is withdrawn from the green cells of the plant, the photosynthetic ability of the cell is injured for a long time, or even permanently. This means that permanent wilting retards, or completely stops, the growth of new organs and, if at seeding or fruiting time, it results in reduced yields of poor quality.

Of course, plants are like every other living thing. They want to survive. In an attempt to withstand dry periods, they close their pores, reduce exposed leaf surface, and even develop hard, woody stems, and layers of waxy substance over leaves. We see this done to perfection in desert plants that are able to live without rain for a year or more. But in vegetables that we put on our table or take to market, dried up leaves, and hard woody stems are undesirable, to say the least. Some plants, such as lettuce, even begin to seed very early when drought conditions are present.

Effect of Chemicals

One of the most important lessons in this whole story is the effect of chemicals on turgidity. If you have a potted plant you *don't want,* and some chemical fertilizer, try this experiment: Put a few spoonfuls of the chemical fertilizer around the base of the plant, or mix it in the soil in the pot, or even just dissolve it in a glass of water and pour it on the soil around the plant. Now water the plant all you want to—even until the soil around it is flooded. In spite of all the water you add, the plant will get more and more droopy, and in a couple of days it will be dead. Water has been sucked out of the plant, not by a hot sun, or a hot, dry wind, but by chemicals!

This simple experiment that so aptly demonstrates what happens to gardeners who use chemical fertilizers in large quantities hoping to get better yields and end up with none or next to none, is based on the pulling power of solutions. The pulling power of the solution of nutrients around the roots of a plant in good soil is around 0.5 to 1.5 atmospheres. Plants are able to suck this solution of water and minerals up because their pulling power is greater. The pulling power of a plant that is drying out may exceed 100 atmospheres. That of some dry seeds is greater than 1,000 atmospheres!

Thus we see that the pulling power of plants for water in soil is very great. Only when there is no more available water or when we have added chemicals that have a greater pulling power for water, do the plants wilt and die.

—Charles Coleman

Protecting Against Drought

"No rain for weeks. Soil dried up. Can't expect anything to grow. Looks like no vegetables from our garden this year!"

Will those be your words in a dry summer? Will you be forced to face defeat after a prolonged dry spell? Isn't there something you can do to protect your growing plants from a scorching summer drought?

The answer is a definite YES! And here's why:

Although you have no control over the amount of rain your garden is going to get in summer, you *can* increase your soil's ability to *hold* any water it receives up to the time that a dry spell begins. To do this, it is absolutely essential you realize that the kind of soil you have largely determines how much and how often it needs water.

For example, Department of Agriculture experts Victor Boswell and Marlowe Thorne state that "the coarser the particles that make up a soil, the less water it will hold under most field conditions. Sandy soils hold only about one-fourth inch equivalent rainfall or irrigation water per foot of depth. Sandy loams commonly hold about three-fourths inch of water per foot; fine sandy loams, about 1.25 inches; silt loams, clay loams, and clays, about 2.5 to 3 inches."

So you can readily see that there is a wide variation in water-holding capacity of soils—enough to make the difference between one-fourth of an inch per foot of soil and 3 inches . . . the difference between garden success or failure during a drought.

To improve soil, keep in mind that large amounts of organic matter will increase the water-holding capacity of sandy soils and sandy loams, as well

Mulching is the best insurance a gardener has against drought damage to his growing plants. Mulching has been found to maintain an even soil temperature, hold down weeds, as well as prevent plants from permanent wilting.

as make it easier for water to soak into heavier-type soils as loams, silt loams and clays.

Sheet composting, green manuring and applying organic fertilizers are the best and fastest ways to increase your garden's humus content.

Now let's examine how subsoil conditions influence the plant-water relationship. Some soils that look all right on the surface are actually covering up a hard, tight layer only a few inches below. It may be practically impossible for roots or water to enter this layer at all. The result is a shallow zone, which prevents normal root development.

"The limited amount of water that can be held by such a shallow layer is soon exhausted in dry weather," observe Boswell and Thorne. "And the plants suffer for water relatively soon after a rain or irrigation. Rock near the surface of the soil is even more unfavorable in this respect. Under such conditions water must be supplied in moderate to small amounts frequently. Heavy watering may virtually drown the plants by overfilling the soil pores and preventing proper aeration of the root zone."

A rather slow, but sure, way of loosening a hardpan is through sheet composting and green manuring. Many organic gardeners have successfully used this method, but it will take at least several years to do the job.

At the other extreme to the hardpan, there are some surface soils that are underlain by very open, deep deposits of sand or gravel. Roots can pass readily into these coarse layers but may obtain little water. Crops on soils having such open subsoils suffer quickly for lack of water just as do those above a tight layer near the surface. Heavy watering of subsoil not only wastes water but leaches some plant nutrients down to levels where the roots fail to reach them. Here again, the best solution is to incorporate large amounts of organic matter into the soil.

Make up a list of the materials available in your neighborhood that you can use as a mulch. Here's a list of the commonly used mulch materials which you can use as a guide in preparing your own list: weeds, native grasses, leaves, leaf mold, compost, stones, grass clippings, packing materials, peat moss, buckwheat hulls, rakings from under trees and shrubs, straw, hay, sawdust, woodshavings, corncobs, coffee grounds, cocoa bean shells, seaweed and vegetable garbage. Next to each item, write down the name of the place where you might be able to obtain it.

Don't put off contacting your sources until July or August when the sun has crisped your soil. As soon as you've prepared your list, get in touch with the persons named; let them know what you want. In some cases, perhaps they'll let you collect the materials at no cost; others might charge you a nominal fee. You can rent a small pick-up truck for several dollars an hour, and collect all the mulching materials you'll need in just a few hours at a cost of approximately $15. That's the cheapest drought insurance you can get!

Here's some advice from expert mulcher Ruth Stout on the best ways to use these materials around your vegetables:

For spinach, lettuce and peas: place 6 to 8 inches of mulch around them. Shade the lettuce if you can.

For beets, carrots, parsnips and kohlrabi: first thin plants; then water thoroughly and put mulch all around them at once, 6 inches deep and between rows. If the mulch is wet, so much the better.

For bush beans: if already planted, thin, water and mulch. If you haven't planted them already, make a drill 4 inches deep; plant the beans sparsely; cover with two inches of soil; water; cover with a board or cardboard and mulch. Remove board as soon as beans sprout.

For corn: if planted already, thin to two plants in a hill instead of customary 3. Water and give it 6 inches of mulch. (If you're running out of mulch, use as many layers of wet cardboard as you can collect. The cardboard is only an emergency measure; it is not, of course, as satisfactory as hay or leaves, because the latter provide valuable nutrients to the soil as they decompose.) Each time you plant corn, advises Miss Stout, soak the seed overnight, make

4-inch drills and cover the seed with two inches of soil. Water thoroughly, put a board over the seed and mulch immediately.

For late cabbage, broccoli, cauliflower, peppers and tomatoes: if not planted yet, put very deep and 4 feet apart instead of 3 and mulch heavily. If peppers and tomatoes aren't in, put them very deep and farther apart than customary. If already planted, water and mulch heavily (6 to 8 inches).

Midsummer Plantings

Even if you live far beyond shouting distance of the balmy South, you can make midseason plantings that will serve up fresh, tasty vegetables all next winter. The only prerequisites are a little advance planning, selection of the right crops and varieties plus some preparation to mulch and protect the late-maturing produce.

For frost-zone gardeners—as far north as Canada's friendly borderline and southward all the way to Florida (where killing frosts are possible anywhere but at the extreme lower tip)—this method can lengthen the growing season surprisingly. More than that, it makes plantings, in midsummer or later, productive and makes possible right-from-the-garden, healthful vegetables on your table through the crisp autumn, frozen winter and even into next year's approaching spring.

Many vegetables that will mature before heavy frost and can be left in the ground through part or all of the winter, can be planted from mid-July through August. Some must have protection; some will survive without it. All benefit from a mulch—and not only for the obvious reasons. If you want to dig up a vegetable on a cold day, a mulch is essential to keep the ground beneath from freezing hard.

In order to get a clear picture of how this idea works in practice—whether you garden in upstate Oregon or downstate Georgia, and especially if it's new to you—here's an outline: Well ahead of midsummer, plan which vegetables you'd like to make late plantings of to carry into fall and winter. (A considerable number of those that fit the program are discussed later.) Next, consult the USDA map contained in Chapter 8 or a reliable local source, for the average date of the first killing frost in your area. Then check on the number of days each of the vegetables needs to reach maturity; count back from the frost date to determine how late each may be seeded.

Once you've made the selections and figured when they should be put in for a safe ripening span in your locality, these vegetables need only the usual cultivation, fertilizing and general care—plus, of course, the doubly helpful mulching with most and some pre-frost protective measures with a few. Remember, the map's frost dates are average; a few days' allowance for an early freeze with the more sensitive vegetables might be a wise precaution; with many stauncher types, particularly cool-weather growers that often improve in light frosts, you can plant up to a week or more later and be moderately safe.

Remember, too, that individual varieties of these vegetables differ in the length of time they require to mature. Some will need several days more than others; some are better late-season planters, have qualities that make them more adapted to ground-storage. Be sure to look over all the varieties of each vegetable suitable for growing in your soil and climate; select those with the most advantageous time requirements, along with other characteristics, for your own late-season garden.

Let's start with carrots. For winter use in Massachusetts, for example, seed should be sown no later than August 1st. You might pull up the pea vines or the first planting of snap beans, then utilize the vacated space. Because of the dry summers that now seem to be the rule, you'll need to water the seedbed before planting, and every day until the seedlings are up and well-established. (This procedure should be followed with all midsummer plantings.) Seed should be sown somewhat deeper, too, and more thickly than in spring.

By October, the carrots will have matured. They are relatively hardy and will stand cold, but just after Thanksgiving when the hard frosts set in, here in Massachusetts, I cover them with a thick mulch of leaves or hay. If the row is marked with stakes visible above a deep fall of snow, you will be able to dig up carrots in February as succulent as any in summer. In fact, they seem sweeter. The carrots stored in dirt in a vegetable cellar, or even in an outdoor pit, are frequently shriveled specimens in comparison and far less juicy and flavorsome. Those kept right in the row are just as easy to get at, too.

Beet seed can be sown now, too. August 15 is not too late for a variety like Crosby's Egyptian or Detroit Dark Red. (Winter Keeper or Long Season beets are bred especially for winter use, but they take a long time to mature and should have been planted in spring.) Follow planting tips outlined above for carrots. In addition, beet seeds are tougher and should be soaked overnight before planting to hasten germination.

The thinnings and baby beets can be enjoyed throughout the fall. When winter comes, mulch. Unlike the carrots, which last until spring (I have dug up perfect carrots in March), the beets may rot in the ground if left beyond January. An explanation: the carrot root penetrates deep into the soil; the beet heaves up, half visible above the surface. A cure: hoe dirt up and around the beets before mulching. Anyway, you'll be getting juicy, flavorsome beets for at least a couple of months of cold weather.

The same remarks apply to turnips as to beets. The tops of foliage turnips seem a bit more hardy, though, than beet greens. I don't count on the latter after the first hard frost.

Lettuce and cabbage seed planted between mid-July and August 1, should form good heads before frost and, if covered lightly, can be kept on for a few weeks longer in the ground. We once had for Thanksgiving dinner a beautiful head of lettuce that had been shielded on cold nights by a bushel basket covered with burlap bags.

Chinese cabbage must be included in a midsummer planting. In fact, unlike lettuce and regular cabbage, it is only a fall and winter vegetable (except where summers are always cool). If the Chinese cabbage comes to maturity during a hot summer, it "bolts" and does not form heads. But if sown during late July, Chinese cabbage will be heading up in the cool days of fall. It can easily withstand several light frosts and even in New England may last well into December, especially if a few forkfuls of hay are tucked in around each plant.

Kale and endive are even more hardy and will survive all but the severest winters. Protect the same as the Chinese cabbage. The plant should nestle in the hay rather than be completely covered by it as the root vegetables are.

If broccoli is desired for late fall use, it would be wiser to set out young plants bought at a nursery rather than to sow seed in midsummer. (The same may be done with Chinese cabbage.) Both take quite a while to mature. To shield the baby plants from the hot August sun, set them in a partly shaded spot, perhaps just within range of a big, bush-type summer squash, or between two rows of tall corn that will have been pulled up before the broccoli matures. Such locations are economical of space, too.

Broccoli has about the same degree of hardiness as cabbage and Chinese cabbage, greater than lettuce. Here in New England, it usually lasts well into December, if protected. In the Middle Atlantic States, I imagine, it might go through the winter. Protect the same as the cabbages.

One of the most valuable of all hardy vegetables—if you like onions—is a bed of the perennial Egyptian onions, sometimes called "treetop" or multiplier onions. For many years I had grown them, knowing they were hardy, of course, but without thinking to put them to winter use. In fact, they served only as spring scallions. One cold day a clump caught my eye—sprouting from a half-frozen compost pile where I had carelessly tossed some superfluous "sets" the August before. Into the potato salad they went, chopped-up green stalks and all. Egyptian onions are a year-round vegetable.

August is the best time to start a bed. Plant the tiny sets that have been growing at the tops of the mature, hollow stalks. Or plant roots from an already established bed, either in smaller clumps or individually. (This second method of propagating can be done later in the fall or in the spring, too.) The first winter the onions may be too small to bother with, if you started with the top sets, except for flavoring purposes. Thereafter, though, they'll reach the size of a small leek. These onions will survive if you don't use a mulch, but it's easier to dig them up if you do.

Last of the winter vegetables, but not at all least, is parsley. It's one of the richest in vitamin A—the nutrient that helps one resist infection. In Massachusetts, parsley is not completely hardy, but it is usually possible to winter over a few plants in a sunny, protected spot if you surround them with a light, loose mulch of hay or straw or even leaves. In the latter case, place the leaves on a rainy day, so they won't blow; and lift the covering occasionally so it won't mat.

As a child, I was once taken to see an old woman who lived in a tumble-down house at the end of a long lane lined with wineglass elms. She was black-haired, though in her eighties, and wore gold hoops in her ears. "The old gypsy woman," people called her, though really her blood was Yankee, with maybe a dash of Indian, and she wandered only over her own fields and woods. Someone remarked on her extraordinary vitality and she replied, "Every day of your life eat something that has just come out of the ground and you'll never be sick."

Probably she was more than half right and actually it can be done.

—Ruth Tirrell

TABLE 36: Time Chart for Mid-Summer Planting

Listed below is the range of days required for maturing by each of nine vegetables suggested for late-season planting. Individual varieties may need the smaller number of days, others between and up to the longer period.

VEGETABLE	DAYS TO MATURITY
Beets	45 to 60
Broccoli	60 to 80
Cabbage	60 to 110
Carrots	60 to 80
Chinese Cabbage	70 to 80
Endive	65 to 90
Kale	55 to 65
Lettuce (leaf type)	40 to 55
Lettuce (head type)	70 to 90
Parsley	70 to 75

In Minnesota, you can plant many vegetable crops at weekly intervals during the early summer to stretch out the harvest period. Three or 4 plantings of corn and snap beans can be made until July 1. Carrots and beets for winter storage can be sown in mid-June. These will be of better quality for winter use than those planted earlier. Cool-season crops like lettuce, spinach, radishes, turnips and kohlrabi can be sown again around August 1 so they will mature during the cool fall weather. Sometimes a fall crop of peas is successful. It should be planted about the middle of July. Chinese cabbage planted around July 20 usually results in fine heads for fall use. Cabbage, broccoli and cauliflower seed can be planted in the garden around June 1. These plants will mature after the weather gets cool in the fall and will usually be of higher quality than those maturing during warm weather.

Further south, plantings of sweet corn and snap beans can be made between date of last killing frost in the spring and July 15. The second planting of corn, for example, can be made when the first planting has 2 or 3 leaves showing. In the case of carrots, beets and cabbage, at least two plantings can be made—one in early spring for summer use and another in July for fall use and storage.

Staking Plants

With certain vegetable plants, it's sometimes necessary to stake them to avoid their drooping or loosening at the roots and to protect them from wind and storm. When tying the plant, be careful not to "choke" it accidentally. Keep the tying material tight around the stake, but leave a wide loop around the plant stem.

It's important to stake any plants before it becomes absolutely necessary, that is, while it is still straight and upright. Plants that become most badly broken and beaten down by wind and rain are usually those already inclined to bend or lie down.

A wide number of materials can be used for both stakes and tying; they vary from branches and twigs to metal rods and from raffia to specially covered wire. Spread the stems out as you stake and don't gather a number of stems together in haphazard fashion.

In the vegetable garden, stakes are always placed before the plants or seeds are in the ground. Supports for tomatoes may be stout stakes, two by two inches; or the plants may be tied to successive wires of a fence as they grow taller. Peas are usually supported on chicken-wire fencing, or they may be supported by twigs pruned from shrubbery. A dual purpose may be served

When staking a tomato plant, the large, soft string should go once around the stake so it won't slip. Stem is then supported by the string so that fruit clusters will develop away from the stake, thus avoiding bruising fruit.

by such a support, if the twigs are freshly pruned from spring-flowering shrubs and immediately thrust into the ground near the peas, because some of the twigs may root, affording a supply of new young shrubs.

Poles for pole beans are usually young sapling trunks, cut during the winter and trimmed for the purpose. They may be set erect at each hill, or 3 poles may be tied together at the top, tripod style, and the seeds planted in the center between them. Poles for beans should be sunk 2 to 3 feet in the ground before seeds are planted.

To make a trellis, space stakes every 10 to 12 feet in the row. Stretch wire between stakes at the top and bottom. Weave string between top and bottom wires to support plants.

Weeding and Cultivating

Mulching is the best method the gardener has in the battle against weeds. It also serves to keep the soil loose and friable. Another way of doing these two jobs—though a much more difficult and time-consuming way—is by cultivation. When mulching isn't practiced, cultivating is necessary to keep weeds out of the garden. Otherwise weeds will rob the vegetable garden of moisture. In drier areas of the country, large numbers of weeds growing in a garden make it necessary to irrigate oftener than where there are no weeds. Another disadvantage of weeds is that they shade vegetables, thereby depriving them of light needed for rapid, strong growth. Underground, weed roots are at work taking plant food from the soil. Since weeds grow faster than most vegetables, they get most of the food materials. Cultivating also aerates the soil, loosening it around plants so that roots can develop more easily.

Another reason for cultivating is that it prevents the soil from packing and cracking on the surface. This happens when soil becomes dry. These cracks permit the sun to evaporate a great deal of moisture from the soil. (Here again, organic gardeners find mulching more effective than cultivating in preserving soil moisture.)

The tool used to do the cultivating job depends upon the size of your garden. Garden tractors and rotary tillers are fine for the larger plot. They

For larger gardens, rotary tillers and garden tractors can easily be used to control weeds and loosen soil between rows of growing plants. The cultivating blades cut off weeds just below soil surface, leave stems on top of soil to provide mulch.

In unmulched gardens, it is necessary to control weeds by cultivating with a hoe or 3 or 4 pronged cultivator as shown above. Soil should be loosened to several inches deep, to destroy weeds and also loosen soil for better root development.

can be easily controlled between rows of growing plants. The cultivating blades cut off weeds just below the surface of the soil, and the stems stay on the surface to provide mulch.

The common hoe, when kept sharp, is one of the best cultivators for the small garden. Nevada's experimental station gives the following advice on its use:

Hoeing should be done thoroughly and in such a manner that all weeds are killed. Don't just scrape the surface, but get under and loosen the soil to a depth of several inches. Be careful not to cut so close to the plants that

Baskets, with plenty of openings for ventilation, give plants set out in midsummer a chance to get a good start without exposure to the sun. Second plantings during July and August are an excellent way to increase fall harvests.

their roots or stems are hurt. For small plants, the cultivation will have to be more carefully done than for big ones, because small plants easily are cut off, dug out or covered up with dirt. As plants grow, the soil should be hoed toward them so that the roots will be well covered and the plants will be in the center of a ridge of soil between the irrigation furrows. The ridge of soil supports the stems and lessens breakage.

Cultivation should be done often enough so that the weeds never get a good start. Don't wait for the weeds to grow big, but do your hoeing while the weeds are still quite small.

All the weeds cannot be killed by hoeing because some of them will come up among the plants. These will have to be pulled out by hand. This can be done as you are hoeing or, if the weeds are quite thick, it may be necessary to hand-weed before hoeing so that the seedlings can be seen. Hand-weeding requires careful work, but often it is necessary.

Always keep the garden free from weeds and in a good condition. Early and frequent cultivation will repay for the required time and effort.

Cultivation must be thorough if success in gardening is to be attained. When the plants are small, the ground should be loosened up close to them but, as they get larger and the roots spread, the tillage must not be deep enough near the plant to cut the roots. The surface of the soil should be kept in a fine, dust-like condition. Do not cultivate after a rain or irrigation while the soil is wet, since this practice will form clods. Wait until the soil is dry enough to crumble in the hands.

After the plants are well started, the garden work will consist chiefly of successive cultivations and irrigations. The cultivations may become less frequent as the plants mature. By midsummer the plants should be large enough so that occasional cultivations will keep out the weeds and maintain the plants in good condition.

Mulching–A Better Solution

As you can see, cultivating by hand or by garden tractor is a lot of work. Mulching in just about every case will accomplish the same purposes as cultivating. In addition, the decaying mulch serves as a constantly rotting compost pile to furnish humus to the soil in your vegetable garden—an important function that no amount of cultivating will accomplish.

Harvesting Vegetables in a Drought

Here is the day-by-day account of Jennie Hutton, a Montana organic gardener, and how she and her family "beat the drought":

To make this a clear picture of a very satisfied gardener, I must present some "before" and "after" sequences.

To begin with the "before," I live in a dry land corner of southeastern Montana with very little average rainfall. Gardens and crops are gambles, to say the least, what with the minimum rainfall, and hot, burning winds that pile up huge thunderheads, usually resulting in an icy shower of hailstones, from pea-size pellets to the large economy egg size.

Add to this sagebrush and alkali, and you have a good picture of what a garden shouldn't be. Yet with a large family like mine, 6 children and two adults, farming is the only life and a garden a necessity. It must be planted and tended whether harvested or not.

My husband plowed the garden and smoothed it with a disk. This was the only preparation. With string and stakes, I marked off the rows and began planting. Big Boston lettuce adjacent to its kin, Great Lakes, comprised the first 4 rows, followed in orderly fashion by Cherry Belle and White Icicle radishes; Sweet Spanish onion sets; Early Alaska and Little Marvel peas; Kentucky Wonder and Golden Wax beans; Detroit Dark Red beets; Copenhagen and Flat Dutch cabbage, started in the house and good-sized plants;

Danvers Half Long carrots, Earliana tomatoes, also started in the house; cucumbers; midget melons; Golden Bantam sweet corn; plus small Sugar pie pumpkins.

All these were in the garden proper, but along one end I set out, one by one, 300 strawberry plants–half Junebearers, half everbearing. My garden in completely, I was more than thrilled by a good soaker of rain. When the ground had dried off sufficiently, I went to work with the hoe, loosening the cementlike ground between the rows. The plants came up fine; cabbage and tomatoes lost their hot caps in due season and everything looked fine.

Flat stones are useful in conserving soil moisture and maintaining an even temperature around growing plants during hot summers. An old custom, stone mulching consists of placing rocks along plant rows so that little soil area is exposed.

Wilting Weather Arrives

We had a good rain on the last two days of May.

June fifteenth and no rain. Hot sun coming out steady. Strawberries look good yet. I pack water by the pail to tomato and cabbage plants Mondays, Wednesdays and Fridays. Strawberries get a drink on Tuesdays and Thursdays. Half of the cabbage left, 100 plants. Three-fourths of the tomatoes, about 40 plants and only 200 strawberry plants. One bucket of water dampens 6 plants, so it is a real chore watering them but I hate to see them die.

June twentieth and no rain. Garden is burning bad. I woke up the other morning and something had eaten all the cabbage off. Tomatoes are struggling hard for survival. *July first* and still the hot dry winds. Hail threat every day, but so far it has gone around us. *July fifteenth,* the same. One mess of peas and vines are gone.

August first, and all I harvested after heat and insects had taken their toll were 8 strawberries, a gallon of tomatoes with large dry-rot spots on all, 3 cucumbers, one mess of peas and sweet corn; no beets, radishes or lettuce for two weeks; carrots the size of threads; one mess of onions, two midget melons and 4 small pumpkins. This, which should feed 8 people through to the next bearing season, wouldn't keep one. I wish it had hailed, at least the agony and hard work would have been shortened. End of the "before" picture.

Decides to Try the Organic Way

Then I heard about the organic method; was this another fad, another pipe dream? Something told me it wasn't. But how would it work in this hot, dry, ever-windy country?

Why not gamble? For 6 of our 10 married years, my gardens had hailed out or burned up. The 4 so-called "good" years had seen average crops, but I had put in a large patch, so I had an abundance. I might as well try it organically, I couldn't lose any more.

As the idea took hold, I began to plan. I had quite a bit of time, as it was November and the snow had already blanketed the ground. My hands itched to begin, as I love to hoe in the early morning with only the birds and me to keep company with the rising sun. It is so peaceful, and everything is damp with sweet-smelling dew. I love it.

April made its showery entrance and I began my grand experiment. On top of the moist garden soil, I spread well-rotted cow manure to a depth of 6 inches, topping it with a shallow layer of poultry manure obtained while readying the brooder house for occupancy. This was all hauled in with a wheelbarrow, loading and unloading all done by me, as my husband was busy in the field. (The only ointment for sore muscles is to envision the results.) I also spread some rotted alfalfa hay I had cleaned out of last winter's feedlot on the top—and hubby plowed the works, disking it smoothly.

Now for the planting. But Mother Nature had an ace up her sleeve. A cold, wet snow fell, and further work was out of the question for 3 weeks. At last the weather settled, but again I had to wait, as 300 strawberry plants arrived and demanded attention, which I gladly gave. Anything to be in that dark, rich-looking soil. It was so soft and had such a nice feel.

My seeds had arrived some time ago and I had really got a lot of them. But I didn't want to waste any of that rich soil I figured I'd have. I had even increased the garden area. I had so much to plant, I was getting very impatient, but the great day finally arrived. I planted the garden all at the same time as the season was far advanced.

Hay Mulch for Berries, Vegetables

Resting my aching back a week later, I decided to use some old alfalfa hay on the strawberries. I took slices off the bale and placed them between the rows all the way until all you could see was green plants and old yellow hay.

I set out 300 cabbages and 200 tomato plants. Friends asked me if I intended to supply the United States, but they knew as well as I how many usually survived. Everything came up on schedule, including the sun. We'd had a good rain on Memorial Day, which we can always predict with pretty good success and another short shower on June sixth. As the plants became good size, I mulched between the rows with more of the baled alfalfa. It gradually grew hotter and hotter, and the wind blew its continuous wail. I watered the cabbage and tomato plants twice, but it was too backbreaking. They were just too many.

June, and the crops in this area are burning badly, but my garden looks wonderful. I haven't seen an insect and weeds are very few. A hard wind blew the hot caps off the tomato and cabbage plants, but cutworms never bothered them.

I made a small compost heap earlier and used it in the yard. What flowers, beautiful beyond description, and they had but little water!

Mulching Makes the Difference

July thirtieth, still no rain. Maybe the mulch did it, but my garden is the best I have ever seen in this part of the country raised without benefit of water. I canned and canned peas but had to give many away, as I couldn't care for them at their peak. I canned *only* 400 pints.

This is August, and the garden is nearly harvested. And what a garden! I canned 300 quarts of tomatoes, 100 quarts of juice, and 50 bottles of catsup. I still had a great surplus for the neighbors. Cabbage the same. Strawberries planted in the spring averaged a pint per plant and were very delicious. The ground between the plants was still moist under the mulch after a record hot summer.

As compared to the year before, I canned 2,000 quarts of vegetables and fruits and gave away plenty of surplus. What a wonderful feeling! My wheelbarrow sure brought me loads of joy. The seeds and varieties were the same ones I had used before, so I know it was the manure and the mulch that did it. My greatest blessing was to see 3 cynical neighbors change their minds—and are now all out for the organic way. I only wish everyone would do the same.

11—Watering and Irrigating

Essential to all life, water carries minerals from the soil to the leaves of plants and serves as raw material in the manufacture of plant food in the leaves. Rains do not always come at the proper time to assure a constant supply of moisture so artificial watering may be needed.

Use a sprinkler or a porous canvas hose. Removing the nozzle and allowing the water to run from the garden hose onto a stone or small board will keep the soil from washing. This procedure will also soak the soil thoroughly. The temperature of the water doesn't matter.

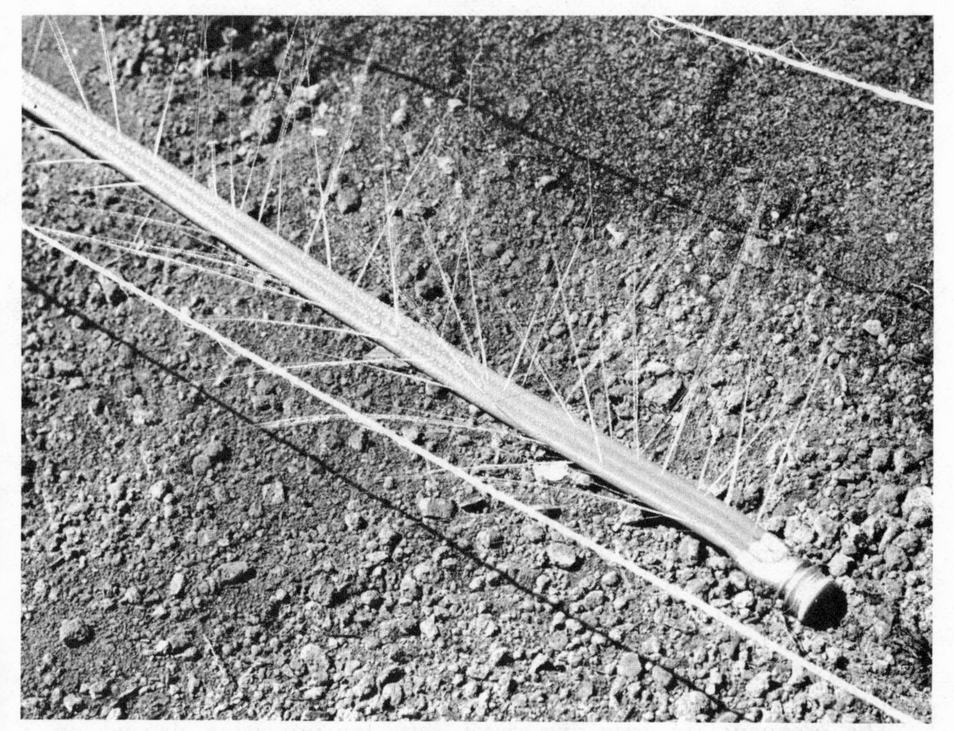

A perforated hose is useful for even distribution of water along the vegetable rows. Watering the soil to a depth of 4 to 6 inches is recommended over frequent, light applications, which tend to cause shallow rooting and poor growth of plants.

In most localities, the vegetable garden does best with a moisture supply equal to about an inch of rain per week during the main growing season. This means about 28,000 gallons of water per acre or 900 gallons weekly on a 30 by 50 foot garden.

Thorough watering of the soil to a depth of 4 to 6 inches weekly is preferable to more frequent light waterings, which encourage shallow rooting. Shallow roots are poor feeders, damage more readily from high winds and suffer more severely in hot weather.

In preparing the garden plot, plan for an adequate water supply and run the garden rows and irrigation furrows properly for watering.

Water in early morning, applying the water gently to the soil, not on the plants.

Irrigation can work for the organic gardener, insuring him against total crop loss and helping him build the land on which he lives. However, irrigation, except in cases of real distress or emergency where complete failure threatens, has certain drawbacks which do not recommend it to the thoughtful gardener. Excessive and continual irrigation must inevitably cause the soil to lose some nitrogen and mineral content through steady leaching. Also, continual watering of the soil overstimulates it into hyperactivity so that, despite a sound composting program, the soil can actually be depleted.

But neither of these considerations is valid when a real drought strikes and the prospect of total crop loss threatens. Irrigation, under such circumstances can mean saving the crop.

Reports from Experiment Stations

Irrigating whenever the "available" water in the upper 24 inches of soil dropped below 50 per cent resulted in significant increases in yield of sweet corn grown for processing in 4 out of 5 years at Cornell's New York State Experiment Station. In one season rainfall was distributed in such a way that irrigation was not necessary.

"The response to irrigation was measured in various ways," explains Dr. M. T. Vittum, Cornell soil scientist. "For one thing, it increased the number and size of marketable ears, thus giving higher gross yields of ears in the husk. Yields of marketable corn averaged 4.9 tons per acre on nonirrigated plots over the 5-year period as compared with 6 tons per acre for the irrigated plots.

"While market growers are interested primarily in ears in the husk, processors are more interested in the yield of kernels which can be cut from the cob," says Vittum. "In our experiment," he continues, "1.8 tons per acre of usable cut corn were produced on the nonirrigated plots and 2.3 tons, or half a ton more, on the irrigated plots."

The irrigated plots were harvested two days later on the average than nonirrigated plots, with little difference in maturity of the corn.

"The economics of irrigating sweet corn for processing depend on a number of things which each grower will have to consider for himself in deciding whether it will pay," states the Station scientist. "Among these are the abundance, location and cost of water, efficiency of the irrigation system, severity of drought, soil physical characteristics, cultural practices, market price and the unpredictable amount and distribution of rainfall during the growing season."

Many vegetable and potato growers wait too long before irrigating, says E. C. Wittmeyer, Ohio State University extension horticulturist.

Many times a small application of water early in the season will pay large dividends at harvest time, the horticulturist points out. The natural tendency is to delay irrigation, hoping it will rain.

The modern grower aims for near perfect stands of plants. This is necessary since production costs are high. Irrigation at the proper time will help insure good stands. Growers putting out plants for late harvest, Wittmeyer says, usually can get a better stand by applying water immediately after planting, unless rain occurs. As little as one-fourth inch of water per acre can mean the difference between a fair stand and an excellent stand.

Potato growers should be ready to apply water if dry weather occurs at tuber-set time, Wittmeyer explains. This is usually the time when blooms occur, but sometimes potatoes do not bloom under Ohio conditions. Dry weather ahead of this time can also interfere greatly with tuber set. Therefore, growers should examine plants and soil carefully to see if water is needed.

Plants differ in their appearance when they need water, Wittmeyer points out. Sweet corn leaves curl, while peppers and spinach wilt. The older leaves at base of tomato, cucumber and muskmelon plants may change color slightly when water is needed.

Some growers are using tensiometers to determine when soil moisture may be reaching the point where crop growth may be restricted. Such procedure, along with actual examination of the soil, is better than visual plant symptoms. By the time visual symptoms appear, yield and quality probably already have been reduced.

Today it is being made increasingly easy for the average home gardener to install and maintain an irrigation system. The national mail-order houses

When furrow-irrigating the garden, washing away of soil can be prevented by placing the end of the hose on a board. Water should always be applied gently to the soil, not on the plants. Mulch cuts down need for repeated applications of water.

and garden supply manufacturers are now making available permanent but portable systems which are comparatively low in cost and very easy to assemble and install right on your property.

The manufacturers are taking full advantage of the new, lightweight metals and are using them to create piping and coupling that are extremely easy to handle and weatherproof.

Aluminum pipes, for example, that are 20 feet long and two inches in diameter, weigh less than 8 pounds; 30-foot pipes weigh less than 12 pounds. Special couplers permit almost immediate snapping-on of one pipe length to the next.

The new nozzles are an important feature of the sprinkler system because they throw a stream of water that approximates a gentle rainfall to a distance of 20 to 40 feet. It is customary to raise these nozzles above the tops of your various crops. Different size riser pipes permit you to do this readily.

The aluminum pipe system can be laid above ground and is easily moved to various points in the garden. However, if you can trench easily, new plastic piping which is extremely lightweight and low in cost is now available. One-inch plastic piping measuring 400 feet weighs only 50 pounds and its manufacturers claim that 500 feet of such pipe can be installed in an hour. Be sure to check if plastic pipe can be left above ground or if it should be covered. If you are interested in the possibilities of using plastic pipe, you will find that adapters, couplings, elbows and joints are also available at a cost that figures in cents rather than dollars.

However, before you send in an order for a system, there are a few things to consider besides the cost. First—do you really need to irrigate? Second—if so, have you enough water available on your own property to do a complete job of watering? Third—do the slope and general lay of your land lend themselves to efficient, successful irrigation? Fourth—will your soil absorb the water without undue loss or wastage? Fifth—how will you combine your composting and mulching with irrigating?

Sources of Water

There are 4 basic requirements which your source of irrigation water must fulfill:

1. It must be adequate at all times, providing enough water for your minimum needs for a single watering; if this is 2,500 or 3,000 gallons, then your source must be able to give this much water in one continuous operation.

2. The cost of providing such a supply of water must fit within your gardening or farming budget.

3. The supply must be legally available, yours to use without restraint or complication.

4. It should be uncontaminated.

Once these obvious requirements are met, the source can be a well, either shallow or deep, a pond or lake, a perennial stream (this means it runs all the time, in dry weather and wet) or a local community water system. The drawback to the last source is plain; communities today restrict the use of water when it is needed most because there isn't enough to go around.

But if you live in the country or have a stream flowing through your property you have a source of water, providing your neighbors have no objection to your storing part of its flow and your rights are legally plain in your community.

A stream or spring-fed pond is an excellent source because the supply should be available in almost any contingency and also because it makes possible a low-cost and easily arranged, comparatively simple irrigation system.

A shallow well is also an excellent source of supply, providing it can deliver a sufficient supply of water in one continuous operation. It should replenish itself during pumping almost as rapidly as the rate of discharge because a horizontal centrifugal pump will not work if the water level falls beyond a point 17 feet below it.

Deep-well irrigation is not for the average homestead because it is a high-cost operation, obviously designed for large-scale irrigating. However, if your deep well is already providing plenty of water for your living needs, you might check if it can also meet your garden irrigation needs. Try to contact the man who drilled it and installed your water system. Keep in mind you will require from 3,000 to 6,000 gallons of water in one continuous operation.

Never try to install a deep-well irrigation system without the advice of an expert.

Irrigation Systems

There are 3 ways the home gardener can irrigate:

1. Overhead sprinkler pipe.
2. Porous hose or perforated plastic pipe laid along the rows.
3. Furrows between the crop rows.

Each has its advantages and drawbacks and each should be carefully considered with regard to the needs and prevailing conditions of the individual garden. The system should be chosen only after such serious consideration.

Overhead Sprinkler

If the pump is the heart of your irrigation system, the pipes, connections, valves and nozzles which convey the precious water to the spot where it is needed are scarcely less vital to success. In the past this paraphernalia or apparatus tended to be bulky, heavy and hard to handle.

Today this is no longer true. The new aluminum equipment can be handled easily and quickly and the new quick coupling devices permit extreme flexibility in pipe-feeding arrangements. If necessary, you can shut off part of your system and relocate it quickly and efficiently while the balance of the system is in operation.

Another improvement is the modern nozzle which throws out a stream of water 1/32 of an inch in diameter from 20 to 40 feet. This approximates a gentle rain and means that your tender young plants won't be beaten down while your soil structure will neither be eroded nor compacted.

The placing of nozzles depends on the size and shape of the area to be covered. In general they should be about 6 feet apart on the feeder line, throwing streams of water 20 or more feet.

As with your pump, discuss this matter with your dealer and local agent and be guided by their advice plus what the manuals and brochures tell you.

Porous Hose or Perforated Pipe

The organic gardener who applies a protective layer of mulch above and around his plants admittedly and understandably faces a problem which does not affect the gardener who does not mulch. It may well be that irrigating with a porous canvas hose or perforated plastic pipe *laid along the rows under the mulch* will be the best way to irrigate your garden, although the hose may tend to mildew and rot because of constant moisture.

Experiments with this ground infiltration system have shown it satisfactory for small fruits, strawberries, and garden vegetables in general. The water should ooze or drip gently right in the row where the hose or pipe is placed. The supply pipe which feeds the row pipes should be on the higher side of the garden with a pressure of 15 to 20 pounds maintained in the porous lines.

The surface method of irrigation has other advantages. Many vegetable crops are susceptible to leaf diseases that can be aggravated by excess moisture on the leaves. Irrigating by pipe or hose and also by furrow between the rows avoids this.

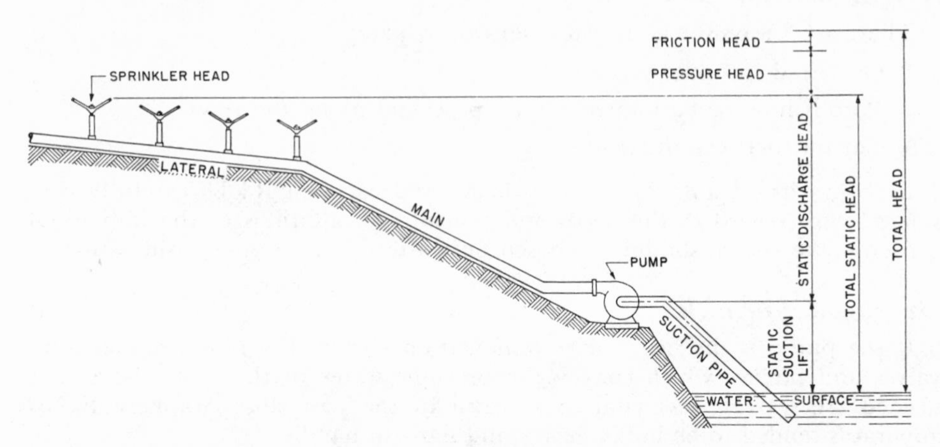

Above is a plan for a home irrigation system, though each irrigation plan will vary according to the layout of the garden. Lifting distances must be taken into consideration when calculating requirements for pump and power unit of the sprinkler system.

For efficient operation, ground slope cannot be too great and the soil should not be too sandy or water will be wasted.

Furrow Irrigation

Effective furrow irrigation can be arranged in the home garden where the slope and soil density conditions are met. The furrow should be made between the rows, permitting the water to soak into the soil and then spread out into the crop root zone between the furrows.

For top efficiency, the soil should hold the water well enough to wet the soil adequately along the entire furrow without flooding or excessive soaking where water is being delivered into the furrow.

It has been found that, where the slope is satisfactory, this form of irrigation is superior to rotary sprinklers where heavy soils prevail. Furrow irrigation causes less packing and crusting of the topsoil.

In the case of heavy soil where water penetration is markedly slow, water may be fed directly into the furrows just where it is needed and at a slow rate which permits a thorough soaking without waste or washing of soil.

TABLE 37: Pump Capacities and Specifications

Type	Highest Well Lift	Gallons Per Hour	Initial Priming	Pump and Motor Location
Shallow Well Piston	22′	250 up	Yes	Over or off well
Deep Well Piston	1,000′	200 up	No	Over well only
Rotary Gear	22′	250 up	Yes	Over or off well
Shallow Well Screw	20′	275 up	Yes	Over or off well
Deep Well Screw	900′	600 up	No	Over well only
Straight Centrifugal (single stage)	15′	50 up	Yes	Near source
Straight Centrifugal (multi-stage)	20′	600 up	Yes	Over or off well
Shallow Well Cent. Jet (single stage)	20′	200 up	Yes	Over or off well
Deep Well Cent. Jet (single stage)	300′	400 up	Yes	Over or off well
Deep Well Cent. Jet (multi-stage)	300′	200 up	Yes	Over or off well
Shallow Well Turbine	15′	100 up	Yes	Near source
Deep Well Turbine	1,000′	900 up	No	Over well only
Submersibles	2,400′	300 up	No	Inside well

Which Pump Should You Use?

There are 4 types of pumps you can use to raise water from your source of supply.

They are:

1. Horizontal centrifugal.
2. Turbine.
3. Propeller.
4. Mixed-flow, a combination of propeller and centrifugal.

Each is designed to do a different kind of pumping job and your choice of pump should depend on the kind of irrigating you will do.

The pump the average homesteader will use should be a low-cost, easily installed unit. The big farms and ranches of the Midwest and Southwest require water in thousands of gallons per minute, and there are big turbine pumps which operate in wells 500 feet deep to bring this volume of water up to the surface.

In practice, however, you will do fine with a 3 or 4 horsepower horizontal centrifugal pump which delivers 50 to 60 gallons a minute or about 3,000 gallons an hour. It will lift this volume of water about 100 feet from the surface of your pond or well. Technically this can be called a head of 100 feet. This pump can be obtained from your garden supply dealer or the mail-order garden and farm supply dealers. It will cost from $200 to $300.

The horizontal centrifugal pump will do a fine job pulling water up out of your shallow well or pond. It must be situated no more than 15 feet above the surface of the water and can be powered by electricity or gasoline—you can take your choice. Study the accompanying chart for general layout and note the distances involved. Keep them in mind when you order your pump and make sure you get the pump that will do the job you require of it.

Water Supply Plus Pump Equal Need

A vegetable garden measuring 60 by 70 feet—that's 4,200 square feet—needs 2,700 gallons of water for adequate irrigation to a depth of 18 inches. That's 2,700 gallons of water *in one continuous pumping.* If your garden measures 100 by 100 feet, you'll require 6,400 gallons of water *in one continuous pumping.* Can your pond or shallow well deliver this volume?

If you place your pump over a shallow well, the top of the water must be *within 17 feet of the pump at all times.* Can your well deliver this amount without dropping so low the pump won't work? Be sure to check this factor before ordering your pump or setting up your irrigation system.

Deep-well turbine pumps are mostly used where the water level is far below the lifting capacity of the centrifugal pump. They are highly efficient and will give long and dependable service. They are also much more expensive and harder to install, inspect and repair. Each is designed to operate at maximum efficiency, rated at a definite discharge for a prearranged distance, based on pump rotation.

Propeller pumps are excellent for lifting rather large amounts of water short distances—10 feet or less. They are well adapted for surface irrigation or drainage.

The combination pump, known as the mixed-flow, will work efficiently lifting large volumes of water up to 25 feet, are at their best working at heads between 10 and 15 feet. Structurally, they combine centrifugal and propeller features.

How Much Will It Cost?

Cost of irrigation can be a determining factor even where the threat of absolute crop failure is real. An efficient sprinkler system can cost $500 to $600 for a 10,000-square-foot garden using an efficient pump which delivers about 5,000 gallons of water in an hour to a point 100 feet away from the source.

Cost of a horizontal centrifugal pump which will do the job is between $250 and $275. Two-inch aluminum piping and laterals and risers which lift the nozzles plus adapters and automatic couplers make up the balance. It should be stressed that the latest piping equipment is extremely lightweight and quickly assembled.

But you can cut costs if you are prepared to work and plan carefully. Again, you should consult a local expert like your county agent and discuss the matter thoroughly with him. The dealer who will supply the pump and piping will also have plenty of reference literature which you should study carefully.

Unless you are very well advised and absolutely sure of all the factors involved, use the pump recommended by the experts who help you with your irrigation plan. Distance of the pump from the top of the water should not be more than 15 feet and the distance it has to push the water uphill to your garden obviously is the second big factor. Using an inefficient or under-powered pump is false economy.

However you can probably cut corners by furrow irrigation, providing the slope and soil favor this. Or you can irrigate by hose along your plant rows, under the mulch. This will save the cost of extra laterals, risers and nozzles. Savings up to one-third can probably be effected safely if you will plan carefully in advance and take full benefit of expert advice. The average home gardener can quite probably have a dependable irrigation system installed for about $300.

This puts the matter squarely up to the individual gardener. If your area is drought prone to the extent that total crop failure can occur, then you should investigate the possibility of home garden irrigation. As stated previously, the manufacturers, dealers and distributors are working overtime to make it easier for you to handle, install and maintain an irrigation system on your comparatively small homestead. Regarding the initial cost, it should be stated in fairness that this sum should be amortized over a 5- or 10-year period because it will work for that time with practically no replacement cost. Thus a $300 system will cost $30 a year for 10 years. Whether or not your crop value warrants this extra insurance cost is a matter only you can decide.

Preventing Frost Damage with Water

Irrigation offers plants protection against cold as well as drought. Many gardeners and farmers have found that using a sprinkler irrigation system when night temperatures drop below 32 degrees protects strawberries and other crops from freezing. (Of course, a heavy straw mulch will also protect plants well.)

One New York strawberry grower used his irrigation system in this way for protection: when the temperature dropped to 34 degrees, he started his pump. By the time freezing began several hours later, his plants were covered with ice an eighth to three-sixteenths of an inch thick. When the temperature dropped still further to 26, the temperature of ice and plants stayed at 32. As a result, no frost damage occurred to all plants under irrigation, while 90 per cent damage occurred to nonirrigated plants.

Measure Soil Moisture

As mentioned previously, it is not healthy to water plants before they need water or to wait too long to start irrigation if the weather is dry. If you water too soon or too often, you can leach precious nutrients out of the soil. If you wait too long to water, damage will be done to the plant from which it may not recover. The best way to find out exactly when to start irrigation is to use a moisture testing device. Then you won't have to rely on your own personal observations, which can be quite wrong. You may very well find yourself saving much water and producing better plants.

There are two types of soil moisture gauges on the market. One type measures the ability of the soil to conduct an electric current between two electrodes. The current is usually supplied by a couple of small batteries, and there is no danger involved in the use of the device. Electric soil moisture checkers can be purchased for prices ranging from about $5.00 to $25.00.

Most irrigated area farmers use the water-pressure type of moisture gauge, a device that is sensitive to varying osmotic pressure in the soil. These are left in one spot for an entire season, and for that reason are probably more accurate on a continuing basis. The "weakest link" in a soil moisture checker is the contact between the measuring device and the soil. If the instrument is left in place all the time, there is less chance for error through faulty contact with the soil.

An osmotic pressure gauge is a plastic tube filled with water, with a porous cone at one end and a pressure gauge at the other. One of the most popular is the Irrometer, made by the T. W. Prosser Company. It sells for over $20 and is made in varying lengths so the moisture content of the soil at different depths can be checked. Most irrigation farmers use a 6-inch and an 18-inch length Irrometer together. Directions supplied with the instruments tell at what readings irrigation should be started.

Irrigation Is Not a Cure-All

Always remember that irrigation will not necessarily mean a bumper harvest for you.

"Irrigation is no cure-all for poor crops," warns William E. Larsen, agricultural engineer at the University of Delaware. "Installation of numerous irrigation systems in recent years indicates that many farmers may be turning to water as a kind of crop insurance against drier-than-normal weather. This may lead to disappointments and poor returns on money invested."

"In normally humid areas," Larsen adds, "irrigation is most likely to be profitable when it is adopted as a final step in a complete soil and water management program on a farm."

Irrigation can bring about spectacular increases in yields, he explains, but only where other conditions are favorable and water is the chief factor in limiting the size of crops. Irrigation will do nothing to solve problems of erosion—it may actually increase the run-off. It may aggravate problems on poorly drained soils or create drainage problems on formerly well-drained soils. Likewise, irrigation alone will do nothing to improve fertility or poor soil conditions. All these things, as well as lack of water, normally limit crop production.

Compost-Irrigation Builds Soils

Thousands of acres in the southwestern United States, which previously were low-yielding, can now be turned into high-producing land. Compost-irrigation is chiefly responsible for this amazing fertility build-up in the soil.

Clinton Ray, whose farm is located near Mesa, Arizona, is one of the many southwestern farmers who has used compost-irrigation successfully. He is using a waste organic material that is readily available in his area—cotton gin trash; by irrigating through the cotton gin trash compost pits and by sheet composting with the same organic material, he's been able to step up productivity rapidly.

Here is how the *Arizona Farmer-Ranchman* recently described Ray's technique:

"The method used by Ray in composting gin trash is to put a pit in the irrigation ditch. During irrigation, the water flows through the stack, carrying plant food and microorganisms into the field. In addition, he side-dresses his crops with the compost, using a cattle feeder truck for the purpose."

Ray believes that the pit can be any size, but should not be made more than 6 feet deep. "When the trash settles," he says, "it resolves to about 4 feet, which is the ideal depth." He has 5 of these compost pits in his irrigation system, each containing 150 tons or more of gin trash.

All of the material is bought from 5 gins located near his farm. Since Ray treats the trash with a bacteria additive, he does not have to turn the compost at any time—the bacteria does the work for him.

Two years ago he was getting a short-staple cotton yield of $1\frac{3}{4}$ bales; last year his yield climbed to $2\frac{4}{5}$ bales. Long-staple yields increased from $\frac{15}{16}$ of a bale to $1\frac{3}{4}$ bales.

Similar increases were made in vegetable production, particularly tomatoes. A few years ago Ray was producing about 300 flats to the acre, and they were poor keepers. Last year he produced 1,000 flats per acre on 7 acres. He says they were better keepers and brought him a premium of 25 cents a flat on the market.

Ray cleared land for this farm out of the desert about 15 years ago, building up a fruit and vegetable farm in addition to straight cropping. His 15 acres of Robin and Redwing peaches have been in for 10 years and are producing well. Three acres of late Royal and early Gold apricots yield well each year.

"I treated my peach orchard with organic material from the first," Ray points out, "and it has been improving every year. I can't say the same for my vegetables and field crops, which I fertilized the usual way. No matter how much I fertilized, yields kept dropping down. I recognized that the soil had a tremendous lack of organic material, so finally I turned to sheet composting with gin trash. This was an improvement, but I didn't get the marked improvement until I treated the trash with bacteria. It gave me improved soil structure. Land, for example, that took 12 hours to sub up during irrigation, now takes only 3."

Idus Gillett of Canutillo, Texas, is another organic farmer who has been practicing compost-irrigation.

The key to his farming success is compost, which he has applied at the rate of 300 tons yearly to his 105 acres, or about 3 tons to the acre. Gillett builds his compost heaps with manure he obtains from railroad livestock cars. He waters the "green" manure heaps down about every two weeks until they create humus by their natural actions. The steps are quickened at times by adding "seed" humus from a compost heap that already has passed through the internal process to rich humus.

This humus is then spread on the land either by dump spreader or through irrigated water. The "pure" humus is fed into the irrigation water and automatically reaches the land in proper proportion. Humus that includes sand, which is often desired, is broadcast by spreader.

The decomposed material is distributed in his irrigation water for two reasons. First, because sheet composting sometimes doesn't work in this semidesert country; second, the water absorbs the salts of the compost, neutralizing much of the alkali present in the soil.

Compost is distributed at every irrigation. Once before planting when Gillett readies his ground for seed and again, 90 days later, about July 1, to mulch the ground around the young plants. A third irrigation is made in mid-August if water is available.

Compost Water for the Garden

Many problem plants and trees can be nursed back to health by treating them with compost water. You can use it on bare spots on your lawn, on trees that have just been transplanted and on indoor plants that need perking up. You can even use it on vegetables in the spring to try to make them mature earlier. Compost water is really good in greenhouses, where finest soil conditions are needed for best results.

It is really no trouble to make compost water on a small scale. For treating house plants or small outdoor areas, all you have to do is fill a sprinkling can half with finished compost and half with water, stir gently 10 or 12 times and pour. Nothing could be easier than that. The compost can be used several times, as one watering will not wash out all of its soluble nutrients. The remaining compost is actually almost as good as new and should be dug into the soil or used as mulch. It takes the action of soil bacteria and plant roots to extract the major value from compost.

If your garden is on a hill (even a slight one will do) you are in an ideal position to use the pit method, because you can let gravity do the job of moving the compost solution from pit to garden and home grounds. Or, if you wish, you can hook up a pump to your garden tractor or tiller and pump the solution wherever it is needed.

The sides and bottom of the tank should be lined with concrete, bricks or blocks, to prevent escape of your solution. If you use gravity for distribution, you will have to construct a drain channel from the base of the pit to your garden. Furrow irrigation can be used to spread the water throughout the garden.

If you are going to use a pump, you will have to build a sump or reservoir at the base of the pit to catch the compost water after it filters through your pile.

Organic Irrigation in Britain

A recent issue of the *New Scientist,* a British publication, carried the following report:

"Organic irrigation by raingun spraying, long familiar in Switzerland, has now come to Britain. Recently a demonstration of one of the schemes being made available here was given on the farm of Mr. J. L. Atherton at Baldwin's Gate in north Staffordshire. In its essentials, organic irrigation is the spraying back on to the land of farmyard manure, suitably diluted with animal urine and waste water from the farm buildings, which would otherwise go down the drains.

"Mr. Atherton's system—which cost him in all some £1,500 ($4,300) to install—centres on a 17,000-gallon cesspit into which inlet pipes from the dairy, piggeries and other farm buildings lead. It takes about a week to fill up. When it is ready for emptying, the liquor is mixed by a line of 5-foot diameter wooden agitator discs, offset to each other: these bring up the sludge which settles to the bottom over the week. A Bücher pump then drives the mixture through 530 yards of aluminium piping to one of two standard Farrow rain guns, which spread about 4,000 gallons an hour over a radius of 100 feet of pasture. They can cope with any lumps and straw and spread the load evenly. The guns can also be used for their normal irrigation work where necessary.

"A number of advantages are claimed for this method in terms of convenience and low operating cost. Perhaps the most telling is the effectiveness with which it disposes of waste detergent from the dairy washing. It is this dairy detergent which gives so much of the unwelcome foam to many of our rivers; the problem has been becoming so serious that a number of river boards are taking action against offending farmers to prevent them emptying into the river. Such an organic irrigation system is a doubly effective answer, for it not only disposes of detergent waste properly, but would also cut down the farming demands on mains water for irrigation uses. Experience so far does not show any harm to pasture from the detergent.

"Though the machinery is modern, organic irrigation as such is nothing new in Britain. Only a few miles from Baldwin's Gate, indeed, it was successfully tried 100 years ago at Hanchurch Farm, near Trentham, where the Duke of Sutherland was an irrigation enthusiast. His equipment was not so remarkably different—a steam engine, pumps, underground iron mains, guttapercha hose and a jet pipe. All this cost £520 and served 83 acres, against Mr. Atherton's total acreage of 180, which is far from being completely served by his present setup. The Duke of a century ago was a believer in large reserves, for he installed settling tanks sufficient for 300 acres."

How Water Fertilizes

Water is a fertilizer. It is a fertilizer that contains an element vital to all plant life. It is a fertilizer that will boost crop yields as much as 81 bushels per acre. That is what it did for University of Illinois farm scientists in a field of corn.

Usually, when we speak of fertilizer we mean organic materials such as compost, manure, bone meal, leaf mold or other plant material. Or, we may mean mineral fertilizers such as nitrogen, phosphorus, potassium, calcium and trace nutrients. But water is a fertilizer equally as important as any organic or mineral fertilizer.

For water contributes the vital element hydrogen, a major constituent of all organic matter.

Whether the organic matter is sugar, starch, protein, fat, vitamins, hormones or enzymes, it contains hydrogen as a part of its basic composition. And this hydrogen comes from water. Growth of plants and production of plant substances is impossible without water. For all organic substances, including all of the food that we eat and that we feed our farm animals, contain hydrogen that has come from water.

In the same sense that organic and mineral substances are fertilizers, water is a fertilizer. For water enters into the life processes of plants and becomes a part of all plant substances.

Scientist F. J. Stevenson, Associate Professor of Soil Biochemistry, University of Illinois, Urbana, says, "There is justification for regarding water as fertilizer. Just as nitrogen, phosphorus and other mineral elements enter directly into growth processes in plants, so does water."

Water is made up of hydrogen and oxygen—two atoms of hydrogen and one of oxygen. It is the hydrogen of water that fertilizes plants through a marvelous process of nature called photosynthesis. Professor Stevenson explains how nature goes about turning water into plant substances and food for animals and man.

"In photosynthesis," Stevenson says, "water is split to yield a hydrogen atom (H) and a hydroxyl group (OH). By a series of complicated reactions in the leaf of the plant, generated by substances called enzymes, the hydrogen is combined with carbon (from carbon dioxide—CO_2) to form cell constituents, such as carbohydrates; a byproduct of the reaction is hydroxyl group (OH), from which oxygen is produced."

Thus water not only is a fertilizer that is vital to the plant, but is a source of the oxygen in the air that we breathe. And the cell constituents and carbohydrates become the food that we eat and that we feed our farm animals.

"The importance of water in the nutrition of plants is apparent," Professor Stevenson points out, "when it is considered that water comprises from 80 to 90 per cent of the total weight of growing plants. The hydrogen derived from water through photosynthesis exists in most of the solid matter remaining after water is removed."

And it is this solid matter that we harvest as crop yields. For University of Illinois scientists, water made a difference of 81 bushels of corn per acre in a special test they ran.

The experiment was to determine the effects of water on crop yield. In the test, one crop of corn was raised in a plot that had the ground covered with thick plates of glass. This was to prevent the soil from losing water through evaporation, and from getting water from rain. This covered plot produced 46 bushels of corn per acre.

Another plot that was left uncovered so that rain could soak into the soil yielded a much larger crop of corn. The natural rainfall of 5.5 inches boosted the yield of corn to 92.4 bushels per acre.

But another plot of corn that had both natural rainfall and 6.5 inches of irrigation water produced 127 bushels of corn per acre! Water made a difference of 81 bushels more of corn per acre over the plot that received no extra water.

A natural reaction to the statement that water boosted the University's yield of corn 81 bushels per acre would be, "Plants do better with plenty of water."

This is true. But why do plants do better? What is it that plants do when they have plenty of water? It's not just that they happen to like water. Water must do something specific towards the production of crop yield by plants. And this specific something is to fertilize the plant with the vital constituent hydrogen that is absolutely required for every bit of tissue and food material produced by the plant.

It is true that water has other functions in the plant, such as transporting nutrients and regulating temperature. But one specific and absolutely vital function of water is to furnish the element hydrogen.

For example, the corn that was raised on the plots by the University of Illinois scientists was just like any other corn. It contained a lot of starch, some protein, some fat, some sugar, vitamins and other organic substances. And scientists analyzing these substances find that every one of them contains hydrogen that comes from water.

Unless water is present to furnish hydrogen, these plant substances and foods cannot be produced. And isn't it just plain common sense that if enough water is available to the plant, the plant will have the hydrogen it needs to produce optimum yields?

In the organic method, moisture is conserved. At the same time, air is allowed to enter the soil and reach plant roots. This is because organic material in the form of compost or green manure acts as a sponge to absorb and hold water until plants need it. And the rough material in organic matter opens up channels down into the soil for air to enter and reach plant roots.

In contrast, soils without organic matter allow water to run off or soak through. Clay soils and hardpan soils, for example, allow valuable water to run off almost like the University of Illinois experimental plot covered with thick sheets of glass. And sand and gravel allow water to soak on through just like a sieve.

On the other hand, organic matter soaks up water during times of plenty and gives it up to plants when they need it most for fertilization and other purposes. This water then goes into the production of food and boosts crop yields tremendously. As in the case of the experiment at the University of Illinois, water boosted the yield of corn almost 300 per cent.

To make sure of getting the full fertilizing effect of water, make sure that plants have sufficient but not too much water. Organic matter can help you control the quantity. Soils should be moist, but not soaked. If necessary, add water to soils by irrigation. A goodly portion of the water is turned directly into the bushel yield that you harvest. This is profit from fertilizing with water.

—Charles H. Coleman

Hints on Watering

To get the best results from the water you give your garden, follow these rules:

1. Don't use dishwater, or any water used for washing clothes, etc. Soap and detergent residues can interfere with essential soil life and aeration; so can bleaches, grease or other impurities usually present in such water. If at all possible, also avoid highly chemicalized city water supplies. Most of these contain too much chlorine and other bacteria-killing ingredients. Artificially fluoridated water is similarly undesirable.

2. Do combine a careful watering program with year-round soil improvement. Organic matter added in the form of compost, turned-under cover crops or plant wastes improves the soil's water-holding capacity; it corrects poor drainage, run-off or puddling caused by hard-packed surface or subsoil;

TABLE 38: Root Depths of Crops

Depths to which the roots of mature crops will exhaust the available water supply when grown in a deep, well-drained soil under average conditions.

Crop	Depth In Feet	Crop	Depth In Feet
Alfalfa	10–15	Ladino clover and grass mix	2
Almonds	6–9	Lettuce	1½
Apricots	6–9	Melons	5
Artichokes	4½	Milo	6
Asparagus	10	Mustard	3½
Beans (dry)	3½	Nuts	4–6
Beans (green)	3	Olives	6–9
Beans (lima)	4	Onions	1
Beets (sugar)	5–6	Orchards, deciduous	6–8
Beets (table)	3	Parsnips	4
Broccoli	2	Peas	3½
Cabbage	2	Peaches	6–9
Cantaloupes	4–6	Peanuts	2
Carrots	3	Pears	6–9
Cauliflower	2	Prunes	6–9
Celery	2	Peppers	3
Chard	3	Potatoes (Irish)	3
Cherries	6–9	Potatoes (sweet)	4–6
Citrus	4–6	Pumpkins	6
Corn (sweet)	3	Radishes	1½
Corn (field)	6	Soybeans	3–4
Cotton	4	Spinach	2
Cucumber	3½	Squash (summer)	3
Eggplant	3	Strawberries	3–4
Figs	5	Sudan Grass	6
Grain and Flax	4	Tobacco	4
Grain Sorghum	4	Tomatoes	6–10
Grapes	8	Turnips	3
Grass Pasture	3–4	Walnuts	12–18
Hops	4–6	Watermelons	6

and it helps water do a better job of getting to plant roots to deliver both food and moisture. Make use of natural ground rock fertilizers for mineral enrichment; skip the quick-soluble synthetics that wash out fast, leave a crust-forming accumulation of caustic salts in the soil.

3. Do mulch for all your garden's worth. Besides helping tremendously to conserve and hold needed moisture through hot, dry spells, a constant organic-matter covering holds down weeds that compete for water and nutrients, cuts direct surface evaporation and helps improve soil structure by decomposing from below. In addition, a mulch tends to keep soil at a more even temperature, prevents harm from sudden weather changes. Whether you irrigate or not, mulching makes a helpful difference—and good sense.

TABLE 39: Guide for Estimating Available Plant Moisture in Soils

Range of moisture in soils	Feel or appearance of soils with varying moisture		
Per cent*	COARSE TEXTURE (Sandy loams and loamy sands)	MEDIUM TEXTURE (Silts, silt loams, loams and very fine sandy loams)	FINE TEXTURE (Silty clay and clay loams)
0-25	Dry, loose, flows through fingers.	Crumbles easily, tends to hold together from hand pressure.	Crumbles readily, will hold together but "balls" with difficulty and breaks easily.
25-50	Looks dry, will not hold together from pressure.	Somewhat crumbly, will hold together in hand with pressure.	Does not crumble, forms readily. Will "ball" with pressure.
50-75	Will form loose ball under pressure, will not hold together with handling.	Forms "ball" readily. Will "slick" slightly with pressure	Forms "ball" readily. Will "ribbon" out between thumb and forefinger. Somewhat "slick" feeling.
75-100	Forms weak ball, breaks eosily, will not "slick."	Forms "ball" easily, fairly pliable, "slicks" readily.	Easily "ribbons" out. Has "slick" feeling.
100-field capacity	No free water appears on soil when squeezed, but wet outline of ball is left on hand. Soil will stick to thumb when rolled between thumb and forefinger.		

*Per cent of plant available water in the soil.

—*Oklahoma State University*

4. Don't apply water haphazardly. Plan your garden's irrigation as carefully as its planting, fertilizing or any other step. Know *when* the plants you're growing need water and time the irrigating. Know *how much* different crops need, too, and control the amount they actually get. Water thoroughly when you do; don't merely sprinkle often; too little can do more damage than none at all. At the same time, avoid overwatering; allow for normal rainfall in your area. USDA research shows vegetable crops should usually be irrigated when the soil surface (one-half to one inch) has dried out. Fruits and vegetables need sufficient water most when fruiting or edible parts are forming. Irrigate at the best hour for your garden—not for personal convenience. Generally, early-morning watering is most effective, provides moisture during the sunny part of the day, and discourages mildew or other conditions brought on by too much dampness in some plants.

According to the Florida Experiment Station, "Vegetable crops are usually irrigated when the soil surface (one-half to one inch) has dried out." They also found that irrigation applied as needed during the blooming period of their citrus crop brought highly significant increases in yield on the 3 varieties of oranges on which the experiment was conducted.

California researchers came up with a very interesting way to tell when trees need water. "When wilting or any other evidence of lack of readily available moisture is hard to detect in the trees themselves, the best method by which the grower can decide when this condition is reached is by watching some of the broad-leaved weeds which may be left as indicator plants in various places in the orchard. Generally, such weeds are deep-rooted enough to indicate by their wilting a lack of readily available water in the soil occupied by the roots of the trees."

Eleazer gives these general rules for when to irrigate some principal crops other than at planting time, when they all need it if moisture is not already in the soil:

Corn: can stand considerable drought early, but needs water just before, during and a while after tasseling time. (Clemson Agricultural College suggests the best indication to irrigate is when the corn wilts by 10 a.m.)

Truck Crops: need sufficient water most when fruiting or edible parts are forming and up to harvest.

Peaches: have two critical periods—first when the pit is hardening and next when they are making their "big swell" before ripening.

12—When to Harvest

VEGETABLES MUST BE picked at the psychological moment, at that stage in their development when they *taste best to you.* This is usually just before they go through chemical changes that convert sugar to starch and fiber to cellulose (wood).

Many vegetables are picked too late.

How to know the right time? Taste them again and again until you recognize the size and appearance at which they reach their peak of flavor and aroma (which has a lot to do with the flavor of things). You will never know what thrills you have been missing until you systematically test your own garden products from the earliest stages of maturity to the latest. They vary in quality and flavor far more than you might believe possible.

EARLY GREENS—The season starts with dandelions and (ten to one) by the time you notice them they are already too large to please a gourmet. Cut them before they are 3 inches long, wash carefully and serve in a tossed, green salad and you have the tastiest of spring dishes. Or cook them gently with a few evergreen bunching onions that have survived the winter in the open. After the flowers appear, both are too tough and bitter for pleasant eating

For sparkling salads there are no better tonics than Upland and water cresses, mustard and Roquette. But be sure to gather them in their "salad days." In their old age they will fairly set your mouth on fire!

ASPARAGUS—Why dig down to cut it with a great, thick, tough white butt that takes so long to cook that the tender green tips are a mush by the time they reach the table? Butts have a strong flavor, too, and if there is any vitamin C in the tops it has cooked away before the butts are done. Don't wait until the stalks spread at the top, either. Cut the young shoots when they are 4 to 6 inches tall. Clean and cook tenderly, then relax and enjoy a most delicious treat.

PEAS—Peas follow hard upon the heels of the asparagus early in June. They are never so good as when eaten raw as you pick them, young and tender, in the garden path. Some of my plantings never get past that stage, though I can sometimes salvage a few handfuls to serve raw in salad. Such peas never find their way to market and none are better than the tiny, smooth-seeded, early varieties like Alaska and the French Petit Pois. Some of the later, commercial varieties have tough skins, are hard to digest at any stage and are comparatively tasteless. Other wrinkled-seeded sorts and the edible-podded are as sweet as the little ones if picked in time. They have more body, are more filling. Quality is recognized at every fair where canned peas are entered in competition. If you look carefully, you will see the blue ribbons on the cans of the youngest, tenderest, greenest peas.

BEANS—Snap beans? Pick them when they are 4 inches long. Sliver and nibble them raw and serve in salad. Slice them once lengthwise and cook gently in just enough water to keep them from burning. Serve in their own liquor for maximum flavor and eat them with a spoon. (Peas, too.)

Lima beans have superb flavor when picked while they are still green, before the pods fill out. Some people feel that they never get their money's worth out of them at that point. Admittedly, they are hard to shell, but you've never tasted limas until you've known them young and tender.

Pick your beans early and often and you will be rewarded with a bonus—the bushes go on bearing crop after crop until frost.

ROOT CROPS—All root crops—radishes, turnips, carrots, beets—are best before they ever attain their largest size. Young and tender, they are all delicious raw and a treat when cooked, gently and just long enough to make them melt in the mouth. Young onions are more delicate in flavor than old ones and far more easily digested.

Salsify and parsnips haven't a chance in the summer garden, but when you go out in earliest spring and pry them out of the half-frozen ground, they can taste like manna from heaven to the winter-weary tongue. After the tops start to grow, however, the roots turn dry and corky and nutrient values go into the seed heads. Young salsify leaves are a pleasant addition to any salad.

POTATOES—Have you ever bored into a potato hill, without disturbing the yet-green tops, and felt around for early tubers about the size of a golf ball? Wash and cook them in their jackets with the last peas of the season or some young beans. Serve with chopped parsley and admit you never had it so good.

LATER GREENS—Lettuce comes in three types, Iceberg for crispness, Looseleaf for vitamin content and Butterhead for delicious flavor. It "shoots" to seed and turns bitter with age. Plant early and often for a constant supply that appeals mostly.

Spinach in its youth is a lively addition to the salad bowl. Later it improves with gentle cooking in a little water. It is not for the table after it has bolted to seed. Celtuce and New Zealand spinach and Swiss chard are reliable cut-and-come-again plants through a long season. The nicest leaves are the middle-sized ones.

Cabbage, if left too long, will crack or split, and it is definitely past its prime when that happens. Savoy is worth growing for its delicate flavor and texture in the raw state. Red cabbage is a rare delicacy when served with a sweet-sour dressing of vinegar, honey and oil. Young cabbage, carrots, onions and turnips cooked together enhance each other's flavor in stew, soup or hash.

Celery is all good at just about every age. The greenest leaves may be tough and bitter but may be used to flavor a multitude of dishes either fresh or dried. The green stalks have the highest vitamin content but the whole plant is rich in iron and even the core of the root is good eating.

Chinese cabbage is prime before the flower stalk develops. Some of the outer leaves are best discarded. Endive and escarole, too. All are at their most delicious combined with grapefruit in the salad bowl.

CORN—Enough has been said about the advisability of getting corn to the table with all speed after picking, but too many gardeners still let it grow until it has passed the milk stage of sweetness and delight. Corn that has reached the dough stage is hard to digest, tasteless and sours when canned.

PEPPERS—Big and green, they are excellent stuffed or raw, but let them turn red for the utmost in flavor—pimento style. Here is an exception to the "pick-early" rule.

EGGPLANT—This is another near exception to the pick-when-young requirement. It changes very little from youth to maturity. Sauté the young fruit, stuff or scallop the older with a chopped onion and a few tomatoes.

TOMATOES—These, thoroughly ripened on the vine until plump and red all over, are altogether different from the same fruit picked with green still showing in the skins. People who buy them in stores never know what they offer to the gardener who grows his own to perfection.

SQUASH-PUMPKINS-CUCUMBERS—Bring them to the table in their tender adolescence. Squash, at its most delicious is Cocozelle, Caserta, no more than 6 inches long. Slice and cook it gently in oil until it is golden brown; add a little water and simmer until done, for a revelation of what is stored in it to tickle your taste buds. Or scoop out the insides and stuff with meat and rice and simmer in tomato sauce until cooked through. If any stuffing is left over, roll it in young grape leaves that have been wilted in boiling water. These have a delicate acid that is an excellent foil for the bland flavor of the squash.

Grate young Alagold pumpkin raw into a salad or into a lime, lemon or orange jelly. Young carrots are grand served in the same way.

Cucumbers, the long-greens are best before the seeds get tough, while they are all green and not too long.

TABLE 40: Harvesting Vegetables

Vegetables	Time of Harvest
Asparagus	Not until third year after planting when spears are 6-10 inches above ground while head is still tight. Harvest only 6 to 8 weeks to allow for sufficient top growth.
Snap beans	Before pods are full size and while seeds are about ¼ developed, or 2 to 3 weeks after first bloom.
Lima beans	When the seeds are green and tender, just before they reach full size and plumpness.
Beets	When 1¼ to 2 inches in diameter.
Broccoli	Before dark green blossom clusters begin to open. Side heads will develop after central head is removed.
Cabbage	When heads are solid and before they split. Splitting can be prevented by cutting or breaking off roots on one side with a spade after a rain.
Carrots	When 1 to 1½ inches in diameter.
Cauliflower	Before heads are ricey, discolored, or blemished. Tie outer leaves above the head when curds are 2 to 3 inches in diameter; heads will be ready in 4 to 12 days after tying.
Sweet corn	When kernels are fully filled out and in the milk stage as determined by the thumbnail test. Use before the kernels get doughy. Silks should be dry and brown, and tips of ears filled tight.
Cucumbers	When fruits are slender and dark green before color becomes lighter. Harvest daily at season's peak. If large cucumbers are allowed to develop and ripen, production will be reduced. For pickles, harvest when fruits have reached the desired size. Pick with a short piece of stem on each fruit.
Eggplant	When fruits are half grown, before color becomes dull.
Kohlrabi	When balls are 2 to 3 inches in diameter.
Muskmelons	When stem easily slips from the fruit, leaving a clean scar.
Onions	For fresh table use, when they are ¼ to 1 inch in diameter. For boiling select when bulbs are about 1½ inches in diameter. For storage, when tops fall over, shrivel at the neck of the bulb, and turn brown. Allow to mature fully but harvest before heavy frost.
Parsnips	Delay harvest until after a sharp frost. Roots may be safely left in ground over winter and used the following spring before growth starts. *They are not poisonous if left in ground over winter.*
Peas	When pods are firm and well-filled, but before the seeds reach their fullest size.
Peppers	When fruits are solid and have almost reached full size. *For red peppers,* allow fruits to become uniformly red.
Potatoes	When tubers are large enough. Tubers continue to grow until vines die. Skin on unripe tubers is thin and easily rubs off. *For storage,* potatoes should be mature and vines dead.
Pumpkins and squash	Summer squash are harvested in early immature stage when skin is soft and before seeds ripen. Winter squash and pumpkin should be well matured on the vine. Skin should be hard and not easily punctured by the thumbnail. Cut fruit off vine with a portion of stem attached. Harvest before heavy frost.
Rutabagas	After exposure to frost but before heavy freeze.
Tomatoes	When fruits are a uniform red, but before they become soft.
Turnips	When 2 to 3 inches in diameter. Larger roots are coarse-textured and bitter.
Watermelon	When the underside of the fruit turns yellow or when snapping the melon with the finger produces a dull, muffled sound instead of a metallic ring.

—Prepared by the University of Minnesota Agricultural Extension Service

OKRA—It turns to wood if it isn't gathered early and often. A nice thickener for soups and stews with other vegetables!

BROCCOLI-CAULIFLOWER-BRUSSELS SPROUTS—Nip them in the bud. Broccoli and purple cauliflower repay early cutting of the main head with numerous side shoots. For the fall garden, after first light frosts, they have no rivals when it comes to bringing the season to a triumphant close.

—*Virginia Conklin*

In harvesting produce from some vegetable plants like peas, beans and cucumbers, take care not to damage plants. Injured plants may dry up partially and stop producing fruit.

According to B. D. Ezell, plant physiologist of the U. S. Department of Agriculture, "maturity is the primary factor affected by the time of harvest and this perhaps is more closely related to quality and production problems such as taste, tenderness, yield, etc., than to nutritional factors. Nutrient content is affected in relation to the amounts of sugar, starch, fiber and to certain vitamins. Young, rapidly growing vegetables are usually richer in vitamin C than the older more mature ones. Other crops such as sweet potatoes may increase in vitamin A values as the season advances.

"High humidity after harvest is important in preserving the appearance as well as vitamin C and probably other vitamins in leafy green vegetables It is of less importance than temperature. Low temperatures are desirable but should not be low enough to cause injury to the product. Low temperatures are also of major importance in retaining quality factors such as taste and appearance."

How Harvesting Can Affect Vitamin Quality

When the housewife goes to market to buy greens for her family she assumes that spinach is spinach regardless of variety, age, conditions of growth and time of harvesting. She takes for granted that a pound of spinach or chard obtainable on one day is equal in quality to a pound on any other day, regardless of time and weather conditions. In the feeding of animals the farmer even assumes that two hours pasturing of his stock in the early morning is the equivalent to two hours in the late afternoon.

Actually these assumptions do not agree with the facts. Definite differences in the amount of starches, sugars, proteins, fats, minerals and of vitamins as well, may be found in plants subjected to different weather conditions, especially at and near the time of harvesting, or even in plants picked at different times of day. Differences, which are particularly noticeable in the leaves, may be found also in plants of different ages.

To increase our knowledge of one of the vitamins, namely vitamin C, studies were made by the United States Public Health Service to determine the effect of age, conditions for growth and time of harvesting upon the quantity of this substance in edible plants. It was assumed that the amount of food in a pound of spinach or peas might depend upon how old the plants were when the vegetables were picked, on the age of the vegetables themselves, on the time of day when they were picked and on whether the weather had been prevailingly cloudy or sunny during their growth, particularly around the time of harvesting. It was found that light has a remarkable effect upon the accumulation of vitamin C. Seedlings sprouted in light contained, after 7 days, more than 4 times as much vitamin C as seedlings of the same age grown in darkness. Plants grown in the greenhouse during May and June in the neighborhood of Washington, D. C., contained twice as much vitamin C as plants grown during December and January. In more northerly latitudes, it might be expected that the differences at the two seasons would be even greater. However, recent tests with tomatoes conducted at the United States Department of Agriculture's Regional Laboratory at Ithaca, New York, yielded differences in vitamin C values in the summer and winter months similar to those which had been found with other types of plants at Washington, D. C.

Fruit from the shaded side of a tree has been shown by other workers to have a lower vitamin C content than that from the sunny side, and even in individual fruits the sunny side has been found to have more than the shaded side. The changes in the amount of vitamin C in a plant under varying conditions of sunlight as compared to shade are noticed first in the leaves, though later differences may be observed in other parts, even in the roots.

Losses of vitamin C at night amounting to as much as 20 per cent of the total quantity, and possibly even more, may occur in some types of plants. Appreciable losses at night occur only when the temperature is high enough to allow growth to take place. Similar losses of the vitamin may occur also during the day but the quantity thus lost is not readily measurable because the vitamin is manufactured more rapidly than it is used. So the net result is an increase in vitamin C. Manufacture at a slow rate occurs at night, but its magnitude is difficult to determine because the vitamin is lost much more quickly that it is made. These facts suggest that the vitamin C is used by the plant in the process of growth. Just what it does with the vitamin is, so far, a secret with the plant. The evidence suggests, however, that it is used for some purpose in the growing regions such as in the tips of the roots and stems and in the development of the young leaves.

As a consequence of its own life processes, therefore, a plant starts the day with a lowered amount of vitamin C. If there then follows a succession of very cloudy days and if the plant is growing rapidly, there tends to be a slow but progressive lowering of the amount of vitamin C. Comparable losses in the sugars and starches of plants under similar conditions have been recognized for a long time. Then comes a bright, sunshiny day. Marked gains in the vitamin are to be observed during the course of the day. Some types of plants may, under these conditions, have more than 25 per cent more vitamin C by late afternoon than at break of day.

An interesting example of this variation in nutritional value of plants as related to time-of-day turned up in an experience in silkworm feeding. In sections of Italy where silkworm production has been an important industry from ancient times, it has been the practice to gather the mulberry leaves, used in feeding the worms, at dusk. These sericulturists have found by experience that leaves gathered at the end of the day tend to yield better results than leaves collected in the morning. Chemical studies of mulberry leaves have revealed why this is true. During the day, under the influence of sunlight, the leaves become enriched in nutritive substances, not only with carbohydrates such as starches and sugars but also with proteins, fats, minerals and, presumably, vitamins, too, since vitamin C, for example, is known to be present in relatively high concentrations in mulberry leaves. Moreover, the protein of young mulberry leaves nearing full size has been found to be superior in quality, quantity and digestibility to that in well-matured leaves.

It seems strange indeed that one should have to turn to the lowly worm for information on the subject of nutrition but actually little is known of the influence of "time-of-day" for collection of food plants, or even of shading, upon their nutritive value to humans and to animals other than silkworms. It is true that variations in protein and non-protein nitrogen have been observed in a number of types of plants harvested in late afternoon and evening in contrast to others collected in the morning. Just as in mulberry leaves, a greater amount of starches and sugars is found in plants kept in sunlight than in those kept in shade, and more also in plants collected in the evening than in those collected in the morning; but nothing was known until recently of the effect of variations in these different conditions on the amounts of any of the vitamins. It remains to be seen whether the amounts of the other vitamins in fruits and vegetables vary as does vitamin C with differences in light intensity, length-of-day and time-of-day for harvesting. It seems probable that if differences occur they won't be so great as those of vitamin C, unless the vitamin in question, like vitamin C, is also used up in the life processes of the plant.

When the time comes to harvest fruits and vegetables, particularly vegetables of the leafy type, due consideration should be given to variations in the amount of light. Present results suggest that for good vitamin C values the harvesting of vegetables should not be done before mid-forenoon, say 10 o'clock, after generally clear weather. It is preferable to harvest, if possible, after a spell of clear weather, or, if it must be done following cloudy days, collection should be made late in the day. Because of the tendency of vegetables, especially those of the leafy type, to lose vitamin C on standing, it would follow that when weather conditions permit, vegetables from the home garden should be freshly picked each day.

—Mary E. Reid,
U. S. Public Health Service

13—Fall and Winter Garden Care

THE ADVENT OF AUTUMN does not mean the end of garden activity for the vegetable grower. Activities include everything from protecting plants from frost damage to fall fertilizing and planting winter cover crops.

When Frost Threatens

You can do something about the weather—especially about the early fall freezes! Watch for that night when the sun goes down in a cloudless sky

—and the wind goes down with it. There is a prickle in the air and radio and television newscasts crackle with warning. Man's ancient enemy, Jack Frost, is on his way—and the product of a summer's labor is threatened with destruction.

Rush ripe, tender crops to shelter where they may be held in storage until they may be dried, frozen, canned or eaten. Such are ripe tomatoes, summer squash, melons, eggplants, cucumbers, peppers, okra.

Where plants are still loaded with immature fruit *cover* them with papers, sacking, blankets. Periods of warm weather after the first freezes will bring many of them to maturity if they are grown under nightly protection but

Large bags are useful in protecting corn plants against damage by frost. Rows of low-growing plants as late lettuce and celery can be protected by drawing the top layer of mulch over them with a rake as a protective cover.

enjoy all the warm sunlight that fall days provide. Rows of late lettuce and celery grown under mulch can be easily covered by drawing the top layer of mulch over them with a rake.

If frost catches you unprepared and the plants are not covered, they may be saved if you will rise before the sun is up the following morning and *water* them with a hose or a watering pot with cold water. Under this treatment the plant cells relax instead of bursting and the plants are good for another span of growth.

Some crops can be rescued by *flooding* with water, as cranberry bogs are flooded every fall.

Orchard crops may be saved by *burning smudge pots or small fires of green wood.* These must be handled carefully, especially in dry weather for fire that gets out of control can do more harm than the frost it combats.

As freeze follows freeze, harvest and store each crop according to its requirements!

Pull *onions,* dry them in the sun and braid the tops so that they may be hung in an airy room until used.

Pull *pepper plants* and *eggplants* by the roots and hang them upside down in a dry attic or cellar.

Green *tomatoes* may be ripened off the vine but they are likely to be very acid. Unless you like them sour, they may not be worth keeping.

Pumpkins and *winter squash* require warm storage for several weeks. Then they can be kept cooler until used.

Celery and *carrots* may be covered with leaves, held down by boards and earth, and dug as wanted until the ground freezes solidly. Just before that happens, dig and store in sand in a cool cellar. *Winter beets* are best boxed in sand somewhat earlier.

Witloof chicory is dug, the old tops cut off just above the crown, and the roots planted close together in a barrel of soil in the cellar. There, new buds will form from February on to provide winter salads.

Move plants of *lettuce, chives, parsley, celery* into the cold frame. With the glass in place and blanketed on very cold nights, they will yield salads and seasoning until Christmas.

Late *melons* may be ripened under gallon jugs with their bottoms removed in the same way that the French and English used their glass *cloches* (bells).

As you survey the devastation of your fall garden take thought as to how you can make it better next year.

Plant hardy, frost-resistant varieties of *vegetables.*

Cabbage, cauliflower, broccoli, Brussels sprouts, collards and *kale* can take a lot of cold and be the better for it. Freezing actually improves their quality and flavor. *Kale, spinach, evergreen bunching onions, parsnips* and *salsify* will survive in the garden all winter and the first 3 send up tender new growth the following spring.

For late season fruits, all too little grown, *quinces* and *persimmons* deserve a place in any well-appointed garden. Persimmons aren't fit to eat until they are well-frozen.

Plan a garden that shrugs off all but the coldest winter weather, one that has *natural defenses against frost.*

Frost strikes on mountain tops and settles in valleys. In between, there is a *thermal belt* of warmer air. A garden situated in such a belt will escape freezing long after those above and below it have been stricken. South, southeast, southwest slopes are best.

Plant near a body of water. Ponds, lakes, streams release heat on cold nights. Watch the steam rise from them and, while it is very hard to measure their effect in British Thermal Units, it can mean just the degree or two of extra warmth that spells safety.

Plant windbreaks. A sheltered field, in late spring and early fall may be as much as 10 degrees warmer than an exposed one. This means that you can grow tender things that otherwise do not survive in your climate belt and even the hardier kinds are the better for not having their roots torn and their tops distorted by the gales.

You cannot lick Jack Frost—but you can keep him at bay for months by taking thought and appropriate action.

—*Virginia Conklin*

Putting a Blanket on Your Garden

Fall is an excellent time to start the year-round mulch garden. Put all available dead leaves on your garden; toss cornstalks on them to keep from blowing. Then get plenty of hay—"spoiled" hay, good hay or salt hay.

An important fall garden activity is gathering materials for the compost pile. Leaves, rich in many plant nutrients, are being placed in a circular bin made of woven wire fence, and will provide humus for the following year's garden.

Decide where your tomatoes will be next year and put cornstalks, cabbage roots, etc. over that area. Shredding the cornstalks will prevent the corn borer from laying eggs in the "joints" of the stalk.

Now spread hay thickly over this refuse. When you plant tomatoes (late May or early June here in Connecticut) your ground will be soft, moist and weedless. Don't be afraid of weed seeds; with a thick mulch they never get a break.

Cover your asparagus with eight inches of hay. Next spring it will come up through the mulch.

The rest of the garden should be more lightly mulched, because next spring, instead of plowing, you will simply pull aside the hay and plant. Therefore, put on your heavy mulch after the seeds have sprouted.

This includes corn. You can save work next spring by marking the corn rows in the fall. Put cottonseed or soybean meal along each row for nitrogen; a little lime if necessary. Mark the rows with a firm stake at each end, leaving only an inch above the ground. If they are taller they may get broken. In the spring you need only rake back the mulch and plant.

When garden crops have been harvested, the foliage can be taken to the compost pile. Above snap beans are gathered after they have finished bearing; their foliage makes excellent material for composting since it is so rich in nitrogen.

Cover Crops Help Vegetable Gardens

Planting a winter cover crop in your vegetable garden in fall can save time, hard work and added expense the following spring, says William M. Brooks, Ohio State University extension horticulturist.

A winter cover crop will greatly reduce the erosion of the soil during the winter and will add organic matter to the soil when it is spaded or plowed under next spring, Brooks points out.

Cover crops should be planted between August 1 and November 1, Brooks says. The time of planting will depend on the vegetables being grown, and the cover crop planted.

Rye grass can be seeded at the time of last cultivation in such vegetables as sweet corn, peppers and eggplant. Rye grass becomes established easily. Rye grass should not be planted after September 15th. Planting after this date results in very limited growth and does not add much organic matter to the soil. Rye grass should be planted at the rate of one to two pounds for 1,000 square feet.

Rye also can be used as a winter cover crop. It should be planted between September 1 and November, at the rate of 3 to 4 pounds per 1,000 square feet.

Winter barley can be used as a winter cover crop in southern Ohio. This should be seeded at the rate of 2½ to 3½ pounds per 1,000 square feet.

Most grass seeds are small and should be planted one-fourth to one-half inch deep. A firm seedbed is necessary when making a summer or fall seeding. For additional cover crops, see the chapter on improving soil structure.

Vegetables for Over-Wintering

The vegetables which are to be left in the garden soil over winter should be mulched to prevent excessive freezing of the soil. With proper mulching, almost any of the fleshy-rooted vegetables may be over-wintered right in the garden. Here is a list of these vegetables:

ASPARAGUS

Asparagus is a wholesome, early spring vegetable. It seems to do best in locations having winters sufficiently cold to freeze the ground to a depth of at least a few inches. The latitude of southern Georgia is about the southern limit of its profitable culture. The plants should be cut off at the ground level in the fall and the tops allowed to remain on the soil as a winter mulch. Since asparagus plants are heavy feeders it is well to apply a liberal amount of compost over the soil and work it in lightly before laying down the winter mulch. It may be necessary to remineralize the soil by adding phosphate rock and potash rock at the rate of ½ pound of each for each square foot of soil. The soil need not be mineralized more often than once every 5 years. If the plants show signs of rust or have been attacked by the asparagus beetle, the tops should be shredded and put into the compost heap and some other plant materials should be used as a winter mulch over the asparagus bed.

HORSERADISH

Horseradish is most commonly grown in some out-of-the-way place in the garden. It is perfectly hardy and is well adapted for growing in the north temperate regions of the United States. It requires no particular winter care. It might be well to stake the plants so that roots can be dug in early spring before new growth begins.

DANDELION

The dandelion is hardy and adapted for growing in almost any garden soil. It is best grown in beds with the plants in rows about 18 inches apart. In colder parts of the country it is desirable to mulch slightly during the winter using leaves, straw, strawy manure or any mulching material which will not pack. Before putting down the winter mulch, apply a thin layer of compost between the rows and work it into the soil lightly. Early the following spring the plants will be ready for use as greens. The leaves may be blanched, if desired, by covering the plants with paper or by placing two boards set in the form of the letter "A" over the row.

KALE

Kale, or borecole, is hardy and lives over winter in latitudes as far north as Pennsylvania and in other locations where similar winter conditions prevail. It is an all-year plant since it is also resistant to heat and may be grown in summer, but its real merit is as a cool-weather green. In northern regions where it lives over winter the last sowing should be about 6 weeks before frost in order that the plants may become well established. It may well follow green beans, potatoes, peas or some other vegetable that has occupied the soil during the summer and early fall. A light winter mulch applied after the ground is frozen will help keep the plants vigorous and insure a good growth in early spring.

SPINACH

Spinach is by far the most popular of the greens. It is a reasonably hardy cool-weather plant that withstands winter conditions throughout most portions of the South. In colder portions of the Southern States it may need some protection during the winter, for, like cabbage and other hardy crops, it is sometimes severely injured or even killed by low temperatures. In most portions of the North spinach is primarily a late fall or early spring crop, but in localities where summer temperatures are mild it may be grown continuously from early spring until late fall. Spinach responds well to compost. A pound of compost mixed with a half pound each of phosphate rock and potash rock makes an excellent fertilizer for the spinach plants. This mixture should be raked into the soil lightly. With some protection, as cold frames, spinach can be grown in winter even in the North. Also, cloches are ideal for keeping spinach and other greens in production throughout all or most of the winter.

CORN SALAD

Corn salad is also known as lamb's lettuce and fetticus. It is a good late winter or early spring vegetable. For an early crop the seed should be planted during the autumn and the plants covered lightly with a winter mulch. In the Southern States the winter mulch is not necessary and the plants are ready for use in February and March. Corn salad is used as a salad in place of lettuce or mixed with lettuce or water cress.

ENDIVE

Endive is another all-year plant as it is a cool-weather plant and is not sensitive to heat. In the South, it is mainly a winter crop. In the North, it is grown during the spring, summer and autumn and is also forced during the winter. Endive is often known on the markets as escarole. For winter use in the North, plants should be removed from the garden with a ball of earth, placed in a cellar or cold frame, where they will not freeze, and tied and blanched as needed.

LETTUCE

Lettuce should be grown in every home garden. It is a cool-weather crop, being as sensitive to heat as any vegetable grown. In the South, lettuce culture is confined to late fall, winter and spring. It thrives in any rich soil, but will not tolerate acid conditions. In preparing the soil, add lime if necessary and a mixture of compost and phosphate rock so that each square foot of soil gets one pound of compost and a half pound of phosphate rock. This mixture should be raked into the soil lightly. In no case should fresh manure be used. Most commercial varieties do well in the South in winter. For northern gardens, sow a variety of winter lettuce which is highly resistant to low temperatures.

PARSLEY

Another winter vegetable is parsley which is hardy to cold but sensitive to heat. If given a little protection it may be carried over winter throughout most of the North. After the plants are established, they thrive in almost any kind of soil. Since the plants are delicate in their early stages of growth, the soil should be mellow and free from clods and stones. Enrich it with compost to which phosphate rock and potash rock have been added. The seeds germinate slowly, but organic matter in the soil will hasten germination greatly.

WATER CRESS

Water cress is one of the few vegetables that can be grown in wet surroundings. In the more moderate portions of the North it grows practically the year round and winter is its best season in the South. It grows best in water from springs in limestone regions. A supply for an entire family may be grown in a small spring-fed brook or a series of shallow pools where the water is about one foot deep. Care must be taken that the water is clean; otherwise the cress is not fit for use. Water cress may be started from seed or from pieces of plants. It pays to prepare the soil by enriching it with a mixture of compost to which has been added pulverized lime, phosphate rock and potash rock. Early spring is the best time to start a water cress planting.

JERUSALEM ARTICHOKE

This plant is really a sunflower, and may become a troublesome weed if care is not taken in its culture. It can be grown in practically all parts of the United States. It is started from pieces of tubers in the same way as potatoes. The tubers are not injured by freezing and may remain in the ground over winter.

SHALLOT

The shallot is a small onion of the Multiplier type. In its requirements it is similar to other onions; and as the bulbs have a more delicate flavor than most onions, the plant is well worth attention in the garden. They seldom form seed and are propagated by means of the small cloves or divisions into which the plant splits during growth. The plant is hardy and may be left in the ground from year to year, but best results are to be had by replanting the smaller ones at the desired time.

—*William H. Eyster, Ph.D.*

Clean Tools Before Storing

Gather all the rakes, hoes, spades and other small tools which you won't be using until spring. Clean them well and paint with oil. Then hang them on the wall. If you have no regular hanging place for small tools, you can make one easily, using peg-board available at most hardware stores.

The care of power tools necessitates a little more work, but it will be rewarded with longer life and better operation. Follow these steps in winterizing power reel and rotary mowers, tillers and shredders:

1. Check the manufacturers' booklets which came with the machines, as they will give very specific directions.
2. Disconnect the lead-in wire to the spark plug to prevent dangerous accidental starting.
3. Drain the gas tank completely.
4. Drain the dirty oil.
5. Clean off all matted clippings, mud and other material from all parts of the machine. Use a wire brush, especially underneath rotary mowers. Paint the underside of the rotary mower with oil.
6. Using a kerosene soaked rag, wipe off excess grease and oil from drive chain, flywheel, axle, wheels, etc.
7. Remove air cleaner and wash upper part with kerosene.
8. Put a few drops of light machine oil on all drive chains.
9. Using a grease gun, grease all lubrication points on the power appliance with heavy lubricating grease.
10. Tighten all nuts, bolts and screws.
11. Clean off rust spots with emery cloth and touch up with good quality paint.
12. Sharpen and clean cutting edges of attachments.
13. Check rubber tires on power units for cuts, nicks and proper air pressure.
14. Refill crankcase with clean oil.

15. Remove spark plug. On your next visit to the gas station, have it tested, cleaned and checked for gap size. Then replace it.

Final Check List at Garden-Closing Time

In late autumn, most gardeners, having harvested their crops, have entirely forgotten about their gardens, which will be something for them to think about again when the seed catalogs roll in next February. Actually, the good gardener should be devoting a lot of time to his garden in autumn.

If you don't believe it, turn to Nature. What is she doing in *her* garden? She is putting it to bed for the winter, blanketing it against the cold and the icy winds. If she has a garden in an open field, she lays down the grasses until they form a deep, matted mass, and she piles the dead stems of the higher plants on top to keep the grasses down. In the forest she gently covers the floor with another layer of leaves and probably drops twigs and limbs on top; for forest trees are self-pruning. Look where you will, you will find that Nature is scrupulous in blanketing the earth against the rigors of the wintertime.

The best time to start making a protective winter covering for your garden is long before the garden year ends. Accumulate all the organic matter that you can find, from residue crop materials, manures, cut weeds, or outside products. Then lay down a thick mulch over the area to be planted to crops next year. All plant residues should be strewn over your garden for winter cover unless they are seriously diseased. Otherwise, gather up the infected plant residues, compost them, scatter the compost on the soil and rake it in.

If you would like to grow a soil-conserving cover crop, cultivate the ground in the rows or rake it over and sow rye grass or some other quick-growing sturdy crop. You can do this between rows even earlier. You can do it at the last cultivation of your corn. Little by little, you can add to the protective carpet, for that is exactly what the rye grass will provide. At the same time, it will be making masses of roots that add humus to the ground.

Be sure to cut all weeds before they go to seed. The old adage says that, "One year's seeding makes 7 years' weeding." Use the cut weeds as mulch. Add any organic matter that will rot—the trimmings of your celery, the outer leaves of your cabbage, the clippings from your lawn, plant stems from your flower beds, the leaves from your trees and so on, provided, of course, that the materials used are not diseased. Corn stalks cut into pieces are excellent to hold down the leaves. Or a little earth can be scattered over them.

Don't ever burn tree leaves. There is nothing you can use for mulch that may be as useful as tree leaves are. Suppose you use carrot tops or bean plants or beet leaves for mulch. How deeply do those plants forage? A few inches only. Whatever plant foods they secure must necessarily come from the top few inches of soil. And in those top few inches much of the original supply of plant food may be exhausted, leached out, washed or blown away. But the tree leaves contain minerals that the roots brought up from deep down in the earth. Even though some of the minerals may be practically gone from the top 10 inches of soil, there may be abundant supplies of these minerals 6 feet below the surface of the ground. The tree roots will bring up some of them. The leaves will contain a part of what is brought up. Spread on your garden and eventually incorporated in your soil, these leaves will help replenish the dwindling supplies of minerals. So you see why tree leaves are especially valuable in your winter mulch or in compost.

Perhaps you cannot collect enough organic material to cover *all* of your garden thoroughly. In that case, you can mulch the areas which you want to be especially rich for your next summer's crops. This naturally suggests that you should plan your next year's garden *this fall.* Then you will know where you will plant this or that vegetable next spring. Beans may grow in very poor soil, but not all crops will. Muskmelons need a shovelful of good rotted manure under each hill. Perhaps you can't get manure. It becomes scarcer each year. But mulch or compost may do the trick for you.

The gardener cannot put back into his garden all that came out of it, because he has to eat the potatoes and tomatoes and beans and corn, etc. But he can return the *equivalent*—and more. His grass clippings may make up for what is lost in the bean crop; his weeds may offset loss through carrot culture, and so on. If the gardener will put back into his soil, by composting, mulching, etc., all that it is possible for him to return to the ground, he may largely or wholly offset the annual loss of plant food. He can also add manure, tree leaves and a host of other enriching products.

From time to time, he may need to sprinkle a little lime on his garden. Yet he should be careful where he puts it. Some plants do not thrive in soil that has been limed. He may need to apply ground phosphate rock, potash rock and ground limestone instead of only lime.

Although your soil is the primary interest in caring for your garden in the autumn, there are many adjuncts to your gardening that also need to be considered now. Tools should be carefully cleaned, oiled, and put away. All broken or damaged implements, etc., should be mended or replaced. Wire trellises should be rolled up and stored in a dry place. Your complement of bean poles, tomato stakes and other similar pieces of equipment should be made ready for spring. For when the spring rush comes, you will find it difficult to do repair jobs. Digging and planting and cultivating will keep you more than busy. If there is anyone who needs to take time by the forelock, it is the gardener. And the best time to do it is in the autumn.

—Lewis E. Theiss

14—Storing the Surplus

THE GARDENER WHO does not find himself, at times, with more vegetables on his hands than his family can consume is a rarity. For all gardeners, it is good economics to preserve such excess for winter use, but for the organic gardener it is especially advantageous to do so, firstly because his vegetables are likely to be better keepers than those grown with commercial fertilizers, secondly because they will be of far better quality and flavor than those bought at the store. On all counts, therefore, he should give serious thought to the problem of winter storage.

Vegetables can be roughly divided into 3 groups:

1. *The quickly perishable* such as green peas, green limas, corn and asparagus and green vegetables such as spinach, chard and lettuce.

2. *The slowly perishable,* in which we include broccoli, cauliflower, late cabbage and onions.

3. *The good keepers,* consisting of potatoes, turnips, beets, carrots and similar root crops, together with pumpkins, late squash and celery.

The 3 groups are not very sharply divided. Under proper conditions, late cabbage and onions will keep for a long time, but with them there is a greater danger of spoilage than with most root crops. Cauliflower and broccoli seldom remain in good condition for more than 2 or 3 weeks, so they almost belong in the quickly perishable group.

Given the proper facilities, the carrying of the good keepers through the winter is a simpler process than either canning or freezing, yet I suppose there are 10 households that freeze or can to one that stores unprocessed roots or other crops. It is common enough to store potatoes but few back yard gardeners seem to provide themselves with a winter supply of turnips, rutabagas, beets, carrots, onions and celery and even fewer trouble to put away pumpkins, squash, late cabbage or sweet potatoes.

Two obstacles seem to account for this: in the first place there is a good deal of uncertainty regarding the conditions needed; in the second, there is a feeling that such conditions are difficult to provide. Let's remove the uncertainty and learn the simple methods of providing the necessary conditions.

—*Leonard Wickenden*

Easiest of All Methods

According to members of Michigan State's Department of Horticulture, storing vegetables is perhaps the easiest and least expensive of all methods of food preservation.

Although each family's needs vary, the amounts listed below are about those needed by the average family of 5 people:

Beets	½ to 1 bushel	Onions	1 to 2 bushels
Carrots	2 to 3 bushels	Cabbage	25 to 35 heads
Turnips and rutabagas	1 to 2 bushels	Squash and pumpkins	20 to 25 fruits
Salsify and parsnips	1 to 2 bushels	Dry beans (navy, lima, soy)	8 to 12 quarts
Potatoes	12 to 20 bushels	Tomatoes (green—mature)	1 to 2 bushels
Celery, Chinese cabbage, Brussels sprouts	Enough for a short time		

The amount of each kind of vegetable to store will depend on your family's tastes and on the amount of canning or freezing you do. If you can carrots and beets, you won't need to store as many of them. By having plenty on hand, however, you can encourage your family to eat more of the healthful, nutritious foods listed. It is better to store too much than too little.

Carrots, beets, parsnips, salsify, rutabagas and turnips must be kept in cool and moist air. A humidity of 90 to 95 per cent is best. Naturally, this high humidity is very hard to achieve in an open basement storage room. Therefore, these vegetables are usually stored in moist sand or leaves so that the humidity can be kept high. If you store carrots in the basement, you can pack them in cans or similar containers with leaves or sand, to keep a high humidity.

TABLE 41: Proper Storage Conditions for Home Garden Vegetables

Cool and moist 32° to 40° F. Humidity 90 to 95%	Cool — moderately moist, 32° to 40° F. Humidity 80 to 90%	Dry and cool (See text)	Dry and warm (See text)
Carrots	Potatoes	Onions	Pumpkins
Beets	Cabbage	Beans	Squash
Parsnips	Cauliflower	Peas	
Salsify	Chinese cabbage	Soybeans	
Rutabagas			
Turnips			
Celery			

Parsnips and salsify are often left in the ground over winter. With a mulch of straw or leaves over them, they will keep very well. However, it is better to dig at least enough for use during the severe part of the winter and to store them where they are easy to get.

The second column of Table 41 lists vegetables that need to be kept in cool and moderately moist air. A humidity of 80 to 90 per cent is suggested. These need not be stored in sand, since you can keep this humidity in the average basement storage room.

Cabbage will keep better if you pull it up by the roots (rather than cut it), then replant the roots in sand. Cabbage stored in the basement tends to "scent up" the house, so you may prefer to store it outdoors by one of the methods suggested later.

The third column lists vegetables that need a cool, dry storage. Do not pull onions until the tops have dried. Then spread them out in a well-ventilated place to dry for a week or 10 days before placing them in storage. A moist root cellar or basement storage room is not a good place to store onions. The attic or a cold, dry room in the basement is best. Do not let them freeze. You can store them in slatted crates, coarse mesh bags, or on shelves in thin layers. Do not place them in deep layers or closed containers, though, or they may heat up and spoil.

Store dry beans, soybeans and peas in closed containers, such as glass jars.

The last column lists vegetables that need to be stored in a dry, warm place. The main ones are pumpkins and squash. Store them at about 40° to 50° F. in a dry room. They keep best if placed on shelves so that they do not touch each other.

You can keep green, mature tomatoes from 6 to 8 weeks if you pick perfect, mature, green fruits before severe frost, then store them at about 55° to 65° F. on shelves or in shallow trays. They will ripen over a period of about two months. Cauliflower and Brussels sprouts, like cabbage, should be pulled and replanted. Store kohlrabi and winter radishes at about 35° F. in moist sand or covered boxes.

A Storage Room Inside the Basement

With a little effort you can build a basement food-storage room that will pay for itself in one year in pared-down food bills if you apply basic storage principles, indicated below.

Essentially the storage room is a place where temperature and humidity are held to the proper level for keeping fruits and produce. This means lower-than-usual household or basement temperatures, ranging from 30 to 40 degrees. After deciding on the size and location where low temperatures can be most easily maintained, the area must be insulated from temperatures prevailing in the rest of the house.

A great many vegetables can be stored in a dry, cool basement. Storage areas should be away from heating pipes; an 8 by 10 foot area is suggested for most families who plan to store vegetables and other foods in the same room.

At least one wall having outside exposure should be used, preferably the one with the least sunlight, on the north side and with a window that is easily reached. The other walls can be made of wood which will do a good job of keeping the storage area cool—providing their construction is tight.

Two types of insulating material can be used: board or loose fill. It is important to keep them dry or their insulating properties will be reduced. So moisture-vapor barriers are used, inside and out, such as damp-proof paper, tar, asphalt.

Board insulation can be nailed to the walls and ceiling. Two thicknesses should be used to prevent leakage through joints.

Loose-fill insulation includes planer shavings, cork dust and minerals. Stagger the studs so they are exposed on one surface only. Sheathing and damp-proofing should be done when the wall space is being filled.

Planer shavings are very satisfactory insulation for side walls and ceiling; they are dry and do not tend to settle. Add hydrated lime to the shavings, from 20 to 40 pounds per cubic yard, to help keep the shavings dry and as a repellent to vermin and rodents. Tamp the shavings as they are filled between studs until a density of 7 to 9 pounds per cubic foot is obtained.

Loose-fill insulation is not used where floor insulating is needed. Use board-type insulation—wood-fiber insulating wall-board to floor insulation. (The floor to side wall jointing is the same as for wall corners.) The insulating board is then mopped with hot tar, and cement or other flooring is laid on top. Where chinks occur at rough wall or floor surfaces, caulking compound, which fills the space and protects the exposed insulating material from dampness, is used.

Temperature of the surrounding basement will determine thickness of the walls. A prevailing temperature of 60 degrees outside the storage room calls for the following thickness of insulating materials:

3 inches: wood-fiber insulating board, cork board, granular cork, fibrous rock, rock wool.

4 to 5 inches: planer shavings.

6 inches: compressed peat.

8 to 9 inches: white pine and soft woods.

Lower basement temperatures will require less insulation.

The walls to close off a corner of the basement are easily constructed with a 2-by-4-inch framework, sheathing on both sides and 3-inch insulation batts between the studding. Leave an opening in one wall for a door. This may be framed with two-by-two-inch studs, faced on each side with quarter-inch plywood, and the center filled with insulation. Fit the door tightly and secure it with a type of latch that holds it firmly closed.

To insulate the ceiling of the storage space, sheathe underneath the ceiling joists and apply 4 inches or more of insulation between the joists, extending the insulation out over the walls of the storage space.

A simple method of cooling the storage is to open the window on cool nights. Light must be excluded from storaged vegetables and fruits, so cover the window with a board. Also, wide, wooden louvres fitted to the outside of the window frame aid in excluding light if the window is opened for ventilation in the daytime. Cover the louvres with copper screening to keep out insects and rodents.

A storage space measuring 5 by 8 feet will hold about 30 bushels. Shelves and bins should be arranged so as to use the space to best advantage, and should be of slatted construction to permit good air circulation. Bottom shelf or bin should be at least 4 inches above floor level, to permit pouring water on floor to maintain humidity.

Organic gardener Harold Lefever of Spring Grove, Pennsylvania, is shown in one of the root cellars on his farm. The Lefever's keep much of their home-grown products in these storage cellars during the winter months.

The two most important points in operating the storage are temperature control and high humidity. Aim at cooling the storage by ventilation to a 35-degree temperature as soon as possible and maintain this during storage period. Hang one reliable thermometer inside and another outside the window for guidance. In early fall, when the nights are cool and the days warm, open ventilators only at night; as weather becomes colder, ventilators may be partially opened for continuous periods.

In cold weather the storage may operate at desired temperature with little attention; in very cold weather it may be necessary to open the storage door slightly for some warm air to enter from basement.

—*Earl W. Gage*

Making an Outside Vegetable Cellar

This story of a Pennsylvania organic gardener, Lewis E. Theiss, illustrates the steps and reasons for an outside storage place:

"We decided to build an outside vegetable cellar because the heat from the basement furnace seemed unbeatable.

"We were guided to this decision by an outside porch, 6 feet wide by 12 long, that had been added to the original house. While the upstairs kitchen door entered into this pantry, it had no cellar beneath it but stood on blocks of stone. After seriously considering all the factors, we decided to dig under it.

"Since the ground was firm and well-compacted, it never caved in and in fact permitted us to dig smooth vertical sides to our new storage room. We built it to fit exactly under the pantry and used the porch floor as a roof, making use of the basement wall on the long side. Eventually we made a small opening in it to permit warm air to enter our storage room when it was needed.

"Building the cellar was purely a spare-moment effort, done after work whenever we had some extra time. Although we completed our excavating in short order without spending a penny on it, we dug it deep enough to give us ample head-room, leaving a sloping excavation for the stairway.

"Next we bought cement, sand and enough foot-wide boards to make forms for pouring our walls. We made an extra-rich concrete to insure smooth walls and did not use stones for ballast.

"As we had to work at the job a little at a time, we mixed our concrete a little at a time. Whenever we had to discontinue our work, we made sure to wet the concrete already laid, when we resumed, to insure proper adhesion of the different layers. It did not take very long to get the walls finished. Constructing the steps was a bit more of a job. We first made the side walls for the steps, then fashioned the little forms for the steps themselves, one at a time.

"When the concrete side walls for the steps reached the desired height, somewhat above the level of the ground, we bedded some sturdy 2-by-4's in the wet concrete to form a frame or embrasure for the outside ground doors, or door. This we made double, so that we could lift up one half and still get down to the cellar. Or we could open both halves so as to have the entire width of the stairway when we wished to take boxes down. To some slight extent this double door arrangement helped to keep the storage room slightly warmer, as less cold air rushed down when only half of the door was opened. We made a second door at the foot of the stairs.

"When all the concrete work was hardened, we took down the forms, smoothed the wall with square-ended masons' trowels, then cleaned the boards as well as we could, and used them clean side up, to make shelves in the new cellar. The bottom shelf was set some two feet above the floor which was also concrete. Above this bottom shelf we set a second shelf something less than two feet higher. These shelves, lining one side of the cellar, gave us 36 linear feet of storage space, including the containers set on the floor. Anticipating that water might seep in after heavy rains, we laid strips of inch boards on the floor to put containers on.

"Here we were in luck. We had put a tin roof on our barn, using tin shingles. These came in boxes that were approximately 12 by 18 inches and perhaps 10 inches deep. These boxes made excellent receptacles for vegetables that were to be stored. We could put them crosswise on our shelves, letting the ends stick out and using approximately 36 of these boxes.

"We never got around to keeping records of the temperatures prevailing in our vegetable cellar. In the coldest weather we opened the door as little as possible and when the outdoor thermometer went to zero, we opened the small hole we had made in the wall between the house basement and our cellar. We have never had anything freeze there, over the years my wife has stored her fruits and vegetables within its walls."

Consider Tile Storage

A tile storage may be just what home gardeners, who produced more fruit and vegetables than their families could eat this summer, need.

W. B. Ward, Purdue University extension horticulturist, says a tile buried upright in the soil is an economical and practical storage for many fruits and vegetables that keep best under cool and moist conditions.

Three bushel baskets of fruit or vegetables can be stored in a tile that is 18 inches in diameter and 30 inches high. A tile 24 by 24 inches will store only two bushel baskets, but about 6½ bushels in bulk.

Ward says that the more porous field tile makes the best storage, although the hard-burned, vitrified or concrete tile may be used.

Best location for the tile storage is a well drained area where a basement can be excavated. The tile should be located away from possible overflow water from down spouts or eaves and where the area will be shaded in summer and winter.

Barrel Root Cellar

A root cellar which is very satisfactory for the home garden is easily made from a large barrel. A strong, well-made one will last many years. Dig a trench a little larger than the barrel so that it can be set into the ground at an angle of about 45 degrees. Before placing the barrel in the trench, drop in a few large stones or bricks to facilitate drainage. Cover the barrel all around with about 6 inches of dirt, then 6 inches of straw or leaves and finally with about two inches of dirt to hold the organic matter in place. After the vegetables are packed in the barrel, place the cover over the top and then pile about a foot of straw, leaves or hay on the cover with a board and a rock to keep the covering in place. This covering is easily removed when it is desired to get into the root cellar.

Surface Piles and Trenches

According to Cornell University experiences, the simplest of the outdoor storages are made on the surface of the ground. A well-drained and convenient spot should be selected. The area to be covered by the stored crop should be covered to a depth of 4 to 8 inches with straw or dry leaves. The vegetables are piled in a cone on this bed. They should be covered with 8 to 12 inches of material similar to that used for the bed. Over this covering, 4 to 6 inches of earth, taken from around the pile, may be placed. Earth to cover the crops is best taken from a shallow trench dug around the pile and at least no closer than 18 inches from the edge of the pile.

Unless the earth is quite permeable, a shallow outlet trench should be dug from the lowest side of this circular trench to insure good drainage.

After the ground is slightly, but not deeply, frozen, a second layer of straw, dry leaves or damaged hay and 8 to 10 inches more of earth are added. The trenches should be widened rather than deepened to get this extra earth.

Crops should not be taken from such a pile when the soil is deeply frozen. If only part of the crops are removed, care must be taken to pack the straw and other litter well into the opening and to cover it carefully with earth.

Trenches are frequently used for the storage of celery, cabbage, Chinese cabbage and root crops. They are dug deep enough so the celery in a standing position does not come to the surface of the ground. Trenches for celery or Chinese cabbage are best when not more than 24 inches wide, with nearly vertical sides to give maximum frost protection by the earth. After the produce is in the trench, slats such as 2 by 4's are laid across the trench to support a tight covering of boards. A strip of roofing paper over the boards helps to keep out water and dirt. As danger of freezing approaches, the boards are covered with straw, hay or leaves and some dirt to hold it in place. Many keep celery this way for use during the Christmas season, the Cornell report points out.

Drying Vegetables

General instructions for drying vegetables, according to Cornell extension workers is to wash, trim and prepare as for serving in slices or strips approximately 1/8 inch thick or in cubes about 3/8 inch on each side. All except onions and herbs, which are used primarily for seasoning, should be blanched or precooked. Blanching may be done either with steam or with hot water.

Blanching with steam saves more nutritive value than does blanching with hot water and in commercial practice has given the best-quality product. A heavy kettle with a close-fitting lid is needed. The vegetable to be blanched is arranged *loosely* in cheesecloth or in a wire basket on a rack above one or two inches of rapidly boiling water. The lid is put on the kettle and a *constant active head of steam* is maintained during the entire blanching period. If food is packed tightly or if the active head of steam is not maintained, uneven or inadequate blanching will result. This causes serious flavor and color losses, as well as some vitamin destruction.

Blanching with boiling water requires a large kettle containing a gallon of boiling water for each pound of vegetable to be blanched at one time. (Two gallons for each pound of leafy vegetable.) The vegetable is put loosely into a cheesecloth or into a wire basket and immersed in the *boiling* water for the required length of time. If the cloth or basket is gently lifted up and down in the water, blanching probably is more even. Water blanching takes approximately two-thirds as long as does steam blanching.

Although steam blanching conserves quality and food value, some persons contend that water blanching is more practical in the home, as there is less danger of its being uneven or inadequate.

Review of General Rules for Storing Vegetables

Store only sound vegetables of good quality. Diseased or injured ones may be used in early fall or preserved in some other way. Harvesting, in most cases,

TABLE 42: Guide for Drying Vegetables

Food	Preparation	Treatment before drying*	Maximum tray loading per square foot	Maximum drying temperature (Fahrenheit) *Degrees*	Characteristics when dried
Beans, lima	Shell	Steam for from 8 to 10 minutes	1 pound	150	Hard, wrinkled
Beans, snap	Wash and cut in 1 inch lengths or Shred French style (cut diagonally or lengthwise)	Steam for from 8 to 10 minutes if in 1-inch lengths or Steam for from 3 to 5 minutes if French style	1 pound	155	Brittle, greenish black
Beets	Wash	Cook until done, for from 45 minutes to 2 hours, depending on size. Slip off the skins and cut the beets in ⅛-inch slices or ⅜-inch cubes	1.5 pounds	150	Brittle
Carrots	Wash, peel or scrape. Cut in slices either lengthwise or crosswise, not more than ⅛ inch thick	Steam for from 5 to 7 minutes	1.5 pounds	160	Brittle
Corn, sweet	Husk	Precook in steam for from 15 to 20 minutes, then cut the corn from the cob	1.5 pounds	150	Dry and brittle
Herbs and celery leaves	Wash, trim and drain	None or steam for from ½ to 1 minute	0.5 pound	150	Crisp
Onions, garlic	Trim, then slice or shred	None or steam for from 1 to 2 minutes	1 pound	160	Brittle, light color
Peas, green	Shell and discard any starchy peas	Steam for from 3 to 5 minutes	1 pound	150	Hard, wrinkled, green
Peppers	Wash, remove the seeds, shred if desired	None or steam for from 1 to 2 minutes	1 pound	160	Leathery to brittle
Pumpkin and winter squash	Cut, remove the seeds, cut in 1- to 2-inch strips, then in slices not more than ¼ inch thick	Steam for from 5 to 7 minutes	1.5 pounds	160	Tough to brittle
Soybeans, green	Blanch the pods in steam for from 5 to 7 minutes, shell	Blanching before shelling is sufficient	1 pound	150	Hard and wrinkled
Spinach and other greens	Trim, discarding coarse stems and ribs, wash and drain	Steam for from 2 to 5 minutes. Pile the leaves loosely in a basket so that steam can reach the center of the mass immediately	0.5 pound	150	Crisp
Turnips and rutabagas	Wash, trim, and slice about ⅛ to ¼ inch thick	Steam for from 6 to 10 minutes	1 pound	150	Brittle

* Blanch vegetables in hot water two-thirds as long as in steam.

—*Cornell Extension Service*

should be delayed as long as possible without danger of freezing. All vegetables to be stored must be handled with great care to avoid cuts and bruises. Carefully remove any excess soil from beets, carrots, celeriac, parsnips, rutabagas, salsify, sweet potatoes, turnips and winter radishes. This may be done either by light rubbing with a soft cloth or glove or by careful washing.

Care must be taken in washing vegetables to avoid injury; let excess water evaporate before the vegetables are stored. Remove tops from root vegetables to within half an inch or so of the crown. Both tops and tap root are commonly removed from rutabagas; tap roots need not be removed from other root vegetables.

Proper temperature is a most important factor. With few exceptions, the most desirable temperature is at or very near 32 degrees—the freezing point of water. Except for potatoes, vegetables are not injured at this temperature. It is difficult, however, to keep the temperature as low as 32 degrees without danger of it going low enough to cause actual freezing during exceedingly cold weather. It is suggested, therefore, that the storage room temperature be kept between *35 and 40 degrees.* Such temperatures cannot be reached and kept except in a room separated from the rest of the basement, reasonably well-insulated and having adequate ventilation.

The size of the basement storage room will vary with the space available and the family needs; 8 by 10 feet is suggested for most families who plan to store both vegetables and other foods in the same room. A storage room, if properly constructed and managed, will be suitable for nearly all foods commonly preserved. Where practical, the storage room should be located either in the northeast or northwest corner of the basement and away from the chimney and heating pipes.

Ability to maintain a desirable temperature range of *35 to 40 degrees* in the basement storage room depends largely upon outside conditions. Day temperatures during both early fall and late spring are likely to be higher than that desired in the storage room. During these critical periods it is essential to close the ventilating system whenever the outside temperature is higher than that in the storage room. When the desired temperature is reached, both timely adjustment of the ventilating system and adequate insulation of the storage room are necessary to keep the desired range. As a general rule, ventilators should be open whenever the outside temperature is lower than that in the storage room and the inside temperature is above 40 degrees. When the temperature in the storage room falls to 35 degrees, the ventilators should be closed. Storage room windows should be adequately screened at all times.

Most vegetables shrivel rapidly unless stored in a moist atmosphere. Shriveling may be prevented: 1. By keeping the air quite moist throughout the storage room. 2. By protecting the vegetables either by wrapping them or putting them in closed containers. 3. By adding moisture directly to the vegetables now and then. If the first method is used, the storage room should have a dirt floor so it can be kept moist by occasional sprinkling; concrete floors may be covered with 4 to 6 inches of soil or sand to help hold moisture. Only those vegetables which require moist conditions may be stored successfully in a room of this type. Vegetables like onions, pumpkins and squash must be stored in some other location where the air is dryer. Moist storage rooms are not well suited for canned foods because can lids and metal cans or other metal containers will rust readily.

A dry storage room is more satisfactory for canned foods and also vegetables which prefer a dry place. Root crops, like carrots, which shrivel easily may be stored in a dry room by adding water directly to the vegetables when needed or by placing them in closed containers. Large crocks, metal cans, tight wooden boxes and barrels are all suitable. Closed containers should be clean, dry and lined with paper before the vegetables are packed; a layer of paper may also be placed between each layer of vegetables. If stored in this way, sand or other materials are not needed to prevent shriveling.

Those who prevent excessive shriveling by adding water directly to the vegetables, commonly store such crops as carrots in crates, boxes or baskets which are kept covered with burlap or a piece of old rug or carpet; water is added by sprinkling as needed and the covering itself is kept moist.

Regardless of the method, stored vegetables should be carefully watched to avoid loss from decay, growth or excessive shriveling. Decaying vegetables should be taken out as soon as noticed. If vegetables start to grow, the temperature is too high. Vegetables which begin to shrivel a great deal should be wrapped, placed in closed containers or sprinkled. In a moist storage room extra moisture may be provided by sprinkling the floor.

—*University of Washington*

15—Controlling Insects and Disease

LOOK TO YOUR SOIL (not a spray bottle) for the best method of insect control. Most organic gardeners report minimum insect damage after they've improved soil fertility and drainage.

"I'm firmly convinced that building up the soil organically and finding the best time and variety to plant is the real solution to the insect problem," writes an El Paso, Texas, gardener. From Pennsylvania comes the comment: "After two years of enriching the soil with compost and manure, there was a noticeable decrease in insect damage."

By concentrating on improving their soils, organic gardeners have proven that they can produce vigorous plants with real resistance to pests and disease. Although organic methods sometimes don't provide 100 per cent control, organic gardeners are content to let the few pests they may find in their gardens have a small share as long as the major portion is left—untainted with poison spray residues—for them. These same gardeners make use of many control measures which do not require the use of pesticides.

A basic control method is to use vegetable varieties which are most resistant to diseases or pests in your area. The list of such resistant varieties is growing constantly. Plant pathologists throughout the nation are helping this phase of insect and disease control by developing healthier vegetables. In many cases, one-third to one-half of the varieties available from many vegetable seed catalogs now are resistant to one or more troubles that had presented a problem to growers in past years.

Rotating crops annually, keeping down weeds and planting crops which are suited to your region and soil also help to eliminate the bug problem. Watchfulness is always important. Check the undersides of leaves and closely examine the ground under plants, flowers and fruit for insects and insect damage.

Don't ignore the value of sanitation either. Eliminate rubbish piles and trash, moving waste organic materials to the compost heap or working into soil. The oldest (and still effective) insect control measure is hand-picking —especially useful against hornworms, squash bugs, harlequin bugs and leaf-feeding caterpillars.

Mulching, of course, has been found effective in protecting plants; tests have shown that heavy mulches reduce root-knot injury to plants by nematodes. But be certain material is not infested with any slugs or other pests.

Incidentally, the question is often asked about mulch attracting insects; I'd like you to know Ruth Stout's observations. After 15 years of continuous, heavy mulching, the grandmother of mulching in America reports: "I see a slug or two now and then on lettuce or cabbage, but I've had no special trouble. . . . I've always found in the many years I've been gardening that however generously insects may help themselves to the products of my labor, they always courteously leave quite a bit for me."

Mantids and Ladybugs

The ladybug (*Hyppodamia convergens*) thrives on a diet of aphids, scale, mealy bugs plus the eggs and larvae of many other harmful bugs. For about 3 dollars, you can get 15 to 20,000—enough for effective control in the average garden. After initial introduction in spring, ladybugs usually remain as long as food supply is sufficient. Another helpful factor is that every female ladybug has from 200 to 500 offspring each season after a single mating. I believe any rose grower who has had trouble with aphids should experiment with using ladybugs as a control.

Praying mantids are other allies of organic growers. When first hatched in spring, the tiny mantids feed on plant lice, flies and many other soft-bodied insects. As they grow to maturity, their diets consist of grasshoppers, chinch bugs and about any pest you can mention. For the past several years, Laurella Andersson of Virginia has been depending on mantids to help keep insect damage under control on her beautifully landscaped grounds. "The exodus of the young mantids in June from the egg cases tied to the trees is a thrilling spectacle. . . . Sometimes we've found that ants will attack the young mantids, so this year we're experimenting with a way to give them protection from such enemies. On the branch of the tree, both above and below the tied-on mantid egg case, I'm painting a broad band of tree tanglefoot. Inside this protected zone, I'll keep the young mantids from the sticky compound by enclosing the egg cases in netting sealed with weatherproof plastic tape. When the mantids are safely hatched, I'll remove the tanglefoot and permit them to roam."

Mrs. Andersson sums up as follows: "Despite the occasional setbacks, I can recommend the breeding of mantids to the gardener who is seriously plagued by insects. They'll help restore the natural balance around your homegrounds —which is, after all, the goal of the organic gardener."

The ladybug thrives on a diet of aphids, mealy bugs plus the eggs and larvae of many other harmful insects. After initial introduction in spring, ladybugs usually remain as long as food supply of insects is sufficient.

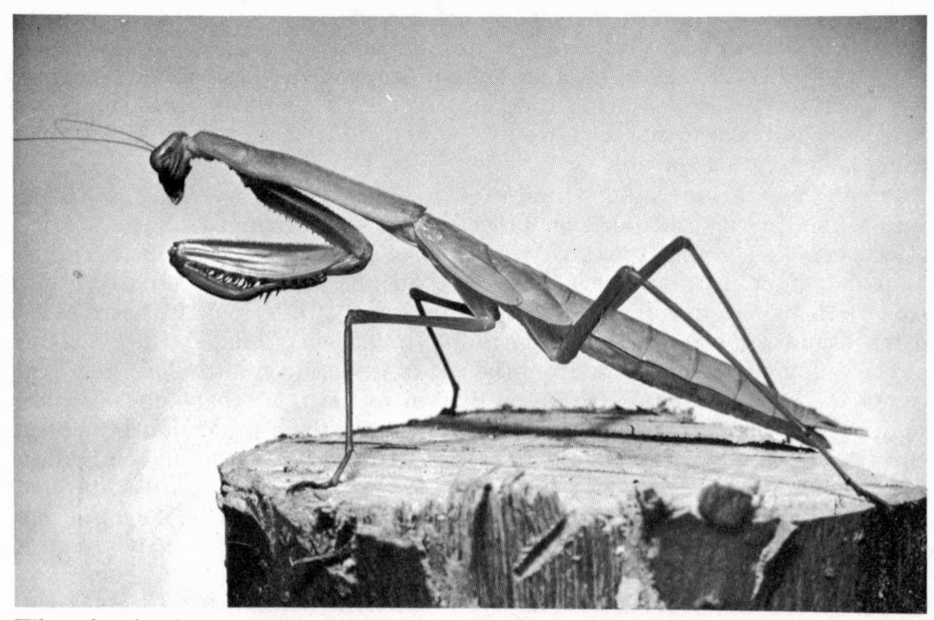

When first hatched in spring, the tiny mantids feed on plant lice, flies and many other soft-bodied insects. As they grow to maturity, their diets consist of grasshoppers, chinch bugs plus many other insects that attack vegetables.

Here are some additional ideas that have been found helpful in controlling insects without poison:

In the small garden, remove insects from plants by brushing them into a pail half full of kerosene; Japanese beetle grubs can be destroyed by milky spore powder, available commercially; mealy bugs can be cleaned off special plants with a cotton swab soaked in alcohol; an inverted cabbage leaf makes a good trap for snails, slugs and any other pests that hide during the day but feast on garden plants at night.

A 3 per cent nontoxic dormant oil spray will protect fruit trees, shrubs and ornamentals from red mite and aphids when applied in winter or early spring; protect plants from cutworms by wrapping a strip of stiff paper around stems. The paper collar should be about one inch below soil and two inches above. A handful of wood ashes will drive away many bugs, such as cucumber beetles; ants can often be controlled by a sprinkling of steamed bone meal; a band of tree tanglefoot also keeps ants and other crawling insects from going up trees.

Some gardeners report favorable results from sprinkling some rye flour over cabbage plants in early morning. Result—flour-drenched cabbage moths and worms "bake" in hot sun. Rotary-tilling crop residues into soil is protection against nematodes.

Grow Your Own Insecticides

The buds, flowers, leaves or roots of some plants can be used as a safe means of insect control. And you don't have to be a chemist to be an insecticide grower; in some cases, merely *raising* the plants in your garden is enough.

For example, marigolds, asters, chrysanthemums and related plants of the aster family are known to drive away some insects. Cosmos, coreopsis and many herbs will also keep insects away. So plant a row of these colorful flowers in your vegetable patch; you'll get pest protection plus some more flowers for cutting.

Many old-fashioned gardens had their herb plot near the kitchen door—not only for convenience but since insecticides were unknown in those days, a sprig kept the house free of "unwelcome visitors." Tansy, growing closely to the kitchen wall, kept out ants. This was one of the first herbs presented to me by a neighbor when I complained about the heavy infestation of ants and the time I had to spend daily cleaning up the area. I placed several tansy plants around the spots, and sure enough they cleared up the infestation.

I recently learned that coriander, one of my favorite seasonings for meat, stews and sausage, contains an oil used in an emulsion spray to kill spider mites and cotton aphids. Anise, another popular herb, has practically the same properties as coriander.

Members of the cucurbit family, as pumpkins and squash, make effective fly repellants. Here's what to do. Nip leaves carefully from strong-growing vines. Crush them and rub on backs and heads of cattle. For another fly (and flea) repellant, gather sprigs of mint leaves; hang them about doorways, place them in the dog kennel and where flies gather.

In India, the Peruvian groundcherry (*Nicandra physalodes*) is the basis for many stories about its effects as a fly repellant, merely by placing some bruised leaves around a room. If planted in quantity near a barn, the animals are said not to be troubled with flies; used in a greenhouse, the white-fly disappears.

Two researchers at the University of California, Peter Ark and James Thompson, have found that garlic is an effective destroyer of many bacteria that damage fruits, vegetables and nuts. In their tests, they used diluted garlic clove juice or powdered garlic extracts. They also discovered that garlic keeps its antibiotic properties for at least 3 years when stored in a dry place.

Grinding Plants

Science Editor Charles Coleman has come up with a unique way of making safe insecticides at home. Here's what he does: "I find some plant in the neighborhood which is not bothered by the pest (the one that is troubling my garden plants). Usually quite a few weeds are resistant. I run these through a meat grinder, or food chopper, saving liquid and residues and add an equal amount of water. I use the ground-material soakings to spray or sprinkle the plants I want to protect. The soakings apparently contain the organic substances that keep the plant from being bothered. As simple as this sounds, it works very well."

This logic is the same as that used by a team of University of Maryland scientists who have isolated a chemical from asparagus plants which, when sprayed on tomato leaves, kills nematodes attacking the tomato plant's roots. Drs. W. R. Jenkins and R. Rhode made their discovery after a statewide nematode-damage survey which revealed that asparagus was affected less by the pests than any other plants. In checking why the plants were resistant, they found that plain asparagus juice killed all types of nematodes found in Maryland. The scientists are now examining other nematode-resistant plants in hopes of isolating additional insecticides.

And home gardeners can do the same thing. For example, when cooking asparagus, save the water, mash up some asparagus ends and pour the mixture around some of your tomatoes. Then keep a check on insect damage between the treated and untreated plants.

Protective Plantings

Many plants have definite likes and dislikes about their immediate neighbors, preferences that will often result in better growth. In some cases, companion plantings will provide added insect protection.

Here is a partial list of vegetables, flowers and herbs that are beneficial to each other when interplanted in the garden:

Repel cabbage butterflies by intercropping with tomatoes, rosemary, sage or peppermint; asparagus beetles dislike tomato plants; aphid damage can be reduced by growing nasturtiums among the fruit trees; soybeans grown as a companion crop with corn shade the bases of corn plants so that they will be avoided by chinch bugs.

These techniques, admittedly unique, are definitely worth experimenting with. The net result of your efforts could be that you'll come up with safe (even edible) homemade "insecticides."

—Eva Wolf

Review of Control Methods

Watchfulness on the grower's part is a most important step in insect control. The home gardener should check the undersides of leaves and closely examine the ground under plants for insects or signs of insect damage.

Hand-picking of insects is probably the oldest method of insect control. It is especially effective against hornworms, squash bugs, harlequin bugs and leaf-feeding caterpillars. Many gardeners simply brush insects into a pail of kerosene.

Clean culture is another important principle and one frequently ignored. Rubbish piles and trash should be eliminated as much as possible and waste organic materials should be put into the compost heap or directly worked into the soil.

A big step in any control program is knowing who the culprit is. With a little practice, you can learn to recognize at a glance the signs and symptoms of common pests. Here's a brief guide:

The various chewing insects make their own patterns. Flea beetles make tiny round perforations; weevils produce rather typical angular openings; beetle larvae (grubs) "skeletonize" leaves, chewing everything but the epidermis and veins.

Sucking insects cause leaves to be yellowish, stippled white or gray. These insects, as well as their brownish eggs or excrement, can often be seen on the underside of foliage. Red spider can be spotted by yellowed leaves that are cobwebby or mealy underneath; whitish streaks mean thrips.

When leaves are curled up, or cupped down, look out for aphids. Deformed leaves may be caused by cyclamen mite; blotches or tunnels by leaf miners; round or conical protrusions by aphids, midges or gall wasps.

The partial collapse and dying of a plant, termed wilt, may result from a number of causes—very often nematodes or grubs.

Insect injury to plants is a result of their attempts to secure food. As insects obtain their food either by sucking out the plant juices or by eating part of the leaf surface, the damage to the plant will vary in appearance.

The injury to plant tissues caused by the feeding of sucking insects is sometimes mistaken for a plant disease. Their needle-like beaks make scarcely visible openings, but the constant removal of the plant's juices soon begins to take effect. Most often, leaves become spotted in color.

The second class of insects are those which damage plants by chewing or eating the leaves. This group includes the common cabbage caterpillar, grasshopper and flea beetles, which eat small holes in the foliage of tomatoes, peppers and potatoes. Some flower garden attackers of the chewing variety are green caterpillars on nasturtiums, thrips on gladiolus and maggots on rosebuds.

Specific Measures

Banding. A suitable band to prevent ants from establishing aphids on the leaves of trees may be made as follows: First put a girdle of cotton around the trunk; over the cotton place another band of roofing paper, which can be secured by the use of small box nails; then with a brush, apply tree tanglefoot over the band of roofing. This prevents crawling insects from going up the tree. Bands also protect trees from cankerworms and codling moth worms.

Bone Meal. Sprinkling steamed bone meal on lawns, flower beds and garden area has also been found to control ants.

Cardboard Collars. To protect against cutworms, place a stiff 3-inch cardboard collar around the stems of plants, allowing it to extend about one inch into soil and protrude two inches above the soil; clear the stem by about one-half inch.

Cultivation. Flea beetles, which attack potatoes, tomatoes, peppers, beets and related crops, can be controlled by frequent cultivation.

Disks. Paper disks encircling tomato plants help protect them from insects.

Flour. During early morning, when dew is heavy on cabbage plants, sprinkle flour on plants (one pint will cover large patch); later in the day, you'll find cabbage moths with dough sticking to their wings, feet and bodies. Acts the same way with worms; they crawl around and then sun bakes dough hard and worms die.

Ladybugs. Available commercially, ladybugs help control aphids, potato beetles.

Mulches. Heavy mulches are effective in reducing injury to plants by nematodes. Make sure mulch materials are not infested with slugs or other pests before spreading around plants.

Oil Sprays. Three per cent dormant oil spray in late winter and early spring protects shrubs, ornamental and fruit trees from aphids; effective in protecting trees against codling moth, mealy bugs and red spider mites. One such spray, Scalecide (manufactured by the B. G. Pratt Company of Paterson, N. J.), contains 97 per cent of a highly refined paraffinic oil.

Milky Spore Disease. Japanese beetle grubs have been destroyed (over 80 per cent) with applications of milky spore disease, an insect pathogen developed about 1933 and available commercially.

Plowing. One control measure against grasshoppers is to plow soil at least 5 inches deep; most species deposit their eggs in ground in late summer and fall; the common injurious species spend 6 to 8 months of the year as eggs in the top 3 inches or so of soil. By plowing deep in the soil, the surface layer—well-compacted by later cultivation—prohibits the hoppers which are hatching from emerging from the eggs. Working the soil also may bring the eggs to the surface, where they are destroyed by the drying action of the sun and wind. Fall tillage is preferable, but spring tillage is sometimes just as effective.

Praying Mantids. Praying mantids (available commercially) have been found most effective in controlling the Mexican bean beetle and tomato hornworms. Suppliers will ship you egg cases of praying mantids, which can be tied to rose bushes and other plants, where they will hatch.

Rock Fertilizers. Many gardeners have found that "dusting" plants with rock phosphate is an effective guard against damage by cucumber beetles and other insects. A spray of rock phosphate in a water solution has also worked well in protecting trees.

Rotary Tilling. Protection against the June beetle, white grubs and nematodes can be obtained from rotary tilling soil in spring. Besides bringing the insects upward, this stirring of the soil discourages weeds, whose roots serve as food for the grubs. The European corn borer has been controlled by tilling the stalk and stubble into the soil. This is done by tilling and cross tilling, which separates the stalk and stubble into small pieces, and then incorporates them with the soil. Borer control depends upon the depth of soil which covers the infested stalks.

Shallow tilling helps control root aphids. Squash vines infested with borers should be tilled into soil (or removed, root and all, in the fall of the year).

Traps. To control earwigs, here is one trap suggested: Take 4 pieces of bamboo, each a foot in length—each piece open at both ends—and tie with nylon yarn into a bundle at both ends. Lightly paint them with a green paint and when dry, put under bushes, against fences and any place where earwigs are likely to gather; leave them there for a few days; early one morning, shake earwigs out of holes into bucket of hot water or kerosene.

Water. Red spider mites can be washed off plants with a stream of water from a pressure tank or ordinary sprayer. (Generally, spiders washed off plants do not return.) A fine spray of water will also keep aphids off roses.

Weeds. Rip out plantain weeds around trees, as aphids often harbor in these weeds. It's good gardening practice to eliminate weeds in garden area, as many insects thrive on them.

Wood Ashes. Cucumber beetles have been repelled by mixing a handful of wood ashes and an equal amount of lime in two gallons of water; then spray both upper and undersides of leaves. A ring of wood ashes around plants was found to be effective against cutworms and slugs.

There are many other methods found successful that experiment-conscious gardeners have devised to fight insects. Some of them may sound a bit unique, but if it works—and is simple enough for you to do—that's what counts.

The main point is that the home gardener should look further than a can of spray if he has some insect trouble . . . he should look all the way to his soil and his garden methods.

Controlling Insects with Birds

We have found that one of the most successful (and cheapest) ways to control *insects* about our gardens and farm is to get myriads of birds to do much of this work for us!

Sixty or more helpful species of birds can be attracted to a place to help with this eradication work in any agricultural area of the United States. These birds come with different appetites. They come all ready to work unceasingly to help. Their "round-the-clock" work brings profit to the gardener and farmer; their varied songs bring much pleasure, too.

How can we know the most helpful birds? How can we attract birds of our choice to come and stay with us? How can we entice their year-round "residency?"

To achieve this and at the same time control our insect population there are 4 items of bird lore, rules or "secrets" we should keep in mind. The first of these is to be found in this paraphrasing statement: "By their bills ye shall know them."

VARIOUS TYPES OF BIRD BILLS

a.—Downy Woodpecker, drumming and drilling; b.—Ruby-Throated Hummingbird, for probing in flowers for nectar and insects; c.—Evening Grosbeak, for cracking and opening seeds; d.—Chimney Swift, degenerate bill; e.—Kingbird, strong and depressed with a notch at the tip. (From AN INTRODUCTION TO ORNITHOLOGY by George J. Wallace.)

WAYS TO ATTRACT BIRDS

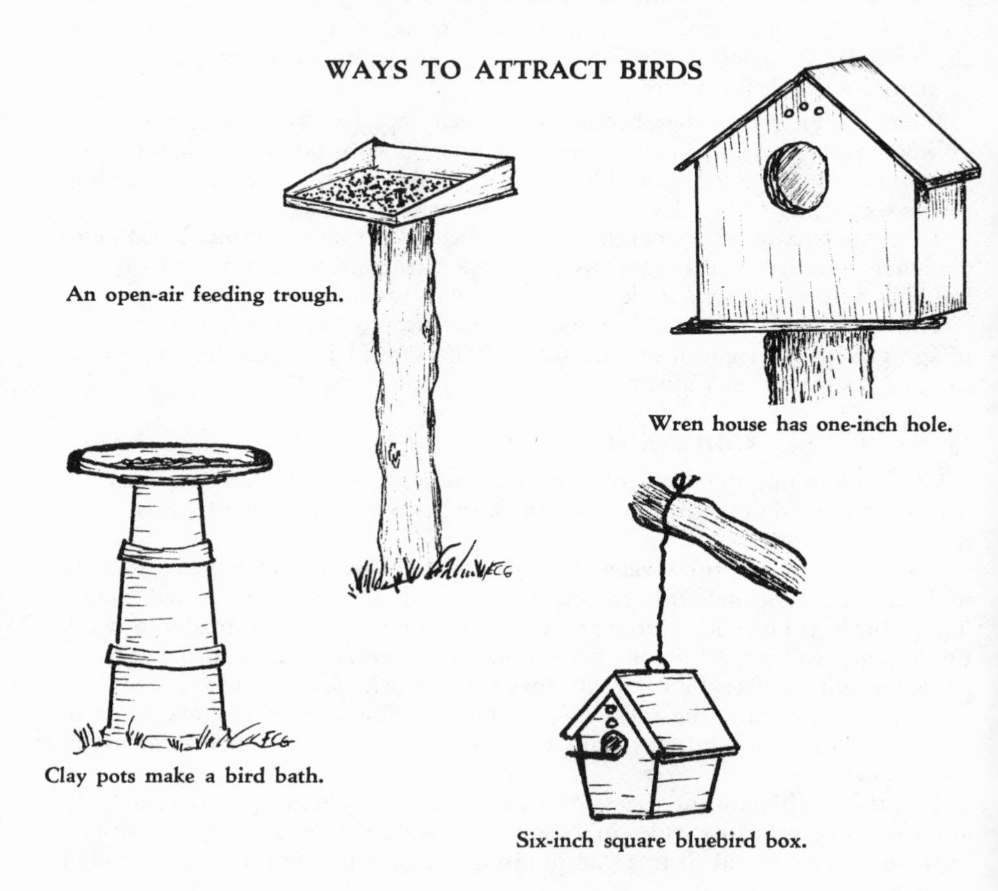

An open-air feeding trough.

Wren house has one-inch hole.

Clay pots make a bird bath.

Six-inch square bluebird box.

Many of us learned long ago to identify birds by their size, coloring and bird calls. But bird bills, no less, will tell us what we want to know about each particular bird. I feel this "secret" has never been sufficiently stressed, although it is highly important to the gardener and farmer who wants to control insects on his place.

For purposes of practical identification, we can classify our bird friends as "vegetarians" and as "heavy meat-eaters." The "vegetarians" are the "all-seed" (or other plant life) eaters. Their *bills* are short and fairly fine. English sparrows, pigeons, quail, bobwhites, mourning doves and a few others are our "all-vegetarian" birds.

But we are more concerned here with the "meat-eaters"—birds which prefer to eat millions of insect life. A few birds prefer an *all-insect* diet. They include barn swallows, swifts, house wrens, gnatcatchers, flycatchers, brown creepers and some of the several species of warblers. Their *bills* are long and straight, or long and curved; or they may be short and whiskered; whip-poorwills and the nighthawk family belong to this group.

The "Mixed-Diet" Birds

Many of our most useful birds enjoy a mixed diet, eating both many kinds of insects as well as seeds and other plant life. "Mixed-diet" birds have *fine, sharp bills.* The finer and smaller the beak, the smaller the insect, the insect egg or plant lice (aphids) such birds are able to reach and eat. In this way, many of the enemies of plant life are destroyed before they hatch; many are destroyed while they are still young and have not had time to grow and devour much of the cultivated crops.

The "Boring" Bills

For further identification of our wonderful trouble-shooters we may give attention to the *long, strong, sharp, "boring" bills.* Flickers, redheaded woodpeckers and downy woodpeckers can make thorough search in deep places for harmful insects and insect eggs. Boring insects make the favorite "banquets" for these birds.

It is important for us to know that the choice of the yellowthroat warblers is cankerworm "cocktails." If troubled with these worms, then invite these warblers to live nearby. Plant lice "luncheons" are day-long treats for the little kinglets, for warblers and for some of the finch family. Hairy caterpillars attract yellow-billed cuckoos only—but how they love them!

Perhaps *beetles* and *caterpillars* head the long list of favorite insect-foods for our birds; it is also possible that there are more of these foods available. Most gardeners and farmers feel that there is a great abundance of these chewers and crawlers on all their crops during all of the growing season. These are eaten "by the millions" by a large majority of our very helpful bird-friends.

Second Secret—"Provide Protection"

In spring and summer provide suitable nesting areas and provide some suitable materials for making nests, so the birds will not have to carry materials far. While birds are very resourceful, a few strings, rags, hair, feathers, always help. Here are a few suggestions.

HAIR is used by house wrens, nut hatches, brown creepers, bluebirds, juncos, Baltimore orioles, indigo buntings, some warblers, some finch and others.

RAGS, TWINE, STRING are used by robins, mockingbirds, Baltimore orioles.

FEATHERS help house wrens, bluebirds, phoebes, titmice, chickadees and others.

UPRIGHT WALLS are preferred by chimney swifts and barn swallows.

OLD TREES are the choices of woodpeckers, flickers and chickadees.

MUD is needed by robins, phoebes, wood thrush, barn swallows, chimney swifts and grackles for nest-making.

BIRD HOUSES help invite some birds. Bluebirds and wrens are attracted to man-made nesting boxes or houses. Certain types of bird houses attract chickadees, woodpeckers, purple martins and the little screech owls. Robins prefer open sheds or shelf-like places as building places.

Most of the other wonderful insect-eating birds prefer to build close to nature, but *more* of them will build *near* places where you want them to be if it is a *safe place* and has plenty of building material nearby.

Birds' Insect Menu

Ants are relished by kinglets, scarlet tanagers, wood thrush, brown creepers, nuthatches, titmice, and barn swallows and others.

Ant eggs are a special delicacy for chickadees, kinglets, gnatcatchers, titmice, nuthatches and brown creepers.

Spiders are said to be the delight of the downy woodpeckers.

Weevils, capable of doing 500 million dollars worth of damage annually, are devoured by the beautiful bluebirds and yellowthroat warblers whenever they are near.

Scale, minute sucking insects, make highly prized food for the little ruby-crowned kinglets, juncos, and native American sparrows (not so-called "English sparrows").

Moths would be the food most welcomed by many birds if the birds were nearby. The scarlet tanagers, phoebes, red-eyed vireos, flycatchers, gnat-catchers and barn swallows find moths of all kinds very palatable.

Millipedes ("thousand-legged worms") bring great joy to the large fox sparrows.

Leaf hoppers are an attraction for gnatcatchers, several warblers and others.

Grasshoppers are eagerly sought by flycatchers, bluebirds, mockingbirds, catbirds, brown thrashers and meadowlarks and some larger birds.

Crickets are relished by scarlet tanagers, blackbirds and grackles.

Mosquitoes are most tempting to the "least" flycatchers and chimney swifts.

Other day-flying insects are regular "bill o' fare" for the flycatchers, gnat-catchers, phoebes, kinglets, barn swallows and others.

Snails bring great satisfaction to the appetites of downy woodpeckers.

Ground insects are "gourmet specials" for towhees and juncos.

You Must Provide Water

The third rule is a very important one: *provide water* for the birds, especially near nesting times, so the parents will not need to leave the eggs or the baby birds later in search of water.

Also, you should provide water in places that are high enough for the safety of young birds when they are learning to fly, and for the older birds when they are weary from long flying.

Offer Substitute Foods

A Delaware farmer finds that crows and some other birds bother his crops less if he scatters a few grains or seeds in plain sight when his crops are peeping through the ground. His wife also keeps the bird feeder near the garden well-filled.

Hungry birds, such as robins, catbirds and others, prefer wild fruits, so cherry trees and strawberry beds are given protection when shadbush, other berries or Russian mulberry trees grow nearby. The latter has a pretty bloom and makes good shade. Raspberry and blackberry fruits will have better protection when mulberry, chokeberry or elders grow close to them.

Grapes are bothered less when wild black cherries, elders or Virginia creeper grows nearby.

The *secret* is that birds *prefer* the strong-tasting wild fruits, berries and other native foods first! They eat the bland-tasting cultivated varieties when the wild foods are not available or not plentiful.

Feeding Supplements

The fourth secret or rule for success in attracting birds is to provide feeding supplements, especially during the cold, hard winter. When you wish to provide a *"year-round" menu* as an added incentive for birds to "come early" and "stay long," here are some tips to remember:

Bread crumbs and kitchen bits are relished by:

Cardinals	Mockingbirds	Juncos
Chickadees	Brown thrashers	and others.
Catbirds	Robins	
Scarlet tanagers	House finches	

Beef suets attract:

Chickadees	Robins	Warblers
Titmice	Flickers	Woodpeckers
Nuthatches	Kinglets	and others.

Orange and apple slices, dried currants, raisins:

Robins	Brown thrashers	Catbirds
Mockingbirds	Some finches	and others
Cedar waxwings		

Cracked corn and sunflower seeds are "specialties" for:

Cardinals	Titmice	Purple finches
Nuthatches	Juncos	Towhees
Cedar waxwings	Grosbeaks	Redwing blackbirds
Chickadees	Goldfinches	and others.

A mixture of millet and hemp seed is prized by:

Cardinals	Chickadees
Nuthatches	Juncos

Good use can be made of dried seeds of melons and pumpkins which have been used earlier in the year.

Dried baked goods are always good; rolled oats are good; crushed egg shells are sometimes greatly relished.

We have learned these *4 rules* to help birds—and to help us have more birds. In turn, they can be more help to mankind near and far:

1. We can know *harmful insects,* then *know the birds which make "special diets"* of those insects.
2. We can provide better *protection* for the birds of today.
3. We can provide *water,* especially near nesting times.
4. We can provide simple *feeding* supplements.

We can begin *now* (at any time of the year) to *invite* as many birds as we wish of this long list of foragers of high-insect consumption.

These are our *4 secrets for success.*

—Emmett C. and Flora Gardner

Plant a Garden to Attract Birds

Make this year's garden a haven for birds. How? Why, it's very simple. You can do it with bird baths, boxes or feeders—or you can even do it with plants alone!

Let's start out with the annuals you're going to plant this spring. Be sure to include sunflowers, cosmos, marigolds, asters or California poppies; these will encourage many weed and seed consumers to remain around your land.

That fence around the outer edge of your property can be a source of food and shelter for birds if you plant a hedge of multiflora or Rosa Rugosa roses. Japanese barberry hedge, bush cherry, bittersweet, Michigan holly or winterberry, highbush cranberry, bush honeysuckle, snowberry and many other kinds of shrubs carry equally good food for birds. A few mulberry trees will also prove more popular than your finest berry crops.

Then, if your well-fed bird helpers still steal an occasional cultivated cherry or berry, it will be small compensation for the enormous job of helping to rid your garden of slugs, caterpillars and assorted bugs and insect larvae.

Keep the birds in mind when you're planting trees or specimen plantings in the yard. Many attractive landscaping trees will supply nesting sites for birds and often food as well. Sugar maple, often called "the aristocrat of all trees," flowering crab, the nut trees such as Chinese chestnut, Russian olive, sand cherry, wild plum and other fruit trees serve a double purpose. Junipers, mountain ash and the hollies are a few other recommended plants for attracting birds.

On a larger scale, plantings of native shrubs are useful along field borders and fence rows. Some farmers in the colder areas of the United States have discovered that planting asparagus and white clover between the last few rows of trees in a shelter belt can keep pheasants from starving in periods of heavy snow.

Many insect-eating birds, such as house wrens, bluebirds, flickers and purple martins, will use bird boxes. Place boxes at moderate elevations—4 feet for bluebirds, 6 feet for house wrens, 6 to 20 feet for flickers and 15 to 20 feet for purple martins—on trees, poles or sides of buildings near the garden.

Other birds, such as the ruby-throated hummingbird, are attracted to feeders of various types—red sugar water for hummingbirds, suet and sunflower seeds for cardinals, sparrows, grosbeaks, woodpeckers and nuthatches. Feeders are especially valuable in winter when the natural supply of food is scarce.

Most birds love to splash in small pools of water. Small, natural-looking pools about two feet in diameter and an inch deep are excellent. If possible the water should be clear, cold and slowly moving. Water which is slowly dripping from a slight elevation often works equally well. Almost every species which is found in the garden will, at times, make use of bird baths.

Spraying and the Law

While few states have passed laws to control damage caused by spraying, there have been many court decisions which have set important precedents. In most cases, the courts have decided that a farmer may be liable for damage for spreading poison or sprays negligently. They generally have found some lack of care on the part of the farmer using the spray or dust. Harold Guither, a researcher in agricultural law, reports in *Farm Chemicals* magazine: "Once damage is established, very little careless conduct is necessary to sustain a finding of negligence."

Here are a few decisions by various state courts which will give you an idea of your rights to recourse from spraying damage:

CALIFORNIA: "No person is permitted by law to use his property in such manner that damage to his neighbor is a foreseeable result."

ARKANSAS: "If one casts a substance into the air which he knows may do damage to others, and which in some circumstance will certainly do so, he is required to know how far the substance will carry or be conveyed through the air and what damage it will do in the path of its journey . . . A rule of strict liability should be applied."

NEW JERSEY: "A farmer might well be considered guilty of 'actionable negligence' if he failed to take the degree of care that the circumstance called for."

LOUISIANA: "The doctrine of strict liability will apply when crops are destroyed by drifting chemical sprays or dusts."

CALIFORNIA: "A farmer who had warned a nearby beekeeper he was going to spray was found not guilty when some of the bees were killed, since the beekeeper could have confined his bees and protected them."

ARIZONA, CALIFORNIA, ARKANSAS: "A farmer cannot escape liability from drift damage by hiring others to do his spraying for him."

SOUTH CAROLINA: "A railroad (or other company) cannot delegate liability for damages to a company it employed to spray weeds and brush along a roadbed. The railroad was forced to pay damages when weedkiller damaged a nearby cotton crop."

IOWA: "Sioux City was sued and forced to pay damages for killing bees and spoiling honey when they sprayed the airport grounds for grasshoppers."

CONNECTICUT: "Norwalk was forced to pay two residents $375 when city spraying killed fish in their private ponds."

Guither lists these instances where the injured person has received payment for at least part of the damages done to his property:

Livestock killed by eating pastures where insecticides containing arsenic had drifted from nearby cotton fields.

Bees killed by insecticides sprayed on nearby vegetable, melon or cotton fields.

Minnows killed in pond near tobacco field that had been sprayed.

Cotton and potatoes damaged by 2,4-D drift from nearby corn field.

The standard method for computing settlements in damage suits is to estimate at the time just before injury the value of the crop at maturity. Crop yields have been estimated on the basis of the undamaged part of the field or neighboring fields of about equal yielding ability.

In the past, countless cases of spray drift damage have not been brought to court because the property owners felt they couldn't win. But now court decisions throughout the country and more factual evidence show that gardeners and farmers do have legal rights for damage caused by spraying neighbors.

The University of Illinois has stated that 2,4-D weedkiller vapors can drift nearly a mile to injure sensitive crops like tomatoes and soybeans. "Slight 2, 4-D damage," reports the Illinois Experiment Station, "can be recognized by prominent veins on the backs of plant leaves, curling of the edges and general distortion of the leaves."

These damage claims are even more significant than the amount of money involved. As more and more people take legal action about spray damage, greater care will be taken by both the spray user and manufacturer. Fear of legal claims can be a most effective weapon in the battle against widespread and indiscriminate spraying.

Controlling Cutworms

You have previously read about the general principles of insect control used by organic gardeners and farmers to combat insects. In forthcoming pages, you will read about recommended measures for controlling insects which attack specific vegetables.

To give you a better idea of the wide range of effective control measures available to organic growers, here is a complete report from a veteran Massachusetts grower, Betty Brinhart, on cutworms and their control:

One of the most destructive of garden and crop pests is the cutworm, and its various species. They feed only at night, cutting down plant after plant. In a matter of days, an entire crop of corn, cotton, wheat, tomatoes or cabbage can be hopelessly destroyed. Thousands of dollars in damages to crops are caused each year in areas where no cutworm control is practiced. The damage occurs so suddenly that the farmer or gardener is completely taken by surprise.

It is difficult to stop these worms once they start on a rampage through the fields and gardens. Control must begin in early fall before the brown and gray moths, or "Millers," lay their eggs.

The Eastern Cutworm

The worm is the larva of the "Miller" moth belonging to the genus *Argotis*. The moths can often be seen flitting around lights on summer evenings.

The female moths lay their eggs during the early part of September. Each moth is capable of producing from 200 to 500 eggs a season. These she cleverly attaches in masses, or singly, to old stalks left standing in the field and garden.

The eggs hatch a few weeks after they are laid. By the end of the month, the larvae move down to the ground and feed upon what vegetation is near at hand. When cold weather sets in, they burrow into the ground and pass the winter as partially grown caterpillars.

Since these moths prefer to lay their eggs in fields and gardens overgrown with weeds, this is where the control should begin. If a field is to be left for grazing the following year, no spring damage will be noticeable. The caterpillars will merely eat their fill of the grass, develop into adult moths and fly away. But, if the field is plowed in spring and planted in corn, the crop will most likely be destroyed by the hungry worms.

By early spring, the cutworms are about fully grown, and feed greedily upon the new shoots of grass. When the field is plowed, their important food supply is suddenly cut off. After several weeks of starvation, these worms will pounce upon the tender shoots of corn that push up through the soil.

Cutworms do their feeding at night, and burrow into the soil to rest during the day. If the earth is carefully examined in the morning around the base of an injured plant, a mound of beaded dirt will be found beside a small hole. If the hole is dug up, the worm will be discovered resting in a curled-up position about an inch below the surface. Destroy it at once to prevent further damage and set a new plant in place of the destroyed one.

Method of Control

If possible, all fields and gardens should be plowed, or disked, by the first of September. This will leave no grass for the moths to lay their eggs upon.

In turn, this means no cutworms in spring. Later, around the last of the month, a cover crop, such as rye, may be planted to hold the soil in place.

If such crops as spinach, late cabbage, pumpkin and squash are still growing, plowing is out of the question. In this case, turn the soil over as soon as the crop is harvested. The cutworms may have hatched by then, but plowing will succeed in cutting off their vital food supply, thus destroying many of the small worms.

In spring, a rotary tiller may be used in the garden to destroy many of the fully-grown worms which survived the fall plowing. To prevent damage from the worms still in the soil at planting time, wrap a strip of heavy, brown paper loosely around the stem of each plant as it is set out. This wrapping should extend at least an inch and one-half below the surface, and an inch above. Around the last of June, when all cutworm danger is past, the paper can easily be removed.

Large fields, that had been in crops or grass last September, should be plowed again in spring. A wise farmer might encourage his flock of chickens to follow in the furrow to snap up any worms that are brought to the surface. Or, he may turn several hogs loose into the plowed field. Hogs are capable of rooting up and eating large quantities of cutworms, grubs and other destructive insects.

As an added protection, make a bait of the following:

Hardwood sawdust	25%	Molasses	50%
Wheat bran	25%	Water	enough to saturate the mixture

Pine sawdust should never be used as it repels the worms. Scatter a handful of this mixture around each plant at dusk, or spread evenly along each field-row of corn. The sweet molasses will attract the worms from the plants. As the cutworms crawl around in the bait, it will cling to their bodies and harden by morning, rendering them helpless. Since they will be unable to burrow back into the ground to hide from birds, the wind and sun, they will soon be destroyed. Although this bait is not 100 per cent effective, it will eliminate an amazingly large number of worms each night.

Army Cutworms

This particular species of cutworm is found chiefly east of the Rocky Mountains. They are caterpillars of various American moths of the *Noctuidae* family known as the *Cirphis Unipuncta*. The scientific name can be used in identifying the adult moth. In Latin "Unus" means "one" and "Punctum" means "a point." Combined, they refer to the single white dot found on each of the front wings of the moth.

The caterpillar is two inches long, green in color, with a yellowish stripe running down its back, and a darker one along each side.

These army worms feed only at night, burrowing into the ground to rest during the day. Although they feed mostly upon grasses, they can do considerable damage to field crops such as corn. They march in hordes like locusts. After they have entirely stripped one field of all vegetation, they move, as one, on to the next. Some farmers, in devastated areas, claim they could actually hear them coming.

Army cutworms cause the most damage during the early part of July, but a second generation can prove just as destructive around the first part of September. In the South, there may occur as many as 6 generations a season, causing a continuous damage to cotton, cabbage, tobacco and tomatoes.

Since the larvae of army cutworms usually hatch in grasslands, they are more difficult to wipe out. Some control, however, may be achieved by keeping such lands plowed or by pasturing the land close.

In early July, when the worms are on the march, the bran bait may be scattered thickly in a strip on the side of the field toward which the army worms are advancing. Apply it just at dusk so it will remain moist all night. A few inches away, between bait and crop, plow or dig a long trench at least two feet deep. What worms are not stopped by the bait will fall into the ditch, thus ending their migration. In the morning, a heavy log may be drawn back and forth through the trench to destroy the worms that have fallen in.

If the land is such that water does not seep into it quickly, another method may be used. Omit the bran bait. After plowing or digging the trench, drill deep post-holes every so many feet in the ditch. Fill the holes with water topped off with at least an inch of coal oil or petroleum. As the worms crawl about, seeking a way out, they will fall into these holes and be destroyed without further action.

Bollworm

Bollworm is the common name for the caterpillar of the noctuid moth, *Heliothis obsoleta* or *armigera*. It is known by several different names, depending upon what type of plant it is feeding on. On cotton, it is called bollworm; on tobacco, it is the tobacco-bud worm; on tomatoes, the tomato-fruit worm, and on corn, the corn-ear worm. But, in all, they are one and the same pest.

The adult reaches a length of 3/4 of an inch, and varies greatly in color, ranging from a yellowish-white to a dull green. The larvae also come in colors. These can be of a bright green, dark brown or deep pink and may be striped, spotted or plain.

Two generations of this insect usually occur in the northern states. In the deep South, there have been as many as 6. The female adult lays her eggs directly upon the plant on which she is feeding. Within a few weeks, the larvae hatch and feed upon the same plant. After 3 weeks of continuous eating, the small worms drop to the ground and pupate in the soil until adulthood.

These caterpillars cause an annual damage of around $50,000,000 to the nation's corn crop and as much as $20,000,000 to the cotton industry. A nation-wide move is on to wipe out this pest. Everyone can help. Bollworms may be controlled by combining the methods used for the Eastern and army cutworms.

To eliminate cutworms and bollworms in your garden and fields, take protective measures before the moths lay their eggs in early fall. If possible, plow all agricultural lands before the beginning of September. If fall crops are still growing, plow as soon as they have been harvested. Plow again in early spring and turn the field or garden over to chickens or hogs for several days before planting. When setting out new plants, wrap each stem with a strip of heavy, brown paper or scatter bran bait around each plant at dusk.

In case of army worms, keep all nearby grasslands cut or pastured. When the insects are on the move, apply bait on the advancing side of the cultivated field and dig or plow deep trenches to prevent them from moving in on your field.

Although your efforts may not rid your land completely of all varieties of cutworms, it will certainly eliminate a large majority of them, in turn, saving you dollars in food crops.

Prevention of Plant Diseases

In plant diseases, as in so many other matters, an ounce of prevention is worth a pound of cure.

Sanitation is of the utmost importance in the prevention of plant diseases. Diseased plant parts, including stems, leaves, and fruits, should be put in the compost heap to be converted into compost.

Crop rotation has been advised for the prevention and control of many plant diseases. If the same kinds of plants are grown in the same soil year after year, parasites are likely to accumulate to the point which makes growing this kind of plant unprofitable.

Rotation of crops and change of location are helpful in all cases and indispensable where root knot and root rots are concerned. Tomatoes, melons, okra and other summer growing crops should not be grown on the same land more often than once in 3 years. The land should be devoted to other crops like corn, grains and early vegetables that are harvested by June. After harvesting, the land should be kept free of weeds.

The small garden can be divided into thirds with summer vegetables susceptible to root diseases grown on a different plot each year.

A living soil, one rich in soil organisms, is usually one having a high organic-matter content and all the nutrient elements in good balance. Such a soil is the best kind of insurance against plant diseases. Tillage of the soil also is an important factor in the control of plant diseases, as it provides conditions favorable to a vigorous growth. Noted soil scientist Selman Waksman and others have shown that in artificial media different bacteria and fungi manufacture substances of varying degrees of antibiotic activity. It is possible that antibiotics are produced in living soils which keep crop plants free from disease. It is a common observation in pot experiments that a great variety of diseases, such as the root rots of cereals, are more destructive in sterilized soil than in comparable nonsterilized soil, both being equally inoculated with the pathogens. In such experiments, if bits of the original soil are introduced into the sterilized soil, the microflora are quickly reestablished, the pathogen checked and the disease controlled.

The physical condition of the soil has an important bearing on the prevention and control of diseases. Important are such factors as temperature,

Above a boy examines cabbage plants affected with the disease known as "yellows," which causes stunted growth and lower leaves to drop off leaving bare stems exposed. Best control measure is to plant resistant varieties.

aeration, moisture and soil reaction. Some disease-producing organisms attack their host plants in soils of certain temperatures, but not when soils have a different and, apparently for the pathogen, unfavorable temperature. An entire tomato crop may be destroyed by wilt caused by Verticillium in wet soil, while the plants will be entirely immune when grown in a well-drained soil. It has long been recognized that potato scab is less prevalent in soils that have a reaction below 5.2 than in soils with a higher *p*H value. The *p*H value becomes less important as the humus content of the soil increases;

the humus acts as an effective buffer in the soil. The physical factors of the soil and air are so interrelated when it comes to their possible effects on pathogens that it is not easy to say just what effect each has separately.

Some Common-sense Measures: Use healthy plants. Many diseases start in young seedlings in greenhouses or plant beds and later cause heavy losses in gardens. One can rarely detect disease at transplanting time. If possible, grow your own plants, or at least purchase plants from a thoroughly reputable grower in whom you have complete confidence.

Eliminate weeds that carry vegetable diseases. Many vegetable diseases spread from weeds to nearby gardens. Cucumber and muskmelon mosaic may spread from milkweed, pokeweed, ground cherry and catnip. Tomato mosaic may come from ground cherries, horse nettle, jimson weed, nightshade, bittersweet and matrimony vine. Many cabbage diseases come from any wild members of the cabbage family, such as wild mustard and shepherd's purse. Weed destruction in the garden, around the garden and along the fences destroys sources of disease-causing organisms.

Make the environment unfavorable to disease occurrence and cultivate, weed and harvest vegetables when foliage is dry. Most disease-causing fungi and bacteria require moisture and some other agent for their spread from plant to plant. Bean blight and anthracnose are easily spread by picking beans when the vines are wet. Any movement of animal or man through wet plants, if disease is present, is certain to spread the causal organisms to healthy plants. Do not take chances.

Practice fall cleanup. Disease-causing organisms in most cases can live through the winter and until the next spring in the old diseased refuse. When plowed under, they rarely cause disease unless brought to the surface through cultivation. If plowing under is not feasible, the raking together of old plant parts and hauling to a refuse heap or to some part of the yard not used for vegetables may save the humus for future use.

According to Oklahoma A. & M. College, two diseases affecting most vegetable crops are root rot and root knot. Here is their description of what the disease looks like and some control suggestions:

Root Rots

(All vegetable crops are affected.)

Appearance: Generally unhealthy appearance of plants. Plants suffer from lack of water and food materials because the root system is partly or wholly destroyed. Leaves may turn yellow, wilt or fall off although no obvious signs of parasites are present. Roots show many rotted and discolored areas.

Control: Plants should be grown under the best possible conditions. Use proper spacing, cultivation, fertilization, crop rotation, field sanitation and weed control to permit and encourage best growth of plants.

Root Knot

(Most all vegetable crops are affected, especially tomatoes, cowpeas, okra, root crops and melons.)

Appearance: Plants show signs of water deficiency, wilt, turn yellow and die. Plants stunted if they live. Roots have many small pinpoint-size to extremely large knots or swellings; these may be rotten. Root system is impaired or destroyed so that the plants cannot obtain water and food materials.

Control: Use a crop rotation system whereby resistant and susceptible crops are alternated on infested soil. The small microscopic worms causing the disease live in the soil over long periods of time so that rotation is a "must." Even a small garden can be rotated. Plant resistant varieties.

Planting Resistant Varieties

There are some varieties of plants which seem to be much more resistant to pathogens than others. Seeds taken from plants grown in your own garden are apt to be better adapted to your conditions than seeds grown elsewhere, especially if they have been grown and selected over a number of years.

Since many garden plants are natives of other countries, it is important to become familiar with the environmental conditions of their native lands. Plants of the desert and open meadow require lots of light, while plants of the woodland prefer some shade.

For each kind of plant, learn whether it prefers an acid or slightly alkaline soil, a northern or southern exposure, a heavy or light soil, a well-drained or wet soil, a long or short day, a dry or humid atmosphere.

A garden planted to one kind of vegetable and a field planted to one kind of crop represent monoculture, a practice which usually is avoided by Nature. Out of cultivation, plants grow in mixed cultures—that is, they grow together to form a more or less complex plant society. Our grandmothers' gardens contained vegetables, flowers, herbs and small fruits, and for this reason, perhaps, were damaged less by pests than present-day vegetable gardens in which are grown only a few kinds of plants and no flowers and herbs.

Many resistant varieties are listed in the section which follows on specific vegetables.

TABLE 43: Causes, Symptoms and Control

The following section has been excerpted from the publication, *Colorado Handbook of Plant Diseases,* prepared by W. J. Henderson, Plant Pathologist at Colorado State University. This report gives an excellent description of the causes and symptoms of plant disease.

Crop	Cause	Symptoms	Control
ASPARAGUS Rust	Caused by the fungus *Puccini asparagi.* Overwinters in winter resting spore stage on disease host tissues or in soil.	Small reddish to brown pustules on stems and foliage. These pustules produce the red summer spores. As the plants reach maturity the same pustules produce the black resting spores or winter spores.	Control by planting Martha Washington or Improved Martha Washington resistant varieties.
BEANS Anthracnose	Caused by the fungus *Colletotrichum lindenmuthianum.* Overwinters on bean refuse in the soil and on infected seed.	Dark sunken spots on seed coat that may extend through to the cotyledons. Brown to black ovals develop on stems of seedlings. They may girdle and kill stems. Veins on lower surface of older leaves show dark-red or dark-purple; in severe cases angular, dead spots on upper surface of leaves. Small elongated reddish-brown borders on pods. Under high humid conditions pinkish masses of spores develop at center of spots.	1. Use 2 to 3 year crop rotation. 2. Plant disease-free seed. 3. Do not work in bean fields when vines are wet.
BEANS Common bacterial blight	Caused by the bacterium *Xanthomonas phaseoli.* Overwinters inside seed and on diseased bean plant refuse in the soil.	*Leaves* first show water-soaked spots which develop into thin, irregular, dead brown spots. These usually have a thin bacterial sheen on surface. Dead tissue of spots falls out leaving a ragged leaf. Defoliation is quite common. *Pods* at first show water-soaked spots, often with reddish borders. These spots may extend through to inside of pod. *Stems* have reddish waxy cankers which may girdle and kill the stems. *Seed:* shriveled with sunken spots, usually brown around the scar.	1. Plant bacterial blight-free seed from Colorado, Idaho and other western states. 2. Practice crop rotation of 2 to 3 years or more.

TABLE 43: Causes, Symptoms and Control (Continued)

Crop	Cause	Symptoms	Control
BEANS Bacterial wilt	Caused by the bacterium *Cornebacterium floccumfaciens.* Overwinters in bean seeds and on diseased bean plant refuse in soil.	Bean seedlings attacked and killed. Leaves become limp, wilt and die. Wilting worst in warm part of the day. If disease is not too far advanced, wilted leaves may temporarily regain vitality in cool of the night. The causal bacterium may gain entrance into seed through the scar and form a yellow mass under the seed coat.	1. Plant bacterial wilt-free seed. Certified seed is recommended. 2. Practice crop rotation. 3. According to Zaumeyer and Thomas of the USDA, none of the commercially grown snap or dry beans are highly resistant to bacterial wilt.
BEANS Rust	Rust is caused by the fungus *Uromyces phaseoli v. tipica.* It overwinters on diseased bean straw in the soil. Carried by wind.	Chocolate colored, dome-like pustules about the size of a pinhead appear chiefly on the lower leaf surface. They may occur on the upper surface of the leaves and on the stems. As many as 2,000 pustules may be crowded on the lower surface of the leaves. Such badly affected leaves turn yellowish, die and drop to the ground.	1. Use a 2 to 3 year or more crop rotation.
BEANS Common mosaic and yellow mosaic	These bean mosaic diseases are caused by different viruses. They are carried by aphids. The virus of common mosaic is carried inside the seed to a large extent. Yellow mosaic virus is not seedborne. Yellow mosaic virus from sweet clover, crimson clover, red clover and infected bean plants is transmitted to healthy bean plants.	*Common mosaic:* Plants from seed infected with common mosaic virus and those which become infected early in seedling stage become severely stunted, set relatively few pods, leaves show dark-green and light-green mosaic mottling, are crinkled and curl downward at edges. Plants that become infected as they approach blooming time or later, either manifest mottling and downward curling of leaves only at the terminal growths, or completely mask the symptoms. Regardless of the degree of symptoms manifested in common mosaic infected plants, they carry virus. *Yellow mosaic:* The contrast between the yellow and green areas of affected leaves is intense and the plants become dwarfed and bunchy.	1. Plant common mosaic-free seed. Certified seed is recommended. 2. Use suitable common mosaic resistant varieties of snap beans which are: Florida Belle, Logan, Pure Gold, Rival, Idaho Refugee, Medal Refugee, Sensation Refugee, Nos. 1066 and 1071, U.S. No. 5 Refugee and Topcrop. The dry bean common mosaic resistant varieties are: Robust, Michelite and Montana No. 1 and Red Mexican. 3. For yellow mosaic, control sweet clover, crimson clover and red clover in and around the bean field or close to fields of these crops.

TABLE 43: Causes, Symptoms and Control (Continued)

Crop	Cause	Symptoms	Control
BEET Leaf spot	Caused by the fungus *Cercospora beticola*. Overwinters on seed balls and affected beet refuse in the soil.	Numerous small (¼ inch) brownish spots with reddish-purple borders appear on leaves and stems. Later they are gray with brown borders. These spots develop fruiting bodies in central portion. Heavy infection causes leaves to become yellow and die, then drop off. Continued heavy infection, dying and dropping of leaves results in pyramiding of the crowns.	1. Use crop rotation of 3 years or more. 2. Plow deeply to cover beet refuse.
CABBAGE Seed rot and damping-off (hotbeds, etc.)	Caused by *Pythium debaryanum* and other soil-inhabiting fungi. Especially bad in hotbeds.	Stems of young seedlings show a water-soaked condition at about the groundline. The plants topple over and die.	
CABBAGE Fusarium yellows	Caused by the soil-borne fungus *Fusarium conglutinans*. May exist in the soil for many years.	Seedlings wilt; older plants stunted, one-sided and yellow leaves; lower leaves drop off leaving bare stem exposed; heads stunted, bitter but usually not rotted; vascular system of stem and leaves turns brown.	*Resistant varieties:* *Early:* Wisconsin Green Acre, Golden Acre, Resistant Detroit, Racine Market. *Mid-season:* Globe, Resistant Glory (kraut type), Wisconsin Bald Head, Jersey Queen.
CANTALOUPE and CUCUMBER Anthracnose	Caused by the fungus *Colletotrichum lagenarium*.	Leaves have round, brown dead spots which may combine and destroy the entire leaf; elongated light brown areas on stems; at first the fruits have sunken, dark-green and water-soaked lesions, later, these become covered with salmon-colored mold growth. The disease is greatly favored by warm moist growing conditions.	1. Use crop rotation of 3 to 4 years in which none of the host plants follow each other.
CANTALOUPE and CUCUMBER Gray mold or Botrytis fruit rot	Caused by the fungus *Botrytis cinera*. Overwinters on affected fruits in soil.	Fruits may rot on the plant, in storage or in transit. Purple color of fruits becomes tan or brown in the infected areas. Rotted areas covered with a dense gray mold growth. Worst when humidity is high.	1. Destroy all affected fruits. 2. Keep fruits as dry as possible. 3. Remove withered blossoms at harvest time.

TABLE 43: Causes, Symptoms and Control (Continued)

Crop	Cause	Symptoms	Control
CANTALOUPE and CUCUMBER Powdery mildew	Caused by the fungus *Erysiphae cichoracearum*. Overwinters on diseased host plant refuse in soil.	Leaves become spotted and eventually covered with white, powdery mold growth. Later the affected leaves become dry and brown. Severely infected plants become yellow with poor quality fruit.	
CANTALOUPE and CUCUMBER Phomopsis blight and fruit rot	Caused by the fungus *Phomopsis vexans* which can live in the soil for 3 or more years. It is also seed-borne.	Seedlings may damp-off at the groundline. Numerous gray to brown spots occur on the leaves. Dark cankers on stems may cause stunting and wilt. A light-brown zonated rot, dotted with numerous pimple-like black fruiting bodies of the fungus. At first the fruit rot is soft, mushy, but later the fruit becomes a shriveled black mummy.	1. Practice a 4 year crop rotation. 2. Destroy all infected fruit and diseased plant refuse.
CANTALOUPE and CUCUMBER Seed rot and damping-off	Caused by the fungus *Pythium debaryanum* and other soil-inhabiting fungi.	Seed rot is a typical semidry rot which decays the seed at time of germination. Damping-off of seedlings is a wet, soft rot of the stems at the groundline. Affected plants topple over and die.	
CARROT Soft rot	Caused by the bacterium *Erwinia caratoo-orus*. Present in the soil. Develops best at high temperatures and high humidity.	Characterized by a soft, shiny, watery, mushy rot of the tap root, either in the field or in storage. It can be readily distinguished from other rots by its offensive sulfurous odor.	1. Store carrots in a cool, dry place. 2. Avoid injury while handling. 3. Do not pack carrots in containers without sufficient ventilation to prevent soft rot.
CELERY Fusarium yellows	Caused by the fungus *Fusarium orthoceras*.	Plants are stunted, usually one-sided growth, and become yellow; vascular strands reddish-brown color, from roots to leaves; affected seedlings die, but older ones live as long as healthy plants.	Try yellows-resistant varieties.

TABLE 43: Causes, Symptoms and Control (Continued)

Crop	Cause	Symptoms	Control
CELERY Seed rot and damping-off	Caused by the fungus *Pythium debaryanum* and other soil-inhabiting fungi.	Seed rot is a typical semidry rot. Damping-off is characterized by a watery soft rot which attacks the young seedlings, stems (petioles) at the groundline. Affected seedlings fall over and dry up.	
CORN (sweet) Seed rot and seedling blight (wire worms and seed corn maggot)	The diseases are caused by the fungi *Fusarium sp.*, *Pythium* spp., *Gibberella sp.*, and *Diplodia sp.*	Seed rot is a typical semidry rot of the seed during germination. Seedling blight is characterized by rotting of roots and crown and yellowing of the leaves. Affected seedlings either die or are greatly retarded in growth.	
LETTUCE Bottom rot	Caused by the fungus *Corticum vagum* which is capable of existing in the soil for several years. Wet, poorly drained soils favor the disease.	Rot begins with bottom leaves on the ground and progresses upward into the head. All of the leaf tissues except the midrib rot. Main stems of affected plants remain solid.	1. Plant lettuce on well-drained soil.
LETTUCE Drop	Caused by *Sclerotinia sclerotiórum*, a fungus which may exist for several years in the soil. Other hosts are cabbage, beans, celery, tomato and cucumber.	Young seedlings have but few leaves which quickly wilt, dry up and die. Older and larger plants may wilt down quickly and resemble a dull green, wet folded rag, or they may show at first a few water-soaked leaves. This water-soaking of lower leaves extends into the stem which then looks like glass or ice when cut across. Stems become brittle and break easily when the affected plants are pushed sideways. Often the affected broken stems quickly show a pink, blood-red or brown discoloration. Eventually the entire affected head becomes an odorless, watery, brown, rotted mass.	1. Crop rotation in which cabbage, beans, celery, tomato and cucumber do not appear. 2. Practice deep plowing to cover the sclerotial bodies of the fungus. 3. Pasturing sheep in lettuce fields after harvest to clean up refuse is advisable.

TABLE 43: Causes, Symptoms and Control (Continued)

Crop	Cause	Symptoms	Control
LETTUCE Downy mildew	Caused by *Bremia lactucae*, a fungus which overwinters on diseased lettuce refuse in soil.	Yellowish or light-green areas appear on the upper surface of older leaves. As the spots enlarge, whitish to gray tufts of mold appear on the upper surface of the spots. Later the spots turn brown. Infected heads often break down with secondary rots, especially in transit.	1. Several strains of Imperial lettuce such as Imperial 410 are highly resistant to many races of downy mildew.
LETTUCE Fusarium yellows	Caused by the soil-borne fungus, *Fusarium sp.*	Leaves become yellow; lower ones drop off. Dark-brown discoloration of the woody vascular tissue or water conducting tubes of the stems and leaf veins.	1. Avoid poorly drained soil. 2. Practice long crop rotations.
LETTUCE Mosaic	Caused by a virus which is carried by aphids. Overwinters in wild lettuce and other perennial weed hosts.	Leaves mottled yellow and green; plants have yellowish cast; usually no head forms.	1. Rogue diseased plants early in season. 2. Control weed hosts in and about the field. 3. Control aphids.
LETTUCE Tip burn	A physiogenic disease believed to be associated with poor growing conditions. Rapid loss of water from the leaves on hot, dry summer days following cloudy rainy weather, or following heavy irrigation appears to favor the disease.	First there appear small, yellowish clear (translucent) areas or spots near leaf margins. As these enlarge and become more numerous the tissues near edge of leaf turn brown and die, thus forming an irregular, brown border along the edge of the leaf. Veins of the affected area of the leaves usually become dark. Often they become infected with soft rot bacteria.	1. Maintain a uniform supply of soil moisture. Excessive irrigation favors development of tip burn. 2. Avoiding too rapid growth reduces tip burn injury. 3. The varieties which have some tolerance to tip burn are: Great Lakes, Imperial 410, Imperial 456, Progress, New York 515 and New York PW 55.

TABLE 43: Causes, Symptoms and Control (Continued)

Crop	Cause	Symptoms	Control
ONION Downy mildew	Caused by the fungus *Peronospora schleideni.* Overwinters on diseased onion plant refuse in the soil.	Best time to examine plants is in the morning while dew is on. At first the affected leaf tissue is grayish, somewhat sunken and water-soaked. Later these areas become covered with a purplish, furry mold growth.	1. Use 3 to 4 year crop rotation. 2. Destroy or plow under deeply all onion refuse.
ONION Neck rot or Botrytis rot	Caused by the fungus *Botrytis alii* and other *Botrytis spp.* which inhabit the soil. These fungi are largely wound parasites. They produce many spores which are carried by wind. Develops most rapidly at 60° to 65° F.	Little or no evidence of the disease up to the time of harvest. Nearly all of it develops in storage. Softening of scales begins at the neck with sunken spongy tissue that breaks down. The disease invades rapidly into one or two, or perhaps all of the fleshy leaves of the bulb causing a water-soaked appearance. Often a gray to brown mold growth (Botrytis) appears on the surface of affected bulbs and between the diseased fleshy leaves. The mold growth produces numerous spores and black kernel-like resting bodies (sclerotia) about ⅛ to ¼ inch in diameter. On red and yellow onions the pigment in diseased tissue is destroyed. On red onion the pigment sometimes becomes a pinkish tint. White varieties are extremely susceptible.	1. Proper curing of bulbs at harvest time is the chief control measure. 2. Good aeration around bulbs in storage crates and rooms. 3. Storage rooms *must be dry* with proper ventilation and temperatures held at 34° to 36° F. 4. Prevent mechanical injury of bulbs, while handling during harvest and storage.
ONION Pink root	Caused by the soil-borne fungus *Pyrenochaeta terrestris.* This organism is able to multiply in the soil and consequently becomes more destructive to onion crops the longer they are produced in the same infested fields.	Affected roots at first show a lead-color, then shrivel and die and take on a distinctly pink color. Affected plants may send out new roots which in turn become diseased and die. Although this procedure may not kill the affected plants, it results in scallions or small bulbs depending upon the extent of infection.	1. Do not plant highly susceptible onion varieties in soils heavily infested with pink root. 2. Plant adaptable pink root-resistant varieties. 3. Transplants or sets from pink root-infested soils should not be used. 4. Good cultural practices and high fertility help.

TABLE 43: Causes, Symptoms and Control (Continued)

Crop	Cause	Symptoms	Control
PEAS Ascochyta blight	Caused by the fungus *Ascochyta pisi.* Overwinters on the seed and diseased plant refuse in the soil.	Small dark-brown spots on stems often in sufficient numbers to girdle and kill the plant. Brownish-gray clots or fruiting bodies of the fungus can be seen in the brown spots on stems. Spots on leaves more or less round with light gray centers and brown borders. On young pods the spots are sunken; on older leaves no detectable depression and the spots are light gray. Brownish fruiting bodies of the fungus appear on dead spots.	1. Use 3 to 4 year or more crop rotation. 2. Use seed from Ascochyta-free crops.
PEAS Bacterial blight	Caused by the bacterium *Phytomonas pisi.* Overwinters in the seed and on diseased pea plant refuse in the soil.	Purplish to nearly black discoloration of stems near groundline, with irregular discolored areas at nodes. Small water-soaked spots appear on leaves first which soon become brown and papery. A thin layer of dried bacteria or sheen of bacteria collect on surface of the spots. Small, water-soaked, yellowish to brown spots appear on pods.	1. Practice a 3 to 4 year crop rotation. 2. Plant bacterial blight-free seed.
PEPPER Mosaic	May be caused by any one of several viruses such as tobacco mosaic, cucumber mosaic, etc. Overwinters in wild perennial host plants.	Plants infected in early season become severely stunted, leaves mottled with light-green and dark-green areas. Few fruits set. Plants that become infected about blossom time either manifest slight mottling by terminal leaves or none at all. Fruits bumpy and are bitter.	1. Eradicate all wild solanaceous perennial host plants in or about the field. Also control wild cucumbers. 2. Control aphid carriers.

TABLE 43: Causes, Symptoms and Control (Continued)

Crop	Cause	Symptoms	Control
POTATO Bacterial ring rot	*Corynebacterium sepedonicum,* a bacterium which overwinters in potato tubers. It is not soil-borne.	Symptoms usually appear when the potato plants are nearly full grown. One or more stems on a plant may wilt and be more or less stunted, whereas the remainder may appear healthy. A characteristic yellowing and dying of the leaf tissues between the larger veins; dead leaves tend to remain on the plants for quite some time. Stems remain rigid and the lower end may show varying degrees of discoloration to complete rotting of the bark, vascular (woody tissues) and pith. Rotting of the stem rarely extends but a few inches above groundline. A milky bacterial exude can be squeezed from lower end of affected stems. Tubers show a faint yellowing to a deep yellow discoloration of the vascular ring. The tissue in tubers separates easily at affected sections of the vascular ring. Secondary rot organisms usually aid in watery soft rot of the more advanced stages of ring rot diseased tubers.	1. Plant healthy seed tubers. 2. Use clean sacks for seed tubers.
POTATO Fusarium wilt	Caused by *Fusarium oxysporium, F. solani,* and *F. eumartii* fungi which exist many years in infected soils.	There are several Fusarium wilts of which 3 are common in Colorado. Each affects the plant in a slightly different manner but in general there is much overlapping of symptoms. A typical symptom is the mid to late season bronzing, yellowing and reddish to purple coloring of the top leaves. The affected plants are somewhat more spindly than normal ones. The foliage becomes quite yellow throughout the entire vine. The woody vascular tissue of stems, roots and leaf petioles becomes brown with some brownish flecking of the pith. The affected plants wilt and may finally die. Tubers show a brownish discoloration of the vascular ring at the stem-end.	1. Plant disease-free seed. 2. Rogue out diseased plants in potato crops intended for seed tuber production.

TABLE 43: Causes, Symptoms and Control (Continued)

Crop	Cause	Symptoms	Control
POTATO Late blight (Psyllid insects)	The disease is caused by the fungus *Phytophthora infestans.* Overwinters in diseased tubers. Spores carried long distances by wind and irrigation water. Cool, wet weather favors its development.	Leaves show dark-green, water-soaked areas which later dry out and die. They remain either a faded green or become light-brown color. In moist cool weather the disease invades the entire plant in a relatively short time. In hot weather infection is either inactivated or greatly retarded in Colorado. Early on cool mornings a sparse, short, white mold (mycelium) growth can be detected, with the aid of a hand lens, on the lower surface of the affected leaves. Infected leaf petioles and stems show dark-brown to nearly black water-soaked lesions. After a short time, under favorable conditions, the entire vine dies. Tubers show irregular, slightly sunken areas, which when cut open and exposed to air, the affected tissue turns a reddish-brown color. These affected areas break down rapidly with a soft rot. This rot has a very disagreeable odor. Tubers may rot either in the field or in storage.	1. *Sanitation:* Destroy all potato cull piles by spreading them out so the tubers will freeze and kill the late blight causal organism. 2. *Storage:* Clean out and disinfect the cellars as in the instance of ring rot of potato. Ventilation to prevent late blight rotting in storage: The cellars should be ventilated such that there will be no condensation of moisture on the tubers or on the ceiling to drop on tubers in bins below. Introduce the air into cellars near the floor to prevent condensation. Keep cellar roof in good repair.
POTATO Verticillium wilt	Caused by the fungus *Verticillium alboatrum,* which is capable of existing in infested soil for many years. Other hosts are pepper, lettuce, celery, eggplant and tomato.	It is very difficult to distinguish the symptoms of verticillium wilt from those of Fusarium wilt. The main woody area, or water and food (vascular) transporting area of the affected stems becomes a reddish-brown. Usually this brown discoloration of the vascular area does not extend as far up the stems and into the leaf petioles as in Fusarium wilt. The leaves roll, become greenish-yellow, wilt, suddenly dry up and die. The vines are somewhat spindly but remain upright except the tips which droop. In general, the yellowing and drying of leaves proceeds from base of the plants upward. Tubers show a brown to nearly black discoloration or net necrosis of the vascular area.	1. Plant healthy seed tubers. 2. Rogue out diseased plants in crops intended for seed production. 3. Pontiac variety is highly resistant to verticillium wilt.

TABLE 43: Causes, Symptoms and Control (Continued)

Crop	Cause	Symptoms	Control
POTATO Witches'-broom	Caused by a virus which overwinters in the tubers. Transmitted by aphid insects.	Characterized by numerous, slender, upright sprouts coming from the seed-piece giving the plant a bushy or broom-like appearance. Leaves are small, velvety and unusually light-green in color. The sprouts are seldom more than 10 inches high.	1. Plant healthy seed tubers.
RADISH Black rot	Caused by the fungus *Aphanomyces raphani* which may exist in soil for many years.	Dark spots develop where lateral roots emerge from the large tap root. These spots enlarge until several of them combine to form a large, black area on the fleshy tap root. Often causes distortion of the fleshy tap root.	Use a 3 to 4 year or longer crop rotation. This will greatly reduce the incidence of infection.
TOMATO Blossom end rot	A physiogenic or non-infectious disease caused by unfavorable growing conditions, due to an irregular supply of available soil moisture.	At first water-soaked spots appear at blossom-end of fruits. Later this increases in size and becomes sunken, flat, brown and leathery. Such fruits are worthless.	Control by keeping the soil in excellent growing condition by frequent light applications of irrigation water. Texto, an excellent green wrap, 70-day variety, is highly resistant to blossom-end rot, as well as resistant to Fusarium wilt.
TOMATO Fusarium wilt	Caused by the fungus *Fusarium lycopersici.* A soil-inhabiting organism.	At first, leaves on one side of stem and also leaflets on one side of the petiole become yellow and wilt. Later all leaves turn yellow, wilt and die. Woody portion (vascular) of stem shows a brown discoloration which may extend from roots into the leaf petioles.	Control by Fusarium wilt resistant varieties: Pritchard (75 days), Rutgers (80 days), Marglobe (79 days), Marbon (68 days), Early Baltimore (78 days), Wiltmaster (80 days), Chesapeak (75 days), Kokomo (80 days). Texto (excellent green wrap, and also resistant to blossom-end rot; about 70 days) and Ohio W-R Brookton.

TABLE 43: Causes, Symptoms and Control (Continued)

Crop	Cause	Symptoms	Control
TOMATO Mosaic	Caused by the tobacco mosaic virus. Overwinters in perennial host plants of the potato family, in diseased dead host tissue, in diseased tobacco leaf tissue used to make chewing tobacco, cigars and cigarettes. Transmitted from diseased to healthy plants by contact of contaminated tools, or hands or clothing of workers and by certain aphis insects and by use of tobacco.	Plants that become infected in *early stages of growth* are stunted. Leaves show light-green and dark-green mottled areas or mosaic pattern which tend to pucker. Very few blossoms set thus resulting in few fruit of inferior quality. Plants that become infected after they approach blooming stage and later, either manifest the light-green and dark-green mottling and puckering *only* on the terminal growth leaves, or completely mask the symptoms. Late infection does not affect the quality and yield of fruit to any extent.	1. Avoid use of tobacco in any form while handling tomato plants, at least up to blooming time. 2. Wash hands with soap and water after handling mosaic-infected plants to prevent transmission of the causal virus on hands to healthy plants, especially until after blooming and fruit-set time. 3. Rogue out all *early* infected plants and burn them. 4. Control weeds in and around tomato crops. 5. Do not plant tomato, potato and pepper crops near one another. 6. Control aphis insects. 7. Do not prune tomato plants with a knife. Pinch off the blossoms instead, without touching any part of the plant but the pruned-off blossom. This is important especially under greenhouse conditions.
TOMATO Verticillium wilt	Caused by the fungus *Verticillium hydromycosis.* Overwinters in soil and may overwinter on seed. Other hosts are pepper, celery and lettuce.	Yellowing of larger basal leaves which later turn brown and die; wilting of tips of branches during day; these may recover at night. Defoliation is common. Branches are uniformly affected, tend to droop or lie prostrate on the ground opening up the crown and exposing the fruit to the sun. Leaves which remain on vine are dull in appearance, and the fruits are small. Woody (vascular) area of the stem becomes brown. This discoloration rarely extends into the leaf petioles as in the case of Fusarium wilt. The plants usually remain alive throughout the season but produce low quality fruit. Difficult to differentiate from symptoms of Fusarium yellows.	1. Use seed from healthy plants. 2. The causal organism is capable of existing for 6 to 7 years after soil has become infected. Crop rotation should therefore be arranged accordingly and having in mind that peppers, eggplant, tomato, celery and potato should not appear, or follow one another in rotation.

—W. J. Henderson, Plant Pathologist, Colorado State University

16—Freezing and Canning Vegetables

Vegetables for Vitamins

ONE OF THE IMPORTANT yardsticks for measuring the nutritional worth of any food is the contribution in terms of *vitamins* that it makes to our diet and our health. Vitamins are organic food substances—that is, substances existing only in living things, plant or animal. Although they exist in foods in minute quantities, they are absolutely necessary for proper growth and the maintenance of health. Plants manufacture their own vitamins. Animals obtain theirs from plants or from other animals that eat plants.

Vitamins are not foods in the sense that carbohydrates, fats and proteins are foods. They are not needed in bulk to build muscle or tissue. However, they are essential, like hormones, in regulating body processes. As in the case of trace minerals (iodine, for instance), the presence or absence of vitamins in very small amounts means the difference between good and bad health. Many diseases and serious conditions in both human beings and animals are directly caused by a specific or combined vitamin deficiency.

The green leaves of plants are the laboratories in which plant vitamins are manufactured. So the green leaves and stalks of plants are full of vitamins. Foods that are seeds (beans, peas, kernels of wheat and corn, etc.) also contain vitamins which the plant has provided to nourish the next generation of plants. The lean meat of animals contains vitamins; the organs (heart, liver, etc.) contain even more, which the animal's digestive system has stored there. Milk and the yolk of eggs contain vitamins which the mother animal provides for her young. Fish store vitamins chiefly in their livers.

Basically there are two different kinds of vitamins—those that can be dissolved in fats and those that dissolve in water. The vitamins found in liver, eggs and butter are fat-soluble. Those in fruits and vegetables are water-soluble. Milk contains both kinds.

The vitamins we know most about are called by a letter and also a chemical name. These are vitamin A (carotene); vitamin B_1 (thiamin), B_2 (riboflavin), B_6 (pyridoxine), the other members of the vitamin B group (biotin, choline, folic acid, inositol, niacin, pantothenic acid, para-aminobenzoic acid, B_{12}); vitamin C (ascorbic acid); the several D vitamins, D_2 (calciferol) and D_3 (7-dehydrocholesterol); vitamin E (tocopherol); vitamins F, K, L_1, L_2, M, P.

Researchers have established approximate estimates of the daily requirements of most of the vitamins for perfect health. These amounts are usually spoken in terms of milligrams. (A milligram is 1/1000 of a gram. A gram is 1/32 of an ounce.) You may also find daily vitamin requirements expressed in terms of International Units, which are each 1/6000 of a milligram. Although there have been cases of overdoses of a vitamin, it seems that the established daily minimum is really a minimum. Actually, from 2 to 4 times that much of any vitamin will produce the most abundant growth and health.

Vegetables—the edible parts of all plants, except the fruits of certain grains, shrubs and trees—are an especially valuable source of many of the vitamins. Yellow and green leafy vegetables, along with tomatoes, contain appreciable amounts of carotene, the plant substance which is changed into vitamin A in the body. Ascorbic acid (vitamin C) is plentiful in tomatoes, peppers and many of the raw leafy vegetables. While potatoes have only a fair amount of ascorbic acid, the quantities in which they are eaten by many people make them a material source of it. Several of the B vitamins, too, are present in a variety of vegetables. The green leafy ones, legumes, peas and potatoes provide some of the needed thiamin, riboflavin and niacin in a well-balanced diet.

Let's not forget that the *way* vegetables are grown has a definite role in the nutritive values—including vitamins—they will contain. As far back as 1939, the United States Department of Agriculture yearbook *Food and Life* stated: "Underneath all agricultural practices there is a guiding principle . . . to carry out this cycle of destruction and construction economically—to see that plants, animals and man utilize raw materials efficiently to build up the products of life and that these products are broken down efficiently into raw

TABLE 44: Vegetables Containing the Largest Amounts of Vitamin A

(Average minimum daily requirement is 5,000 units)

Vegetables	International Units of Vitamin A
Asparagus, fresh	1,000 in 12 stalks
Beans, snap	630 to 2,000 in 1 cup, cooked
Beet greens	6,700 in ½ cup, cooked
Broccoli	3,500 in 1 cup, cooked
Carrots, fresh	12,000 in 1 cup, cooked
Celery cabbage	9,000 in 1 cup
Collards	6,870 in 1 cup, cooked
Dandelion greens	13,650 in 1 cup, cooked
Endive (escarole)	10,000 to 15,000 in 1 head
Kale	7,540 in ½ cup, cooked
Lettuce, green	4,000 to 5,000 in 6 large leaves
Parsley	5,000 to 30,000 in 100 sprigs
Peas, split	1,680 in 1 pound
Peppers, green	3,000 in 2 peppers
Peppers, red	2,000 in 2 peppers
Pumpkin	1,200 to 3,400 in 1 cup, cooked
Spinach, fresh	9,420 in ½ cup, cooked
Spinach, canned	5,500 in ½ cup
Squash, winter	4,950 in ½ cup, cooked
Sweet potatoes	7,700 in 1 medium potato, baked
Tomatoes, fresh	1,100 in 1 medium tomato
Turnip greens	9,540 in ½ cup, cooked
Watercress	4,000 in 1 bunch

TABLE 45: Vegetables Containing Largest Amounts of B-Complex Vitamins (Continued next page)

Vegetable	Milligrams of		
	Thiamin (B_1) (M. D. R. 1.8 mg.)	Riboflavin (B_2) (M. D. R. 2.7 mg.)	Pyridoxine (B_6) (M. D. R. 1.8 mg.)
ARTICHOKES, Jerusalem			
ASPARAGUS, fresh	.16 in 12 stalks		
BEANS, dried lima	.60 in ½ cup, cooked	.24 to .75 in ½ cup	.55 in ½ cup
BEANS, dried soy		.31 in ½ cup	
BEANS, green			
BEETS			.11 in ½ cup
BEET TOPS		.17 to .30 in ½ cup	
BROCCOLI			
CABBAGE			.29 in ½ cup
CARROTS			
CAULIFLOWER	.09 in ¼ small head		
COLLARDS	.19 in ½ cup, cooked	.22 in ½ cup, steamed	
CORN, yellow	1.9 in 1 lb.	0.5 in 1 lb.	
DANDELION, greens	.19 in 1 cup, steamed		
ENDIVE (escarole)		.20 in 1 head	
KALE			
MUSHROOMS			
MUSTARD GREENS		.20 to .37 in 1 cup, steamed	
PEAS, green, fresh	.36 in 1 cup, cooked	.18 to .21 in 1 cup, steamed	1.9 in 1 cup
POTATOES, Irish			.16 in 1 medium potato
POTATOES, sweet			.32 in 1 medium potato
SPINACH		.24 to .30 in ½ cup, steamed	
TOMATOES, fresh			
TURNIPS			.10 in ½ cup
TURNIP, greens	.10 in ½ cup, steamed	.35 to .56 in ½ cup	

TABLE 45: Vegetables Containing Largest Amounts of B-Complex Vitamins

Milligrams of				
Choline (M. D. R. not established)	Folic Acid (Micrograms) (M. D. R. not established)	Inositol (M. D. R. not established)	Niacin (M. D. R. 18-20 mg.)	Pantothenic Acid (M. D. R. not established)
				.40 in 4 artichokes
130 in 12 stalks	120 in 12 stalks			
	330 in ½ cup	170 in ½ cup	9.6 in 1 lb.	.83 in ½ cup
340 in ½ cup				
340 in 1 cup	71 in 1 cup			
8 in ½ cup	42 in ½ cup, diced	21 in ½ cup		
	25 in ½ cup			
	90 to 110 in 1 cup			1.4 in 1 cup
250 in 1 cup		95 in 1 cup		
95 in 1 cup	97 in 1 cup	48 in 1 cup		.2 in ½ cup
	44 in ¼ small head	95 in ¼ head		.92 in ¼ head
			7.8 in 1 lb.	2.3 in 1 lb.
	62 to 75 in 1 head			
	100 in 1 cup			.30 in 1 cup
	98 in ½ cup, cooked	17 in 7 mushrooms	6.0 in 7 mushrooms	1.7 in 7 mushrooms
252 in 1 cup				
260 in 1 cup	22 in 1 cup	80 in ½ cup		.60 in 1 cup
105 in 1 med. potato	140 in 1 small potato	29 in 1 med. potato		65 in 1 med. potato
35 in 1 med. potato		66 in 1 med. potato		.95 in 1 med. potato
240 in ½ cup	225 to 280 in ½ cup	27 in ½ cup		
	12 to 14 in 1 sm. tomato	46 in 1 sm. tomato		
94 in ½ cup		46 in ½ cup		
245 in ½ cup				

materials that can be used again . . . By the proper use of fertilizers and other cultural practices it might be possible to insure the production of plant and animal products of better-than-average nutritive value for human beings."

In selecting vegetables, planning and varying those you serve at meals, it's important to remember that fresh vegetables offer higher overall nutrient values than either canned or frozen and in season are usually less expensive. Within economic reason, it is preferable that as much of any family's vegetable dietary as possible be of the fresh variety. Of course, the sooner they are eaten after picking or purchase, the more that vitamin and other food values are retained. Careful storing and refrigeration of those that must be held is essential for vitamin retention.

TABLE 46: Vegetables Containing the Largest Amounts of Vitamin C

(Average minimum daily requirement is 75 milligrams)

Vegetables	Milligrams of Vitamin C
Asparagus	20 in 8 stalks
Beans, green lima	42 in ½ cup
Beet greens, cooked	50 in ½ cup
Broccoli, flower	65 in ¾ cup
Broccoli, leaf	90 in ¾ cup
Brussels sprouts	130 in ¾ cup
Cabbage, raw	50 in 1 cup
Chard, Swiss, cooked	37 in ½ cup
Collards, cooked	70 in ½ cup
Dandelion greens, cooked	100 in 1 cup
Kale, cooked	96 in ¾ cup
Kohlrabi	50 in ½ cup
Leeks	25 in ½ cup
Mustard greens, cooked	125 in ½ cup
Parsley	70 in ½ cup
Parsnips	40 in ½ cup
Peas, fresh cooked	20 in 1 cup
Peppers, green	125 in 1 medium pepper
Peppers, pimento	100 in 1 medium pepper
Potatoes, sweet	25 in 1 medium potato
Potatoes, white, baked	20 in 1 medium potato
Potatoes, white, raw	33 in 1 medium potato
Radishes	25 in 15 large radishes
Rutabagas	26 in ¾ cup
Spinach, cooked	30 in ½ cup
Tomatoes, fresh	25 in 1 medium tomato
Turnips, cooked	22 in ½ cup
Turnips, raw	30 in 1 medium turnip
Turnip tops, cooked	130 in ½ cup
Water Cress	54 in 1 average bunch

TABLE 47: Vegetables Containing the Largest Amounts of Vitamin E

(Minimum daily requirements have not been established)

Vegetables	Milligrams of Vitamin E
Beans, dry navy	3.60 in ½ cup steamed
Carrots	.45 in 1 cup
Celery	.48 in 1 cup
Lettuce	.50 in 6 large leaves
Onions	.26 in 2 medium raw onions
Peas, green	2.10 in 1 cup
Potatoes, white	.06 in 1 medium potato
Potatoes, sweet	4.0 in 1 medium potato
Tomatoes	.36 in 1 small tomato
Turnip greens	2.30 in ½ cup steamed

Blanching Vegetables for Home Freezing

Besides providing a number of advantages in the keeping and eating qualities, blanching has been shown in recent nutritional research to retain more of the actual food value—vitamin C in particular—of several popular vegetables.

The blanching idea isn't new. Methods for scalding or steaming fresh produce in preparation for freezing were introduced about 30 years ago. Since then, however, food specialists have been discovering more about the unrealized and subtle effects of using this pre-cold-storage process. They have found, for example, that blanching makes certain enzymes inactive which would otherwise cause unnatural colors and disagreeable flavors and odors to develop while the foods remain frozen. Then, too, they've found that blanched vegetables are somewhat softened, so that they can be packed more easily and solidly into freezer containers.

First and foremost reason for blanching, of course, is to help frozen produce keep better. Next to eating them straight from the garden, it is the best way to retain good taste, appealing color and texture in vegetables stored for future use. What freezing—and the right preparation for it—does is literally to hold on to as much natural "freshness" as possible.

Now there's news of another benefit in the blanch-freeze method, a benefit that is tremendously important to those who actually eat the frozen vegetables. Recent experiments show ascorbic acid—or, as it's more commonly known, vitamin C—is retained in much greater amounts in many of those blanched before freezing. Some held 2, 3 and even 4 times more of this elusive element through periods ranging up to 9 months. Aside from maintaining better quality, this nutritive advantage over vegetables frozen unblanched (or treated in other ways, such as with sulfur dioxide gas) is significant.

Tests Show Difference

Concentrated research on food freezing, blanching and quality and nutrient retention has been in progress at the University of Illinois College of Agriculture for several years. The Foods and Nutrition Division of their Home Economics Department, under the direction of Dr. Frances O. Van Duyne, has a continuing project in "Methods of Home Freezing." The University's Experiment Station quarterly, *Illinois Research,* recently reported extensive experiments on the blanching questions.

Five vegetables—broccoli, peas, snap beans, spinach and corn—were picked fresh, at optimum maturity for freezing, and processed promptly. Several lots of each were used. Part of these were given preliminary preparation, blanching, cooling, packaging, freezing and freezer storage according to standard directions. The rest were packaged and frozen without blanching.

The green vegetables were compared for ascorbic acid retention after freezer storage periods of 1, 3, 6 and 9 months. In one instance (broccoli after one month) the amount was equal. In all the others, analyses showed *more vitamin C was retained in the blanched vegetables than in those frozen unblanched,* at every period of testing.

After 3 months, for example, blanched broccoli had held 64 per cent of its raw ascorbic acid content, compared to 57 per cent for the unblanched samples. At 6 months, the difference had widened to 60 to 40 per cent; and by 9 months, 54 to 36 per cent.

Peas revealed even greater losses where blanching had been omitted. After a month, blanched specimens tested 70 per cent vitamin C retention, against 63 per cent for the unblanched. By the third month, it was 75 to 55; 6 months, 71 to 36; and 9 months after freezing 70 per cent to 37 per cent.

With snap beans, a larger disparity was found right at the start. Blanched beans had held 85 per cent of their initial ascorbic acid after a single month in the freezer, while those unblanched had dropped back to 58 per cent. In 3 months it was 83 to 44; at 6, 64 to 15; and at 9 months the blanched beans still had 43 per cent—and the unblanched just 3 per cent.

Spinach was tested after two weeks' freezer storage, and showed a 52 to 28 per cent advantage for blanching. Corn, which is low in vitamin C to begin with, was not tested for retention, but was included in the cooking and palatability comparisons since it is one of the more popular vegetables for home freezing.

Other Qualities Affected

In addition to checking for ascorbic acid variation in the blanched and unblanched produce, the Illinois research sought to compare what is called *palatability,* too. This includes such factors as appearance, color, texture, flavor (or off-flavor) and general acceptability. "During the entire 9 months," states the report, "scores of blanched samples were 4.0 to 4.4, corresponding to ratings of good to high good. After cooking, the blanched vegetables were bright green and tender, and had a good flavor. No off flavors were noted."

On the other hand, unblanched samples that had been stored for only one month received general acceptability scores of 2.4 to 2.9, which corresponds to ratings between poor and fair. "Strong off-flavors developed and unblanched products lost color during the first month in the freezer."

Blanching Time Chart

Vegetables	Minutes Water	Minutes Steam
Asparagus	2–3	4
Beans, green snap	2–3	3
Beans, wax	2–3	3
Beans, lima	2	3
Beets (until tender)	25–50	—
Broccoli (split stalks)	3	5
Brussels sprouts	3–4	5
Carrots (small, whole)	4–5	5
Carrots (diced or sliced)	2	4
Cauliflower	3	4
Corn on the cob	7–11	—
Corn (cut after blanching and cooling)	4	—
Kale	2	—
Peas, green	1½	1½
Spinach	2	—
Squash, summer	3	4
Squash, winter	20	—

Over the longer storage periods, the vegetables that were not blanched deteriorated still further, some of them so much that they were considered inedible. Most became faded, dull or gray; all became tough or fibrous; and some, broccoli especially, developed an objectionable hay-like flavor.

Tests with the frozen corn disclosed palatability ratings between good and very good for the blanched samples, while none of the unblanched was considered even fair. Although the appearance and color of the unblanched corn had held up, its flavor had become disagreeable and there was deterioration in texture.

What You Can Do

What do these research results mean to the home gardener? Just this: to keep the largest possible amounts of vitamin C in the vegetables you freeze and to keep them tasty and appetizing, it pays to blanch carefully before freezing. Whether you want to store your own organic surplus for a healthful supply next winter or if you'd like to be sure of putting up enough naturally grown produce bought throughout this season, here are some tips on freezing:

1. Line up everything needed for blanching and freezing *first*. Nothing counts more than speed in holding on to freshness, taste and nutritive value. Plan a family operation deep-freeze; have all hands on deck to help quickly, and arrange equipment, containers, etc., in advance for a smooth production.

2. Pick young tender vegetables for freezer storage; freezing doesn't improve poor-quality produce. As a rule, it is better to choose slightly immature produce over any that is fully ripe; avoid bruised, damaged or overripe

To insure a year-round supply of vitamins and minerals in your diet from your organic vegetable garden, you will want to process your excess produce. Home freezers are becoming more and more popular, but canning is still more commonly practiced.

vegetables. Harvest in early morning. Try to include some of tastiest early-season crops; don't wait only for later ones.

3. Blanch with care and without delay. Vegetables should be thoroughly cleaned, edible parts cut into pieces if desired, then heated to stop or slow down enzyme action. For scalding, use at least a gallon of water to each pound of vegetable, preheated to boiling point in a covered kettle or utensil (preferably stainless steel, glass or earthenware). Steaming is better for some vegetables, helps retain more nutritive value. Use a wire-mesh holder or cheesecloth bag over one inch of boiling water in an 8-quart pot; same arrangement, to hold one pound at a time, is handy for plunging vegetables into boiling water. Start timing as soon as basket or bag is immersed or set in place for steaming. (See CHART for scalding or steaming times required for different vegetables.)

4. Cool quickly to stop cooking at the right point. Plunge blanched vegetables into cold water (below 60 degrees); ice water or cold running water will do best. Change water often; then drain well, using absorbent paper toweling.

5. Package at once in suitable containers. Glass jars require 1 to 1½ inches headspace; paper containers call for leaving ½ inch headspace, except for

vegetables like asparagus and broccoli that pack loosely and need no extra room. Work out air pockets gently and seal tightly.

6. Label all frozen food packages; indicate vegetable, date of freezing, variety, etc. Serve in logical order—remember food value and appeal are gradually lowered by long storage. Maximum freezer periods for most vegetables is 8 to 12 months. Except for spinach and corn on the cob, cook without thawing. Avoid overcooking.

Teamed with organic growing and careful variety choice, the blanch-freeze idea can help bring a wide selection of truly taste-delighting frozen freshness to your table.

Quality of Vegetables for Home Freezing

These helpful pointers on quality control for the home freezing of vegetables are offered by Richard Hopp, Assistant Horticulturist and Susan B. Merrow, Assistant Nutritionist of the Vermont Agricultural Experiment Station:

The most important factor in obtaining a frozen pack of high quality is the condition of the fresh product at the time of processing. Freezing is primarily a process aimed at preserving the fresh characteristics as long as possible. Properly done, it is an excellent method of preservation since it maintains very satisfactorily the initial color, texture and flavor of the fresh product as it comes from the garden. At the same time the original nutritive value is retained to a high degree. Remember, however, that quick freezing cannot improve the quality of the fresh product and, therefore, you should consider only the finest quality of fresh vegetables good enough for freezing. Variety selected, stage of maturity at the time of harvest and handling between harvesting and processing are the chief factors responsible for the condition of the fresh product.

Stage of Maturity

The home gardener has the great advantage that he can pick and process his vegetables when their quality is at its best. This means he can harvest them at the proper stage of maturity and as short a time as possible before using.

In most cases you find the finest flavor when vegetables are still young: peas and corn while they taste sweet and not starchy; snap beans while the pods are tender and fleshy before the beans inside the pods get plump; summer squash while their skins are still tender. Carrots and beets, too, have a particularly delicious flavor when they are young.

It is not always economical for the commercial grower to harvest a crop at the same early stage of development at which a home gardener can well afford to pick his crop. It is, therefore, difficult to buy most vegetables in a store if you expect to produce a superior frozen pack. The surest way to obtain fresh produce of high quality for freezing is to grow it yourself and pick it yourself, or at least obtain it garden fresh direct from a grower.

Vegetables which do not mature their entire crop at the same time require repeated pickings to ensure a uniform, high quality. This may mean processing a few packages at a time. By planning your garden to allow for succession plantings you can extend the harvesting season and provide a continued supply of fresh produce at a desirable stage of maturity. A poorly planned garden often results in an oversupply of certain vegetables. Then not all of the crop can be used at the prime stage of development. As maturity advances, vegetables increase in size and fiber content. Changes in color and nutritive value take place. In peas, corn and other crops, sugar content, after reaching a maximum, may decrease while starch increases, giving the particular produce a characteristic starchy flavor. Overmature leafy vegetables are tough because they contain too much fiber.

Unfortunately, too often the overmature portion of the crop is used for canning or freezing. This is poor policy. As a general rule, for most satisfactory results, vegetables for freezing should be harvested when slightly younger than desired for eating fresh.

In canning, the prolonged cooking process sometimes improves texture and flavor of fully matured vegetables. The same results cannot be expected when overmature vegetables are frozen.

Handling After Harvest

The next important factor influencing the condition of a vegetable at the time of processing is its handling after harvesting, during the time between picking and freezing. Vegetables should be prepared for freezing and put into the freezer immediately after harvest.

A few examples may serve to explain why speed is so essential, and why delay interferes with the quality of the final product. Physiological and chemical changes continue after harvesting. The rate of these changes depends on the temperature of the product. In the case of corn and peas, for example, at high temperatures the sugar contained in these crops changes rapidly to starch. At a temperature of 85 degrees Fahrenheit sweet corn loses half its sugar in 24 hours. Asparagus, too, undergoes marked changes in chemical composition after it is cut. These changes are chiefly a loss in sugar and an increase in fiber; they are most pronounced during the first 24 hours and at high temperatures. Losses in sugar and increases in starch and fiber obviously affect the quality of the product. Furthermore, the vitamin content may be seriously reduced when processing is not done immediately after harvesting. A delay of 24 hours may cause a loss of 50 per cent of the vitamin C content and smaller losses of other vitamins.

If you cannot avoid a delay in processing, cool vegetables immediately after harvesting. Do not keep them at room temperature or, even worse, expose them to the sun. The quickest way of cooling them is to immerse them in ice water. After draining, keep them at low temperatures, preferably between 32 and 40 degrees Fahrenheit. Covering the produce with cracked ice is another means of cooling and thereby slowing down the loss of quality. These aids, however, do not replace the need for prompt processing.

A crop of superior quality improperly handled will give disappointing results. On the other hand, an inferior raw product can result only in an inferior frozen product even though the handling is carried out with the greatest of care.

In other words, you cannot expect to get anything better out of the freezer than you put in and a good quality fresh product can be ruined by poor handling.

What Vegetables Can Be Frozen?

Generally speaking, vegetables best adapted for freezing are those that are usually cooked before serving. These include asparagus, snap beans and lima beans, beets and beet greens, cauliflower, broccoli, Brussels sprouts, peas, carrots, kohlrabi, rhubarb, squash, sweet corn, spinach and other vegetable greens. These are discussed in detail on the following pages. Vegetables that are usually eaten raw, such as celery, cabbage, cucumbers, lettuce, onions, radishes and tomatoes, are least suited for freezing.

TABLE 48: Selection and Preparation of Vegetables for Freezing

Vegetable	How to select	How to prepare	Scalding time in boiling water
Asparagus	Use young, rapidly grown stalks. Freeze same day harvested. Discard short stubby white stalks. Cut stalks 5- to 10-inch lengths in garden.	Wash in several waters. Sort for size. Eliminate tough portion of stalk. Cut into ¾- to 1-inch lengths or to fit container.	Small stalks, 2 min. Medium stalks, 3 min. Large stalks, 4 min. Cool.
Beans — Lima	Pick when ready for table use and pods slightly rounded and bright green. Discard overmature and discolored pods or use in dry form.	Wash. Scald 4 minutes before shelling; cool in cold water; drain. Shell with pea sheller or by hand. Rinse shelled beans in cold water.	Scald in pods 4 min. No additional scalding necessary. Cool.
Beans — Shelled Green	Pick when pods are well filled but beans are still green and tender.	Shell beans. Wash.	1 minute. Cool.
Beans — Snap	Pick when pods desired length but before seeds mature. Pods must be tender. Discard moldy or imperfect pods.	Wash in cold water. Rinse and drain. Snip ends and cut into desired lengths or leave whole.	3 to 4 min. depending on maturity. Cool.
Beans — Vegetable Soybeans	Pick when pods fairly well rounded, but while still bright green. Yellowish pods indicate overmaturity. Two or three days too long in garden result in overmaturity.	Wash. Scald 5 minutes before shelling; cool in cold water; drain. Shell with pea sheller or by hand. Rinse shelled beans in cold water.	Scald in pod 5 min. No additional scalding necessary. Cool.
Beets	Harvest while tender and mild flavored. Beets are preferred canned.	Wash. Leave on ½ inch of tops. Cook whole until tender. Skin and cut as desired.	No further heating necessary. Cool.
Broccoli and Brussels Sprouts	Select well-formed heads free from insect infestation. Buds showing yellow flowers are too mature.	Wash, peel and trim. Split broccoli lengthwise into pieces not more than 1½ inches across. Remove outer leaves from Brussels sprouts.	Broccoli, 3 min. Brussels sprouts, 3-5 min. depending on size.
Carrots	Harvest while tender and of mild flavor. Root cellar storage is preferred to freezing.	Top, wash and peel. Small tender carrots may be frozen whole. Others, cut into ¼-inch cubes or slices or Frenched.	2 min. for small cubes; 3 min. for slightly larger pieces. 5 min. for whole. Cool.
Cauliflower	Select well formed heads free from blemishes.	Wash. Break into flowerlets. Peel and split stems. Soak in salt water 30 minutes.	3 min. May add acid to scalding water to keep color white.

TABLE 48. Selection and Preparation of Vegetables for Freezing (Continued)

Vegetable	How to select	How to prepare	Scalding time in boiling water
Corn — Sweet	Select tender, ripe ears in full milk stage. Do not use in dough stage. Discard smutted and badly worm-infested ears. Hard cutting ears indicate old corn.	Husk, de-silk, wash, sort and trim. After scalding cut off kernels. For corn on cob, cut off tips of ears and square up basal ends.	For cut corn, 4 min. Corn on cob — small ears, 7 min., medium ears, 9 min., large ears, 11 min.
Eggplant	Select firm, heavy fruit of uniform dark purple color. Harvest while seeds are tender.	Wash, peel, cut into ⅓ to ½-inch slices or cubes. Dip in solution of 1 T. lemon juice to 1 qt. water. Dip again in lemon juice solution after heating and cooling.	4 minutes. Cool.
Kohlrabi	Harvest while tender and of mild flavor. Avoid overmature products.	Wash and trim off trunk. Slice or dice in ½-inch pieces or smaller.	1 to 2 min. depending on size of cubes or slices. Cool.
Mushrooms	Select firm, tender mushrooms, small to medium size.	Wash, cut off lower part of stems. Cut large mushrooms into pieces. Add 1⅓ t. ascorbic acid to 1 gallon scalding water.	Small whole — 2 min. Large whole — 4 min. Slices — 2 min.
Okra	Select young tender pods.	Wash, cut off stems so as not to rupture seed cells. Freeze whole or slice crosswise after scalding.	Small pods — 2 min. Large pods — 3 min. Cool.
Peas	Pick when ready for table use as seeds become plump and pods roundish. Pack same day harvested. Sugar and quality are lost rapidly at room temperature.	Wash peas, before and after shelling. Discard immature and tough peas.	1½ min. Cool.
Peppers — Sweet and Pimentos	Select deep green or red color, glossy skin, thick flesh, and tender.	Wash, halve, remove seed. Slice or dice as preferred.	Scalding not necessary but makes packing easier. Sliced — 2 min. Halves — 3 min. Cool.
Pumpkin and Winter Squash Summer Squash	Harvest when fully colored, firm and when shell becomes hard. Use only fully matured pumpkin and squash, ripened on healthy vines. Select summer squash before rind becomes hard.	Wash, pare, cut into small pieces. Cook completely. Do not add seasoning. Cool in air or float pan in cold water to cool. Slice summer squash ½ inch thick.	Cook before packing. Scald summer squash 4 min. Zucchinis 2 to 3 min. Cool.
Spinach and Other Greens	Harvest while tender, before plants become extremely large. Cut before seed stalks appear. Harvest entire spinach plant. Use only tender center leaves from old kale and mustard plants. Select turnip leaves from young plants.	Wash through several changes of water. Trim off leaves from center stalk. Trim off large midribs and leaf stems. Discard insect-eaten or injured leaves.	2 min. Stir to prevent leaves matting together. Cool.
Sweet Potatoes	Use only high quality, smooth, firm sweet potatoes.	Wash. Cook in water until soft. Cool, remove skins. Pack whole, sliced or mashed. To each 3 cups pulp mix 2 T. lemon or orange juice or dip whole or sliced sweet potatoes in ½ cup lemon juice to 1 qt. water.	

TABLE 49: Approximate Yield of Frozen Vegetables from Fresh Vegetables

VEGETABLE	FRESH, AS PURCHASED OR PICKED	FROZEN
Asparagus	1 crate (12 2-lb. bunches) 1 to 1½ lb.	15 to 22 pt. 1 pt.
Beans, lima (in pods)	1 bu. (32 lb.) 2 to 2½ lb.	12 to 16 pt. 1 pt.
Beans, snap, green, and wax	1 bu. (30 lb.) ⅔ to 1 lb.	30 to 45 pt. 1 pt.
Beet greens	15 lb. 1 to 1½ lb.	10 to 15 pt. 1 pt.
Beets (without tops)	1 bu. (52 lb.) 1¼ to 1½ lb.	35 to 42 pt. 1 pt.
Broccoli	1 crate (25 lb.) 1 lb.	24 pt. 1 pt.
Brussels sprouts	4 quart boxes 1 lb.	6 pt. 1 pt.
Carrots (without tops)	1 bu. (50 lb.) 1¼ to 1½ lb.	32 to 40 pt. 1 pt.
Cauliflower	2 medium heads 1⅓ lb.	3 pt. 1 pt.
Chard	1 bu. (12 lb.) 1 to 1½ lb.	8 to 12 pt. 1 pt.
Collards	1 bu. (12 lb.) 1 to 1½ lb.	8 to 12 pt. 1 pt.
Corn, sweet (in husks)	1 bu. (35 lb.) 2 to 2½ lb.	14 to 17 pt. 1 pt.
Eggplant	1 lb.	1 pt.
Kale	1 bu. (18 lb.) 1 to 1½ lb.	12 to 18 pt. 1 pt.
Mustard greens	1 bu. (12 lb.) 1 to 1½ lb.	8 to 12 pt. 1 pt.
Peas	1 bu. (30 lb.) 2 to 2½ lb.	12 to 15 pt. 1 pt.
Peppers, green	⅔ lb. (3 peppers)	1 pt.
Pumpkin	3 lb.	2 pt.
Spinach	1 bu. (18 lb.) 1 to 1½ lb.	12 to 18 pt. 1 pt.
Squash, summer	1 bu. (40 lb.) 1 to 1¼ lb.	32 to 40 pt. 1 pt.
Squash, winter	3 lb.	2 pt.
Sweet potatoes	⅔ lb.	1 pt.

—U. S. Department of Agriculture

Canning Vegetables

Many of the principles concerning the quality of vegetables apply to canning as well as freezing. You should always choose young, tender vegetables, sorting them for size and maturity so that they will cook evenly. Always can vegetables quickly while they are still fresh—preferably on the day you harvest them. This is especially important for vegetables, since the delay of even a few hours means change in flavor and the growth of many resistant bacteria.

The following information on canning vegetables is excerpted from Circular 749 of the University of Illinois College of Agriculture:

Types of Canners

Steam-Pressure Canner. For safe use of the steam-pressure canner, clean the safety valve and petcock openings by drawing a string or narrow strip of cloth through them. A dial pressure gauge should be checked each year before the canning season. See your home adviser, dealer or manufacturer about checking it. A weighted gauge needs only to be thoroughly cleaned.

Wash the canner kettle well before using it. Wipe the cover with a damp, clean cloth—don't put it in water.

When using the canner, follow the manufacturer's directions. At end of processing time, be sure to let the pressure return to zero before opening the canner.

Pressure Saucepan. A pressure saucepan having an accurate indicator or gauge for controlling pressure at 10 pounds (240° F.) may be used for processing vegetables in pint jars or No. 2 cans.

Water-Bath Canner. Any large vessel will do for a boiling-water bath canner if it meets these requirements: It should be deep enough to have at least one inch of water over the top of the jars and an inch or two of extra space for boiling. It should have a snug-fitting cover. And there should be a rack to keep the jars from touching the bottom.

If your steam-pressure canner is deep enough, you can use it as a water bath. Set the cover in place without fastening it. Be sure to have the petcock wide open, so that steam escapes and no pressure is built up.

Containers

Glass Jars. For processing foods in a boiling water bath or a steam-pressure cooker, use only jars made especially for canning. Be sure all jars are in good condition, clean, and hot before packing food in them.

There are four types of closure for glass jars. Be sure to follow the sealing directions that come with each type closure. Some general suggestions are given in the following paragraphs.

Mason Top. If porcelain lining is cracked, broken, or loose, or if there is even a slight dent at the seal edge, discard the cover. Opening these jars by thrusting a knife blade into the rubber and prying ruins many good covers. Each time you use a jar, have a new rubber ring of the right size.

Three-piece Cap fits a deep-thread jar with or without a shoulder. The metal band holds cap in place during processing and cooling. Remove it when contents of jar are cold, usually after 24 hours. Use a new rubber ring each time.

Two-piece Cap. Use the metal lid only once. The metal band is needed only during processing and cooling. Do not screw it farther after taking jar from canner. Remove band after contents of jar are cold, usually after 24 hours.

Glass Top with Wire Clamp. Wire clamps must be tight enough to click when the longer one is snapped into place on the cover. Unless cracked or nicked, the glass cover may be used again and again, but a new rubber ring is needed each time.

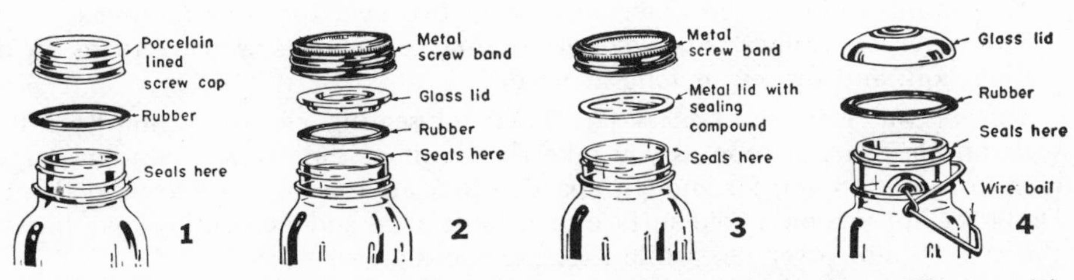

Four types of closures commonly used in canning vegetables are shown above. They are (1) mason top; (2) three-piece cap; (3) two-piece cap; and (4) glass top with wire clamp. (Courtesy of U. S. Department of Agriculture.)

Tin Cans. Use only perfect cans and lids and gaskets. Wash cans in clean water and drain upside down. Do not wash lids, as washing may damage the gasket. If lids are dusty or dirty, wipe with a damp cloth just before putting them on the cans.

Tin cans need a sealer. Be sure the sealer you use is properly adjusted. To test the sealer, put a little water in a can, seal it, then submerge the can in boiling water for a few minutes. If air bubbles rise from the can, the seal is not tight and you will need to adjust the sealer, following manufacturer's directions.

Exhausting Tin Cans

The temperature of food in tin cans must be 170 degrees Fahrenheit or higher when the cans are sealed. Heating food drives out air and helps prevent loss of color and flavor. Sealing hot also keeps cans from bulging and seams from breaking. To get the right sealing temperature, pack food hot or heat it in the open cans (exhausting).

To exhaust tin cans, place the open, filled cans in a large kettle with boiling water about two inches below the tops of the cans. Cover kettle, bring water back to boiling, boil until food reaches 170 degrees Fahrenheit (about 10 minutes). To be sure the food is heated enough, test the temperature with a thermometer, placing the bulb in the center of the can.

Remove cans from water one at a time. Replace any liquid spilled from cans by filling them with boiling packing liquid or water. Place clean lid on filled can. Seal at once and process.

Cans don't have to be exhausted if temperature of food is 170 degrees or higher when cans are filled. If you don't exhaust, work out air bubbles after filling cans. Add more boiling liquid if needed to fill cans to top. Seal at once and process.

Processing in Boiling-Water Bath

High-acid foods like fruit, tomatoes and pickled vegetables may be safely processed in a boiling-water bath.

General Directions. Place filled glass jars or tin cans in canner. If you've used a cold pack in glass jars, the water in the canner should be hot but not boiling. For hot pack in glass jars or for tin cans, have the water boiling. If necessary, after you have put the jars or cans in the canner, add enough boiling water to bring the water an inch or two over tops of containers.

Put cover on canner. Start counting time as soon as water returns to a rolling boil, and process as long as needed.

Out of the Canner. Unless jars have self-sealing closures, complete the seals on *glass jars* as soon as you take them out of the canner. Set jars on a rack, top side up and far enough apart so that air can circulate around them. Don't set hot jars on a cold surface or in a draft as sudden cooling may break the jar. Do not cover jars while they are cooling.

Cool *tin cans* quickly in cold water, using as many changes of water as necessary. Remove tin cans from the cooling water while they are slightly warm so they will dry in the air. If you must stack the cans, stagger them so that air can circulate around them.

Processing in the Steam-Pressure Canner

It is not safe to use a boiling-water bath for vegetables other than pickled ones or tomatoes. The temperature of boiling water is not high enough to destroy spoilage organisms in low-acid foods in a reasonable length of time. By using a steam-pressure canner in good condition, however, you can obtain a temperature of 240 degrees Fahrenheit at 10 pounds pressure—which is high enough to kill the organisms if the foods are processed for the required length of time.

Salting vegetables is not necessary. The small amount of salt used in canning does not prevent spoilage. To can unsalted vegetables use the same processing times recommended for salted vegetables.

In using the pressure canner, be sure to follow the manufacturer's directions. Some general pointers are given below:

General Directions. Put 2 or 3 inches of hot water in the bottom of the cooker. Set filled glass jars on rack so steam can circulate around each one. Tin cans may be staggered without a rack between layers. Fasten cover securely so that no steam escapes except at the open petcock or weighted gauge opening. Allow steam to escape from opening for 10 minutes so all air is driven out of canner. Then close petcock or put on weighted gauge and let pressure rise to 10 pounds.

Start counting time as soon as 10 pounds pressure is reached and process for the required time. Keep pressure as uniform as possible by regulating heat under canner. At end of processing time, slide canner away from heat.

Out of the canner. If you've used *glass jars,* let canner stand until pressure returns to zero. Wait a minute or two, then slowly open the petcock or remove weighted gauge. Unfasten cover and tilt far side up so steam escapes away from you. As you take jars from canner, complete the seals if jars are not the self-sealing type. Set jars upright on a rack, placing them far enough apart so that air can circulate around all of them. Don't slow down cooling by covering jars.

For *tin cans,* release steam in canner at end of processing time by slowly opening the petcock or taking off weighted gauge. When no more steam escapes, remove canner cover. Cool tin cans in cold water, changing water often enough to cool them quickly. Take cans out of cooling water while still slightly warm, so that they will air-dry. If you must stack cans, stagger them so that air can get around them.

Don't Destroy Vitamins in Cooking

The principal weakness in American cooking lies in the preparation of vegetables. As they are usually cooked by boiling, 50 to 90 per cent of many nutrients and much of their flavor is lost before they reach the table. These losses are largely avoidable.

Three factors are important in evaluating prepared vegetables. First, they must provide nutrients; second, they must have a pleasing taste; and third, they must be eye-appealing. The homemaker must learn to prepare foods which incorporate these factors, as nothing could be more tasteless, less appealing or less useful than eating waterlogged, overcooked, boiled vegetables.

First, a word about raw vegetables. Actually enough could not be said for the merits of eating fresh raw vegetables right out of your garden. There is nothing that could supply more vitamins and minerals than a fresh garden salad. Don't think you are limited to vegetables commonly used in salads such as lettuce, cabbage, tomatoes, radishes, onions, celery, parsley, green peppers, cucumbers and carrots. This gives you a good list, but add to it by using raw shredded beets, chopped spinach, chopped cauliflower, chard, Brussels sprouts, kohlrabi, and nothing is sweeter than a fresh raw turnip just pulled from the garden. Mix some of these raw vegetables in with your favorite vegetable salad.

Ideally your vegetables should be gathered immediately before using, but, if you must gather them some time before, wash and dry them immediately. Chill and put them into a dark place—your refrigerator—as quickly as possible to stop enzyme action. Make your salad just before serving.

Beyond a shadow of a doubt, the best way to insure your health with a daily dosage of vitalizing minerals and vitamins is to see to it that you eat a good part of your vegetables raw.

Vitamin Losses

Vegetables to be cooked should be handled in the same manner as salad vegetables. Gather immediately before using or else wash, dry and store in a cool place. If they are left at room temperature and in the light, much folic acid, vitamin B_2 and 50 per cent or more of the vitamin C in most fresh vegetables can be lost in a few hours.

The greatest loss of vitamins and minerals occurs in soaking and boiling vegetables. Vitamins C, P and many B vitamins pass out of the vegetables and dissolve in water very quickly. Various studies made of nutrient losses have shown that when whole vegetables are boiled 4 minutes, 20 to 45 per cent of the total mineral content and 75 per cent of the sugars they contain pass into the water. Since vegetables are often soaked before and during cooking for longer periods, as much as 75 to 100 per cent of the sugars, minerals and water-soluble vitamins are often lost. The color of the water left after certain vegetables are washed slowly or boiled indicates how quickly these substances can pass out of vegetables, even though unpeeled. The losses are hastened when vegetables are soaked after being peeled, chopped, sliced or shredded and especially after the cell walls are softened by cooking.

Save Flavor

Flavor is lost when nutritive value is lost. Vegetables contain aromatic oils which give them their characteristic flavor. These oils are not true oils and they readily dissolve in water. Vegetables also contain minerals and sugars which add to their flavor. Minerals add a saltiness to the taste, and sugars cause them to be sweet. If you want your vegetables to be delicious, as well as nutritious, do not let them soak during washing and cook them by methods other than boiling.

If vegetables are not soaked, the saving of vitamins C and B_2 during preparation and cooking largely depends on preventing enzyme action and eliminating oxygen and light. Enzymes are known to be inactive when cold and are destroyed by heat. It is important to exclude as much oxygen and light as possible as vitamin C is destroyed only in the presence of oxygen and vitamin B_2, only in light. When preparing vegetables try not to let them reach room temperature and, if you prepare vegetables prior to cooking them, put them back into the refrigerator to keep them chilled. It is important to heat vegetables as quickly as possible to destroy enzyme action; therefore, preheat the oven or the utensil to be used so it is ready by the time your vegetables are prepared. Lift the lid as few times as possible during cooking. To insure flavor and the most nutrients, cook vegetables for the shortest time necessary to make them tender.

Salt draws moisture out of substances; consequently when a vegetable is salted at the beginning of cooking, its juices, carrying vitamins, minerals, sugars and flavors, are drawn out. Do not use baking soda to intensify color in vegetables. This will still amplify vitamin and mineral losses. When you prepare and cook vegetables to preserve nutrients and flavor, their natural colors are preserved also. Green vegetables contain plant acids which react chemically during cooking with the coloring matter. If the vegetable is cooked quickly and not overcooked, little acid is freed and the bright color you desire is preserved.

Don't Peel Vegetables

Another great cause of lost nutrients is the peeling of vegetables. The minerals in most root vegetables are concentrated under the skin and are discarded when the vegetable is peeled. Leave the skin on whenever it is practical. A good rule to follow is peel vegetables only when the skin is tough, bitter or too uneven to be thoroughly cleaned.

A good cooking method should meet the following requirements: the initial heating must be rapid to destroy enzymes; to prevent loss of vitamin C, contact with oxygen should be avoided by leaving the vegetable unpeeled, by covering cut surfaces with oil or by displacing the oxygen in the utensil with steam; the vegetable should be cooked in the shortest time necessary and all of the liquid which touches the vegetable should be used to prevent loss of other nutrients.

There is no objection to boiling vegetables if the water is rapidly boiling before the vegetable is dropped into it; if it is quickly reheated to boiling; if the vegetable is not overcooked; and if *all* the cooking liquid is used. But how often is all this liquid used?

Waterless Method Best

The best method of cooking vegetables is probably the waterless method. Fresh vegetables contain 70 to 95 per cent water, which is sufficient for cooking them if the heat is controlled so that no steam escapes. The utensil must have a tight-fitting lid, and the heat should be evenly distributed to the sides and the lid. In this way the vegetable cooks by heat coming from all directions. Vegetables can be cooked without any added water by this method, but a tablespoon or two should be put into the preheated utensil in order to replace oxygen by steam. The important thing to remember in this method is to keep the heat low after the first few minutes, so that no steam escapes. Waterless-cooking equipment pays for itself in saving fuel, and what is more important it pays for itself a hundredfold in promoting health.

Steaming in the pressure cooker is another good method, provided the cooking time is watched. Length of time for cooking each vegetable is included in the directions with the equipment. Only a few tablespoons of water need be used and the remaining liquid should be served with the vegetable. As soon as the cooking time is completed, the utensil should be cooled immediately. There is some controversy over this method. Some believe this method causes too much intense heat, thus destroying nutrients. There is no definite proof one way or the other so it is up to the individual's discretion whether to use this method or not.

Broiling Techniques

Another good method of cooking vegetables and one which almost everyone would have equipment for, is to cook your vegetables in the top of a double boiler. A few tablespoons of water are put into the top of the double boiler and brought to a boil; the vegetables are added and steamed over high direct heat until the enzymes are destroyed. The vegetables are then placed over boiling water, in the bottom part of the boiler, to finish cooking.

Strange as it may seem, broiling is a good method of cooking vegetables. The important factors of a good cooking method are utilized in that the initial heating is rapid, no moisture is needed and the vegetables cook quickly. If the cooking time is relatively long, the vegetable should be brushed or tossed with oil to prevent loss of vitamin C. It is important to keep the heat moderate or low after the vegetable is once heated through; otherwise you may end up with a shriveled and unattractive vegetable.

Baking Superior to Boiling

Baking is superior to boiling as a method of cooking vegetables, but vitamin C can be destroyed because of slow initial heating and long cooking. Vegetables to be baked should first be quickly steamed or put under the broiler until they are heated through and the enzymes destroyed. Then they should be transferred to the preheated oven to finish cooking. When vegetables are baked in a casserole, the liquid used should be hot, the casserole and the oven preheated and a lid put on to hold in steam and prevent contact with oxygen.

To insure a year-round supply of vitamins and minerals in your diet, you will want to process your excess produce from your garden for winter use.

17—Marketing Organic Foods

In one sense, of course, every organic vegetable garden more than pays for itself, since the money spent on it is returned several times over in the value of the superior food it produces, to say nothing of the pleasure of all the activities connected with it. In many localities, however, it is possible to make the garden pay for itself in cash, as well as in other ways. Where this can be done, all the usual advantages of gardening can be enjoyed free of charge. In other words, you can actually get paid to do what you would be happy to pay to do. All you need are a few near neighbors and a little more space than you use to grow vegetables for yourself. Your neighbors presumably buy vegetables and there is no reason why they should not buy some from you, particularly since yours have a freshness and quality they cannot obtain elsewhere.

Now, as a result of extensive tests over a period of some years, I have concluded that ways of making the home garden pay are of two main types, and that the gardener will do well to choose one type or the other according to his own circumstances.

Briefly, the first type consists of doing the minimum of work, and concentrating on a single specialty crop of high value. The second type includes methods of production and sale that involve a good deal of time and effort. They are best suited to gardeners who have retired from other work and who find the extra activity more enjoyable than not. Let us consider these alternatives in more detail.

The gardener who is regularly employed in a business or profession, and who can harvest crops only in the evening and on weekends, can make his garden pay for itself by growing a specialty crop. The only real difficulty is in choosing the right crop to grow, and success depends almost entirely on your choice. Let us see what sort of product it should be.

1. It should be one that everyone particularly likes, so it can be sold without special effort. Your own favorite vegetable may not have such general appeal.

2. It should be easy to grow, with the minimum expenditure of time and labor. Crops that need thinning, weeding or much other attention are unsatisfactory.

3. It should be equally easy to harvest. Crops like peas and beans will not do at all. Strawberries, raspberries or blueberries are not much better.

4. It should have high value for the space it occupies. Radishes, cucumbers, squashes, carrots, beets and the like do not qualify.

5. The whole crop should be ready for sale over a short period. Since your time is limited, you want a specialty that can be marketed in a few evenings or over a weekend.

6. The quality of the product should depend primarily on its freshness, so your potential customers will recognize that nothing equally good can be bought elsewhere.

7. Your specialty should be a crop that is not grown by every neighbor with a dozen square feet of land at his disposal. In some areas, for instance, nearly everyone grows his own tomatoes and, though few may grow anything else, tomatoes would be the worst possible specialty there.

As you consider these criteria for a specialty crop, you may find it hard to think of anything that meets all of them adequately. If so, you are not alone. In many years of experience, I have found just one crop that is ideally suited to the purpose, and that is sweet corn. Fresh sweet corn is a delicacy that nearly everybody appreciates and one that comparatively few home gardeners grow.

Using some of the year-round system of mulching, in combination with close planting and manure or other organic fertilizer, an enormous crop of corn can be grown in a very small space. Virtually nothing has to be done to the crop between planting and harvesting, and the harvesting itself is extremely quick and easy. At prices of 60 to 80 cents a dozen, 10 dollars worth of sweet corn can be gathered and sold in an hour or two. In an area as small as 12 feet by 25 you can easily grow more than 30 dozen ears of corn, which should sell for at least 20 dollars. This is actually a low figure. I have always done better and should expect any good gardener to.

Now for a few suggestions on production and sale.

1. If you live in an area that is not liable to late frosts in the spring, you will do best to concentrate your efforts on a single crop of corn and to aim at having it ready to sell before any is available in the stores. In this way you will have a local monopoly at a time when demand is greatest, and you will be regarded as a public benefactor in selling it at premium prices.

2. If you cannot beat the market with your corn, it is well to grow two crops rather than one. Your early corn will thus be ready when your customers have had just enough of the store product to see how much better yours is

and you can inform all of them that you will have even finer corn for them a few weeks later.

3. A few telephone calls should locate all the customers you can handle. If possible, get them to come after their corn in person and let them watch you harvest it. Customers who get their corn from you just before dinner, and cook it immediately, are going to come back for more. It will probably be the best they have ever eaten.

4. Husk all the corn as you harvest it, removing the silk cleanly, and snapping off any tips that are not filled out. This small trick takes very little time, and it enables you to see that every buyer gets perfect ears ready to cook. Explain to your customers that corn retains its flavor best if husked at once. This is not only good psychology, it also happens to be true.

5. Ask your customers what stage of maturity they prefer and try to give them what they like best. Since you inspect all the ears, this is easy enough to do and your buyers get corn that is virtually custom made.

6. Add an extra ear to each dozen without calling attention to the fact. Add two, if some of the ears are a trifle small. Customers who think your prices are high will change their minds when they discover this. Oddly enough, the extra corn is more appealing than a lower price would be.

7. Impress on everyone the importance of getting your corn to the table without delay. A customer who cannot do this should be told to get it into the refrigerator as soon as possible. Quality is what he is paying for and he will appreciate your care in seeing that he gets it.

A gardener with time on his hands can use some of it to advantage by simply growing more of all his favorite crops and keeping his neighbors supplied with a variety of vegetables throughout the season. He will have early peas for them in the spring and Brussels sprouts after the autumn frosts. In between, he will have such delicacies as baby carrots and beets, leaf lettuce, scallions, early cabbage and vine-ripened melons. He will have at least one or two items to offer each day of a quality that cannot be matched by any store anywhere. How much money he makes naturally depends on the scale of the operation; it may be anywhere from 20 dollars to 100 or more. Whatever the size of the enterprise, it should easily pay the whole money cost of the garden that helps to support it.

Although every gardener must develop his own program in the light of local conditions, experience suggests that a number of general principles are likely to be applicable everywhere:

1. It is better to supply all the vegetables required by a few neighbors than to supply only part of those needed by a large number. If you are able to count on a steady, if modest, demand for your products, you can arrange your plantings on an appropriate plan, instead of growing crops haphazardly and then trying to find buyers for them. By dealing steadily with the same customers, you learn their preferences and can satisfy them better as time goes on. You can also keep your customers informed concerning crops shortly to be ready, they can tell you their needs for the near future.

2. Remember that the most noticeable superiority of your vegetables over those available elsewhere is their freshness and prime condition, and be sure that your customers get the full benefit of this. Explain the importance of getting peas and sweet corn served within minutes of harvesting and arrange in advance for delivery of such perishable items just before mealtime. Encourage buying for immediate needs and discourage buying ahead for storage. Pass along hints on preparation and serving that will preserve the full flavor of your vegetables.

3. Keep your garden looking neat and attractive and invite your customers to make some of their own selections on the spot. If they form the habit of calling in person for their vegetables, you can discuss the garden with them and let them watch you at "work." When you are pulling a bunch of carrots for a customer, and he sees you casually discarding one or two with slight imperfections, he appreciates what he is getting better than he otherwise could.

4. Be sure that everything you sell looks as good as it really is. Since nearly all vegetables in the stores look much better than they are, this means that yours must present a specially attractive appearance. Every gardener knows that a slightly crooked carrot or a cracked cabbage is exactly as good to eat as any other, and it is tempting to convey this intelligence to a customer along with the item in question. It is better, however, to save such things for your own use. When you must supply something short of the handsomest specimens, make the price a little lower than usual. Be firm, but unostentatious about this and it will leave fewer doubts about your standards than would any amount of conversation.

5. Do not be afraid to charge fair prices for your products. They are much superior to any available in chain stores or supermarkets, and they are likewise better than anything offered by roadside stands. It is the rare quality, rather than the cheapness of your vegetables that you should stress and that your customers are going to recognize. If any of your neighbors is unable to see how much better your products are than those he can buy elsewhere, ,
do not want him for a customer at all.

A good way to determine your prices is to check occasionally with the stores at which your customers trade and charge the nearest price above this figure that is evenly divisible by 5. In other words, if snap beans are 27 cents a pound at the local stores, you charge 30 cents; if they are 33 cents, you charge 35, and so on. This makes bookkeeping easy, saves trouble in making change, and keeps your prices in line with market conditions.

6. Keep accurate records of cash expense and receipts. An excellent device for doing this is a low-priced 5-year diary in which you enter purchases and sales each day, together with any other data you wish. A useful technique is to keep totals up to date as you go, circling these figures, or distinguishing them from the rest in some other simple way. A 5-year diary provides just the right amount of space for daily records, and has the special advantage of enabling you to compare the current year's operations with those of preceding years from day to day.

7. Do not undertake this method of making your garden pay for itself unless you are sure that you have ample time for it. If you are to supply one or

more families with vegetables every day, you must be on hand to take orders and make deliveries whenever your customers choose to do business. To be sure, you can, by prearrangement with them, take a day or two off from time to time. But unless you mean to be at home as a rule for other reasons, you should not try to keep your neighbors steadily supplied with vegetables. Another way of making your garden pay will work better for you.

Although gardeners with limited time at their disposal are unlikely to be drawn into selling much besides their specialty crops, those who have plenty of time may be tempted to add the corn specialty to their other methods of making the garden pay. Since this effort can easily lead to trouble, it may be well to insert a word of warning here.

The difficulty in trying to combine the alternative ways of making your garden pay is that you must keep your markets separated, which is very hard to do. As soon as your numerous corn customers discover that you are supplying not only corn but all kinds of other vegetables to a few families steadily, they are going to want more than corn from you. If you refuse to sell them other vegetables, you are likely to lose valuable good will. On the other hand, if you try to sell what you can spare, these new customers are not going to be satisfied. At best, they will feel that they are getting what is left over after others have had their pick and this will be perfectly true. At least, it had better be true. If you became sufficiently demoralized, you might try to please your new customers at the expense of the rest and make a mess of the whole thing.

My advice, then, is to choose the one alternative that fits your own situation, and make the most of it. If you do, I am sure you will find that making your garden pay its way financially is easy. You will also find that your garden is paying better than ever in enjoyment, quite apart from the money you make.

—Richard V. Clemence

Stress "Organically Grown"

There are a great many people who want to obtain organically grown vegetables and fruit. You can let the people in your area know that you have organic produce for sale.

Remember to stress the words, "organically grown," and put additional small signs explaining the value of chemical-free produce on the stand itself. Small cards bearing the same message and given out with each purchase are good advertising too. Some roadside sellers keep a register book on the stand and encourage customers to write their names and addresses in it for possible mail-order business in the future.

75 Crops on 120 Acres

How can small farmers stay in business? Walter and Fred Fischer of Fullerton, Maryland, have come up with one solution: *raising as much and as many different vegetables as possible.* The family farm consists of some 80 acres plus an additional 40 acres which they rent. Close to 50 acres are in vegetables, with the balance going for grain, hay, cattle, hogs and chickens.

As might be expected, growing over 75 different vegetables in a season is

The roadside stand of Frank Miceli, a California organic grower, is pictured above. Always remember to stress the words "organically-grown" on signs; some growers have cards printed explaining the value of chemical-free produce for distribution to customers.

a lot more time-consuming and sometimes more confusing than specializing in only a few crops. But this variety is also more profitable.

In late spring, the Fischers set up two stands along the road in front of their farm. Their children ranging in age from 11 to 16, do most of the work connected with the stand. In the last few years, a regular trade has been built up, so most of the produce is sold at the stands. When there's a surplus, the younger brother, Fred, loads the truck and manages to sell the balance of the vegetables in the nearby suburbs of Baltimore.

Generally speaking, the Fischers try to price their organically grown vegetables, beef and eggs somewhat lower than similar foods on sale at retail stores. They don't believe in charging premium prices. Potatoes and carrots are sold at $3.00 a bushel; turnips, beets and rutabagas range from $1.50 to $2.00 a bushel. And as for flavor and quality, Walter Fischer challenges: "We dare the chemical people to say they can raise better crops."

Besides their vegetables, the two brothers raise all their own grain—wheat, barley, oats and corn. They stone grind and sell their own whole wheat and rye flour. Their poultry set-up is working out well. This spring out of 200 pullets and 100 straight-run, they lost only 3 chicks. There is a steady demand for their eggs.

To make a diversified operation of this kind practical, you have to be willing to test different ideas and techniques all the time. And the Fischer brothers are always willing to do that. Sometimes the ideas work well, sometimes not, but they keep right on trying.

Raising several crops on the same piece of land during one growing season is important for economy's sake. The Fischers try to do this whenever they can with crops such as radishes, spinach and soybeans. As a result, they often get two, sometimes 3, crops from a single area.

If they are planting a large area with the same crop, they use the farm tractor with a two-row seeder. When cultivating a small section, they use one of two garden tillers they have.

Often many different varieties of a vegetable are grown to satisfy both the tastes of their customers as well as to vary the time of harvesting.

Getting early crops in spring is important, and it's worth some extra effort to encourage fast maturing. For example, in a one-half acre rhubarb planting, bushel baskets were laid over about half of the plants when they first appeared above ground. This was done to make the leaves come up straighter and the fruit develop earlier.

The rotation schedule followed is a simple one. Soil-depleting crops, such as corn and vegetables thriving best in rich soil, as tomatoes, are rotated each year. They often grow the same vegetables in several different fields as insurance against a poor crop in one section of the farm but not in another.

A large root cellar capable of storing close to 1,000 bushels of carrots, beets, potatoes, rutabagas and other vegetables permits them to store vegetables for sale over the winter months. Fischer is sure that the keeping quality of these vegetables has improved since he stopped using nitrate of soda and other artificial fertilizers. Now his vegetables keep firm and little, if any, rotting takes place.

Only in the peak time of harvesting is any outside help hired. Aside from that, the two Fischers with some help from their children do all the work on their 120-acre farm. They have to buy practically nothing to feed their families. Meats, dairy products, fresh vegetables—all are produced right on the farm.

While there's a lot of work to be done and it keeps them busy all the time, they're confident that they'll be able to maintain their small family farm on the plan they've developed. And that's in addition to the pride they have in raising top-quality crops.

Raising Herbs for a Living

When I became an herb grower 7 years ago, it was a big move from being a general merchandise jobber. The change of occupation was not my choice, but was forced on me through illness after I was 40. Now that my health is restored, I would not trade places with anyone. The life is independent and the business is free from pressure.

My first attempt to sell herbs was to advertise them in the local paper. They sold well in the beginning, but when all the herb enthusiasts were supplied, we ran out of customers. In order to continue in business we decided to sell through mail order by advertising in magazines with a national circulation. We chose garden and farm publications because their readers would be potential customers for our plants and seeds.

It was with high hopes in the fall of 1953 that we launched our venture. The start was slow and it took several months before we were able to show a profit. The first year we were in business we call "the year of the creditors" because there were so many of them. To cut expenses we planted a large garden, grew our own fruit and raised all of the eggs and meat that we needed. A couple of beehives produced all of the sweetening that our family of 4 could use.

From a start of less than a quarter of an acre of herbs we now grow over two acres of them. Herbs are simple to grow and require only average fertility which we provide through barnyard manures and cover cropping. Plants forced into quick growth through the use of commercial fertilizers are inferior because they are low in aromatic oils. Herbs are comparatively disease free and the few insects in our fields that may prove injurious are kept under control by a flock of black Cochin Bantams.

The 30 different kinds of plants that we grow to sell are the perennial varieties of culinary, aromatic and decorative herbs. We supply only the seeds of the annual herbs because the plants do not ship well. Due to the considerable publicity that herb cookery has received in recent years, the culinary varieties such as sage, rosemary, thyme and tarragon have become our best sellers.

Most of the propagation of the herbs is done through plant divisions in the fall. A few varieties such as sage, rosemary and the thymes are increased through summer cuttings rooted in sand. The small plants are kept ready for shipping by setting them out in raised beds filled with a mixture of peat moss and sand. Herbs left over after the shipping season are planted in the field to fill out rows and replace old overgrown plants.

Seed growing starts in the middle of spring and continues until late summer. The seeds are harvested as soon as they have matured and turned color. They are then sifted free of impurities, placed on sheets in a well-ventilated room and thoroughly dried. Our State Department of Agriculture demands that seeds packaged for sale must meet minimum germination requirements.

In the last couple of years we have expanded the herb business into producing packaged dried culinary herbs, herb teas and potted herb plants for indoor use. We even sell surplus honey from the several hives we now keep for plant pollination. These products, sold locally to retail stores, help diversify our business and have proven a good source of income.

The culinary herbs raised for selling to grocery stores are sweet basil, garden sage, thyme and summer savory. These plants are easily grown and require little equipment to process and dry for packaging. Harvesting the herbs consists of cutting two-thirds of the top foliage. In late summer the same plants will have grown sufficiently to produce a second crop.

The herbs are dried indoors by placing a thin layer of foliage on screen racks with a good circulation of air. In drying they must never be exposed to direct sunlight or it will cause plant discoloration and loss of flavor. The dried culinary herbs are packaged in small transparent polyethylene bags and stapled on display cards. Sales increased when printed instructions and recipes were enclosed in the packages. We have recently developed an all-herb seasoning for Italian spaghetti which is becoming very popular.

The plants used for teas are sage, catnip, lemon balm and lemon verbena.

They are harvested and dried the same way as the culinary herbs. Herb teas are a specialty food that command a good price and for that reason we package them in fancy glass jars with attractive labels. A small printed pamphlet is tied to the top of each jar with instructions for herb-tea making and the story of their health-giving benefits.

There is a growing interest in potted herbs and in the fall of the year we sell several hundred of them to garden stores and florists. Chives, marjoram and sweet basil, sage, rosemary and parsley make excellent indoor plants for winter use. They are decorative and provide fresh seasoning material for cooking. Last year we had a local wood-worker make us some indoor window boxes and offered the herbs in groups with the window box. Each pot carried the name of the herb and its culinary uses.

There are many other products made from herbs that can be readily sold. Gift shops and specialty stores are good outlets for lavender sachets, tarragon vinegar, camomile hair rinses and bergamot after-shave lotion. Herbs make excellent confections and no sweet tooth can resist candied angelica stalks or honeyed borage blossoms. We recently heard of an elderly woman who grows catnip for making catnip mice. She sells hundreds of them by mail to pet shops.

Herb growing is not a "get-rich-quick business," but a good living can be made on acreage raising them if a grower processes and markets his own plants. Raising herbs is an ideal part-time business for retired and semiretired people who want to supplement their income. This may be the small outdoor enterprise that will help solve your problems. At least it did for us.

—N. P. Nichols

Roadside Stands

Many growers have found that a roadside stand is the most profitable method for selling surplus produce. However, care must be taken in locating and building the stand, displaying the produce itself and publicizing. Clifford R. Eckstrom, Maine Extension Economist in Marketing, makes these comments:

Location of the Stand

In deciding where to locate your roadside stand, you should consider how much traffic the highway carries. You may have little choice as to the highway where you place your stand, yet the type of traffic will largely determine how much business you can do.

A location near a city or a summer home area is much to be desired. Repeat buying by established customers is important in maintaining a good volume of business.

You should carefully consider the location of the stand beside the highway. The best place, both for business and for safety reasons, is where you can see quite a distance both ways. Locate your stand, if you can, away from hills, or curves, or anything that may obstruct the vision of motorists coming in either direction. The approach should be open enough so that the driver can see the stand in time to slow down gradually. He should also have ample

An attractive roadside stand can be a most profitable method of selling surplus produce. The stand should be located beside a highway, and should be well-kept. Signs should be neat and easy to read, may also quote prices of major products for sale.

vision both up and down the highway for making a left hand turn to get into your driveway.

You should have ample parking space around the stand for the motorists who stop. You can do that by placing the stand back from the highway so cars can park well off the road in front of the stand, or you may provide parking space at the sides. Anything you can do in locating the stand to make it safer and easier for motorists to stop will encourage better business.

The appearance of the surroundings of your stand are also important. Well-kept buildings or farmyards will add much to the general appearance and attract customers. Poorly kept surroundings will drive them away.

It is also wise, when you can, to locate the stand near the field or orchard where the produce is raised. When the motorist can see the garden, it is good advertising as to the freshness of the produce handled.

Try to locate your roadside stand in a shady spot. Shade not only provides comfort and an added inducement for customers to stop, but it also gives protection to the produce so that it will retain top appearance and quality for a longer time.

The Stand Itself

The kind of stand you build and the size of investment that is warranted will vary with the produce to be handled and the amount of business that can be expected. If you plan to operate for just a few weeks while one or two items of produce are in season, an elaborate stand is neither warranted nor necessary. If however, the stand is to be opened early in the season to sell early summer vegetables and to be kept going through the fall to market late crops, a more elaborate building may be warranted.

A successful business does not of necessity require a large investment in a building. If you have a limited amount of produce to market, a table or bench in a shaded spot may be adequate. If you plan to operate for a longer period and will handle a greater variety of products, a more permanent stand should be considered. This need not be elaborate, but should have a roof and walls for protection in foul weather. It should also provide plenty of shelf space for the display of produce. The construction or location of the stand should provide shade and protection for the display shelves.

Neatness in construction and general surroundings of a stand is good advertising for the business. A little landscaping and care in tidying the surrounding area will do much to attract the eye of the passing motorist and encourage him to stop at your place of business.

Grading and Display of Produce

Proper grading and display of produce are important. Produce should be so graded and displayed that the customer can see exactly what he is getting when he buys. Misrepresented produce may make a first sale but dissatisfied customers will not bring repeat trade. Produce should be the same in quality throughout the lot which is offered for sale. It should not be "deaconed" with the best quality on top of the package and poorer quality underneath.

Where volume of produce handled permits, it may be well to have two grades for sale and offer the lower grade at a lower price. There is usually a market for the lower quality when it is sold at the right price and in the long run you should be money ahead by selling it as a separate grade rather than mixing it in with the top quality produce.

A pleasing display of produce is a good attraction for increased sales. Volume display is good. A customer is less likely to buy what appears to be the odds and ends left from a lot than to buy from a well-stocked stand. Volume and variety both attract the eye.

It's a good idea to make provision for sprinkling water over the produce on the display rack. This will do much to add to the eye appeal as well as to improve the keeping quality of the produce on display.

Often produce is not completely sold out the day it is picked and must be held over till the next day. Produce thus held over will retain much of its quality and eye appeal if handled properly. A good sprinkling with water and holding it in a basement or other cool place will do much to retain quality during a short holdover. Packing produce in a crate or barrel with crushed ice will keep it crisp and fresh for quite a long time.

Pricing the Produce

Because direct marketing at roadside stands avoids many of the handling and distribution costs it should result in a saving. This saving is made possible by the efforts of both grower and customer. The grower contributes by performing certain retail functions. The customer contributes by buying the produce where it is grown. If both parties contribute, both should share in the savings.

A stand operator should not price himself out of the market. He deserves, and should expect, a price somewhat above the farm wholsale price. On the other hand the customer deserves, and should expect, a price somewhat below the city retail price. When the price is such as to benefit both, the business is most apt to continue.

Market reports may be used as a guide in establishing prices. These are available from your state experiment station. They list the prevailing wholesale prices in produce markets in different cities in the state and many market operators keep posted on them and then set their stand prices somewhat above these wholesale levels.

Many stand operators keep posted on the prevailing retail prices in town and then set their prices somewhat below those figures. Prices charged by other roadside stands along the highway are also watched and operators try not to get far out of line with these prevailing prices.

Customers stop at roadside stands because they are looking for fresh produce of good quality and expect to get it at a price somewhat lower than city retail prices. It has been estimated that approximately 15 per cent of the customers who patronize roadside stands are mainly quality conscious. They want quality produce at any price. Then about 70 per cent of the customers are both price and quality conscious. This is the largest and the most important group to which you will cater. They want quality produce but they are also conscious of price. The remaining 15 per cent are mainly price conscious. People in this group are interested mainly in a bargain price, and they are willing to take lower quality at what they consider to be the right price.

Signs

Your sign is the potential customer's first contact with your business, and every effort should be made to make this contact appealing enough to induce him to stop.

A good identifying sign is always an asset for your place of business just as are signs that advertise at least the major products which are for sale. Signs need not be professional but they should be neat and easy to read.

Often it is a good idea to quote the prices on your signs advertising your major products for sale. Many customers like to know the price before they stop. If your price is more than they wish to pay it will save you time and trouble if they do not have to stop to find out. You will then have more time for those customers who are more likely to make a purchase.

Signs should not be long and wordy because the motorist as he drives past has time only for a quick glance. Too many signs are distracting. It is desirable to place identifying signs beside the highway so that motorists approaching from either direction have time to slow down gradually to a stop before they pass the stand. Motorists are reluctant to have to stop quickly as well as to back up if they pass the place of business before they have a chance to stop.

You will get along best with your customers if you advertise only those items that you actually have for sale. If any item is sold out, the sign advertising it should be taken down or covered.

Other Advertising

Some operators of roadside stands have found that it pays dividends to carry their advertising beyond the use of signs along the highway. This may include newspaper or radio advertising as well as marking the stand name on bags and other containers in which produce is sold.

Operators sometimes keep a record of repeat customers to whom they may send notices when produce is coming on the market.

You will find that good quality produce at a fair price makes satisfied customers. This is the most effective of all advertising. Satisfied customers will stop again and will bring new customers to your place of business. There is no substitute for advertising of this type.

—Clifford R. Eckstrom

Personal Experiences

The following experiences of Paul Keene and W. Hamilton Noyes will give you additional information on the methods used by organic growers in marketing their foods. Mr. Keene of Penns Creek, Pennsylvania concentrates on mail-order selling, while Mr. Noyes of Woodstock, Illinois sells direct to his customers. Here are their stories:

Walnut Acres

Walnut acres is a farm located in Penns Creek, a tiny village in central Pennsylvania, not located in a suburban area or on any bus, rail or air lines. Because we are situated away from any large city or centers of population, there is no use of thinking of marketing organic produce in any way other than by mail order at present. Of course, the idea of marketing such produce at all was slow in appearing and developing. It had to start in a very small way and grow naturally—organically if you like.

Today we sell the grain produce from our 160 organically tilled acres, plus a great variety of other natural food products which we cannot raise or prepare here. About 97 per cent of this is shipped out by parcel post, express and freight. The other 3 per cent is sold directly to customers who come from near and far to see the farm and the mill and store.

In the mill, we grind grains into all kinds of flour and cereal. The grinding is done every day, fresh to order only. This procedure holds up orders for a day or two but it guarantees absolute freshness, which is of great importance. It also prevents insects from getting into the ground grain as nothing stands around waiting to be sold.

Another way in which we overcome the insect problem is through the use of refrigeration. After harvest our grains are thoroughly cleaned and then run into large clean refrigerated bins. There are no insects, rats or mice in these bins. The grain and grain products are never treated in any way with fumigants, preservatives and the like, for none are necessary. The Food and Drug inspectors are always highly pleased when they come here for they do not have to work very hard!

Perhaps the most difficult thing to obtain, and the most highly to be desired, is the confidence of one's cutomers. An organic farmer who markets his own produce must be absolutely honest both with himself and with his customers. There can be no easy, get-rich-quick shortcuts. Organic foods must be of highest quality. One shipment of second-rate goods can destroy much good will which has been slow in building up.

Just as organic farmers lean over backward in returning to the soil more than they take out, so they must be willing to go the second mile with their customers. No effort in courtesy, kindness, understanding and desire to please is too great. If this way of farming is really a way of life, the same principles will follow through in all the farmer's dealings. One's business advertises itself if it is sincerely carried out.

To sum up, we have really had very few problems. It seems that vision, sincerity, cleanliness, friendliness and hard work just about guarantee success. The organic market is here and it is growing. People are tired of being fooled. Each farmer must work out his own solution in his own way, but it can be done!

—Paul Keene

Top Farm

In order to sell your product, I feel it is best for the farmer or grower to sell direct to the consumer and thus avoid any of the natural foods becoming contaminated through mishandling, spraying at point of sale or delays, thereby losing the freshness as well, of the product. In trying to find a market in stores for organic foods within the area, I found it most discouraging. Because of this, I built a store on my farm where the customers can come out and make their purchases in confidence. My old customers know that I have spent 8 long years in building up the soil on this 160-acre farm to a purely organic state, and there is no question as to the quality of the produce I have to offer. We have found that the customers are usually willing to pay for quality, especially when they realize the present-day foods do not contain the nutrients they should. The buyer cannot get full value for his money in the stores.

Satisfied customers inform their neighbors and relatives, and this sets up a chain reaction which of course is not to be ignored.

If a customer is new and unfamiliar with organic foods as against processed foods, it has been found advantageous to spend a little time in explaining the worthiness of purchasing organic foods and also in an explanation of how the food is grown and produced for sale as well as its contents in minerals and vitamins. It helps in selling the new customer on the new approach to better living. In other words, a short lesson in nutrition helps to convince the customer he is being "short-changed" in buying foodless foods.

Some of our customers drive out to the store coming as far as 150 to 200 miles and they buy in quantity. Where there is a car-trip pool among neighbors or friends, they can take turns in driving out for the needs of several families and obtain foods more frequently. We also found it necessary to educate the people on ways and means of preparing the natural foods so they do not destroy the vital elements which are in them—so methods and recipes are suggested.

—W. Hamilton Noyes

18—Growing Vegetables in the Greenhouse

A BIG ADVANTAGE of a greenhouse lies in its ability to produce fresh vegetables for the organic gardener's winter table. Lettuce, radishes, Swiss chard, kale and scallions do very well in the cool greenhouse. So do carrots, cauliflower, peas, red and green cabbage and beets, if you have the extra space they need. Tomatoes and cucumbers need more room too, and a night temperature of 60 degrees or higher. Many growers prefer to start their tomatoes from seed in January and thus have tomatoes two months before their outdoor plants begin to bear.

Slips may be taken from garden-grown tomato plants in mid-August and rooted in tubs for the greenhouse. These plants may be staked and pruned to one main stalk to save space. They will begin to bear at about the time when frost nips garden plants and will sometimes continue to bear until next year's garden plants begin to ripen their fruit. Italian plum tomatoes are especially good for this purpose.

Herbs are another good crop. A few pots or boxes of rue, sage, mint, marjoram, parsley, chives and the like will provide garnishes for winter meals.

Cucumbers and melons are only practical as a summer crop in the small house for they need space and higher temperatures. It is grand to have the large, juicy melons that can be grown to perfection under glass in the summer and cucumber with seed so soft there hardly seem to be any at all.

One of the most profitable things done in a greenhouse is to raise plants for setting out in your summer garden. They can be grown by the thousands in a home greenhouse. Husky plants for tomatoes, cabbage, cauliflower, lettuce, melons, broccoli, celery and peppers, as well as all of the bedding plants for the flower garden. Seed is inexpensive and the plants are much finer than those offered for sale and may be raised in strains not available on the market.

In the Organic Gardening Experimental Farm greenhouse there are no benches. Instead, the plants are set directly in the soil which is considered better with long season crops. The entire soil area is treated with a mulch of leaf mold, manure and rock powders in July and this application is rototilled in during the early part of September.

Many greenhouse owners have found that less than 10 hours a week can maintain a 16- by 20-foot greenhouse thusly: 15 minutes a day for watering, a few minutes a week for preparing compost, keeping the house clean, etc., the rest in spare time or on weekends for actual work with plants. Of course, the more time you put into it, the greater will be your pleasure and profit.

Early fall is a good time to build a greenhouse. Most plants can be started during the winter, but tomatoes, lettuce and cauliflower need an early start, which you can give by sowing them in a garden frame while the greenhouse is being erected. Husky plants before frost is your aim.

What size and style of greenhouse is best for you? Judging from the enthusiastic reports of owners, no matter what size you build, it will soon become too small. Your activities will grow as fast as the fun you have and much faster than your space.

Certain basic considerations, however, are important. If you decide to add a charming sunny room to your home by building an attached greenhouse, it must be on the southerly side, preferably with a southeast exposure. (Heating is easier in an attached greenhouse—you can utilize your home heating system.) If you want a free-standing greenhouse, put it where you can see and enjoy it fully. In the North, experts recommend running a greenhouse on a line 10 degrees northeast and southwest to take full advantage of meager winter sunlight.

A well-planned greenhouse can supply lettuce, radishes, carrots, cauliflowers, peas, cabbages and many other vegetables throughout the winter. In the greenhouse, vegetables can be grown directly in soil, as well as in benches.

The simplest kind of greenhouse is, of course, the cold frame, a box covered with window sash sloping to the south. Make it a little bigger, add a foot or so of heat-producing manure, sawdust and compost and you have a hotbed. Both will grow good crops of vegetables in cold climates. Dig a pit or aisle to stand in, add perhaps heat and electric light and there's a pit greenhouse. All of these styles can be attached to a building, such as your home or garage, or free-standing. You can make canvas covers that roll up on poles for cold night protection.

For those who don't want the trouble of building a greenhouse from scratch, the new aluminum prefabs are great. Completely rustproof and requiring no putty or painting, their gleaming beauty is an asset to any home. Some have weather-tight, fuel-saving, rubber channels to hold their large panes of glass—glass strong enough to withstand the heaviest hail or wind storm. They come in a large selection of sizes, in attached and free-standing styles and straight and curved eaves. They can be set up easily with hand tools.

You want a strictly organic soil, open and rich in humus. It should feel, look and smell good, with bounteous life in it so that seedlings can take advantage of the warm, moisture-laden air to give that luxuriant "tropical" growth characteristic of greenhouse plants. The best soil is made by composting manure and sod, with some rock phosphate added. Any good compost, however, is fine. It should be mixed in equal parts with garden loam and sand or leaf mold.

Bone meal, wood ashes, dried blood and the like, plus lime, may be needed for many plants. Some growers change the soil in their boxes or benches after a few crops, but the use of organic fertilizers generally makes this unnecessary. Soil from earthworm boxes, say some growers, produces marvelous plant growth. Soil changes, plus cleanliness and disposing of infested plants, will lick what few insect pests and diseases may appear in a greenhouse.

All Year Greenhouse in Maine

Another example of how a greenhouse can produce food throughout the year (even when the temperature goes down to 45 degrees) is offered by Helen K. Nearing of Maine. Here is her story:

Most folks plant their garden in the spring, weed it occasionally as it comes to fruition and, after the crops are in, let the empty garden gather weeds till the following year.

The Orientals, especially the Japanese and the Chinese, plant a continuous garden. If they pull out radishes, they put in beets or lettuce or some other crop which, when it comes out, makes way for other seeds. They refertilize between crops, but the gardens are never empty—not a row. They sow and harvest and sow again from spring to fall.

We have practiced this way of gardening for years—our fall garden being as full of young green things as a usual spring garden. Lately we have tried to extend our season by gardening when the snows are a couple of feet high and our thermometers drop to 25 degrees below zero.

We put up two 6-foot high, 12-inch wide, stone walls on the north and east corners of our herb garden and against them built a small, simple, unheated, lean-to greenhouse, which makes possible our winter garden. The roof consists of 16 by 24 inch double-strength window glass laid on 2 by 4 inch cedar rafters, 8 storm windows for front wall, with a glassed door for entrance. Two parallel beds supported by 2 by 8 inch planks run lengthwise, with an 18-inch path between. The beds are 15 inches high and rest on the earth beneath.

The greenhouse is full of tomato and pepper or melon plants all summer. These and other plants are carried fresh and green into the fall when their brother plants outside are frozen stiff.

Through most of the winter we plant, transplant and harvest, in this small space, lettuce, radishes, mustard and cress, chives and parsley. When winter days or nights get as cold as 10 or 15 degrees, we leave the ground unstirred but cut vibrant succulent lettuce that has revived on sunny days from its limpness during the freeze. In the fall we transplant mature cos, Chinese cabbage, escarole and celery into one of the beds and eat them till almost Christmas. Parsley and chive plants carry all through the winter.

On September 7th, we planted Oakleaf, White Boston, Salad Bowl and Blackseeded Simpson lettuce in our open garden for future transplanting into the unheated greenhouse.

The first week in October we transplanted about 100 of the little lettuce seedlings, shut the greenhouse door, barred it and left October 10th for a 6 months' tour, arriving back home April 20th. During this time, no watering or cultivation was done and the outside temperatures went to 25 degrees below zero.

When we got back home April 20th, our first look-around was to the greenhouse. It was bright with the pale green of masses of mature and healthy lettuce plants. Some of the Salad Bowl variety were a foot across and a foot high. None had gone to seed and few were overripe. We counted 67 heads that were immediately edible and a few younger plants just coming along. There were 17 fine sturdy parsley plants, also transplanted from the outside garden in October. They carried an abundant crop of curly green leaves, with a particularly fine flavor. Our 5 chive plants were growing nicely.

These plants must have frozen stiff in the winter nights and sunless days. Yet here they were, revived and healthy. Condensation dripping from the inside of the glass, plus capillarity from the underlying soil, took care of the watering. The plants were vigorous and upstanding, though the ground was quite dry the day we returned home. No one had opened the greenhouse door for over 6 months and they went unwatered and unheated (except for the sun) during that whole time.

We have tried this same scheme of winter gardening in Vermont, where the temperature used to go to 45 degrees below zero. It succeeded as well.

Coldhouse Winter Gardening

The word "hothouse" was formerly used to describe a greenhouse, but it wouldn't do for ours. "Coldhouse" would be a better term, for we have operated it for the past two winters without heat of any kind except from the sun. There is no stove and we don't use fermenting manure to produce heat. The greenhouse is in the middle of an exposed field and we don't cover anything at night or any other time. Yet the greenhouse supplies us with lettuce, kale, parsley and other greens all winter here in New Jersey. There are peas, red and green cabbage, carrots and beets growing there right now in January. In fact, the carrots and beets were planted in December and germinated on schedule.

The inside temperature goes below the freezing point almost every night. It has been down to 20 degrees, 15 degrees and even 13 degrees Fahrenheit, the lowest point I have noticed in the morning, without harming the plants. The inside temperature is usually within 3 degrees of the outside temperature at night.

How do I account for this?

Frankly, it's a mystery to me and the only explanation I have to offer is that the organic method is followed. Last year we took the stove out in the fall because oil was too expensive and in the middle of the winter planted lettuce and parsley in the benches, using topsoil, sand and compost. The seeds were slow to germinate and grow, but we did have some fresh greens for the table.

Last winter we tore out all the benches, made an 8-inch wall with cinder blocks and filled the planting area this deep with a mixture of half-rotted cow manure picked up from the pasture, half-finished leaf mold and topsoil. This was all run through a grinding machine and when it went into the greenhouse looked like dark brown humus. We took the precaution this time to sow our seeds in early October so that the plants were large and vigorous by the time cold weather came. The kale was transplanted from the garden.

Between the rows of plants we placed flat stones to soak up the heat by day and give it off by night, as well as to retain moisture and this, coupled with the fact that the planting bed is slightly below ground level, probably accounts for the protection of the plants against destruction by below-freezing temperatures. Naturally, we would not attempt to grow tropical plants under these conditions.

The greens are picked 2 or 3 times a week and keep right on growing, as we take only the largest leaves. They are watered every week or 10 days, if we remember.

The greenhouse is a 3-section one on a cinder-block foundation. Neighbors are growing all-winter greens with even simpler equipment—cold frames and compost—and have done so for years, even in winters when the thermometer dropped well below zero. A friend of ours had heard about the cold greenhouse and came up on a bitter winter day to take a look. She said, "I simply couldn't believe it until I saw it."

As an experiment, I tried covering one section of lettuce with a large window—in other words, this was a cold frame within a greenhouse. The lettuce grew more rapidly with this added protection but somehow its health was altered, for aphids began to appear. To combat this invasion I simply removed the extra glass, and the aphids promptly froze to death. The lettuce, no longer coddled, regained its health.

Fresh greens in winter are a problem to the organic gardener, for he knows that any he buys in the store have been raised with commercial fertilizer and liberally sprayed with poison. We found raising our own to be easy and without cost in our organic "coldhouse."

—Alden Stahr

A Home-Made Plastic Greenhouse

If you have many times wished for your own little greenhouse, but had to forget about it because of the price, you may well be interested in a small plastic one that costs around $30. With some used lumber, the cost will be even less.

The one we've just built here in Decatur, Indiana is 6 feet square and can be heated on cold nights by a 250-watt heat bulb (the type used in brooder houses).

The construction of the framework is quite simple. The studding sills and plates are all made from 6-foot 2 by 3's. The bench frames, 8 corner braces and rafters are also made from 2 by 3's. The ridge pole is a 6-foot 1 by 6. There is a ½ by 1-inch strip inside the door frame.

The plastic (6 feet wide, 4 millimeters thick) was fastened on with heavy staples applied by a staple gun. We started by fastening to the bottom of one side, up over the top and down the other side. When the back and front were covered, we tacked thin narrow strips of scrap wood over the rows of staples; this helps to keep the plastic from tearing loose at the staples when the wind causes it to flap a little. Soil was packed around the outside bottom of the greenhouse to shut off any wind that might blow in underneath.

Since there is no foundation on which to fasten the greenhouse, we drove 4 iron rods into the ground inside each of the 4 corners. These are 18 inches long with 3-inch ells bent on the top ends which rest firmly upon the bottom sills when driven into the ground as far as possible.

We placed the greenhouse with the door facing the east—that is the direction from which we get the least wind. For ventilation, a small window was placed in the back, or the west side. I can open it all the way or just a little bit by fastening it with a small chain.

This cross ventilation is quite necessary, for when the spring days start getting warmer, the temperature inside will shoot up to over 100 degrees. That's one thing you have to remember if you want to go away in the morning —always open the window or door a little just in case you don't get back as soon as you expected, or else you may come back to a 120-degree house and some rather "cooked" plants.

In February, we checked the temperature for 3 consecutive days. At noon on the first day, the temperature outside was 24 degrees and inside the greenhouse, it was 80 degrees since the sun was shining; the second day, outside 34 degrees, inside 64 (cloudy); the third day, outside 19 degrees, inside 81 (very sunny). At night, it gets almost as cold inside as outside, so you can see heat is only needed then.

The bench is 36 inches from the ground and forms a U shape, as it runs from the door (which is two feet wide), down one side, across the back and up the other side to the door. That makes a space to stand 2 feet by 4 feet which is covered with gravel. The bench will hold 10 standard-sized (21 by 14 inches) flats. If I ever want to, I can add another shelf about 24 inches above the bench.

Since the plastic covers the greenhouse frame from the ground up, I've made a planting bed on the ground directly below the U-shaped bench. Some old bricks were hunted up and laid to form a border for this bed. A heating cable is placed in the soil and here is where I start most of my plants from seeds.

I like to use a thin layer of vermiculite over the soil, plant the seeds, and cover with a little more vermiculite. Having sprinkled the soil so that it is properly damp before planting, I then cut pieces of scrap plastic and lay it over the planted seeds. I do not take it off until the seeds have sprouted. You've noticed, no doubt, that different kinds of seeds require a different number of days to germinate, so instead of covering the whole seedbed with one long piece of plastic, I cut pieces just the size of each different little bed of seeds. As the earlier ones come up, the plastic is removed and the later-germinating ones are left covered till they are ready to come up.

The plants in a plastic house need not be watered as often as those in a glass house, for the plastic tends to hold moisture inside better than glass. I can run a hose from the faucet at the back of the house to the greenhouse very easily, so watering is no problem.

We intend to cover the framework with panels of lath (held in place temporarily by wiring) and this will serve as a lath-house during the summer.

Sometime during Indian summer, the lath panels will be removed and stored away and the framework will be covered again with plastic. I'm going to cover the ground bed in the greenhouse with plastic-covered sash and grow lettuce and keep clumps of endive, escarole and parsley all winter. This idea of having a cold frame inside the greenhouse gives sort of a double-insulation effect. Late last fall, I placed a small cold frame out in the garden over a row of late-sown Oakleaf lettuce and covered it with a glass sash. The lettuce stayed green all winter and even grew a little.

—Lois Hebble

Build a Basement Window Greenhouse

Have you ever wished for a greenhouse that is more than just a hotbed, yet doesn't require too much of an outlay in cash? So did I, until the day I made myself the little greenhouse, the instructions for which are given below.

Preparations should be made during the fall so it will be ready when you are—early in the spring or late winter. After the frame has been made, set the edges down at least 3 or 4 inches, using the earth removed to bank with on the outside. Push the frame up against the house as tightly as you can with a basement window inside the frame. Foam rubber strips provide insulation against the cold without driving a single nail into your house. A window with southern exposure is best.

I have found that planting seeds in flats, then putting the flats in the frame works out best. Upturned peach crates or boards laid across bricks keep the flats off the cold ground. When you're ready to use the greenhouse, remove the basement window and set the flats in. The heat from your basement will keep it warm. If the mercury takes a nose dive, as it does all too often here in Minnesota, it takes only a few minutes to lift out the flats and replace the window for a day or two until the cold snap is past.

We have early sweet corn, cabbage and tomatoes, thanks to my little greenhouse. Last summer we enjoyed long southern watermelons instead of the little round early ones recommended for this far north. If my plans do not exactly fit your needs, it is easy to alter length, width, or height so that they will fit your early gardening requirements.

Directions

1. Nail the ends to the front piece. If necessary, use inside braces. By that I mean a two-inch square piece $1\frac{1}{2}$ feet long nailed inside each corner.

2. Nail the 5-foot-11-inch board to the *inside* of the ends at the top.

3. Attach plastic to top rim with brads. Nail two inches on top, using as few nails as possible.

4. Cut foam rubber into two-inch wide strips and nail them along the side that will be against the house.

5. Dig a trench 3 or 4 inches deep and settle the frame into it. Bank the surplus earth up around it.

6. After the frame fits snugly against the house, drive either pointed wood or metal rods down to hold it.

7. From the basement, look into the frame. Plug any air spaces around the frame with left-over bits of the foam rubber.

8. Lay boards, or weighted cardboard or canvas across the plastic top of the frame to keep the snow off until you are ready to use it.

Materials Needed

1. Two yards and 4 inches of heavy clear plastic at least 36 inches wide.
2. Eighteen feet of 1 by 2 inch strips. One board 1 inch by 6 inches by 5 feet 11 inches.
3. One by 3 foot piece of foam rubber at least one inch thick.
4. Piece of ½ inch plywood that measures 6 feet by 4½.
5. One box of medium-sized brads to attach plastic.
6. One-half pound of shingle nails.
7. Two stakes.

Cost

These figures are approximate. Most of the materials used, with the exception of the rubber and plastic, will likely be found in almost any workshop.

1. Foam rubber	$1.50
2. Plastic	3.00
3. Brads	.15
4. Shingle nails	.10
5. Plywood remnant	2.50
6. Strips	1.50
	$9.75

—Ethelyn Pearson

Growing Early Vegetable Plants Under Glass

W. B. Nissley of the Pennsylvania Agricultural Extension Service offers this advice on growing young vegetables in a greenhouse:

Ordinary topsoil from the field or garden is not suitable for starting early plants under glass. Such soil becomes hard, dries out rapidly and is not very fertile. Plant-growing soils, therefore, usually are mixtures consisting of one-third topsoil, one-third well-rotted manure or compost and one-third sharp sand; or one-half topsoil and one-half well-rotted manure; or one-half topsoil, one-quarter well-rotted manure and one-quarter sand. These ingredients may be screened separately and later thoroughly mixed. In the case of compost the ingredients are screened as a mixture. The important thing is to have at least one-quarter and not more than one-half organic matter in the mixture. Such soils are friable, will not bake or crust, will hold moisture longer and produce a good root system.

TABLE 50: Seed-Sowing Dates in Small Greenhouses*

Vegetables	North Pa.	South Pa.	Temperature Day	Night
Cabbage (early) Cauliflower Lettuce Kohlrabi Onions Beets Early broccoli	Feb. 15 to Mar. 1	Feb. 1 to Feb. 15	60—70°	45°
Celery	Mar. 1-10	Feb. 10-15	65—75°	55° min.
Tomatoes Peppers Eggplant	Mar. 15-20	Mar. 1-10	75—90°	60°
Squash Cucumbers Melons	May 1-10	Apr. 1-10	85—100°	60°

*In larger houses with better heating facilities, seed-sowing dates are a little earlier than in sash houses. Seed-sowing dates for the hot bed are three weeks later than those recommended for the sash greenhouse.
—*W. B. Nissley*

Seed Sowing

Two methods are used in sowing seed; broadcasting and sowing in rows. The row method is preferable especially for amateurs. Space the rows two inches apart and make the furrows a quarter-inch deep for all vegetables except celery, which should be barely covered. (Cover the seed with a quarter inch of screened soil, or coarse white sand if damping-off is troublesome.) For cabbage, cauliflower and lettuce sow about 12 to 15 seeds per inch of row. For tomatoes, peppers and eggplants sow 8 to 10 seeds per inch of row.

Cover the flats with wet burlap and water thoroughly through the burlap to prevent washing. Remove the flats to a warm (70° to 80° F.) place for germination. They may be stacked on top of each other to germinate, but should be brought to full light after germination begins and the temperature lowered 10 to 15 degrees.

The day and night temperatures will fluctuate greatly from those given in Table 50. Night temperatures near the freezing point will occur and on clear days the temperature may reach 100° or more without any serious damage if ventilation is provided. Yet the temperatures given should be maintained as nearly as possible for best results.

When to Transplant

Young seedlings are usually transplanted when the plants are 1½ to 2 inches tall. Since the seed is sown thickly, it is necessary to transplant the seedlings while very young or they will become spindling and undesirable. Cabbage, broccoli, lettuce, cauliflower and kohlrabi are usually set 1½ by 1½ inches apart in 2½ or 3-inch deep flats. This is the only transplanting they receive, except setting in the field. Onions are not transplanted at all until the field setting. Celery usually is set one by one inch or 1½ by 1½ inches apart. Tomatoes, peppers and eggplants usually are set two by two inches apart. If choice plants are desired, a second transplanting to other flats, pots or veneer bands is desirable. Remember that excessive heat, over-watering and crowding result in slender, spindling undesirable plants. Good plants grow slowly, are dark green and are short and stocky. Moderate temperature and judicious watering and ventilation produce stocky plants.

Watering

Water during the forenoon on sunny days; this permits the soil to become warm during the day and gives the plants a chance to dry off. Plants seldom need watering on cloudy days. When watering, do it thoroughly so that the water reaches the bottom of the flat. Do not sprinkle. Frequent light watering is conducive to damping-off.

Ventilation

Ventilation can be controlled best in greenhouses. Hotbeds and cold frames are more difficult to ventilate, especially during cold, cloudy periods. Excessive condensation on the glass stimulates damping-off and plants are often lost. In greenhouses often it is advisable on cloudy days to do a little more firing and ventilate slightly to reduce humidity. Always ventilate greenhouses and hotbeds or cold frames on sunny days. Never guess at temperature. Inexpensive thermometers should be placed in greenhouses, hotbeds and cold frames.

19—Year-Round Schedule of Garden Activity

Below is a month-by-month calendar to serve you as a check list for *when* to get busy in your vegetable garden. It is divided into 4 major geographical sections—North and East, the South, the Southwest and the Northwest.

The North and East

January

Now is the time to begin planning the vegetable garden. Rhubarb can be forced quickly in the basement at this time of year, as may roots of French endive and several other vegetables.

Start growing mushrooms this month. With spare room in a cellar, shed, barn or under the greenhouse benches, this enjoyable venture can be profitable.

February

During the month, in hotbeds, cold frames, greenhouses and indoors, sow seeds of early cabbage, cauliflower, celery, tomatoes, radishes, peppers. This pleasant pre-spring chore will bring fine rewards in spring and will save the expense of buying many nursery plants.

Garden tools need a thorough cleaning and mending before the hectic spring season starts. Repair cold frames and hotbeds—they should be ready for heavy use next month.

March

Sow asparagus seed or set out roots, and give the bed a good application of pulverized phosphate and potash rock. Cabbage and cauliflower seed may be sown now in sheltered spot, if not already in hotbed. Also set out horse-radish, onions and rhubarb. Dig any parsnips which have overwintered in ground before the tops get too high. Early potatoes may also be planted and in hotbeds, sow seed of garden sage and tomatoes.

April

Many vegetables may be planted now . . . which ones, of course, depend upon the specific varieties and the particular location. Among sure bets are the very hardies, such as asparagus, horseradish, parsnip, rhubarb, salsify and winter onion. Varieties recommended for sowing or planting around April 10th include some varieties of asparagus, beets, cabbage, lettuce, mustard and parsley. Some peas may also be sown now. If tomatoes were sown last month, they may be transplanted into beds or boxes. Transplant cabbage plants for an early crop. Transplant onion plants from hotbeds to open ground.

May

Beans, corn, cucumbers, melons, potatoes and squash may be safely planted in open ground in mid-May, assuming the ground is not too cold. If the season is mild, early corn may be planted during the first week. Add plenty of compost and a pound each of pulverized phosphate rock and potash rock to every 10 square feet of soil. Harden off tomato plants and plan to set them out the beginning of June. Work in phosphate and potash rock; tomatoes require little nitrogen.

June

Plant seed for late cabbage, cauliflower and broccoli first week in June at the latest, to be transplanted in mid-July. Sweet corn planted early this month will make rapid growth, and could be followed by a mid-month planting. Plant beans, beets, carrots, kohlrabi and turnips for a late crop, digging in one-half pound each of phosphate and potash rock for each 10 square feet. Thin root crops planted earlier. They need ample room. Use the tops of beets and turnips for salads. Generally, they contain more vitamins than the more popular root parts, and they taste good, too. Peppers should be kept well watered, but not kept too moist. If the fruits have not set, overwatering may produce too much leaf growth.

July

Fill up empty spaces in the garden with second plantings. Beets, carrots, beans, chard, endive, kale, kohlrabi, lettuce, broccoli, Brussels sprouts, late cabbage, cauliflower, late beets, celery, turnips, radishes, spinach and others may be planted now. Check the number of frostless days left. Dig in plenty of compost for each planting. Mulch to preserve moisture, keep down weeds and protect soil from temperature changes. A mulch will also keep ripened vegetables and fruits from rotting on the ground.

August

Make sure the mulch is in place, especially for plants like tomatoes, cucumbers and melons, whose fruits come into contact with the ground. There's still time for second plantings: turnips, spinach, lettuce, beets, radishes, chard, endive, kale, mustard, winter onions and late kohlrabi. Check the date of maturity on the seed packet against the average date of first frost.

September

There's still time to plant radishes, corn salad, mustard, spinach (for overwintering) and turnips. Root crops can be left in the ground until serious freezes start. Parsnips must be exposed to a slight frost to become best tasting. Asparagus bed should be mulched. In the cold frame, you may sow parsley, lettuce, spinach and chives.

Parsley and other herbs for winter-growing indoors will do better if potted this month and left in a cold frame or cool cellar until December. The compost pile should be building up with waste materials from the garden. In case of a dry Indian summer, be sure to keep the heap moist. Put up some screened soil for potting mixtures for winter and spring flats and for house plants.

October

Harvest all remaining vegetation before frost. Tomatoes, squashes, pumpkins, peppers and ornamental gourds are damaged by the first frost. Give rhubarb a liberal application of manure, compost or kelp.

Close the season with a thorough cleanup to destroy weeds and control pests and diseases. Compost weeds and discarded plants. Shred woody stalks before composting. Clean out cold frames, line with dry leaves, hay or straw. Close sash and cover with manure or earth.

November

Now is a good time to incorporate compost and green matter into the soil. It will blend in with the soil over winter. Rotary till the material into the garden soil and let it rough surfaced until spring.

Gather up and add leaves to compost pile . . . or shred up on location with one of the new rotary mowers. A few passes of the mower will shred leaves fine enough to sift in between the blades of grass. Turn under the last of the vegetable remains. Test soil. If acid, add a layer of lime. If lacking in nutrients, prepare to get some organic materials, to add now or in spring. Remove nests of tent caterpillars and cocoons attached to branches with a stiff brush or broom. Save the egg masses of the praying mantis. Learn to distinguish between the cocoons of both. Get winter protective mulching materials together now. Hay, buckwheat hulls, peat and leaves are good. Mix some potting soil now, before the ground freezes.

Heap manure on rhubarb when the weather gets cold to stay.

December

Sow radish and cauliflower in frames. Mulch strawberry plants as soon as the ground freezes. Good tools well cared for will give you lots of pride and satisfaction. They will also lessen labor. Keep them shiny and sharp during the season. Remember, the place to use a tool is in the garden but when not in use each tool should have a place of its own to make it easy to find. Some tools such as rakes, hoes, and shovels can be hung on the wall on nails. Outline the place for the tool and perhaps paint the space they occupy.

After the tools have been thoroughly cleaned the wooden parts should be painted; bright colors will make them easier to find. All metal parts, especially the shiny parts, should be cleaned with kerosene and then coated with oil to prevent rust. Moving parts such as the wheels and bearings on the hand cultivator or lawn mower should be packed with a good grade of cup grease.

The South

January

Plant Bermuda onion plants, onion sets, cabbage plants, rape, mustard, turnips and carrots. Garden peas may be planted in mid-South, using only the smooth-seeded varieties for early planting. In lower South, many more vegetables may be planted. Gardens well-fortified with humus will grow crops that withstand a spell of cold weather. Rhubarb does well only in the higher altitudes of the upper South. Plant now.

February

Many vegetables can be seeded now, including beets, broccoli, cabbage, carrot, cauliflower, chard, endive, kale, kohlrabi, lettuce, leek, onion, parsley, peas, potatoes, radish, spinach, turnip. Start plants of asparagus, beets, cabbage, cauliflower, leek, lettuce, onion and rhubarb. Check the local frost date. Cucumbers respond to heavily composted soils, doing away with fusarium wilt and being free from insect pests. Make a hole the depth of a shovel and a foot or more in diameter and fill with compost. Mix fine soil and compost deep enough to cover the seed. Allow 3 plants to the hill.

March

Set out tomatoes, eggplant and peppers along Gulf Coast and in Florida when danger of frost is past. In upper South the hardiest of vegetables may be planted: beets, kale, carrots, kohlrabi, mustard, smooth varieties of English peas, radishes, spinach and turnips. In Middle South, add collards, black eye peas, lettuce, parsley, Swiss chard and Irish potatoes. In lower South, English peas, Irish potatoes and onion sets may still be planted. Also, all kinds of bush and pole beans, limas or butterbeans, squash, cucumbers, cantaloupe, pumpkin and watermelons.

April

In sections where an Easter freeze is usual, plant tomatoes, peppers, eggplants, cauliflower, onions, celery, sweet potatoes and cabbage plants immediately after the freeze. In frost-free sections, plant any time during the month. Make sure soil is workable before planting.

May

For better health, more tasty vegetables and economical living, grow ample garden produce for the family. Planting must be done immediately for summer crops. Be generous with compost, and when plants are 4 to 5 inches high, apply a mulch at least 3 inches deep. Use old sawdust, hay, leaves, pine needles or rocks to hold moisture, lessen weed growth and cut down on labor. Practically all of the vegetable seeds may be planted in May, including heat-loving ones, such as okra, squash, watermelons, cucumbers, cantaloupes, summer spinach and second plantings of corn. Set tomato plants 4 to 6 inches deep. Pinch out suckers that develop between the main stalk and the branches while they are quite small. This will prevent them from utilizing plant food and making too much foliage. Cut okra pods before they mature, so the plants will continue to bear all season. Plant in wide rows with stalks every 2 or 3 feet. Use compost liberally. Cauliflower, celery, cucumbers, pumpkins, cowpeas, gourds, spinach and crowder peas may be planted now; also vegetable plants, such as sweet potato slips, tomatoes, eggplants, collards, peppers and cabbage for fall crops.

June

In harvested rows, plants of tomato, eggplant, sweet potato and pepper can be set out in the middle and upper South. Sow seed of bush and pole lima beans, pole snap beans, crowder peas, black eye peas, pumpkin, cantaloupe, watermelon and sweet corn. In the lower South, try planting collard, eggplant, mustard, okra, peanuts, peppers, rutabaga, New Zealand spinach, sunflowers, turnips, sweet potatoes, pole snap beans and pole lima beans.

Take care of the compost heap this month. Water if it has dried out and keep adding material. Warm weather is breakdown time, but the bacteria cannot do their work in dry material.

July

The compost heap needs plenty of attention this month. For best results, grind all material with a shredder, or a rotary mower and keep the heap moist to let bacteria do their work faster.

Keep tomatoes well watered and staked. Use soft cloth cords to tie them—a strong wind will break stalks against hard cords. Lots of vegetables can replace those which are being harvested now.

August

By the middle or latter part of the month, start the winter vegetables in seedbeds for early September planting. The different cabbages and cauliflowers, Italian broccoli, Bermuda onions, Fordhook chard, parsley and winter lettuce are all good bets. In many places the fall crop of Irish potatoes can be planted in thoroughly irrigated rows. Dig in plenty of compost before planting. It's also time to plant tomatoes, winter squash, drought-resistant Black-eye beans, radishes, rutabagas, Chinese cabbage, cabbage, broccoli, Brussels sprouts, chard, kale, leaf lettuce, fall peas, salsify, late spinach, turnips, bush and pole beans, sweet corn and pumpkins.

September

Plant as many greens this month as needed. They like rich soil, so apply plenty of compost. September is usually a dry month, so keep mulching to preserve moisture. In the garden planting, it is a good idea to include a few rows of sweet potatoes for early spring use. They should be far enough along to stand a slight frost if it comes in early winter. If a hard frost is expected, cover the plants with hay or grass or, if available, burlap bags. Beets, Brussels sprouts, carrots, Swiss chard, cress, corn salad, kale, spinach, radishes, rape, turnips, leeks, mustard and lettuce may be planted now. In Florida and the warmer regions of the South, tomatoes and many frost-tender vegetables may be planted.

October

A dozen or more vegetables may still be planted. They are beets, carrots, radish, cress, leaf lettuce, leek, turnips, mustard, Swiss chard, kale, kohlrabi and onion sets. In the lower South, cabbage, Chinese cabbage, head lettuce, parsley, broccoli and collards. Gather matured tomatoes just before frost. Wrap in paper and store in a cool dark place on shelves or in wire baskets. Bring them out into a warm room two days before using. Store rutabagas and turnips soon after cold weather. Cut off tops to half-inch of root. Place on well-drained ground and cover with soil to prevent freezing. Thin fall plantings of carrots, beets and turnips when they are 3 to 4 inches tall. After frost, cut asparagus tops and put them on compost heap. Apply heavy feeding of well-rotted stable manure or compost and a good mulch.

November

Put in an asparagus bed now. Place it on one side of the garden so it will be out of the way of annual vegetables. Asparagus crowns may be planted in rows or beds, 18 to 24 inches apart, in rich soil. If fed properly and not cut until the third year, they will produce for as many as 50 years to come. Plant head lettuce in cold frame and transplant (to another cold frame or hotbed) 12 inches apart when large enough. Other vegetables which may be planted now include cabbage, endive, kale, lettuce, radishes, onions, spinach, beets, Chinese cabbage, turnips, kohlrabi, mustard and parsley. Check catalog for varieties suitable for your region.

Clean and store garden implements before putting away. Oil applied to wooden parts will prevent cracking and rotting. Collect and save leaves. They can be used either on the compost pile, shred right on location with a rotary mower, or as leaf mold.

December

Now is a good time to test soil. Transform findings into organic formulas, using raw ground phosphate rock and granite dust or greensand. After the powders have been broadcast, spade up the ground. Allow it to lie rough over the winter unless you plant a cover crop. Broken ground will warm up more quickly in spring.

In parts of the South strawberries may still be planted. Don't overlook this opportunity for good eating. Many vegetables may be planted now, depending upon your particular location. Check into onion sets, carrots, English peas, beets, cabbage, lettuce, greens and radishes.

The Southwest

January

Plant perennial vegetables, artichokes, asparagus, rhubarb, and horseradish while the roots are dormant. In warm sandy soil, potatoes may be planted. Spread well-prepared compost over the area before planting. Also plant warm weather vegetables—tomatoes, eggplant and peppers—if your area has a mild, frostless belt.

Make the last turning of sheet-composted material preparatory to spading under for the February planting. Mulch winter vegetables with unsifted compost. It will keep down winter weeds, conserve heat and provide nourishment.

February

Humus is lacking in the sun-baked soils so spread plenty of compost in rows before planting. Start new bins or fill old ones immediately. Add neither lime nor ashes to compost, for Southwest soils are generally high enough in *p*H. Make good use of the lush growth of winter weeds before they go to seed. They may be spaded under along with other cover crops. The waste from winter crops, which are now waning, furnishes more good material. Save lawn clippings for mulch. Spade in cover crops and sheet composting materials early in the month.

By the middle of February, the time has arrived for planting the early spring garden. Plant the first crop of Irish potatoes in well-composted soil; the second crop to follow in July. Set out plants of cabbage, Italian sprouting broccoli, kale, chard, leaf lettuce and onion seedlings and sets. Sow seeds of beets, carrots, mustard, kohlrabi, parsley, turnips, Swiss chard and radishes.

March

Make a late planting of the hardy vegetables: onion sets and seedlings, beets, carrots, kohlrabi, chard, lettuce, broccoli, savoy cabbage. Watch heated cold frames carefully and transfer peppers, eggplant and tomatoes to plant bands or cartons made from milk containers. Discard all weaklings. With a warm location in a sheltered area, use hot caps on tomatoes, eggplant and peppers—planting directly a few seeds in a place and selecting the strongest one. Plant the last week in March. By the end of March, plant bush beans and summer squash. Holes for squash return double the yield if they are carefully prepared beforehand.

Spade in cover crops if they are planted on contoured sections. Watch night temperatures. Do not rush to plant the heat-loving plants when the night temperatures drop below 50 degrees. Beware either a very hot spell or a sudden frost in March.

April

April is a midseason planting month. Conserve every bit of moisture by spading in compost and mulching all plantings. Return all waste material to the soil, especially the abundant greens of the spring weeds and lawn clippings.

Continue to plant vegetables this month. In the area with the last killing frost occurring between April 1 and April 20, the warm weather crops should be planted in April and May. These include New Zealand spinach, tomatoes, cantaloupes, etc. But root crops and other hardy vegetables may still be sown. The more hardy ones, including beets, broccoli, cabbage, leaf lettuce, radish, carrots, turnips, should be planted in April for the first time in regions where the last killing frost may occur between April 20 and May 10. In the cooler parts, only asparagus, onions, spinach, radishes, parsley and Chinese cabbage are sown out by the end of April, but the warm weather crops must be started early in the month inside, especially cucumbers, peppers and tomatoes, so that they can be set out in late May or early June. If vegetable garden space is limited, plant chives and parsley as an edging to flower borders. Leaf garlic can be used in the bulb border, as well as the handsome rhubarb chard.

May

In carefully prepared rows plant okra, bush squash, bush and pole beans, lima beans and corn. Mulch heavily. For small home gardens use as many vertical growers as possible. Chinese and Armenian cucumbers, pole beans and tomatoes should be trained to stakes. Plant summer cover crops of the drought-resistant legumes. Tepary and Black-eye beans give a food crop and can be plowed under later as a soil renewer. Plant the leaf lettuces, which do not get bitter in summer, in a little shade. The heat-loving vegetables—lima beans, summer squash, peppers, eggplant, tomatoes—should be planted in deeply composted holes. Water cress can be planted in a shallow tub or discarded sink.

The importance of compost and mulches cannot be overstressed for the season to come. Continue to add to the compost heap, and gather mulch materials because they hold soil moisture and improve soil tilth.

June

Plant in deep, well-prepared holes, well-established plants of tomatoes, eggplant and peppers. Put compost in the bottom of the holes and sink water down before planting. In deep, well-prepared soil, plant okra, lima beans, bush and pole beans and corn. Mulch heavily and water with soil soakers. Plant summer cover crops of drought-resistant legumes such as tepary and Black-eye beans. Plowed under later as a green manure, they help renew the soil. Put lettuce for summer in partial shade. Sow such annual herbs as dill, summer savory and sweet basil in your herb beds. Keep parsley, evergreen garlic and sorrel growing for your salad bowl. Water sweet potatoes and melons, using a soil soaker and mulch heavily afterwards.

July

July is a crucial "mulch month" in the Southwest. Many gardeners have an aversion to sawdust as a mulch, but experiments have found it to be excellent for protecting soil and for breaking down slowly into humus. Keep the compost pile moist during the dry spell and keep adding materials. Spend lots of time out in the garden and on the patio this month.

Continue setting out plants of eggplant, tomatoes and peppers. This is also the time to fill bare spots in the garden with a variety of vegetables, including okra, summer squash, string beans, hybrid sweet corn, spinach, kohlrabi, chard, beets, broccoli, carrots, cauliflower, endive, lettuce, radishes, turnips and potatoes. Conditions vary from area to area, so check seed packets for maturity dates against the number of frostless days left in your area.

August

Start winter vegetables in seedbeds in the latter part of the month. The variety is endless; radishes, rutabagas, Black-eye beans, cabbage, broccoli, Brussels sprouts, chard, kale, leaf lettuce, fall peas, late potatoes, salsify, late spinach, tomatoes, turnips, bush and pole beans, sweet corn, pumpkins and winter squash. In many areas, the fall crop of Irish potatoes can be planted in thoroughly irrigated rows. When summer crops are harvested, pull out old plants and weeds and add them to the compost heap. The internal action of the heap should provide enough heat to kill weed seeds. For better results, shred materials before composting and keep the heap moist.

September

This month marks the beginning of the "second spring" of the Southwest. Planting of winter vegetables should be rushed as soon as the soil is warm and night temperatures are favorable. Keep an eye on your thermometer and, when the night temperature drops down to 40 and lower, the dormant period has set in. In September, when the soil is thoroughly warmed up and the night temperatures are favorable, rush plantings of winter vegetables.

Sow seeds of beets, carrots, chard, early onions, winter radishes, Chinese cabbage, mustard, turnips. Canadian Wonder bush beans and Pencil pod wax beans will mature about the middle of November if planted the first week of September. In water-saturated and compost-enriched rows, set out stocky plants of cabbage, cauliflower, Italian sprouting broccoli.

Mulch every planting if possible. Use lawn clippings, leaf mold from the home grounds.

October

Make the most of favorable planting weather while the soil is warm. Watch the thermometer for night temperatures. When they drop to 50 degrees and lower, slow up on garden operations. "Shooting to seed" in early spring is easily prevented by this precaution.

Continue setting plants of such winter vegetables as cabbage, cauliflower, kale, kohlrabi, Italian sprouting broccoli, celery, seedling onions, Romaine, leaf and head lettuce. Chinese edible pod peas can be trellised like sweet peas. Telephone peas and Windsor beans are excellent winter crops.

Plan to sheet compost any unused areas of the garden this fall. Use waste raw materials; in our section "turkey poult" is good with dry grass, straw, spoiled hay and vegetable refuse, which can be layered, watered first, soaked by the winter rain and spaded or disked early in the spring.

November

Sheet compost for early spring planting. Add manure for nitrogen insurance. Have the soil tested now for *p*H and analyzed to find out what minerals are lacking and add mineral rock and other soil amendments. Use pulverized seaweeds if obtainable locally. Make compost by the quick "dry method" to keep *p*H down so that the minerals which are in the soil can be absorbed.

Sow Telephone peas, Chinese edible pod peas, Windsor beans. In sunny and light soils, put in more onion sets and set plants of leaf lettuce.

December

Things are virtually at a standstill in the Southwest. Food lockers and pantries are chock-full, but let's not forget about next season. Cover crops may still be planted.

Asparagus, artichokes and rhubarb can be set as soon as available. Make new strawberry beds and renew old ones. Other vegetables which may still be planted in parts of the Southwest are beets, chard, cucumber, broccoli, Brussels sprouts, cantaloupes, carrots, leaf lettuce, muskmelon, green onions, parsley, peppers, potatoes, radishes, rutabaga, spinach, tomatoes and turnips. Divide and reset garden herbs. Use chives as edgings, along with parsley. Divide your chives while they are dormant.

The Northwest

January

Take a tour of the garden this month. Make sure all mulches are in place. Replace plants which have heaved out of the ground. Eliminate standing puddles with down-grade ditches. On one of the nicer days, turn the compost heap. Add a layer of soil on top, keep moist and protect from rains.

February

Seed onions, early potatoes and carrots can be seeded this month if the soil is sandy. If soil does not crumble and fall apart when a handful is squeezed, it is too wet. Wait until a later month to make the garden planting. In colder parts of the Northwest, cold-frame or hotbed seedings should be made of many vegetables, including peppers, tomatoes, early cabbage, cauliflower, celery and radishes. Gardeners who seeded the ground to rye and vetch for green manure should turn it over now and prepare at least that strip which is to be seeded to early peas. Washington's Birthday is the date on which the peas are sowed in the Pacific Northwest.

March

Warm, sunny days near the coast may be tempting, but avoid planting until the latter part of the month, when seeds will be fairly safe from rotting in the wet ground.

Plant some lettuce as a border for flower beds. It fills the gaps in the spring and will be harvested before the space is needed. East of the Cascades, plant only cabbage, lettuce and onions in the vegetable garden this month. West of the Cascades, the following may also be planted: peas, carrots, spinach, turnips, celery, radishes, early potatoes, cauliflower, rhubarb and asparagus roots. Wait for the soil to ready itself for digging. If a spadeful breaks apart when hit with the spade, it is ready. If not, wait until later in the month. Near the middle of the month, put in another planting of peas. The first one should be made around Washington's Birthday. Place a 2-by-4 on the ground, and plant the first planting on one side of it, the second on the other, using the same strings for both plants.

April

Never plant seeds in poorly prepared soil. See to it that compost is well worked in, the seedbed raked over, the surface pulverized. Lime may be needed, particularly if your soil is low-lying and wet. Don't, however, add lime without testing to see if it is needed.

In the West, approximate dates for seeding or setting out plants are: asparagus and early potatoes until April 8; spinach, peas, onions, Swiss chard, beets and lettuce, during the first half of the month. Strangely enough, parsnips and salsify, both among the hardiest crops, are recommended for April 25 only. Sometimes, not coolness but light conditions may require a late or early planting, since too early planting may produce seed formation with some crops. In all cases, check catalogs and seed packets, as dates also vary with different varieties. In general, if the ground is friable and ready for use, these vegetables may go out in the open: Cabbage and broccoli plants, seeds of turnips, beets, lettuce, carrots, chard, potatoes, spinach, parsnips, radish and salsify. Partially fill the seed furrow with compost containing a small amount of ground steer manure to furnish a little warmth. Plant and cover as usual.

The winter compost pile should be ready to yield some returns now. Turn the top layer over into a new heap and sift the well-rotted compost out of the remainder. Place the residue on the new heap for further decomposition. Take advantage of the lush new growth of grass and use the clippings for mulching.

May

Prepare the vegetable garden well with standard organic methods. High humus content of soil helps keep plants supplied with moisture during summer, mulch keeps down weeds, and compost and other natural fertilizers supply needed food. All plants may be set out now—cabbage, cucumbers, cauliflower, tomatoes. East of the Cascades, be sure the frost date has passed. Stop cutting asparagus the latter part of May.

June

Continue mulching to control weeds. A little effort now will be doubly rewarded next month. The warm weather is just right for bacterial action in the compost heap. Give the bacteria a chance by keeping the heap moist. Try to use rain water, if possible.

It is late, but several vegetables may still be planted, such as sweet corn, carrots, pole beans and heat-resistant varieties of lettuce. Prepare a seedbed and plant seeds, for transplanting later, of winter crops such as cabbage, broccoli, chard and Brussels sprouts. Tomatoes should be set out when they flower. Set paper bags about them and stake down the bags to prevent blowing away. Put sawdust or grass cuttings around the plants. Drop root vegetables into a bucket of water after harvesting them. Ready them for the kitchen by washing and removing the tops. Contents of the bucket and the tops may then be put on the compost pile. Don't throw out the turnip and beet tops, though; they're loaded with vitamins.

July

Spring-applied mulch should be doing a good job in keeping down weeds this month. If it isn't, the material may be too loose. New straw, although it is a fairly effective mulch, isn't much good in smothering weeds. Better for this purpose are rotted or ground straw, ground corncobs, buckwheat hulls, sawdust, grass clippings or any other fine-textured material. The mulch will also improve results tremendously during the present dry season, especially east of the Cascades.

In those empty spaces in the garden sow kale and kohlrabi. Also plant beets, carrots, rutabagas, bush beans and a fall crop of peas. Also think of radish, head lettuce, St. Valentine broccoli, cauliflower, beets, cabbage, endive and leaf lettuce. Check the number of frostless days left against maturity date and hardiness of individual varieties.

August

Make enough second plantings now to insure food until spring. Try fall and late varieties of artichokes, beets, cabbage, carrots, parsley, peas, radishes, spinach and turnips. Check maturing time of individual varieties against the date of the first fall frost.

September

Keep carrots and beets in moist sand. Allow an inch or two of the tops to remain. Too much moisture will cause them to put out feeders, so be careful. Cucumbers should now be producing. Keep them well-watered and pick them every day. Cut ends of squash to within leaf of a fruit. Vines need moisture now. Keep mulching. Set out plants of late cabbage, cauliflower, Brussels sprouts and broccoli. Rhubarb clumps may be divided now. Set at ground level, leaving one or two eyes per root. Cover mulch of straw manure about 5 inches. Plant spinach in a sunny spot before the 15th. It should be ready in October or November. Plant radishes now in time for Thanksgiving garnishes.

Now is the time to add plenty of material to the compost heap . . . garden wastes, lawn clippings, weeds, any organic material. Prepare some potting soil for the winter. Sifted compost mixed with rich loam makes an excellent soil to use during winter months, when rains and snows make it difficult to secure. Sand or peat may be added according to the structure of your soil.

October

October is the harvest month. Canning, freezing and storage activities are at a peak now. The less time lost in getting fruits and vegetables from garden to freezer, the less vitamins are lost. Vitamin C is particularly vulnerable to this danger period.

October is a good time to divide rhubarb. Even the smallest garden can find room for this coarse, yet beautiful plant. It is a heavy feeder and likes lots of manure.

Clean out all plant residues and throw them on the compost pile. Now is the time when the heap should be built up for next year's use. Grass clippings, leaves and any other organic matter can help. The difference between a compost pile and a pile of green matter, is the presence of manure and layers of soil put between each application of green matter. Soil and manure will help the pile to decompose much more quickly. Another trick is to shred up all green matter with the rotary mower before applying to the heap. Add nutrients to the heap, with rock powders to each layer of soil. The result will be a beautiful combination soil conditioner and fertilizer.

November

Earth up celery for the winter. Replant winter varieties in deep frames. Heap manure around rhubarb plants. Where possible, plant vegetable garden in fall to expose hibernating insects to freezing. Plant peas in well-drained soil. Leave a ridge of soil to the south of the row to shade the row of small plants from sudden temperature extremities. Onion sets put out now bring the earliest green onions. The size of harvest may be small, but none the less rewarding. Onion seeds can be sown now and will germinate at the first sign of spring.

December

The garden should have been put to bed last month. Remember to keep the compost heap covered to protect it from the heavy Northwest rains.

Section 2.

20—Organic Vegetable Culture: A Complete Listing

ARTICHOKE

ARTICHOKE, globe. *Cynara scolymus* or *C. cardunculus,* of the family *Compositae.* Thistle-like perennial herb, cultivated form of the cardoon. Native in its wild state to the Mediterranean, southern Europe and central Asia. Also called bur-artichoke. Known in present form since the fifteenth century. Edible portions include the flower bud, a compact cluster of scales, and the bottom or heart to which they attach. These are eaten steamed usually with melted butter. Blanched base of scales, the tenderest portion, probably contains little or no vitamin A or C.

Propagation

Artichoke seed tends to revert to parent thistles, so beds usually started from roots or rooted suckers of old plants. Varieties include Italian, French and Chinese; the "Green Globe," an Italian artichoke, being the only one planted in this country.

Soil

While climate is a more important factor in production of tender buds than soil, artichokes are heavy feeders requiring large amounts of nitrogen and moist but well-drained soil. Only portion of the United States where commercial planting is profitable is along the California coast between Los Angeles and San Francisco, where best beds are on slopes overlooking the Pacific. For year-round production winter temperatures should be above freezing, summers should be cool and foggy. Winter crops may also be obtained in the Gulf states, and summer crops along the North Atlantic as far as Massachusetts. In cold areas, plants are protected from winter weather with a basket covering, which keeps moisture from crowns, and a thick mulch of manure over all.

Compost or manure should be deeply dug into beds before planting, especially if bed is expected to last 4 years. Roots or suckers should be planted 6 inches deep, 3 to 6 feet apart in rows 4 to 8 feet apart. Good plants yield 15 to 20 buds each. In cold sections, plant in spring; in warm sections, plant in fall or winter. New beds may be started annually with suckers from old plants, if desired.

Cultivation

Light mulch may be spread between plants to conserve moisture and keep down weeds. Heavy mulch of manure should be spread over plants each year after cutting back, and an additional side dressing of organic fertilizer should be applied at the rate of 3 pounds per hundred feet when crowns begin new growth. Soil around roots must be kept moist or buds will be tough. If irrigation is necessary, water must not be permitted to stand around crowns, which are subject to fungus disease.

Artichoke plume moth, aphids, and botrytis disease are most serious enemies. Plume moth lays eggs in buds, and larvae tunnel foliage and base of buds. If permitted to hatch, entire planting may become infected. Eggs may be destroyed by hand, or infected parts cut and destroyed. Aphids may be controlled by spraying with a strong stream of water. Botrytis sometimes results after warm summer rain; seriously affected plants should be cut and removed.

Harvesting

Buds are cut with 1½-inch of stem attached before blossoms begin to open, in cold areas in late summer; in frost-free sections, harvest from early fall throughout winter. Unexpected frost in warm areas will destroy developing buds, putting back harvest two or three weeks. Bud-bearing stem is cut back to ground after bud is harvested. Globe artichokes are usually eaten fresh, but may be frozen whole, or hearts may be canned or pickled.

Artichoke, Jerusalem

Tuber of a native American perennial sunflower, *Helianthus tuberosus,* of the

Compositae family. The name is thought to be a corruption of the Italian name, *girasoli articocco,* meaning sunflower artichoke. One of the few plants cultivated by the Indians, the Jerusalem artichoke was introduced to Europe by some of the early explorers and became more popular there than it has been here. Plants grow 4 to 8 feet tall, with coarse hairy foliage which may be used as forage, and yellow flowers 3 to 4 inches in diameter. Seed is seldom formed, but the root system spreads far and wide from the main stalk, making tubers which produce new plants the following year.

Jerusalem artichokes are valuable in the diet of diabetics because they contain no starch and store sugar in the form of levulose. They are one of the few vegetables which supply pantothenic acid, one of the B vitamins, about .10 milligrams being supplied by a single tuber weighing 50 grams. In addition, a tuber yields 200 units of vitamin A, .075 milligrams of vitamin B_1, .015 milligrams of vitamin B_2, 10 milligrams of vitamin C, 20 milligrams of calcium, 47 milligrams of phosphorus, .4 milligrams of iron, 1 gram of protein and a total of 32 calories. When cooked, the tuber has a sweetish, nutlike flavor. It may be sliced raw in salads, when its flavor is very much like the crisp water-chestnuts used by the Chinese.

Range and Soil

Jerusalem artichokes may be grown throughout the United States, but they do best in the northern two-thirds of the country. They should be tried only with caution in gardens of the Gulf states and in the arid Southwest, where they may require too much irrigation to make them worth growing. They need about as much rainfall as sweet corn. A season of 125 growing days is needed to mature a good crop.

Being large, vigorously growing plants, they need abundant nutrients as well as moisture to make good yields. Best crops are grown in rich, sandy loam. They should not be grown in heavy soil. The bed should be isolated from the garden, because spreading roots make volunteers which easily become weeds, difficult to eradicate.

Grown commercially along the California coast, artichokes are heavy feeders requiring large amounts of nitrogen and moist but well-drained soil.

Seed

Whole tubers or pieces of tubers weighing at least two ounces make the best seed. Smaller pieces make smaller plants, and yields are lower. About two quarts of tubers planted in an area of approximately 25 square feet will produce all the tubers needed by one family.

Planting

Jerusalem artichoke may be planted in fall when tubers are dug, or may be planted as early in the spring as the soil can be worked. Spring frost will not injure the planting. Tubers are placed 4 inches deep and two feet apart in rows which are 3 to 4 feet apart. Or they may be planted 4 feet apart in each direction in areas where the growing season is very long and moist, and soil is rich alluvial river-bottom loam.

Culture

Weeds should be kept down by a light mulch which will not disintegrate until leaves grow over bed. After the first year's crop has covered the bed with foliage, growth will be too vigorous for competition from smaller plants, and weeds are no problem. Tops provide mulch for the second year crop.

Harvest

After frost blackens foliage, cut tops back and leave them to mulch the bed. Additional mulch in the form of leaves and garden trash may be added in areas where the ground freezes very hard. Tubers may be dug any time after the first frost, and additional tubers may be dug throughout the winter and into spring, until growth starts. Because Jerusalem artichokes have a very thin coating, with no skin like that on a potato, they do not keep well after digging, so no more than a couple of weeks' supply should be dug at one time. Seed for new beds to be started in spring should be left in the ground through the winter.

If spreading plants become a weed problem, they should be dug up after top growth has been made, but before August. New tubers form from August through the fall, and will spread the plants further from their original beds. It is best to confine them within a barrier, such as a wall, walk, drive or roadway.

Tasty Tuber

"My nomination for the get-to-know vegetable of the year is the Jerusalem artichoke. Usually missing from gardens and seed catalogues, the Jerusalem artichoke bears decorative flowers and produces tasty tubers. Some farms have a stand of this sunflower-like plant growing wild, its nutritious edibility unappreciated.

"I know of no plant easier to grow. Artichokes seem to have few if any enemies. That they are amazingly sturdy is testified by our planting here in central western Jersey. When two foot tall, a friend pulled off the above-ground parts as weeds. Undaunted, the roots put forth sprouts again which attained a height of 3 feet. Then the horse in an adjoining pasture reached over and devoured the tops. Again new blooms persisted, only to be run down by a truck seeking a short-cut along the fence. Yet in the fall we harvested a nice crop of artichokes, in addition to earlier bouquets of the flowers.

"After growth is stopped in autumn by a killing frost, the tops can be cut off and used for compost or as a mulch. As Jerusalem artichokes have tender skins and are not hardy keepers, I dig a supply for a month or two at the most. These tubers need to be kept moist in storage or they will go limp and perhaps rot."

—Marilyn Neuhauser

ASPARAGUS

ASPARAGUS. *Asparagus officinalis,* family *Liliaceae.* Originally native to seashore and river banks of southern Europe, the Crimea and Siberia, now naturalized in all parts of the world. Cultivated as early as 200 B.C. by the Romans. A vigorous-growing fernlike perennial herb, attaining a height of 4 to 10 feet, with strong root systems spreading as much as 6 feet horizontally and 6 to 8 feet deep. There are male and female plants; the female bears red berries in late summer. Edible portion is the young unbranched shoots harvested in early spring. Food value of 8 green stalks: 1100 units vitamin A; .360 milligrams vitamin B_1; .065 milligrams vitamin B_2; 20 milligrams vitamin C; 21 milligrams calcium; 40 milligrams phosphorus; 1 milligram iron; 2 grams protein; 20 calories.

Range and Soil

Grows best where cool temperatures prevail during growing season, where summers are not extremely hot, as in southern and Gulf states, and where winters are cold enough to give the plants a dormant season. Winter injury from cold in the United States, except Alaska, results only in fields where tops are removed.

Bed should be well-drained, in full sun. For best results, soil should be rich with *p*H 6 to 7, preferably light, deep, loose, free of large rocks which cause spears to grow crooked, and free of perennial weeds. In heavy soils, shallow planting is recommended. Soil may be prepared the fall before planting, deeply cultivated, enriched especially with phosphorus and potash rock, such as rock phosphate and granite dust, at the rate of 5 to 10 pounds per 100 square feet. It is important to dig these deeply into the bed before planting to make nutrients more readily available. A nitrogenous green manure crop may be planted in fall and dug under in spring before planting asparagus.

Buying Seed

Beds may be started from seed or purchased roots. Variety recommended for the entire United States is Mary Washington, a rust-resistant strain. Also rust-resistant is Martha Washington, recommended for New England and southern states. In addition, Paradise for central states; Viking for the Great Lakes region; F-1 hybrids for North Central states; and Waltham Washington for New England.

One ounce of seed will produce 10,000 plants if started inside at 80 to 90 degrees; if started in the field, about one-tenth that number may be expected to germinate. One packet will sow 100 foot row. Seed produces a crop 1 to 3 years later than roots, which may be purchased 1, 2, or 3 years old. Two-year roots are recommended for home gardeners who wish to cut a small crop a year after planting. Roots are spaced 18 inches apart in rows 5 feet apart; 70 plants to the 100 foot row will yield about 30 lbs. of asparagus when bed is established.

Planting

Thick mulch is recommended at all times on asparagus, not only to keep down weeds but to maintain the high soil-moisture content necessary for best production. Unless bed is to be perpetually mulched, roots must be planted deep enough to be entirely below range of cultivator. If bed will be kept free of weeds with mulch, deep planting is unnecessary.

Many growers use a trench method as follows:

Dig a trench 12 inches deep and 12 inches wide. Fill the bottom 4 inches with rich compost, rotted manure or a mixture of topsoil and dried manure. Place roots at 18 inch intervals on conical mounds, with roots spread around mound and crown an inch below soil surface. Fill trench and firm soil over roots. If deeper root planting is desired, trench must be made 15 inches deep, plants must be set with crowns 6 inches below soil surface and filled only to one inch over the crown. When shoots begin to come up, soil is added to cover them until entire trench is filled.

Seed may be started indoors in flats kept moist and warm until germination. When 2 to 3 inches high, transplant to stand 3 inches apart. Set outside 6 to 9 inches apart first year. Second spring, set them in permanent bed, treated as purchased roots. Or sow outside one inch deep as early as possible; thin to 3 inches when well up and to 6 to 9 inches when about 12 inches high. If seed is sown in warm weather when soil is dry, it may be sown 1½ inches deep.

Cultivation

Unless bed is mulched, weeds should be kept down by thorough hoeing or cultivating early in spring before shoots begin to sprout. After the harvest, a light top growth of weeds is thought to help protect soil from drying sun. Small annual weeds will not compete with deep roots or tall tops, and may be worked under the following spring to provide humus.

Mulch, if used, should be 4 to 6 inches deep, preferably of a granular or chopped material which has a neutral *p*H and a high nitrogen content. If wood chips, sawdust or ground corn cobs are used, they should be enriched with blood meal or cottonseed meal to provide extra nitrogen. Leaves are best shredded before use. Cocoa shells, pea or lima bean pods and grass clippings make excellent asparagus mulch. Oak leaves, pine needles, peat moss and coffee grounds are too acid.

Winter mulch should be a 3-inch application of manure, plus the plant tops which are allowed to stand until spring. Tops should never be cut because they continue to store food in roots for spring shoots after they are apparently dead. In spring they should be crushed down over the bed and allowed to disintegrate.

In addition to the winter manuring, an application of organic fertilizer should be dug in around each plant after harvest. This feeding should consist of 3 parts granite dust or greensand, 1 part dried blood or 2 parts cottonseed meal and 1 part bone meal, applied at the rate of 3½ pounds per 100 square feet.

Because of the depth to which its roots go, asparagus should never need to be irrigated in the home garden, especially if kept under a heavy mulch. But being a gross feeder, asparagus needs plenty of moisture, so sandy or gravel soils should be avoided because they dry out too quickly.

Enemies

Most important pests are the asparagus beetle and its larvae and the 12-spotted beetle. Adults winter in trash around the bed and appear early on new shoots. If shoots are picked early every day, the beetles can be caught by hand and killed, because they are unable to fly in the cool of the morning. Whether picked or not, the new shoots should be examined and beetles removed daily for perfect control. A time-tested control method is to turn poultry loose in the bed. They eat both adults and larvae.

Asparagus rust is a fungus which may attack and kill an entire plantation in areas where mists and heavy dews are prevalent. In such sections Mary Washington, a rust-resistant variety, should be planted.

Harvest

Harvest begins when plants are 3 years old. If two-year roots are planted, a limited cutting lasting not more than two weeks can be made the following spring. The next year, when the plants are 4

Asparagus shoots should be cut when 4 to 8 inches high, before scales on tips begin to open. Begin harvesting when plants are 3 years old.

years old (and from that date on) cutting may continue from 4 to 6 weeks. Last cutting date varies in different localities, but probably the latest is in New England, where harvest stops on Fourth of July. Large thick stalks only should be cut or pulled. Thin stalks should be allowed to grow. Production of thin stalks means that the food reserve in the roots has been exhausted and the tops should be permitted to grow to replenish the supply. When one stalk is permitted to develop side branches, shoot production from that root will be inhibited. So until cutting is to be discontinued, all thick shoots should be pulled, even though they are not usable for any reason. Early morning is the best time to harvest new shoots, which become tougher and drier in the heat of the day.

Shoots are pulled when 4 to 8 inches high, before scales on tips begin to open. They may be cut with a sharp knife two inches below the soil surface—a method which often results in injury to neighboring shoots below the surface—or they may be broken off by hand. Spears should be stored in the refrigerator until used or prepared immediately for canning or freezing.

Asparagus season may be prolonged by a couple of weeks if mulch is removed from half the bed early in spring, to permit the soil to warm up. As soon as the soil is thoroughly warmed, mulch should be returned. Production will start in a bed treated in this manner approximately two weeks earlier than in the bed on which mulch is undisturbed.

Saving Seed

If certain plants in a bed produce unusually well, it may be worth while to save seed from them to propagate the strain. To prepare seed, the plant is cut and hung in a well-ventilated place to dry. Berries are then removed, soaked in water until soft enough to break outer shells and are mashed. They are then washed under running water to float off the shells. Seeds are spread on a board in the sun to dry for a day, then further dried in an attic for about a week. They may then be stored until planting time. Best berries to save are those which develop early in the season and also those which grow deep in the plant, rather than on the tips of shoots or on the lowest branches. But unless fertilizing plant can also be selected and seed hand-fertilized, plants grown from seed will not all come true to the parent plant.

Don't Work So Hard for Asparagus

Ruth Stout, the famous exponent of the year-round mulch system of gardening, describes here her method of growing asparagus:

The old-timers used to dig a trench 2 or 3 feet deep and a foot or two in width, then fill it almost to the top with dirt and manure and put in the roots. Nowadays the authorities are backing down on this elaborate procedure and are advising much shallower trenches.

They tell you to dig a trench about a foot deep and 10 inches wide, fill it almost to the top with very rich soil (if you can get manure, that's fine). Follow what they tell you about planting and making your first cutting and the length of time to cut each season. Beyond these things, my earnest desire is that you ignore the experts.

Some of them will tell you to plant a cover crop of soybeans and dig it in. This is unnecessary work. Others will say that manure should be spread over your bed every fall. Well, if you have some manure handy, which didn't cost you anything, go ahead and use it for this, but your bed doesn't need it.

Some gardeners put salt on their asparagus bed to keep out weeds and I'm told that this is effective, but there's a better and simpler way to outwit weeds.

You may be told that you should cut the stalks in the fall, and some authorities even advise that these be taken off and burned. I suggest that you do neither of those things; just leave the stalks where they are. Like everything else, they will die when their time comes, so let them rest in peace. They will help mulch the bed, too.

Let's just skip the fantastic idea of making mounds over the asparagus in order to bleach it. That's for the birds—and some Europeans, who were brought up on white asparagus and haven't seen the light. Nowadays, health-conscious people urge us to eat green-colored foods, the greener the better. Assuming that this is a beneficial thing to do, isn't it wonderful that for once the thing that's good for us is less work than that which isn't so good?

I've read that it's desirable to mulch

an asparagus bed lightly in late autumn to protect the crowns, but we are also told to remove the covering in the spring and cultivate the soil. There is, however, no reason given as to why we should go to this trouble, and my guess is that the experts don't *know* why. It can't be to soften the soil or to kill weeds, because under a proper mulch the earth is always soft and there aren't any weeds.

So what *should* you do? I imagine you will follow whatever method sounds most sensible and reasonable to you and I wouldn't presume to advise you, but I'm going to tell you what results I get with *my* system.

I have two 50-foot rows of asparagus, one of which was planted 30 years ago in the old-fashioned way, before I knew better; that is, in a very deep, wide trench. The other was put in two years later, in a shallow trench. The two rows are doing equally well.

For the first 14 of my gardening years, I covered my asparagus each fall with manure, cultivated it each spring, weeded it all summer long. Then, one fine day in April, I got the bright idea of abandoning plowing and hoeing and weeding. I covered my plot with hay, left it there, added more now and then and for the past 16 years the work in my garden has consisted of replenishing the mulch here and there and planting and picking my wonderful produce.

This of course includes the asparagus bed and, as I said above, I can't for the life of me figure out why anyone should think that the hay should be removed in the spring and the bed cultivated. The tips will come up right through the hay, so why disturb it?

It's true that mulch prevents the soil from warming up in the spring as rapidly as it would if the ground was bare, which means that your season will start a little later than it otherwise would and also means, of course, that it lasts further into the summer.

For me all this is an advantage, because we get occasional frosts in May, and sometimes even in June; these kill any exposed asparagus stalks, so I'm glad to have the crop somewhat delayed.

If, however, you are in a hurry for any reason—if, for instance, you sell asparagus and want it to mature early, when it brings top prices—it isn't much trouble to pull the mulch back and leave it off until the ground warms up, then return it.

Or you might push the mulch back on only a part of your bed, which would give you a longer season. You would in this way be cutting one section a week or two earlier than the rest and you could cut the second section a week or so after you've stopped cutting the first.

In general, leaves are a good mulch, but loose hay is the best for asparagus; leaves, or hay that has been baled, may prevent the sprouts from coming through. Straw is all right. However, J. A. Eliot, of New Jersey, an asparagus expert, believes that hay is the best mulch of all; he says that for nutritive value it is superior to manure. And his reasoning is that part of the nutrients in hay, which is fed to horses and cows, go to build up the body of the animals and to make milk; manure is the residue. But a rotting hay mulch still has all the nutrients left in it.

Now a word about picking asparagus. People can't seem to get away from that slow business of cutting it with a sharp knife or a two-pronged asparagus cutter, just below the surface of the ground. For my money, that method has 4 things wrong with it: it takes quite a little time; one is likely to injure a nearby shoot which doesn't yet show above the ground; the stalk-ends are dirty; and the tough part has to be cut off and disposed of.

My system is much simpler: I walk down the row and snap off any stalk which has matured and, since I break it where it is tender, there's nothing to be cut off afterward. And the stalks are so clean that all they need is a quick rinse under cold running water.

The amount of money my method of growing this vegetable will save you depends on how much you have to pay for hay and how much you would spend for fertilizer if you grew asparagus the old-fashioned way. But I am sure my system will save you a tremendous amount of time and energy.

—***Ruth Stout***

BEAN

BEAN. *Phaseolus vulgaris* (climbing) and *P. nanus* (bush). Family *Leguminosae*. Common bean, haricot or kidney bean, including all common garden, snap, string and stringless beans whose immature pods are cooked and served or whose seed is dried and cooked. Also includes shell or butter beans—the larger seeded varieties except sievas or limas, whose seed is used either green or dried. About 200 types; 1,500 varieties.

Most beans are native to America where early explorers found many varieties being cultivated by Indians. Ancient Old World beans were believed to belong to another genus, *Dolichos*.

Climbing or pole bean vines, sometimes grown trailing in dry climates, achieve a length of 8 to 10 feet. Bush beans grow 15 to 24 inches high. Roots generally shallow, with feeder roots near surface. Growing in root nodules are colonies of nitrogen-fixing bacteria, which convert atmospheric nitrogen into nitrates, an important step in improving soil fertility. Edible pod (snap) beans, may be either green or yellow (wax) podded. Beans grown for shelling may have red-spotted pods and white seed. Both snap beans and shell beans may be dried to provide beans for baking or special varieties of field beans may be planted for the purpose. Plantings of drying beans are not recommended for home gardeners unless space is unlimited.

One hundred grams of cooked snap beans (3/4 cup) provides the following: 950 units of vitamin A, .060 milligrams of vitamin B_1, .100 milligrams of vitamin B_2, 8 milligrams of vitamin C, 55 milligrams of calcium, 50 milligrams of phosphorus, 1.1 milligrams of iron, 2 grams of protein and 43 calories. Wax beans contain much less vitamin A. Dried navy beans contain little vitamin A and vitamin B_2, no vitamin C, but about 3 times as much vitamin B_1, phosphorus, iron and protein per 100 grams.

Range and Soil

Can be grown in any part of the country where there are 3 frost-free months. In the South and Southwest, beans are not grown during the hot summer months because flowers are often blasted by extreme heat and pollen injured by day temperatures above 95 degrees. Flowers also drop in rainy weather. Plantings are made in fall, winter and spring in extreme South.

Beans prefer slightly acid soil, *p*H 5.8 to 6.5, sandy or clay loam, but will thrive in any except heavy cold, clayey soil. Plants will mature faster in sandy than clay soils.

Fair to high humus content improves crop. High nitrogen content not desirable, because it tends to promote foliage rather than pod growth. Wax and pole beans need richer soil than green bush beans. If manure is used, it is best dug in the previous fall or at latest two weeks before planting. Green manure should be turned under one month before planting.

Climbing or pole bean vines, sometimes grow 8 to 10 feet long; beans are best at north end of garden or along border fence.

Beans add nitrogen to the soil. Since they may be considered a soil-enriching crop, they should be rotated yearly with soil-depleting crops. Climbing or pole beans are best at north end of garden, or along boundary fence. All varieties need full sun.

Seed and Varieties

Bush bean seed planted one pound per 100 foot row. Pole beans planted 1/2 pound per 100 foot row. Yield from either type about two bushels per 100 feet.

Most recommended green bush bean varieties are Top Crop, Contender, Wade, Improved Tendergreen, Stringless Black Valentine, Bountiful. Cherokee, Pencil Pod and Brittle are most recommended bush wax beans. Kentucky Wonder is by far the most popular green pole bean, with Blue Lake second. Most recommended shell beans are Shelleasy, Red Shellout and (except in northernmost areas) French Horticultural. Varieties especially recommended for growing in the south are Ideal Market and Alabama No. 1, the latter resistant to root-knot nematodes. Varieties highest in vitamins: vitamin A, Early Bountiful; vitamin B_1, Burpee Stringless; vitamin C, Bountiful, Burpee Stringless, Idaho No. 1, Great Northern, Strider, Stringless Black Valentine and VBL 46, a hybrid. Low in vitamin C are Blue Lake, Georgian, Logan Imperial and U. S. No. 1 Refugee.

Planting

Beans are very tender, killed by a slight touch of frost. Earliest plantings cannot be made before the last expected frost. Optimum air temperatures for germination and growth are 65 to 85 degrees. Seed germinates in 4 to 7 days. Bush beans mature in 6 to 8 weeks, pole beans in 75 to 80 days, depending on temperatures. Bush varieties can be picked in successively smaller amounts for about 4 weeks after maturity. Successive sowings may be made to extend the season, the last planting 10 weeks before the first expected frost. Pole varieties may be harvested from maturity until frost.

When planting in soil where legumes have not previously grown, it is advisable to inoculate seed with a commercial preparation of dried nitrogen-fixing bacteria. After legumes have been rotated through the entire garden, the practice may be discontinued, since the bacteria remain in the soil.

Bush beans are sown in a drill one inch deep, seed spaced two inches apart, in rows 15 to 18 inches apart. After germination they are thinned to 4 to 6 inches. Closer spacing promotes fungus diseases. Poles 6 to 8 feet tall are set before pole varieties are planted. Rough-surfaced poles, such as stripped saplings with bark, are better than smooth poles. Poles may be set separately or in tripods, tied at the tops. Or posts with horizontal wires 6 inches and 5 feet above ground may be strung, with strings at one foot intervals for support. Tripods are spaced 4 feet from center to center. Individual poles should be spaced 3 feet apart. Six or eight seeds are planted at the foot of each pole, thinned to 3 or 4 best plants.

Beans planted in cold wet soil germinate badly, if at all. Seed rots quickly, the cotyledons which should be pushed out of the soil by the developing root shriveling and turning brown if the temperature is too low for quick germination. Some protection may be afforded if early seed is soaked in water until swollen, then coated with oil before planting. Because the small root must lift two large cotyledons, germination may be hindered by soil which cakes over the seed. This can be avoided by covering seed with sifted compost. Also, woodchucks relish newly sprouted cotyledons. If they are known

Six to 8 foot high poles should be set before planting climbing beans. Rough-surfaced poles, such as stripped saplings with bark, are better than smooth poles.

to be near garden, it is advisable to protect seedlings until a few leaves develop.

Seed planted late in season when soil has dried out should be planted 1½ inches deep, the drill well-flooded before planting. Additional irrigation should be unnecessary if soil is mulched between rows.

Cultivation

Beds should be kept weed-free, either by mulch drawn up to base of plants or by shallow cultivation. Mulch will also preserve moisture. On unmulched beds, 12 to 15 inches of rainfall or irrigation is needed in 60 days for a full crop. If beans are irrigated with an overhead sprinkler or hose, it is essential that foliage dry before nightfall.

A light, top dressing of finished compost can be applied to bean plants. Bean plants add nitrogen to the soil; foliage should be added to compost heap.

Side-dressings of fertilizer should be unnecessary if the soil contains enough phosphorus and potash when beans are planted. At time of planting, 3 pounds of one of the following per 100 feet may be raked into the row: equal parts of bone meal and greensand; or, equal parts of tankage, wood ashes and granite dust. The latter formula should be used only when *p*H is low, since it tends to alkalize the soil.

Enemies

Most common bean pests are Mexican bean beetle and two fungus diseases, mosaic and anthracnose.

Mexican bean beetle is a brown, spotted beetle, slightly larger than a ladybug. Usually appears on the young plants by the time the first blossoms appear. Early plantings are freer of the pest than late-season sowings, so it is advisable to sow heaviest crop for canning and freezing early. If first beetles are hand-picked and destroyed, ravages of larvae, which are yellow and wooly, may be minimized. Small clusters of yellowish or brownish eggs on under sides of leaves should also be destroyed when beans are being harvested. Larvae chew and skeletonize leaves and later in the season also attack beans.

Fungus diseases are carried in seed. By purchasing seed grown in Idaho or California, where the diseases are infrequent, they may be avoided to some extent. Anthracnose and bacterial blight both spread when plants are wet. To prevent dissemination of spores, do not touch or walk among plants while they are wet from rain or dew. Spores of mosaic disease are harbored in clover. Avoid planting beans near old clover sod.

Harvesting

Snap beans should be picked when pods are almost full-grown, but do not yet show the outlines of enclosed seeds; tips of pods should be soft; pods should snap easily and should be stringless.

Shell beans should be picked when forms of beans are clearly discernible within the pods, but pods have not yet begun to turn limp or dry.

Beans grown especially for drying are left on the plants until almost dry. Plants are cut and hung until pods are brittle, then are threshed. Seed must be allowed to dry thoroughly before storing. If weevil eggs are present in seeds, the organic gardener may destroy them by heating dried seed to 145 degrees for ten minutes (temperature of seed, not of oven) or may freeze dried beans, inhibiting growth of weevils.

Excess crops of snap beans may be treated like dried beans. Pole beans dry better in most climates than bush beans, being hung in mid-air where there is less danger of mildew or sprouting in the pods. In a rainy season, it is advisable to pick dried pods promptly and shell the seed at once to obtain the best dried beans from the home garden.

Seed which is heat-treated cannot be used to plant next year's crop. However, except for Idaho and California residents, gardeners are advised to purchase seed each year for disease-free crops.

Bean Sprouts

Shoots from germinating legume seed, principally from soybean, *Glycine soja;* green or golden gram, *Phaseolus aureus;* or from *P. mungo,* the urd, mungo or black gram. Bean sprouts constitute one of the chief fresh vegetables in the diet of large Oriental populations. Edible portion is the hypocotyl, or sprout, with or without attached cotyledons, used when the sprout is about 1/8 inch in diameter and 3 inches long, but before roots or leaves develop.

Used raw in salads, they have the highest vitamin content. Cooking need be only very brief, with little or no water, in the Chinese manner. If water is used, it is incorporated into a gravy or sauce to be served with them. Sprouts may be frozen, needing two minutes' steaming.

Sprouts are produced commercially under carefully controlled temperature, moisture and atmospheric conditions. The process can be adapted to home use as follows:

Soak beans 8 to 10 hours in water at room temperature, when they should have swollen to triple their volume. Place in a wide-mouthed jar, such as a mason jar and tie a cloth over the mouth. To drain stand inverted on a rack in a dark corner where the temperature remains near 70 degrees day and night. If soybeans are used, temperature may be as much as 10 degrees higher. Every 4 hours beans must be thoroughly rinsed in water at 70 degrees, to wash away putrefactive bacteria and molds and to change air in jar. Full development of sprouts should take place within 4½ to 5½ days.

The amount of oxygen in sprouting container and number of washings affect not only the putrefactive bacteria, but also the amount of vitamin C developed by sprouts. Ascorbic acid content is highest when not too much oxygen is present during sprouting; however, at home it is almost impossible to avoid putrefaction unless air is changed at each washing. Highest concentration of ascorbic acid occurs when sprouts are about two days old.

A comparison of the nutrients contained in one cup of sprouts from soybeans and mung beans follows:

	Soybeans	*Mung beans*
Calcium	48 mg.	29 mg.
Phosphorus	67 mg.	59 mg.
Iron	1 mg.	.8 mg.
Vitamin A	180 Int'l Units	10 Int'l Units
Vitamin B		
Thiamin	.23 mg.	.07 mg.
Riboflavin	.20 mg.	.09 mg.
Niacin	.8 mg.	.5 mg.
Vitamin C	13 mg.	15 mg.

BEET

BEET. *Beta vulgaris,* family *Chenopodiaceae.* This biennial herb is native to Europe and North Africa. Though used as a vegetable by the Romans in the third century A.D., improved varieties were not known until the middle of the sixteenth century.

Planted from seed first year, beets develop thick roots consisting of alternate bands of storage and conducting tissues, with tap roots 2½ to 3 feet long; leaves grow from the crown on short stems to 15 inches high. Second year growth is 2 to 4 feet high. Seed is produced the second year in fruits which are planted whole.

Edible portions are storage roots and leaves. Roots can be cooked, leaves cooked or eaten raw in salads. Nutrients in ½ cup cooked root: 50 units vitamin A, .041 milligrams of vitamin B_1, .037 milligrams of vitamin B_2, 8 milligrams of vitamin C, 28 milligrams of calcium, 42 milligrams of phosphorus, 2.8 milligrams of iron, 2 grams of protein and 40 calories. In ½ cup cooked greens, nutrients are 22,000 units of vitamin A, .100 milligrams of vitamin B_1, .500 milligrams of vitamin B_2, 50 milligrams of vitamin C, 94 milligrams of calcium, 40 milligrams of phosphorus, 3.2 milligrams of iron, 2 grams of protein and 28 calories.

Range and Soil

May be grown in all parts of the United States, from spring to fall in northern sections; fall, winter and spring in extreme south. Roots resent hot weather, become tough and stringy when hot and dry. Any good well-drained garden soil in full sun, well-enriched with humus and nutrients. Will tolerate *p*H 5.8 through 7.5, but optimum 6. to 7. Will not grow on acid soil. Soil should be deeply prepared with rotted, not fresh, manure or compost. Stones should be raked out of top 4 inches to ensure well-formed roots.

Seed

Varieties most recommended for all parts of the country are Detroit Dark Red, maturing in 65 days, Early Wonder, maturing in 60 days and Crosby's Egyptian, maturing in 56 days. For late planting, Winter Keeper which matures in 80 days is the recommended variety. One ounce of seed will plant 80 to 100 foot row, a packet, 20 foot row. A 100 foot row yields 150 to 200 pounds roots, plus tops and thinnings.

Planting

Beets may be sown in garden one month before last expected spring frost. Season may be anticipated by planting seed in cold frame or hotbed 6 weeks earlier and transferring to garden when first seeds are planted. For best continuous yield, seed may be sown in 10 foot rows, an additional 10 feet planted every 3 or 4 weeks until 90 days before the first expected fall frost. Earliest beets may be followed by late cabbage, broccoli or lettuce. Late beets may be planted in rows used for early lettuce or peas.

Beet seed balls consist of two to six seeds in a dry fruit husk which must be softened before shoots can emerge. Germination can be hastened by soaking seed for 12 hours before planting. Early plantings may be made in drills one-half to one inch deep, depending upon soil moisture. Place seeds one to two inches apart, rows 12 inches apart. As soil becomes warmer and drier, succeeding plantings should be deeper until summer plantings are two inches deep. When days are hot enough to dry soil quickly, seed should be covered with a light mulch or with burlap until first shoots appear. Germination should begin 10 to 12 days after planting.

Beets can be thinned to stand 4 to 6 inches apart before they become crowded in the row, so that they will develop properly.

Cultivation

Seedlings should be thinned when 6 inches tall to stand 4 to 6 inches apart. Young thinnings may be used in salad or cooked whole. At same time, a side-dressing of nitrogenous fertilizer—cottonseed meal or blood meal—may be applied at the rate of 5 pounds per 100 foot row.

Beets should be heavily mulched, with mulch pulled close around plants, both to preserve moisture which they must have for best quality and to keep out weeds, which cut yield. If season is very dry, plants must be irrigated to keep top 8 inches of soil moist.

Enemies

Principal pests are spinach flea beetle, leaf spot and leaf miner. None of these seriously affects plants on clean, healthy soil. Rotation of crops is sufficient to discourage infestation.

Harvest

Early beets are pulled when 1½ to 2 inches in diameter. Tops are cut 1½ inches above root, leaving stub of stems to prevent bleeding during cooking. Late beets may be left in the ground all winter in areas where soil does not freeze solid. In cold areas they are dug before the first severe freeze and stored in cold moist cellars at 33 to 40 degrees or in root pits. Baby beets may be canned whole, frozen or pickled.

Beet seed is seldom saved from roots grown in the home garden. If the gardener wishes to preserve an especially fine strain of beets for which commercial seed is not available, he must carry over roots from one year to the next to plant them the second spring. In order to do this, roots must be dug early in fall before crown is exposed to a severe frost. Roots are stored in a frost-proof cellar, planted the second year. Stems grow as high as 4 feet. When seed is ripe, tops are cut and dried in an attic or dry barn. When thoroughly dry, seed is removed and saved for following spring's planting.

Declining Sugar Content in Beets

The *San Francisco Examiner* of September 21, 1958 and the *Wall Street Journal* of November 19, 1958, discuss the alarming reduction of sugar in beets grown in California. The sugar content of a beet is important because the beet is used by the sugar refiners as their raw material in the manufacture of sugar. In 1938 in California the average content of the beet was 18 per cent. In 1958 it had dropped to 14½ per cent and its downward trend was accelerating alarmingly. In 1959, it had gone down a full one per cent, whereas in the previous 20 years the total decline was 3½ per cent.

Dr. A. F. Yeager of New Hampshire University's Experiment Station exhibits the "Sweetheart," a new variety of table beet that is 50 per cent sweeter than other kinds.

Since the growers are paid for their beets on the basis of their sugar content, the reduction in the sugar is costing farmers millions of dollars. Some authorities think that something is happening to the seed. This may be true, because there has been evidence of the dying out of the variety of various plants, that is, loss of potency in the seed, due to the overuse of chemical fertilizers. But the article in the *Wall Street Journal* speaks of a comparison made by two farmers using the same seed and, where one obtained a 14 per cent average of sugar, the other one, a Mr. Grainger, got only

10 per cent and he stated this is the last time he will grow beets.

The *Wall Street Journal* mentions the case of a Mr. Nielsen and his brother who raise beets on two adjacent pieces of land. Carl Nielsen got only 13 per cent sugar while his brother, John, got 17 per cent. The authorities don't seem to know exactly why this should be, although they have vague theories. My suggestion would be *to determine the amount of organic matter in the soil of these 4 farmers.* It will probably be found that *where there is more organic matter in the soil, there will be more sugar in the beets.*

The case of Mr. Grainger is illustrative of what these farmers are up against. "I'll be lucky to break even this year," he grumbled. Ten years ago his beets averaged 17 per cent sugar. This last season they contained about 12 per cent. He is getting 30 per cent more crop yield, but due to the fact that costs have increased, his net income has declined. In other words, excessive amounts of chemical fertilizers raise the yield, but reduce the quality of the crop.

The sugar refineries measure the sugar content of the beets but do not concern themselves with their vitamin and mineral make-up, because that does not reflect itself in any way in profit and loss. The fact that the beet-pulp residues that are fed to cattle might contain less nutrient value is no affair of theirs.

Authorities have attributed the vanishing sugar to the fluctuating climate and also to the heavy use of nitrogen fertilizers. The *Wall Street Journal* says, *"Proper climate and nitrogen can be crucial to sugar beet development, most producers agree. Too little nitrogen in the ground brings scrawny, undergrown beets; an excess and the beets use too much of their own sugar in growing, draining supplies normally stored in their roots."*

Dr. Albert Ulrich, a plant physiologist of the University of California, has said, *"In many cases, it's probably a case of overfertilization, but in others it may be something else.* There's no definite pattern on the one hand and yet, on the other there's the definite pattern of decreasing sugar content." Could there be a pattern of constantly dwindling organic matter in the soil? It is a known fact that the *use of nitrogenous fertilizers bring about a reduction of organic matter in the soil. Would it not be better to depend on the safe amount of nitrogen in organic matter?* There may be somewhat less total production of beets but there would be profits. Research has shown that the overuse of nitrogen fertilizer has reduced the sugar content of other fruit crops—apples, for instance.

BROCCOLI

BROCCOLI. *Brassica oleracea botrytis cymosa.* Family *Cruciferae.* Broccoli is believed to be the *cyma* referred to by the Romans, who knew it as the flowering shoots developed by cabbage plants when they were allowed to stand two years in the field. Sprouting broccoli, also known as Italian asparagus broccoli, has been grown widely in American gardens only since 1900. Another kind, the heading type, looks so much like cauliflower that it is sometimes marketed for such. This takes a long period to develop however.

Plants grow two feet high and almost as wide, root systems spreading quickly both laterally and vertically. A plant 8 to 10 inches high may have a root 2½ feet deep. Aborted green flower heads develop at top of central stalk; after cutting, additional smaller heads develop from leaf axils.

Edible portions include more tender leaves, fleshy stalks and flower heads. Stalks are peeled and eaten as finger salads; leaves and flower heads can be cooked. Nutrients obtained from 100 grams (¾ cup) broccoli flower: vitamin A, 6,000 units; vitamin B_1, .120 milligrams; vitamin B_2, .350 milligrams; vitamin C, 65 milligrams; calcium, 64 milligrams; phosphorus, 105 milligrams; iron, 1.3 milligrams; protein, 2 grams; calories, 35. From an equal quantity of broccoli leaves: vitamin A, 30,000 units; vitamin B_1, .120 milligrams; vitamin B_2, .687 milligrams; vitamin C, 90 milligrams; calcium, 262 milligrams; phosphorus, 67 milligrams; iron 2.3 milligrams; protein, 3 grams; calories, 35. From the same amount of peeled stem: vitamin A, 2,000 units; vitamin B_2, .187 milligrams; calcium, 83 milligrams; phosphorus, 35 milligrams; iron, 1.1 milligrams; protein, 2 grams; calories, 35.

Range and Soil

Broccoli requires cool nights and a steady supply of moisture. In hot

Spring plants of sprouting broccoli can be started inside in flats and set out when seedlings are 5 to 6 inches tall; plants should stand 15 to 18 inches apart in 3-foot rows.

weather, the blossoms "rice," separating into individual florets. Except along the seacoast or on the northern portions of the Great Lakes, summer crops are unsatisfactory. A small spring crop may be grown, started indoors 2 to 3 months before outdoor planting time, which should take place at least two months before the onset of hot weather. Fall crops are started outdoors in midsummer.

Optimum *p*H is 5.5 to 6.5. Soil should be rich in moisture-holding humus and in nutrients, for rapid growth. No members of the cabbage family should have been grown on the area during the past 3 years. Calcium requirements are high, and compost should be incorporated into the soil at the rate of one pound per plant. The equivalent of one inch of rainfall per week is needed for best production.

Seed

One packet of seed should provide 150 to 200 plants. Yield is approximately 45 heads per 100 feet of row.

Most recommended varieties are Waltham 29, DeCicco, Calabrese, Green Mountain, Green Sprouting and Italian Green Sprouting. Waltham 29, the most popular variety, is thought to be better in fall than in spring. Two varieties especially rich in vitamin C are Freezer and Propagane.

Planting

Spring plants are started indoors in flats and planted outside as soon as soil can be worked and plants are 5 to 6 inches tall. Plants should stand 15 to 18 inches apart in rows 3 feet apart. Fall plants may be started in the garden in July and transplanted to permanent positions in August. Discard any plants showing blackened or wiry sections on base of stem when transplanting. Set slightly deeper in permanent position. Protect seedlings against cutworm with paper collars.

To prolong fall harvest, sow seed in midsummer where fall crop will stand. Transplant excess seedlings to new row, leaving some in position where seed was sown. Those not transplanted will mature two to four weeks earlier than transplants.

Cultivation

Plants should be well mulched to preserve moisture. Side-dressings of a nitrogenous fertilizer, such as cottonseed meal or blood meal, may be dug in when plants have been set in garden for two weeks and again when heads begin to form.

Enemies

Like all members of the cabbage family, broccoli may suffer from mosaic, blackrot, blackleg, clubroot, yellows, damping-off and ring-spot. Most damaging insect pest is cabbage worm, green larvae of a white butterfly.

Some of the fungus diseases may be avoided by purchase of only American western-grown seed. Yellows and clubroot are prevalent in the South. For gardeners of other sections, avoidance of southern plants is advised. Clubroot is a disease which is difficult to eradicate from infected soil, but raising the *p*H to neutral (*p*H 7) sometimes helps.

If cabbage worm infestation is heavy, the organic gardener might consider dusting with a 1 per cent rotenone preparation. Dusting must be done early in the day, while plants are wet with dew and the dust will cling to the under sides of the leaves. Rotenone, an extract of derris-root, is thought to be nontoxic to humans.

Harvest

Flower clusters should be cut with a 4 to 6 inch stalk before flowers break open. No color should show. After the first large central cluster is cut, clusters of smaller size will form in leaf axils. As long as clusters are cut promptly, they will continue to form until hot weather or heavy frost stops production.

Broccoli is best preserved by freezing. Stalks should be washed briefly, dried and refrigerated immediately after being picked to preserve the largest amount of vitamin.

BRUSSELS SPROUTS

Brussels Sprouts. *Brassica oleracea bullata gemmifera*. Family *Cruciferae*. A development from wild cabbage, native on both sides of the English Channel and on the coast of Denmark. It is said to have been developed and grown before the sixteenth century near Brussels, Belgium, but there is no definite proof of its existence as a variety distinct from cabbage until the nineteenth century. Brussels sprouts is a slow-growing biennial from 2 to 4 feet high, depending on variety. Roots are shallow, seldom deeper than 2½ feet. Tuft of cabbage-like leaves borne at the top of leggy stalk, lower leaves spaced further apart. Edible portion—the sprouts or buds one to two inches in diameter formed in the axils of the lower leaves, those at the bottom maturing first. It is eaten cooked.

A slow-growing biennial from 2 to 4 feet high, Brussels sprouts should be harvested when they are full-sized and fairly firm, picking lowest sprouts first.

Nutrients present in 100 grams (¾ cup) cooked Brussels sprouts: 400 units vitamin A, .180 milligrams vitamin B_1, .090 milligrams vitamin B_2, 130 milligrams vitamin C, 27 milligrams calcium, 121 milligrams phosphorus, 2.1 milligrams iron, 4 grams protein and 55 calories.

Soil and Range

Sprouts are best if matured in cool weather and are improved by a slight touch of frost. Cannot be grown in frost-free regions. Commercial crops are grown along central California coast and on Long Island. Best product is obtained on sunny days followed by nights of light frost. Needs rich, deep sandy loam, with good drainage both at surface and around roots. Uses large quantities of nitrogen during slow growth. Soil acidity as for cabbage—*p*H 6 to 7.

Seed

Germinates at 45 degrees or above. A packet of seed should give 150 plants. Each plant yields about one quart of sprouts. Most recommended varieties are Catskill and Long Island Improved, both dwarfs growing no more than two feet tall and Half Dwarf Improved, a taller variety which is raised by California growers from home-grown seed.

Planting

Dates for starting seed vary greatly through the country. As for late cabbage, seed is started in greenhouses or in special seedbeds and transplanted to garden. Seedlings require 50 to 75 days to grow to transplanting size—7 to 8 inches tall. About 90 to 110 days are required to mature plants after they are in the garden. Plants grow best at every stage in cool weather. If possible, they should be planted to reach maturity and to start bearing at about the time of the first fall frosts.

On the cool foggy portions of the California coast, seed may be started outside from February to April and crops harvested beginning in the latter part of August. In most of the rest of the country, plants are set out in the garden about the first of August to begin bearing around Oct. 15. Space plants 18 inches apart in rows 3 feet apart. A spadeful of compost is dug into the hole prepared for each plant. A few leaves should be pinched off from the bottom when the plant is set. Cutworm collars should be used on all spring-set plants. Brussels sprouts may follow early carrots, beets or corn in the garden.

Cultivation

Like all cabbage plants, Brussels sprouts need plenty of moisture, so a thick mulch is beneficial. A side-dressing of nitrogenous fertilizer such as cottonseed meal or blood meal may be applied two weeks after plants are set out and again when sprouts begin to form. Lower leaves should be removed when buds sprout, to make room for the sprouts and to concentrate all the food delivered to that portion of the plant into the sprouts. Plants should be staked when they are about one foot high.

Enemies

Same as Broccoli, which see.

Harvest

Pick when sprouts are full size and fairly firm, before they grow tough or yellow. Lowest sprouts should be picked first, even though they do not usually make as firm heads as later ones. To pick, first remove leaf below sprout, then twist sprout off stalk.

Harvest in freeze-free areas may continue in the garden for 5 or 6 months. Where the temperature goes well below freezing during the winter months, plants may be lifted just before the first hard freeze and kept in a protected cold frame, an unheated cellar or an outbuilding where they will not freeze hard. Except in the coldest areas, this can be done after December 1. Sprouts which are budding when plants are lifted will mature in storage, providing an extra month or two of harvest. Soil about the roots of lifted plants should be kept moist.

Where plants are carried through the winter outside, seed may be grown from the best plants for the following year's planting. Choose a plant with high yield and healthy foliage. Protect it from winds and from birds. Split the outer leaves of the sprouts with a knife, to permit seed stalk to emerge. One plant should provide 4 ounces of seed, enough to raise more than 10,000 plants.

CABBAGE

Cabbage. *Brassica oleracea capitata,* member of the mustard family, *Cruciferae.* Native to the seacoasts of southern Europe from Greece to Britain, along the northwest coast of France and up to Denmark. Cabbage has been cultivated since before recorded history. It is known to have been grown by the Egyptians, by the early Greeks and by the ancient Celts.

Rich, moist, well-drained loam is best for all members of the cabbage family; early cabbages require more soil nutrients than the later crop.

It seems always to have lent itself to ease in culture. Experiments done in England show that the wild cabbage currently found growing on the coast of Wales can be cultivated, the open-leafed varieties becoming kale, the incurved-leaf plants becoming heart cabbage and the many-flowered types becoming cauliflower.

A hardy biennial, cabbage can be obtained in many types of head—conical, round, flat, pointed—with smooth or crinkled leaves (Savoy cabbage), green or red. All except the most coarse outer leaves are eaten, either cooked or raw. Considered one of the "strong-flavored" vegetables by tradition, but recent nutritional experiments prove that strong flavor results from the breakdown of sulfur compounds, which does not happen, unless cabbage is overcooked, cooked at too high temperature or in too much water. Contact for prolonged periods with water in washing or in cooking may also hasten breakdown of the sulfur compounds. Vitamin C losses in cooking cabbage are cut to a minimum by the same procedures which prevent the development of strong flavors.

Vitamin content of early and late cabbage, fresh or stored, has been found to vary greatly. Also affecting the nutritional value is the growing season, the variety planted and the maturity of the head. Immature heads with green leaves contain more vitamin C than mature heads from which outer leaves have been stripped or heads blanched white. But cabbage which has been stored for several weeks or months will have larger quantities of C in the inner leaves and in the midribs of the outer leaves. Salted cabbage, or sauerkraut, also loses much of its nutritional value.

Following are average contents of 100 grams of raw green cabbage, raw white inner leaves and sauerkraut. Vitamin A: green, 160 units; inside leaves, none; sauerkraut, 20 units. Vitamin B_1: green, .090 milligrams; inside leaves, .078 milligrams; sauerkraut, .008 milligrams. Vitamin B_2: green, .150; inside leaves, .075;

sauerkraut, .370. Vitamin C: green, 50 to 100 milligrams; inside leaves, 27 to 50 milligrams; sauerkraut, 5 milligrams. Calcium: green, 429 milligrams; inside leaves, 46 milligrams; sauerkraut, 45 milligrams. Phosphorus: green, 72 milligrams; inside leaves, 34 milligrams; sauerkraut, 29 milligrams. Iron: green, 2.8 milligrams; inside leaves, .2 milligrams; sauerkraut, .3 milligrams. In all cases the total amount of protein in 100 grams is about 2 grams; each supplies about 28 calories.

Range and Soil

Cabbages may be grown in all parts of the United States. In the South, plantings are made in fall and winter; in the North, most cabbage is grown in the fall, extending into the early frosty months. Spring crops can be grown where long cool springs prevail; summer crops succeed only where the summers are cool and moist. Well-hardened cabbages will stand frosts down to 20 degrees, some varieties even as low as 15 degrees, over short periods. Where temperatures drop and stay just below freezing for a month or more, plants will bolt and form seed upon warming up.

Rich moist loam is best for all members of the cabbage family. Early cabbages require more soil nutrients than late crops. All types need abundant moisture, but beds must be well-drained to minimize fungus disease. Soil should contain large quantities of humus. Manure may be spread over the entire bed to a depth of 3 inches and dug in at least two weeks before planting. If the manure is fresh, the earlier it is incorporated in the soil, the better will be the results. If manure is not available, compost at the rate of one pound per plant may be dug in when transplanting; or the following fertilizer may be applied at the rate of 8 pounds per 100 feet of row: two parts cottonseed meal, one part colloidal phosphate, two parts granite dust.

Cabbage should be given a position in full sun, in soil that has not grown any member of the cabbage family for at least 3 years. Early cabbage may occupy rows to be used after harvest for late corn, beets or beans. Late cabbage can follow early plantings of lettuce, peas or carrots.

Seed

One packet of seed will provide more than enough plants for 100 feet of row. Yield of early cabbage, 100 foot row, is about 100 pounds. Late cabbage, which is allowed to stand longer and achieve greater maturity, may supply 175 pounds in the same space.

Of all the hundreds of varieties that may be grown the following are especially recommended. Those marked with an asterisk are especially high in vitamin C.

Early Cabbage

Early Jersey Wakefield, *Golden Acre, Badger Market, Marion Market, Detroit, *Allhead Early, *Babyhead Special, *Early Copenhagen, *Early Select No. 99.

Midseason

*Midseason Market, Copenhagen Market, Bonanza, *Glory.

Late

Danish Ballhead, *Wisconsin All-Season, Penn State Ballhead.

Red Cabbage

Red Acre, Mammoth Red Rock.

Savoy Cabbage

Chieftain, Drumhead Savoy.

Planting

Early cabbage is planted indoors 4 to 6 weeks before it is transplanted to the garden, which can be done after the last severe freeze. Properly hardened-off in a cold frame, seedlings will stand light frosts. Seed should be started in a cool greenhouse or grown in a cool room, where night temperatures are no higher than 50 to 60 degrees.

Seed for late cabbage may be started directly in one end of the row where the planting will stand and may be transplanted to fill the row when plants are 4 to 6 inches tall. Or it may be sown in a special seedbed or in flats kept outside. In the North, late cabbages should be set in the garden no later than August 1.

Early cabbage is planted 18 inches apart. Late plantings need more room—about two feet each. Rows are spaced 30 to 36 inches apart. Seedlings should be placed lower in the soil than they were in their flats. Plant deep enough to cover stems to the first leaf axil. Protect early plantings with cutworm collars.

Culture

Soil in which cabbage is grown must be kept constantly moist to assure well-shaped, firm heads. A heavy mulch will help, but when rain fails, cabbages must be regularly irrigated. If the plants are allowed to dry out and subsequently receive a normal watering, heads may crack open. Late fall drying winds, when the heads are large, may have the same effect. To prevent cracking, break some of the feeder roots near the surface by twisting the head or by plunging a trowel into the soil next to the plant.

Side-dressings of nitrogenous fertilizer, cottonseed meal or blood meal, may be used on early cabbage two to three weeks after they are set out.

Enemies

Besides the pests which beset other members of the cabbage family (see BROCCOLI), cabbage maggot is a serious threat to the cabbage crop in some areas. The maggot is the larvae of a fly which lays her eggs on the soil near the stem. Larvae riddle the stem, killing young plants. When eggs can be found in time and destroyed by hand the plants are safe. But as soon as the larvae crawl underground, the plants are often lost. Tar-paper collars may be fitted to the plants when they are set out, to prevent the fly from laying her eggs close to the stem. To make a collar, cut a circular piece of tar-paper about 4 inches in diameter. Punch a hole in the center with a spike or nail-set and slash the paper from one edge to the center hole. Place the collar around the newly planted seedling and close the slash with a stone. A maggot deterrent is a strongly alkaline area around plants, which may be achieved with an application of a heaping tablespoonful of wood ashes dug into the soil around the stem. Or when setting out August plantings, a mixture of 4 parts of wood ashes, 1 part each of ground limestone, rock phosphate and bone meal may be stirred into the soil at the rate of two cups within a two-foot radius of each plant.

Cabbages are sensitive to nutrient deficiencies in the soil and show their hunger in symptoms which may be mistaken for disease. Potash deficiency, for instance, is apparent when outer leaves become first yellow, then dry and brittle around the margins. Heads fail to firm up. Similar brittle areas appearing in spots between the veins indicate a need for magnesium, which can be supplied by an application of dolomitic limestone. When spots appear on leaves and in the core which look water-soaked, the plants need a trace of boron in the soil.

Harvest

Early cabbage should be picked before it is too mature. Heads are cut and used as needed.

Late varieties are improved by a slight touch of frost, though they may be used as soon as heads begin to firm up. Before the first hard freeze (usually, except in the coldest areas, around December 1) plants may be dug and stored for the winter. They are best stored outside or in an outbuilding and should never be kept in the house. Firm heads are pulled with their roots and stacked upside down in a protected corner of the yard or garage on a bed of straw. A foot-deep layer of straw is then laid over the pile and cabbages will be sufficiently protected for the entire winter.

CARDOON

CARDOON. *Cynara cardunculus,* a thistle-like member of the daisy family, *Compositae*. Native in its wild form in southern Europe and central Asia. One of the most esteemed vegetables of the Greeks and Romans, more recently grown principally in France and Spain and to a smaller extent in England. Closely related to globe artichoke, with similar horticultural requirements.

Plants are 3 to 6 feet tall, with deeply cleft leaves, mostly spiny, deeply felted, the flower heads purple and thistle-like. Edible portion is the blanched leaf stalk, which is used in salads or cooked in soups or stews. Its flavor is somewhat bitter. In Spain, an infusion is made from the down on the leaves which is used like rennet in cheese-making.

Range and Soil

Like artichokes, cardoons are tender and may be grown throughout the United States if treated like tender annuals. Soil should be well-drained, not too acid and enriched with compost or well-rotted manure.

Planting

In the South, seed is planted outside when the soil becomes warm. In the

North, it is started indoors in flats and treated about the same as tomato plants. When the soil has warmed and the nights are settled, seedlings may be planted in the garden. Plants should stand 3 feet apart in each direction.

Culture

Soil must be kept moist at all times or stalks will become hollow and pithy. Mulch heavily and irrigate in dry weather.

As fall approaches, plants are blanched starting at least 30 days before the first expected frosts. Leaves are pulled up close to the main stem, the plant is wrapped in heavy paper and tied.

Harvest

Leaves are stripped from the midrib and stalks are used for salad or cooking. If required for winter meals, plants may be lifted with a large ball of soil and replanted close together in a cool dark cellar, where they will blanch themselves. In moving plants, care must be taken not to cover crown with soil, which will rot the crown and kill the plant.

CARROT

CARROT. *Daucus carota,* belonging to the family *Umbelliferae,* called the carrot family. A biennial herb growing, the first year, a heavy yellow or orange root with foot-high feathery foliage sprouting in a circle from its crown, feeding roots going 25 to 30 inches deep in sandy or gravel loam. During its second year, the root sends up a stem two to four feet high which bears flowers and seed.

Native to Europe and adjoining portions of Asia, carrots are now cultivated throughout the world. They are believed to have been cultivated by the Romans and Greeks, who describe red and yellow varieties; however, a confusion in their literature makes it difficult to understand whether they are describing carrots or parsnips. It is known that present-day garden carrots may be obtained by a few years' cultivation and selection of the wild types.

Roots may be long and tapered or short and stubby, depending upon variety. In general, the varieties taking the longest time to mature are long and tapered; those maturing in a shorter time are short and stubby.

The fleshy edible root consists of an outer skin, a pithy central core and an outer core where sugar and carotene are stored. Best varieties have the largest proportion of outer core. Carrots may be eaten raw or cooked. Carotene, which becomes vitamin A upon being absorbed by the digestive system, is enclosed in cellulose cell walls in the root. Since digestion cannot break down cellulose, the body cannot make use of the carotene unless the raw carrots are juiced, shredded very fine or are softened by cooking.

Nutrients obtained from one-half cup of diced, cooked carrots (100 grams) are as follows: vitamin A, 4,500 units; vitamin B_1, .070 milligrams; vitamin B_2, .075 milligrams; vitamin C, 5 milligrams; calcium, 45 milligrams; phosphorus, 41 milligrams; iron, .6 milligrams; protein, 1 gram; 30 calories.

Range and Soil

Carrots may be grown in all portions of the United States. In the northern areas where ground freezes hard in winter, they are planted as early in spring as soil can be worked and successive sowings are made until midsummer. In the South and on the Pacific coast, they are grown as fall and winter crops. Optimum soil temperatures are between 65 and 70 degrees. Best atmospheric temperatures are 60 to 70 degrees. Above these temperatures, the roots become shorter and stubbier. Below 60 degrees, roots become longer and poor in color, which means correspondingly poor in vitamin A.

Best soil is sandy or open loam with a large humus content. Heavy clay soils should be lightened with humus for best root development. Soil should be slightly acid, *p*H 5.5 to 6.5, loosened at least 9 inches deep and free of lumps and stones. Early carrots require very rich soil; later varieties require somewhat less fertilizer.

If manure is to be used in the carrot bed, it is best dug in the previous fall. Fresh manure should never be used—it makes the carrots rough-skinned and liquid since it stimulates root branching. In addition to manure, rock phosphate at the rate of 1½ pounds per 100 square feet may be dug in the previous fall. If rotted manure is not available, the following fertilizer may be dug into the bed in spring at the rate of 7 pounds per 100 square feet: 1 part dried blood, 1 part rock phosphate, 4 parts wood ashes. Or a two-inch layer of compost may be spread over the bed after it has been dug and prepared.

Seed

One packet of seed will sow 30 feet of row; one ounce will sow 100 feet. Carrots mature in 65 to 85 days from seed, depending on variety sown. Yield of mature main-crop planting may be about 125 pounds per 100 feet.

Most recommended varieties of early, short stocky carrots are Red Cored Chantenay, Nantes and Danvers Half-Long. These mature in about 65 days. Tendersweet, Imperator and Gold Pak are best long-tapered, slower-maturing varieties, recommended for late planting and winter storing. Danvers Half-Long has an unusually high percentage of carotene. Tendersweet contains larger quantities of vitamin C than most carrots. On heavier soils, the shortest roots are better than long tapered ones, which are more likely to be deformed in clay soils.

Planting

Carrot seed is slow to germinate and germination is uneven under the best of conditions. First shoots are small and weak, so soil that bakes hard on the surface must be made open and porous for seedlings. This can be accomplished if seed is planted in a drill and covered with sifted compost. Seed should be sown comparatively thick—about half a dozen seeds to the inch—and thinned later. Rows should be spaced 12 to 16 inches apart. Early sowings, while spring soil is cool and moist, should be ½ inch deep. Later sowings, when soil dries in the late spring and summer heat, should be one inch deep, in drills that have been thoroughly soaked before planting. Seed for late sowings, which may be difficult to germinate because of dryness, will come up 4 or 5 days earlier if seed is pregerminated before sowing. Spread the seed in a thin layer between two sheets of wet blotting paper and store in refrigerator. As soon as white root tips break out of the seed, it is ready to be planted. Mix the damp seed with a little dry sand to make it easier to sow evenly.

Because young carrots are much more tender and nutritious than those that have remained in the ground too long, a succession of plantings are needed to provide them throughout the growing season and late crops for winter storage. Plantings may be made every 3 weeks from very early spring, before the frosts are over, until 2½ months before the first expected fall frosts. If first seeds are pregerminated and sown under glass cloches in the garden, they may be brought to maturity early enough so that tomato and pepper plants may be planted in their rows to succeed them. Brussels sprouts and broccoli, which are also rich feeders in spring, may be interplanted among slightly later plantings. For the last sowings, lettuce seed planted in the same drills will sprout early and provide shade for the slower carrot seeds.

Culture

When seedlings are 2 to 3 inches high, they should be thinned to stand half an inch apart. Thinning should be done while the soil is damp, in order not to disturb soil around the roots of remaining plants. When the young carrots are ½ inch in diameter, they should be thinned again to stand 1½ inches apart. These thinnings may be used for salad, or may be cooked whole.

At the same time as the latter thinning takes place, soil should be brought up to cover the shoulders of the carrots. Exposure to the sun turns the shoulders green.

If they are heavily mulched, carrots should need watering only during long dry spells. Soil should be flooded, because mature feeder roots go deep.

Enemies

Larvae of the carrot rust fly, a tiny, yellowish white maggot, burrows in carrot roots to make tunnels. If the depredations are slight, the tunnels can be cut out and the carrots used for cooking. If the infestation becomes serious and cannot be controlled by rotating the crop, skip the early plantings one year and do not plant until early summer, when the grubs will have died of starvation. If this treatment is followed, celery planting must also be held up in the same year, because otherwise the maggots will live happily on celery roots while they wait for carrots. Main crops planted after June 1 will escape the larvae if harvested in early September, before the second brood are hatched.

Damage from wireworms is usually not very great, but it can be controlled by a sprinkling of wood ashes along the row.

Best soil for carrots is sandy or open loam which has a large humus content, has been loosened to about 9 inches deep, and is free of stones.

Bacterial soft rot, which turns the center of the carrot soft, may occur either in the field or in storage. Rotation of rows usually prevents soft rot in the garden. To prevent it in storage, thoroughly dry the outsides of roots in the sun before storing.

A deficiency of manganese or boron in the soil can turn the carrot core black. Both rock phosphate and ground limestone usually contain these trace minerals.

Harvest

Early carrots may be pulled and used any time after they are ½ inch in diameter. They are best flavored and most tender if they are not allowed to grow to more than one inch in diameter.

Late carrots are dug soon after the first fall frosts, when the ground is dry. Tops are cut off leaving about an inch of stem. They can be stored in baskets in a cool humid cellar. If the atmosphere of the storage room is dry, they should be buried in sand Temperature for best storage should be just above freezing. Outside storage pits, if above freezing, may be used. To make a pit, dig a hole 12 inches deep, and place in it a wooden box or crate. Line the bottom with 3 inches of straw and place carrots on the straw. Fill the interstices with sand and cover with a 6-inch layer of straw. Invert a second box over the filled one. Cover the whole with a 6-inch layer of straw, then a 6-inch layer of soil. Protected in this way, carrots will remain crisp and fresh all winter.

Fresh-Dug Carrots at Snow-Melting Time

Leftover snow lay in sheltered corners of the yard when my husband spaded the garden last spring and found two carrots—fresh and crisp and sweet—that had been overlooked the previous fall. When you consider that the ground had been frozen to a depth of 3 feet and that our northern Minnesota wilderness temperature had reached 45 degrees below zero, the healthy roots took on the appearance of a minor miracle.

Our next harvest of Improved Long Orange was so bountiful that, last October, after we had exhausted our supply of sand and containers suitable for winter carrot storage, two rows were still not dug. We decided to try to hold them over in the ground. After all, if two carrots could survive, why not two rows?

We clipped the tops about a half-inch above the roots and covered the rows with a wide, 3-inch layer of coarse sawdust, which we hoped would be a protection against extreme cold snaps that might come before the snow. The glittering, white blanket was 3 feet deep by December first and did not begin to melt until the latter part of April. As soon as the garden area was bare, we scraped away some of the sawdust. Bright and green, young carrot leaves peeped out.

The thought of tender carrot strips was mouth-watering, because our stored supply was long gone and it does not pay to have them mailed to us from the village in winter as they stand a good chance of being hopelessly frozen on their 45-mile trip. But the ground was still ice-hard; not even a pick would break it up! Regretfully, we replaced the sawdust as a deterrent to strong sprouting which might result in seed-stalk formation and inedible roots.

In mid-May, the ground was fit to dig for this year's carrots and, on the day we planted, we ate the first of those held over from last year! On June 15, when leaves were beginning to poke up through the hindering sawdust layer, we ate the last of the roots, which were still perfectly firm, and replanted the two rows.

Our sand-stored carrots, which are watered regularly and kept at an above-freezing temperature, put out many tiny rootlets and show considerable sprouting. By March, their texture is less delicate and their flavor less sweet, but the moist storage prevents their wilting. Apparently the dryness and very cold storage temperature of those left in the ground held back growth so that there was no deterioration.

Why don't you experiment with a few of your own carrots, if your garden is in an area where the ground freezes to at least a depth of a few inches? If your snow cover is not dependable, a thick layer of any close mulch which extends somewhat beyond the rows should prevent either too hard freezing or frost heaving. If this works for you, you will find that the sight of those bright orange roots, fresh from the bedraggled earth of early spring, is the sort of thrill no vegetable gardener should miss.

—Helen Hoover

CAULIFLOWER

CAULIFLOWER. *Brassica oleracea botrytis cymosa,* family *Cruciferae.* Cauliflower and broccoli were lumped together in one genus and species by Linnaeus' classification and have been so classified ever since. For many years, the plant we now call cauliflower was called cauliflower broccoli and our broccoli was known as asparagus or sprouting broccoli.

A biennial herb, cauliflower is somewhat more tender than most of the *Brassicas.* Under the right conditions of soil, temperature and moisture, it forms a large leafy plant with cabbage or kale-type leaves. Just before maturity a small flower, at first enclosed in curled leaves, appears at its heart. This flower, or "button," is protected from sun by covering with its own leaves and matures in 5 days to two weeks as a white compact flower head. Varieties which make purple heads do not need to be covered or blanched. Roots go deep into the soil, 2 to 3 feet in mature plants.

Florets of the cauliflower head may be used in salads or may be cooked. They are a rich source of calcium in the diet. Nutrients to be obtained from a 100 gram portion (3/4 cup) of cooked cauliflower are: 90 units vitamin A, .085 milligrams vitamin B_1, .090 milligrams vitamin B_2, 75 milligrams vitamin C, 122 milligrams calcium, 60 milligrams phosphorus, .9 milligrams iron, 2 grams protein and 25 calories.

Range and Soil

Like cabbage, cauliflower does best in a cool, humid climate. But unlike cabbage, cauliflower is tender to frosts and sensitive to rough handling. It needs very fertile soil, heavy enough to hold large quantities of moisture and well supplied with humus. Because it cannot be planted out as early as cabbage, it will not make heads where hot weather comes in late spring. At least 65 days of cool but frost-free weather are required after setting out to mature a spring crop. In most of the country only fall crops can be grown.

On the Pacific coast "winter cauliflower," which takes 5 months to mature, is planted in fall and harvested early the following spring.

Though cauliflower demands moist soil, it must have good drainage as well. Spring crops should have even richer loam than that needed by fall crops. Both need large quantities of nitrogen and potash. Well-rotted manure is the best fertilizer and should be dug into the bed the previous

When the flower heads of cauliflower become egg-sized, the large, inner leaves should be drawn over and tied to protect head from the sun.

season. Optimum *p*H is 6.0 to 7.0. Since it needs large quantities of calcium, if the soil is acid a layer of limestone at the rate of 10 pounds per 100 square feet may be raked into the soil. Rotted compost may be substituted for manure.

Seed

One packet of seed, or 70 plants, will plant 100 feet of row. Yield from 100 feet is about 50 heads of cauliflower. Seed germinates in 7 to 9 days; plants mature in 60 to 150 days after setting out, depending on weather and variety. Most recommended varieties are Snowball, Super-Snowball and Snowball Imperial for white; Purple Head and Royal Purple for purple heads.

Planting

Seed for early cauliflower is planted in flats indoors about two weeks later than seeds for spring cabbage. Soil temperature of seed flats should be about 70 degrees. From the time they germinate until they are set out, special care should be taken to see that they are not set back. If stunted at this stage, heads will be small and inferior. They must be well-supplied with moisture; temperatures should be fairly uniform; they should be well-ventilated and, when transplanted from flat to cold frame, they should be protected from the sun for a few days. Roots may be pruned in the flat a week before transplanting by running a knife through the soil between plants, thus cutting back the longest roots. When they are transplanted, care should be exerted to keep a ball of soil around seedling roots.

Seedlings are set in garden when about 6 inches high, at the age of 4 to 6 weeks, 20 inches apart in rows which are 30 inches apart. Spring crops should be set out to mature before hot weather; fall crops must mature before the first frosts. Early carrots or onions may be planted in the rows with fall cauliflower. If they follow a planting of legumes such as snap beans or peas, they will benefit by the extra nitrogen in the soil.

Culture

A heavy mulch will help preserve soil moisture and keep soil temperatures down in hot weather. Cauliflower planted in well-prepared beds should need almost no further care except watering in dry weather until head begins to form. While heads are maturing the plants require more moisture than at any other time. Even though rainfall is normal, it is advisable to soak the soil around each plant when leaves are tied up for blanching.

Enemies

See Broccoli, Cabbage, both of which have the same pests and diseases as cauliflower. Black leaf spot of cabbage is responsible for brown spots which sometimes appear in cauliflower heads. Best control for this fungus disease is the purchase of seeds grown in western states where the disease does not occur.

Harvest

When flower head begins to break out of the curling protective leaves and before it gets bigger than egg-size, enough of the large inner leaves to protect the head from sun must be brought over the button and tied loosely. Or the midribs of the leaves may be snapped on the outside, so that the leaves will flop over the head, to form a cover. Ribs must be snapped high enough to give the head room to develop without pushing through the sun shade. Outer leaves should be left intact, to make food which is used in heading. After covering, the heads must be closely watched. As soon as a solid flower develops, it is ready to harvest and must be cut immediately or it will deteriorate. Heads do not become larger, only looser and eventually branched or "riced" when allowed to stand too long. Flavor and color both suffer in an overgrown head.

Stem should be cut well below the head, if cauliflower is to be used immediately. If it is to be stored (it will keep about one month in a cool cellar), plants are pulled up by the roots.

CELERIAC

CELERIAC. *Apium graveolens rapaceum.* Family, *Umbelliferae.* Also called knob celery and turnip-rooted celery. Developed from the same wild plants as celery but selected for the fleshy tuber or root which grows to 4 inches in diameter. Well-grown, the tuber is irregularly shaped, solid, tender and delicately flavored. It has been cultivated for food since about the end of the sixteenth century.

Since the root alone is used, cooked or raw, no vitamin A, vitamin B_1 or vitamin B_2 is derived from celeriac and a small amount of vitamin C–2 milligrams per ½ cup portion or 100 grams. It is richest in calcium and phosphorus, providing 47 milligrams of calcium and 71 of phosphorus per 100 grams; .8 milligrams of iron; 3 grams of protein; and a total of 38 calories.

Culture

Celeriac needs exactly the same treatment as celery, for best growth. (See CELERY.) It is not quite so demanding in matters of moisture and fertilizers, however, and does not need blanching. It may be grown successfully where celery may fail. If celery is grown only for flavor, rather than for nutritional value, celeriac may be an acceptable substitute. Principal variety is Giant Prague. For fall crops, which may be stored outside in the ground where winters are mild or in sand-filled boxes in a cool cellar, planting schedule is about the same as for fall celery.

CELERY

CELERY. *Apium graveolens,* a member of the carrot family, *Umbelliferae.* Obsolete name for celery or a closely allied plant is smallage. An annual or biennial marsh plant, native to Europe from Sweden south through Africa to Egypt and Abyssinia and to India on the east. Isolated specimens of wild celery have also been found in California and New Zealand. It was known and described by the ancients, but not cultivated except for medicinal purposes until the seventeenth century.

Growing 12 to 18 inches tall, celery has a shallow root system which penetrates the ground no more than 8 inches in any direction. Edible parts are the stalks and leaves, which are eaten raw or as flavoring in soups and stews and the seeds, which are used for flavoring. Two types are commonly grown, the self-blanching or golden and the green (Utah or Pascal) varieties.

Comparative vitamin contents of the blanched and the green stalks are given below. In 4 stalks (weighing 100 grams) of blanched celery there are about 20 units of vitamin A; in the same amount of green celery, 640 units. In each variety there are .030 milligrams vitamin B_1. Blanched has .015 milligrams of vitamin B_2 compared to .045 in green. Vitamin C content is almost the same, 5 milligrams in blanched to 7 in green. Green stalks contain 98 milligrams of calcium, compared to 78 in blanched and .8 milligrams of iron to .5 in blanched. Each has 46 milligrams of phosphorus, about 1 gram of protein and a total of nineteen calories.

Range and Soil

Cool even-temperatured areas where climate is moderated by large bodies of water grow the best celery. A long, cool spring of almost 4 months without frost is needed to mature an early summer crop. In the North, fall crops are grown almost exclusively, most in the Great Lakes region. The Pacific coast, Florida and sections of the South grow winter crops. Muck land where the water table is within 20 to 28 inches of the surface is ideal for celery, a marsh plant. In the garden where muck soil is not available, a soil loaded with humus is necessary to keep moisture content high enough for crisp fast-growing stalks. Optimum *p*H is 6.0, but celery will tolerate soil with any rating between 5.2 and 7.5.

Muck soil, where the water table is within two feet of the soil surface, is ideal for celery, actually a marsh plant.

Because of the short roots which must supply plant nutrients, very rich soil in the immediate vicinity of the planting row must be provided. It is almost impossible to make the soil too rich. If well-rotted stable manure is available, it can be dug into an area 14 inches wide and 9 inches deep at the rate of 90 per cent manure to 10 per cent loam. Five pounds of bone meal per 100 square feet should be added to this. To achieve this soil mixture, 100 pounds of manure are needed in a 100 foot row—about 5 full wheelbarrows. If rotted manure is not available good compost or leaf mold may be susbtituted, with 50 pounds of dried chicken manure added per 100 feet of row.

Seed

Since seed may carry blight, one of the few diseases which attack celery, it is advisable to buy blight-resistant strains from a reliable dealer. One packet of seed will provide more than enough plants for a 100 foot row. Yield from 100 foot row is about 200 plants. Seed germinates slowly, taking between 10 and 15 days in soil whose temperature is 63 to 68 degrees.

Especially recommended green varieties are the various Utah strains or Summer Pascal. Golden Self-Blanching is the most popular variety for bleaching.

Planting

Where spring crops can be grown, seed is started in January or February indoors for planting in garden 10 weeks later. Seeds for fall or winter crops are usually planted outside, but in flats or cold frames rather than directly in rows where it will grow. Five to six months should be allowed to mature celery from seeding to harvest.

A mixture of two parts garden loam, one part sand and one part peat moss or leaf mold is best for starting seed. Scatter seed thinly on the surface of the flat and cover with a fine sifting of clean sand. Flats must be well watered from the bottom and never allowed to dry out. If seed is started in cold frame, care must be taken not to expose seedlings to temperatures below 50 degrees for an extended period. If they are too cold, they may bolt without forming stalks. Indoor temperatures should be kept between 60 and 70 degrees during the day and between 50 and 60 degrees at night. Flats should be shaded, indoors or out, until seedlings are up and appear to be well-rooted. Shade should be gradually withdrawn to permit them to harden to sunlight. Care should be taken to provide good ventilation even when plants are shielded from direct sun, because celery plants are weak when newly sprouted and easily succumb to damping-off.

Thin seedlings to stand one inch apart in flat. When they are 3 to 4 inches high, they should be transplanted to a second flat or cold frame containing rich loam, where they are spaced 3 inches apart. Shade until growth again begins. About 10 weeks after sowing, seedlings should be 4 to 5 inches tall and ready to plant in the garden.

If plants are to be blanched, they should be planted in a trench 3 or 4 inches deep and a foot wide, with soil at the bottom of the trench prepared as described above. Green varieties may be planted in rows prepared flush with garden soil level. Plants are set very slightly below depth to which they grew in flats and care should be taken not to pull soil inside the leaf stalks. Space seedlings 6 inches apart in rows 24 inches apart. Shade from the sun and water well after transplanting. Best celery is grown from seedlings that have suffered the smallest possible amount of shock in transplanting.

Culture

To preserve an even amount of soil moisture in the garden, a heavy mulch is needed around celery plants even though the soil is full of humus. Best for this purpose is a mulch of ground corncobs, shredded leaves mixed with old sawdust or shredded hay or straw. About 1.5 inches of rainfall or irrigation per week is recommended.

For most luxuriant growth 3 applications of manure water should be given, the first one week after planting in garden and two more at 10 day intervals.

If celery is to be blanched, soil from the edges of the trench in which it stands is gradually pulled down around plants as they grow taller, being careful each time not to get soil into centers of plants.

As cold weather approaches, it is advisable to keep a supply of dry straw near the bed for cover on frosty nights. Straw spread loosely over the row will protect it from the first light frosts. When weather is mild, the cover should be removed.

Enemies

Greatest losses from damping-off of seedlings occur during hot, humid, rainy summer weather or when temperature above 75 degrees takes a sudden drop in fall. Best protection from the summer heat is a shade of muslin stretched over the bed but held high enough to permit ventilation below. Flats should always be watered by being placed in a container of water which raises the level no closer to the surface than 3 inches. Seed at the top of the flat will be kept damp by capillarity.

Virus diseases which attack celery can be spread from weeds surrounding the garden to the celery by aphids. If weeds in the vicinity are kept mowed, danger of infection is decreased.

Larvae from the carrot rust fly sometimes attack celery. For control, see CARROTS.

Harvest

Self-blanching celery must be blanched 2 to 3 weeks before harvest in warm weather or 3 weeks to a month before harvest in cool weather. In warm weather, recommended blanching method is either by means of paper collars or boards. In cool weather, soil may be banked around plants.

Heavy building paper or thickly folded newspaper may be used to make paper collars. To apply them, gather the leaves and stems close around the plants and remove any suckers. Wrap collar around stems, allowing 4 to 6 inches of leaves to protrude at top. Tie firmly and bank up soil against paper at the base to exclude light from below.

Twelve-inch boards may be used for summer blanching if they are available. Place the boards next to the plants on each side, working them into the soil to a depth of an inch at the base. Stakes are driven into the soil on the outside of the boards, to prevent them from falling away from the plants. Six-inch cross braces may be nailed on the tops at intervals to hold the boards free of the leaves.

Boards may be used to hold soil against stalks in cool weather or the soil, if sufficiently moist, may hold by itself. It is hilled up against plants to permit only the top leaves to emerge, after leaves have been drawn up close to the center of the plant. This method of blanching is recommended for celery which will be used after frosts and the plants may be left hilled in this manner for the entire winter where soil does not freeze hard.

To harvest, cut the plant root immediately below the soil surface if celery is to be used immediately. If for winter storage, plants are lifted with roots. Except in the coldest areas, celery may be left in the garden and covered with a deep layer of leaves or straw. Where temperatures go far below freezing, plants may be stored in deep outdoor pits where they are covered with straw and soil and kept dry with mats. They are lifted with their roots for such storage and are placed close together in pits 12 or 18 inches deep. If small mats or wooden covers are placed over the top of the leaf and soil cover, small portions of the pit may be opened at a time and a few plants at a time removed to indoor refrigeration. If a cool cellar with temperatures just above 32 degrees is available, celery may be kept in boxes packed with plants placed close together and covered to maintain humidity. Plants treated in either of the two latter ways will blanch themselves during the winter.

CHAYOTE

CHAYOTE. *Sechium edule,* a member of the gourd family, the *Cucurbitaceae.* A perennial vine native to the West Indies, but cultivated in tropical America, Madeira and the West Indies. The name used by English-speaking countries for the fruit is derived from the word the Aztecs used for it, *chayotli.* It is also known as Christophine, chuchu, vegetable pear and mirliton.

Almost every part of the chayote vine is edible—the pale green fruit which is a 4-inch squash having one edible seed; the young leaves which can be used in salads or cooked greens; the old leaves which are used as forage; and the tuberous root which is similar to sweet potato.

Range and Soil

Chayote vines are perennial only where the ground does not freeze to a depth greater than an inch or two. It is grown in California and Florida as a perennial. In areas further north, it may be grown as an annual, but the vine takes a long time to mature, so it can be grown only in regions where there is a very long growing season.

Soil requirements are very similar to those of other members of the gourd or cucumber family. Rich moist soil is best, with a *p*H around neutral, from 6.0 to 8.0. Where the vine is perennial, heavy annual applications of an organic fertilizer should be made during the growing season. A mixture of cottonseed meal, wood ashes and rock phosphate can be dug into the top layer of soil within a radius of 3 feet of the plant. Where it is grown as an annual, several forkfuls of rotted manure should be dug into the soil 6 inches to one foot below where the seed is to be planted.

Planting

Vines are usually started by planting a whole fruit where the vine is to grow. It is placed in the ground with the stem end just protruding. It is advisable to start several vines, although each may become very large, because some plants have both male and female flowers, but other vines have flowers of only one sex.

Vines are spaced 10 feet apart each way. A support should be provided to keep the fruit off the ground.

Harvest

Fruits may be produced throughout the year on vines which are well established. As many as 100 fruits may be yielded in a single year by one plant.

After the vines are two years old, small tubers are produced around the main tuber. They may be dug and used at any time. New leaves may be harvested year round from the growing tips of the branches.

CHICORY

CHICORY. *Cichorium intybus.* Also called succory and blue dandelion, a member of the daisy family, *Compositae.* Native to Europe and the Orient. The wild plant is still used in much of Europe, where leaves of the seedlings are gathered in spring for salad. It has been cultivated since the early part of the seventeenth century for large roots which are either dried, roasted and ground to use as a coffee substitute or adulterant or are stored fresh and forced in winter to provide witloof or barbe de capucin, both used for salad. Fleshy roots are also sometimes cooked like turnip. The plant has been little changed by cultivation. Red midrib varieties of the wild chicory are the prototype of the Magdeburg and red Italian cultivated chicories. They are closely related to endive, which is *C. endivia.* Witloof is often referred to as French endive.

Chicory is a hardy plant which grows

2 to 4 feet high, has blue flowers which open in the morning and close with the sun. They somewhat resemble bachelor buttons in appearance. The root is thick and tough, going deep into the soil. It has escaped from cultivation and grows wild through large parts of the United States.

Seed

Several varieties of seed are available, each intended for a specific use. Witloof or French endive seed is used to grow roots for winter forcing. Seed for Large Rooted Magdeburg provides roots for drying and roasting, as a coffee substitute. Roots of Magdeburg may also be cooked and served as a vegetable or young tops may be used for salad. Asparagus chicory, also called *Cichoria Catalogna,* is the Italian dandelion of which leaves and young seed stalks are used for salad. A packet of seed provides 300 to 500 roots. Yield depends upon use to which they are put. A packet used to grow roots for witloof will provide 200 forcing roots plus salad greens from the thinnings.

Range and Soil

Chicory may be grown throughout the United States. It is a better crop for the northern sections, however, than for the southern. It will grow in any soil with a *p*H of 5.5 to 6.5, but the bed should be deeply prepared to provide straight roots that may be easily lifted in fall. Roots will be stronger for the addition of some rotted manure or compost to the bed.

Planting

Seed is sown is spring approximately 3½ months before the first killing fall frosts, usually about June 1. Seed is sown one-half inch deep in rows two feet apart. If grown for roots, the plants may stand 3 inches apart throughout their growing season. If roots are intended for forcing, they are thinned to stand 6 inches apart in the row and young seedlings pulled in the thinning are used for salad. They should be mulched to keep down weeds and preserve moisture.

Harvest

Roots are dug after the first killing frost has killed the tops. If they are intended for cooking, they may be stored in a cool cellar with high humidity or buried in sand in a cellar or pit until needed.

Witloof

Forced salad shoots which are compact, with cos-type heads are called witloof or French endive. After roots are dug in the fall, tops are cut to within an inch of crowns and the roots are shortened to a uniform length of 9 inches. They are then buried upright, close together, in damp sand or soil with a 6-inch layer of sand over their crowns. They may be stored in this condition until two or three weeks before they will be needed for salad, if the soil temperature is kept down near the freezing point. When they are to be forced, the roots in their containers are brought into a warm cellar where the temperature is about 60 degrees. They are watered once a week, to keep them moist but not wet. Within two or three weeks, blanched heads—5 or 6 inches high—may be cut for salad. The roots will then be of no further use.

Barbe de Capucin

The French method of forcing chicory roots produces loose pink and white leaves which are very tender, but essentially the same as witloof. To produce these, roots are planted in a triangular or pyramidal pile of soil in the cellar where they are to be forced. A 3-inch layer of soil is laid down on the floor where the temperature can be maintained at 60 degrees. On this bed the trimmed roots are laid horizontally, their stem ends pointing outward. They are covered with 3 inches of soil and a second layer of roots is placed on top of the soil, with crowns back about 3 or 4 inches from the first layer. The layering is repeated until roots are used and pile is 2 or 3 feet high. Keep the soil moist and in 2 or 3 weeks leaves may be cut from the outside of the pile. If the roots are not disturbed, several cuttings may be made.

CHINESE CABBAGE

Chinese Cabbage. Common name applied to two species of the genus *Brassica* of the *Cruciferae,* the mustard family. *B. chinensis,* known to the Chinese as *pak-choi* or *bok-choy,* is a leafy vegetable resembling chard. *B. pekinensis* is the more widely known heading Chinese cabbage, called *pe-tsai* in China.

So far as can be discovered, Chinese cabbage has been cultivated for more than 1500 years in the Orient, probably developed from wild mustard as the *B. oleracea* varieties in Europe were developed from wild cabbage. It has found slow acceptance in Europe and in America, having been first introduced in Paris in 1837.

B. pekinensis grows to a height of 16 to 20 inches, with flaring green outer leaves which, when stripped off, reveal a slender tapering cylindrical head. *B. chinensis* grows to about the same height, but does not form heads. Leaves are green and slender, with a crisp white thick midrib. Size varies with maturity and richness of soil.

Entire plant is edible, though outer leaves grow somewhat tough on mature plants. Leaves are cooked or used raw in salad. In China seeds are ground to make mustard. Food value of one cup of raw Chinese cabbage weighing 100 grams is as follows: vitamin A, 5,000 units; vitamin B_1, .036 milligrams; vitamin B_2, .462 milligrams; vitamin C, 50 milligrams; calcium, 400 milligrams; phosphorus, 72 milligrams; iron, 2.5 milligrams; protein, 2 grams; and 30 calories.

Range and Soil

Like other members of the cabbage family, Chinese cabbage demands cool, moist weather for best growth. It can be grown in fall in the northern parts of the United States and in winter in the areas where there are no heavy winter freezes. It will stand a little light frost, but not as much as cabbage.

Any soil base—clay or sand—can grow Chinese cabbage if it is very well enriched with humus and soil nutrients. To prepare the bed, spread either a 3-inch layer of compost or a 6-inch layer of well-rotted manure over the row and lightly fork it into the top layer of soil. Soil should not be acid, about 6. to 7. *p*H. If rating is lower, prepare bed the previous year by raking in a half-inch layer of ground limestone. Chinese cabbage is planted at the same time as late cabbage, so it may follow early plantings of peas or snap beans. Since it needs a very rich soil, it is best not to plant it where early crops have depleted the soil.

Seed

One packet of seed of the *pe-tsai* varieties will plant 100 feet of row and will yield 65 to 100 mature heads, plus salad greens from thinnings. Some varieties of the *bok-choy* type are so prolific that literally tons of greens may be cut from one ounce of seed broadcast over an area 40 by 60 feet. For the heading type, the most recommended variety is Michihli, which replaces Chihli in most seed catalogues as being more dependable. Wong Bok is a shorter-headed variety grown in the South.

Chinese cabbage needs plenty of moisture, especially when plants are very young. As soon as plants are tall enough, mulch heavily to eliminate need for irrigation.

Planting

Unlike cabbage, Chinese cabbage is not started in seedbeds or in flats in the greenhouse for early spring planting unless the springs are very long and cool. Seedlings resent being transplanted and plants do not head up well when their growth is interrupted. Also, at the first sign of hot summer weather, Chinese cabbage goes to seed.

In most parts of the country seed is started in the garden bed where it will mature. It is planted any time between July 1 and August 15. That will give the plants about 80 days to mature before the first frost. In warm climates where there is no heavy winter freeze, successive sowings may be made starting one month before the first cool weather and continuing monthly until 80 days before the weather is expected to grow hot in spring.

After soil has been thoroughly enriched, a one-inch trench is dug and half filled with sifted compost. Seeds are planted about 3 inches apart and covered with another half inch of sifted compost. Soil is firmed and well-watered. When plants are 3 inches high, they are thinned to stand 6 to 8 inches apart and thinnings are used for salad. When they are 6 to 8 inches high, they are again thinned to 12 to 16 inches apart. The second thinnings may be used for cooked greens.

Culture

Most important factor in growing good Chinese cabbage is a soil so rich in humus that it will hold moisture well. If it is properly enriched, heavily mulched and timed to mature in the rainy fall season, it should need little additional irrigation, except when plants are very young. If the weather during the summer months after planting is hot and dry, thorough watering may be necessary every day to keep topsoil moist. After plants are tall enough so that mulch may be pulled up around them, irrigation should be no longer necessary.

A side-dressing of dried manure dug in during the early fall may be beneficial.

Enemies

Few insect pests bother Chinese cabbage, though it is subject to the same enemies and diseases as the rest of the cabbage family. (*See* BROCCOLI, CABBAGE.)

Harvest

Cut as needed, after heads form. After first light frost, pull up plants with roots attached. Trim outer leaves and stack on a bed of straw in a frost-free corner of an outbuilding. Cover with a foot-thick layer of straw. Stored this way, the heads will keep for a month or more.

It is possible to grow seed from old roots left in the ground over the winter, if the ground does not freeze too hard. After cutting head from the stalk, cover stalks with a thick layer of mulch. Uncover in spring and, if the stalk has lived over the winter, it will immediately blossom and form seed.

COLLARDS

COLLARDS. *Brassica oleracea acephala.* A tall form of kale or open-headed cabbage. Kale and collards were doubtless among the earliest cultivated forms of cabbage, known to the Greeks and Romans. Where collards are hardy, plants sometimes grow to a height of 3 or 4 feet. A biennial herb with coarse cabbage-type leaves growing in a tuft at the top of a stout stem. Edible portion of the plant includes the immature leaves, eaten in salad or cooked.

Collards are very rich in vitamins A and C and in minerals. A 100 gram portion (½ cup) of cooked collards provides 6,300 units vitamin A, .130 milligrams vitamin B_1, 70 milligrams vitamin C, 207 milligrams calcium, 75 milligrams phosphorus, 3.4 milligrams iron, 3 grams protein and a total of 41 calories.

Range and Soil

Collards are winter-hardy in regions having the first frosts about October 30, roughly from Virginia southward. In the cotton belt, they are grown as biennials. They stand summer heat better than cabbage and are grown in its place where a cabbage crop would fail.

Any well-drained soil with a *p*H between 5.5 and 6.5 may be enriched to grow good collards. Well-rotted manure or good compost may be dug into the row before planting or the following organic fertilizer may be substituted, at the rate of 8 pounds per 100 feet: 2 parts cottonseed meal, 1 part colloidal phosphate, 2 parts granite dust.

Seed

One ounce of collard seed will produce 2,000 plants. One packet will produce more than enough for one family. Ten or twelve mature plants will provide all the collard greens needed by the average family for a winter's use. Seed germinates easily and plants mature in about 80 days from seed.

Most recommended variety is Vates, which does not grow as tall as some of the others. Also recommended are Georgia and Louisiana Sweet. Seed for cabbage collard, which is a cross between Wakefield Cabbage and Georgia Collard, produces loose heads which are claimed to have the hardiness of collards and the flavor and quality of cabbage. Heads of this variety can be left on plants all winter, and used as needed.

Planting

Seed may be sown in the row where they are to grow either in spring or in fall in the mild areas. In the North, seed is sown only in spring. If desired, it may be treated like cabbage and sown early indoors, then transplanted to the garden as soon as the soil can be worked. In the garden, seed is sown thinly in drills one-half inch deep. As seedlings develop, they are thinned to stand first 3 inches, then 6, then 18 or 24 inches apart. Thinnings may be cooked as greens or used for salad. Rows are spaced 3 feet apart.

Culture

Rapid growth makes tender collards of good flavor and quality. Although they will stand some drought, they are best if the soil is kept mulched around their roots to keep them moist.

As the first leaves are harvested, the stem begins to grow tall, with the tuft of leaves at its top. A stake or other support will keep the plant upright, so that it will not blow over in high wind.

Enemies

Collards suffer from the same pests and diseases as other *Brassicas.* (For their control, see BROCCOLI, CABBAGE.)

Harvest

Leaves are picked for cooking when they are almost full size but have not yet begun to turn tough or woody. For salad, half grown leaves or young plants are used. The main stem is allowed to stand to produce a further crop. In the North, leaves may be harvested until the first hard freeze, sometimes up to Christmas. It is thought that light frost improves the flavor.

CORN

CORN. *Zea mays.* A member of *Graminae,* the grass family. Native to tropical America. It is believed to have been cultivated by the Aztecs as early as the eighth century. It was the Indians' chief staple of diet and figures in their folklore. The first corn seen by Europeans is believed to have been that carried back by Columbus in 1492. Corn has never become an important crop in Europe because summers there are not hot or sunny enough to grow it well. In the United States, about one-third of all plowed land is planted to corn. Most of this is *Z. mays dentata,* dent corn, or *Z. mays indurata,* flint corn, both grown for stock feed, silage, ground meal and many commercial purposes.

Sweet corn, *Z. mays rugosa,* was not known in New England among the early colonists, although the Pilgrims planted maize during their first few years. It was not until 1779 that the sweet variety now grown in home gardens was first introduced to the northern colonies. It is now grown in all of the states of the continental United States and can be grown anywhere where there is a 70 to 80 day growing season with plenty of heat. Two types are grown, white or yellow, the yellow preferred recently since the discovery that it contains more vitamins than white.

Nutrients that may be obtained from a medium-sized ear of yellow corn weighing 100 grams include the following: 860 units vitamin A, .209 milligrams vitamin B_1, .055 milligrams vitamin B_2, 8 milligrams vitamin C, 8 milligrams calcium, 103 milligrams phosphorus, .4 milligrams iron, 3 grams protein and 90 calories.

Range and Soil

Corn can be grown almost all over this country and in almost any soil, though its preference is a deep mellow rich loam with plenty of moisture-holding humus. The soil must be well-drained and the *p*H 6.0 to 7.0. More acid soils must be limed to bring up the *p*H rating. The hotter the climate, between 40 and 90 degrees, the faster the crop will mature and also the quicker the ears will reach and pass their prime.

Corn is a deep-rooted plant, though it does not have a single deep taproot. First roots developed by germinating seed are temporary. Within a few days, the first set of permanent roots grows out of the stalk above the seed, developing out and downward on all sides. Succeeding sets continue to sprout above each other on the stalk, each becoming more horizontal than the preceding, until when the tassel begins to form, a set is thrown out above the surface of the ground for the purpose of bracing the tall stalk. By the end of the season, the entire area of the corn patch is penetrated with feeder roots, many of which go deep into the subsoil.

Because the roots are deep and the top grows rapidly producing large leaves, corn must have moisture and nutrients well incorporated through the soil. Best fertilizer is a one-inch layer of fresh stable manure applied over the entire bed the fall before planting. During the winter and early spring, rain and melting snow will carry the leached-out nutrients from the manure deep into the soil. When the soil is prepared for planting, the manure residue is dug in to raise its humus content. If manure is not available, a one-inch layer of compost may be turned under before planting. Or 7½ pounds of wood ashes and 3 pounds of the following mixture may be used, if humus content of the soil is already high: 2 parts cottonseed meal, 1 part colloidal phosphate and 2 parts granite dust.

Seed

Open-pollinated varieties of sweet corn are sweeter and contain more protein than hybrid varieties. They are usually designated as "standard" or "open-pollinated" in seed catalogs, while the hybrids are called either "crosses" or "hybrids." In warm weather, seed will germinate in 5 to 8 days and corn matures in 65 to 100 days, depending on weather and variety. One-fourth pound of seed will plant 100 foot row. Yield from 100 foot row is about 48 ears or more, depending on fertility and growing conditions.

Most popular early variety is North Star; second earliest, Golden Beauty; and midseason, Golden Cross Bantam. All these are hybrids. They represent the 3 types which may be planted at the same time in spring to mature in succession, giving ears for the table in about 65, 70 and 84 days after planting.

Planting

For a long corn season, several varieties maturing at different dates may be planted all at the same time, or a succession of sowings of one of the early varieties may be made at 2 or 3 week intervals until midsummer. The last planting should be made about 80 days before the first fall frost, even though the variety should mature in 65 days. Cooling weather as the days grow shorter will delay its ripening.

First plantings may be made on about the date of the last spring frost, though if the soil is heavy with clay, seed will germinate badly until two or three weeks later. When planted in cold soil the seed is attacked by molds and may rot before it sprouts.

In heavy soil or in early spring when the soil is thoroughly moist, seed should be planted no more than 1½ inches deep. Succeeding plantings may be deeper, until a light soil planted in July may have seed buried 3 or 4 inches deep. Roots sprouting above the seed will establish themselves one inch below the soil surface, whatever the depth of the planting. Space seed 3 or 4 inches apart and thin tall varieties to stand 12 inches apart, shorter varieties 8 inches. Or, if tall varieties are to be hilled (soil pulled up around the maturing stalks for support), the hills are spaced 18 inches apart, with 6 seeds planted in each, thinned later to 3. Rows are 2½ feet apart.

Corn must be planted in blocks no less than 3 rows wide, rather than in single rows, to facilitate pollination. If 4 rows each 50 feet long are to be planted with 4 successive sowings, each row should be planted one-fourth of its length at each sowing, rather than one complete row at a time. Fine pollen forms in the tassels, sifts down on breezes to silk of surrounding plants, which in turn carries the pollen to the kernels. A plant completely surrounded by pollinating plants has a much greater chance of being well pollinated than one with neighbors only on two sides.

Culture

Corn needs plenty of water throughout its life because of its large leaf surface and rapid growth. Because it is not planted until late in spring or even until summer, the soil moisture should be conserved. The bed should be prepared early and a mulch spread over it to protect it from drying spring winds. At planting time, the mulch on the rows can be pulled back to permit opening of drills and, after seed is sown, an inch of light mulch may be drawn back over the row. New corn sprouts can penetrate an inch of mulch, but would be stifled by heavier cover. When the new sprouts are 8 inches tall, a 6-inch mulch may be drawn up to the plant and left there throughout its growth. The old idea that corn must have sun on its roots has been proven faulty by gardeners who mulch. The practice is now recommended widely and not only by organic gardeners.

Beginning 10 days before tasseling and continuing through the flowering period,

Mrs. Ann Dovell, an organic gardener from Vista, California, stands beside the 15-foot corn she grew—an indication of the results to expect from organic gardening methods.

corn needs plenty of water. If rainfall has been scanty and soil seems dry under the mulch when this period occurs in late-maturing corn, the crop will benefit by irrigation. After flowering is finished, the ears will form more quickly without too much water.

Side shoots or suckers forming at the base of the stalk have been found by trial to have no deleterious effect on the crop —in fact, it is now believed that the extra leaves may benefit the ears by supplying extra food. The practice of removing suckers has also been discontinued through most of the country.

When stalks are knee-high, they may be side-dressed with a mixture of one part rotted manure and one part cottonseed or soya meal, if desired.

Enemies

Purchase of disease-resistant seed, crop rotation, early turning under of garden trash or manure and removal of old stalks in the fall, all help to prevent loss from insects and diseases of corn.

The corn earworm, which bores into the tips of growing ears, should be destroyed as affected ears are picked. If permitted to fall to the ground, he will pupate there during the winter and emerge in spring as a moth to lay eggs for more worms in next year's crop.

Corn smut, a fungus disease which causes kernels and sometimes tassels to grow monstrous grey boils, should be removed before the boils burst, and affected plants should be destroyed. The black smut inside the boils contains fungus spores which spread the disease.

Harvest

Corn should be harvested as soon as ears are mature, especially in hot weather when it may pass its prime in only one or two days. When the silk is brown and dry, the tips of the ears are rounded but not hard and kernels exude milk when dented with a fingernail, the corn is ready to pick. Past its prime, the kernels get tough, the milk turns doughy and starchy and the corn loses its sweetness.

Corn should be cooked immediately after picking for best quality. If cooking is delayed by so much as an hour, part of the sugar in the kernels will have turned to starch. When it is necessary for any reason to delay cooking, ears should be chilled immediately after they are picked and stored in the coldest part of the refrigerator in their husks. Corn for canning or freezing should be rushed to the pot just the same as corn for the table.

Remove old stalks from the garden immediately after harvest and shred them, if possible, into the compost bin. Do not leave them standing in the garden over the winter to harbor pests or diseases.

Corn from open-pollinated seed may be saved for seeding the following year's crop. Good solid ears should be selected from among the first which mature on plants of good habit. They should be left on the stalk until they are completely dry. After they are picked, they should be stored in a dry, well-ventilated place until required for seed. It is advisable to test the viability of saved seed, to avoid disappointments and delay in starting the next year's crop.

Popcorn

Zea mays everta, popcorn is a variety of corn grown especially for its small ears of pointed kernels which explode when heated. Cultivation of popcorn is believed to be as old as cultivation of maize. Most popular variety is Purdue.

Popcorn culture is essentially the same as sweet corn growing, except that it takes longer to mature. Ears are allowed to dry on the stalks and are harvested as sweet corn seed. Except for seed saved for the next year, popcorn can be husked after picking and kernels stored in glass jars until required.

How Shallow Mulch Doubled a Corn Crop

Jim McNeilus is one farmer who isn't satisfied with things as they are and doesn't rest easily unless he's conducting an experiment of one kind or another. Last spring some of his neighbors said he was crazy when they saw him supervising the spreading of straw on a portion of his corn land. This year they can safely say he's twice as crazy, for he has piled up what is undoubtedly the largest stack of straw in Floyd County, Iowa, comprising about 2,400 bales, laid in at 17 cents per bale.

McNeilus, who farms about 3 miles northwest of Marble Rock, is crazy like a fox, for the straw—according to a yield check—doubled corn output on one of the two 2½ acre strips upon which it was applied.

The germ of the idea to mulch corn came from a friend, Paul Bucklen, who had good initial results two years prior when he mulched his garden. Such an operation is now standard recommendation for home gardens; mulching benefits all vegetable and fruit crops. One of Bucklen's cabbages, uncracked, weighed 13½ pounds. Why, McNeilus asked himself, can't mulching work with corn as well?

To test the idea he split his corn acreage into several 2½ acre strips, with the mulch to go on the poorest two, which were spread over the top of two slight knolls. Two kernels were power checked every 16 inches—17,820 stalks per acre. The strips to receive the straw were cultivated but once and then the straw was applied at a rate of 50 bales to the acre.

When in place and somewhat compacted it averaged about 1½ inches in depth.

During the summer Bucklen checked the stalks for growth and condition and discovered that the mulched areas had average stalk height of around 10 inches more than the unmulched corn on the better land.

Additionally, the unmulched corn was cultivated once or twice more and sprayed to control weeds. It had a tendency to fire while that with the straw remained moist all summer, never baked on top—as is customary in midsummer heat—and the earthworms and soil organisms had ideal conditions in which to work.

The corn was picked by Philip Parcher and son Earl and cribbed by Jim Rex. When McNeilus and Bucklen computed the yield they found corn on the unmulched strips averaged 70 bushels per acre. On one of the mulched strips, which stood well, the yield came to 140 bushels per acre and on the other, where many stalks had been caught by a wind and leveled, the average was 120 bushels per acre.

However, on the latter strip the downed ears were lying on straw, off the soil and pastured cattle easily cleaned them up, so they weren't lost. The main difference between the test areas is that the ears on the mulched strips had much deeper, drier and better colored kernels; the ears were longer and the corn was of better quality.

McNeilus secured the yield increase without commercial fertilizer, which he has never used. The only applications on his land have been of lime.

Despite the fact that it cost him $25 per acre to get the doubled yield, he considers the transaction had an extra profit, the labor not needed for the land that would have been needed to secure the same total crop.

McNeilus feels that if mulching can be a basis for doubling yield, small farms can be worked to yield a profit, something which he says is fast disappearing. "The little farmer," McNeilus says, "is forced to pay taxes to support programs for the large farmer, which are ruinous for the fellow with but a few acres."

Next year McNeilus plans to spread 65 bales of straw per acre on 18 acres, all the land he will plant to corn. Rather than plow down and take quite a nitrogen loss when the straw and stalks rot in the soil, he's going to chance disking, side raking the trash to the side while the corn is planted and gone over with a rotary hoe after it germinates and then putting the trash back in addition to the 65 bales of straw per acre.

And there will be two more experiments. He plans to sow rye on one 2½ acre strip after the corn is in. Then it will be mulched. From this he anticipates green feed in the fall after the corn is picked and green manure to plow down the following spring. On another strip he will be doubling his stalk stand, placing the rows but 20 inches apart for a per acre plant total of 35,640. Harvesting will be done either by hand or with a combine.

Although standard farming practices would call for the addition of nitrogen, McNeilus may not be so far off base with his mulch-trash combination as it may at first seem apparent. For our Iowa virgin soils had been enriched for thousands of years by lush grasses, which fell and decayed upon the surface each year. Man's plow and clean cultivated crops have been the major culprits in the loss of fertility. Mulch can possibly bring back prairies conditions for corn which after all, is merely a stalk of grass with muscles.

—Robert Fischer

Reprinted from the Charles City Press, Charles City, Iowa.

CORN SALAD

Valerianella olitoria, the only edible member of the family *Valerianaceae.* Also called lamb's lettuce and fetticus. An annual native to Europe, North Africa and Asia Minor. It has been in use as a salad green for centuries, but was not brought into the garden for cultivation until about the sixteenth century. Leaves are large and round or oval, grey-green, and tender, with a very mild flavor. Best mixed with a more pungent green, like cress in salad or mustard in cooked greens.

Culture

Corn salad cannot stand hot weather, so it must be planted very early in spring or late in the fall. One-fourth pound of seed will sow a 100 foot row. Seed is sown in drills one-half inch deep in rows one foot apart and thinned to stand 3 inches apart. Six weeks after sowing, leaves may be picked for salad. They should be picked, not cut.

If seed is broadcast late in fall, a few plants will grow large enough to use before the ground freezes hard. Any seedlings too small for use may be lightly mulched at about the same time when the straw is laid on the strawberries. Many of them will live through the winter and will spring up early to provide the first spring garden salad. In southern states, the plants do not need a winter cover.

COWPEA

COWPEA. *Vigna sinensis,* a legume, also called southern pea. A native of central Africa, the cowpea has spread to practically all warm tillable areas of the earth. It is a multirooted legume grown for food, hay, grazing and soil-building.

Range and Soil

With sufficient warm weather, cowpeas grow wherever corn will grow. In the United States, it grows in the southeastern portion roughly from central Pennsylvania to Kansas, though it has been grown as far north as Michigan. It is grown in this zone as a secondary crop with corn, as a main crop and as a soil builder, especially in areas where cotton and tobacco have depleted the soil.

With suitable climate, cowpeas grow on practically any type of well-drained soil, preferring a *p*H around 5.5 to 6.5. They withstand drought well and make forage at time of year when it is most needed. As soil conditioners, they make sandy ground more compact and heavy clays more friable through added humus. Their roots reactivate dead soils with nitrogen-fixing bacteria, which grow in nodules.

Culture

Cowpeas are sown broadcast where they are intended as a soil-building crop. One ounce of seed provides ground cover for 25 to 30 square feet. In protein the straw runs 3.4 per cent, higher than soybean, oat, rice or barley straw. It is used as litter, mulch and feed, being preferred by some growers to the hay.

As food, cowpeas are served like snap beans, shell beans or are dried and cooked like dried beans. Dried, they have 19.4 per cent protein, 54.5 per cent carbohydrate, and 1.1 per cent fat. They are especially rich in vitamin B_1. Sprouts, ready to use after 24 hours germination, are high in vitamin C. For garden culture, see BEANS.

CRESS

CRESS. This name is applied to several members of the mustard family, *Cruciferae.* Cress, garden cress, peppergrass or pepper cress is *Lepidium sativum,* an annual salad plant said by Xenophon in 400 B.C. to have been eaten by the Persians before they became acquainted with bread.

Upland cress (as distinct from water cress) is a name sometimes used erroneously for garden cress. Correct alternate names for upland cress are winter cress, rocket or yellow rocket. Latin name, *Barbarea verna.* Upland cress is a hardy biennial or perennial native to Europe and temperate Asia.

Water cress, *Nasturtium officinale,* is a bog or aquatic plant. (See WATER CRESS.)

Garden cress or peppergrass is a cool-weather plant which grows quickly and passes its prime early. It has small deep green curly leaves with a pungent taste, high in vitamin A. It should be sown thickly in one-fourth inch drills 10 to 12 inches apart in earliest spring. Successive plantings may be made at weekly intervals to mature 40 days later, as long as growing weather is cool. In hot weather, the plants acquire a strong flavor and quickly go to seed. Leaves may be cut for salad within 4 weeks of planting. If they are not cut too close, several cuttings may be made from the same plants.

Upland cress is hardy in northern climates and will not grow where winters are too warm. It is sown in early spring in rich soil and first leaves may be gathered 50 days later. Thereafter, the leaves are picked in early spring, even late winter, before flower stems develop. The leaves have a somewhat bitter flavor. In Sweden, they are cooked like kale. In England and Scotland, they are used for salads.

CUCUMBER

CUCUMBER. *Cucumis sativus,* a member of the gourd family. *Cucurbitaceae.* A trailing or climbing annual vine thought to have originated in Asia and Egypt, but because it has been cultivated for more than 3,000 years, its original habitat is not certain. Now grown throughout the world, it is a rapid grower spreading as much as 6 feet in all directions in a season. Flowers are male and female on the same plant, the male greatly outnumbering the female. Fruits average about 10 to 12 inches, but in some varieties achieve a length of 2 to 3 feet.

Cucumbers are more than 95 per cent water, but contain a fair amount of vitamins A and C. One medium cucumber weighing 100 grams contains 35 units vitamin A, .060 milligrams vitamin B_1, .054 milligrams vitamin B_2, 12 milligrams vitamin C, 10 milligrams calcium, 21 milligrams phosphorus, .3 milligrams iron and yields 1 gram protein from a total of 15 calories.

Range and Soil

A warm-season crop easily injured by frost, cucumbers may nevertheless be grown almost anywhere in the United States since they mature quickly. Light loam, neither too sandy nor too clayey, is best, if it is well provided with humus. If possible, grow cucumbers where legumes were grown the year before. Soil *p*H should be 5.5 to 6.8, but they will tolerate a slight margin above that range.

Where soil is heavy and wet they may be grown in raised beds or hills and supported on fences or trellises to keep the fruit off the wet ground where it might rot. They need a strong support, because when fruit develops the vine is heavy. Eight-foot poles spaced 4 feet apart and supporting 6-foot 20-gauge poultry wire is recommended. Ends of the fence should be braced with guy wires with turnbuckles.

Soil in which cucumbers are to be planted should be well enriched, preferably with manure at least 4 months old, plus a rich supply of organic fertilizer containing a nitrogen source, such as cottonseed meal or dried blood, with rock phosphate or bone meal and tobacco stems, wood ashes or greensand. The NPK formula for the fertilizer mixture should be about 5-10-5.

Deep stirring of the soil is important for cucumbers, not only to assist their deep roots but also to improve drainage. While they need plenty of water, they will not tolerate water standing around their roots.

Seed

A packet of seed will plant 20 hills, about twice the amount required by the average family. The seed germinates in the field in 8 to 10 days and fruit matures 60 to 75 days later.

Disease and blight-resistant seeds are best to buy, since cucumbers are subject to a number of fungus diseases. Most popular varieties for slicing are Marketer and Burpee Hybrid. For pickling, National Pickling, Ohio and Wisconsin are all recommended. China cucumbers are interesting to grow, if they can be supported on a fence or trellis. They grow to 20 inches in length, while their diameter never goes much over two inches. If grown on the ground they are inclined to curl up in doughnut or horseshoe shape.

Planting

The soil should be warm and danger of frost past when cucumber seed is planted outside. They may be planted in rows or in hills. If the soil is well drained, "hills" are nominal—merely a group of seeds planted together in soil level with the ground. If soil is wet, hills may be raised 8 or 10 inches above ground level. The usual method is to plant in level hills, 4 to 6 feet apart each way.

For each hill, dig out an area 12 to 15 inches in diameter to a depth of 6 to 8 inches. Spread a couple of inches of rotted manure or compost over the bottom of the hole. Replace the topsoil and firm. Sprinkle 8 or 10 seeds over the area and cover them with an inch of soil. When seedlings are 6 inches tall, select the 3 strongest vines and pinch out the

By growing cucumbers on a fence or trellis, you can reduce the space ordinarily needed for growing this crop, by 80 per cent; poultry wire makes a good support for plants.

others. Do not try to pull them up, because their roots will probably be entwined with the seedlings that are to remain.

If planting in a row, the row should be 6 feet from its neighbors on each side. Dig out a trench and treat the soil as for planting in hills. Cucumbers grown in rows are thinned to one plant every 12 inches. The vines should be trained to grow to alternate sides of the row. If they are to be trained on a fence, the seeds should be placed about 6 inches from the fence, on the south side if the fence can be run from east to west. Soil should be prepared and firmed in the trench before the fence posts are put in place, in order not to weaken their position. No matter how they are planted, care should be taken not to allow the seeds to come into direct contact with the manure, which may cause the seedling roots to rot.

For the earliest cucumbers, seeds can be started indoors under heat, but they must have special handling. If arrangements can be made to heat the seedbed to 80 to 85 degrees, either by placing the flat over a heater or by use of electric wires, seedlings will appear above ground in 30 to 36 hours. Approximately 72 hours after sowing they are ready to be lifted into 5-inch flower pots, 7 to 10 seedlings being planted in each pot. Pots are then placed in a sunny window where the daytime temperature is 70 to 75 degrees and night temperature is not lower than 60 to 65 degrees.

Three or 4 weeks later, when plants have acquired 3 or 4 true leaves, they will be ready to set in outdoor frames or in the garden under protection. Seedlings should have been thinned to 3 to the pot by this time. If they are to go in frames, the frames must be large enough to hold full-grown vines at maturity. Or they may be set in the garden in hills and protected with glass cloches until weather is warm. Before setting out, the plants are well-watered. They are removed from pots with the ball of earth around their roots intact and set directly in the soil. Or they may be grown indoors in pots of pressed peat moss, which can be set in the ground with the plants. In order to keep new vines coming along all summer, a few seeds may be sown in each hill beside the potted seedlings when they are planted in the garden.

Culture

Being composed of such a large percentage of water, cucumbers are among the thirstiest garden vegetables. Not only do they need a large amount of moisture-holding humus around their roots, but they must have a thick mulch to protect the soil moisture and possibly an extra source of water besides. During a drought, they should be thoroughly watered each week. In areas where summer rainfall is always slight, they may need individual water tanks in each hill.

To provide a constant source of water, dig a hole in the center of each hill (before planting) large enough to accommodate a 6-inch flower pot, sunk to the rim. Firm the soil around the pot and drop a little clayey soil in the bottom, enough to fill the drainage hole. Plant seeds 2 or 3 inches away from the pot's rim. The pot should be filled daily with water and, if the hole at the bottom is properly plugged, the water will seep away slowly through the drainage hole and through the pores of the unglazed clay, keeping the hill constantly moist.

When plants are about a foot high, they may be side-dressed with rotted manure or with organic fertilizer. If staked or grown on a fence, after the main stem of the vine reaches the top of the support, its tip should be pinched back, to encourage side branching.

Harvest

Picking may begin when cucumbers are an inch long if they are to be pickled. Or they may be allowed to grow almost to maturity if they are to be sliced. To remove them from the vines, cut the stems with shears about an inch from the cucumbers. As soon as the fruit begins to form, it must be carefully watched. Fully mature cucumbers seem to grow overnight when they get started. If any are missed and turn yellow on the vines, they should be removed and composted. If they are left to grow, they will slow or stop the vine's production.

Enemies

If disease-resistant strains are planted, the worst enemies of cucumbers are the yellow and black striped cucumber beetles. Adult beetles feed on the leaves and spread wilt; larvae bore into stems and roots. A spray that can be used for very bad infestation may be made by mixing a handful of wood ashes and a handful of hydrated lime in two gallons of water. Both sides of the leaves should be sprayed thoroughly with this mixture.

Marigolds planted near cucumbers are said to repel beetles. If the cucumbers are planted on a fence, the marigolds may be planted on the other side of the fence. Or if in hills, plant rows of marigolds between the hills. The odor of the foliage is believed to be responsible for chasing beetles away.

Cucumber on a Trellis

If you've kept yourself from growing cucumbers because you lack space in your garden, cheer up. By growing cucumbers on a fence or trellis, any gardener can grow and reap a bountiful harvest in less than one-fifth the room ordinarily required. If you have space enough for another row of carrots or beets, you have room enough for a row of cucumbers!

There is no specific climbing cucumber. All cucumbers have a tendency to climb. Just choose your favorite all-around cucumber and go ahead. We have found it wise to plant two varieties—one just for slicing and one for pickling. If you wish, one variety will take care of both needs.

The material that cucumbers like best to climb upon is 20-gauge poultry wire. The thin wire provides good anchorage for the springy tentacles. The fence should be at least 6 feet tall. If you're unable to get wire that wide, overlap two narrower strips. Since the vines produce mostly on their tips as they grow taller, it is important to give them plenty of height in order to prolong the producing season.

Before planting, place 7 or 8-foot poles every 4 feet along the row. Hammer them securely into the ground, then nail the fence to them, stretching it tightly between the poles. Do not space the poles any further than 4 feet apart, for the vines become quite heavy with fruit by midsummer and will cause the fence to sag if sufficient support is not present. Be sure to place a guide wire on either end of the fence to prevent sagging in the middle.

If possible, run the fence from east to west. This will keep all the vines growing on the south side. If the fence must be run the other way, try to keep the vines growing on one side for easier picking by removing the vine tips, now and then, from the holes should they grow through.

—*Betty Brinhart*

CUSHAW

Cushaw. *Cucurbita moschata.* Also called Canada crookneck squash, winter crookneck squash or winter crookneck pumpkin. Erroneously called pumpkin, due to its having the same kind of stem, swollen at the point of attachment. (See Squash.)

DANDELION

Dandelion. *Taraxacum officinale,* member of the daisy family, *Compositae,* a perennial with deep taproot, milky juice and only ray flowers in the head. As a wild plant, dandelions have long been known and used for medicinal purposes and for food. It is only during the past century that gardeners, mainly French, have cultivated them. The cultivated varieties developed in that time have a very superior flavor and are more tender than the wild types. Parts used for food are principally the stemless leaves, which are cooked like spinach, used from the garden for salad or forced in winter. Flower heads of the wild varieties are also used to make wine and the roots may be dried and used as a coffee substitute.

Cooked dandelion greens yield an enormous quantity of vitamins A and C and are rich in B vitamins and minerals. In a 100 gram portion, approximately ½ cup of cooked greens, there are 15,000 units of vitamin A, .190 milligrams of vitamin B_1, .270 milligrams of vitamin B_2, 100 milligrams of vitamin C, 84 milligrams of calcium, 35 milligrams of phosphorus, 6 milligrams of iron, 3 grams of protein and about 45 calories.

Range and Soil

Dandelions will grow in any temperate climate where spinach or kale will grow and are so common as to be noxious weeds. They will grow in any garden in the United States except in the hottest parts of the South. In the warmer areas, they may be grown as a winter crop; in the North, they are grown for spring harvest. Any good garden soil with a *p*H rating between 6.0 and 8.0 will grow them.

Seed

One packet of seed will sow a 15 foot row; one ounce will sow 100 feet. Greens mature in about 95 days from seed. Most commonly sold in this country are the Thick-Leaved or Arlington Thick-Leaved varieties. Also recommended are French Broad-Leaved and Common French.

Planting

Seed may be sown in spring or until midsummer to start plants for the following winter or spring. Plant in drills one-half inch deep in rows 18 inches apart and thin plants to stand one foot apart in the rows. Cover the plants during the winter with a straw mulch in the North. In the South, they need no protection.

Harvest

A light crop of leaves may be cut the first fall from seed planted in spring. Or the plants may be allowed to grow until spring for the first cutting. Leaves are cut like spinach for cooked greens or salad.

Blanched greens lose many of their vitamins, but they are very tender and less bitter than the dark green leaves. They may be blanched by tying the leaves together over the plant, and covering the whole with a flower pot until the leaves turn light green. Or the plants may be buried in sand.

Dandelion roots may also be dug in fall and forced like *barbe de capucin*. For the forcing method, see CHICORY.

Raising Tame Dandelions

Being a dandelion greens' lover and regretting that they last only a couple of weeks in early spring in the wild state, I bought seeds from a nursery some 3 or 4 years ago and planted them in the garden. They are still there and in the same rows where they were planted, as I don't plow or till up the garden any more.

Each spring I loosen the soil and plant around them, then harvest the greens all spring, summer and fall. Besides, I always freeze the winter's supply out of probably 25 plants in all. Last summer they self-seeded a few plants, which I'll take up and put in the row this spring.

All summer, I dump the dry garbage and wastes right on top of the leaf and straw mulches, whereas the smelly type of garbage is hidden underneath. My earthworm population is sufficient to take care of such practices—and a good thing too, as I don't have the physical strength to do much composting.

The garden has had generous applications of rock phosphate, seaweed, wood ashes. When the soil became covered with moss from too many oak leaves, we put fresh marl over the entire surface, which corrected this sour condition.

The dandelions, being untemperamental and undemanding, get no extra care whatever and last summer they grew to a terrific size. I don't know whether the seaweed, the marl, the leaf mulches, the rock dusts or what accounted for this outlandish growth. They were tender and good-flavored all summer, whereas before, during the hot, dry weather, they were tough and I left them alone until the fall rains caused new, tender growth, after which I cooked them all autumn.

I dig the root vegetables from under the mulch all winter. The dandelions and cress will be much earlier this spring for having had this coverlet of leaves and straw through the zero weather. I uncover them by degrees and will probably find a mess of dandelions ready under the mulch in early April, possibly in March.

—Gertrude Springer

DASHEEN

DASHEEN. *Colocasia esculenta,* a tropical herb of the arum family, *Araceae*. Also called taro, kalo and eddo. It is grown in tropical America, Jamaica, Japan, south China, Tahiti, the Philippines and Hawaii. In many of the places where it is cultivated, it is the chief item of diet. A paste called poi is made from the root.

Dasheen is a caladium-like plant, closely related to the ornamentals called elephant's-ears. Its large leaves grow on 6-foot stems from tubers. Roots are shallow. Edible portions are the tubers and the young shoots which may be gathered in spring or forced in winter.

Tubers are more starchy than potatoes or sweet potatoes and contain more protein. They are somewhat sweet and nut-like in flavor. They contain about half as much sugar as sweet potato. The shoots, which are cut and used like asparagus, have a flavor something like mushrooms. Leaves are sometimes cooked in soup, where they almost dissolve and thicken it.

Range and Soil

Seven months of hot weather are required to mature a crop of dasheen. In the United States, it can be grown successfully only in Florida and the Gulf states. Where only 6 months of growing weather are hot enough, the tubers may be sprouted inside and transferred to the garden when the weather is settled.

Best soil for dasheen is rich, moist, but well drained. Silted creek or river-bottom enriched with an organic fertilizer with an NPK formula of 4-6-10 is ideal.

Seed and Planting

Dasheen is planted outside two weeks before the last killing frost. Best variety for planting in the United States has proved to be the Trinidad dasheen. Tubers weighing 2 to 5 ounces are used for seed. They are planted 2 to 3 inches deep, 24 to 30 inches apart in rows that are 3½ to 4 feet apart. Weeds should be kept down and soil kept moist until leaves are tall and large enough to shade the ground.

Harvest

Leaves are cut back and removed first to make harvest easier. Roots are then lifted in the same manner as sweet potatoes before the first fall frost. The corms are divided and cleaned and stored for use very much like sweet potatoes, but they do not require a curing period. Largest corms should be used first because they do not keep as well as the small ones.

Forcing

To force shoots for winter use, plant corms near the surface of a sand bed in a warm cellar or greenhouse. A tent-like arrangement of heavy burlap may be placed over the bed if blanched shoots are required. Shoots are cut and used as they attain a height of about 6 inches.

EGGPLANT

EGGPLANT. *Solanum melongena.* A perennial or sub-shrub native to the tropics, member of the potato family, *Solanaceae.* It has been cultivated for many centuries in the southern Asiatic countries, but was not known in Europe until the Middle Ages and then it was grown only as an ornamental. It is now used widely through southern Europe and Asia Minor as food. It has not been very popular in the United States, though it is gradually gaining wider acceptance.

In areas where temperatures drop even for a short period below freezing, eggplant is grown as a tender annual. As an annual, the variety grown most widely in the United States attains a height of two to three feet, with attractive grey-green foliage, large lavender blossoms and purple-black glossy fruit which may grow 10 to 12 inches long. Elsewhere varieties with smaller fruits are grown and color of the fruit varies from black, through purple, green, yellow and white. Smaller fruited varieties are thought to have much better quality by growers in India.

Nutritional value of eggplant is not high, though it is used as a substitute for meat in some countries, probably due to its flavor. In ½ cup of cooked eggplant weighing 100 grams, there is only one gram of protein. The same quantity yields 70 units of vitamin A, .042 milligrams of vitamin B_1, .036 milligrams of vitamin B_2, 10 milligrams of vitamin C, 11 milligrams of calcium, 31 milligrams of phosphorus, .5 milligrams of iron and a total of 15 calories.

Range and Soil

A relatively long hot growing season is needed for good production. Five months of hot weather, preferably with day temperatures from 80 to 90 degrees and night temperatures 70 to 80 degrees. In the North, plants may be started in a warm greenhouse and moved to the garden after the soil is thoroughly warm. Plants are stunted and made unproductive by long periods of chilly weather. Climate that will grow sweet potatoes is good for eggplant.

Eggplant is a heavy feeder and requires a deep, rich, porous soil that is reasonably moist but well-drained. Sandy loam is better than clay. It should have plenty of humus and a *p*H of 6. to 7. Best fertilizer is thoroughly rotted manure or compost, dug in at the rate of two pounds per plant. If either manure or compost is fibrous, it should be dug into the soil the previous fall. An organic fertilizer with an NPK formula approximating 5-10-5 may also be added, but it should not have tobacco stems in it. Tobacco stems, though a fine source of potash, are said to kill any member of the *Solanaceae* family.

In Asiatic countries, eggplant is sometimes grown in a tub as a patio ornamental and the fruit is eaten. Handled in this way, it may easily be protected against bad weather. Since only a few well-grown plants supply the average family's needs, it might be grown in the same way by the home gardener in northern states.

Seed

One packet of seed will start 40 to 60 plants, enough for a 100 foot row. Each plant, when successfully grown, will supply 3 to 8 fruits. If stunted for any reason, they may not bear at all. Seed germinates in 10 to 12 days and Black Beauty matures in 85 to 90 days after plants are set out. New Hampshire, a small-fruited variety, matures fruit in 75 days and requires a foot less room in the garden than other varieties.

Planting

Seed is started indoors 8 to 10 weeks before plants may be set out. It is started in flats or pots in light mellow soil one-fourth to one-half inch deep, and kept at a soil temperature of 70 to 75 degrees until germination. In two weeks or a little more, the seedlings should be 2 to 3 inches high and ready to transplant to plant bands, clay pots or to flats or beds where they can stand 4 or 5 inches apart. Rich, friable soil should be used in the second pots or flats.

After seedlings emerge, the temperature of the air should be about 65 to 70 degrees in daytime, 50 to 55 degrees at night. Temperature control, watering and ventilation are all very important in growing good plants. Insects may cause stunting of seedlings and must be removed if plants are to produce well. Any stunting or slowing of plant growth will cause seedlings to become woody, when they should be discarded.

If seedlings are grown in flats, they should be "blocked out" a week before setting out in the garden. To do this, run a sharp knife through the soil midway between plants, cutting roots. Plants should be kept slightly shaded and well-watered after this operation, which cuts the longer roots and causes a mass of fine feeder roots to grow at the base of the stem. When these plants are transferred to the garden, the entire block of soil should be moved with each one, leaving roots undisturbed. This prevents the setback which can be fatal to good production.

Seedlings should not be planted outside until mean daily temperatures are up to 65 to 70 degrees unless special provision can be made to protect them. They may be set out one or two weeks earlier under hot caps or glass cloches. A good general rule is to plant when new oak leaves are fully grown. Plantings will do best in a situation protected from north and west winds. Seedlings should be protected by cutworm collars when they are set out and should be spaced 2½ feet apart in rows 3 feet apart.

Culture

Soil around eggplant should be mulched early in the spring to preserve moisture. When the plants are set in the garden, mulch may be pushed aside to uncover a spot just large enough for the plant and its cover. A deep mulch will protect the small seedlings from wind damage, which might injure the tender plants.

As each plant is set in the garden, it may be treated to a cup of liquid manure to help settle its roots. Four weeks later, a regular bi-weekly manure-water feeding should begin. About 3 cups of manure water should be poured around (but not touching) the stem and another quart should be poured over the adjacent mulch.

Eggplant needs a lot of moisture, especially while fruit is setting. In periods of drought, especially if the temperature is high, the bed should be well irrigated weekly.

Enemies

A fungus, *Phomopsis vexans,* may attack stem, leaves or fruit in hot wet weather, causing rot to appear. In areas where this is a problem, fungus-resistant strains may be planted. Florida Beauty and Florida Market were both developed by the Florida Agricultural Station to combat this disease.

Flea beetles sometimes attack seedlings both before and after they have been set in the garden and are capable of causing enough damage to stunt the plants. Protection against them is rotenone dust, which should be discontinued before blossoming begins.

Harvest

Eggplants may be picked for use when they are one-third to fully grown. As soon as a high gloss appears on the skin they may be harvested. The young fruits have more tender flesh and smaller seeds than nearly mature ones. More fruit per plant will develop when they are picked young. Stems are tough and should be cut with a sharp knife to avoid breaking branches.

Last eggplants should be picked before the first frost. They may be stored for several months on a shelf in a cool cellar if they are handled very carefully to avoid bruising.

Eggplant in New England

Every summer now, for the past 5 years, my family and I have found great

Massachusetts gardener J. Francis Michajluk has found that eggplants can be grown in colder regions if given an early start, plenty of manure, and a good watering once a week during drought periods.

pleasure in visiting organic gardeners throughout Connecticut, Vermont, Massachusetts and New Hampshire. These gardens are about the best we have ever seen anywhere, the products the tastiest and the gardeners the friendliest. But what puzzles me about most of these gardeners is that few of them grow eggplant.

Last July, while visiting a gardener in Vermont, I asked her if her family liked eggplant. She said they loved it, but, because organically grown eggplant was hard to come by, they did not enjoy it too often. When I asked her why she did not grow her own, she looked as though I had asked her to bring down the moon.

"You know darned well our seasons are too short," she replied. The look she gave me said quite plainly that a woman of my position should know as much.

But there are ways of stretching the growing season for many fruits and vegetables which enable us to grow a wider variety in our gardens each year. Eggplant is one vegetable that must have its growing season lengthened in the colder regions of the country. If the seeds are started indoors early enough in spring, eggplant can be grown with good success throughout the northern part of the United States regardless of the length of the growing season.

Back in Illinois, my father was a great one for growing eggplant and we all loved it. Later, when I married and moved East, I missed the organically grown eggplant. But I was unable to grow my own, and those I bought at the store were spongy and tasteless.

In Illinois the weather warms up early enough in spring so that all my father had to do was sow the seed right in the cold frame. But, up here in the East, the ground is often still frozen during the first week of April.

My father decided there were two things I must do in order to have any success with eggplant in Massachusetts. The first was to soak the seeds before planting to hasten germination, and second, to start them indoors at least 8 to 9 weeks before transplanting time.

The following spring I put his advice to work. The results were very satisfactory, to say the least. We harvested far more large eggplants than we could possibly use. This method of starting plants early proved to me, and to my many gardening friends, that eggplant can be grown with success in the colder regions if a little beforehand preparation is practiced.

During the last week of March, I begin by getting my seed flat ready for the eggplant seeds. I use any sized flat I might have on hand, as long as it is 5 inches deep. (You may use a large flower pot if many plants are not desired.) Into the flat I put 4 inches of a soil mixture prepared the fall before. It consists of one part garden loam, one part compost, one of leaf mold, some peat moss and two parts of sand. On top of this I pour ½-inch of sterile vermiculite, then water the entire flat down very well.

About the second of April, I soak the seeds in warm water overnight to soften the shells, then sow them thinly in furrows made in the vermiculite. After the seeds are planted, I sprinkle a thin layer of vermiculite over the rows, then moisten it with a fine spray.

Because eggplant seeds germinate best in a warm, moist atmosphere, I like to keep the flat on my kitchen sink until sprouting occurs. (No top glass is necessary when vermiculite is used in the flat.)

As soon as the majority of seeds have sprouted, and the seedlings have sent their long taproots into the soil mixture below, I move the flat to a sunny window. There the plants remain, growing very well with no fertilization until the weather warms up enough to transplant them into the cold frame. If you do not have a cold frame, it is wise to transplant the two-inch seedlings into other flats, or back into the same flat after the vermiculite has been removed. Transplanting creates strong root systems so important to good production.

After transplanting the young seedlings into the cold frame, I begin to prepare the section of the garden where they will later grow. My father had discovered that, even though eggplants thrive vigorously on cow manure, they will not grow well in a plot freshly fortified with it. I follow his method of turning in manure as soon as the ground can be worked in spring, then seeding that plot down in lettuce, peas, kohlrabi or any other vegetable that matures early. These rows are spaced far enough apart to allow room for the eggplant transplants later.

By the first week in June, my eggplants are large enough for moving out into the prepared plot. I usually wait for a slight drizzle before I transplant them to lessen shock and always take a ball of earth along with the roots. Preventing shock is quite important. The less transplanting shock these plants suffer, the quicker they will catch on and resume growth. This, of course, means an earlier harvest.

When setting out the young plants, I allow 3 feet of space between plants and rows. This gives the mature plant a foot of space on all sides for good air circulation, and also enables the sun to ripen the fruit earlier. When planted too near, the dense shade prolongs ripening and cuts down on the amount of fruit each plant sets.

After all of the early crop has been harvested from that plot, I loosen the soil slightly in the aisles, hill each plant separately, then mulch the entire area heavily with grass clippings, dried grass or straw. If the weather is dry, I water down the bed once a week, applying at least one inch of water at a time. Eggplants must have moist soil around their roots at all times so that growth may proceed normally. Drought will retard growth, thus delaying maturity.

Even though I apply manure in early spring, I find an additional feeding in July necessary. This steps up growth and hastens maturity. It also aids the plants in setting more fruit. I use an all-organic fertilizer of a 5-10-10 formula, about the middle of July. I push back the mulch, work a handful of fertilizer into the soil over the root area, then water it in well with the hose. If I am sure a heavy rain is coming, I fertilize and let the rain carry it down to the roots. If the mulch has grown thin by this time, I replenish it to conserve moisture through the hot days of August.

Eggplant fruits may be harvested and eaten any time after they take on a good, purple color. They are a lot like cucumbers in that they may be eaten at any stage.

As the cold weather descends on your garden, keep an eye on the temperature. Eggplants can withstand a light frost with no damage to their foliage or fruits. But, should a heavy frost threaten, cover several plants that are still bearing heavily with burlap bags. If thus protected, they will go on producing wholesome eggplant well into late fall.

The most important things to remember when growing eggplant in the colder regions are to give them an early start, plenty of manure and a good watering once a week during drought.

—Betty Brinhart

ENDIVE

ENDIVE. *Cichorium endivia.* A hardy annual or biennial herb of the daisy family, *Compositae,* grown as a main salad crop in the fall. Native to the East Indies and possibly also to Sicily, it was known early to the Egyptians and Greeks as a salad herb. Two varieties have been developed, the narrow-leaved with curly, finely cut leaves, usually called endive, and the broad-leaved, popularly known as escarole.

Endive is very rich in vitamin A, yielding as much 23,000 units from ¾ cup of the green leaves, .075 milligrams of vitamin B_1, .250 milligrams of vitamin B_2, 7 milligrams of vitamin C, 29 milligrams of calcium, 27 milligrams of phosphorus, 1.5 milligrams of iron, 1 gram of protein and a total of 20 calories.

Range and Soil

Endive may be grown throughout the United States, forming a fall crop in the North and fall, winter and spring crop in the South. It is hardy and will stand light frosts without protection; with light straw mulch over the plants on cold nights it will continue to yield in the North until a hard freeze. Its requirements are similar to those of lettuce, but it is more heat-resistant, and less likely to bolt if started too early in the summer.

Soil requirements, too, are similar to those for lettuce. Any moderately rich garden soil will grow good endive. If a heavy feeder was grown in the row earlier in the season, a two-inch layer of compost should be incorporated in the soil before plants are set out.

Seed

A packet of seed will sow 15 feet, one ounce will sow a 100 foot row and will yield 100 fully mature plants, plus thinnings. Matures in 90 to 95 days from seed. If started in a seedbed or flat and seedlings are transferred in the fall to the garden, a packet of seed will provide plants enough for one family. Recommended varieties are Green Curled endive or Full Heart Batavian escarole.

Planting

Seed is usually sown in flats or partly shaded seedbeds any time after midsummer and succession plantings may be made at intervals of 2 or 3 weeks up to 90 days before the first hard freeze. Seedlings are transferred to the garden when they are about 4 weeks old and protected for a few days from hot sun. If slugs are a problem, remove the cover as soon as possible, because tender endive seedlings are their favorite food.

Seedlings are set 12 inches apart in rows 18 inches apart. If seed is planted directly in rows, it should be thinned to stand 12 inches apart and thinnings may be used for salads.

Blanching

Blanched leaves contain much less vitamin A than green ones do. They are also more crisp, tender, and less bitter in flavor.

Plants should be almost fully grown when they are blanched—that is, about 12 to 15 inches in diameter. Outer leaves are pulled together and tied above the plant's heart, either with soft string or with strips of cloth. The operation should be performed when the plants are perfectly dry and water should not be allowed to get into the hearts subsequently or it may cause rot. If they do get wet, plants should be opened to dry, then retied to continue blanching. Blanching may also be done with two wide boards, set at an angle to meet over the plants. About 3 weeks are required to remove vitamins, color and bitterness from the inner leaves.

Harvest

Just before the first hard freeze in northern gardens, endive may be lifted with a large ball of soil, placed in a tub or pail and stored in a cool cellar free from frost. Plants should be inspected and slugs or other pests removed before storing. Soil may be kept almost, but not quite dry and plants will be usable for several months.

ESCAROLE

ESCAROLE. *See* ENDIVE.

FENNEL

FENNEL. *Foeniculum vulgare.* An anise-flavored perennial or biennial herb of the carrot family, *Umbelliferae.* Native to temperate Europe and Asia, but growing wild in South America and possibly other places, as an escape from gardens. Fennel was much esteemed by the Romans, who used it to flavor most of their dishes. It was used through the middle ages for food and medicine.

Bitter fennel, the wild type, is a perennial growing to 6 feet tall, with feathery leaves, coarse stalks and aromatic seeds. Only the seeds are used, principally to flavor liquors.

Sweet fennel is the cultivated type and, if grown as a perennial, will revert in 4 or 5 years to the wild state. Two varieties are grown, usually as annuals. *F. vulgare dulce,* called Florence fennel or finnochio, has leaves two to three feet high arising from a thickened stalk or bulbous enlargement made by the overlapping thickened stems. Leaves of finnochio are used for flavoring sauces, stems are eaten like celery and the bulb is eaten raw in salads or cooked, usually blanched. The other sweet variety is *F. vulgare piperitum,* called carosella. There is no bulb in this variety, but the young stalks are picked before flowering to use like celery.

Range and Soil

Fennel can be grown as an annual throughout the United States. It can be grown as a perennial in the warmer sections. Frost damages its top, but seed will form the second year where the top dies back.

Good rich garden soil is best, and the tenderest stalks are produced when it is treated like celery and grown in very rich soil. It responds to the same treatment as celery, but requires much less trouble.

Seed

One ounce of seed will sow a 100 foot row and yield 200 plants. Fennel is ready to use in 60 to 110 days after sowing, depending on variety and use to which it will be put. Good celery-like stalks are ready in about 90 days.

Planting

Seed may be planted in early spring for summer use, but it is often planted in late spring to mature in early fall. Sow in rows prepared with two inches of compost or rotted manure, in drills one-half inch deep and 18 inches apart. Thin seedlings to stand 6 inches apart. When plants are a foot high, draw the soil up around the bulbs to blanch them.

Harvest

When the bulb reaches a diameter of two inches, the plant is ready for use. Leaf stalks should be used before they become tough and stringy, their outer skin peeled.

FENUGREEK

FENUGREEK. *Trigonella foenum-graecum.* A hardy annual of the bean family, *Leguminosae,* native in Europe and the Orient. It has a tough fibrous stem, long taproot, three-parted leaf and small clover-like flower. The seed, a rectangular wheat-size pea, is enclosed in a beaked pod. In Egypt, where it is called helbeh, the leaves are eaten raw or the seeds are mixed with honey to make a conserve. In India, seeds are used as a condiment and leaves as a potherb. Throughout the Near East and the Orient, the seeds are used to brew a tea. In this country, fenugreek is sometimes planted in the warmer regions, as in California, as a winter annual cover crop, to provide a heavy green-manure crop when it is plowed in before spring planting.

When required for seed, fenugreek may be planted in the garden in spring in any moderately good garden soil, with a fairly high phosphorus content and a *p*H rating between 6.0 and 7.0. Seed is sown in drills one-half inch deep and plants are thinned to stand 3 inches apart. It will quickly go to seed in the hot weather.

If planted for its leaves, fenugreek should be planted in late fall. Seed will germinate well even in very cool soil and plants are hardy at temperatures down to 16 degrees Fahrenheit.

FETTICUS

FETTICUS. *See* CORN SALAD.

GARLIC

GARLIC. *Allium sativum.* A flat-leaved onion of the lily family native to western Tartary, now grown throughout the world wherever the climate is mild. It was grown by the Greeks and Romans and is mentioned in the Bible. Plants grow to two feet, have pinkish flower heads and roots that penetrate to two feet into the subsoil. It is a perennial growing from top sets or division of bulbs, but seldom makes seeds.

Range and Soil

Mild temperatures are best for garlic, which is winter-hardy along the Pacific coast through Oregon. It is grown commercially in California and Texas, but can be grown in the northern states as a summer crop. It is sensitive to frequent rains. Rich sandy loam with plenty of humus is preferred, or any light friable soil. Organic matter should be deeply incorporated for the plant's use in hot weather.

Seed

Actual seed is seldom planted, but garlic sets, the "cloves" into which bulbs can be divided, are referred to as seed. Usually bulbs have about 10 cloves each, which are planted separately. One-half pound of seed will plant a 100 foot row. There are two types, the early which is white, prolific, of poorer keeping quality and the late which is pink, not quite so prolific and matures about two to three weeks later. Recommended varieties are Creole and Italian.

Planting

Plant cloves two inches deep, spaced 3 to 6 inches in rows that are 12 to 16 inches apart. In California and Oregon, crops are planted in October and January and vegetation continues through the cool weather, making large plants and large bulbs. In the North, garlic is planted in early spring.

Culture

Garlic is treated much the same as onions. It should be mulched to ensure a continuous supply of moisture, and irrigated during its growing period if the soil becomes dry to an appreciable depth. If soil is very rich the bulbs will be larger, but also the tops will be stronger and may not fall over at maturity. If this happens, they are kicked or bent over to slow growth and mature the bulbs. When they fall or are bent, irrigation is stopped.

Harvest

When the tops are dry and bent to the ground, garlic is ready to pull. If a quantity is to be cured, it may be stacked in the field with tops covering the bulbs from the direct rays of the sun. After they are thoroughly dry, tops and roots are removed with shears, leaving one inch of top and one-half inch of root on the bulb. If the quantity is small, tops may be bunched and tied or may be braided and the strings hung in a cool, well-ventilated room to dry and store.

GROUND CHERRY

GROUND CHERRY. Common name given to a number of species of the genus *Physalis,* of the nightshade family, *Solanaceae. P. angulata* and *P. peruviana* are ground cherries native to the tropics. *P. alkekengi,* winter cherry, is native to Europe and Japan. *P. obscura, P. philadelphica, P. pubescens, P. virginiana* and *P. viscosa* are all North American ground cherries. All the above varieties are annual or perennial herbs bearing edible berries enclosed in papery pods.

The species *P. pubescens* is sometimes grown in American gardens under the name of ground cherry or husk tomato. Its fruits are small, round and yellow, borne in papery husks and are very sweet. It is used in salads and to make preserves, jams and pies.

Most species of *physalis* need a long, warm growing season, so seeds should be started indoors in the North and planted outside when the soil is warm. *P. pubescens* ripens fruit in about 75 days after being set in the garden. Culture is the same as for TOMATO, which see.

HORSERADISH

HORSERADISH. Latin name assigned by Linnaeus, *Radicula armoracia;* also called *Cochlearia armoracia* and more commonly *Armoracia rusticana.* A hardy perennial of the mustard family, *Cruciferae,* related to cabbage, kale, cauliflower, radishes and turnips. Originally native in Europe from the Caspian through Russia and Poland to Finland. Leaves and roots were both used by the Germans during the Middle Ages, but were not used throughout the other European countries until much later. Horseradish roots penetrate 5 to 10 feet below the surface of the soil, the large taproot sending out innumerable vertical and horizontal side roots. Top grows usually no more than 2 to 3 feet tall and seed is seldom formed. Root is grated and used as a condiment.

Range and Soil

Horseradish needs a long growing season in a temperate climate, but will not grow in the South where soil does not freeze in winter. Grows best in the rich, deep, moist loam of river-bottom gardens, but will do well in any soil where there is not too much sand nor too hard a subsoil. Soil should be deeply prepared, with manure dug in to a depth of two feet 2 or 3 months before planting or the previous fall. Near neutral *p*H rating is best, between 6.0 and 8.0. To avoid common pests of the cabbage family, choose a site removed from the other members of the family.

Seed

Very little seed is produced by horseradish and that little is not viable very long, so new plantings are made from division of crowns or root cuttings. A crown cutting is made by slicing the main root lengthwise leaving a few crown buds on each piece. Root cuttings are pieces 6 to 8 inches long of older roots as thick as a pencil. These are cut square at the top, slanting at the bottom, to help the gardener to know which way to plant them.

Planting

Crown cuttings are planted with tops slightly below the surface of the soil.

To plant root cuttings, dig a trench 4 to 5 inches deep in previously prepared soil. Place cuttings 18 inches apart in the trench, with 3 feet between trenches. Each cutting should be placed at an angle, with its top near the soil surface and its bottom end on the bottom of the trench. Replace soil carefully in order not to disturb position of roots. To develop the strongest plants, the roots should be dug up carefully by hand when leaves begin to sprout. It will be found that leaves are sprouting from several points along the cuttings. All sprouts except the strongest clusters near the top are removed and roots are re-covered in soil.

Culture

Commercial growers who must grow straight roots for market dig up their horseradish when plants are about half grown, strip off the lateral roots and then replant them. This may be done by the home gardener, or he may leave his roots to grow branched and crooked if their shape makes no difference to him. It will not change their flavor.

After the first year, the horseradish bed should be top-dressed each spring with manure.

Harvest

Horseradish makes mainly top growth and deep feeder root growth during the summer months. The main root does not begin to thicken until cool fall weather sets in. Usually roots are not dug before September and they may be dug from then on until ground is frozen too hard. Main roots may be dug before a hard freeze and stored in root storage pits or in a cool cellar. If care is taken to remove only the main roots at harvest time, the side roots will sprout in the spring to make new plants. Or the whole root may be dug, and the side roots stored and saved for cuttings to plant in the spring.

KALE

KALE. *Brassica oleracea acephala.* Also called cole and borecole. A nonheading member of the cabbage genus of the mustard family, *Cruciferae.* This is probably the form in which cabbage was known to the Greeks and Romans. Kale is closely related to collards and similar in growth except that it lacks the tall stem. Seed for 2 different types are offered by most seed houses—Scotch and Siberian. Scotch kale has curled and crumpled foliage that is gray-green in color. Siberian kale is less curly and its color is blue-green.

Kale is seldom eaten raw because its leaves are rather coarse for salad. But even cooked, it is a mine of vitamins and minerals. One-half cup of cooked kale, a portion weighing about 100 grams, contains 20,000 units of vitamin A, .189 milligrams of vitamin B_1, .570 milligrams of vitamin B_2, 96 milligrams of vitamin C, 195 milligrams of calcium, 67 milligrams of phosphorus, 2.5 milligrams of iron and yields 4 grams of protein out of 45 calories. It contains 10 times as much vitamin A as an equivalent amount of green lettuce, 2 to 3 times as much vitamin C as an equivalent weight of orange juice, more calcium than an equivalent weight of whole fresh milk and more of vitamins B_1 and B_2 than a slice of whole wheat bread.

Range and Soil

Kale can be grown throughout the United States and even in Alaska. It is a cool-season crop which takes only 55 to 65 days to mature. In the southern portion of the country, where the first fall frost does not come before the end of October, it can be grown throughout the winter. Soil should be fertile sandy or clay loam rich in humus and should have a *p*H of 6. to 8. It needs plenty of calcium, so it should be planted in soil which has recently had an application of ground limestone.

Seed

One packet of seed will sow a 30-foot row, more than enough for the average family. Best varieties from a horticultural standpoint are Dwarf Blue Curled Scotch and Green Curled Scotch. From the nutritional angle, Bloomsdale Double Curled is very high in vitamin C and Blue Scotch is richest in vitamin A.

Planting

Plant seed one-half inch deep in rows 18 inches apart. Thin seedlings after they have grown large enough to provide a meal from the thinnings. Mature plants of the large varieties should stand 12 inches apart, of the dwarf varieties, 6 inches. Seed is sown from early autumn to late winter in the southern United States. In the northern portions, it may be sown any time from June 1 until 6 weeks before the first frost and it will yield until the first hard freeze. In the South, seed sown in February or March will provide greens for 2 years in all except the hottest summer months.

Best flavor and tenderest leaves are the product of quick growth, which can only be attained in the richest soil. To prepare a bed which will bear well over a long period, dig a trench 6 inches deep and 6 inches wide, line the bottom with crushed limestone and fill to within an inch of the top with compost or rotted manure. Water well. Add enough of the same mixture to bring the level up to planting depth, that is, one-half inch below the surface. Sprinkle in seeds—broadcast if they are being planted in fall after weeds have done most of their growing—and cover with another half inch of sifted compost or manure. If compost or manure are not available and the soil contains plenty of humus, the following organic fertilizer may be added to the soil in the trench at the rate of 7 pounds per 100 feet: 3 parts basic slag, 2 parts wood ashes and 2 parts leather dust or dried blood.

Culture

A thick mulch drawn close to the plants will help to preserve soil warmth in late fall and to keep the leaves growing after everything else in the garden is finished.

If a bed slows down in its bearing, except in hot weather when it might be expected to slow, a side-dressing of manure water will help to give it new life. About 1½ cups of manure water per plant may be poured along the row, not too close to the stems. The treatment may be repeated in 2 weeks.

Harvest

Large outer leaves are cut before they are too mature and inner smaller leaves are left to grow to full size. If the leaves are to be added to a salad, they should be smaller and younger than those which are picked for cooking. Like many other members of the cabbage genus, kale is considered better after it has had a touch of frost and least desirable in midsummer, when leaves are less flavorful, more tough and stringy.

In cold areas where kale cannot be wintered outside, several flats filled with seedlings in early fall will start to bear at about the time when the rest of the garden is frostbitten, if they are kept in a sunny window in an unheated room.

KOHLRABI

KOHLRABI. *Brassica oleracea caulorapa.* A biennial herb, cousin of cabbage and cauliflower, with a swollen bulb-like main stem from which leaves arise. Kohlrabi was not mentioned in horticultural literature before the sixteenth century, though it may have been known much earlier as a somewhat less developed variant of the cabbages. It is grown for the bulbous stem, which can be eaten cooked or raw in salads.

Kohlrabi is rich in calcium and has a moderate amount of vitamin C. In one-half cup of the cooked vegetable weighing 100 grams, there are .030 milligrams of vitamin B_1, .130 milligrams of vitamin B_2, 50 milligrams of vitamin C, 195 milligrams of calcium, 60 milligrams of phosphorus, .7 milligrams of iron, 2 grams of protein and a total of 32 calories.

Soil and Range

Kohlrabi, like most of the cabbages, is a cool-weather vegetable. It is grown in early spring and in fall in the northern part of the United States and through the winter in the South. In hot summer weather, it becomes woody and strongly flavored.

A rich, loamy soil with plenty of humus will yield the tenderest bulbs. It must have ample moisture to grow rapidly and steadily. A check to its growth, through drought or heat, spoils its tender texture.

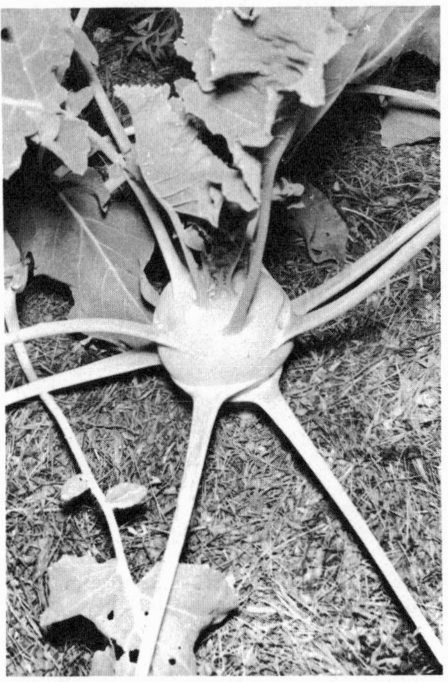

Like other members of the cabbage family, kohlrabi is a cool weather vegetable grown in early spring and fall in the North and throughout the winter in the South.

Soil in which kohlrabi is to be grown should be sprinkled liberally with lime to bring the *p*H up between slightly acid *p*H 6., and neutral *p*H 7. After raking in lime, spread a 3-inch layer of rotted manure or compost along the row, and dig it in lightly. Raw manure may be used for kohlrabi, if it is free of cabbage waste.

Seed

One packet of seed will plant a 40 foot row and yield 80 bulbs. Plants mature in 55 to 65 days. Recommended as the best variety is Early White Vienna, with Purple Vienna a second choice.

Planting

Being hardy, kohlrabi may be planted as soon as the frost is out of the ground in spring. Successive plantings of a few feet of the row at a time should be made until two months before the hottest summer weather. Fall crops may be planted 60 days before the first frost.

Plant seed one-half inch deep in rows 15 to 18 inches apart. Thin the seedlings to stand 6 inches apart. If allowed to grow too close together, kohlrabi is hard to harvest.

Culture

Moisture is of the greatest importance in feeding kohlrabi for best growth. A thick mulch should be drawn up to the seedlings as soon as they are tall enough, and soil beneath the mulch should be kept constantly moist. Fall plantings may need weekly irrigation, if rainfall is not abundant.

Harvest

Do not let kohlrabi grow larger than two to two and a half inches in diameter. Larger bulbs lose their delicate flavor and become tough. If the bulbs are not needed when they become full-size, they should be picked and stored until they can be used. Fall-grown kohlrabi may be stored in the same way as cabbage. (*See* CABBAGE.)

If the bulbs are close together when mature, cut them off below the bulb, leaving the roots in the soil. Roots are usually tangled and, when one is pulled, its neighbor may lose enough roots to slow growth.

LAMB'S LETTUCE

LAMB'S LETTUCE. *See* CORN SALAD.

LEEK

LEEK. *Allium porrum.* A biennial herb of the onion genus of the lily family, native to Switzerland and also found growing wild in Algeria. It was cultivated by the Greeks and was so much esteemed by the Romans that Nero is said to have eaten it regularly for several days each month to clear his voice. It is a hardy bulbless onion with flat, rather than round, hollow leaves. Its thick stem, which might be considered an elongated soft bulb, is blanched and used in soups or salads where a delicate onion flavor is required.

Leeks have a little more nutritive value than onions, although their principal use is to provide flavor rather than sustenance. In one-half cup, there are 20 units of vitamin A, .150 milligrams of vitamin B_1, 24 milligrams of vitamin C, 58 milligrams of calcium, 56 milligrams of phosphorus, .6 milligrams of iron, 2 grams of protein and 40 calories.

Range and Soil

Leeks may be grown throughout the United States. They grow best in a rich loam with a *p*H between 6. and 8. and containing an abundance of humus and of nitrogen. They need plenty of moisture, especially early in their growth, but must have very good drainage.

In milder climates, leeks may be left outdoors all winter and harvested as needed, if they are well-banked with soil.

Seed

One packet of seed will sow 25 feet, one ounce, 100 feet. A packet will yield 50 mature plants plus thinnings, which may be used in salads. In 85 days, leeks are as large as scallions and can be used like them. They are mature in about 130 days. Recommended variety is American Flag.

Planting

Seed may be started indoors in February for a summer crop, or may be sown directly in the garden in early spring to mature in the fall. Early sowings are made one-half inch deep in flats or pots, the seed sprinkled rather thickly. When seedlings are 3 inches high, they should be transplanted to stand 1½ inches apart in a flat. By the middle of April or early in May, they should be 5 inches tall and ready to plant outside.

In the garden, dig a trench 6 inches deep and 6 inches wide. Line the bottom with an inch of rotted compost or manure. Do not use fresh manure. Place the seedlings in the trench 6 inches apart and plant them deeper than they grew in the flat. If soil is not too well drained, omit the trench and plant on the level of the garden. Seed started outside may be planted either in a trench or on the surface and thinned to 6 inches between plants. Rows should be 18 inches apart.

Culture

When leeks begin to attain their full height, draw soil from the sides of the trench around the stems. By the time they are mature, they should be banked with soil to the full height of the stem, to blanch for delicate flavor.

Harvest

In mild climates, leeks may be left outside all winter and pulled as needed, if they are well-banked. Where the winters are severe, they should be dug up before the ground freezes and stored, heeled-in, in a cool cellar.

LENTIL

Lentil. *Lens esculenta,* an annual herb of the pea family, *Leguminocae.* Native to the Orient, but have been cultivated since ancient times in Egypt, Europe and Asia. Lentils have been found in Egyptian tombs that date back to 2400 B.C. Egypt was known in the ancient world for the fine quality of its lentils.

The lentil plant is a straggling vine 12 to 18 inches in length, having small lavender or white flowers that are followed by flat pods containing two seeds, which may be small and brown, grey or yellow or larger and pea-shaped. The seeds are dried and used like dried beans.

Lentils are rich in B vitamins. One-half cup, cooked, contains .378 milligrams of vitamin B_1, .390 milligrams of vitamin B_2, 200 units of vitamin A, 20 milligrams of calcium, 77 milligrams of phosphorus, 1.7 milligrams of iron, 9 grams of protein and a total of 115 calories.

Range and Soil

Lentils are a cool-season, hardy crop that will stand light frosts. They may be grown throughout the United States, in the winter in the South and in early spring in the North. Best soil is a sandy loam with a *p*H from 6. to 8.

Like other legumes, lentil roots harbor nitrogen-fixing bacteria, which enrich the soil. However, lentils do not provide as high a total yield as most garden vegetables, so they are recommended for growing in home gardens only when space is unlimited. Homesteaders with several acres at their disposal can use lentils as a rotation crop with some of the small grains, such as wheat or may use it as an early crop which will be followed by late corn.

Planting and Culture

Lentils are planted in early spring and are treated in the same way as peas. (See Peas.)

Harvest

Seeds are dried in the pods before being threshed or shelled. If seed is saved from year to year for planting, it will be more viable if left in the pods throughout the winter.

LETTUCE

Lettuce. *Lactuca sativa,* the cultivated form developed from *L. scariola,* wild prickly lettuce which is native to Europe and the Orient. Other wild forms are *L. virosa* and *L. canadensis.* Grown and used by the Greeks and Persians at least as early as 500 B.C. Lettuce is a hardy annual, shallow-rooted and growing best in cool weather. Five types have been developed: loose leaf or nonheading; butterhead type, with soft leaves grown in a closed head; cabbage head type, with closed head of crisp leaves; cos or romaine, with a tall or elongated head; and stem lettuce, or celtuce, grown for the thick stem rather than the leaves. Of the 5, the cabbage head type is most sought after for its flavor and tenderness, but is most difficult to grow. Stem lettuce or celtuce is seldom planted.

Green leaves on the outside of heading lettuces and the leaves of the leaf type contain more vitamins than the white inner leaves of heads. In green leaves the following may be obtained from a 100 gram portion: 2,000 units of vitamin A, .075 milligrams of vitamin B_1, .150 milligrams of vitamin B_2, 5 milligrams of vitamin C, 17 milligrams of calcium, 40 milligrams of phosphorus, .5 milligrams of iron, 1 gram of protein and a total of 10 calories. In the same amount of white lettuce there are only 125 units of vitamin A, .051 milligrams of vitamin B_1, .062 milligrams of vitamin B_2, 5 milligrams of vitamin C, 17 milligrams of calcium, 40 milligrams of phosphorus and .5 milligrams of iron.

Range and Soil

Being a cool-weather plant, lettuce is grown best in early spring or fall in the North and in fall, winter and spring in the South. Strains have been developed, especially of leafy varieties, which will withstand the heat of an average northern summer and there are a few growing tricks which help to keep beds cool. But in general, it is not advisable to plant lettuce that should make its heads during hot weather.

Rich soil which retains moisture, yet has good drainage, will grow the best lettuce. Early types do best in sandy loam; later varieties in muck or clay loam. Enough lime should be added to bring the *p*H to 6. to 7., that is, almost neutral. Soil should be aerated 6 to 8 inches deep. If available, one pound of well-rotted compost or manure should be dug into each square foot, with one-half pound of rock phosphate or bone meal. If soil is already rich in humus, the following organic fertilizer may be substituted, if dug in at the rate of 5 pounds per 100 feet: one part dried blood, one part rock phosphate and 4 parts wood ashes.

Seed

One packet of seed will sow a 100 foot row. Seed germinates in 6 to 8 days and sprouts best at 60 to 65 degrees. At 80 degrees and above, it will sometimes not germinate at all. If it is necessary to sow summer plantings while the weather is hot, best results are obtained by pregermination in the refrigerator. Seed is sprinkled on a layer of wet blotter or cheesecloth, covered with another layer of the same material and stored for 4 to 6 days in the refrigerator. It is then partly dried, or is mixed with dry sand and planted immediately. Lettuce seed is viable for many years if it is kept in a cool dry place. It may lose its viability when kept warm or moist.

Best varieties for the home gardener vary with the soil, elevation, climate and rainfall. Most popular leaf varieties are the Grand Rapids strains, Black Seeded Simpson and Salad Bowl. Great Lakes is a popular iceberg or cabbage head variety; Bibb and White Boston lead among the butterheads. Paris White is a slow-bolting variety of cos but Dark Green, as the name implies, has more of the vitamin-rich green color and its head is more open at the top, so fewer leaves are blanched. Recommended for an unusual amount of vitamin C are Black Seeded Simpson, Romaine and Tennis Ball. Lower than average in vitamin C are Salamander, White Boston, California Iceberg and Mignonette.

Planting

Cabbage head and butterhead varieties require 80 to 90 days of cool growing to head well. If the weather turns hot before they have headed, the heads will never form. In order to give them their full growing season before hot summer months and in order to obtain an early lettuce crop, seed is usually sown indoors in flats, or outside in hotbeds or cold frames.

Seed is planted one-fourth inch deep in flats and kept cool and moist until it comes up. Start to thin seedlings as soon as they are large enough to handle. Leaves of adjacent plants should never touch throughout the entire growing period of lettuce for the best plants. Lettuce seedlings have shallow roots and are easily transplanted, but care should be taken at each transplanting to keep the main tap-root growing vertically. Seedlings should be transplanted when two inches high to a flat or cold frame in which they will stand two inches apart. They may be transplanted again before setting out or hardened off and set into the garden immediately from the second flat, if the last frost is past.

Seedlings for early spring planting in the garden may be started in fall except in the most severe climates. Seed is broadcast in a seedbed at about the time of the first frost and will make small plants before the ground freezes. As the weather grows colder, straw, leaves and twigs are heaped higher and higher over the seedlings. Early in spring, the cover is removed and the soil will warm quickly and growth again resume. As soon as this happens, the hardened seedlings are ready to set in their permanent rows.

If a long cool growing period may be expected in spring, seed may be planted directly in rows instead of indoors or in seedbeds. It may be planted in pinches spaced 6 inches apart or may be sprinkled thinly down the row and covered with one-fourth inch of soil. Rows should be two feet apart. Plants are thinned as they grow, the first thinnings being transplanted, the later ones used in salads. Final spacing should be 10 to 16 inches apart, depending on variety grown.

To provide a constant supply of lettuce through the growing season, several varieties may be planted at the same time or successive plantings of one variety may be made. Early in spring where the cool season will last 3 months, for instance, the first planting can be made as soon as soil can be worked and will consist of a row of 6-inch seedlings started indoors two months earlier; another row of 3-inch seedlings started 6 weeks earlier; a row of seed of the same variety; and a row or two of seed for varieties that mature in shorter or longer periods. A month or 6 weeks before the hot weather starts, a row each of romaine and of leaf lettuce of the Grand Rapids strain may be planted, to mature during the early days of summer. Three weeks later another row of Grand Rapids can be planted, to carry the supply through the hottest months. In midsummer seed is started in shaded beds for fall heading, to mature before the first frost.

Culture

Heading lettuce must be kept shaded in late spring if heads have not formed before the weather turns hot. A canopy of netting or double cheesecloth may be spread two feet above the bed or a slatted cover supported at the same height to provide partial shade. Where summers are extremely hot, lettuce may be grown

Some California growers have found that growing lettuce and other salad greens under netting protects plants from drying up; netting reduces heat of sun by 35 to 45 per cent.

by the same device, if light reaching the plants is cut 35 to 45 per cent. An accurate measure of the light may be made with a photographic light meter. Shading material should be raised high enough above the plants to permit the air to circulate freely below and should be removable for overhead sprinkling or for rain.

Lettuce must be kept constantly moist for best growth. In early spring, a mulch drawn up to the seedlings when they are 6 inches high is usually sufficient to protect soil moisture. But during dry weather, the soil should be irrigated if it becomes dry below the surface. After irrigation, it should be wet enough to ball when a handful is squeezed. Too much moisture, on the other hand, can make cabbage-head type plants puffy instead of firm and compact.

Enemies

Lettuce rot, which affects first the lowest leaves in contact with the soil and then spreads through the plant, can be avoided by spreading clean sand over the surface of the soil. Fungus and bacterial diseases are best avoided by rotating the crop. Do not plant lettuce in the same rows two years in succession and do not plant where endive, chicory or dandelions have just grown. These are all subject to the same diseases.

Insects which cause the greatest damage in new lettuce plantings are cutworms and slugs. Cutworms may be outwitted if loose collars of stiff paper are put on the seedlings when they are planted. The collars are made about one to one and a half inches in diameter and are wide enough so that half an inch may be buried in the soil and an inch or more remain as a barricade above ground. If slugs are a problem, limestone or wood ashes may be sprinkled over the soil around the plants to discourage them. In a wet season when slugs are very bad, do not use small baskets or boxes to cover newly planted seedlings. It is better to shade them with a canopy held a foot or more above the bed. Slugs find small baskets convenient shelters during the hot part of the day and may sometimes be found by the dozen inside inverted berry boxes.

Harvest

Pick lettuce in the early morning hours to preserve the crispness it acquired during the cool of the night. Immediately after being picked, it should be washed thoroughly but as briefly as possible and dried immediately to prevent loss of vitamins. If stored in a closed container in the refrigerator, it will retain its crispness for days.

All varieties of lettuce, including the Boston-type shown above, need moisture for best growth. Mulch is always recommended.

Loose leaf types should be cut off at ground level and the roots left in the soil. Later in the season the roots will send up new leaves for a second crop.

Growing Lettuce Under Netting

If you want crisp, green, flavorful lettuce all summer long – grow it under netting.

After 10 years' experience as a gardener on the Malibu coast of Southern California, I have found this is the best way to keep salad greens coming to the table during the hot, dry days of midsummer and the early fall.

The netting does not have to be expensive—it can be cheesecloth or old corded fish netting.

Stretched across old slats, odds and ends of old fencing and poles, netting cuts the heat of the sun by as much as 35 to 45 per cent. If you use cheesecloth, double it over. To make sure my netting will do the job, I measure the light *above and then below* the netting with a photographic light meter. It tells me how much light I am cutting out and guides me when I am rigging my nets.

When its readings tell me I have cut the direct burning sunlight down to the above limits, I know my netting will protect the lettuce and keep it from wilting and drying out. Of course the soil must be kept moist and fertile all the time it is growing successive crops of lettuce during the hot summer days. But you will find that your nets positively reduce soil evaporation and plant transpiration.

Making the growing bed is not hard. Again, use whatever comes handiest and cheapest. I have used old railroad ties for frames, raising the walls to the height of one foot, thus providing protection from side surface winds and drying-out. Use old slats to make hinged bed frames with the netting tacked and draped directly to them.

The soil should be good, fertile topsoil with plenty of humus and, if you are specializing mostly in lettuce, with a neutral to a slightly alkaline *p*H rating. I work with soil that is about 6 inches deep in the beds but I have seen such soil produce sunflowers measuring more than 10 inches across!

Lettuce in shaded beds transplants easily, so I frequently sow 3 feet thickly of a favored variety in a single row. I permit this to grow densely to a height of two inches and then transplant directly to a final growing site. Here I do not set them too closely, allowing 5 to 6 inches between plants. I cut parsley back almost to the ground after it reaches its prime and then permit it to regrow for future cutting.

A word of warning: you will find that not all truck garden vegetables do well under netting, in the shade. You must be selective. Tomatoes, for example, do not thrive under netting, but make a heavy vine growth producing few tomatoes. They require a full sun, heavy mulch and a deep, moist soil to really yield.

However, growing under netting will give you repeated crops of midsummer lettuce—enough to supply your friends as well as yourself with plenty of crisp-leaved, flavorful, salad greens, all packed with health-giving vitamins because they were grown naturally—the organic way.

—Gordon L'Allemand

My Favorite Lettuce Variety

I keep my garden here in Ohio under a year-round mulch but for lettuce I rake the mulch off a spot about 5 feet square and then I dig rotted manure or rich compost into the soil. Since my soil is on the acid side, I sprinkle lime over the top and rake it in thoroughly before planting. I thin the plants until they stand about 6 inches apart each way. Then I mulch the bed with half rotted compost to keep the ground moist and help feed the plants.

Since I can count on lettuce to cut for weeks, I make only 3 plantings each season, 2 months apart. One during the first week in April, another the first week in June, and the last the first week in August. I spade the ground for the April planting in October or November for my ground is very heavy and needs the mellowing effect of winter's freezing and thawing spells.

Seedling lettuce can stand considerable frost, but I have found that mature lettuce needs protection. I place short stakes here and there in the bed so that the lettuce won't be smashed and then I cover the whole bed with a plastic cover, sealing the edges with stones or clods of dirt. With this protection, I can have lettuce for many more weeks than I would otherwise.

My favorite variety is Oakleaf. It's not a lettuce for the truck gardener; it won't keep well enough to stand around in a grocery store for days at a time. It's strictly for the home gardener who likes to pick his lettuce 5 minutes before a meal is to be served, wash it and rush it to the table. Of course, if the weather is hot, it's best to gather the lettuce early in the morning before the sun wilts it, wash it and wrap it in wax paper. Store it in the refrigerator until you want to serve it later in the day.

Its leaves of medium green have a tenderness and a wonderful flavor that non-gardeners never have a chance to savor. The delicate appearance of this lettuce is very deceptive for it can stand both heat and cold. All summer long I cut Oakleaf plants over and over again and it never turns bitter nor does it bolt to seed.

Some of the seed catalogues list Oakleaf as being ready in 38 days but by eating the thinnings as soon as they reach an inch high I actually start eating it in about 20 days.

—Lucille Shade

Growing Lettuce in Summer

My efforts in growing lettuce during summer have met with great success in which I would like others to share. Here in New Jersey, I plant lettuce in a bed, not in rows. A row is satisfactory for spring or fall, but summer lettuce should be grown in a makeshift cold frame: 4 boards standing on edge around the bed. I make a screen of mason lath, or it could be made of mosquito netting, so that the lettuce is shaded on hot days. I water the lettuce as needed and find that Black Seeded Simpson does fine, especially if it has a light mulch of sheep manure amongst the heads. The screen should be removed during damp weather or otherwise the lettuce would mold. Seeds should be sown in pinches 8 inches apart and the seedlings thinned by cutting away all but one when they are a week old. By using every second head when they are half grown, I finally end up with heads 16 inches apart and they almost touch each other, so good is their growth.

—J. A. Eliot

LIMA BEANS

LIMA BEANS. *Phaseolus lunatus* or *P. limensis.* An annual herb, sometimes a vine, of the pea family, *Leguminosae.* Lima beans are native to tropical America. They are believed to have been cultivated first in the vicinity of Peru, where they have been found in ancient tombs, and they are also found growing wild near old Indian garden sites in Florida. Early and wild forms had seeds of several colors, sometimes mottled. Re-

Lima beans should be harvested when the beans are clearly visible by the swellings on the pods; pole limas yield continually until frost.

cent cultivated types have large or medium-sized white or green seeds, growing 3 to 5 in thick green pods on vines or low bushy plants, referred to as pole or bush limas, respectively.

Lima beans are used green or are dried, but are always cooked, since they contain a fairly large amount of raw starch. Green limas are rich in vitamin A and contain a moderate amount of vitamin C, both of which are lost in drying. On the other hand, dried limas contain more protein, though it is of an incomplete type which must be supplemented by other amino acids before it can be used as protein in the body. In one-half cup of green limas, cooked, there are 900 units of vitamin A, .225 milligrams of vitamin B_1, .250 milligrams of vitamin B_2, 42 milligrams of vitamin C, 21 milligrams of calcium, 130 milligrams of phosphorus, .9 milligrams of iron, 7 grams of protein and 116 calories. The same quantity of cooked dried limas contains .300 milligrams of vitamin B_1, .250 milligrams of vitamin B_2, 72 milligrams of calcium, 2.9 milligrams of iron, 8 grams of protein and a total of 129 calories.

Range and Soil

Lima beans may be grown throughout the United States, though the northernmost sections can grow only the quickly maturing varieties, and even those must be started indoors. To grow a crop outside requires warm to hot days, and warm nights over a period of about 3 months. Seed cannot be planted until the soil has thoroughly warmed up—at least 2 weeks after the last spring frost. The earliest fall frosts will kill the vines.

Any soil except heavy wet clay will grow lima beans. Best soils are sandy loam or very light clay loam, which warm up fast in spring. Fertility of the soil is less important than sunshine, though a small amount of compost dug in before planting will improve the plants. The usual practice is to rotate the lima rows in the garden to follow another crop which was heavily manured, such as celery. Soil should be on the acid, rather than alkaline side, with a *p*H of 5.5 to 6.5.

Seed

Bush varieties mature in 65 to 75 days, require a pound of seed to sow a 100 foot row and yield about 2 bushels to 100 feet. Pole varieties mature in 85 to 95 days, require one-half pound of seed to 100 feet, and yield more than the bush types, the amount depending on the length of the growing season. Bush limas should be planted where seasons are short. Pole limas may just get to the bearing stage in the northern sections when they are cut down by frost. Where it is possible to mature them, however, pole limas are better, because the suspended pods are freer from fungus disease, which can infect pods which rest on the ground.

Henderson Bush Lima is a popular variety, having thin green seeds, though it is not too rich in vitamin C. A thick-seeded variety which is widely grown is Fordhook 247. Burpee and Cangreen are both bush limas rich in vitamins and are recommended for that reason. Jackson Wonder is short on vitamin A and is not recommended.

King of the Garden, the most popular pole lima, is preferred because of its abundant yield, though it is low in vitamin A, as is Giant Podded. Ideal is another pole type low in vitamin C and not recommended. Best pole bean for vitamin content is Burpee's Best, rich in both vitamins A and C. The small seeded Carolina or Sieva bean, sometimes called the butter bean, is high in vitamin C.

Planting

Limas may be planted outside any time after the soil is thoroughly warm. If they are planted too early, the seed will rot. They are more sensitive to cold than kidney beans and should not be put in the ground until about 2 weeks later than the first snap beans. If the season is cold, or if the area has too few hot days to mature a crop, seed may be planted inside in paper pots which can be put into the garden bed without disturbing the roots. One seed is planted in each pot for this purpose.

Bush limas are planted 3 inches apart in rows 30 inches apart and seedlings are thinned to stand 6 to 8 inches apart. Pole limas planted along a fence are put in the ground 6 inches apart and thinned to 12 inches. On poles, several seeds are planted in each hill and thinned to one or two to a pole. Seed should be placed 1½ inches deep and, if possible, with the eye downward. Poles should be placed in the round before seed is planted. The poles should be rough-surfaced and 10 feet long, 3 feet of which is firmly buried underground.

Culture

Mulch limas when the plants are 4 inches high. Culture and enemies are the same as for snap beans. (See BEANS.)

Harvest

Limas should be picked when the beans are clearly discernible through the swellings on the pods. Do not allow them to become too mature, or they will be tough. Bush limas may be picked several times in the season and then they will be finished. Pole limas yield continually until frost.

To dry limas, permit the pods to remain on the plants until they are dry. Pull up the plants and hang up to dry. When they are thoroughly dry, the beans may be picked from the vines by hand or they may be threshed by walking on the vines.

Lima beans are best preserved for winter by freezing or they may be canned.

MARTYNIA

MARTYNIA. *Martynia proboscidea, Proboscidea jussieui,* or *P. louisianica.* Also called unicorn plant, devil's claw and proboscis flower. A sprawling annual herb, the only cultivated member of the *Martyniaceae* family. Martynia is native to southwestern North America, but is now naturalized throughout the southeastern states. It is cultivated in southern gardens for the fruit which is picked green to make pickles similar to cucumber pickles. The fruit is a capsule about 1½ inches in diameter and 3 inches long, tapering to a slender curving horn or beak. It is downy when green and becomes woody when completely mature.

Martynia is a tender plant, and should be grown like tomatoes. It needs plenty of room, since it is spreading in habit and should be allowed 5 feet in each direction.

MUSHROOM

MUSHROOM. Cultivated mushrooms are fungi of the *Agaricus campestris* species, white, with thick cap, short stem and usually light-colored gills. They are grown commercially in this country in underground caves, cellars or specially built mushroom sheds. In France they are grown mostly in abandoned underground quarries.

Being plants without chlorophyll, mushrooms do not supply vitamin A and have very little vitamin C. But they are rich in the B vitamins—folic acid, riboflavin, biotin, thiamin and pantothenic acid; they contain a fair amount of calcium and also some of the trace elements, such as copper and manganese. In ¾ cup of cooked mushrooms, there are 4 grams of protein and only 36 calories.

BUTTON AND OPEN CUP STAGES OF MUSHROOM DEVELOPMENT

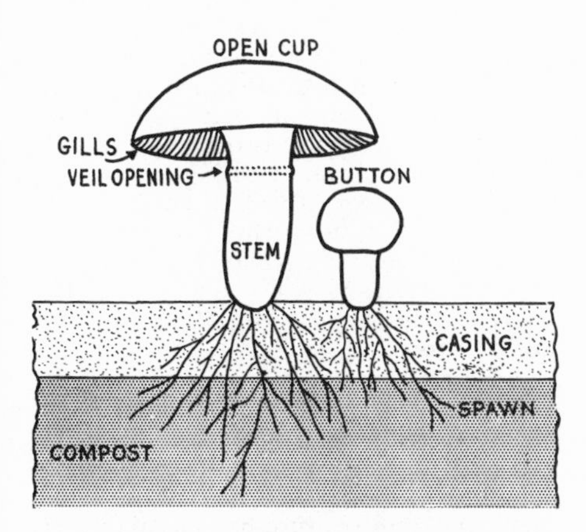

Range and Soil

Mushrooms can be cultivated in any section of the country where the temperature drops below 60 degrees and stays down for 3 or more months and where the atmosphere is not too arid. They cannot be grown successfully in the South where low temperatures last for no more than a month or two in the winter, nor can they be grown in the dry Southwest.

In their wild state, mushrooms spring up in shady moist spots where the soil is rich in leaf mold. Such spots in odd corners of the garden may be enriched with manure for their culture, though the crop will not be certain, depending as it does on so many variables of heat and moisture. Shady moist areas of the lawn are better than border or shrubbery areas, but they cannot be relied upon to provide a crop over a very long period. To raise a profitable crop, the home gardener can force mushrooms indoors in odd corners of the cellar and obtain a more certain return for his investment of time than if he plants where he is at the mercy of the weather.

Commercial Culture

The planting medium used by commercial growers consists of a rich mixture of horse manure and wheat straw which has been stacked in piles at least 4 feet high, moistened and turned or aerated until it has heated to 135 to 150 degrees and cooled back to 76 degrees. The manure must be entirely free of DDT or other disinfectants or sprays. As soon as the pile has cooled down, the manure is spread in trays 6 inches deep and stacked in houses which are heated to 58 to 65 degrees. When the tray has cooled to 70 degrees, it is inoculated with

Many commercial mushroom growers use a planting mixture of horse manure and wheat straw which has been composted.

mushroom spawn, a two-inch piece being buried an inch deep every 10 inches. For the next 3 weeks, it is watered regularly with a fine spray of tepid water, keeping the medium moist and crumbly but not wet enough to drip when squeezed in the hand. By the end of 3 weeks, the mushroom roots—a fine network of white mycelia—should have permeated the entire bed.

Now, an inch of garden loam, free of stones, weed seed, artificial fertilizer or spray residues, is spread over the entire bed. Virgin soil taken from below freshly stripped sod is best for this purpose. The bed is kept moist and at about 58 degrees for the next 6 to 7 weeks, when the first white dots, which will develop into mushrooms, appear. These spring up in clumps over the spots where the spawn was planted. The mushrooms develop quickly and are picked when full grown. Successive waves, or "flushes," of mushrooms come up in the same spot at about weekly intervals over a period of 3 to 5 months after the bearing commences. It has been found that beds kept at 45 to 50 degrees will bear for 4 to 5 months; beds kept at 60 to 65 degrees bear for about 3 months. The total crop of the cooler bed is slightly higher. Limits for growth are 45 to 68 degrees. Temperatures below 45 by a few degrees over a short period will not injure the bed, but temperatures of 70 degrees or more for even a short time will cause permanent damage.

As the last flush dies out of any given spot in the bed, the compost should be removed from that spot and replaced with clean soil. This will prevent the spread of disease from the rotting portions of stems that are left after harvest. The spent compost is no good for further mushroom growing, but is fine rich compost for the spring garden. It is sold by many commercial growers for that purpose.

Home Culture

Not many home gardeners are equipped to grow mushrooms in the way they are grown commercially. Adaptations of the method may be made by the individual to suit his needs.

For instance, horse manure of the proper quality and quantity is available in few gardens. A heap 4 by 4 by 4 is the minimum for proper heating. Chicken manure may be substituted or an organic compost pile built to contain the same plant nutrients may be prepared. This might get its bulk and most of its humus from 250 pounds of ground corn cobs and straw, mixed 50-50. The mixture should be dampened to prepare it for heating before other ingredients are added. It should be wet thoroughly and trampled to compact it, the process repeated once or twice in the next 4 or 5 days. Then, after the following has been added, the pile is mixed and turned: 18 pounds of tankage, 15 pounds of greensand or 30 pounds of granite dust, 25 pounds of moist acid leaf mold or peat moss, 40 pounds of whole grain mill sweepings or 25 pounds of brewers' or distillers' grains. Water should be added as the pile is turned, as much as it will absorb without run-off. The pile should heat rapidly and, if it is outdoors, should be protected from chilling winds to promote heating. At the end of one week it should be turned and again a week later. By this time, if the *p*H is not over 8.2, it is ready for use. If the *p*H is still high, turn and allow to stand until it drops.

Trays to contain the compost may be built like flats, but should be 6 to 8 inches deep. An appropriate growing shed will probably not be found on most homesteads, but a corner of the cellar where the temperature is not above 60 degrees should not be hard to find. If the atmosphere in the cellar is not moist enough, wet burlap may be stretched above the flats, raised a few inches to permit circulation of air.

Mushroom spawn may be purchased from most seed dealers. It usually comes in a brick composed of a sterilized growing medium, like tobacco stems, permeated with germinated spores. If a pure culture of spawn is available, it is better than bricks.

The type generally sold is white and produces fine fat white mushrooms with a good flavor under proper conditions. A brown type is occasionally offered for sale which is also good and has a very distinctive flavor. The brown mushrooms are more hardy and will take a little more abuse than white ones, so they may be more practical for the home gardener. One-half pound of spawn, cut into pieces the size of half an egg and planted at 10 inch intervals, will inoculate 35 square feet. Yield should be one to six pounds per square foot, if conditions are all right.

If the home gardener wants to take a chance and use short cuts, he can fill his planting trays with a mixture of 2/3 fresh horse manure without the straw and 1/3 soil, mix thoroughly and plant immediately. This mixture he will not want to keep in his cellar, because of the odor, but in mild climates he may be able to find a shed with the right temperature. Or he may put the mixture into a cold frame, cover it with 6 inches of straw or with heavy mats and grow his mushrooms outside. The frame should be in a sheltered position, not exposed to the heat of the sun or the cold winter winds.

A crop of mushrooms planted in a lawn in early spring may make good underground growth during the summer, if conditions are right, and send up mushrooms in September and October. The spawn may be placed directly under the sod in two-inch chunks and manure or compost raked into the grass over the bed. Or a two-by-two foot hole may be dug under the sod, some of the soil replaced with rich manure or compost, the spawn planted and sod replaced. The gardener who uses this method may risk the price of the spawn on a bad season, or he may be rewarded as much as 15 months later by a crop of mushrooms. The bed may bear for several years, once it has started.

Mushrooms Under the Kitchen Sink

Ever hear of growing mushrooms *under the kitchen sink?* I once met an elderly gentleman who did just that.

Living in a tiny New York City apartment, he nevertheless wanted to grow some organic food of his own. Herbs did fine on his window sills and he even grew radishes, lettuce and peppers in a couple of big window boxes. But his happiest experiment was with mushrooms. He found they thrived in trays under his kitchen sink, in the bottom of a large dish closet and in a spare bathroom cupboard!

The moral is this: mushrooms are a lot easier to grow than most people think. Provided their requirements of darkness, humidity and temperature are met, they take very little care and are great fun to grow—and even greater fun to eat.

Growing mushrooms is a kind of "reverse gardening." Being fungi instead of plants, mushrooms have no chlorophyll. Thus they are grown in darkness, having no need for sunlight to synthesize their foods. Instead, they rely on the richness of the compost on which their dusty black spores sprout and grow. A mushroom farmer wears a miner's lamp on his head, rather than a straw hat.

Homesteaders have found mushrooms an excellent spare-time crop to raise in barns, root houses, stables and chicken houses. According to the Ohio Experiment Station, growing mushrooms in winter in a house that is used to raise broilers in summer can be a high income producer. Poultry litter makes very good mushroom compost, giving higher yields than regular stable manure compost. This is probably due to its high nitrogen and vitamin B content.

You can, of course, buy prepared trays, complete with compost and planted spawn. Put in a cool, damp spot, they will bear a pretty good crop.

But if you have room in your cellar or other suitable building, you can start from scratch and grow a really big crop. The most important consideration is temperature.

The ideal air temperature is 60 degrees, but it can vary 5 degrees in either direction. Relative humidity must be high, as near 80 to 85 per cent as possible. An inexpensive hygrometer will enable you to check this regularly.

Make tests with a thermometer and hygrometer in various spots to find those that maintain the right conditions day and night. Remember that there may be a difference of 10 degrees or more at various levels in the same spot. There should be no drafts, but absolute darkness is not essential, as long as the light is very dim.

Usually, due to the difficulty of keeping the air cool in summer, mushroom growing is limited to the winter. But if you are lucky enough to find a spot that has the right temperature all year, you can grow crop after crop and get 6 or more pounds per square foot of bed.

You may find it most practical to make permanent box trays like greenhouse benches, setting them in tiers about a foot above one another. Or you can construct boxes without bottoms to stand on the floor, or close off a corner or end of the room with a single board on the floor. The beds should be about 6 inches deep.

Now for the compost. Fresh, strawy horse manure is the standby of commercial growers. It is piled up 4 feet high and turned, shaken and watered every 4 days. Dried brewers' grains, cottonseed meal or other nitrogenous materials, plus rock fertilizers, are generally added to enrich it. In 2 to 3 weeks, when the temperature is down to about 75 degrees, the compost is put into the beds and planted.

Sometimes sawdust, shavings or ground cobs that have been used for bedding animals is composted instead of horse manure. Or you can make your compost with ground cobs or sawdust without manure:

Mix 100 pounds of the cobs or sawdust with an equal amount of straw, adding about 20 pounds each of tankage (or similar organic nitrogen fertilizer), leaf mold, greensand and granite dust and whole grains. Mix and tamp the pile well, water thoroughly and turn it in 5 days, then again in a week. Put it into the beds and plant when the temperature drops to 75.

Spawn generally comes in brick form. Break it into pieces a little smaller than a golf ball and plant it one to two inches deep, 10 inches apart in both directions. Keep the beds moist but not wet; use a gentle spray of tepid, never cold water.

It takes about 3 weeks for the filaments or threads, called mycelium, to spread out completely over the bed. Then you "case" the beds, covering them with an inch of clean soil, free from any chemical residues. In another 3 weeks, tiny pinheads, the first "flush" or "break," will begin appearing on top of this.

If you maintain correct temperature and humidity from casing time on, you'll have no trouble with diseases or pests. A layer or two of moist burlap will help keep the proper humidity over the beds if the air tends to become dry.

In about 10 days, you can start harvesting your crop. But don't hurry the picking. Mushrooms are at their most flavorful when the bottom of the cap has broken away from the stem and spread out almost flat. Never leave them after this, as they stay good only a short time.

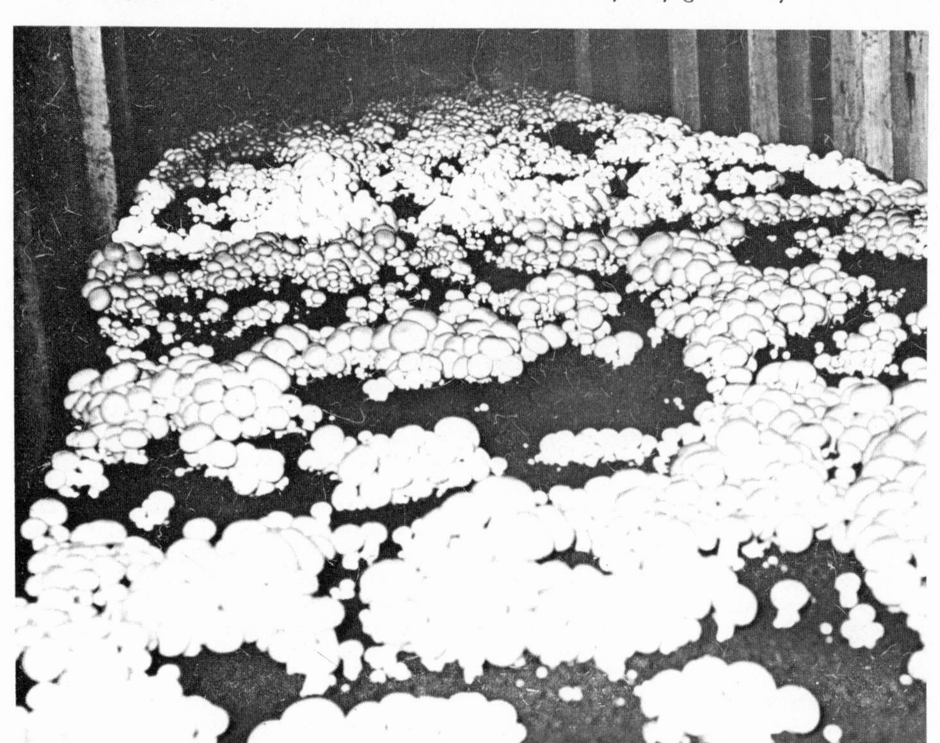

When harvesting mushrooms, carefully press the soil around the base of the stem, cutting off the stem with a sharp knife.

Pulling up a mushroom may injure others just breaking through. So, carefully press down on the soil around the base of the stem with one hand, using the other to cut off the stem with a sharp knife or twist it off. Try to pick as fast as they become ripe, daily if possible, to make your beds yield for two months or more.

After harvesting each flush, it's a good idea to clean out the remaining mushroom tissue beneath it to prevent decay. Fill the hole with fresh soil. After the final flush, remove the compost and give the bed a good scrubbing and sunning.

The spent compost, although no longer good for growing mushrooms, is fine for the garden. Apply it directly to the soil or mix it with heap or sheet-composted materials. Thus it will do double duty, giving you as excellent summer crops as your winter mushroom crop.

Some people like to eat mushrooms raw as soon as they pick them. But they will keep fresh in the refrigerator for quite a while or you can blanch them by steaming or boiling and freeze them. Better yet, dry your crop on a sheet of cheesecloth strung between convenient supports. They will dry in 3 to 5 days in a well-ventilated room and keep for a year or more.

Some gardeners grow mushrooms in a cold frame. They make the compost and plant the spawn the usual way in the frame, then cover the bed with a 6-inch layer of straw or mats. If your frame is on the north side of your house and the weather isn't too hot, you may get a pretty respectable crop.

It's even possible to naturalize mushrooms in a lawn or field. Simply plant 2-inch squares of spawn 2 inches deep. Try to pick a spot that stays fairly damp and is high in humus. If moisture conditions are right, you should get a fairly good crop in late summer and for several summers thereafter. Mushrooms grown in the light, incidentally, are quite high in vitamin D.

One last tip: never pick wild mushrooms without first arming yourself with a mushroom identification booklet (from your state experiment station). Those buttons you see popping up merrily from leaf mold or sod after a night's rain may be gourmet's delights like the puffball or inky cap or one of the many deadly species like the very common amanita.

—Thomas Powell

A Tier-Shelf Bed for Mushrooms

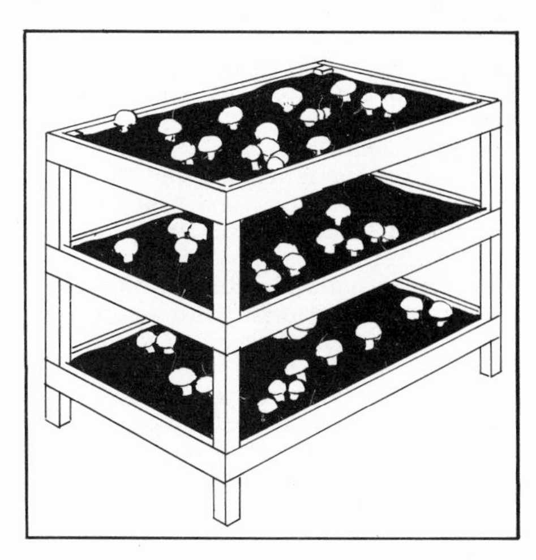

In a short time, you can build these tier-shelf beds which provide excellent growing places for your mushroom crop. "Why didn't I try this before?" you'll say when you discover the convenience of this practical bed arrangement.

Space-saving tier-shelf beds are easy to fill, empty, clean and disinfect.

Construct the uprights and cross-bearers of rot-resisting cedar, redwood or cypress, 3 by 4 inches for the uprights and 1 by 4 inches for the cross-bearers. Or you may use 1½ inch tubular steel. The supports must be strong.

The bed-boards should be 3 feet wide, 2¼ feet deep and 6 inches high. You can cut the bed-boards up to 5½ feet if you have an open path on each side. Use rot-resisting woods.

Place the bottom bed 6 to 8 inches away from the floor. Leave 18 inches between bed-boards and allow for from 24 to 30 inches between the top shelf bed and ceiling. The air will circulate easily. Place away from drafts and sunlight.

Fill the beds with 5 inches of good-quality horse manure and wheat straw compost. The final fermentation (peak heat) will destroy insect and fungus enemies. You may encourage it with artificial heat.

When the beds cool below 80 degrees and the *p*H is between 8.2 and 7.7, you are ready to plant the spawn. Purchase dry spawn of the prolific and disease-resistant brown variety from a commercial grower. Plant spawn the size of a walnut every 8 to 10 inches forming a diamond pattern one inch below the surface. Fill in the holes with compost.

From 7 to 14 days after spawning, check to be sure a blue-grey color is "running" through the compost. Spread one inch of organically sterilized loam over the compost.

Keep the humidity between 70 and 80 and provide fans for good ventilation.

Pinheads will appear from 10 to 20 days after casing. Spray the beds lightly with water, maintain the humidity at 95 per cent and regulate the temperature between 60 and 65 degrees Fahrenheit.

Within 6 to 8 weeks, mushroom clusters will appear. When the veils are stretched and they are ripe, harvest by grasping the stem and twisting. Cut out mushroom butts and fill the holes with ten parts casing soil to one part lime.

Rest periods and flushes will alternate for from 2 to 5 months before the medium is worn out.

Protect your new crop from enemies by sterilizing the tier-shelf beds, walls, floor and tools.

—Kathleen Porteous

MUSTARD

MUSTARD. *Brassica nigra,* black mustard and *B. juncea,* Chinese or Indian mustard. Black mustard is the mustard of the ancients, now found wild throughout Europe and is naturalized in the United States. It is grown here and in most of Europe for its seed, which is ground and used as a condiment. Young plants were used here at one time for salads, but are now largely supplanted by *B. juncea,* used particularly in the South for cooked mustard greens and salad. Mustard is a half-hardy annual herb, quick-growing and quickly running to seed.

One of the greens richest in vitamins and minerals, mustard ranks almost as high as kale nutritionally and its flavor and texture are preferred by many. One-half cup of cooked greens yields 11,000 units of vitamin A, .138 milligrams of vitamin B_1, .450 milligrams of vitamin B_2, 125 milligrams of vitamin C, 291 milligrams of calcium, 84 milligrams of phosphorus, 9.1 milligrams of iron, 2 grams of protein and a total of only 25 calories.

Range and Soil

Mustard can be grown in any part of the United States in cool weather. Its growth is so quick that even where winters are short, a crop of mild-flavored mustard may be brought to maturity. In hot weather, the flavor becomes strong. Sandy loam or any other light loam will grow a good crop. A light application of compost or manure supplies enough soil nutrients.

Seed

In the spring, mustard big enough for salad can be picked 30 days after planting. In fall, the growth is a little slower. One ounce of seed will sow 100 feet. Yield depends upon the size to which it is grown and the number of cuttings, as well as the length of the cool season. Most popular varieties are Tendergreen, the fastest growing and Southern Curled, which takes 2 weeks longer to mature.

Planting

Successive plantings are made every 10 days in early spring, starting as soon as the ground can be worked and ending one month before hot weather. In fall, the plantings are repeated about every 2 weeks until 6 weeks before winter weather is expected.

Sow the seed thinly one-half inch deep in rows 12 inches apart and thin plants before they get too crowded to stand 4 to 6 inches apart. Thinnings may be used for salad.

Harvest

When the leaves are 4 inches high, they are tender and right for salad. For cooked greens, they may be permitted to grow to 6 inches. They should be cut with shears. If the shearing is not too close, a planting will yield several cuttings.

Mustard makes a good cool greenhouse crop, since it is almost free of disease and insect pests and grows rapidly. It will grow well in a sheltered cold frame in the winter or on an unheated enclosed sunporch.

NEW ZEALAND SPINACH

NEW ZEALAND SPINACH. *Tetragonia expansa,* of the *Aizoaceae* family, native in New Caledonia, China, Japan, Chile, Australia and New Zealand, where Captain Cook's round-the-world expedition found it in the eighteenth century. It is a tender annual herb with a spreading habit, sprawling to a diameter of 4 or 5 feet and growing a foot or two high. The leaves of some varieties are encrusted with white dots, like the ice plants to which it is closely related. Leaves are brittle and tender when young and taste something like spinach when cooked. Like spinach, it is rich in vitamins and minerals.

Range and Soil

New Zealand spinach can be grown in weather too hot for spinach and is usually used as a between-season crop to supply greens between the spring and fall spinach plantings. Although it is heat-resistant, it will not grow in the hot weather of the extreme South without protection from the full rays of the sun. Best soil is a deeply-prepared loam, with a *p*H between 6. and 7. New Zealand spinach likes lime and does best in a plot which has been well manured.

Seed

One packet of seed will plant 25 feet, one ounce 100 feet. Seed germinates slowly and should be soaked in hot water for 3 to 4 hours or in cold water for 24 hours before planting. Plants mature 70 days after germination.

Planting

Plant seed in hills 3 to 4 feet apart or in rows spaced 4 feet apart. In hills, plant 3 seeds to the hill and thin later to one plant. In rows, thin plants to 12 to 18 inches apart. Being a tender plant, New Zealand spinach should not be planted outside until all danger of frost is past. Seed may be started indoors and transplanted to the garden when the soil is warm.

Culture

Like SPINACH, which see.

Harvest

Pick the tender stem tips about 3 or 4 inches long. The young leaves have better flavor than the older ones. If tips are picked, more branches will be put out by the plant and it may be harvested continually until frost.

OKRA

OKRA. *Hibiscus esculentus*. A tall ornamental tropical annual of the mallow family, *Malvaceae*. Also called gumbo. Okra is native to Asia or Africa or both; authorities disagree on the subject. It is certainly used in both places and is found growing wild along the Nile. It is a tropical plant growing to 3 feet in the dwarf form or 4 to 5 feet in the taller varieties. The flower is a typical mallow, yellow with a red center. Pods are used as a vegetable or in soups and meat dishes. Young but fully developed seed may be shelled and cooked like peas. Dried seeds may be used as a coffee substitute. The pods are sometimes dried and used as flavoring for soups or are pickled.

Okra pods cooked fresh have a fair amount of vitamin A and are rich in minerals. In one-half cup, a 100 gram portion, there are 440 units of vitamin A, .126 milligrams of vitamin B_1, 17 milligrams of vitamin C, 72 milligrams of calcium, 62 milligrams of phosphorus, 2.1 milligrams of iron, 2 grams of protein and a total of 24 calories.

Range and Soil

Okra can be grown where vine crops, like tomatoes, squash or cucumbers, thrive.

Okra pods are ready to harvest several days after flowers fall and should be picked promptly, before they become woody.

In very warm areas, two crops a year are grown. In temperate climates, it is grown as a summer crop. It needs a light loam, well-drained, but with plenty of moisture-holding humus, since the pods are 90 per cent water. The soil should be moderately rich. Poor soils should have manure dug in, but rich soils will need no special treatment. Too much nutrient in the soil results in much foliage and few pods. Approximately neutral *p*H is best—anything between 6.0 and 8.0.

Seed

One packet of seed will sow 15 feet of row, which is more than enough for the average family. If pods are picked at least every other day, 3 or 4 plants will supply more than a family can use. Germination is quick when the soil is warm. First pods may be picked in 55 to 65 days after planting. The plants yield until frost. Total yield depends upon the length of the season. Most popular varieties are Clemson Spineless and Dwarf Green.

Planting

Because the flowers are ornamental, okra is sometimes used in the flower border. Seed should not be planted outside until the soil is thoroughly warm because it rots quickly in wet cold soil. It is hard to transplant, but may be started indoors in flower pots a month before the outdoor planting time. Dwarf plants should stand 30 inches from their neighbors; tall ones need 3 to 4 feet. In rows, space the rows 3 feet apart and thin seeds to 12 to 18 inches apart.

Enemies

Cabbage loopers, corn earworms and green stink bugs are the chief pests to be found on okra. They may be hand-picked or somewhat discouraged with a dusting of rotenone. Because of the nature of the pods, it is advisable to remove all pods, mature or immature, before dusting.

Harvest

Pods are ready to pick a few days after flowers fall and should be picked promptly while young, before they become woody. They should be soft and the seed no more than half-grown if the pods are to be used whole. Picking should be done every other day, whether the pods are needed for the kitchen or not, in order to keep the plants producing until frost. Whether they are picked or left on the plants, the pods quickly become woody after they have reached their prime, so they should be used or frozen, canned or dried at once. To dry them, slice them or leave the small pods whole and string them. Hang the strings in a well-ventilated place out of the sun until the okra is dry.

ONIONS

ONIONS. *Allium cepa.* Onions have been cultivated for so long that it is difficult to trace their origin. It is believed that they may have originated in Asia Minor or India. They were used in early Egypt, in Greece and are mentioned in the early books of the Bible. A genus of the lily family, some varieties have attractive flowers and are sometimes planted in the flower border as well as in the vegetable garden. *A. cepa* is a shallow-rooted hollow-leaved plant averaging 18 inches tall when mature, with lavender or white flower heads or sometimes heads of small bulbils rather than flowers.

Three main types of onions are cultivated today. The American onions, with large yellow bulbs, are strong in flavor, medium size and keep well in storage. European or foreign varieties, which include Bermuda, white Portugals, Spanish and Italian, are usually milder in flavor, sometimes very large and do not store as well as American varieties. Egyptian onions are perennial garden herbs which form bulbils or sets in place of flowers. Their sets plant themselves when the tops bend over in midsummer and they supply early spring green onions. Egyptian onions are also called top onions, multiplier onions and potato onions. Any type—American, European, or Egyptian—can be used for bunching when picked at the small, bulbless stage.

Except in their vitamin A and C content, green onions and dry onions yield about the same nutrients. In 100 gm. of either are to be found: .042 milligrams of vitamin B_1, .125 milligrams of vitamin B_2, 41 milligrams of calcium, 47 milligrams of phosphorus, .4 milligrams of iron, 1 gram of protein and 42 to 45 calories. In addition, green onions supply 60 units of vitamin A and 7 milligrams of vitamin C, while dry onions have no vitamin A and only 2 milligrams of vitamin C.

Range and Soil

Onions may be grown at some time in the year in any part of the United States. When young, they are fairly resistant to frost. Those grown from seeds prefer cool weather for developing foliage and hot weather to mature large bulbs. In consequence, seed is planted in fall or winter in the southern, central, southwestern and western states. In the northern states, the seed is usually started indoors and transplants set in the garden as early as the soil can be prepared.

Best soil is a friable sandy loam, silty loam or peat. Heavy clay, coarse sand and gravelly soils are to be avoided. Onions need a large amount of moisture, with a supply constantly on tap in the top 8 inches of soil, but they must also have good drainage. Since roots are comparatively shallow, they must have abundant nutrients within a small area. To supply the needed humus, a heavy application of manure should be made the previous fall, if the manure is fresh or before planting, if it is well-rotted. Rock phosphate and rock potash should also be supplied. Or the following may be dug into each 100 feet of row: two pounds rotted manure, a sprinkling of wood ashes and 4 pounds 3½-3½-3½ organic fertilizer. The fertilizer can contain 3 parts granite dust, 1 part dried blood, 1 part bone meal and 5 parts seaweed or the equivalent in other materials. Soil should be slightly acid, *p*H 5.5 to 6.5.

Seed

Onions may be grown from seeds or from sets or transplants may be purchased from most seed men. Spring-sown seeds are usually successful only in the North. In the South and in the central states, the young plants too often encounter hot weather before they are large enough and the plantings fail. To obtain the earliest crops, sets or transplants (either home-grown or purchased) are set out in spring; in warm sections, seed can be started in fall.

Sets are small bulbs, bulblets, or bulbils. They may be bulbs formed by grow-

When plants are 10 to 12 inches tall, onions should be mulched to preserve moisture necessary for forming bulbs.

ing seedlings in crowded rows, which will grow to normal size when set out with plenty of room. Or they may be bulblets, the small bulbs which form beside large ones. Or they may be bulbils, the flower-growth on top onions. The latter are set out in fall in North as well as South to make early green onions. The advantage of using sets instead of seed is that they are less work and they mature faster. The disadvantage is that sets are obtainable in only a few varieties, while many types can be grown from seed.

One ounce of onion seed will sow 100 feet of row or will grow enough transplants for 400 feet of row. Yield from 100 feet, transplants or sets, is 1½ bushels of large bulbs. If they are picked for bunching, yield will be 400 onions. If seed is sown outdoors in the row, yield will be 1½ bushels of large bulbs per 100 feet, plus 200 bunching onions. Two quarts of onion sets will plant 100 feet.

Seed germinates best at 65 degrees, but will germinate at any temperature between 45 and 85 if it is fresh. Onion seed does not remain viable long, and should not be planted when more than one year old. Germination occurs in 8 or 10 days; bunching onions mature in 55 days, dry onions in 100 to 130 days.

Below are some of the varieties which are popular, with some of their characteristics:

TABLE 51: Onion Varieties and Characteristics

Variety	Maturity	Size	Color	Flavor	Storage period
Sweet Spanish	Late	Large	White or yellow	Pungent	Medium
Yellow Globe	Late	Medium	Yellow	Pungent	Long
Ebenezer	Early	Medium	Yellow	Mild	Long
Elite	Late	Large	Yellow	Pungent	Long
Bermuda	Early	Medium	Yellow or white	Mild	Short
White Portugal	Early	Medium	White	Mild	Medium
Italian Red	Midseason	Large	Red	Mild	Short

Planting

Sow seeds indoors 10 weeks before the outdoor planting season to obtain transplants which are one-fourth to three-eighths inch in diameter. Seedlings larger than three-eighths inch should not be set out, because they may send up seed stalks instead of forming bulbs.

Sets are planted in the garden two inches deep and two inches apart in rows 8 to 18 inches apart.

Seed sown directly in the garden is planted one-half inch to one inch deep, depending upon the lightness of the soil. In a very light soil, plant deeper to prevent drying. Seed is sown rather thinly, and when seedlings are 3 inches tall, they are thinned to stand 3 inches apart in the row. Transplants may be set in the garden 3 inches apart or may be spaced 4 to 6 inches apart, if they are not to be used green and depending upon the expected size of the mature bulbs.

When seedlings are 8 inches tall, before they have begun to form bulbs, they may be pulled for use in salads. Earlier thinnings may also be used for salad or used like chives in cooking.

Culture

After they are 10 to 12 inches tall, onions should have their mulch drawn up close to the plants to preserve moisture they will need while forming bulbs. If the soil is permitted to dry out at this time, "splits" may result—onions which form two bulbs on one root.

A side-dressing of chicken manure, spread close to but not touching the plants, may be applied one month after the transplants are set in the garden and the application repeated once or twice at one month intervals.

Enemies

Onions have very few enemies in northern gardens, but in the South they are not so fortunate. Onion thrips, which attack the foliage, may be washed off with a strong spray of water and they seldom return. Onion maggot, from a fly which lays its eggs at the base of the plant, can cause rot in the bulb. All affected onions should be pulled and destroyed before the maggots mature.

Onions like a variety of trace minerals and some deficiencies appear to be diseases. Copper deficiency, for instance, causes poor color and thin outer scales, which reduce the keeping qualities. A liberal use of rock fertilizers will overcome trace mineral deficiencies.

Harvest

Green onions should be pulled before bulbs begin to form. Bulbs become mature after the tops have begun to fall over, due to drying in the neck region of the leaves. In warm weather, when early onions are maturing, bulbs should be pulled when 25 per cent of the tops are down. In cooler weather, harvest when 50 per cent are down. In the Northeast, all the tops should be down at harvest time. There are a few exceptions to this—the thick-necks or "scullions." Leaves of scullions start to die back from the tips and the necks do not close, no matter how long they stand. These should be separated from the rest of the crop and used first, because they do not keep well.

In the South, or in hot weather, onions should be dried in the field for a few hours, then spread out in dry, well-ventilated sheds to cure for two weeks before storing.

In the fall, or in cooler sections, onions should be pulled and left one or two days in the garden beside their rows to cure. Tops should be laid over the bulbs, to prevent damage from the sun.

At the end of the curing period, tops are removed an inch above the bulb and the onions are stored in a cool cellar in well-ventilated containers. Tops should no longer be green at the end of the curing process, but shriveled and dry. Small onions may be stored by braiding their tops and hanging the braids in a cool cellar. The weight of larger onions will break the tops so they cannot be braided.

OYSTER PLANT

OYSTER PLANT. *See* SALSIFY.

PARSNIP

PARSNIP. *Pastinaca sativa* or *Peucedanum sativa.* Native to Europe and North America, parsnips were grown and used by the Greeks and Romans. A close relative was also used by the American Indians. They belong to the same family as carrots and celery, the *Umbelliferae.*

Parsnips are slow-germinating, slow-growing biennials which develop roots as much as two feet long, 3 to 4 inches in diameter at the shoulder. In their second year, a 3 to 5 foot stem shoots up on which greenish white flowers bloom. The roots contain about 18 per cent carbohydrate which is in the form of starch during the growing season, but changes, after a few weeks' exposure to near-freezing temperatures, into sugar. They have a strong distinctive flavor greatly relished or greatly disliked, according to individual taste.

One-half cup of cooked parsnips, a 100 gram portion, contains 100 units of vitamin A, .120 milligrams of vitamin B_1, 40 milligrams of vitamin C, 60 milligrams of calcium, 76 milligrams of phosphorus, 1.7 milligrams of iron, 2 grams of protein and a total of 65 calories.

Range and Soil

Parsnips can be grown throughout the United States, but are not often planted in the South or West, because the temperatures do not fall low enough to develop their best flavor. When they are planted in the southern states, they are grown through the winter and harvested before the weather becomes warm.

Soil must be deeply prepared to accommodate the deep parsnip taproot, and should be light in texture to avoid pronging. It need not be very rich, because of the slow growing season, but is improved by a layer of mature compost or well-rotted manure dug in at least a foot deep. Fresh manure stimulates the root to divide and roots may also be malformed if stones are left in the rows. Soil should be about neutral, with a *p*H rating of 6.0 to 8.0.

Seed

Parsnip seed germinates slowly and not too well, even when fresh. It should never be kept from year to year, because it loses its viability quickly. A packet of seed will sow 20 feet, an ounce 100 feet. Yield from 100 feet is about two bushels. Two to three weeks are needed for germination; roots mature in 3 to 4 months after they are planted. By far the most popular variety is Hollow Crown.

Planting

As soon as the soil is warm, sow seed rather thickly in drills one-half inch deep and cover with sifted compost. The seed is very fine so it may be mixed with compost, if desired, to facilitate sowing. If seed is soaked in water overnight, it will germinate a little sooner. Spray with a fine stream after planting, and cover the seedbed with burlap or a plank to preserve moisture. Or sow radishes thinly in the row to keep the crust broken and to mark the drill. Radish leaves develop quickly and will shade the weaker parsnip seeds and prevent their drying out. Thin seedlings to stand 3 inches apart in rows 18 inches apart.

Culture

Young parsnip seedlings need plenty of moisture until they are big enough to send their roots deep in the soil. Mulch them early, and they should be able to find enough of the spring residue of rainfall in the soil, except in the driest seasons, to supply their needs.

Enemies

Parsnips, carrots and celery all have the same enemies and diseases. (*See* CARROTS, CELERY.)

Harvest

Parsnips may be pulled when they are large enough to cook, but they improve greatly in flavor if left in the ground for a few weeks at near-freezing temperatures. They are perfectly hardy and may be left in the garden all winter. Where the season is mild enough to permit digging in midwinter, they can be dug up as needed. Where the ground freezes hard, they may be mulched to prevent alternate freezing and thawing (which spoils their flavor) and used in spring before tops begin to grow. Or they may be dug and stored in a root pit, covered first with soil, then with straw and leaves and finally with a waterproof cover. They cannot be stored indoors except in a very cold cellar and then, they must be covered with moist soil to prevent shriveling. If left in the ground until spring, roots should be dug before the tops begin to develop and stored in a cold place.

PEA

PEA. *Pisum sativum,* garden or English peas; variety *humile,* dwarf garden peas; variety *macrocarpon,* edible pod peas; variety *arvense,* field peas.

Peas have been known from ancient times, and were thought to have originated in Europe and Asia. Garden peas are believed to have been developed from *P. sativum arvense,* the grey field peas now used for food in India and Egypt, but in this country only for forage or a green manure crop.

Two types of peas are planted in gardens, those having smooth and those having wrinkled seed. The former are more hardy and earlier, but are not as sweet. Wrinkled peas are usually planted as the main crop, with a few smooth planted to provide the first few meals in spring.

Tall peas are annual vines climbing by means of tendrils to a height of 3 to 6 feet. The dwarf variety grows to one and a half to two feet, is somewhat earlier, takes less space and is less trouble to grow, though it is not so prolific as the tall variety and its season is shorter. Edible pod peas give greater yields of peas, which are cooked in their pods like snap beans.

Peas are rich in vitamin A and the B vitamins. In a half cup, cooked, there are 1,500 units of vitamin A, .390 milligrams of vitamin B_1, .250 milligrams of vitamin B_2, 20 milligrams of vitamin C, 28 milligrams of calcium, 127 milligrams of phosphorus, 2 milligrams of iron, 7 grams of protein and about 100 calories.

Range and Soil

Peas are a cool weather crop. The young plants are not injured by light frosts, but the blossoms and pods are, so they are grown as a very early spring crop in the North and as a fall crop only in the South and Southwest. Any soil will grow good peas, but the season in which they will produce best is determined by the type of soil. A light sandy loam will yield the earliest crop; heavy clay soil which is cold and wet will yield a later crop. Varieties should be chosen with reference to the soil in which they will be planted. Plenty of humus is needed in the soil to hold moisture. A plot previously manured for a richly feeding crop, such as broccoli or celery, is best. If possible, no peas should have grown on it for 5 years.

Peas are often the first crop planted in spring, sometimes even planted in fall just before a hard freeze.

Peas do not need much nitrogen, but like a soil rich in potash and phosphorus. Soil *p*H should be just below neutral, about 6.5.

Seed

One packet of seed will sow 15 feet, a pound 100 feet. Usually 2 to 3 pounds should be planted by the average family for the table and for canning or freezing. Yield from 100 feet of garden peas is one bushel, of edible pod peas, two bushels. Seed germinates in 8 to 10 days, pods mature in 50 to 80 days after planting. It is best to sow several varieties which will mature in different periods, planting them all as early as possible, rather than to sow the same kind in several successive sowings. The small plants grow better in the very cool weather of early spring than later when the soil has warmed up.

Most popular varieties are: very early smooth-seeded, Alaska; early dwarf, Laxton's Progress, Little Marvel; early tall, Freezonian, Thomas Laxton, World's Record; tall late, Alderman and Lincoln.

Planting

Peas should be just about the first thing a gardener plants in spring. They may be planted in fall just before a hard freeze and will come up in the spring, though there is no advantage to fall planting unless spring planting may be delayed for some reason. Plant them just as soon as the soil can be prepared, if possible 6 weeks before the last spring frost.

If soil is dry at planting time, the seed may be soaked overnight. Or the soil may be flooded with a hose after the drills, two inches deep, are opened and before the seeds are planted. If the soil is wet, do not soak the seed or it may rot. Seed should be inoculated with nitrogen-fixing bacteria which will grow in nodules on the roots and provide a supply of nitrates for the plants.

Peas may be planted in double or even quadruple rows, in drills 8 inches apart, with 3 feet between row centers. Peas which will climb on a fence or trellis are usually planted in double rows, one on each side of the support. Dwarf varieties may be planted in 4 rows, two to be picked from each side. If the dwarf varieties will be supported by a thick mulch on each side, they will be easier to mulch in single rows.

Space between rows of peas may be planted with spinach, lettuce, green onions, radishes or any small crop which will mature in two months and be out of the ground before the peas grow tall or begin to sprawl.

Seed should be sown about 12 to the foot in each drill. When seedlings are two inches high, dwarf peas are thinned to 3 inches apart, tall varieties to 4 inches apart.

Even dwarf peas may need support to keep the pods off the ground. If they are well mulched with straw, they will not be damaged by sprawling on the ground except in very wet weather. But if they are not mulched, they should be supported. Twigs of privet, birch or of any well-branched shrub or tree may be pushed into the ground next to the rows immediately after seeds are planted. If they are put in later, they may damage roots.

Supports for tall peas may be chicken wire, 4 feet wide and supported by 6 foot posts of which two feet are underground. Or they may be strings wound around two wires which are strung 6 inches and 4 feet above ground. A horizontal support as well as a vertical one to which the tendrils cling will help tall varieties to weather strong spring winds. To strengthen them, when the vines are about one foot tall, fasten a stout string to the end post 10 inches above the ground, carry it outside the vines to the next post, and wind it around that and so on down the row. As the vines grow, reinforce them at each foot mark in the same way.

Culture

If peas are planted early enough in the spring, the normal spring rainfall should supply enough moisture, especially if they are well mulched. However, if the spring is dry, they will need irrigation at about the time when the blossoms fall. It is very important to give them enough water while they are filling out the pods.

Enemies

Insect pests do not bother peas much in northern gardens. In the South, a weevil lay eggs on the pods and the larvae bore in to feed on the seeds. If the peas are wanted for seed, they may be heated to 120 to 130 degrees for 5 or 6 hours to kill the weevils. The heat treatment will not damage the seed.

Powdery mildew may attack peas in a wet season. Nothing can be done to protect the plants from frequent rains, which bring on the attacks, but overhead sprinkling can be done early in the day to ward off the disease. If mildew has infested a planting, dig under all remnants of the vines as soon as the last peas have been picked to prevent the mildew from perpetuating itself until the next year.

Peas are also subject to a variety of blights and wilts, but carefully selected and grown seed from a reliable dealer should be free of these diseases. In areas where they are prevalent, buy resistant strains.

Harvest

Peas should be picked as soon as pods seem well-filled, but before they begin to lose color. Immature peas, very small and soft, are considered a great delicacy by some people, but the low yield when pods are picked too early makes such harvesting impractical.

Peas should be cooked or processed immediately after picking to preserve vitamin and sugar content. Within two hours after they are picked, the sugar begins to change into starch. If they must be kept for a few hours, they should be quickly chilled in the ice-tray compartment and then stored in the refrigerator until time to cook them. If it is necessary to wash them, wash before shelling, not after. For best flavor and greatest nutritional value, peas should never touch water.

Best way to preserve peas for winter use is by freezing. They may also be canned, but the canned product cannot compare with the frozen one.

PEA, BLACK-EYE

PEA, BLACK-EYE. *Dolichos sphaerospermus.* This is a native of the West Indies and is used in Jamaica and through the southern states. It is a hot-weather crop, resembling snap beans horticulturally rather than garden peas. It is grown throughout the warm areas of the United States in preference to garden peas, which are not too successful there. Black-eye or Crowder peas are heat-loving plants and not well adapted to the seasons of the northern states.

The pods may be used like snap beans or they may be shelled and peas and pods cooked together. They are also sometimes dried and used like dried beans, but they are somewhat sweeter.

Seed

One-half pound of seed will plant a 100 foot row. Popular varieties are California Black-Eye, Extra Early Black-Eye, Large Black-Eye, Brown Crowder, Cream Crowder, Conch and Cream Lady.

Planting

Black-eye peas are planted about 6 weeks after the last frost, when the soil is warming up. Seed is sown 1½ inches deep in rows that are 2½ feet apart and plants are thinned to stand 3 inches apart.

Harvest

Black-eye peas are picked after the peas are well developed and can easily be removed from the pods, but before the pods begin to dry out.

Peas (Sugar) in the Pod

Ever start out with a dishpan full of freshly picked peas and wind up with a dishpan full of empty pods that looked good enough to eat? It seemed almost a shame not to, they were so crisp and fresh and juicy, especially if they came from your own garden. One look at the few handfuls of peas in the pot, all that remained of what looked like a big harvest, and you couldn't help wishing all the more that you could cook the pods, too. But even though Nature does gift-wrap the pea you just can't eat the wrappings.

Not, that is, unless it happens to be that most delectable legume, the sugar-pod pea, the *mange-tous,* or "eat-all" as the French call it. It's not a bad name, for that is exactly what you do, eat all, pods included. It's an old favorite in many country gardens, but most people who must depend on city markets for their fresh vegetables have never even heard of edible-podded peas.

Sugar pod peas produce most abundantly and, not only is there no "waste" to your crop, in the form of the usually discarded pods (although you do add them to your compost pile, of course) but each plant bears so heavily that a pan full can be picked in no time.

It's a good idea to select one of the cooler spots in your garden for the peas. Avoid a southern exposure. In the case of spring plantings, if warm weather arrives prematurely, the vines will bust out in bloom and you'll have to hurry and pick a nice early crop, but it will be a smaller one than you'd have at the end of a *longer* cold spring. Now, by careful timing the fall planting can escape this hazard. Just be sure, however, to consult the frost schedule for your region before planting and allow about 65 days for your plants to mature. The cool of the fall can be favorable to pea culture provided care has been taken to plant far enough ahead of the frost date line.

Once the flowers begin to fade you must keep a close watch to catch your peas at the peak of perfection.

The picking must be done with great sensitivity, a fine feeling for that just right moment. A day later and the flavor is overblown. In the case of the sugar pea there is a much wider time margin here. The fact that sugar pods can be eaten long before the peas inside have fully developed makes them available considerably earlier and for a longer period of time than ordinary peas.

Incidentally, the tenderness of the pods of the sugar pea is due to the absence of that parchment lining present in the pods of all other kinds of peas.

Sugar pod peas are planted just like other kinds of peas. Being so prolific they are a good space-saver, something to consider now, with all the slow growers like corn, tomatoes and melons taking up so much room in the garden.

—*Clee Williams*

PEANUT

Peanut. *Arachis hypogaea.* Peanuts are legumes which are native to tropical America. Seed has been found in early Peruvian tombs indicating that they were cultivated by the Indians. They are grown principally in America and in China at present.

Vines are sometimes planted in northern gardens, but produce a dependable crop mainly in the South. Plants grow 12 to 18 inches tall and have two sets of flowers, yellow showy pea-like flowers which are sterile and inconspicuous fertile flowers growing in the lower leaf axils. After the lower flowers are fertilized, they send out long peduncles which bend over to bury their tips in the soil. The peanuts, their seed, form underground on the tips of the peduncles.

Peanuts are extremely rich in all the B vitamins, containing more pantothenic acid than any other food except liver. They are also rich in vitamin E, 100 grams yielding from 26 to 36 milligrams. In 20 grams, about 18 nuts, there are 70 units of vitamin A, .225 milligrams of vitamin B_1, .110 milligrams of vitamin B_2, 15 milligrams of calcium, 73 milligrams of phosphorus, .4 milligrams of iron, 5 grams of protein and a total of 110 calories.

Range and Soil

Peanuts are grown commercially from Virginia south along the Atlantic, in the Gulf states, Texas and Oklahoma. They can be grown farther north in favorable seasons, but they require special care and handling. With luck, a crop can be gathered from a planting as far north as Massachusetts.

Peanuts require a growing season of 4 to 5 months, with fairly hot weather and a rainfall of about 20 inches evenly distributed through that period. They do not need rich soil, but an acid soil with a *p*H of 5.0 to 6.0 is essential. A loose soil is necessary to permit the fruiting stalks to penetrate the ground. Plenty of humus provided by mature compost is the best means of opening up a crusty or heavy soil to accommodate them. In the North, a warm sandy soil in a sheltered place with a southern exposure will give peanuts the best chance of bearing a good crop.

Seed

Whole peanuts may be planted in their hulls if the shells are thin. Or they may be shelled, and nuts planted separately. One packet will sow 15 to 25 feet, one pound 100 to 200 feet of row, depending upon whether or not they are hulled.

If weather is hot, seed germinates in 7 days. If there is no sun and weather is cold, it may take two weeks for them to appear. Two varieties are commonly planted in the United States, the Spanish red and the Mexican brown nuts. Spanish are larger and better flavored, the Mexican small and prolific. Usually the Mexican ones are sold salted whole with their skins left on; Spanish nuts are skinned and heated in oil.

Seed to be planted in northern gardens will give much better results if it was raised in the North. If possible, buy from a northern grower or save seed from your own plantings.

Planting

In the South, peanuts are planted after the last frost when the soil has warmed up. In the North, in order to stretch the short growing season, they are planted at about the last frost date. Hulled nuts are sown 3 to 6 inches apart; those in hulls 8 inches apart. Rows are spaced 30 inches apart. In the South, seeds are planted as much as 4 inches deep; in the North no more than 1½ inches deep.

Culture

When plants are about a foot high, draw soil up around them in the way that it is hilled around potatoes. Raising the soil level helps the fruiting peduncles to bury their tips more quickly and to start forming seed sooner. After the plants are hilled, mulch between rows with straw.

Enemies

Few insects or diseases bother peanut vines. Corn earworms may attack the foliage late in the season, but they will not affect the crop, so they may be ignored.

Harvest

In the South, peanut vines are dug before frost and are hung in open sheds to cure. To determine the time for pulling, examine the inside of the shells. They should show color and veins and the foliage should begin to turn yellow.

Peanuts grown in northern gardens are left in the ground until mid-October before they are dug. The leaves may be killed earlier by frost but food stored in the stems will feed the nuts that are still developing underground.

Vines should be dried by hanging in a well-ventilated attic in the North or an open shed in the South or the nuts may be spread on a piece of wire screening to dry. They should cure for about two months before they are ready to roast.

Roast peanuts for 20 minutes in the oven at 300 degrees. Remove them from the oven and allow them to cool, to finish the roasting process. Homemade peanut butter may be made by putting the nuts through a food chopper several times, using the finest knife.

Peanut shells and vines make excellent mulch or compost. They contain a large percentage of nitrogen and a moderate amount of phosphorus and potash as well. On an asparagus bed, the shells make a fine mulch which the spears have no difficulty in penetrating.

Select large yellow pods which respond with a good rattle when shaken to save for next year's seed. Nuts should be left in the shells until planting time and should be stored in a dry place where they will not mildew.

Growing Peanuts in the North

If you're going to plant peanuts in northern states, be careful of your seed variety. L. W. Martin, organic farmer from Wheatland, Iowa, has this to say:

"If you start with southern grown seed

and grow them this far north, you undoubtedly will run into the same trouble I had. In the beginning, I didn't even get my seed back.

"But now, I've gotten to the point where peanuts are as sure a crop as corn. I have seed in its twelfth year—grown organically—and it is acclimated to Iowa."

Here is how Iowa farmer Martin plants his peanut crop.

"I've found that plowing under sweet clover ground is best, when the clover is about 18 inches high. Here this would be around May 10th. The soil should be a rich, black sandy loam and mellow.

"You can plant the peanuts in the shell or out, whichever you prefer. I gummed out an old corn planter butt drop plate to fit a medium-sized peanut and use the full row width. I plow them, using standard equipment, until the peanuts start to show yellow blossoms."

Experts advise not planting peanuts on the same land oftener than once in 3 or 4 years. The rotation should include at least two soil-building crops—one a winter cover crop. Many growers use cow-peas or velvet beans. For winter cover crops, crimson clover, giant red clover, alfalfa, rye and barley are popular.

Planting is generally done between April 10 and May 10. However, no planting should be done until the soil is fairly warm. The crop should be harvested before the vines are killed by frost. To determine when to harvest, check the foliage for a slight yellowing and examine the pods. If the peas are full-grown and the inside of the shells has begun to color and show darkened veins, they can be assumed ready for harvesting.

Controlling weeds is a major factor in peanut production. For best results, start early—many weeds can be killed just before and just after the peanuts come up. In addition, cultivate shallow at all times. Most weeds germinate in the top one-half inch of the soil.

Many commercial growers have been forced to stop using Benzene hexachloride and other pesticides because of the off-flavors produced in the peanuts.

Peanut vines can be dried by hanging in a well-ventilated attic or the nuts can be spread on a piece of wire screening to dry.

In fact, growers are warned not to raise peanuts on land that has been treated with questionable materials. This gives an advantage to organic farmers who have not resorted to using these chemicals.

"Peanut seed should be relatively dry before being placed in storage," advises the USDA, "about 8 per cent moisture or less for unshelled seed, 6 per cent or less for shelled seed. Peanut growers who require only a few hundred pounds of seed often use clean second-hand 50-gallon drums for storage. One end of the drum is cut out close to the rim and smoothed down; a square of plywood is laid on top of the drum as a cover. Ventilation may be provided by punching numerous small holes in the sides of the drum. The peanuts may be stored either in bulk or in sacks. Each drum will hold approximately 100 pounds of unshelled peanuts."

Make Your Own Peanut Butter

There are many good reasons for making your own peanut butter. For one thing, you won't be adding hydrogenated oil as is done commercially to prevent separation. Hydrogenation is the process which causes natural peanut oil to harden, converting the oil from a healthful, unsaturated fat to an unwholesome, saturated one.

It's the content of high-cholesterol saturated fatty acids (in contrast to the *essential* fats and oils present in a natural product) that medical specialists and nutritional research have been emphasizing as a major contributing factor in several serious disorders. Among these are arteriosclerosis (hardening of the arteries), blood circulation disturbances and heart trouble complications.

Just roast your peanuts (preferably grown organically) in their shells in a 300 to 320 degree oven for about 20 to 25 minutes, turning occasionally. Then run them through the kitchen meat grinder 3 or 4 times for a fine, smooth butter. Of course, the first turn will be somewhat of a "grind" and requires a bit of effort (good exercise, though). After that, they'll pour through easily.

Large-scale manufacturers roast *preshelled* peanuts at 350 to 400 degrees for 40 to 50 minutes. From a nutritional point of view, this is overdoing it in the worst way. The added exposure, longer roasting and higher temperatures are not only unnecessary, they quickly destroy most of the high vitamin B content of peanuts. An average roasting leaves but 12 to 20 per cent of these important elements.

The rule, then, is to roast as lightly as possible. You might even try *not roasting at all.* The taste and texture won't be the same, but you'd have a much better peanut butter *healthwise.* The heavy commercial roasting is done mostly so that the peanut butter will have an attractive *color.*

PEPPER

PEPPER: *Capiscum frutescens,* a genus of the potato family, *Solanaceae.* Variety *C. frutescens grossum,* the square and blocky fruit, includes sweet bell peppers, the most popular of the green peppers. *C. frutescens longum* is a long tapered variety which includes many of the hot peppers such as cayenne and chilis. *Piper nigrum,* whose seeds are ground to give us black or white pepper, is of another family altogether.

Peppers are natives of the tropics, where they grow on woody perennial shrubs 6 to 8 feet high. They are grown in temperate latitudes as tender annuals, which make a growth of 2 to 3 feet during a summer and bear pod-like berries with thick rinds which may be eaten green or may be allowed to ripen to bright red or yellow, depending upon variety. The wild peppers are hot and pungent; cultivated ones are often sweet rather than hot. Hot varieties are spicy while green, but become much hotter upon changing color.

Vitamin C content of peppers is also greatly increased—in fact, almost doubled—when the color changes. At the same time, the vitamin A content is somewhat decreased. One hundred grams of green peppers provide 700 units of vitamin A, .025 milligrams of vitamin B_1, .025 milligrams of vitamin B_2, 125 milligrams of vitamin C, 12 milligrams of calcium, 28 milligrams of phosphorus, .4 milligrams of iron, 1 gram of protein and 25 calories. In the same amount of red pimentos, there are 500 units of vitamin A and 200 milligrams of vitamin C.

Range and Soil

Climate requirements of peppers are about the same as those of eggplant and tomatoes, except that peppers are somewhat more hardy. They will stand a little more cold weather than tomatoes or eggplant, though they produce best at moderately hot temperatures. At blossom time, they need a mean temperature of between 65 to 80 degrees to set fruit. But if the weather is too hot and dry, they may drop the small fruits and wait for

Green peppers are mature and ready for picking when they feel firm and have become heavy. Cut stems one-half inch from pepper cap when picking.

more temperate weather to bear. A heavy crop of foliage over the fruit helps to protect it from the sun, so that less fruit is dropped by plants with good foliage than by those with fewer leaves. For that reason soil, which was at one time thought to be best if rather thin, is now believed to be best for peppers when it is quite rich.

Peppers will grow faster on soil that is light and sandy, so in northernmost portions of the United States, a crop may be matured on light soil where it might not ripen on heavier soil. In most sections of the country, a slightly clayey, rich soil with plenty of humus will produce most heavily. Bell peppers like a lighter soil than hot peppers. Both do well in average garden soil enriched with compost and with 4 pounds of the following mixture dug in on a 100 foot row: 1 part rock phosphate, 3 parts greensand, 2 parts wood ashes. Tobacco stems, which are a good source of potash for any other family, will kill peppers if used near them. Greensand and wood ashes should supply all the potash needed.

Seed

One packet of seed will start 50 plants. A dozen plants of any one variety will more than supply the needs of the average family. A 100 foot row will yield about 4 bushels of peppers, depending upon the length of the growing season and the variety planted. Peppers mature in 14 to 18 weeks from seed.

Most popular varieties of hot peppers are Long Red Cayenne, Red Chili, Tobasco and Anaheim Chili. Good sweet pepper varieties are California Wonder, Yolo Wonder, Ruby King, Vinedale, Sunnybrook, Merrimac Wonder. Sunnybrook and World Beater are not as high as average in vitamins, however, while Neapolitan, which is not raised very often, is very high in both vitamin A and vitamin C.

Planting

Seed should be started indoors in the North or in outdoor, protected seedbeds in the South about 6 to 8 weeks before the last expected frost. Plants will produce best if their growth is uninterrupted, so the best way to plant seed for family use is in flower pots, 6 or 8 seeds to the pot, thinned later to the one strongest seedling. Two weeks after the last frost, when the soil has warmed up, the plants are set in the garden 12 to 18 inches apart in rows that are 3 feet apart. If both hot and sweet peppers are being grown, separate them by the length of the garden, otherwise they will cross-pollinate and surprise you with hot peppers where nothing but sweet should grow.

Plants should be somewhat deeper in the garden than they were in the pots. Plant them deeply enough to bring the soil level within an inch or two of the lowest leaves. Protect each seedling with a cutworm collar of stiff paper and give it a cup of manure water to start it off.

To obtain the earliest crop of peppers, plants may be set in the garden 2 or 3 weeks earlier if they are protected with glass cloches and surrounded by tar paper squares. A square of tar paper 18 by 18 inches, with a 5-inch hole in its center, is laid over the spot where the pepper is to be planted. Edges of the paper should be weighted with stones to prevent its being blown away. The pepper is planted through the 5-inch hole, and given its drink of manure water and its cutworm collar. Then a gallon clear-glass jug, its bottom removed, is set over the plant. The jug should not be closed by its cap, but left open for ventilation. The tar paper gathers all the day's sunshine to warm the soil and prevents it from cooling at night. The glass jar permits the sun to reach the seedling, but keeps out chill from early spring breezes. By the time the plant has filled the jar, the weather should have warmed enough so that the protection can safely be removed and an ordinary straw mulch substituted.

Culture

In their early stages peppers need plenty of water, and should be irrigated regularly if rainful is scanty. Later in the season their roots have gone deeper and they will stand a little more drought than tomatoes. A good straw mulch around the plants should be enough to protect them from summer drought, unless it is extreme.

Enemies

Damping-off in cool weather is to be expected when seedlings are started outside, unless the seedbed is well ventilated. Cutworms also menace the young seedlings but can be warded off by stiff paper collars thrust into the soil in a circle about an inch from the young stems.

Of the diseases to which peppers are subject, mosaic and anthracnose are the most destructive. Mosaic disease attacks the leaves, causing them to become mottled and will eventually kill the plant. Affected plants should be pulled out and destroyed. Anthracnose is a fungus disease which causes soft spots in the peppers. It is spread by spores which are disseminated during wet weather. Avoid touching the plants while they are wet and keep the pepper and bean plantings separate in the garden, since both vegetables are affected by anthracnose.

Harvest

Green peppers are mature when they feel firm to the touch, and when they become heavy. If they are left on the plants beyond that point, they will begin to turn color and become red or yellow, depending on variety. Red peppers should have a full rich color before they are picked.

Cut the stems one-half inch from the pepper cap when picking them. Branches are brittle, and will break sooner than the stems if they are not cut. To store peppers, pick them green at about the time of the first frost, put them in ventilated baskets and keep them in a room as close to 32 degrees as possible and as humid as it can be kept. If the humidity is 98 per cent, the peppers will keep for about 40 days.

Hot peppers may be dried and used throughout the winter for flavoring. Shortly before the first frost, pull up the plants and hang them upside down in a well-ventilated place to dry. When the pods are dry they may be tied in long strings and hung in the kitchen until used.

PEPPER GRASS

PEPPER GRASS. *See* CRESS.

PHYSALIS

PHYSALIS. *See* GROUND CHERRY.

PIEPLANT

PIEPLANT. *See* RHUBARB.

POTATO

POTATO. *Solanum tuberosum*. A member of the *Solanaceae,* the family which includes tomatoes, peppers, tobacco and, in the flower garden, petunias. Potatoes are natives of southern Chile and were grown by the Incas before the early explorers carried them back to Europe.

A shallow-rooted plant, the potato grows from sprouts arising from its tubers, seldom making seed which can be planted. The new tubers are formed on roots arising above the original tuber from the stem. When exposed to light above ground, the tubers turn green and develop a toxic substance called solanine.

Potatoes are composed principally of starch, but they contain a fair to large amount of vitamin C, depending on variety and supply minerals as well. The average medium-sized baked potato weighing 100 grams will yield .200 milligrams of vitamin B_1, .075 milligrams of vitamin B_2, 20 milligrams of vitamin C, 13 milligrams of calcium, 53 milligrams of phosphorus, 1.5 milligrams of iron, 3 grams of protein and 92 calories.

Range and Soil

White potatoes were formerly grown only on cool soils. In southern gardens the term "potato" meant only sweet potato. New potato varieties and growing methods have made it possible for southern growers as far south as Florida to harvest a good crop of early potatoes before the summer heat. One of the most important factors in growing them in hot climates is a heavy mulch spread early to preserve the soil's winter coolness. Commercial growers in the South now supply the nation with early new potatoes, while main crop storage potatoes are grown in the northernmost states.

Any good garden soil that is not too heavy or wet can grow good potatoes. Preparation for the crop should begin the year before they are planted. Best crop to follow with potatoes is a legume, either beans, peas or a plowed-under cover crop of soybeans or clover. Potatoes should not follow tomatoes, which harbor many of the potato diseases. Soil should not be limed within a year of potato planting, nor have wood ashes dug in. Lime in the soil encourages scab, one of the most persistent of potatoes' 50 or 60 diseases. If manure is to be used for potatoes, it is best dug in in the fall before they are planted. Compost may be used, but only that made without the addition of lime.

An acid soil is best for potatoes, though they will bear well in any soil with a *p*H of 4.8 to 6.5. A *p*H of 5.6 inhibits the spread of scab; below 5.0 destroys it. If it is impossible to make the soil as acid as that, the next best thing is to plant a scab-resistant variety, such as Russet Burbanks. Legumes turned under several months before planting, help to make the soil acid as well as provide humus and soil nutrients.

A soil rich in potash tends to produce mealy potatoes. With a shortage of potash, they are more apt to be soggy after cooking. But potash must not be supplied from tobacco stems which should not be used on any crop belonging to its own family. Granite dust or greensand are best potash sources. Five pounds of the following fertilizer on a 100 foot row is recommended: two parts cottonseed meal; one part phosphate rock; two parts granite dust.

Seed

Seed potatoes certified free of disease are available from seed dealers. Unless the gardener has been able to save some of his own which he knows to be disease-free, certified potatoes are safest to plant. Potatoes purchased for cooking should not be used, because their background can never be known. Different varieties prosper in different areas of the country and the best recommendation is that provided by the experience of local gardeners. Below are some of the popular varieties, with some of their characteristics.

TABLE 52: Popular Varieties of Potatoes

Variety	Season	Resistant to	Vitamin C
Irish Cobbler	Early	Wart, mild mosaic	Medium
Kennebec	Late	Late blight, mild mosaic, net necrosis	
Katahdin	Late	Net necrosis, mild mosaic, wart, brown rot	High
Triumph	Early		Medium
Cherokee	Medium early	Late blight, common scab	
Sebago	Late	Yellow dwarf	Medium
Chippewa	Midseason	Mild mosaic	Low
Green Mountain	Late	Wart	
Houma	Midseason	Net necrosis, mild mosaic, wart	High

Seven pounds of seed should plant a 100 foot row and yield 2 to 4 bushels if they are all allowed to mature. If they are dug early for summer meals, the yield will be less.

Seed potatoes smaller than an egg should be discarded and tips where eyes are crowded should not be used. To speed development, seed may be "greened" before planting. New potatoes can sometimes be harvested in as little as 6 or 7 weeks after greened sets are planted. To prepare them, 6 weeks before planting spread the seed potatoes in a single layer, uncut, in a light warm shed or on a porch where the temperature is about 60 degrees and the light is bright but diffused. If they are greened in the sun, the process will take about 2 to 3 weeks, but the seed may be damaged by too much heat. The potatoes are turned at intervals to green evenly on all sides. By the end of the period, short stocky sprouts about one-fourth inch long should have developed from the eyes. Any potatoes with long spindly sprouts should be discarded. Diseased and blind sets are easily spotted after greening and the development of a large number of weak, minor shoots is discouraged.

Many planters prefer to plant whole potatoes rather than cut sets. Whether whole or cut sets are planted, each piece should be about the size of an egg to provide food for the sprouts until roots form and should contain one or two eyes —never more than 3. Seed should be cut several days before planting to give the cut faces a chance to harden, thus preventing seed rot.

Planting

Early potatoes are planted as early as the soil can be worked in spring, usually 5 to 6 weeks before the last frost. The late or storage crop is planted early enough so that it will mature 3 to 4 weeks before the first fall frost.

Potatoes may be planted in rows or in hills, but rows are easier. Tubers are planted 6 to 12 inches apart in rows 2 to 3 feet apart and are covered with at least 3 inches of soil—more if soil is sandy.

Several organic gardeners report bumper crops from potatoes planted either on top of a thick layer of partly decomposed leaves, on top of the remnants of last year's straw mulch or even on top of enriched, but bare, rocky soil, covered only with a foot of straw. These potatoes, they say, are almost entirely pest-free and need no further care until the harvest, which consists of turning back the straw and picking up the potatoes.

Culture

When shoots are 9 inches above the soil, hilling should begin. Soil is drawn up around the plants, leaving 6 inches of stem exposed above ground. The hill should be fairly broad, because its purpose is to cover deeply the developing tubers, so that they will not be greened by the light or sun-scalded. Hilling should be repeated every 3 weeks until flowering. A deep straw mulch, spread between plants and drawn up to the stems, would serve the same purpose. It would also preserve the soil moisture, which is essential to the young tubers.

If the weather is dry, potatoes may need irrigating until the flowering stage. After they begin to bloom, water may be gradually withheld. After the tops become yellowed, they should be gradually allowed to dry out completely.

Harvest

For summer use, potatoes may be dug any time after the blossoms form. Tiny new potatoes no more than an inch in diameter are delicious and contain a very large amount of vitamin C. A few whole plants may be carefully lifted with a fork or individual tubers may be dug, the gardener working carefully with his fingers to find them in a loose soil. If they are dug in this way, the plant may be left to produce the rest of its crop undisturbed.

For storage, potatoes should be completely mature, with skin that has become tough enough so that it cannot be rubbed off. The majority of the tops should have yellowed before this stage and the potatoes may be left in the soil for several weeks after the first frost. If the weather turns warm and wet, however, they should be dug immediately, to avoid the chance of starting a second growth.

Best time to harvest is when the soil is very dry. Tubers are lifted carefully with a fork and are allowed to dry for an hour or two on top of the soil. They should then be stored immediately in a cool dark cellar, frost-free and away from all light. Best temperature for storage is 36 to 40 degrees, and the atmosphere should be humid.

Growing Potatoes Above Ground

Many more gardeners now grow potatoes *above ground.* The reasons are simple—less work, better potatoes.

Veteran organic grower Alden Stahr of New Jersey reports that his plants had from 24 to 36 scab-free potatoes on them when grown as follows:

In April he prepares a patch by turning under the winter rye and allowing it to decompose before planting. Next he sticks a whole potato (it's generally agreed that cutting weakens the plants) in the furrow every 6 inches. A shovelful of compost was placed over each potato before covering with soil. Stahr's final step was to spread a foot of hay over the patch.

Comments Stahr: "The taste of our compost-and-mulch-grown potatoes is incomparably better than that of store potatoes and we grew them the lazy way."

You can also grow potatoes without furrows, without any digging at all. This above-ground method calls for placing the potatoes right on *top* of the previous year's mulch.

Housewife Lois Hebble of Decatur, Illinois, actually prepares her potato patch in fall when she selects a wide strip of her garden and piles leaves, weeds and other garden wastes there. By spring this material has packed down and is fairly decomposed. All she does then is lay the seed potatoes right on top, finally covering them with 12 to 14 inches of straw. If any tubers stick through, she adds more mulch.

Says Mrs. Hebble: "The first time I tried planting potatoes in leaves, I made the mistake of pulling the leaves back and sticking the potatoes down under them next to the soil. The spring rains were heavy and the leaves kept the potatoes too wet and some rotted. So that is why I now plant on top of the leaves and cover with a good heavy mulch at planting time. When it's time to dig the potatoes, I just pull the mulch aside and harvest.

For winter storage, potatoes (as Irish cobblers above) should be completely mature, with skin that has become tough enough so that it cannot be rubbed off.

They are the best-tasting, smoothest and largest I have ever grown. I might add that so far I have never seen a single potato bug on the potatoes I grow this way."

Professor Richard Clemence of Wellesley College in Massachusetts likes the above-ground method too, since he gets a heavy potato yield and can "inspect the vines from both sides occasionally, taking care of a rare potato bug or a bunch of eggs that the ladybugs have missed." Professor Clemence makes double rows, 14 inches apart on top of the old mulch, with the seed the same distance apart in the rows.

After laying the small whole potatoes in straight rows, he covers the rows with 6 or 8 inches of hay. When the blossoms fall, Clemence begins to harvest new potatoes. To do this early picking, he moves the hay carefully aside, separates some small potatoes an inch or two in diameter from the stems and gently replaces the hay. (He rates Irish Cobblers best to eat this way.)

Aside from saving much work and giving plants increased protection from insects and disease, the deep-mulch planting method also offers the gardener a good chance to rotate his potato crop wisely. Unless potato rows are rotated, plant food will be rapidly used up. Second and third year crops are more susceptible to insect attack.

Any of the legumes, such as peas or beans, are excellent to precede or follow potatoes; leaf vegetables or cucumbers that have received large applications of compost or manure are also recommended. (Potatoes should never follow tomatoes and vice versa.)

A workable plan is to select several rows of your garden in fall for potatoes the following spring. These rows can become your deep-mulch section—where you can pile leaves, weeds, etc. By changing this spot each year, you'll be building up your soil's fertility with these materials—a different section annually—and, of course, rotating the potato patch.

Tests have revealed that potatoes take in nutrients from the soil much faster in the early stages of growth than when the plant is older. Thus if the deep mulch is well decayed and other plant foods have been added to it, the young plants immediately have an available supply of food and growth is rapid. It's a good practice to apply compost and some well-rotted manure to the surface mulch in fall, so that the potato plants will be sure to have a rich, steady supply of nutrients.

If you don't have enough mulch material, try seeding a green manure crop on the part of your garden scheduled for potatoes. Soybeans are excellent to grow before potatoes. In fact, many expert potato growers recommend growing soybeans for several years as a way to "cleanse" soil that is heavily infested with scab. Scab, they discovered, often thrives in neutral or slightly alkaline soil and soybeans, turned under as green manure, tend to acidify the land. It has long been an accepted fact that soybeans and other green manures will help control scab by rapidly decaying and encouraging the development of beneficial soil organisms.

Incidentally, if you've found that the potatoes you've grown become soggy when cooked, a lack of soil potash may be the trouble. To correct it, add about a pound of greensand or granite dust to each 5 square feet of potato row.

Potatoes, Poison and Humus

Here's an experience from Vere L. Griffen of Key West, Florida on growing potatoes:

"In 1943 I took a notion to be a farmer.

Many gardeners grow potatoes above ground, placing compost over each potato before covering with soil and hay mulch.

I bought an 80-acre farm, intending to raise navy beans, but thought I better plant a couple acres of potatoes, as my father had raised them as his main cash crop when I was a boy, so I thought I knew how to raise 'taters.

"Certified seed was high priced, so I got seed from any farmer that had any for sale. They proved to be scabby and blighted, but I figured the formaldehyde treatment would kill all the germs anyway, so I cut off all bad pieces and cut up the best parts for seed. I soaked them in the recommended strength solution—twice as long as recommended—to be sure all the germs were dead. They got planted with 300 pounds per acre of 2-12-6 and grew well until fair-size tubers were forming and then, they blighted *bad*. In the meantime, they had to be sprayed several times to keep the Colorado potato beetle from eating the tops all off.

"The blighted and scabby parts I just tossed into an old straw stack bottom and covered them with some of the rotten straw to hide them.

"During the summer, I noticed there were some good dark-green potato vines in the old straw stack and there seemed to be no bugs on them. Occasionally, there would be a very few old bugs, but I never saw any young bugs. No spray was ever used on these.

"Well, after all the harvesting was done, one day I wondered if any potatoes had grown in that rotten straw, so I got a fork and uncovered the most perfect, smooth potatoes you ever saw. There was not a trace of blight or scab. I was dumbfounded, to say the least.

"Maybe this incident proves that enough organic material in the soil kills germs and somehow keeps bugs off better than disinfectants and poison sprays!"

PUMPKIN

PUMPKIN. *Cucurbita pepo*. A tropical annual vine of undetermined origin. Pumpkins were known to the Greeks and Romans, who considered them excessively large cucumbers. They were also grown by the Indians in their corn patches, according to Captain John Smith.

Pumpkins and squashes have become so crossed that it is difficult to tell which is which and many of the vegetables we considered squash in the garden or kitchen are actually pumpkins. For the sake of clarity here, we give directions for growing only the commonly recognized jack-o'-lantern fruits. Cushaws, summer squash and the others which are botanically pumpkins are treated under SQUASH.

The principal botanical differences between pumpkins and squashes are in their skins and stems. Pumpkins have soft rinds which never become very hard, even

Pumpkins grow on large, sprawling vines preferring light-textured soil containing large amounts of humus.

when mature and cured. Their stems are hard, woody and furrowed. True squash have hard rind and soft spongy stems.

Pumpkin is stronger in flavor than squash, and the coarser flesh makes it less popular as a vegetable. It is more often used strained in pies. It is very rich in vitamin A, one-half cup containing 2,500 units. The same amount yields .056 milligrams of vitamin B_1, .057 milligrams of vitamin B_2, 8 milligrams of vitamin C, 23 milligrams of calcium, 50 milligrams of phosphorus, .9 milligrams of iron, 1 gram of protein and 27 calories.

Pumpkins grow on large, sprawling vines. They are monoecious—that is, they have two different flowers, male and female, both on the same plant. Because of this, they must be insect-pollinated and beehives in the vicinity of the pumpkin patch will help to set a larger quantity of fruit.

Range and Soil

Pumpkins can be grown throughout the country wherever there are 4 months of comparatively hot weather. They like clear, cloudless skies and hot days to mature fruit, but do not need hot nights as do watermelons and cantaloupes.

Rich light soil with a large amount of humus is best for pumpkins. The very large exhibition fruits are grown in pure compost or spent mushroom manure. For family-sized fruit, a spadeful of rotted manure or compost in each hill will provide enough nutrients. Slightly acid to slightly alkaline soil is best, with a *p*H rating between 6.0 and 8.0.

Seed

One packet of seed will sow 5 or 6 hills, plenty for the ordinary family. One ounce will sow 25 hills. Seed germinates in 8 to 10 days. Fruit matures in about 4 months.

Two types of pumpkins are offered by seedmen. The Connecticut Field Group includes the largest varieties, some weighing as much as 200 pounds. They are very sweet and have a deep orange rind. The Cheese Group are smaller with creamy to yellow skins. Best variety for pies is Small Sugar.

Planting

In northern states where the growing season is short, pumpkin seed may be planted in individual pots in the greenhouse or hotbed and transferred to the garden when the weather is warm and the soil has warmed up. The plants dislike transplanting, so they should be treated with care not to disturb their roots.

In the garden, sow seeds 2 to 4 weeks after the last frost. In rows, sow one or two seeds per foot and thin to 2 to 3 feet apart, in rows spaced 6 feet apart. In hills, plant 2 or 3 seeds in each hill, and thin to one plant, spacing the hills 4 to 8 feet apart each way. Seeds should be planted one inch deep.

Like the Indians, many farmers plant their pumpkins among the corn to save space. This may be done in the home vegetable garden or the pumpkins may be planted along the southern edge of the garden and allowed to stray out over the grass, unless they will be in the way there. They take up so much space that they are not a profitable crop for a very small garden unless they can be doubled up with another planting.

Culture

Pumpkins produce best in mulch, because they require a lot of moisture to mature large fruit.

After enough fruit has been set to mature before frost, pinch back extra foliage growth and remove fruits set too late to mature. This will throw the nutrients into the previously set fruit.

Enemies

Bacterial wilt, anthracnose and downy mildew all attack pumpkins, but they do

Many organic gardeners, such as young Stephen Carsten, of Guilford, New York, have been surprised by the large pumpkins which "spring up" in the compost heap.

little damage. More trouble is caused by squash bug and squash vine borer. When a whole branch of the vine dies off while the rest of the vine is healthy, cut the stem where the wilt begins and look for a borer. He will be on his way back to the main stem. If he can be caught and killed before he comes out again, future generations will die with him.

Squash bugs are long-legged, grey, elongated pentagons which may be found and caught on the top sides of the leaves in the full hot sun of noon. In cooler parts of the day, they are more likely to be among the mulch or dead leaves below the vine and are harder to catch. Their eggs are brown clusters of dots which grow darker brown as they prepare, in a couple of days, to hatch. They should be crushed by hand and the mature bugs should be knocked into a can of kerosene. If the first few bugs can be caught in the first hot weather, many hundreds will be spared the planting later on.

Harvest

After the first frost, cut the stems about one inch from the pumpkins with a sharp knife and allow the fruit to cure for 2 or 3 weeks in the field. Then, handling it very carefully because of its tender skin, remove the fruit to a cool, dry cellar, where the temperature is about 50 degrees. Pumpkins do not last as long in storage as squash, but they will last for several months. Or they may be steamed and the strained pulp canned or frozen.

RADISH

Radish. *Raphanus sativus.* A member of the mustard family, *Cruciferae,* and cousin to all the cabbage tribe, the radish has a widely diversified and long history. Its origin may have been anywhere from Japan to the Mediterranean. Present forms of the radish are not directly traceable to any one ancestor, though wild radish, *R. raphanistrum,* is widely spread through Europe and northeastern North America.

Varieties of radishes have been grown for many centuries in China and Japan. They were also cultivated in ancient Egypt, Greece and Rome. Every part of the radish plant has been used at one time or another and some varieties without the familiar small red root are still grown in some countries. One variety is grown in China for the oil in its seed. Egyptians grow a radish for its green tops. Another variety, the rat-tail, has edible seed pods which are pickled. Some varieties have roots which grow enormous and are cooked like turnips. In Japan, almost every dish contains or is served with some type of radish.

In this country 3 types are grown for their roots. Spring radishes are the small red globes, those grown in summer are white and long and the fall radishes may be black or white, long or round.

More useful as an appetite stimulant than for nutriment, radishes actually supply only a small amount of vitamin C and some minerals. In 15 large roots weighing a total of 100 grams there are .030 milligrams vitamin B_1, .054 milligrams vitamin B_2, 25 milligrams vitamin C, 21 milligrams calcium, 29 milligrams phosphorus, .9 milligrams iron, 1 gram protein and 22 calories.

Range and Soil

Radishes may be grown anywhere in the United States, if planted at the right time. They prefer cool weather, though some varieties have been developed to withstand summer heat. In the South, they are grown almost solely as a winter crop. In the North, they are planted from earliest spring until late spring, then again after midsummer for fall.

Most varieties of radishes must be grown rapidly for best quality. At the same time, their feeder roots are only 2 to 8 inches long. Consequently their moisture and nutrients must be concentrated in a rich band near the row.

A large percentage of humus in the soil will grow the best radishes. Rich sandy loam is best for early varieties; rich clay loam is best for late ones. Soil *p*H should be 6. to 8. Soil may be enriched with well-rotted manure, never fresh manure, or compost.

Seed

One packet of seed will sow 20 to 30 feet of row, depending on variety. One ounce will sow about 100 feet. Early varieties mature in 20 to 30 days; summer varieties in 35 to 40 days; fall or winter varieties in 55 to 60 days.

Best seed is large. To be sure of growing good roots, sift the seed before planting and discard the smaller seeds.

Most popular early varieties are Scarlet Globe and Cherry Belle. White Icicle is the best summer variety. For fall, Long Black Spanish, Round Black Spanish or White Chinese or Celestial are preferred.

Planting

Early radishes are frequently sown in the rows with carrots, parsnips or parsley to mark the rows until the slower growing *Umbelliferae* come up. Treated this way, they will supply some early roots, but the crop will all mature at about the same time, since they are all planted at about the same season, unless different types are planted.

Radishes are very hardy and a few may be planted as early as the soil can be worked. The usual method is to plant early, and to make successive sowings of about 5 to 8 feet of row every 10 days until the weather begins to warm up. Then the variety is changed to summer radishes, which are sown every 10 days until the weather begins to get hot. Fall radishes are planted after the hottest of the summer weather is past and will mature in 60 days, which may be partly after the first frost.

Make a drill with a corner of the hoe, about 1½ inches deep. Dribble an inch of sifted compost into the drill and sow the seeds, 15 to the inch, on the compost. Cover with another half-inch of compost. Water well and keep moist continually until mature. Early radishes should be sown thinly enough so that they will not need to be thinned. Fall radishes, especially the large varieties, should be thinned to stand 4 to 6 inches apart. Rows for early varieties may be as close together as 4 inches; late rows should stand 12 inches apart.

Culture

Rapidly grown roots are firm, crisp and mild in flavor. Slow-growing roots are woody and tough. Radishes planted too late, when the days get long, go to seed quickly.

To grow good early radishes, water often and mulch just as soon as the leaves are a couple of inches tall. Even a thin mulch around the short foliage will help keep the soil cool.

Enemies

If the roots are grown fast enough, they do not suffer from pests to any great extent. If allowed to dry out, the leaves may be troubled with leaf hoppers. Root maggot, which also attacks others of the cabbage family, sometimes bores into a few roots of late radishes. Wood ashes raked in along the row helps to discourage them, but best insurance against maggot is to grow radishes in soil that has not had a member of the cabbage family in 3 years.

Harvest

Early and summer radishes should be pulled promptly when they attain their full growth. A few days past their prime may spoil the crop.

Winter radishes may be left in the ground until after frost, then dug and stored in a root pit or in damp sand in a cool cellar. They will keep through the winter if properly stored.

RHUBARB

RHUBARB. *Rheum rhaponticum.* Also called pieplant. A cool-climate perennial of the *Polygonaceae,* or buckwheat, family. Native to southern Siberia and the Volga region, rhubarb had been cultivated in the Asiatic countries for many centuries before it was introduced to Europe in about 1600. Varieties were listed by horticulturists in America by 1800.

Rhubarb's large red-veined leaves spring from a crown several inches below the soil level, from which a deep network of roots penetrate several feet into the subsoil, some of them enlarged to store food for the early spring growth of stalks. The plants sometimes make flowers, but if seed develops at all, it is not to be trusted to propagate plants true to the parent stock. Edible portions are the thick reddish leaf stalks. The leaves themselves must never be eaten. They contain amounts of oxalic acid and its salts which may be fatally poisonous.

Rhubarb is a good source of vitamin A and contains a lot of calcium, most of which is not available for nutritional purposes, because it is combined with the oxalic acid in insoluble compounds. In one-half cup of rhubarb, a 100 gram portion, there are 650 units of vitamin A, .024 milligrams of vitamin B_1, .024 milligrams of vitamin B_2, 12 milligrams of vitamin C, 48 milligrams of calcium, 18 milligrams of phosphorus, .5 milligrams of iron, 1 gram of protein and 20 calories.

Range and Soil

Rhubarb is one of the few vegetables which can be grown in our northernmost state, Alaska, but cannot be successfully grown in the South in the Gulf states. It is most productive in the northern third of the United States. In the South, if it grows at all, the stalks are likely to be entirely green, instead of pinkish or red, except at elevations where the temperature range is cool. Best growth is where the summer mean temperature

Rhubarb grows best in the northern third of the United States, its large, red-veined leaves springing from a crown several inches below the soil level.

does not rise above 75 degrees or the winter mean temperature above 40 degrees. It needs a good cold period of dormancy, preferably at freezing temperatures. Special varieties needing less dormancy are being developed for parts of California.

Sand, peat or clay will grow rhubarb as long as they are well drained. Best soil for forcing is a heavy loam which will cling to the roots when they are dug. For early rhubarb, sandy loam is best. But whatever the soil, it must be well supplied with humus and nutrients. It is almost impossible to give rhubarb too much fertilizer or to dig the fertilizer in too deeply. Soil is best if slightly acid—*p*H 5.5 to 6.5.

Seed

Some seedmen offer rhubarb seed, but since it is not always to be trusted to come true to type and since it takes much longer to develop plants to the production point, it is advisable to plant roots instead.

Most popular variety through most of the country is MacDonald, followed by Chipman's Canada Red, Valentine and Victoria. Cherry or Giant Cherry are recommended for growing in California and may be tried in other states where the temperatures are too high in winter for a complete freeze-up.

To provide root cuttings, the crowns are dug and divided into as many pieces as will give one or more good eyes each. If the crowns have roots attached when they are planted, they will grow into producing plants more quickly than if the roots are stripped. One or two roots per family member should provide enough rhubarb. Yield is about 10 to 20 stalks per plant.

Planting

Starting new roots or dividing old ones may be done in early spring in the northern areas. In regions where there are long cool falls, the job should be done in fall.

Cuttings should be planted at an end of the garden devoted to perennials, such as asparagus, winter onions and perennial herbs. The bed, when established, will last for as much as 10 years. Commercial plantings are dug and renewed every 4 years, but in the home garden if enough nutrients and mulch are supplied, the plants should bear well for much longer periods. Only when the stalks become small and crowded should the plants be dug and divided.

Roots are usually planted 3 feet apart in rows 4 feet apart. To prepare the soil, dig out a hole deep enough to hold a bushel basket. Remove the topsoil and reserve; take the subsoil out and discard it. In the bottom of the hole, place one-half bushel of manure. Over that replace the topsoil mixed with enough compost to fill the hole to ground level. Place the root in the center of the prepared spot, covered with 3 inches of soil. Firm the soil well over the newly planted root.

When dividing old clumps, dig the soil away carefully from over the fleshy crown. Cut through the crown, leaving half where the old plant stood, its roots undisturbed. The other portion, with as many roots attached as possible, should be lifted and, if it has many eyes, divided to provide one or more eyes per piece.

Culture

Thick stalks result from continuous heavy feeding of rhubarb. In order to keep the soil up to the rich standard with which the plants were started, a thick mulch of strawy manure should be spread over the bed each winter after the ground freezes. Winter rains and melting snows will leach much of the nutrient out of the manure directly to the plant roots. In spring the residue should be raked aside to uncover the plants to the sun, so that the soil may warm quickly to start an early growth. After stalks have developed, the manure, with an addition of straw mulch, may be pulled close up around the plant, to preserve the soil moisture. Plants should not be allowed to dry out, even when the bearing period is past. They need a continuous supply of moisture to enable them to store up nutrients for their early growth next spring. They should be watered regularly during periods of drought.

Flower stalks should be removed as soon as they develop. They are unlikely to develop seeds, but if they do, they will take more of the plant's vigor than the dubious offspring are worth.

Enemies

Rhubarb is usually healthy, especially when fed and watered sufficiently. However, it may be a victim of foot rot, which causes stalks to rot at the base and the leaves to topple over. The only treatment for this is to remove affected plants immediately and to expose the soil in which they were grown to the full rays of the sun for a few weeks.

Most widespread insect pest is rhubarb curculio, a rusty snout beetle about three-quarters of an inch long. He bores into stalks, crowns and roots and also makes forays on wild dock which may be growing nearby. He may be seen easily and hand-picked from the rhubarb. All wild dock should be dug out and burned in July, when the beetle will have laid its eggs in it.

Harvest

The first year after rhubarb is planted it should be allowed to keep all its stalks. Harvest should not begin until the plants have had a full year, or a year plus a winter, in the ground.

For an early harvest of rhubarb, force plants by covering in early spring with bushel baskets and mound with manure and soil.

To harvest, pull the stalks, do not cut them. Grasp the stalk near its base and twist away from the plant. It will separate easily from the crown. Harvest only the large stalks—do not take thin or short ones. The stalk, not including the leaf, should be at least 10 inches long and one inch thick. Shorter stalked leaves and thin stalks may be left to feed the roots for the next year.

Harvest may continue for about two months or until the peas are ready to pick. After that time, the stalks will become inferior in texture and perhaps hollow.

For an early harvest, the plants may be covered in very early spring with bushel baskets which are mounded over with fresh manure. The heat from the manure will raise the temperature of the soil and stalks will be forced early under their warm cover.

Forcing

Crowns may be lifted from the garden and brought indoors to force, in order to supply the family with fresh tender rhubarb in midwinter.

Two-year-old crowns are best for forcing. The stalks should not be harvested in spring from plants that will be lifted for forcing in fall. After frost has killed their leaves, the crowns are lifted with as much soil as will cling to the roots. Only large strong crowns should be used. Plants are placed in tubs or boxes and soil packed around them, allowing a square foot of soil surface for each plant. The boxes are well-watered and left in the open or in unheated sheds until the soil in the boxes is well frozen. They should have at least 6 weeks and may be permitted as much more as there is time for, before being brought indoors.

Boxes are then brought into a dark cellar where the temperature is 55 to 60 degrees. In a short time, small leaves will begin to form and stalks which are pleasantly pink and tender will grow. Humidity in the vicinity of the forced plants should be kept high, if necessary with tubs of water standing near them. Boxes of roots should be kept moist, but not wet, after they are thawed. Yield from forced roots is usually from 2 to 6 pounds per plant.

RICE

RICE. *Oryza sativa,* one of the members of the grass family, *Graminae.* Thought to be of Asiatic origin, rice has been cultivated in China since 3000 B.C. It was well after explorers had traveled around the world that it was introduced to Italian gardens, but some of the earliest settlers in this country planted rice as early as 1647. North and South Carolina for many years grew some of the finest rice in the world, but their production has recently declined, while that in Louisiana has increased to supply their market.

Rice grows to a height of 3 to 4 feet, normally with yellow or brown seed, though there are a few red varieties. White rice is the brown product with the vitamin-rich outer coating removed by polishing. Converted rice is brown rice which has been treated by a steaming process, during which the water-soluble vitamins flow throughout the grain, so that when the brown husk is later removed there is less vitamin loss than there is in ordinary polished rice. Rice in the husk, as it is harvested and planted, is called "paddy."

Three-quarters of a cup of cooked brown rice contains 20 units of vitamin A, .190 milligrams vitamin B_1, .075 milligrams vitamin B_2, 22 milligrams calcium, 112 milligrams phosphorus, 1.6 milligrams iron, 4 grams protein and 117 calories. The same quantity of cooked polished rice contains no vitamin A or B vitamins, only 3 milligrams calcium, 33 milligrams phosphorus, .2 milligrams iron, 2 grams protein, but still the same 117 calories, mostly pure starch.

Range and Soil

Rice needs a long, hot growing season. It can be grown only in the South on level ground which may be flooded for a large part of the summer either by a diverted stream or by water from a drilled well. Soil well-enriched with manure is best, with a *p*H near neutral, about 6. to 7.

Seed

An American variety called "Gold Seed" is thought to be superior to any grown elsewhere. Other varieties are Honduras and Japanese.

Culture

After the field is drained in fall, it is fertilized and prepared for spring planting. In March or April, seed is broadcast and then harrowed in like oats.

When plants are up a few inches, water is gradually returned to the field until, when they have reached their full growth, the water is 8 inches deep. Best irrigation is with slowly moving, not stagnant water. When the rice is ready to harvest, the water is drained from the field and harvest proceeds as for any other grain.

RICE, WILD

RICE, WILD. *Zizania aquatica.* A grass of entirely different origin and genus from ordinary cultivated rice. Also called Indian, water or Canada rice. It is native to northern America and eastern Asia. Single stems of wild rice grow to 10 feet in height, the blossom being a panicle two feet long. The seed is about one-half inch long, dark grey or purplish, enclosed in a husk an inch long. Wild rice is a self-seeding annual which will reappear year after year, once a bed is started under the right growing conditions. Nutritive value is about the same as whole wheat, that is, high in the B vitamins, protein and with some oil. Because of the difficulty of harvesting it, wild rice has become an expensive delicacy in this country, used principally as a stuffing or accompaniment of wild fowl. It was originally used as food by the Indians of the Great Lakes and Mississippi valley region. In China, where it is cultivated, the solid base of the stems is considered a choice vegetable.

Range and Soil

In the United States, wild rice grows from the Delaware River north in the east, through the Great Lakes region, down the Mississippi as far as Louisiana and in the Northwest. It could be grown in almost any portion of the United States under the proper planting conditions. It prefers quiet but pure water 5 inches to 3 feet deep with a current of not more than one mile per hour. The water level must not fluctuate more than 6 inches as tidal backwater or flood, otherwise the seed will wash out. Acid water or run-off from chemical fertilizers can kill it. Best *p*H in the water is 6. to 7. It may be planted along stream edges, ponds, flood plains of rivers or in marshes where water constantly stands. It makes a handsome decorative plant along landscaped waterways and attracts wild birds.

Seed

Hulled seed sold in stores for cooking will not germinate. Seed must be purchased in hulls, preferably moist, and should be kept moist until planted. If dry, it should be soaked in a sack 24 hours before planting, so that it will be water-logged enough to sink when broadcast.

Seed purchased dry may not sprout until the second year after planting. One seed may produce several stems, but the yield depends upon wind conditions when it is ripe, the number of birds around and the method of harvesting. It is usually not possible to harvest more than half the crop, the other half planting itself to grow the following year.

Planting

Scatter wet seed at the rate of a bushel per acre or a large handful on a plot 6 by 6 feet. It will sink quickly if wet enough. The best time to plant is just before the ice forms in winter or just after it breaks up in spring. If planted too early, it may be eaten by migrating wild fowl.

Culture

Wild rice has been getting along by itself for a long time and will do it again wherever you plant it. If it likes the surroundings, it will grow well and bear seed. In the winter, the ice will freeze and lift the old plants out of their moorings, allowing them to drop back as a kind of underwater mulch when the thaws come. Any seed left in the heads will plant itself and grow the following spring.

Harvest

Indians in the Great Lakes region harvest wild rice by running their canoes into a planting, bending the stalks over the canoe and beating the seed out of the heads. Seed ripens over a period of about 10 days in September in that region, so each portion of the rice bed must be visited several times in the 10 days.

After the rice is harvested, it is still incased in its hulls. It is dried for several days in the sun, then heated gently in a metal tub over an open fire, stirred like popcorn, until the hulls are dry and brittle. Threshing removes the hulls, but leaves them mixed with the grain. The threshing is done either by flailing the kernels or by tramping on them. Then it is winnowed by shaking or pouring it from one vessel to another in a wind, which will blow away the chaff and leave the kernels.

ROCKET

ROCKET. *See* CRESS.

ROMAINE

ROMAINE. *See* LETTUCE.

RUTABAGA

RUTABAGA. *Brassica napobrassica,* or *B. campestris.* Also called Swede or Swedish turnip and winter turnip. A hardy biennial herb native to the Scandinavian peninsula, Russia and Siberia. It is thought that it may have developed from a cross between rape and turnip. It is a member of the cabbage family somewhat similar to kohlrabi, but its root, rather than its stem, is enlarged. It is similar to but hardier than turnip and takes 4 to 6 weeks longer to mature.

Rutabagas quickly develop an extensive and finely branched root system. At 3 weeks, rootlets go down to a depth of two feet; at 41 days they are 3 feet below the surface. Leaves are smooth and blue-green, like typical cabbage family leaves, rather than hairy like turnips. Also, rutabagas have more of a neck from which the leaves spring than does turnip.

Rutabagas are slightly higher in nutritional value than turnips. In a 100 gram portion, about three-quarters of a cup, there are 25 units vitamin A, .075 milligrams vitamin B_1, .120 milligrams vitamin B_2, 26 milligrams vitamin C, 74 milligrams calcium, 56 milligrams phosphorus, .7 milligrams iron, 1 gram protein and 36 calories.

Range and Soil

Rutabagas can be grown throughout the country, but they are not a good crop for the South, where turnips grow much better. They may be grown in the southern states in fall, winter and spring, but they really thrive only north of a line drawn from New York City to Chicago, and directly west from there. In this section, they may be grown in early spring, but are usually planted in early summer and grown through the fall as a storage crop, for which they are excellent. They are hardy to frost, but cannot stand hard freezing.

Well-drained humus loam is best for rutabagas, especially with a clay subsoil. Heavy clay in the topsoil can cause branching of the roots. Soil may be slightly acid, with a *p*H of 5.5 to 7. For a spring crop, which must be grown rapidly before the hot weather arrives, the soil should be well-enriched with manure which is turned under in fall, so that the nutrients will leach deep into the soil. A fall crop needs less fertilizer—in fact, if the soil has been enriched for a spring crop grown in the same rows, it will need no further fertilizer when the rutabagas are planted in late June or early July. Rutabagas should not be planted in soil that has had a crop of any member of the cabbage family, including radishes, in the past 3 years.

Seed

One packet of seed sows 50 feet of row, and should yield about one bushel of rutabagas. The seed is fine and may need to be mixed with sand if it is to be sown thinly. Germination is quick and average time to maturity is 90 days from planting. Most popular varieties are American Purple Top, Improved Long Island, Alta Sweet, Laurentian and Macomber.

Planting

Seed may be planted from one-fourth inch to one inch deep, depending upon how dry the soil is at planting time. In very dry weather, flood the row after opening the drill and plant more deeply. In spring, for the early crop, plant less deeply. Rows should be spaced 18 to 24 inches apart, and plants should be thinned when they are 4 inches high to stand 8 to 12 inches apart in the rows. Plant early rutabagas as early as the ground can be opened; late ones 90 days before the first fall frost.

Culture

Because of their very deep root system, rutabagas do not suffer from drought unless the summer is extremely dry. A straw mulch between rows is all that they require in moderately dry weather.

Enemies

Pests and diseases of rutabagas are much the same as the ones which attack the other members of the family. (*See* BROCCOLI, CABBAGE.)

Harvest

Pull rutabagas after the first frost and store in a root cellar or root pit, where the temperature is between 32 and 40 degrees. They will not keep if they are frozen. If they are stored in a dry cellar, bury them in moist sand.

SAFFLOWER

SAFFLOWER. *Carthamus tinctorius.* Also called false saffron. A coarse annual member of the daisy family, *Compositae,* with seeds like miniature sunflower seeds. Some varieties are spiny. It is cultivated in India, China and other parts of Asia for the oil extracted from the seeds which is used there in cooking and for oil lamps. In Spain and the Levant it is grown for its orange flowers which yield a drug and a dye used for food coloring and for rouge making. It is grown in this country in the western part of the Plains states principally for the oil which has industrial uses. A secondary product is the meal left after extracting oil from seeds, a protein supplement fed to lambs and beef cattle to fatten them. Average oil yield is 22 to 36 per cent of the seed, by weight.

Recent investigations have discovered that safflower oil contains large amounts of linoleic acid, a vitally important fatty acid and it is also a good source of vitamin E.

Planting and Culture

Safflower needs a growing season of 120 days, but is quite hardy and may be planted early in spring. Seed is sown one inch deep in rows 20 to 24 inches apart, at the rate of 15 to 30 pounds per acre. It is harvested when 98 to 100 per cent of the heads are dry and thresh easily by hand.

SALSIFY

SALSIFY. *Tragopogon porrifolius.* Also called oyster plant and vegetable oyster. A biennial root-vegetable native in the Mediterranean area. Related species were grown by the Romans, but this one was not cultivated until the Middle Ages. Spanish oyster plant, *Scolymus hispanicus,* which is very similar, has an older history in Europe, but is not much grown in the United States. Black salsify, *Scorzonera hispanica,* is the central European equivalent, with a black-skinned somewhat slimy root and leaves which are used for salad. It is a perennial, but difficult to establish.

The salsify grown here resembles parsnips, but the white taproots are more slender and their flavor, when well-developed by freezing, resembles that of oysters. Roots grow to 1½ inches in diameter and 8 inches or more in length. They are found very seldom in the market, probably because the roots shrivel quickly when they are out of the ground for any length of time. Flower stalks which develop the second year may be cut and cooked like asparagus if they are harvested before they become hard. Like parsnips, salsify is useful because it may be left in the ground all winter and dug as needed or the final harvest may be made in early spring.

Salsify does not supply many vitamins, having only 7 milligrams of vitamin C in 100 grams and no vitamin A or vitamin B, but it does contain a fair to large amount of minerals. In two roots, an average portion, there are 60 milligrams of calcium, 53 milligrams of phosphorus, 1.2 milligrams of iron and 3 grams of protein out of 78 calories.

Range and Soil

At least 120 growing days are needed to form a root of edible thickness, so salsify cannot be grown where the seasons are very short. But being hardy, it may be planted before the frosts are over in the spring, and will continue growing on into the fall, if days are warm. Best roots come from a deep, light, rich, mellow soil, with a *p*H near neutral, about 6. to 8. Manure should not be used unless it is very old, because it may cause pronging of the roots. Well-rotted compost, dug in to a depth of one foot, is best. Where seasons are no longer than the minimum, dig a trench a foot deep and a foot wide and fill it with compost.

Seed

Salsify seed does not remain viable long, so it should be purchased fresh each year. Sow it thickly for best results. It is better to thin the plants than to have to plant over again because the seed did not germinate well enough to fill out the row. One ounce of seed will plant 50 feet, one packet 10 to 20. Fifty feet of row should supply the average family. Most popular variety is Mammoth Sandwich Island.

Planting

Sow as early in spring as the soil can be prepared. Seed should be planted one inch deep in rows 12 to 15 inches apart. When the plants are 3 inches high, thin to stand 4 inches apart.

Harvest

Heavy frost or freezing of the roots improves the flavor of salsify. Dig as needed as long as the ground is open. If roots will be wanted for winter meals, dig up enough to last through the coldest months just before the ground freezes and store them in an outdoor root pit or in a box of damp sand in a cool cellar. Mulch the roots left in the garden, so that they will not alternately freeze and thaw during the winter. Dig in spring when the soil is thawed and store the remaining roots in a dark cellar before they sprout.

SCALLION

SCALLION. A popular name for bunching onions, onions pulled before they have developed bulbs and eaten raw. The name is a corruption of the Latin name of the species of shallot, *Allium ascalonicum*. The name is also applied to shallots, young leeks and hardy white onions grown as perennials. Special varieties of hardy onions which do not form bulbs have been developed and the culture of this type of scallion is described below. (*See also* LEEK, ONION, SHALLOT.)

Range and Soil

Scallions, like onions, can be grown throughout the United States. They like a cool season for developing. Soil should be well-enriched with mature compost, because the roots are shallow and they need abundant moisture in the top layer of soil. Optimum *p*H is slightly acid, from 5.6 to 7. Since the plants will remain in the ground through the winter, rows of scallions should be planted near the perennial vegetables.

Seed

One packet of seed will sow a 25 foot row. Yield varies, depending upon spacing and culture. Scallions may be pulled in 60 days from seed. Hardy White Bunching, Japanese Bunching and Long White Bunching are popular varieties.

Planting

Fresh seed should be planted thickly and covered with one-half inch of sifted compost. Onion seed does not remain viable long and should be purchased fresh for each planting. Rows may be 12 to 18 inches apart. If they are given enough room, some of the varieties will form several stalks per plant. Seed may be planted very early in spring for a crop which will mature in late spring or it may be planted about the time of the first frost to germinate and form roots in fall. A winter protection of straw and leaves should be given the fall planting, until very early spring. Seed planted in fall will provide the first scallions of spring.

Harvesting

Scallions are pulled when they are 8 to 10 inches high and half an inch thick. The outer leaves are peeled off and the inner leaves with the white stems are used.

SEA KALE

SEA KALE. *Crambe maritima*. Also called scurvy grass. A seacoast perennial native in Europe along the sandy shores of the North Sea, the Atlantic and the Mediterranean. Sea kale was known to the Romans, who carried barrels of it, preserved, on long sea voyages. It was known but apparently not cultivated in England until the eighteenth century, well after the colonists had begun to grow it in America.

Sea kale is a vigorous perennial vegetable much like, in cultural needs, rhubarb. It sends up leaf shoots early in spring which may be blanched and eaten like asparagus. Later in the season, the leaves, blue-green and cabbage-like in appearance, make food which is stored in the roots for the following spring's crop. The white flower cluster grows 3 feet high.

Range and Soil

Sea kale can probably be grown throughout the United States, but does especially well in cool or temperate sections which duplicate its seashore origin. It sends its roots deep, so it needs a deeply prepared soil with manure or other moisture-holding humus dug in to a depth of two feet. It is a heavy feeder with needs similar to RHUBARB, which see.

Seed

Plants may be started from seed, but they will not yield a crop until their third year. Root cuttings, if obtainable, will yield a crop the spring after they are planted.

Planting

Sea kale should be planted in the perennial section of the vegetable garden, where it will yield a crop each spring for 6 or more years if it is not disturbed. Plants should be spaced 3 to 4 feet apart in a rich bed.

Culture

As soon as the first shoots appear in spring, the plants should be mounded over with soil to a depth of about a foot or should be covered with light-proof boxes or tubs. The shoots must be blanched, even though they lose vitamins thereby, or they will become tough.

Harvest

Cut shoots when they reach a height of 12 inches. Cutting may continue until the leaves begin to open, when they should be allowed to grow to full size.

Blanching materials should be removed at this time and the plants should be encouraged to make their full growth.

Leaves are not cut down until late fall. After the soil has frozen, they should be mulched with a deep layer of strawy manure.

SHALLOT

SHALLOT. *Allium ascalonicum.* A kind of multiplier onion which does not occur wild in nature and is thought to have been a variant of *A. cepa,* the common onion. Shallots are mild in flavor, though they resemble garlic in every other way. They develop greyish pointed, angular bulbs which are joined at the base but are not enclosed in a sheath like cloves of garlic. They are often used green in spring and summer as very mild-flavored scallions or they may be allowed to mature their bulbs for winter use.

Range and Soil

Shallots will grow anywhere in the United States, but are grown in the South as a fall-planted winter crop, while they are grown in the North either as a perennial, with plants left in the ground from year to year, or from sets planted in spring and lifted in fall.

Seed

Shallots are usually planted from sets, divisions of the loosely joined bulb cluster. One clove planted in spring will make a bulb cluster by fall. Best variety is the Louisiana shallot, though seedmen usually offer only Jersey or False shallot. Small onions are sometimes offered as shallots, but the difference may easily be seen in the color and shape.

Planting and Culture

Shallots should be treated like onions grown from sets. (*See* ONIONS.) Soil requirements are the same as for onions. Although the sets may be planted in the perennial vegetable garden and left from year to year, since they are completely hardy, they will do best if lifted when mature and replanted the following year.

Harvest

Young shallots may be pulled while green and used like scallions, top as well as bulb being eaten. In fall before the first frost, the mature bulbs are pulled and allowed to dry. Tops are then removed and the bulbs are stored in well-ventilated baskets in a cool cellar. They keep like onions and may be held until the following spring.

SORREL

SORREL. *Rumex acetosa,* sorrel or sour dock; also *R. scutatus,* known as garden sorrel, a more acid plant. Members of the buckwheat family, *Polygonaceae,* the sorrels were at one time popular in Europe, where they are native. They reached the height of their popularity in Elizabethan England, but are grown now mainly in French gardens. Wild dock, a first cousin, is a naturalized garden weed in this country.

Sorrel is a perennial cultivated more in the South than in the North, though it is quite hardy. It has long thin bright green leaves with a bittersweet flavor which are used in salads or are cooked like spinach. Flower stalks in some varieties grow to 3 feet.

Culture

Sorrel should be planted in the perennial vegetable garden, in rich soil with a *p*H of 6. to 7. One ounce of seed will sow 100 feet of row. Two kinds are offered by the seed dealers, a broad-leaved and a narrow-leaved.

Sow the seed one-half inch deep in rows 12 to 18 inches apart and thin the plants to stand 3 inches apart. Leaves may be picked for cooking or salad within 60 days from planting. Flower stems should be cut off before they develop, otherwise the leaves become tough and acid. Plants will bear for three or more years, but clumps should be divided and reset every 4 years.

SOYBEAN

SOYBEAN. *Glycine max* or *G. soja*. Native to tropical Asia, soybeans have been cultivated there since before 2800 B.C. They are widely used through all of India, China and Japan, but were not known in the Western countries until Perry's expedition brought back seeds in 1854. Since that time their use has gradually spread and only recently have we begun to realize the many uses to which soybeans can be put.

Two types of soybeans are now cultivated in this country, out of the hundred or more varieties of the Orient. Field soybeans are used to make hay, for seed, for green manure and in combination with sorghum, for silage. The seed is a rich source of oil which has industrial as well as culinary uses. Vegetable soybeans are used green like peas or lima beans; dried, like navy beans, or are sprouted like Mung beans. Soybean protein is more complete than the protein found in any other vegetable. Soy flour is a valuable protein additive to baked products and is used for diabetics because it is low in carbohydrates. Soy milk and cheese are useful for babies and children suffering from cow's milk allergies. Soy meal, like cottonseed meal, is a rich source of nitrogen when used as a fertilizer.

The vegetable soybean plant is a sprawling, bush legume similar to bush snap beans. Above, Wilbert Walker, a Pennsylvania organic gardener, is shown in his garden soybean patch.

Soy flour is richer in B vitamins than any flour except peanut flour and wheat germ. In one cup weighing 113 grams, there are 37 grams of protein, 200 milligrams of calcium, 7.4 milligrams of iron, .65 milligrams of vitamin B_1, .37 milligrams of vitamin B_2 and 8.2 milligrams of niacin. Dried cooked soybeans are also rich in the B vitamins, containing, in one-half cup, .525 milligrams vitamin B_1, .300 milligrams vitamin B_2 in addition to 10 units of vitamin A, 104 milligrams calcium, 300 milligrams phosphorus, 4 milligrams iron, 20 grams protein and a total of 108 calories. When soybeans are sprouted, they develop a supply of vitamin C which increases, rather than decreases, upon refrigeration.

The vegetable soybean plant is a sprawling bush legume similar to bush snap bean or lima bean. It grows taller the longer it takes to mature and sprawls over more ground. Its root system is shallow compared to some of the other green manure crops such as clover and it will return less nitrogen to the soil than they will unless the whole plant is plowed under. It draws heavily upon the soil's supply of lime, but will return nitrogen if the residue of the vine is left in the field and turned under or used as mulch.

Range and Soil

Being a tropical native, the soybean does not do too well in the North of the United States, except in isolated areas. It needs at least 100 days of warm growing weather to mature a crop. In the Corn Belt (that is, in conditions similar to those in Iowa) a good crop of early and late vegetable soybeans may be grown. North of that latitude, it is advisable to plant only the early varieties.

Soil required for growing soybeans should be about the same as for snap beans, except that more lime is needed. Any good garden soil, provided it has some humus and a *p*H of 6. to 7. will grow soybeans. If fertilizer is added, it should be richer in phosphorus and potash than in nitrogen. The following is recommended, spread at the rate of 3 pounds per 100 feet of row: one part tankage, one part wood ashes, one part granite dust.

Seed

One packet of seed will plant 25 feet of row; one pound 150 feet. Yield is about 30 pounds per 100 feet, of which one-half is beans, one-half pods. The yield is about the same as peas, better than lima beans.

Tests at Iowa State College have shown that the best way to plant soybeans is to plant several varieties, maturing in different periods, at the same time. Judging by gustatory as well as horticultural standards, Iowa State recommends Sac, maturing in 9 weeks, Kanro, maturing in 10 weeks and Jogun, maturing in 12 weeks.

The protein in soybeans is considered more "complete" than the protein found in any other vegetable. Soy flour and soy milk are valuable dietary products.

These may be expected to supply green beans for the table during the late summer.

Planting

Soybeans are somewhat more hardy than snap beans and may be planted a little earlier. The best rule is to plant when apple trees are in full bloom. Seed should be inoculated with nitrogen-fixing bacteria before planting.

Plant seed one inch deep, early types with two feet between rows, late types in rows 3 feet apart. Thin plants to 3 to 4 inches apart.

If soybeans are to be planted for green manure, a late-maturing variety should be grown because it will make more top. Seed should be planted solid and the crop turned under when seeds are half-formed. A few weeks should be allowed to elapse to permit the beans to rot before the following crop is planted.

Enemies

Soybeans are not bothered by Mexican bean beetle and will be a welcome addition to any garden in an area where the bean beetle infestation is bad. They are not bothered by other bugs, either, but they are one of the favorite foods of rabbits. In the country, it is wise to protect them with fencing or wire covers; otherwise the rabbits may reap the crop.

Brown spot, bacterial blight and downy mildew are principal diseases which attack soybeans. Best protection against their recurrence is to turn under the residues of the vines deep enough so that the spores cannot spread and to rotate soybeans with crops from other families.

Harvest

Green beans should be picked when the pods are two-thirds to fully mature, but before they start to turn yellow. The period for their harvest is short, lasting only a week or 10 days. To shell the beans, which are rather difficult to remove from their pods, parboil them for 5 minutes in the pods. Parboil a few at a time, so that they will still be hot, but the heat will not penetrate to the seed when they are shelled.

Dry soybeans must be allowed to dry on the vines, but must be picked immediately after the beans are dry and while the stems are still green, otherwise the shells will shatter and the seed be lost.

For sprouting soybeans, see BEAN SPROUTS, under the description on BEANS.

SPINACH

SPINACH. *Spinacia oleracea.* A member of the goosefoot family, *Chenopodiaceae,* which also includes beets and Swiss chard. Spinach is almost universally cultivated. It was known in China in the eighth century and was later introduced to Europe through Spain, probably in the

For quick growth, spinach should have a fertile loam soil, with a *p*H between 6. and 6.7. Bloomsdale variety is shown above.

twelfth century, when it is mentioned as being used in the diet of monks on fast days.

Properly prepared, spinach is rich in vitamins and minerals, though not the source of calcium that it was once believed to be. It contains a good quantity of calcium, 78 milligrams in 100 grams to be exact, but the oxalic acid which it also contains joins the calcium in cooking to form calcium oxalate, which cannot be absorbed by the digestive system. In one-half cup of cooked spinach, there are 11,000 units of vitamin A, .090 milligrams of vitamin B_1, .312 milligrams of vitamin B_2, 30 milligrams of vitamin C, 46 milligrams of phosphorus, 2.5 milligrams of iron, 2 grams of protein, 2 grams of carbohydrate and 25 calories. Some varieties contain larger quantities of vitamin C.

Range and Soil

Spinach needs cool weather and it must also have short days for best production. When the mean temperature rises above 60 degrees during the first 6 weeks of its development, it is likely to bolt. In northern states, it is grown as a very early spring crop, with a later fall crop also planted in sections where blight is not a problem. In the South, it is planted in winter to produce a very early spring crop or planted in fall and wintered over, if the winters are mild. In the North, late fall sowings can be started in uncovered cold frames, the plants covered with straw after they have made a good growth. A few weeks before they are needed for the table, sashes may be placed over the frames and the spinach will finish growing, even though the soil is frozen.

Almost as important as the proper growing temperature is the correct soil acidity. Spinach will grow only if the *p*H is between 6. and 6.7. If the acidity is greater than *p*H 6., lime must be added to bring it up to that point. But if it gets too much lime, spinach will suffer from a lack of manganese, its leaves becoming pointed and pale yellow.

To make its growth quickly, spinach should have a fertile loam rich in humus and nitrogen. Well-rotted manure should be applied at the rate of one pound of stable manure per square foot or one-half pound of chicken manure per square foot. If fresh manure is used, it should be dug into the rows in the fall. In addition, the following organic fertilizer (or equivalent) can be added at the rate of 2½ pounds per 100 feet: 1 part dried blood, 1 part phosphate rock and 3 parts wood ashes.

Seed

Spinach seed should be purchased fresh each year, because its viability does not last long. One packet of seed is enough for 25 feet, one ounce for 75 feet of row. Yield from 100 feet is about 3 bushels of whole plants. Seed germinates in 5 to 9 days and plants mature in 40 to 50 days. Seeds for two types of spinach are sold, smooth and crinkly leaved. Nobel and King of Denmark are typical smooth varieties. Bloomsdale Savoy is a crumpled leaf type. Some varieties are called "long standing," indicating that they resist bolting in hot weather.

Most popular varieties are Bloomsdale Long Standing, America, Giant Nobel, Virginia Savoy, Old Dominion and Viking. Recommended for their superior vitamin C content are Broad Flanders, Viking and Virginia Savoy. Princess Juliana is lower than average in vitamin C. All of the popular varieties are recommended for freezing except Giant Nobel. Old Dominion and Virginia Savoy bolt easily and should be sown only for the coolest seasons. (New Zealand spinach, which withstands heat, is not spinach at all. *See* NEW ZEALAND SPINACH.)

Planting

Spinach seed should be planted three-quarters of an inch deep in rows 12 to 15 inches apart and plants thinned when they have two true leaves to stand 4 to 6 inches apart. Or seed may be planted in double rows 8 inches apart, with 20 inches between row centers. If seed is fresh one seed per inch should ensure a good stand.

Several plantings may be made spring and fall to provide a continuous supply of spinach through the cool seasons. Seed may be sown in fall-prepared rows before the soil is thawed, for the first planting, and successive plantings made every 10 days until mid-April in the North. Fall plantings may be made up to the time the ground freezes, with the last seeds or seedlings covered with straw and leaves, to remain out all winter. In spring the covering is gradually removed to allow the spinach to grow and provide the spring's earliest greens. Last spinach of the spring should be out of the ground in time for the main crop of corn to occupy its space. Earlier plantings will yield their rows to string beans or limas.

First fall plantings may coincide with a late summer hot spell, when it will be difficult to germinate spinach seed in the garden. In that case, seed may be pre-germinated in the refrigerator. Sprinkle the seed over a damp blotter, cover with another one and store in the coolest part of the refrigerator for a week or less. In that time, small breaks in the seed should show where the tiny roots will first appear. Mix the seed with dry sand to make it easier to handle and sow in drills which have been flooded to make them moist. Firm soil well over them.

Culture

Mulch spinach with grass clippings, straw or ground corn cobs to preserve moisture. Do not use acid material, such as peat moss or acid sawdust, which will lower the soil's *p*H. The mulch should not be applied until the leaves have made a good growth and, if slugs are abundant, should not be drawn up too close to the plants. If the season is dry, water the spinach every 3 or 4 days when it does not receive one-half inch of rain.

Enemies

Spinach damps-off in cold wet weather to some extent and enough seed should be sown to supply extra plants to fill the gaps. Another drawback to its preference for cool moist days is the blue mold which may attack the plants. When this is discovered, plants large enough should be harvested immediately; those too small to pick should be given an extra feeding of fertilizer to try to stimulate their growth of new leaves to replace the ones lost to the mold. Another disease to which it is subject is fusarium wilt. Resistant strains are available and should be used in sections where this is a problem. Strains have also been developed which are resistant to mosaic and mildew.

Harvest

Spinach is usually harvested by cutting the whole plant, which is the end of that plant. But it may also be gathered, if the gardener has the time and inclination, by cutting the outside leaves and allowing the small center leaves to remain, in the same way that Swiss chard is picked. The drawback to this method is that the crinkled leaves are difficult to cut away without damaging the plant. Plants are considered mature when about 6 or more leaves have grown to a length of 7 inches.

In order to preserve the largest possible amount of the vitamins and minerals in spinach, the leaves should be washed as quickly as possible, without soaking and dried by whirling in a salad basket. Water left on the leaves dissolves vitamin C, sugars and minerals out of it and spoils the flavor of the cooked product. To cook, heat it as quickly as possible in a small amount of water, then reduce the heat and cover, to allow it to steam for about 5 minutes in its own juices, until leaves wilt.

SQUASH

SQUASH. *Cucurbita pepo, C. maxima, C. moschata* and *C. mixta.* These species of the genus *Cucurbita* include all squash, pumpkins, cushaws and gourds and have become so mixed and interwoven that it is almost impossible to sort them out. Technically, most summer squash are really pumpkins, but they are pumpkins only to the botanist. And since their culture is the same, we treat here all cucurbits which are commonly called squash.

It is thought that squash originated in North and South America, but there is no definite proof. The name is derived from a corruption of the Indian name for gourds, *ascutasquash.* They are very tender annuals, probably originally tropical vines.

Two types of squash are cultivated, summer squash, which grows quickly and is eaten immature and winter squash, which takes the whole season to mature and is hard-shelled enough to store well during the winter. Many of the summer squash may be grown in "bush" form, that is, the vines are very short and compact. Most winter squash grow on long straggling vines which may spread as much as 20 feet.

Summer squash, like cucumber, is composed largely of water, but contains a good quantity of vitamin A. In one-half cup, cooked, there are 1,000 units of vitamin A, .040 milligrams of vitamin B_1, .050 milligrams of vitamin B_2, 3 milligrams of vitamin C, 18 milligrams of calcium, 15 milligrams of phosphorus, .3 milligrams of iron, 1 gram of protein and 15 calories.

Winter squash contains about 4 times as much vitamin A as summer squash and about 3 times as many calories, most of which are in sugar. Otherwise the two types have about the same nutritional value.

Range and Soil

Summer squash may be grown anywhere in the United States where there are 60 to 75 hot growing days. Winter squash, with the exception of Acorn or

A new edible-seeded variety developed at New Hampshire University, sweetnut squash matures in 85 to 90 days. Seeds contain 35 per cent protein.

Table Queen, takes 90 to 110 hot growing days. Squash is not quite as sensitive to cold as muskmelons, but it needs hot days even though nights are cooler.

Rich light soil, preferably sandy so that it will warm up quickly, and with a *p*H of 6. to 8., will grow good squash. Although they are tolerant to a slight amount of acid or alkali, yield will be best from soil with a *p*H between 6. and 6.5. The squash bed should be well-drained, but supplied with a uniform amount of moisture with sufficient humus to hold it. Two or three spadefuls of rotted manure or compost should be dug into each hill. In very heavy clay soils, the same amount of sharp sand should also be added.

Bush squash may be planted in a row in any part of the garden, but spreading vines should be placed so that they will not interfere with smaller row crops. Like pumpkins, squash may be planted among the corn and allowed to run there. Investigations at Michigan State College seem to indicate that yield is better when they are planted in the corn. If space is small in the garden itself, the larger vines may be planted along the edge of the cultivated area and trained to run out over grass or weeds. The vines do not resent being gently lifted and aimed in the direction in which they are intended to grow.

Seed

Squash seed remains viable up to 4 years, which is fortunate because it takes only a small part of a packet of most varieties to supply the average family. One or two hills of most summer squash will supply not only the family, but also all friends who can be induced to take them. More winter squash may be planted, because they will keep well through most of the winter in storage or may be canned to use through the fall, winter and spring. Summer squash is usually eaten as soon as picked—it is not very good either canned or frozen.

One packet of seed will sow 8 to 10 hills. Yield from that amount may be as much as 500 pounds, depending on variety, maturity when harvested and fertility of the soil.

Summer varieties are crookneck or straightneck, both yellow; coccozelle or zucchini, both green; and white scallop or patty pan squash. Winter varieties are butternut, buttercup, Hubbard, acorn, Table Queen, banana, Boston marrow, Delicious and cushaws. Most of the winter types come in yellow or green rinds, the flesh being about the same in both cases.

Planting

Bush squash should be spaced 4 to 5 feet between hills. Running vines need 10 to 12 feet—if planted in rows, the rows should be spaced 6 feet from their nearest neighbors and plants should be 6 feet apart. Plant 6 seeds in each hill in a circle about a foot in diameter and cover with one inch of soil. Thin each hill to two best plants. If the season is short, plants may be started indoors no more than one month before they can be put in the garden, after the last frost. Squash roots resent being disturbed, so seeds should be sown one to each paper pot for starting inside and the plants should be set two to a hill outdoors. If plants started this way produce flowers immediately after being set out, the first flowers should be removed; otherwise the plants may not set any further fruit.

Culture

Soil should be cultivated around squash until vines begin to run, after which they should be well mulched with grass or leaves. By midsummer, the winter squash vines will have set all the fruit they will be able to mature. After that time, flowers should be removed, so that all the strength will go into the fruit already set.

Organic grower Frank Miceli proudly displays a variety of giant squash grown in his large California garden.

Enemies

Squash vine borer pierces the young vines and hollows the stem, causing the leaves to go limp and die. When a whole section of vine dies, look for a small pile of sawdust along the stem and with a sharp knife slit the stem until you come to the borer. If stems are buried under a small amount of soil as they grow, roots will be sent down from many leaf nodes and borer damage will be less.

Squash beetle is a grey bug about an inch long which appears at midday to lay clusters of shiny brown eggs on both sides of the leaves. At night, the beetles take refuge under dried leaves or mulch under the vines. If a board is placed beneath the vine, the beetles may be caught and killed there in early morning or they may be hand-picked from the leaves later in the day. Egg clusters can be destroyed on the leaves; the young beetles which swarm out of the clusters which are missed can be killed with a dusting of pyrethrum.

Striped beetle is a danger not only for the damage he does himself but because he spreads mosaic. If this pest is prevalent, plants may be protected until the runners are 5 feet long with cheesecloth tents, held above the plants with stakes, boxes or barrel staves, the edges covered with soil.

Harvest

When summer squash is allowed to mature, the seeds become large and tough, the skin leathery, and the flavor loses its delicacy. Long varieties should be picked before they are 8 inches long and may be used when they are scarcely more than fingers. Round patty pans may be used any time from 1 inch in diameter to 4 inches. Skin should be soft enough to break easily when pressed with a fingernail. All squash should be picked when at the edible stage, whether it can be used or not, in order to keep the vines bearing well. When a few are permitted to grow very large and to mature, production will slow down or stop altogether.

Winter squash should be left on the vine until frost threatens. Two weeks before frost, cut the squash from the vines with a sharp knife and leave them out to cure in the sun. This process helps toughen the skins so that they will keep better. If frost suddenly overtakes the squash before they are cured, they should be brought into the house before they are frosted and held for several days in a room whose temperature is about 70 degrees before being stored. Stems should always be cut, not broken, with a piece left on the squash measuring at least an inch to avoid stem-end rot.

Handle winter squash with great care to avoid bruises. Wash and dry the skins carefully and store them in a single layer on shelves in a dry cellar where the temperature is about 50 degrees. Examine them every week or two for molds developing on the skins, which may become deep if allowed to remain. When molds are found, wipe them off with an oily cloth. Treated this way, squash will keep for at least 4 months.

Save Room for Hubbard Squash

In recent years, more and more home gardeners have turned to raising Hubbard squash instead of pumpkin. Its growing popularity is due to the many ways it may be prepared for the table. It can be creamed, baked, sautéed, made into mouth-watering cupcakes, pies and soup, only to mention a few. One of the greatest advantages in its favor is its tendency to remain stored in a cool place over a long period of time without rotting. This means no canning for the housewife. And fresh squash may be enjoyed far into the winter.

The middle of May is about the best time to plant Hubbard squash in New England. In the Middle West, it may be planted as early as the last of April. Light soil, rich in organic substances, is best for growing this vegetable. Choose a well-drained and sunny location, allowing at least 6 square feet for every plant. The vines grow very thick and long and do not like to be overcrowded.

While preparing the soil, soak the seeds in warm water for at least 3 days. Their shells will become soft and the center will swell. This will insure rapid growth, more productive seeds and prevent damp rot which happens often to hard-shelled seeds.

Before plowing or spading the planting area, scatter it thickly with good, aged chicken manure. This makes the plants grow like wild. After the plot has been raked smooth, make mounds 6 feet apart in all directions. In the top of each mound, then, make a hole as big as your fist. Into this, put one handful of dried or aged cow manure. Cover this with at least one inch of soil and plant 6 seeds. Cover them lightly with approximately an inch of ground.

To give the plants a good start, pour a cupful of liquid fertilizer into each mound. This may be made by soaking a bag filled with cow manure or sheep manure in a barrel of water. Keep a barrel of this good fertilizer on hand for future use.

When the plants are 4 inches high, save the ~~two~~ strongest plants in each mound and pull out the rest. The more plants that are allowed to grow, the smaller will be the squash.

Once a week, regardless of rain, water the plants with the liquid fertilizer. Hubbard squash are heavy feeders. In order to produce prize-winning squash, the plants must be fed regularly. Cultivate frequently among the mounds. But care should be taken not to get too near the roots of any one plant. Though they grow to be vigorous plants, they are sensitive in the early stage. A severed root system will result in a dead plant.

When the vines begin to creep, stop cultivating and mulch heavily around the root systems. Grass or leaves may be used. This will discourage insects who love to bore into the hollow stem and suck the juices of the plant. It will also help to keep the soil light and moist around the roots.

After the plant has sent out all the vines it intends to, snip off all except two. Choose the best vines and cut out the weaker ones. This will leave you 4 vines to a mound. The less vines you have, the bigger your squash will grow.

Later, after the small squash have formed on the vines, snip off all except one from each vine. That will leave 4 squash to mature from each mound. Too many squash left on the vines will result in small fruit. It is far better to have one large squash than two small ones. You will get more pulp and the flavor will be far better.

If you find your plants running together as the vines grow longer, you may lift them up gently and circle them back to their home mound. They do not mind being moved. But care should be taken not to snap the crisp vine, or disconnect the heavy squash.

You need not harvest the grown squash until after frost kills the plants. Their shells are hard and the light frost will not harm them. Store them in a cool basement or cellar for future use. Some may be canned if you wish to save time when making pie, but it is not necessary. The squash will keep for months and retain their good flavor.

When planting, grow yourself a good-sized patch to insure delicious eating next winter. Around Thanksgiving, you will agree with me that nothing can beat a good, spicy squash pie.

—Betty Brinhart

SUNFLOWER

SUNFLOWER. *Helianthus annuus,* the giant common sunflower is one of the largest daisies of the family *Compositae.* It is native to North and South America. In the Plains States and the Northwest, it grows wild, sometimes becoming a noxious weed. Sunflowers were grown by the Indians for food before white men explored the western states. The tall

Leo Wenner of Seattle, Washington, stands alongside his 9-foot sunflower plant. The flower is 24 inches in diameter, with leaves averaging 16 by 18 inches.

flowers were sometimes planted in fields of maize, their seeds ground to a meal. The Indians mixed this meal with marrow to make a cake which, for nutritional value, we probably could not beat.

Sunflowers are now grown in many countries, notably Russia, Egypt and India, where they are used for food and for the oil in their seeds. In Canada, large-scale plantings are made for seed and for forage provided by the fast-growing stalks and leaves. In many European countries as well as here, the seed is munched as a delicacy, like peanuts. Seed is also used as feed, both for poultry and, ground to a meal, for livestock. The meal is one of the richest sources of protein and fat than can be used for fattening cattle. Sunflower seed, ground into a flour, can be added to recipes for biscuits, muffins and bread to enrich their protein content.

In addition to these uses, people from various countries have invented other purposes to which the plant may be put. The yellow petals, for instance, may be crushed to provide a buttery yellow dye. Dried leaves have been used in Germany as tobacco. The whole seed, roasted, may be used as a coffee substitute. The stalks, when treated like flax, produce an excellent silky fiber. Bees kept near sunflowers make an abundance of delicious honey. And wild birds flock to feeding stations stocked with sunflower seed.

But most important to organic growers is the food value of sunflower seeds, eaten as they are hulled. In addition to their 22 per cent protein and 30 per cent oil content, the seeds are rich in the B vitamins, calcium, phosphorus and iron and also supply vitamins A, C, D, E and K. The oil contains a large proportion of linoleic acid and lecithin and has no linolenic acid, which means that it does not turn rancid quickly. One hundred grams of sunflower seeds contain 1.96 milligrams of vitmain B_1, .23 milligrams of vitamin B_2, 5.8 milligrams of niacin and 7.2 milligrams of iron.

Range and Soil

Sunflowers can be grown wherever corn will grow. A total of 120 to 140 growing days are needed to mature seed heads, but sunflowers will stand more frost than corn, both in the young stage and when seeds are drying. When ripening seeds are near maturity, even heavy frosts will not damage them; young plants before they develop more than 4 leaves will stand the last light frosts of spring.

Soil should be fertile and deeply prepared, with about the same specifications as for corn. Sunflowers will tolerate a soil with a *p*H of 6. to 8., but do best with the *p*H at 6.7. If soils are slightly sandy, they will be warm earlier in spring and will give the seedlings a better start than if they are heavy. Soil on which water stands should not be used.

In the home garden, soil containing a large quantity of humus may be enriched for sunflowers with 3 pounds per 100 feet of a mixture of two parts cottonseed meal, one part colloidal phosphate and two parts granite dust. In addition, 7½ pounds of wood ashes should be dug in per 100 feet. If the humus content of the soil is not high, rotted manure or compost should be added.

Seed

Seed is sown in large plantings at the rate of 4 pounds per acre. A packet of seed will be plenty for the small gardener with limited space. Most seedmen offer only the Mammoth Russian variety, which grows to a height of 8 to 12 feet, with heads 9 to 15 inches in diameter. These flowers need staking and must be hand-picked; which is not a disadvantage to the small grower but makes them impractical for large plantings.

Varieties of smaller sunflowers, suitable for growers who use machinery for planting, cultivating and harvesting have been developed and some of the best are listed below:

Sunrise is a small-seeded dwarf type of uniform growth. Seed is high in oil but yield comparatively low.

Advance, another dwarf type, produces higher yield and is first choice as an oil-seed type. It is a hybird, so only first generation seed should be planted each year.

Arrowhead is an earlier dwarf type with seed that is larger but lower in oil than the foregoing. Recommended for areas where Sunrise or Advance cannot be brought to maturity.

Jupiter has black seed of high oil content, but heads sometimes shatter in the field and it is only semidwarf, making it too tall for successful combining.

Manchurian is a late, tall, large-seeded type. Seed is low in oil but yields are good in areas where it can be brought to maturity.

Planting

Sunflowers may be planted two weeks before the last frost date. Planting should not be delayed, because every week lost in growing time decreases the yield. Sow seed where it will remain one-half inch deep in rows 3 feet apart and thin plants to stand two feet apart in the rows. If smaller varieties are grown, the plants may be spaced as little as one foot apart in the row.

Sunflowers, with their large heavy leaves, are sometimes used to smother weeds and to clean fields of undesirable undergrowth. When they are planted for this purpose, the field should be thoroughly disked before planting to give the sunflowers an even start with the weeds, and plants should be spaced a foot apart in rows one foot apart.

In the home garden, the row of sunflowers should be on the north end of the garden to prevent them from shading smaller plants. When they are well-spaced, they may be used as poles for beans to climb upon, although their heads will need support when they begin to ripen. The very heavy heads of the Mammoth Russian sunflowers usually grown in the home garden may bend to the ground if not supported, and field mice or rats will empty them of seed before they can be harvested.

Enemies

Chief enemy of sunflowers is the stem borer, a grub which tunnels the main stem, killing all growth above. When the top of a plant droops for no apparent reason, look for the borer's point of entrance and slit the stem both ways with a pointed knife until he is uncovered and destroyed. Best preventive measure is to clean up all old stems each fall, shredding and composting them to destroy borer eggs.

Harvest

The back of a thoroughly ripe and dry sunflower head is brown and dry, with no trace of green left in it. All seeds in such a head will have matured, but the trouble with leaving it in the field until that point is reached is that the birds may have harvested the crop for you, or the head may have shattered and dropped many of its seeds to the ground.

In the home garden, sunflower heads may be covered with cheesecloth to keep the birds away. Or the heads may be cut off when the seeds are large enough and they may be allowed to dry elsewhere. If they are cut with a foot or two of stem attached, they may be hung in a dry, well-ventilated attic to finish drying. They may also be cut and spread on boards on the ground, protected with a wire screening from mice and rats, to dry in the sun for a couple of weeks. Heads should not be piled on top of each other or seed will become moldy or rot.

When the heads are dry, the home

gardener may brush seeds out with a stiff brush, a fish-scaler or a currycomb or he may remove them by rubbing them over a screen of half-inch hardware cloth stretched over a box. If some of the seed is still moist, it may be spread out to complete its drying after being removed from the heads. Dried seed should be stored in small containers and should be stirred once or twice a week to prevent mustiness. Stored in large bins, it may heat up and some of the vitamins will be lost.

Kernels are separated from the husks by cracking like nuts, or by blanching. The seeds should be shelled only as required for eating or cooking, because they will lose their vitamins if exposed to the air for a long time.

Growers of large sunflower plantings who have planted the smaller varieties will find it possible to harvest them with a combine after they have dried thoroughly in the field. Seeds may be removed by a thrasher set for soybeans. Large quantities of seeds can be shelled with a grain hammer mill with the screens removed. The shells are broken by the mill, but left mixed with seed. If they are then submerged in water, the shells will float off and the seed be left at the bottom of the container. Seed must then be spread out to dry in the sun or may be dried by an air blower.

SWEET POTATO

SWEET POTATO. *Ipomoea batatas,* a tropical American native of the morning glory family, *Convolvulaceae*. Indians in South and Central America grew sweet potatoes in a variety of colors, from white through yellow, orange, red and brown, long before Columbus discovered them on one of his voyages. Later discoverers carried them on round-the-world voyages and left some on Pacific islands on their way, so that they may now be found growing in the South Pacific, Japan, China, New Zealand, India, Egypt and a few even in the South of Europe.

Grown in the tropics, sweet potatoes make a long-trailing vine with pale lavender, morning-glory type blossoms which seldom make seed. Grown in cooler climates, the vines seldom blossom, but can be induced to produce their tubers in warm sheltered positions.

Sweet potatoes are primarily a carbohydrate food and are a rich source of vitamin A. In a medium-sized potato weighing 100 grams there are 3600 units of vitamin A, .155 milligrams of vitamin B_1, .150 milligrams of vitamin B_2, 25 milligrams of vitamin C, 19 milligrams of calcium, 45 milligrams of phosphorus, .9 milligrams of iron, 3 grams of protein and 130 calories.

Range and Soil

Commercial plantings of sweet potatoes are confined to warm states, from the southern tip of New Jersey southward. In sandy areas, however, as in Connecticut and Michigan, it is possible to grow some varieties with success and, in favored positions in many northern gardens, a few sweet potatoes may be produced. For best production, they should have 175 frost-free days, mostly warm and with warm nights. Soil temperature of 70 to 85 degrees is ideal. At 60 degrees the vines stop growing, and held at 50 degrees, they will die.

Plantings of sweet potatoes can be made when soil warms up enough, about 10 days or more after the last frost.

Sandy loam with a clay subsoil is considered best for well-shaped tubers. Clayey soil develops poor yields of misshapen roots. The soil should be enriched with phosphorus and potash, but should be rather lean in nitrogen. Three pounds per 100 feet of the following is recommended: 1 part tankage, 1 part wood ashes, 1 part granite dust. Too high a nitrogen content makes lush foliage growth, but with long slender roots which mature late. Heavy potash fertilizing tends to make the roots short and chunky.

If the soil was given a heavy feeding with manure or compost the previous year, it will not need more for sweet potatoes. But if the humus content is running low, a furrow may be dug where the planting will be made and an inch of either well-rotted manure or mature compost laid in it. This should be done at least a week and preferably a month before planting. Except where the soil is very light sand, a ridge should be made above the buried manure, which is 8 to 10 inches high and a foot wide. Sweet potatoes are planted in this ridge.

Sweet potatoes have a preference for slightly acid soil, doing best when the *p*H is between 5.2 and 6.7, with some varieties liking the lower range and some the higher range.

Seed

If the home gardener wishes to plant a particular variety of sweet potato, he must buy plants from growers. If he is willing to take a chance, he may start his own plants from potatoes purchased in the market.

Sweet potatoes are not planted from potato cuttings, as are white potatoes, but from sprouts started indoors or in heated frames outside. These sprouts are called slips or draws. Good slips are 6 to 9 inches long, with well-developed leaves and a good cluster of roots on the under side and are free of disease. A plant with 4 to 6 joints will grow well. Smaller ones should be discarded.

Slips are planted a foot apart, so 100 good slips will plant a 100-foot row and yield should be about two bushels.

Slips may be started in the kitchen window, if the home gardener wants to try a few sweet potatoes to see whether they will grow in his area. Half a dozen potatoes should suffice for growing slips. Potatoes chosen should be of moderate size and not too slender. They should be half-submerged in a pan of water and kept in a light window at about 75 degrees for a month before planting time. Or they may be laid horizontally in pots of sand, covered with two inches of sand and kept warm and moist in a light window. Rooted slips are broken from the parent potato with a twisting motion when they are wanted for planting.

Varieties that may be grown with some success in the northern states include 3 which were developed in Oklahoma—Nemagold, which is nematode-resistant, All Gold, maturing early and extremely high in vitamins A and C and Red Gold, which is wilt resistant. Porto Rico and Nancy Hall are popular varieties in the warmer areas. A new variety, Apache—developed by the California and New Mexico experiment stations—is about one-third richer in vitamin A than most sweet potatoes.

Planting

Slips are planted a foot apart in the tops of ridges, with rows spaced 3½ feet apart. The planting should be made at the end of the garden, if possible, so that vines which stray will not disturb other row plants. Planting should be done as soon as soil is really warm, about 10 days or more after the last frost.

Purchased plants should be plunged into the soil slightly lower than they were grown in cold frames. Home-grown shoots should be planted 4 to 6 inches deep. They should be watered as soon as they are planted to settle the roots and should be watered again when the vines begin to run. From that time on, they may be permitted to dry out somewhat, because roots will go deep enough to find the small amount of moisture that they need. Sweet potatoes are more drought-resistant than almost any other vegetable.

Harvest

First sweets may be dug when they are large enough to cook. The only way to determine this is to dig them up and look at them.

If wanted for winter storage, in the South they are permitted to stay in the ground until the foliage turns yellow, when the tubers will be mature. In the North, they will stay in the ground until the first frost blackens the tips of the vines. When this happens, the vines should be cut off immediately and the tubers dug the same day.

Digging should be slow and careful, because the skins of sweet potatoes are very tender until they are cured and any bruise received when they are dug will form rot in storage. Tubers may be allowed to dry in the sun, but should not be left out overnight if the nights are cold or damp.

Curing

To toughen their skins and heal cuts, sweet potatoes must be cured, that is, held at a temperature of about 85 degrees where the humidity is about 90 per cent for 10 days. If they are not cured, they will not keep more than a few weeks in storage. To extend these few weeks, they may be placed, uncured, in moist sand, which will preserve them for a slightly longer period.

At the end of the curing process, the temperatures are gradually reduced to about 55 degrees and the sweet potatoes are stored in a dry, well ventilated place at that temperature for as long as 4 months. During long storage, their flavor improves, some of their starch turning to sugar.

SWISS CHARD

SWISS CHARD. *Beta vulgaris cicla*. Swiss chard is the beet as it was cultivated by the Greeks and Romans, long before the thick fleshy root we now know as beet was developed. Use of chard spread to China in the Middle Ages and is now almost universal. It is found growing wild in the Canary Islands and around the Mediterranean.

Swiss chard has a tremendously developed root system when it is mature, though the root is not thickened at its top like beet root. Within two months after it is planted, it has developed a taproot 3½ feet long, with much-branched laterals which spread out in a diameter of 5 feet. Fully mature, the roots cover an area 7 feet in diameter and go 6 to 7 feet into the soil. If kept over the winter, in its second year innumerable small roots develop in all directions from the first year's root. These extensive roots are valuable in helping to open up a heavy subsoil and materially aid in its aeration. When the roots have died, they add humus at great depths.

Chard is grown primarily for its leaves, which are used during hot weather to supplant spinach. Heavy midribs of the leaves are sometimes cooked separately like asparagus. The green leaves are a very rich source of A, B and C vitamins and of minerals. One-half cup, cooked, contains 15,000 units of vitamin A, .450 milligrams of vitamin B_1, .165 milligrams of vitamin B_2, 37 milligrams of vitamin C, 150 milligrams of calcium, 3.1 milligrams of iron, 50 milligrams of phosphorus, 2 grams of protein and only 25 calories.

Range and Soil

Swiss chard may be grown all over the United States, though in the warmest regions it may need to be replanted halfway through the season because the leaves begin to dwindle and run out.

The outer leaves of this Swiss chard are ready for harvest; plants average around a foot high. Leaves should not be allowed to become coarse and overgrown.

Any soil in which lettuce thrives will grow good Swiss chard. An addition of a small amount of compost or well-rotted manure is all that is needed in most garden soils. A ridge of compost 3 by 3 inches down the row may be dug in before planting. Chard is more tolerant of acidity than beets, but a *p*H of 6. to 7. is optimum.

Seed

A packet of seed will sow 15 to 25 feet of row; an ounce is enough for 50 to 100 feet. A packet will supply most families unless a large quantity will be frozen. Fifty-five days after planting the first leaves should be large enough for cutting.

Planting

Swiss chard may be planted quite early, because it is hardy to the last spring frosts as well as to the summer's heat. In California and other places where curly top is prevalent, it should be planted as early as possible to avoid the damage from that disease. Most recommended variety is Lucullus, with Fordhook Giant a close second.

Germination will be hastened if the seed is soaked 24 hours before planting. Plant seeds 4 inches apart and one inch deep in rows that are 18 inches apart. After the plants are a few inches tall, thin them to stand 8 to 12 inches apart. Thinnings may be cooked or used in salads.

Harvest

When the largest leaves are 7 inches high, Swiss chard cutting may begin. With a sharp knife, remove the large outer leaves and allow the inside small ones to remain. Harvest should continue at intervals of 3 days to one week all through the summer and fall. If the leaves cut cannot be used in the kitchen, they will be very welcome in the poultry pen. Chickens are fond of both stems and leaves of Swiss chard.

In all but the coldest areas, a mulch of straw and leaves over the plants will keep the roots alive until spring, when leaves will sprout to provide early greens. In milder winters and in the warmer sections, leaves may be cut throughout the winter.

TAMPALA

Tampala. *Amaranthus gangeticus.* Also called amaranthus. A tropical herb from the Orient widely used there in many forms as a potherb or salad green. Tampala corresponds roughly to the variety *viridis* which is used in Bengal. A red variety, a white one and a giant form are cultivated in India and China. Several closely allied forms are grown here as ornamentals because of their colored foliage. Chinese name for tampala is "salty vegetable," because of the naturally salty tang it has when cooked. It is thought to be very rich in minerals, which produce the tang, but no analysis is available. Its flavor is more bland than that of spinach, indicating a lower acid content.

Tampala is a warm-weather crop which is sown after the ground is warm in spring. Seed is planted one-fourth inch deep and plants are thinned when about 8 inches high to stand 18 inches apart in the row. Harvest of leaves which are cut from the branches may begin 6 to 8 weeks after planting. If largest leaves are kept cut, it will continue to produce through the hot weather and into the fall.

TARO

Taro. *See* Dasheen.

TOMATO

Tomato. *Lycopersicon esculentum.* A tropical perennial and one of the most important members of the horticulturally and commercially valuable nightshade family, *Solanaceae.* First cousins in the upper hierarchy of importance are potatoes and tobacco and the family also includes eggplant, green and red peppers in the vegetable garden, petunia, browallia and salpiglossis in the flower garden.

Tomatoes are native to tropical America, having been found by the early explorers in both Mexico and Peru, where they were cultivated and eaten by the Indians. The name comes from the Mexican Indian name, *tomatl.* Contrary to stories about them, tomatoes were grown for food in Europe as early as the sixteenth century, though they were not popular in some countries. They were introduced to the oriental countries in the eighteenth century and Thomas Jefferson grew them in his garden in 1781. Gradually their use spread north from Virginia, probably following the development of a forcing technique for early production. The first varieties cultivated were ribbed, some of them red, others yellow. Smooth round tomatoes were developed in the nineteenth century. Cherry tomatoes, now grown as a novelty and for pickling, were grown in the sixteenth century; pear-shaped tomatoes, also known as plum tomatoes, were a nineteenth century development.

Although a perennial in the tropics, tomatoes are grown as annuals in the temperate zones and varieties have been developed which will bear fruit as little as 65 days after planting. They still have a preference for hot weather, though, and do their best growing in temperatures ranging from 70 to 90 degrees. Most varieties will not set fruit if night temperatures drop below 60 degrees or day temperatures rise above 94 degrees over an extended period. Once set, however, the fruit ripens most quickly if the temperature drops to between 55 to 65 degrees.

Most cultivated tomatoes make vines which will continue to grow taller the longer the growing season. A few varieties have been developed which make their growth and then stop growing while the fruit sets and ripens. These are called "determinate" vines; the long-growing ones are "indeterminate." Both types have extensive root systems which spread approximately to the same extent as their tops. If the seeds are planted in the field, they develop deep tap roots in addition to the roots which spread laterally near the surface from the main stem and those sent down by any branches which touch the ground. Plants started under glass and later transferred to the field do not develop tap roots so deep.

Tomatoes are among the few vegetables whose vitamin content is almost as great after cooking as when they are raw. With the exception of vitamins A and B_1, the loss through cooking is very little. Fresh tomatoes, in a 100 gram portion, contain 1500 units of vitamin A, .110 milligrams vitamin B_1, .050 milligrams vitamin B_2, 25 milligrams vitamin C, 11 milligrams calcium, 29 milligrams phosphorus, .4 milligrams iron, 1 gram protein and 20 calories. The same weight in tomato juice, a little more than one-third cup, contains 1,125 units vitamin A, .081 milligrams vitamin B_1, .520 milligrams vitamin B_2, 20 milligrams vitamin C, 10 milligrams calcium, 29 milligrams phosphorus, .5 milligrams iron, 1 gram of protein and 25 calories. Increase of some mineral quantities may be due to the evaporation of water in cooking, which concentrates the solids.

The above vitamin content is based on analysis of the average field-grown tomato. If tomatoes are grown in partial shade, they lose in vitamin C content. And special varieties are being developed which were selected for their high vitamin A and vitamin C content. Caro-Red, just coming into the market, has a vitamin A content almost 10 times that of the older varieties. A medium-sized Doublerich tomato weighing 100 grams, contains 56 milligrams of vitamin C, more than twice the amount in the old varieties.

Range and Soil

Tomatoes will produce a good crop wherever they have 3 to 4 warm growing months with weather that is clear and fairly dry, where they have uniform moisture and can be protected from hot drying winds. Plants do not increase in size above 95 degrees and the fruit does not color properly when night temperatures stay above 85 degrees. High temperatures accompanied by high humidity favor foliage diseases. On the other hand, the plants are sensitive to frost and will usually be killed by temperatures as low as 32 degrees.

Almost any good garden soil will grow tomatoes. Sandy loam will produce earliest; heavy clay loam will produce well when late tomatoes growing large vines are planted on it. Tomatoes like a soil that is just below neutral in acidity, with a *p*H between 6.0 and 7.0.

Earliest and heaviest crops will be produced when the nitrogen content of the soil is not too high, but the phosphorus content has been given a boost. Too much nitrogen produces abundant foliage and delays ripening. Nitrogen deficiency in tomatoes is rare, but when it occurs the top leaves of the plant turn yellow and the stems become a deep purple. Unless these symptoms show, addition of nitrogen is not necessary after plants are started.

To prepare the soil for tomatoes, an application of one-half pound of mature compost per plant should provide humus enough when dug in where the plant will stand. If compost is not available, a trowelful of dried manure may be substituted. In addition, the following formula should be dug in at the rate of 7 pounds per 100 feet of row: 2 parts dried blood, 1 part phosphate rock and 4 parts wood ashes. A trowelful of bone meal mixed into the soil at the bottom of each hole or dug into the hill before seed is sown, will provide the extra phosphorus which speeds ripening.

Tomatoes should not be planted in soil that has recently had an application of tobacco stems, nor should they be planted too near to a walnut tree's roots. Either of these will kill the plants. They are also very sensitive to the effects of illuminating gas and seedlings being grown on the kitchen window sill may do very badly if there is a gas leak in the room.

It is best to avoid a position in the garden where other members of the family—peppers, potatoes or eggplant—have recently been grown, since tomatoes are subject to the same diseases. And finally, choose a spot for the tomatoes where they will get a minimum of 8 hours sunshine a day.

Seed

One packet of seeds will usually sow enough plants for 100 feet of row, that is, 26 plants if they are to be unstaked or 51 plants if they will be pruned and trained to supports. New varieties however, are sometimes sold in packets containing smaller amounts of seed. The seed catalogs usually specify the number of seeds per packet in this case.

Some seedmen sell hot-water treated seed which is more disease-free than untreated seed. Also, some varieties are disease resistant. Where bacterial or fungus disease is prevalent, these seeds should be planted.

Seed germinates in 8 to 10 days at 70 degrees. Plants start to yield 65 to 100 days after planting. It takes 45 to 50 days to produce ripe fruit from a blossom. Yield per 100 feet of row varies widely, with method of growing and length of season, from 4 to 15 bushels. A dozen unstaked plants will provide all the tomatoes the average family can use on the table and another 12 to 18 will provide enough for canning. Tomatoes do not freeze well.

For early tomatoes and a long season, plant several varieties. Below are some of the most popular, together with some of their characteristics.

Planting

Because they take so long to reach full production, most tomatoes are started under glass, either in flats in the house, in cold frames or in greenhouses. Plants 7 to 9 weeks old are planted in the garden about two weeks after the last spring frost. The gardener may purchase plants if he will plant only a few, but his choice of varieties will be limited unless he starts the seeds himself.

Seed should be planted no more than 6 or 7 weeks before the last expected spring frost. Plants allowed to grow too long indoors do not yield as well as those planted outside promptly. Soil mixture for seeding should be two parts topsoil, one part sand and one part compost. Seeds should be planted one inch apart in pots or flats and one-fourth inch deep. Three times as many seeds should be planted as the number of plants required. Keep seed pots in a sunny window where the temperature is constant at 70 degrees. When seeds have germinated, temperature may be 60 degrees at night.

Two weeks after seeding, the first small plants, usually just beginning to develop second leaves, should be ready for transplanting. They may be transferred to another flat or may be planted in plant bands or individual containers. Soil mixture should be the same as in the seed pots and each plant should be given a space 3 by 3 inches. Sun is the most important requirement of the seedlings at this time. Plants should be kept slightly on the dry side, but they must have at

TABLE 53: Most Popular Varieties of Tomatoes

Variety	Days to Maturity	Plant type	Size of fruit	Shape
Fireball	65	Determinate	Medium	Globe
Marglobe	77	Indeterminate	Medium large	Deep globe
Moreton hybrid	70	Indeterminate	Large	Globe
Pritchard	76	Determinate	Medium	Globe
Rutgers	86	Indeterminate	Large	Globe
San Marzano	90	Indeterminate	Small	Pear
Sioux	70	Determinate	Medium	Flattened globe
Stokesdale	75	Indeterminate	Medium	Globe
Urbana	78	Determinate	Medium	Globe

least 6 hours' sun per day to make stocky growth. By the time they are 7 to 9 weeks old, the plants should be 8 to 10 inches tall and with a leaf spread as wide as the height. Leaves should be dark green and set close together; stems should be thick and dark green without hard or knobby joints; texture should be neither soft and sappy nor hard and woody.

During the last two weeks before they are set in the garden, the seedlings should be hardened-off outdoors in the daytime. They may be set in a covered cold frame where the sash can be kept on during the nights and on cold days or they may be moved indoors at night. When the plants will stand up under a hot sun for a full day without showing signs of drooping, they are hardened off enough to plant in the garden.

Spacing in the garden will depend upon how the plants are to be grown. If they will be pruned to one stem and tied to a stake, they may be planted 18 to 24 inches apart in rows 3 feet apart. If they will be permitted to develop 3 stems but will be tied upright, they will need 24 to 30 inches between plants. If they will be permitted to sprawl on the ground and will not be pruned, they should have 3 to 4 feet each way, depending upon variety.

In placing the plants in the ground, dig a hole large enough to bury the seedling almost up to the lowest leaves. Roots will develop along the stem that is covered with soil. If leggy plants must be used, dig a trench long enough to accommodate the stems, placing the roots about 6 inches below the surface and allowing the stem to lie horizontally with only the leafy portion of the plant above ground. After planting, water well with manure water or water in which rich compost has soaked overnight.

Roots should be disturbed as little as possible during the transplanting. If pots were used, the whole root ball should be planted just as it is removed from the pot, without spreading the roots. If the plants were grown in flats, they should have been blocked out a week before transplanting by running a sharp knife through the soil half-way between plants. This cuts through the long spreading roots and forces the plant to develop small feeder roots close to the main stem.

After the plants are in the ground, place a stiff paper collar loosely around each stem, extending an inch into the soil and rising an inch above the surface. This will foil the hungry cutworms in the neighborhood. If the sun is bright on the next few days after transplanting, shade from a bushel basket inverted over the plant will give it a chance to become established more quickly.

You can have an abundant harvest of tomatoes and save space in your garden by training plants to grow up a trellis.

In areas where the temperatures are high enough in spring or in northern areas where early plants are started under glass and later ones may be allowed a long period outside, seeds may be started directly in the garden. There are many advantages to planting directly outside, even in the North where seed for early varieties must be used to get a crop which will have a long enough season.

Plants started in the garden develop deeper taproots when they do not have to be transplanted several times, so they are better able to stand dry weather later in the season. And because the roots go deeper, the fruit is less likely to crack in wet weather and will suffer less from blossom end rot. Also, plants started outside will stand considerably more cold than those which are softer from having been started in a warm place and then artificially hardened off. Plants started in the field are less likely to succumb to disease than those started indoors and they are stronger than those purchased from a grower who has had to ship them a distance.

When planting outside, plant 5 to 6 seeds to a hill in hills spaced as the plants would be spaced. Seed should be covered with an inch of sifted compost. When the seedlings are 6 inches high, all but the strongest plant should be removed.

Culture

Thick mulch of straw or grass clippings under tomato plants will help to keep the fruit clean, if the plants are allowed to sprawl on the ground and will keep the moisture content of the soil more uniform, thus preventing much of the blossom end rot. Mulch should be drawn up to the stems when the plants are a foot high. During periods of drought, tomatoes should be watered once a week.

Supports

Tomato plants permitted to sprawl on the ground freely without pruning will set and mature more fruit than plants that are pruned, but the fruit will be smaller and much of it may be lost by rotting on damp ground, by damage from slugs and other insects and by becoming overripe while it is hidden under a dense mass of foliage.

Supports of all kinds have been used for tomatoes and all of them are good if they keep the fruit off the ground. They may be simple racks 18 inches high, which surround the plants with a sturdy frame over which unpruned branches may be trained. They may be simple stakes with about the dimensions of a broomstick, sunk a foot into the ground, to which a vine pruned to one stem may be tied. Or a group of 3 or 4 such stakes may lean together and be fastened at the top like pole bean tripods, with a plant trained to each stake. If plants are to be pruned to two stems, two stakes will be needed; 3 stems need 3 stakes.

Various trellis or fence arrangements have been devised on which tomatoes may be trained. The simplest is a wire fence with two or three strands of stout wire strung between heavy 8-foot posts, two feet of which are under ground. The ends of the fence should be kept taut with guy wires and turnbuckles. Posts should be spaced 8 to 10 feet apart and 3 or 4 plants are placed between each pair of posts. Ties are fastened to the horizontal wires and dropped to the growing plant below, training it to grow upright.

A double row may be planted, one on each side of the fence if the top wire is lower and plants are placed a foot or two out from the fence. Heavy cords or laths are strung from the horizontal top of the fence to a peg at the base of each plant and the vines are trained along this support. Plants pruned to one stem may be spaced 24 inches apart on each side of such a trellis and staggered so that their tops will be spaced a foot apart when they reach the top wire.

Stakes which will be sunk in the ground near the plants should be put in place when the plants are set out in order not to damage the roots later on.

Material used for tying vines should be soft and broad in order to prevent cutting the stems. Best for the purpose is raffia or cloth torn into one-inch strips. If these are not available, soft twine may be used. Wire and stiff twine should be avoided.

Pruning

Allowed to grow unpruned, tomatoes will make branches at each leaf node and sometimes fruit cluster stems will extend to form additional branches.

Whether the vines are to be trained or not, these fruit cluster branches should be pinched off just beyond the cluster, in order to throw all the nourishment from that stem into the fruit.

When vines are to be trained to one stem, all side branches should be pinched back after they have made one set of leaves. Experiments have shown that more and better fruit will result from permitting the first pair of leaves on each branch to develop before the growing tip is removed. Suckers arising from the root after the main stem has made a good growth should also be removed.

If 2 or 3 stems will be permitted to grow, the branches on the first leaf nodes are allowed to develop. Each branch is

(Top) When growing climbing tomatoes, begin suckering plants as soon as they bloom. Cut away all vines except the main one. (Bottom) Tie vines to stakes often as they continue to grow taller, to prevent wind from breaking off loose vines.

then tied to its own stake and treated in the same way as the parent stem.

Be sure that fruit clusters are directed away from the stakes and strings, even if the stems must be bent or twisted. Fruit which sets in contact with the support may suffer contact injury and leave itself open to the entrance of disease.

Once pruning has begun, it must be repeated every week throughout the growing season. If it is started and then abandoned, tops of the plants will develop heavily and many suckers will take the strength from the plants without providing much fruit.

Six weeks before the first expected frost, nip out all growing tips including that on the main stem. This will stop vine development and permit the plant's nourishment to concentrate on maturing fruit. New blossoms may also be pruned out after this date, since fruit set late will not be large enough for use.

Enemies

Organically grown tomatoes that are well-mulched and are grown from disease-free seed should have very little trouble with pests and diseases. Cutworms may attack them, but they may be protected by paper collars. Tomato hornworm sometimes appears and feasts off the foliage, removing all the soft leafy parts from the stems. One tomato hornworm, which grows as big as a little finger, can do a lot of damage in a short time. When the foliage has been eaten, look for him lying close to the stem at one end or the other of his trail of chewed leaves. If his back is covered with a cluster of small white bodies that look like eggs, leave him alone. He will die from the parasites that are on him and they will live to prey on other hornworms.

Where fungus and bacterial disease is prevalent, all wild nightshade vines in the vicinity of the garden should be rooted out. They are subject to the same diseases and will carry them over from one year to the next.

Cracks that develop in the tops of tomatoes are caused by a sudden supply of water reaching fruit that has been growing slowly, its skin hardening, on a limited supply. Thick mulch and attention to the soil moisture can prevent much of the top cracking.

The same attention will prevent blossom end rot, which shows itself in both green and ripe fruit by a brown, bruised portion on the underside of the fruit. This is caused by the drying out of the fruit, which starts at the end furthest from the stem. A constant supply of moisture can prevent blossom end rot, but too much moisture can also cause it. If the roots are swamped by saturated soil, the smaller feeder roots will rot and die. This will result in lack of sufficient water reaching the fruit just as surely as drought will do it.

Early Tomatoes

Because no tomatoes have the fine flavor of home-grown, vine-ripened ones, each gardener tries to start the season as early as possible and make it last as long as possible. A number of tricks to outwit the seasons have been devised, each gardener having one of his own.

Most of the early tomato schemes are based upon one of two principles or a combination of them: they protect the plants from cold encountered by early planting outside or they warm up the outside to accommodate the plants.

Plants may be protected when they are put into the garden by any number of mechanical devices. They may be covered by bushel baskets—the oldest protection of all. The basket may have one stave in three replaced by plastic to let in light. They may be covered with plastic hot caps which allow the light to go through, but keep out frost and cold winds. They may be covered with gallon glass jugs which have had their bottoms removed, thus putting each plant in an individual cold frame. They may be surrounded by very deep mulch with a small depression left for each plant, the top of the depression being covered by a pane of glass.

Plants may be protected from the cold by conditioning the plants to stand the cold, even to like it. One cold treatment recommended by Michigan State College is to grow seedlings for 3 weeks after the seed leaves have unfolded at night temperatures of 50 to 55 degrees. Plants grown in this way may be expected to stand transplanting better, to have stronger side shoots and to blossom earlier with the blossoms set closer to the ground.

Tomato seedlings started in outdoor cold frames will be tougher than those started indoors and may be set into the garden bed earlier. A cold frame covered with cloth rather than sash will permit more of the cold air to circulate among the plants and harden them still more.

Soil may be warmed up early to receive tomato seedlings. This may be done by pushing back any mulch left from last year to permit the sun to shine directly on the soil. The sun-warming process will be speeded if black roofing paper, which absorbs all the sun's rays, is laid over the ground to help warm it by day and blanket it at night against loss of the day's warmth. When the soil is warm enough, plants may be placed in the ground through a hole in the center of the paper, which is left as mulch.

Or the soil may be warmed from below by addition of organic matter which will heat mildly in fermenting. Such a method is described below:

Dig a hole 18 inches deep for each plant. Fill it with 3 inches of chopped corncobs or stalks, two inches of cow, chicken or rotted horse manure and 4 inches of mature compost. Foot-high plants are set upon this bed, which is now 9 inches deep and all but the top pair of leaves pinched off. Rich soil is used to finish filling the hole. Plants are mulched after making some growth.

Plants which will bloom and bear in shorter than usual time after being set in the garden may be grown from branches of plants started indoors. Tomato seed would have to be started about two months earlier than usual for this method and they would need a very light place to grow, preferably a greenhouse. Seedlings should be planted in 6-inch pots and allowed to grow in the light and to develop side branches. When the branches are 6 to 8 inches long, they are removed from the plants and rooted in sand. The branches, not the parent plant, are set in the garden after they are well-rooted. Such cuttings begin to blossom and set fruit much earlier than seedlings of comparable size.

Harvest

Vine-ripened tomatoes have more vitamin C than tomatoes picked before they are quite ripe. Allow the fruit to ripen fully on the vine. Peak of ripeness usually comes about 6 days after the first color shows. Technically, they are ripe when the plant sugar stops moving into the fruit, a criterion which the amateur grower can judge only from flavor.

To pick a tomato, do not pull it directly away from the plant, because it may tear the branch loose at its joint if it is not quite ready to part with its stem. Twist the fruit on its stem, tipping it sideways as you twist and the stem will part with the fruit or break off close to it.

When frost threatens in fall, all fruit near ripeness or of mature size should be picked and stored where it will ripen. If the tomatoes are wrapped separately in newspaper or are stored in a dark, cool place, they will not ripen too fast. A few may be brought into a warm sunny window sill to ripen as they are needed. The season may thus be prolonged for 6 to 8 weeks, though the tomatoes ripened indoors will have less flavor and less nutritional value than those ripened outside on the vine.

Fencing Tomato Plants

Last year, we grew our tomatoes along a line enclosed with wire fencing and we picked more tomatoes than we ever did before. What's more, we did it with a lot less bending and trouble. No secret growing technique is involved. Any gardener or farmer who's willing to invest a few dollars in wire fencing can step up tomato production—just as we did.

When we set out to take the backache out of tomato picking, we went on a search for hog-netting. (Use other cheap netting materials if they are more available where you live.) We were able to buy 40 rods at $10 for old fencing that was still perfect for our use. The junk yard yielded us one roll at no cost and we had some on hand. All this was stacked by the garden gate in readiness before planting time.

Furrows 6 feet apart were made for setting out plants. We set the plants 18 inches apart in the rows, much closer than we had ever planted before. This is one of the secrets of this system—your plants close together so they hold each other up and do not slump down within the enclosure.

After several weeks, when the plants were large enough, we put on a light mulch of pine needles. Any mulching material would do.

Now, you are ready to put up the netting. Don't try to substitute chicken netting. It is too light weight. The heavy vines crush it down.

We used short wooden fence posts that we had available for the end posts. Steel posts were driven approximately one rod (5½ yards) apart down each side of the row with the posts 18 inches apart across the row. This is a point we wish to emphasize. The posts must be set no farther than 18 inches apart across the row. We set some farther apart than this in our first trial plot and the plants, heavy with fruits, slumped down within the enclosure. Steel posts cost approximately $1.00 each, but would last during the gardening years of a retired person or persons such as we. We also used a few old two-by-fours that my husband had purchased.

The netting was rolled out so that the larger mesh would be at the bottom to aid in reaching through and picking tomatoes that are low on the vine. The netting was fastened to the posts 6 inches above the ground. Most hog-netting is 26 inches high and fastening it 6 inches above the ground makes the enclosure 32 inches high. It needs to be this high for tomato vines.

We literally made our soil. Our garden plot was originally a light sandy soil, but it was in a little sheltered valley with hills on the west and east and subirrigated to some extent. We wanted it for our garden. It has been built into a fertile soil with rotted manure from a feed lot and partially rotted alfalfa bales from the base of stacks scattered and plowed in. Last year's tomato plot was formerly a strawberry bed. We plowed in a great deal of organic matter—the mass of strawberry plants and also the straw mulch between the rows and that which had worked in among the plants from the winter cover. This was done in June the year before, after the strawberries were through bearing.

I mentioned to a garden visitor that we had 650 tomato plants in our plot. "Oh, you must have a lot of green worms to pick," she said with a distasteful squirm. When I told her we had killed only two, she looked at me unbelieving.

We even surprised ourselves. We hadn't realized until we checked our garden records for the last 5 years that the tomatoes ripened a great deal earlier. We found we were 10 days to 3 weeks earlier than previous years in having our first ripe tomatoes. I think all people, including myself, drool more waiting for the first tomato to ripen than over any other garden vegetable. To those who sell, the earlier ripening always means a better price. Our telephone is on a party line with 10 other people. One of my neighbors says the requests for early tomatoes is like a broken record. When she hears ours ring and listens, all she hears is tomatoes, tomatoes, tomatoes.

We grow our own plants in a hotbed with horse manure used to provide the heat. It is difficult to find, but we were able to get it from a racing stable. Raising our own plants is an important factor to us. We like to have the kinds we want and perhaps try new ones that have just been put on the market.

We feel conservative estimates of our gain over other systems of tomato growing would be about 30 per cent! We used to lose 10 per cent or more of our tomatoes from rotten bottoms. What a joy it was to reach for a tomato and know it was going to be usable. We save another 5 per cent formerly lost from sun scald. We gained 10 per cent by getting more of the tomatoes ripened before frost than ever before. The fact that we were also early would be counted in this percentage. The reason, I'm sure, was that they were suspended with the light and air circulating through them. The tomatoes were much larger last year, which should give us another 5 per cent.

Furthermore, with fenced-in tomatoes, the plants shade the ground and help to conserve moisture. The ripening is more even, not one green side and one ripe. If you are an older person, perhaps the most important advantage is that you don't have to lift a heavy vine in search of its fruits.

There is no pruning or tying as in the case of staked tomatoes. Some tendrils sneak through the netting, but to the gardener with one row, this wouldn't present a problem. On an evening walk through the garden, as all good gardeners take, if he notices such a culprit, pulling him back within the enclosure is an easy task. We didn't find it a major problem in our large plot.

Taking the fence down and stacking the posts is an after-frost job. They come down faster than they go up. The convenience during the summer far overshadows the work of erecting the fences and taking them down.

—Anoma Hoffmeister

50-to-the-Plant Tomatoes

When visitors saw Everett Malone's long rows of thrifty, heavily bearing tomato vines last summer, many of them asked him wonderingly, "What new miracle plant food do you use?"

Actually, these abundant tomatoes were grown in almost pure compost. It was the first time Mr. Malone had tried this process in a truck gardening project. These vines were far more prolific and the fruit of much better quality and flavor than any he had previously grown.

Everett Malone, a Kansas truck farmer, grew 50-to-the-plant tomatoes by setting them in almost pure compost.

The Malone farm is in Chautauqua County, Kansas. It consists of 320 acres, 100 acres of which are planted in rotation in wheat, alfalfa, oats, barley and milo. There are about a dozen head of beef cattle and a number of hogs on the farm. Two of Malone's sons raise purebred hogs.

Although he had used conservation practices in terracing, making ponds, rotating crops, manuring and green manuring, Mr. Malone previously had not given special attention to a truck gardening project. It was last winter, during a lull in routine farm work, that he started.

In late winter, Malone began making compost, which consisted mainly of sawdust from a nearby sawmill and chicken manure containing much litter. He added small amounts of bone meal, agricultural limestone and rock phosphate, then watered the heap well. When it became heated, he turned the heap, repeating the turning process several times.

After it had ceased heating, it had decomposed into rich black compost.

A furrow was then turned in a badly eroded plot near the house, exposing almost pure shale. Into this furrow 100 plants were set in the pure compost at a distance of about 3 feet from each other. A handful of a seaweed product which is said to prolong bearing and forestall frost damage, plus a handful of pulverized granite, were added in the planting of each vine. The plants were heavily mulched with sawdust, the spaces between them were also filled with sawdust and the plants were then watered well. Another 200 vines were planted in the same manner on a slope near a pond where the soil was better to begin with.

Very little time was spent on the patches, as the plants required no hoeing. The vines were pruned and tied to stakes. These stakes were of rough lumber with crossbars connecting them. As the vines grew and bore fruit, the crossbars were needed for additional support, as each plant often had more than 50 large tomatoes on it at one time.

The Malones offered the tomatoes for sale from door to door in Wichita residential sections. Their customers invariably were delighted with the quality of the fruit, which they said was unequaled by any which they had bought previously from others.

By late Summer other gardeners' tomato plants had stopped bearing, but the Malones' were still bearing heavily and they continued to do so right into fall when frost finally killed them. (Mr. Malone found that, in addition to bearing until the first heavy frost, his vines were still undamaged after those in other gardens had blackened and died.) Thus, there was a great demand for the Malones' fruit at that time. Many of the best large tomatoes went to stores for table use; the smaller ones were sold by the bushel for canning. The total sales of extra tomatoes—the Malones used many themselves—amounted to over $350.

Lois M. Smith

Rack Method for Tomatoes

Here in Illinois, tomatoes grown on racks are superior to those left on the ground. They do not rot and I get a better yield. They also bear longer and fruits are more easily picked. By this method I can also set plants closer.

I set plants in a row 18 inches apart. When they're growing well, I put up racks 8 feet long and 4 feet high, made of one-by-two-inch slats spaced a foot apart. Two of the upright slats are nailed a foot from the end and extend a foot below so they can be set in the ground. I lean the rack slightly and brace with stakes driven into the ground at an angle and nailed to the rack. I use pieces of old steel posts that have holes in them, mak-

An inverted V-rack can be used as support for tomato plants to increase yields and permit easier harvesting.

ing them easy to nail. Wooden stakes may also be used.

Racks and stakes must be set deep enough to keep them from blowing over. An inverted V-rack may be used and plants set on both sides, but single racks make hoeing and picking easier. When spring weather is quite warm I plant some seeds of Stone or Firesteel in the garden and reset when big enough. Those grow very fast and will bear late or until frost when the others are through. I have never had any trouble with disease, wilt or pests on my tomatoes.

As soon as side shoots appear on the plants, I prune them off, leaving the main stem and top which I tie to the rack. I usually prune off side shoots 3 or 4 times, training and tying main stems to the rack as growth progresses. I then let the plants go, tying as necessary.

When freezing weather threatens in the fall I gather all the fruits that show a whitish tint, spread them out on a table in the basement and cover with paper or cloth. These fruits will ripen nicely. If too hot they will all ripen too soon. Temperature in an ordinary basement is about right. I have tomatoes for Thanksgiving and one time kept some in the refrigerator until Christmas after they had ripened. They are not as good a quality as those ripened in the garden, but I consider them superior to any shipped in.

—*Charles W. Norris*

TURNIPS—TURNIP GREENS

TURNIPS—TURNIP GREENS. *Brassica rapa.* Turnips are biennials grown as fast-growing annuals in cool periods throughout the United States. Native to the cool portions of Russia, Siberia and the Scandinavian peninsula, they were grown early by the Romans as far south as France in the first century A.D. The Romans knew two varieties, one grown for its roots, the other for its leaves. Today we grow 3 varieties, one for its leaves, a white and a yellow-rooted variety.

Early turnips can be planted as soon as ground can be dug in spring. Grass clippings shown make a good mulch for turnip plants.

Like rutabagas, turnips quickly develop a deep and vast root system. The principal differences between rutabagas and yellow turnips are the attachment of the leaves, which in rutabagas make a neck or crown and in turnips are attached directly in a rosette to the root and in the length of the growing season. Rutabagas require a much longer period to mature than do turnips.

Turnips or turnip tops may be eaten raw in salads or cooked. The tops are an excellent source of A and C vitamins and of calcium. One-half cup of cooked turnips contains .062 milligrams vitamin B_1, .062 milligrams vitamin B_2, 22 milligrams vitamin C, 56 milligrams calcium, 47 milligrams phosphorus, .5 milligrams iron, 1 gram protein and 33 calories. The roots contain no vitamin A. One-half cup of cooked greens, however, contains 11,000 units vitamin A, .060 milligrams vitamin B_1, .450 milligrams vitamin B_2, 130 milligrams vitamin C, 347 milligrams calcium, 49 milligrams phosphorus, 3.4 milligrams iron, 2 grams protein and only 28 calories.

Range and Soil

Turnips grow quickly and well during the cool weather, but become hard, woody and may go to seed during hot weather. They are grown as early spring or fall crops in the North and are grown during spring, fall and winter in the South. Fall crops may be left in the ground all winter and dug as needed in any intermediate states where the soil does not freeze hard.

Rapid growth is essential for good root quality and also to mature a crop and get it out of the ground before hot weather sets in. For that reason, soil must be well-enriched for the spring planting, but may be somewhat less rich for turnips grown in fall, when haste is not so important.

Loose, friable soil, well-supplied with humus, will grow the best spring turnips. Manure may be dug in during the fall and permitted to rot over the winter or rotted manure in a one-inch layer may be turned under in early spring as soon as the ground can be worked. In addition, 3 pounds of the following organic fertilizer should be worked into 100 feet of row for the spring crop: 2 parts cottonseed meal, 1 part colloidal phosphate, 2 parts granite dust. Soil acidity should be around neutral, with a *p*H between 6.0 and 8.0. Soil for fall crops may not need special treatment if turnips are planted in rows that were well-enriched for the crops planted in spring.

Seed

One packet of seed sows 50 feet of row; one ounce is enough for 250 feet. Seed is very small and special care must be taken not to plant too thickly or too deep. Tops are ready for the pot in 40 days; roots mature in 35 to 60 days, depending upon variety. Varieties grown for tops may have worthless roots, which is the case with Seven Top. If the roots are not needed, this is the variety to plant. But if they are grown for both greens and roots, Shogoin is a better variety. Most popular variety for the fall root crop of white turnips is Purple Top White Globe. Golden Ball is a recommended yellow type.

Planting

Early turnips are planted as soon as the ground can be dug in spring or even earlier, if the rows were prepared in fall. Seed is covered with one-fourth inch of fine compost in the spring planting. Fall crops are planted two months before the first expected frost and seed is covered with one-half inch of soil to help it germinate in the hot dry summer. Rows should be spaced 18 to 24 inches apart and plants should be thinned to stand 3 or 4 inches apart in the rows.

Enemies

Same as for CABBAGE and BROCCOLI, which see.

Harvest

Pull and use the spring planting when roots are 2 to 3 inches in diameter. Fall or winter roots may be left in the ground until after the first light frost, but should be dug before the ground freezes. Tops should be removed and roots stored in a root pit or cool cellar at just above 32 degrees.

UPLAND CRESS

UPLAND CRESS. *See* CRESS.

VEGETABLE MARROW

VEGETABLE MARROW. A variety of SQUASH, which see.

WATER CRESS

WATER CRESS. *Nasturtium amphibium, Nasturtium officinale,* or *Roripa nasturtium-aquaticum.* A hardy European perennial, native to the north temperate regions, whose native habitat is in streams and along their boggy banks. Water cress is used in salads and, in a lightly cooked form, as flavoring in soups, stuffings and sauces. It is pungent and peppery and contains some important nutrients.

A small bunch of water cress, about ¾ cup weighing 25 grams, contains 1250 units of vitamin A, .030 milligrams of vitamin B_1, .090 milligrams of vitamin B_2, 15 milligrams of vitamin C, 40 milligrams of calcium, 11 milligrams of phosphorus, .8 milligrams of iron and supplies only 6 calories. It is a very rich source of iodine as well, if it grows in an area where iodine is present in the soil.

Range and Soil

Growing naturally, water cress can be found in black muck in and along streams whose water is pure and cold. It grows better in the northern states than in the South, except at elevations where cold streams are to be found in the southern states. Unless supplied by a cold spring, it will grow best in spring and fall in the North. But under favorable growing conditions, the plants in the North may become a foot or more tall during the summer, instead of the 3 or 4 inch growth it makes when growing periods are limited by heat and warm water or freezing weather.

Commercially, water cress is grown in areas which may be flooded after plants are set out in muckland. Somewhat similar conditions may be made for it in the home garden, if the water source is unlimited.

Cress grows along a stream in soil that is rich in woods mold and it grows best in limestone regions. Sifted leaf mold or sifted compost, with limestone enough to bring the *p*H up to between 6.0 and 8.0 or running water with a similar *p*H is necessary.

Culture

Plants may be started indoors before they are planted out in a stream or seed may be sown in the soggy bank of a stream outside. If they are planted in a stream, the purity of the water should be ascertained first.

To start seed inside, sow it in a pot of sifted humus and place the container in a pan of water on a sunny window sill. Water in the pan should be changed daily. When the plants are large enough to handle, they may be transplanted to a specially prepared cold frame, to individual pots or to a stream.

Pots of water cress may be matured if they are grown standing in pans of water that are kept cool and are renewed daily. They may be grown in a sunny window, in a greenhouse or in a pan of water outside kept cool by being sunk in the ground. If the outdoor pan is kept under a water faucet, the water may be allowed to trickle into it during the hottest part of the day, to keep the plants cool.

A small cold frame which is near a water source can be used to grow water cress after the seedlings are started indoors. The soil in the frame should consist of at least 50 per cent humus and should be well limed. Soak the soil thoroughly before setting out the plants, which are spaced 8 inches apart. After they are set out, the plants are watered every other day, the soil being thoroughly soaked each time. This type of culture is successful only in the cooler areas and then only early in spring. Water cress is fairly hardy, so with the protection of the cold frame, it may be set outside a month before the last frost.

If seed for water cress is not obtainable (the supply is sometimes short) cress may be grown from bunches purchased in the stores. Buy it as fresh as possible, with no yellowed leaves. Plant the stems with their roots directly in pots standing in water or allow the roots to develop by placing the bunch in a glass of water which is changed daily. Do not submerge all the leaves. A few must be left above water, here or when it is planted in a stream, to supply the plant with carbon dioxide.

WHEAT

WHEAT. *Triticum aestivum* or *T. sativum*. A cultivated grass of the family *Gramineae* which does not occur in the wild state now. It is believed to have originated in Asia, where closely allied grasses are found growing wild. Wheat is thought to have been cultivated by prehistoric man, some specimens having been found in the Stone Age relics in Switzerland. It is certain that its culture predates historical records in Egypt, China and Greece, where the earliest paintings show wheat in full cultivation. It is grown all over the world and in all climates except the humid tropics, the Arctic and Antarctic.

The wheat plant grows to about 4 feet in height, with half-inch wide, grass-like leaves 12 to 16 inches long and develops a head about 4 inches long which may or may not have awns (a beard). There are two general types grown in the United States—one that loses its chaff and another type of grain to which the chaff clings in the threshing process. The former is considered our true wheat, while the latter is used for stock feed and includes emmer, spelts and einkorn.

The free-threshing type comprises several groups which vary in their climatic adaptation, planting time and use, as follows:

Hard red winter wheat is seeded in fall and can stand colder winters than any other class. It is grown from Texas north through Nebraska, with a little grown farther north under favorable conditions. It is used for bread flour, having a high gluten (wheat protein) content which ranges from 12 to 14 per cent.

Soft red winter wheat is less tolerant of extremely low temperatures and requires more moisture than the hard red type. It is also sown in fall and is grown in the eastern part of the country, as well as along the Pacific coast. It is used for pastry flour, containing only 10.5 to 11.5 per cent gluten.

Hard red spring wheat is seeded in spring where the winters are too cold for the winter wheats. It is grown in the coldest states and up into Canada. It contains 14 to 15.5 per cent protein, more than any wheat except durum, whose gluten is of a quality which cannot be made into light bread.

Durum wheat grows in the same northern areas as hard red spring and may contain up to 17 per cent protein. Its gluten is extremely elastic and it is used commercially to manufacture spaghetti. The home gardener who wants to raise a small quantity of wheat for his own bread and does not care whether the bread is tall and light or compact and heavy in the loaf, will have a more nourishing product if he raises durum.

White club wheat or *common white* is the starchiest variety of all, and is raised entirely for pastry. It contains less than 10 per cent protein, sometimes as little as 8.5 per cent. It is grown in the warmer, more moist regions, especially in the far West.

Whole wheat and wheat germ are among the few foods which contain all of the B vitamins. Flours sold in the stores which have had the wheat germ removed are often sold "enriched," that is, with one or two of the B vitamins returned. Even whole wheat flours, as usually sold, have the perishable wheat germ removed. Unless a flour specifically states that it is 100 per cent wheat and that it is perishable unless refrigerated, it probably lacks some of the valuable B vitamins and wheat germ.

For that reason home gardeners may wish to grow, thresh and mill small quantities of flour for their own bread. The directions given below will assist a gardener who is not a farmer and who does not have heavy farm machinery available.

Range and Soil

No part of the United States is too hot for wheat, but some sections may be too humid. When too much humidity is supplied, the foliage proliferates, but the grain refuses to make heads. Wheat is grown in this hemisphere as far north as 60 degrees North Latitude, a line running through the southern part of Alaska. It will grow wherever there is an annual rainfall between 30 and 60 inches and where the average temperature for two months before ripening is equal to 68 degrees. Where the temperatures are higher, a greater amount of moisture can be supplied. At lower temperatures, the lower ranges of rainfall are best.

Soil acidity should be close to neutral—*p*H 6.0 to 7.0. If lower, ground limestone should be harrowed in after the soil has been plowed. For a planting of winter wheat, the soil should be prepared in late summer or early fall. A half acre, well-fertilized and cared for, should grow from 10 to 15 bushels of wheat. On a plot that size, if the *p*H is 5.5 or below, work in 1,000 pounds of limestone, and in addition, 500 pounds of phosphate rock, 500 pounds of greensand or granite dust and as much rotted manure or mature compost as is available. Allow the field to lie for several weeks and after weeds have sprouted, harrow or rotary till it again. If there is time before planting, the process may be repeated again. The more often weeds are eradicated in this manner in fall, the fewer there will be at harvest time.

Seed

Certified seed grain should be purchased, to get the cleanest stand of wheat with fewest diseases. Seed should be broadcast over the area as evenly as possible at the rate of two bushels per acre. It will germinate best if the temperature is between 75 and 88 degrees, but is able to germinate at any temperature between 40 and 108 degrees. Seed does not germinate properly unless it has had at least one month's rest after ripening.

After the first seed has been purchased, it is best to save seed from one year to the next. Seed saved should be carefully cleaned of weed seeds and diseased grain. Better strains for the particular situation and area will be gradually developed from seed grown there than may be purchased from fields that may be in entirely different situations, if not in different parts of the country.

Planting

Winter wheat is planted after the Hessian fly free date, which differs in various parts of the country. Obtain the date for your area from your county agent. In the mid-North, the date is around September 15.

Broadcast the seed as evenly as possible and rake or harrow it under. It should grow 3 to 6 inches before frost. Toward spring, top-seed the field with clover in order to grow a soil-developing cover crop after the wheat is harvested.

Harvest

When stems and heads are golden and the grain cannot be dented by a thumbnail, the wheat is ready to harvest.

If a neighbor can help with a combine, much work will be saved. If not, the home gardener may need to resort to the old methods of harvesting. A sicklebar mower on a small tractor will cut the grain, but by running over it afterwards, it may shatter the heads so that the wheat will be lost on the ground. A sickle or scythe provides the best method of mowing. Stems should be cut close to the ground when the wheat is entirely dry—after 11 A.M. when the dew is off the field, on a sunny day. After it is cut, it should be tied in bunches and stacked in sheaves, heads up. It is allowed to cure in the weather outdoors for several weeks. Then, when it is as dry as possible, it is stored in a well-ventilated, dry, covered place for 1 to 5 months to sweat before threshing. If the sweating process is entirely completed, the threshed grain will store better without heating up. If it was cut and threshed by combine, it must be stored carefully, or it may spoil when it heats.

Threshing may done without machinery in one of two ways. A handful of stems may be beaten against the inside of a barrel and the grain and chaff will fall to the bottom of the barrel. Or the whole grain, on its stems, may be laid on a concrete or hard dirt floor and flailed. A flail can be made out of a broom handle, with about a foot cut off and tied loosely to one end with wire or a leather thong. This loose end is used to beat the grain. When the process is finished, the stems may be picked up from the floor and used as straw mulch in the garden or may be put on the compost heap. Left on the floor will be the grain and chaff.

Next the grain must be winnowed to remove the chaff. This can be accomplished by shoveling the grain from one pile to another, tossing it in the air, on a windy day. The chaff will blow away and the wheat remain. Or it may be poured from one container to another in front of a strong electric fan. All that remains now is to pick out weed seeds and store until it is needed for bread. The grinding should be done just before baking, if possible, because the wheat germ will keep best if it is not exposed to the air. After being ground, the flour should be kept refrigerated until used. Mills may be purchased which will grind the grain by electricity, a much better method than hand-grinding, which is a long and arduous process.

WILD RICE

WILD RICE. *See* RICE, WILD.

YAM

YAM. *Dioscorea,* various species. True yams are the fleshy tubers of tropical and subtropical vines seldom grown or eaten in any part of the United States. The only place where some of the food species might be grown is in the southern tip of Florida. The name is erroneously applied to sweet potatoes in southeastern United States, though sweet potatoes do not even belong to the same family. Yams provide an important part of the diet in many of the tropical countries where they are grown. Their culture centers in the South Pacific islands, but has spread to Japan, China, Australia, India, Africa, the West Indies and South America.

Yams are easy to grow in any place where they have a 12-month growing season with plenty of heat. Their tubers may be planted any time of the year in warm sandy soil. Some species produce their tubers above ground in leaf axils; others produce them so far under ground that they are difficult to dig. A few of the most important species are listed below.

D. aculeata, the birch-rind yam, grown in the West Indies, Indian archipelago and the Pacific islands. It is sweetish and regarded by some as one of the finest esculent roots on the globe. Grown in Jamaica, it has a slightly bitter taste.

D. alata, the white yam grown in all the tropical countries, one of the most widely grown species.

D. divarticata, the Chinese yam, known in the United States as the cinnamon vine and grown as an ornamental but not for food. In China and the Philippines the tubers are eaten.

D. fasciculata is also grown in the Philippines as well as in the vicinity of Calcutta and closely resembles some of our potatoes.

D. sativa is grown in South America and sometimes in the tropical South of the United States. Its roots are flattish, sometimes pronged and resemble sweet potatoes.

ZUCCHINI

ZUCCHINI. *See* SQUASH.

Section
3.

21—Planning the Home Fruit Garden

THE IDEAL FRUIT garden or home orchard should contain many kinds of fruits, represented by a number of varieties which ripen one after another through as long a period as possible. It should be planned to provide the best fruits from a nutritional as well as from a gastronomic standpoint. A fresh supply of just-picked fruit should be available from early spring to late fall and stored or frozen garden fruit should carry through the winter.

The home gardener need not sacrifice vitamins and flavor for marketable eye-appeal and shipping qualities, as the commercial grower is forced by competition to do. The gardener can grow for his own use delicate fruits which never appear in the markets because they are too perishable or new fruits to which the fruit-buying public has not yet been educated. Whatever the size of his garden—whether it occupies a window sill, a suburban lot or an acre or more in the country—the organic fruit gardener can supply his family with health-giving desserts of a quality that cannot be purchased in the stores. But whether he can grow all of the fruit they need depends not only on the size of his garden, but also upon the time which he can give to its care. A large fruit garden takes a lot of time and care, especially for the organic gardener who eschews quick bug-killing sprays in favor of poison-free fruits which can be eaten directly from the garden.

Fruits purchased in the markets have been sprayed between 5 and 9 times each while they hung on the trees. If the fruits were stored for any length of time, they may also have been gassed at regular monthly intervals to preserve them. Advertisements by the companies who manufacture poison sprays assure the gardener that all residues disappear within a few days. But warning after warning comes from state agricultural stations, as well as from other experts who write for the commercial growers, that they must be sure to eliminate sprays during the last month or more of field growth *if* their products are expected to cross state lines. The Federal government can seize shipments of fruits bearing more than a certain residue of poison when the crop crosses a state line. But it is powerless if the poisoned fruit is sold within the state. Moreover, not every shipment over the border is examined. So even if you do not buy fruit grown within your own state, you are not fully protected by the government. If you want full protection, you must buy organically grown fruit or grow it yourself. In the following pages, we have tried to assemble the best possible directions for growing fruit of every kind and in every section of the country, without the use of poisons.

Nutritional Value

Fruits contain considerable amounts of several vitamins; they are major contributors of most essential minerals; and their role in aiding digestion, body development and tone is particularly important. But beyond the ordinary nutritional statistics, fruits have some unique and delightful functions in the well-rounded diet that make them deserve far more consideration —and use—by those who do the family food planning *and* the family gardening.

More than any other food group, fruits introduce a fascinating variety of color, taste and texture to meals and snacks. Fruits have a very low caloric content and, with the exception of avocados and olives, a practically negligible amount of fat.

Fresh fruits, because of the cellulose and organic acids they contain, have a natural laxative effect, constantly aiding the passage of food along the gastro-intestinal tract. What's more, they yield an alkaline residue or ash that neutralizes the acid residue formed by meat, eggs and other protein-rich foods. Because of this, fruits are exceptionally important in maintaining acid-base neutrality—that balance so essential to good health, nutrition and digestion.

Most of us know the citrus fruits are especially valuable for their ascorbic acid (vitamin C) content. Along with these go berries and melons as fairly good suppliers. All fresh fruits, in fact, make some contribution of this highly perishable vitamin that needs constant replenishing.

Another nutritive factor found in fruits is carotene, which our bodies convert to vitamin A. Such fruits as apricots, peaches, cantaloupes and bananas supply appreciable amounts. And although tree crops in general offer only small quantities of the B-complex vitamins, several of the dried fruits and citrus varieties contribute some thiamine (B_1) to the diet. Speaking of dried fruits, make sure those you select are *sun-dried,* not sulfured. There's an important difference in the overall health value.

On the mineral side of the ledger, fruits shine even more brightly. Their potassium content is quite high and is usually combined with those vital organic acids. Large amounts of calcium are present in the dried fruits and somewhat more moderately in oranges, raspberries and strawberries. Iron makes an appearance in dates, figs, bananas, peaches, prunes, raisins and apricots. In the last 4, it is found in quite high amounts—and, what is another important consideration, mostly in an *available* form.

Since fruits have a high water content and low proportions of fat and protein, they represent especially good sources of food iron and other minerals because they can be added to the diet without replacing other foods or making the total calorie consumption too great. Nutritive values of specific fruits are given under the separate fruit entries.

While fresh fruits, as emphasized before, are to be preferred (any processing or other treatment lowers the nutritional value), they may of course be cooked and prepared in numerous ways to lend further variety to our eating and to make them suitable for infant or invalid feeding. What the cooking—either stewing, steaming or baking—accomplishes is a softening of the cellulose so that they are more readily digested. It's also one way to add to their keeping quality and, second to freezing, it is the best solution to a sizable surplus.

Making a Plan

First consideration in planning the home fruit garden or orchard is the requirements of the family. The list of fruits to be grown, within limitations of climate, should be dictated by the family's preferences. Plan to grow as much of the family's favorite fruit as space and time will permit. Only if both extra space and time allow, include in the plan marginal fruits which may not appeal to all the members of the family.

TABLE 54: Nutritional Values of Common Fresh Fruits

Fresh Fruit	Measure	Calories (Energy Value)	Protein gm.	Fat gm.	Calcium mg.	Iron mg.	Vitamin A Internat. Units	Thiamine (B_1) mg.	Riboflavin (B_2) mg.	Niacin mg.	Vitamin C (Ascorbic Acid) mg.
Apples	1 medium	76	.4	.5	8	.4	120	.05	.04	.2	6
Apricots	three	54	1.1	.1	17	.5	2990	.03	.05	.9	7
Avocados	½ peeled	279	1.9	30.1	11	.7	330	.07	.15	1.3	18
Bananas	1 large	119	1.6	.3	11	.8	570	.06	.06	1.0	13
Blackberries	1 cup	82	1.7	1.4	46	1.3	280	.05	.06	.5	30
Blueberries	1 cup	85	.8	.8	22	1.1	400	.04	.03	.4	23
Cherries	1 cup	65	1.2	.5	19	.4	660	.05	.06	.4	9
Currants	1 cup	60	1.3	.2	40	1.0	130	.04	0	0	40
Dates (fresh or dried)	1 cup, pitted	505	3.9	1.1	128	3.7	100	.16	.17	3.9	0
Figs	4 large	79	1.4	.4	54	.6	80	.06	.05	.5	2
Grapefruit	½ medium	75	.9	.4	41	.4	20	.07	.04	.4	76
Grapes (American)	1 cup, raw	84	1.7	1.7	20	.7	90	.07	.05	.3	5
Guavas	one	49	.7	.4	21	.5	180	.05	.03	.8	212
Lemons	1 medium	20	.6	.4	25	.4	0	.03	tr.	.1	31
Limes	1 medium	18	.4	.1	21	.3	0	.02	tr.	.1	14
Loganberries	1 cup	90	1.4	.9	50	1.7	280	.04	.10	.4	34
Mangoes	1 medium	87	.9	.3	12	.3	8380	.08	.07	1.2	55
Oranges	1 medium	70	1.4	.3	51	.6	290	.12	.04	.4	77
Papayas	1 cup	71	1.1	.2	36	.5	3190	.06	.07	.5	102
Peaches	1 medium	46	.5	.1	8	.6	880	.02	.05	.9	8
Pears	1 med., peeled	95	1.1	.6	20	.5	30	.03	.06	.2	6
Persimmons	1 med., seedless	95	1.0	.5	7	.4	3270	.06	.05	tr.	13
Pineapple	1 cup, diced	74	.6	.3	22	.4	180	.12	.04	.3	33
Plums	1 medium	29	.4	.1	10	.3	200	.04	.02	.3	3
Prunes (dried)	4 med., unsulfured	73	.6	.2	15	1.1	510	.03	.04	.5	1
Raisins (dried)	1 cup, unsulfured	429	3.7	.8	125	5.3	80	.24	.13	.8	tr.
Raspberries (black)	1 cup	74	1.5	2.1	54	1.2	0	.03	.09	.4	32
Raspberries (red)	1 cup	70	1.5	.5	49	1.1	160	.03	.08	.4	29
Strawberries	1 cup	54	1.2	.7	42	1.2	90	.04	.10	.4	89
Tangerines	1 medium	35	.6	.2	27	.3	340	.06	.02	.2	25

Of prime importance in successful fruit growing according to William Ackerman, is the selection of nursery stock, procurement of varieties that are best adapted to your particular climate and soil and the proper arrangement, spacing and planting of these varieties.

Fruit trees may be planted in the spring or fall. In the milder parts of the United States where minimum temperatures are not likely to go below 0 degrees Fahrenheit, they may be planted in the late fall before the ground is frozen. Peaches, apricots and other tender fruits often suffer considerable winter injury in more severe climates and should therefore always be planted in the spring. Spring-planted trees should be set as early as the ground will permit.

In an article written for an English organic gardening magazine, Mr. Ackerman gives the following advice:

In buying nursery trees, it is safest to purchase from a firm of known honesty and reputation. There is some advantage in giving preference to reliable local nurseries, where transportation charges are less and there is less likelihood of delay in transit with consequent drying out or freezing of the plants. However, these points should not outweigh the advantages of receiving good trees wherever they may be obtained.

Nursery catalogues usually list several grades of trees according to height and caliber of tree. The age of trees is also stated as one-year-old, two-years-old, etc. It pays to buy the better grade larger plants of a given age.

Most large nurseries dig their nursery trees in, in the fall and store them over winter. Any reliable nursery will have sufficient sales to unload its stock without lowering the sales prices appreciably. Therefore, beware of so-called "bargains." Such trees may have been winter injured, dried out from warm storage or have frosted roots. It is often difficult to detect these injuries until it is too late.

When you receive your trees from the nursery, be sure they do not dry out at any time before planting. If not planted at the time they are received, the trees should be heeled in. Select a place that is protected from excessive exposure and where the soil is well drained. Dig a trench deep enough to accommodate all the roots, throwing the soil in a mound along the south side of the trench. The boxes or bales in which the trees were shipped should be opened or heating may occur within the package and the trees injured. Cut the bundles and separate the trees from each other somewhat, making sure that the varieties are separated so they may be distinguished later. The trees should be placed close together in the trench row and slanted toward the south to prevent sun injury. Fill the trench with soil being sure all the roots are covered. Trees heeled in this fashion may be held for a considerable length of time before they are permanently planted in the orchard.

Planning the Orchard

Assuming that the location of the orchard has been determined, a plan must be decided on for the arrangement of the trees. Orchard trees are generally planted according to some definite system. The 3 most common arrangements of trees in an orchard are the square, the quincunx or diagonal and the hexagon or triangle. Where the land is rugged and steep a fourth system, the contour arrangement, is best.

F. M. Green and A. M. Binkley of the Colorado Experiment Station describe these 4 common arrangements as follows:

1. The square plan is the usual arrangement and makes for easy orchard operation.

2. The quincunx arrangement is the square plan with an additional tree planted in the center (a good method for using filler trees).

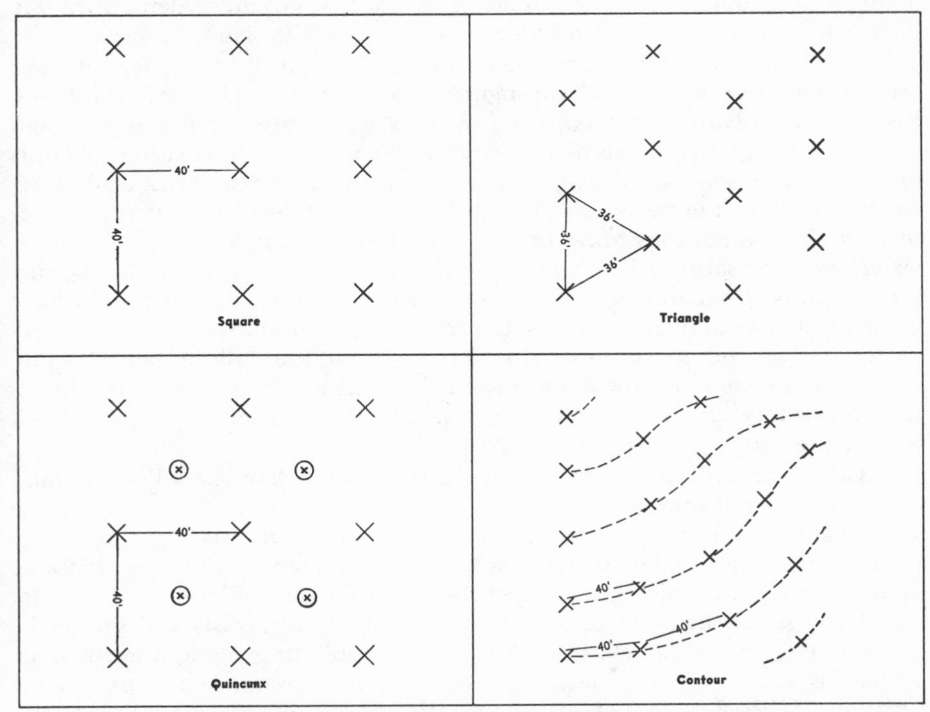

The 4 common arrangements which are used in laying out fruit orchards include the square, triangle, quincunx and contour. The square plan is the usual arrangement as it provides for easiest orchard operation. The contour plan is used primarily on slopes to reduce erosion.

3. In the triangle plan, all trees are equidistant. This arrangement allows for 15 per cent more trees per acre than the square system.

4. In the contour arrangement, trees are planted along contour lines. This plan reduces erosion and makes it possible to grow an orchard on hilly ground.

Trees in a small home orchard may be arranged according to one of these systems if a large enough number will be planted. Usually, however, no more than half a dozen of any one kind of tree is required to supply a family's needs and a larger planting requires more attention than the part-time gardener can allot to it. It is better to plant fewer trees which are within the gardener's capacity than to plant so many that he is overwhelmed with work that he cannot get done.

Where the home orchard occupies a suburban lot, its capacity and arrangement is dictated by the amount of space in the lot. The home orchard in the country allows greater latitude in the planning. The choice of site is most important to the success of many fruits.

Location

The site of the home fruit planting is usually determined by circumstances. M. B. Hoffman of the Cornell Extension Service says that nearness to the house is desirable, but if the building is on low ground where frosts are likely to occur it is better, if possible, to plant on higher land.

Fruit plants differ from crop plants in that they are perennials and their root systems live in the soil throughout the year, Mr. Hoffman continues. Because fruit plants start their top growth early, a correspondingly early and vigorous root growth is needed to supply the tops with increasing amounts of water. The soil must have, in addition to available water, an adequate supply of air so the roots can "breathe" and thus release the energy that is used by them when they grow and absorb water and nutrients. A soil that is waterlogged or saturated during the spring does not contain the air needed for the proper functioning of the roots of fruit plants. The soil should have enough internal drainage to allow the roots to start unimpaired activity early in the spring and to continue this activity until leaf fall in late autumn. Otherwise, maximum growth and production will not be realized. In obtaining satisfactory performance of fruit plants there is no substitute for a well-drained soil.

A satisfactory site for an orchard is slightly higher than the adjoining land and has good soil and air drainage. Cold or cool air functions in much the same manner as water. Air in the absence of wind currents has a tendency to move from the higher to the lower levels. Because of this, the desirable orchard site is one which has adjacent low land for the cold air to drain into. Air drainage can be good insurance against late spring frosts and should be a deciding factor in the selection of an orchard site. In general, a north slope will delay blooming and maturity, whereas a south slope will hasten these processes. The effect of a slope, to a certain extent, may be partially or completely overcome by local conditions such as topography or nearby bodies of water. Low areas are usually subject to winter-killing and frost damage unless protected by air currents from adjoining canyons.

The orchard may be located on new land, old crop land or old abandoned orchard soils which have previously proven satisfactory for fruit growing. If an old orchard site is to be replanted, the land should be cropped for several seasons or a green manure crop may be grown and turned under before the new orchard is set. Turn the green manure crop under in the fall and set the young trees the following spring. It is also a good idea to plant a different kind of fruit on an old orchard site than was previously grown there. Trees set in new holes can also be expected to do better than those planted where old trees have been recently removed.

The contour system is the most feasible arrangement for planting an orchard on a hillside. In contour planting on sloping land, all trees in any one row are in the soil at the same elevation. Obviously, the tree rows will not be equally spaced in all parts of the orchard.

The first step in laying out a contour planting is to decide upon a minimum allowable interval between rows. This distance will depend on the kind and varieties of fruit trees planted. The first contour line should be laid out at the highest elevation. From a given point, the line is projected on the contour in both directions to the limits of the area to be planted, by using an engineer's level and rod. Then proceed to the steepest slope in the orchard below the first line and establish a point at the minimum distance between rows below the first line. A line is then projected in both directions from this point in a similar manner as that in establishing the first line. If at any place in the orchard the distance between two adjacent lines becomes twice the minimum interval, a new line is laid out on the contour between them.

The planting distance must be determined principally by the size the trees are likely to attain at maturity. This is largely dependent upon the type of fruit, the variety, fertility and depth of the soil and amount of rainfall or available moisture. The trees must be sufficiently far apart to allow the sun to hit the lower branches if fruit of satisfactory quality is to be grown on the lower parts of the trees.

Although the matter has never been definitely settled the following distances are most frequently used:

TABLE 55: Planting Distances, Yields and Bearing Age of Fruits

Fruit	Min. Distance Between Plants (Feet)	Approx. Yield per Plant (Bushels)	Bearing Age (Years)
Apple—standard	35	8	6-10
Apple—semidwarf	20	2	4-6
Apple—dwarf	12	½	2-3
Pear—standard	25	3	5-8
Pear—dwarf	12	½	3-4
Peach	20	4	3-4
Plum	20	2	4-5
Quince	15	1	5-6
Cherry—sour	20	60 qts.	4-5
Cherry—sweet	25	75 qts.	5-7
Grape	8	8 lb.	3-4
Raspberry	3	1 qt.	2-3
Blackberry	3	1 qt.	2-3
Currant	3	3 qts.	2-3
Blueberry	4	2 qts.	3-4
Strawberry	2	1 qt.	1

—Virginia Polytechnic Institute

The variation in the planting distances for each type of fruit is largely dependent upon the variety. This can be illustrated in the example of apples. Vigorous apple varieties, such as Baldwin, McIntosh and York Imperial, planted 25 feet apart in fertile soil, may crowd when only about 15 years old; planted 30 feet apart they will usually crowd when about 20 years old. Less vigorous trees such as Rome Beauty, Winesap and Golden Delicious, planted 25 feet apart, would not crowd seriously until 18 or 20 years old.

Dwarf Fruit Trees

Dwarf trees, usually growing only from 5 to 7 feet tall, are truly Tom Thumbs compared to the 35 feet or more their regular variety relatives reach. To the busy part-time gardener, this means a number of important things: They are considerably simpler to prepare soil for, plant, fertilize and mulch. In each of these operations, there's much less land area, material and tree itself to handle and give attention to.

Dwarf trees usually begin fruiting far sooner than standard varieties—most by their second year. Standard trees take up to 10 years to start bearing and

PLAN FOR HALF-ACRE ORCHARD

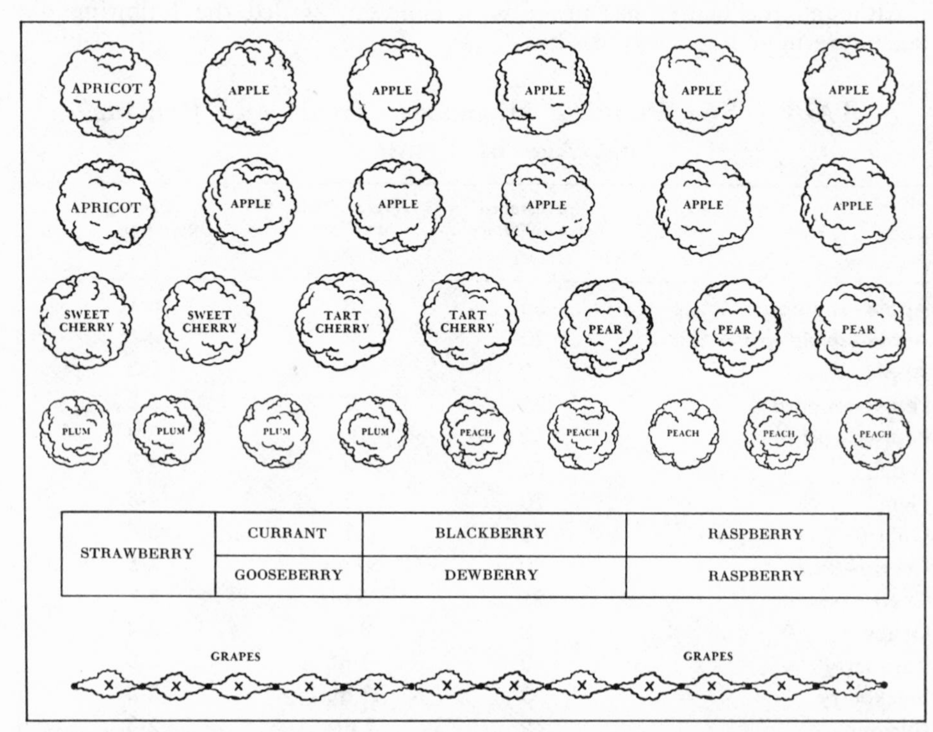

A half-acre is required for this orchard, which provides space for standard size apple trees. The whole orchard covers an area 130 by 167 feet. Grapes are trained to a wire fence on the south side of the orchard. If a quick fruit crop in the first years of orcharding is required, spaces between the larger trees at the top of the orchard might be interplanted with dwarf trees which would be removed when the standard apples begin to mature.

usually a minimum of 5. Dwarf apples, pears, peaches, plums, nectarines, apricots and sweet cherries begin yielding within a couple of years, producing well by the time they're 3-year-olds.

The fruit from dwarf varieties is equal in size to that of standard types. As a matter of fact, because these trees can be given individual care more readily, the fruit they bear is frequently *larger* than on the full-scale kind and often is of better quality, color and taste. To top it off, they are obviously easier to harvest and there's less waste or spoilage from fallen fruit. All in all, dwarf trees produce more fruit faster in any-sized area than any other method.

The 3 general types of dwarfs are:

1. Trees grafted on dwarf rootstocks.
2. Those with dwarf interstocks.
3. A true dwarf variety.

Dr. Robert A. Norton of the Logan, Utah Agricultural Experiment Station defines the differences between these as follows:

The dwarf rootstock. Most of the dwarf apples, pears and other fruits

PLAN FOR 100′ x 125′ ORCHARD

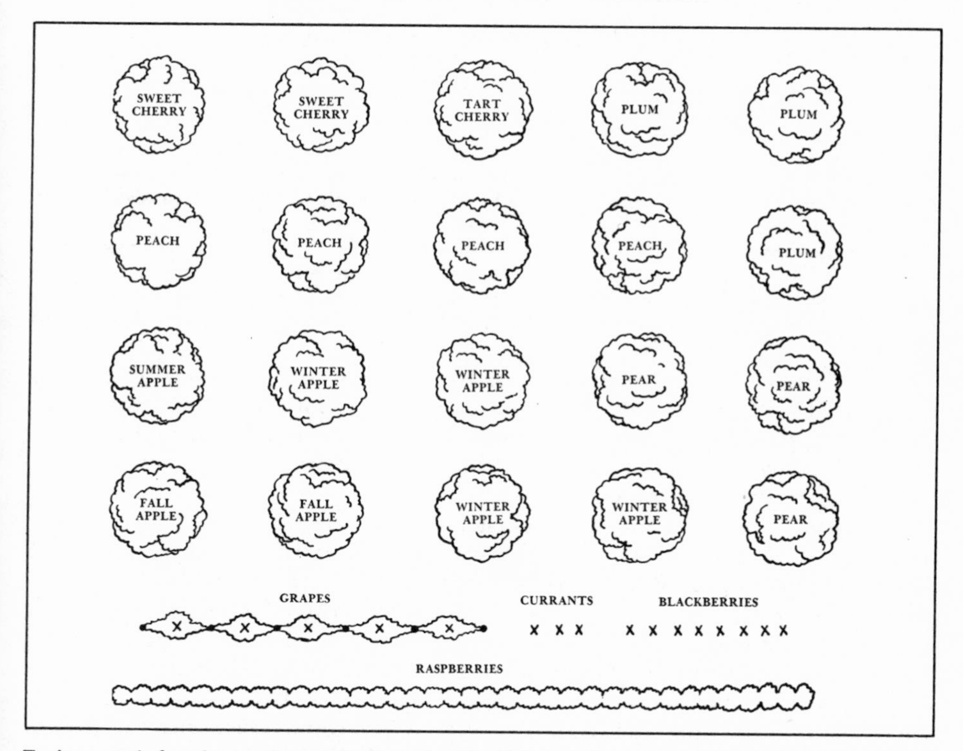

Fruit enough for the year's supply for a family of 5 may be grown in an orchard 100 x 125 feet if dwarf apple and pear trees are interplanted with standard size peaches and plums. This plan makes provision for the longest possible extension of the season by mixing varieties, and also allows for cross-pollination where necessary.

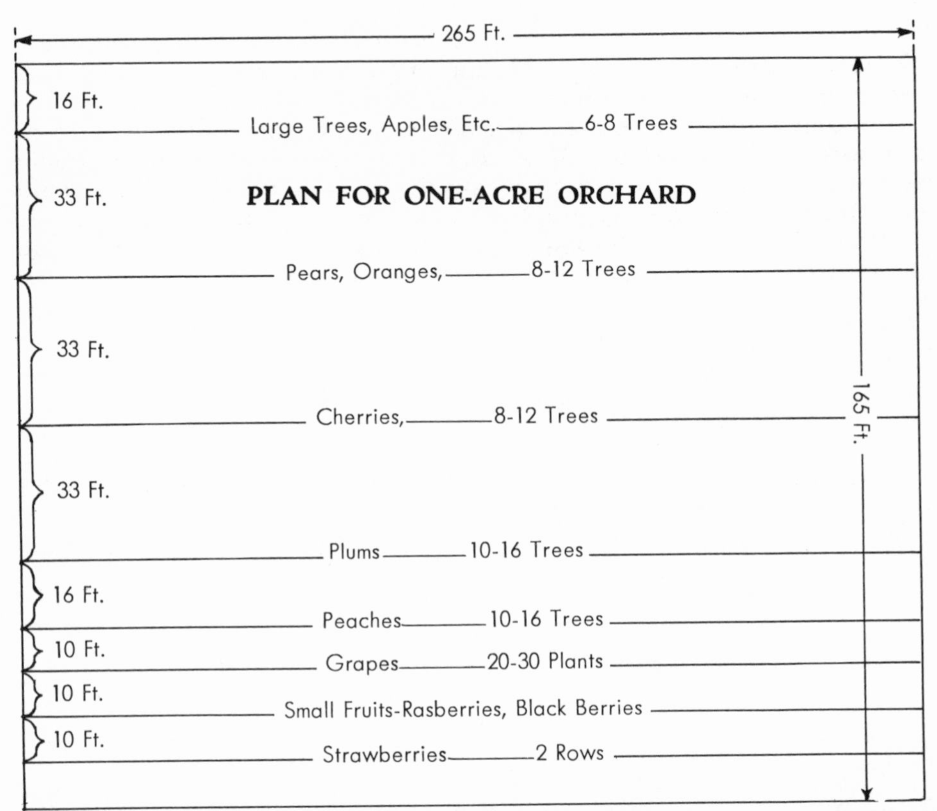

A fruit orchard covering an entire acre should provide enough fruit for a good-sized family, with some left over for sale. All trees in the plan given above are standard size. While they are small, the space between the larger trees might be interplanted with dwarf trees, small fruits, or a row or two of vegetables. Remember, though, when interplanting, that the primary object of the orchard plan is to grow trees that will bear well for a generation or more. Do not sacrifice the ultimate good of the trees to temporary benefit from interplantings.

which originate in the Northwest, Dr. Norton says, are produced by grafting the desired variety on a selected rootstock. The rootstocks are actually old semiwild apple varieties from Europe which have been selected and classified by horticulturists of England's East Malling research station according to the degree of dwarfing produced.

The most common clones for rootstocks are the Malling IX, VII and II. Trees on Malling IX roots are extremely small, usually not exceeding 6 to 8 feet in height. Malling VII produces a semidwarf tree, similar in size to a peach tree, while trees on Malling II are about three-fourths the size of a standard apple tree. Trees on Malling IX rootstock come into bearing at an early age and are considered an ideal type for the small property owner. These trees require staking since they have a weak, brittle root system; thus they are generally not planted commercially. The commercial possibilities of Malling VII and II are being studied in Utah and in many areas of the country. They are ideally suited to the backyard orchard where only a 15 to 18 foot spacing between trees is needed. Except in extremely windy areas the semidwarf apples do not need artificial support.

The dwarf interstock. In the dwarf interstock type growth is controlled by the weak root system. To attain better anchorage in the ground and more resistance to wind the dwarf interstock was developed. In this type a vigorous, well-anchored rootstock is used. The dwarf types are grafted or budded on this stock, either IX, VII or VIII (Clark dwarf). The desired variety is budded or grafted on to the dwarf interstock about 4 to 12 inches above the original graft union. The degree of growth control depends upon the type and length of interstock. A swelling of the interstock is a common trademark of this type of tree. The double grafting necessary to produce dwarf interstock trees accounts for their higher cost.

Although the interstock method has been limited mainly to apples, other kinds of fruit trees have been produced in this manner. To develop a dwarf pear tree, the Angers variety of quince is used as a rootstock. It is grafted to a compatible variety of pear such as Hardy or Old Home. The desired variety, Bartlett, for example, is in turn grafted to the interstock. Bartlett and many other popular pear varieties are not directly compatible with the quince, thus the use of the interstock.

We are finding that with many other kinds of fruits, through the right combination of rootstock, interstock and variety, we can obtain numerous variations in size, shape and bearing potential.

The true dwarf variety. A few cases of true genetic dwarfing have been found in desirable fruit varieties. The Delcon apple, which resembles Delicious, is naturally a semidwarf variety, that is, the tree size is smaller while the fruit is of normal size. Several dwarf and semidwarf red tart or sour cherry varieties have been introduced recently including Dwarfrich, Meteor and North Star. The new "spur-type" strains of Delicious apples are semidwarf, similar in size to trees on Malling II rootstock. This type of dwarfing is rare, however.

Most fruit trees can be "dwarfed." Apple growth control can be attained either through a root or interstock of one of the East Malling and, more recently, the Malling-Merton clones. The latter are resistant to woolly apple aphid. One true dwarf variety, Delcon, has proved suitable for planting in the home garden.

A series of 3 types of quince, named Quince A, B and C, may be used for pears, each being grafted to a compatible interstock of Hardy or Old Home. Quince A, an Angers type, is the best of the series, producing a semidwarf tree, while trees on Quince C are extremely dwarfed.

For commercial production, dwarf forms of peaches, plums, apricots and nectarines are not needed. Standard size trees come into bearing at from 3 to 5 years of age and can be controlled in size by pruning. To satisfy the demands of the backyard orchardist, several dwarfing rootstocks have been used with varying degrees of success. St. Julien A and *Prunus tomentosa* have been used for many types of peaches, apricots and plums and are especially good for plums. The western sand cherry, *Prunus besseyi,* is compatible with most plums and is fairly successful with peach and apricot, although it suckers profusely.

No completely satisfactory dwarfing stocks have been found for the sweet and sour cherry, though the need is great from a commercial as well as from the home gardener's standpoint. Some promising combinations of root and interstock are now being tested in several nurseries and experiment stations throughout the country.

One vital note regarding how to plant dwarf fruit trees: They will stay dwarf only if planted correctly. Because the dwarf is produced by grafting, it must be planted with the graft-union above ground. This is a knob at the base of the trunk and can be seen by a change in the color of the bark. Be sure this union is kept above ground. If it is placed below, the upper part of the tree (standard variety) will form its own roots—and, instead of growing 6 to 8 feet, it will shoot up 35 or 40!

Plant dwarfs in early spring if you live where winters are severe; in late fall or early spring in more moderate climates. Place them 10 to 12 feet apart each way or 6 to 8 feet apart in rows with 15 feet between them. Pack good topsoil firmly around the roots.

Not all varieties in either standard or dwarf trees are self-fertile. In order to set fruit, some must be cross-pollinated by another variety planted nearby. With apple or pear trees it's best to plant two dwarfs of different named varieties to assure cross-pollination and get more fruit. Peaches, nectarines and apricots are usually self-fertile, as are plums which set a better crop, however, when several varieties fertilize each other.

Sour cherries, by the way, have some special advantages that make them doubly attractive for dwarf fruit planting by home gardeners. They're hardier, more disease and insect-resistant than any of the others. They are self-fruitful, do not require cross-pollination. And they're early bearing, yield fruit in June or by mid-July, starting with the third year.

Not much pruning is needed with the dwarfs. When planting dwarf apples, the lowest branch should be about 6 to 12 inches above the ground. Prune back any on the trunk below that height. After that, simply watch for water sprouts or suckers, trim off broken ends and let the tree take its natural shape.

Against a wall or trellis, the ornamental or "espalier" dwarf fruit tree is an attractive variation. Popular in Europe where they achieve a double purpose in providing food and decoration, the espaliers are trained by special pruning, bending and tying into flat, interesting designs.

Another recommendation is that excess fruits (anything more than one fruit along each 6 inches of a branch) be thinned by picking off the small fruits. If not thinned, say the experts, dwarfs may set more fruit than they can carry and develop to good size and quality and may not bloom the following year.

As with all fruit trees, mulching is ideal for dwarfs. Mulch with straw, grass clippings or other organic material and make it deep enough around each tree to smother grass and weeds. Keep mulch a few inches away from the trunk to avoid attracting bark and root-damaging mice or rabbits. Protect trees from rodents or other animal pests by applying a shield of fine wire mesh (hardware cloth) around the trunk base.

There are several varieties of most dwarf fruits available in just about every part of the country. Weather extremes and soil conditions can, of course, make a difference. Your state agricultural experiment station or county agent can advise you on the adaptability of a specific dwarf type or variety to your locality. For the most part, those fruits that grow well as standard trees in the area may also be grown as dwarfs.

For a better selection of hardy, satisfying and new varieties, check with the nurseries, with the latest catalogs and advertisements. Here are just a few dwarf fruit tree varieties from which you can choose. (In addition, some of the nation's larger nurseries and mail-order suppliers have developed many of their own named varieties that are excellent choices.)

Apples. Baldwin, Cortland, McIntosh, Red Delicious, Northern Spy, Yellow Delicious, Rome Beauty, R. I. Greening, Jonathan, Stayman Winesap, Gravenstein, Redwin Spy, Spitzenberg, Wealthy, Yellow Transparent.

Pears. Bartlett, Seckel, Kieffer, Clapp's Favorite, Duchess, Anjou, Fame, Lincoln, Flemish Beauty, Red Bartlett, Gorham, Easter, Sheldon.

Peaches. Elberta, Golden Jubilee, Red Haven, Eclipse, Rochester, Valiant, Crawfordy, Red Bird, Hiley, J. H. Hale, Belle of Georgia.

Plums. Italian Prune, Burbank, Stanley, Damson, Abundance, German Prune, Reine Claude, Red June.

Apricots. Alexander, Early Golden, Moorpak, Perfection, Riland.

Nectarines. Boston, Hunter, Napier, Newton, Sure Crop, Red Roman.

Choosing a Site for a Citrus Grove

Experience has shown that in selecting a location for a citrus grove, cold protection is still of primary importance, according to Fred P. Lawrence of the Florida Extension Service.

"Cold protection cannot be defined by broad zones within the state, but rather must be determined by local conditions.

"The first consideration is topography. Rolling lands with gentle slopes and no depressions are preferred because of air drainage. Cold air, heavier than warm air, drains downhill. Any preferable location for citrus should have an adequate outlet provided by adjacent low areas into which cold air can drain. The site should never be trapped by surrounding higher areas that could prevent natural drainage away from the grove.

"In addition to favorable topography, nearness to water is also an important factor in cold protection. The heat radiated from large bodies of water tends to moderate cold air moving across them. Cold fronts usually approach the state from the northwest, so locations on the south and southeast borders of lakes are more desirable. Large, deep lakes afford maximum protection. However, the number of lakes within an area is important, for one lake, even if fairly large, may fail to give the protection that can be had from a number of smaller *deep lakes.* Some degree of protection is afforded in coastal areas where the warm waters of the gulf and ocean have a moderating effect, but there are areas where this does not extend inland to an appreciable distance."

Using Filler Trees

According to Colorado's Green and Binkley, filler or temporary trees are sometimes used in an attempt to obtain greater yields and income while trees are young. In some instances, fruits of earlier bearing types, such as peaches among apples, are used. When apples are used as fillers among apples, temporary trees are planted close together in the rows. Main objection here is that many growers fail to remove filler trees before serious damage from crowding occurs to permanent plantings.

Also, because of the larger spacing between the rows, the planted space is not equally divided between trees; this complicates orchard operations.

The best system for planting filler trees is the quincunx, in which the filler tree is set in the center of the square formed by 4 permanent trees. Early bearing varieties such as Rome Beauty, Golden Delicious and Jonathan are the most suitable varieties of apples to plant as fillers. Inasmuch as the useful life of filler trees is short, the same varieties on semidwarf stocks could be used, though the initial cost of the trees is high.

Fruit on a Suburban Lot

The gardener with only a small city lot can have a homegrown fruit crop if he plants trees and shrubs for harvest as well as for shade, ornament, screening and fencing. Few city lots have room for more than one of the standard size apple or sweet cherry trees. But if dwarfs are planted, they take up little more room than a magnolia or forsythia. Fruit trees espaliered against buildings or at the property line yield larger and better fruits than those allowed to branch naturally. Used as foundation plantings, fruit bushes are just as decorative and a great deal more useful than privet or barberry.

The bush cherry is perhaps the most likely fruit to replace many small hedges and foundation plantings. This variety can be planted in close rows and can be trimmed to almost any height desired. Blossoms are beautiful in spring, and the silvery green foliage is always attractive, but the real payoff comes when quarts of succulent cherries are picked in late summer for canning, pies and preserves. Bush cherries are also adaptable for singular plantings. Use this just as you would any ornamental of comparable size. (Up to 5 feet.)

The "wild" hedges, those which should not be trimmed, include the blueberry, raspberry and *Rosa rugosa*. They may best be used for spots where larger, rambling hedges would ordinarily be used, such as along property lines, farm pastures and along roadways. These plants generally grow with amazing speed and are perfect for shutting out undesirable views or for secluding a patio from a roadway or other houses. Raspberries generally grow about 6 feet high and should be planted about 3 feet apart. A trellis can be used to help train canes, prevent wind damage and keep fruit off the ground. Blueberries grow 6 to 10 feet high, and for commercial purposes are planted 5 feet apart. You may, however, wish to sacrifice some of the fruit surface to gain a more dense hedge. If so, plant about 3 to 4 feet apart.

The *Rosa rugosa* is famous for its "hips," or fruits that form on its branches after the blossoms fall. Rose hips are nature's most compact source of Vitamin C. A quarter pound of rose hips contains as much vitamin C as 120 oranges. The plant itself will grow to a height of 5 or 6 feet and, when planted about two feet apart, will grow into an attractive, thick hedge, with beautiful blossoms appearing all summer long.

Another family of plants which are both attractive and fruit bearing are the dwarf trees previously described. Full-sized apples, pears, cherries and other fruits can be enjoyed all summer in the space which otherwise might have been devoted to ornamental shrubs. Most of these trees are very attractive, especially at spring blossom time and the fruit they bear is equal in all respects to that of full-sized trees. Dwarfs usually range from 6 to 15 feet in height. In Europe, dwarf apples are commonly seen used as hedgerows.

Strawberries can be used as a border for the flower bed. The plants are very attractive and provide another example of how you can combine landscaping with fruit growing. One expert says of the strawberry plant, "it does make an extremely attractive low hedge or border and when a vigorous upright growing variety such as Empire is used it forms a very attractive picture, both in bloom and in fruiting."

In the section devoted to cultural directions, many fruits are recommended to a northern slope, away from early spring warmth or the hot summer midday sun. The city lot gardener can translate this recommendation to mean the north side of his house or garage. For instance, apricot is every bit as hardy as the peach, but blooms so early in the spring that the late frosts often nip the flowers. Planting on the north side of buildings holds them back in spring so that they bloom after danger of freezing the flowers is past. They attain the size of a dogwood tree or a large lilac or Japonica plant. A natural hybrid dwarf apricot will grow no larger than a large Scarba Deutzia.

On the fence side of your garage or as a vine on the back porch, plant grapes. Plant Caco for a very fine sweet red, Seneca for a sweet fine amber and a seedless Concord for an easy-to-eat good-quality black grape. Grapes take very little extra space and pay very large and delightful dividends.

On the fence, you can also plant black or purple raspberries. These are tip-rooting and do not sucker from the root to be a pest in the little garden. Of the blacks, try Morrison and, of the reds, plant Sodus. This is a large, rich-flavored berry which is a cross between a red and a black variety. It is more fruitful than the older kinds and almost entirely blightproof.

If you live in the Middle West and not further north than New York, you can grow papaw, a very nice and interesting fruit, on a dwarf tree. This fruit is unfamiliar to most people, but is well worth a place in your garden. The ripe fruit looks much like a potato and the aromatic pulp is like a banana custard. It is sometimes called a custard apple, but differs from the tropical fruit having the same name.

If you live in the cotton belt, you can plant Garber or pineapple pears. They are inferior in quality to pears grown in the North, but are very good for cooking. In place of apples, plant Oriental persimmons. This is a delightful fruit, about the size and color of an orange. The trees grow to about the same size as the Anoka apple recommended for the northern garden.

In the coastal area, from Charleston and Savannah around to the Gulf try growing figs in place of plums grown in the North. Celeste and Brown Turkey are good varieties. They are easy to grow and set fruit the first year they are planted. In this area you can also plant Satsuma oranges which grow no larger than a good-sized forsythia bush. These oranges are very hardy and can endure considerable frost without being damaged.

This southern area will also grow loquats and Chinese date. These fruits are not very familiar even to the southern gardeners, but are worthwhile and please those who try them. The trees are dwarf and easy to grow.

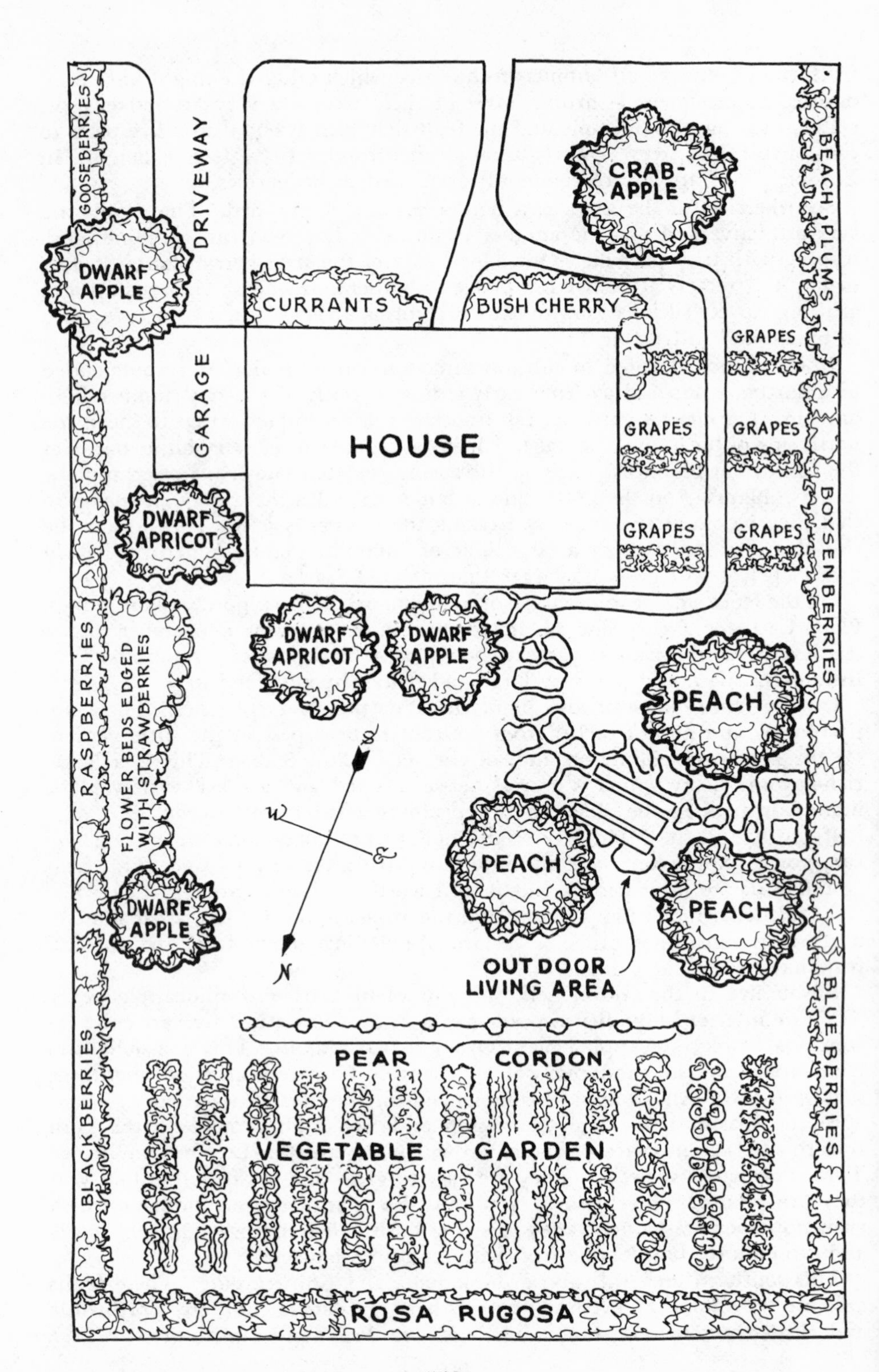
GOOSEBERRIES
DRIVEWAY
CRAB-APPLE
BEACH PLUMS
DWARF APPLE
CURRANTS
BUSH CHERRY
GRAPES
GARAGE
HOUSE
GRAPES
GRAPES
GRAPES
GRAPES
DWARF APRICOT
BOYSENBERRIES
DWARF APRICOT
DWARF APPLE
RASPBERRIES
FLOWER BEDS EDGED WITH STRAWBERRIES
PEACH
S
W
E
PEACH
DWARF APPLE
PEACH
N
OUTDOOR LIVING AREA
BLUE BERRIES
BLACKBERRIES
PEAR CORDON
VEGETABLE GARDEN
ROSA RUGOSA

Three-in-One Tree

If you are trying to grow more varieties of fruit than you can squeeze into a city lot, even by planting dwarfs, consider planting a 3-in-1 tree. This is not a single tree bearing 3 fruits, but 3 trees planted in one hole, each one allotted a third of the bearing circumference.

The 3 trees in each hole should be of the same family: apple, peach, pear, plum or cherry. The best idea is to combine an early, midseason and late variety of a single fruit. Then the same "tree" will yield for a number of months. Because this type of planting makes sense only in cramped quarters on a small lot, you will want to use dwarf trees.

Here's how to plant the 3-in-1 tree:

1. Dig a triangular hole 18 inches deep and 3 or more feet in each direction. Break up the subsoil, and add to it plenty of weathered compost, peat moss or leaf mold.

2. Get dormant, bare-root trees. Prune off most of the roots from one side of each tree—the side that is going to face the other trees. Also prune away any damaged roots.

3. Prune all branches on the same side that you pruned the roots. Cut back the main trunk of each tree to a healthy lateral branch about two feet above the ground.

4. Place a tree in each corner of the hole—trunks about 18 inches apart. The unpruned roots and branches face out from the center.

5. Firm the soil around the roots gently but thoroughly. Flood the partially-filled hole to help settle the mixture of compost, subsoil and topsoil. Add more soil until the hole is filled, but leave a basin around the young trees which will tend to hold water.

Essentially, these steps jibe with standard planting practice—except for possible overcrowding. Your trees will need careful inside pruning each year to admit some sunlight and air, but actually most of the fruit will be borne on the outer branches of the trees.

Window Sill Fruit

Growing plants indoors has several advantages, the most important being that it can be done all year 'round. Plants, to thrive indoors, must like year-round warm temperatures.

In general, tropical plants do best in a loose, rich soil. You can start with garden soil and add some sharp sand (to provide good drainage) and then lots and lots of good compost or some other humus material. If you are not an organic gardener yet and don't know what compost is or where to get it, then scrape up the top inch or two of leaf mold from the woods or fields; better yet, look around your neighborhood to see who keeps a big pile of "trash" (properly cared for, that will be compost) and beg some for your own use. But unless the potting soil is loaded with humus material of some kind, it will not retain enough moisture for proper plant growth.

Don't ever expose your plants to scorching noonday sun; coming through the glass, it may burn their leaves. And the "Saturday night bath" routine

The Creeping-fig is a natural climber and suitable for indoor planting where space is limited. While small, the Creeping-fig has tiny, heart-shaped leaves; when fully mature and ready to bear fruit, it sends out very large leaves.

is wonderful for plants, since it removes accumulated dust, washes off any undesirable insects that may be trying to establish a beachhead and the moisture soaking into the stems encourages new growth. If it isn't too burdensome, tepid showers oftener than once a week are very beneficial when the plants are in active growth. And do give *all* your plants as much humidity in the air as you can manage.

Many of the fruits listed in the chapter on exotics are suitable for window sill or greenhouse culture.

The most commonly grown window fruit is the Ponderosa lemon. Its giant-sized fruits are delicious and one lemon will make enough ade for a host of thirsty friends or furnish enough juice for a dozen pies. The Ponderosa bears fruit and flowers simultaneously, which makes the plant interesting as well as attractive. The fruits last extremely well on the plant and may be allowed to remain on it for months without danger of deterioration.

In addition to the Ponderosa lemon, which is readily available from many nurseries, several growers are now offering other citrus fruits for the indoor garden. Otaheite orange is a dwarf plant of great beauty, although its small fruits are not as useful as those of the Ponderosa lemon. The leaves are glossy and the fragrant blossoms are pink-tinted, waxy white. Plants in 5-inch pots may carry as many as a dozen oranges.

The dwarf tangerine not only has glossy foliage and fragrant flowers but the brilliantly colored fruit is very good to eat. Kumquats, limes and other small citrus fruits are equally attractive and useful. If you know you should get more vitamin C, but you don't care for prepared juices, pick whatever you like from your own window garden.

Figs are no novelty in the indoor garden. Fig plants begin to bear very early, although more fruit is obtained from larger plants, naturally. If you find it inconvenient to handle the large plants indoors, grow them in tubs, in the basement during the winter and on the porch or patio during the summer.

If you want to grow an edible fig, but lack the space for the standard types, try Creeping fig. This is naturally a climber, and support should be furnished, to which its aerial rootlets can cling. While small, the Creeping fig has tiny, heart-shaped leaves, but when the plant is fully mature and ready to bear fruit, it sends out very large leaves.

Dwarf pomegranates are beautiful plants, evergreen, glossy-leaved and with fragrant flowers. These plants have edible fruits, as do the Strawberry guavas.

If you want to grow another "strawberry," try the Strawberry tree *(Arbutus unedo)*, a handsome shrub with red or white flowers and small, scarlet, warty fruits.

Ladyfinger bananas are edible and the plants are attractive. Many other varieties are equally decorative as foliage specimens and some mature at 3 feet.

22—Selecting Fruit Varieties

A WIDE RANGE OF fruits may be grown in home gardens in almost all sections of the United States and the same fruits can be found in a garden in Texas that might be found in a New York state garden. But if the New York garden were dug bodily and transferred to Texas, few of the plants would yield good crops. This is because of temperature, of course, but also because of such factors as length of days, humidity or aridity, length of growing season between the last spring frosts and the first fall frosts, the amount and persistence of winter cold, annual rainfall, depth of snows and many other considerations. State Agricultural Experimental Stations throughout the country carry on an extensive program of trials in which varieties of many fruits are planted, to find the best ones for the various areas. Few fruit varieties grow equally well in all parts of the country. A variety which may bear bumper crops in Texas may be suitable for few other places. It is most important, then, to plant the varieties recommended to your section, taking into consideration any special needs of the particular spot on which your own home orchard will stand.

A few of the weather factors may be modified slightly by cultural practices in a home garden, but they cannot be changed to any great extent. In an arid area, irrigation will supply moisture to the roots of the trees, but a plant which definitely must have the moist atmosphere of a jungle cannot be made to grow in a desert, however much it is watered. Within certain limits, frost injury may be avoided or postponed. In the citrus areas, it is not usually practical for the home gardener to heat his orchard with smudge, especially in suburban areas. But he can prevent some damage by use of an overhead sprinkler on cold nights. Plantings further north can be protected from early or late frosts by special protective coverings, either temporary, as in the case of staw on the strawberry blossoms, or permanent, in specially constructed cold frames. But these expedients should be considered as emergency measures. An orchard should be planned to need very little emergency treatment, if it is to be at all practical.

In general, the following lists of fruits and nuts issued by the United States Department of Agriculture are suitable for planting in designated areas of the country. This list does not attempt to include all of the possible fruits that may be grown in each section, but is intended only as a general guide.

Southeastern and Central Southern States

Under most conditions in this region the best fruits and nuts for the home garden are grapes (muscadine), pecans, figs, strawberries, blackberries (trailing), blueberries (rabbiteye varieties), pears, blackberries (erect), bunch (American) grapes, peaches, plums, apples and raspberries. Under the more subtropical conditions, several citrus fruits—guavas, oriental persimmons, feijoas, loquats, pomegranates, papayas—may be grown. In certain locations black walnuts and Chinese chestnuts may well be included.

Muscadine grapes are adapted to the greatest number of locations and conditions, except in the more northern districts, where the bunch grapes are better adapted. The muscadines produce heavily and furnish fresh fruit over a long period as well as fruit for jelly, preserves and beverages.

Pecans are very widely used as, and are well adapted for, shade trees for the home and yard. The fig also is well suited to most of this region. It should never be cultivated, but it should be planted near a building or in a part of the yard that is kept in grass; otherwise it is soon killed by root knot nematodes.

Strawberries are also well adapted to this region and are the first fruit to ripen.

The Young and Boysen trailing blackberries succeed except in central and southern Florida and in the high mountains. They grow vigorously and produce an abundance of high-flavored fruit one year after planting.

Strawberries, trailing blackberries, figs and grapes cover the season from April or May till frost in most of this region. Larger gardens that include blueberries, pecans, pears, peaches, plums and other fruits will furnish a greater variety of fresh fruit during much of the year.

Northeastern and North Central States

The best fruits for the home garden are strawberries, raspberries, sour cherries, grapes, plums, pears, sweet cherries, blackberries and apples. Under some conditions peaches, cherry-plum hybrids and blueberries may be grown. Currants and gooseberries, which succeed well in all parts of the region, should be planted wherever quarantine regulations permit, that is, where white pines are not important. In certain locations black walnuts, Chinese chestnuts, hickories and filberts may well be included. Sour cherries succeed in all but the coldest part of the region.

Strawberries are adapted to the greatest number of locations and conditions in this region. They are the first fruit to ripen, are of fine flavor and are very high in vitamin C content. Even when frozen, strawberries keep most of their vitamin C content for many months. Strawberries should be a part of almost every garden. The everbearing strawberries Superfection and Gem can be grown in all districts of this region.

Red and purple raspberries can be grown in all districts of this region and black raspberries except in northern Minnesota. Usually it is best not to grow both red and black varieties in the same garden, for cultivated and wild red raspberries often have a virus disease that spreads to black raspberries and quickly kills them. Either of the red raspberries Taylor or Latham (not both) should be grown in Northern New England. Red raspberries are relatively high in vitamin C.

Strawberries, raspberries, plums and grapes cover the season from June until frost. More than one variety of some fruits may be grown to extend the season. Thus, Howard 17 (Premier), Catskill and Gem will furnish strawberries for most of the summer.

In all except the coldest portions, more nuts, which are high in food value, may well be planted. The newer named varieties are better than wild seedlings. Good varieties of black walnuts are the Thomas and Ohio, of filberts the Bixby and Buchanan and of Chinese chestnuts the Carr and Hobson. The planting distances are 40 feet apart for black walnuts, 30 feet for chestnuts and 15 feet for filberts. Filberts can be grown wherever peaches are hardy. The black walnut may be used as a shade tree, especially in the milder parts of the region.

Northern Plains States

These are the coldest parts of the states, except Alaska, which we list separately, and require the hardiest fruits. It is most important in this area to plant on northeast slopes. Where this is not possible, the plantings should have windbreak protection from the north, west and south. The following fruits may be grown here: apples and crab apples, plums and sand cherry hybrids, raspberries, gooseberries, currants, grapes and strawberries. These must be on hardy rootstocks to meet the climate requirements of the coldest areas. Native fruits which do well in the home garden are Juneberry, Missouri currant, wild plum, buffalo berry and high bush cranberry.

Southern Plains States

Under most conditions in the southern part of the region the best fruits and nuts for the home garden are grapes (muscadine), pecans, figs, dewberries, strawberries, blackberries, bunch grapes, peaches and plums. Under the more subtropical conditions, several citrus fruits, guavas, oriental persimmons, feijoas, loquats, pomegranates and many other fruits may be grown. In certain locations black walnuts, Chinese chestnuts and filberts may well be included. In the northern part of the region, strawberries, bunch grapes

(American), sour cherries, plum and cherry hybrids, plums, peaches and apples are most widely adapted.

Muscadine grapes are adapted to the greatest number of locations and conditions, except in the more northern districts, where the bunch grapes are better adapted. The muscadines produce heavily without spraying and furnish fresh fruit for over a long period, as well as fruit for jellies, preserves and beverages.

Pecans are very widely used as shade trees for the home and yard and are well adapted for this purpose. The nuts are high in food value. The fig also is well adapted to the southern half of the region; it does best when not cultivated and should be planted near a building or in a part of the yard that is kept in grass; otherwise the tree is soon killed by root knot nematodes.

Strawberries are also well adapted to this region. They are the first fruit to ripen, are of fine flavor and except for citrus fruits are highest in vitamin C content of any fruits that can be grown in this region. Even when frozen, strawberries keep their high vitamin C content for many months. Therefore, strawberries should be a part of almost every garden.

The Young and Boysen dewberries succeed in southeastern Kansas, eastern Oklahoma and most of Texas. Their high flavor, productiveness and vigorous growth enable one to obtain an abundance of high-flavored fruit one year after planting.

Strawberries, dewberries, figs and grapes cover the season from April or May till frost in most of the region. Larger gardens, which include pecans, cherries, peaches, plums and other fruits, will furnish a greater variety of fresh fruit during much of the year.

Pacific Coast and Arizona

Under most conditions in this region the best fruits and nuts for the home garden are grapes, strawberries, Young or Boysen dewberries, red raspberries, filberts or Persian (English) walnuts, plums and prunes, cherries, pears, peaches, apricots and apples. In restricted locations in California other fruits, including figs, olives, avocados and citrus fruits, can be grown. Fortunately, in most parts of this region fruit plants are free from many diseases and some insects that attack unsprayed trees in the more humid regions.

In western Oregon the bunch (American) grapes and in California the European grapes are well adapted to the home garden.

Strawberries are also well adapted to gardens in this region. They are the first fruit to ripen, are of fine flavor and, with the exception of oranges, are the highest in vitamin C content of any fruits that can be grown in this region. Even when frozen, strawberries keep their high vitamin C content for many months. Strawberries, therefore, should be a part of every garden. The everbearing strawberries Rockhill and Gem can be grown in eastern Oregon and Washington where the season is short. Other varieties may be grown in other districts.

The Persian walnut is widely used as a shade tree about the home, especially in western Oregon and California; also, the sweet cherry, so well adapted to this region, is often used for this purpose.

Peaches, plums, prunes and apricots produce abundantly under irrigation in the warm interior valleys of California and in favorable locations in Oregon and Washington. A few trees of these fruits will supply ample quantities for home use. Most of the stone fruits, apples and pears are not well adapted to locations in southern California and Arizona where the winters are relatively warm. Apples do well in the home gardens in Washington and Oregon and in the cooler parts of northwestern California and along the foothills, but not in the river valleys of central or southern California.

Alaska

Many small fruits will grow in Alaska, although few of the varieties familiar to new settlers from their former homes are adapted to Alaskan conditions. Special attention must be given to planting only varieties recommended for the area. Under those conditions, strawberries, raspberries, currants and gooseberries all succeed. The culture of tree fruits is limited to southeastern and southern areas. Even here the only fruits which reliably produce a harvest are yellow transparent apples and a few crab apples, plus some of the smaller cherries. Bush cherries and Juneberries may be grown as far north as the Tanana Valley.

Pollination and Varieties

For the fruits of apples, pears, sweet cherries and some plums to set and develop, there must be cross pollination between two varieties of the same fruit. Certain apple varieties, such as Gravenstein and Rhode Island Greening, are not only self-unfruitful but produce poor pollen which is ineffective in setting fruit of other varieties. To be sure of adequate cross pollination, at least 3 varieties of apples should be planted; while any two of the recommended varieties of pears, sweet cherries and plums should prove satisfactory for this purpose. The Montmorency cherry, most listed varieties of peaches, and the small fruits are sufficiently self-fruitful to set satisfactory crops with their own pollen.

W. H. Griggs of the California Extension Service makes the following distinctions:

Pollination refers to the transfer of pollen from the anthers to the stigma of a flower. Self-pollination refers to the transfer of pollen from the anthers of a flower of one variety to the stigma of a flower of the same variety. Self-unfruitful, in the strict sense, means the inability of a variety to produce fruit with viable seeds following self-pollination, while self-fruitfulness involves the production of fruits with viable seeds. Cross-pollination refers to the transfer of pollen from a flower of one variety to a flower of a different variety. If one variety, when used as the pollinizer of another, is unable to produce sufficient fruits for a commercial crop, the two varieties are considered cross-unfruitful toward each other; if fruits are produced, cross-fruitful. Most of the fruit types display rather definite patterns of self- and cross-unfruitfulness among the different varieties.

The reasons for nonbearing, from a pollination standpoint, are not all known. Some of them, aside from weather conditions, are incompatibility, imperfection or degeneraton of sex organs, slow growth of the pollen tube and premature or delayed pollination.

The orchardist has a pollination problem with all almond and sweet cherry varieties, with most apple, plum and pear varieties and with the Calimyrna fig. Under certain conditions, cross-pollination will also increase the production of some varieties of walnuts, filberts, pecans, chestnuts and olives. In general, apricots, peaches and nectarines set well with their own pollen. The Riland and Perfection apricots, however, are both self-incompatible. The J. H. Hale, Hal-berta, June Elberta, Candoka and Alamar peaches are also exceptions since they have nonviable pollen and are, therefore, self-unfruitful.

With fruits having a pollination problem, the grower must provide facilities for cross-pollination. In selecting the pollinizers he should consider the following factors:

1. Time and coincidence of bloom.
2. Amount of pollen produced.
3. Germinability (viability) of pollen.
4. Desirability of fruit of pollinizer.
5. Regularity of blossoming of the pollinizer used. Varieties that do not blossom at the same time cannot cross-pollinate each other.

Two factors determine the time of bloom:

1. Breaking of the rest.
2. Temperatures suitable for growth of particular species and varieties.

To break the rest period, deciduous fruit trees require some chilling; the amount varies with the different species and varieties. In the Northwest and the East, where there is extended cold weather during dormancy, the rest period is completely broken. This may not occur in most areas of California and Florida, where the climate is milder. As a result, blossoming is delayed and bloom is generally extended over a longer period, thus avoiding early spring frosts and favoring pollination and fruit setting. Less cold than normal may, however, retard blossoming excessively, so that certain varieties may be delayed too long for cross-pollination—for example, Bartlett and Winter Nelis pears. In seasons following normally cold winters, these varieties would blossom practically together. In the East and Northwest (where the rest period is completely broken), the Bartlett, which starts growth at a lower temperature, blossoms ahead of the Winter Nelis.

After a warm winter many peach, nectarine, apricot, plum and prune varieties will drop large percentages of their flower buds before the blossoming period. Dropping usually begins the latter part of January and continues through the blossoming period.

In orchards lacking varieties for cross-pollination, the situation can be corrected by grafting. Some prefer to graft a pollinizer branch into each tree, while others graft over (topwork) entire trees. The single-branch method has not proved satisfactory in many orchards because growers have not selected large enough branches in the best locations and have not given them proper care. As a result, the branches were choked out. It is more practical to graft over entire trees, rather than a single small branch. Grafting over every third tree in every third row will usually provide a sufficient number of pollinizers.

Whole trees offer a distinct advantage in keeping the fruit separate during harvest. It requires 3 or 4 years for grafts to furnish enough blossoms for cross-pollination.

Until provision can be made for permanent cross-pollination, the introduction of bouquets of bloom taken from suitable pollinizing varieties is probably the most effective and economical way of insuring a fruit set in orchards lacking varieties for cross-pollination. A bucket of blossoming branches for each tree, hung on the leeward side, is usually adequate, although the common practice is to place large blossoming branches in barrels of water on the ground. The bouquets must be kept fresh, and an abundant supply of pollinizing insects is necessary for the success of this method.

23—Preparing Soil for Fruit Plantings

Obtaining good soil balance in the home fruit garden is not just a matter of luck, according to orchardist William Ackerman. It requires some basic factual knowledge and a good deal of common sense.

First you must know the general soil requirements of the fruits you choose to grow. Studying their needs will help you to decide what you want to grow. After that, you must know the composition of your own soil and just what specific improvements are needed to make it adequate for your selection of fruits. To obtain this information, you should take a representative soil sample and have it analyzed. You need well-qualified recommendations to guide you in improving your soil for ideal balance. Then plant good stock and continue to treat it properly and you can expect plentiful, nutritious yields.

For real disappointment, of course, you can neglect such steps. Just plant poor stock in unknown soil and ignore its needs from then on. Or pour on indiscriminately every other "miracle" plant food and "wonder" spray that comes on the market. If such mismanagement seems fantastic to you, that's a good sign. It shows you are off to the right start, fortified by common sense.

The ideal soil for all home fruits is not a sand, not a heavy clay, but a medium loam. Sandy soils require additions of plenty of both fine and coarse humus to give them a loamy texture. Clay soils need coarse peat, ground corncobs, rice or buckwheat hulls and sometimes sand.

The depth of orchard soils should be 5 feet or more. The depth of the soil is determined by underlying rock; impervious hardpan, sand or gravel; or lack of aeration in the soil. Root penetration may be stopped by any one of these and the depth at which the penetration is stopped is the depth of the usable soil. This depth can be determined, except that limited by aeration, by borings made with a soil tube. In general, soils at higher altitudes may be underlaid with heavy lime deposits, while those in and around river bottoms are usually deep but may have a high water table.

No fruit tree, shrub, vine, cane or plant likes to have wet feet. Poor drainage causes root rot and unfavorable plant food balance. Sometimes deep and hearty digging improves the situation; but where drainage is very poor, tile lines must be laid or beds dug out and underlaid with coarse gravel. Rapid leaching, the opposite of poor drainage, occurs in soils that are too light or sandy. Such loss can be counteracted by working in humus materials and by using slowly available plant foods.

Subsoiling or blasting before planting may be desirable under special conditions where particular kinds of hardpan are present. These exist when the hardpan may be broken up economically by these methods and it will not resume its original impervious condition upon being wetted again and where the soil is pervious and fertile below.

For most soils, 3 to 5 per cent organic matter is ideal. A soil with an analysis of less than 2 per cent is poor. Adequate humus is essential for soils to keep plant foods available and yet hold them against leaching. Soil humus also retains moisture and creates good physical structure.

Basic humus build up materials are manures (fresh or dried), compost, leaf mold, mushroom soil, and peat. Cover crops and mulching material also supply humus, as they are turned under.

In the days of our fathers plowing the land was considered a "must." Plowing filled a double purpose; it got the cover crops under and it loosened and aerated the soil, making it easier for the young trees to send out their roots.

Today the average homesteader is working in a smaller space and, unless he has commercial aspirations, is more modest in his ambitions as a tree-grower. But, it is axiomatic that an organic gardener will grow a cover crop to protect and enrich his soil, particularly when he is growing a tree crop.

The putting in of a cover crop on the larger plot will pay dividends far beyond the time and trouble invested. In addition the earnest and hard-working gardener of today has a whole host of power tools to ease his task for him.

Outstanding among these is the light but potent rotary tiller which can turn a cover crop under in addition to performing other chores.

The advantages conferred by the propagation and subsequent turning under of cover crops are many. The cover crop serves a double purpose; it protects the soil from the elements and it improves its nutritional qualities. Kinds of cover include the various clovers, vetches and leguminous crops such as mixed beans. All of these will, when turned back into the soil, add valuable nutrients as they break down and will also promote better soil structure through the addition of humus.

The function and benefits of the cover crop may be itemized as follows:

1. Directly improve the physical condition of the soil.
2. Prevents hard soils from puddling or cementing.
3. It holds the rain and snow until they have time to soak into the land.
4. Serves as an added protection against frost.
5. Catches and holds some of the leaching nitrates.
6. Renders plant-foods available.

A last word on the cover crop—plant it one full year before you put your trees into the soil. That's right; try to plan in this case a year ahead. This means putting in a winter cover of vetch or clover and then turning it under in the spring. Permit it to age in the soil until September or early October when you plant your trees.

The reasons for this are sound and, if you follow this plan, it can save you a lot of trouble. First: don't put raw manure of any kind directly into the soil when you plant the trees. By plowing under in the spring and waiting 3 to 4 months you give the green manures time to age and weather. Second: you may stimulate late tree growth which means that your trees will carry tender, sensitive wood into the chilling frosts of winter. It is almost certain beyond all doubt that this wood will be killed and the young tree itself seriously weakened if not killed outright.

The humus requirements of your trees may be met if heavy cover crops are plowed in, or additional material may be needed, Mr. Ackerman says.

Apples, peaches, pears, plums, cherries need medium soil humus. The best materials for improvement are compost, leaf mold, manure (rotted or dried) or mushroom soil, well worked in.

Grapes, blackberries, raspberries need plenty of soil humus, plus spring applications of nitrogen-rich organic materials around their bases.

Strawberries do not demand high soil humus, but want plenty of coarse mulch material around the plants, especially in winter.

Blueberries demand high soil humus, the more the better, as long as the soil is not soggy. They also need a heavy mulch of peat, acid leaf mold or woods humus.

An underlimed soil does not mean just one that is too acid. It also means one deficient in calcium and magnesium, which are both essential plant foods. Working in the proper amount of ground limestone or marl will bring soil to a weak acid reaction. At such a reaction, microörganisms do their best work in breaking down humus into plant foods and the intricate organo-chemical balance of plant foods in the soil is most efficient.

Overliming must be avoided. There is nothing so difficult to counteract as too much lime in the soil, which ties up plant foods and trace elements and discourages biological action. The right amount of lime to use depends on a soil analysis.

Apples, peaches, pears, plums, cherries need a well-limed soil, not only for their own sakes but to benefit cover crops, if any; and to encourage decomposition of mulch material.

Grapes, blackberries, raspberries need a fairly well-limed soil, medium to slightly acid.

Strawberries are not as particular about lime, but will not thrive on a strongly acid soil.

Blueberries are the exception to the rule. They must have a strongly acid soil and will fail on a well-limed soil. They get along with less calcium and magnesium than most plants. Peat, oak leaf mold or woods humus are often needed to make blueberry soils more acid. Concentrated organic soil acidifiers can be used to advantage. Aluminum sulfate should never be used to acidify soil.

All plant tissues and all animal proteins contain nitrogen. There is nitrogen in the air around us, in most waters and in the soil. The problem is how to make it regularly available to plant roots. Stunted growth and pale leaves are signs of a lack of nitrogen in the soil. Too lush growth, at the expense of flowering or fruit, is a sign of nitrogen oversupply. Good humus in the soil is the basic nitrogen provider. Coarse mulches like peat, straw, corncobs or sawdust use up nitrogen as they decompose; while rich mulches like manure, compost or peat-plus-tankage provide more nitrogen than they need for their own decomposition. Legumes, when properly inoculated, take nitrogen out of the air and add it to the soil.

Manure, compost, leaf mold, mushroom soil and sludge are bulky organic materials carrying nitrogen. Cottonseed meal, dried blood, tankage, fish meal, hoof and horn scraps and tobacco stems are more concentrated nitrogenous materials. Those of the first group are best worked into the soil or in some cases used as a mulch. Those of the second group are more often used as side-dressings, spread on the surface and worked in or combined with coarse mulching materials.

Apples have a low nitrogen demand when young, a fairly high nitrogen demand when mature. A legume cover crop helps supply nitrogen to mature trees; or a generous mulch of hay, grass clippings or ground corncobs, plus cottonseed meal, dried blood, tankage, fish meal or chicken manure.

Pears require less nitrogen than apples. There is danger of oversupplying. Frequently, the cottonseed meal, etc., should be omitted, even for bearing trees.

Peaches, plums, cherries have a generally high plant food demand, including the need for generously applied nitrogen when mature. Thick cover crops and coarse mulches should be avoided, unless supplemental nitrogen is added.

Grapes are usually nitrogen-hungry when in leaf. In addition to being mulched they should have plenty of concentrated nitrogenous material during the growing season, such as dried blood, tankage, fish meal or chicken manure.

Blackberries, raspberries want good manuring or addition of rich nitrogenous material in spring. But they should have no high-nitrogen material in summer or fall, for danger of encouraging weak cane growth.

Strawberries should have a medium supply of nitrogen, through coarse mulch fortified with moderate concentrated nitrogenous material in spring, repeated in August.

Blueberries need plenty of spring nitrogen when of bearing age. In addition to a thick mulch of acid material, they should have cottonseed meal, tankage, dried blood or fish meal as soon as the weather turns warm.

Phosphate is not only essential to leaf and stem growth, it is a vital part of all flowers, fruits and seeds. Clay soils and soils poor in humus have a high ability to absorb phosphate. The appetite of the soil particles for phosphate must be satisfied before any is available to plant roots. Increasing phosphate availability is another reason for having adequate humus in soils. Proper liming is also important for making phosphate available. Too little lime causes formation of unavailable iron and aluminum phosphates; too much lime creates insoluble tri-calcium phosphates.

Good compost and manure contain some phosphate; but for many of our soils, more than this is needed. Where soil analysis guides, plenty of rock or colloidal phosphate should be worked in before planting. Bone meal supplies phosphate, but not as efficiently as the mineral materials. It also has an alkalizing effect and should not be used on heavily limed soils.

Finely ground granite rock, glauconite and greensand are suppliers of potassium. In many areas, there is not enough mineral potassium in the soil to supply thriving crops or trees. It, too, must be provided by man. These potassium bearing rock materials provide that essential nutrient in the same slow, steady way.

An average application of phosphate rock would be one-half ton per acre; of potash rock three-quarters ton. Both should be placed fairly deep in the soil, near the feeding root zone.

What about those trace elements? The lack of only a few pounds per acre of iron, manganese, boron, copper or zinc in the soil can cause failure to an orchard. Soil analysis tells when these are lacking. In a well-managed orchard, there are ordinarily no trace element deficiencies, because mineral-rich materials have been provided. Organic matter brought in for the orchard should have been grown on productive land, not on worn-out soils. That includes manure, compost, leaf mold, mulching material and other animal and plant by-products. Most phosphate and potash rocks contain valuable impurities in the form of trace elements. Thus, there is a double reason for including them in the soil building program.

A soil rich in humus holds both major plant foods and trace elements in readily available form, and at the same time holds them against being leached out. Humus may be incorporated in the soil before planting and maintained by a liberal mulching program.

For more information on improving the fertility of your soil, see the chapters on this subject in the vegetable section.

Cover Cropping

A. F. Vierheller, extension horticulturist at the University of Maryland, offers this advice for preparing soil prior to planting fruit trees:

Unless the soil is fertile, it is well to spend a year or more in its preparation or improvement before planting the trees. For home plantings, where only a few trees are to be planted and where soil is not of the best, it is advisable to replace the soil by filling good soil around the tree as it is being planted. If good soil is not available, the use of sand, well-rotted manure or compost, *thoroughly* mixed with the soil will be of great benefit. For the commercial orchard, a winter cover crop of rye 1½ bushels per acre, or better, a mixture of rye 3 pecks and winter vetch 20 pounds per acre may be used, sowing the seed about August 20th to September 10th. Recent research at University of Maryland Field Laboratory, Hancock, indicates that *either* of the following mixtures will be cheap and effective for winter cover cropping:

MIXTURE #1

Seed Mixture per Acre

8 lb. crimson clover
8 lb. winter vetch
6 lb. winter rye
2 lb. golden millet

MIXTURE #2

Seed Mixture per Acre

8 lb. crimson clover
8 lb. winter vetch
8 lb. winter rye
8 lb. Wong barley

Mixture No. 1 should be sown between August 20th and September 1st but if later sowing is necessary, Mixture No. 2 is suggested, up to September 10 as millet must make its growth before frost and barley is substituted for millet in Mixture No. 2. Fertilize the cover crop well. Heavy growth of cover crop is desired as this adds bulk of organic material to the soil. Then next spring, when the rye and crimson clover begin to develop woody growth (usually in early May), the cover crop should be either plowed under or disked rather thoroughly, to incorporate the organic material with the soil. If fall planting is to be done, a summer cover crop of soybeans or cowpeas may be grown, seeding them in June and disking them under when the pods begin to harden.

Importance of Improving Soil

Some idea of the possibilities of organically built-up soil may be derived from the following account, by A. P. Thomson, of his experiences at the Golden Acres Apple Orchard, Front Royal, Virginia.

In 1944, after a long period of service with the Navy as a diesel engineer, A. P. Thomson bought a 45-acre farm in the colorful Shenandoah Valley. The land had been literally "mined" for nearly 250 years. Once rich and high-yielding, its soil had lost practically every vestige of fertility through careless tillage. Still, Thomson thought of his being the fifth generation to farm this land bordered by the Blue Ridge and Allegheny mountain ranges—and became downright determined that he would revive it.

How did he accomplish it? Actually, by concentrating on a fundamental truth Thomson himself has put this way: "An apple cannot give you the living elements and delight that you need and want if it comes from a tree itself deficient because of the deficiencies of the soil."

And so, before a single apple tree was set out, a vigorous program of rebuilding the soil was begun. In the fall of 1945, Thomson turned under 5 years' accumulated growth of lespedeza, broom sage, foxtail, berry vines and assorted weeds. Over 15 tons of organic matter per acre were composted and incorporated in the soil. Several hundred tons of manure were applied. More than 100 tons of corn fodder were composted by the Indore method, developed by Sir Albert Howard, and also added to the soil. Now, even the tree prunings are shredded and allowed to lay as compost.

In addition, Thomson applied 500 pounds per acre of ground phosphate rock, then sowed a mixture of alfalfa, sweet and red clovers, alsike, some brome, rye grass and oats—all as a further soil-enriching cover crop. He soon found, though, that the heavy loam soil was too acid and so in the fall of 1946 broadcast 315 tons of limestone rock. By the following spring, growth was luxuriant, much of it over lanky Thomson's head. It took hold on the farm's bare galls and it completly stopped erosion. In October, he disked under the heavy cover, planned to use brome, oats and alfalfa as a new cover for his orchard's middle rows and was ready to plant apple trees by December.

When it came to choosing varieties, Thomson selected first the Golden Delicious, a variety especially suited to his locality. Among its advantages are that it is a very high quality apple when grown well; it's a heavy bearer and a good pollinator. Another feature is that it attains an appetizing golden yellow color as it ripens. On the other hand, it hasn't found much commercial orcharding popularity because it requires a long season to mature and is one of the most winter-tender varieties. Furthermore, it is particularly susceptible to injury from chemical sprays and from mice.

Just the same, Thomson wanted to grow this variety. He wasn't going to use any poison sprays—and he was certain that the organic methods he was following would overcome the other difficulties and provide him good yields of a high-quality fruit that should be popular. Choosing this apple suggested a good name, Golden Acres Orchard. It's a fitting designation and one that aptly describes the glowing farm today. In addition, he planted Red York Imperial, another variety well suited to the area. Both this and the Golden Delicious are superb for eating raw and cooking.

The trees at Golden Acres were spaced 38 feet by 40 feet apart, contrasting with the 15 feet to 22 feet spacing usual in commercial orchards. Thus, Thomson's trees are far enough apart to prevent touching at maturity, whereas the average grove becomes like a jungle. Although production per acre can't be as large, the whole orchard is benefited because the trees and fruit get better air circulation and more sunlight and because room is provided for those enriching cover crops.

Thomson's orchard management includes another unusual feature that

A. P. Thomson, organic orchardist of Front Royal, Virginia, stands alongside one of his trees which has benefited from the rock mulch "fortress method." Around each tree in a circle 5 or 6 feet in diameter and over 6 inches deep, Thomson places 500 pounds of ground rock fertilizer.

he calls the "fortress method." Around each tree, in a circle 5 or 6 feet in diameter and over 6 inches deep, he places approximately 500 pounds of half-inch dolomitic rock. What's this do? Thomson cites 4 major advantages:

1. It gives the trees greater anchorage against strong winds, a protection otherwise lacking where they are widely spaced.

2. It absorbs a large amount of heat from the spring sun, creating convection currents during frosty nights to provide some protection for bud and bloom.

3. It discourages mice from burrowing in around the trunks and damaging or killing the trees. (The area has an abundance of the pine and meadow mouse.)

4. As the rock weathers, it becomes a source of calcium and magnesium for the trees' continuing requirements.

Sawdust mulching is also practiced. Thomson spreads a ton per acre each yeaı between the rows. He uses aged Blue Ridge Mountain hardwood sawdust, selected because the rock and virgin soil of these mountains affords the trees many valuable trace elements. And, he notes, the sawdust takes up excess nitrogen produced by the legumes. In late fall, a winter hay mulch of 200 pounds per tree is applied—only to the drip line.

Currently, the entire orchard is sheet composted 4 or 5 times a year with the legumes and grasses grown between rows, not only to maintain but steadily increase the soil's fertility and structure. A rotary cutter is used to macerate this growth so that decomposition can take place quickly and a rich humus released for the trees to draw upon.

Then, too, earthworms aren't forgotten, nor their tremendous contribution to any soil-building program. Right from the start, Thomson began a sizable worm-breeding project, added them to all 45 acres and used them in compost and mulch piles. With some initial help from earthworm-specialist Dr. Thomas Barrett, the farm's earthworm population has grown to over 5 million per acre—all working constantly to enrich and aerate the soil.

A pointed sidelight on what these earthworms can do is found in one 60-year-old Pippin apple tree originally at Golden Acres. In 1950, Thomson discovered this tree was dying. He added a million earthworms to its soil. By last year it had not only regained health, but yielded 51 bushels of fine apples.

As for insect control, always a major problem in apple growing, Thomson relies on some very effective, natural measures, plus an army of natural allies. In February, he applies a miscible oil dormant spray which checks scale, mites and developing eggs of many orchard pests. For severe oubreaks of codling moth, an extract prepared from the root of a British West Indies tropical plant has been used experimentally and found helpful. Called Ryania, this spray is very mild and leaves no residue.

In addition, there's a host of predators on hand. Ladybugs and praying mantids do their share, and thousands of birds, encouraged by the multiflora rose hedges around Golden Acres, willingly pitch right in. Those hedges, by the way, also draw swarms of welcome bees who contribute to the pollinating job throughout the orchard. When some Japanese beetles showed up in 1954, Thomson put 4 teaspoonfuls of milky spore disease powder around each tree and hasn't had any difficulty with these troublemakers since.

Now, what about the harvest, the apples themselves? Without resorting to chemical fertilizers, hormones, defoliants or poisonous sprays (most orchardists pour on DDT, arsenic, lead, copper, methoxychlor—to mention just a few), the Thomson orchard yields are as high as 30 bushels per tree of top-quality, really healthful apples. These crisp, good-looking fruits can be eaten—nutritionally rich skins and all—*without* the cumulative dangers of all those toxic and artificial concoctions and *with* all the natural flavor and food value put into them by good soil.

24—Planting Fruits

FRUIT TREES, VINES and bushes are planted at various times of the year, depending upon the area. In the coldest sections, approximately those where the temperature goes below 0 degrees Fahrenheit in the winter, fruits are generally planted in spring. In sections where the winter temperatures are less severe fruits are set in fall or early winter. Early setting permits the roots to begin to take hold and grow before the buds begin to expand in spring and to demand moisture. However, young fruit plants are more easily winter-injured than older ones, so early spring planting is recommended where winter injury is a problem.

Whether they are planted in spring or fall, deciduous plants should be entirely dormant when they are planted. It is safest to assume that they will be dormant one month after the first killing frost in fall until one month before the last killing frost in spring. Evergreen trees and shrubs, especially those grown in the frost-free districts, should be planted between growth flushes. Citrus trees are best planted in California between March and May. In Florida and the Islands, they are usually planted in the winter months, between the middle of November and the middle of February. This schedule takes advantage of the winter or spring rains, which help to wash the soil in around the roots and to settle the young plants.

Now, if you want your trees to grow surely and steadily, putting out sound wood from the first year on, there are 5 basic conditions which must be met. We are not recommending short cuts; instead we are passing on to you tried and true methods that have worked with other tree-growers and that will work for you—if you follow them.

These 5 guide posts are all of equal importance and each calls for serious thought and hard work. If you cut down on any one of them, you will hold up the entire program. Here they are:

1. Advance preparation of the soil.
2. Careful planting technique.
3. Feeding the young tree.
4. Fungi growing in the soil.
5. Pruning of roots and branches.

FOUR STEPS IN PLANTING A TREE

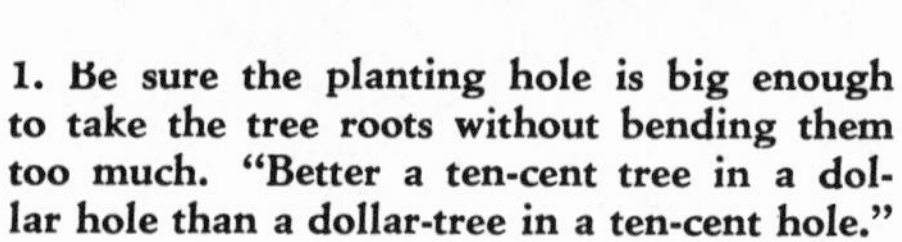

1. Be sure the planting hole is big enough to take the tree roots without bending them too much. "Better a ten-cent tree in a dollar hole than a dollar-tree in a ten-cent hole."

2. Use good quality topsoil in filling hole to 3 inches of the top. Do not fill hole all the way but be sure to leave a depression for the water from rain or irrigation to collect.

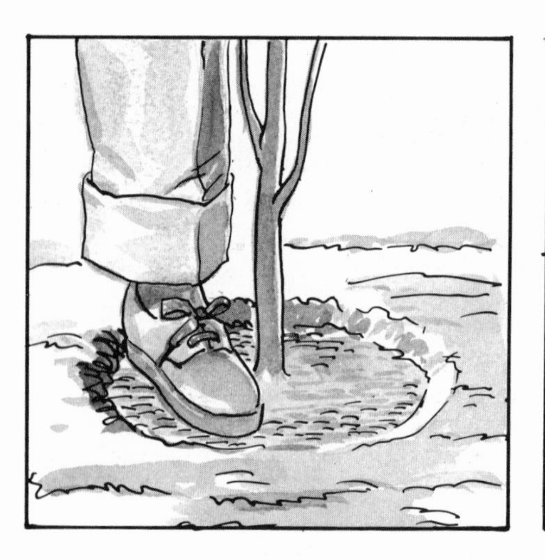

3. Tamp the earth gently but very firmly around the roots, leaving no air pockets. Disturb the roots as little as possible when you are working the soil in around them.

4. Set a strong stake next to the tree to keep it from falling, also to help it grow straight. Take care to make the loop tight around the stake, but loose around the tree.

These are common-sense guides that have helped a lot of people solve that problem of sustained tree growth.

Plan Before Planting

Plan the entire operation of tree-planting carefully and in advance. Know what you are going to do and what and how much of everything you will need. This will enable you to work swiftly once you get started and you won't have to cut corners in order to finish your job on time.

There are 5 basic steps which must be taken to achieve successful planting of a tree. More or less in order, they are:

1. Digging an adequate or even more-than-adequate hole.
2. Providing a mellow soil with plenty of humus.
3. Supplying plenty of organic nutrients to help your young tree get started.
4. Giving the tree a sound planting mechanically, making sure it stands straight, is planted deep enough and is adequately pruned, trimmed, braced and protected.
5. Following through with a program of watchful maintenance and after-care.

Techniques will vary slightly to considerably depending upon soil condition, climate, time of year and the individual tree. But the above rules will, in their essentials, prevail where and whenever trees are planted. Break them and you break them at your own risk.

Digging the Hole

This is not as easy as it sounds and too many of us can fail on this very first step. So, make your hole big enough to accommodate your young tree with room to spare! It should fe at least one foot wider and 6 inches deeper than the spread of the roots. Most tree-planters recommend a minimal 3-foot hole—3 feet wide by 3 feet deep.

When digging the hole, be sure to pile the topsoil separately from the subsoil. Later, when you start filling up, it will be the first soil to go back into the hole, mixed with compost, sphagnum moss, rock minerals. In general, when digging the hole, be guided by that sage old maxim: "Better a 50-cent tree in a 5-dollar hole than a 5-dollar tree in a 50-cent hole."

The problem of soil drainage can be important to the success of your tree program. If you are working in a stratum of tight clay you should seriously consider placing 4-inch tile piping leading from the bottom of the hole. Writing on this aspect of tree planting, H. W. Gilbert and A. L. Leiser of Purdue University advise that:

"The most critical need for a transplanted tree (assuming it is alive and vigorous) is water. An apparently healthy tree can die as readily from too much water as from not enough water. When water is needed, man can apply it to the tree.

"One problem when planting is to determine how to adjust to the local soil condition. A tight, poorly-drained clay soil can become either too wet or too dry under normal climatic conditions. A light well-drained soil can become too dry, but not too wet except under abnormal flood conditions.

If little or no topsoil is present, care must be taken not to increase the drainage problem.

"A pocket of high quality loamy, friable soil with relatively high organic content placed in a hole dug in tight soil or subsoil will hold free water in prolonged wet weather and can kill the tree for lack of air. In such cases a 4-inch tile line to a drainage outlet should be installed. If this cannot be done, it would be better to plant the tree in the soil removed from the hole or with no more than 20 per cent peat moss thoroughly mixed with this soil. On the contrary, very light sandy, well-drained soils can benefit by one-third peat moss and one-third loamy or topsoil mixed with one-third original soil.

If trees cannot be planted immediately, plant them in shallow trench, making sure the roots are covered with damp soil. The sketch above, prepared by the Vermont Experiment Station, shows how, "heeling-in," will protect roots from drying out.

"There are other variables than the kind of tree selected for the location, the kind of soil found when digging the hole and the variation of water-holding capacity of soil types, but suffice to say here that when the water relationship is favorable a healthy plant should grow."

Advance preparation of the soil is necessary not only because you are enriching the soil but also because you break it up, which eases the subsequent growth and spreading of the tender, young rootlets. Phil Arena, highly successful organic orchardist, of Escondido, California, comments thus on this aspect of tree growth: "Roots penetrate according to the depth of the topsoil. But the fibrous, or feeder roots feed at the level where nourishment lies. If you furnish plenty of organic matter, they will thrive much better."

Transplanting of any tree involves injury and shock. It's your job to keep these to a minimum and you can—if you know what you are doing and then do it quickly. Tender roots are damaged, the entire organism is suffering from undue loss of water through the exposed areas.

The roots of your tree should be kept moist from the time they are dug until they are safely planted. Carry the trees to the planting site with the roots wrapped in burlap bags. Do not try to carry too many at one time. Keep the roots of all trees except the one you are planting covered with wet burlap.

Don't worry, however, if you have to keep the trees waiting. Just keep them in a cool place and don't unpack them. They'll stay moist that way. But if you must wait a week or so—plant them in a shallow trench, making sure the roots are covered with damp soil.

Use Topsoil to Fill

Use a mixture of rich topsoil, finished compost or well-rotted manure and rock fertilizers when filling the hole. This gives your tree a real chance to get started. Later when the roots extend out into the poorer surrounding soil, it will not matter so much. But it is obvious that lack of fertility during the critical starting period can result in loss of the tree.

Remember, the hole must be large enough to take the roots of the tree without bending them too much. Make a mound of the topsoil at the bottom of the hole and set your tree on this mound with the side roots running down its sides.

Be sure not to leave air pockets when tamping soil over and around the roots. You cannot have these if the tree is to flourish and produce. Pack the soil firmly about the roots, being careful not to injure them. You can use a stick with a rounded end for poking and tamping the earth into place.

Tramping the soil down, layer by layer, is also recommended. The experts also say you can't tramp hard or long enough.

Here is another way to avoid air pockets. Fill the hole with earth to within 6 inches of the top. Then pour water up to the rim. Hold the tree trunk and rock and jar the tree gently as the earth settles with the water. After this use more topsoil to fill the hole.

Change root positions as little as possible while planting the tree. Try to keep as they were before transplanting. Hold the side roots up as the soil is placed around the tree. Let go of the roots one by one as the hole fills up.

Set the tree deep enough in the hole to cover it adequately without filling the hole all the way to the top. Leave a space of 3 to 4 inches to act as a basin for holding water during the first critical years. Later it will gradually fill up by itself. You must water the trees—thoroughly—as soon as they are planted. We mean thoroughly water them—several pails to the tree. Soak the ground so that the water remains on top for several minutes. This thorough soaking once again helps settle the soil around the roots. So add more topsoil if necessary. But don't fill up the hole; remember to leave 3 to 4 inches for water storage.

Water the trees once a week after growth starts in the spring. Do a thorough job. If enough rain falls to fill the depressions about each tree you can omit watering that week. Watch the topsoil around the tree and replenish it if and when necessary.

But—water once a week only. Trees will not grow in a swamp. Standing water in the soil keeps the air out. The proper amount of air in the soil is just as important to the health of the tree as the moisture.

The water that should always be in the soil is the film that adheres to each soil particle. As trees and other plants take up this film of moisture from each soil particle, it is replaced through capillary action by water deeper in the soil.

It is this water that dissolves the plant-food elements and holds them in solution for the tree roots to absorb. This is the only way that trees and other plants can take food up and out of the soil.

Composting the Tree

Start right from the very beginning. Use compost to start and nourish your tree. We cannot recommend commercial, chemical fertilizers which do not add humus to the soil. You must add organic matter to the soil if you want productive, disease-free trees. Soil that is well supplied with humus retains moisture and has good drainage and aeration.

As humus decomposes, it releases a continuous supply of plant food in contrast to the "flash" action produced by chemical fertilizers. It encourages the existence of beneficial bacteria and earthworms. It fights erosion and overcompactness of the soil.

Every tree grower must have a compost pile. Use your compost as a mulch around each tree. Your trees need nitrogen, phosphorus and potash and your compost pile should contain these nutrients in good proportion. Take a series of soil tests in the orchard. It will tell you what you need and save you trouble and worry later.

White Planting Method

When organic expert Herbert Clarence White of Paradise, California, plants a tree, he doesn't even glance at the little instruction sheet that the nursery sent with the stock. He proceeds to plant the tree using an unusual method handed down to him by his grandmother years ago. Grandma White's method has worked so well for Herbert over the years that he has used it to plant hundreds—possibly even thousands—of trees. He has seen fruit trees planted by Grandma White's method show 3 or 4 feet of new growth in a year, and start bearing crops in only a couple of seasons. His method requires a lot of work and a lot of raw material in the form of compost, peat and topsoil, but he claims (and others have observed) that the results amply justify the investment in time and material.

You start out by digging a hole 3 feet wide and 3 feet deep in which to plant your young fruit tree. Separate the topsoil from the subsoil that is dug from the planting hole. In the bottom of the hole place a couple of pieces of 4-inch drain tile and plug up the ends with stones. Fill up the bottom foot of the hole with a mixture of equal parts of topsoil, peat moss and finished compost, plus about 5 pounds of phosphate rock or colloidal phosphate.

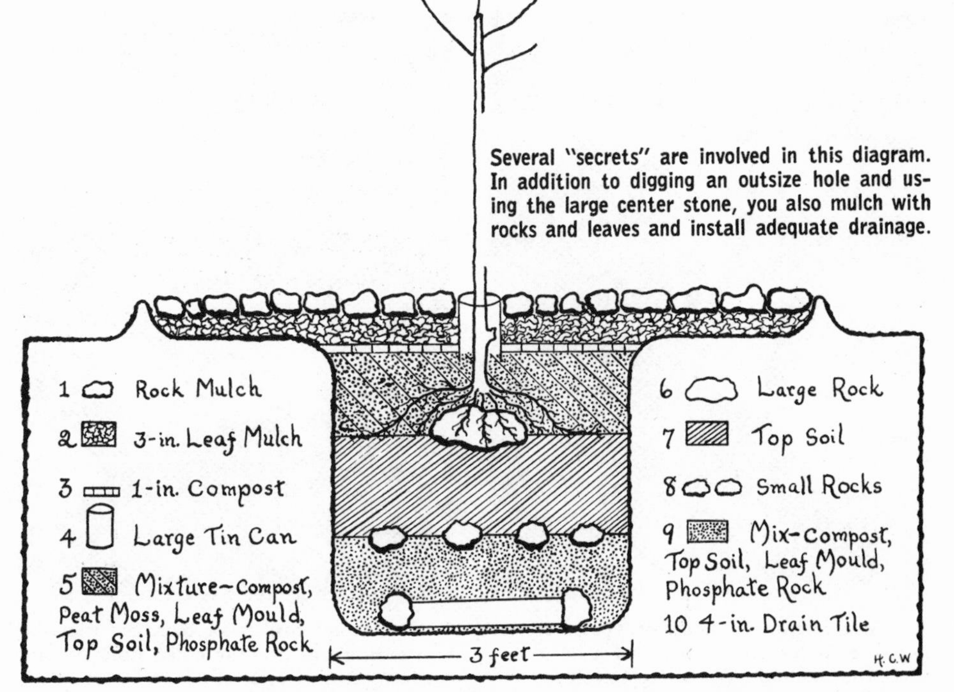

Above is California orchardist Herbert Clarence White's diagram for planting a tree.

On top of that mixture place a layer of small rocks. The next one-foot layer consists of pure topsoil. Now put into the hole a large stone. Spread the roots of the tree over that stone, then fill the rest of the hole with the compost-topsoil-peat-phosphate-rock mixture. As mulch over the planting, place one inch of compost, 3 inches of leaves, plus a layer of stones if desired. White also advises putting 250 to 500 earthworms in the top compost layer, and adds this postscript to the description of his method:

"Does all this sound too weird and grotesque? Too utterly fantastic? If so, far be it from me to try and convince you. But if you are just a wee bit interested in watching a miracle, just try it out on one little tree—following the planting plan as indicated in the diagram carefully—and it will be hard for you to believe your own eyes when that baby tree starts growing."

The planting board is of value to set the trees in line after the field has been staked out. This board is 3 to 4 feet long with a notch at each end and another at one edge in the exact center. Before digging the hole for the tree the board is so placed that the stake, showing where the tree will be, fills in the center notch. A stake is then placed at each end of the board after which the center stake and the board can be removed. After the hole is dug, the planting board is placed over the two remaining stakes in the original position. With the tree trunk in the center notch, the alignment of the original staking will be retained. The tips of the notches in the board should be in line with each other and the board must be used in one position only.

PLANTING YOUR FRUIT TREES ORGANICALLY

1. Dig a good-sized hole for your tree, being sure to loosen hard-packed soil to give roots the best opportunity to develop.

2. Cut off any broken or damaged roots before setting tree into the hole. Root pruning encourages development of feeder roots.

3. Work in large amounts of organic materials such as compost, leaf mold or rotted manure when filling the hole. This will provide the necessary nutrients for the tree.

4. After placing tree on mound of topsoil, partially fill hole with soil and humus mixture. Watering during this step helps settle the soil closely around roots.

5. Pack the soil down firmly about the roots to avoid air pockets which can prevent proper feeding of tree, keeping it from making good growth.

6. Soak the ground thoroughly so that the water remains on top of the soil for several minutes; thorough soaking once again settles the soil around the roots.

7. It is often necessary to prune back some of the branches of young trees in order to make sure that the branch system is in proper balance with the roots.

8. Always provide plenty of mulch for the young tree. Above, the basin has been filled with organic material and is slightly depressed to act as a basin for water.

It is not advisable to dig the tree holes very much in advance of planting. Exposure of the soil to the weather will tend to harden the sides of the holes, especially in certain types of soil, so that the tree roots will not easily penetrate the soil.

During all operations incident to transplanting trees, care should be taken to prevent the roots from drying out, especially in the case of one-year-old trees, the roots of which are fibrous and very tender. The trees should be kept in a barrel of water or covered with moist earth or sacking at all times until finally planted.

Prune the roots of the young tree only where necessary to remove broken and damaged roots or to head back some that are excessively long. Should a tree be so badly scarred or damaged that there is grave doubt of its survival, the wise thing to do is to discard it entirely.

See that the soil is loosened in the bottom of the hole. Set the tree at approximately the same depth it grew in the nursery, never more than two inches deeper. Dwarf apple trees on Malling IX and dwarf pear trees on quince roots must, however, always be planted with the union between the variety and the rootstock slightly above ground level. If this is not done, the trees, particularly apple trees on Malling IX, may form their own roots on the variety portion of the stem above the graft union and in such cases the dwarfing effect of the rootstock will disappear and the trees will grow to a large size.

Experiences in Planting

The sound planting technique, described here by E. Hamilton Fairley, of California, is painstaking and thorough. When planting avocado trees he advises us to "dig a hole twice as large as the ball of tree roots or larger. A hole 3 feet wide and deep is worth the extra trouble because it will give the new, tender roots a soft run for some time to come.

"In the bottom of the hole place two shovelfuls of well-rotted compost mixed with the same quantity of good topsoil which may be a good, sandy loam. Add enough topsoil to bring the top of the ball level with the ground. Place the tree on this in the center of the hole and fill in with good soil in which some compost is mixed. This will put humus in the soil.

"Firm the mixture around the ball of roots as the filling proceeds. When the hole is almost filled, have a gentle stream of water from the hose run in and settle the soil so there will be no air pockets.

"Let the water run long enough so it reaches down below the roots, then fill with more soil to ground level. Now make a basin around the tree which will hold water. Water thoroughly once or twice a week until the newly planted tree is established."

Jonathan Hughes of Visalia, California, has sent a report of 3 Lombardy poplar trees that grew 19 feet in one year. Mr. Hughes achieved that phenomenal rate of growth by planting 3 trees together in a hole 7 feet in diameter and 4 feet deep—filling the hole with topsoil mixed with equal parts of leaf mold, compost, rotted manure and peat. After planting, the trees were mulched with leaves. They are now, 4 years after planting, more than 80 feet high.

The moral of the above reports strikes home. Taking time and trouble pays off when you are growing trees. And it is obvious that a good start is useless unless your care and feeding of the young tree are just as painstaking and thorough.

This is how California avocado tree-grower Gus Albrecht ensures sound, sustained tree growth. Once again there are no short cuts; this system is for the grower who wants results and is willing to work to get them.

In 1951 Mr. Albrecht started a 300-tree grove on a slope, preparing an extra-large basin around each tree—10 feet wide. Since they were on a slope, he had to build up a small terrace for each tree, leveling around it to maintain an even flow of water.

Next he spread minerals, colloidal phosphate, compost and mulches out from the trunk to the dike of each basin. Almost 500 pounds of nutrients were spread around each tree.

Response on the part of the trees was immediate and gratifying and can be chalked up to the credit of the big basin idea. Sixteen months after planting, the trees were 9 feet high and 9 feet across—something of a record in California.

These excellent results can definitely be ascribed to the abundant variety of food that was made permanently available to each tree. The animal, fungal and bacterial inhabitants of the soil under those trees were stimulated to prepare choice food for the young, tender feeder roots all the way out to the rim of those 10-foot basins.

The feeder roots were then stimulated to spread out beyond the confines of their original 3-foot holes in search of this food. It was this extraordinary feeder-root development which in turn made possible the spectacular growth of the trees.

Action of Soil Fungi

You have been planning and working thus far with visible and tangible factors. But if your tree is to grow well and soundly you must now take into consideration a factor which involves the life of soil fungi—and even the scientists don't know too much about this aspect of the life cycle.

These fungi live in the soil in close harmony and cooperation with the feeding roots of trees. They are called mycorhiza and they are independent organisms with an existence of their own. They are to be found in great profusion in the soil of our dense forest lands and many soil scientists credit them with the vigorous growth of our forest trees.

This teamwork is called mycorhizal association. When it is absent, the tree suffers. Its leaves turn yellow out of season and its growth is spindly. Meanwhile and close at hand "trees aided by the presence of the fungus are strong and healthy with much larger and more vigorous root systems covered with celium."

This celium or mycelium is the underground vegetative part of the fungus. It "envelops the roots of the tree and, acting as root-hairs, absorbs nutrients from the soil and passes them on to the tree." Many of our forest trees cannot even grow without proper mycorhizal association, the study of which is "in its infancy but may in time revolutionize our ideas of forestry and many of our fertilizing practices."

You must have mycorhiza in the soil if you want your trees to grow properly. How can you make sure of their presence in your soil? Quite simply. Duplicate as much as possible forest soil conditions in your growing sites. We have hinted at this before but we stress it now. Make your tree soil as fertile as the soil in the forest.

Also, spread a heavy mulch as a ground cover in your tree basin. Remember how the forest floor is covered with leaves all year? This leaf blanket not only is a source of compost but it also keeps the temperature of the soil constant. The sun is never able to bake the ground to a hardpan and the winter frosts never freeze the ground as hard and deep as your fields and pastures. It is believed that this consistently even temperature is what mycorhiza and other benevolent soil organisms need in order to function effectively.

Another, and recent, report on this aspect of tree growth reveals that spent hops and dried blood, when mixed well, appear to stimulate the growth of mycorhiza. It is also known, beyond all doubt, that spent hops is an excellent mulch for trees. It is on the acid side with a *p*H factor of 4.5 and it is recommended for all wood-bearing plants. You can get a free supply from your local brewery and it will really help your trees grow.

Admittedly, there is something mysterious about the growth of trees. Most growth occurs within a very short space of time—6 to 8 weeks in the spring—and the rate of growth is absolutely phenomenal in this brief period. Obviously this burst of visible energy is the result of previous preparation which stores energy all year and releases it only when everything is right and ready for it.

A tree needs food in order to grow. This food comes from 3 sources: the soil, the sun and the air. The vast amount of carbon which makes a tree comes through its leaves as carbon dioxide. The roots absorb potassium, calcium, magnesium, iron, sulfur and phosphorus—all suspended in life-giving water.

These minerals reach the leaves where, combining with the carbon, hydrogen and oxygen, they are converted into starch by the process known as photosynthesis.

The power responsible for this transformation comes from the sun. The sunlight-energy is absorbed by the leaves and it is responsible for the conversion of mineral and gas molecules into the more complex molecules of soluble starch.

This synthesized product goes into the sap which is circulated throughout the tree by the cambium layer. It is this food that creates the new wood growth you want so badly. This is the food which extends the length of the twigs you can see and the rootlets under the soil which, although unseen, are equally important to the life and growth of the tree.

Planting Bush and Vine Fruits

The soil preparation needed for small bush and vine fruits is similar to that needed for trees, except that in some cases the roots will not penetrate so deeply under ground. Roots on the plants, when they arrive from the nursery, are not as long as those on the trees.

For setting grapes and bush fruits it may be convenient to open a furrow with a light plow. The plants can then be set in the bottom of this furrow and the earth drawn about their roots and pressed firmly.

Somewhat more care is necessary in setting strawberry plants. It is important to have the crown of the plant even with the surface of the ground so that the terminal bud, or growing point, is just above the surface. The plant is likely to rot if planted too deep and, if planted too shallow, the crown and roots may dry out. When the ground is well prepared, strawberries can be set rapidly and satisfactorily with a garden trowel. As with other plants, it is necessary to firm the soil about the roots.

Young trees as grown in the crowded nursery have not developed a tough resistant bark. The practice of planting the trees as shallow as possible fre-

Soil where strawberries and other small fruits are to be grown should be prepared carefully. Above, a strawberry plant is set out in early spring, the thumb and forefinger clasping plant at the crown which is to be placed at ground level.

quently places the bark that was below the soil line in the nursery above the ground, thus exposing very tender tissues to sunlight. Unless the trees are protected from direct sunlight, sunscalding frequently results on the southwest side of the trunks. Such injured spots on the trunk provide an entrance place for disease organisms and as a result many trees are killed. To prevent sunscald some sort of protection should be placed around the trunk. Protection can be made of newspaper mats, heavy magazine paper or any cheap material that will last for two or more years until the bark on the trunk becomes thick and resistant to sunscald. Do not wrap too tightly as the material may bind and girdle the tree when the diameter of the trunk increases. Whitewashing the trunk is sometimes resorted to, but protectors seem to be more effective in preventing sunscald.

A paint formula which has been effective in preventing sunscald is as follows: Mix 6 pounds of whiting, 1 pound of casein and 1/3 pint of raw linseed oil with enough water to make a thick paste and allow it to stand for 24 hours. The following day, thin it to the consistency of buttermilk and paint it on the tree trunks or larger limbs.

The same protection should be used against winter injury as for sunscald, since the protection will prevent excessive fluctuations in temperature of the bark when the sunlight is reflected from the snow onto the bark. This reflection raises the temperature of the bark considerably above that of the air and when cold night temperatures follow, killing of the bark in that area may result.

The root system of Malling IX is brittle and trees on this stock may break off when the tops become large enough to offer resistance to a strong wind. This can be prevented by providing supports in the form of a stake or post for each tree. A piece of two-inch galvanized gas pipe makes a permanent and inconspicuous support when driven into the soil. The support should be about 4 or 5 inches from the trunk and should extend several inches above the lowest scaffold limbs. A durable piece of hemp cord or a heavy wire covered with a section of garden hose serves to fasten the trunk to the support. The tie should always be loose enough to prevent binding or girdling as the trunk increases in circumference.

After the tree is planted, the top should be headed back to correspond with reduction in the root system. In transplanting, a considerable portion of the roots has been lost and the water-absorbing capacity of the plant has been materially reduced. If the tree top is left at the original size, the root system will be unable to provide sufficient moisture to replace that lost by transpiration and the tree will gradually die back and in some cases may completely die. Generally, cutting the tree tops back to a height of 18 to 24 inches will be found satisfactory. Trees cut back at planting as directed generally attain a larger size in less time than similar ones not cut back.

As soon as the young tree is planted, a protection against mice and rabbits should be installed. If the trunk is well-wrapped to protect it from sunscald, this may be protection enough against the small rodents who gnaw the bark. But a more certain protection against rabbits is chicken wire, especially in areas where deep snow enables them to reach the tender bark higher up in the tree. Use one-inch mesh chicken wire or hardware cloth. Cut roll 36 inches wide lengthwise, into two 18-inch strips. Then cut these strips into pieces 14 inches long. Roll and bend the strip around the trunk of the tree so that the long way is up and down the trunk and the edges overlap. Twist a small wire loosely about the center to prevent the strip from unrolling. Pushing the lower edges well down into the soil will also help in protecting the tree against meadow mice.

25—Fertilizing Fruits

A TREE IN THE FOREST feeds itself. Every leaf, blossom, fruit and branch is returned to the soil. Nothing is wasted.

A cultivated fruit tree is not so fortunate. Each year its produce is removed, never to return to its native soil. Somehow this continual taking-away must be balanced.

Man insists that the fruit tree produce abundantly, not merely enough to ensure its own reproduction. As a result, he must keep the fruit tree supplied with plant foods and trace elements and must restore the organic material he sends away to market. He expects the tree to provide yields unnaturally high in quantity and quality.

Thus man has a duty to treat his fruit trees fairly, and see that they don't go hungry on the job.

To maintain good growth on young trees, the fertilizer program must be planned to keep a moderate but adequate level of nutrition available throughout the growing season, but not in excessive amounts at any one time. Young fruit trees respond quickly to nutrition deficiencies. Symptoms of deficiencies may appear any time during the growing season on young trees. If they are observed in early development they can be corrected and the young tree will respond to the corrective measure quickly. In contrast, bearing trees permitted to become deficient require an application the following season to correct that deficiency.

Tree Nutrition and pH

The *p*H of the soil is most important in tree nutrition. Plants are able to obtain some of the necessary nutrients only when the soil's grip on them is unlocked by a certain amount of acidity. Each plant has its own *p*H requirement which we have tried to indicate throughout the text. Before trees or bush fruits are planted the *p*H should be adjusted for the particular needs of the fruits. But after the orchard is established, this acidity level should be checked frequently, because many of the orchard practices change the reading from time to time.

Orchards which have been treated in the past with chemical fertilizers may have a surprisingly low *p*H. This is brought about by the excess sulfur in the soil, accumulated from the various sulfates and the sulfur spray materials. When the subsoil becomes too acid some trees, especially apples, become shallow-rooted, languish and perhaps eventually die. This could have been prevented if the *p*H of the soil had been kept at a consistent level. If trees or bushes show signs of nutritional deficiencies, first check the soil's *p*H. It is possible that there is plenty of the needed nutrient in the soil, but because the *p*H is wrong, the plants cannot get at it.

According to research reports, orchard soils in the Northeast are naturally acid or sour, but this condition is being intensified by the sulfur deposits from spraying and dusting. For example, in a Massachusetts orchard in which sulfur has been used liberally for 15 years, the soil *p*H was found to be 3.8 in some places, an extremely acid condition which prevents the growth of the usual orchard grasses and clovers. Of course, it would take many years before all of the soil reached by the tree roots could become acid, which is fortunate, because tests have shown that an acidity of *p*H 4.0 will cause serious injury and death to trees.

Liming is the logical answer, of course, for growers who find that their orchard soils have become very acid. One to two tons of limestone per acre is considered a good application. Many growers find that poor trees are seldom found in orchards where the soil will support a good ground cover, indicating sufficient lime content. For a full discussion on the use of lime, see the chapter on fertilizing vegetable soils.

When soil analysis shows the need, the orchard soil should have the required amount of ground limestone or marl disked in, before the trees are set out. If the subsoil is unusually acid, part of this lime should be disked in before plowing. Where trees are already growing, lime can be spread and cultivated in before seeding a cover crop. Where the cover crop is not to be disturbed, lime must be top-dressed. Limestone only moves downward through the soil at about two inches per year; thus, it is of more immediate value when well worked in.

If the soil is too alkaline, sometimes woods humus, leaf mold or peat moss can be worked in to counteract the condition. All are strongly acid materials. Bone meal and wood ashes, being alkaline, should be avoided in such cases. An important rule for alkaline soils is to make every effort to build up organic matter, so that toxic salts are absorbed.

Feeding Trees

With reaction under control, tree feeding materials come next in consideration. Choosing the right ones depends on the species and age of the trees and on the particular soil. Young fruit trees are moderate feeders and have a relatively low nitrogen demand. The basic build-up of the orchard soil keeps them going for several years. Bearing trees naturally have higher demands. They especially need a regular supply of nitrogen. Peach trees are heavier feeders than apple, often taking up twice as much plant food; cherry and pear lie somewhere between in their demands.

No two soils need just the same plant food applications to gain ideal nutrient balance. In some areas, phosphate may be particularly low; in others, potassium may be the limiting element; or poor humus and lack of suitable cover crops may cause a shortage of nitrogen. Soils on adjacent farms can differ greatly, even soils on adjacent fields. Varying geological origin decides much of this. Man's past management—or mismanagement—often decides just as much. It is because of these many differences that soil analysis is important. From a properly taken soil sample, a good laboratory can prescribe definite kinds and amounts of soil improving and plant feeding materials and tell how to apply them to the ground.

Phosphate rock and colloidal phosphate are vital feeding materials wherever phosphate must be added to the soil; and that means almost everywhere. Basic slag also may be used, but preferably not on alkaline soils, since it contains much lime. A slow but steady supply of that essential plant food is provided to the roots. The fine particles are gradually decomposed by natural organic acids in the soil. At no one time is much of the material in soluble form. One adequate application of such a mineral phosphate lasts for 5 to 10 years in tree feeding.

Similarly, finely ground granite rock, glauconite and greensand are suppliers of potassium. In many areas, there is not enough mineral potassium in the soil to supply thriving crops or trees. It, too, must be provided by man. These potassium bearing rock materials provide that essential nutrient in the same slow, steady way.

Compost can be worked deeply into the soil around old fruit trees by making holes a foot apart all around the tree and packing these with compost or other organic fertilizer. In this way, nutrients are placed near tree roots.

An average application of phosphate rock would be one half ton per acre; of potash rock three-quarters ton. For individual trees, apply about 15 to 100 pounds per tree, depending upon size and age. Spread rock powders to edge of drip line and work into soil. Both should be placed fairly deep in the soil, near the feeding root zone. Like limestone, these rock products move downward through the soil at only about two inches per year. Thus, proper plowing under or disking in deeply is far more beneficial than merely top-dressing. The time to create the proper phosphate and potassium balance in an orchard is before setting the trees, or at least before establishing a heavy cover crop.

Lime, phosphate and potassium being generally accounted for, what else is needed? There is that all-important gadabout, nitrogen. Good humus in the soil means good nitrogen. But seldom do our soils have enough humus to provide all the nitrogen that fruit trees require. Help is needed, through continually building up humus and supplying nitrogenous materials. Manures and composts and organic wastes are the great standbys, used at the rate of 10 to 15 tons per acre. Of course they supply many nutrients, but are most famous for their nitrogen value. They can be spread on the surface, but are better worked in moderately, for nitrogen, unlike most other plant foods, likes to travel. Through decomposition, it soon becomes soluble and starts on its journey downward through the soil. Good conservation measures are needed, to see that it does not travel laterally as well—into the nearest creek or river, along with a generous part of the topsoil. That means, contouring, cover cropping, mulching.

There are many other useful materials, some especially important for their nitrogen-supplying ability. Manures and composts also supply phosphate, potassium, calcium, magnesium and many trace elements. Leaf mold ranks with compost, usually having an acidifying effect to be considered as well. Mushroom soil and sewage sludge are low-concentration organic materials, well worth using if transportation and handling costs are reasonable. Ten to 20 tons per acre would be normal application. Ground tobacco stems are a source of nitrogen and potassium and carry some other nutrients. Dried, blood, cottonseed meal, castor pomace and tankage, applied at one-half to one ton per acre, supply moderate slowly available nitrogen and some other nutrients. Bone meal acts mainly like phosphate rock, also providing a little nitrogen. Wood ashes are high in potassium, and also have a liming effect.

Mulching is often useful around the trees. It helps hold moisture and prevents leaching and erosion. It encourages microbiological activity in the soil. Chopped legume hay makes a good mulch, particularly because of its high nitrogen content. Straw, ground corncobs or sawdust may be used, but should be supplemented by a sprinkling of high nitrogenous material like dried blood, tankage or rich plant wastes. A practical way to use these coarse mulch materials is in the barn or barnyard, as absorbents for animal manure. Grass clippings should not be used alone, since they tend to form a mat. Liquid or composted garbage is good and can be mixed with coarser mulch materials. Dried and washed seaweed makes a useful mulch and is often rich in trace elements.

The beginning orchardist may find the principles of fruit tree nutrition and soil balance seemingly involved. With experience, this impression should disappear. His first move toward experience is to find out what his soils need, proceeding then to build up his land and feed his trees accordingly. A competent fruit grower does not flounder around in ignorance or guesswork.

—Edwin Harrington

Applying Compost to Trees

Compost should be applied under each tree. Start about 2 to 3 feet away from the trunk and go to about a foot beyond the drip-line at the end of the branches. How thick shall it be applied? If you are going to apply it every year, a half inch to an inch will do. First cultivate under the tree to work the grass mat into the soil; then work in the compost, keeping it in the upper two inches. It is a good practice then to apply a mulch of old hay or other green matter. A layer of compost about 3 or 4 inches thick would be sufficient for 3 or 4 years.

Where there are poisons in the soil from many years of spraying, a 3- or 4-inch layer of compost worked into the soil will tend to counteract somewhat their harmful effects. To save time, the compost, instead of being made in a separate place and then hauled to each tree, can be made right under the tree. Thus it acts as a mulch also. The reason it is called the *ring method* is

Most fruit trees should receive annual applications of organic fertilizers. As shown above, material should be applied to area from trunk to drip-line of tree, and worked into the top 2 or 3 inches of soil. This procedure will supply sufficient nutrients for best growth.

that since you start from about 3 feet away from the trunk, the material looks like a ring. Apply the raw materials under the tree as if you were making compost, but instead of making the heap 5 feet high, make it only about two feet high. To hasten the formation of compost, a large quantity of earthworms can be placed in the material.

Another good plan for fertilizing old fruit trees is to auger holes a foot apart all around the tree and pack these with compost or other natural fertilizer.

Nitrogen for Fruit Trees

Available from various organic sources, nitrogen may be used in the form of 2¼ pounds of dried blood or 4½ pounds of cottonseed meal or about 15 pounds of compost to the tree, if the winter is not too severe. It should be applied several times in the spring when the frost is well out of the ground and a good 3 weeks before the time when the trees will be in blossom. The actual amount of nitrogen given in the form of dried blood or cottonseed meal, or any of the other basic materials from which nitrogen is released, will depend on the type of mulch which has been used on the young trees, as well as on the age of the trees, plus the nature of the soil.

One-half the amount recommended above may well be sufficient for young trees grown in a cover crop which is later used as a mulch around their base. For instance if apple trees are grown in sod and mulched with nonlegume hay, obviously they will obtain little nitrogen under those circumstances and the dosage of dried blood or other material will have to be increased. Similarly, with each recurring season, the amount would have to be increased depending on the condition of the trees, reaching a maximum application of two pounds of nitrogen when the trees are 7 or 8 years old.

Irrespective of the amount of nitrogenous material, it should be applied in a circle about 3 feet wide under the outer extremities of the branch spread.

A deficiency of nitrogen will show up in the tree by the leaves being small and yellowish; the remedy has already been suggested. If the foliage rolls and scorches, though, that indicates a lack of potassium in the soil. A liberal mulch of compost or a clover mulch to which lime has been added, mixed with from 400 to 500 pounds of potash rock applied to the acre, will adjust the potassium deficiency.

—Milton Johnson

Fertilizing Citrus Trees

Similar to other fruit trees, citrus trees require well-drained, well-aerated and fairly fertile soils. Excellent drainage is particularly important. Therefore cover cropping, sheet composting and the addition of organic fertilizers are vital to the success of every citrus grower, if soil structure is to be improved.

The time cover crops are grown usually coincides with the period of heaviest rainfall. In California practically all of the rain occurs during the winter and so, cover crops that grow in cool weather are started in the fall just before the rains stop. In Florida, most of the rain takes place in the summer; cover crops are started in the spring and worked into the soil in fall—usually just before the harvest season.

To get an idea of how to fertilize your citrus trees, let's take a look at how Will Kinney of Vista, California, handles his 24-acre orange grove. When he was first building up the soil, Kinney used 20 tons each of natural limestone and gypsum, plus 30 tons of a rock phosphate mixture that included many trace elements such as cobalt, zinc, boron, molybdenum, iron and magnesium. Comments orange grower Kinney: "Seventy tons of this material on 24 acres is a very heavy application. However, this sheet composting program will be continued, with some changes, for several more years to bring the grove into maximum production of maximum food value fruit.

"Each month throughout the year, the permanent cover crop is chopped at the rate of 6,000 pounds per acre. These chopped plants return a high-nutrient green manure residue to the soil. I have been using grasses and clovers for cover crops and they produce an abundance of organic matter for

The Golden Delicious apple leaves above are suffering from a magnesium deficiency, indicated by the development of flesh-colored patches of dead tissue (not necessarily restricted to leaf edges). Apply dolomitic limestone or rock phosphate to correct deficiency.

our soils. In addition, I've been able to add hundreds of tons of cow manure, which is an excellent organic nitrogen carrier."

Recently more and more reports have been issued about the decline in citrus production, especially in Florida. To a great extent, the cause seems to be excessive use of concentrated chemical fertilizers. In general, California growers have been making better use of organic nitrogen fertilizers, such as manure, alfalfa hay and straw, than orchardists in Florida, where only a small percentage of the nitrogen is derived from organic sources.

Importance of Trace Elements

It is becoming increasingly evident that many minor elements play a large part in the health of every plant. The following is taken from the USDA Yearbook for 1938, which was devoted to the relation of soils to plants.

"It is now generally admitted that for normal development, plants require the following chemical elements in suitable compounds: carbon, hydrogen, oxygen, nitrogen, phosphorus, potassium, calcium, magnesium, sulfur, iron, manganese, boron, copper and zinc. Animals may require all of these and in addition sodium, chlorine, iodine, and cobalt. The quantities necessary for the normal development of plants or animals vary considerably. Those required in small quantities include iron, manganese, boron, copper, zinc, iodine and cobalt. The members of this last group are the ones that commonly become toxic in slightly larger quantities than those required. Another group of elements that are toxic at relatively low concentrations includes aluminum, barium, chromium, fluorine, lead, selenium and thallium."

Dr. William Eyster points out that the deleterious effects of the poisons which find their way into the soil from the use of sprays and dusts in attempts to control insects, pests and diseases have already reached the danger point in all crops that are being grown intensively, as orchards of fruit trees, tobacco, potatoes and vegetables grown in monoculture as beans, tomatoes and spinach. Often sprays are used in an effort to control "diseases" which are caused by deficiency of one or more essential elements in the soil. The growers of these crops have been using excessive amounts of both highly soluble acid fertilizers and poisonous sprays so that the soil is no longer capable of providing the plants with their nutrients in proper balance.

While the continuous use of chemical fertilizers tends to deplete the essential elements not supplied to the soil, the use of stable manure, leaf mold, wood ashes and peat tends to conserve them. On dairy farms, a large part of all elements is returned to the soil and the secondary elements contained in such concentrates as are purchased from the outside are therefore actually added to it. Leaf litter, leaf mold and wood ashes contain many of the elements taken from the forest soil in proportions desirable for the nourishment of the trees. The undesirable ones have been largely eliminated. Furthermore, the secondary elements in leaf mold, particularly manganese, are in a very available form. In long-continued experiments at Woburn and Rothamsted in England, it has been found that stable manure has maintained the fertility of the soil over much longer periods than has the use of chemical fertilizers containing nitrogen, potash, phosphorus and sulfur. The numerous chemical elements contained in the manure are undoubtedly an important factor in this observed maintenance of fertility.

But sufficient manure may not be available to the gardener to take care of all his needs. Composted organic materials of all kinds may be substituted, particularly if they are enriched with natural rock supplements.

Susceptibility of plants to certain fungi and insects must be regarded as a deficiency of one or more essential elements. In controlling diseases and insects, prevention by adding the deficient elements to the soil is by far the more natural and more economical measure to be taken.

When fruit trees or bushes show signs of ill health which cannot be caused by insects, check the following list of deficiency symptoms before deciding that they are suffering from virus or another type of disease:

Nitrogen Deficiency.—Examine the leaves on the old branches. With a lack of nitrogen, these older leaves turn a yellowish-green, working toward the tips. You may also notice reddish or reddish-purple discolorations. If nothing is done to relieve the deficiency, leaves become very small and the twigs slender and hard.

Phosphorus Deficiency.—The young twigs develop a ghost-like hue; stems show purple coloring; leaves are abnormally small and dark green. Old leaves become mottled with light and dark green areas. Occasionally bronzed leaves will show up on mature branches.

Potash Deficiency.—The key to potash deficiency is purplish discoloration and scorching of leaf edges. The dead spots will be found on mature leaves, but under continued deficiency, even very young leaves are affected. Peach foliage often becomes crinkled and twigs are unusually slender.

Magnesium Deficiency.—The large, old leaves will display flesh-colored patches of dead tissue, not restricted to the leaf edges. Watch for dropping of leaves, first on old branches, then on twigs of the current season. Defoliation may be so severe that only tufts or "rosettes" of thin, small leaves are left.

Zinc Deficiency.—Both zinc and magnesium deficiencies are very much alike. Each of them can cause rosettes of leaves in the advanced stage. But without zinc, crinkled leaves are common, which are also chlorotic (rather washed-out yellow in color). In peach trees this is very true. With citrus fruit, very small, smooth fruit and pointed leaves are the symptoms. There may also be striking contrasts in leaf patterns—dark green veins and yellow tissue.

Calcium Deficiency.—Calcium and boron shortage will show up first on young twigs rather than the mature branches. Dead areas are noticeable on the young, tender leaves at the tips and margins without calcium. Later, the twigs will die back and roots are injured.

Boron Deficiency.—Immediately coming to mind should be internal cork of apples, which is the commonest boron deficiency problem. Early in the season, hard, brown spots with defiinite margins form inside the fruit. As the season progresses, the spots soften, become larger and lose their definite outline. The leaves may be entirely unaffected.

In other cases, the young leaves can become very thick and brittle, then cause dieback of twigs. Some trees may also form wrinkled, chlorotic leaves.

Iron and Aluminum Deficiency.—With an overdose of lime comes an unavailability of certain minerals like iron and aluminum. These minerals may be right in the soil, but are held insoluble when the acidity is low. Look for yellow leaves with brown patches and loss of flavor in the fruit.

To correct these deficiencies:

Nitrogen—apply cottonseed meal, dried blood, tankage, raw bone meal, fish wastes, legume hay or one of the organic nitrogen commercial products now on the market.

Phosphorus—raw or colloidal phosphate rock, bone meal, fish wastes, guano or raw sugar wastes.

Potash—granite dust, glauconite marl (greensand), wood ashes, seaweed or orange rinds.

Magnesium—dolomitic limestone or raw phosphate rock.

Zinc—raw phosphate rock.

Calcium—raw pulverized limestone.

Boron—raw phosphate rock (avoid lime, add acid organic matter like peat moss, sawdust or ground oak leaves).

Iron and Aluminum—glauconite marl (avoid lime, use the acid organic matter recommended under boron).

Results of Organic Fertilizing

Dennis Seale, a California organic grower, root-feeds his trees with the compost twice yearly, pulling the dirt away from the roots for a depth of 6 or 8 inches and in an area of about 6 feet around the tree. He fills in this ring with finished compost and covers it over with the soil again, watering it down well.

Among his prize-bearing fruit trees are an Everbearing lemon which produces around 400 pounds of lemons per year, some of which measure 9 and 10 inches around with a tape measure. Also winners are an Elephant Heart plum, a Burbank HalBerta Giant peach and his Golden Delicious and Double-Red Delicious apples.

Another California fruit grower, Herbert White, reports the following:

"Fig trees planted in late April, loaded with luscious fruit in early October. Raspberry vines, planted in late July, with fruit-laden terminals 3 months after planting. Grapes planted in May bearing well-formed bunches of nectar-filled fruit of the vine 5 months after planting. Lemon trees planted in July, pushing out beautiful 3-foot terminals covered with glorious foliage and, believe it or not—lemons the size of plums already forming on the branches!

"Charlotte and Joe Merrik of Sierra Madre in suburban Los Angeles, California, had a few square yards of land on a steep, formidable looking hillside.

Joe Merrick of Sierra Madre, California, stands next to a flourishing peach tree in his organic garden. Planted in April, this tree measured 8 feet in height with 65 inches of terminal growth only 6 months after planting.

The soil? Just about coarse decomposed granite. And yet they wanted a fruit garden—and worked the right way to achieve it!

"Given impetus by an interest in better foods, the Merriks sought aid and materials for converting their 65 by 75 feet back yard into a productive miniature orchard. Joe did the initial work of excavating steep banks, building retaining walls and leveling terraces. Then, despite the liberal gifts of fine ornamentals and a variety of deciduous and citrus fruit trees from a family friend, they paid no heed to his advice to use the standard commercial chemical fertilizers, sprays and so forth. Instead, the Merriks turned to the author whose suggestions for organic care were eagerly followed. Soil amendments—organic, that is—were secured and applied carefully. These included compost, leaf mold, trace minerals, peat moss, raw phosphate rock, earthworms, topsoil and stones I had recommended for mulching and drainage. The result? In less than 6 months they have 24 varieties of fruits and berries, an herb garden, a compost pit, a lily pond, an earthworm culture bed and a vegetable garden."

26—Mulching and Watering

MULCHING MEANS THE introduction of organic matter into the orchard and its placement upon the soil as a cover, more especially under the tree and under the outer branches and less near the trunk. This method is almost ideal from the standpoint of maintaining the organic matter, preventing erosion where sloping land exists and in maintaining the fertility of the soil. It also helps greatly in unifying the moisture content and its temperature. Both of these results are important. Moreover, the mulch insures the penetration of rainfall and hinders the washing of the soil and its solutes. When the mulch material is hauled in, it does not compete while growing with the fruit trees in their need for moisture and nutrients in the orchard. Under this practice the mulch should be sufficient to suppress the growth of weeds and grass under the trees. If the mulch or manure is ample to cover all the ground between trees, so much the better, as there is then more water and nutrients for the trees. The orchard tree also gets the benefit of the introduced product and in addition the benefit from the soil just as before the application of the mulch.

Two objections have been made to the mulching practice. One is the fire hazard in a dry season. Another is the rodent problem as the mulch seems to afford shelter and breeding medium for mice. But with proper precaution these hazards are not insurmountable, in fact they are overcome in a well-ordered orchard.

The mulching material may have many sources according to its nature. Swale hay is good and generally inexpensive to get; leaves from the woods or from city street sweepings serve well. Sawdust is sometimes handy and generally effective. However, it tends to make the soil acid. Lime should be mixed with a sawdust mulch. Straw or waste hay have been used. Bluegrass is much used. It is, however, less desirable for dry places. Pea vine is also excellent. Buckwheat or other straw will do. Orchard grass is excellent and is sometimes grown in the orchard between the rows of trees and is cut just before seeding and placed under the branches. Buckwheat hulls and cottonseed meal make fair mulches where available at low total cost including haulage and distribution.

Many years' experience with orchard grass and street leaves have shown them good mulching products. Orchard grass *(Dactylis glomerata)* is a per-

This California orange grove is being mulched with alfalfa. Trees under mulch have been found to come into bearing earlier than trees grown under clean cultivation.

manent grass. It makes a good growth and forms fine sod even though tufted and yields well year after year. It was used at the Vermont Experiment Station for 20 years and was as good after two decades as at the beginning. It was grown between the rows and when cut spread mostly under the trees. In years of abundant rain two cuttings were made—one in mid-June, another in mid-August. Orchard grass grows well in slightly acid soil, is very hardy and stands dry summer weather.

Leaves were used for several years. They were available in large quantities as the orchard was within the city limits and the street department delivered them to the orchard free of charge as the nearest dumping place for the city trucks. For several years leaves were placed one foot deep under the branches and packed to some extent by tramping; but often after the first good rain, the material packed itself and stayed where placed. Leaves were then used for 3 consecutive years and they were satisfactory. After the first year, the leaves decay at the bottom of the mulch fast enough to permit annual applications to be put on top. They settle fast.

Orchardists in Door County, Wisconsin, experimented with a number of mulches under their young orchard trees and found them all helpful, according to Charles F. Swingle. Their mulch sources included sawdust, shavings, cedar bark peelings, marsh hay, apple pomace, cherry pits, manure, cornstalks, paper bags, leaves and stones. These were all found to be useful, especially when used under small trees which had not grown dense enough to cast a good shade.

Mulching mature orchards has proven advantageous in a number of cases cited by William Ackerman. He points out that mulch reduces to some extent the amount of bruising of the windfalls and fruit which falls during the picking operation. This fruit is also cleaner than where clean cultivation is practiced. Trees under mulch have been known to come into bearing somewhat earlier than trees under clean cultivation and sod. Also, the color of the fruit is somewhat improved.

Some of the results performed by agricultural investigators have been quite revealing. For example, yields from cherry trees have been increased 100 per cent in experiments at Michigan State College. It was done by mulching the trees with straw or legume hay, in tests by Dr. A. L. Kenworthy, horticulturist. The increase came on trees receiving only mulch and no fertilizer, as compared with trees which received complete fertilizer but no mulch.

It is reported from New Zealand that excellent results have been secured by mulching citrus trees, as indicated by the dark green foliage and the smooth clear skin of the fruit. Since mulching has been undertaken, the quantities of artificial manures have been reduced by nearly half. The practice is to grow clovers or grasses in adjacent fields and place the cut materials in the late summer around the trees out to the limit of the branch spread.

One of the problems in a large mulch orchard is to secure sufficient litter. There are 3 common sources of such material. It may be grown within the orchard, produced elsewhere and hauled into the orchard or purchased from an outside source.

Mulching materials that may be utilized include a wide variety of substances. The cut plant material of a large group of field and cover crops are available. Summer cover crops that may be planted in the orchard and cut for mulching material include soybeans, cowpeas, millet, sudan grass, buckwheat, lespedeza, crotalaria and sesbania. Orchard winter cover crops that may be cut for mulching purposes include rye, wheat, vetch, crimson clover, sweet clover, alfalfa and kudzu beans. Besides these materials, many other substances are often available, including weeds, sawdust, seaweed, peat moss, peat, whole or ground corncobs, stones, rotted wood, leaves, pine needles, waste products such as apple pomace, pea vines, brewery and canning wastes, etc. When one considers the vast store of organic materials nature provides, it should not be difficult to procure sufficient mulching material.

Although mulching was originally developed as a cultural practice for apples and pears, experiments in several states have shown that peach and other stone fruit trees can be profitably grown in sod areas if adequate mulch is provided about the trees. For a long time it was considered that the only way to grow stone fruits profitably was through clean cultivation. The results of more recent experiments have disproved this. The peach and stone fruit problem was the last stronghold of the clean cultivation advocates. Now this too has broken down and with it has evolved the practicality of growing stone fruits on rolling terrain where it was previously considered unadvisable.

Some experimenters advise that mulches should not be placed around the tree on dry soil. This often makes it difficult to get the next irrigation application through the mulch material in sufficient quantity to wet the soil to a two foot depth for the young tree. The tree then suffers from lack of adequate moisture. It is usually best to apply the mulch early in the spring or immediately following a good irrigation.

Another old cry in opposition to mulching has been the depletion of nitrogen from the soil because it is used up by the soil organisms in their effort to break down the organic materials. Because the mulch material is not mixed with the surface soil itself, but exists only as a cover, decomposition is not rapid and takes place primarily at the interfacial layer where soil and mulch are in contact.

Temporary nitrogen deficiency does not occur when mulches are used properly, but for the individual who does not go overboard in the use of

If mulch is heavy enough, it will conserve moisture by killing weeds and reducing water loss through evaporation. Besides keeping water in the soil, a straw mulch will also hold about 3 times its weight in water.

low-nitrogen materials, there is little danger. If straw, corncobs and sawdust, which are low in nitrogen content, are mixed with other mulching materials such as legume residues, felt or wool wastes, hoof and horn meal, cottonseed meal, castor pomace, etc., which are high in nitrogen content, little difficulty should be encountered with nitrogen deficiency.

The use of mulch in preference to cover crops, especially near young trees, is advocated universally. Between their first and fourth years, fruit trees are especially sensitive to competition from other vegetation. At that time, the roots of the trees are limited, both in horizontal and vertical spread. Competing growth within 4 feet of the trunks in the form of weeds, grasses or even cover crop legumes such as alfalfa can prevent the young trees from making normal growth. One pound of dry weeds or grass requires little in the form of nutrient, but 500 to 800 pounds of water must pass through the leaves of the plants to produce that amount, according to figures issued by Oregon State College. If a cover crop is used, therefore, it should be turned under in early spring before growth starts. This precludes the use of permanent cover crops in a young orchard, which would be damaged by being turned under too often. Such material can be used only in established orchards and there it is advisable to mow it at intervals through the season and to rake the mowed material under the trees where it can act as mulch.

An orchard cover sod recommended by the Ohio Agriculture Experimental Station in Wooster, Ohio, is made up of seed in the following proportions: 3 pounds timothy, ½ pound ladino clover, 5 pounds bird's-foot trefoil and 3 pounds bluegrass per acre. The seed may be sown in August, but March or April are preferred in the sections with climate similar to Ohio's. After it is established, a process which takes 2 or 3 years, the trefoil provides heavy mulch for the trees. The growth should be clipped 2 or 3 times each season. This practice conserves moisture, but it must be emphasized that it conserves it only under the mulch.

In studies made in California, Veihmeyer and Hendrickson found that an orchard with a cover crop of alfalfa needed one and a half times as much water as when the orchard was under clean cultivation.

Perennial Mulch

During the summer months of the past few years, there has been a serious shortage of rainfall. Adding to the problem of water shortage only about one-third of the rainfall enters the soil and not all of this is available to plants. Since it is possible to have a water deficit, or at least an amount that is deficient for good growth of our fruit trees, some cultural system that may help on this shortage should be considered.

Many times water deficits are not noticed on fruit trees as the leaves of most orchard trees are stiff and do not drop easily. The adverse effects of water shortage may show up in smaller yields, smaller fruit, poorer quality and stunted trees. All of these shortcomings may not show up the first year, but if water shortage continues over a few years all these problems will make their appearance.

A perennial mulch system should be considered for conserving larger amounts of rainfall and soil moisture and at the same time protect the soil against

Wood chip mulch on young raspberry plants makes an excellent covering, conserving moisture for the development of berries. Sawdust and wood chips are especially useful around berry plants which prefer an acid soil.

erosion. The perennial mulch system of culture is one in which the mulch extends from about one foot from the tree trunk to the tips of the branches and is heavy enough (6 to 8 inches deep) to keep down weed growth under the trees.

Mature trees require approximately 200 to 300 pounds of mulch material the first year, plus an additional 100 pounds yearly, to receive the benefits of a good mulch system. The mulching material may consist of wheat straw, oat straw, hay, corncobs, sawdust or weeds. Some mulching materials may create critical soil acidity and care must be taken to correct this.

The system of perennial mulch has been found to give less fluctuation of soil moisture, particularly during dry periods. An additional benefit is the ability of the soil to absorb water more easily because it becomes loose and friable under the mulch.

If the mulch is heavy enough, as stated earlier, it will conserve moisture by killing weeds and removing this competition for water. A certain amount of loss due to evaporation will also be prevented.

Besides keeping water in the soil, a straw mulch will also hold approximately 3 times its weight in water. Many live roots work into the mulch and this may be beneficial during a critical water shortage.

The system of perennial mulch not only helps during a drought period, but also has the advantage of:

1. Less fluctuation in soil temperature.
2. Higher organic matter content.
3. More availability of many nutrients.
4. Serve as a cushion for dropped fruit.

—*Stephen Patronsky*

After the leaves are formed, Veihmeyer and Hendrickson found, the trees begin to draw upon the soil moisture and continue to do so until it is reduced to slightly beyond the PWP (the soil-moisture condition at which plants cannot obtain water readily). After irrigating, this process is repeated. In practice, of course, the orchard in an arid part of the country should be irrigated before the trees wilt.

The number of times this cycle of events takes place during the growing season depends upon the size of the trees, the climatic conditions and the kind and depth of the soil.

The total amount of water that comparable trees will use will not be greater on a clay soil than it is on a sandy soil if both are fertile and have readily available water at all times. Usually on sandy soils, however, water must be applied more frequently and in smaller amounts than on clay soils.

With the coming of warm weather, the readily available water is quickly used by the trees. This should be replenished.

The orchardist should be able to judge when his trees need water because of his close association with them and his daily observation of their condition.

When wilting or any other evidence of lack of readily available moisture is hard to detect in the trees themselves, the best method by which the grower can decide when this condition is reached is by watching some of the broad-leaved weeds which may be left as indicator plants in various places in the orchard. Generally, such weeds are deep-rooted enough to indicate by their wilting a lack of readily available water in the soil occupied by the roots of the trees.

It should be pointed out that evergreen trees use water later in the fall and earlier in the spring than deciduous trees. They even use some moisture on clear warm days during the winter.

Studies in the irrigation of citrus groves show that an average of not more than 5 per cent of the moisture used was taken from the 5th foot of soil, which indicates that most of the roots were above this depth. In fact, in soils less than 3 feet in depth 50 to 60 per cent of the roots probably are in the first foot of soil or below the cultivated layer.

On the other hand, walnut trees on a fairly uniform soil extracted the moisture from a depth of 12 feet or more.

If possible, the soil should be wetted at each irrigation to the depth in which most of the roots lie even though the lower layers still contain some readily available moisture. It is less expensive to wet this depth at this time than later.

Wetting the soil to a depth of 5 or 6 feet will usually be sufficient with

A great many organic orchardists use hay and other cut plant material to mulch their trees. Experiments at many agricultural stations have proven that fruit trees benefit from such a mulch.

most deciduous trees and to a shallower depth with citrus trees. If there is an impervious layer within the depth mentioned, just enough water should be used to wet the soil above this layer.

General Guide to Irrigation

The New Mexico Experiment Station offers this advice on irrigating orchards:

Orchard irrigation is something which can be discussed only generally. The water requirements of an orchard will vary from season to season or within any particular growing season because of weather conditions, cover crops and soil types. No specific recommendations can be made, therefore, about the number and frequency of irrigations.

The need for irrigation can best be determined by actual examination of the soil at a depth of 2 to 3 feet. A shovel can be used but a soil auger is preferable and much quicker. Experimental data indicate that fruit trees need about 6 to 10 per cent available moisture in the soil. A little experience will enable you to tell fairly accurately just by looking at your trees if less than this amount is present.

A check for soil moisture which can be used in the field is to weigh out 100 ounces of soil and dry it by direct exposure to sunlight for a day. The loss of weight of the sample in ounces will be the percentage of available or free water in the soil. If the loss in weight is less than 6 to 10 ounces, water is needed and should be applied. Orchards should always be irrigated when water is needed, regardless of the season of the year.

Always take care in irrigating fruit trees. Late or heavy irrigations in young orchards may delay maturity and permit winter injury to immature wood. Wide fluctuations in the moisture content of the soil should be avoided. On the other hand, excessive soil moisture is also to be avoided, since it may cause the fruit to crack at harvest time, may impair quality or reduce the season of storage.

Winter injury in many parts of the country may be due to lack of adequate soil moisture rather than to low temperatures, although the latter is usually blamed. The damage often results from the cold, dry air drawing moisture from the trees, which cannot be replaced from a dry soil. A late fall or early winter irrigation, after the trees are dormant, will generally reduce winter injury to a minimum.

Orchard soils, when irrigated, should be wet down to a depth of 4 to 5 feet. The time required for water to penetrate to this depth will vary with individual orchards and soils. A steel rod or soil auger can be used to check the depth of water penetration. In most soils, a rod can be pushed into the soil easily as far as the water has penetrated.

27—Aftercare of Fruits

When the gardener starts a new orchard, he is able to shape his trees, shrubs and vines from the beginning to accord with what he believes to be their best condition. But a home gardener who has just purchased an old home may find established plantings begun by someone else and perhaps neglected or trained as the gardener would not have done it himself. Making over an old planting poses problems not encountered in a planting which has had steady care.

Trees which have been neglected usually will have grown too tall to be practical and they may be full of dead and diseased wood. Bush fruits unpruned for more than one season become choked with old canes. Grape vines left unpruned for only one year sprawl and proliferate to such an extent that their crop may be reduced to almost nothing.

Old neglected trees in fairly healthy condition, with few dead branches, may need only light pruning to open the tops. If the tops are very dense, this should not all be done in one year or a crop of water sprouts will result. Moderate top thinning over the course of 2 or 3 years will result in better crops.

If the old tree is weak, with large amounts of dead wood, heavy pruning and removal of the dying branches in the first year will stimulate new growth and fruit production. When weak branches are cut back to side branches, there is not much danger of forming water sprouts and the old spurs which remain may be stimulated to bear.

This apricot was near death from gummosis, a dreaded trunk disease of fruits, when California gardener Regina Hughes Jones started a sustained compost-treatment program. Within a year the tree had recovered completely, showed no signs of the brown rot infection, and became productive.

A tree which has grown so tall that the jobs of pruning and harvesting have become difficult should be cut back gradually. If the top is dehorned radically, water sprouts will fill the entire top and choke out any fruiting growth. Instead, cut back only one or two tall branches each year, cutting back each branch half-way the first year and the balance later. If the tree has a very tall, dense top, this may take 4 or 5 years to complete. If many branches are cut at once, especially from the top of the tree, not only is too much undesirable growth started, but the upper branches may suffer serious sunscald.

Bramble fruits may be brought under control the first year if all the branched canes (these are the two-year canes) are removed. If the bushes have spread by suckering or tip-layering, the best way to tame the wilderness is to cut paths at intervals with pruning shears and then to drive a tractor, rotary tiller or even a wheel-hoe through the paths to kill off excess plants. Directions for pruning specific brambles are given under their separate entries.

Concord-type and vinifera grapes increase their wood almost 10-fold each year unless they are drastically pruned. Each leaf node on a grape vine becomes, in its second year, a branch. In its third year, the branches branch and the vine is soon a complete shambles of small, fruitless shoots, instead of a few orderly branches on which each node produces a bunch of grapes.

As the branches on the grape multiply, the fruiting portion becomes further and further from the root each year, unless the vine is cut back to force latent buds. A neglected vine may need to be cut back in 2 or 3 stages. The first year most of the wood may be removed, leaving no more than 4 branches which may be tied to the trellis. This drastic cutting may force some buds along the trunk, where they may be selected for arms two years later. If necessary, suckers from the roots may be used to replace a top which has gone out of control, but the cutting back to the sucker should not be done all in one year.

A neglected strawberry patch is one of the most difficult of fruit plantings to restore. The best treatment is usually to dig any vigorous small runners early in spring and replant them in a newly prepared bed. This is the only possible treatment if numerous perennial weeds have become established in the old bed.

Repair Pruning

Pruning out damaged, diseased or winterkilled branches gives rise to entirely different problems from those created in the training of a tree. Most emergency work done on injured tissue should not be postponed until the regular dormant pruning time, but should be done as soon as the injury takes place. The only exception to this rule is the treatment of winterkill or frost injury, especially in tender evergreens such as oranges.

Repair work of branches broken in a summer storm should be prompt, to prevent further damage from decay or infection. Breaks in the branches should be cut back to a point where a smooth wound may be left, similar to those made in pruning, so that rapid healing may take place. Occasionally splits may be repaired by a temporary bracing of the branch, followed by an inarched or bracing graft to strengthen the broken member.

Surgery in the case of disease should be prompt and thorough. As soon as an infection from bacteria or fungus is discovered, the best cure is to cut back the diseased portion of the branch well below the infected area. Cut back far enough so that only completely healthy wood remains and make the cut just above a lateral bud. Remove any infected material at once from the orchard and burn it. Before making cuts in healthy trees with the same tools, be sure to disinfect them. Surgery in the case of virus disease does no good. A virus infects the circulatory system of the plant and is carried throughout its tissue. If the infection is bad, the best procedure is to remove the infected plant before the virus spreads to healthy trees.

In the treatment of frozen trees or frostbitten evergreens, great care should be exercised. If the treatment is too drastic, the injury may be accentuated rather than reduced. It is best to wait until growth has started before treating winterkill, because in some cases the unfolding of buds is delayed, but not entirely prevented by cold. In the case of citrus trees, the frost injury should be pruned out gradually, if it is extensive. In very bad cases, the work should be spaced out over the course of 2 to 3 years.

Bark splitting is found in some trees, especially cherry, following a winter of alternate thaws and freezes. If the bark is tacked back in place as soon as possible, much of the damage may be healed over. The bark may not unite with the wood immediately, but sometimes after 2 or 3 cambial layers have been laid down the fusion will take place.

Less winter injury is suffered by unpruned than by recently pruned trees. If the pruning is delayed until late winter, not only can small amounts of winterkill be trimmed back, but less of it will occur. This is true of red raspberries and grapes as well as of fruit trees.

Maintenance

M. B. Cummings, an official nursery inspector in Vermont, sums up the problems of orchard maintenance as primarily one problem—that of maintaining the organic content of the soil. Humus, says Mr. Cummings, is used up and needs annually to be replenished. To the fruit grower, whether large or small his operations, nothing in practice can exceed in importance the nature and content of his soil. The more successful he is in this matter, the greater his success in growing good fruit and doing it economically. One may do more or less pruning, more or less fruit thinning, etc., but he should not be stinting in supplying organic material for it is the very life and the dynamic center of the land. It is the basis of healthy, vigorous and productive fruit trees.

A tree in the orchard is somewhat like a tree in the woods, a large permanent specimen, but its needs are much greater, for in addition to wood production it should yield an annual crop of fruit. This gives the orchard tree a double function in contrast to a single one for the forest tree. A fruit tree needs ample growth of wood for the formation of new branches, to develop fruit spurs and fruit buds. These must precede fruit development. The growing and maturing of a crop of fruit makes a heavy draft on the stored and accumulating food materials. A forest tree gets its litter of leaves, which may be ample, but an orchard needs offside and additional material for its double duty, wood and fruit. Dead vegetation helps the land.

In the beginning of fruit growing, years ago, trees were grown in grassland—sometimes in pastures. Grass was allowed to grow, was cut as hay and hauled away. This method deprived the trees of needed organic matter and the results were generally meager and the crop irregular in the early days. Then there was a long period of clean cultivation with or without cover crops, preferably the former. After this stage there was another period in which the practice called for the application of manure. The latter is still a good practice, in fact one of the best. And finally the mulch system which is the best practice of all. And yet only a few practitioners know its value.

Making Fruit Trees Bear

Occasionally there is a tree or shrub which has had all the attention that a loving gardener can give it, has been fed on the best organic fertilizers and seems free of deficiencies, insects and diseases, but still it does not bear fruit. Before deciding that it is hopeless and must go, check the following:

Is it of bearing age? Some trees take much longer to arrive at their bearing age than others. Some apples, like Northern Spy, may need 10 or more years to mature. Most pear trees bear little fruit before the age of 6 or 7; sweet cherries,

A straw mulch will help the growth of established trees, often resulting in greater harvests; workers in the apple orchard above are spreading this material. To avoid chance of rodent injury, keep mulch about a foot away from trunk of tree.

5 to 7; some nut trees much longer. Or is the tree too old? It may have been neglected for long periods and a careful and patient course of rejuvenation may be necessary to bring it back.

Was the weather in spring right for fruit set? Rainy weather during pollination can wash out all the pollen and prevent fruit set. A slight touch of frost can kill many blossoms. Even if there was no frost, some trees cannot set fruit at temperatures under 40 to 42 degrees. When the weather is very windy during blossoming or when the temperature is low the bees are inactive, so the blossoms may not be fertilized. Perhaps intense winter cold killed the reproductive organs of the blossoms and, though they opened, they were sterile. Or the cold killed the entire flower buds during the winter and the tree could not blossom.

Are the plants self-pollinating? Some kinds of fruits need a second variety, blooming at the same time, for successful pollination. Or the plants may be dioecious—male on one plant and female on another. Check the entries under separate fruits to determine whether your tree needs a mate.

Has the tree been properly pruned? Are you sure that you have not removed fruiting spurs in mistake for water sprouts or that the trees were not too severely pruned?

Has the soil under the mulch been permitted to dry out? Dry weather can make fruit drop prematurely. Mulch can conserve moisture, but it cannot make it. In extremely arid areas or in prolonged drought even a mulched tree may need water.

Have you used too lavish a hand when shoveling fertilizer? Excessive feeding of some trees and shrubs can cause lush vegetative growth at the expense of fruiting.

Did the tree bear an extremely heavy crop last year? Perhaps the fruit should have been thinned to prevent the tree from using all its strength to ripen last year's crop. If the tree has become an alternate bearer because its crops are too heavy in some years, the cure is to thin the fruit and to remove some of the fruiting wood.

If none of these suggestions seems applicable to the case, it is still possible that the tree may be brought into bearing by one or two artificial treatments. Either one of these is designed to reduce the flow of sap to the branches and thus to increase the production of fruit buds.

The first method is that of root pruning, similar to the treatment given to espaliered trees. To accomplish this, the whole tree may be dug out and replanted in its former position. In the course of digging, the small roots at the ends of the deeper ones are inevitably removed and the tree gets less from the soil. If the tree is one with a tap root, this may be severely pruned back or removed entirely.

Root pruning can be done in a more limited way by digging a trench around the tree, two feet deep and half way between the trunk and the tips of the branches. Either of these root jobs should be done in late autumn, after the leaves have fallen.

The second treatment (it should not be tried at the same time as the first) is bark ringing. To do this, early in spring remove two half-inch strips of bark, each one going half-way around the trunk, one 6 inches above the other and overlapping by an inch. Cover the cuts with adhesive tape. Bark ringing has the effect of cutting down on the flow of sap to the branches, which set more fruit buds in consequence. Any treatment which has the effect of disturbing the flow through the bark will accomplish the same purpose. For example, if a branch is bent so that the bark is strained on the outside and cramped on the inside of the bend, the result is similar. If the tip of the branch is tied down to the trunk, it may be made to fruit unless it is too shaded. This method, called circling, cannot be used on trees whose wood is brittle.

Treating a Sick Tree

Herbert White's plan for planting a tree has already been given. Here is this California tree grower's advice for bringing trees "back to life":

With correct planting procedures and good "aftercare," this California organic gardener's peach trees grew 10½ feet tall with a 9-foot spread, 7 months after planting.

"In treating a sick tree or one which has failed year after year to produce a crop, I use the same formula as my Grandmother White used in planting a baby tree. In other words I make a 'blend' or mixture consisting of: ½ yard (12 bushels) topsoil; 2 bushels compost (completely decomposed); 2 bushels leaf mold (completely broken-down); 20 pounds colloidal phosphate (or rock phosphate); and 20 pounds of granite dust. The above formula will be sufficient for a young tree of from 2 to 6 years of age that has shown little vitality and below average growth. This enriched earth will be used to fill the 20-inch deep holes that encircle the young tree. About 8 to 10 holes can be made with an auger; these 8-inch diameter holes can be equally spaced along drip-line of tree.

"The next step in 'treating' our sick tree is to level off the land around the trunk a little beyond the drip-line of the branches, and to build a sturdy dike just outside the ring of 20-inch holes. Ten pounds of each of the two above-mentioned minerals should then be spread from the trunk of the tree out to the rim of the basin, and worked lightly into the soil with a rake. One inch of compost and leaf mold (mixed) should then be added to the area within the basin and spread out evenly.

"If domesticated earthworms are to be used in this 'health-building' program, they should be spread over the compost (from 500 to 1,000) and covered with a 3-inch leaf mulch. A little corn meal or coffee grounds spread on the compost before spreading the leaves would help to give the worms a good start in their new environment."

Regina Hughes Jones of Visalia, California had an apricot tree infected with gummosis, a gummy exudation on the trunks and branches caused by disease. Here is Mrs. Jones' story:

A few years ago, we bought an acre ranch on the outskirts of town here in Visalia, California. A young orchard ran in single file down each side and across the back of the acreage. The first year we lost 3 trees on the south; the soil was strongly alkali. Two of them were plum trees and early in the season a gummy substance began to form on the trunks and limbs. At this time we did not know too much about gummosis or what caused it; however, I noticed that the trees seemed to be slowly dying of what appeared to be rot.

One of the transplanted trees was a healthy, young apricot that had begun to bear. We had moved it in December. Early in March the buds on it began to swell and soon burst into an abundance of bloom. But alas, our gratification was short-lived. Scarcely had the blooms dropped, when I noticed the same sticky gum forming on the trunk and branches. I was heartsick. It was the only apricot tree we had. It was also a very fine variety. I rushed to the telephone and called our neighborhood nursery and asked if they could tell me what was causing the formation of the gum and if there was any way I could save the tree.

The owner of the nursery came to the telephone and promptly stated that there was no known cure for gummosis or, in other words, the brown rot that was causing the formation of the gum. He added that as far as they could determine, it was caused by something in the soil, or the condition of it. He said I might as well dig the tree up and cast it into the fire.

I set my jaw. "I won't give up—not yet," I said to myself.

I picked up a hoe, cut all the grass and weeds growing nearby and loosened up the hard soil around the tree. This done, I turned to a large pile of fertilizer on my garden plot. It consisted of sand, sawdust and steer manure, scrapings from big cattle trucks. I dumped two wheelbarrow loads of it around the tree, covering a circle of 3 feet in circumference. As the hot summer days passed, I kept lovingly pulling the gum from the dying tree; adding fresh compost and grass clippings; cutting away the limbs as they died. By autumn, the soil was soft and loamy, alive with earthworms, but the trunk and the main limb of the tree was all that was left. All winter, it protruded from the mellow soil like a lonely sentinel.

Then one day in February I decided that I might as well take it up and plant another one. I bought another tree and the next day, with shovel and new tree in hand, went out to carry out my decision. A few feet from the sick tree, I stopped—and gazed at it as if I were seeing things.

Across the fence, my neighbor was working in her garden. I looked at her, and cried out with delight, "Look! I guess I have proved that trees, as well as people, are what they eat!"

What she saw was the one brown limb shooting forth green branches in every direction.

That was less than two years ago, and the tree attained more than its former size and is entirely free of the gum.

28—Pruning and Training Fruit Trees

BEFORE BEGINNING an explanation of the art of pruning, here are some of the terms that will be used, referring to the growth and shaping of woody plants:

Collar. The point where the roots join the above-ground portion of the plant.

Trunk. The portion attached to the roots above ground. It extends as high as the first branches.

Whip. A young unbranched shoot of a woody plant, usually the first year's growth from a graft or bud.

Leader or *Head.* The extension of the trunk above the first branches. The height of the head is the distance from the ground to the first branch.

Framework. The branch structure of a tree, which gives it shape and strength.

Scaffold. Primary scaffold branches are those which arise directly from the leader. Secondary scaffold branches arise from the primary scaffold branches. Together with the leader, the scaffolds make up the framework.

Lateral. A branch or shoot arising from a side bud on another branch or shoot.

Bud. An unexpanded flower or vegetative shoot.

Shoot. A stem with its leaves. The shoot may be terminal or lateral. During the growing season a new green shoot is called the current year's growth. The following year it becomes one-year wood, the next year, two-year wood, etc.

Spur. A short shoot, usually of less than 4 inches growth in a season and usually arising from a lateral bud.

Sucker. A vigorous shoot arising below the ground from the roots or (after deep transplanting) from the buried portion of the trunk.

Water sprout. A vigorous shoot arising above ground from the trunk or older branches. These originate usually from latent buds.

Heading back. Cutting back the tip of a shoot, twig or branch, usually to a lateral.

Thinning out. Removal of an entire shoot, twig or branch from positions lateral to the main branch or leader.

Reasons for Pruning

Fruiting plants are cut or pruned for one of several reasons. They are cut to increase the crop, to improve the quality of the fruit, to shape the plant for greater convenience of culture or for decorative purposes or to repair damage.

At planting time the plant is pruned to equalize the evaporation surface of the leaves with capacity of the roots, which suffer some pruning whenever the plant is disturbed. Immediately after planting the young tree is trained by means of the smallest amount of pruning possible that serves to form a strong shape capable of bearing fruit without breakage.

To prune any tree, shrub or vine one must understand the characteristics of that particular plant. It is not enough to know the peculiarities of the family or even of the genus. Each variety has its own growth pattern which must be taken into account by its trainer. Some of the characteristics which the pruner needs to know are: the reaction of the plant to heading back; the number of small branches formed by the particular variety; whether the flower and fruit buds are located on shoots or spurs and where on the branch they grow; the position on the lateral, spur or shoot of the fruit of the best quality.

Fruit trees are pruned to increase yield, improve fruit quality, repair damage or shape the tree. The tree above is being headed back in winter with lopping shears.

It is also necessary to understand a few fundamental principles of the reaction of all plants to cutting. Most important is this: all pruning has a dwarfing effect and reduces fruit production. It is true that cutting is sometimes done for the purpose of stimulating growth, which seems contradictory to the principle just stated. When a piece is cut from a plant, the root system and the plant's circulatory system both find themselves providing more food than the curtailed part can use. Consequently the plant is stimulated to repair this unbalance by replacing the part. The greatest growth increase is always in the vicinity of the cut. But the total amount of growth achieved is always less than the plant would have made if the removed portion is added to the amount of new growth. An unpruned plant always grows faster than a pruned one and, because it develops faster, it comes into bearing sooner.

The dwarfing effects of pruning are greater when a large amount of stored food is in the part removed and the plant suffers greatest shock when its food storehouse is disturbed. For that reason, summer pruning has a greater dwarfing effect than pruning in the dormant season. In summer and during the growing season large quantities of carbohydrates are located in the leaves, stems and branches. During the winter these supplies recede to the roots. They are at the greatest ebb in the upper part of the plant in late winter, just before the sap begins to rise to swell the new buds. Cutting done at this time will have the least dwarfing effect on the plant. Also, wounds made in February, March and April heal most rapidly. Evergreen plants, whose leaves remain on the plant all winter, do not go through a period of complete dormancy, so their balance is much more affected by pruning at any time than that of deciduous plants pruned while dormant. They are usually trimmed when growth has slowed or stopped, between growth flushes, but most evergreens are pruned very little.

When young fruit trees are severely pruned, their coming into bearing is delayed, sometimes by several years. A large amount of heading back delays flowering much more than if the same amount of wood were removed by thinning out cuts. Heading back also dwarfs the tree more than thinning out. After the plant begins to bear, pruning tends to increase the percentage of flowers which will set fruit, but it decreases the total yield.

The quality and quantity of some fruit is more affected by pruning than others. The degree to which they are affected depends upon the location of fruiting buds on the trees. A tree whose fruits are borne on terminal shoots will have an increased crop if it presents a larger outside shell, or bearing surface, to the sun. If the fruits are borne on lateral fruiting spurs on the branches, the sun must be able to penetrate into the center of the tree, both to color the fruit and to stimulate the formation of new fruiting spurs. Thinning out will increase the fruiting wood on the latter tree; heading back on the former.

There are usually more branches on a plant—on its sides or crowns—than can possibly survive; a certain percentage will die through the struggle for existence. Nature eliminates and selects by a slow action; controlled pruning does this quickly and thus may be defined basically as a thinning process.

There is no reason why cutting off limbs and branches should injure the tree or plant. If practiced with the correct knowledge, common sense, good sharp tools and some muscular effort, pruning is a benefit.

Plant Growth and Pruning

However, before you can acquire a correct knowledge of pruning, you should understand the process of plant growth. Here is a description given by Gustav Wittrock in *The Pruning Book.*

The tissue that is the seat of new growth is called meristematic tissue. This new growth is either primary or secondary. Primary meristematic tissue is found on the extremes of the plant; in the aerial part of the plant it is located in the buds, either terminal or lateral; and in the subterranean part of the plant it is found at the ends or tips of the roots.

Since the bud encloses this embryonic growing tissue, it is the seat of the growing point for height or elongation of the plant and is actually a protective enclosure of primary meristematic tissue. Millions of lateral buds remain dormant in an average tree, but nature has this form of preparedness constantly ready to assure growth, in case it is necessary to replace leaves, branches and stems.

Terminal buds are normally the most active in plant species and are apt to continue growth when the growing period is favorable. But if this terminal bud is injured in any way, the immediate lateral bud will be stimulated to carry on and become the leader.

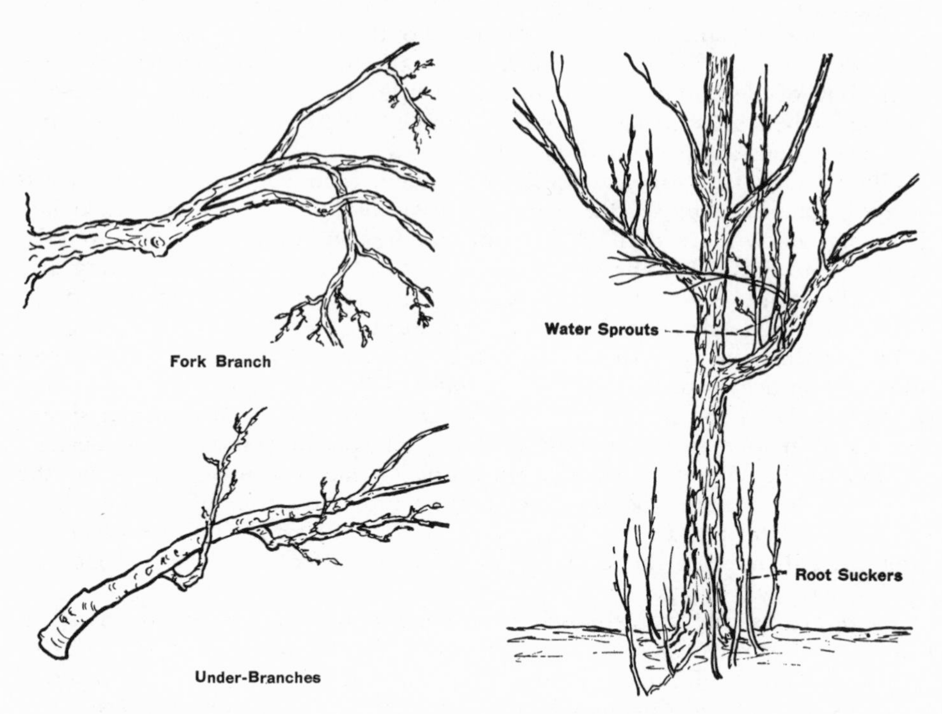

These are different forms of tree growth which should be among the first to be pruned.

Therefore when pruning, one must recognize these delicate areas and remember that by cutting back, active terminal buds are severed. Dormant lateral buds must then carry on the function of elongating the branch; in young trees, extreme care should be taken not to cut through the lateral bud, as this leads to direct exposure. If too long a stub is left above the bud, the stub is also subject to drying up and decaying, thus injuring the bud. The ideal cut is one-eighth to one-quarter of an inch above the bud, which then has a chance to survive.

Secondary growing tissue is found between the bark and wood of roots, twigs, branchlets, stems and trunks. This tissue is called the cambium layer and consists of a thin, slimy layer of minute brick-shaped cells. These microscopic cells are almost rectangular in outline, thin-walled and easily torn, especially during the growing season.

The function of the cambium layer is to produce and increase the diameter of the stem, branch or trunk and to heal wounds. All secondary wood and bark originates from this active growing layer or cylinder of living, dividing cells. If a limb or stem is girded by a wire, the wire will eventually strangle the cambium layer and kill the branch or tree.

Any injury to this important layer is always serious. Whenever the bark is punctured to the cambium layer, the sweetish, living substance inside oozes out, thereby attracting insects and fungi. These then feed on this substance and penetrate the injured bark tissue, establishing a foothold within the plant.

After acquiring this rudimentary knowledge of how a plant grows, you still must bear in mind that no tree or shrub should be pruned until you know *why* you are pruning. First diagnose your tree; note the dead and injured limbs and branches; then decide what additional healthy branches should be removed, if balance is necessary or if the green limbs removed are going to be a serious handicap for the health of the tree. You should have a good reason for removing any branch and you should thoroughly understand the consequences of such a removal.

Not more than 2 or 3 large healthy branches should be removed in one season, especially from large trees, cautions Mr. Wittrock. It has been shown experimentally that the greatest fault is not that too much pruning is done, but that too much is done *at one time*. Excessive pruning during any one outing not only makes unsightly scars and exposes too many areas to fungi, but it also induces the development of a great deal of suckers.

Making the Cut

Now let's suppose after duly considering all of the above-mentioned "don'ts" you have decided which limbs you want to cut from the tree. There is only one way to do it and that is the safest and surest way for the remaining scar to heal. You're not particularly interested in the cut limb, but you are concerned about the future of the tree you have left.

Very often the proper cut is not the easiest one. The shortest cut through a limb is at right angles to its diameter, but this will leave a stump even though the cut may be started at the crotch of the trunk. This stump is a hindrance to the healing of the wound.

Here's what actually happens to a stump left on a trunk. Sap will not flow out into a stump 6 inches long, as a general rule; hence there will be no healing. If the stump is shorter, a certain amount of healing will occur. If the stump is less than one inch long, complete healing or a considerable amount of it can be expected.

The proper cut of a limb should be parallel to the trunk of a tree. This rule applies to all trees in general, whether shade, ornamental, timber or fruit. This cut may result in the largest one in terms of diameter, but it will be the easiest and best cut to heal.

If the limb is more than two inches thick, 3 cuts should be made. The first cut is the undercut, from 6 inches to one foot from the trunk of the tree. Then the uppercut, a few inches further out on the limb from the undercut, is the next step; this cut is made through the limb or until the branch snaps off by its own weight. By this method there will be no tearing-back of the bark to the trunk, which is one of the dangers in pruning.

After the weight of the limb has been removed by the preliminary cuts, the final cut is made. If the stump is thick and heavy, from 4 to more inches in diameter, a careful pruner will make a fourth cut on the under side of the stump at the point where the final cut will come through; this is done to break through the bark and to ensure against ripping.

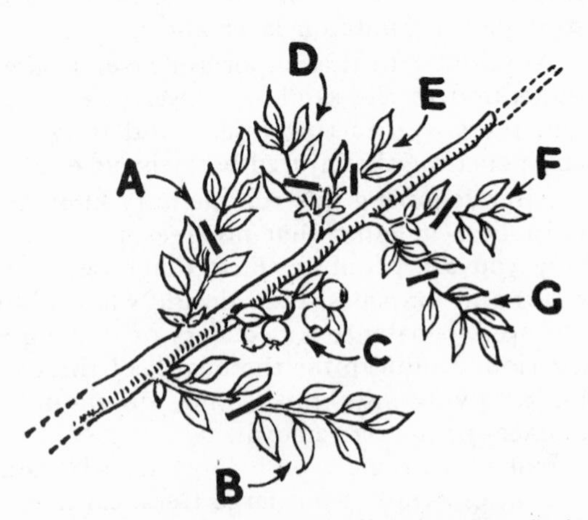

Summer pruning: note the main single laterals (A and B) are cut back to 3 leaves at the base of the lateral. Cut the twin or double laterals (D, E, F, G) back to one leaf above the smaller pair of leaves. "C" shows the fruiting spur.

Another method is to tie up the stump to prevent it from falling and tearing the bark when the final cut is made. All these precautions may seem a trifle exaggerated, but they should be observed in removing limbs to ensure not only the best chance for the tree to survive, but the safety of the pruner as well.

In the general form of a tree, many branches develop that are destined to die because of their location, the way they are attached to other branches and their relation to the general form of the tree. These branches are the ones chosen by the pruner to thin out the fruit trees. Always prune twigs and branches at the point where they start, making a clean cut and leaving no stump. These weak branches may be water sprouts, horizontal forks of equal thickness and branches growing in a downward direction from the main branch. All of these should be removed.

Water sprouts usually rise within a tree from adventitious buds. These buds develop spontaneously, usually near wounds, but they can arise from normal bark, particularly if the tree has been pruned severely. They grow very fast the first year, are soft and flexible and do not cause serious harm at first. But if they remain for the 2nd and 3rd year or more, they thicken in diameter, become stiffer and rub other branches, bruising them as they poke through to the top of the tree seeking light.

The second type of weak branch is the fork. When pruning the head, remember not to stand on a limb of the fork, even though it may be 3 to 4 inches in diameter. It is best almost always to remove one of the limbs. If that would make too big a gap in the form of the tree, you might try putting braces a few feet from the fork, with bolts firmly screwed through the center of each limb.

The underbranch, 3rd type of weak branch, often appears in the primary development of the tree. At first they are not usually serious because of their flexibility, but as they grow older and become a part of the framework of the tree, they are the weaker branches. They also develop from a water sprout origin, the adventitious bud arising from the underside of an older limb. The young sprout growth either elongates horizontally to an opening in the crown of a tree or twists around the limb to grow upward or vertically toward the top of a tree.

Some trees and many bush fruits send up suckers from buds which form on the roots. This is common especially with red raspberries and with a number of tropical fruits. Unless they are controlled, the planting may become a wilderness in a few years. Suckers can be permanently removed only by digging out the soil to the point where they spring from the root and removing the sucker flush with the root. Make a clean cut, leaving no stump. In sections where foot-rot is a problem, the cuts should be painted for protection and left uncovered until the wound has callused. In other sections, the soil may be replaced immediately.

When you are ready to do the actual cutting, make certain that it is all done with sharp tools. In the case of saw cuts, it is often advisable to smooth the surface with a plane or sharp knife, for it is believed that a clean cut heals faster than a jagged one. For small branches, a pruning knife skillfully used is probably the best tool, since its use involves a minimum of injury to plant tissues.

Pruning shears, if kept sharp and carefully used, can also do an excellent job, causing little damage to surrounding tissues. Long-handled lopping shears are also used successfully on a large scale in pruning shrubs where the branches to be removed are up to one or one and one-half inches in diameter.

Thus far we have discussed pruning from the standpoint of achieving the largest possible crop from the plant. To do this, the plant must be pruned as lightly as possible and must be allowed to grow as large as it will grow within limits of the convenience of the gardener's operations. But if space in the garden is at a premium, it may be necessary to prune deliberately for a dwarfing effect upon the tree. The following method of pruning for size of tree is recommended by D. Macer Wright in his book, *Dwarf Pyramid Fruit Culture*.

Dwarf fruit trees in the United States are usually grown in a "bush" shape—a polite way of saying that they are more or less left to find their own form naturally. Intensive orcharding with dwarf trees requires the use of the "pyramid" form, a conical shape that allows sun and air to reach and pass through the closely planted trees. Shaping and pruning the trees is done at two seasons—winter and summer.

Winter pruning helps develop the general form of the tree, but in intensive fruit growing, summer pruning achieves great importance. It is important to understand that winter pruning of a healthy tree encourages it to put out more growth the next season. Summer pruning has the exact opposite effect. You can easily see how an ignorance of that rule can lead to trouble in growing dwarf fruit trees, especially when they are planted close together in an intensive system. An intensive orchard that is left to grow unchecked can be the hardest of any type to bring back under control.

Summer pruning holds back the growth of the tree by encouraging formation of fruit spurs instead of new lateral shoots. As you can appreciate, it is a time-consuming project and is, in fact, the major limitation to the size of an intensive orchard. Mr. Wright claims that 10 acres is the absolute maximum that even a fairly large orcharding company can undertake and 5 acres are sufficient for a man and wife and helper to care for as a full-time operation.

Mr. Wright uses the Modified Lorette System for summer pruning, and feels that a rather simple formula can describe how it is done. When lateral shoots have matured for at least half their length (usually by mid-July) cut back each lateral growth to 3 leaves. The small twigs that are found growing from the lateral branches after those branches are cut back to 3 leaves are themselves cut back to only one leaf. That sounds complicated, but perhaps the accompanying drawing will help clear it up for you.

It is important to continue pruning the lower branches during the winter to encourage a continuing growth of branches on which fruit spurs can form. Fruit spurs on old branches gradually weaken and cease producing.

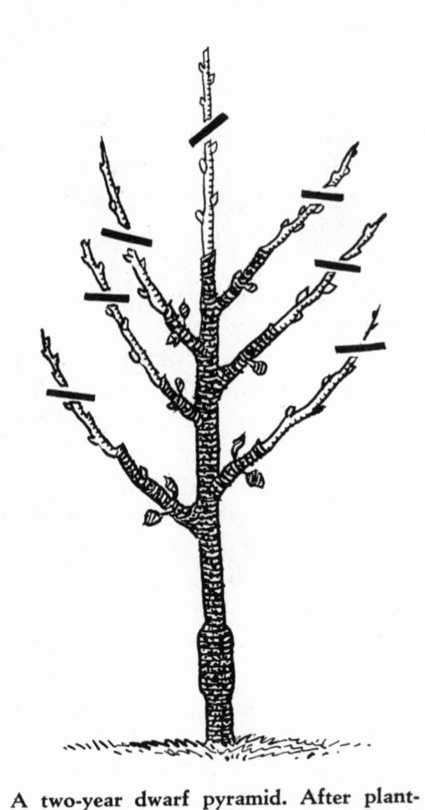

A two-year dwarf pyramid. After planting, cut the central leader to leave 9 inches of the current season's growth and the side branches at 6 inches. Repeat each winter with lessening severity. Note that the pyramidal form permits a maximum of light and air to reach tree.

The pyramid shape is not practical for a standard-size tree, because it results, roughly, in the old type of "central-leader" tree, which is too tall for convenience Several other methods of training deciduous trees have been developed, of which two are most often used. The following description of them, and the steps needed to develop them, are given by F. M. Green and A. M. Binkley in a bulletin issued by the Colorado Agricultural Experiment Station, Fort Collins, Colorado. They recommend the use of either the "vase-type" or "open-center" system or the "modified-central-leader" system, also called the "delayed-open-center."

The open-center is the type originally found in most western fruit plantings. Properly handled, it is reasonably satisfactory, Green and Binkley say. It provides a low, open, spreading tree which is easy to establish and maintain. However, the open-center method has one serious defect. Since all the branches come out from the trunk at practically the same point, within a space of 6 to 12 inches, the tree is too often structurally weak.

Three to 5 main branches should form the framework of the vase-type tree. One-year-old trees which are 4 to 6 feet high are usually used. After the trees are planted, they are cut back to a height of 18 to 24 inches from the ground. The buds below the cut then develop into strong branches which form the framework of main branches. Generally 5 to 6 branches start to form; you should allow all of them to grow the first year.

At the beginning of the second season in the orchard, the number should be reduced to from 3 to not over 5 branches, evenly distributed around the trunk. If the remaining branches make an excessively willowy growth longer than 18 inches, cut them back lightly. This will reduce the tendency toward excessive bending. Continue thinning of superfluous secondary branches. The tree will develop into a vase form without a central column or trunk. The absence of a central trunk distributes not only wood growth but also fruiting wood over a wider area lower on the tree.

If pruning is carefully done during the first 4 or 5 years, there should be little occasion for heavy pruning after that time providing the trees are pruned lightly annually. In annual pruning, thin out superfluous shoots where too abundant and shorten lateral growth to side branches where too long.

Modified-central-leader-type is a combination of the old "central-leader" tree, in which the leader or main trunk was permitted to develop without checking its height and the "vase-type" tree. The objective of this training method is to obtain the mechanical strength of the first type and the desirable open head of the second. It differs from the central-leader-type in that the leader is removed *after* the main or scaffold branches have been selected. It differs from the open-center or vase-type *only* in the placing of the main scaffold branches.

Instead of coming out from the trunk in one limited area, the main or scaffold branches are spaced at intervals ranging from a few inches to 12 inches apart up and down the trunk. Usually, only 1 to 3 of the main branches may be selected and established the first season: the balance are established by the end of the third year. When the desired number of main branches, usually 4 or 5, are established, the leader is cut out above the highest main branch. The best practice is to delay this removal until the topmost main branch has become stocky enough to continue growing out from the tree at a slight angle. The resulting tree is low-headed, open and spreading. Because of the better placement of the main branches along the short trunk, this type has the distinct advantage of greater mechanical strength.

The modified-central-leader tree is somewhat more difficult to obtain than the vase-type previously outlined. If one-year-old whips are used, a heading back or cutting back of the top of the tree is all that is needed when the trees are set in the orchard. This tends to compensate for the loss of roots incidental to transplanting. If the young tree is from 4 to 7 feet tall, cut it back to 24 to 30 inches from the ground. Higher heading makes it difficult to obtain branches low on the trunk. If the tree makes good growth the first year in the orchard, the grower may be able to select several of the main branches at the beginning of the second year.

The first step in this pruning is to select the terminal branch which is to continue the development of the trunk. Choose a vigorous shoot and the one most centrally located. Then select as many suitably placed side shoots for main branches as are available. Be sure to space them up and down the trunk not less than 6 to 9 inches apart with the lowest one at a height of 24 to 36 inches from the ground level. Locate them around the trunk in such a manner that no branch is directly above another. Keep the length of the side shoots that are to become the main scaffold limbs shorter than the terminal shoots. In certain sections where sunscald is serious, locate the lowest branch on the southwest side of the tree.

Two other methods of obtaining a modified-leader head are known under the names of "disbudding" and "de-shooting." In the first method, tall, one-year whips are planted and the trunk is left full length. Four or 5 buds, depending upon the number of scaffold limbs, are left properly spaced around and up and down the trunk. All other buds are shaved off the trunk. There is considerable risk that some of the buds left will be lost, too.

With the second method, the tree is either left full length at planting time or is cut back to 36 inches in height. In the case of branched trees, the side branches are stubbed back to one or two buds. When the lateral shoots have developed to a length of 8 to 12 inches in early summer, select the desired number of scaffold branches which are properly spaced along the trunk. Then cut away all other shoots. Remove any water sprouts that develop later in the season.

Two-year-old nursery trees should be pruned when set in the orchard as described for year-old whips which have been growing in the orchard for one year. At the beginning of the third year, the scaffold branches selected the previous years should be maintained. Others desirably placed are selected. If the 4 or 5 branches are now available in the proper positions, cut off the terminal or leader just above the last branch selected. If the top scaffold tends to grow upright, delay cutting the leader until another season.

The modified-leader form will often produce a lower tree than the vase-type because the branches tend to grow out at a broader angle. Consequently, although the branches may be rather long, the tree itself will be lower. This effect is obtained by retaining the central leader for a period of 2 to 4 years after the main branches have been selected or until they have become fixed. It is cut back sufficiently each season to side laterals to prevent its outgrowing the side branches.

The number of main branches left may vary from 3 to 7. A happy medium would be 4 or 5, since in trees with only 3 remaining branches the loss of one might seriously cripple the tree. With too many branches, the tree eventually may become a mass of long bare poles with a tuft of smaller branches at the outer ends. The vertical spacing will also vary with the number of limbs. With a smaller number, the spacing could be greater (up to one foot apart when only 4 branches are left). The entire number of main branches may not be obtained in the first or even second year of growth in the orchard. However, by the third year the head should be well-established and then the central leader can be removed.

The intermediate period in the life of a fruit tree is the time after the head is established but before the bearing begins. During this time, pruning should be confined to removal of excess branches in order to maintain the framework already established and to keep the balance between the main branches. Remember that branches that may be removed at this time with shears may require saw cuts for removal in later years. But do not prune the tree too heavily.

Until the main branches have reached approximately the desired height for the mature tree, they need not be shortened except to maintain the necessary balance with neighboring branches. Where shortening is necessary, prune by cutting to a side branch.

Trunk diameter will increase at a more rapid rate if no heading-back cuts are made. The top of the tree at this age may be left somewhat dense because the weight of the first crops of fruit will spread the main leaders and open the top.

After the trees have begun to bear fruit, continue pruning to keep the tree at a convenient height. The main consideration now is to encourage and maintain the fruit-bearing structure, which in the case of apple and pear are the fruiting spurs. There are two methods of pruning in use for the bearing tree, heading back and thinning out.

Heading back is the annual cutting back of one-third to two-thirds of the yearly growth. Sometimes this cutting is done into 2- and 3-year-old wood. In such cases stubs may be left which may not heal readily. This type of pruning often results in the production of a large number of new shoots within a small area of the tree and only a small number of fruit spurs. On any kind of fruit tree this will result in a dense top with the growth concentrated at the outer ends of the branches. Heading back such varieties as Jonathan, Ben Davis and Winesap, which tend to produce fruit from axillary buds on the new growth, will result in the removal of part of the crop.

Thinning Out

Thinning out removes entire twigs or branches back to the points of origin along main or secondary branches. This type of pruning produces new shoots that are spread all over the tree; many of the shoots may produce fruit spurs in their second and third seasons from lateral buds. This will tend to cause annual bearing and the production of larger crops.

Pruning for fruit production should consist primarily of thinning out rather than heading back, except where it is necessary to subordinate one-year-old wood that is growing excessively rank. If you desire to shorten branches, the cuts should be made to lateral branches which are in a position to assume the lead in place of the one removed. Such cuts may be made in 2- and 3-year-old wood or older and the severity or amount of cutting will vary with the vitality of the tree and the amount of growth it is making.

Increase in vigor following pruning is more pronounced in old trees than in young ones. Excessive pruning of the top may reduce the root system to a greater extent than the top. Trees growing on poor soils require heavier pruning to maintain vigor.

Fruit thinning as a means of improving the quality and quantity of the crop is recommended by M. B. Hoffman in a pamphlet issued by the Cornell Extension Service.

Young trees, says Mr. Hoffman, seldom set heavy enough to warrant thinning of the fruit; but after the bearing habit is well established, some attention should be given to this practice. This is especially true with peaches and some varieties of plums, prunes and apples. When the set is excessive, proper thinning results in larger, better colored and higher quality fruits. Thinning should be done early, from 2 to 3 weeks after bloom if possible.

Where fruits are clustered, all but one fruit in each cluster should be removed. Peach fruits should be from 4 to 8 inches apart, using the wider spacing for the early varieties. Plums and prunes should be thinned so that the fruits are 4 to 5 inches apart. The small, insect and disease injured fruit should be removed first.

An excessive set on apples prevents flower-bud formation for the next season and results in alternate bearing or a heavy crop of small sized fruits one year and no crop the next. This is a common occurrence with Golden Delicious and some other varieties. It can be prevented by early thinning. Apples should be spaced from 4 to 6 inches apart, which means one fruit on every third or fourth spur. Thinning actually requires very little time and the improved size, quality and repeat bloom are well worth the effort.

The grapevine is pruned to control the quantity and quality of both wood growth and fruit production. In pruning, it is the aim to leave enough of the best fruiting wood to obtain maximum production of good clusters and to remove the surplus to prevent overbearing and its consequent inferior wood growth. This is *pruning* proper. Pruning is also employed to make well-proportioned vines that are easily managed and kept within definite bounds. This is more correctly termed *training*. Both operations are often referred to as pruning, but the two terms should not be confused. Various methods of training grapes are described in the section dealing with their culture.

For bramble fruits, the two-year-old fruiting canes should be removed as soon as the harvest is completed. Of no further use, these tend only to spread disease to the new shoots. The old canes should be cut off close to the ground.

Black raspberries, purple canes and blackberries respond to summer pinching of the new shoots. The tip ends of the shoots are pinched off when they have reached a height of from 18 to 24 inches, which is usually in early June. When the shoot is pinched back, it stops growing at the end and the buds on the side push out to form lateral branches. Plants treated in this way are lower and more self-supporting.

Red raspberries produce suckers freely and for this reason are commonly grown in hedge rows about 2½ feet in width. Sprouts appearing in the row middles should be destroyed, else the patch will become a thicket of weak canes. The new shoots of red raspberries should not be summer-pinched, because laterals forced by this treatment are subject to winter killing. Before growth starts in early spring, the unbranched canes are cut back to a height of about 3 to 4 feet. The weakest canes should be removed. In plantings where cane growth is exceptionally vigorous, red raspberries may be kept in hills and supported by a stake. This permits leaving vigorous canes taller in order to take advantage of their greater fruiting capacity. It also simplifies picking.

In the spring, black raspberries, purple canes, and blackberries are pruned so as to shorten the lateral branches which developed as a result of summer pinching. The laterals on blackcaps and purple canes are cut back to 5 or 6 inches, while those on blackberries are left from 10 to 12 inches. The weak, spindling canes are cut out entirely.

The pruning of currants and gooseberries is governed by the fact that 2- and 3-year-old wood is the most fruitful. In pruning, therefore, branches older than 4 years are cut out at the ground, as are the weakest of the young shoots.

The fruit of the blueberry is produced on wood of the previous season's growth. Strong shoots from the base of the plant or vigorous laterals from the older canes produce the best berries. Little or no pruning is required the first 2 or 3 years, but older bushes, if not properly pruned, overbear and produce large quantities of small, worthless berries. Bearing branches close to the ground should be removed as well as dead and broken branches. Dense areas of weak, twiggy growth should be thinned by cutting back to vigorous side shoots. Occasionally, old, weak canes which produce no strong laterals are cut out in favor of younger shoots that will grow more vigorous fruiting branches.

Supports

When grapes, espaliered trees and some bush fruits are trained, they must be tied to supports to keep them under control and to lift the fruit off the ground. If the supports serve only that purpose, they are usually built by attaching wires to fence posts. If they are decorative in purpose, they may take the form of trellises or arbors, neither of which is important to this book, but belongs to a study of the decorative garden.

Decay-resistant wood used for fence posts lasts much longer than posts made from tree thinnings or from casual lumber. Cedar or redwood are the longest-lived post materials. Posts used for raspberries need not be as sturdy as those used for grapes, because they will be called upon to support smaller loads.

End posts for grape supports must be large, deeply buried and well-braced against the pull of the crop load. The smaller end of the end post should be at least 5 inches in diameter, and the post should be 9 feet or more in length. Three to 4 feet of the end post is sunk into the ground, leaving 5 to 6 feet above ground. The end posts may be braced on the outside, beyond the row of vines, or on the inside, between the end post and the second post.

An exterior brace may be attached to a 3-foot length of log or to a rock which is buried in the soil 3 feet from the end of the trellis and on a line with it.

A galvanized steel wire of at least number 9 gauge is attached to the anchor rock or post and run to the end post a foot from the top. The wire is fastened securely and tightened before the anchor is buried.

If space does not permit of an exterior brace, the end post may be braced between the end of the trellis and the first line post. Set a rail or a 2 by 4 of decay-resistant wood diagonally between the top of the end post and the bottom of the line post. Tie a wire around the two posts on the other diagonal, forming an X with the bracing post and the wire. Twist the wire until it is tight.

Line posts on grape trellises, or posts used to support raspberries and other brambles, need not be more than 7 to 8 feet long and 4 inches in diameter at the small end. They should be set two or two and a half feet into the ground and may be spaced from 16 to 24 feet apart.

If the Munson trellis system is used for grapes, crossarms are needed. These may be made of two-foot lengths of 2 by 4's, either of Douglas fir or of a decay-resistant wood. The crossarms are fastened to the posts with 3/8 or 1/2 inch bolts. They are attached to the outsides of the end posts.

The Munson trellis requires 3 wires run the length of the trellis. Number 9 wire is best, though number 11 may be substituted on the two top wires. Number 9 or 11 may be used for a Kniffen trellis. Wire is usually sold by the pound. In one pound of number 9 wire there are about 15 to 17 feet; in one pound of number 11 wire about 25 to 27 feet. Lighter wires run as follows: number 12, 34 feet per pound; number 16, 102 feet per pound.

Wires should be attached loosely to the trellises, with provision made for pulling them tight as the season demands. They may be run through holes drilled through the posts or through grooves in the tops of the line supports. They are usually threaded through holes in the end posts and a tightening device is fastened to the wire's end. The device may consist of a block of wood 1 1/2 inches square and about 8 inches long. Bore a 1/4 inch hole through the block, fasten the wire through the hole and wrap it around the block. The wire may be tightened by turning the block with a wrench. Or the device may be made out of a 1/2-inch steel bar, bent to L-shape. Drill the hole through the bar, and fasten and tighten the wire as with the block. A small reel with a ratchet may also be used. Whatever the device, the wire should be loosened before it contracts with the cold in winter and may be tightened when the summer's heat again expands it.

Special Training

Dwarf fruit trees are sometimes trained to grow in special shapes, such as the pyramid described earlier, to suit them to a particular need in the garden. This type of training is seldom encountered in this country except in gardens operated by professionals who devote a large amount of time to it. In Europe some of the old gardens contain fine old espaliered trees trained to cover the walls of houses, wreathing the windows with orderly flat branchings, or fanning out against garden walls. Elaborate branching schemes and grafted shapes are the ultimate achievement of the most formal of these old-world gardens, achievements impossible for the busy home gardener who takes care of the vegetable garden, the lawn and the flower beds as well as the fruit plantings. But for the 6-handed gardener who cannot keep busy otherwise, we offer these brief notes on the charming art of espalier, as described by a number of old-world gardeners.

Espalier is actually a method for training fruit trees to grow in a shape best suited to the grower. More specifically, the objective is to train the trees in a manner similar to vines or grapes, so that the branches grow flat along a

Espalier is a method for training fruit trees so that the branches grow flat along a wall or trellis; thus espalier trees develop two-dimensional growth, growing up and across a wall.

wall or trellis. Therefore an espalier tree becomes a two-dimensional growth; it grows up and across the wall but not out from it.

The advantages from growing trees this way are:

1. *Space-saving.* If your grounds are limited, you probably will like to keep a good portion of your yard clear. By planting your trees against your house, you can have your fruit and still have an open area too.

2. *Quality fruit.* Actually, if you practice training and growing trees espalier fashion, you produce fewer but better fruit because each gets more light and air and is less likely to suffer abrupt temperature changes.

3. *Individual charm.* You will bring a real individuality to your home and grounds by training your trees to grow on its walls. With the passage of time and the development of tree forms, your home will assume a flavor and charm which will be a source of real pride to you—not to mention a source of nutritional harvests.

Many home gardeners disregard espalier, thinking it impossible, since tree branches normally thrust out in every direction from the tree trunk. But by bending the succulent little twigs in the direction desired and fastening them, you get the tree to assume the shape you wish. As described below, you will see that much trimming and care is necessary, but some of this is also often necessary to do for any fruit tree you are shaping.

Coming from the French word, epaulet, which refers to a shoulder strap, espalier got its name because of the way the branches go out from the trunk almost at right angles or shoulder-like. It's based on the principle of sap flow control. In practice, this means that you may have to prune or kill off stronger branches in order to help along weaker ones, so that the tree has a more even shape.

All trees grown espalier fashion should be dwarf stock, which bears fruit within 3 years or less after planting. Apples, pears, peaches, plums, nectarines and cherries are dwarf fruits suitable for espalier.

You can plant espaliers against sides of buildings, walls or fences; below or in between windows; or even along driveways or between properties as a hedge. These trees can be trained to grow horizontally to cover low or medium high walls; upright as a screen or for high walls; or fan-shaped to stretch out over broad wall areas.

When planting espaliers, use the same methods described for dwarf fruit trees. The soil should be moderately open and well-drained; the hole should be big enough to take the tree roots without bending them. So cultivate the soil deeply, making certain to loosen the soil at the bottom of the hole.

If you plant trees close to the foundation of your house, you may run up against a heavy, hard-pan subsoil—probably full of bricks, stones and maybe a few pipes left by the builders. Clean them out and work in compost, peat or other humus material, so that the roots of your tree won't be cut off from water, air and nutrients.

Both tree roots and top should be pruned at time of planting. Use good quality topsoil in filling hole to a few inches of the top. By leaving a slight depression around the tree, rain water can collect there. Tamp the earth gently but firmly around the roots leaving no air pockets; next water and mulch 2 to 3 feet around the tree and several inches deep.

Keep in mind that dwarf fruit trees should always be planted with the graft-union above ground. The graft-union is an onion-shaped knob at the base of the trunk and can be recognized by a change in the color of the bark. This graft serves as a guide for how deep to plant the tree and generally should be about an inch above the soil surface.

Eastern, northern or western exposures are thought to be better for espaliered trees than southern. Not only is the southern exposure too hot in the middle of summer, especially when the heat is caught and reflected from a stone wall, but trees in this position may start into growth too early in spring. This is especially dangerous for the fruits which may be inclined to bloom too early for their own good, such as apricots and peaches. When planting against a wall, it's best to allow about 18 inches root space.

The final shape of the tree depends on the equal flow of sap throughout the branches. You may have to hold back certain portions that are naturally stronger and faster-growing and encourage the slower and weaker branches.

There are several ways to do this:

1. Prune the strong branches short but allow the weaker to grow long.

2. Kill off the useless buds on the strong parts as soon as possible. Do this as late as possible with the weaker parts.

3. Tie up the stronger branches very early and close to the wall but delay doing this to the weaker branches.

4. Pick off some of the leaves on the stronger side.

5. Leave as much fruit as possible on the stronger branches but pick all fruit off the weaker ones.

6. Keep the strong side close to the wall and keep the weak part away.

7. If necessary, even place a covering over the strong side to deprive it of light.

So much for trying to keep the development of your trees equally in balance for all parts. Here are a few more common-sense principles to keep in mind:

1. Branches cut short develop more vigorously than those left long.

2. Sap has a tendency to flow to the end of the branch, making the terminal bud develop more than the lateral buds.

3. You'll get more buds and less wood if you slow up the circulation of the tree's sap.

In general, espalier growing is considered more an art than a science. Results cannot be predicted with unvarying certainty; you will learn more from getting out and doing than from study. However, there are two basic principles which should guide you in your tree-training and growing.

1. The side shoots—not the leaders—need constant summer pruning. Pinching back these shoots induces the formation of fruit spurs, fruit buds and, finally, fruit.

2. Leaders, the principal extensions of the main trunk, are headed back in the dormant or late winter season, if and when necessary.

The term espalier, now applied to the form of rigidly trained trees, actually has a more limited meaning. It applies specifically to the framework to which free-standing flat fruit tree arrangements are trained. Thus far we have talked about trees trained to walls or buildings. These may be shaped to the same forms if they are trained to trellises of wire, supported by posts, anywhere in the yard. They may be trained to form a slender hedge which is used to screen one portion of the yard, as in our plan for the suburban garden. Because the tree is so drastically trimmed at frequent intervals, such a hedge or divider is thinner and occupies less space than, say, a privet hedge. An espaliered cordon may be used to border a driveway; in a series of fan-shaped forms or vertical arms it may be trained to a picket fence. An oblique, free-standing espalier makes a good screening for the corner of the lot which contains the service portion or the compost and manure heaps.

Before launching on a detailed description of the various forms to which espaliers are commonly trained, here are the points necessary to successful espalier training, as outlined by another expert, Arthur C. Bobb, of the Connecticut Agricultural Extension Service:

1. The training of trees into ornamental forms is largely a mechanical process. Trees are shaped while they are growing. Young shoots are twisted

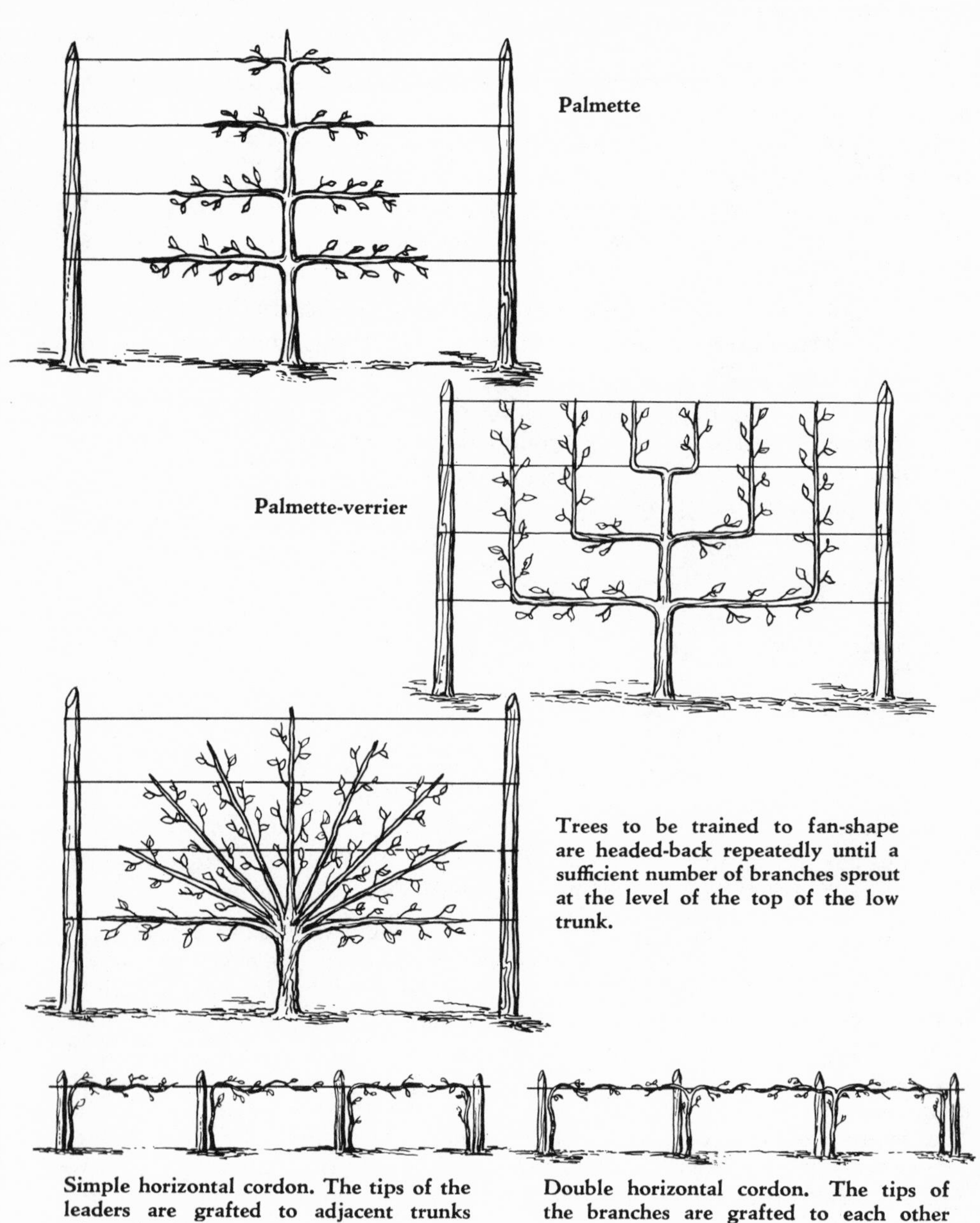

Palmette

Palmette-verrier

Trees to be trained to fan-shape are headed-back repeatedly until a sufficient number of branches sprout at the level of the top of the low trunk.

Simple horizontal cordon. The tips of the leaders are grafted to adjacent trunks when they meet.

Double horizontal cordon. The tips of the branches are grafted to each other when they meet.

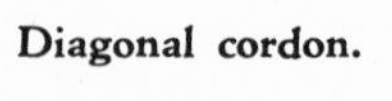

Diagonal cordon.

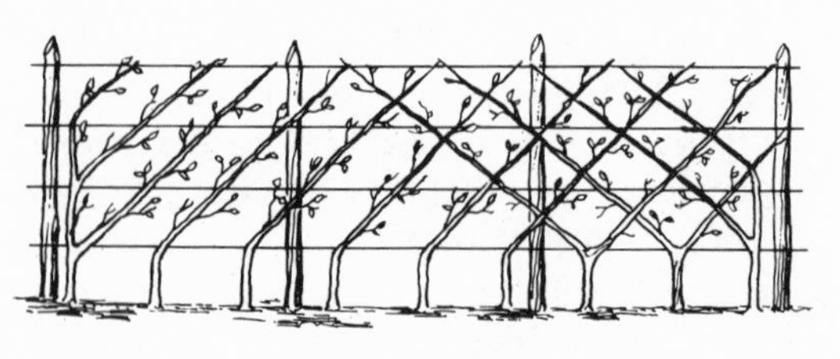

Double diagonal cordon.

and bent into the desired positions and are tied into place until the stems become hardened. The most important rule to remember is that constant attention must be given the shoots while they are growing. Mistakes are corrected with difficulty after an undesirable form has been allowed to harden.

2. Trees are severely headed in. By "heading in" we refer to shortening of the leaders. Such shortening is usually given at the spring pruning, while the trees are dormant.

3. Summer pruning is essential.

4. Side shoots need pinching during the growing season. Leaders are allowed to grow unchecked throughout the season. Summer pinching has the purpose of encouraging the formation of fruit spurs.

5. Four successive examinations of the trees are usually required. The first pinching falls about 3 weeks after the trees have started into growth. The next one comes 10 days later, the next one 10 days later again and the fourth pruning two weeks after the third. The bulk of this pruning can be done with the thumb nail and forefinger.

6. Root pruning is sometimes advisable. It should be done late in the fall. They may even be dug out completely and set right back in place. This is a special practice not to be indulged in too freely.

7. A certain equilibrium between vegetative growth and fruit bearing should be established. If there is a proper balance between summer pruning and winter pruning, combined with proper control of vegetation and fertilization, then the balance between vegetation and fruitage can be maintained.

Supports for espaliered trees planted as hedges, freestanding or against walls or buildings are essentially the same. For end posts, drive 6-foot lengths of two-inch galvanized pipe into the soil a foot deep. If the trees are to grow more than 5 feet tall, the pipes must be correspondingly longer. The plants are tied to a series of wires, for which 14-gauge is sufficiently strong for most trees. The wire is stretched between the posts and fastened by twisting it at the end. The bottom wire should be the height of the first serial branch of the tree, usually 12 to 14 inches above the ground. All higher wires are spaced 10 inches apart. If 3 or more plants are to be supported, additional posts should be used for every two trees. End posts should be braced against the weight.

If the trees are to be planted in front of a wall or building, the support is the same, with one exception. The top of each post is fastened to the wall by means of an L-shaped pipe, an angle threaded to the post and attached to the wall by means of a flange. The flange is held with screws or, if it is a stone or brick wall, with expansion screws. Plants should be set about 10 inches from the wall; the support never less than 6 inches.

Espalier shapes may be trained vertically, the leader being pinched back at intervals until it produces side branches. If two opposite side branches are trained horizontally at each such interval, the resulting shape is a simple palmette. If the ends of the side branches are subsequently trained to turn upward, the form becomes a palmette-verrier. To turn the ends of the branches at the exact point where it must form a right angle, a vertical stake must be supplied for each branch. Trained against a picket fence, a palmette-verrier can be tied to the alternate pickets and shaped.

A simple horizontal cordon is formed by turning the leaders at the desired height and tying them to horizontal wires. Apple trees are often used for this form. They are planted two, three or four feet apart, depending upon variety and rootstock. No side branches are permitted to form until the leaders meet. The tip of one leader is then grafted to the trunk of the next tree, until the end of the row is reached, when the two leaders at the end are grafted together. When the grafts have grown firmly together, the trees make a solid fence.

Diagonal cordons may be formed by bending the leaders of apple, peach or pear trees to make a 45° angle to the bottom wire and continuing with subsequent trees to form parallel diagonals. If two leaders are permitted to grow from each trunk and are trained in opposite directions as diagonals, a diamond pattern may be made. Two leaders may also be started in the horizontal cordon, to form a so-called double cordon.

The success of an espalier, when it comes into bearing, depends upon the regularity of the fruiting spurs. When a spur is lacking in a desired position, a latent bud may be forced into growth if a slit is made through the bark one-half inch above it during the late winter. Slits are also sometimes cut around the base of fruit-bearing branches at blossoming time, with the result that the fruits set often become larger and ripen better.

European gardeners employ root pruning to inhibit vegetative growth, once the espalier has achieved its full size. One method is to uncover the roots at the base of the trunk and to leave them uncovered for most of the summer. Or they may be uncovered, part of the roots cut away and the soil replaced. If the trees are lifted at the end of the summer and then replaced in their same positions, another form of root pruning is accomplished.

29—Propagating Fruit

FRUIT AND NUT PLANTS are propagated in one of two ways—from seed or by vegetative means. Usually they cannot be satisfactorily grown from seed. Our highly developed varieties do not come true to the parents, but show their mixed heritage by reverting to one or another of their ancestors. Fruit varieties that come true to seed might be developed, but this would take a special effort on the part of breeders and since more practical means of propagation are at hand, it would be uneconomical.

Starting a New Plant Vegetatively

Starting a new plant vegetatively means simply growing roots on a portion of a parent plant whose characteristics are desirable. This might be done by rooting cuttings, but most fruit and nut trees do not make roots easily. A few, notably quince, fig, pomegranate, grape, olive, currant and gooseberry can be rooted by planting portions of a branch or stem. This may be done by severing the piece from the plant first and then treating it in such a way that it will form roots, in which case the new plant is a rooted cutting. Or if a portion of a root can be severed from the plant and can be made to send up a new top, it is a root cutting. If the root forms buds underground and sends up a shoot from the bud to start a new plant, or sucker, beside the old one, the new plant may be divided from the old and it is said to have been propagated by suckering.

Some plants will form roots on any stem or branch which is bent down to the ground and buried. A new plant formed in this way is said to be layered. If the tip of the branch is used for rooting, it is tip-layered. If the roots are formed on the branch above ground, in a special sac, it is air-layered, a process also known as marcottage.

Pears, apples, cherries, peaches, almonds, walnuts and most plums are among the plants that are difficult to propagate from cuttings or layerings. Such plants are usually propagated by first growing seedlings and then budding or grafting the desired varieties to the ready-made roots. Sometimes the tops are grafted on to roots, called rootstocks, grown from their own seeds. But more often they are grafted to other varieties chosen for special hardiness or disease resistance or for the vigor or limitation of size which they transmit to the new top.

Seeds from most deciduous trees require a certain amount of winter chilling before they will germinate. The chilling requirements of the seeds of fruits usually corresponds roughly to the chilling required by the parent trees for good fruit setting. The following periods of cold, between 32 and 40 degrees, are needed by seeds before they can be expected to germinate: almond and apricot, 20 to 30 days; apple and pear, 60 to 90 days; cherry, peach, plum and walnut, 90 to 120 days.

Before planting these seeds for rootstock, they must be exposed to chilling temperatures. This is done by a process known as stratification. Stratification means to *layer* the seed in sand, sawdust or peat moss and then expose it to freezing and thawing for a period of time sufficient to duplicate natural winter conditions. When starting out to germinate tree seeds in the late winter or early spring, the stratification or wintering process can be speeded up by placing the seeds, mixed with moist sand, in a deep freeze for a few days, then removing them for a few days. This freezing and thawing process is repeated for a week or two before the seeds are planted.

When more time is available, the stratification box is placed in a refrigerator or country root cellar for 30 to 120 days, depending on the type of seed being germinated. If a refrigerator is used, it is advisable to place the box on a bottom shelf with a thick cardboard under it to absorb moisture.

When the ground is fit for working in early spring, take the seeds out of the stratification container and sow them in rows in good garden soil that is as free of weed seeds as possible. Small seeds are covered with only a little earth and large seeds are planted deeper—usually about an inch below the surface. Plant the seeds at least a foot apart in the row and make the

STARTING TREES FROM SEEDS

1. Basic materials include a handy wood flat which permits compartmentation for different seeds. Small, separate boxes may also be used.

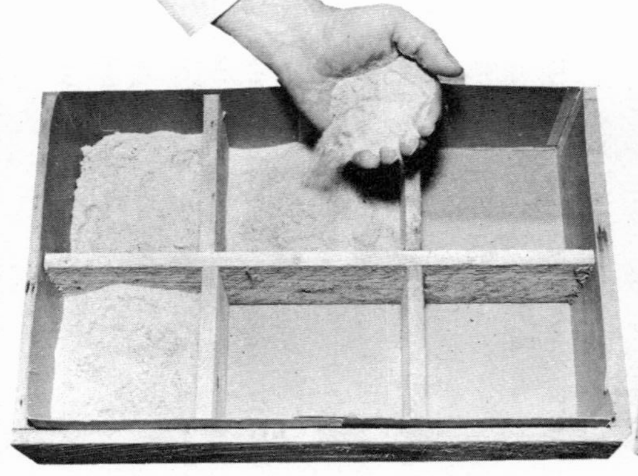

2. After lining the box with cardboard, fill the compartments with clean sand or peat moss to serve as layering medium for various seeds.

3. Seeding the compartment is simple but must be completed one section at a time, including covering seed and placing marker.

4. Use plenty of sand or peat moss when covering seeds. Do not tamp down too hard and make sure sand or peat is always moist.

5. Identify each compartment with a wooden tag which bears the date of layering as well as variety plainly marked with moisture-proof ink.

6. If you have no root cellar, place in refrigerator. Set the flat on a thick, absorbent layer of cardboard with large pan underneath

rows wide enough to allow a tiller or mower to pass between them. If only a small number of seeds are sown, they can be grown in a large flat and transplanted later, but direct garden planting is usually best for the amateur.

Most fruit trees and some nut trees don't come true from seed. So if seedlings are grown they must later be budded or grafted to known varieties. For that reason, it is best to start out your tree growing program with nut trees or ornamentals. Walnuts are often grown from seed.

Seedlings are transplanted after one or two years' growth. The purpose of transplanting is to encourage the trees to put out stronger roots and to give them the room they need. After 2 to 4 years in the nursery, they can be set in permanent locations.

Rootstock seeds most often planted directly in the nursery row are apricot, almond, peach, walnut, pecan, myrobalan plum and mahaleb cherry. Those usually grown for a year in a seedbed before being set out in the nursery are apple, pear, mazzard cherry and sometimes myrobalan plum and mahaleb cherry.

For the home gardener who wants to plant only a few trees to try his hand at grafting or budding, the simplest plan is to plant the seeds in a small nursery row in a corner of the garden in the fall. The wintering outside in the ground, where they will grow, will produce the same effect on the seeds as stratification. In the spring the seeds will come up and, after they have achieved a reasonable size, they may be transplanted to the spot in the garden where the mature tree will some day stand. After the first year the rootstock seedling may be budded or grafted right in place and it will be ready to proceed with the business of developing a bearing top a full year earlier than the tree which is transplanted 2 or 3 times, from seedbed, to nursery, to garden.

Budding and grafting are vegetative methods used to transfer a desirable top to the roots of questionable seedlings. Karl D. Brase at the New York Agricultural Experimental Station emphasizes that it is essential that the plants joined together in this way are closely related and compatible and that the cambium of the scion or bud is placed in contact with that of the rootstock.

In fruit trees the cambium is found between the bark and wood. It consists of a thin layer of cells from which new bark and wood cells originate. In the spring and summer, the bark readily separates from the wood in the cambium region, some cambium adhering to the lifted bark and some remaining on the wood. In winter the position is not so clearly defined, but close examination of a shoot will reveal its position at the junction of the soft bark and the hard solid wood.

Budding

The budding operation can be carried out when the bark of the rootstock easily parts from the wood and when sufficiently developed buds of the desired fruit variety are obtainable.

Depending on the region, budding can be carried out from June until September, although the month of August is the time when budding is mostly done in the northeastern United States.

The best time to bud some fruit stocks is dependent on certain growth characteristics of the stock. Fruit plants that complete their growth earlier in the growing season must be budded first. *Prunus Besseyi, P. domestica,* pear, apple, mazzard cherry seedlings and quince stocks are best budded in the order given from mid-July to mid-August.

Of the several types of buds, the T-bud is probably the most widely used. It is usually done in the summer after buds of the current season have matured and bud wood is cut as needed. The leaves on the bud-stick should be trimmed away as soon as it is cut from the tree. This helps to prevent drying out of the buds, as leaves left on the stick will cause it to dry out much faster. A short portion of the leaf stem is usually left on to serve as a handle to the bud when it is being inserted, but has no other influence on the success of the operation. As soon as the leaves are trimmed off, the bud-sticks should be wrapped in moist burlap or other material that will keep them from drying before being used, for once they become dry, they are worthless.

At the budding point the stock is wiped clean of soil particles and a T-shaped cut is made through the bark but not into the wood. The upper transverse cut is made first about one-third around the stock, followed by a vertical cut upwards to meet the transverse cut. As it reaches the transverse cut, a twist of the knife blade raises the edges of the bark just enough without tearing so that the bud may be easily inserted.

After completing the T-shaped incision on the stock, the bud is cut with a shield of bark by holding the bud-stick by the upper end with the lower end away from the body. The knife is placed half an inch below the first suitable bud and by a shallow slicing movement is passed beneath the bud approaching the surface an inch above it.

The shield bearing the bud must be cut fairly thin, but not so thin that the soft growing tissue between the bark and the wood is injured. With fruit plants one may leave the thin strip of wood that was cut with the shield, while with certain ornamental plants, such as roses, it is best to remove all the wood. This can be accomplished by cutting a thick shield and severing the bark by a cross cut an inch above the bud.

Grasping the shield firmly between the thumb and forefinger, it can be carefully lifted from the wood without tearing the bud. After cutting, the bud is held by the leaf stem between finger and thumb and is inserted into the T-shaped incision on the stock.

With rootstocks of sufficient diameter, in good growing condition and with the bark separating readily, the insertion of the bud is easy. When properly inserted, the bud should be at least three-quarters of an inch below the transverse cut.

To assure firm cambium contact and to prevent drying out of the severed bark, it is necessary that the cut be wrapped with a rubber budding strip. Waxing is not necessary. If the bud has been properly inserted, the wrapping may either be done downward or upward. The important point is that the wrapping is tied and that the bud itself is not covered by the rubber band. The rubber band is held in place by placing the free end back under the last turn. Gaps may be left between each turn.

Although other tying materials may be used, such as raffia or woolen yarn, the rubber budding strips have the advantage of expanding with growth of the rootstock and after exposure of a month to the sun will rot and fall off. By this time a good union between bud and stock has taken place. In the Northeast rubber budding strips of .010 gauge are the better ones to use, whereas in more southern regions strips of .016 gauge should be used.

The first indication that the bud has united with the stock is the dropping off of the leaf stem. Shrivelled adhering leaf-stems often indicate failure. In such a case, provided the bark still separates readily from the wood, a new bud may be inserted in a new position on the stock.

The bud is ready to be forced into growth in about two weeks if the budding is done in the early summer or it may be left until the next spring if budded in the late summer or early fall. To force the bud, cut the stock off just above the bud. In hot summer weather it is usually cut an inch or two above the bud and after the bud has grown a few inches it is cut again just above the bud. If the bud is not forced until the next spring it is just as well to cut it immediately above it at first. In either case one should keep all growth removed from the stock except the bud.

Grafting

The chief purpose of grafting, as of budding, is to multiply varieties that cannot be propagated by cuttings or layers. Grafting differs from budding in two ways. First, it can be done when the two components to be grafted, the stock and the scion, are still dormant. Second, instead of a single bud, as in budding, a short section (scion) of a one-year shoot that may contain only a single bud or several buds is used.

Grafting involves, as does budding, the bringing together of the cambium layers of two different individuals that will unite as one. Growth of both stock and scion in the cambium region results in an interlacing of tissues and "union." A clean straight cut, careful matching of cambium regions, snug fitting and fastening and careful protection from drying of the tissues both before and after the grafting operations all contribute to successful grafts.

Whip or Tongue Grafting

Whip or tongue grafting is used in grafting a scion on a stock that is of similar diameter. The stock may be a one-year-old seedling, a short piece of a root, a rooted or nonrooted cutting, a seedling on which a previously inserted bud failed to grow or a 1- to 3-year-old tree that is to be top worked to some other variety. The technique is used in grafting young stock plants that are either in the nursery row or that have been dug to be retransplanted after being grafted. In the latter case, the term bench grafting is used, i.e., the grafting operation is carried out indoors and the operator has his grafting materials on the work bench. Bench grafts are stored in the same manner as scion wood until planting can take place in the spring. The whip graft is easily made and is one of the most commonly used because the several uniting edges of this graft form a strong union. It is suitable for apple, apricot, cherry, pear, plum, prune and grape. Quince and peach, however, do not unite readily by whip grafting and budding during summer is the preferred method for these two fruits.

A whip graft is essentially the joining of two parts, the rootstock and scion, each cut off diagonally to fit together. But the woody center of each piece, which is almost the whole diameter of the pieces to be joined, will not grow together. Only the cambium layer, a very thin skin, will make a firm joint. In order to give the pieces more of a purchase on each other, both rootstock and scion are split with a cut almost parallel to the diagonal cut and the two splits are fitted together. This not only makes a firmer joint, but also permits more of the cambium layer of each piece to come in contact with the cambium layer on the other.

Properly cut, the rootstock and scion should fit very firmly without splitting or bulging. If one is slightly larger than the other, it is essential to fit the cambium layers together on one side, rather than to center them and miss a jointure on both sides. After they are joined, they should be firmly wrapped with nurseryman's tape or with narrow strips of waxed cloth. The binding material must be watched and cut before it tightens too much on the expanding union.

When one wishes to graft on a stock that is large (one to several inches in diameter), the cleft is one type of graft that can be used.

First, cut the stock to be grafted, leaving a smooth cut. If it is a large limb which may split, it is best to cut it first a few inches above the desired position. Saw from the lower side of the limb until the saw begins to bind and then finish from above. This will prevent splitting of the stub that remains. If the cut is not smooth, saw it off again a short distance below the original cut.

Now take a grafting tool or some other sharp instrument and split the stock. Place the grafting tool or some type of wedge in this split to hold it open enough to take the scion, but avoid splitting too far as it will not hold the scion tightly and will leave a larger injury than is necessary. Now take a 5- to 6-inch scion and trim the lower edge into a wedge shape. Trim both sides the same, with one edge thinner than the other. Place the scion in the stock with the cambiums united and with the thick edge of the scion outside and then remove the grafting tool. The stock should now be holding the scions tightly in place. This is necessary to bring the cambium of the scion in close contact with the cambium of the stock. Remember that the cambium is the line of union between bark and wood and do not be confused by the fact that the bark on the stock is often several times as thick as that on the scion. A good way to insure contact of the cambium of the stock and scion is to place the cambium of the scion just outside the cambium of the stock at the top of the stock and just inside it at the lower end of the scion. In this way the cambium will cross in such a manner as to insure contact for a short distance at least.

It is not necessary to wrap this type of graft, as the pressure of the stock will hold the scions in place unless the split is made too far down the stock or the stock is too small. The cleft graft should be covered with a good grafting wax. Be sure that the top of the scion is covered with wax to prevent drying out until it has had time to make a union with the stock and the entire cut surface of the stock, including as far down the sides as it has been split, should also be covered.

Two very important points to remember in making this type of graft are to see that cambiums of stock and scion are in contact and that all exposed cut surfaces are covered with wax.

Preparation of an inlay graft is described by J. H. Waring and M. T. Hilborn of Maine Extension Service. The branch is cut off as for the cleft graft but is not split. The scion is prepared by cutting from a point opposite a bud and 1½ inches from the base, first sharply toward the heart, then parallel with the grain of the wood to the base end. The base is then finished by a cut to make it slightly wedge-shaped and the sides of the cut face are trued and the cambium exposed by slightly shaving both edges.

Then the cut surface of the scion is placed against the stock and outlined on the bark with the knife. A nicety that is liked by some men is, after thus marking the full length of the sides of the cut surface of the scion, to make the cross cut a bit short of that length so that, when the bark is removed and the scion placed, it may be forced down under a lip of bark. Scions must be tacked, using two very thin and preferably flat-headed nails ½ to ¾ inch long in each scion. Cuts are then protected by coating with grafting wax.

Subsequent care of inlay grafts is essentially the same as for cleft grafts. It is well to remember, when finally the stub is to be cut slanting away from the one scion that is retained, that there are small nails to be avoided.

Bridge Grafting

Bridge grafting is used to repair and improve trees completely or partially girdled near the ground level by rodents or suffering from low temperature or disease injuries. Such injured trees may die unless the injured areas of the trunk are bridged over by living tissue to reconnect the fruit-bearing top with the root system.

Ragged edges and loose bark must be cut away cleanly in early spring. When the injury extends to the roots, the earth must be removed from the base of the tree and from the larger roots until sound bark is uncovered.

There are various ways of inserting the scion. When dealing with younger trees having relatively thin bark, a short slit is made through the healthy bark above and below the area to be bridged over. A dormant scion is cut wedge shape on both ends, but the cut on one side is only about half as long as on the opposite side. The wedge-shaped ends are then inserted under the bark at the point it was slit so that the longer cut surface is in contact with the wood of the tree. Scions are placed about 3 inches apart and should be from 3 to 4 inches longer than the space to be bridged. The scions should be long enough to bow outward from the trunk when in place. Each scion is secured by driving a small nail through each end of the scion and the loosened bark. That part of the scion inserted under the bark is thoroughly waxed over.

With old trees having very thick bark the inlay graft instead of the bark graft is preferred. Instead of making a slit through the bark a 3- to 4-inch long strip of bark as wide as the scion to be used is separated from the trunk. The scion is cut flat on one side on both ends and fitted into the place where the bark strip has been separated. Where the inlay graft is used, care must again be taken to cut the scion long enough to form a slight bow when in place. The properly placed scions are held firmly to the trunk by driving two small nails through each place of attachment. The ends of the scion must be covered with wax, but it is not necessary to cover all of the bridge portion of the trunk.

In instances of extensive injury to trunk and root system it becomes necessary to inarch seedling or young trees into the tree to be kept alive.

One-year-old seedlings or budlings are planted beside the tree and are then grafted into the tree trunk. Depending on the age and thickness of the bark, the seedling can be attached to the tree trunk by the bark or inlay graft. Again each seedling is nailed in place and all cut surfaces are waxed. Shoots that appear on the inarched seedlings during the following growing season are best left but must be somewhat suppressed by pinching back their tips. After the inarched plants have securely united all shoot growth can be removed. Inarching as well as bridge grafting is most successfully done in early spring as soon as the bark of the injured trees separates readily from the wood.

Two different grafting waxes can be used: a soft wax that can be applied by hand; and a hard wax that must be melted when applied. Grafting waxes also often contain some pigment, such as lampblack, to give them a color different from the tree. The black color absorbs sun rays which merge minute cracks formed in the wax cover during cold weather. Using colored wax also helps to determine quickly if all grafting wounds are covered.

A soft wax may be made from the following ingredients:

Resin	4 pounds
Beeswax	2 pounds
Tallow	1 pound
Lampblack	1 ounce

Melt the first 3 ingredients together and add the lampblack. In order to obtain a softer wax, the same amount of linseed oil may be substituted for the tallow.

A hard wax that must be applied when melted can be made from the following ingredients:

Resin	5 pounds
Beeswax	1 pound
Raw linseed oil	3/4 pint
Lampblack	1/2 pound

Melt the resin, add the beeswax and melt, add linseed oil and remove from heat and stir in lampblack little by little. Since the materials used are inflammable proper precautions must be taken in the preparation of grafting waxes.

The grafting and budding techniques described in the foregoing paragraphs are of use to the home gardener principally as means of repairing damage or of grafting pollinating branches to home orchards. It is usually uneconomical for the home gardener to try to propagate his own trees, since he needs so few of them and he must wait so much longer for them to start to produce. But a few of the small fruits, such as grapes, bramble berries, gooseberries and currants start to produce crops within a few years after planting. These may profitably be propagated in the home garden by simple vegetative methods which do not involve surgery.

Layering

Layering is an operation by which shoots still attached to an established mother plant are made to root.

The simplest method is tip layering, employed to obtain young blackberry, dewberry and black raspberry plants by covering the ends of the young canes with soil during late summer. The covered portion will develop roots and the rooted tip, when separated from the mother plant the following spring, will serve as the new plant.

Another method is simple layering that consists of bending and pegging down branches and covering part of each branch with soil, but leaving the tips of the branch uncovered. This method is successful with grapes and filberts.

There are two methods of air layering. The first method is used in England. Select a twig or branch about the size of a pencil in circumference and, with a sharp knife, make an incision in the branch center and continue the cut upwards for about two inches. This cut should be made just below a dormant bud situated not more than 12 inches from the tip of the branch. Now, carefully bend open the incision or open it with the aid of your knife and keep it open until you have inserted in the incision, a tiny pebble or stone, or better still a small piece of sphagnum or other moss. The object of this is to prevent Nature from healing the wound by contact.

You now envelop this incision with a good-sized handful of sphagnum moss, crushed straw or hay, all of which must be moistened—preferably in rain or pond water—to the consistency of a squeezed-out sponge. Next, cover the moss with a sac of transparent plastic or polyethylene, completely enclosing the moistened moss and securing each end of the shroud with string, tape or such things as "tyems" or "knotems." In 4 or 5 weeks, depending on the type of plant you are air layering, roots should develop.

If you are anxious to take a peep to see how Nature is progressing, you may carefully untie the shroud, but be sure to again properly wrap to prevent drought.

The importance of the shrouds and their proper application to the location of the incision or removal of bark is so important that it must be emphasized. It is the shroud which prevents the rapid evaporation of the essential moisture and once you have properly moistened the enveloping materials in which the roots actually will form, the materials will thereafter be kept properly moist and there will be no need to add additional moisture at any time the roots are forming.

Here is the second method: it is not necessary to be too precise about the size of the plant stem or branch that you wish to make produce roots. I have actual rooted specimens varying from one inch to three inches in circumference. The length of the branch may also be approximately from 6 inches to a couple of feet long. In any case, you circle the bark and cambium with a cut from your sharp knife and down to the actual hard wood. In the case of the smaller branches, another similar cut is made one inch away from the first cut.

AIR LAYERING STEPS

1. Cut or girdle the outer layer of the bark just below bud. Width of girdle depends on size of branch; from one-half inch to one and one-half inches. Scrape the girdled area, removing all tissue, so area will not heal and roots will grow.

2. Place sphagnum moss or compost in a plastic cover, wrapping it about the cutting and tying securely. Be sure to leave the top open a bit to catch rain. Softwood grows roots faster than a hardwood which takes 10 to 16 weeks. Check area for roots in 8 weeks.

3. If there is no rain, the cutting will need water once a week. Six ounces is enough during a dry spell. If plants root by July 1, they will be ready to set out in early September. They will then be well established before the heavy winter frosts.

Allow a space of up to one and one-half inches between cuts in the larger branches. Now, entirely remove all the bark and cambium—down to the white wood stem—between the two cuts made with your knife. You have entirely severed the sap flow, the life line, from the mother plant. If the branch is left like this and uncovered, this branch ahead of the wound will die. To complete this operation, all you have to do now is cover the wound with sphagnum moss, moistened, and tie it with the shroud of your choice. It should look like the bulge of a sweet potato with tapered ends. Roots may be

Above is an excellent example of air layering being done on a pear tree at the Robbins Gardens in Florida. The rooting method is damp, rich earth, wrapped in aluminum foil; rooting period has been 8 weeks.

formed in from one to two month's time. When roots are well-established and several inches long, the branch may be severed by a saw or clean cutting shears, immediately below the bottom end of the shroud, which is conveniently cut to about 8 inches in length.

The successfully rooted branches are now on their own and entirely separated from the mother plant. In starting life on their own, they must be handled with some care until they gain experience in this life. Remove the shroud and soak the roots in tepid water for an hour. Discard the moss or other rooting media. The newly formed roots must now supply life-giving liquids to the plant. To give them a hand, it is necessary to reduce the volume. Therefore, prune the entire branch at least one-third of its stature.

These rooted branches may now be set out in suitable sized pots or planted directly into the garden soil. They must be kept damp and shaded for several months; then transfer them to the permanent quarters that you have selected.

—Philip Wells

Mound or stool layering is often used to propagate gooseberries. The process is described by C. J. Hansen and H. T. Hartmann of the California Agricultural Experiment Station.

Before growth starts in the spring, the mother plants are cut back close to the ground. Soil is mounded around the bases of the new shoots as they grow during the spring, but the tips are not covered. The mound is built up to a height of 8 to 12 inches. Roots arise from the bases of the new shoots, growing into the mound of soil. If the mounds are kept moist through the summer the shoots are usually well-rooted by winter when the soil is removed and the rooted shoots are cut off. These are often lined out in the nursery row the second spring and grown an additional year to produce stronger plants. Mound layering is used in the commercial production of gooseberries, currants, quinces and certain vegetatively propagated plum and apple rootstocks.

Plants that will root when stooled or layered also will often root when a piece of the mother plant, such as a stem, a root or even a leaf attached to a short stem portion, is placed under conditions favorable for root development. The important factor that makes rooting possible is the condition of the material used and the time the cuttings are planted.

Cuttings

Hardwood cuttings consist of sections of the stem usually taken from the previous season's growth although with some species, such as the fig and olive, wood up to 2 or 3 years old can be used. For deciduous species, the cuttings are made after the leaves drop, from late fall to early spring. The length of the cuttings varies with the species, ranging from 5 to 15 inches. Each cutting should include at least two nodes. Roots ordinarily arise in the vicinity of the lower node, or nodes, while the shoot originates from one of the uppermost buds. In species with short internodes it is immaterial where the cuts are made in relation to the nodes, but if the internodes are long, the basal cut should be about ½ inch below a node and the top cut about 1 inch above a node.

A number of varieties of fruit species are propagated commercially by hardwood cuttings and are thus grown "on their own roots." No graft or bud union is involved. These are varieties of the grape, currant, gooseberry, fig, quince, pomegranate and olive. In addition, a number of rootstocks are propagated in this manner, to be budded or grafted later to a named fruit variety. These include certain plum rootstocks, such as Myrobalan 29 and Marianna; nematode- and phylloxera-resistant grape rootstocks; types of the quince used as a dwarfing stock for pears; and certain clonal apple rootstocks.

There are several methods of handling hardwood cuttings of deciduous fruit species. The cutting material may be gathered in the fall, made into cuttings and planted immediately. Better rooting of some plants is more often obtained with fall planting than with spring planting, especially in regions with mild winters. In areas with severe winters, cold damage may occur to fall-planted cuttings and freezing and thawing of the soil may cause the cuttings to heave out of the ground. In addition, rodent injury during winter may be a problem. Even in areas where winters are mild, weed growth in the nursery during winter may require expensive hand weeding in the spring.

Another method commonly used is to prepare the cuttings in late fall or early winter and store them in callusing beds, for planting in the nursery row in early spring. The bundles of cuttings may be packed in some moist material, such as sawdust, shingle tow or peat moss, in large boxes and kept in an unheated, outdoor building or a cool cellar until planting time. Temperatures of 40 to 60° F. are desirable.

As a rule, cuttings are so planted that only one bud is above the ground level, and are placed 2 to 3 inches apart in the nursery row.

A semihardwood cutting is a type of stem cutting used in propagating certain kinds of broad-leaved evergreens. The olive is propagated commercially in this manner. Other fruit plants which can be propagated by semihardwood cuttings, although not commercially, include certain citrus species, macadamia and some avocados.

These are leafy cuttings which must be rooted under conditions of high humidity, such as glass-enclosed propagating cases or mist propagating beds.

A cold frame can be useful for starting young blueberry plants as shown above. Hardwood cuttings for deciduous species are made after the leaves drop, from late fall to early spring; each cutting should include at least two nodes.

Semihardwood cuttings can ordinarily be rooted at any time of year, but for those plants which have distinct flushes of growth, it is best to take the cuttings following a period of growth activity after the wood has become partially matured.

Root cuttings are small sections of roots which, when planted, produce a new shoot system and a renewal of growth of the existing root piece. Success is more likely with root cuttings if they are taken from young plants, not over two or three years old. Plants which produce suckers from the roots ordinarily can be propagated readily by root cuttings. Fruit species started in this manner include the red raspberry, trailing and upright blackberries, quince and some apple rootstocks.

The cuttings can be made from roots of various sizes, even very small roots. They are cut into sections 2 to 4 inches long and may be planted either horizontally, about two inches deep in the nursery row or vertically with the top just below the surface of the soil. Red raspberry and blackberry root cuttings are ordinarily planted horizontally. In vertical planting, the end of the root piece which was nearest the crown of the plant should always be up. The best time to take the cuttings is in late winter or early spring before growth starts. They may be made and planted immediately or stored in a cool, moist callusing bed until they are planted in the spring.

Red raspberries, upright blackberries and the Stockton Morello cherry (a rootstock for cherries), are propagated by suckers. The suckers, together with some roots, are removed from the parent plant during the winter.

30—Protecting Fruits Against Insects and Disease

THE LIST OF INSECTS, fungi, viruses and bacteria which can and often do attack fruits is almost endless. Experiences of organic gardeners who have put into practice programs without chemicals seems to indicate that this list may possibly be eliminated by organic methods. But whether they can be eliminated or not, it can be stated with certainty that by far the greater number of diseases and insects can be brought under control without chemicals.

Commercial orchards in the United States have not yet reached the point in their mass destruction of helpful insects, soil bacteria and earthworms where they are forced to realize their folly. But in small, isolated instances the realization of what they are doing seems to have been forced upon them. Citrus growers in both California and Florida are having borne in upon them the fact that they have killed the soil flora in their orchards and have left themselves open to a battle with nematodes and other difficulties. Cotton growers know that they have poisoned not only the bees that they need, but even the honey produced by their bees. One experimental station after another, throughout the country, recommends organic culture of the soil "wherever it is practical."

A mass program is under way in Nova Scotia to change over from chemical insect and disease control to what their Department of Agriculture calls "natural control." A. David Crowe relates that in the early 1940's, as the officer-in-charge of entomological research in Nova Scotia, Dr. A. D. Pickett began wondering where they were going in the spray program. Bud moth, codling moth, gray-banded leaf roller, oystershell scale, rosy aphid and red mite were all serious pests—and seemed to be getting a little bit worse each year.

Dr. Pickett decided to initiate work to see if spray programs could be devised which would avoid this worsening of the pest control problem. Very quickly it was found that oystershell scale would be controlled by its natural enemies if the grower avoided the use of sulfur sprays.

This was encouraging, particularly as new fungicides were appearing which gave the promise of interfering less with the predaceous and parasitic fauna of the orchard. The entomological staff continued their research along these lines and by 1950 they had pushed their experiments to the point where they felt they could recommend to a number of growers a trial program in which very few if any insecticides or miticides were used.

Fortunately, most growers who tried this approach were able to get satisfactory results and usage of the natural control program expanded. New materials and techniques have assisted this development.

Today the large bulk of the apples grown in Nova Scotia are grown under this program. Along with changes in the program have come changes in terminology. The title of natural control has been replaced by the more accurate term of modified control. And it is likely that integrated control would be an even more desirable term.

Growers on the modified control program actually do use chemicals for insect control, but only when necessary. This is the crux of the problem. When is an insecticidal spray necessary? Can a grower just sit by and take a chance on extensive damage while waiting to see if nature will do the job for him? Fortunately, growers have now reached the point where they do not need to go through this nerve-wracking process.

There are reasons for this. In the first place, once an orchard is adapted to a modified program, the changes in insect and mite populations proceed at a much slower pace. The only increase in populations is that which is over and above natural control agencies, whereas under a full chemical program there is the full expansive potential of the pest to contend with.

It would be wrong to leave the impression that there are no problems. There are. We need more chemicals that are specific for a single pest, says Dr. Pickett, the way ryania will control codling moth and not interfere with the natural control of mites. Species which were not economically important before have a way of developing.

Prior to the adoption of the modified control program, random samples of all apple orchards showed an average of about 15 per cent of the fruit was being directly damaged by insects. This figure has declined steadily with the general adoption of the modified control program and has leveled off at an average of about 4 per cent for the past 3 years. In addition, mites were a widespread pest whereas now they are only an economic problem in orchards where growers have used wide spectrum insecticides.

How would the Nova Scotia program work elsewhere? It wouldn't. But there are indications that a New York modified control program, for example, would work as well in New York as the Nova Scotia program does in Nova Scotia. Modified spray programs in the United States and British Columbia have not been successful to date, due, no doubt, to a more favorable climate than in Nova Scotia for insect development.

Dormant Oil Spray

Early spring is the only time the organic gardener has recourse to that much overused implement, the sprayer. And he uses it then only to apply an entirely safe, nonpoisonous spray, not DDT, parathion or any of the other deadly concoctions in the chemical arsenal.

Used properly, a 3 per cent miscible oil dormant spray is marvelously effective against a host of chewing and sucking insects, organic gardeners report. Aphids, red spider, thrips, mealybugs, whiteflies, pear psylla, all

A dormant oil spray can be applied to orchard trees before any of the buds open in late winter. Insects which have hatched from eggs laid on plants the previous fall, can be readily destroyed; the egg shells and protective covering become more porous then.

kinds of scale insects and mites fall before it. The eggs of codling moth, oriental fruit moth, various leaf rollers and cankerworms are destroyed.

A dormant spray is applied to orchard trees before any of the buds open. Some gardeners make it a practice to use it on all dormant trees, shrubs and evergreens every spring, but this is rarely necessary if the plants have been organically grown for a number of years. Fruit trees, however, have many enemies and dormant spraying should be a regular practice for them, along with a strict program of sanitation.

In early spring insects that hatch from eggs laid on plants the previous fall can be readily killed because the shells of the eggs and the protective covering of hibernating scales become softer and more porous at this time. The dormant spray penetrates and makes a tight, continuous film over these, literally suffocating the organism to death.

It will, of course, form a similar film over leaves and injure them, which is why it is applied only while the trees are in a leafless state. Citrus trees, which do not shed their leaves, are given a very dilute spray, usually made with "white oils," highly refined oils that present the least chance of foliage injury.

Dormant oil sprays have a residual effect, too. An oil film covering the plant interferes with the successful establishment of any insects that may hatch for several days after spraying.

Stock preparations of miscible oil sprays are sold by all garden supply stores, with instructions for dilution and use. You can also make your own, using a gallon of light grade oil and a pound of fish-oil soap (an emulsifier) to each half gallon. These ingredients are mixed together, brought to a boil and poured back and forth from one container to another until emulsified (thoroughly blended). Since all oil emulsions tend to separate into oil and water again, the mixture should be used as soon as possible after it is prepared. Dilute it with 20 or more times its volume of water for use.

The grower who has always used chemical sprays must not expect to drop them all at once. In converting from chemical to organic methods, or in combatting an especially severe infestation, the use of rotenone or pyrethrum dusts or Ryania (which are plant extractives) may be necessary. Either should be applied only where a particularly serious condition is present or where a conversion is being made. They should be discontinued completely as soon as moderate control has been attained. Natural resistance and plant strength gained through constant organic treatment make the use of violent chemical poisons needlessly dangerous and costly.

Sanitation

The commercial fruit grower today seldom practices the orchard sanitation techniques that were so vital in years gone by when sprays were not as deadly as they are today. Today the spray gun is basically his only weapon against insects.

If you want to grow fruit organically, you must do more than just stop spraying. You must make things difficult for harmful insects by cleaning out their hiding places. You will be surprised at the fine insect control you can achieve by "house cleaning" your orchard just as you would house clean your home.

Sanitation is quite simple and maybe that's why we so often neglect it. Take that perennial pest, the codling moth. A few simple steps will make things mighty uncomfortable for him.

Thoroughly scraping off all loose, rough bark in the spring will destroy almost half the overwintering codling moth larvae on a tree. (The progeny of a single pair of these can damage 150 bushels of apples.) You can make a good scraper from an old hoe, a piece of saw blade or a mower section or a floor scraper. Use a gentle, even movement. The tool should not be too sharp and don't scrape down to live bark, which might result in damage to the tree.

Put a sheet or canvas under the tree to catch all the scrapings and immediately put them in a fresh compost pile. Scraping, by the way, will permit more thorough coverage with the dormant oil spray you apply to control scale insects, aphids, red mite and other pests.

Next, apply tanglefoot, or wrap and tack a 2- to 4-inch-wide band of corrugated paper or burlap around the trunk. This will trap a great many of the remaining larvae. But don't use chemically treated bands, which can injure the tree.

Old decaying wounds are a favorite habitat for codling moth larvae and other destructive pests. The cavities resulting from poor pruning cuts, split trunks and broken branches are filled with punky wood where they hibernate.

Prune off all dead, split or broken branches and stubs cleanly at the point of origin, beveling the edges to encourage healing. Compost or burn all the prunings promptly—dumping them into a gully or brush heap means the larvae can continue to develop and eventually return to the tree as adult moths.

All running cavities and stubs that have rotted back into the tree should be cleaned out with a gouge chisel to solid live wood. This can be done in almost any season, except in wet weather. A straight chisel is best for finishing the edges of a cavity, trimming them back to live bark and bringing the cavity to a point at top and bottom. Cut all edges so they are at right angles to the bark to promote rapid healing.

All cuts over two inches in diameter should be painted with tree paint. It's not a good idea to fill cavities with cement—it may look neater, but you won't be able to see if more decay shows up in the cavity later.

Cutting out cankers and diseased limbs during the winter and rubbing off suckers and watersprouts while they are still very small, will go a long way towards eliminating bacterial diseases like fire blight. Bitter rot, blister canker, black rot and blotch fungi also live in old cankers and attack the tree through new, open wounds. Remove the bark on old cankers to healthy tissue and paint them and all new wounds promptly.

Certain pests require specific treatments. Black knot fungus of plum trees must be eradicated by cutting out the knotted tissue, removing it for an inch beyond the boundary of the knot. The plum curculio beetle, which attacks apples and stone fruits, can be shaken off the tree into a sheet by jarring it or hammering the branch with a mallet.

Tree hoppers live on weeds, so keep your fruit plantings weed-free. Apple

SIX STEPS TO ORCHARD SANITATION

1. Scrape bark.

4. Paint cuts.

2. Band with tanglefoot.

5. Keep mulch from trees.

3. Prune trees correctly.

6. Pick up dropped fruit.

scab fungi overwinter on leaves on the ground—make sure these are incorporated well into your mulches.

A mulch should be of fairly fine material: whole cobs, cornstalks and large weeds can harbor larvae and disease organisms. And keep it at least two feet from the trunk—not only rodents, but also such pests as the apple flea weevil, like to live in the cover there. A wire mesh guard, veneer bands or paper wrapped around the trunks of young trees will prevent mice and rabbits from girdling them.

Whitewashing the trunks of fruit trees is of some value, especially in certain regions of the country. The advantages of the practice are as follows: It is claimed that whitewashing reduces insects and disease organisms that burrow in the bark tissues; insect eggs are also supposedly killed by the whitewashing substance. Whitewashing prevents some scald and heat injury of the bark tissues of very young trees. This is especially true in the sunny regions of the country. The general practice of whitewashing is to make an application over the trunk and main supportings of the tree.

A wire probe or knife is usually necessary to dig baby borers out of the trunk near the ground. You may have to hoe away the soil a few inches deep to be sure of getting those that have bored into the tree below ground level. Look for tiny holes, soft spots in the bark and masses of amber, gummy material. Keeping the ground clean around the trunk will let the birds find the eggs and young worms before they start boring. (We need more birdhouses in our gardens and orchards!) Tanglefoot, building paper or a loose band of tin filled with tobacco dust around the trunk will also prevent borer damage.

During the growing season, pick off all wormy or diseased fruit as early as possible to reduce sources of infection. And pick up all dropped fruit every 2 or 3 days. Put the dried, shriveled "mummies" in the compost heap. The rest can be used for fruit dishes or jelly or fed to livestock.

Keep your garden or orchard clear of rubbish, for old baskets, sacks and decaying prunings can harbor a surprising number of destructive organisms. Also, don't mow any tree sprouts or underbrush. The ragged stubs left by the mower are excellent hibernating places. Always grub them out. Weeds and fallen bark should likewise be promptly removed to the compost pile and buried therein.

Here's another point orchardists may slip up on: crates stored in an open packing shed provide a fine breeding ground for codling moths and other pests. Sheds should always be enclosed, or screened with heavy cloth in spring and early summer to prevent the adult insects from escaping to the orchard. A pan of kerosene in the shed, with a light left burning over it, will kill many of the bugs.

With berry fruits, sanitation usually consists merely of removing diseased or pest-attacked parts of the plants promptly. Watch particularly for swelling of the canes or wilted tips, signs of borer damage. If you have any wild brambles growing nearby, keep an eye on them, too and destroy infested parts immediately.

One final sanitation tip: if you grow sweet clover in your orchard, always clip it at the proper time. The rank stem growth that results when it is allowed to reach maximum size the second year is a favored place for codling moth larvae to spin.

Plant Resistant Varieties

One way to outwit various diseases is to plant resistant varieties. The following varieties, plus others named in the sections under various fruits, are recommended for their resistant qualities:

Apple: Baldwin, moderately scab-resistant; Black Twig, cedar-rust resistant; Golden Delicious, moderately resistant to quince rust and scab; Grimes Golden, cedar-rust resistant and very resistant to scab; Ingram, very resistant to scab; Jonathan, Lowland Raspberry, Maiden Blush, Mann, Oldenburg, Ortley, Red Astrachan, Wagener and Yellow Transparent, moderately resistant to scab; Stayman and Winesap, cedar-rust resistant; York Imperial, moderately resistant to quince rust and very resistant to scab.

Brambles: Eldorado blackberries, very resistant to orange rust; Milton and Newburgh raspberries, moderately resistant to mosaic; Sodus raspberry, moderately resistant to all disease.

Blueberries: Harding and Stanley, stunt-disease resistant; Jersey, moderately resistant to stunt disease.

Grapes: Kendaia, downy-mildew resistant; Seibel 1000 (a wine grape) generally disease resistant.

Pears: French and Mazzard rootstocks both resistant to armillaria root rot.

Strawberries: Blakemore, Brightmore, Howard 17, Klondike, all resistant to viruses; Catskill, Dresden, Fairmore, Fairfax, Howard 17, Klonmore, Midland, Missionary, Red Star, Southland and Starbright, resistant to leaf spot; Fairland, Marshall, Md-683, Pathfinder, Sparkle, Temple, US 3203, US 3205, US 3374, all resistant to red stele root rot.

Watermelon: Blacklee, Black Kleckley, Blue Ribbon, Dixie Hybrid, Georgia Wilt Resistant, Kleckley Hybrid, Klondike R7, Leesburg and Miles, all wilt resistant; Black Kleckley, Congo, Dixie Hybrid, Kleckley Hybrid and Missouri Queen, anthracnose resistant.

Here are recommendations from agricultural experiment stations of the more resistant fruits and nut trees in these states spotted through the country:

Alabama: Pecans, black walnuts, hickory, Chinese chestnuts, filberts, pears, cultivated crab apples, mulberries, May haws, black haws, jujube, persimmon (both American and Oriental), bronze Elaeagnus, blueberry, pomegranate, papaws, strawberries, Sapa plums, compass cherries, Oka cherries, figs and muscadine grapes.

Varieties of pears are Seckel, LeConte, Garber, Kieffer, Baldwin and Oriental.

California: Varieties of the following kinds of fruit will do reasonably well if given normal garden care without the use of sprays for disease and insect control: blueberry, Carissa, chestnuts, feijoa, fig, jujube, macadamia, olive, papaya, pecan, persimmon, pistachio, pomegranate, quince, strawberry tree (*Arbutus unedo*), and black walnut.

Not all of these species are immune to insect or disease problems, but most of them will do quite well with minimum attention.

Florida: A few fruit trees that will do reasonably well without chemical means of controlling insects and disease are: banana, pineapple, pear, Stewart pecan, Japanese persimmon, sapote, sea grape, cattley guava, mulberry and loquats.

New Hampshire: Pears as a whole are very much less subject to insects and disease than are apples. The varieties we recommend are Clapp's Favorite for early, Gorham for midseason and Bosc for late.

New Mexico: The Arkansas Black Apple is a vigorous, disease-tolerant tree which produces a firm, thick-skinned apple. The growing season requirements for the Arkansas Black, however, are long and only in southern New Mexico will it reach proper maturity.

Tennessee: Pears and strawberries (resistant varieties) come as near being free from insect and disease attack as any fruits. The Orient pear is practically immune to fire blight, a bacterial disease and in addition is resistant to leaf spots. Ayres and Morgan are two new varieties that are very resistant to these diseases. Orient and Ayres have sterile pollen and must be cross-pollinated by another variety that blossoms at the same time. Many growers in Tennessee have managed these new varieties for 7 or 8 years without a drop of spray chemicals.

Home plantings of strawberries, using resistant varieties, rarely if ever receive any spray chemicals. Tennessee Beauty, Blakemore and Tennessee Supreme are commonly planted in Tennessee.

Utah: In this area the native American plum, *Prunus Munsoniana,* or Potawattamie plum and the Stanley and Italian prune are not bothered too much by insects. The apricot and sour and sweet cherry are grown commonly without spraying.

Virginia: The York apple is more resistant to scab. In our home fruit planting recommendations we also recommend the planting of early season maturing varieties such as Lodi, Williams Early Red, Summer Champion and Rambo. It is our feeling that these early season maturing varieties will become ripe before the insect and disease problems are too serious.

West Virginia: The Seckel pear is one of the most satisfactory varieties to plant for a supply of home fruit. As you know, it thrives relatively well under neglect and is reasonably resistant to fire blight. Codling moth will not attack pears as much as it will apples and the pear is fairly resistant to scab and other foliage diseases. The principal problem with pears in this country happens to be spring frost because of their early blooming habits. Other fruits which we sometimes recommend for home planting are Lodi, Yellow Transparent, Wealthy, and Rome for the apple varieties. Keiffer and Winter Nellis are often recommended along with the Seckel of the better pear varieties. Among cherries we like the Montmorency and Windsor, which is a sweet cherry.

One more word about fruit trees and disease: Many states have a program of inspection of nursery stock. In these cases, stock is certified disease-free by the inspectors and should be so labeled. If you live in a state with an inspection program, insist upon certified stock when you buy your fruit plants.

Rodent Protection

At the approach of winter, fruit growers should see that their trees are properly protected against mice and rabbits. Injury may occur in both sod and clean-cultivated orchards, but is most frequent in orchards where the sod system of soil management is practiced, especially in orchards that have received additional mulch. It is far better to prevent girdling than to try to save trees by bridge grafting after the damage is done.

Several methods are recommended by T. A. Merrill in a pamphlet published by Michigan State Extension Service to be used to protect trees from injury by mice and rabbits. They are:

1. Placing mechanical protectors around the tree trunks.

2. Cleaning the sod and weeds away from the tree trunks and mounding with soil, cinders or gravel.

3. Pruning the trees (at least partially) in the fall and leaving the prunings on the ground until spring.

4. Painting trunks of trees with repellent.

Several types of mechanical protectors may be used to protect trees from mouse and rabbit injury. Some of the more satisfactory are made of wire netting having 3 to 4 meshes to the inch, wood veneer strips and waterproof paper. These protectors should cover the trunk of the tree to a height of 15 to 20 inches and have been employed for this purpose and given satisfactory results. If wire netting is used it should be pushed into the soil an inch or two to hold it firmly in place after removing weeds and grass from the trunk of the tree. The wire should be cut in strips of sufficient width so they will not have to be removed for several years. Wood veneer strips and paper, however, should be placed around the trunks every fall and removed the following spring. Tar paper should not be used as it may cause injury to the trunk. The wire netting is the most satisfactory and cheapest over a period of years.

Cleaning the sod and weeds away from the trunks of the trees and mounding around the trunks with cinders, fine gravel, sand or earth are helpful.

In the case of danger from underground injury, removing the soil to a distance of 8 to 10 inches from the trunk of the tree and 6 to 8 inches deep and filling in with coarse cinders, has been found to be a satisfactory means of protecting the roots from being girdled by mice. However, caution should be taken to work fine soil well among the cinders to prevent injury from winter freezing.

Many fruit growers have observed that when trees are pruned in the fall of the year and the prunings have been allowed to remain on the ground until spring, mice and rabbits will feed upon those relatively tender branches in preference to the tree trunks. Consequently some growers are making a practice of pruning a few branches from every tree in the fall. In many cases this method has been effective, but growers are cautioned not to rely upon it alone.

A tree coating consisting of a mixture of rosin and denatured ethyl alcohol has given satisfactory results as a repellent for rabbits. This combination, however, is not satisfactory against the ravages of mice.

The above coating or repellent is made by dissolving 7 pounds of finely powdered rosin in one gallon of denatured commercial ethyl alcohol. So-called "antifreeze" alcohol may be used if it does not contain methyl alcohol. Methyl alcohol (wood alcohol or methanol) does not dissolve rosin and, therefore, should not be used. These proportions are slightly more than one part of rosin to one part of alcohol by weight. A good method of mixing is to add the rosin to the alcohol in a container with a cover tight enough to allow shaking and prevent evaporation. If the container is kept in a warm place and shaken occasionally, the rosin will dissolve more rapidly. *Do not apply any heat.* To heat the solution is not only dangerous but may evaporate enough alcohol to alter the composition of the mixture. Handled in this manner, it usually takes about 24 hours for the rosin to dissolve. It is best if only rather small quantities are mixed at a time—just what will be used in 2 or 3 days' time.

Water causes a white precipitate to be formed in this solution. If much of this precipitate is present, it will greatly alter the consistency of the repellent or even seriously interfere with its application. To avoid contamination of the reserve stock, a smaller container should be used in the orchard.

This material can be applied to the trunks of the trees and lower branches as long as the bark is dry and any time when the weather will permit working out of doors. Best results can be expected when the temperature is above freezing. Early November is a good time to make application. It should be repeated each year.

Trees treated with this rosin-alcohol repellent always turn white in the next snow or rain. This, however, does not alter the effectiveness of the repellent.

31—Harvesting and Storing Fruits

THE HOME FRUIT grower can harvest his fruit 15 minutes before it is served. This fact alone makes it worthwhile to grow your own fruits. Commercial growers are forced to pick their crops when an approximation of ripeness is attained—an approximation that is likewise an approximation of flavor and quality. Until you have picked fully ripe fruit and eaten it still warm from the sun, you can have no idea of the flavor possibilities locked up in the greenish product that comes from the store.

Fruit reaches its peak of juicy, sweet perfection only when it comes to full maturity on the plant. At this point it undergoes subtle chemical changes that convert acids to sweets, the leaf-green of the skin mellows to a yellow-green, changing to yellow, red, blue and purple in different species. It reaches maximum weight and the seeds that it contains become ripe, ready to leave the mother plant and shift for themselves.

Many fruits are picked too early. They are rushed to market while they are still firm and green for fear that they will soften and spoil before they reach the consumer. To make them palatable they are sweetened and stewed and volatile aromas and vitamins disappear in steam. Volume is sacrificed, for they are at their plumpest and juiciest only when they hang on the parent plant to the end of the ripening process. Practice patience, watch your pomes and berries swell and redden and reap your reward in flavor and lusciousness beyond anything you have ever imagined.

Strawberries—Do you like your strawberries with green noses? Have you ever tasted fresh home-grown or wild ones, ripe and red to the tips?

Cherries—Spread a sheet or canvas under your cherry tree when the cherries are red and shake the branches with a long pole. Eat those that drop easily; leave the rest for another day—and another day—until all have puffed up to their top size and peak in flavor.

Currants—Black, white or red, let them hang on the bush until the last berry on the stem is plump and ready to fall. Some will hang on for the greater part of the summer, growing mellower and mellower. Try some of them as juice as well as for jam.

Happiest time of year for the fruit grower—harvesting is a thrill for the entire family. Fruit reaches peak of juicy, sweet perfection only when it comes to full maturity on plant.

Gooseberries—Should be golden-green or russet-red and soft before they are brought to the table. They are best raw but make excellent jam.

Raspberries—Red or purple, gold or black, these are the jewels of the mid-summer garden. They should pick easily and go into small baskets to avoid crushing the delicate flesh. Serve them promptly.

Blackberries—Boysenberries—You don't know what they taste like until they fall off the canes when you look at them.

Elderberries—Mulberries—These should be royal purple or the subtle, delicate flavor may escape you completely.

Blueberries—Sometimes it seems as though the best blueberries are the ones the children and the birds pass up during the summer. Linger by the bushes on a drowsy September day and strip the fruit still hanging from the twigs and find your reward in a mild, tender berry that seems only remotely related to the tart, rather tough-skinned fruit you have known for the last two months.

Peaches—If you have known only market peaches, you have never known peaches. They are all too often yellow-fleshed, half-ripe, half-developed. The peach, par excellence, is white in flesh, free-stone and with a red in the skin that strikes right through the flesh to the pit. You must stand under the tree and eat it warm with sunshine, skin and all, for it should not have been sprayed.

Nectarines—Apricots—These are scarce and rather hard to have in the Northeast because frost so often catches the flowers but even when you see a crop only once in 3, 4 or 5 years, that crop can be so delicious that they are well worth growing. When they just can't get any bigger or more colorful without bursting, then, and only then, you come to the rescue of the tree and pick them.

Pears—You must watch these; they *can* hang on the tree until they are overripe. For prime quality, gather them when the leaf-green in their skins turns to buttery yellow-green. Then spread them out in a cool, dark room and eat them as they soften. Properly stored, some varieties will keep well into the winter. There are none sweeter or more dependable than the little Seckel. It is russet-brown when ripe and ready for picking.

Melons—The gathering of them is a fine art. They must be ready to leave the parent vine—a trifle reluctantly, perhaps, but the parting is definitely in order. The seeds should be mature, the fruits as heavy as they ever will be and, again, that subtle change has taken place in the color of the rind. In case of doubt, test as you would a cheese: cut a plug and examine and taste it.

Grapes—Grapes are never so good as when they grow, spread out, over a stone wall in sun. When they drop at a touch, then bring them to the table and let their perfume fill the room. Heat from the wall will help avert the cold, and early varieties, just as good eating as the late ones, may be selected for planting in exposed positions. Try the Interlaken Seedless and the exquisite Delaware for some of the daintiest morsels nature has to offer.

Apples—Too often they are spoiled by early picking before the leaf-green has left the skin and by bruising and storing under conditions that inhibit the development of their best flavor. They vary amazingly, too, according to where they are grown. Little ones from the Northeast may surpass huge, highly-colored samples of the same variety grown on the West Coast.

Red apples should be picked when they have a good red color for their variety and when the ground color is yellow-green. At such time the fruit has matured enough for storing, but it will be more satisfactory for eating out of hand if it ripens for a few days at room temperature. Apples picked too green will shrivel and lose flavor.

Figs—These are chancy things for the trees are tender and need winter protection in the northern states, but there is no describing the lusciousness of ripe figs eaten fresh from your own tree. Warm, melting, sugary, they put the best efforts of the candymakers to shame.

Quinces—These, like rhubarb, never mellow enough to be eaten raw, but let them ripen until the bloom is lavender-gray on the golden cheek, stew them gently until soft, sweeten with clover honey and enjoy an intriguing flavor found nowhere else. Something for the connoisseur of fruits and foods.

Persimmons—Where you really pay for your impatience is when you try to eat a green persimmon. These must freeze, on or off the trees until it would seem that they must perish with the cold. Without that treatment they can tie your mouth up in bowknots.

—Virginia Conklin

Storage

How successfully fruits keep in storage depends largely on:

1. The quality of the produce. Fruits should be free from decay, disease, insect injury or mechanical injury caused by handling. The inclusion of one diseased or damaged specimen in the storage may start decay that will rapidly spread to destroy the entire lot.
2. The selection of the proper varieties.
3. Harvesting at the proper stage of maturity.
4. The temperature and the humidity of the storage space.

Placing vegetables or fruits in storages, either pits or basement rooms, before cold weather starts in the fall is a frequent cause of early spoilage. One of the most difficult steps in the successful common storage of both fruits and vegetables is to keep them in prime condition from the time of maturity until the night temperature is low enough to cool the storage place.

Cornell Extension Service recommends that the so-called winter varieties of apples, such as Northern Spy and Baldwin be used for home storage. Other varieties with a storage life, such as McIntosh and Cortland, are satisfactory if removed from storage before they become overripe.

For a long-time storage, apples must be picked at the proper degree of maturity. Apples harvested when too green are subject to a number of storage disorders, such as scald and bitter pit. If picked when overmature, they quickly become overripe in storage.

Apples must be handled carefully at all times. Bruising and skin cuts not only make the fruit more subject to decay but actually make the fruit live faster and hence deteriorate faster in storage.

Varieties that are highly subject to scald, such as Cortland and Rhode Island Greening, stored much after Thanksgiving, probably should be mixed with shredded oiled paper at the rate of one-half pound per bushel or else wrapped with oiled paper.

For long-time storage, the temperature should be as close to 32 degrees F. as possible and the relative humidity from 80 to 85 per cent.

While apples and pears often are successfully stored in the vegetable storage, they sometimes become tainted by the odors of such vegetables as potatoes. Wrapping apples and pears or packing them in maple leaves in barrels often will prevent absorption of such odors.

Varieties of pears such as Bartlett can be satisfactorily stored for only a month or so. Other varieties, such as Winter Nelis, keep for several months. Kieffer, while not a high-quality pear, often keeps for fairly long periods in home storage.

Pears should be harvested in a condition that would seem to the amateur to be immature. If allowed to begin to "yellow" on the tree, pear fruits develop hard gritty cells in the flesh. They should be harvested when the dark green of the skin just begins to fade to a yellowish green and the fruit begins to separate more or less readily from the tree.

Pears ordinarily do not ripen satisfactorily at storage temperatures as do apples. For highest eating quality they should be removed from storage while they are still comparatively hard and green, and ripened at room temperature with a high relative humidity. Pears often keep somewhat better in home storage if wrapped with newspaper or other paper. This fruit, like apples, should be stored at a temperature as close to 32 degrees F. as possible and with a high relative humidity.

The old-time cool, moist, dirt-floor cellar was an ideal fruit storehouse. But even modern hot, dry basements can be used if you follow this plan:

Build a small, shelf-equipped storage room away from the furnace, using building paper and matched boards to cover the partition studs both inside and out. This will leave a 4-inch insulating space between the boards. Concrete or cinder blocks are also good. Put in a heavy, tight door. Two windows, shaded to keep out light, are necessary to provide free air crculation (to dissipate the gases given off by the fruit, which hasten ripening). Ventilate regularly but carefully. The ideal storage temperature for most fruits is 32 degrees—apples and Bartlett pears will ripen 3 times as fast at 40 degrees as at 32; neither will freeze until it drops below 30 degrees.

Humidity should be between 80 and 90 per cent. To insure this, sprinkle the floor whenever the humidity gets too low. A 3-inch layer of gravel will evaporate much more water than a bare floor because it has a greater surface area. An inexpensive instrument called a psychrometer will tell you both humidity and temperature. Most late varieties of apples, as well as some pears, will keep until April or longer in your storage room or in a well-insulated garage or shed. Always pick them while still very hard.

A two-foot-deep garden pit lined with leaves and straw and having a board cover will store apples and pears successfully where winters are cold but not too severe. Top it with several inches of soil and a heavy mulch. Or you can use a barrel buried in a mound of soil and covered with a deep mulch of straw and leaves.

The importance of storing fruits in a cold room quickly after harvest is pointed up by these figures supplied by William Ackerman. All varieties of apples have been found to soften very rapidly at warm temperatures. Softening proceeds about twice as fast at 70 degrees F. as at 50 degrees, while at 50 degrees it is almost twice as fast as at 40 degrees which in turn is fully twice as fast as at 32 degrees. Fruit held at 32 degrees will ripen about 25 per cent faster than that held at 30 degrees.

The effect of delays in placing McIntosh apples in storage was studied at the New Hampshire Experiment Station. It was found that a delay of 5 days in the harvesting, grading, packing and transportation operations did not materially lower the storage quality, if the fruit was marketed by March 1. However, a delay of 10 days reduced the storage life of the fruit 8 to 10 weeks and 20 days caused it to become overripe and possess no storage value.

Under storage conditions the commercial life of Bartlett pears was limited to 30 days at a temperature of 43 degrees F.; 60 days at 36 degrees; and 100 days at 31 degrees.

The most usual temperature recommended for apples is 31 to 32 degrees F. The temperature at which apples freeze varies with varieties but is always well below 30 degrees F. Naturally, the lower the storage temperature the more expensive it is to maintain. When it is not desired to keep fruits in storage for the maximum length of time, somewhat higher temperatures than those described may be used. Actually, the problem of disease and physiological disturbances are less at temperatures of 35 to 36 degrees F. than at lower temperatures.

Relative humidity refers to the per cent of saturation of the air with water at any given temperature. When the air contains all the moisture it can hold the relative humidity is 100 per cent. The ability of air to hold moisture is directly proportional to the temperature. That is, the higher the temperature, the more water it can hold. If the temperature of the air is lowered the holding capacity decreases and condensation or "sweating" takes place.

The most practical relative humidity for most fruits is about 85 to 90 per cent. Fruits held in dry atmospheres tend to shrivel and wilt, greatly reducing their value. Too great a humidity favors blue mold and the growth of other fungus organisms.

Fruit in storage gives off gaseous substances, such as ethylene and volatile esters, the continued presence of which greatly speeds up the ripening processes of the fruit. It is extremely important that a rather free circulation of air, which will prevent the accumulation of these gases, takes place in storage.

Fruit Processing

Many fruits are too soft and perishable for storage in a cellar or even in a refrigerator for more than a very short time. Also, a certain percentage of fruit from each harvest is defective in some way that precludes long storage in a cold cellar. Another method of preserving such fruit must be used. Most commonly the culls from the harvest are used for jellies, jams or juices. The perfect fruits are used for canning, freezing or drying.

Drying

Commercially dried fruits contain large amounts of sulfur which is used to process the fruits before they are dried as a preventive for discoloration. Sulfured fruits should always be avoided, even though the unsulfured product is tougher, because it must be dried to a lower water content to improve its keeping qualities.

Where the atmosphere is arid in the fall, drying may be accomplished outdoors in the sun. Where heavy dews and frequent rains are the rule at harvest time, drying must be done indoors, in the oven, in a specially constructed drier or in a dry attic.

Fruit should be perfect for drying. Blemished or bruised fruits will not keep as well and may turn a whole tray of drying fruit bad. Wash and drain the perfect fruit and remove all inedible portions. The smaller the pieces to be dried, the briefer will be the drying time. Slice or quarter apples, peaches, pears or other large fruits. Place the pieces on clean trays or racks which have ventilated bottoms made of wire mesh or narrow wooden slats. The layer of fruit should be no more than one piece deep.

If drying is to be done outdoors, place the trays on raised racks above ground in a comparatively dust-free location. Turn the fruit often and cover the racks at night to prevent the dew from settling on the fruit. Be sure to exclude animals of all kinds from the drying yard.

Drying by artificial heat is to be preferred if the heat can be closely regulated. Quick drying is better than slow, but the heat must not be sufficient to cook the fruit.

TABLE 56: For Drying

Product	Drying time	Temperature
Apples	4-6 hours	110°-150°
Apricots (let stand 20 minutes in boiling water)	4-6 hours	110°-150°
Cherries	2-4 hours	110°-150°
Peaches	4-6 hours	110°-150°
Plums (treat same as apricot)	4-6 hours	110°-150°
Rhubarb	6-8 hours	110°-130°

Drying is finished when the fruit feels dry on the outside but slightly soft inside. It should not be brittle, nor should it be possible to squeeze out any juice.

After the drying is finished, store the fruit in glass or cardboard containers. For 4 successive days stir the contents thoroughly each day to bring the drier particles in contact with some that are more moist. In this way the moisture content will be evenly distributed. If, at the end of the 4 days, the fruit seems too moist, return it to the drier for further treatment. When well dried and conditioned, the fruit should be stored in a cool place. It is best to examine it occasionally, to discover molds if they appear. The danger of molds is prevented if the dried product can be stored at freezing temperatures or below.

Canning

Small fruits, overripe fruits and defective fruits may be used to make jams and jellies or they may be used in a much more healthful way in fruit purées or juices. Jams, jellies and other sweet preserves contain such large quantities of sweeteners (whether honey or sugar) that any nutritive value of the fruit is overshadowed by the concentrated sweets which supply nothing but calories.

The same fruits which might be used for jams and jellies can be more profitably stored for the family's winter use in the form of fruit juices and purées,

Nowadays, a wide selection of modern equipment—including some astonishingly quick and efficient blenders, liquefiers, juicers and mechanical presses—makes on-the-spot juice extraction of any fruits or vegetables supersimple. But if you do not own a mechanical juicer, you can use home-improvised equipment and methods to "put up" any fruit juices; you do not have to use automatic appliances of this type. You *will* need to follow a careful procedure to prepare juice to be stored—which differs from squeezing or readying any that is used right away, of course.

Nutritionist Adelle Davis, in her popular text *Let's Cook It Right,* provides one of the best ways to prepare fruit juices for preserving. Her directions are uncomplicated and the method is one that seeks to retain as much vitamin value as possible. On this point, she explains: "In extracting juices to be canned, such as tomato or grape juice, far less loss of vitamin C occurs if the raw fruits are first quickly cooked and the enzymes (responsible for vitamin loss) destroyed before the juice is extracted."

Miss Davis adds that any fruit for juicing is to be cooked *only* until tender. "After the enzymes are destroyed, the shorter the cooking time, the less the nutritive loss." Sour fruits, she points out, such as plums, green apples and rhubarb, sustain little loss of vitamin C because the acids they contain inhibit enzyme action.

The following "recipe" can be used in making juice or sauces (or both) from these fruits: apricots, apples, plums, berries (all kinds), grapes, cherries, peaches, pears, nectarines and red or yellow tomatoes.

Use ripe fruit; it may be smaller or riper than suitable for other canning. Do not peel, core or cut apples or tomatoes except to remove damaged spots; do not remove stems from strawberries or stones from small apricots or cherries. Add only enough water at the beginning of cooking to prevent sticking.

Set fruit in a large utensil (stainless steel, pyrex or enamelware) over high heat. Use a large knife and cut through many fruits at one time, bringing knife from one side to the other, as if cutting a pan of fudge. Mash slightly to squeeze out juice.

Cover utensil; keep heat high and cook until tender, stirring occasionally to ensure even heating. Then press fruit through a cone-shaped colander. Collect thin juice in one utensil; change utensils and keep thick purée separate.

Bring juice to a rolling boil. Set clean jar in a pan of warm water to prevent breaking; pour boiling juice into jar to ⅛ inch of top; wipe edge of glass, adjust rubber and lid and seal.

Invert jar to sterilize lid. Let stand in a place free from draft. Do not move jars until cool. If screw top is used, remove ring after 24 hours.

Use the purée for making fruit sauces; or bring to a rolling boil, stirring frequently; pour into clean jars and seal.

If clear berry, plum or grape juice is desired, strain through a cloth after passing through the colander. And if you have rose hips available, add them at the beginning of the cooking, advises Miss Davis, to enhance both flavor and vitamin value.

Once you've readied any fruit juices for canning, you can freeze them if you prefer—and if your storage space permits. Prepare for canning (as just outlined), then chill thoroughly, pack into suitable containers, freeze and store.

To keep vitamin and mineral content as high as possible in the finished product, remember a few always-true pointers on retaining food values:

1. Pick fruits when fully ripe; nutritive levels are lower in immature tree or vine produce.

2. Use promptly; if fruit is to be processed for canning or freezing, get to the project with as little delay as feasible; refrigerate harvested fruit right after picking—until eaten or prepared.

3. Wash quickly, just before use; don't soak or let it stand in water. (A quick rinsing is usually ample for spray-free organic fruit.)

4. Avoid cutting, peeling, chopping, etc., ahead of time—or at all if unnecessary for preparation. The less surface area of fruit or vegetable that is exposed to air (oxidation) the smaller the vitamin loss.

5. Cook rapidly when required for canning; apply high heat for fastest process, do not overcook; use a *minimum* of water; avoid aluminum or copper-inner-surfaced utensils.

6. Store carefully; keep from exposure to heat or light; if for frozen storage, quick freeze.

And as for fruit juices themselves, keep another hint or two in mind: All fruits contain 3 *natural* sugars (glucose, fructose, sucrose) and rarely need sweetening. Get your family to like foods and beverages as nature flavors them. Add honey when actully needed—never white sugar or synthetic sweeteners.

If a large supply of perfect fruit is available, some of it may be canned whole by the cold-pack method. The same rules apply here as given above for preserving vitamins and nutrients in the fruits to be canned. Wash the fruit briefly, prepare it quickly and process it promptly to get the best nutritive value from it.

Fruits may be canned in one of two ways—by cooking before or after they are packed in the containers. If they are cooked by the open-kettle method, more vitamins are lost through oxidation, though the product may contain little or no water, since the juicier fruits may be cooked in their own juices. Cold-pack canning requires that the raw fruits packed in the jars be covered with a syrup made of water and sugar or honey.

For your health's sake, use honey and use it sparingly—rather than refined sugar in canning fruits. General recommendations include cooking gently in the heaviest syrup and stopping just before the fruit is tender. Then put it into the hot sterilized jars, filling them to the top and screwing on the lid until almost tight. The jars are then placed on a rack in the kettle, covered with water to at least an inch over their tops and processed in a rolling boil for 10 to 15 minutes before being removed and sealed tightly.

Freezing

Fruits lend themselves to freezing better, in most cases, than do vegetables, because fruits do not need to be blanched or cooked before they are frozen. It is true that certain changes take place in the texture of frozen fruits which is similar to the changes in cooked fruits. A few fruits, notably pears, watermelon, papayas, mangoes and avocados cannot be frozen very satisfactorily—some not at all. But wherever it is possible to freeze them, fruits can be made to retain more nutritive value and flavor when frozen than by any other method of preservation and the process usually takes only about half as much labor as canning.

Studies published by Ohio State University show that frozen foods retain their food value the same as fresh foods.

Vitamin A is practically all retained in fruits and vegetables during processing and freezing.

Vitamins B, C and G (riboflavin) are water soluble and tend to suffer some loss during the blanching and cooling of vegetables before freezing and in the leakage or drip after defrosting.

Vitamin C of cooked frozen or fresh foods is about the same. Vitamin C of fruits is well retained since fruits are not blanched, but vegetables lose variable amounts.

Changes brought about in the freezing of foods are admirably explained in a pamphlet by Faith Fenton, Geraldine Bryant, Carey D. Miller and Kathryn Orr published by the University of Hawaii. They explain the changes in texture as follows: It is known that a uniform distribution of the ice crystals in the fruit aids in the retenton of the texture. Texture change is largely due to the separation of water from the fruit tissues in the form of ice. Once separated, most of the water cannot be reabsorbed by the tissue.

The cellular breakdown or softening, which is characterized by loss in crispness and turgidity, is similar to the effect of cooking. Some fruits, such as papaya, are more subject to softening than are other fruits. Some varieties of a given fruit are also more subject to loss of texture than are other varieties.

One of the most effective ways of minimizing loss of texture in frozen fruits is to serve them before they are completely thawed and while some ice crystals remain.

Spoilage of foods, in general, is caused largely by enzyme activity, by the growth and activity of microörganisms and by decomposition of certain of the constituents through reactions with each other, with the oxygen of the air or with container walls. In freezing, the tissues are softend and the constituents are allowed to mix with each other. Furthermore, the natural resistance of the living plant tissue to the action of microörganisms has been broken down and the enzyme systems, which aid in the ripening, respiratory and other life processes, have been interfered with.

Microbial spoilage: In sound fruits, microörganisms are present only on the surface. As stated previously, fruits ripened on the tree or vine should be prepared and frozen immediately. If they must be held, refrigerate them.

The rate of growth of all microörganisms decreases as the temperature is lowered. Reducing the temperature to 40 degrees F. as quickly as possible will markedly retard the growth and activity of microörganisms and reduce the danger of fermentation and spoilage of the fruit during storage and thawing.

Precooling the fruit is important for obtaining excellent products. This may be done by refrigeration or, in the case of freshly picked berries and small fruits, by washing them in water containing ice.

Microörganisms are retarded in growth and activity during freezer-storage and many are destroyed. They are not all destroyed, however and those remaining may increase in number and activity during thawing and during holding of the fruit after thawing. In frozen fruit, the softening of the plant tissue permits a more rapid rate of spoilage after thawing than in the corresponding fresh fruit.

The type of spoilage naturally occurring in frozen fruits is fermentation. The changes are similar to those occurring in making fermented beverages and the products are not likely to be harmful.

Enzymatic activity: Changes in color, taste, aroma and ascorbic acid (vitamin C) content of fruits during freezing and thawing are caused mainly by oxidative enzymes. The discoloration of fruits and the resultant loss of characteristic flavor and the production of off-flavor are caused largely by oxidative enzymes. No browning will occur in fruit tissues until practically all the ascorbic acid has been oxidized to dehydroascorbic acid.

Low temperature retards, but does not entirely prevent, the activity of enzymes. To obtain high-quality products:

1. Select varieties which do not change readily in color, flavor and texture.

2. Freeze mature fruit. Immature fruit is usually higher in tannins and other constituents involved in darkening; some contain compounds which become bitter during freezer storage and thawing.

3. Handle fruit and fruit products quickly during preparation for freezing, packaging and partial thawing and serving to minimize exposure to air. Cut directly into the syrup any fruit that is likely to discolor.

Non-enzymatic activity: Even though the enzymes may have been inactivated, auto-oxidation (self-oxidation) may occur. It is difficult to control but it can be minimized by the exclusion of oxygen and by the addition of sugar or syrup. Off-flavors and off-colors may be caused by auto-oxidation.

Sometimes a white fondant-like material, which after thawing appears as white patches, forms on the surface of frozen fruits during storage. This is due to the crystallization of sucrose.

Fruits are usually frozen in one of two ways—dry, or floated in a sweet syrup. The usual directions found in most articles on the subject recommend freezing fruit, when dry, by mixing it with dry sugar or in syrup, by floating it in a sugar and water syrup. Nutritionists interested in natural foods recommend the substitution of honey for sugar in freezing fruits. In general, recipes for freezing with sugar may be converted to honey recipes by substituting one-fourth to one-half the amount of honey for sugar. This means that 2 to 8 tablespoonfuls of honey are mixed with one pint of fruit dry, where the sugar recipe calls for one-half to one cup of sugar. A honey syrup suitable for use with most fruits may be made by mixing 3 parts of honey with 5 parts of warm water and chilling before using. If desired, two tablespoonfuls of lemon juice may added, as an antioxidant.

One precaution must be taken into consideration in the use of honey for freezing fruits. Lucille Fisman, in an article in ORGANIC GARDENING AND FARMING, warned that, "Only the mild-flavored, high-quality honeys are suitable for this purpose. In our locality this is the spring or early summer honey which is locust or clover honey. I have found that the honey harvested in the late summer or in the early fall usually has a more pronounced flavor which will overpower the delicate fruit flavors. In fact, a few years ago I had to discard some fruit I had prepared with fall honey because even to us the honey flavor was too dominant. So you see it is important to find a reliable source of good, mild-flavored honey, but you will be pleasantly rewarded for any trouble you may take in order to obtain the proper honey for freezing fruits."

For the actual containers of fruits preserved by quick freeze methods, any may be used that will exclude air and prevent contamination and loss of moisture. Ice cream containers, waxed inside, are ideal. If tins cans with friction lids or glass jars are utilized, don't fill them to within more than an inch of the top as that space would allow room for the normal expansion of the food without accidents. Fruits which are to be packaged dry should be placed in moisture-proof cellophane bags, heat-sealed with a hot iron, at point close to the contents as possible so that there's a minimum of air in the package, or waxed containers could be used. It's advisable to overwrap cellophane packages with stockinette to prevent the plastic material either breaking or tearing.

Instruction for freezing specific fruits are given in the accompanying table. We have substituted honey for sugar in our recommendations for processing. Since the sweetening quality of different honeys differs, the processor should use his own taste to guide him in the final decision about quantities to be used.

TABLE 57: Freezing Fruits

Product	Varieties to freeze	Preparation	Packing
Apples	Yellow Transparent Early Cooper Lodi Duchess Wealthy Golden Delicious Jonathan McIntosh Spitzenburg Winesap Yellow Newton Rome Beauty Northern Spy Cortland Gravenstein	Slices: Peel, core and slice. Sauce: Grind whole apples, including skins, in blender. Juice: Wash fruit, crush, press out juice.	Pack dry or mix with 2 to 4 tablespoonfuls of honey mixed with 2 tablespoonfuls lemon juice. Add honey and lemon juice to taste. Add lemon juice to taste.

TABLE 57: Freezing Fruits (Continued)

Product	Varieties to freeze	Preparation	Packing
Apricots	Moorpark Tilton Royal Blenheim Chinese Riland	Blanch and peel or leave skins on. Cut in half and remove pits. Add a few pits to each container for flavor.	Trickle over honey thinned with warm water. Or pack in honey syrup.
Blackberries	Eldorado Lucretia Ebony King Wild berries Brainerd Evergreen Himalaya	If organically grown, pick out leaves and debris but do not wash.	Pack dry or trickle small amount of honey into bag. Seal and turn end for end until mixed.
Blueberries	Herbert Coville Dixi Atlantic Pioneer Cabot Concord Jersey June Rancocas Katherine Rubel Wild berries	If organically grown, do not wash. Pick out stems and leaves. If wild, scald for one minute to prevent toughening of skins.	Pack dry or with small amount of honey added.
Cantaloupe	Honey Rock Hale's Best Hearts of Gold	Cut flesh in slices, cubes or balls. Texture best when served partially frozen.	Add honey and lemon if desired. For salads, add whole seedless grapes.
Carambola		Wash and slice. The tough rind is not softened by freezing, so it is best used as a garnish.	Pack in honey syrup.
Cherries sour	Montmorency Morello Early Richmond Late Duke	Wash, chill, stem and pit.	Add a small amount of pure honey. Mix and pack.
Cherries sweet	Bing Lambert Black Tartarian Deacon Black Republican Royal Ann Schmidt	Wash, chill, stem and pit.	Add lemon juice or honey. Light varieties need more lemon than dark varieties.
Bush cherries	Keyapaha Oahe Teepee Wampum	Sort, wash, pit or pack whole.	Add honey to taste.
Coconut	Any variety	Drain out milk, remove hull and skin, grate or grind in meat chopper.	Pack dry. Will keep well for one year.
Cranberries	Early Black Howes McFarlin	Sort, wash and drain. Raw relish: Grind 1 orange with 1lb. berries. Add 1 cup crushed pineapple.	Pack dry. Add honey to taste.

TABLE 57: Freezing Fruits (Continued)

Product	Varieties to freeze	Preparation	Packing
Figs	Celeste Brown Turkey Magnolia Lemon Ischia California varieties	Whole or slices: Wash, sort, cut off stems, peel, leave whole or slice. Crushed: Wash and coarsely grind figs.	Cover with thin syrup. Add honey if desired.
Gooseberries	Red Lake Champion Houghton Glendale Pixwell	Sort, remove stems and blossom ends. Wash.	Pack dry or in honey syrup.
Grapefruit	Marsh Seedless Ruby Any seedless variety	Sections: Peel by cutting through white membrane. Remove sections from membrane. Juice: Squeeze juice, trying to avoid mixing in oil from rind.	Pack dry or add honey to taste. Pack and freeze quickly.
Grapes	Concord Lutie Lucille Any good eating varieties	Whole: Wash and stem. Leave seedless whole, cut in half and remove seeds from others. Juice: Crush and heat to 140° in top of double boiler. Extract juice.	Pack dry or in thin syrup. Thaw and strain before serving.
Guavas		Purée: Remove seedy portion and strain to remove seeds. Shells: Pare and halve. Slice if desired.	Sweeten with honey. Cover with thin syrup.
Lychees		Leave about ¼ inch stem on fruit. Wash.	Pack dry.
Mulberries		Stem. Wash if necessary.	Pack dry or in honey syrup.
Oranges	Any good juice varieties	Squeeze juice. Work quickly, but avoid mixing oil from rind with juice.	Pack and freeze quickly.
Passion Fruit		Wash fruit. Cut in half. Remove pulp and seeds with spoon. Extract juice through a cheesecloth.	Add honey to taste.
Peaches	Halehaven Elberta J. H. Hale Rio Oso Gem Champion Redhaven Sunhigh Triogem Golden Jubilee Georgia Belle Afterglow Eclipse Erly-Red-Fre Raritan Rose Gold Medal	Use only peaches ripe enough so that skins may be pulled off without blanching. Wash, skin, pit, freeze in halves or slices. Prepare only fruit enough for one package at a time, to avoid discoloration.	Add honey mixed with a small amount of lemon juice, if desired.

TABLE 57: Freezing Fruits (Continued)

Product	Varieties to freeze	Preparation	Packing
Persimmons	Japanese Native varieties	Sort, wash, slice and freeze or press through a purée sieve.	Add 2 tablespoonfuls of lemon juice per pint. Sweeten to taste with honey.
Pineapple	Smooth Cayenne	Use only fruit ripened on plant. Pare, trim, core and slice or cut in wedges. Juice: Cut pared fruit into 8 or more pieces. Squeeze in a poi cloth or flour sack.	Pack in own juice or in thin syrup. Freeze without adding sweetening.
Plums and Prunes	Methley Italian President Agen Sugar Imperial Epineuse Abundance Damson Toka Wantea Oka Stanley Reine Claude	If freestone, wash, pit, halve or quarter. If clingstone, crush slightly, heat just to boiling, cool and press through purée sieve.	Add 2 tablespoonfuls lemon juice per pint. Sweeten with honey or add honey syrup.
Raspberries red	Washington Indian Summer Cuthbert Latham Newburgh Lloyd George Chief Taylor Herbert	Clean and remove stems.	Pack dry or fill bags and trickle in 2 tablespoonfuls of honey per bag. Seal and turn end for end to mix.
black purple	Cumberland Bristol Dundee Munger Columbian Ruddy Sodus		
Rhubarb	Victoria Wine MacDonald Ruby Canada Red Valentine	Wash. Cut in 1-inch pieces.	Pack dry.
Soursop		Peel and cut lengthwise through the center. Remove and discard seeds. Force through purée sieve.	Sweeten to taste with honey.

TABLE 57: Freezing Fruits (Continued)

Product	Varieties to freeze	Preparation	Packing
Strawberries	Evermore Dunlap Gem Brightmore Dorsett Catskill Fairfax Improved Oregon Marshall Narcissa Redheart Rockhill Blakemore Pocahontas Howard 17 Robinson Sparkle Temple Klondyke	Wash, hull, slice, cut in half or freeze whole.	Sweeten to taste with honey or use honey syrup.
Surinam cherries		Prepare and freeze quickly. Wash and sort. Remove stems, blossom ends and pits.	Pack in honey syrup.

Section 4.

32—Organic Fruit Culture: A Complete Listing

ACEROLA

ACEROLA. *Malpighia glabra,* one of a dozen or more members of the tropical family *Malpighiaceae* whose fruits are edible. The acerola, or Barbados cherry, is grown widely as a dooryard fruit in Jamaica and other tropical American countries. It was found growing in the West Indies by the early Spanish explorers, but was not introduced to Florida until 1880. It is extremely tender and can only be grown in the frost-free areas of central and southern Florida and a small portion of southern Texas.

Fruit of the acerola is a round, berry-like drupe, bright red to crimson in color, about the size and shape of a cherry. Its seeds are contained in a 3-part stone, 3 angular pits lying together to look like a single stone. The plant is a shrub growing from 5 to 15 feet tall, but it may be pruned to a short trunk to give it a tree-like growth. Leaves are dark green, oval and glossy—evergreen in well-grown specimens. Its deep roots adapt it to poor soils, but it needs abundant moisture and good drainage. Some varieties are almost everbearing, from April through November bearing pinkish flowers which attract the bees and develop in 3 to 4 weeks into ripe fruits. Acerolas are thin-skinned and delicate, with a flavor more like crab apple or raspberry than like cherry.

Since the 1946 publication of results of experiments by two Puerto Ricans, Asenjo and Guzman, acerola has been the subject of much interest because of its extremely high vitamin content. One fruit of the acerola contains more than the minimum daily requirement of vitamin C, yielding a higher percentage of ascorbic acid than any fruit except *Rosa rugosa* rose hips. Fruits grown in Florida have been found to contain 1,030 to 3,309 milligrams of vitamin C per 100 grams of fruit. The riper the fruit, the less C it contains, until it reaches the lower figure quoted above. The larger quantities are found in green fruits, less than half-ripe. Also, more vitamin C is found in fruit harvested in August than in that which ripens in June and larger quantities in fruit ripened in the sun than that ripened in the shade.

Processed acerola products, though they contain less vitamin C than the fresh fruit, still contain phenomenal amounts in comparison with the same products prepared from other fruits. Canned juice, for instance, will retain 80 per cent of the ascorbic acid content of the fresh fruit for a year if it is stored at 45 degrees. When stored at room temperature, it loses 54 to 82 per cent of the vitamin C. However, it also loses flavor and color, no matter how it is stored. For that reason, frozen juices and pulps are preferred. Even jellies prepared from acerola contain as much as 500 milligrams of vitamin C per 100 grams, an amazing amount to survive the cooking necessary to make the jelly.

Planning the Planting

Several varieties, developed from clones of the best producers and those with the largest amounts of ascorbic acid, are now in process of cultivation in Puerto Rico and Florida. Two of these, now ready for distribution, seem outstanding.

Puerto Rican B-17 is a tart variety which gives high yields. Its ascorbic acid content averages 1,325 to 2,250 milligrams per 100 grams of juice. Growth habit of the bush is dense and drooping.

Florida Sweet Barbados cherry produces a large bright red fruit with a skin that is thicker than average, so the fruits keep better than most acerolas. It has a semisweet fruit, with apple-like flavor. Vitamin C content is about 1,500 to 2,000 milligrams per 100 grams of juice. The tree is open and upright in growth.

For the home garden, one or two plants are sufficient to provide plenty of fruit, especially of the varieties which have a long bearing season. A 7-year-old tree of the Florida Sweet variety may yield up to 170 pounds of fruit in a single year. Two-year-old trees will bear flowers and a small amount of fruit; in their third and fourth year they begin to bear heavily.

Soil

Acerolas will grow in either alkaline or acid soil, though if the *p*H is as low as 5.4, they will benefit from an application of limestone. Sand, marl, or clay have all been used to grow healthy specimens. More important than the composition of the soil is its drainage. Roots of the acerola will not stand flooding for more than a very short time, so the soil must be thoroughly drained to a good depth. Plantings in Florida have been entirely killed by floods which stood about their roots for two days.

Planting

Six to 12 month old plants may be set out between April 1st and the end of June in Florida. If they are planted alone, as specimens in a lawn or shrubbery border, they should be given a space which can be expanded to 12 to 15 feet square when they are mature. In orchards they may be planted in hedges, being spaced 6 to 8 feet apart or they may be interplanted among avocados, mangoes, lychees or other plants that stand 25 to 40 feet apart. Planted 2 to 4 feet apart in a border, they make a very effective hedge and can be kept in shape with a small amount of trimming.

Maintenance

A heavy mulch of grass, hay, straw, leaves, wood shavings, sawdust or similar materials will help to preserve the moisture needed by acerolas and will also tend to control root-knot nematodes. Water is not so necessary during the dormant season, from November to April, but during the early spring months when Florida gardens are dry, the plants need irrigation.

Not much phosphorus is needed by acerolas, but they require fairly large amounts of nitrogen and potash. Fertilizer should be applied 3 times during the year—in May just after growth starts, in July during the summer rains and after the fall pruning in September. A total of one ounce of pure nitrogen and one of potash should be applied each year for each year of the tree's age, up to 10 years, after which the trees should receive a total of 10 ounces of each per year. For example, to feed this amount of nitrogen and potash, it would be necessary to give each tree 6 ounces of blood meal and 14 ounces of wood ashes or tobacco stems or their equivalent, for each year of its age. Spread the fertilizer on the ground under the tree, covering an area one to two feet beyond the drip line of the branches.

Pruning

When they are well established, acerolas benefit by being pruned every year or two. Bushes with a dense growth habit should be opened out to permit the sun and air to reach their centers. The upright varieties may be headed back to prevent the tallest branches from growing up out of reach.

Pruning is best done in September or October when the trees will have a chance to make some growth before entering their dormant stage. An application of fertilizer after pruning will help to stimulate this growth. When pruning is done in early spring, yield the following year is definitely decreased.

Propagation

Acerolas may be propagated by air layering, by hardwood cuttings, by grafting or from seed. For the home gardener who wants only a few plants, either layered or rooted cuttings will provide the most dependable means of propagation. Varieties grown from seed may not give the most desirable fruit or growth habit.

An air layer is made on a stem whose bark is girdled, a ring of bark being removed. The cut surface is covered with damp sphagnum moss which is sealed in place with a sheet of vinyl plastic film. After 6 weeks, roots should have formed within the plastic bag and the shoot is ready to be severed from the parent plant. Pot it in a quart to a gallon size container and allow it to grow for several months to become established before setting out.

Cuttings are made from shoots 5 to 10 inches long, one-fourth to one-half inch thick, bearing 2 to 3 leaves. They may be rooted in sand, peat moss, vermiculite, crushed granite or wood shavings or a combination of several of these. Cuttings must be kept moist and shaded and will benefit from a frequent spraying to retard evaporation. Within two months the cuttings should develop enough roots to permit transplanting to containers.

A bushy tree native to tropical and subtropical regions of the Western Hemisphere, acerola produces a cherry-like fruit rich in vitamin C.

To start seedlings, clean off all flesh from the stones and permit them to dry. Plant the whole stones without removing the seed from the pit. The seed is very delicate and is almost always injured when an attempt is made to open the pit. Germination may be poor from fruits from some plants, because not all embryos are viable and a 50 per cent failure may result from planting these pits. Plant the seed in flats and transplant when seedlings are 2 to 3 inches tall to containers in which they may develop before being set out.

Grafting is usually used to start plants where root-knot nematodes are a problem. Scions of *M. glabra* are sometimes grafted to *M. suberosa,* which is fairly resistant. Cleft graft or side-veneer graft may be used. Grafting is also sometimes used to propagate desirable clones which are hard to root from cuttings.

Enemies

Root-knot nematodes are the chief enemies of acerolas when they are grown in soil lacking in organic matter. Best preventive is the maintenance of a deep year-round mulch, which encourages the growth of a fungus which preys on the nematodes. Where the condition is very bad, it may be necessary to graft acerola to a resistant root.

A few scales also attack the fruits in some areas. A dormant spray of oil emulsion may be used to check the condition if it becomes serious.

Harvest

During the fruiting period, acerolas should be picked every other day. The fruit ripens quickly and may be picked when slightly green or half ripe for best ascorbic acid content. Being very delicate, it should be used as soon as possible after picking.

ACTINIDIA KOLOMITKA

This is one of a family of woody vines grown chiefly in Asia. The kolomitka is hardy through the central section of the

United States, and as far north as Long Island on the Atlantic seaboard. Its fruit, a seedy, sweet, greenish-yellow berry about an inch long, is about 15 to 20 times as rich in vitamin C as the lemon, which itself is considered an excellent vitamin C source.

The kolomitka vine grows 6 to 10 feet high and bears fragrant white flowers, male and female on separate vines. Plants bearing each kind of flower must be grown together for fruit production. The vines are quite ornamental, some having leaves which are variegated with blotches of pink or white.

Yields are good and grow larger as the plants grow older. On an 8-year-old plant, a crop of 55 pounds of fruit may be harvested in a single season. Older plants may bear as much as 80 pounds. The fruit has a slight flavor of pineapple which is most discernible when it is dried, like raisins. It may also be eaten fresh, preserved or cooked and puréed before canning. In the fresh or cooked state it tastes like gooseberry.

The vines have attractive foliage which makes them desirable as ornamentals in the home garden. They are especially useful as screens or they may be used to cover trellises or arbors.

APPLE

Apple. *Malus pumila,* sometimes classified as *Pyrus malus,* is probably the most important food member of the rose family, *Rosaceae*. Apples are native to temperate sections of Europe and Asia and have been used for food since prehistoric times. Carbonized apples have been found in the remnants of the Swiss lake dwellings. Though the best apples have never been grown in the warm climates, they were cultivated and treasured by both the Phoenicians and Greeks. The Romans, when they penetrated north of the Alps, expressed amazement that there were a few isolated spots where apples were unknown. By the seventeenth century, a catalog of apple varieties included more than 500 kinds. Today the number is well into the thousands.

Apples are grown as one of the principal fruit crops in almost every state in the United States. A soil and climate to suit them may be found in all but the hottest sections. Almost any soil except very sandy or mucky will produce a crop. Long hot summers, warm winters providing too short a dormant period and winter temperatures which consistently drop below 20 degrees F. below zero, are the only unfavorable weather conditions.

Varieties have been developed which will grow in each section of the country. A few versatile apples, such as Golden Delicious, will grow from Florida to Minnesota. Others, like Red Astrachan, are suited only to the North; or like Yellow Newtown, thrive best in the Central States.

A gardener who chooses his varieties with care and plants with the thought in mind that he is planting for himself, his children and his grandchildren, will find the work and time investment involved in starting his home orchard small compared to the return. Apple trees are planted to last at least 35 years and their most productive years are usually from the 10th to the 30th. During these years a productive standard size tree may be expected to yield 7 to 10 bushels per year. A dwarf tree may yield 1 to 5 bushels per year.

Different varieties of apples provide varying amounts of food value, their supply of vitamins, minerals and sugar depending not only upon the soil on which they are grown, but also upon the amount of color in their skins, the amount of sunlight which they can absorb on the tree which in turn depends upon the openness of its growth habit and probably also upon the length of time which they need to mature. Of the varieties which we list, Yellow Newtown, Baldwin and Northern Spy contain 15 to 20 milligrams of vitamin C per 100 gram apple, a small apple. McIntosh, on the other hand, contains only about 4 milligrams of vitamin C in an apple of the same size. On the average, about 90 milligrams of vitamin A, .036 milligrams vitamin B_1, .050 milligrams vitamin B_2, 7 milligrams calcium, 12 milligrams phosphorus and 64 calories will be found in the same small apple.

Most of the vitamin C content of an apple is just under the skin. Consequently, small apples will be richer in C than large ones, because the proportion of skin to flesh is greater. Of course, if the apples have been so thoroughly sprayed as to make eating the skin unsafe, then the skin, vitamin C and most of the minerals must be pared off and thrown away.

TABLE 58: Apple Varieties

Listed in Order of Their Popularity and General Usefulness

Variety	Section for which recommended	Fruit — season and description	Tree — characteristics	Use and keeping quality
Delicious (Includes Red Delicious, Starking, Richared)	New England Great Lakes Northern plains Southern plains Mountain States Appalachians Southeast Pacific States	Winter fruit, medium to large size, sweet, striped red to solid red.	Not fully hardy. Narrow weak crotches. Susceptible to scab. Self-sterile but excellent for pollinating other varieties. Upright habit.	Excellent for eating raw. Poor for cooking. Keeps to March but deteriorates in storage.
Golden Delicious	New England Great Lakes Northern plains Southern plains Appalachians Corn Belt Southeast Pacific States	Winter, ripens late and requires long season on tree. Medium to large bright yellow fruit. Mild subacid flavor.	More vigorous growth and better branching habits than Delicious. Good pollinator. Moderately hardy. Moderately resistant to scab, susceptible to cedar rust.	Excellent for both cooking and eating. Keeps to March but may shrivel in storage.
Winesap (Includes Stayman Winesap)	New England Great Lakes Northern plains Southern plains Appalachians Corn Belt Southeast Pacific States	Early blooming, fall ripening. Medium size, red. Moderate amount of vitamin C. Juicy, subacid.	Hardy. Trees bear when young. Susceptible to scab. Pollination poor. Resistant to cedar rust.	Good for eating, fair for cooking. Good keeper.
Jonathan (Includes JonaRed)	New England Great Lakes Northern plains Southern plains Corn Belt Mountain States Pacific States	Winter. Small to medium size. Red. Fruit subject to Jonathan spot but moderately resistant to scab. Subacid flavor.	Tree late in bearing and long-lived. Not fully hardy. Good for pollination. Susceptible to cedar rust and fire blight.	Very good for both cooking and eating. Keeps to February. Resistant to bruising.

TABLE 58: Apple Varieties (Continued)

Listed in Order of Their Popularity and General Usefulness

Variety	Section for which recommended	Fruit — season and description	Tree — characteristics	Use and keeping quality
Yellow Transparent	New England Great Lakes Northern plains Southern plains Southeast Pacific States Alaska	Extra early. Medium size, pale yellow fruit. Tart to subacid. Mealy when overripe.	Most versatile and hardy variety. Bears when very young. Moderately resistant to scab.	Flavor tart for eating, but good for cooking. Bruises easily.
Lodi (Developed from Yellow Transparent)	New England Great Lakes Northern plains Southern plains Corn Belt Appalachians Southeast	Early. Large yellow fruit. Ripens 5 to 7 days after Transparent. Acid.	Almost but not quite as hardy as Transparent.	Excellent for cooking and eating. Less susceptible to bruising and keeps better than Transparent.
McIntosh	New England Great Lakes Northern plains Corn Belt Mountain States Pacific Coast	Fall. Medium size, red-striped fruit. Low in vitamin C. Subacid, aromatic and juicy.	Moderately hardy. Annual bearer. Trees bear when young. Susceptible to scab. Very spreading habit.	Very good for eating, fair for cooking. Bruises easily. Keeps to January.
Wealthy	New England Great Lakes Northern plains Mountain States Corn Belt Pacific States	Fall. Medium size fruit, smaller on old trees. Green, striped with red. Crisp flesh, subacid.	Bears when young. Hardy, but susceptible to fire blight, cedar rust. Tends to make weak crotches and to bear biennially.	Good for cooking and eating. Poor keeper.
Cortland	New England Great Lakes Northern plains Corn Belt	Fall to winter. Medium size, red fruit. Subacid.	Moderately hardy. Spreading habit.	Good cooking and eating. Keeps to January.

TABLE 58: Apple Varieties (Continued)

Listed in Order of Their Popularity and General Usefulness

Variety	Section for which recommended	Fruit — season and description	Tree — characteristics	Use and keeping quality
Minjon	New England Great Lakes Northern plains Mountain States Corn Belt	Fall. Small to medium red fruit.	Very hardy.	Very good for cooking, fair for eating. Keeps to mid-December.
Melba	New England Great Lakes Northern plains Corn belt	Early. Medium size fruit, red over yellow. Aromatic, subacid.	Hardy.	Excellent for eating, fair for cooking. Bruises easily.
Oriole	New England Great Lakes Northern plains Corn Belt	Summer, early. Large green-yellow fruit with red blush.	Hardy, but susceptible to fire blight.	Excellent for cooking and eating. Bruises easily.
Rome Beauty	New England Great Lakes Mountain States Appalachians Pacific States	Late blooming, late ripening winter apple. Large red fruit. Subacid.	Not fully hardy. Bear when young, small, short-lived trees. Good pollinators. Susceptible to scab, powdery mildew and cedar rust.	Excellent baker. Not an eating apple. Keeps to March.
Victory	New England Great Lakes Northern plains Corn Belt	Late winter apple, medium size, part red, with aromatic, slightly acid flavor.	Similar to McIntosh, but more hardy.	Excellent for cooking and eating. Keeps to March.
Fireside	New England Great Lakes Northern plains Corn Belt	Late winter apple, large, sweet, red over yellow.	Hardy.	Excellent for eating, fair for cooking. Keeps to May.
Beacon	New England Great Lakes Northern plains Mountain States	Early medium size, red, juicy, subacid.	Very hardy but susceptible to cedar rust.	Good cooking, fair eating, keeps 4 weeks. Bruises easily.

TABLE 58: Apple Varieties (Continued)

Listed in Order of Their Popularity and General Usefulness

Variety	Section for which recommended	Fruit — season and description	Tree — characteristics	Use and keeping quality
Grimes Golden	Northern plains Appalachians Corn Belt	Fall, yellow, medium size, aromatic, subacid.	Resistant to scab and cedar rust. Good pollinator. Vigorous medium size tree, bears when young. Buy double-worked stock to avoid phytophthora collar rot.	Very good for cooking and eating. Keeps to January.
Duchess	New England Great Lakes Northern plains	Fall. Medium size, tart, red stripes on yellow.	Hardy.	Good for cooking. Tart for eating. Bruises easily.
Northern Spy	New England Great Lakes	Winter. Large striped fruit, excellent subacid flavor, rich in vitamin C.	Upright, hardy, healthy. Biennial bearer, some not bearing until 12 to 15 years old. Susceptible to scab, bitter pit.	Good all purpose fruit. Keeps to March. Resists bruising.
Mantet	New England Great Lakes Northern plains	Early summer. Yellow, lightly striped, medium size.	Hardy.	Excellent flavor, good cooking, excellent keeper.
Hibernal	New England Great Lakes Northern plains	Fall. Medium size, acid, green with dull red stripes.	Very hardy.	Good for cooking. Tart for eating. Keeps 1 month.
Northwestern Greening	New England Great Lakes Northern plains	Large, late, green or yellow, juicy, slight aroma.	Not fully hardy. Resistant to fire blight, cedar rust.	Fair eating. Good cooking. Keeps to May.
Haralson	New England Great Lakes Northern plains	Late winter, medium size, red, tart.	Hardy. Biennial bearing.	Good eating. Excellent cooking. Keeps to March.
Prairie Spy	New England Great Lakes Northern plains	Late winter. Large, red, mild flavor.	Hardy.	Excellent cooking, eating. Long keeping.

TABLE 58: Apple Varieties (Continued)

Listed in Order of Their Popularity and General Usefulness

Variety	Section for which recommended	Fruit — season and description	Tree — characteristics	Use and keeping quality
Gravenstein	New England Great Lakes Pacific Mountains	Summer. Large, green to orange, striped red. Aromatic, acid.	Subject to winter injury in coldest areas. Biennial bearing, poor pollinator.	Excellent for cooking. Keeps to November.
Red Astrachan	New England Great Lakes Pacific Mountains	Summer. Medium size, red, subacid.	Moderately resistant to scab, large, productive, short-lived tree.	Good cooking and eating. Poor keeper.
Patten Greening	New England Great Lakes Northern plains	Medium to large, fall, yellow-green with red blush.	Hardy.	Good for cooking. Keeps to December.
Red June	Pacific coast Southeast	Early, yellow-red to crimson, sprightly flavor.	Susceptible to scab.	Good for eating.
Yellow Newtown	Pacific States Appalachians	Medium to large, very late, green-yellow with slight blush, juicy, firm, rich in vitamin C.	Not very hardy or productive. Susceptible to scab.	Excellent cooking, eating. Long keeping.
Early Harvest	Southeast	Early, yellow, medium size, acid to subacid.	Vigorous, spreading habit.	Good cooking and eating.
Baldwin	New England Great Lakes	Late, large, red over yellow, sweet, rich in vitamin C.	Fairly hardy, but not entirely. Moderately resistant to scab. Poor pollinator.	Very good cooking. Fair eating. Keeps to January.
York Imperial	Appalachians Corn Belt	Late, medium to large, red over yellow, subacid, low in vitamin C.	Good pollinator, not very hardy, scab resistant but susceptible to cedar rust.	Very good cooking and eating. Keeps to April.

Apples stored at 36 degrees lose none of their vitamin C in storage. If the temperature of the storage room rises to 45 degrees, some vitamin will be lost. Recommended storage temperature is usually from just above 32 to 35 degrees.

Ordering Stock

The home orchard designed to provide fruit for the family may consist of a dozen trees, standard size, of different varieties, or of one or two dwarf trees, depending upon the space and time that the gardener can give it.

All apple trees offered by nurserymen are grafted or budded on seedling roots or root pieces. The types of roots limit the tree's mature size. Dwarf trees are usually grafted on Malling stocks, many of which have been tried and a few chosen for their special qualities. If the trees will be planted in a small home plot, they will fit the scheme best if they are dwarfs. Trees grafted to dwarf rootstocks also have the advantage of bearing while young and of being easy to prune and harvest without special equipment.

A standard apple tree requires a space about 40 by 45 feet. A dwarf tree grafted on Malling IX stock needs 12 by 12 feet. Some half-dwarf trees, grafted on Malling VII stock, may be available to fill a space 20 by 20 feet.

All trees should be allowed enough room so that when mature their branch tips will not meet. Roots spread slightly further than the branches and they should not need to compete for available food.

During the first few years of the apple tree's life in the garden, temporary plantings may be made between trees in the space which the trees will eventually occupy. This planting should always be kept at least 5 feet from the young tree on all sides and is best if it consists of vegetables or fruits which have confined root systems. It should also be arranged that these plants will not interfere with sun, rain, or air which the young tree needs. Extra nitrogen should be fed to interplanted young trees.

Best fruit setting will be obtained from two or more varieties of apples planted together. Some varieties are self-sterile, others have very weak pollen. Even those varieties with good pollen will be most successfully fertilized if they are planted near a different variety which blooms at the same time. When a variety with known weak pollen is planted, two others with good viable pollen should be planted near, to pollinate the tree with weak pollen and to pollinate each other.

When choosing varieties to plant, it is well to remember that a tree which would produce its crop in the fall if planted in New England may mature fruit in late summer in the Appalachians or in midsummer in the Southeast. Particularly favorable sites within a section will have the same effect of speeding the crop to an early maturity. On the other hand, a site which warms up slowly in spring, if it cannot be avoided, may have the opposite effect, so that an apple which matures late elsewhere in the vicinity may never reach maturity in the cold orchard.

Nurseries commonly offer trees that are 1, 2, or 3 years old. Best for the home gardener is a large one-year-old whip or a medium sized two-year tree. After 4 or 5 years, the one-year whip will have produced as large a tree as the older, more expensive one and, if properly pruned, the tree will be stronger and better shaped.

Orchard Site

If there is a choice of sites, the home gardener should plant his trees on a well-drained plot where the soil consists of sandy to heavy clay loam, with a permeable but not sandy or gravelly subsoil. An elevated exposure is good, though a slight elevation next to a large, shallow valley is better than a high elevation next to a steep, narrow valley. Bottom land, where the spring frosts are cradled long after the heights are drained of cold air, is to be avoided.

Best soil is sandy loam over clay subsoil. A sandy loam may be enriched with as much organic matter as the gardener can spare. The organic matter will absorb water, while the excess drains away through the sand. If a hole in a dense clay soil is enriched with large quantities of organic matter, it will become a basin of wet muck in rainy weather and the trees may drown in the mud before the water drains away. Better, in a heavy clay, to leave the soil around the roots heavy, so that it will absorb no more water than the surrounding area. Best of all is to make a tile drainage system at the bottom of the root hole to carry off excess water and then to add organic matter to the soil around the roots.

Apple trees thrive on soils on which common cereals and potatoes grow well. They do not necessarily do well on corn land. Soil should be almost neutral, with a *p*H of 5.5 to 6.5. In the lower range, it is best to add limestone. Lime also has an inhibiting effect on some of the larvae of apple pests which winter in the soil. Nitrogen content of the soil should be moderate, but potash and phosphorus may be comparatively low.

Apple trees do not thrive as well when planted in grass lawn as when they are planted where a heavy straw mulch can be maintained. If they must be planted in the grass of a home lawn, a 6-foot circle around the base of the young tree should be left clean-cultivated. If possible, plant early summer varieties away from the house and place only the late varieties near the living area. Apples which fall from the trees interfere with lawn mowing and attract fruit flies which can be a pest indoors.

When planting dwarf trees, choose a site where they will be protected from strong winds. Dwarfs have shallow root systems and should have a strong support to hold them upright during their first few years. Their shallow roots also make them more sensitive to drought than the larger trees and they need more frequent watering.

Planting Trees

Apple trees may be planted in either spring or fall. But whether the job is done early or late, it must be done at a time when the leaves are naturally off the tree and before the buds start to swell.

Before planting cut back any broken roots until the cut shows vigorous tissue. Any extremely long roots may be shortened to bring them within reasonable distance of the rest of the root system. Dip the roots into a thin mud mixture to protect them while planting. Dig a hole big enough to accommodate the roots, when they are fanned out, without crowding or cramping their spread. Set the tree two inches deeper than it was growing in the nursery and incline it slightly toward the direction of the prevailing wind. If the tree is grafted on dwarf stock, take care to keep the graft above soil level. When the graft is covered with soil the top stock may root and the tree will grow to the size of a standard tree. If the tree is two years old or more, turn the most heavily branched side toward the wind.

Fill the hole carefully, leaving no air pockets. Topsoil should be placed in the bottom of the hole, around the roots and the subsoil used to fill the hole to an inch below soil level. Puddle well with plenty of water, shake the tree to settle it and add a little more soil when the water has drained away. Protection against mice and rabbits in the form of an 18-inch collar of hardware cloth should be placed around the trunk immediately. Next spread 3 bushels of rotted manure around the tree in a 6-foot circle, and mulch the same area with 8 to 10 inches of hay or straw, leaving bare a circle one foot in diameter around the trunk.

Maintenance

Experiments have been carried out by the agricultural stations in several states to discover the best treatment for young fruit trees and orchards. Trials in the New Mexico Experimental Station showed that trees grown under mulch, as described above, grew much larger and more healthy than those grown in sod, in clean cultivation, with chemical weed killers or with a nitrogenous fertilizer. Further experiments carried out by the Beltsville, Maryland, station showed that a mulch of alfalfa hay produced better trees than a wheat straw mulch. A heavy mulch of alfalfa, renewed every year, provides enough nitrogen for best growth of the young tree. If it is mulched with straw, the tree should be given a yearly feeding in early spring of a nitrogenous fertilizer, such as cottonseed meal, dried blood, sludge or manure.

Renewed application of nitrogen in some form is necessary every year for best apple production. The fertilizer should be applied 6 to 8 weeks before the tree blooms. It should never be applied as late as June, because of the danger of stimulating a late growth which will not harden sufficiently before the following winter.

If the nitrogen is to be supplied to the tree from mulch, it will be necessary to administer supplemental nitrogen for the first 2 or 3 years after the mulch is laid down. Experiments show that a period of about 2 to 3 years are needed to break down a mulch and convert it into a form that can be used by the tree as food.

Nitrogen should be supplied, for minimum maintenance, at the following rate for each year of the tree's age: activated sludge, 12 ounces; bloodmeal, 4 ounces; cottonseed meal, 8 ounces; manure, 8 pounds. To encourage biennial trees to bear annually and to produce larger and heavier crops on all trees, that amount may be doubled. Fertilizer should be spread under the mulch in a band starting one foot from the trunk and extending out to the branch tips or a little beyond.

If trees are heavily fertilized with nitrogen, the color of the apples may suffer. It has been found that with large applications of fertilizer, the color remains more green in the ripe apples, although the food value seems to equal that in more highly colored fruit.

When large quantities of nitrogen are used, it is sometimes necessary to feed additional potash to the tree. After a heavy nitrogen feeding, watch for signs of potassium deficiency, curling and spotting of leaves and, in extreme cases, a burnt appearance near the stems. These signs should never appear in a tree fed with heavy mulch, because the green matter contains a balanced diet of nitrogen and potash. But a feeding of nitrogenous fertilizer may also necessitate the application of wood ashes, bone meal or potash rock. These will make potash available to the tree roots in varying lengths of time, the wood ashes leaching out most quickly and the potash rock moving most slowly through the soil. Old orchards almost always need potash. Heavy applications of wood ashes or bone meal are best, though it may take two years for the food to reach the deep roots of mature trees.

The mulch which feeds the apple tree also protects its water supply to such a degree that, in all but the most arid sections, irrigation is unnecessary. However, there are parts of the Southwest and Western states where there is not enough rainfall, yet apples are so successfully grown commercially as to form a major crop. Orchardists in these sections irrigate their trees at least once every 30 days during the growing season, with the last irrigation timed for mid-August. It has been found that trees are less susceptible to winter injury if water is withheld during the fall months. In some areas where the soil is very sandy or where winter winds may become unseasonably warm or drying, it may be necessary to irrigate once or twice during the dormant season. Though evaporation at that time is at a minimum, there is always some evaporation going on, even when the leaves have dropped. And roots may become dangerously dry where the drainage is good and the winter rainfall insufficient.

Pruning

A one-year whip, 5 feet long, is ideal for planting and training in the home garden and will develop within 4 or 5 years to a tree that is ready to bear.

Experiments have shown that too severe cutting back of the top is unnecessary if the roots are in good condition. After it has been planted, the whip should be cut back to about 44 inches. In June or later, when the laterals have grown several inches long, select 4 strong ones for the framework of the tree. They should be well spaced from top to bottom of the whip and should be distributed well around the circumference of the tree. Remove all other laterals, rubbing them off with the fingers and permit the 4 selected shoots to grow. During the first summer it will probably be necessary to remove shoots at two week intervals several times, until the selected shoots have made a strong growth and have outstripped all competition. These 4 shoots will become the primary laterals in the mature tree. The topmost lateral will become the leader.

Late the next winter, prune the 4 laterals to unequal lengths, leaving the leader longest. No further pruning should be done the second year. In later prunings, the primary laterals will be favored instead of the leader, to prevent the tree from becoming too tall. This means that when there is conflict between secondary branches, the small branches should be removed from the leader, rather than from the laterals. In pruning varieties which have a naturally drooping habit, prune to buds that point upward. Trees of erect habit should be pruned to buds that point downward.

Pruning should be light after the first year or two, when the main framework should be making strong growth. After the pattern for the tree is set, the principal pruning will be confined to removing branches that cross, those that close in the center of the tree, shutting out sun and air and those that are damaged or killed by wind, cold or disease. As the tree approaches bearing age it develops short stubby twigs with rounded "'cluster" buds at the tip. These are fruiting spurs which must not be removed or damaged. Each spur may bear fruit for 10 years or more. Many blossoms will be produced on each spur and several fruits may set. However, most of them will drop in the so-called "June drop," which is the tree's own way of pruning its crop to a size that it can nourish.

Renovating Old Trees

An old tree that has been neglected may sometimes be restored to production if it is still fairly vigorous in its growth. If the trunk or main branches are badly rotted or if the top is one-fourth dead from disease or winter injury, it will not be worthwhile trying to save it. Such a tree should be taken out. To leave it in its neglected state is to provide a haven for many orchard insects and diseases.

If the old tree needs much pruning, the job should be done during a period of 2 or 3 years rather than all at once. Too severe pruning of the top in one year may result in a large crop of suckers the following summer or even the death of some of the roots.

Wherever possible, in pruning an old tree, cut out the older wood and leave the young vigorous branches. Remove all suckers not necessary to replace the top. Prune out interlacing branches to open the center of the tree to light and air.

Trunks and large branches of most old trees are covered with a thick layer of crumbly, dead bark. This dry cork provides a hiding place and breeding ground for many apple tree enemies. As much dead bark as possible should be scraped off, the scrapings gathered in a tarpaulin spread beneath the tree and burned. A hoe, wielded with care to avoid digging into the live bark below, is the best tool for this job.

In early spring, soil under the old tree should be broken up with a fork and a two-inch layer of compost spread over the soil to the tips of the branches. This should be lightly worked in with a fork or rake. A heavy application of cotton-

Apples are grown throughout the United States, the number of varieties ranging into the thousands. A standard-size tree yields 7 to 10 bushels per year; a dwarf yields 1 to 5.

seed meal, dried blood or sludge should be spread over the surface—25 to 35 pounds, depending upon the size of the tree. Finally, spread a 10-inch hay mulch over the soil. Each spring for the next two years this treatment should be repeated. Within 3 years, the old tree should be bearing well.

Propagation

Few home gardeners find it worthwhile to grow their own trees from seedling roots and grafted tops. However, a gardener with apple trees may find a knowledge of grafting useful at times. A planting which contains no viable pollinating varieties, for instance, may be made much more productive if a few branches of good pollinators are grafted to one or more trees. Occasionally winter damage may be minimized by a few grafts, applied in early spring before rot has set in. In extreme cases, a whole tree may be grafted to change its variety, over the course of a few years, with the least loss of production.

Whip grafting or cleft grafting are most useful for repairing winter damage. (See Chapter 29 on propagating fruits.) Either of these two methods or budding may be used to graft on pollinating branches. Budding, which is performed in midsummer, is perhaps to be preferred, because the buds may be selected and marked in spring during the blossoming period, to ensure their blooming at the same time as the tree to which they will be grafted.

Enemies

Some varieties are less attractive to insects than others, possibly because of their growing vigor. Ben Davis, an old stand-by now seldom planted, was thought to be able to withstand insects. The Ben Davis tree was used to develop Cortland, which has many of its parent's virtues. Late varieties, such as Winesap, Jonathan and York Imperial, are less attractive to apple maggot, a pest which destroys many fruits on earlier varieties. It is well to separate early and late varieties as far as possible, to prevent the insects which infest the early trees from spreading to the later, more resistant ones.

One spray which an organic gardener can use on his trees without poisoning his crop or his soil is the dormant oil spray, applied in late winter before the buds begin to swell. Because it is applied so early, the oil spray material will all have evaporated before the leaves or blossoms open and will not infect the food. To be effective, an oil spray must coat every part of the twigs, branches and trunk of the tree. It is effective against many insect eggs which are under crevices in the bark and against scales, aphids and mites. Preparations sold as miscible oil sprays are emulsified preparations which mix easily with water.

To combat all the other pests, for which some commercial orchardists use as many as 20 to 30 applications of spray materials, an organic gardener must rely upon the robust good health of his trees, plus a few mechanical aids and strict sanitary measures. Dropped fruit, whether it is fallen because of disease or because of the normal "June drop," should be removed to the compost heap immediately. A close watch must be kept on the trees at all times for the first signs of an infestation. Many times if the first few pests are found and killed, a much greater infestation may be avoided.

Trees should be examined carefully in June, after the fruit has set and any deformed or infected fruit should be removed immediately. This will leave more of the tree's nourishment for the perfect fruits. Also, deformed fruit invites attack from insects.

Codling moth is the principal enemy of apple trees in many sections. Larvae of the moth develop from eggs laid in the blossom end of the apples in June, work their way up the core and out to the skin. After leaving their trail through the apples, they make their way to the rough bark where they spin their cocoons in crevices. In June the moths emerge to lay a new generation of eggs.

A nineteenth century method of combatting codling moth, before the advent of DDT, was to build bonfires in the orchards on June evenings. The moths were supposedly attracted to the flames and consumed. Whether or not it was effective against codling moth, the bonfires must have provided many pleasant June evenings.

A mechanical control which will help to cut down on next year's moths is a band of corrugated paper fastened around main branches and the trunk. The larvae spin their cocoons inside the corrugations during the summer and then the paper is removed and burned.

Powdery mildew may damage many

trees in a season when the rainfall is heavy and the nights are cool. Usually the mildew attacks only the leaves and may be destroyed if the fallen leaves are carefully collected in fall. Occasionally, however, some of the twigs and branches may be affected. The best time to detect this infection is in fall, just before the leaves drop. Examine the leaves for the telltale whitish coating of mildew. Where it is found, the twigs and branches should also be examined. Any badly infected ones should be removed.

Scab, a fungus which makes olive-brown round velvety spots on the fruit and leaves, may cause heavy damage in some areas. Many resistant varieties are available and should be planted when possible. The best weapon against scab in an old tree is strict cleanliness. Infected fruit should be picked and removed; leaves and fallen apples should be promptly raked up. Dormant oil spray may also help to inhibit the fungus.

Apple fly is another source of apple "worms." Maggots of the apple fly are gray-yellow and tunnel into the fruit during the summer. The greatest numbers of flies are to be caught in the trees between July 15th and August 15th, though sometimes they remain in large numbers well into the fall. Traps to catch the flies are the best method of combatting them.

A trap may be made from a wide-mouthed jar, with a coarse-meshed hardware cloth cover. A string may be attached to a wire around the jar's neck and the trap hung on the sunny side of the tree. The sun should strike the trap. Bait for the traps may be made as follows: Mix 1 part molasses, 9 parts water and one tablespoonful yeast per gallon. Allow the mixture to ferment 48 hours. Fill the jars half-full and hang in the trees.

Cedar rust and fire blight are diseases which plague some sections of the country. Varieties that are resistant to one or both of these fungi have been developed. Gardeners who live in areas where the diseases are endemic should plant resistant trees, rather than relying upon poison sprays which make the fruits as unfit for human as for fungus consumption.

Experiments performed at Michigan State Agricultural Experiment Station have shown that fungicides applied to the trees within 10 days following full bloom —a period when the orchardists are busiest spraying—can inhibit fruit setting. Even when pollination conditions are most favorable, that is, when the temperature remains above 42 degrees for 7 days after full bloom and when the trees get plenty of sunlight, fruit setting can be slowed or stopped if fungicides are applied. So work is constantly going forward in the development of resistant trees and new resistant varieties are constantly being evolved. The home gardener should consult with his county agent before ordering trees, to find out which diseases are most prevalent in his section and which apple varieties are most recommended.

Harvest and Storage

Summer apples are picked when ripe, usually during a 2 or 3 week period for each tree and are used or preserved at once. Few summer apples can be kept for more than a few days in storage. Fall and winter varieties should be picked at peak ripeness for best storage.

Fruit which is to be stored must be picked with the greatest care to avoid bruising or breaking the skin. No attempt should be made to store imperfect or bruised fruit.

When picking apples, be careful not to break off the fruiting spur. Fruit may be borne on the same spur year after year if it is undamaged. Pick by cupping each apple in the hand with the thumb against the stem. Lift the fruit up and sideways, twisting it slightly. Pick with the stems, which should be left in apples which are to be stored. If the stems are removed, a break in the skin is left which will let in bacteria and cause rot.

Apples may be stored for 2 to 3 weeks at 60 to 70 degrees, but they will lose some of their vitamins at that temperature. Best storage is just above 32 degrees, in a moist atmosphere. Apples should never be permitted to freeze in storage and the temperature should be fairly even. They will keep best if stored in shallow baskets or trays containing a half bushel or less. Weight of a large number of apples may cause bruising in the lower layers of deep containers and consequent rotting.

If the apples are used to make cider, they should be washed thoroughly before pressing. Cider will keep longest without pasteurization or the addition of chemical preservatives if the equipment is sterile and only clean, sound apples are used. After pressing, the liquid should be filtered to remove dirt and pulp. Best filters are those made of diatomaceous silica.

APRICOT

APRICOT. *Prunus armeniaca,* a member of the stone-fruit genus of the rose family, *Rosaceae*. Apricots are thought to have originated in the Caucasus. Early reports of them come from Armenia, Arabia and the higher portions of central Asia, centering around the land of the healthy Hunzas, who have grown them for more than 1,500 years. In fact, they have become so much a part of Hunza culture that they have given rise to a unique custom there. Apricot trees are inherited by the Hunza women, independently of their brothers and regardless of the ownership of the land on which the trees stand. The climate in the Hunza territory is very similar to that in some parts of our own Rocky Mountain area, where apricots grow well.

Perhaps one of the reasons for Hunza health is the large vitamin content of their apricots. Dried and fresh fruits both contain large quantities of vitamin A, fair quantities of the B vitamins, and some calcium and iron. In 3 fresh apricots weighing 100 grams, there are 7,500 units of vitamin A, .033 milligrams of vitamin B_1, .100 milligrams vitamin B_2, 4 milligrams vitamin C, 13 milligrams calcium, 24 milligrams phosphorus, .6 milligrams iron and 70 calories, all from sugar. In dried apricots the nutrients are concentrated to provide 13,700 units of vitamin A in 100 grams, .96 milligrams vitamin B_1, .5 milligrams vitamin B_2, 16 milligrams calcium, 30 milligrams phosphorus, .8 milligrams iron and 102 calories.

Apricot trees have rather stringent climate requirements which make them difficult to grow in many areas. Though they are among the earliest trees to bloom, their flowers and fruit buds are very sensitive to cold. On the other hand, the trees will not set fruit properly unless they are chilled by freezing weather during their dormant period.

Conditions suitable to good yields are difficult to find on the Atlantic seaboard, where most of the area is subject to alternate freezing and thawing or to late frosts. Better success attends their culture in the Great Lakes area and the northern plains, where spring comes late but is less unsettled and in some Pacific and mountain sections where spring and fall are long and moderate and winters frigid enough to chill the trees.

Heat requirements for ripening the fruit are not so difficult to meet. Apricots seem to ripen when a certain number of heat units have been applied, whether the heat is intense and supplied over a short period or more moderate and of longer duration. However, temperatures of more than 100 degrees during the ripening period often result in pit burn, which may damage much of the fruit.

Choosing a Site

In any area where late frosts are a possibility, apricot trees should be planted where they will have a northern exposure. This will delay the blossoming period and may save the crop from failure in many years.

Deep, fertile, well-drained soil of fine texture is best for apricots. Loam to clay loam are preferred to sandy soils which tend to warm up early. Shallow hardpan should be avoided.

If possible, apricots should be planted in a part of the home grounds remote from the vegetable garden and strawberry patch. Tomatoes, potatoes, Persian melons and strawberries all harbor verticillium wilt, which causes "blackheart" in apricot trees.

Ordering Stock

Apricot trees are sold on 3 different rootstocks, apricot, peach and myrobalan plum, each of which has merits. The type chosen should conform to local conditions.

Trees grown on their own roots or grafted to apricot rootstocks succeed better in moderately heavy soil and are more resistant to peach twig borer than those on peach or plum stock. Also, apricot stock is immune to root-knot nematode.

Apricot trees on peach stock sometimes lack vigor and are somewhat more susceptible to crown gall. Where the rootstock is congenial, however, the fruit is larger and matures earlier. Peach stock is better adapted to light soils.

On clay soils, trees grafted to myrobalan plum do best, because the plum roots are less likely to be damaged by excess water. Trees on plum stock are somewhat resistant to oak root fungus, but are more subject to bacterial canker.

Sometimes apricot scions make poor unions with plum roots and the trees may be broken over in strong winds. The gardener who is in doubt about which rootstock to order should consult with his county agent about the diseases and conditions in his own area.

Trees should be healthy one-year stock of medium size, that is, the trunks should be approximately one-half inch in diameter two inches above the buds. These will make better growth than older trees and may be trained from the first year.

One self-pollinating apricot tree will yield 200 to 250 pounds of fruit in a good year. If fruit is to be dried, 5 pounds of fresh fruit will yield one pound dried.

Planting

Apricot trees should be planted in early spring before the buds begin to swell. In California they are planted between the middle of January and the first of March. Throughout the rest of the country they should be planted as soon as the soil can be prepared.

A mature apricot tree needs space to spread over a circle 25 to 30 feet in diameter. During its first 5 years it may be given less space, but it should never be crowded because its roots spread out in a larger circle than its top.

Immoderately long roots should be shortened before planting and broken roots removed. Make the hole large enough to hold all the roots without bending, and plant the tree to the depth at which it was grown in the nursery. Settle the soil about the roots with water and thoroughly wet the soil after planting. Wrap building paper loosely around the trunk immediately after the tree is planted to protect it from mice, rabbits and sunburn. Spread a one-inch layer of

TABLE 59: Apricot Varieties

All varieties listed below are self-pollinating except where noted.

Variety	Section	Fruit characteristics	Tree characteristics
Moorpark	Corn Belt Mountain States Northern plains Pacific States	Large, soft, fine flavor. Best used fresh, not good for drying or canning. Late.	Sometimes a shy bearer.
Early Golden	Northern plains	Large. Early.	Reliable.
Alexander	Northern plains New England Great Lakes	Small. Not as good as western varieties. Freestone.	Very hardy, good in cold areas.
Tilton	Pacific States	Large, light color, bland flavor. Used for drying. Late midseason.	Prolific but tends to be biennial.
Riland	Mountain States	Thick meaty fruit with rich flavor, tends to crack if large. Moderately early. Not good for canning or drying.	Viable pollen but not self-pollinating. Upright habit.
Royal	Pacific States Southern plains	Medium size, very good flavor. Good for drying, canning. Midseason.	Productive.
Blenheim	Pacific States	Large. Very good for drying and canning. Late midseason.	Productive.
Newcastle	Pacific States	Good flavor but soft coarse texture. Early.	Succeeds in mild winters, but subject to brown rot and gummosis.

well-rotted manure mixed with phosphate rock and ground granite over the soil in a 3-foot circle and then mulch.

The top of the newly planted tree should be cut back to balance the top with the roots, many of which were lost when it was dug. The top of the tree should be no taller than 24 to 30 inches after pruning. Select 3 branches for the two main scaffold branches and the leader. They should be well distributed around the trunk and should be no closer to each other than 6 or 8 inches. Cut these branches back to 4 to 6 inches long. Remove all other side branches, leaving a small stub of each in order not to damage the buds at their bases, which should be permitted to grow.

Pruning

During the first summer encourage the 3 scaffold branches to make the strongest growth on the tree by pinching back other side shoots after they are 3 to 4 inches long. If the first summer's pruning is carefully done, the tree will have made good progress toward building a healthy, strong framework. If the tree was not pruned during its first summer, these scaffold branches would need to be selected and started at the first dormant pruning.

The first winter pruning should further develop the tree's shape. The leader should be the strongest branch at this point. All 3 scaffold branches should be shortened at the first dormant pruning to about 2 to 3 feet long, but if the bottom two threaten to overtake the leader, do not shorten the leader. Prune the other two to make them assume a secondary position. If the laterals are permitted to overtake the leader, weak crotches will result. If the leader is strong, the laterals must reach out horizontally to get the sun and their crotches will be more nearly at right angles to the trunk.

During the second summer very little pruning should be necessary. Sometimes whorls of small branches tend to form below the point where the leader or a branch was cut back. Where this happens, remove all but 2 or 3 of the branches from the whorl.

Secondary laterals are selected at the second dormant pruning. A number of small side branches will have developed from the two main lateral branches and the leader. Five to 7 of these, strong and well spaced around the trunk, are selected to form the secondaries. All other secondary lateral branches are removed, but the small spurs which grow less than 6 inches long from branches and trunk are allowed to remain. Make sure of good secondary crotches by heading back the secondary branches to less than the growth of the main laterals. The leader should still be given preference over all the laterals. Where a main lateral is becoming too strong, it should be headed all the way back to one of its laterals.

After the third season, the young apricot tree will begin to bear. Pruning after this becomes a matter of cutting out unnecessary branches, water sprouts and dead wood and keeping the center open.

When the tree attains the desired height, stop favoring the leader and permit the laterals to catch up. The height of the tree should be determined, within reasonable limits, by the height which can be comfortably reached for pruning and harvest. It is not possible to make a dwarf tree of a standard tree by pruning. But the tree may be made to stop its upward growth to accommodate itself to work from a 10 or 12 foot ladder. When standing on the next to the top step, a worker can reach the top of a 16 to 18 foot tree. Once established, this height should be maintained.

From this point forward, the tree will be pruned to permit sun to reach the lower branches. Most fruit is found on wood two years old or older. Fruiting spurs are present in dormant buds on the old wood, but they will not develop unless they are exposed to the sun. The top must be kept sufficiently open to allow the sun to penetrate to wood of the older branches. Under these conditions old spurs remain alive and new shoots arise each year. If the top is allowed to become too dense the old wood is shaded and new shoots cannot arise.

This pruning to keep the center open will eventually result in a top that will begin to spread, vase-like. When this happens, select wood that is growing into the center and prune back drooping wood.

A pruning job may be judged the following year by the number of new spurs growing in the lower branches. If the growth is sparse, cut out more of the top. If too prolific, leave more growth in the top.

Maintenance

Apricot trees need plenty of water early in the summer, when the fruit buds are forming. If there is a drought, the tree should be given water enough to moisten the ground 5 to 6 feet deep, to the point where most of the roots may be found.

Fertilizer should be given to the tree in moderate quantities and only when the need is apparent. A nitrogen deficiency will show itself in yellowing leaves, fruit that ripens unusually early and is small and firmer than normal. When this occurs, rotted manure should be spread below the branches at the rate of one bushel per square yard. However, it should not be administered unless the tree shows a need. Too much nitrogen may result in fruit which is too soft and suffers from pit burn even in mild summers.

In general, sufficient nitrogen should be supplied if the tree is well-mulched with hay or a leguminous mulching material. It needs about as much potash and phosphorus as is supplied by the general garden broadcast every 2 or 3 years and should thrive if its temperature requirements are met.

In some years the trees may set too heavy a crop of fruit. When this happens, the fruit should be thinned in order to preserve the strength of the tree and to increase the size of the individual apricots. Also, when too much fruit is set, there is more danger of brown rot. When the green fruit attains the size of a thumbnail, thin it to stand 3 to 4 inches apart. Tight clusters of fruit should be broken up or malformed apricots will result.

Propagation

New trees are seldom started by individual growers, because the techniques are more suited to a nursery. In general, the process of propagating is as follows:

Seeds of the rootstock are planted in early spring, and are grown in the nursery until midsummer or later. While the bark still slips easily the seedlings are budded just above the soil line with the

Though among the earliest trees to bloom, the flowers and fruit buds of the apricot tree are very sensitive to cold. Trees should be planted in early spring before buds begin to swell.

desired apricot scion. The following spring the seedling is cut back to the bud, which is then allowed to take over. When an especially vigorous peach stock is used, it is sometimes budded in June and the top is cut back in late summer.

Enemies

Brown rot is a fungus which covers the blossoms with a mass of gray spores. The blossoms wither and die and the fungus works back through the twigs which are also blighted. When the fungus attacks ripening fruit the fruit becomes brown, rotted and covered with fungus. Best preventive is to go over the tree in early spring looking for the telltale gray tufts on the twigs which will produce the fungus spores. Remove all infected twigs before the blossoms are in bud.

Blackheart is caused by verticillium fungus. Foliage wilts on the first hot days of summer and an examination of the wood shows black streaks an inch below the surface. Sometimes the tree will recover from a blackheart infection, but sometimes it will die. No cure is known for the disease beyond the vigor of the tree itself, but to avoid infection, do not plant the tree where tomatoes will be its close neighbors.

Crown gall is the name for swellings which occur where bark has been damaged and bacteria given entry. The protuberances sometimes occur at the soil line, when cultivating tools have been carelessly used or they may appear on branches damaged by ladders. Remove the gall with a sharp, clean knife and paint the wound with a tree surgeon's paint to prevent further infection.

Sometimes skin of ripening fruit will crack for no apparent reason and give entry to bacteria which may rot the fruit. Usually the cause of cracking is sunburn. This affects the skins more in hot, humid weather than when the season is cool and dry. When the fruit is thinned, particular attention should be given to that growing on the exposed tips of branches. A few leaves which shade the fruit part of the day will help to protect it from sunburn. Where these leaves are lacking at the tips of the branches, it is better to remove the fruit.

Like peaches, apricots are subject to pit burn when they ripen in very hot weather. This turns the area around the pit of the half-ripe fruit soft and brown, as if the pit were heated. A thick mulch under the tree tends to keep the soil cool, which in turn will keep the whole tree cooler than the surrounding garden. When the heat is very bad during the ripening period, it may help to sprinkle the mulch with water in the warmest part of the day.

Fruit buds may be shed from an apricot tree in the warmer sections if the tree did not go through a cold enough winter. A thorough chilling for an extended period is necessary to soften the buds enough so that they can develop the following year.

Cankerworms or measuring worms sometimes attack apricots and chew the edges of the leaves. These worms do not have wings in their adult egg-laying stage, so they may be caught as they climb the trunk in a sticky substance like Tanglefoot.

Branch and twig borers, as well as shothole borers, breed in dead and diseased wood. Remove all prunings from the vicinity of the tree as soon as the operation is finished and cut out all dead or diseased wood from the tree as soon as it is spotted.

Harvest and Drying

Apricots do not increase their supply of sugar after they are picked, so they should be left on the tree until fully ripe. For drying, they should be ripe, but still firm.

Best process for preserving apricots is sundrying. Freestone fruit should be used for this purpose and it should be ripe but not mushy.

Split the fruit in two and remove the pits. Lay the halves on trays in a single layer with the cut side up, exposed to the sun. Leave the trays out in the sun for 1 to 3 days, until the cut side is well healed. Then stack the trays in a well-ventilated shed to finish the drying before the fruit is stored. Five pounds of fruit, fresh, will yield one pound of dried fruit.

AVOCADO

Avocado. *Persea americana,* a subtropical evergreen tree of the laurel family, *Laurineae*. Also mistakenly called alligator pear, a name which the California avocado growers discourage because it leads strangers to the fruit to expect a sweet, rather than a nut-like flavor.

Avocados are native to tropical America, where they have been cultivated for centuries. The word is derived from the

Spanish *aguacate,* which in turn was derived from the Aztec *ahuacatl.* Like its name, the fruit was first the product of Indian culture and later was adopted by the Spanish explorers. It is grown in the United States in southern Florida, California and Hawaii.

Three races of avocados are grown under cultivation. They stem from those native to the West Indies, which are the most tender varieties ripening in summer and fall, with large fruit, rind leathery and not more than 1/16 inch thick; the Guatemalan, somewhat less tender trees with large fruit which ripens in winter and spring, its rind thicker and woody in texture; and the Mexican race, the most hardy, ripening its somewhat smaller fruits from late spring to early autumn. Hawaii grows West Indian and Guatemalan avocados almost exclusively; Florida grows mostly Guatemalan and Mexican; and in California the varieties are principally Mexican or hybrids, with a few of the Guatemalan varieties grown in the southern coastal areas.

Avocado trees are tall and handsome and make beautiful shade trees in the home grounds. In the wild state, they sometimes grow 50 to 100 feet tall, but under cultivation they are held to a maximum of 30 feet. They may be counted on to yield good crops of fruit year-round in areas where the temperatures never dip below 32 degrees, if two or more varieties are planted. In the colder sections, Mexican trees which will stand temperatures briefly as cold as 25 degrees will give good summer and fall crops. They do best in a soil with plenty of humus to hold moisture, of which they require an abundance, at the same time that they need good drainage. Neutral soils are best, with a *p*H of 6. to 8.

Avocados are more closely allied to vegetables in the diet than they are to fruits. They are always eaten raw, either as an appetizer or in a salad. Cooked, they become bitter and they cannot be frozen successfully. Their flavor is rich, buttery and somewhat nut-like. Because of their large oil content—this varies from 7 to 23 per cent with different varieties—they are usually served with an acid dressing or with lemon or lime juice. They are very rich in minerals which make up 1½ per cent of their total weight. They contain 9 vitamins, are low in carbohydrates and 93 per cent of their oil is in unsaturated fatty acids, which are entirely digestible. They also contain vitamin E.

Amounts of vitamins and minerals vary greatly in different types of avocados. In 1/3 cup of cubes, for instance, there may be 300 units of vitamin A, or as many as 2,000 units, depending upon whether the fruits are of the summer or winter type. In the same quantity, there may be .04 to .10 milligrams vitamin B_1, .06 to .17 milligrams vitamin B_2, .8 to 1.9 milligrams niacin, 2 to 10 milligrams vitamin C, 35 milligrams calcium, 35 milligrams phosphorus, 5.4 milligrams iron and 480 to 980 calories.

Ordering Stock

Some varieties of avocados are self-pollinating; others have flowers which require another variety for pollination. Two different types of flowers are to be found on some of the different trees, which makes for a complication in finding the right types to plant together. It is advisable to buy stock from a reliable nurseryman who can recommend the best plants for the area, especially since many of the trees sold are grafted upon seedling rootstocks from unnamed varieties.

Usually used for rootstocks are Mexican varieties, which lend some of their hardiness to the trees. More important, however, is to get a tree which was budded to a seedling grown from a disease-free parent. A reliable nursery keeps in close touch with the parent trees to be sure that disease is not carried in the seeds to rootstocks or in twigs to scions.

In the following list, the varieties which are not self-pollinating are listed as *a* or *b* and, in order to pollinate, an *a* and a *b* which bloom simultaneously must be planted together. Areas where these are recommended are also noted.

Pollock. b. Florida, only in the southern tip. Very tender West Indian variety.

Gottfried. a. Florida or California, a Mexican type hardy to 25 degrees.

Lula. a. Florida. A hybrid hardy to 28 degrees.

Taylor. a. A very tender West Indian type. South of Florida.

Waldin. a. Also West Indian and tender. Florida.

Nabal. Florida or California. Fine flavor. Ripens June to September. Hardy to 29 degrees.

Fuerte. b. Fine quality hybrid grown

commercially in California. Ripens November to June. Spreading tree, hardy to 28 degrees.

Hass. California. Heavy yields of purple-black fruit with fine flavor from May to October. Hardy to 29 degrees.

Beardslee. Many varieties under this name are grown in Hawaii, all developed from the same parents. Guatemalan origin, ripen winter and spring.

Anaheim. California. Upright tree bears heavy crops of large fruits May to August. Hardy to 30 degrees.

Duke. Recommended for the colder interior valleys in California because it is hardy to 25 degrees.

Puebla. a. A small tree which bears large fruits November to January and is frost-hardy to 28 degrees. California.

Edranol. California. Upright tree bears large fruit with rich flavor May through August. Hardy to 29 degrees.

Schmidt. b. A Guatemalan type recommended for Florida. Hardy to 30 degrees.

Avocado yields vary greatly with variety and maturity. One large well-grown tree in California has a record of more than 1,300 pounds in one year.

Site

In the home grounds an avocado tree should be planted where it will have uninterrupted sun all day long. If there is a choice of elevation, a slope where the air has free movement is best. Avocado trees, even though they may be able to stand a brief dip of the temperature, must warm up again as soon as possible. If they are on a slope where the cold air can drain away readily, they will be able to warm up earlier in the morning following a frost. Also, a large sunny area near the tree, such as an unshaded expanse of lawn, will retain more heat on cold nights and thus prevent chill. Trees of most types require a space about 30 feet in diameter when fully mature. Some of the smaller or upright trees may be grown in a space no more than 20 feet in diameter. Soil must be well-drained, because avocados resent standing water around their roots.

Planting

Trees may be planted out any time between November and May, though March or April, after frost danger is past, is best.

In order to give the young roots a soft medium in which to spread, a hole at least twice as big as the root ball is recommended. Dig the hole 3 feet in diameter, if possible, and as deep. In the bottom spread 2 or 3 shovelfuls of rich compost and cover it with topsoil up to the level of the lowest roots. Set the tree with the bud union two inches above soil level, to give it a chance to settle in, and with the bud turned away from the sun. Fill in around roots with a mixture of topsoil and compost, washing the soil in with a stream of water to be sure there are no air pockets. Provide light shade for the newly planted tree until it becomes established.

Make a basin two inches deep around the trunk for watering. A newly set tree should be watered thoroughly once a week until it is established. After it has grown 8 or 9 inches, waterings may be reduced to one every two weeks. Flood thoroughly with the hose at every watering, giving it 45 to 60 minutes. If the tree starts to grow rapidly, it should be tied to a stake until the trunk is strong and woody.

In cold areas, give young trees winter protection until they are 3 years old. This may be done with burlap, cornstalks or by mounding the soil around their trunks.

Maintenance

Successful avocado growers each have their own recipes for producing strong, healthy trees, but their methods have one thing in common: they provide a rich organic diet with plenty of humus.

Trees which are strong and healthy are better able to withstand drops in the temperature in winter, they are seldom troubled by insects or diseases and they yield heavy crops. Because they are evergreen, avocados need water all year. In winter seasons when the rainfall is scanty, it should be supplemented by frequent irrigations. A 4-inch mulch of strawy manure, leaves or compost should be maintained year-round. The mulch should cover the soil under the entire spread of the branches except for a ring 6 inches wide which should be left bare around the trunk.

Trees should be fed from early spring until August. At that time stimulus to new growth should be withheld in any section where frost is a possibility. If the mulch is coarse, it should be pulled back and the nutrients spread under it. Phos-

Avocado trees make beautiful shade trees on the home grounds. The young tree above is laden with fruit, and has been given support to protect the limbs from wind breakage.

phorus and nitrogen should be supplied in about the following amounts: in spring a one-inch layer of rabbit or poultry manure may be spread over the entire root area; or, every 6 weeks from early spring to August one trowel of bloodmeal and two trowels of bone meal may be fed to each tree.

Pruning

Trunks and branches of avocado trees are very sensitive to sunburn. For that reason, little pruning should be done, especially to the young tree. All leaves are necessary to shade the wood.

As the tree grows full and mature, little need be cut from it except to shape it or to remove dead wood. Any branches which droop to the ground may be shortened to protect the fruit. The only time when extensive pruning may be necessary is after frost damage, but no cutting should be done until midsummer following the damage. By that time, it will be possible to spot branches which have put forth no new growth and to remove them.

Propagation

Avocado trees are always grafted, because they cannot be trusted to come true from seed. For hardiness, rootstocks are usually from seeds of Mexican varieties. Most commonly used are Ganter, Duke, Mexicola or Topatopa. Seed should be taken from known trees and windfalls, which may be infected by fungus diseases on the ground, should be avoided.

A slice 1/8 inch thick cut from top and bottom of the large seeds speeds its germination. The seed should be buried in sand, its flat side down and pointed end up, with 1/4 inch of sand over its top. The sand must be kept constantly moist until germination take place, in 30 days in warm weather or 60 to 90 days in cooler weather. When the seedling has developed half a dozen leaves it is ready to transplant to a nursery row.

Seedlings must be dug carefully to avoid damage to the cotyledons, which still supply much of the nourishment to the young plants. Holes should be deep enough to allow the roots to spread without bending or the young plants will be stunted. Seedlings may be set 14 inches apart in the nursery.

By the following spring seedling trunks should be 3/8 to 1/2 inch thick and ready for budding. Bud grafts are placed on the side away from the sun, to prevent drying out. After the bud begins to grow, the seedling should be cut back to a 10-inch stub. As the bud lengthens, it should be loosely tied to the stub for support. After it has put out its second set of leaves, it should be tied to a 4-foot stake until the trunk is 3/4 inch thick. When the young tree is 3 to 5 feet tall, it is ready to transplant to its permanent position.

A few varieties may be grown from cuttings, but a special technique is required which is usually beyond the means of the home gardener. Zutano, Ganter and Fuertes are among the varieties which can be made to root. Cuttings 3 to 6 inches long, each with 3 or 4 leaves, are used. They are planted in special cutting boxes which are supplied with bottom heat to 70 degrees, and in which the humidity can be held at 100 per cent. Two to 4 months are needed to make enough roots for transplanting. The cuttings are transferred to pots and the humidity gradually reduced before they are ready for planting outside.

Enemies

Insects and diseases affect avocados very little when a good mulch is maintained. Animals are more of a problem to the orchardists, though they may be less troublesome in the home grounds. In areas where irrigation ditches must be maintained, gophers sometimes tunnel in the ditches and destroy the system. One organic grower has imported owls to prey upon the gophers. He supplies artificial owl houses in his orchards and the wise birds accept his food and shelter gratefully. Snails under the trees may also present a problem, but one which can be solved easily if chickens or ducks are allowed to forage there occasionally.

Harvest

After oil develops in the fruit and it becomes soft to the touch it is ripe and ready to use. Fruit which is to be held for a short time may be picked while it is still firm, but it should have developed enough oil before picking so that when mellowed it will not have a rubbery texture.

After a frost, mature fruit should be picked as soon as possible. Severely frosted fruit, that which shows brown water marks on the skins, will not keep. But if the frost has damaged only the stems, the fruit may be stored for a short time.

BANANA

BANANA. *Musa paradisiaca sapientum.* The banana is one of the oldest known fruits. Some idea of its antiquity may be derived from some of the myths about it. One old belief is linked to its early name, "Adam's Fig," which has it that the fig leaves used by Adam and Eve were actually banana leaves. In support of this, an early writer reported that in Ceylon in the fourteenth century bananas were called figs by the natives. Another myth relates that the serpent in the Garden of Eden was hidden inside a bunch of bananas and to this day one of the banana varieties is called Apple.

A wide range of banana varieties are grown throughout the tropics. Many of them are used for cooking and are called plantains, only those which are eaten raw being called bananas. Fruits of some are as big as a man's arm. In one variety the stem, rather than the fruit, is cooked and eaten. Those sold in markets throughout the United States are mostly of the variety called Gros Michel in tropical America or Bluefields in Hawaii. They grow on plants which are up to 25 feet tall and are extremely tender. In the southern tip of Florida, a few of the more hardy types can be grown, but Gros Michel is not adaptable to the Florida climate.

Banana plants are rapid-growing herbs 5 to 25 feet tall. Their stalks or trunks are succulent, being actually composed of compressed layers of leaf sheaths. Flower clusters force their way up through this tightly packed stalk after it reaches its full growth and emerge at the top to droop over the outside of the leafy crown. Flowers on the cluster are all perfect, but stamens, pistils or both may be abortive. The flowers at the base of the cluster are female and set fruit without

Banana plants grow rapidly, ranging from 5 to 25 feet tall. "Hands" of 5 to 20 bananas develop in whorls around the base of the fruiting stalk.

pollination. Next flowers have both stamens and ovaries, but both abort and the flowers drop. Flowers at the tip of the cluster have only stamens, so they cannot set fruit, but in some varieties the flowers persist. "Hands" of 5 to 20 bananas set in whorls around the base of the fruiting stalk, the end being cut off when the bunch is cut from the plant.

After bearing once, the stalks die back to the plant's true stem, which is an underground rhizome. New suckers on which further fruit is borne are constantly rising from a healthy, productive rhizome. On a plant that is growing well, a sucker bears within 12 to 18 months after its emergence from the soil, but the length of time required to develop fruit may vary with soil and climate.

Bananas of the plantain type and green bananas are used for cooking. Until the dessert varieties are ripe they are composed principally of raw starch which must be cooked to be digestible. In their all-yellow state with green tips, the fruit should be cooked. Fully ripe bananas are brown-flecked, all brown or red, according to variety. Only these are ready to eat raw.

Varieties which are used only for cooking are, in general, richer in vitamins than those which are eaten raw. In one medium-sized banana of the cooking variety, there are 420 units vitamin A; .04 milligrams vitamin B_1; .06 milligrams vitamin B_2; .6 milligrams niacin; 12 milligrams ascorbic acid; and about 100 calories. In a dessert variety the amounts are approximately as follows: vitamin A, 120 units; vitamin B_1, .03 milligrams; vitamin B_2, .05 milligrams; niacin, .6 milligrams; vitamin C, 9 milligrams; and 85 calories.

Site

Banana plants may be grown as far north as the Georgia border and a few are grown as ornaments in California. They will fruit, though, only where frost almost never touches them. If the tops are badly frosted, they may die to the soil but the roots are not killed unless the cold is severe. New suckers will arise from the roots if the rhizomes are not damaged. However, if the leaves are damaged by frost, the fruiting clusters will never appear.

Bananas may be grown in any soil except sand or limestone, but they prefer a porous, moist, well-drained soil with a large percentage of humus. Tall varieties may be badly damaged unless they are protected from the wind. Also, they resent traces of salt in the soil or in ocean spray. Soil should be as nearly neutral as possible.

Planting

A hole 30 by 36 inches, 18 inches deep should be prepared for planting either rhizomes or suckers. Cavendish may be planted 8 to 10 feet apart; others should stand no closer than 12 to 15 feet. Plant the sucker or rhizome a foot deep, and fill the hole with a mixture of topsoil and compost or rotted manure.

While the plant is young, remove all but one sucker, which should be allowed to bear its fruit and be cut back before another sucker is permitted to grow. Older plants may be allowed to develop one new sucker every 3 months. A stool will grow and bear well for 4 to 6 years, after which it should be dug out, the soil enriched and new suckers or rhizome pieces planted. After its first fruit has ripened, the plant may be allowed to grow 3 to 5 suckers at a time, depending upon its vigor. All others should be cut out.

Bananas are gross feeders. Because of their heavy growth they need plenty of fertilizer and a large amount of moisture. They do best where the rainfall averages 60 to 100 inches per year. In areas where it is less than that, they will need frequent watering.

A heavy rich mulch should be maintained under the plants at all times. This may be rotted manure, compost or a mixture of manure and leaves or grass.

Propagation

Because they contain no viable seeds, bananas must be propagated by separating the suckers from the parent plants or by making cuttings of the rhizomes. Suckers 2 to 8 months old are used and are moved in March or April. Leaves are removed from the suckers except on

TABLE 60: Banana Varieties

Variety	Section	Fruit characteristics	Plant characteristics	Use
Lady Finger or Hart's Choice	Southern Florida	Yellow, 4 to 6 inches long.	20 ft. tall. Fairly tender.	Eating raw.
Cavendish or Chinese	South to Central Florida Hawaii	Yellow, 6 to 8 inches long, thin skinned but good flavor.	5 to 7 ft. tall. Wind-resistant, hardy.	Eating raw.
Apple or Brazilian	South Florida	Small to medium, tart flavor, firm flesh.	Fairly hardy, prolific.	Better cooked than raw.
Orinoco	South Florida	Thick fruits 6 inches long, inferior flavor.	Small bunches.	Cooking.
Bluefields or Gros Michel	Hawaii	Largest eating variety, flavor mild but good.	25 ft. tall, tender.	Best raw, but should be cooked if green.
Red Cuban or Red Spanish	Hawaii	Medium size, red skin, pungent flavor.	Tender.	Not recommended for cooking.
Maimaoli	Hawaii	Large, tart. Yellow skin, creamy flesh.	Tender	Best baked or boiled.

Cavendish, which may be permitted to retain their leaves.

Seven to 10-pound cuttings of the rhizome are removed, each with two buds, by cutting with a spade or mattock. Rhizome cuttings are replanted immediately. Suckers may be held for a few days, exposed to the air, or they may be planted immediately.

Harvest

Bananas which are cut 7 to 14 days before ripening may be hung in a cool shady place to develop their flavor and sugar. Their nutritional value will be the same as that of tree-ripened fruit.

After the bunches are cut from the plant, the ends of the stalk are trimmed and the bananas are held at room temperature until thoroughly ripe. Stalks of the plant are cut back and are chopped into small pieces which are added to the mulch around the roots.

BARBADOS CHERRY

BARBADOS CHERRY. *See* ACEROLA.

BLACKBERRY

BLACKBERRY. A large variety of berries of the genus *Rubus,* of the *Rosaceae* family, are grouped together as blackberries and, since they are horticulturally similar, we list them together here. Included in the genus are species which grow wild in every state in the United States, including Alaska and Hawaii. Consequently, cultivated species can be grown anywhere in the country.

Most of our cultivated varieties have been developed from native blackberries within the last hundred years, with the exception of the varieties Evergreen and Himalaya, both of which are imports. The wild species include the following: *R. arcticus,* the crimson or Arctic bramble, which grows in Alaska, Labrador and the Scandinavian peninsula. *R. canadensis* is the trailing blackberry or dewberry, thought to be one of the parents of some of the California trailing hybrids. *R. hawaiensis* is a small red species native to Hawaii. *R. ursinus* and *R. villosus* are the Latin names for the Western and Eastern wild blackberries. In addition, we grow *R. phoenicolasus,* the Asiatic species known as Japanese wineberry and *R. loganobaccus,* the hybrid loganberry.

Most of the cultivated blackberries grown in the northern sections of the country are bush species, growing very tall and upright with long, vicious thorns. Trailing vines, called trailing blackberries or dewberries, are more tender as a rule and are winter-killed when grown in the North. Some of the trailing sorts are now thornless, having been developed from chimeras, or canes of mixed parentage. The thornless varieties can be propagated only from tip-rootings, the root suckers invariably reverting to thorns.

Blackberries are adaptable to many types of soil, but they flourish best in a moderately light loam containing some clay and a large amount of humus. Roots penetrate to a depth of 3 feet. Soil must be well-drained to that depth to support healthy growth, but the water table should not be too much deeper, because during the hot summer months the bushes need their largest amounts of water.

Moisture should be supplied in good quantities while the fruit is ripening and the young canes are making their growth, from early summer into early fall. If the summers are too dry, the berries become small and seedy and the young canes do not make their growth early enough to harden off before cold weather approaches and may suffer winter injury as a consequence.

In the home garden, blackberries should be grown where they are sheltered from drying winter winds, though they produce better on a northern slope which is cooler at ripening time than on a southern slope. They may be planted in half-shade, in a corner where they will have half a day of sun during the hottest months. The best fruit is produced where the bushes are protected from the sun at midday. The life expectancy of a planting which is given a favorable position may be as much as 15 to 20 years.

Blackberries are a fairly good source of vitamin A in the diet, although some varieties contain much more than others. A small serving, about ¾ of a cupful, contains on the average about 300 units of vitamin A; .025 milligrams vitamin B_1; .030 milligrams vitamin B_2; 3 milligrams vitamin C; 32 milligrams calcium; 32 milligrams phosphorus; .9 milligrams iron; and 52 calories, all supplied in sugar. Some varieties are much sweeter than others; also, the amount of sugar supplied increases sharply depending upon the ripeness of the berries.

Ordering Stock

Blackberries may be planted in several ways and their culture varies with the method of planting. Upright or bush varieties may be planted singly, in hills, in hedges or in rows which will be trained to trellises. Trailing varieties are seldom grown prostrate, but are usually trained to trellises, wires or stakes. The number of plants which can be used in a home garden depends upon the method of culture.

Bush berries may be planted 3 to 4 feet apart in hedges; if the same varieties are trained to stakes or planted in hills, they should be spaced 6 feet apart. Trailing varieties on a stake need 6 feet in each direction; on a fence or wire, they should be planted 10 to 12 feet apart in rows spaced 6 feet apart.

Usually a planting of 12 to 18 bushes of two or more varieties is sufficient to supply a family of 4 or 5 people. Yields vary with the method of training and the variety planted, but may be from 10 to 15 pounds per plant.

Blackberries pollinate fairly well, but if the planting includes two or more varieties, the pollination is always improved. When ordering the plants, be sure to get certified stock. This means, in the case of bramble fruits, that the stock was examined by state inspectors and found to be "apparently free of insects and diseases."

Blackberry plants are supplied bare-rooted by the dealers, but the roots should be protected by damp sphagnum or peat moss or by being enclosed in plastic bags. They should not be allowed to dry out.

Planting

In warm sections of the country where vines will make some growth during the winter months, blackberries are planted in late fall after the plants are dormant. In all other sections, they should be planted as early in spring as the soil can be worked, but always before the buds start to swell.

Soil should be deeply prepared, with as much humus worked in as possible. A top-dressing of two ounces of bone meal per square yard should be raked into the surface and the soil *p*H should be tested and corrected, if necessary, with ground limestone. Trailing varieties usually need more soil acidity than bush species, tolerating a *p*H as low as 5, but not much higher than 6.0. Bush varieties prefer a soil with a *p*H of 6.0 to 8.0.

If plants are to be trained to stakes, a trellis or a fence, the supports should be placed before the bushes are planted. Do not place the young plants too deep—the cane buds should be no more than two inches underground. Firm the soil well around the roots and tie the shoots, if there are any long enough, to the wires or supports. Plants should be prevented from bearing the first year, to strengthen their root systems.

Pruning and Training

Established plantings are pruned each spring, after harvest and sometimes also at intervals through the summer until early fall.

Fruiting canes are cut back and removed from all varieties except Himalaya and Evergreen immediately after harvest. Most varieties have biennial canes on perennial roots—the canes die back after fruiting. Himalaya and Evergreen have perennial canes—they bear repeatedly on the same canes.

In order to do an intelligent pruning job, it is necessary to study the bearing habits of each variety. Some bush varieties bear best in the middle of the canes, some toward the tips and some near the base. Most bush blackberries should have their young canes pinched back when they reach a height of 30 inches to 3 feet, though some may be permitted to grow to 5 feet. After the canes are pinched back, they will throw out lateral fruiting branches. These may be cut back in the spring before growth starts, usually to about one-half their length.

Trailing varieties are usually pruned when they are tied to their supports. This may be done in the late summer, after harvest if the section has mild winters; where winters are severe it is usually not done until spring.

In addition to pruning each cane, the young canes should be thinned on each plant to no more than the 10 strongest canes, and in some cases no more than 6 should be allowed to develop. Larger berries and stronger plants will result from a good job of pruning.

Bush blackberries may be trained to grow within a wire framework, a method which requires constant attention during the growing season, but pays off in ease of harvest. Supports for this type of training consist of heavy posts set 10 to 16 feet apart, 1½ feet below ground and 4 to 5 feet above, with crossarms 18 to 24 inches long at the top and half way between top and ground. Wires are strung between the tips of the crossarms, making a box-like framework within which the young canes are trained to grow. Varieties making strong canes need no further support; those which are semirecumbent may need to be tied to the wires.

Trellises for trailing vines may consist of one or two wires strung between posts set 12 to 20 feet apart. If smaller supports are placed between the posts, the spacing may be wider. Wires used to support the vines must be strong, because all the weight of the canes and fruit will hang upon them. It is well, also, to arrange a mechanism for tightening the individual wires as the burden on them grows during the summer. Main posts should be 6 inches in diameter sunk 1½ to 2 feet in the ground and rising 5 to 6 feet above. Similar posts are used as stakes for individual plants, if the vines are trained to a stake instead of a wire.

Usually the staked plants are given a crossarm a few inches below the top of the post. The young canes, as they sprout, are trained to run along the ground during their first summer and may be held together with wire wickets, like croquet wickets. After the fruiting canes have been cut out when the harvest is finished, the young canes may be trained up to take their place. Or if the winters are too cold, the training may be done in spring, though the old canes are still cut back after the harvest.

To train the young canes to the stake, twist them together slightly to form a rope and wind them loosely around the post, bringing them up in front of the crossarm, behind the top of the post, in front of the other end of the crossarm and then down. The ends should be woven inside the rising rope, and any excess growth cut off.

If it is possible to winter the vines in an upright position without winter injury, fall training is preferable, because less damage to the fruiting laterals will occur at that time. However, where the climate is uncertain, it is best to leave the young canes on the ground and to cover them with straw after the soil freezes. As early in spring as seems safe, the covering should be removed and the canes trained upright before the buds expand.

When vines are trained to a horizontal wire, the young canes are usually trained to run along the ground under the wire in both directions from the root. This training must be done at regular intervals of no more than two weeks during the growing season. This method gives the plants two "ropes" of canes each. When the time comes to train them to the wires, one rope is carried along the wire in each direction, being tied loosely at intervals or intertwined with the rope from the neighboring vine so that they will not shift in the wind. The objection to this system is that a great weight must all be supported by one wire.

When two wires are used to make a trellis, the top wire may be supported on the tops of 4½ to 5 foot posts and the lower one may be strung two feet lower. In the so-called 4-arm system, the longest canes are divided in two ropes and carried up to the top wire, half being trained in each direction. Shorter canes are carried along the lower wire in both directions, the ends of the canes being intertwined with their neighbors, as in the one-wire system.

Where all the canes are long and the plants are set somewhat closer together all the canes may be gathered into two ropes which are carried to the top wire and trained in one of several ways.

The weave or wreath system is used where the canes are very supple, but cannot be used where they are more stiff and likely to be damaged at the bend. Two or 3 canes are taken together from the rope and wound around the two wires, finishing with a reverse turn which locks the ends in place. This method makes a solid wall of fruiting branches between the wires after the leaves are open and makes the best use of the trellis, but damage to the canes is possible if greatest care is not exercised.

In the loop system, canes from one root are divided into two ropes, each one being carried around the top wire once or twice, then brought down to the lower wire and turned back toward the plant with one or two twists. Thus a loop is

METHODS OF TRAINING BLACKBERRY PLANTS

1. Trained to a stake.

2. Two-arm system trained to one wire.

3. Wreath or weave system—2 wires.

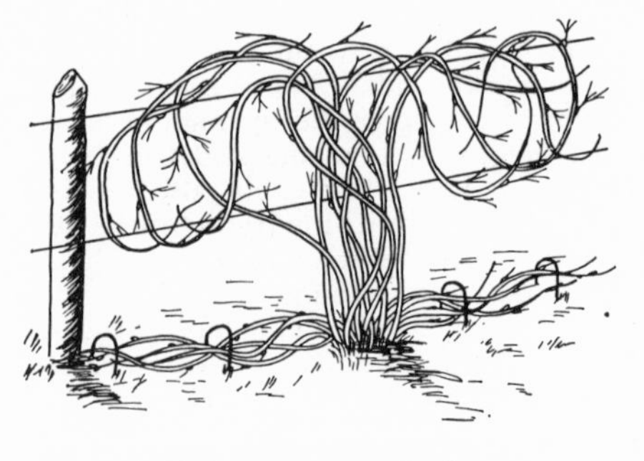

4. Loop system on 2 wires.

5. Semi-loop system.

6. Four-arm system.

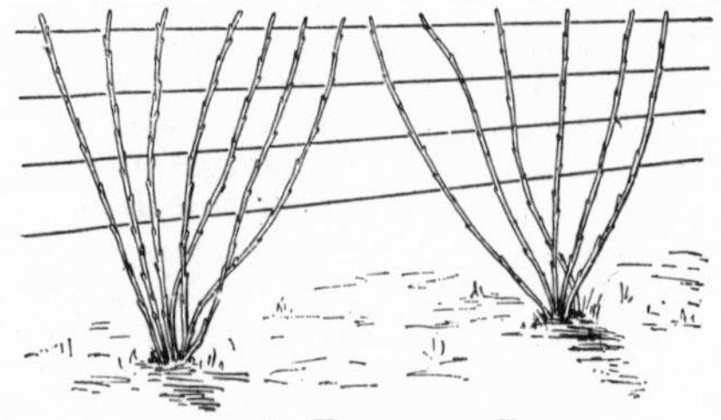

7. Fan system.

formed in each direction from the root and the excess is trimmed off each cane where it returns to the center, whence it started. A variation of this system is to loop the ends of the canes around the lower wire going in the direction of the neighboring plant, instead of turning it back toward its own center. Ends of the canes of the two neighboring vines then are interwoven, as in the 2-arm or 4-arm systems or they may be cut off where they meet and tied.

With very vigorous growers, such as Brainerd, canes may be pinched off when they reach the cross wires and strong laterals will develop at right angles to the parent canes. The laterals are trained along the wires as in the 4-arm system. This method eliminates one bend in the canes and the likelihood of damage at that point.

To achieve the largest yield in a small space, vines may be planted 3 or 4 feet apart if they are trained in a fan pattern to a 3- to 5-wire trellis. This system works well only where summer training is possible, because neighboring vines become too much entangled unless the canes can be tied as they grow.

To train in a fan, bring up 2 or 3 canes at a time and tie them to each wire. Cut off the canes when they reach the top wire. Canes should be spread out fan-shape on the wires until the outside canes approach the neighboring plant. After that, additional canes are pruned out, or are tied between the first ones that were tied. This system produces very high yields, but takes much more labor in training than any of the others.

Taming a Brier Patch

If you find yourself in possession of an overgrown planting or of a wild blackberry patch, it is possible to subdue it to make harvest possible without the use of wires or trellises.

First map out rows 7 feet apart, each row 18 inches wide, through the jungle. For the next step, stout leather gloves, long pruning shears and a good deal of determination are necessary. Remove all

Blackberries will grow in many soil types, thriving best in a moderately light loam containing a large amount of humus. Roots penetrate to a depth of 3 feet.

canes growing between the projected rows. After they have been cut off to the ground, cut back canes on remaining bushes to no more than 3 to 5 feet. Remove all dead and weak canes and prune to leave strong canes growing no closer together than 8 to 12 inches.

If the soil around the roots seems dry and lacking in humus, prune more severely until the soil can be built up to support more growth. In spring, when the soil is thawed, cover it with a thick layer of strawy manure. During the first growing season or two after cultivation of the patch is begun, it will be necessary to grub out roots of the bushes that were cut back to make aisles. All roots must be removed wherever a shoot appears—the shoot must be pulled up, not cut back. Each shoot cut back encourages two more to sprout from the same spot. A thick mulch will help to discourage growth, but will not stop it entirely.

Fertilizer

Blackberries use nitrogen in large quantities, with potash and phosphorus in smaller amounts. They also need large amounts of humus to maintain soil moisture during the driest months of the year. Demands made by different varieties vary, with loganberries requiring more than most of the others.

The best way to supply the various foods, and at the same time the humus, is with thick mulches of strawy manure, leaf mold, lawn clippings, sea weed, peat moss, compost or spent hops. Of these, well-rotted manure is preferred.

An application of 50 to 100 pounds per 1,000 square feet of rotted manure may be applied in the late fall after growth has stopped and the nitrogen will leach down with winter rains or melting snow to be ready for spring growth. This amount should supply one to two pounds of pure nitrogen, enough for 15 to 18 vines. The manure may also be applied in early spring, but never after the first of May or it will stimulate late soft growth of canes which will not ripen before the next winter's cold hits them.

If potash and phosphorus are supplied separately, they should be applied to the plants to reach the roots in late summer and early fall. Both of these substances hasten ripening; they will help the canes to harden against the coming winter. Where winter injury is a problem, an application of wood ashes or bone meal in late summer may help.

Propagation

Bush blackberries are usually propagated by root cuttings. Trailing varieties are more often propagated by rooting the tips of the canes. Where thornless varieties are to be propagated, they will come true to the parent plant only if they are tip-rooted.

To make root cuttings, take up roots one-fourth to one-half inch in diameter in late fall or early winter. Cut the root into pieces 2 to 4 inches long and bury in sand or light loam in a cold frame, greenhouse or (if the weather is mild in the area) in outdoor beds. In spring, after shoots develop, take up the cuttings and set them 3 to 4 inches apart in a nursery row.

Allow the young bushes to grow in the nursery the first summer. After the leaves drop in fall, bend the canes down and cover them with 3 to 4 inches of soil or a heavy mulch, if the winters are cold. Remove the cover early the next spring, before the buds begin to swell, and plant the bushes in their permanent positions.

Cane tips are rooted in early fall, after the canes develop a characteristic "rat-tail" tip. Canes which are to be used for rooting are pinched back at the beginning of the summer to force several canes to develop below the pinch. Each well-grown cane may be used for rooting.

When they are ready to make tiproots, the ends of the canes thicken up and the leaves become small. The canes should then be bent down, and the tips inserted at least two inches into a shovel cut in the soil, entering it at a 45 degree angle. Firm the soil against the tip. After several weeks have passed, roots should have formed and the old cane may be cut away from the tip. In spring, the plantlet is moved to the nursery, where it is treated in the same manner as a root-cutting sprout.

Suckers may also be used for propagating blackberries, but the number which can be dug in a single year will not increase the size of the planting very much. If a large number of new bushes are wanted, the root cutting or tip rooting methods are better.

Enemies

Blackberries are usually very healthy when grown with enough food, moisture and proper pruning. A few diseases, however, should be watched for and rooted out as soon as they appear.

Orange rust is one of these. When plants show a tendency to throw out spindling shoots with narrow leaves, look for bright orange spores under the leaves. If they are there, the only remedy is to dig out and destroy the infected plants, including all suckers, before the infection spreads. Orange rust is a systemic disease which spreads through the canes and not one must be permitted to remain. If there are any wild berry plantings in the neighborhood, they, too, should be inspected for the disease. There is no gain in rooting out all the infected members of your own planting if you allow the disease full sway over the fence.

Another fungus disease is the one which makes double blossoms from which no fruit is set. It is called, appropriately enough, double blossom. If any blossoms with crinkled multiple petals appear among the healthy blossoms of your planting, the best remedy is to inspect all buds before they open. Buds of double blossom may be identified by their chubby contours. The buds are not only fleshier, but also are redder than normal blooms. Pick off and destroy all such buds before they open and release spores which they enclose.

Anthracnose, which infects a number of garden subjects, is also the enemy of blackberries. Symptoms of the disease are small ash-gray spots on canes near their base and also on the leaves and fruit. Best preventive is to avoid planting other anthracnose-prone fruits and vegetables near the blackberries. This includes beans, cucumbers, grapes, raspberries and sweet peas. Also, do not work on the bushes or pick the fruit while they are wet.

Various virus diseases may infect blackberries, but in a thrifty planting they seldom do. However, if you find single plants with mottled yellow leaves in a healthy, dark green row of blackberry bushes, you might justifiably suspect virus. Dig out and destroy all such plants. If the whole row presents the same mottled yellow aspect, the chances are that the planting is suffering from a shortage of nitrogen. If the season is not too far advanced, a heavy application of blood meal or cottonseed meal will bring back the healthy green color to the leaves.

An injury which seldom troubles organically grown bushes is crown gall. This is a tumorous growth on the plant crown made by a bacterial infection which enters a bark abrasion on the root or crown just under the surface of the soil. Such abrasions are usually caused by cuts made by a hoe or harrow. If the bushes are well mulched and not cultivated, their roots need never suffer from crown gall.

Spider mite is another enemy which should not trouble mulched bushes, except in seasons of extreme drought. If leaves are dull and pale, with yellow dots on the upper surfaces and webbing beneath, they may have an infestation of spider mite. These pests are more prevalent when the weather is dry. If the infestation becomes bad, provide irrigation and the mites will disappear. Healthy plants with enough moisture do not suffer from them.

Best preventive for all blackberry diseases and pests is a clean berry patch. Remove all old canes as soon as they are through bearing, and cart them away from the growing bushes to the compost heap. Any leaves which drop out of season are suspect and should also be removed at once. When pruning, cut the canes off all the way back to the soil level. Do not leave untidy stubs for insects to feast on. Pick all the berries, even the few that turn small and dry toward the end of the season. With these precautions, an organically grown blackberry patch should never be troubled by enemies.

Harvest

Blackberries are not ripe as soon as they turn black. They must be allowed an extra day or two to sweeten. A ripe berry will fall off the bush into the hand.

Pick berries into shallow containers which will allow the fruit plenty of room to spread out without being weighed down by too many layers above. When the containers are full, put them in the shade until the picking is finished. If they are left too long in the sun, they lose color and turn sour. Berries are very tender—they should be picked in the cool of the morning, handled as little as possible and refrigerated or processed as soon as they can be taken indoors. Berries that are not sprayed seldom need washing, unless they are exposed to roadside dust. If they must be washed, dip them briefly into still, cold water and drain immediately.

TABLE 61: Blackberry Varieties

Variety	Suited for these sections	Fruit characteristics	Bush characteristics	Cold endurance
Evergreen	North Pacific coast Appalachian	Black. Subject to attacks by red berry mite.	Trailing, thornless. Productive. Perennial canes.	Hardy to +5°.
Marion	North Pacific coast	Medium large, firm black. Excellent flavor.	Trailing, canes few and long, arched fruiting laterals. Thorny. Resistant to verticillium.	Hardy to +5°.
Logan	Pacific coast	Medium to large, dark red, medium firm. Subacid to acid.	Trailing. Thornless or thorny. Vigorous. Productive.	Hardy to +5°. (Fruit and leaves both shrink in hot weather.)
Eldorado	New England Appalachian Southeast Corn Belt Northern plains Southern plains Mountain States Pacific coast	Black, medium size, medium firm, sweet. Midseason.	Bush, productive. Prune canes back to one-half to increase size of fruit. Resistant to orange rust.	Hardy to —15°.
Snyder	New England Northern plains	Black.	Bush. Pinch back canes 30 to 36 inches. Shorten laterals to 7 to 10 buds in spring.	Hardy to —25°.
Early Harvest	Delta States	Very early. Firm, small berries, black.	Bush. Vigorous. Subject to rust.	Tender.
Lawton	Southeast Delta Southern plains South Pacific coast	Early to midseason. Large, soft, black. Sweet.	Bush, productive. Fruits near base of lower laterals. Prune cane back moderately, shorten lateral one-half.	Hardy to —5°.
Boysen (Nectar)	Pacific Southeast Delta Southern plains	Early. Large, firm, dark red. Tart.	Trailing, occasionally thornless, not dependably so.	Hardy to +5°.

TABLE 61: Blackberry Varieties (Continued)

Variety	Suited for these sections	Fruit characteristics	Bush characteristics	Cold endurance
Brainerd	Pacific Southeast Southern Corn Belt	Midseason to late. Large, black, firm.	Semi-upright, but needs support. Thorny, vigorous, very productive.	Moderately hardy.
Cascade	Pacific coastal areas	Dark red, slightly smaller than Logan. Soft. Early.	Trailing. Especially recommended for home gardens.	Hardy to +5°.
Youngberry	Pacific States Delta Southern plains	Large, dark red, sweet, excellent flavor. Later than Boysen.	Vigorous, trailing, thorny. Not a heavy yielder.	Subject to winter injury.
Bailey	New England Appalachian Corn Belt	Midseason. Large, firm, black. Subacid.	Bush. Vigorous. Tall. Productive.	Hardy to —15°.
Hedrick	New England	Early. Large, medium firm. Black.	Tall vigorous bush. Susceptible to orange rust.	Hardy to —15°.
Lucretia	Appalachian Delta Southern plains Mountain States Corn Belt	Black. Large. Firm. Tart.	Trailing. Moderately vigorous and productive. Winter protection needed in North.	Hardy to —5°.
Japanese Wineberry	Southern New England Appalachian	Amber to scarlet, soft, tart.	Semiupright, soft spines, very decorative.	Hardy to —15°.

To freeze blackberries that have been picked at peak ripeness, simply pour them into freezer containers as soon as they are brought in from the garden and put them in the freezer. They are sweet enough for a delicious midwinter dessert and their garden freshness is best preserved in this way.

BLUEBERRY

BLUEBERRY. Several species of the genus *Vaccinium* having similar fruits are called, collectively, blueberries. The same genus of the heath family, *Ericaceae,* embraces bilberries, whortleberries, cranberries and huckleberries, with which blueberries are sometimes confused. In many parts of the country the names huckleberry and blueberry are used interchangeably and in some places they are applied to the wrong fruits exclusively. The name huckleberry is properly applied only to members of the *Gaylussacia* genus, also the heath family, whose fruit, which has 10 definite stones, is a gritty object much less suitable for cultivation than the small-seeded blueberry.

Until 1909 there were no garden varieties of blueberries. Many wild ones were picked every year in the more favored sections of this continent, in such remotely separated places as Alaska, Oregon, Michigan, New Jersey and Labrador. A few grew wild through some parts of the Appalachians and up into New England. In the South, the rabbiteye blueberry was picked from the swamps in Georgia and parts of Florida.

In 1909, Dr. F. V. Coville of the United States Department of Agriculture, with his experiments with blueberries, started work on the cultivated fruit which is newest to our gardens. Attempts had been made sporadically to improve blueberries, but progress was not very satisfactory until Dr. Coville pointed out that the key to their culture was acid soil. All of the fine modern strains are the result of his experiments.

Most of the modern varieties are developments of the high-bush blueberry, *V. corymbosum,* the low-bush blueberry, *V. lamarckii,* and the rabbiteye blueberry, *V. ashei.* The evergreen species, *V. ovatum,* which is native on the Pacific coast, is the blueberry used for greens by the florist.

In the wild, blueberries grow where there is loose, acid soil with abundant year-round moisture, usually supplied by a water table no more than two feet below the surface. Where they seem to grow in swampy bogs, close examination shows that the blueberry plants are seated on hummocks which raise their delicate roots above the standing water. Where they are perched on mountain sides which may be dry on the surface, usually it will be found that their roots reach down to lateral underground seepage along the surface of bedrock. This profuse supply of water is necessary because they feed directly through fine roots with no root hairs. The bushes grow from underground rhizomes which give rise below ground to a mass of very fine roots and to stems above ground.

As might be expected from their natural habitat, blueberries are very winter-hardy, though the more recently developed varieties are becoming less so. When properly hardened, they can stand temperatures down to 25 to 30 degrees below zero, though some stems may be killed below 20 degrees below zero. Even their blossoms are tough, being able to withstand freezes down to 23 degrees. On the other hand, they demand a period of winter chilling and, except in the rabbiteye varieties, will not bear properly without it. To produce well, blueberries must have 650 to 850 hours in their dormant period of temperatures which are below 40 degrees—approximately 100 cold nights. They ripen their fruits 60 to 90 days after blooming and then need an additional 40 to 50 days before the first fall frost in which to mature buds for the following year's fruit. They are the latest berries to ripen, starting to bear when the brambles are almost finished and their season may be extended until September if late varieties are planted, along with early and midseason types.

Blueberries contain small quantities of most of the vitamins and minerals, being richest in vitamin A. In one cup, raw, there are 21 grams of sugar, 22 milligrams calcium, 1.1 milligrams iron, 400 units vitamin A, .04 milligrams vitamin B_1, .03 milligrams vitamin B_2, .4 milligrams niacin and 23 milligrams vitamin C.

Ordering Stock

Blueberries are slow to mature. They do not bear a full crop until 6 to 8 years after planting, though they may be permitted to ripen a small amount of fruit in their fourth year. A fully mature bush,

TABLE 62: Blueberry Varieties

Variety	Section for which recommended	Fruit characteristics	Bush characteristics	Comments
Earliblue	New England Great Lakes Appalachians	Very early. Large, firm, subacid. Good quality.	Vigorous, productive. Upright bush. Hardy to —25°.	New and very satisfactory.
Rubel	New England Pacific Mountains	Late. Small to medium size berry. Firm. Subacid.	Wild bush brought into cultivation. Very hardy. Upright habit. Resistant to canker. Vigorous. Productive.	Very good for muffins.
Jersey	New England Appalachians Pacific	Late. Large berries, large loose clusters. Acid until ripe.	Vigorous, productive. Subject to canker. Hardy to —30°. Upright.	Old variety but good.
Bluecrop	Appalachians New England	Midseason. Big berries, firm, subacid, good quality.	Vigorous, productive. Fruit should be heavily pruned to maintain size. Hardy to —25°. Upright.	New.
Berkeley	Appalachians New England	Large berries, large fruit clusters. Midseason. Subacid.	Vigorous, productive. Open spreading habit. Less hardy than average.	New.
Burlington	Southeast Pacific Appalachians New England	Latest variety. Fruit cluster fairly compact. Medium size fruit.	Moderately productive, moderately vigorous. Upright. Hardy. Grows slowly. Resistant to canker.	Old. Superseded by better late varieties.
Coville	New England Appalachians	Very late. Big berries of very high quality. Open spreading habit. Too late for cold areas.	Vigorous, productive. Open and spreading. Subject to canker.	New.
Ivanhoe	Appalachians	Midseason. Very large firm berries. Subacid.	Vigorous, productive. Large upright bush.	New.
Rancocas	Pacific Appalachians	Early midseason. Good firm berries, medium size. Subacid.	Moderately vigorous, productive, hardy. Upright. Resistant to canker.	Old variety now superseded.
Callaway	Delta States Southern plains Southeast	Early. Medium size. Excellent quality.	Rabbiteye variety. Tall, vigorous and productive. Plant 8 feet apart. Hardy only to 5°.	New.
Croatan	Delta States Southeast	Early. Large, dark, firm berries. Fair quality.	Vigorous. Canker resistant.	

TABLE 62: Blueberry Varieties (Continued)

Variety	Section for which recommended	Fruit characteristics	Bush characteristics	Comments
Scammell	Pacific States Appalachians Southern plains Southeast	Midseason. Medium size. Fair quality.	Subject to canker.	
Stanley	Appalachians Southern plains	Midseason. Medium to small fruit. Firm. Sweet.	Vigorous. Moderately productive. Upright. Hardy.	Old, but very good for pies.
June	Pacific States Southern plains	Very early. Fair quality. Medium size.	Not vigorous. Moderately productive. Upright. Resistant to canker. Subject to leaf spot.	Old variety now superseded.
Concord	Pacific	Late midseason. Medium size, firm, good quality. Subacid.	Vigorous, productive. Upright, hardy to —25°. Spreading habit.	Old, but very good for freezing.
Dixi	Pacific	Late. Very large but size and quality diminish after peak production.	Very tall, moderately productive. Less hardy than average.	Old. Superseded.
Pemberton	Pacific	Late. Large, but size and quality diminish after peak production.	Very hardy, vigorous and productive. Upright.	Old.
Weymouth	Pacific	Very early. Medium to large. Firm, but poor quality berry.	Medium vigor. Moderately productive. Spreading. Resistant to canker.	Old. Superseded.
Atlantic	Pacific States Northeast	Late. Firm large berries, acid till ripe.	Vigorous. Productive. Hardy to —30°. Open, spreading. Subject to canker.	Old variety.
Cabot	Pacific	Medium size, early, subacid, medium firm.	Moderately productive, not vigorous. Low bush, open habit. Moderately hardy.	Old variety now superseded.
Herbert	New England	Late. Very large, firm, very good quality. Subacid.	Very productive. Upright, vigorous.	New and good.
Pioneer	New England	Midseason. Medium firm, with good mild flavor.	Medium vigor, productive. Low, spreading. Hardy. Resistant to canker.	Old. Superseded.
Blueray	New England Great Lakes	Very good flavor, large. Midseason.	Hardy to —25°. Tall vigorous bush.	New.

one between its 10th and 15th years, may yield as much as 14 to 15 quarts of fruit, when it is well-grown.

Since they grow so slowly, one-year bushes are too small to plant, having just come out of the nursery beds. Order at least two-year-olds, which will be 8 to 18 inches high, or 3-year-olds, which may measure up to 24 inches. After the first 3 years they grow faster, but it is not advisable to buy older plants. It is difficult to dig complete root systems after the third year and they may suffer too much shock from transplanting. Also, an older bush is probably a cull, left in the nursery because it was not large enough at 3 years to dig. Bushes should never be dug bare-root, but always with a ball of soil.

It is best to buy blueberries from a specialist who will have a larger selection of named varieties. With all of the modern developments, it is foolish to buy unnamed varieties which are probably older types now superseded by new improvements. Blueberries are self-fertile, but experience with the new varieties has shown that the berries are larger, more abundant and earlier if they are cross-pollinated. At least two varieties should be planted for best results.

Site

Except in very hot and arid sections, blueberries should be given a spot where they will have full sun. A very little shade may be permitted them where summers are hot; a little more where the air is dry.

Because they are so close to their wild forebears, blueberries need to be given conditions as near as possible to those which they enjoyed in nature. This makes them ideal subjects for the organic garden. They thrive on large quantities of humus, thick cool layers of mulch over the soil, natural fertilizers and simple composts and very little fussing. They resent worked out mineral soils, chemical fertilizers and spraying.

Best soil is a light loam with large quantities of acid humus and sand which has never been near a seashore, without salt or alkali. Heavy clay loams or dense clay will not grow healthy crops. The fine roots are not able to penetrate too dense a medium and the heavy clay loams are usually not acid enough, nor is it possible to permanently acidify them. Soil with a natural *p*H of 4. to 5. is very good, with *p*H 4.5 best. Blueberry roots are not able to utilize nitrates, but must have their nitrogen in the form of ammonia. Under *p*H 5.2 the soil organisms which convert nitrates to ammonia flourish. When the *p*H rises to 5.5, organisms which carry on the opposite function begin to operate. As the *p*H rises, the plants begin to suffer from malnutrition. In a limestone area, water usually has a *p*H of 6. or more. Even if the soil is made acid by mixing acid humus with it, the water will eventually raise the *p*H again unless an acid mulch is maintained. There also seems to be some indication that the presence of calcium in the soil, even though the *p*H is lowered, is injurious to the plants.

Where the soil is well-drained and loose, blueberries may be grown successfully in the home grounds in areas where the soil can be properly prepared for them, even though the *p*H of the garden in general is higher than their requirement. Acid materials should be worked into the place where they will be planted at least 6 months to a year before they are set out. These materials may be in the form of pine, fir or oak sawdust, acid peat moss, oak leaves or pine needles and wood chips.

Where the soil tests higher than 5.5, it is advisable to grow the bushes in containers such as metal washtubs or half oil drums, sunk into the ground within one inch of the rims. Four drainage holes should be bored into the bottom of each container, which should be filled with acid leaf mold or peat mixed with topsoil, half and half. One plant is set in each tub. This treatment prevents seepage water from the surrounding area from reaching the roots and alkalizing the soil.

Planting

Bushes should be planted as early in the spring as the soil can be worked without packing. In California, planting is done in late fall; in some of the warm areas, it is done in late winter. Where the winters are not too extreme and two-year plants are being set, they may be planted in fall if they are protected against winter heaving. This may be done by mounding soil high around the stems or providing them with a very thick mulch.

Plants are set 4 to 8 feet apart in the row, depending upon variety. Rows are

When harvesting blueberries in the home garden, allow them to ripen until they are slightly soft and the flavor is rich and sweet.

usually 8 feet apart. Before planting, spade peat moss or other organic matter into the top 6 inches so that the soil is half topsoil, half humus. This mixture should cover a two-foot circle where the plant will stand. Set the bush so that the root-ball is one to two inches below the surface of the soil and lightly pack the soil around it. Give it a 4- to 6-inch mulch of well-rotted sawdust immediately.

Trim back branches one-half when planting, to equalize the top and the roots. Remove all fruit buds, short twigs and dead wood. Keep the soil constantly moist under the mulch, especially the first year.

Maintenance

Each spring blueberries should be given an application of acid compost (compost made without lime) and the mulch should be renewed to bring it up to a depth of 6 inches. As the bush grows, the area mulched should be increased until the mature bush has a circle of mulch 6 feet in diameter. Where the soil has naturally a low *p*H, the mulch may be composed of straw, hay, corncobs or leaves. If the *p*H must be lowered, a mulch of pine needles or oak leaves is best. If sawdust is used, it should be composted with manure first or it should be aged for at least a year.

Pruning

The principal purpose of pruning blueberries is to regulate the crop size in order to maintain plant vigor and fruit quality. Improved varieties tend to overbear and an excessive crop can check growth almost completely, reducing future crops, as well as reducing the size of the berries.

Until they are 4 years old, bushes should not be allowed to bear at all. Blossom clusters may be stripped off as they appear or may be removed in the dormant period. Their buds are identified by their plumpness, in contrast with the small pointed leaf buds.

As the bushes become large enough to

bear a crop, only part of the fruiting wood should be removed. The amount to leave depends upon the bush and must be learned by experience. A good rule of thumb is to leave one fruiting bud for each 3 inches of new shoot. It must be remembered that each bud will produce 5 to 8 berries. To crowd more than that number into a space of 3 inches would seriously reduce their size.

In addition, the number of canes which are allowed to grow should be limited to one for each year of the age of the plant, up to 6 or 8 canes maximum. As the bushes grow older and have an extensive root system, they will send up new vigorous canes taller than the bush. These may be used as additional or replacement canes. They should be headed back to the height of the bush to encourage laterals, which usually fruit the following year. After the bush has been growing in the garden for 3 or 4 years, some of the lower, more spreading branches may be cut out and replaced by taller canes which make harvesting easier.

Pruning out the small, weak laterals can be a heavy chore when the bushes become full. If it is done in the winter, wear a pair of heavy leather gloves and rub the mittened hands along the branches. The spindly, weak twigs, made brittle by the cold, will break off easily, leaving only the stronger fruiting wood.

Propagation

Blueberries of the improved varieties are propagated by hardwood cuttings, started in spring in special frames or propagating beds. They are difficult to root and take experience and skill. Since the bushes bear for a long time, it is probably better to buy new ones than to try to raise them.

It may be easier for the amateur to propagate the few extra bushes. He will want to increase his planting by layering than by rooted cuttings. Make a cut half way through one of the lower, drooping branches, on its underside. Pin the cut side down and mound a little of the mulch over it. Allow it to remain for a year before it is severed from the parent bush.

Enemies

Tent caterpillars are the most damaging insects which infest blueberry plantations. Their eggs are laid on twigs and should be removed if they can be found during the winter. If not, they begin to spin their webs in early spring. As soon as they start, the webs and all their inhabitants must be removed and destroyed. It is also helpful to remove all the wild cherry trees in the vicinity because wild cherries are the caterpillars' favorite host.

Mummy berry is caused by a fungus which infects the berries and sometimes travels back into twigs. Its spread is implemented by wet weather and poor air circulation. If blueberries are planted where the air circulation is bad, leaves and fruit cannot dry quickly after a rain and the fungus flourishes. A healthy specimen grown where the air currents move freely is usually resistant to the fungus.

Canker which causes reddish-brown to black lesions in the stems kills the buds in the vicinity of the canker and sometimes girdles the stem and kills the whole cane. As soon as dead buds are discovered, cut the stem back to the canker and remove it. Also watch for blossoms which turn brown before setting fruit. This condition is often caused by botrytis blight. Twigs where it occurs should be removed.

Harvest

Berries turn blue several days to a week before they have developed their full quota of sugar. For the market, they are picked as soon as they develop a blue color, so many of the berries in the stores are tasteless. But in the home garden they may be permitted to ripen until they are slightly soft and the flavor is rich and sweet.

Do not pick the berries by pulling them from the stem. Instead, roll them gently with the finger tips cupping the hands below the stem. Those that come off with a gentle touch are ripe. Those that do not should be permitted to ripen further. After picking, they should be refrigerated until processed or used.

BOYSENBERRY

BOYSENBERRY. *See* BLACKBERRY.

BUFFALO BERRY

BUFFALO BERRY. *Shepherdia argentea,* a small thorny tree or large shrub of the oleaster family, *Elaeagnaceae.* Buffalo berries are small, currant-like fruits borne in great profusion on silvery, spiny trees 10 to 18 feet high. They may be grown in the most extreme climates and on wind-swept sites where little else will thrive. The species *S. argentea* is native to the northwestern states. Another even hardier species, *S. canadensis,* whose less flavorful berries are produced on a shorter, spineless shrub, is grown as far north as the Arctic circle.

Both species are dioecious, that is, the male and female flowers are borne on separate plants. In order to produce fruit, both kinds must be planted. They are easily propagated from seed.

S. argentea produces red to yellow berries with a very acid flavor which becomes sweet after the first frost. The berries are dried by the Indians for winter fruit. They are also used to make jelly. Fruit of the *canadensis* species is acid and bitter until frost-bitten, when it becomes rather sweet and insipid. It is used to make a drink which ferments very quickly and is refreshing in hot weather.

CALAMONDIN

CALAMONDIN. Whether the calamondin is a hybrid mandarin orange or a separate species of citrus fruit, *Citrus mitis,* is a matter for dispute among the experts. It is a small red-orange fruit with a loose skin like the mandarins, few seeds and very acid but flavorful. It is used for the same purposes as lime. The tree on which it grows is useful as an ornamental and is one of the most hardy of all citrus trees, especially on trifoliate orange roots. It is a useful dooryard tree in sections that are too cold for lemons or limes. Culture is similar to that of oranges. (*See* ORANGE.)

CANTALOUPE

CANTALOUPE. *Cucumis melo* is one of the melon genera of the family *Cucurbitaceae* which includes a number of varieties, one of which is cantaloupe. The real cantaloupes belong to the *C. melo* variety *cantalupensis,* which is the European cantaloupe not grown in this country. Our melons belong to two other varieties, *C. melo reticulatus,* the netted or nutmeg melons and *C. melo inodorus,* which includes the so-called winter melons, casabas and honeydews. But since we have become accustomed to calling all our melons cantaloupes, however incorrectly, we will use that name here.

Cantaloupes are the fruits of tender annual vines thought to have originated in India. In this country they must be grown on warm soil, preferably sandy loam, which is enriched to force quick growth and maturity during the warm weather. They must have plenty of sunshine and heat in order to ripen properly, but the heat is not necessarily applied while they are ripening—some of the winter melons prefer a cool period for ripening.

Melons supply vitamins and minerals in varying amounts, depending upon variety. The netted, orange-fleshed cantaloupes are richest in vitamin A; honeydews supply more vitamin C. In half a small cantaloupe there are about 900 units of vitamin A, .090 milligrams vitamin B_1, .100 milligrams vitamin B_2, 50 milligrams vitamin C, 32 milligrams calcium, 30 milligrams phosphorus, .5 milligrams iron and about 44 calories. One-fourth of a honeydew, weighing the same amount as the half cantaloupe, contains only 100 units of vitamin A, 90 milligrams of vitamin C and 35 calories.

Range and Soil

The *reticulatus* melons are grown throughout the United States, in any section where 80 to 100 days are not only frost-free but are also warm, with warm nights. Casabas and honeydews need 105 to 125 days and can be grown only in warm regions. In most sections these conditions can be found only where there is sandy to light clay loam. Melons do not thrive on heavy clay, adobe, muck or peat. Near neutral soil is necessary; cantaloupes will grow in soil with a *p*H of 6. to 8., but the less acidity, the better. Ground limestone should be added to any soil which tests below 7.

The larger the variety, the more nutrients cantaloupes must find in the soil to make a proper development. The best way to supply their food is in the form of well-rotted manure or compost. If a plentiful supply is at hand, the manure may be spread on the entire melon patch at the rate of 75 pounds over an area large enough for 20 hills. If the supply is limited, it may be dug into the hills themselves at the rate of one to two pounds to the hill—a good spadeful. Either manure or compost will provide the humus which melons must have to

TABLE 63: Cantaloupe Varieties

Variety	No. days to maturity	Fruit	Vine
Hale's Best—#36	90	Small, round, with heavy netting. High quality. Salmon flesh.	Susceptible to powdery mildew.
Hale's Best—Jumbo	88	Large, round, with heavy netting. Flesh salmon-orange.	
Honey Rock	85 to 88 days	Small to medium size, grey rind, orange flesh, sparse netting. Good.	
Hearts of Gold	85 to 90 days	Small to medium, green netted rind, orange flesh. Sweet, juicy.	
Hale's Best—#45	95 to 110 days	Small, oval, netted, shallow ribs. Sweet, excellent flavor.	Powdery mildew resistant.
Delicious #51	86	Medium size, round, coarse netting, grey rind, orange flesh.	Resistant to fusarium wilt.
Harvest Queen	95	Medium size, oval, sparse netting, orange flesh.	Resistant to fusarium wilt.
Iroquois	90	Medium to large round melon with green rind, coarse netting, orange flesh.	Fusarium wilt resistant.
Honeydew	110 to 125 days	Heavy round to oval melon with cream-white skin, no ribbing, sweet green flesh.	Stem remains attached at maturity.
Melogold (Honey Ball)	90 to 105 days	Small, cream to tan smooth skin, pink-orange flesh, very sweet.	Very prolific, vigorous. Very susceptible to powdery mildew.
Golden Beauty Casaba	110 to 125 days	Globular fruit, pointed at stem end, skin yellow with longitudinal wrinkles. Thick sweet white flesh.	Large, vigorous. Intolerant of heat at harvest time. Very susceptible to powdery mildew.
Crenshaw Casaba	110 to 125 days	Smooth pear-shaped fruit, with thin pale yellow to tan skin. Flesh very sweet, light salmon-orange.	Susceptible to powdery mildew.
Persian	105 to 120 days	Large melon with dark green netted skin, orange flesh, sweet.	Vines vigorous but subject to powdery mildew.

hold in readiness as much moisture as they require.

Ordering Seed

Most important in ordering seed is to find a variety which will mature in the number of days in which you have hot weather. A week, or at most two weeks, may be cut from the number of days to maturity if the seed is started indoors. But when the nights begin to turn cool in fall, even though there are no frosts, melons will not ripen properly. Their season should finish while days are still 80 degrees or more and nights above 50.

One ounce of seed will plant 20 hills. The number which may be harvested from a patch of that size depends entirely upon the conditions. It is possible, under the best combinations of soil and climate, to take 16 melons from a hill. It is also possible to take no more than one, though at that rate the returns do not justify the space that they occupy in the garden.

The melon patch should be given a sunny spot, one protected from cold spring or fall winds. It must be well-drained and situated where the foliage will have free air circulation to dry it after a rain. Contrary to old beliefs, cucumbers and squash do not cross-pollinate melons and there is no need to keep them apart in the garden.

Planting

In cool areas melon seeds may be started in peat pots or bands indoors or in the cold frame 10 to 30 days before the period for planting outside. The young plants must be put into the garden before they have developed more than 2 or 3 leaves, because the roots will not transplant well when they are too much developed.

Time to plant outside is when the soil temperature has risen to 50 degrees or more, when the nights are getting warm and the days are near 80 degrees. Melon seeds will not germinate under 50 degrees, but will rot in the soil if it is too cold. In some places, particularly in California, planting is done under hot caps. This is helpful if the soil temperature is high enough, but if not the seeds will germinate no better than if no hot caps are used. When seed is started inside, however, hot caps will help them to adjust to the outdoor conditions when they are transplanted.

Melons are usually planted in "hills" spaced 4 feet apart in each direction. If manure or compost is to be dug into each hill, remove a spadeful of soil and replace it with a generous forkful of manure which is well-decomposed. Replace the soil over the manure, mound it gently and plant the seeds in a 12-inch circle, 8 to 10 seeds to the hill. Cover the seed with ½ inch of fine soil and mulch the space between the hills with a thick layer of straw, hay or leaves.

The cantaloupe patch should be located in a sunny spot, protected from cold spring and fall winds. Melon seeds are planted in "hills" spaced 4 feet apart.

SAVING CANTALOUPE SEEDS

1. Cut the fruit in half, using colander to hold seeds scraped out of melon.

2. Separate the seeds from the pulp with a knife. Seeds should be stored carefully away from rodents.

3. These dried seeds from a home-grown cantaloupe can be planted in next year's garden.

When the seeds germinate, thin plants to 4 to 5 to each hill. Final thinning should not be done until the young vines have 4 or 5 leaves each, when all but two plants are removed from each hill. If hot caps are used, they may be safely removed by this time.

Melons need abundant moisture from the time they come up until they are almost full-grown. Water should be withheld during the period when they are ripening, for best flavor. If the soil contains plenty of humus and is thickly mulched, normal rainfall in most places will supply all the moisture necessary. But in periods of drought or where soil humus has not yet been built up, additional moisture may be necessary.

In dry areas a constant source of water may be supplied from a 5-inch clay flower pot sunk in the middle of each hill and which is filled each morning with water. If a clay plug is pushed into the hole in the bottom of the pot, the water should seep out into the soil slowly during the day, providing a moist root area several inches below the soil surface.

If too many melons set on the vines, especially if they continue to set late in the season, some of them may be removed to improve the size and quality of those which remain. Usually, fruit which is set after mid-July will not have time to ripen before cool weather. If this is removed, the food which would have gone into its development will be used by the remaining melons.

Enemies

Downy and powdery mildew, fusarium wilt and striped cucumber beetles which spread the wilt, are the melons' principal enemies.

Mildews attack the leaves of the plants in warm, wet weather, sometimes damaging them enough to kill the foliage. Fruit can be protected to some extent from mildew when the weather is wet, if a wooden shingle or small board is placed under each melon. Varieties subject to mildew are best grown in arid climates

Fusarium wilt causes the vines suddenly to wither away, sometimes almost overnight, and for no apparent reason, when they have been flourishing. This disease is caused by a fungus which is spread by the yellow and black striped cucumber beetles. If the plants are protected from the beetles, they will be protected from wilt. Hot caps over the hills when the seedlings are young help to protect the tender shoots from beetles, which attack them sometimes as fast as they germinate. Or tents of cheesecloth or mosquito netting may be spread over the hills until the plants achieve a large enough growth so that they are tough enough not to tempt the beetles.

Harvest

Most varieties of melons are ripe when their stems begin to separate from the fruit. Usually, when a crack shows all around the stem, the melon is ready to pick. If it is lifted and twisted slightly, it may come free of the vine. Or it may be necessary to press against the stem to free it. A few varieties do not leave their stems and must be judged for ripeness by their color or softness.

If melons are kept in a warm room for one or two days after picking, they develop more flavor. Their aroma at the stem end indicates to an experienced sniffer just when the fruit is ready to use. Cantaloupes may be stored at 50 to 55 degrees in a humid place for 1 to 3 weeks after harvest. Honeydews may be kept as much as 3 or 4 weeks; casabas can be stored from 4 to 8 weeks.

CAROB

Carob. *Ceratonia siliqua.* Carob is one of the borderline edible plants of which the seed and pod are eaten, yet it is not a nut, though it grows on a tree; it has no succulent flesh so it is not a fruit; and it grows on a woody plant so it is not a vegetable in the commonly accepted dictionary sense. We class it here with fruits, as being the closest category. But it does not really matter what we call it—it has been used by man since before the era of dictionaries or classified botanies.

Carob is also known as St.-John's-bread, because it is believed that John the Baptist lived on the husk of the carob during his sojourn in the wilderness. It is an evergreen tree indigenous to the Mediterranean shores, from Spain to Asia Minor and south to the coast of Africa. Carob belongs to the legumes, but it lacks the characteristic pea-like flower. The fruit is a pod 4 to 12 inches long, containing seeds surrounded by a sweet, nutritious pulp. The pods contain 40 per cent sugar and 6 per cent protein. They are ground into a flour which can be used in baked products, giving them a chocolate flavor; coarse-ground they make an excellent breakfast cereal; cooked down with water, the pods make a sweet honey-like syrup; and in the Mediterranean, the pods are used as fodder.

Carob trees need a warm climate with 14 to 24 inches of rain per year, none of it falling in autumn while the pods are ripening. In Europe they grow well near the sea. In California, where most of the American carob crop is produced, they must not be grown in the fog areas near the coast. When grown in such sections, the ripening pods turn moldy and are quickly spoiled by worms. A short rain during the fall months, if followed by a drying wind, does not do much harm to the crop. But after the pods have sweetened in September and October, they quickly ferment if they are not kept comparatively dry. Harvest is in October and November.

Culture

Carob trees are grown from seed, which should be planted as soon as the pods have ripened. Seeds are usually started in flats indoors and are transplanted, when their true leaves begin to form, to containers 12 inches deep. These may be carried along inside or planted out when frost is past. When the seedling trunks attain a thickness of 3/8 inch, they are budded with scions from desirable trees.

Trees should be allowed a space 30 feet in diameter in the garden. Best soil is a deep heavy loam with good drainage. They require almost no care after they are tall enough to surmount nearby vegetation. Bearing, which begins about the sixth year, may be expected to produce 5 pounds of fruit the first year and increase steadily until the crop is near 100 pounds by the twelfth year.

Dry carob pods may be shaken from the tree in fall and gathered from the ground. They are spread in the sun for a few days after harvest to complete the drying, after which they may be stored in a rodent-proof container almost indefinitely.

CHERRY

Cherry. The two principal species of the stone fruit genus, *Prunus,* from which our cultivated cherries are derived are *P. avium* and *P. cerasus.* Though they are distinctly different species and

TABLE 64: Sour Cherry Varieties

Variety	Section for which recommended	Fruit characteristics	Tree characteristics	Cold endurance
Early Richmond (amarelle)	Appalachians Corn Belt Southern plains Northeast Northern plains (southern portion)	Early. Small. Fair quality.	Fairly productive. Good on dwarf stock.	Hardy where winters average 10° to 20°.
Montmorency (amarelle)	Mountain States Appalachians Corn Belt Southern plains Southeast Pacific coast Northern plains Northeast	Midseason. Large fruit, bright red, tart.	Vigorous. High yield on good soil. Large trees.	Hardy with 10° to 25° winter average.
English Morello	Mountain States Northern plains Northeast	Late. Almost black, medium size, high sugar content but predominantly acid.	Small, spreading. Low yield, susceptible to leaf spot, but vigorous.	Hardy with winter average at 10° to 20°.
North Star (hybrid)	Northern plains Northeast	Cross between morello and Serbian Pie—both sour.	Small tree.	Very hardy in wood and bud.

TABLE 65: Dukes—Hybrid Sour and Sweet Cherries

Variety	Section for which recommended	Fruit characteristics	Tree characteristics	Cold endurance
May Duke	Appalachians Southern plains Southeast	Earliest hybrid, dark red, subacid, rich flavor.	Vigorous, erect, grows to 25 feet.	About as hardy as peach.
Royal Duke	Appalachians Southern plains	Midseason. Dark red, very large, sprightly flavor.	Large, vigorous, productive.	Hardy as peach.
Late Duke	Appalachians Southern plains Corn Belt Mountain States	Late. Dark red, medium to large.	Vigorous, productive. Breaks fairly easily in high winds.	Hardier than two above—blossoms later.
Reine Hortense	Appalachians Southern plains	Midseason. Light red. Poor keeper. Sweet.	Vigorous but brittle. Sometimes unproductive.	Hardy as peach.
Brassington	Corn Belt	Red, medium size, good for pies. Early.	Sometimes lack vigor. Low yield. Breaks easily.	

TABLE 66: Sweet Cherry Varieties

Variety	Section for which recommended	Fruit characteristics	Tree characteristics	Cold endurance
Black Tartarian	Southeast Pacific States Appalachians Northeast (southern part)	Black, juicy, rich. Small to medium. Early.	Vigorous. Productive. Good pollinator. Susceptible to brown rot.	Less hardy than Napoleon.
Governor Wood	Southeast Corn Belt	Yellow. Small. Very good. Hangs well on tree. Early.	Healthy. Vigorous. Sometimes but not always productive.	Fruit buds fairly hardy.
Napoleon (Royal Ann)	Northeast Appalachians Corn Belt Pacific States	Yellow, tinged red. Large. Sweet. Firm.	Vigorous, productive. Large, open top, spreading. Self-sterile.	Fairly hardy fruit buds.
Windsor	Northeast Appalachians	Purple-red. Sweet, good flavor. Small, does not crack easily. Medium late.	Good pollinator. Spreading, vigorous. Resistant to brown rot.	Hardiest sweet cherry.
Schmidt	Northeast Appalachians	Black. Large. Medium late.	Hardy. Productive. Late coming into full bearing.	Relatively hardy in changeable weather.
Bing	Pacific States Southeast Northeast	Dark red, large, aromatic. Very good. Midseason. Crack easily.	Large, spreading. Self-unfertile. Better producer in West than in East.	Fairly tender in fruit bud and blossom.
Lambert	Mountain States Northeast Pacific States	Midseason. Purple-red, large, firm, sweet, aromatic. Cracks in rainy weather.	Vigorous. Strong. Productive. Upright. Self-unfertile.	Fairly hardy.
Seneca	Appalachians	Very early. Purple-black. Medium size. Rich flavor.	Vigorous, productive. Tall. Susceptible to brown rot.	Not very hardy.
Giant	Corn Belt	Dark purple to black. Smaller than Bing, but better flavor.	Vigorous, productive.	Fairly hardy.
Deacon	Mountain States Pacific States	Black. Early midseason. Medium to large. Sweet. Fairly subject to cracking.	Large, upright, moderately spreading. Vigorous.	Not hardy.

of differing origins, they are nearly enough allied so that they may be crossed to produce hybrids and they may both be grafted to the same rootstocks.

P. avium is the sweet cherry species, native to Europe and the warmer sections of the Caucasus. It includes wild cherries, mazzards and bird cherries. Several cultivated strains have come out of *P. avium,* notably the mazzard seedlings, used as rootstocks for the best sweet and sour cherry trees; the Hearts or Geans, sweet, soft-fleshed juicy varieties such as Black Tartarian; and the Biggareaus, such as Royal Ann, equally sweet but firmer of flesh and less juicy. In nature *P. avium* trees are tall, growing sometimes as much as 100 feet. They are sensitive to both heat and cold, disliking long, hot summers and cold winters, especially if the freezing weather is broken by warm spells that suddenly turn cold again.

Pie cherries or sour cherries belong to the species *P. cerasus,* a more hardy type with smaller trees and a wider range, native to many parts of Europe and the Orient. Two definite strains of sour cherries have been developed, the amarelles and the morellos. Amarelles are small trees with pale red fruit which has colorless juice and is moderately acid. Fruit of the morellos is more definitely acid, even when fully ripe. They are dark red, with red juice and the morello trees grow almost like tall, lilac-size shrubs.

A group of hybrids which are between sweet and sour have been developed, the group being designated Dukes after the first-named hybrid, which was called Médoc and corrupted to May Duke. Flavor of the Dukes is less sour than any of the pie cherries, and the habit of the trees varies, sometimes resembling one parent and sometimes the other.

Several other species are grown where conditions do not favor sweet or sour cherries, but most of them are not much improved by cultivation. In Alaska, portions of Canada and the northwestern United States *P. pumila,* the sand cherry and *P. virginiana,* the chokecherry, are grown for their astringent fruits. These were used by the Indians who dried them for winter fruit. They are too sour for anything except jelly or jam and are cultivated only where few other fruits can be grown.

P. Besseyi, also a hardy species, has been slightly improved to produce some of the modern bush cherries. These may be grown in some of the northern plains states where temperatures are too low in winter for even the hardiest sour cherries. Fruit of the bush cherry is black-purple and tart, the bushes growing 4 to 5 feet tall. Two varieties, Hanson and Brooks, are recommended for climates as severe as Alaska's; 3 others are being developed for the colder portions of the United States—South Dakota Amber, Black Beauty and Sioux.

Sour, sweet or hybrid cherries all seem to have about the same nutritional value. In one cup of any of them, raw, there are about 650 units of vitamin A, though this varies with the color of the fruit, the lighter ones containing less, the darker more; there are also .05 milligrams vitamin B_1, .06 milligrams vitamin B_2, .4 milligrams niacin, 9 milligrams vitamin C, 19 milligrams calcium, .4 milligrams iron and about 65 calories.

Within the last few years, experiments have been made with cherry juice as a diet for arthritics. When their malady is gouty arthritis, patients on the cherry diet seem to have improved. It is less efficacious for the rheumatic type of arthritis.

Ordering Stock

Successful cherry culture is more dependent on climate than on almost any other factor. It is very important, then, to order stock of a variety that suits the climate where it is to grow. Sweet cherries succeed where peaches are grown, but are just as tender as the peaches. Sour cherries may be grown in the same areas as some of the less hardy apple varieties, such as McIntosh and Northern Spy. Dukes are variable, depending upon their parents. Cherries are less adaptable to unfavorable climates than many fruits and should not be attempted in sections where they have a bad record.

Cherries are grafted to either of two rootstocks, mazzard or mahaleb. Mazzard is a wild cherry which grows to 40 feet or more, with a strong, sturdy trunk. Mahaleb is a bush-like tree, shorter, with a thick top and slender branches. Where soil conditions are perfect, trees grafted to mahaleb do almost as well as those on mazzard stock. But where there is any deficiency in soil or climate to overcome, trees on mazzard stock grow very much better. Mazzard-grafted trees are longer

The sour cherry is not only beautiful in springtime, but it is an early and bountiful yielder as well. Given proper care, it will produce quarts of fruit, good for eating and preserving.

lived and sturdier, but on the other hand, trees grafted to mahaleb begin to bear while they are younger. Mazzard grafts are usually more expensive because they are harder to make in the nursery and the trees are not as large at one or two years as those on mahaleb, so they do not have so much sales appeal and nurserymen are reluctant to attempt to grow them. They may be harder to find, but where they can be purchased, mazzard trees will be worth their extra cost.

Trees are usually sold at the end of their first or second year after grafting, as one- or two-year olds. They should be 4 to 5 feet tall, with trunks nearly ¾ inch thick. If they are two-year olds, they should have developed some strong lateral branches. Roots should be well protected from drying by damp moss and burlap.

Cherries come into bearing 3 to 5 years after being set out in their permanent places, but do not attain their maximum growth and bearing for 10 to 20 years. A mature sweet cherry tree may yield as much as 3 bushels of fruit; an amarelle, perhaps 2½ bushels; and a morello does well to yield two bushels.

Most sour cherry trees are self-fertile. Sweet cherries usually are not and some varieties of sweet cherry, such as Bing, Lambert and Napoleon, are intersterile. These 3 must be planted with good pollinators, such as Windsor or Black Tartarian, but care must be taken to get two varieties which bloom at the same time for cross-pollination. The advice of a good nurseryman is invaluable when choosing varieties to plant together. Hybrid cherries are, of course, self-sterile. They must be planted with sour cherry varieties for pollination.

If the home garden is too small for more than one cherry tree, it is possible to graft pollinating branches to a self-sterile variety while it is young. (This job can also be done when the tree is older, but it will take at least 4 years for the grafted branch to bloom and be effective.) Scions for bud grafts should be available from the nursery where the tree is purchased. It is advisable to wait for the young tree's second spring, to allow it to start growing, before attempting to make the graft.

Site

Like other fruits with tender blossoms and fruit buds, cherries profit from being planted on a slope where the cold air will drain away on frosty nights in late spring. If there is a choice, plant them on the side of the slope, rather than in a pocket at the bottom.

Soil on the hillside is also likely to have better soil drainage than that in the hollow and drainage is very important to cherries. On a loose, well-drained soil with no clay hardpan near the surface, cherries will send their roots down 6 feet or more. At this depth, they are fairly safe from spring droughts, which tend to produce small fruits. Where a solid clay hardpan approaches the surface, drainage is bad and the trees will not flourish. Moreover, their roots will remain in the upper few feet of soil where they will dry out and suffer with the first dry weather.

Sweet cherries like a dry, warm, gravelly, friable loam. Sour cherries grow well on fertile clay loams or clay soils more retentive of moisture, though they prefer lighter soils, too, and thrive sometimes on almost pure sand. Bush cherries will grow in a variety of soils, from sandy to clayey loams, if they are enriched with compost. Nearly neutral soil with a *p*H from 6. to 8. is preferred by the cultivated varieties, but some of the chokecherries and sand cherries like a little more acid, with a *p*H of 5. to 6.

Planting

Except where winters are severe, cherry trees should be planted in fall. Since severe winters are not suitable for cherries in most cases, this means that almost all cherries are best fall-planted. Experiments have shown that a greater percentage of success has attended fall plantings as far north as New York state, Michigan and Oregon. Buds of the cherry start to develop very early in spring—usually before it is possible to work on the soil. It is probably for this reason that many spring plantings fail. If it is impossible to do the planting in the fall, the trees should be heeled in where they will be kept cold until late into the spring, to prevent the buds from swelling.

Make the holes large enough to contain all of the roots without bending. Soil should be loosened as deeply as possible before the tree is set. Long, straggling roots may be cut back and injured roots should be removed. Set the tree a little deeper in the soil than it was in the nursery. Spread the roots well and fill in with topsoil around them. As the soil is thrown into the hole, tamp it well around the roots, taking care to leave no air pockets. After a few shovelfuls have been thrown in, raise and lower the tree slightly to help settle the soil. Take care to protect the roots at all times from the sun. If the job can be done in early morning or late evening there will be less chance of damaging the roots by drying them out.

All branches that form narrow angles with the trunk should be removed immediately when the tree is set. If the young tree is a two-year-old, it may be pruned immediately to its main scaffold branches or the work may be done in early spring after the most severe winter weather is past.

If planting is done in the fall, a thick mulch should be applied at once to protect the tree from heaving. First wrap the trunk loosely with building paper or with wire hardware cloth, to keep the mice and rabbits away and then mulch with straw or hay to a depth of 6 inches.

Maintenance

Because cherries bear their crop so early in the season, they seldom need more moisture than is supplied by the winter and spring precipitation. This is especially true if a good mulch is maintained around the trees. It is conceivable, however, that a dry winter and spring could leave the soil without sufficient moisture, especially if it is sandy. In that case, the trees will need additional water while the fruits are ripening. Cherries increase rapidly in size during the last few days before they ripen and at this time the trees must have plenty of water deep around their roots.

Cherries need less fertilizer than many tree fruits, but they will be more healthy if nutrients are supplied to them at regular intervals. They seldom suffer from a deficiency of anything except nitrogen. This should be given them at the rate of one ounce of pure nitrogen per year of the tree's age until it reaches its mature size, when the feedings should remain constant. In terms of organic fertilizers, one ounce of pure nitrogen will be found in 5 pounds of poultry manure, one pound of cottonseed meal or one-half pound of bloodmeal.

Pruning

Sweet cherries grow tall and upright, if left unpruned; sour cherries and hybrids are more bushy and spreading. Trees of both varieties are usually trained to the modified-leader system, though sweet cherries must be allowed to make a taller growth than is necessary with sour.

Cherries are given their first dormant pruning as two-year-olds. This may be the winter that they are planted or it may be the following year, if whips were planted. First, all branches with narrow angles to the trunk and all weak branches should be removed. Of the remaining strong laterals with good crotches, select 3 or 4 radiating in different directions and at least 4 to 6 inches apart up and down the trunk. Do not leave two branches at the same height. The topmost branch will be the leader. Prune back all the other branches to make them shorter than the leader. This will make them stretch out laterally to the light and will help them to maintain right angled crotches, which will be stronger than crotches which make acute angles.

In the following winters, any scaffold branches that become too vigorous and tend to crowd out others should be headed back to a lateral which grows outward or upward and some of the laterals should be removed from it. Two or 3 more scaffold branches, which will be secondaries, are selected on the upper part of the trunk, so that the mature tree will have about 6 to 8 strong scaffold branches. After 2 or 3 years, when the tree has filled out its proper form, the leader is cut back to a lateral.

Only light, corrective pruning is needed for the next 4 or 5 years, to remove crossing branches, to keep the center of the tree open and to maintain a balance between the scaffold limbs.

When the tree reaches its mature size weak branches should be pruned from the center and top and side branches should be thinned. This will open it up to permit sunlight to reach the lower branches, necessary to keep up the production of fruiting wood in the lower part of the tree. If the top is permitted to become too bushy, fruit will develop only in the upper branches and half the tree's potential will be lost. Some heading back of the upright branches at the top may also be necessary to maintain the tree's height.

Sweet cherry trees that have been allowed to become too tall may sometimes be cut back severely to reduce their height. This forces the development of new fruiting wood on the lower branches, where it can be reached for harvest.

Propagation

As has been explained earlier, cherries should be grafted to mazzard or mahaleb rootstocks for the best trees. Cherry seedlings which may pop up where birds have dropped pits may revert to an ancestor of the cherry from which the pit came. The roots may or may not be suitable rootstocks for grafting and it may be a waste of time to bother with them. It is better to buy trees for the home grounds, especially if time and space are limited.

Enemies

Sour cherries are more resistant to insects and fungi than any other stone fruits. Sweet cherries are slightly less resistant, but they are attacked by a few more pests. In this category, unfortunately, must be classed birds. At fruiting time the single cherry tree standing in the home grounds is likely to be stripped of its fruit as it ripens. Less damage is noticed in a large orchard, because there are so many more fruits per capita to the bird population.

Two possibilities are presented to the home cherrygrower. He may cover his entire tree, when the fruit first starts to color, with a netting of some sort that will keep the birds away. Or he may plant mulberry trees alongside when he plants cherries. Given their choice, birds will eat mulberries in preference to cherries and will do much less damage to a cherry tree with mulberry neighbors.

Two insects damage cherry trees and are not so easy to cope with. Tent caterpillars are likely to develop their webs in spring and, if not removed immediately, they will eat every leaf in reach before they drop to the ground. These are particularly damaging on the wild types of cherries. In fact, the home gardener who has wild cherries growing in his neighborhood could often save much trouble in his home orchard by destroying the wild trees.

"Wormy" cherries are the work of the larvae of plum curculio, or snout beetle. The adult curculio is about one-fourth inch long, brown mottled with gray and with 4 humps on its back. It feeds upon the blossoms to some extent and also upon the fruit, where it makes a crescent-shaped slit to lay its eggs in. The developing larvae burrow into the fruit, causing it to drop to the ground, where the larvae leave the fruit and dig into the soil to pupate.

Curculios have a habit of playing possum which makes it possible to catch them in wholesale lots. When they are disturbed, they pull in their legs and drop to the ground. This makes it possible to catch them by the "jarring" method. It takes a lot of work, but if the curculio population can be decimated in the home garden it is worthwhile. Each morning, starting with blossoming time and continuing for 6 weeks, spread a tarpaulin under the tree. Jar the tree with blows on the trunk. Curculios will fall to the tarpaulin, where they may be squashed.

The only fungus which causes much damage to cherries is brown rot. This attacks the ripening fruit and rots it, traveling back into the twigs. It is also sometimes present on the blossoms. Best cure is to keep a careful watch and remove any infected fruit as it appears, also removing and destroying any that falls to the ground. If the rot has entered the fruit and completely rotted it on the tree, the chances are that it has traveled back into the twigs, which must also be removed.

Harvest

Picking cherries with stems requires skill and experience, if it is to be accomplished without damaging the fruiting twigs. The picker grasps the stems of several fruits in a cluster and strips them from the spur with a slight backward twist.

In the home garden, when the cherries are picked just before they will be used, it is better to pick them without stems. This, of course, will leave a break in the fruit where the stem is removed and, if the cherries are not used at once, bacteria will enter the break and start to spoil the fruit. However, if they can be used at once, the trees will be better for having the cherries picked this way.

If sour cherry trees are covered with netting, the fruit may be permitted to hang as much as 2 to 3 weeks after it is ripe. The longer it hangs, the sweeter it becomes. Fruit from an amarelle becomes sweet enough to be eaten out of hand if it is left on the tree to develop all its sugar.

CITRANGE, CITRANGEDIN, CITRANGEQUAT

CITRANGE, CITRANGEDIN, CITRANGEQUAT. *See* ORANGE, TRIFOLIATE.

CITRON

CITRON. The citrus fruit with the oldest history is the citron, *Citrus medica,* its Latin name derived from that of ancient Media. Citrons were known to the Greeks and Romans, who knew nothing about our modern citrus fruits. Culture of the citron was introduced to the American continent soon after its discovery, but though it was grown in Florida for a time it has been almost entirely allowed to die out there. The reason for its lack of popularity as a commercial fruit as for a home garden fruit is its extremely limited use. Citron is grown entirely for the peel, which is candied and used in baked products. A very few fruits are all that could be consumed in a year by the most avid citron fanciers. For that reason, the tree has little interest for the home gardener. Added to this is the fact that citron is one of the tenderest citrus fruits, which can be grown only in the warmer lemon districts. It has least resistance to low temperatures of any of the *Citrus* genus.

The citron is a 6- to 9-inch oblong warty yellow fruit with a thick rind, white except on the outside, sparse pulp, acid bitter and scanty juice. It grows on a 10-foot shrub or tree with a short trunk and straggling, thorny branches. Its flowers are small and often unisexual. Two varieties are grown from time to time in Florida and California. They are Lemon, an everbearing tree when it grows in favorable areas; and Orange, with a cone-shaped orange-colored fruit which is less bitter than the common citron.

CITRUS FRUITS

Citrus Fruits. Citrus, a genus of the family *Rutaceae,* which includes all the most valuable tropical and subtropical fruits—orange, grapefruit, lemon and lime, plus a host of hybrid and lesser species.

The citrus fruits, with the exception of citron, are comparatively recent additions to the list of Western comestibles. Citron seems to have been known to the Romans, but oranges were not introduced to Europe until the Crusades and the Moorish invasion of Spain. Their origin is now obscure, but they are thought to have been brought to Arabia from India and thence westward.

Citrus trees are broad-leaved evergreens, all more or less intolerant of frost and cold weather. Limes are the tenderest, being killed by a whiff of frost. Kumquats are the hardiest. In between, the oranges can be made more or less hardy depending upon bud variety and the rootstocks to which they are grafted. Trees grown on rough lemon stock are more tender; those grown on trifoliate stock are more hardy.

Citrus trees are normally spiny, though most of the more highly cultivated varieties have lost their spines. Flowers are usually white and waxy and many are very fragrant. Flowers of some oranges and some grapefruit have no ovules or viable pollen, but will set fruit without pollination. These fruits are seedless where ovules do not exist; they may have a few seeds where ovules exist but pollen is not viable when trees with viable pollen are planted nearby. Fruit will set on any citrus tree without pollen, so pollination is no problem, as it is with apples.

Citrus fruits have another sexual peculiarity in that they will give rise to new plants in the fruit, but from the tissue that lies outside the seed or ovum. These plants, which arise from the *nucellus,* the tissue of the ovule or seed capsule, are called "nucellar" plants. They are often much more hardy than seedlings and many of them are being raised in experimental stations in both Florida and California, where they are watched for disease-resistant qualities. It is believed that some of the fungus diseases which plague the citrus industry may be eliminated by the use of nucellar trees.

Citrus fruit thrives in very hot tropical climates, but the fruit is of better quality where the temperatures do not climb too high. Semitropical areas produce the best fruit. Where dry summer winds blow, the orchards must be planted with windbreaks to produce good crops. On the other hand, areas where the summer climate is excessively humid are infested with a variety of fungus diseases.

Citrus-growing sections of Florida and California have quite different problems, based upon the amount of annual rainfall and their aridity. In California, Arizona and Texas the citrus districts are mainly desert in character and the principal problems are wind and irrigation. In Florida, there is plenty of rainfall in most areas, but the problem is one of drainage and fungus disease. In California, many of the areas where citrus might grow have soil that is alkaline from centuries of evaporation without run-off. Florida gardeners have a problem much easier to solve, that of draining their gardens. It is easier to raise the *p*H of sour soil, after draining it, by the addition of lime, than it is to lower the *p*H to between 5.5 and 6.2, optimum range for citrus.

In the past 10 or 15 years, a decline has been noted by citrus growers both in California and Florida. In California where irrigation is necessary in almost all orchards, the soil alkalinity constantly builds until it is impossible to grow cover crops between the trees. Chemical fertilizers have been heaped on, with counter-chemicals to correct alkaline conditions and then counter-counter-chemicals to correct the damage done by the acidifiers. The soil is so dead from applications of chemical fertilizers, weed killers, soil fumigants, insecticides and fungicides, that the size and quality of fruit has declined to an alarming degree. The same series of conditions have begun to appear in Florida, where the process is slower only because less irrigation is needed.

A few growers in Polk County, Florida, have experimented during the last 10 years with organic treatment of their orchards. They have found it a great success, according to a report in Organic Gardening and Farming magazine of February, 1956.

Organically treated orchards where formerly no cover crops could be grown, now have annual crops of legumes growing as much as 5 feet high. Trees are healthy and injurious insects are almost lacking. The change has been brought about in one orchard, owned by George Brenzel, by the addition of natural rock minerals, such as pulverized granite and basalt, greensand and organic humus to the soil while at the same time he eliminated chemical fertilizers, sprays and insecticides. In 4 years the Brenzel renovated orchard produced 3 times the fruit crop that it formerly produced, with fruit that contained far more minerals, juice and sugar than the Florida average.

A similar renaissance is reported from a California orchard at Redlands, owned by the Champion brothers. The Champions bought a herd of Herefords, in order to supply their trees with manure instead of chemical fertilizers. They report that in 3 years their crop has tripled, they have larger fruit and no pest problems. The response of citrus fruit to organic gardening methods is encouraging for the home gardener who wishes to grow healthful fruit without poison sprays or chemical fertilizers.

Many of the presently known citrus species are the result of cultivation. The following are the most important members of the genus:

C. aurantifolia, the limes, whose very acid fruits are usually picked and used green, though some of them become yellow when fully ripe. This species is more frequently used in this country than the sweet lime, *C. limetta.*

C. Aurantium, the sour or Seville orange, is grown mostly in Spain for making marmalade. It is not grown in this country for its fruit, which is sour with bitter "rag" (the white tissue between sections), but the roots are used for stocks to which many top varieties are budded.

C. limonia, lemons. Beside being grown for fruit, lemon roots of the rough-lemon variety are also used for budding oranges.

C. maxima, shaddock, is also called pummelo or pompelmous. The fruit is something like grapefruit, but pear-shaped, very large, coarse and inferior in quality.

C. medica, citron, has fruit that grows 7 to 10 inches long, melon-shaped and yellow. Its aromatic skin is candied and sold for use in baked products.

C. mitis is the calamondin or Panama orange. Its fruits are small, being about one inch in diameter, with loose skins and acid pulp.

C. nobilis, king orange, is grown principally in the two varieties *deliciosa* and *unshiu.* This species is characterized by sweet pulp, loose skin and sections which can be easily separated, like those found in Temples. The variety *deliciosa* includes tangerines or mandarin oranges, with smooth skins. Variety *unshiu* is the Satsuma orange, somewhat flatter and with a more pebbly skin. Temples are thought to be a hybrid of tangerine and sweet orange.

C. sinensis is the common or sweet orange. All the commercial oranges are developments of this fruit. Many varieties are not even named, being sports developed by one grower or by one neighborhood in the Florida or California orange-growing country. In some cases, growers develop whole orchards of an unnamed variety by budding their own young trees from a single parent tree.

C. taitensis, the Otaheite or Tahiti orange, is a miniature tree grown for ornament only, usually in pots. Its fruit is small and insipid in flavor.

C. trifoliata, the hardy or trifoliate orange, is an important species, though its fruit is acid and dry. The roots of *trifoliata* are used widely for grafting stock, producing a more hardy, healthy and somewhat dwarfed orange tree.

Kumquats belong to another genus, the *Fortunella,* though they are occasionally classified as *C. japonica* and are considered citrus fruits. Kumquats are small, being about 1 to 1½ inches in diameter, but their flavor is sweet and aromatic. Three species are grown, *F. crassifolia,* Meiwa kumquat; *F. japonica,* the round kumquat; and *F. margarita,* the oval or Nagami kumquat. Kumquats are the most hardy of the citrus fruits, some being able to withstand temperatures down to 15 degrees.

For further information about growing citrus fruits, see the following: LIME, LEMON, ORANGE, GRAPEFRUIT, PUMMELO, SHADDOCK, CITRON, CALAMONDIN, TANGERINE and KUMQUAT.

CRAB APPLE

CRAB APPLE. Several species of the genus *Malus,* the rose family *Rosaceae,* are included under the popular name of

TABLE 67: Crab Apple Varieties

Variety	Sections for which recommended	Fruit characteristics	Tree characteristics	Cold endurance
Whitney	Mountain States Northern plains Northeast Pacific States	Large, red striped, early. Juicy and subacid, but becomes mealy quickly after ripening.	Sometimes a shy bearer.	Fairly hardy.
Dolgo	Southern plains Mountain States Northeast Pacific States	Small and ornamental bright red fruit. Early. Very juicy, sprightly flavor.	Upright. Prolific.	Hardy.
Transcendant	Pacific States	Red over yellow, medium to large, crisp, juicy, subacid, astringent. Early midseason.	Thrifty and productive but susceptible to blight.	Hardy.
Chestnut	Northeast Pacific States	Yellow with red stripes. Midseason. Crisp, juicy, spicy. Keeps two months.		
Hyslop	Northeast Pacific States	Large. Purplish-red. Midseason. Firm. Astringent. Stores three months.	Somewhat biennial.	
Red Siberian	Alaska Pacific States	Red. Firm, astringent. Late midseason.		Very hardy.
Yellow Siberian	Pacific States	Yellow. Firm, astringent. Late midseason.		
Red River	Pacific States Northeast	Red. Late. Firm, crisp, juicy, aromatic.		
Piotosh	Pacific States	Red over yellow. Very late. Crisp, hard, juicy, pleasant flavor.		

TABLE 67: Crab Apple Varieties (Continued)

Variety	Sections for which recommended	Fruit characteristics	Tree characteristics	Cold endurance
Young America	Northeast	Large, bright red. Early midseason.	Crops ripen unevenly, sometimes yield small.	
Sylvia	Alaska Northern plains	Earliest crab. Yellow, medium size. Should be picked slightly green—becomes mealy when ripe.	Upright. Resistant to blight.	Hardy.
Sweet Russet	Northern plains	Early. Large. Green marked with russet. Good for eating.	Fairly productive.	Fairly hardy.
Florence	Southern plains Northern plains Northeast	Early midseason. Medium to large, red-splashed. Subacid but astringent. Mealy when ripe.	Blight-resistant. Bears when young.	Hardy. Blossoms withstand frost fairly well.
Robin	Northern plains	Midseason. Medium size, firm, juicy, subacid. Yellow with red blush.	Spreading. Prolific. Blight resistant.	Fairly hardy.
Virginia	Mountain States Northeast Northern plains	Late. Large. Red blushed and striped. Juicy. Fair for eating, good for cooking.	Thrifty, with strong structure.	Hardy.

crab apple, which has no exact meaning botanically. Any small apple used more for jelly-making than for eating might be called a crab, or crab apple. Most varieties cultivated as edible crab apples are members of the species *M. baccata,* the Siberian crab. Several hundred varieties of *M. floribunda,* the flowering crab, have been developed for ornamental purposes, their fruits edible but usually too small to be of any value. Half a dozen species of crab apples are native to the northern North American continent, some growing wild as far north as Alaska. Some cultivated varieties are hybrids of these, crossed with natives of China or Siberia. The trees are useful for their fruits, flowers and also may serve as windbreaks, being very hardy.

M. coronaria, the American crab, or sweet crab, grows wild through the states east of the Rockies, from New York to Missouri. In places where apple and cedar rust are prevalent the American crab may serve as host to the disease and should be destroyed.

Culture, pruning and training of crab apple trees is the same as that of apples, which see.

CRANBERRY

CRANBERRY. *Vaccinium macrocarpon,* a member of the heath family, *Ericaceae,* is a native of temperate American bogland from Virginia to Wisconsin and across to the Pacific coast. Indians used cranberries before the colonists settled the East, but no cultivation was attempted until about 100 years ago, when plantations were started on Cape Cod.

Very few home gardeners are able to supply the conditions essential to success with cranberries. Climate plays a large part in their culture. They must have winters not too cold, summers not too hot and springs that warm up fairly early. They require acid soil, whose *pH* is 3.2 to 4.5. Commercially they are grown on peat or muck land with a hardpan at least 15 inches below the surface and a layer of grey sand 2 to 3 inches deep on top. They must be grown on flat land which can be flooded with acid water whose *p*H matches that of the soil and which can be drained to a depth of 15 inches at the will of the gardener. Unless the home gardener has these conditions practically at his doorstep, it does not pay to put in the work and preparation necessary to grow cranberries, a fruit which, at best, has very small nutritional value. Cranberries supply a very small amount of vitamins: 30 units of vitamin A and 6 milligrams of vitamin C in ¾ cup of sauce. In addition, they contain 13 milligrams calcium, 11 milligrams phosphorus and .4 milligrams iron in the same amount.

New Jersey, Cape Cod, Wisconsin and Oregon are the principal cranberry areas in the United States. In all 4 places they are grown near the lake or ocean, where the large bodies of water modify the temperature extremes summer and winter.

Briefly, their culture is as follows: A flat site, which a slow stream of water can be made to flood, is provided with a dam or flood gates above and below. The proposed cranberry bed may be flooded for two years or it may be prepared in the fall before planting. Destroy all perennial weeds and invert sod over the entire bed to provide a weed-free base for the cranberries.

Ditches must be dug through the bed to a depth of a foot to 15 inches and must be connected with the stream above the bed and the flood gates below. Three inches of clean sand should be spread over the entire bed before it is planted. Every 3 years thereafter, an additional half-inch of sand is spread during the winter. If the water freezes, the sand may be spread on the ice, to drop down into the bed as the ice melts. If the plant tips are buried, they should be pulled above the surface of the sand with a rake.

Cranberries are started from cuttings, which are planted in early spring immediately after the first layer of sand is applied. Two or three cuttings 5 to 10 inches long are pushed into the sand almost to the peat layer 6 to 18 inches apart. In Wisconsin the growers use the closer planting distance; in the East they space the cuttings farther apart. The closer together they are planted, the sooner the ground will be covered and the less weeding will need to be done. Tops of the cuttings need to protrude no more than one inch above the soil level.

Keep the water table high the first year, but not quite up to the level of the cuttings. Cranberry roots will not grow under water, though the plants may be flooded and kept submerged for long periods without suffering damage. The first

Cranberries are harvested by hand-picking or by combing the plants with scoops. Scooping is an art that requires just the right rolling motion to comb the ripe berries off the vines without damaging the tender vines themselves.

year a few upright stems will sprout from the cuttings. In the second year a few runners will develop. After the third year the soil will be more uniformly covered and a few berries may be produced. Full production will not start until after the bed is 4 years old. By that time the soil will be covered with trailing vines having upright branches 4 inches high, about one-third of which will bear one or two berries each. On good soil 400 uprights can be counted on a square foot of mature cranberry bog. A good bog, well cared for, can last as long as 60 years. Yield is as high as 100 barrels per acre under the best conditions.

Water should be brought in to flood the bed for one or two days after planting, then drained almost to the bottoms of the ditches. During the balance of the first year, the bed should be flooded only one or two days at a time, whenever it is necessary to moisten the sand or to control insects.

When the soil begins to freeze in fall the bed should be flooded deep enough to cover the plants and the water should be kept at that level all winter. Excess must be drained away after heavy thaws or rains to prevent heaving. In the latitude of Massachusetts, the beds are drained about May 5th, though flooding may again be necessary if there are late May frosts. Usually 2 to 3 inches of water under the vines in late spring will prevent frost damage. Some growers drain their beds in March to give the plants a breather, then flood them again in April until May. Winter buds will endure a temperature down to 25 degrees. After the buds start to swell, they will stand temperatures down to 20 degrees until they blossom. While blossoming, the plants cannot be entirely flooded without washing out the crop.

Beds are flooded in June as insect protection, but if the water is warm it cannot be left over the bed for more than 24 hours without spreading fungi. Principal enemy is a virus disease called "false

TABLE 68: Cranberry Varieties

Variety	Where grown	Fruit	Vine	Season
McFarlin	Massachusetts	Good keeper. Attractive berry with some bloom.	Resists false blossom. Forms many runners.	
Early Black	Massachusetts	Dark red. Excellent flavor.	Grows rapidly. Resists fungus diseases and to some extent false blossom.	Early.
Beckwith (McFarlin x Early Black)	New Jersey	Large. Good flavor.	Resists rot, false blossom.	Late.
Bennett	Wisconsin	Good keeper.	Subject to false blossom.	Late.
Centennial	Massachusetts	Extra fine flavor. Large.	Very subject to false blossom.	
Howes	Massachusetts	Red. Very high in pectin. Good keeper.	Subject to false blossom.	Late.
Searls	Wisconsin	Large. Dark. Fair keeper.	Subject to fruit rot in East. Highly productive in Wisconsin.	Midseason.

blossom," which is transmitted by the blunt-nosed leafhopper. Flooding late in May or early June also helps control the black and yellow-headed cranberry worm and the cranberry fruit worm.

Cranberries cannot be given too rich an organic diet, because the decomposition of the material under water robs the plants of necessary oxygen. Unless the leaves get enough oxygen, they are unable to manufacture carbohydrates and the fruits are small. Once in 3 or 4 years, the beds should be given an application of bloodmeal, phosphate rock and a granite dust low in lime.

Cranberries are harvested by hand-picking or by combing the beds with scoops. A cranberry scoop resembles a wooden dust pan with a large pocket above the handle and with its edge slitted to form a comb. The scoop is run through the vines in the same direction each year, to minimize damage. The fruit will keep for several months if stored at 36 to 40 degrees.

CRANBERRY BUSH

Cranberry Bush. *Viburnum trilobum,* or *V. americanum.* A 6- to 8-foot shrub of the honeysuckle family, *Caprifoliaceae,* which bears bright red berries in midsummer. Also called highbush cranberry, the bush bears edible fruit which is inferior to real cranberries, but makes acceptable jelly. Cranberry bush is grown chiefly for ornamental purposes. It is extremely hardy and can be grown in the northern plains in situations where few other fruits will grow. For that reason, it is cultivated there as a fruit bush, but is grown elsewhere only as part of a shrubbery border. Its red berries persist into the winter.

CURRANT

Currant. *Ribes odoratum,* the buffalo currant, *R. nigrum,* black currant, *R. rubrum,* the red currant and *R. sativum,* the common currant, are first cousins of gooseberries and members of the *saxifrage* family. All are edible currants more or less frequently cultivated. The buffalo currant gives us our domestic yellow and white varieties. Black currants are grown in England where they are esteemed for jams. Red currants are cultivated more often in Europe than in this country, because the common currant varieties have been developed into larger and more popular fruits.

TABLE 69: Currant Varieties

Variety	Section for which recommended	Fruit characteristics	Bush characteristics
Red Lake	Corn Belt Pacific States Alaska Northern plains Great Lakes Northeast	Large red berry, subacid, good quality. Long, well-filled clusters. Late.	Upright. Very hardy. Vigorous. Productive.
Wilder	Corn Belt Alaska Northern plains Great Lakes Northeast	Large, dark red berry. Mildly subacid. Midseason. Large compact clusters.	Very large bush, upright. Set 6 feet apart. Long-lived, but not hardy in upper Mississippi Valley.
Cascade	Pacific States Northern plains Great Lakes Northeast	Dark red medium size berry in compact clusters.	
Perfection	Alaska Pacific States Northern plains	Large bright red berries. Best dessert variety. Long compact clusters. Midseason. Berries scald in hot weather when allowed to hang after ripening.	Small spreading bush. Heavy yields. Canes break easily.
White Grape	Corn Belt	Large pale yellow berries, mild flavor.	Very productive.
Fay	Alaska Pacific States	Large dark red acid berries. Early to midseason.	Small spreading bush. Canes break easily.
Cherry	Pacific States	Deep red large acid berry that becomes small on old bush. Midseason.	Spreading. Not very productive.
Prince Albert	Alaska	Medium to large red berry. Large clusters. Late.	Large upright bush. Productive. Disease resistant.
Diploma	Northern plains Great Lakes Northeast	Glossy, very large, bright red berries. Subacid.	Upright, slightly spreading. Canes brittle.
Filler	Northeast	Large, similar to Fay.	Upright.
London Market	Northern plains Great Lakes Corn Belt	Medium to large deep red berry, acid. Compact clusters. Midseason to late.	Upright. Resistant to borers, disease, white pine blister.
Long Bunched Holland	Mountain States Pacific States	Acid, light red berries of medium size in long clusters. Very late.	Upright. Vigorous. Moderately productive.

Excepting the black ones, all currants grow on tidy little bushes 2 to 4 feet high, spreading or upright according to variety. They are very hardy, neat in the garden and make pleasant additions to the shrubbery border. Their only fault is that their foliage is inclined to drop early in the fall.

Currants are hardier than any of the fruit trees, bearing good yields where winters are as cold as Alaska's. They like cool summers with plenty of moisture from spring through August. Their leaves are rather soft and do not stand heat or drying winds, but even in the Plains States they are hardy and will produce good crops where they are given some shelter from wind and from the hottest summer sun. They find almost exactly their optimum requirements around the Great Lakes and in the Hudson River Valley. Some areas of the Rockies would give them the same conditions but their culture in that section is limited by their susceptibility to white pine blister rust, to which they are an alternate host. In fact, many of the states where white pine is an important resource prohibit the planting of black currants and provide that any currant or gooseberry bushes infected by white pine blister fungus may be seized by the authorities. To be safe, consult a nurseryman or county agent about local laws before planting either red or white currants or gooseberries. The United States Department of Agriculture discourages the planting of black currants anywhere in the country, even where it is not unlawful, because black currant bushes are so much more susceptible to the fungus.

Currants are a fairly good source of vitamin C, but unfortunately the species richest in that vitamin is the forbidden *R. nigrum*. The varieties which can be eaten raw will provide more vitamins than the ones which must be cooked for pies or jelly.

Ordering Stock

All currants are self-fertile, so there is no pollination problem. One species may be planted without a second for cross-pollination. Best stock is two years old, that is, in its second year from cuttings, because it will begin to produce the following year. One-year-old stock will need another year before beginning to yield. A well-grown, mature bush may bear up to 3 quarts of fruit.

Site

Fertile heavy clay soil is better for currants than a light sandy soil, which may be hot. Currant bushes root near the surface, so a soil which holds its moisture well is better than one that is too dry. But that does not mean that currants will stand wet feet—they must have good drainage, too. Clay soil which is opened up with a good proportion of humus is best, with a *p*H of 6. to 8.

Where the summers are rather hot, currants may be planted where they will have partial shade. The north side of a building is usually coolest, because it will afford shade in the hottest part of the day. Or the bushes may be interplanted with apple trees in the home orchard, though they should be planted in the middle of the space between the trees, rather than under them where they would not get enough moisture. Having fairly hardy fruit buds, they may also be planted toward the bottoms of slopes where the higher sites are reserved for fruits more susceptible to the effects of bad air drainage.

Planting

Prepare soil for currants in the same way that it would be prepared for a vegetable crop; that is, dig in a supply of the rock fertilizers and humus to a depth of 6 to 8 inches. If the site has previously been cultivated for another crop, there will be less trouble with perennial weeds near the bushes.

Currant buds start their growth early in spring. The bushes should be set before this growth starts. If the winters are not too severe, set the bushes in fall after they have become dormant. This gives the roots a chance to make some growth before freezing soil heaves them up. Where winters are very cold, the plants may be set in spring, but the work should be done before growth starts.

Bushes are set 4 to 6 feet apart, depending upon variety. If the roots are dry when the plants are received, they may be soaked in water for a few hours before they are planted. Trim off any broken or damaged roots and cut back any that are too long. Set them slightly lower than they were set in the nursery. If they tend to have a trunk, set them deep enough so that the bottom branches are covered with soil at their bases. Cut the top back to 6 to 12 inches, depending upon the strength of the root system. Firm the soil well around the roots and spread a mulch in a 3-foot circle around each bush, leaving the canes bare where they emerge from the soil. Mice like the tender shoots on currant bushes and must not be encouraged by a mulch too close to the canes.

Currants are among the fruits which show a decided preference for organic rather than chemical fertilizers. Even gardeners who pour on the chemicals through most of their garden find that manure, wood ashes and bone meal are the only nutrients on which currant bushes thrive. Fertilizer should be applied in late winter, just before growth starts. Since the roots are near the surface they will be able to use the nutrients with a minimum of leaching. Rotted manure or compost should be spread in a layer an inch deep over a 3-foot circle around each bush and one ounce of wood ashes and two ounces of bone meal should be raked into the same space.

Pruning

Currants, like cherries, are borne in greatest profusion on wood that is 2 or 3 years old. After that age the size and quantity of the fruit declines. During the first year, a cane should make a strong, straight, unbranched growth. In the second year, it sends out 3 or 4 lateral branches which are 6 to 12 inches long. In the third year more numerous and shorter laterals grow from the main cane and from one or two laterals.

Ideal pruning limits the bush to 3 or 4 canes of each of these ages. Older canes are cut out when they reach 4 years old. A two-year old bush, when pruned for the first time, will be permitted to keep no more than 6 to 8 shoots, half new growth and half those that it had when it was planted. Each year thereafter the new shoots are cut out to leave no more than 3 or 4 strong ones which have a desirable growth habit. On spreading bushes, the more upright canes should be selected. On upright bushes, the canes which spread slightly to open up the top would be preferred. Beginning in the second or third year in the garden, the oldest canes are also cut out each year. All canes removed are cut back as far as possible into the ground, to keep the base of the bush clean and free of stumps. Canes are almost never headed back, unless they are injured. The best time for pruning is in early winter.

Old bushes may be renovated by cutting back all except 3 or 4 strong canes the first year. This will force the growth of many new shoots from which a selection of a half dozen the second year will make it possible to cut out the remaining old canes. By the third year the bushes will be filled out and will be ready for the regular pruning schedule. Some authorities recommend lifting the bushes after 8 to 10 years and dividing them to prevent the canes from growing too close together.

Propagation

Currants are usually propagated from cuttings of wood of the current season. These may be taken in fall and wintered outside, buried in bundles upside down in the garden. Or the cuttings may be made early in spring before growth starts and set out immediately.

Cuttings should be 8 to 12 inches long. They are planted in the nursery row 3 inches apart with no more than two buds above ground. Some shade may be provided until growth starts and the row should be kept moist. The new bushes are left in the nursery bed until the following spring or they may be left until the second spring before they are set out.

New bushes may also be started from the old ones by layering, if the variety is of spreading habit. A low branch is pulled down to the ground and pinned there, its bark slit on the underside at the point of contact. Soil should be heaped over the point where the slit was made. The tip of the branch should be allowed to rise above the soil, and should be headed back slightly to encourage branching. At the end of a year enough roots should have formed from the slit to permit the new bush to be severed from the parent plant. It may then be moved to the nursery, it may be left where it is for another year or it may be planted in its permanent position.

Currant bushes may be propagated by mounding, as are gooseberry bushes. For that method, *see* GOOSEBERRY.

Enemies

White pine blister rust, while it attacks almost all currants, is so much less to be feared on red currants than on black that the red varieties seem almost resistant by contrast. However, they are susceptible to some degree and for that reason should not be planted within 900 feet of white or 5-needled pine. This is the uttermost limit to which the spores may be carried by the wind from them to pine trees. The rust is a fungus which kills the pine trees. It cannot spread from pine to pine without an intermediate stage on either currant or gooseberry bushes—no other alternate host will do. On the bushes the fungus produces orange-colored pustules under the leaves. In this stage, the spores are very hardy and may be carried miles by the wind to infect other bushes, though this spore will not infect the trees. Bushes which show signs of the rust should be destroyed.

Currant maggot, the maggot of the currant fruit fly, is a small "worm" which attacks the fruit. Clean conditions around the bushes when the fruit is ripening will help to prevent much damage from this particular pest. Any fruit that ripens prematurely should be inspected for worm-holes and destroyed if they are present.

Anthracnose and crown gall are two diseases to which the bushes are subject. As in other cases, anthracnose is best combatted by preventing its spread, which may occur when the bushes are worked upon while they are wet. When crown gall attacks, if the lesions at the soil level are not too severe, the bush may survive them. In any case, the infection is one which might occur from ordinary soil bacteria and is not spread by its occurrence in one plant. The bush might as well stay where it is as long as it continues to produce. If the attack becomes crippling, the bush must be lifted and replaced.

Cane borers can become serious if they are not sought out and destroyed. Whenever a cane shows wilting leaves at its tip, cut it back until the borer is discovered, and destroy him before he emerges.

Harvest

Currants may be left on the bushes for 4 to 6 weeks after they are ripe. The longer they hang, the sweeter they become. At the same time, however, they lose their pectin content. If they are to be used for jelly they should be picked a little green.

Pick the berries when they are dry. Grasp the stems, not the berries and pull each cluster off the bush. Berries may also be stripped from the stems as they hang, but the job becomes a messy one and the fruit will not keep long when it is picked that way.

Dried currants are not made from the *Ribes* fruits, but are small viniferous grapes.

DATE

DATE. *Phoenix dactylifera,* the date palm has supported desert peoples with its fruit as far back as history records. A native of tropical Africa and Arabia, it has been imported into the subtropical arid portions of the United States, where it is grown with some success, despite the drawback of temperatures that are less than torrid. First plantations were attempted in Florida, where the highest temperatures prevail, but they were less successful than those now flourishing in the less hot but more arid portions of Arizona and California.

Grown primarily in California and Arizona, dates develop on a tall, feathery palm tree. Flowers are male on one plant, female on another.

Dates grow on a tall feathery palm tree which forms suckers at its base. Flowers are male on one plant, female on another. Roots penetrate deep into the earth and spread much farther than the top of the tree. Date palms will grow in any section where the mean temperatures for 6 months of the year are 75 to 90 degrees, and where winter temperatures never go below 20. Dry air is essential while the crop is ripening. Even night dews at that time are injurious.

Dried dates are a concentrated food, containing large quantities of sugar, some vitamins A and B, calcium and iron. Fifteen medium size dates provide about 350 calories and contain 155 units of vitamin A, .060 milligrams vitamin B_1, .054 milligrams vitamin B_2, 70 milligrams calcium, 56 milligrams phosphorus and 3.5 milligrams iron.

Culture

Three varieties of dates are grown in the United States, the Deglet Noor, Khadrawy, and Halawy. They are most productive where they are planted in soil with a deep water table, but where they can be regularly and frequently irrigated.

Trees are propagated from suckers removed from the base of young plants. Removal of suckers should be a gradual process, not more than 3 being taken from one plant in a season. Suckers which grow high on the trunks of older trees are not used for propagation. The suckers should be allowed to reach an age of 3 to 6 years before they are severed from the parent tree. Sex of the sucker will be the same as that of the parent plant.

The suckers are set out 30 feet apart in their permanent positions and are well headed back. Female plants begin to bear in 5 or 6 years and yield 100 to 200 pounds of fruit each year after they are 10 to 15 years old. One male tree must be provided for each 300 females, but the pollination will not take place without an assist from the grower. In Egypt, the custom is to make a hole in the female bud and insert a piece of the pollen-producing inflorescence. In this country, it is more often accomplished by tying a piece of the male flower beside the female cluster. Pollination may take place any time within 4 to 6 days after the female flower opens.

When the fruits are ½ to ¾ inch long they are thinned out, for better size and quality. They are then covered with paper protectors to keep them dry. During the fruiting period, water is supplied to the trees every week to 10 days, depending upon the temperature. As the fruit approaches maturity, water is withheld or supplied at less frequent intervals. In dry weather the dates may be left on the trees until they are thoroughly ripe. If the weather becomes wet, they must be picked before any rain touches them and the ripening process is finished in warm drying sheds.

DEWBERRIES

DEWBERRIES. *See* BLACKBERRIES.

ELDERBERRY

ELDERBERRY. *Sambucus canadensis,* the Canadian elderberry, is most commonly cultivated for its berries throughout the northeastern United States, where it is also found growing wild. Several other species of *Sambucus* may be used for food, some of them native to the North American continent. *S. caerulea,* a tree-like species with light blue edible berries, is common on the West Coast. *S. mexicana,* which also grows in the West, has the most delicious berries of all the varieties, said to taste something like blackberries. Europe and parts of Asia cultivate *S. nigra,* whose berries are used to make wine and to color port wine.

The Canadian elderberry is a 6- to 10-foot shrub which can be very ornamental in the proper surroundings. In spring it bears large flat clusters of greenish-white blossoms, some of which are a foot or more in diameter. These give way, in summer, to heavy blue-black berry clusters, berries which contain more vitamin C than any other northern garden subject except rose hips and black currants.

Both flowers and berries may be used as food or to make wine. Flower clusters are sometimes dipped in batter and fried, to make crisp dessert-type fritters. The berries are used, often in combination with apples, for pie with a distinctive flavor; they make a jelly with little pectin but a nut-like flavor; and they are used to brew a thick, heavy wine. Wine made from the flowers is said to be quite delicate. Birds are very fond of the berries.

Culture

The Canadian elderberry is a hardy shrub which grows well in climates with cold winters. Its native habitat is the moist, partially shaded woodland along a stream or at the edge of a bog. It is not restricted in its choice of soils, but is seldom found in dry places. It likes a soil which is nearly neutral, with a *p*H of 6. to 8.

In the home garden, the best place for elder is near a pond or stream, in the dampest situation available. It should be given plenty of space because at maturity it may spread to a diameter of

The elderberry is a hardy shrub which grows well in climates with cold winters. In spring, it bears large flat clusters of greenish-white blossoms which later give way to blue-black berries.

8 to 12 feet. A few cultivated varieties are available, the best of them being Adams. So far as is known, there is no problem in pollination, but if a planting does not seem to bear as heavily as it might, a wild bush may be planted near it for cross-pollination.

Elderberries can be propagated by cuttings or division. The plants send up new shoots from the base each year and some of these may be dug up and separated from the parent plant to make new plants. Plants may also be started from seeds, but named varieties may not come true from seedlings.

Fruit is borne on shoots 1 to 3 years old. During the first year, the green shoots attain the height of the mature bush and in the second spring they bloom. During the second season the pithy stems become hard and woody and send out a few lateral branches which bear fruit the following year. In the winter following its 3rd year, the shoot should be cut back to the ground, because its fruit clusters will be smaller and its wood more brittle the following year.

Elderberries are not troubled with many insects or diseases. The currant borer sometimes makes its way into the hollow stems and if discovered should be destroyed. Elderberry leaves are believed in some countries to be repellent to flies and it may be that this property repels other insects as well.

FEIJOA

FEIJOA. *Feijoa sellowiana* is not actually one of the guavas, though it is closely related to them and is often called pineapple guava. The feijoa is a native of South America, a member of the evergreen myrtle family which includes several spices of culinary importance as well as the guavas and pomegranates. Like the guavas, the feijoa is grown only in subtropical climates, though it will stand a little more cold than the true guavas. They may be grown in Florida and California and also along the Gulf, being able to withstand a temperature down to 14 degrees for short periods.

The feijoa is a shrubby plant which attains a height of up to 15 feet. Branches arise from the ground level and their total spread is about equal to their height. Purplish-white flowers which bloom in April have striking brushes of crimson stamens, which make the plant a desirable ornamental in the home grounds. The thick flower petals are sweet, and may be used in salads. The fruit, which is round to pear-shaped, is not very showy. It ripens to a gray-green color in August or September and drops to the ground when mature. Its flavor is a cross between strawberry and pineapple, according to some growers. The flesh is white, with numerous seeds embedded in it. It may be eaten fresh or used to make jelly.

Propagation is mainly by seeds, which must be planted as soon as they are removed from the fruit. Seedlings may not always resemble the parent fruits. Seeds germinate in 2 to 3 weeks. A few varieties have been developed, but to be sure of propagating them they must be grafted, air-layered or started from root or woody cuttings. Named varieties include Andre, Superba, Coolidge and Choice. Sometimes the blossoms are not self-fruitful, in which case two or more varieties should be planted.

Feijoas may be planted at any time of the year if they are purchased in pots from the nursery. They are very tolerant about the type of soil in which they grow, but they like good drainage and full sun. If they are to stand alone as specimen plants, they should be set 15 to 20 feet apart. A fine, dense hedge may be grown from plants set 6 feet apart.

Dig the hole about two feet in diameter for each plant and stir into the soil at the bottom a generous spadeful of well-rotted sheep manure or compost. Set the plant a little higher than it was grown in the nursery, so that when it has finished settling it will be at its former height. Fill in the hole with a mixture of topsoil, compost or manure, mixed with a cupful of bone meal or rock phosphate. Water the soil thoroughly after the tree is set and make a 4-inch deep basin over the roots for future waterings. Mulch with a heavy cover of hay or straw, keeping the mulch 6 inches away from the main trunk or branches.

Every 6 to 8 weeks during the growing season, which starts late in February, the feijoa should be fed. A scattering of rotted manure, scratched into the soil over the roots, an application of fish emulsion or a dusting of cottonseed meal or bloodmeal will help the bush grow rapidly and to bear fruit in its second year. Before the fall rain starts, dust bone meal or rock phosphate under the branch-spread. The soil should be kept moist at all times. Feijoas and guavas need more moisture than do citrus fruits. In dry climates the young plant should be watered every week until it starts to take hold; after the new green growth shows, the waterings may be reduced to one every two weeks.

Very little pruning is necessary to keep the bushes shapely, unless they grow taller than is desirable. If so, they may be headed back from time to time and the resulting new growth will bear larger and better fruit.

FIG

FIG. *Ficus carica* has been grown in Syria, Persia and the Mediterranean countries since time immemorial, though its fruiting processes have only recently been thoroughly understood. Figs played an important part in ancient mythology and religion. The sweet fruits, easily dried and stored, were the mainstay of many Mediterranean diets.

In order to understand fig culture, the gardener should know something about the construction of its strange fruit. A fig is a collection of many fruits which grow, not in a flat umbel or from a cluster of stems as many fruits do, but inside of a fleshy receptacle. The outer wall of the fig is actually a stem with a specialized shape. It forms a cavity, in some cases almost closed, in which sprout a number of tiny flowers, some male and some female. In the Smyrna fig, this cavity is almost closed at the bottom end—the end which, in apples or pears, would correspond to the blossom end. And since Smyrnas bear only female flowers, it would seem that they would be doomed to falling, infertile, from the tree every year, for lack of pollination. Which is exactly what does happen without the aid of a specialized wasp, the fig wasp, or *Blastophaga psenes,* which has been busy fertilizing this type of fig for many centuries in the Old World. It was not until this was understood and the wasp was imported to this country, that Smyrna figs were successfully grown here.

The fig wasp lays its eggs inside the fruiting portion of the fig. The wasps are raised on a wild variety called the caprifig, whose fruits are inedible but contain male as well as female flowers. When the female wasps mature, they break their way out of the caprifigs and seek new fruits in which to lay their eggs. As they emerge, their bodies are coated with pollen. The fig grower profits by this habit to pollinate his female fig trees. When the female trees are ready for pollination their flowers last from 4 to 8 days—the gardener hangs 20 to 30 fruiting receptacles from the caprifig in his tree. Wasps from this number of caprifigs are capable of pollinating the whole tree.

But climate which favors the culture of fig trees is not always congenial to the fig wasp. In this country the trees may be grown where the temperature does not go below 20 degrees for extended periods. While in its completely dormant state, the wood will sometimes even survive temperatures as low as 10 to 15 degrees, though after the buds start to swell, they will be damaged at 28 to 30 degrees. The wasps, unfortunately, cannot stand this

TABLE 70: Fig Varieties

Variety	Section for which recommended	Fruit	Tree
Magnolia	Southern plains	Bronze fruit of mediocre quality when eaten fresh. Used for canning and jam. Needs no caprification. Sours badly if matured on tree.	Can be pruned back vigorously and will bear crop on new wood.
Celeste	Southeast Delta States Southern plains	High quality bronze and violet fruit, closed eye prevents souring. Fruit borne on new wood. Needs no caprification.	Slightly more cold-hardy than others, but will not bear in first season after freezing to the ground.
Brown Turkey (Everbearing)	Pacific States Southeast Southern plains	Bronze, medium size, needs no caprification. Sours more easily than Celeste, less than Magnolia.	Prune back upper branches. Rampant grower.
Kadota	Pacific States Southeast	Medium size, yellow fruit. Souring prevented by honey in eye. Good quality. Needs caprification.	Prolific. Needs light pruning to keep center open.
Green Ischia	Delta States Southeast	Green fruit of not too high quality. Needs no caprification.	Not as productive as most figs.
Mission	Pacific States	Black fruit of excellent quality. Needs no caprification.	Grows into a fine shade tree in California and bears two crops in the central portion of state.

cold. They flourish in California in districts where the temperatures are always mild, but cannot be raised in Texas. Summers in Texas and some southeast and delta states are hot enough and long enough for fig culture, but the winters are too cold for the wasps, so the Smyrna figs cannot be grown. Instead, two groups of figs—the common figs and those of the San Pedro type, are grown where winters will not support the wasps.

The common fig produces its fruit asexually and any pollination or seed development which follows is not necessary to the production of fruit. Trees of the San Pedro group produce two crops, the first asexually and the second which must be fertilized by pollen from the caprifig. These two types of trees are recommended for home gardeners, even those who live in the mild climates where the wasps can grow. Caprification is a job for a specialist. Moreover, caprifigs are susceptible to brown rot and the disease is often carried by the wasps to fruit-bearing trees. Many California growers are now producing their wasps under laboratory conditions, to prevent the spread of brown rot.

Trees of the fig are rapid-growing, like many tropical plants. Under favorable conditions, cuttings of certain varieties may be raised to tree size and may bear a crop in the first year.

Though the plants survive winter cold to some extent, their wood is sometimes damaged without any outward sign. Crops in the following year may be very limited or completely lacking. Tops are sometimes killed back to the roots. On some varieties, new wood which sprouts after such a freeze may bear fruit the following year.

Fresh figs contain small amounts of vitamins A, B_1 B_2 and calcium. These amounts diminish in the dried fruit, but the sugar content increases. Thus, in two large fresh figs there are 50 units of vitamin A, .037 milligrams vitamin B_1, .030 milligrams vitamin B_2, 26 milligrams calcium and about 42 calories. In two small dried figs the vitamin A is reduced to 15 units, vitamin B_1 to .015 milligrams, but the calories go up to more than 100.

Site

Figs grow best on sandy or clay loam well supplied with humus and plant nutrients. They will tolerate high lime soils and also higher concentrations of salts than most fruits. In a cool area they may be planted near a building for warmth and winter protection, but they should have full sun. The soil must be well supplied with moisture and must also be well-drained. If drainage is at all dubious, raise the planting bed above the soil level by making a mound or terrace in which to plant the tree.

Planting

Space required by trees depends to some extent on the variety, but also upon climate. If the trees are often killed back, they will not make large growth. But if the growth persists from year to year, the trees need more space. One-year-old number one plants should be purchased and planted in early spring. Space required is from 14 to 25 feet.

In warm areas, set the tree one to two inches lower in the soil than it grew in the nursery. Where there is danger of freezing back to the ground, set the short trunk portion of the tree entirely under soil level, so that 3 to 6 leaders may be grown as a bush.

Maintenance

Figs sometimes refuse to bear fruit when they are given too much fertilizer, so it is best to allow the tree to go its own way as long as it is bearing satisfactorily. If growth and production fall off, a light application of rotted manure or compost should be spread under the tree in early spring.

If water is supplied in hot dry weather, it will lengthen the period of production. A thick mulch under the trees is also helpful, not only to preserve moisture,

Like many tropical plants, fig trees grow rapidly. Under favorable conditions, cuttings of certain varieties may bear a crop in the first year.

but also to keep the soil around the roots cool.

Fig trees should be pruned very little, except to remove dead wood, to open the top for sun and air and to stimulate new growth. Fruit is borne in the leaf axils. The asexual fruit of the San Pedro type is borne only on new growth of the previous year. This must not be headed back or a diminution of the crop will result.

Propagation

New trees may be easily produced in the home garden from cuttings. Pieces of 2- to 3-year-old wood no more than ¾ inch in diameter are taken in winter or early spring and cut into 8 to 10 inch lengths, the cut being made through a node. These are set in a nursery bed or they may be planted where the tree is to have its permanent home. They are set into the soil with only the tip exposed and the soil is kept moist, but not wet. In the nursery, cuttings are spaced two feet apart. The following spring they are dug and moved to their permanent positions.

Enemies

Cotton root rot attacks the figs in some areas and no cure is known for the disease. Trees attacked by the rot suddenly wilt and die. Best preventive is to avoid soil on which cotton has been grown.

Some fruits which have eyes that are open wide enough to admit insects suffer a disease known as souring. Insects carry bacteria and yeasts into the ripe fruit and fermentation follows which spoils the fruit flavor. Two preventives are possible. Either plant a variety which does not have open eyes, such as Celeste or Kadota, or harvest the fruit before it is fully ripe. Sanitary measures which also help, include removing all fallen or soured fruits from the trees.

GOOSEBERRY

GOOSEBERRY. Two species of the *Ribes* genus of the *saxifrage* family have been developed into the modern varieties of gooseberry. In England, where gooseberries are much more popular than they are in this country, *R. Grossularia* which is native to Europe has been developed into large-fruited bushes, some of which bear berries almost as big as hen's eggs. The native American *R. hirtellum* naturally finds the American climate more to its taste and has been developed into a number of varieties which may be grown in all of our cooler states. A few of the European varieties have been adapted to this country, but they succeed best where the summers are coolest, as along the New England and northern Pacific coasts and around the Great Lakes. In addition to their dislike of heat, European varieties are also more susceptible to mildew, one of the gooseberries' most destructive diseases.

Susceptibility to mildew is often a result of the gooseberry bush's tendency to form a dense top. Like currants, gooseberries send up a crop of new shoots from the ground each year. In the second year, these shoots form many more laterals than currants do and by the third year each cane has almost the dimension of a bush in itself, if it is not pruned. Roots are shallow, but spreading. Gooseberries send out long feeder roots laterally and for that reason must be planted young and left in their positions after planting. They are not moved easily when mature. Also because of their shallow roots, they prefer a cool heavy soil with a large percentage of humus to a sandy or even a light clay loam. Where the summers are rather warm, they should be given some shade. Gooseberries will tolerate more shade than any other fruit in the home garden. The fruit is improved if it matures during a long, cool period.

Berries of the gooseberry may be red, yellow or green when mature. The varieties which remain green (actually they are whitish when fully mature) are the most sour; those which are red are the sweetest when ripe and can be eaten out of hand. The green and yellow varieties are used mostly for pies, cooked desserts and jams or jellies, with some sort of sweetening added.

Not including the food value of the added sweetening, gooseberries have a low caloric value, supplying only about 37 calories in a ¾ cup serving. They also supply some vitamin A, the amount varying with variety, but averaging about 150 units to a serving; also .150 milligrams vitamin B_1, 25 milligrams vitamin C, 40 milligrams calcium, 50 milligrams phosphorus, .4 milligrams iron and 1 gram of protein in 100 grams.

TABLE 71: Gooseberry Varieties

Variety	Section for which recommended	Fruit characteristics	Bush characteristics
Fredonia	New England coast North Pacific coast	Late. Medium to large red berry. European.	Open habit.
Carrie	Great Lakes (In New England it resembles Houghton)	Small to medium size berry, red when ripe. American.	Fairly mildew-free bush with a few short thorns. Very productive.
Downing	Pacific coast Corn Belt New England	Large pale green berry. Midseason. Good for canning. American.	Seldom has mildew. Very productive. Aphid-resistant. Large bush.
Houghton	Pacific coast Northeast	Small dark red berry. American.	Large productive bush, drooping habit. Somewhat susceptible to mildew and aphids.
Josselyn (Red Jacket)	New England Great Lakes	Large red-green berry. Early. American.	Productive. Mildews when grown further south.
Oregon (Oregon Champion)	Pacific coast New England Alaska	Large green berry. Late. American.	Large productive bush. Mildew resistant.
Poorman	Corn Belt Great Lakes New England Mountain States	Large berry, red when ripe. Midseason. Best variety for eating raw. American.	Vigorous, productive bush with a few short thorns. Large bush.
Pixwell	Alaska Pacific coast Northern plains New England	Large berry, pink when ripe. American.	Hardy, vigorous, productive bush with less thorns than many varieties.
Chautauqua	Appalachians Pacific coast	Very large greenish-yellow sour berry. European.	Small bush, not as hardy as American varieties.
Como	New England Great Lakes	Medium size berry, green, tinged yellow. American.	Hardy.

Ordering Stock

American varieties are most productive, hardiest and of the best quality when grown in the United States. Though the European berries are larger, they do not produce their best crops here. Except in special situations, American bushes will be most successful in American gardens.

One-year stock should be purchased in fall and planted any time after the leaves have dropped. Gooseberries bloom so early in spring that it is almost impossible to transplant them at that time. The first warm days, before the ground has warmed up at all, will see them flowering. Since they require a clay soil which warms up late, the soil is almost never in condition to be worked early enough for spring planting.

Site

Gooseberry bushes are strong enough to flourish in a wind-swept position, even when they act as a windbreak at the top of the garden. They should not be planted in air pockets, not because of the cold air, but because they need good air circulation to prevent mildew. They will stand some shade, particularly if the climate in summer is hot, but they should stand free under a large spreading tree, rather than in a corner or nook in the shrubbery border. Where shade is needed, they are sometimes interplanted between rows of grapes, on a slope with good air drainage. They grow lower than the grapes and are partly shaded by the vines. Like currants, gooseberries prefer a soil with a *p*H of 6. to 8.

Planting

Before planting gooseberries, incorporate plenty of potash into the site, as well as humus. The bushes should not be fed too heavily with nitrogen, which encourages heavy green growth which in turn promotes mildew. If rotted manure is used to supply humus, it is better to use stable manure than poultry manure. Even better is a material like sawdust, seaweed or spent hops which is not over-rich in nitrogen.

Plants may be set 4 to 6 feet apart in the row, depending upon variety. European varieties usually make smaller bushes—some of the American bushes are larger and spreading. Rows should be spaced 6 to 8 feet apart.

Trim off any broken roots before planting and cut back the tops to 6 inches. The bushes should be set slightly deeper than they were in the nursery. If the form of the bush tends to a main trunk, plant it so that the junction of trunk and branches are all below the soil surface.

Maintenance

Because of their surface roots, gooseberries are grateful for a thick straw mulch. But be careful to leave an area of bare soil near the bushes, to discourage field mice. If the mulch is of a material which supplies some nitrogen, additional feedings each year should consist of only bone meal. This should be spread at the rate of two ounces per square yard.

Pruning

Gooseberries are pruned very much like currants. The mature bush should consist of 2 to 3 canes which sprouted during each of the past 3 years. After planting, during the first dormant pruning all but two of the strongest canes are cut back. The next year these are supplemented by 2 or 3 from the following year's crop. In the third year 2 or 3 more young shoots are left, the rest pruned out. Beginning either in the third or the fourth year the oldest wood is cut out at each pruning. In the East, 3-year canes are usually removed. In the West, these are left for another season. In addition to removing new canes, each 2- or 3-year cane must have its laterals thinned each year and low-hanging branches should be removed.

Bushes which are shaded need more severe pruning than those in full sun. Unless their heads are kept relatively open, they are susceptible to mildew.

Pruning should be done in late fall to midwinter. Spring pruning may find blossoms already opening on the bushes.

Propagation

Gooseberries may be propagated by layering or from cuttings in the same way that currants are propagated, but they are more reluctant to make roots than currants. They are most often increased by mounding, which is a special type of layering.

The plant from which mound layers will be taken should be cut back heavily before it begins growth in spring. By midsummer it will have sent up numerous new shoots and the bases of the shoots should have begun to harden.

Earth is mounded over the base of the plant halfway to the tips of the shoots. By autumn the shoots will have sent out roots. Those with strong roots may be severed from the parent plant immediately and set in the nursery. The balance should be left on the parent bush until the following fall. In the nursery they are set two feet apart and grown for one or two years. When they have made a strong root growth they are ready to be set in their permanent positions.

Enemies

Like currants, gooseberries are susceptible to white pine blister. Their enemies are very similar to those of the currants, but they are more susceptible to mildew. For other pests, *see* CURRANTS.

Harvest

Unlike strawberries or cherries whose ripening is spread over a long period, gooseberries of one variety all ripen at the same time, so that they may all be picked on one day. This is fortunate, because the bushes of most varieties are so thorny that the usual methods of picking would be pretty painful.

Usually the picker wears stout leather gloves and strips the branches of their fruit by running his hand along the whole branch, catching the berries in a large open container, such as a bushel basket. Many small pieces of leaf and twig are stripped with the fruit by this method. They may be separated if the fruit is rolled down a gentle incline, such as may be made with a long piece of cardboard. The twigs and leaves are mostly left on the incline and moderately clean fruit rolls to the bottom. Since the fruit must be picked over to remove the petals and stems before it is used, this residue of debris is not too important.

When gooseberries have attained their full size, they may be used for cooking, even though they are not fully ripe. The longer they are left on the bush, until they are fully colored, the sweeter they become.

GRAPE

GRAPE. One of the fruits universally described in all the ancient literatures is the grape. Some type of grape is native to almost every portion of the temperate and many of the subtropical portions of the world. Grapes, and the wine made from them, have been important in man's diet since earliest times.

Some variety of grape may be found to grow in any state in the United States. Many wild types were found here by the earliest European explorers, some of which we still grow in cultivated varieties, in hybrids or as rootstocks on which are grafted Old World varieties which cannot be grown here on their own roots.

Early settlers on the eastern seaboard tried vainly for many years to grow the grapes that they brought with them from Europe, only to find that they could not be grown except in greenhouses or under the most artificial conditions. It was only when they gave up the attempt to grow European grapes and turned to the native species that American grape growing became successful. The production of grapes on European vines, which can now be grown in California and a few portions of the hot, arid Southwest, became possible only after many of the varieties were grafted to native American roots.

Grapes now grown in this country comprise 3 species of the genus *Vitis.* They are: *V. vinifera,* the European type, whose skins do not separate easily from the berry, grapes which require a long, hot season to mature; *V. labrusca,* called Eastern, American or fox grapes—the Concord type, with slip-skins and a comparatively short ripening period and the fruit of much hardier vines than the other two species; and *V. rotundifolia,* the Muscadine or southern fox grapes, also called Scuppernongs. The Muscadines are native in the southeastern portion of the country, south of the Potomac River, where Eastern grapes are not successful because of the many fungus diseases endemic there. Muscadines are slip-skin grapes, growing on comparatively tender vines, the bunches usually containing no more than 2 or 3 berries. Most of them are dioecious, that is, the plants bear either male or female flowers, but not both on one plant. Plantings, to produce fruit, must contain at least one male plant for each 25 females. A few varieties with perfect flowers are being developed.

Grape vines are deep-rooted when they find soil that suits them, the roots sometimes penetrating 8 feet into the ground. Roots usually develop in two whorls: an upper layer which spreads laterally and a lower one which penetrates diagonally to vertically. Vines spread 6 to 8 feet in each direction in the Eastern type, 12 to 20 feet in the Muscadine and, in the viniferas, may be pruned short or made to cover an arbor.

Fruit is borne on the current season's growth. From each of the canes made last season, a bud at each leaf node sends out a new lateral branch in the spring. Latent in this bud is a shoot which will have leaves at the first node or two, followed by flowers, fruits and leaves at the following nodes. Beyond the fruit, the shoot continues to grow and make leaves. At the end of the season, this shoot is known as a one-year cane. It develops a fibrous bark and will send out, the following year, its own crop of shoots, each with flowers and leaves. Grapes do not bloom until fairly late in spring, but the young shoots are very tender and may be damaged by frost before the flowers have opened. As soon as the buds begin to grow they become vulnerable to cold. Grape culture is based upon these peculiarities of the plants and successful planting, pruning and training is based upon a knowledge of them.

Grape vines in the home garden may form a part of the landscape picture, at the same time providing dessert for the

TABLE 72: Grape Varieties

EASTERN GRAPES

These grapes are similar to Concords grown in all sections where winters are frigid. They are more or less hardy, have skins which slip off easily and grow in fairly large bunches, unless otherwise noted.

Variety	Section for which recommended	Fruit characteristics	Vine characteristics	Cold endurance
Concord	New England Great Lakes Northern plains Corn Belt Southern plains Appalachians Pacific coast	Black, aromatic. Late. Used for dessert, juice, wine, jelly. Seedless sport available.	Grows well where many others fail. Requires 170-day growing season, 10 to 12 ft. spread.	Hardy to —15°.
Fredonia	Corn Belt Southern plains Northern plains Great Lakes Northeast Appalachians	Early. Black. Medium size fruit—compact clusters. Susceptible to mildew.	Vigorous. Productive.	Hardy to —15°.
Delaware	Southeast Appalachians Pacific coast Southern plains New England	Red. Small berries, small clusters. Used for white wine. Midseason.	Lacks vigor on own roots. Moderately productive. Small vines—need 6 to 8 ft.	Hardy to —5°.
Niagara	Corn Belt Southern plains Southeast Appalachians Pacific coast Northern plains New England	Sweet white medium to large berry. Large clusters. Midseason.	Moderately susceptible to disease. Productive.	Hardy to —5°.
Elvira	New England	White delicate berries. Used for wine.	Vigorous. Productive.	Hardy to —15°.

TABLE 72: Grape Varieties (Continued)

Variety	Section for which recommended	Fruit characteristics	Vine characteristics	Cold endurance
Catawba	New England Southern plains Northern plains	Red. Late. Used for wine, juice.	Productive. Moderately vigorous, but subject to fungus disease. Need 10 to 12 ft.	Hardy to —15°.
Interlaken Seedless	New England	White seedless. Very early. Sweet, meaty, crisp.	Very low temperatures may injure wood.	Hardy to —5°.
Schuyler	New England	Black. Very early. Nonslip skin.	Require short pruning.	Moderately hardy.
Van Buren	New England	Early. Black. Concord type.	Flower clusters and fruit susceptible to downy mildew.	Hardy to —5°.
Portland	New England Appalachians Southeast Corn Belt	Amber. Very early. Berries and bunches medium size.	Moderately vigorous and productive. Harvest as soon as ripe. Bunches shatter.	Hardy to —5°.
Ontario	New England Appalachians Corn Belt Great Lakes	Green-white. Very early. Small berry.	Vigorous. Productive.	Hardy to —15°.
Seneca	New England Southern plains	White. Nonslip skin. Firm, tender. Early.	Susceptible to mildew.	Hardy to —5°.
Kendaia	New England Corn Belt	Black. Early. Concord type.	Very productive.	Hardy to —15°.
Buffalo	New England Corn Belt	Midseason. Black.	Vigorous. Productive.	Hardy to —5°.
Steuben	New England Great Lakes Northern plains Corn Belt	Blue-black. Late. Sweet and spicy. Large bunches. Store well.	Vigorous. Productive.	Hardy to —15°.

TABLE 72: Grape Varieties (Continued)

Variety	Section for which recommended	Fruit characteristics	Vine characteristics	Cold endurance
Golden Muscat	Appalachians Pacific coast	Gold-yellow. Large bunches. Late midseason.	Vigorous. Tends to overproduce some years, weakening vines.	Not hardy below 0°.
Campbell	Pacific coast New England Southern plains	Early. Black. Large clusters.	Require two varieties of pollination. Should be grafted. Needs fertile soil.	Hardy to —5°.
Brockton	Southeast Southern plains	Midseason. White. Medium size.	Productive.	Hardy to —5°.
Beacon	Southern plains	Midseason. Black. Large clusters.	Productive.	Hardy to —5°.
Bailey	Southern plains	Midseason. Black. Large clusters.	Productive.	Hardy to —5°.
Extra	Delta States Southeast Southern plains	Black. Midseason. Compact bunches.	Productive.	Hardy to +5°.
Moore Early	New England Great Lakes Corn Belt	Moderately large. Black. Early.	Fairly productive, vigorous. Needs fertile soil and heavy pruning.	Hardy to —15°.
Worden	Northern plains New England Great Lakes Appalachians	Black. A week earlier than Concord. Sweet. Berries split in wet weather.	Needs another variety for pollination. Moderately tolerant to mildew. Recommended for home gardens.	Hardy to —15°.
Athens	Appalachians Southeast	Black. Early. Large loose clusters.	Vigorous. High yields on heavy soils.	Hardy to +5°.
Brighton	New England Appalachians Corn Belt	Dark red. Midseason. Sweet. High quality.	Yield medium to good. Needs another variety for pollination. Self-sterile.	Hardy to —5°. Kills to ground in severe cold.

TABLE 72: Grape Varieties (Continued)

Variety	Section for which recommended	Fruit characteristics	Vine characteristics	Cold endurance
Caco	Appalachians Southern plains Corn Belt	Late midseason. Light red. Small bunches, large berries.	Moderately productive.	Hardy to —5°.
Sheridan	Appalachians	Late. Black. Medium size berry, compact bunch.		Hardy to —5°.
Alpha and Beta	Northern plains New England	Small berry and cluster. Too sour for eating—can be used for jelly or juice.	Two varieties of wild grape grown where winters are too severe for other varieties.	Completely hardy to —25°. May be left on trellis all winter.
Blue Jay	Northern plains New England	Blue-black. May be eaten when fully ripe.	Needs second variety for pollination.	Hardy to —15° but needs protection in exposed sites.
Red Amber	Northern plains New England	Red. Sweet. Medium size berry and bunch.		Hardy to —15° but needs protection in cold areas.
Moonbeam	New England Northern plains	Green. Large berries, small bunches. Mild flavor.		Hardy to —5°.
North Dakota #11	New England Northern plains	Green. Moderate size. Too sour for eating. Jelly resembles crab apple flavor.	Productive.	Hardy to —15°.
Lucille	Corn Belt New England	Red table grape. Fair quality.	Vigorous. Productive.	Hardy to —15°.
Isabella	Hawaii	Medium size. Black-blue with bloom.		

TABLE 72: Grape Varieties (Continued)

MUSCADINE GRAPES

These resemble the Eastern grapes, but are more resistant to the fungi diseases and are longer-lived when grown in the Southeast than the Concord group. Bunches are very small, usually consisting of one or two berries. Hardy only to 5 degrees. Grown southward from the Potomac River. Vigorous vines, spread 12 to 20 feet.

Variety	Description
Burgaw	Midseason. Reddish black. Perfect flowered. Fair quality.
Creek	Late. Reddish purple, with thin skin. Good quality.
Dulcet	Midseason. Reddish purple. Good quality.
Higgins	Midseason. Bronze. Large berry. Good quality.
Hunt	Early. Black. Skin thin. Excellent quality.
Scuppernong	A wild variety, early, green to bronze, quality very good. Rich in vitamin C.
Tarheel	Midseason. Small berry, black, thin skinned. Perfect flowers.
Topsail	Midseason. Green to light bronze. Skin medium thick, tough. Good.
Wallace	Midseason to late. Yellow to bronze, Sweet, thin skinned. Perfect flowered.
Yuga	Late. Reddish amber, thin skin. Very good.

VINIFERA GRAPES

Vines in this group are mostly tender below 15 degrees, a few will tolerate temperatures down to 5 degrees. They need a long season of hot weather for ripening and fairly arid conditions, such as are found only in some of the southwestern states and the interior valleys in California. Most do not ripen well in California coastal fogs. They are all nonslip skin berries in large bunches.

Variety	Description
Ribier	Black, very large berries in medium size bunches. Early midseason.
Red Malaga	Large red berries. Adapted to hottest areas. Midseason.
Flame Tokay	Similar to Malaga, but will ripen in cooler valleys.
Olivette Blanche (Lady Finger)	White, elongated berries. Very good eating.
Thompson Seedless	White, medium to small berries, large bunches. Very sweet. Early. Used for raisins.
Muscat	Large white berries. Sweet and flavorful. Used for raisins.
Emperor	Very late. Suitable for hot valleys. Large red berries and clusters. Keep well.

table. Although they are deciduous and of little use as a cover in the winter, they can be used in outdoor living areas as a summer screen to mask service fixtures or to give shade. Training the vines against a building is not recommended, because the air drainage in such a position is usually bad—consequently in early spring more frost damage may be suffered there. But open arbors which shade garden paths or patios may be covered, in warm areas, with a canopy of grapes. Muscadines are especially useful for this purpose. Two vines have been known to cover completely a 50-foot arbor.

Fresh grapes supply a small amount of vitamins A and C, some of the B vitamins and some minerals. Grape juice loses all its vitamins A and C and some of its B vitamins. Raisins, probably because they are concentrated in drying, provide a much larger proportion of everything except vitamin C in equivalent weights. Most of the food value of grapes is in sugar, of which they contain a fairly large amount. A small bunch of fresh grapes supplies 80 calories, all sugar; one-half cup of grape juice contains 60 calories, and one-fourth cup of raisins contains 105 calories.

Ordering Stock

Thousands of grape varieties are available and, among these, many more combinations may be made with grafted rootstocks. Gardeners in most areas of the country have a wide choice of varieties, most of which will grow on any well-drained soil and produce well as long as their other requirements are met.

Most important in selection of a variety is to plant a grape suited to the climate and growing season. In the accompanying chart, Eastern grapes are listed as "late" if their growing season is about as long as the season needed to mature Concord grapes, that is, 170 frost-free days. If the number of days between his last spring frost and his first fall frost is 170 or more, the gardener can expect grapes listed as late to have time to mature in his garden, assuming that their other needs are met. If he has less than 170 frost-free days, he should plant a mid-season or early type instead. Certain favored spots in any area may have a couple of weeks more of frost-free weather than the norm for the section as a whole. Such a sanctuary may sometimes be found on a southern slope, where the air drainage is good and the soil contains enough gravel or stone to retain the sun's heat, or it may be found near the shore of a large body of water.

When ordering stock, choose one-year plants in preference to two-year. The stock must be healthy to make sufficient growth in one year to plant out and a one-year vine will come into bearing as quickly as a two-year vine, provided the root system is good. Before being set out, the roots will be cut back to 6 to 8 inches each, so there is no advantage in buying long roots. Look for a bushy clump of roots rather than long, straggling roots.

Most varieties of grapes are self-fruitful except where noted on the variety chart. Some varieties, however, need a second type for pollination. Grapes are pollinated by the wind, rather than by insects. It is necessary, then, to plant the pollinating varieties near each other in order to set fruit. Maximum distance is 50 feet; closer setting is desirable.

A vigorous mature vine may be expected to yield 15 or more pounds of fruit, when properly pruned. Eight or 10 vines are usually ample for the needs of the average-sized family. Under favorable conditions and with proper care, vines will produce for at least a generation and sometimes as long as 50 to 60 years.

Preparing a Site

A deeply drained corner of the garden with full sun, good air circulation and warm soil will grow the best grapes. If the site has been cultivated for 2 or 3 years before the vines are installed, less trouble with deeply rooted perennial weeds may be expected. Where there is a choice, a south slope with good air drainage will give the longest frost-free period for maturing the fruit and developing its full quota of sugar.

Whether the site has been used before or not, it should be dug or plowed deeply the fall before grapes are set out. Granite dust and rock phosphate should be liberally applied and dug into the soil in an 8-foot circle around the spot where each vine will be placed. These supplies of phosphorus and potash will gradually become available in the soil for many years. Grapes will tolerate some acidity, but prefer a soil that is nearly neutral, with a *p*H between 6. and 8. If the site is more acid, bring it up with an application of dolomitic limestone.

Planting

Vines may be planted either in spring or fall. The usual rule is to plant in fall only where there is no danger of winter injury and to plant in early spring through the rest of the country. Plants should be set out before their buds begin to swell, because the buds are easily damaged after growth starts.

Spacing of vines depends upon the variety planted, the system of training and pruning to be followed and the type of support. When they are planted in rows in the garden, an aisle 9 to 10 feet wide should be left between rows to facilitate work and harvest and to permit plenty of sun to reach all parts of the plants. Varieties which produce only a moderate growth in a season, such as Delaware, may be planted 6 feet apart.

The more rampant growers need 8, 10 or even 12 feet.

Roots and tops should be pruned before the vines are planted. Roots should all be shortened to no more than 6 inches, in order to encourage formation of feeder roots near the trunk. Remove all but the two strongest canes in the top and cut those two back to two buds each. Dig the hole one foot in diameter and spread the roots carefully when planting, being careful to spread the lower whorl, tamp it in and then spread the upper whorl over it. Set the vine with the lowest bud level with the soil surface. Flood the hole with water to wash the soil in around the roots.

A trellis need not be provided for vines until their second year, but they will need some sort of support the first season. A stake 6 to 7 feet long and 1½ inches in diameter or a pole like a bean pole, will be adequate for the first summer's training. Set a stake next to each cane before growth starts. Also spread fertilizer in the form of compost or rotted manure over the soil in a 3-foot circle in a layer an inch deep. Mulch with gravel, straw or a coarse material such as cocoa shells or ground corn cobs.

Four shoots should develop the first year from the two buds left on each of the two spurs. When the shoots are 2 or 3 inches long, remove all but the strongest one. This shoot should be encouraged to make a strong growth; it will form the trunk of the vine. Tie it at intervals to the stake and allow it to develop as many leaves and grow as tall as it will in one season. By the end of the summer it should be at least 30 inches tall and preferably 5 or 6 feet.

If the new vine was planted below an existing trellis, a soft twine tied to the horizontal wires may be dropped to the vine and tied to one of the stubs left after the excess shoots are removed. As the trunk grows, twist it loosely around the twine to keep it upright, and tie it every 8 to 10 inches with soft jute twine

Grape vines in the home garden can be an attractive part of the landscape picture, at the same time providing a nutritious dessert. A mulch of alfalfa hay is valuable for grapes.

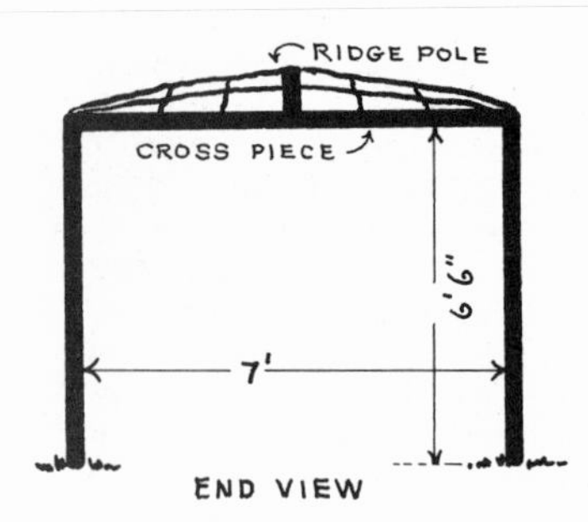

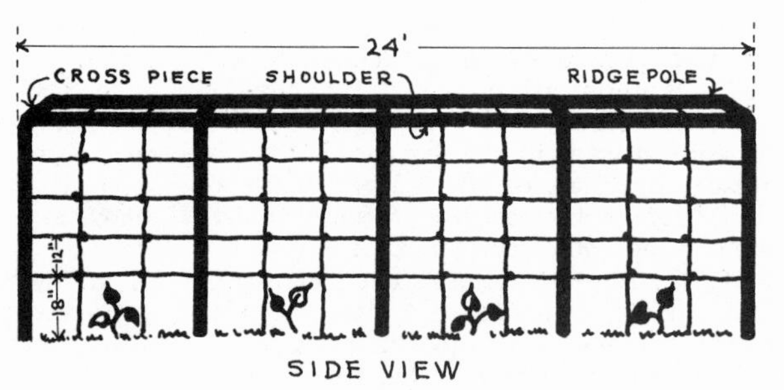

or raffia. Make all knots loose to avoid girdling the shoot and handle the vine carefully, because the tender shoot is easily damaged.

Pruning and Training

Pruning should be done in late winter or early spring before buds start to swell. If the job is done late enough in winter, all winter-killed wood may be spotted and cut out, and only good canes left. Late pruning may result in "bleeding" from the cut ends when the sap begins to run, but this apparently does the vines no harm.

The trellis type should be decided upon before the young vine is pruned the first time and it should be erected before the second season's growth starts. A number of training systems may be used where the climate is not too cold, though one or two are best adapted to each type of grape. Vinifera grapes are usually trained in the field to spur pruning, but in the home garden they may be trained to a stake with canes permitted to make an umbrella, they may be trained to the Munson system or they may be trained to an arbor. Muscadine grapes are usually trained to an arbor, to cover a building or a garden eyesore, or they may be trained in the field to the permanent 4-arm system. Eastern grapes are usually pruned to the 2-, 4- or 6-arm Kniffin system or to the Munson system. Of these, the 4-arm Kniffin seems most popular and is easiest to train.

Posts for a trellis will last longest if they are made of redwood or cedar, though steel posts are being used more and more. End posts should be 7 o 7½ feet long, 2½ feet set into the soil. Intermediate posts may be a foot shorter, with only 18 inches set below ground. Posts should be set 16 to 18 feet apart, spaced so that 2 or 3 vines are supported between each pair of posts.

Wires may be attached to the posts in different positions, depending upon the system of training. Wire is usually number 9 or 10, to be strong enough to support the vines when they become mature and carry a heavy load of fruit. Some means of tightening the wire in hot weather and when it carries its greatest load should be devised when it is installed. Also, the end posts should be throughly braced against the direction of greatest strain.

Wires for the permanent 4-arm system or the 4-arm Kniffin system are set 2½ and 5 feet above ground. The two-arm Kniffin is not recommended because it places too much weight on one wire, allows the vines too little spread to catch the light and air and makes pruning difficult. Vines, when growing on the Kniffin system, form a more or less vertical wall in a plane at right angles to the ground.

The Munson system trains the vines in an umbrella pattern, in a plane which is horizontal but elevated 4 to 5 feet above ground. One wire is stretched between posts 4½ feet above ground and the trunk is trained up to that wire. Crosspieces two feet long are set above the wire on each post and two more wires are stretched to connect the ends of the crosspieces. As the shoots grow from the main trunk, they are trained to the raised outer wires. The advantage of this system is in more light reaching all the leaves, with the fruit hanging below and shaded from sunscald, and also in ease of harvest. However, the trellis is more expensive to construct and more time must be spent training the vines.

Home arbors are usually constructed no more than 7 feet high and 8 to 12 feet across. They may be as long as desired. Vines are planted 8 to 10 feet apart along the sides of the arbor and trunks are carried to the top edge, where they are pruned to the cane-renewal method, the canes being arranged over the top of the arbor.

For extremely cold areas, the best way to train vines is in the fan system. Wires are arranged as for the 4-arm Kniffin or they may be increased to 3 wires in the same space. The trunk is kept very short—usually no more than one foot high. Four or 5 canes pruned by the cane-renewal method are allowed to develop each year. Canes can be most easily protected in this type of system, either by being covered on the trellis or by being bent to the ground and covered with soil through the winter.

Grapes are fully ripe when they become aromatic, sweet, and the stem of the bunch begins to show brown areas. Clip bunches from vines with sharp shears.

During the first winter after the vines were planted, they will be pruned back to the trunk, which is tied to the top wire in the Kniffin system, to the center wire in the Munson or to the lower wire in the fan system. Side shoots, if any, are removed. If the trunk is less than 30 inches tall, cut it back to two buds and start it over again, handling it as if it were in its first year.

In its second summer the vine will send out horizontal shoots from each leaf node along the trunk. These should be allowed to grow, since each leaf manufacturing food contributes to the vigor of the young vine.

The second winter's pruning will begin to shape the vine in its final form. This may be done by pruning it to the cane-renewal system, as for the Kniffin trellis, fan or arbor, or by pruning it to the spur system. The spur system may be trained to the Munson trellis, or it is sometimes used in California vineyards with no trellis at all, the canes being allowed to form an umbrella and trail, from the top of the trunk, toward the ground. Muscadine grapes, which grow in small clusters, do not need vigorous yearly pruning because they will not commit horticultural suicide by overproducing, as will the other vines. They may be trained to 4 permanent arms, with a small amount of yearly pruning to stimulate production or they may be trained over an arbor or other support and trimmed very little each year.

Whatever the system, canes must be selected for growth the second year and all others removed from the vine. The best canes to leave are those intermediate in growth—neither too spindly nor too rampant. Best fruit is produced on canes which are about the thickness of a lead pencil and with the internode between the fifth and sixth buds from the trunk measuring 6 to 7 inches.

Pruning to the cane-renewal system tends to produce the best bunches of grapes, which are to be found at about the sixth or seventh nodes of the canes, but the total yield is slightly less than the yield on vines pruned to the spur system. Typical of the cane-renewal type of pruning is the 4-arm Kniffin system, the most popular method.

Four canes are selected for the 4-arm Kniffin and are tied to the wires, one traveling in each direction from the trunk on each of two wires. These canes are shortened to a predetermined number of total buds, depending upon the age, vigor and productive capacity of the vine. Pruning the second winter should leave no more than 12 buds total to each vine. In addition, 4 spurs of one bud each are left, one beside each cane. These spurs will grow the canes which will be tied to the trellis the following winter.

As the vine grows older, more buds are left each year on the 4 canes which are tied to the wires. Many methods have been devised of computing the number of buds which may safely be left for best production. These involve, in some cases, pruning one vine and weighing the canes removed, then from this weight judging the number of buds to be left. (This implies, of course, that the pruner knows how to prune the first vine properly.) For Concord vines, 30 buds may be left for the first pound of pruned wood, with 10 additional buds left for each additional pound.

Best rule for the novice is to assume that his vines will not come into full production until they are 5 or 6 years old, and to leave a little more wood each year until maximum production is reached. If a vine may be expected to yield 15 pounds of grapes in one-fourth pound bunches and an average of two bunches are produced at each bud, then at top production the vine should be permitted no more than 30 buds. Success of the previous year's pruning may be judged when the vine is ready for pruning the following year. If the canes from buds which were permitted to remain on the vine are spindly, then too many buds were left. If growth of the young canes is too rampant, then too few buds were left and a larger number may be left at the next pruning.

In spur pruning the canes selected are all cut back to spurs. These are used both for renewal (growing shoots for the next year's spurs) and fruit production. On a spur-pruned vine the best canes at the top of the trunk are cut back to two buds each. In the young vine, 4 spurs are left the second winter, 6 or 7 the next year and, with each winter, the number is increased until the buds on the spurs equal the total number of buds which may be grown on a mature vine. Most satisfactory support for a vine pruned this way is the Munson trellis, which holds the young canes in a horizontal plane as they develop.

Total number of buds which may be left on mature vines vary from 30 to 35 on the less vigorous to 70 to 90, which are recommended in the northern plains on the most vigorous vines grown there. Concord and Niagara are typical of the vigorous vines which may be expected to bear heavy growth and yields. Delaware, Moore Early and Catawba are less vigorous.

After the vines are pruned they should be tied to the trellis wires. This should be done when the weather has warmed somewhat, because when they are frozen the canes are brittle and more easily damaged. But it should be done before growth starts, in order to avoid injury to the new shoots.

Renovating Old Vines

When they are pruned regularly and properly, grape vines will produce their best fruit and will bear a crop each year. When they are not pruned for a year or two, the wood multiplies on the vines with such rapidity that only a few years' neglect will reduce the crop to small bunches of miserable berries borne, sometimes, only biennially.

A vine which has run wild for several years will present a complicated tangle of canes and, sometimes, more than one trunk. If the growth has become very prolific, it should not all be cut back in one year, or the result will be the production of too-rampant canes which will not fruit. The more overgrown the vine, the slower will be the rejuvenation process.

If the vine was originally intended to grow in the 4-arm Kniffin system, each of the arms may be pruned moderately each year for 4 years, with one arm entirely reconstructed each year. If the vine was spur-pruned, the spurs will have to be cut back gradually and any adventitious growth from the trunk (that is, buds forming on old wood where they are not expected) should be encouraged. The difficulty with old spur-pruned vines is that the fruiting spurs grow farther and farther from the trunk with successive years. First all dead or weak wood should be removed. Next shorten the long trunk and branches to leave 8 to 10 good canes. Shorten half of the canes to 6 to 8 buds; the others to one or two. The following year thin the new canes and provide renewal spurs. Thereafter the pruning may follow the regular system.

When the trunk of the vine is very old or if it is decayed, it may be cut off at the ground. New shoots will arise from the crown and one may be selected to form a new trunk. The vine will be trained as if it were a new planting, but the development should be much faster.

Maintenance

Grapes are adapted to rather dry climates, and need less water than most of the small fruits. They are known to grow on islands where the rainfall is almost nil and where the moisture in the air from mist-laden clouds never falls as rain. But of course there is some evaporation from the leaves and, where the air is very arid, a certain amount of irrigation is necessary. Most of the water consumed by the vines should be supplied in early spring while the shoots are developing and again in midsummer before the canes begin to harden off for the winter. Vines do not need water while ripening fruit and too much rain in fall is actually injurious.

Because of their deep root systems the vines are protected naturally from short periods of drought, so they do not need mulch for moisture protection, as do many other small fruits. However, mulch which keeps down weeds and prevents their theft of nutrients from the vines, mulch which actually supplies nutrients itself, mulch which prevents erosion around the roots, is very valuable. Like most fruits, grapes need large quantities of nitrogen. A mulch of good alfalfa hay which supplies nitrogen as it breaks down can help to keep the vines thrifty.

Where the growing season is barely long enough to mature a crop of grapes of the variety planted, it is frequently possible to lengthen the season by a week or more by raking off the mulch for a brief period in early spring. Mulch over the soil retards warming from below and vines will not begin to grow until their roots are warmed. Rake the mulch back when the end of the frost season approaches and allow the soil to warm thoroughly. As soon as it is warm, replace the mulch.

Starting with the first year after planting, the vines should be given a feeding each spring. In the first season, the feeding roots will extend in a circle no more than 3 feet in diameter. Each succeeding year the circle will increase until it reaches a diameter of 8 feet. This circle should be spread with fertilizer, in order to reach the feeder roots. A mature vine should receive the equivalent of one bushel of well-rotted manure plus two pounds of granite dust each year. This should be applied in early spring under the mulch, and the mulch raked back to cover it.

Propagation

Grape vines are easily propagated from cuttings or layering. Select vines which have borne heavily for propagation. In California, where many vinifera grapes are subject to attacks from phylloxera, the vines must be grafted to resistant stocks. Eastern grapes are more easily propagated in the home garden.

If only a few new vines are wanted they may be produced by layering. Bend down the growing canes until they touch the ground. Make a cut through one or two buds on the under sides of the stem. Bury the cut nodes under 3 or 4 inches of soil. In a few months, after roots have formed from the node, the new plants may be severed from the canes and set out.

Where a larger number of vines are needed, they may more easily be grown from cuttings. These are made from pruned canes, taken in winter. For the cuttings, select well-matured canes neither too spindly nor too vigorous, about the thickness of a lead pencil. Each piece to be rooted should include 2 or 3 nodes which are 6 to 8 inches apart. The basal cut should be made just below a node; the top cut one inch above the second or third node above. Cuttings will not grow if by chance they are placed upside down in the soil. So in order to mark the top, make the top cut slanting and the bottom one straight.

After making the cuttings, tie them in bundles and store them in the field. Turn the bundles upside down and place them in holes in the ground. Cover, with soil at least two inches deep over the top end of the bundle. This will help to develop a callus on the bottom end of the pieces. Dig up the bundle early in spring, as soon as the nursery row can be prepared. Place the cuttings 3 to 4 inches apart in the row, with the top bud just above the soil surface. Allow it to grow as vigorously as it will throughout the summer. By the following year it will be a one-year plant and will be ready to be set out in its permanent position.

Enemies

Most common grape disease is black rot, which makes dried mummies of the grapes as they hang on the vine. This disease is prevalent wherever grapes are grown. The only cure for black rot is prevention—remove all bunches as soon as the fungus affects them and keep mummified berries cleaned up before the disease can spread.

Root aphids, or phylloxera, occur in various parts of the country, but principally in California. Some rootstocks have been found to be resistant to phylloxera and, in areas where their damage is great, only vines grafted to resistant rootstocks should be planted.

Japanese beetle caused a great amount of damage for years in the eastern part of the country, skeletonizing leaves in large vineyards. Beetles have been largely brought under control now by biological means. Inoculation of the soil with milky spore disease in areas where the beetle is still prevalent will cut his depredations.

On the whole, grapes are not much bothered by disease or insects, especially when the vines are well-fed and well-pruned. Where an old planting has been neglected, proper pruning and sanitary measures will sometimes serve to control the pests as well as the vines.

Harvest and Storage

Grapes do not improve much in aroma or flavor after they are picked, so they should be allowed to become fully ripe on the vines. Indications of fully ripe fruit are: it is aromatic, it is sweet, and the stem of the bunch begins to show brown areas.

Clip the bunches from the vines with sharp shears when both the grapes and vines are dry. Handle them by the stems, rather than by the berries. Place the fruit in a shallow basket and keep the picked bunches out of the sun.

If the berries are to be stored, they should be picked in the coolest part of the day. Choose only firm varieties for storage and store only perfect fruit. Cool the fruit to 50 degrees as soon as possible after picking, spread out in single layers. Allow to remain in this condition until the stems shrivel slightly. Then place in trays no more than 4 inches deep in a cellar which is slightly humid and where the temperature can be kept at about 40 degrees. Stored in this way, the fresh fruit can be kept for several months.

In hot and arid areas, grapes may be dried for winter use as raisins or currants. Vinifera grapes are used for this purpose. Currants are a special variety—the Zantes—which are very small, seedless and are used for this purpose alone. Seedless raisins are dried Thompson Seedless; seeded raisins are dried Muscats. Fully ripe grapes are spread on wire trays and are left out in the sun. The grapes are turned frequently and are protected from dew and rain with tarpaulins. Drying can be successfully accomplished only where the sun is hot enough to do the job quickly. However, it is well worth the trouble for gardeners who are able to do it.

Making a Grape Arbor

One organic gardener, John Mortimer, gives this advice for building a grape arbor:

All the material needed for this structure is a few cedar poles about 9 feet long and 4 inches in average diameter, a coil of fence wire, a quantity of boards or 2-by-4's, some nails and staples and a number of grape plants. If you can find used lumber which is suitable, your edifice will cost surprisingly little.

Your structure must be of adequate dimensions. If you care for comfort, you must have freedom of movement. This is the essential in which most arbors are lacking. They are usually constructed for the purpose of yielding much fruit and little human comfort. In a cramped-up arbor, to which no one is attracted, straying branches soon make the interior so impenetrable that most of the fruit cannot be reached.

An ideal size is 7 feet wide by 6½ feet high by 24 feet long. Such a structure will produce abundant fruit and allow complete freedom from a shut-in-cooped-up feeling. If coolness is your main object, both ends should be left open to the breeze. If you desire greater privacy, the ends may be closed, with one central entrance or a side opening at each end. Still more privacy is obtainable by making a "T," "U" or "L" shaped arbor and, if your yard is of irregular contour, something like this may inspire an unusual or artistic design.

Actual building of the arbor requires no great ingenuity. Posts should be sunk about 2½ feet into the ground, opposite each other in pairs, with a 2-by-4 or similar piece connecting their tops. These crosspieces support a center ridge pole or board, which may be composed of several parts. The pairs of posts should be 6 feet apart, connected also with 2-by-4's from top to top along each side.

Along the ridge pole, at two-foot intervals in each section, strands stapled in the middle reach to the ground on both sides. These wires are also stapled to the shoulder 2-by-4's, then looped alternatelv around each of the horizontal wires. Along the top, strands run from end to end, one foot apart, looping the cross wires and stapled to the crosspieces.

The use of wire in place of wood wherever possible will make your fruit easier to pick and your arbor less subject to rot. Life of the wood may be extended considerably by applying a coat of paint or preservative.

GRAPEFRUIT

GRAPEFRUIT. The *Citrus paradisi* is an anomalous fruit whose origin is obscure. So far as is known the seeds were brought

Fruits of the grapefruit hang from the inner branches of the tree, usually in clusters of 3 to 6, like enormous grapes.

to the West Indies at an early date, but they were believed to be shaddock seeds, a different citrus species. At any rate, grapefruit or pomelos grew from them.

The correct name for the species, pomelo, has never been popular in this country, where the fruit is known as grapefruit, possibly from its habit of growth. The fruits hang from the inner branches of the tree in clusters of 3 to 6, like enormous few-berried bunches of grapes. In this, as in the shapes of the individual fruits, they differ from shaddocks.

The first Florida planting of grapefruit was made in 1823, though the fruit was not developed until after 1885. It is interesting to note that some of the trees from the original 1823 planting were still growing and bearing in 1925, some of them 60 feet tall, with an equal spread.

In frost endurance, grapefruit trees are intermediate between sweet and sour oranges. They are less affected by sudden heat and are more resistant to wind damage than other citrus fruits. The trees require a high total amount of heat to produce their best quality fruit—at the average temperatures at which oranges and lemons thrive the fruit is sour, bitter and late ripening. At least 9 to 10 months are needed to mature fruit and, in some sections, they take two seasons. The best grapefruit areas are in central and southern Florida, the Rio Grande district in Texas and the mild-winter areas of Arizona and California.

Fresh grapefruit is a good source of vitamin C and contains some A and B vitamins as well. In one half grapefruit there are, on the average, 20 units of vitamin A, .070 milligrams of vitamin B_1, .060 milligrams of vitamin B_2, 45 milligrams of vitamin C, 21 milligrams of calcium, 20 milligrams of phosphorus, .2 milligrams of iron and only 36 calories.

Grapefruit, like the orange, presents no pollination problem, because fruit will set without fertilization and is often seedless. The trees are shallow-rooted, but they must have good drainage about 4 feet down, the depth of their penetration. Grafted one-year stock is best for planting. Except that they need more fertilizer, they are grown the same as oranges. (*See* ORANGES.)

GUAVA

GUAVA. One of the subtropics' stepchildren is the guava, a fruit which bears flavorful fruit full of vitamin C, but which has been little cultivated until fairly recently. The common guava, *Psidium Guajava,* is native to the American tropics, though it has become widely distributed through all the warm areas around the globe. Its seedlings grow wild in profusion in Hawaii, southern Florida and in many other places, but the fruits are seldom for sale in the markets. The common guava is sometimes called lemon guava, pear guava or apple guava—names arising from its flavor, size or shape. The seedlings show a great variation, some being round, some oblong, their flesh

TABLE 73: Grapefruit Varieties

The only grapefruit grown successfully in California is Marsh, with the standard grey-white pulp. Pink Marsh is not grown there because the pink fruits do not color well in California.

Variety	Season	Fruit characteristics	Best rootstock	Flavor
Duncan	Late	Medium size, light yellow. 50 seeds.	Trifoliate	Superior—sweet and acid.
Marsh	Florida—Feb., March. California—Nov. to June	Medium size, light yellow, seeds 0 to 6, can be held late on trees.	Rough lemon	Good quality. Pronounced flavor but not too bitter.
McCarty (Indian River)	Jan. to March	Large, very light yellow. 49 to 59 seeds.	Rough lemon	Excellent quality. Bitter and sweet. Juicy.
Royal	Medium early	Small, orange yellow, 40 seeds.	Sour orange	Sweet—bitterness lacking. Can be eaten like an orange.
Triumph	Medium early	Heavy, medium size, 37 seeds.	Sour orange	Moderately bitter. Acid and sweet.
May	Nov. to Feb.	Small to medium, 44 seeds.		Flavor excellent.
Marsh Pink	Feb., March	Medium size, seedless, pink flesh.		Good quality.
Ruby (USDA Red)	Feb., March	Deepest red variety.	Rough lemon	Good flavor, not too bitter.

white, yellow, pink or red, skin green to yellow, few to many seeds, sweet to acid, mild to pungent. In size the fruits range from one ounce to one pound. Desirable wildlings cannot be propagated by seed, because too many strains have gone into their composition. In order to obtain plants with identical fruit, a vegetative means of reproduction is necessary.

Strawberry guava, *P. Cattleianum,* or Cattley guava, is more often cultivated than the common strains. It is a species with dark red flesh or, in one variety, *lucidum,* pale yellow when ripe. Cattley guavas are small round fruits which are sweeter and more delicately flavored than the common species. They are very good when eaten fresh and their juice is pleasing as well as extremely rich in vitamin C.

Both the common and the Cattley guavas are excellent sources of vitamin C, either when eaten fresh or when a juice is extracted by cooking them briefly with a small amount of water. Samples of this juice prepared under laboratory conditions have yielded 70 to 130 milligrams of vitamin C in 100 grams of juice. Guava pulp and juice, canned, averages about 66 milligrams of vitamin C to 100 grams. In addition to ascorbic acid, guavas contain small amounts of vitamin A, calcium and vitamins B_1 and B_2.

Guavas are members of the evergreen myrtle family. They are large tree-like shrubs or small trees which branch freely almost to the ground. Under favorable conditions, when they are pruned to tree form, they may attain a height of 30 feet, with an almost equal spread.

Common guavas are slightly more cold-sensitive than Cattleys. Young trees of the common type are killed by temperatures that dip to 29 degrees for a few hours, but older trees are fairly safe for short periods to 26 degrees. Cattley guava will stand a temperature down to 22 degrees for a short time when the plants are mature. Both species are fast-growing plants and, within 2 or 3 years of being frozen to the ground, the new growth which sprouts again from the roots will be bearing fruit.

Guavas grow on any well-drained loam, muck or sandy soils which are supplied with enough moisture. They require more water for their growth than citrus trees do. They are also very tolerant of acidity and alkalinity, doing well in all the circumneutral soils and even tolerating a *p*H down to 4.5 or up to 8.2. At either extreme, their need for nitrogen increases.

Nurserymen in Florida are beginning to raise a few named varieties of guava which were selected and named by the University of Miami Experimental Farm. Some of those released at the present time are Red Indian, Ruby, Supreme, Miami Red and Miami White. Work is still proceeding on many other desirable clones.

Planting

Best time to set out young guava plants is just before the rainy season; in Florida in late spring, in California in the fall. If the bushes or trees are to stand alone, they should be allowed 25 to 30 feet. Used as a hedge, they may be spaced 10 to 15 feet apart.

Given the right start, guavas will make rapid growth and will be ready to bear fruit in two years after planting. In order to manufacture wood enough, and leaves enough to nourish the wood and roots, they must be well-fed when they are planted. Dig a hole two feet in diameter for each plant. In the bottom of the hole dig in one or two shovelfuls of rich, well-rotted compost or sheep manure. Set the tree about one inch higher than it was planted in the nursery, in order to maintain its nursery height after the soil has settled. Fill the hole around the roots with a rich mixture of topsoil, rotted manure, compost or other organic fertilizer, plus one cupful of either bone meal or rock phosphate. Settle the soil with a liberal watering and make a 4-inch watering basin above the roots. Mulch the surrounding soil with a 6-inch layer of straw, hay or leaves.

Keep the soil moist with weekly waterings until the young plant shows some new growth. After that, the waterings may be spaced further apart until the mature trees are watered on alternate weeks. This should be sufficient moisture, unless the soil is very sandy or the season very hot and dry. While ripening, the fruits require a little more water, but too much at this time tends to make watery fruit or cracked skins.

Fertilizer should be applied every 6 to 8 weeks during the growing season. This may be in the form of rotted manures or cottonseed meal, blood meal, sludge or fish emulsion. No statistics are available on the amount needed, but it is difficult to be too generous. Before the rainy season bone meal or rock phosphate should be raked into the soil.

Pruning

If a tree form is desired, lower branches should be cut back on the young plant during its first year. After the tree has been shaped, no further pruning should be done for several years. The largest and best fruit is borne on wood that is 1 to 3 years old. After the tree is in full bearing it may be pruned once in 2 or 3 years to thin and head back the top. This will decrease the crop during the current year, but it will stimulate new growth and increase both quantity and quality of crops in the next couple of years.

Propagation

Guava seedlings are very easy to start, though the quality of their fruit is always a matter of luck. Seed should be planted as soon as possible after it is removed from the fruit. Sow it one-fourth inch deep in a flat filled with a mixture of vermiculite and peat moss. After the true leaves appear, transplant to a pot of loam and keep it partly shaded and moist until it is large enough to set out.

Good specimens may be multiplied by vegetative methods which will ensure the promulgation of their quality. It is possible to propagate them by cuttings and by grafting, but guava requires special treatment so both of those methods are difficult for the amateur. The easiest way to increase stock for the home garden is by root cuttings, a method which does not always work, but may be repeated if it does not work the first time. To make the cuttings, plunge a sharp spade into the soil under a mature bush about two feet from the main trunk where it will sever a few of the surface roots. Keep the spot moderately moist during the next two months. Sprouts should arise from the cut ends of the severed roots. When these have attained a large enough size they may be transplanted to their permanent places or to pots.

Guavas may also be asexually increased by air layering. Select a branch which is not more than one-half inch in diameter for the purpose. Girdle it by removing a ¾ inch strip of bark completely around it, down to the wood. Bind tightly over the wound a 4-inch ball of damp sphagnum moss and cover the sphagnum with a wrapping of clear polyethylene. Bind the polyethylene to the limb with rubber bands on each side to make an airtight sac around the sphagnum moss, so that it cannot dry out.

In 3 to 5 weeks a half dozen roots should be visible through the polyethylene. If they are not, remove the wrapping and examine the bark. If it has made a callus bridge across the girdle, regirdle it and wrap it up again. In a few more weeks the roots should show.

When the moss seems well impregnated with roots, sever the cutting, wrapped roots and all, from the bush. Remove the polyethylene but not the sphagnum. Pot the rooted cutting and keep it partly shaded until it is large enough to set out.

Enemies

Guavas are almost entirely free of insects and diseases, but a few of the endemic Florida pests will attack them if they have a chance. Crown or root rot may sometimes destroy a planting which is made on soil which still contains tree stumps and old roots, particularly if the trees were oak. It is best to plant guavas where the land has been cleared long enough for all of the old roots to have decomposed.

Nematodes sometimes attack and kill wild plantings of guava. They should not menace any planting fed and mulched with organic matter, however, because the fungus which kills the nematodes is present in the decomposing organic fertilizer.

Harvest and Storage

Guavas ripen during a period of about 6 weeks in the summer. If they are to be used for jelly or juice, they may be picked before they are quite ripe. They contain such a large percentage of pectin that a pound of fruit will make more than 3 pounds of jelly.

For dessert, fresh fruit should be picked when it is entirely ripe. It may be stored for brief periods in the refrigerator. If it is to be kept longer, it should be made into juice, jelly, pulp or preserves and canned or frozen. Or the outer portion, minus seeds, may be cut into strips and frozen for use in winter desserts and salads.

HIGHBUSH CRANBERRY

HIGHBUSH CRANBERRY. *See* CRANBERRY BUSH.

HIPPOPHAË RHAMNOIDES

HIPPOPHAË RHAMNOIDES. This large shrub or small tree, sometimes called sea buckthorn, bears small yellow-orange to red berries which are very potent sources of vitamin C. The shrub is hardy through most of the United States and can be grown in almost any type of soil. Plants are dioecious, bearing male flowers on one bush and female on another. In order to produce a crop of fruit, both male and female plants must be grown. Propagation is by seeds, cuttings or layering.

JUJUBE

JUJUBE. Several species of the genus *Zizyphus* are grown in the United States, the least desirable being the Texas buckthorn or Texas jujube, *Z. obtusifolia,* the only native American jujube. Fruit of the buckthorn is not more than a third of an inch long and is rather tasteless, though it is eaten and relished by the Mexicans. Chinese jujubes, *Z. jujuba,* with fruits growing to two inches in diameter, or Indian jujubes, *Z. mauritiana,* with slightly smaller fruits, are more palatable and more often cultivated.

Chinese jujubes are most successful in hot arid sections of the country. Their fruits are tender to frost and require long hot summers to mature. In southern Florida, they are plagued by a number of fungus diseases which flourish there during the humid summers.

The tree is upright, growing to 20 or 30 feet under favorable conditions, with deep green, shining, deciduous foliage. Some trees have sharp slender thorns arising from the leaf nodes, where the fruits also grow.

Jujubes are stone fruits, in appearance somewhat like olives, in flavor more like dates. The Chinese species grows one to two inches in diameter, round to oblong and it ripens to a deep brown from July to September. The fruit may be eaten fresh, but is more often candied, dried or preserved. Its flesh is crisp, white and subacid. Varieties recommended by the United States Department of Agriculture are Mu Shing Hong, Lang, Sui Men, and Li. Grafted plants of these varieties may be purchased. Seeds are easily propagated, but they will not come true to the parent fruit.

Indian or Malay jujube is a small vigorous evergreen tree whose spreading branches are covered with small thorns. Its leaves are coated on their undersides with a white matting, from which the plant is sometimes named the cottony jujube. Indian jujubes are somewhat more hardy than Chinese and are also more resistant to the Florida fungus diseases. The fruits are slightly smaller and show wide variations when grown from seeds. The best varieties have a mild subacid flavor and may be eaten fresh. Some others have astringent pulp which may become disagreeably slippery when eaten.

JUNEBERRY

JUNEBERRY. This name, along with serviceberry, shadbush and shad-blow, is applied to various species of *Amelanchier* which are grown as ornamentals, and sometimes for fruit. Many of the species are native to this country and were used by the Indians for food. They are hardy, some species growing as far north as Alaska.

Juneberries are shrubs or small trees of the rose family, their fruits similar to miniature red to purplish-black apples with bony centers. The so-called berries may be eaten fresh, dried or used for jelly. Trees and shrubs are quite showy in May when they bear clusters of white blossoms on the tips of the branches.

Best species to plant for fruits are *A. sanguinea,* a slender 6- to 8-foot bush which is hardy and tolerates drought; *A. alnifolia,* a slender shrub or tree hardy in the northern Great Lakes area, and bearing the largest fruit of the genus; *A. stolonifera,* a twiggy 4-foot shrub which spreads by suckers; and *A. ovalis,* variety Saskatoon, which is recommended for Alaska.

None of the juneberries are particular about soil, but they like a *p*H just under neutral and do best on a dryish limestone soil. Propagation is by seeds, layering and by suckers. Juneberries are alternate hosts to some juniper rusts which cause leaf spots, yellowing and dropping leaves in spring. Where this is present, it may be checked by removing the junipers in the immediate neighborhood or, if they are valuable, the juneberries.

KUMQUAT

KUMQUAT. Though very similar to the citrus fruits, kumquats actually belong

TABLE 74: Kumquat Varieties

Variety	Season	Fruit	Rootstock	Tree
Nagami *(Fortunella margarita)*	Oct. to Jan.	Oblong to pear shaped. 1 inch thick.	Trifoliate or rough lemon.	Thornless shrub or small tree.
Neiwa *(F. crassifolia)*	Late fall	Round, sweet. 1½ inches thick.	Trifoliate or rough lemon.	8 to 15 feet, thorny.
Marumi *(F. japonica)*	Late fall	Round, sweet, 1¼ inches thick.	Sweet orange.	7 to 10 feet, many branches.

to a different genus, the *Fortunellas,* which are so closely allied to *Citrus* that crosses between the two can be made, the same rootstocks can be used and the little orange-colored kumquat fruits look like miniature oranges.

Kumquat trees are very decorative and make handsome ornamentals in the home garden, bearing at the same time a good crop of fruit. They grow no more than 10 or 12 feet high, are very full and dense to the ground, with glossy foliage and bright orange fruit. In addition, kumquats have the advantage of being hardier than almost any other citrus fruit, especially when grafted on trifoliate orange roots. They may be grown as far north as the northern borders of Florida and are sometimes found even in Georgia and South Carolina. In the North they are grown as pot plants which are as attractive but more useful than the Otaheite orange. Where especially dwarfed specimens are needed, they are grown either on trifoliate orange or on *Severinia buxifolia* rootstock.

Fruits of the kumquats are seldom more than 1½ inches in diameter and are eaten without peeling. The skins have an agreeably spicy flavor and the pulp is sweet and juicy. No analyses are available, but the indications are that kumquats are somewhat less rich in vitamin C than oranges. They are used for making marmalade and preserves, for which they are especially suited because they contain large amounts of pectin.

Culture of kumquats is very much like that of oranges. (*See* ORANGE.)

The 3 varieties grown most frequently represent 3 species of *Fortunella. F. margarita* is the Nagami kumquat; *F. japonica* is the Marumi; and *F. crassifolia* is the Neiwa variety. The citrangequat and the limequat are both hybrids.

LEMON

LEMON. *Citrus Limon* or *C. limonia.* Lemons were introduced to Europe at about the same time as oranges, and were brought to America by some of the early Spanish explorers. They have never been as widely planted as oranges because their use is more restricted, most varieties being almost as acid as limes. A few varieties of sweet lemons have been developed, but their acceptance has been hindered by their inferiority to some of of the other sweet citrus fruits.

Like most of its cousins, the lemon was first grown in Florida in this country, but diseases in that region, added to a heavy freeze in 1894 which wiped out most of the orchards, discouraged growers there. More lemons are grown in California than in Florida at present, though it is possible to succeed with them in central and southern Florida as well as in southern California. A gum disease which attacks them in Florida makes the trees unsightly in the dooryard, but an abundant crop may be produced in spite of the disease. Lemons are more tender than oranges or grapefruit, but more hardy than limes. Winter temperatures of 28 to 30 degrees will make them shed their young fruits; 26 to 28 degrees kills the new growth; and, below 26 degrees the whole tree may be injured. Immature lemons have a very tender skin which is injured by sunburn, by being wind-tossed and by diseases which flourish during damp summer months.

Lemons are a good source of vitamin C, but because of their acidity they are seldom eaten in large enough quantities to provide any considerable amount of either vitamins or minerals. A quarter of a cup of lemon juice contains 25 milligrams vitamin C, .024 milligrams vitamin B_1, .002 milligrams vitamin B_2, 11 milligrams calcium, 6 milligrams phosphorus, .3 milligrams iron and 20 calories.

Lemon trees grow 10 to 20 feet high, with an open head. They prefer heavy soil, but it must be well-drained. Trees are almost everbearing under favorable conditions, bearing flowers, young fruit and ripe fruit all at the same time. For this reason cold during the winter months is able to wipe out a considerable portion of the tree's yearly yield.

In planting, set grafts high above the soil surface and space trees 24 to 30 feet apart. The trees need more pruning than most citrus fruits because they must be kept compact to avoid sun and wind damage. Straggling branches should be shortened at any time during the year to inward-growing laterals. Except for these few differences, lemons are grown just as oranges are. (*See* ORANGE.)

When buying trees in Florida, it is important to know something about the stock from which they were budded. A reliable dealer who has access to disease-free trees can make recommendations about best varieties to grow locally. The same care must be used in California in order to get good trees, but for different reasons. In California the Eureka and Lisbon varieties have become so intermingled that not even the growers can always be sure which they are growing. It is best to buy trees there from an orchardist known to produce good fruit.

For anyone who wants a source of vitamin C without the accompanying acid, the Dorshapo sweet lemon is the best variety to plant. It grows on a medium size tree which has many of the characteristics of Eureka.

Several lemon hybrids are grown as curiosities, though Meyer, a cross between lemon and orange, is valuable for its

This Meyer lemon bush is protected during the winter by a frame and a double thickness of fish net. Grown along the southern California coast, this bush has yielded up to 18 pounds of lemons at a single harvest.

TABLE 75: Lemon Varieties

All lemon varieties are everbearing, some with a tendency to a peak season in fall.

Variety	Fruit characteristics	Best rootstock	Tree characteristics
Lisbon	Large, juicy, excellent flavor, 1 to 5 seeds.	Fla.—rough lemon Calif.—sour orange.	Large, dense, vigorous, thorny. More cold-resistant than Eureka.
Eureka	Medium size, juicy, acid. Seeds mostly abortive.	Does not do well on sour orange. Fla.—sweet orange. Calif.—Sampson tangelo.	Good dooryard tree, not as vigorous as Lisbon. Open, spreading, few thorns. Nucellar varieties best.
Genoa	Medium size, few seeds, acid clear and strong. Juicy. Keeps well.	Rough lemon or sour orange.	Dwarf, thornless.
Villa Franca	Medium large, good flavor, 30 seeds.	Calif.—sour orange. Fla.—rough lemon.	Few thorns. Vigorous, productive.
Meyer	Hybrid. Golden yellow, thin peel, good flavor, 10 seeds.	Fla.—rough lemon. Calif.—sour orange.	Good home garden tree. More disease-free and hardier than others. Stand temperatures to 15°.

hardiness. Two very large lemons of indifferent quality are Ponderosa and Chinese, both hybrids with lemon and citron parents. A cross between lemon and lime, the Perrine, is being experimented with on Florida farms.

Lemons may be pot-grown in the North and, after they have reached a fair size, they begin to yield a few lemons each year. They are usually grown in a greenhouse or a sunny window through the winter and summered outside in the garden. With frequent watering, an occasional feeding of bone meal and sheep manure and plenty of sun, some of the pot-grown trees have been known to live for nearly 100 years. An old specimen will yield up to 50 lemons in a year.

Enemies

Lemon trees grown on sour orange stock are particularly susceptible to citrus scab, which is the principal lemon disease in Florida. The scab is caused by a fungus which attacks the new growth and young fruits of citrus trees. It seems to attack the growing parts of the treetop only while the shoots are very young and tender. When the tissues have become developed and hardened, it cannot affect them. Leaves more than one-half inch wide and fruits more than three-quarters of an inch in diameter are immune. The scab develops its worst attacks during cool weather when the foliage and fruit are growing slowly, and it has time to get a foothold. Dampness also favors the disease.

Scab is not a disease of dry, arid sections of the country. California plantings never suffer from it, though the disease has doubtless had many chances to reach the California orchards in stock from Florida. For that reason, lemons can be grown in California with greater success than they can in Florida.

Scab fungus is carried over from year to year in infected twigs and leaves. It does not live over from season to season in ripe fruit. The fungus finds most congenial conditions in sour orange sprouts which may arise from the rootstock, and from water sprouts from rough lemon roots. These should be destroyed as soon as they appear in any citrus tree. Also, where the scab has appeared, all infected growth should be pruned out of the tree before new growth starts in early spring. The strictest sanitation should be observed in the vicinity of trees which have been infected, and all fallen twigs and leaves should be removed and burned.

Making Lemon Trees Bear

If you really want your fruit trees to bear, give them plenty of compost and protection from the wind. These two rules solved a problem for Frank L. Coe, retired Los Angeles police captain, who had several Meyer lemon trees . . . but no lemons.

Soon after moving into his home along the Malibu coast in California, Coe discovered that the strong winter winds prevented lemon trees as well as other plants from bearing. So, 4 years ago, he put in rows of Myoporum trees that grew into efficient wind-breaks around his cliffside home. Within this protected enclosure, he planted his organic gardens, set up his compost bins and put in fruit trees—including the Meyer lemon.

The only drawback was—the lemon tree still didn't bear. Protected from the driving Pacific winds, it managed to live but showed no blossoms and—no lemons. In desperation, Mr. Coe transplanted it from one place to another, hoping it would find a congenial site. But he hoped in vain.

As Mr. Coe reports it, things changed when "my compost bins began turning out wheelbarrow loads of rich compost made of leaves, grass and loads of manures. So I dug up that poor little Meyer lemon bush and planted it for the last time in a hole deep with rich compost.

"You never in your life saw such a transformation. That Meyer lemon bush throve. Today it bears the year around. We raise all we need from this one bush and have plenty to give to friends."

The Meyer lemon, bush type, is excellent for the small garden outdoors, also for greenhouse and window-growing in the colder climates. Coming from the more rigorous zone of temperate China, it is frost-resistant down to 10 and 15 degrees, which is outstanding for a citrus plant. Outside of California and Florida, it should winter indoors and will grow freely in a sunny window. The flowers are delightfully fragrant and the fruit is large and juicy. As a house plant, the Meyer lemon has an undeniable charm.

It should be noted here that drying winter winds, common along the California coast, have made it necessary for us to put frames around the Meyers and to cover them with two layers of discarded fish net or some equally protecting material. Under this shelter the plants bear all winter long and, we have found, they need no protection for the rest of the year. We have also found that wind protection from a nearby wall or building eliminates the need for additional shielding.

We feed the Meyers from time to time by raking sifted, ground pigeon, chicken or stable manure into the soil, starting from the middle of each bush and working out to the drip lines. With such care, these prolific little bushes respond bountifully.

One such lemon bush will easily supply a family with plenty left over for neighbors. We have found that they also bear well left in containers and do very nicely in colder climates in large pots near the window during the winter season.

Since these bush lemons love full sunshine, pick sunny spots in your yard or garden for planting. Outdoors the bushes average 5 feet tall by 8 feet wide when fully grown. Being evergreens, the leaves are a bright green the year round while the pinkish-white blossoms perfume the entire house with their happy fragrance. You will find buds, flowers, all sizes of fruit, both green and ripe, the year round on the Meyer lemon bush.

Providing, advises Mr. Coe, you give it plenty of good, rich, organic compost and take care to shelter it from the cold, driving winter winds.

—Gordon L'Allemand

LIME

LIME. The most tender of the citrus fruits are the limes, both the acid limes, *Citrus aurantifolia,* and the sweet limes, *C. limetta.* These are both semitropical fruits and can be grown successfully only in the southern tip of Florida and on the Keys. They are not troubled by the citrus scab which is rife in Florida lemon groves, so they are better adapted to Florida's humid summers.

The sweet limes are of little importance, except as a curiosity, being of rather insipid flavor. Acid limes, however, have a distinctive flavor of their own which make them useful in desserts and cool drinks. They have much the same uses as lemon, and may be substituted for it where they are grown in the dooryard. They contain more acid than lemons and more sugar. In substituting, about 2/3 of the quantity of lemon juice is needed.

Lime juice does not contain quite as

much vitamin C as lemon juice, but it contains more vitamin A. In one-fourth cup of juice there are about 65 units of vitamin A, 18 milligrams vitamin C, 28 milligrams calcium, 17 milligrams phosphorus and 20 calories, all sugar.

Two types of limes are grown in Florida, the Mexican, grown mainly on the Keys, and Persian limes. The Mexican limes are actually not a variety, but a group of varieties from seedlings in which there may be wide variations. In general, they are more tender to cold than the Persian limes, being damaged by temperatures of 28 to 30 degrees. They are also sensitive to sudden heat waves, and they must be protected from wind damage to their sensitive rinds. The trees are almost everbearing, so a drop in winter temperatures may injure blossoms, young fruit or fruit approaching maturity.

The young, sweet lime trees above growing in California greatly benefited from the rock mulch placed around the trunk.

Limes are extremely tolerant of soils, requiring only that the soil be well-drained to a depth of 3 or 4 feet, which is sufficient for their shallow roots. In areas where the rainfall is heavy, bud joints should be set high. Mexican trees are smaller than Persians, the former requiring a space about 15 by 15 feet, the latter 20 by 20. Culture is very similar to that of ORANGES, which see.

Limes are best when grafted on rough lemon or sour orange roots. The heaviest harvest may be expected in early summer, though they are everbearing under favorable conditions. Color of the fruits is lemon yellow when entirely ripe, with green flesh having a characteristic flavor, which is stronger in the Persian varieties than in Mexican. Most tender are the Mexican and Tahiti or Persians, with a little more hardiness in the Bearss variety.

LIMEQUAT

LIMEQUAT. The hybrid limequat is a fruit more hardy than its parent lime, but not quite as hardy as its other parent, the kumquat. Plants resulting from this cross may be grown in areas too cold for limes, and they are also more disease resistant. Their fruits are used as a substitute for lime, but they are recommended only for areas where the winters are not warm enough to grow the parent fruit, because they are not as good.

The only important variety available for planting in Florida is the Eustis, which is slightly more cold-resistant than sweet orange. It is a cross between a Mexican lime and Marumi kumquat. Its light yellow fruits are small, with 5 to 12 seeds and juicy, greenish pulp. The trees bear a succession of crops that make them almost everbearing.

LINGONBERRY

LINGONBERRY. *Vaccinium vitis-idaea,* the lingonberry, is also known as the mountain cranberry, which is a very apt description. It belongs to the same genus of the heath family as cranberry and blueberry, and it resembles the former except in some of its requirements.

A very hardy plant, lingonberry grows through the colder areas of the American continent as far north as Alaska. It makes an excellent dark green ground cover under the acid-loving shrubbery border, and its small bright red berries in fall are decorative as well as edible. Like the cranberry, it grows only a few inches high and is best propagated by cuttings. Its fruit may be used in the same ways as cranberries, which it resembles in flavor.

Lingonberries like a gritty soil, such as might be found in the mountain crannies where they are native. They grow well in full sun or part shade, but they will not grow in any except an acid soil whose pH is 4. to 5. When their requirements are met they will form a solid mat which shades the ground and provides its own mulch.

Because they are so small (they are like miniature cranberries) and are borne sparsely on the plant, lingonberries are back-breaking to harvest. Scandinavian cooks make good use of their excellent flavor in a pancake sauce, but unless one is Scandinavian and has a yearning for pancakes, he will find the lingonberry not worth the work necessary for its culture. In sections of the country where other fruits can be grown, it is not advisable to try lingonberry.

LOGANBERRY

LOGANBERRY. *See* BLACKBERRY.

MANGO

MANGO. The tropical apple is the mango, widely grown through all the world's tropics. It is a native of India, as is indicated by its Latin name, *Mangifera indica,* and is a member of the sumac family. Though many named and unnamed varieties are grown in Hawaii, the only other states with suitable climate are the southern tips of Florida and California. Mangoes are very sensitive to the slightest amount of frost.

The mango is a medium-sized fruit varying from 2 to 4 inches wide by 3 to 7 inches long. Like the avocado, it has a skin which is strong and thick enough to be pulled from the flesh when it is ripe. As the fruit matures its green skin changes to purplish-red, deep crimson or yellow with red spots. Color of the flesh varies from light lemon to deep apricot. In the best varieties it is smooth and juicy with an excellent flavor somewhat reminiscent of peaches. In some of the worst varieties, the flesh is full of fibers and the flavor is strong of turpentine. Where the turpentine varieties are not grown, good mangoes may be propagated directly from seeds.

Mango trees are tall and handsome and are sometimes planted as shade trees and ornamentals, even where the climate will not often permit the fruit to ripen. The trees sometimes attain a height of 90 feet where conditions are favorable, though in the home garden they may be held down to half that by careful pruning. They start to bear fruit when they are 5 to 7 years old. They thrive on poor sandy soil, but produce better fruits with some fertility. Too much nitrogen, however, may cause excess growth at the expense of fruit.

Though quantities differ in different varieties, mangoes are a good to excellent source of vitamin A; fair to good source of vitamin C; and they contain some thiamine, niacin and riboflavin. Most ascorbic acid is contained in green fruit, but the quantity steadily declines until only a moderate amount remains in the fully ripe fruit. This vitamin C is not lost when the green fruits are cooked if they are then refrigerated and held no more than one week, or are canned or frozen.

Grown throughout the tropics, mango trees are tall and handsome, sometimes planted as ornamentals. Fruit is medium-sized, has a strong skin as the avocado.

In India all types of mango are eaten, even the most turpentine-flavored. In times of famine even the large drupelike seed is cooked and eaten. Mangoes are used there and in Hawaii as dessert fruits, when fresh, or are cooked in chutney, preserves and desserts of all kinds.

Varieties recommended for planting in the far South are Haden, which is high in vitamin A and supplies a fair amount of vitamin C; Kent, notable for its smooth texture; Zill, Brooks and Cambodiana. Pirie is the variety most often planted in Hawaii. Also recommended there for trial or home planting are Ah Ping, Gouveia, Joe Welch, Kent, Zill, Julie, Ono, Edwards, Fairchild, Georgiana and Smith.

Mangoes may be propagated from seed, but if a desirable variety is at hand, it can be multiplied with certainty only by grafting, excepting only Julie of those mentioned above. Julie is one of the polyembryonic mangoes which may be depended upon to propagate from seed. Side veneer, side-tongue and whip grafts are most successful.

Trees are usually planted during the rainy season and spaced 35 to 40 feet apart. Pruning is confined to removal of dead wood and low-growing branches which trail on the ground. Any necessary cutting should be done just before a new flush of growth starts and pruning just after harvest is preferred.

Because they make several growth flushes during the year, mangoes should have a constant source of food at their roots. Several applications each year of rotted manure, compost or a mixture of organic fertilizers should be spread over the root area, the amount increasing with the growth of the tree. We do not have specific quantitative recommendations, but the trees need heavy feeding with nitrogen, phosphorus and potassium.

The most serious disease of mangoes is anthracnose. In wet areas and areas where there are heavy rainfalls, resistant varieties should be planted. Especially resistant are Paris and Fairchild.

MULBERRY

Mulberry. An Asian plant, of which one species is native to the American continent, the mulberry has spread to many parts of the world. Mulberry is the common name for a whole genus, the *Morus,* most important members of the mulberry family, the *Moraceae.* Most of the mulberries are hardy in the northern sections of the country, with the exception of the black mulberry, *M. nigra,* which is grown in the South and in Hawaii. The form of the tree varies from the 50 to 60 foot white or red mulberries to the shrub, *M. multicaulis,* a white species grown in France for silkworm culture. Flowers of the mulberry are dioecious, male and female flowers being borne in catkins on separate trees. The fruit resembles blackberry in form, but is usually sweet and insipid in flavor. In Hawaii, black mulberries have become almost seedless and are used with spices or lemon juice to make desserts, jams and cold drinks.

Mulberries vary according to species, but those analyzed have proven to be fair sources of vitamin A, riboflavin and niacin and good sources of vitamin C. They are very attractive to wild birds and, when planted near cherries or other fruits which the birds might otherwise molest, they are more effective than scarecrows. Mulberry trees are sometimes planted in chicken yards, because they provide a good summer food for the poultry. Pigs also like them.

Best fruit is produced on the following mulberry varieties: Black English, Downing, Hicks and New American, most of them cultivated forms of white or red mulberry. Trees should be given about 30 feet each way. They grow well in any moderately good garden soil with a *p*H of 6. to 8. and are seldom attacked by insects or diseases.

Mulberries are very soft and perishable, so they are usually picked as they are needed. Trees spread rapidly after one or two have been planted, because the birds drop seeds far and wide. Usually a few trees in a hedgerow or in the back of the garden will be enough to provide mulberries throughout the countryside.

OLIVE

Olive. The common olive, *Olea europaea,* figures in the Bible and in tales of Homer, but in Biblical and Homeric times, it was already an ancient fruit. Evidences found in ruins on the island of Crete show that olives were grown there as far back as 3500 B.C. It is thought that they originated in the hot portions of the Mediterranean coast, but it is difficult at this late date to trace their birthplace.

The first olives were introduced to California in 1769 by the Spanish missionary Junipero Serra, who brought seeds to the mission in San Diego. Some of the trees planted at that time are still growing. They are the forebears of all of California's Mission olives. Though these may seem to be venerable trees, they are youngsters compared to some

TABLE 76: Olive Varieties

Variety	Harvest	Fruit	Tree	Disease resistant
Mission (Spanish)	Oct. and Nov.	Small to medium. Use—oil and pickling.	Tall unless pruned. Tends to be biennial. Old trees hardy to 8°.	Susceptible to peacock spot.
Baroumi (Tunisian)	Oct.	Medium to large. Use—pickling.	Spreading. Injured by low temperatures. Productive in interior Calif. valleys, not so productive in S. Calif.	Susceptible to olive knot.
Manzanillo (Spanish)	Oct.	Small to medium. Use—oil and pickling.	Low, spreading, productive. Not biennial. Susceptible to low temperature injury.	Susceptible to olive knot.
Sevillano (Spanish)	Oct.	Very large fruit used only for pickling. Oil content relatively low.	Low spreading trees. Productive.	Susceptible to olive knot, shot berry, split pit and soft nose. Resistant to peacock spot.
Ascolano (Italian)	Sept. and Oct.	Large fruit. Use—pickling ripe. (Shrivels when pickled green).	Not as productive as others. Trees tender even when mature.	Resistant to olive knot and peacock spot.

of the Mediterranean specimens, which are known to be as much as 1500 years old. Olives are among the longest-lived fruit trees, but their most productive period is during their youth—before they are 50 years old.

Olives are successfully grown only between the latitudes of 30 to 45 degrees N., and within the same southern latitudes. Although they require a great deal of heat, they will not grow in the tropics below 30 degrees N. because they also require a certain amount of winter cold. For satisfactory yield, they should be planted only where the mean January temperature is between 45 and 50 degrees, but the thermometer never goes below 10 degrees. When, to these requirements, we add their need for hot dry weather during the summer months and in late spring when the fruit is setting, we are left with very few sections of the United States where olives may be grown. Although the trees are grown as ornamentals down into Florida and in the Delta States, they produce a crop of fruit dependably only in the hot interior valleys in northern California, in a few spots in southern California and in the south of Arizona.

The olive is a 25 to 70 foot evergreen with silvery willow-like leaves that last on the tree for 2 or 3 years. They shed in spring. They are related to forsythia, and their panicles of yellowish-white flowers are borne in the spring from leaf nodes, like the forsythia flowers. Flowers and fruit form mostly on one-year wood, though a few dormant buds develop each year on 2- or 3-year wood. Buds begin to swell early in March; blossoms are open between May 1 and June 1 and fruit, if it has had sufficient heat during the summer, ripens between October and January. For pickling, it is picked in a green-ripe stage usually in October. For oil, it is left on the tree until midwinter to develop its full quota of oil, even though the fruits shrivel and freeze. Oil content of California olives is 18 to 25 per cent by weight, when they are fully ripe.

Olive oil does not contain vitamin E, as does the oil extracted from various kinds of seeds. It contains about 15 per cent linoleic acid. One tablespoon of olive oil yields 125 calories, which is 25 more than in the same amount of cottonseed or corn oil. Green olives contain a very small amount of vitamin A and some calcium, amounts unimportant because so few of them are included in the total of the diet.

Site

Olives are tolerant of soils, but an open sandy or clay loam is best for them. They are also moderately tolerant of salt or alkali, but do better without either. Their roots do not penetrate very deep, so they are at home on rocky hillsides in shallow pockets, but they grow equally well in deep fertile valley soil. In the home garden, it is best to plant them in a spot removed as far as possible from the vegetable garden, because they are subject to vercillium wilt which affects tomatoes and some other vegetables.

Ordering Stock

Most trees sold by nurserymen are grafted on seedling olive rootstock. While it is possible to start trees from cuttings, some growers believe that seedling roots are better under stress from windstorms than roots developed from cuttings. Grafted trees are sold a year or more after being budded or grafted, at which time their roots are about 3 years old.

Trees are sold with bare roots or potted in gallon cans. If they are purchased bare-root, the roots should be well protected with damp sphagnum or peat moss. They should never be dry. Bearing will start 4 to 5 years after planting. Full production should not be expected until the trees are 12 to 20 years old.

Most olive varieties are self-fruitful, but growers have found that in some years a better crop is set when two or more varieties are planted together. The reason for this, as well as the reasons for many other habits of the olive tree, is still the olive's secret.

Planting

Trees should be planted where they will have a free area 35 to 40 feet in diameter when they have matured, and where they have full access to sunlight from morning to night. In the home ground, this means that they must not be placed too close to a building or to another tree.

Planting is best done in December, January or February. It should be postponed until a day when the soil is not too wet or heavy for being worked. If the tree is bare-root, prune any straggling-long roots and cut away any broken pieces. Bare-root plants should have all branches removed when they are set, and should be headed back to no more than 24 to 30 inches.

Dig a hole large enough to hold the roots without crowding or bending them. Fill in around them carefully, using a stream of water to wash in the soil and to puddle air pockets. Work the tree up and down slightly to settle the soil as the hole is filled. Trees should be set at the same depth as they were in the nursery. As soon as they are planted, the trunks should be protected from sunburn. This may be done with a coat of white paint, or they may be wrapped with several layers of newspaper or a thickness of roofing paper.

The young trees need ample water to keep them growing well. If the spring following their start is dry, water them often enough to keep the soil under them moist, but not wet. No fertilizer should be given them until bearing starts.

Training and Pruning

Premium size fruit is the result of sufficient irrigation combined with intelligent pruning and thinning of the fruit. Most important pruning is that done during the first 3 years, when the tree is shaped to provide plenty of outside fruiting wood which will receive sun enough to mature good fruit.

First pruning is performed at the end of the tree's first growing season, if it was planted bare-root, or at planting, if it was potted. Select 3 to 5 strong branches well spaced both around the trunk and from top to bottom of the tree. These will be the main scaffold branches. Remove all other branches, as well as any suckers that arise from the base of the plant.

About 3 more prunings should be performed during the next growing season. At this time remove the small sprouts which start up to replace branches which were pruned out, as well as any suckers. All unwanted branches should be removed before they grow large enough to compete with the scaffold branches or develop many leaves.

Pruning during the second, third and fourth dormant seasons should be confined to removal of broken limbs, branches which cross or sucker-like growth which may fill up the center of the tree. Lateral branches should not be removed when they make a consistently outward growth.

Fruit should be produced during the tree's fifth season in the garden. It is borne laterally on one-year wood. The bearing area constitutes a 2- to 3-foot shell over the entire surface of the tree. After bearing starts, prune each year to remove any twiggy growth or spindling branches which sprout inside this shell. A certain amount of wood cut out each year stimulates growth of new wood which will bear the following year's fruit.

When the tree has grown as tall as can be comfortably harvested, head back the main upright branches each year to stop its upward growth. But remember that trees of varieties whose habit is tall cannot be kept within the same bounds as the naturally lower-growing trees. If a low tree is desired, it is better to plant one with that type growth than to try to confine a tall tree to a spreading habit.

In some seasons a very large crop of fruit sets, for no very apparent reason. If all of it is left on the tree to mature, the fruit will be small and will sap the tree's strength to such a degree that, the following year, no fruit can be matured and the tree will become a biennial bearer. To prevent this, fruit should be thinned late in June after the normal June drop. This is accomplished by running the hand along each fruiting limb and removing all but 3 to 5 olives from each foot of branch.

Dormant pruning is usually done in the winter months. In a planting where olive knot is present, all pruning should be done in the hot dry months of midsummer, when the heat will quickly dry and callus all wounds before infection can spread.

Fertilizing

Too large a crop is sometimes set when the trees have received too much nitrogen. Since this is not only a waste of good fertilizer, but of thinning time as well, it is better to fertilize olive trees less rather than more than they need. If new growth is less than 4 inches a year, fertilizer applications should be increased. But if ample new growth is put out each year and the foliage remains dark green and sleek, the tree is receiving enough fertilizer. A record kept from year to year of the amount applied will help to establish the proper feeding schedule for trees in the home garden.

Mature trees need approximately one to two pounds of pure nitrogen per year. In terms of organic fertilizers, this amounts to 12 to 24 pounds of tankage; 8 to 16 pounds of dried blood; 12 to 25 pounds of cottonseed meal; or 40 to 80 pounds of dry poultry manure. Other manures may also be used, in appropriate amounts, but the other sources of nitrogen are preferred, when they are available. Fruits of manure-fed trees are more subject to an ailment known as soft nose than those whose nitrogen comes in other forms.

Propagation

Olive trees will grow from seeds or cuttings or special varieties are grafted on seedling roots. When grown from seeds, they may not resemble the parent trees. New trees from cuttings, suckers or graftings resemble the parent tree.

Softwood cuttings are made in September from wood of the current season's growth. Plunge the cuttings in pots of sand and peat moss and cover with a glass jar. Bottom heat is helpful, but not essential. Roots form in 10 to 15 weeks. Hardwood cuttings may also be made from wood pruned out in the winter, but it is much more difficult to root the hard wood than soft, especially if the soft wood can be taken from water-sprout type growth.

Seedlings used to graft desirable varieties are started from seeds of fully ripe Redding Picholine or Chemilali olives. They will sprout more quickly if the pits are cracked gently or chipped at the tip end before planting. The seed should be planted as soon as the fruit is ripe and may take as much as 6 months to germinate. If a tree of a desirable rootstock variety is available, it may be easier to hunt under its branches for seedlings which are volunteers. Buds from the top variety are grafted to the roots and, no more than 2 to 3 weeks after grafting, the tops of the seedlings are cut back. Or it is sometimes possible to start new trees by whip-grafting scions of the desired variety to pieces of root taken from mature trees. The root pieces must be allowed to callus before the graft is made.

Enemies

Most of the olive's enemies may be eliminated by sanitary measures. If the tree is watched and any infections removed promptly, a tree in the home garden should not be troubled by many pests.

Olive knot causes swellings on trunk, limbs or leaves. As soon as such a swelling is discovered, it should be removed and the wound, if extensive, painted with tree-surgeon's paint.

Peacock spot is a fungus infection which causes dark green spots on the leaves, which yellow and drop off. Similar spots are found on infected fruit. Any infected foliage should be removed as soon as found. If leaves or fruit drop under the tree, they should be gathered and burned. Peacock spot attacks the tree in periods of cold and damp.

Split pit is the result of allowing the trees to get too dry while the fruit is expanding, then watering them. The preventive is obvious—keep the soil moderately moist at all times.

The spotty fruit resulting from an ailment known as shot berry is thought to be caused by a delayed form of pistil abortion, in which the aborted fruit fails to drop from the tree in June. There is no known cure for this trouble.

Renovating Old Trees

Very old trees which are healthy but are not bearing well may be renovated by top grafting. To accomplish this, all branches except one, known as the "nurse branch" are removed by being cut off a foot or two from the point where they join the main trunk. Within 3 to 4 hours of cutting the branches should be grafted. Two to 4 scions are bark-grafted to each stub, spacing them no farther apart than 3 or 4 inches to ensure thorough healing of the stub. Scion wood should be taken from a tree with desirable fruit and bearing characteristics. It should be one-fourth to one-half inch thick, from one- or two-year-old wood. Remove leaves as soon as the scions are cut and keep them moist. Each scion should be 5 to 6 inches long with two nodes (4 buds) above the stub. Immediately after grafting cover all surfaces with wax.

If the tree is exposed to strong winds, any grafts that make vigorous growth the first year should be staked or tied in some way to lessen the strain at the joint. Many suckers and water sprouts will form the first year. These should be removed from the stub, but they may be allowed to grow on the trunk, where their leaves will provide shade and nourishment to the roots. During the first year or two after grafting the nurse branch is removed and its stub grafted. Grafts should be pruned lightly, the best one on each branch favored to replace the branch, the others allowed to remain until the stub is healed. After it is thoroughly healed over, the excess grafts are completely removed. The tree should begin to bear again 3 years after it was grafted.

Whenever heavy pruning or cutting, as in this operation, are performed, nitrogen should be withheld for 2 or 3 years, to discourage excess sucker and water sprout growth.

Harvest and Storage

Olives are picked for pickling either in the straw-green stage, when partly colored, or when fully colored. Usually only the largest fruit is picked for pickling, the smaller fruit being allowed to remain on the tree until fully ripe for oil.

Commercially pickled olives are usually treated with lye to remove their bitterness before they are pickled. But they may also be prepared by a lactic acid fermentation process, similar to that used in making dill pickles, or by salting.

Oil is pressed from completely ripe olives by expensive hydraulic presses in the factories, but hand presses are still in use in some of the less mechanized countries where olive oil is an important item in the diet. The olives, sometimes with, sometimes without their pits are ground fine before pressing. The oil which comes from the first pressing is clear and pure and is called, commercially, virgin olive oil. It is separated from its water content by settling or filtering. Several other pressings may be made, and the residue of the oil is then removed by a sulfur solvent in the commercial mills. This residue oil is referred to as olive foots or as sulfur olive oil. After the first pressing, the caked olive is reground before subsequent pressings. The oil pressed in the second and later tries is stronger flavored and more colored than the virgin oil, and it does not keep as well, because of the fatty acids contained in it.

ORANGE

ORANGE. Most of our oranges are varieties of the species *Citrus sinensis,* the common or sweet orange, mixed in some cases with strains of the Satsuma or mandarin species. So many hybrids and sports have been involved in the development of modern orange varieties that it would be impossible to trace most of their lineages or to go back to their original parents. Nor would it be desirable to go back. Wild oranges, found growing in some parts of the tropics, are miserable sour or insipid little things, compared to the large, fragrant fruits that we can grow today. (*See* CITRUS FRUITS.)

Sweet oranges are hardier than their cousins, the limes and the lemons, but are less hardy than grapefruit, sour oranges and kumquats. Where a slight touch of frost will kill a lime tree, oranges can stand temperatures below 30 degrees without too much injury. Fruit and the young growth are damaged when the thermometer goes down to 26 or 25 degrees and severe tree injury is suffered at 20 degrees.

The only areas of continental United States where temperatures remain high enough year-round for orange culture are in the south and central portions of Florida, the California coast from Los Angeles south and a small strip of the southwest corner of Arizona. Oranges grown outside these sections must be very hardy varieties or must have artificial winter protection or heating. A few valleys in the northern interior of California and some portions of the southern Texas Rio Grande area provide conditions which make the latter type of culture possible.

Total heat applied during the growing season to an orange tree also influences the yield. In some of the northern California areas, summers are much hotter than they are on the southern California coast and oranges ripen much earlier in the interior. Valencia oranges in the south do not ripen their crops until the second summer after blooming; consequently, the trees there seem to be everbearing, with oranges from two blooms developing at the same time. In the hot interior orchards, the Valencias ripen their fruit by November of the first year after bloom.

Oranges are one of the most dependable sources of vitamin C. When they are taken directly from the tree to the table they provide much more vitamin than the same oranges would yield if stored before using, because they have not suffered from enzyme action. It is for this reason

TABLE 77: Orange Varieties

SPANISH ORANGES

These are mostly seedling orange varieties from Florida groves. They are more successful in Florida than in California. Trees are large, vigorous and very productive, with strong foliage. Fruit is coarse-grained, large, seedy, but of good quality.

Variety	Season	Fruit characteristics	Best rootstock	Flavor
Homosassa	Dec. to Feb.	Medium to large, yellow-orange, 20 to 24 seeds.	Trifoliate	Sprightly, rich.
Indian River	Dec. to Feb.	Medium to large, bright orange, 17 to 20 seeds.	All standard stocks	Rich, excellent.
Magnum Bonum	Dec. to Feb.	Large, orange-yellow, juicy, 13 seeds.	Trifoliate	Sweet, rich
Nonpariel	Dec., Jan.	Medium to large, yellow to orange, 11 to 15 seeds.		Rich, vinous.
Parson Brown	Oct., Nov.	Medium to large, yellow-orange, 10 to 19 seeds.	All standard stocks	Good if picked early.

MEDITERRANEAN ORANGES

Although a few originated in Florida, these are considered of southern European origin. Trees are standard size or half-dwarf with abundant foliage, heavy, rich and vinous fruit, similar to blood oranges except in color. Succeed in California or Florida.

Variety	Season	Fruit characteristics	Best rootstock	Flavor
Hamlin	Oct., Nov. and later.	Medium size, golden yellow to orange-red, 28 seeds.	Any except rough lemon, on which it lacks flavor.	Acidity and sweetness well blended.
Jaffa	Dec. to Feb.	Medium to large, orange, 9 seeds.	Sweet orange.	Rich, juicy.
Lue Gim-Gong	June to Sept.	Large, deep orange, 4 to 8 seeds, best quality.	Own or trifoliate (Hardiest sweet orange).	Sweet, subacid.
Pineapple	Jan., Feb., sometimes later.	Medium to large, orange, 13 to 23 seeds.	Sweet or trifoliate orange.	Rich, sprightly, resembles pineapple.
Valencia	Mar. to June	Medium to large, golden orange, 6 seeds.	Trifoliate or rough lemon.	Sprightly, vinous.

TABLE 77: Orange Varieties (Continued)

BLOOD ORANGES

Quality of these oranges is excellent though the size is often small. Trees are dwarfish and compact, fruit has reddish rind and red flesh when fully ripe.

Variety	Season	Fruit characteristics	Best rootstock	Flavor
Ruby	Feb., Mar.	Small to large, orange to reddish, 11 seeds.	Sweet orange	Sweet.
St. Michael	Jan., Feb.	Medium size, yellow to orange-red, seeds.		Sweetness and acid well blended.

NAVEL ORANGES

Florida navel oranges are not usually very productive. Greater success attends their culture in California, where most of the navels are grown. Trees are almost thornless dwarfish, with rounded tops, pollen usually lacking, umbilical mark more or less developed. They are technically or actually seedless.

Variety	Season	Fruit characteristics	Best rootstock	Flavor
Summerfield		Medium to large, deep orange, juicy.	Bears well in Florida.	Excellent quality.
Washington	Dec. to Feb.	Large, orange. Juice slack in stem end of some.	Trifoliate, some on rough lemon in Florida.	Sweet rich.

MANDARIN ORANGES

In this group are all the loose-skinned oranges which peel and divide easily. Most of them are grown in Florida, but they are suitable to California climates as well, some varieties being among the hardiest of the citrus fruits.

Variety	Season	Fruit characteristics	Best rootstock	Flavor
China Tangerine	Nov., Dec.	Medium size, dark orange, 15-25 seeds.	Cleopatra mandarin or trifoliate orange.	Distinctive tangerine flavor, vinous.
Cleopatra (Spice Tangerine)	Jan., Feb.	Small, dark orange, juicy, 20 seeds.	Own or trifoliate.	Vinous.
Dancy Tangerine	Dec., Jan.	Medium size, orange-red, 7 to 20 seeds.	Cleopatra, sour orange or trifoliate.	Rich, sprightly.
Oneco	Jan. to Mar.	Medium to large, deep orange-yellow, 12 to 14 seeds.		Excellent flavor.
Owari Satsuma	Oct., Nov.	Size variable. Orange-yellow, juicy. Seedless or 1 to 4 seeds.	Very hardy on trifoliate stock.	Agreeable, sprightly.
King (King of Siam)	Mar., April	Large, deep orange, 18 to 20 seeds.	Cleopatra mandarin or sweet orange.	Agreeable, sweet to acid.
Temple	Jan. to April	Medium to large, deep orange-red. About 20 seeds.	Cleopatra mandarin is closest, sour orange next.	Spicy, rich, sweet.

that the unique virtue of citrus fruit is important to home gardeners—it can be stored on the tree for months after ripening before it is picked. Valencia oranges are especially good for this purpose. They may be held for 6 months or more before the juices from the fruit begin to recede and run back into the tree.

A fair amount of vitamin A also is supplied by oranges and they contain calcium as well. Whole oranges or those liquefied in a blender are preferred to those juiced, because some of the valuable nutrients are contained in the "rag," the white tissue surrounding the sections, which is discarded after juice is extracted.

A medium sized orange weighing 100 grams contains about the following amounts of nutrients: 190 units of vitamin A, .090 milligrams vitamin B_1, .075 milligrams vitamin B_2, 50 milligrams vitamin C, 44 milligrams calcium, 18 milligrams phosphorus, .4 milligrams iron and about 50 calories, all supplied by sugar. Organically grown oranges from a Florida farm were analyzed several years ago and were found to contain more nutrients than the average grown by other farms in the state by the following amounts: 14.8 per cent more minerals; 37 per cent more juice; and 67 per cent more sugar. Vitamin analysis was not made.

Ordering Stock

Four orange trees, two of navel and two of Valencia varieties, will keep a family supplied with citrus fruit year-round. Navel oranges bear from New Year's to May. Valencias may be picked during almost all of the other months. Trees should be purchased and planted when growth is slowed by cool weather, usually in January or February.

Whenever possible, the home gardener should go to the nursery to select his tree before it is dug. Purchasing in this way, he is sure that the tree he is buying is not a cull from a lot that has been sorted, and the best quality trees sold first. Also, he will be able to find out whether the soil in which it was grown was infected with fungus disease by gauging the general health of the planting.

Almost all orange trees are grafted. A healthy young tree should have the budded joint at least 6 inches above soil level, and the thickness of the trunk above the joint should be almost equal in diameter to the thickness of the rootstock below. If the trunk is definitely thinner, it may mean that the rootstock or bud was diseased or that the root was budded twice. The graft should show a smooth joining, well-healed. Bark on the trunk and rootstock should be smooth, clean and yellowish, not rough and greying. It is desirable to buy trees that were grown in one flush without interruption. Setbacks in the growing period are registered by ridges or hardening rings around the trunk and main branches. Leaves should be large and healthy.

It is better to buy a large one-year tree than a tree the same size that took two years to achieve the same amount of growth. Only a healthy tree can grow rapidly enough to be large in one year. A one-year-old tree can be distinguished by the leaves that are still growing out of the trunk below the main branches. These leaves drop off after the first year. However, they may also drop during a period of intense heat, so the leaf test is not infallible. Thickness of the trunk of a one-year-old tree one inch above the union of bud and rootstock should be three-eighths to three-quarters of an inch.

In California almost all trees are sold balled, that is, they are dug with a ball of soil around the roots which is held there by a burlap wrapping. California atmosphere being very dry, this is the only way that roots can be kept moist for any length of time. In Florida, where the atmosphere is more humid, trees are sometimes sold with bare roots.

A balled tree should have a root ball at least 14 to 18 inches long. Soil inside the ball must not be broken away from the roots, due to drying or jarring the ball. Trees sold with bare roots must be kept moist with a covering of damp peat or sphagnum moss until planting. They should not be held in pails of water, because such treatment makes them more susceptible to a number of fungus diseases.

Some rootstocks are more disease-resistant than others. Also, rootstocks may tolerate more or less cold and more or less drought, depending upon variety. They should also be compatible with the tops that are grafted to them. Different combinations of root and top are better for different sections of the orange-growing country. Before buying, deter-

The Valencia orange trees shown growing above in California produce abundant foliage and heavy fruit. Bearing season ranges from March to July.

mine which is your section and what varieties will do best there.

In California, budwood trees are registered by the state if they are free of virus disease. When buying in that state, demand registered trees.

Trees budded from nucellar seedlings are being offered by some nurseries and, so far as can be judged at present, they are much more free of the common citrus diseases than those budded from grafted parents. Nucellar sprouts arise from the *nucellus,* asexual tissue surrounding the embryo, rather than from the embryo itself. Trees grown from such shoots are propagated asexually, just as cuttings are, and inherit only the characteristics of the mother tree. Nucellar trees are more vigorous, upright and thorny than the parent. Trees budded from nucellar varieties have the same characteristics as the budwood parent and will come into production in the same period.

The rootstock to which the top is budded is of utmost importance to the mature tree. A purchaser has no way of knowing what the rootstock is except from what he is told by the nurseryman, so it is important to buy trees from reliable dealers. Six varieties are used for stocks for almost all orange trees. They are:

Sweet orange, best on sandy, dry soil but not suitable for heavy, clay soil because it is susceptible to foot rot. This stock also tends to produce smaller fruits.

Sour orange, which is used mainly for grapefruit and sweet orange varieties. It is hardy and disease-resistant, but in California recently it has been subject to the "quick decline" which has mysteriously attacked some orchards. For that reason it is no longer recommended. It has roots which go deeper than most citrus roots, and so, it is not suitable to areas in Florida where the water table is near the surface.

Grapefruit rootstock is used more often for lemons than for oranges and is not too productive. Lemons on grapefruit stock are also susceptible to quick decline.

Rough lemon is most suitable to high,

dry, sandy soil where fruit is harvested early in the spring. It is vigorous, giving quick-growing trees, but they are likely to be short-lived. They mature early, but fruit tends to be coarse and acid. Also, it is susceptible to cold and to foot rot.

Cleopatra is a good rootstock for mandarin orange varieties including tangerines and is compatible with most citrus varieties. It is tolerant to quick decline and can be used on moderately heavy soils. It is resistant to fungus disease.

Trifoliate rootstocks are particularly suitable to home plantings because they tend to produce dwarfed trees and large fruit of good quality. Trifoliate rooted trees are more hardy and resistant to cold, to foot rot and to citrus nematodes, and they do not seem to be subject to quick decline. However, they are susceptible to scaly butt. They should not be used in regions never affected by frost or in tropical areas.

Preparing the Site

Orange trees should be planted in full sun and, if possible, where they are protected from strong drying winds. A sunny position on the leeward side of a building where the air can circulate freely is ideal. Where irrigation is necessary, it is better if the tree can be irrigated independently of lawns and other portions of the garden.

The planting site should be well-drained to a depth of 3 or 4 feet. If the water level is nearer to the surface than that, or if water tends to stand on the soil in rainy seasons, the tree should be planted on a mound which is 12 feet in diameter at the base, 8 feet in diameter across the top and 18 to 36 inches high.

Sandy loam is the best soil for oranges, though a clay loam lightened with organic matter is good in warm areas. Where heavy clay soil is used, trees tend to be short-lived. Soil should be slightly acid, with a *p*H between 5. and 7. Optimum is 5.5 to 6.2. Chlorosis occurs at 7.5 and above.

A large orange tree needs a space about 25 by 25 feet. Navel trees are somewhat smaller than Valencias. On dwarfing rootstock, oranges require no more than 20 by 20 feet.

Before planting it is well to incorporate coarse phosphate rock and granite dust into the soil. These substances break down slowly and move slowly through the soil. If they are near the roots, they will give the young tree a better start.

Planting

A planting board is a useful tool which enables the gardener to place his tree accurately. It can be made of scrap wood of any width, but should be 4 feet long. A notch 2 to 4 inches wide is cut in each end and another in the center of one side.

Lay the board on the ground with the center notch placed where the tree is to stand. Drive a stake into the ground in each of the end notches and remove the board.

Dig the hole large enough to contain the tree roots without crowding. If the tree is balled, dig it a foot wider than the diameter of the ball and not quite as deep. Remove the topsoil first and lay it aside.

If the tree roots are bare, cut away any damaged roots until the cut shows live undamaged tissue. If any of the roots are very much longer than the average, shorten them to bring them to a reasonable length.

Replace the planting board, fitting the end notches to the stakes. Place the tree in the hole, with the trunk resting in the notch in the center of the board. The soil line on the trunk should be two inches higher than soil level—in other words, the tree will be set a little higher in the garden than it grew in the nursery. This should bring the juncture of the main roots and the trunk just about even with the top of the ground.

Fill the space around the tree roots with soil, returning the top soil first, so that it will occupy the bottom of the hole. If the tree is balled, fill the hole two-thirds full while the burlap is still around the ball. Then untie the cord and fold the burlap back as far as possible. Finish filling the hole, leaving the burlap in the hole to rot away. If the roots of the tree are bare, fill in carefully around them, taking care to leave no air pockets.

When the hole is filled within two inches of the top, run water into it to wash the soil down among the roots or around the ball. Shake the tree several times, while filling and watering the hole, to help settle the soil. Allow the water to drain away, then fill the hole to ground level.

Now make a basin 3 or 4 inches deep

and two feet in diameter for watering. The sides of the basin should slant toward the ball. Water the tree thoroughly, to dampen the soil all the way through the ball.

The trunk of the young tree should be protected from bark injury and sunburn. Several thicknesses of newspaper or a layer of building paper, tied loosely, will keep mice away and prevent injury until the tree has recovered from the shock of transplanting. If the weather is cold the trunk should have a wrapper against frost. In Florida, this protection is provided by drawing the soil up to the lowest branches, making a bank 18 to 24 inches high. In California, a wrapping of cornstalks around the trunk provides similar protection. This precaution should be repeated each year until the tree is at least 4 years old and is acclimated.

Irrigation

During the first two weeks after it is transplanted, the young tree should be kept constantly moist. If the soil is sandy, it may be necessary to water it 3 times a week. A clay loam will need water only once a week. During the remainder of its first growing season in the garden, it should be watered once every week to 10 days.

By the time the tree's second season starts, roots should have penetrated somewhat deeper into the soil and waterings can be cut to one every 2 to 3 weeks, with 10 to 20 gallons of water applied to each tree. By its third year, the Florida tree will be on its own, except during periods of extreme drought. When the soil is getting dry, 25 to 30 gallons per tree should be administered. California, in the irrigated districts, supplies water every 15 to 30 days to all trees after the second year, the frequency depending on soil and temperatures. Soil should never be allowed to become completely dry in the root area. On the other hand, when water is supplied, it should never be flooded into the tree's soil in quantities that will penetrate deeper than the immediate root area. If too much water is applied, soil nutrients wash down beyond the reach of the tree's shallow roots. Also, trees should not be watered daily, if they are planted in lawns that need this much water, or foot rot may result.

Soil should not be allowed to dry out completely in the orchard in winter, when rainfall is scanty. Because they are evergreen, orange trees lose water by evaporation all year round. This must be supplied by the soil even in the most dormant periods or damage to the tree will result.

Fertilizer Program

In Florida, where orange groves have been forced for years on heavy applications of chemical NPK fertilizers, growers now realize that their trees are starved for trace minerals. As a result they are adding sulfates of magnesium, manganese, copper and other minerals which they believe are lacking. The fertilizer formulae are becoming more and more complicated, as the chemists add new ingredients to try to bring the soil back to normal. A few growers have switched to organic fertilizers, realizing that these would give them the most complicated formulae of all, and their orchards have responded with startling results.

Before planting, it is well to incorporate plenty of slowly-available coarse granite dust or greensand and phosphate rock into the soil. If soil is very acid, dolomitic limestone, to bring up the *p*H to 6. will add more trace minerals.

In addition, yearly applications of an organic source of nitrogen is necessary. If this is administered in the form of manures, alfalfa hay or tankage, the trace minerals as well as additional phosphorus and potash will be supplied at the same time.

Orange trees need yearly applications of nitrogen in the following amounts: trees under 10 years old should have the equivalent of 2/3 pound per year of the tree's age of a 6 per cent nitrogen carrier. Over 10 years, they should receive one pound, up to 30 pounds per year. Old trees should be given 30 pounds. To supply the equivalent of one pound of 6 per cent nitrogen carrier organically, use 3 pounds poultry manure; 6 pounds dairy manure; 3/4 pound cottonseed meal; 6 ounces dried blood; 1½ pounds sludge; or 3 pounds good alfalfa hay.

Since the organic fertilizers depend upon bacteria in the soil to break down their complicated nitrogenous compounds into a form which the trees can use, additional nitrogen must be supplied in the beginning to feed the bacteria. After the cycle is well under way, the bacteria in the soil die and their stored nitrogen becomes available and supplies this need for additional nitrogen. But during the first year or two of a newly begun organic program, amounts of nitrogenous fertilizer should be more generous than the quantities given above. Also, it would be well to provide this additional amount in a form which is more quickly available than strawy manure or hay. Blood meal, cottonseed meal or sludge will break down more quickly, to speed the fermentation process in the soil.

Pruning

Orange trees should be pruned no more than is necessary to remove dead or diseased tissue or to shape the young tree. Removal of water sprouts may be necessary in older trees to prevent clogging of the center of the tree, but little more pruning should be done. Very old trees may be stimulated to new growth if some light pruning of the oldest wood is done, but for the most part a productive tree should be left alone. All its leaves are necessary to the feeding of the fruit.

When the tree is planted, 4 or 5 vigorous shoots should be selected to form main branches. This selection is usually made in the nursery and the trees as purchased should have the first scaffold branches started. All that the gardener must do is remove succulent sprouts which appear on the trunk below the scaffold limbs, while they are small and green. These can be removed with the fingers—pruning shears should be unnecessary if they are removed when young.

If trees are injured by frost, slight damage should be attended to immediately. As soon as a few damaged shoots are noticeable, they should be removed. But if the damage is fairly severe, the tree should be permitted to recover from the shock of the injury before any pruning is attempted. Six months to a year may be permitted to pass before pruning trees very badly injured by frost. In any case, the trees should not be cut back enough to stimulate new growth unless danger of further cold has passed.

Normal citrus trees produce vigorous shoots that tend to become tangled with the older branches. This growth can usually be bent down and pulled to the outside, improving the structure of the tree. When old trees are cut to stimulate them to new production, only the shoots which arise from the trunks and main branches on the inside, where they cannot be pulled out into the sun, should be cut. If large amounts of such wood need cutting, the operation should be spaced out over a period of 2 to 3 years.

Citrus trees are not pruned, as apple trees are, to open out their tops. Sometimes, however, the upper branches grow so much longer than the lower ones, that the lowest growth is shaded. When this occurs in a planting containing several trees, the best solution is to remove one or more trees to permit the sun to shine from the sides. The condition cannot be corrected by pruning the branches.

Size and quality of oranges is not improved by pruning, so the work is never done for this reason. Lemons, on the other hand, are improved by an annual pruning, so some annual trimming of lemon trees is advised.

Enemies

Experiences of growers who have changed from their chemical fertilizer and spraying programs to the organic method seem to bear out the theory that insects and fungi do not attack healthy plants. The fact that these organic orchards are free of many of the pests which plague other growers may also be due partly to the fact that the beneficial organisms are not killed off when the trees are not sprayed. These biological pest controls include, on citrus trees, several benevolent fungi as well as the lady beetles which are helpful in controlling aphids and mites on other crops.

Two fungi whose host is the citrus leaf are helpful in the control of white flies. Red aschersonia kills the immature flies. It forms pink or red pustules one-eighth inch in diameter or less on the undersides of the leaves. It is easily seen and is so colorful, that it looks like a menace. Actually it is harmless to the tree.

Brown white-fly fungus also attacks the white fly. It makes cinnamon or brownish pustules, about one-eighth inch in diameter on the undersides of the leaves, which look very much like Florida red scale, but lack the characteristic raised center of the scale.

Sooty mold, while not beneficial, is less dangerous in itself than annoying. Actually, it is a symptom rather than a disease. It is a mold which grows in the honeydew excreted by some scales and aphids on the fruit and foliage. When the scales and aphids are destroyed by lady beetles and by the general good health of a well-grown tree, the sooty mold also disappears.

Nematodes in the soil around the roots have brought about the death of some fine citrus orchards. It is generally recognized that the best control of nematodes is through the fostering of fungi which are their natural enemies. Soil which is so full of chemicals that all microscopic life in it is killed, makes a happy hunting ground for nematodes. Restoration of decomposing organic matter to the soil brings fungi which prey on the nematodes. Trees grown with plenty of organic fertilizers and mulches are never troubled by this pest.

Mulch under a citrus tree should be kept at least 8 inches from the base of the tree in order not to foster foot rot, also called brown rot gummosis. This is a disease of the bark which is favored by long periods of wet weather. Patches of the bark become infected at the soil line and, in extreme cases, the tree is girdled and dies. Best method of combating foot rot is by sanitary measures. Keep all refuse away from the base of the tree and remove the soil down to the juncture of the trunk and the first roots. Keep the mulch pulled well away from the trunk and remove any leaves that fall into the area. Do not allow irrigation water to stand around the foot of the tree.

Harvest and Storage

Oranges with tight skins, such as navels and Valencias, may be picked by pulling away from the stem with a slight twist. Those with loose skins, such as mandarins, Temples and tangerines, are more safely picked with clippers, which clip the stem near the juncture with the fruit. Usually clipping is done in two stages—the fruit is cut from the tree with one-half inch of stem attached and then, the stub is cut close to the button. In this way the danger of puncturing adjacent fruit in the basket is eliminated.

Color is not always a sign of maturity in oranges. Their skins contain a mixture of pigments: green, orange and yellow. In fall the green predominates until the weather turns cool enough to check their growth, when the green fades out and the other pigments predominate. But in spring when growth begins, green pigment will again appear in perfectly ripe fruit as it hangs on the tree.

Oranges may be stored for 6 to 10 weeks at very low temperatures, under refrigeration. Florida oranges can be kept 8 to 10 weeks at 30 to 32 degrees with 85 to 90 per cent humidity. California oranges can be kept 6 to 8 weeks at 35 to 37 degrees, but are subject to rind disorders at lower temperatures.

ORANGE, MANDARIN

ORANGE, MANDARIN. Much confusion exists in the classification of orange varieties and hybrids, but most authorities agree that the species *Citrus nobilis* (sometimes called *C. reticulata)* should be separate from the sweet orange species, *C. sinensis,* principally on the basis of difference of fruits and seeds. Mandarins have at times been called kid-glove oranges (because they might be eaten with kid gloves without soiling the gloves), Muscat apples, King oranges and tangerines, Satsumas belong to the group. In a recent classification, the mandarin name is given to the whole *C. nobilis* species and is applied especially to a group of fruits of the variety *deliciosa,* which also includes tangerines. *C. nobilis Unshui* is made to include all the Satsumas and their variations, while King and Calamondin are considered hybrids.

Mandarin oranges originated, as might be inferred from their name, in China. They were introduced to England in 1805 and made their way to Louisiana by 1850. Probably the reason for the name is that the mandarin orange is considered an aristocrat of the citrus fruits, just as the mandarins of China were once the aristocrat class. Fruits of mandarin trees are, in general, smaller than those of the sweet orange, though some of the varieties have been developed into large oranges. Cotyledons of the seeds are usually green, rather than white, the skins are loose and the sections separate easily.

Mandarin trees are not as productive as sweet orange trees and they require most careful culture, in order to produce their fruit best, both in size and quality. Some of the mandarins, especially Satsumas on trifoliate orange stock, are among the hardiest citrus fruits in regard to cold-endurance. If well-protected while they are young, the trees will acclimate themselves to any of the Gulf States. Culture of most varieties of mandarins has been more extensive in Florida than in California, though some varieties are equally suited to the California climate.

For complete information about mandarin varieties and culture, *see* ORANGE.

ORANGE, OTAHEITE

ORANGE, OTAHEITE. This is a true dwarf citrus fruit, *Citrus taitensis,* which is grown as a pot plant in many northern greenhouses, but is not much cultivated in the orange areas of the country because of its insipid flavor. The Otaheite orange should be considered primarily an ornamental plant.

Propagation is by seeds or grafting. It may be grown in any well-drained, mellow soil, with frequent watering and full sun. Its fragrant pink and white flowers are followed by two-inch oranges which are very decorative and hang on the plant for a long time.

ORANGE, TRIFOLIATE

ORANGE, TRIFOLIATE. This is not a member of the *Citrus* genus to which the true oranges belong, but it is closely related to them. *Poncirus trifoliata* is the correct botanical name of trifoliate orange, and is applied to a tree with deciduous leaves and a hairy, inedible fruit. The plants are sometimes grown for ornamental purposes, but more important are the uses of trifoliate orange and its hybrids as rootstocks for many of the citrus fruits. Trifoliate stock imparts to the trees grafted on it a cold-hardiness which is very important to citrus fruits. Some of the hybrids formed by crossing *P. trifoliata, Citrus sinensis* (sweet orange) and a kumquat, one of the *Fortunella* species, have proved superior to *P. trifoliata* for rootstocks. Common names for some of these hybrids are citrange, citrangedin, and citrangequat.

PAPAW

PAPAW. Or pawpaw, *Asimina triloba,* is a native of the more temperate regions of the United States east of the Mississippi River and, while not evenly distributed throughout that area, it may be grown anywhere within it, as well as in other portions of the country with similar climate. In the north, its habitat extends to the central portion of Michigan and New York state, in the south, to within 100 miles of the Gulf, though it is sometimes also found in northern Florida.

The papaw tree is long-lived, though not very large when mature. It grows usually to about 20 to 30 feet, with a trunk 10 to 12 inches in diameter. Specimens have been known to live to 100 years and others to have borne fruit for at least 60 years. Due to its habit of throwing up suckers from its roots, the papaw is often found in thickets where the growth of the individual plants is shrubby and not exceeding 10 to 15 feet. Plantings are often mixed with mulberries or locust.

With its smooth, glossy leaves, which are 6 to 10 inches long, the papaw makes a decorative tree for the home grounds. Two-inch blossoms, which are first green and hidden in the leaf axils, later change to purple, red and yellow. The fruits develop slowly, ripening from the end of August until some time after frost in November. They are brown when ripe, resembling potatoes in appearance. Two types have been observed, though they are not named varieties, except locally. One type has smallish fruits 2 to 3 inches long whose flesh is white and watery, with a disagreeable flavor. The other type is larger-fruited, growing to about 6 inches long and 3 inches thick. Seeds, some of which are an inch long, are arranged in two parallel rows down the center of the fruit. Flesh of the larger type is cream-yellow, soft and banana-flavored. Its consistency when ripe is reminiscent of custard, from which it is sometimes called a custard apple.

Papaws are rarely cultivated. They are usually gathered from trees which grow wild in rich, alluvial, river-bottom soil or in moderately moist and lightly shaded locations. Accordingly, when they are planted in the home grounds they should be given strong, rich, well-drained soil in a position where they do not get a full day's sun. Soil which supports a growth of black ash, May apple or ginseng is known to be congenial. From this it might be inferred that soil *p*H should be near neutral, from 6. to 8., though actual tests have not been made. In nature the trees are often found growing under tall, open trees which gives them some light shade. In the home garden they may be planted near or under such trees as persimmons.

Vegetative propagation, which ensures the perpetuation of a desirable variety, is difficult with papaws. They are very difficult to graft and cuttings of the top wood root only with the greatest amount of care. It is possible, though not known with certainty, that root cuttings may be made to sprout. The best means of vegetative propagation is by digging and moving suckers from a papaw thicket. The suckers should be small—preferably less than two years old—and several times as many should be dug as will be required, because the percentage of success in moving them is small.

Propagation by seed is more easily accomplished, but the trees which come from seed may differ considerably from their parent tree. Since it takes 6 to 8 years to mature the first fruit from seedlings, seed should be planted only with the expectation of possible disappointment. Suckers are sometimes brought to bearing within 3 years, so the investment of time is not so great when they are planted.

Seed must be planted as soon as the fruit is ripe in fall, or it may be stratified over the winter and planted in spring. Sometimes whole fruits are planted, and seedlings thinned after they have germinated. Individual seeds should be planted one to two inches deep and mulched to keep the soil moist. They will not germinate until the following July or August. Greatest success seems to attend shaded seedbeds, with some sort of shade provided for the seedlings especially when they are small. Growth during the first two years is slow, the plants achieving a height usually no greater than a foot by the end of their second year. After that they grow more quickly.

Like many plants that have not been much cultivated, papaws are not often troubled with diseases or insects. Trees are not too productive, bearing half a bushel to one bushel when they are 15 to 20 feet tall, but the fruit they bear is almost always perfect.

Best flavor is thought to develop after the fruit has been picked, if it is kept cool. When the papaws begin to soften, pick and store them in the refrigerator for 2 to 3 weeks. They will keep best when picked at the half-ripe stage, rather than when they have entirely ripened on the tree.

PAPAYA

Papaya. Originally a native of the American tropics, the papaya, *Carica Papaya,* has spread to many of the Pacific islands, including Hawaii. It is sometimes called papaw, or pawpaw, but this name belongs to an entirely different fruit, *Asimina triloba,* which is not even distantly related to papaya.

Papayas are often called tree melons, a term which describes their large fruits, some weighing as much as 20 pounds, which hang in clusters from a palm-like tree. The fruit is green to orange when ripe, with thin skin, yellow to salmon-pink flesh, one to two inches thick, inside of which is a cavity containing a mass of small black seeds enclosed in a gelatinous membrane. Texture of the fruit is soft and melting when ripe, with a flavor that has been described as similar to peach, cantaloupe, strawberry or just papaya. However it is described, most people agree that it is delicious. Best flavor is thought to be that of the smaller fruits of the only variety which has been developed, called Solo, Hawaiian Solo or Blue Solo. Larger fruits grow on some of the native trees and they may or may not have the quality or flavor of the Solo fruits.

The papaya tree is small, old specimens growing no more than 25 feet tall. It bears fruit, in its first year in the garden at a height of 6 to 10 feet. Many growers allow the trees only 2 or 3 seasons before they tear them out and replant, because the first few years' crops are best. If permitted to continue, the tree will bear smaller fruits each year and the quality is thought to deteriorate. The papaya tree has a straight palm-like trunk with large deeply lobed leaves on long stems. Leaves, stems and unripe fruit exude a milky white latex.

Three types of blossoms are found on papaya trees—male, which are funnel shaped and grow in long panicles; female; and bisexual, which are 5-petaled blossoms which occur in the axils of the leaves. Male trees bear no fruit. Bisexual trees yield cylindrical fruits with small seed cavities. Fruits of the female trees are rounder. Plantings containing bisexual trees have no need of male trees, though the male trees are more active pollinators. Where there are no bisexual trees present, one male tree must be left in the planting to every 20 to 25 female trees.

Papayas are very tender to frost damage. A temperature of 31 degrees will damage the plants, but mature plants can stand lower temperatures for several hours before they are killed. However, exposure to lower temperatures may prevent bearing during the following season. Unless they can be protected from cold, trees should not be planted north of the frost-free areas.

Any good garden soil will grow healthy papayas, but the more fertile the soil, the larger will be the yield. The trees must have good drainage, and will not tolerate water standing around their roots for as long as 48 hours. Soil should be deeply prepared before the trees are set and, when fertilizer is applied, it should be incorporated into the soil around the mature trees a foot or more beyond the spread of the branches, because the roots are not only deep, but are also spreading. Growth is very rapid and crops are heavy. A tree may yield a good crop in as little as 10 months after planting. A good tree will bear 75 pounds of fruit per year and some have been known to bear 300 pounds in a single year. In order to nourish so much growth, fertilizer applications must be heavy, frequent and organic. Papaya will not tolerate chemicals.

Trees in this papaya grove in Florida are less than two years old. Paper bags cover melons which have been hand-pollinated for special seeds.

The papaya has a reputation, partly deserved but greatly exaggerated, of being able to tenderize tough cuts of meat. It is true that green fruits contain a small amount of papain, a protein-splitting enzyme which is capable of breaking down tough tissue to some extent. However, it is not true, as the Chinese once believed, that a chicken hung in a papaya tree would become tender. And a ripe papaya contains little or no papain.

The fruit is a good source of vitamins A and C and provides some B vitamins as well. A piece of a Solo papaya weighing about 100 grams can supply 1,750 units vitamin A, 84 milligrams vitamin C, .02 milligrams vitamin B_1, .05 milligrams vitamin B_2, .4 milligrams niacin and a fair amount of calcium. In papaya, the vitamin C content increases as the fruit ripens. When the skin is dark green and the flesh yellow, the fruit contains only 60 to 70 per cent of the vitamin C that it will contain when fully ripe.

Because of its increased vitamin content, the fruit is most often used when ripe. It is served raw in salads or desserts, baked, or in sherbets. However, it is not often frozen as a means of preservation, because it loses flavor and texture on being thawed. Green papayas may be cooked and served as a vegetable.

Planting

Papayas are grown from seeds, usually seeds taken from known plantations where the trees are isolated from wild or undesirable varieties. In Florida, the seeds are planted early enough indoors so that the plants are ready to be set out in March or April, and they will then produce fruit by the next November or December. If they are not ready to set out until July, they will not fruit until the following year.

Seeds are usually sown in quart cans. The roots do not like to be disturbed, so the bottoms are cut almost free of the cans, but are left attached until the plants are ready to remove. Then the bottom of the can is bent back, and the plant with its soil ball is pushed out through the opening in the bottom.

Five seeds should be sown in each can. After they have germinated, the 3 strongest are allowed to remain and two are removed. Plants are set out 8 feet apart if they will be renewed yearly; 10 to 12 feet apart if they are to remain for longer periods. When they bloom and the sexes of the plants have been determined, male plants are all removed if a sufficient number of bisexual plants are present. If possible, a female and a bisexual plant should be permitted to grow in each hill. If there are not enough bisexual plants for pollination, one male should be left for each 20 females.

Dig the hole large enough to receive the plants and then dig in a generous amount of compost or manure. Chicken manure, tobacco stems or sludge may be incorporated into the soil before planting at the rate of 5 to 10 pounds to each hill. Set the plants in the garden at the same height as they are growing in the pots. Mulch the soil thoroughly around each hill as soon as the trees are planted.

Maintenance

Papayas must be kept moist at all times, but never wet. When the blossoms have opened, the plants must be watered regularly or they will not set fruit. On dry soils and those which drain very well, water must be supplied every 5 or 6 days during the dry weather.

Starting two weeks after planting, a regular biweekly fertilizing is necessary to nourish the fast-growing plants. This may be administered in large amounts of well-rotted compost or it may be in the form of nitrogen-rich fertilizers such as fish meal, cottonseed meal or bloodmeal. When the plants are small, the amounts should be approximately as follows, for each plant: 1½ ounces of bloodmeal; 2 ounces fish meal; or 3 ounces of cottonseed meal. As they grow, the amounts should be increased gradually, until after 6 months the trees are receiving 4 times that amount of fertilizer every other week. Spread the fertilizer carefully over the entire root area.

PASSIONFRUIT

PASSIONFRUIT. *Passiflora edulis,* called passionfruit, purple granadilla or sometimes erroneously water-lemon, is a robust evergreen climber native to Brazil. Its most popular name, passionfruit, was given to it by the early Spanish explorers. It is believed that the flower, which has a multitude of ray-like filaments, suggested to them the symbolic rays used by medieval and renaissance artists to represent Christ's passion, hence the name.

Passion vine is very tender in its first year, but after that it is hardy down to 28 degrees. It is grown extensively in Australia and Hawaii for its fruit. Very few commercial plantings now exist in the United States, though it is grown in some California and Florida dooryards. It may be grown in any climate which is suitable for lemons or limes.

The vine has dark green leaves, usually 3-lobed, and it climbs by means of tendrils. Flowers are about two inches wide, white, with a purple and white crown. The fruit is plum-shaped, 2 to 3 inches long, purple, with a leathery, thin, tough shell, numerous flattened edible seeds surrounded by soft yellowish juicy pulp

The fruit of the passion flower is about the size of a plum, and is considered a delicacy because of its unique flavor.

which is acid to subacid and has a delicious aroma. The pulp is spooned from the shell or it is squeezed in a cloth, to extract the juice. A yellow variety of the passionfruit widely grown in Hawaii has more acid in its juice than the purple variety.

As a source of vitamin A, the yellow passionfruit is superior to purple. They are both good sources of vitamin C and supply small amounts of calcium, niacin and riboflavin. One-half cup of juice from yellow passionfruit contains 2,400 units of vitamin A and 20 milligrams of vitamin C. In the same quantity of juice from the purple variety there are 700 units of vitamin A and about 33 milligrams of vitamin C.

Passion vines prefer a well-drained soil composed of fibrous loam, and enriched with leaf mold. They grow best in an almost neutral soil, with a *p*H of 6. to 8. When growth is active, they are improved by a feeding with liquid manure.

Vines are short-lived under the best of conditions and must be replaced after about 8 years. In their fourth to their fifth years they produce the largest amounts of fruit, in some cases up to 40 pounds per vine.

Propagation is by seeds, which germinate in 2 to 3 weeks; by layering; or by cuttings of mature wood.

In Florida the passion vine is attacked by crown rot and nematodes. California plantings are not so much troubled by these enemies, especially when they are grown on an organically rich but dryish soil.

Passionfruit may be canned or frozen and its juice may also be preserved by either of these means, though it should not be kept more than 6 to 8 months or it will lose its color and flavor.

PEACH

PEACH. *Amygdalus persica* is the Latin name now given to the peach, formerly classified as *Prunus persica,* in the genus which contains all the other stone fruits. Because of slight botanical differences, peach and almond are now classed separately in the genus *Amygdalus.*

Peaches originated somewhere in the Orient, probably in southern China or perhaps Persia, and were introduced to the West at about the beginning of the Christian Era. Their cultivation spread rapidly and they were among the first fruits planted in this country when it was settled. In fact, so many peach trees were planted by the colonists that many escaped and became naturalized in the South. Peaches are still among the most popular of American fruits, ranking fourth among commercial crops and being exceeded only by apples, oranges and grapes.

Very few places in the country are unsuitable for some variety of peaches, but very few states are actually suitable to peaches grown as nature intended them to grow. Excepting in the dozen or so southern states where they now grow wild, peaches must be coddled or adapted to garden culture. In some areas—those where there are fewer than 650 hours below 45 degrees, or where the temperatures regularly drop 15 degrees below zero during the winter—peaches cannot be grown at all. But these areas are very few. Hawaii does not have enough winter cold. Northern New England, Alaska and the northern fringe of the Plains States are too cold. Peaches may be made to grow almost anywhere else.

Buds of immature trees, even the most hardy varieties, will stand less cold than buds on mature trees. Elbertas and Golden Jubilee, two widely planted varieties, suffer bud damage at 12 degrees below zero. Some of the hardier varieties can take temperatures to 18 below, though they should not be expected to make crops every year where such cold is common. After a warm period in spring, which has started growth in the buds, damage may be suffered at a much higher temperature, 10 below sometimes causing as much damage as 20 below earlier in the winter. During the blossoming period 25 to 30 degrees can kill the entire crop. The amount of cold which any one planting can stand may vary from year to year, depending upon conditions during the previous growing season and upon such circumstances as fertility of the soil and the rainfall. Any factor which induces late growth in the trees may make them more subject to winter injury. Any general weakening caused by disease, or even by overbearing, may make a hardy variety sensitive to cold.

Summer temperatures are less important to peaches than those of the winter, though they have some bearing on the length of time taken to mature fruits. Areas where the mean temperature during the summer months is around 75 degrees produce the best quality fruit. But lack of sufficient summer heat only slows the ripening process. Lack of sufficient winter chilling can prevent setting of the crop. Unless the trees have a certain number of hours of winter cold—750 to 1,150 hours below 45 degrees, according to variety—their rest period is not complete and they will not set fruit. Also, within certain limits, trees require less chilling in an area where the temperatures go below 45 degrees and stay there, than where they drop below that at night and rise during the day.

Soil influences the hardiness of peaches more than that of almost any other fruit. A warm sandy or gravelly soil may support a fine crop of peaches at temperatures that would winter-kill trees planted in clay. Elevation may also have much to do with climate in an orchard. Experiments carried out in New Hampshire have shown that a difference of 40 feet in elevation may make as much as 15 degrees difference in winter temperatures. All of these factors should be taken into account when home plantings of peaches are made.

Peaches and nectarines are botanically the same, except that nectarines lack the fuzz that covers peach skins. Nectarines have been known to grow from peach pits, and peaches from nectarine pits, without cross-pollination. The two have even been known to grow on the same tree. No explanation for the variants is known. Culture of the two is essentially the same, but for some reason nectarines succeed better west of the Rockies than in the central and eastern part of the country. So far as their treatment is concerned, they differ from peaches only in being more susceptible to damage from curculio and other tunneling larvae. Nectarines may be yellow or white-fleshed, like peaches, and either freestone or cling. Their flavor is individual and is more relished by some than peach flavor.

Yellow peaches are rich in vitamin A, of which they retain about half when they are cooked, canned or frozen. White peaches contain a very small amount of vitamin A, and their other nutrients are about equal to those supplied by yellow varieties. In 100 grams of yellow peaches, raw, there are, on the average, 1,000 units vitamin A, .025 milligrams vitamin B_1, .065 milligrams vitamin B_2, 9 milligrams vitamin C, 10 milligrams calcium, 19 milligrams phosphorus, .3 milligrams iron and about 50 calories. The same quantity of white peaches contains only 100 units of vitamin A.

Ordering Stock

It is very important to plant a variety of peach suited to the section of the country in which you live. Thousands of varieties have been developed and new ones are constantly coming on the market, each one suited to a special climate. Since climate is one of the most important factors in peach growing, selection of the variety suited to the climate is the most important job that the gardener can do for his peaches. It is essential to know, to choose a variety, the temperature fluctuations on the particular piece of ground on which the trees will be planted, as well as the average temperatures for the region. A cold northern hillside in a warm area may be suited to the same variety as a warm, sandy river bank in a cold area. Early-blooming varieties should not be planted where the site warms up before the neighborhood, nor late-ripening varieties in areas of early fall frosts.

Most varieties are self-fruitful, with the exception only of J. H. Hale among those listed on the accompanying chart. And since most trees set more fruit than may be allowed to ripen, there is no advantage to providing cross-pollination. Where shy bearers are planted, a second variety planted nearby may increase yields.

The average yield per tree is about 3 to 4 bushels, though exceptional cases have been known where a large old tree produced as much as 16 bushels. Bearing starts 3 to 5 years after planting and 2 to 3 more years are required to bring a tree to full production. The average life of a planting is 10 to 12 years, though, in a few isolated cases, orchards have been known to flourish for 25 years or more. Because of the short life expectancy, it is wise to plant replacements for peach trees every 5 years. Best trees to plant are one-year stock 3 to 5 feet tall, the trunks at least one-half inch thick.

Site

Where a temperature-moderating factor is available, the peach orchard should be planted to take advantage of it. The best such device is a hillside which slopes down to a fairly large body of water,

TABLE 78: Peach Varieties

Where available, the number of days between blossoming and ripening are given. Crops can succeed only where this number is well below the total number of frost-free days for the area. Blooming, described as early or late, can only be comparative because dates differ in various sections. Hours of winter chilling given represent the minimum number of hours required at 45 degrees or lower.

Variety	Section for which recommended	Fruit characteristics	Tree characteristics	Wintering
Afterglow	Southeast	Late blooming, 149 days to mature. Large yellow freestone peach of good quality.	Upright habit. Moderately vigorous. Good yields.	750 hrs. chilling.
Ambergem	New England Great Lakes Appalachians	Small to medium yellow clingstone of high quality for canning. Midseason to late.	Moderately heavy yield.	Flower buds hardy.
Anza	Southeast Pacific States	Yellow freestone of good quality. Early to midseason.	Upright, moderately vigorous.	800 hrs. chilling.
Blake	Southeast Appalachians	Large to medium yellow freestone with excellent flavor. Late.	Vigorous, productive.	Flower buds hardy.
Bonita	Southeast Pacific States	Medium to large yellow freestone. Flavor sweet, sometimes slightly bitter. Early.	Upright, vigorous, very productive.	800 hrs. chilling.
Cardinal	Southeast Pacific States Appalachians	Late blooming, 89 days to maturity. Medium size yellow cling, good flavor. Slightly tart and astringent.	Moderately vigorous, spreading. Productive. Many twins. Thin early and heavily.	900 hrs. chilling.
Carman	Southern plains Appalachians	Small to medium white semi-cling. Fair quality. Early.	Light to medium yield. Replaced by Golden Jubilee.	
Coronet	Southeast Appalachians Southern plains	Early blooming. 99 days to mature. Yellow cling, medium large, sweet mild flavor.	Spreading, vigorous, productive. Susceptible to bacteriosis.	750 hrs. chilling. Early bloom subject to frost damage.
Dixigem	Southern plains Southeast Delta States Appalachians	Late blooming, 97 days to mature. Medium size yellow, almost freestone, sweet, slightly flat, high quality fruit good for canning and freezing.	Spreading, vigorous, very good yields.	850 hrs. chilling. Moderately hardy.

TABLE 78: Peach Varieties (Continued)

Variety	Section for which recommended	Fruit characteristics	Tree characteristics	Wintering
Dixired	Southern plains Delta States Southeast Appalachians	Late blooming. 80 days to maturity. Medium to large yellow, semi-cling, firm, sweet mild flavor.	Upright, vigorous. Reliably productive but yields moderate.	950 hrs. chilling. Less hardy than Elberta.
Elberta	Appalachians Southern plains Southeast Delta States Pacific coast Mountain States New England Great Lakes	Early blooming. 140 days to maturity. Large, yellow, freestone, strong flavor, slightly astringent. Below average in quality.	Productive, widely adapted to many soils and climates. Upright, vigorous, fair to good yields.	850 hrs. chilling. Tender to very low temperatures. Flower buds are tender.
Erly-Red-Fre	Southern plains Appalachians Southeast	Large white semi-cling to freestone, firm, good quality and flavor, early.	Reliably but moderately productive. Tendency to produce split pits.	850 hrs. chilling. Flower buds hardy.
Erly Vee	New England Great Lakes	Small yellow semi-cling with tender skin. Juicy, good flavor. Early.	Requires early, heavy thinning. Vigorous, productive.	Hardy.
Fairhaven	Northern plains New England Great Lakes	Late blooming, 107 days to mature. Medium size yellow freestone, sweet coarse flesh, good flavor. Good for freezing	Spreading, vigorous, productive. Good yields. Recommended for home gardens.	850 hrs. chilling. Hardy.
Georgia Belle	Southeast Delta States Appalachians Southern plains	Medium size white freestone. High quality. Midseason to late.	Vigorous, productive.	850 hrs. chilling. Not very hardy but buds hardy.
Golden Blush	Pacific States Southern plains	Early blooming, 144 days to mature. Medium size yellow freestone, slightly astringent.	Spreading, vigorous. Good yields. Slightly susceptible to bacterial leaf spot.	850 hrs. chilling.
Goldeneast	Southern plains	Late blooming. 125 days to mature. Large white freestone, sweet, good quality.	Upright, vigorous. Light to medium yields.	1,050 hrs. chilling.

TABLE 78: Peach Varieties (Continued)

Variety	Section for which recommended	Fruit characteristics	Tree characteristics	Wintering
Golden Globe	New England Great Lakes	Large yellow freestone, juicy, good flavor. Midseason to late.	Productive, vigorous.	Medium hardy.
Golden Jubilee	Corn Belt Appalachians New England Great Lakes	Early blooming, 99 days to mature. Large yellow freestone, juicy, sweet to slightly acid, strong flavor. Low in vitamin C. Ripens fast at high temperatures.	Moderately vigorous, reliably productive of fair yields. Spreading. Does not require much thinning.	850 hrs. chilling. Hardy.
Golden State	Southeast Pacific coast	Large, yellow, freestone, coarse flesh, fair flavor. Early	Vigorous. Upright.	800 hrs. chilling.
Halehaven	Appalachians Mountain States Corn Belt Southeast Southern plains Pacific coast Northern plains New England Great Lakes	Late blooming, 115 days to mature. Medium to large freestone, yellow, sweet to slightly tart, fine flavor, flesh slightly stringy in some areas, but usually high quality.	Recommended for home gardens in almost all areas. Large, upright to spreading tree wtih strong scaffolds. Fair yields.	850 hrs. chilling. Fruit and leaf buds very hardy.
Hermosa	Southeast Pacific States	Medium size white freestone. Sweet, mild. Midseason.	Spreading. Moderately vigorous, productive.	800 hrs. chilling.
Hiland	Southeast	Early blooming, 87 days to mature. Small to medium yellow clingstone. Fair quality.	Spreading, vigorous. Light yields. Susceptible to brown rot.	750 hrs. chilling.
Hiley	Southeast Southern plains	Medium to large white freestone. Good quality. Good for canning or freezing. Midseason.	Medium yields.	750 hrs. chilling.
Jerseyland	Southeast Corn Belt	Late blooming, 93 days to mature. Medium to large, rather coarse, slightly acid, good flavor, yellow, semi-cling.	Spreading. Vigorous, very good yields.	850 hrs. chilling. Hardy.
J. H. Hale	Pacific States Southeast Appalachians	Late blooming, 132 days to mature. Large sweet yellow freestone of high quality.	Upright, rather weak tree, less vigorous than Elberta. Light yields. Pollen not viable—needs pollinator.	850 hrs. chilling. Short-lived in cool areas.

TABLE 78: Peach Varieties (Continued)

Variety	Section for which recommended	Fruit characteristics	Tree characteristics	Wintering
July Elberta (Burbank)	Southeast Southern plains Mountain States Pacific coast	Late blooming, 112 days to mature. Medium size yellow freestone, sweet, medium to good quality.	Spreading, not too vigorous. Tends to overbear, must be thinned.	750 hrs. chilling. Moderately hardy.
Kalhaven	New England	Late blooming, 131 days to mature. Medium size sweet freestone.	Spreading, vigorous, good yields. Requires heavy thinning.	950 hrs. chilling.
Laterose	Southeast Appalachians New England	Early blooming, 148 days to mature. Greenish-white sweet freestone. Best late white peach.	Spreading, vigorous, productive.	800 hrs. chilling.
Lizzie	Great Lakes	Large yellow freestone of good quality. Very late.	Medium to light yield. Vigorous.	Flower buds tender.
Loring	Appalachians Southeast Southern plains	Early blooming, 145 days to mature. Large sweet yellow freestone. Good flavor.	Upright, vigorous, productive.	850 hrs. chilling.
Maygold	Southeast	Early blooming, 98 days to mature. Medium size yellow clingstone. Slightly acid.	Moderately vigorous, not very productive. Spreading.	650 hrs. chilling.
Nectar	Corn Belt Appalachians	Late blooming, 125 days to mature. Large sweet white freestone, good quality.	Upright, vigorous. Light to medium yields.	1,050 hrs. chilling.
Newday	Southeast Southern plains	Early blooming, 107 days to mature. Medium size yellow, slightly acid, mostly freestone.	Upright, vigorous, fairly productive. Somewhat susceptible to bacterial leaf spot.	750 hrs. chilling.
Oriole	New England Northern plains	Late blooming, 94 days to mature. Medium size yellow semi-cling with good flavor, slightly acid.	Upright, productive, vigorous. Medium yield. High percentage of split pits.	850 hrs. chilling. Very hardy.
Prairie Dawn	Corn Belt New England Great Lakes	Late blooming, 90 days to maturity. Medium size sweet yellow cling with coarse flesh but good flavor.	Spreading, vigorous, very productive.	850 hrs. chilling. Very hardy.
Prenda	Southeast Pacific coast	Medium to large white freestone, juicy, sweet, mild. Early.	Upright, vigorous, productive, subject to preharvest drop.	750 hrs. chilling.
Ramona	Southeast Pacific coast	Medium-large yellow cling. Very good for canning. Late.	Vigorous. Upright. Tends to produce many twin fruits.	800 hrs. chilling.

TABLE 78: Peach Varieties (Continued)

Variety	Section for which recommended	Fruit characteristics	Tree characteristics	Wintering
Ranger	Southeast Southern plains	Late blooming, 101 days to mature. Medium to large sweet yellow freestone with good flavor. Good for canning, freezing.	Upright, vigorous, very good yields. Resistant to bacteriosis.	950 hrs. chilling.
Raritan Rose	Southeast Southern plains Northern plains Great Lakes New England	Late blooming, 96 days to mature. Large white firm freestone, sweet, good flavor.	Upright, vigorous, productive.	950 hrs. chilling. Medium hardy.
Redcap	Southeast Southern plains	Early blooming, 92 days to mature. Medium size yellow cling, slightly acid.	Upright, moderately vigorous, good yields.	750 hrs. chilling.
Redhaven	Mountain States Appalachians Corn Belt Southern plains Southeast Northern plains New England Great Lakes	Late blooming, 94 days to mature medium size yellow freestone. Sweet, pleasant flavor. Oxidizes slowly, so good for freezing, canning.	Spreading, moderately vigorous, good yields. Good for home gardens. Thin early and heavily. Resistant to bacterial leaf spot.	850 hrs. chilling. Flower buds moderately hardy.
Redskin	Southeast Southern plains New England Great Lakes	Early blooming, 133 days to mature. Medium size yellow freestone, slightly astringent.	Spreading, moderately vigorous, very good yields.	800 hrs. chilling. Hardy.
Richaven	Southeast Appalachians New England	Large yellow freestone. High quality for canning and freezing. Midseason.	Vigorous, productive.	
Rio Oso Gem	Pacific States Southeast Appalachians	Late blooming, 140 days to mature. Medium size yellow freestone. Sweet, good quality.	Upright trees lacking in vigor, medium to light yields. Susceptible to mildew.	850 hrs. chilling. Not very hardy but flower buds moderately hardy.
Rubidoux	Pacific States	Late blooming, 126 days to mature. Medium size yellow freestone, sweet, good flavor, but somewhat fibrous.	Vigorous, moderately spreading, productive.	850 hrs. chilling.

TABLE 78: Peach Varieties (Continued)

Variety	Section for which recommended	Fruit characteristics	Tree characteristics	Wintering
Shipper's Late Red	Appalachians Southeast	Late blooming, 134 days to mature. Large yellow freestone, slightly astringent, good quality.	Spreading, vigorous, productive.	850 hrs. chilling. Moderately hardy.
South Haven	New England Great Lakes	Early blooming, 106 days to mature. Medium to large yellow freestone. High quality, slightly astringent.	Upright, vigorous, medium yields.	900 hrs. chilling.
Southland	Southeast Appalachians New England Great Lakes	Early blooming, 106 to 112 days to mature. Medium to large yellow freestone, coarse but good flavor.	Upright, vigorous, productive. Recommended for home gardens.	750 hrs. chilling. Not as hardy as Elberta.
Sullivan Early Elberta	Corn Belt Southern plains Southeast Mountain States	Early blooming, 122 days to mature. Large yellow freestone, slightly astringent.	Spreading, vigorous, good yields. More susceptible to bacterial leaf spot than Elberta.	850 hrs. chilling. As hardy as Elberta.
Sun Glo	Pacific States	Late blooming, 126 days to mature. Large yellow freestone, sweet, juicy, rich flavor.	Spreading to upright, vigorous, productive. Somewhat subject to preharvest drop.	900 hrs. chilling.
Sunhaven	Southeast Appalachians New England	Large yellow freestone. Good flavor. Early.	Vigorous, productive.	
Sunhigh	Southeast Appalachians New England Great Lakes	Early blooming. 110 days to mature. Large yellow freestone. Good flavor, high quality, good canning, freezing.	Upright, vigorous, high yields that are sometimes reduced by winter injury. Subject to bacterial leaf spot.	750 hrs. chilling. Less hardy than Elberta.
Triogem	Southeast Southern plains Appalachians Pacific coast	Late blooming, 102 days to mature. Medium to large, yellow, mostly freestone. Juicy, sweet, good flavor. Good canning, freezing.	Spreading. Reliable large yields. Should be thinned early. Fruit hangs well after ripening.	850 hrs. chilling. Medium hardy. Flower buds tender.
Ventura	Pacific coast	Very early blooming. 122 days to mature. Yellow, slightly acid, freestone. Fair quality.	Upright, moderately vigorous, very susceptible to leaf spot.	750 hrs. chilling.

TABLE 79: Nectarine Varieties

Variety	Section for which recommended	Fruit characteristics	Tree characteristics	Wintering
Cavalier	Appalachians Southern plains	Early blooming, 150 days to mature. Medium size, yellow, sweet aromatic flavor, freestone.	Spreading, vigorous, productive.	950 hrs. chilling. Blossoms hardy.
Early Flame	Appalachians	Late flowering, 97 days to mature. Small to medium, yellow, clingstone. Sweet, excellent flavor.	Spreading, moderately vigorous, shy bearer.	950 hrs. chilling.
Fuzzless Berta	New England	Late blooming, 144 days to mature. Large, greenish yellow, freestone, sweet, strong flavor.	Productive, spreading, vigorous.	1,150 hrs. chilling. About as hardy as Elberta.
Garden State	Appalachians New England	Late blooming, 129 days to mature, medium to large yellow freestone. Rich sweet flavor.	Upright, vigorous, shy bearer. Susceptible to bacterial leaf spot.	950 hrs. chilling. Less hardy than Elberta. Blossoms tender to frost.
John Rivers	New England	Small white clingstone. Good flavor. Early.	Productive.	Hardy.
Lexington	Appalachians	Medium size, yellow, good quality. Mid-season.	Vigorous, productive.	Moderately resistant to frost.
Nectacrest	New England Southern plains	Late blooming, 150 days to mature. Medium size greenish-white freestone. Sweet, slightly flat.	Spreading, moderately vigorous. Free of bacterial leaf spot.	850 hrs. chilling.
Nectarose	New England	Early blooming, 145 days to mature. Medium size white freestone. Sweet, rich flavor.	Upright, vigorous, productive.	850 hrs. chilling. Hardy.
Redbud	Appalachians	White, medium size, early.	Moderately vigorous, heavy fruit sets.	Buds medium hardy.
Redchief	Appalachians	Early blooming, 154 days to mature. Large greenish-white freestone, sweet, mild flavor.	Vigorous, upright, low yields.	1,050 hrs. chilling.

such as a river or lake. The bed made by the flow of water provides a natural air drainage, where unseasonal currents of cold air can drain away from the trees. Moreover, the water warms and cools more slowly than the air to prevent early thaws or late freezes, and also to moderate early fall drops in the temperature.

Where there is no such natural protection, trees should be planted on a site which warms up slowly in spring, and one which is least subject to winter thaw and freezes. The northern slope of a hill usually provides this type of protection.

Best peach soil is sandy or gravelly loam with a large percentage of organic matter, underlaid by a clay subsoil which is no closer than 18 inches to the surface. Peach trees are less tolerant, than almost any other fruit trees, of water standing among their roots, so the soil must be very well drained. Near neutral *p*H is required, between 6. and 8. Where the *p*H is below 5.5, regular liming of the soil near the tree will result in larger yields.

Soil should be dug or plowed as for any other planting before the trees are set. Plenty of organic matter may be dug in, but it should consist of material such as peat moss or straw which is not too rich in nitrogen. Phosphate rock and granite dust may be incorporated deeply into the area where roots will ultimately spread, because these minerals do not leach rapidly through the soil. But nitrogen should not be fed in large quantities to the young trees, because it tends to make them too vigorous and more subject to winter injury.

Planting

Except in the areas where winters are very mild, peaches are better planted in spring than in fall. Only in states where the planting could be done in midwinter is it safe to set them out during the fall or winter months. Plant in spring instead, as early as the ground can be prepared, and on a day when the soil is not too wet.

Mature trees will need an area 20 to 24 feet in diameter, in full sun. Dig the hole large enough to hold the roots without crowding or bending them. Prune off broken roots and cut back any that are very long. Set the tree so that the bud union is at or near the surface of the soil. As soon as they are planted, the small trees should have their trunks loosely wrapped with burlap or building paper to protect them from rabbits, mice and sunburn. Tie the wrapping loosely, so that the strings do not bind the bark. The wrapping and strings should be inspected at regular intervals after planting, to be sure that mice do not take up their residences inside it or that the young tree is not outgrowing the ties.

Pruning

Peach trees do not transplant well, so their tops must be pruned severely when they are set. Head back the top to a height of 24 to 30 inches, making the cut just above a branch. All branches should be cut back to one-inch stubs, to reduce leaf-surface evaporation. They should not be cut flush with the trunk, because that would damage the buds from which new branches will grow. Training the tree may start during the first summer after it is planted. Summer pruning, if carried out consistently the first year, can form the main scaffold of the tree from the start, making it unnecessary to remove large amounts of the first year growth the following winter, and thus hastening the tree's development.

When new branches sprout beside the stubs which were allowed to remain, select 3 strong ones which are well spaced between a point a foot above ground to the top of the trunk, and also well distributed around the trunk's circumference. Rub off all the shoots except these 3. During the summer, continue to rub off new growth from other parts of the trunk, before sprouts are able to put forth more than two leaves. The earlier a sprout is removed, the less the tree loses by its removal.

One of two tree shapes may be formed during the succeeding years, and the tree's own habit should be consulted when the choice between them is made. Peach trees are naturally either upright or spreading in habit, depending upon their variety. An upright tree may be trained to spread, but with difficulty. It is better to allow it to grow a taller trunk and to form a round top than to try to force it into the vase shape naturally made by a spreading tree.

A tall trunk is made by favoring the leader—the topmost branch. If it is to be favored, the leader is allowed to grow taller and stronger and to develop more laterals than the other two branches. When it has attained a height of about

Peach trees should be planted in spring as early as soil can be prepared; best soil for peach trees is sandy or gravelly loam containing a large percentage of humus.

8 feet, it should be headed back, and the other two primary branches allowed to catch up with it.

If the tree is to be trained to an open form, the 3 primary branches should be kept as nearly equal as possible. An equal number of secondary branches should be allowed to grow from each of them, and they should be headed back to about the same length each year. This training is begun the first summer and continued at each pruning during the second and third dormant periods.

Little pruning should be necessary after the first year until the trees begin to bear. Mainly, pruning at this time is intended only to prevent the tree from losing its intended shape, and to keep too-vigorous growth within bounds. During the first 3 to 4 years, terminal growth should be about 20 inches per year. If it is much more than that, the excess should be cut back just above a vigorous lateral. After bearing starts, terminal growth should be about 6 to 10 inches a year. If it is much more, the tree is growing too fast and probably receiving too much fertilizer. If less, it may need more fertilizer, more laterals pruned out, or it may have been allowed to set too much fruit.

When thinning laterals, remove any which arise within 15 inches of the trunk, any which grow toward the center of the tree, or toward the ground. Favor those which grow outward and upward on a spreading tree or those which grow outward on an upright tree. On mature trees, remove enough laterals each year to keep the top open, so that sun may penetrate to the center of the tree. This will stimulate the growth of new fruiting wood low on the branches. If the leaves make a dense shade in the tree's center, fruiting wood (which is one-year wood) will be entirely confined to the surface of the tree and will get further and further away from the trunk each year. The best fruit is produced on terminal growth that is no more than 14 to 16 inches long or laterals 8 to 10 inches long, and usually it is toward the base of such new wood. It is possible to stimulate new wood to keep it growing each year while the fruit is forming by cutting back all shoots to remove half of their length.

Old trees may be renovated to stimulate new fruiting growth by cutting back terminal branches into wood that is 2 or 3 years old. This is done with least loss of yield if it is accomplished in easy stages, part of the wood being removed each year over 2 or 3 years.

In warm areas, pruning may be done at any time during the dormant period. Where winters are cold, it should be deferred until growth starts in spring, so that all winter-killed growth may be spotted and cut out. In extreme cases, pruning may even be done while trees are in bloom. Where winter damage has been heavy, very little pruning should be done. Trees which have lost much of their wood need as many leaves as possible to help them recover. Winter-killed buds may be identified by their appearance. Their centers, then the whole buds, have a water-soaked look.

Maintenance

Where peaches are grown in their natural habitat—in a climate which is mild and only sufficiently cold in winter to supply their chilling needs—the trees may be fertilized and treated like any other fruit trees. But where they are grown in spite of the weather, they must be treated much more carefully in the matter of feeding.

In cold areas, peach trees must be encouraged to make all their growth early in the summer, and to slow down and stop growing by the end of July, so that the new wood will harden off against the coming winter. In order to do this, nitrogen feedings must be heavy and quickly available in the very early spring, but must gradually slacken and almost stop in summer. This necessitates feeding with a fertilizer that can be controlled. Manure is fine food for any plant, but it is available over an extended period which may be the death of a peach tree whose new growth continues too late in the season. The same thing applies to compost. Where climates are mild enough so that winter damage is rare, rotted manure or compost may be used freely on peach trees. But in cold areas they will be better served by being fed with bloodmeal, cottonseed meal, tankage or fish meal.

Compost or manure, when used, should be spread in late fall to midwinter, so that the nutrients will have leached into the soil by the time the trees start to make their growth in spring. The more concentrated nitrogen-bearers are spread in very early spring. Fertilizer of all kinds should be spread under the branch drip line, which is about where the feeding roots are located. Nitrogen should be supplied at the rate of one ounce of pure nitrogen per year for each year of the tree's age. One ounce is found in 12 pounds of dairy manure, 4 pounds of poultry or sheep manure, 10 ounces of fish meal, 12 ounces of tankage, 8 ounces of bloodmeal or 14 ounces of cottonseed meal. These amounts are increased each year until the mature tree is receiving the equivalent of a pound of pure nitrogen, or 16 times the amounts listed above. This should be sufficient for trees grown under straw mulch, or with clean cultivation and cover crops. In sod, the amounts must all be doubled.

Fully tree-ripened peaches have more sugar and less acid than fruit picked half-ripe. When flesh gives way to a light pressure of the thumb, peaches are ready to pick.

Peaches do not require large amounts of phosphorus or potassium, though their potassium needs are the higher of the two. If the soil is adequately supplied with these elements when the trees are planted, yearly light applications of rock phosphate and granite dust may be made each fall thereafter.

Cover crops planted in July in an orchard help to slow down the new growth and to prepare the tree for winter. If the tree is mulched, the mulch should be applied in early spring. In cold areas, this may be very strawy manure, clean straw, ground corncobs, cocoa shells, peat moss or other cellulose wastes. Mulch should never be drawn up to the tree trunks, because of fungus and rodent damage. A clear circle at least two feet in diameter should be maintained with clean culture around the base of the tree.

Propagation

Peach trees are all grafted, usually budded on peach stock. Wild stock was used exclusively at one time, but now a number of varieties—mostly unnamed—are used instead. Where apricots flourish, peaches may be grown on apricot stock.

Peach pits must be planted in fall or stratified. When the seeds sprout in spring, transplant them to a nursery row where they may be set 3 feet apart. Buds should be grafted in August. Six weeks after they are budded, the top should be cut back about half. As the buds grow, remove more and more of the rootstock top, until by winter it is reduced to a stub that stands no more than one inch above the graft. Allow the budded tree to grow one year in the nursery before setting it in its permanent place.

Thinning

Overbearing weakens peach trees so that they are much more subject to winter injury. Also, fruit size suffers when too much fruit is allowed to mature, and flavor may also suffer Almost all varieties need some thinning, some much more than others.

Thinning should be done as soon as the June drop has slowed down. Usually the rule for thinning is to leave one peach to each 4 to 6 inches of fruiting wood. Or, calculating it by leaf growth, one peach may be matured to every 50 to 75 leaves.

Most accurate thinning can be done if the peaches are removed by hand, working along each branch in turn. But an approximate job may be done by tapping the branches two feet from the tip with a piece of rubber hose attached to the end of a broomstick. A little practice with the truncheon method is needed, but it is much faster than hand thinning.

Bracing

As the crop nears maturity, branches sometimes bend dangerously low and may break under the weight of the fruit if they are not supported. Young trees may be braced by tying light ropes to opposite branches through the center of the tree. Older trees need more than this support, and may even need heavy wires run through old garden hose where it girdles the branches. Props may also be placed under the branches to brace them from below. One-by-three inch pine, forked at the top and pointed at the bottom, is best for this job. Props and braces should be removed after the fruit has been picked.

Enemies

Peach tree borers, plum curculios, brown rot, bacterial leaf spot and yellows are the most dangerous enemies of peaches. For treatment of plum curculio, *see* CHERRY. Treatment of brown rot is described under APRICOT.

Peach tree borers are larvae of a moth with clear wings that is often mistaken for a wasp. Eggs are laid in the crotches of the tree and in the trash around the base of the trunk in two generations a year which hatch in June and September. The larvae bore into the bark, allowing the tree's sap to escape in a gummy exudation. The gum should not be removed, because it seals the holes. Look for the eggs, which are small clusters of light to dark brown granular flecks, before the hatching date and destroy them. Also, keep trash away from the base of the tree.

Greatest damage is suffered by neglected trees from the disease called bacterial leaf spot, which causes brown spots which eventually fall out of the leaves, as well as sunken spots on the fruit. Trees which are kept in vigorous growing condition seldom suffer from this disease.

Yellows is a contagious and virulent disease which strikes without warning. When it has been definitely diagnosed, the tree should be destroyed before the disease can spread. First symptoms are premature ripening of the fruits which are covered with red blotches, their flesh streaked with red. Next winter buds begin to open prematurely before the winter starts. In a year or two, leaves yellow and drop before the autumn, and then the tree will die. There is no known cure for the disease.

Harvest

Peaches are ripe when the green color in the skin changes to yellow. Fully tree-ripe fruit has more sugar and less acid than fruit picked half-ripe. When the flesh gives way to a light pressure of the thumb, the peaches are ready to pick.

Remove the fruit from the tree by tipping and twisting it sideways. A direct pull makes a bruise and, once bruised, the fruit spoils rapidly. If it is picked carefully, ripe fruit may be stored from several days to two weeks in a cool cellar. When it is stored, it should be sorted every other day to remove fruit which is beginning to soften.

Peaches may be frozen by peeling and slicing them and putting them immediately into bags, excluding all air. Their flavor may be preserved through the whole winter in this way.

PEAR

PEAR. Most of our pears are derived from *Pyrus communis,* the European pear, though a few have strains of the Japanese pear, *P. pyrifolia,* a pear that is hardier and more resistant to fire blight, but of poor quality. Pears have been grown in the cool temperate zones of Europe and Asia since prehistoric times, but the greatest developments in their culture have come within the last 200 years.

In America, pears were planted in the original 13 colonies as soon as they were settled. Father Junipero Serra planted pear seeds in the California missions very early, and some of his trees are still growing there. The boom in California pear growing came with the Gold Rush, when farmers whose gold was mined from their acreage planted large orchards to supply the newly rich miners with delicacies. California is still the largest pear-growing area in the United States. In the California valleys where climate ranges from torrid to cold in a season, pears do better than they do anywhere else in the country. Outside of this area and excluding isolated islands, climate is suitable for pear growing only in a band extending from the Great Lakes eastward to New England. New varieties are being developed which are suited to other areas, but the older types grown in the South and West have much less desirable fruit than is produced under optimum conditions.

Generally speaking, pears flourish where both apples and peaches may be grown. Their cold endurance is below that of peaches—they will withstand freezes down to 20 below, and occasionally 30 below—but their buds suffer winter injury at those temperatures. Their southern limits are set by their need of winter chilling, like that of peaches. Where winters are too warm, they do not have sufficient rest and flower buds die without opening. Leaf buds sometimes remain dormant for a whole year, until they are chilled the following winter. When the chilling is almost, but not quite, adequate, blooming is delayed, sometimes until after the bloom on the pollinating species is finished, so fruit is not set. Pears blossom at about the same time as sour cherries, but their blossoms are more cold-resistant.

Harvest pears when they have reached full size and skins begin turning a light green. Stems will separate easily from fruit spur when pears are gently lifted.

Pears are more tolerant of unlikely soils than most fruits. Their greatest enemy, fire blight, flourishes in growth that is too lush. Consequently, the trees are grown on the lean side. They root deeply and seem to prefer heavy rather than sandy soil, though the clay must be well-drained. Just as the roots go deep down, so the tops reach high up, unless pruned to lower growth. Some varieties, while they are not bent down by weight of fruit, almost resemble Lombardy poplars. They cannot be pruned too much or headed back too severely, or their growth will be stimulated to make blight-susceptible wood. In sandy soils, pears grow too rapidly. The bark sometimes splits from the swift growth and the trees die young.

Soil should be nearly neutral for best growth, with a *p*H of about 6.5. Limestone should be mixed with soil wherever the *p*H is below that. Pears thrive on frequent applications of lime. They contain tannin and all trees containing tannin in large quantities need plenty of lime.

Pears are decorative in the garden and they play a more decorative than useful role in the diet. They actually supply very little of any important nutrients. One medium size pear weighing 100 grams contains 17 units of vitamin A, .030 milligrams vitamin B_1, .060 milligrams vitamin B_2, 4 milligrams vitamin C, 15 milligrams calcium, 18 milligrams phosphorus, .3 milligrams iron and yields 60 calories. At one time, pears were thought to have a medicinal value. If they have, the reason for it has not yet been discovered.

Ordering Stock

Most pears are self-sterile; that is, though their blossoms are perfect, containing both male and female cells, they do not self-pollinate and will not set fruit, unless a tree of another variety is present. Trees should be planted fairly close together to cross-pollinate, because they bloom early when the bees do not make long flights.

With a few exceptions, pear trees will cross-pollinate if their periods of bloom coincide. One notable exception to this rule is the Bartlett-Seckel combination, which will not cross-pollinate. In the accompanying chart, blooming periods are described as early, midseason and late. Actually, the seasons are close together and may overlap by several days, so that early blooming trees usually can be used to cross-pollinate midseason blooms, or midseason to pollinate late. But in some seasons there is too much spread between early and late bloom for dependable cross-pollination.

Best trees to order are one-year whips 4 to 6 feet tall. Trunks should be at least one-half inch thick with healthy bark not pitted by hail or insects. A good combination of varieties, one which may be grown on a small home lot, is Bartlett, Seckel and Anjou. Their blooming dates coincide, cross-pollination is possible through the medium of Anjou, and they supply a crop from early midseason through the end of the season and may be stored into the winter. Trees of all varieties may be dwarfed by being grafted to dwarfing rootstocks, though they make a brittle junction.

All pear trees are grafted. Best rootstock is Old Home, which may be grown for 3 to 4 years before being top-worked to the desired variety. This makes an expensive tree, but in quality it may pay for the added expense in the home grounds. Dwarf trees are occasionally available double-worked with Old Home or Hardy grafted to quince rootstock, and the desired variety to the intermediate. The purpose of such grafts is to limit the damage which can be done by fire blight, and to make a sturdier joint between the rootstock and top. Winter Nelis is good rootstock, producing strong and vigorous standard-size trees, but it is seldom available. Most often presented for sale are trees grafted to *P. serotina,* a Japanese pear. These trees are more blight-resistant than those grafted to French pear stocks, but their fruit is subject to an ailment called black end or hard end, which does not afflict that of other trees.

Yields which may be expected from a mature standard-size tree average about 4 to 5 bushels. A dwarf tree will produce ½ to 1½ bushels.

Site

Because they bloom early, and also because they should be protected from the hottest sun, pear trees do best on northern or eastern slopes. They should be planted where the air can move freely around them, to dry them quickly after a rain or heavy dew. Windbreaks, hedgerows or buildings which protect them from the breezes should be shunned. Fire blight is a fungus disease which is fostered by a hot, damp atmosphere. Con-

TABLE 80: Pear Varieties

Variety	Section for which recommended	Fruit characteristics	Tree characteristics	Blooming season
Anjou	Pacific coast Mountain States Great Lakes New England	Medium to large, smooth green skin with faint blush. White sweet flesh, fine flavor. Late. Stores well.	Large tree, sometimes self-sterile. Not a reliable producer.	Midseason.
Bartlett	Mountain States Pacific coast Northern plains Great Lakes New England Corn Belt	Large green to yellow, medium quality and flavor. Most popular commercial pear, many better for home garden. Medium early. Keeps.	Adaptable to many soils, climates, conditions. Bears early, heavily, regularly. Upright, medium to large tree. Moderately vigorous. Subject to fire blight. About hardy as peach.	Midseason. Will not cross-pollinate Seckel. Best pollinators are Anjou, Bosc, Beierschmitt, Ewart.
Beierschmitt	Corn Belt	Medium size, juicy, sweet, good flavor. Medium early.	More blight-resistant than Bartlett.	Midseason.
Bosc	New England Pacific coast Great Lakes Corn Belt	Long, tapering neck. Russet skin. Juicy, rich aroma, fine flavor. Late. Stores well.	Large vigorous tree needs 25 by 25 ft. Hard to propagate and train. Sometimes self-sterile. Late coming into bearing. Susceptible to fire blight. Tender to cold.	Late.
Carrick	Appalachians	Medium size, russet, good quality when fully ripe. Cream colored flesh, sweet, crisp, juicy. Late.	Large, vigorous, spreading tree. Productive. Fire blight-resistant.	Medium early.
Clapp Favorite	Pacific coast New England Great Lakes Corn Belt	Resembles Bartlett. Softens quickly in center after harvest. Early.	Productive, vigorous. Very hardy. Sometimes self-sterile. Susceptible to fire blight.	Late.
Comice	Pacific coast	Large, roundish, green-yellow to yellow, with delicate blush, russet spots. Best quality and flavor of all. Midseason.	Soil and climate must be just right. Bears irregularly. Subject to fire blight. Self-fertile.	Midseason.
Dana Hovey (Winter Seckel)	Pacific coast	Small, pale yellow and russet. Flesh sweet, juicy, aromatic. Keeps well.	Very good home garden variety.	

TABLE 80: Pear Varieties (Continued)

Variety	Section for which recommended	Fruit characteristics	Tree characteristics	Blooming season
Easter	Pacific coast	Large, round, green with russet patches. Very good dessert pear. Late. Keeps very well.	Sometimes self-fertile.	Early.
Ewart	Corn Belt	Medium to large, green-yellow to yellow. Flesh soft, juicy. Good dessert quality. Stores. Midseason.	More blight-resistant than Bartlett. Productive, but tend to be biennial.	Midseason.
Flemish Beauty	Northern plains New England Great Lakes	Large, yellow with red blush. Juicy, sweet, spicy, good quality. Should ripen on tree. Midseason.	Very hardy, vigorous, productive. Very susceptible to fire blight and pear scab; fruit sometimes cracks badly.	Early.
Hardy	Pacific coast	Large, slightly russet. Sweet, juicy, aromatic. Ripens just after Bartlett.	Sturdy, vigorous, heavy bearer. Self-fertile.	Late.
Kieffer	Appalachians Northern plains Southern plains New England Great Lakes Southeast Delta States	Low in quality. Fruit hard and woody unless properly mellowed, even then not good dessert quality. Must be picked before ripe, stored at 60 to 65° for 2 to 3 weeks.	Least susceptible to fire blight of all pears. Will grow in climates where few other pears succeed. Sometimes self-fertile.	Early.
Madeleine	Pacific coast	Small. Best very early pear, ripens 3 to 4 weeks before Bartlett. Pale green-yellow, flesh white, juicy, delicate.	Recommend for home gardens. Self-sterile.	
Maxine	New England Great Lakes Corn Belt	Large, bright yellow. Sweet mild flavor. Flesh coarse.	Productive. Vigorous. Blight-resistant.	Midseason.
Morgan	Appalachians	Medium to large, yellow, rose blush. Subacid, good texture, flavor. Midseason.	Large, upright, vigorous, fire blight-resistant. Productive under favorable conditions.	Very late.

TABLE 80: Pear Varieties (Continued)

Variety	Section for which recommended	Fruit characteristics	Tree characteristics	Blooming season
Seckel	Mountain States Corn Belt Appalachians Southern plains Pacific coast Northern plains New England Great Lakes	Small, flavor excellent. Very good for cooking. Late.	Very vigorous, productive. Immune to fire blight. Very hardy. Trees come into bearing late. Sometimes self-sterile. Very good for home gardens because seldom bothered by insects.	Midseason to late. Will not pollinate Bartlett.
Tyson	New England	Medium size, yellow, flesh granular around core. Sweet, juicy, good quality. Early.	Productive. Vigorous. Good for home orchard. Somewhat blight-resistant. Comes into bearing late.	Midseason.
Vermont Beauty	New England	Small. Excellent quality. Late.	Very hardy and productive.	
Winter Nelis	Pacific coast Mountain States New England Great Lakes	Medium size, roundish. Sometimes russet. Sweet, juicy, good flavor. Very late. Stores well.	Vigorous but hard to train. Somewhat resistant to fire blight. Self-sterile.	Late.

sequently, one of the best preventives is good air drainage of the site.

Planting

In areas where the winter temperatures go no lower than zero, pears may be planted in the fall. They are more safely planted in very early spring where the winters are colder than that. Whether planted in spring or fall, the trees should be entirely dormant when they are set out.

Standard size trees should be allowed a space 20 by 20 feet. Dwarf trees need only 12 to 15 feet.

If a one-year whip is planted, the top should be cut back to 3 to 3½ feet when it is set out. Roots which are broken or are too long and straggling may also be pruned back to a point where they are fairly uniform with the others.

Dig the hole large enough to contain all the roots without crowding or bending them. Set the tree at the height at which it grew in the nursery, unless it is a dwarf, in which case it may be set a little higher, to ensure that the top will not form roots. A standard top on dwarf roots may make roots of its own and, if it does, the tree will grow to standard size. Do not use nitrogen fertilizer when the young trees are planted. Manure or compost, or any other nitrogenous fertilizer, stimulates heavy vegetative growth. Under any other circumstances, this would be welcome, but it is to be strictly avoided on pear trees, because it invites fire blight. Limestone, rock phosphate or granite dust may be used to dig into the hole, but not nitrogenous stimulants.

Pruning

An unbranched one-year whip should be headed back to 3 or 3½ feet before the first season's growth. If the new tree is a branched two-year old, the scaffold limbs and central leader should be selected and the remaining branches removed. A long leader should be headed back, leaving no more than 18 to 24 inches of the previous year's growth.

Pruning for the first few years has one main purpose—to select 5 to 8 scaffold limbs spaced 4 to 12 inches apart and distributed uniformly around the trunk. Growth of the scaffolds should be balanced, with too-vigorous growth headed back. When it is 4 or 5 years old, the tree may begin to reach up too high. To maintain the mature trees at a maximum of 15 to 18 feet, a light heading back and thinning out of the higher limbs, especially those in the center of the tree, should be done at this time. If the upward growth is checked at the proper time, it will not be necessary later to make large cuts which stimulate heavy growth and increase the susceptibility to blight.

Heavy pruning at any time may cause the tree to grow too rapidly and increase its blight susceptibility. After the tree's main shape has been trained, little cutting should be necessary until after bearing starts. During the fourth year and after, remove all sucker growth but leave small twiggy growth along the branches. Fruiting spurs, which occur singly or in clusters, produce fruit for 7 or 8 years.

Trees should make some terminal growth each year. If this growth slows or stops entirely, the tree needs more drastic pruning to bring it back. If a few cuts back into 2- or 3-year wood are made, the tree may be stimulated to continue its growth. However, sometimes (especially in trees grafted to *P. serotina*) growth stops, the bark becomes thick and hard and the tree takes on a stunted look. When this occurs, scrape the bark thoroughly and deeply, almost to the live layer, and wash it down once each month with a kerosene emulsion. A year of such treatment should serve to stimulate the tree to new growth.

Old, neglected pear trees may be brought back to usefulness if a 3-year pruning program is instituted to control their growth. Cut back top limbs, one third each year, for 3 years. Little additional pruning should be done in these 3 years. After the tree has recovered from being topped, any necessary alterations of the lower limbs may be made, a little each year, until the tree has taken its desired shape.

Maintenance

Some varieties of pears require more fertilizer than others do. Seckel pears like a rich soil, but even though they are given more than the others, they should not be fed heavy manures. After they have grown in the garden for 2 or 3 years, they may be lightly mulched with strawy manure mixed with wood ashes. Most pears should be given only a mulch of a bale of straw per tree each year, unless they show definite signs of nitrogen deficiency, which shows itself in yellowing of the leaves, especially in the growing tips of the branches. When this happens, substitute strawy manure for the straw mulch in early spring.

Pears should be supplied with water throughout the year, when they are grown in arid areas, but their requirements are much more elastic than those of many fruit trees. They are equally able to withstand too much and too little water, but they do not produce their best crops unless a reasonable amount of moisture is maintained in the soil at all times.

Propagation

Some of the best pear rootstocks are difficult to propagate from seed. Winter Nelis, for example, is more easily started by layering than from seed. The Japanese rootstocks, which have many undesirable traits, are easier to grow from seed. More satisfactory trees are grown on stocks from imported French pear seeds.

Seed may be planted in the nursery in fall, or it may be stratified to soften the hard coat which surrounds the embryo. Seeds should be covered with one-half inch of loose soil, and a thin mulch spread over the row to help maintain uniform moisture.

When the seedlings are 3 inches tall, they should be moved to nursery rows where they are planted 3 feet apart. In the Northeast, these seedlings are permitted to grow until the next spring before being budded. In Pacific coast states, they may attain a size which can be budded by midsummer. Trunks should be about one-fourth inch thick for budding.

About a week before budding, remove all leaves and twigs within 4 to 6 inches of the ground. Select budwood from thrifty, large buds of the current year's growth on bearing trees. Union should take place in about 10 days. The following spring, the stock should be cut back to just above the bud, to force all the growth into the latter, forming the top of the budded variety. At the end of this growing season the young trees are known as one-year whips on 3-year roots, and are ready to be transplanted to their permanent places.

Enemies

So much has been said about fire blight that it must be quite apparent that this fungus is the pear's most deadly enemy. Very few trees are completely immune, and those that are usually are inferior in quality of fruit.

Fire blight attacks leaves, flowers, fruit, branches and trunks, making the portions infected blackish as if they had been scorched. On the trunk the infection

The Kieffer pear tree shown above is less susceptible to fire-blight of all pear varieties and will grow in climates where few other pears succeed.

looks like a canker. There is no known cure for the blight except surgery. Trees should be inspected for blight every 2 or 3 days from blooming to midsummer. When it is found, the infected portions should be cut out, using sterilized instruments. The cut should be made at least 6, and preferably 12 inches back toward the roots. All material removed should be burned.

Pears are not bothered by many other diseases or insects, but occasionally pear scab, pear slug, psylla, curculio or codling moth may attack them.

Pear scab appears as a velvety olive-green spot on the fruit, becoming black and scabby at maturity. On the leaves the scab makes black spots. The disease is favored by warm damp weather which also fosters blight. Remove any leaves or fruit infected with scab, and keep the area under the tree free of fallen leaves and fruit.

Pear slugs are larvae of a small fly.

They appear on the tops of the leaves—small dark-green slug-like creatures who eat the green portions of the leaves and leave the veins. They may be killed, when found, by dribbling limestone over them.

Psylla are insects whose honeydew becomes black and sooty, insects that attack the blossoms and prevent fruit setting. The best preventive measure is a thorough dormant oil spray in early spring.

For information about curculio, *see* CHERRY. Codling moth is described under APPLE.

Harvest

During the last few weeks before ripening, stony granules are formed through the flesh of many pears to spoil their smooth texture. If the pears are picked before these particles develop, their quality will be much better. Picking is done when the pears have reached their full size and the skins begin to change to a lighter green. Seeds will just be starting to turn brown at this stage, and the stems separate easily from the fruit spur when the pears are gently lifted.

When they are picked at this stage, many varieties may be stored for as long as 2 to 3 months, if the temperature of the storeroom is near freezing and the humidity is high. As they are wanted for the table, they may be brought into a room where the temperature is 60 to 65 degrees for ripening. If they are to be held for any length of time, the fruits should be separately wrapped in soft paper before storage.

PERSIMMON

PERSIMMON. Various species of the genus *Diospyros* of the ebony family are known as persimmons. The only one native to this country is *D. virginiana,* the hardiest of the genus. Since the Perry expedition to Japan, *D. kaki,* the so-called Japanese persimmon, which is really a native of China, has been grown in our southern states. *D. lotus,* also an Oriental import, is known as the date plum, and is of interest here only because it is frequently used as a rootstock, to which choice persimmon varieties are grafted.

All of the persimmons are handsome ornamental trees, growing to 40 feet or more, with glossy, dark green leaves which are shed in early autumn. Their flowers are inconspicuous, but their fruits are colorful through the fall and early winter. Wood from persimmon trees is very hard and fine and is prized for wood carving. Leaves are sometimes picked when young and tender, fermented and dried, and used like black tea.

Three types of blossoms are borne on the trees, sometimes all on one tree, sometimes on separate trees. Blossoms may be male, female or bisexual. Fruit will set without pollination on some trees, resulting in seedless fruit. Other trees will not set fruit unless pollination occurs and the fruits are more seedy as a result.

Fruit of the American persimmon is small, averaging about one inch in diameter. Japanese fruits are larger, tomato-shaped, up to 4 inches in diameter and 3 inches deep, and orange. Fruits of most varieties are very astringent when green, but become sweet when fully ripe. Though they may be frosted in the process of ripening, the frost is not an essential factor in sugar formation.

Japanese persimmons are an excellent source of vitamins A and C. (No analysis of the American species is available.) In one large persimmon weighing 150 grams there are 1,600 units of vitamin A, 40 milligrams vitamin C, 22 milligrams calcium and a total of 116 calories, mostly in the form of sugar.

Culture

Wild or seedling persimmons should not be planted in the home garden, because their fruits are often worthless. Grafted trees of named varieties should be purchased.

In the northern areas up to Rhode Island and across to the Great Lakes and Iowa, the American persimmon is hardy. Most recommended variety is Early Golden, followed by Penland, Killen, Garretson, William, Golden Supreme, Lena, Grayville and Owen. Japanese varieties are able to withstand winter temperatures down to 10 degrees, which confines them to an area south of the Ohio River and to the Pacific coast. Most frequently grown in this country among Japanese varieties are Tanenashi, which is self-fertile; Hachiya, Fuyu, Gailey and Fuyugaki, which must be cross-pollinated.

Persimmons must be purchased as very young trees, because they have a very long tap root which cannot be transplanted when the tree is older. The trees should be allowed a space 18 to 20 feet in diameter in soil that is well-drained. They will grow in any soil, but respond well to fertility. The young trees should be set about an inch deeper than they grew in the nursery, and should be mulched well to conserve the soil moisture. Fertilizer may be supplied by a mulch of strawy manure applied in the early spring.

The trees throw up a quantity of suckers which must be removed each year in order to prevent formation of a thicket. These young trees cannot usually be transplanted successfully to increase the garden stock. If transplanting is attempted, almost all, if not all of the top must be cut back, and a large section of root must be dug with each sucker. It is easier to increase a planting by starting seedlings which may be grafted in mid-May of the following year. Grafting cannot be done as early on persimmons as on many fruit trees, because they start their growth late in the season.

Persimmon trees require very little pruning. The only trimming that should be done is to remove broken branches—the wood is fairly brittle and breaks easily—and to open up the top occasionally.

Harvest

When they are thoroughly ripe, persimmons are very soft and cannot well be handled. They are usually picked while still a little firm, and are finished ripening in a warm room. When they are completely ripe they are very sweet. Japanese varieties contain so much sugar that they may be dried and eaten like figs.

American persimmons may be harvested just before they are ripe, also, or they may be left hanging on the tree into the winter months. Fruit left hanging through January, even though it is frozen on the tree, retains its flavor when picked and thawed. Or the ripe fruit may be blanched, peeled, wrapped individually and frozen for winter use.

PINEAPPLE

PINEAPPLE. *Ananas comosus* or *A. sativa,* the pineapple is a member of the bromelia family, known in the North principally for house plants with stiff rosettes of sword-like spiny leaves. In the tropics many of the bromelias are air plants, perching in trees and leading a parasitic existence. Pineapple, however, is strictly a terrestrial member of the family. A perennial herb native to South America, the pineapple was discovered with surprised delight by the early Spanish explorers. So impressed with its aromatic fruit were they, that they took samples of it around the globe with them and dropped off shoots to be planted in many of the countries which they touched. At present, pineapples are grown throughout the tropics. In the United States, they can be grown in central and southern Florida and in Hawaii. No other states have consistently warm enough climate for them.

A pineapple plant consists of a typcial bromelia rosette of stiff, sharp-pointed, prickle-edged leaves. From the center

TABLE 81: Pineapple Varieties

Variety	Season	Fruit characteristics	Comment
Red Spanish	May-June	Medium to large, coarse and juicy, white flesh, sweet and spicy.	Fruit inferior to others.
Natal Queen	June-July	Small, crisp, tender, very sweet and rich, with yellow flesh.	Produces many suckers. Should be replanted every 3 years.
Pernambuco	June-Aug.	Small to medium, very sweet, tender, juicy. White flesh.	Very good table quality.
Smooth Cayenne	July-Sept.	Large, sweet but tart, tender, juicy. Slightly yellow flesh.	Widely grown for canning.
Abakka	July-Sept.	Medium to large, sweet, tender, juicy. Yellow flesh.	Subject to black heart.

rises a stem, sometimes 4 feet high. Near its head it swells into the fruit, which consists of a spike of flowers which coalesce into one mass with the stem, and are surmounted by a tuft of leaves. The fruit portion, juicy and fleshy, is actually the fruit stem. In primitive varieties, seed may sometimes be found in the pineapple "eyes," the remnants in the mature fruit of the separate flowers.

Fresh pineapple is a good source of vitamin C, and contains a small amount of other vitamins as well. In a 100 gram portion (approximately 2/3 cup) there are about 30 units of vitamin A, .100 milligrams vitamin B_1, .025 milligrams vitamin B_2, 38 milligrams vitamin C, 8 milligrams calcium, 26 milligrams phosphorus, .2 milligrams iron and 57 calories, all in sugar. One cupful of juice, canned, contains about twice as much vitamin A, calcium and phosphorus, but the same amount of vitamins B and C.

Culture

Pineapple plants will live through exposure to temperatures as low as 28 degrees, but they do not grow at anything under 55. If they are exposed to 40 degrees or lower for any protracted period they suffer from an ailment called heart rot. On the other hand, they do not prosper above 95 degrees, either. Optimum temperature range is 70 to 90 degrees. Rainfall requirements are more flexible—anything between 20 to 80 inches of rainfall per year is satisfactory, especially when the soil is well-drained. Roots penetrate about a foot into the surface, and the top 12 to 18 inches must be well-drained. Pineapples require an acid soil, with *p*H between 5.5 and 6. Best soil is sandy, well supplied with organic matter.

Pineapple is propagated by cuttings known as planting pieces, which may be taken from 1 of 3 positions on the parent plant. The tuft of leaves growing out of the top of the fruit may be used as a planting piece and is quite acceptable for propagation. More desirable is a ratoon—a sucker which arises in the lower leaf axils which can be allowed to make larger growth before being removed and planted separately. Third choice for propagation is the slip or sucker which arises just below the fruit from the fruit itself or from the stalk. Although this slip makes an acceptable planting piece, it may give rise to offspring which make too many suckers from this position, crowding and disfiguring the fruit. Planting pieces may be of any size, but the larger they are, the sooner they will produce fruit.

In Hawaii, planting pieces are allowed to dry for about a week after they are taken from the plants before they are put in the ground. In Florida they are planted at once, because they are likely to decay if allowed to wait. Planting may be done any time from early spring to September, but the best time is in early spring, to produce a winter crop. Soil should be deeply prepared to remove all weeds.

A pineapple plant consists of a typical bromelia rosette of stiff, sharp-pointed, prickly leaves. A stem rises from the center which later develops into the fruit.

In Florida, pineapples are often planted in 4- or 5-row beds which are set between two drainage ditches. Plants are spaced 12 to 18 inches apart, with 18 to 21 inches between rows. Where drainage is not necessary, the beds are made 3 to 4 feet wide, with 3 foot aisles between.

Do not place the planting piece too deeply. The tip must be raised sufficiently so that sand cannot sift into the bud, which would injure or kill it. To prevent this, the tip may be filled with bulky material such as cottonseed meal, tobacco stems or castor pomace. If the bud should become clogged with sand, wash it out with a gentle stream of water.

During its first few months the pineapple plant has very few roots. At this time, the fertilizer, which may consist of any of the 3 substances mentioned above, should be applied very close to the stem or should be sprinkled in the lower leaf axils. A very good all-round fertilizer for pineapples is composted seaweed, which contains just about the right proportions of all of the necessary nutrients. As the plants grow larger, their roots will fill the spaces between them, and the fertilizer should be spread over the entire bed. Fertilizer application should be made monthly from early spring until harvest, when it should be omitted, but applications should be resumed when the harvest is over.

At least two crops of fruit should be obtained from a planting before it is dug and replanted. Only one fruit is produced from each stem. Subsequent crops are produced from suckers which develop from the original plant. It is not necessary to dig and replace the plants as long as fruit production is satisfactory. When a new planting is to be made, all old material should be turned under to provide organic material, and to reduce the danger of propagating the mealy bug population.

Mealy bug is the most important pineapple enemy. If the planting is small, watch for the first appearance of the bugs. They may be killed by applying alcohol directly to them, either with a small paint brush or with a cotton swab. Rubbing alcohol serves this purpose quite well. If they are caught early, before they have multiplied on the plants, they are not difficult to control.

Red spider sometimes attacks the young plants, weakening them. These insects, sometimes referred to as pineapple mites, may be controlled by dusting the plants with tobacco dust, which, when it washes into the soil, acts as a fertilizer.

Pineapple Plantation in Florida

Pineapples are grown organically on one of the largest pineapple farms in the United States outside Hawaii, Plantation Paradise, near Lake Placid, Florida, operated by Lucille and Jake Emminger.

It is not large as plantations go, being only 35 acres with 10 in pineapples, yet it is one of the largest in Florida and one of the few remaining commercial pineapple plantings in the state.

First came clearing of the virgin land which was done by hand-grubbing of brush and palmettos. Lucille worked with her husband on this job. "It was sure hard work," she recalls, "but I loved every minute of it."

Most of their fruit is sold to tourists who take it with them or have it shipped to "the folks back home."

The pineapple is a unique crop, requiring from 12 to 16 months to produce fruit. A single fruit grows on a sturdy stock that grows from a plant which is started by setting out a slip, sucker, ratoon or crown (top) of a mature fruit.

A slip is a shoot growing out from the lower part of the fruit. Suckers emerge from the plant stem between the ground and the fruit, while ratoons emerge from underground.

Planting is best done in April, although it can be done the year around. Emminger follows the latter practice in order to have fruit ripening all year. However, the best fruit with the largest yield is produced from June to November.

The planting stock is set in 4-foot beds with 3, 4 or 5 rows per bed depending on the variety.

Soil on the Emminger plantation is sand with some organic matter from the native growth on the land. This is supplemented with a mulch of tobacco stems applied in September or October at the rate of 2½ to 3 tons per acre. In the spring, castor pomace is applied by hand to the leaves of the plants at the rate of 700 to 800 pounds per acre. No other fertilizing is done.

Pineapples require well-drained soil and, although Plantation Paradise is equipped with an irrigation system, it is not often used. "If we go 5 or 6 weeks without rain, I turn on the water," Jake explained, "and that is usually during the winter."

The irrigation system also serves as a protection against frost. When the weather bureau issues a frost warning, Jake turns on the sprinklers and lets them run until the danger is past. During the freeze of December 12th and 13th, 1957, Emminger's plants were encrusted with ice for 39 hours.

Following that freeze Lucille developed

a recipe for pineapple preserves which enabled them to salvage some of the frozen fruit. In fact she made 3,000 one-pound jars of preserves from freeze-salvaged fruit. Today she makes preserves with off-grade fruit and sells it at their roadside fruit and souvenir shop. This shop, established in 1953 when the first plantings were made, was a lifesaver for the Emmingers during the months it took the plantation to recover from freeze damage.

Another innovation of Lucille's is the use of stunted fruits in flower arrangements. The little pineapples, only an inch or two in diameter, yet perfectly formed and on stiff stems, grow in the old beds which have begun to "run out." These little fruits have become popular items with garden club members from as far as Miami, 150 miles away. They will stand shipping as far as will the mature fruit.

About 150 varieties of pineapples are grown throughout the world, of which 12 are to be found at Plantation Paradise with 5 making up most of the plantings. Emminger's favorites are Natal Queen, from South Africa, and Smooth Cayenne, a major Hawaiian variety.

—Charles Stookey

PLUM

PLUM. The genus *Prunus,* which includes apricots, cherries and almonds, also embraces a number of species which are all known as plums. Though there is a certain similarity between all of the *Prunus* group, principally in the appearance and stones of the fruits, there is as much difference between the culture of some of the different kinds of plums as there is between the culture of cherries and apricots, and about as much similarity.

Because of the wide range of plum types, there are plums to grow in almost every state in the Union. Of greatest interest, because of their quality, are the European plums, *P. domestica.* These grow on rather large trees and are better suited to the more temperate zones. They accustom themselves better to heavy than to light soils. They bloom rather late, so are less subject to damage from late frosts than some of the others. Included in this group are the prune plums. A prune is any plum which contains enough sugar to preserve it without fermentation while it is drying when its pit is not removed. (When pits are removed and plums subsequently dried, they are called dried plums, not prunes.)

Japanese plums are the second most important group, placed first by some gardeners because of their many virtues. Some varieties of the Japanese plum, *P. salicina,* and the hybrid varieties which have been developed by crossing it with European, are of very high quality. Some are very large, but somewhat lacking in flavor. The trees have a wider soil-tolerance than European—they are at home in sandy or clayey loams. They are immune to curculio and a few less important pests, but are quite susceptible to brown rot. They bloom early and many times their blossoms are killed by frosts unless they are particularly fortunately situated. Japanese plums are almost entirely dessert fruits, or they may be canned.

Damson plums, *P. institia,* are also from Europe. They are more hardy than Europeans and more easily adapted to sand. Their fruits are used almost entirely for jams and jellies, being too tart for anything else.

Half a dozen native American species complete the plum roster. These are plums which are usually hardier and more drought-resistant than many of the more highly developed varieties, and are adapted for growing both north and south of the usual plum areas. *P. subcordata* and *P. nigra* are very hardy, even in the coldest areas. Their fruits are small and tart, though some varieties may be used for dessert. They are usually not grown where better quality plums are hardy. *P. maritima* is the Eastern beach plum, which grows wild in the vicinity of Cape Cod, but may also be cultivated in any sandy garden. Its fruit makes some of the best plum jelly. *P. pumila* is the sand cherry which has been hybridized to produce hardy bush-like trees with fruit which is less tart than most native plums. Some of the sand cherry group are closer to cherries than to plums. Their trees are bush-like in growth, are drought-resistant and are short-lived.

Like peaches, pears and apples, plums need more or less winter chilling. Most trees are hardy in any section where apples may be grown. They are also able to stand hot summers and drought pretty well, some varieties more than others.

The European species root very deeply and must have a deep, fertile soil. Japanese varieties are more shallow-rooted and their roots spread more widely. But, for both types, the soil must be well-drained and hardpan should be well below the surface. They respond well to a large amount of moisture-retentive humus in the soil, whether it is sandy or clay, and do best with a *p*H of 6. to 8.

Plums contain small amounts of A and B vitamins. Prunes are much richer in vitamin A. Neither contains much vitamin C. In 3 medium size plums, there are 130 units of vitamin A, .120 milligrams vitamin B_1, .056 milligrams vitamin B_2, 5 milligrams vitamin C, 20 milligrams calcium, 27 milligrams phosphorus, .5 milligrams iron and 80 calories, mostly sugar. Six medium size prunes, weighing half as much, contain 1,500 units of vitamin A, .075 milligrams vitamin B_1, .325 milligrams vitamin B_2, 4 milligrams vitamin C, 27 milligrams calcium, 57 milligrams phosphorus, 1.5 milligrams iron and 173 calories.

Ordering Stock

The number of plum varieties which are self-fertile is very few. Most trees, even when they are self-fertile, bear better when a second variety is planted nearby, for cross-pollination. This is more true of the Japanese than the European varieties. In addition to being self-sterile, many of the Japanese plums have very weak pollen which is not good for pollination of other trees. A garden which has half a dozen plum trees, all different, will have better crops than one where one or two kinds are planted.

In order to cross-pollinate successfully, the trees must bloom almost simultaneously. However, European plums which bloom early may be used to pollinate some of the later-blooming Japanese varieties, if the spread between their blooming periods is not too great. When choosing varieties for cross-pollination, it is important, too, not to choose biennial bearers which may produce blossoms only in alternate years.

Best growth may be expected from one-year trees of medium size, that is, about 4 to 6 feet tall, with trunks not less than one-half inch in diameter. This means that the top has grown one full year in the nursery after the top of the rootstock was removed. The roots (all plums are grafted) will be 2 to 3 years old.

Rootstocks are usually myrobalan plum, peach, almond or occasionally apricot. Myrobalan roots make hardy and long-lived trees which are not rampant growers, but are resistant to nematodes. They like a deep, fertile soil and tolerate clay. Peach stock is used in the early-ripening areas, and thrives better than the others in poor, shallow or light soil. Its moisture requirements are correspondingly high. Almond needs a deep, fertile and well-drained soil, but is most drought-resistant of the 3. The trees on almond stock grow less rapidly than those on peach, but are longer-lived. Apricot stock is not recommended, because it does not make a satisfactory union with plum.

Planting

Plums may be planted in fall where winters are comparatively mild. The trees are stronger where fall planting is possible, and will go into their first growing season in the garden more ready for whatever setbacks may come, if their roots have had a chance to become established before leaves, demanding moisture, open in spring. Where winters are very cold—usually where they are below zero—spring planting is advisable. Trees should be completely dormant when planted, and the operation should be carried out very early, before buds begin to swell.

Where plums are to be planted in a home lawn, a circle at least 6 feet in diameter should be cleared of all grass and other roots, and the soil should be stirred to a depth of 10 to 12 inches. A good supply of phosphorus and potash may be dug in at the same time. If these take the form of phosphate rock and granite dust, they will slowly become available to the young tree over the course of several years.

Plum trees need a space 18 to 24 feet in diameter after the trees are fully mature. When planting Japanese plums, try to give them a situation where the spring growth will not start too early, so that their blossoms will be delayed until the last of the spring frosts are past. Damsons and native species take less room than the large and spreading European and Japanese types.

Make the hole large enough to contain the roots without cramping. Cut off any damaged roots, and cut back slightly any

TABLE 82: Plum Varieties

JAPANESE AND JAPANESE HYBRIDS

Variety	Section for which recommended	Fruit characteristics	Tree characteristics
Burbank	Corn Belt New England	Medium size, purple to dark red, clingstone. Flesh firm, juicy, sweet. Skin thin, tough. Fair quality. Midseason.	Low, wide, spreading. So productive it must be thinned severely. Self-sterile and will not pollinate Shiro. Hardy. Somewhat resistant to brown rot.
Formosa	New England Pacific coast	Large yellow-green, cling to semifreestone, sweet, excellent dessert plum. Early.	Large, vigorous. Biennial bearer. Hardy. Comes into bearing late. Self-sterile. Blooms so early that blossoms sometimes frosted. Not a good pollinator.
Santa Rosa	Southern plains Southeast Delta States Pacific coast Mountain States New England	Large, dark red clingstone. Skin medium thick. Flesh dark red. Good keeper. Midseason.	Large, upright but compact. Vigorous. Productive. Light crops in alternate years. Needs cross-pollination.
Shiro	Delta States Southeast Appalachians New England	Medium size, light yellow, thin, tough skin. Very juicy, sweet, mild flavor.	Large, vigorous, upright-spreading, productive, hardy.
Beauty	Pacific States New England	Medium size red clingstone. Sweet, good flavor. Early.	Vigorous. Open-spreading habit. Somewhat tender. Biennial bearer. Fruit needs thinning. Self-fertile.
Abundance	New England Great Lakes	Medium size red clingstone. Thin, tough skin. Good flavor. Fair quality. Early.	Vigorous, vase-shaped, too productive (fruit must be thinned.) Hardy. Self-sterile.
Duarte	Pacific coast Mountain States	Large, red, dark red flesh. Early.	Self-sterile. Productive when cross-pollinated.
Satsuma	Pacific States	Medium to large, red, purplish-red flesh. Good flavor. High dessert quality.	Early blooming. Vigorous, but light producer until mature. Not hardy in North. Self-sterile.
America	Southern plains New England	Medium size golden yellow, firm yellow flesh, fair quality.	Bears very early.
La Crescent	New England Northern plains	Medium size, yellow, thin skin, sweet. Very high quality. Midseason.	Young trees somewhat tender. Older trees hardy. Erratic bearer.
Monitor	Corn Belt Mountain States Northern plains	Very large, bronze-red, clingstone. Tart in skin and around pit.	Pollinates all except Underwood. Medium size tree, moderately hardy, fairly productive.
Waneta	Northern plains New England	Very large, deep red, firm, clingstone. Fair quality. Midseason.	Prolific. Moderately hardy. Sometimes tends to overbear.
Kaga Hanska	Northern plains Southern plains	Similar varieties, both firm, fine flavor from a common apricot-plum parent. Midseason.	Suffer from drought. Only moderately hardy.
Elliot	Northern plains New England	Red, firm, good for cooking. Good keeper. Midseason to late.	Hardy. Good yields

TABLE 82: Plum Varieties (Continued)

JAPANESE AND JAPANESE HYBRIDS (Continued)

Variety	Section for which recommended	Fruit characteristics	Tree characteristics
Omaha	Mountain States	Large, mottled red and yellow. Very sweet and juicy. Late.	Very hardy, productive.
Superior	Corn Belt Northern plains New England Great Lakes	Large, red, firm. Good quality. Early.	Vigorous. Fairly heavy bearer.
Tecumseh	New England Great Lakes Northern plains	Large, red, thin skin, firm, sweet, juicy. Early.	Hardy and prolific.
Radisson	New England Northern plains	Large, red, sweet, firm, juicy, excellent quality. Early.	Moderately hardy but early bloom sometimes killed by frost. Not heavy yielder.
Pembina	New England Northern plains	Large. Red. Good quality. Early.	Moderately hardy but blooms early. Thrives best in North.
Underwood	Mountain States Corn Belt Northern plains New England	Large, red, good quality. Midseason.	Hardy. Not a good pollinator for Monitor.
Ember	New England Northern plains	Large, mottled red and yellow. Firm flesh, good quality. Midseason.	Moderately hardy and productive.
Fiebing	New England Great Lakes	Large, good quality. Early to midseason.	Very hardy. Spreading.
Methley	Hawaii (5,000 ft. elev.) Southeast Delta States Southern plains Corn Belt Appalachians	Small to medium, dark red, red flesh, clingstone, sweet, good flavor but bitter skin. Good dessert but turns bitter when cooked or frozen.	Tender. Self-fertile.
		EUROPEAN PLUMS	
Yellow Gage	New England	Golden yellow with blush. Excellent quality. Tender skin. Freestone. Early.	
Green Gage (Reine Claude)	Southern Plains New England Corn Belt	Yellow-green to golden yellow. Semi-clingstone. Excellent quality but tends to crack. Late.	Trees short-lived, large, not too hardy. Self-sterile.
Jefferson	New England Pacific coast	Greenish to bronze-yellow. Medium size, semi-clingstone. Best quality. Good for canning.	Vigorous, productive. Late coming into bearing. Self-sterile.
Italian Prune (Fellenburg)	Mountain States New England North Pacific coast	Medium size, purple-black, freestone. Good keeper. Hangs well. Midseason to late.	Vigorous, hardy, productive under favorable conditions. Spreading to upright. Susceptible to pests and sensitive to drought. Self-sterile.
De Montfort	New England	Medium size dark purple freestone, tough, thick skin, subacid.	Vigorous, spreading, open. Heavy annual bearer. Self-sterile.

TABLE 82: Plum Varieties (Continued)

EUROPEAN PLUMS (Continued)

Variety	Section for which recommended	Fruit characteristics	Tree characteristics
Stanley	Mountain States Corn Belt Southeast Northern plains New England	Large, dark blue with heavy bloom, freestone, sweet, moderately juicy, tender, good quality. Midseason.	Early and annual bearer. Productive, hardy, good pollinator, vigorous, upright-spreading, resistant to low temperatures and to frost while in bloom. Self-fertile.
Bradshaw	New England	Large, blue-purple. Good quality. Early.	Self-sterile.
Mt. Royal	Northern plains New England Great Lakes	Medium size, blue, juicy, mild, excellent quality. Good for canning.	Medium vigor. Hardy. Annual bearer.
French Prune (Prune d'Agen)	California	Medium size, purple skin, yellow flesh. Very rich, sweet. Tough skin. Late.	Strong, vigorous, regular crops. Self-fertile. Needs deep, fertile soil for large fruit.
Imperial (Imperial Epineuse)	Pacific coast	Large, red-purple, yellow-green flesh. Very good flavor. Midseason.	Self-sterile. Vigorous, but unless well-pollinated, a shy bearer. Requires deep, fertile soil.
Sugar	Pacific coast	Large, purple-red with yellow flesh, large pit. Tender skin. Sweet. Early.	Tends to be biennial. Wood brittle. Blooms late, is self-fertile. Prolific—needs thinning. Open, spreading.
Grand Duke	Pacific coast	Large, firm, high sugar content but not very good for dessert. Late.	Requires cross-pollination.
Tragedy	Pacific coast	Medium size. Very good dessert plum. Early.	Self-sterile.
Shropshire Damson	Appalachians Southern plains Corn Belt New England	Purple-black with heavy bloom. Clingstone. Skin thin and tender. Tart, good for jam or jelly, not dessert.	Large, vigorous, hardy, vase-shaped tree. Bears heavy annual crops. Self-fertile.
French Damson	New England	Black, larger than Shropshire. Freestone some seasons.	Hardy. Productive. Self-fertile.
		NATIVE PLUMS	
Assiniboin	Northern plains	Large, red, sweet, juicy. Early.	Hardy, but not too vigorous or productive.
Wolf	Northern plains	Small to medium, freestone. Early. Good quality.	Hardy, productive.
Terry	Northern plains Northern New England	Small to medium. Yellow blotched with red. Semi-freestone. Fair quality. Midseason.	Very hardy, prolific. Good pollinator.
Mandan 48	Northern plains	Large, red, juicy. Sweet. Good quality. Midseason.	Hardy. Productive. Strong framework.
Teton	Northern plains	Medium to large, red, sweet, clingstone. Fair dessert quality. Late.	Very hardy, thrifty, prolific. Good pollinator. Wood tends to split.

TABLE 82: Plum Varieties (Continued)

SAND-CHERRY HYBRIDS

Variety	Section for which recommended	Fruit characteristics	Tree characteristics
Opata	Mountain States Southern plains Northern plains New England	Small to medium, deep blue, green flesh, small pit, clingstone, fair for dessert, good for cooking. Very early.	Drought-resistant, fairly hardy, Very prolific. Tends to overbear. Susceptible to brown rot.
Sapa	Mountain States Southern plains Northern plains New England	Small to medium, deep blue, dark red to black flesh, very good quality, very good for cooking. Early to midseason.	Not as hardy as Opata and not as prolific, but much better fruit. Susceptible to brown rot.
Oka	Northern plains New England	Small to medium, dark blue, red flesh, sweet, juicy, small pit, good quality, early.	Only moderately hardy. Susceptible to brown rot.
Compass	Southern plains Northern plains New England	Small, red and yellow, clingstone, fair for dessert, good for cooking. Midseason.	Very hardy, drought-resistant. Heavy annual bearer. Good pollinator for above three varieties.

that are overlong. Set the tree at the depth at which it grew in the nursery, or slightly deeper. Firm the soil well as it is filled in around the roots and water it well before adding the last 2 or 3 inches. As soon as the young tree is planted, it is well to give it some protection from sunburn, which may injure its bark. Either paint the trunk with white paint or wrap it loosely in several thicknesses of roofing paper.

Maintenance

Best fertilizer for plum trees is stable manure which can be used in the strawy state as a mulch as soon as the trees are planted. Spread the manure in a layer 3 to 4 inches deep over the entire area which was dug for the tree. As the tree grows, the root area will increase so that feeding roots are at least as far from the trunk as the drip line of the branches. Manure should be spread to this point early each spring, although if the program is a continuous one, the date at which the manure is applied does not make very much difference, so long as there is always rotting manure available over the roots. A circle one to two feet in diameter around the base of the tree should be left bare at all times, in order not to encourage rodents to build their nests where they may nibble on the bark.

If manure is not available, the mulch may be composed of straw, hay, sawdust or ground corncobs. With these materials, a nitrogenous fertilizer should be applied early each spring, at the rate of two ounces of pure nitrogen for each year of the young tree's age. When the tree reaches bearing age, at 4 to 5 years, the fertilizer allotment should be stabilized at 8 to 12 ounces of nitrogen per year per tree, depending on size. Two ounces of pure nitrogen are contained in the following amounts of organic fertilizing materials: one pound two ounces of fish meal; one pound 8 ounces of tankage; one pound of dried blood; two pounds of cottonseed meal.

When trees arrive at the full bearing age, they are often too heavily loaded with fruit for the strength of their somewhat brittle wood. Even when fruit is thinned to lighten the load, they sometimes need added strength. This may be supplied by a system of props and braces, which are applied before the crop becomes too heavy.

Where the tree is grown with a central leader, a heavy ring may be attached high on the leader, and wires strung from it to the smaller branches that need support. Large branches opposite each other through the tree's axis may be directly wired together, to counterbalance each other. Some crops are so heavy that even this support is not enough to prevent breakage. In the heavy years, props of one-by-two inch wood, forked at the top end, may be set under the heavily laden branches to take part of the load.

Pruning

Tops of newly planted trees should be cut back severely to balance the tops with roots which have been curtailed by digging. If the tree is a Japanese with upright growth habit, it should be shortened to stand no taller than 30 inches after planting. A tree with more spreading habit may be permitted to stand 36 to 42 inches tall. Make the cut at the top of the whip just above a budding lateral.

If the young tree has already developed some lateral branches, 3 may be selected to form the scaffold, and all others completely removed. The 3 laterals which remain should be cut back to a length of 6 inches each. Laterals should be chosen which make good right angle crotches from the trunk, start no lower than 18 inches from the ground, and are spaced no closer than 6 inches apart up the trunk. Also, if possible, they should point in 3 different compass directions from the trunk. If there are no laterals on the tree when it is planted, these may be selected during the tree's first summer in the garden. At that time, shoots which are not to become a part of the tree's mature scaffold may be rubbed off before they become large enough to bear more than two leaves. By constant attention to the new tree, the gardener may save it from making much unnecessary growth, which would be removed at the first dormant pruning, and he may encourage faster growth of the main framework of the tree.

After the first year, pruning of plum trees is confined mainly to encouraging the main scaffold limbs to throw out useful laterals, to keeping the head of the tree open, so that sun will reach lower branches and their fruiting spurs, and to removing dead or damaged wood. On trees of upright habit, the most upright-growing branches should be headed back when they approach the limits of ladderwork. Trees inclined to a more spreading form, on the other hand, should have their horizontal branches headed back to upright laterals when they threaten to become too long.

After the trees come into bearing, pruning should be limited to removal of dense top wood, damaged wood and crossing branches. If there is little of this to remove, some of the more rampant new growth should be cut out each year to encourage production of new fruiting spurs and to stimulate some new growth during the coming year. A tree should never be allowed to come to a standstill in its growth—it should make a little new growth each year, but not too much. This is especially important in the Japanese varieties whose fruit is grown on new wood. With plum trees, it is better to prune too little than too much. The trees are very easily thrown out of balance, and may be forced to make undesirably dense tops with too much pruning.

Many varieties of plums tend to overbear when the season is favorable. Not only does the size and quality of the fruit suffer when too much is allowed to mature, but the weight of so much fruit may break branches, or even split the trunk of the tree. Moreover, where the fruits touch along the bough they make snug nests for brown-rot fungus, which spreads from fruit to fruit right along the branch. To avoid this, when the crop is still heavy after the normal June drop, thin the fruit to stand far enough apart so that it will not touch when it matures. Fruits should be permitted to remain no closer than 1 to 3 inches apart when thinned.

Propagation

Plum trees are usually purchased, because appropriate rootstocks must be grown from seed, which is not easy to find. However, for the gardener who wishes to try his hand at grafting, the procedure is much the same as for many other grafted trees.

Rootstock characteristics have already been described in the paragraphs concerning stock. Rootstock seedlings are grown for one full year in the nursery. In the summer or fall of their second year they are budded with the desired variety of top. Early in the following spring, the rootstock top is cut back to a short stub, and the bud sprout encouraged to make a strong top growth. That fall or early in the following spring, the young tree is ready to transplant to its permanent home.

Enemies

Plum trees are all susceptible to heart rot which is the result of careless pruning practice. When pruning a plum, be very careful to cut all limbs as short as possible. Stubs which are too long to heal over permit the entry of bacteria to the heart wood, and heart rot results.

Most of the other plum enemies have already been covered under other fruits. For San Jose scale and dormant oil spray, *see* APPLE. Brown rot is described under APRICOT. For curculio, *see* CHERRY. Peach tree borers and their prevention are to be found under PEACH. The only other important deterrent to a good plum crop may be the bird population, which flocks around the plum trees almost as enthusiastically as around the cherries. Our recommendation, if you have the room, is to plant enough mulberries close by to distract the birds. Because one thing that they love more than plums, and even more than cherries, is a treeful of ripe mulberries.

Harvest

For canning or jelly-making, plums may be picked as soon as they have developed their bloom. At this point they are slightly soft, but still retain some of their tartness and firmness. For dessert, they should be allowed to become fully ripe on the tree. At this time many of them will be very soft, so that they cannot be handled excessively, but then, they have their most delicious flavor and are full of natural sweetness.

Prune varieties are usually distinguished by the fact that they will hang on the tree long after they are fully ripe, and develop more sugar as they hang. Home prune drying is usually practical only where the climate is very dry at harvest time. Where early fall rains or heavy dews may be expected, the crop may be damaged or completely ruined before it is dry.

In commercial orchards, prune trays are placed beneath the trees when the prunes are ready to fall, and the trees are shaken. The prunes drop into the trays, which are then moved to the drying yard and are kept in the sun. Fruit must be raised off the ground, and the trays should be made with wire-mesh bottoms, to allow full circulation of air. Fruit is stirred regularly to expose all sides to the sun. In the home garden, small amounts of fruit may be protected from dew or rain by moving it indoors, or covering it with tarpaulins.

In northern climates and those with wet fall weather, prunes may be dried in a specially constructed drier, or they may even be done in an ordinary oven if it can be closely regulated. After removing the fruit from the tree it should be spread in drying trays which will fit the oven. Fruit should be no more than one layer deep in each tray, and a little space must be left between the fruits to permit the air to circulate. It is then placed in the oven, which has been set for 110 to 150 degrees. The fruit must be stirred frequently. Total oven drying time is usually between 4 and 6 hours. If the plums are soaked in hot water for 20 minutes before oven drying, their skins are made more permeable and they dry more quickly. However, some loss of nutrients takes place during the soaking. Drying is complete when the fruit feels tough and leathery, but is still soft inside. If it is stored for 2 or 3 weeks in an airtight container, a more even distribution of the moisture takes place, and the outside becomes less tough. At the end of that period, it may be frozen, if desired, or stored in a cool, dry cupboard.

PLUMCOT

PLUMCOT. Very few gardeners have taken up the hybrid plum crossed with apricot which was originated by Burbank about 1901. The fruit has a fuzzy skin like an apricot, but its flesh is dark red and more acid. Recommended for Pacific coast gardens are Rutland, the original Burbank hybrid, Sharpe and Stanford. Other varieties are Apex, Corona, Silver and Triumph. Some of these are hardy through the apricot section; that is, the more moderate climates from the Appalachians north to the southern Great Lakes and southern New England. Culture is similar to that of APRICOTS, which see.

POMEGRANATE

POMEGRANATE. A unique fruit, the pomegranate is one of the two members of the *Punicaceae* family, the other member being a very unimportant fruit with similar characteristics. The *Punica granatum,* or pomegranate, was thought, by the ancient Greeks and Persians, to have some mystic connection with procreation because of its numerous and prominent seeds.

The pomegranate grows on a small tree or shrub, 10 to 20 feet tall, and is tender to winter cold. It can be grown outdoors along the eastern coast as far north as Maryland, in Florida and the Delta States, in the warmest parts of the southern plains and Southwest, on the Pacific coast, and in hot interior California valleys. Grown where the winters are cold in contrast to summer heat, the tree is deciduous. On the California coast, where the summers and winters are both mild, it is sometimes evergreen. In its dormant, leafless state it can stand a considerable amount of cold, but while budding or in leaf it is very tender.

Pomegranate plants may be trained to tree-shape, though this requires frequent attention to removal of suckers. Where it is grown for fruit, a tree-form is more desirable. When they are grown for decoration, they may be permitted to make a more shrubby growth. Their glossy, green leaves make them desirable ornamentals, especially when they blossom in spring, and are covered with large orange hibiscus-like flowers, or when they bear their dark red fruits in fall. In the more humid regions, as in Florida, both foliage and fruit may be covered with an unsightly fungus, making them less useful in the shrub border.

The only variety recommended for fruit in all of the sections is Wonderful, which has dark red fruit, red juicy flesh, and becomes quite sweet when thoroughly ripe. Several double varieties are offered for decorative purposes, but they do not bear fruit. Best quality fruit is produced where the summers are very hot and dry, but where the trees may be irrigated throughout the summer and fall.

Trees are propagated by hardwood cuttings taken during the dormant period. The cuttings should be 10 to 12 inches long. They are planted in nursery rows with only the top 3 inches above ground. They may be transplanted to their permanent position in spring after they have rooted, or they may be grown for a year in the nursery before transplanting.

Trees in the garden require a little more water than citrus fruits. If the water supply is uneven, fruit tends to split before it ripens in fall. They are also nitrogen-consumers, needing about one-half pound of pure nitrogen per year per tree. This may be supplied in the form of a thick mulch of good dairy manure, or it may be concentrated in 5 pounds of fish meal, 6 pounds of tankage, 4 pounds of dried blood or 7 pounds of cottonseed meal. If supplied in manure, the mulch should be renewed early each spring. The more concentrated feedings may be made in two installments, one in winter and another in spring.

Pruning should be confined to the removal of suckers, removal of dead wood and cutting back interfering branches each winter. This light annual pruning is necessary to maintain production of large, good quality fruit.

Fruit is picked in fall after it has changed color, and will ripen as it is held in cold storage. Or it may be permitted to ripen on the tree, provided it does not split.

POMELO

POMELO. *See* GRAPEFRUIT.

PUMMELO

PUMMELO. *See* SHADDOCK.

QUINCE

QUINCE. *Cydonia oblonga,* formerly classified with apples and pears as *Pyrus cydonia,* quinces are natives of Asia and have been known to the Western world since the time of the Greeks. Until many of our other fruits were so greatly improved within the last hundred years, quinces were widely grown as a great garden delicacy. Their popularity waned as other fruits were developed in more desirable table varieties, but they are still grown by those who esteem their distinctive and delicate flavor.

Quinces can be grown only in temperate climates, where they have about the same requirements as peaches. They grow on shallow-rooted large shrubs or small shrubby trees. The fruit is something like a lumpy, woolly pear, colored greenish yellow until fully ripe in late fall, when the skin becomes yellow and the flesh creamy to apricot-colored. Texture of the flesh is very woody until ripe and cooked, when it becomes much like the more firm pears. It contains a large amount of pectin, and is prized for making delicately-scented jelly. It is a fair source of calcium and vitamins A and C.

Culture

Quince trees may be propagated by layering, division, root grafting or budding, but the easiest method is by making cuttings of the current year's growth from the numerous suckers that are thrown up from the roots. Cuttings 8 inches long should be buried to two-thirds of their length in a nursery row in late summer. After a year many of them will have rooted and will be ready to plant in their permanent positions. Quince bushes or trees are slow-growing, and do not come into bearing for at least 4 years after they are set out. Heavy crops should not be expected until they are 8 or 9 years old, but they will then continue to bear for 25 years or more.

Quinces grow best on heavy clay soil that is well-drained. They do not need large amounts of nitrogenous fertilizers, but should be given ample applications of rock phosphate and potash. Their shallow roots appreciate a thick straw mulch which preserves the soil moisture.

Young plants may be pruned to tree-shape if desired, or they may be permitted to sucker freely and form bushes. In the tree form it is easier to spot and control borers; a bush provides plenty of extra branches to continue growth if fire blight kills off one or two. Unless the quince is to be grown as a tree, it needs little pruning. Fruits are borne on new growth of the current year, so a small amount of annual thinning helps to stimulate new shoots. If it is grown as a bush, some of the less robust branches may be removed from ground level each year.

Enemies of the quince are almost identical with those of the PEAR, which see. In addition, it is attacked by the quince curculio, which appears in midsummer. Treatment for curculio is described under CHERRY.

Harvest

Quinces hang well on the bush, and may be permitted to remain until after the first fall frost. If they are to be stored, they may be picked a few weeks earlier, packed in shallow baskets and stored in a cool, moist cellar. If the fruit is gently handled, it will keep for 2 to 3 months in storage.

RASPBERRY

RASPBERRY. A number of cultivated species of the genus *Rubus* are grown in the northern two-thirds of the United States, many of them more hardy in cold sections than any other fruits.

Most prized for their flavor and quality are the red raspberries, developed from *R. strigosus,* the American, and *R. idaeus,* the European red raspberries. Yellow and white varieties, which are mutations of the reds, are sometimes grown as novelties, though their quality does not compare to the true red varieties. The red raspberries are distinguished by their stiff, erect canes which, in rampant growers, may attain a height of 8 to 10 feet. They propagate by means of suckers which arise from the roots.

Black raspberries, *R. occidentalis,* are native Americans, a little less hardy than the red varieties. Their canes are graceful and curved, touching the ground with their tips before fall comes. As the days begin to shorten the tips of black raspberry canes thicken and then send out roots, which in another season will become a new bush.

A cross between the red and black raspberry varieties has produced *R. neglectus,* a purple species, with fruit which is less attractive in appearance and more tart, but makes excellent juice or jam. Some of the purple varieties are propagated by suckers, but most form plants from cane tips like the black varieties.

R. parvifolius, a trailing blackcap, is sometimes planted in the western states to halt soil erosion. The trailing species is not cultivated, but is used in its wild state for its ability to cover the ground quickly, and to form a dense mat of roots which hold the soil and vines which provide wild life shelter. Fruit is small, seedy, but with a good flavor when prepared for jams or jellies with most of the seeds removed.

Raspberries may be found to plant in

TABLE 83: Raspberry Varieties

RED RASPBERRIES

Variety	Section for which recommended	Berry characteristics	Plant characteristics
Latham	Appalachians Corn Belt Pacific coast Alaska Northern plains New England Great Lakes	Large, dark red, mild flavor, medium firm, fair quality. Late.	Hardy, vigorous, productive. Mosaic-resistant. Tall, slender canes. Suckers freely. Reliable yields.
Canby	Pacific coast	Medium to large, bright red, mildly acid, soft to medium firm. Midseason.	Thornless. Productive in loose, deep, well-drained soil. Hardy. Canes large, straight.
Cuthbert	Pacific coast	Dark red, medium firm, medium size, intense flavor.	Light yields. Tender to cold.
Newburgh	Pacific coast Corn Belt New England Great Lakes	Light red, very mild, large, medium firm, crumbles if not fully ripe. Fair quality. Midseason.	Productive. Not hardy on heavy soils. Resistant to root rot and "crumbling."
Sumner	Pacific coast	Medium size, medium dark red, firm, sweet, intense flavor. Midseason.	Adapted to heavy soils, even when not well-drained. Hardy. Productive. Resists root rot, mildew, rust. Medium size straight canes.
Taylor	Pacific coast Appalachians New England	Medium size, medium firm, mild flavor, bright red, good quality, late.	Not consistent yielder, but vigorous. Susceptible to mosaic. Large, strong canes.
Washington	Alaska Pacific coast	Small to medium, dark red, intense flavor, soft. Late.	Vigorous in light soil. Hardy. Susceptible to red rust, anthracnose.
Willamette	Pacific coast New England Great Lakes	Large, firm, dark red, acid, mild flavor, midseason.	Will adapt to any well-drained soils. Productive. Hardy. Medium to large canes. Subject to root rot.
Ranere (St. Regis)	Pacific coast New England	Small to medium, dark red, soft, good quality.	Fall-bearing.
Indian Summer	Corn Belt New England Alaska	Good quality but crumbly. Dark red, medium size, late summer.	Vigorous, hardy. Fall-bearing.
September	Appalachians Corn Belt Alaska Northern plains New England	Medium size, round, firm, bright red, fair to good quality. Midfall.	Hardy, reliable fall crops.
Viking	New England	Bright red, large, firm. Good quality. Midseason.	Resistant to spur-blight. Large, vigorous. Subject to mosaic.
Chief	Northern plains Alaska New England Corn Belt Pacific coast	Small, bright red, good quality. Early.	Very hardy. Resistant to mosaic. Erect, vigorous canes. Productive.

TABLE 83: Raspberry Varieties (Continued)

RED RASPBERRIES (Continued)

Variety	Section for which recommended	Berry characteristics	Plant characteristics
Durham	Northern plains New England	Firm, good quality, bright red, medium size. Midfall.	For fall crop, cut canes back to ground in spring. Old canes not reliably winter-hardy.
Sunrise	Alaska Appalachians Corn Belt New England	Bright red, medium size, medium firm, good quality. Very early.	Hardy, productive. Sometimes suckers too freely.
Puyallup	Pacific coast	Large, medium red, sweet. Fairly firm. Midseason.	On good soil, hardy and productive. Subject to root rot and mildew. Medium to large canes.
Madawaska	Northern plains	Large, fair quality. Inclined to crumble. Early.	Short, erect, hardy.
Dixie	Southeast Delta States	Tart, soft.	Latham cross, suited to South.
Early Red	New England Great Lakes	Early. Good quality.	Productive.
		BLACK RASPBERRIES	
Cumberland	Appalachians Southern plains Southeast Pacific coast Northern plains New England	Glossy, large, firm. Very good quality. Late.	Not reliably hardy in coldest areas. Subject to anthracnose.
Logan	Corn Belt New England	Glossy, large, firm, very good quality. Early.	Fair yields. Subject to anthracnose.
Black Hawk	Corn Belt Northern plains New England	Large, firm. Late.	Hardy. Strong, vigorous canes.
Bristol	Corn Belt Pacific coast Appalachians New England	Large, excellent quality. Midseason.	Subject to anthracnose.
Morrison	Appalachians New England Corn Belt	Very large. Fair quality. Late.	Good yields. Subject to anthracnose.
Plum Farmer	Pacific coast Appalachians	Medium size, fairly firm. Good quality. Midseason.	Vigorous, productive. Subject to orange rust, mosaic.
Kansas	Northern plains	Medium size, glossy, firm, mildly acid. Good quality.	Very productive. Resistant to mosaic.
		PURPLE RASPBERRIES	
Marion	New England	Light red-purple, attractive. Very large, firm, good quality. Late.	
Sodus	Corn Belt Southeast Appalachians New England	Very large, attractive, firm, good quality. Late.	Large vigorous canes, to 10 ft. Needs protection in cold areas.
Ruddy	Pacific coast New England	Reddish-purple, small, very soft. Flavor like black raspberries.	Produces from suckers. Hardy, productive.
		YELLOW RASPBERRIES	
Goldenwest	Pacific coast	Light yellow to slightly pink. Good flavor.	Vigorous. Grown as novelty.
Amber	New England	Large, high quality.	Hardy.

almost any soil, if it is well-drained. One or two varieties are even indifferent to drainage. In general, black raspberries prefer heavier soils than the red ones do, and best results come from sandy to clayey loams containing large amounts of humus. They are also indifferent to soil acidity, and will tolerate any *p*H from 5. to 6.5. Bumper crops come from deep, fertile soils, and those into which the gardener pours a never-ending stream of rich organic material. Rotted manures are acceptable to raspberries, but rotted vegetable matter seems to produce healthier bushes.

In choosing a site for a raspberry patch on the home grounds, it is best to plant on a sloping piece of ground, preferably a northern slope, with full sun in summer and possibly shade from distant trees in winter. The planting should be at least 50 feet from shade trees and 30 feet from fruit trees. If possible, do not plant raspberries where potatoes or tomatoes have recently grown. All 3 are subject to verticillium wilt, which can be carried over in the soil.

There is little difference between red and black raspberries nutritionally, except that black raspberries lack vitamin A, which the red ones contain in small amounts. In half a cup of red raspberries, there are about 260 units of vitamin A, .021 milligrams vitamin B_1, .04 milligrams vitamin B_2, 41 milligrams calcium, 38 milligrams phosphorus, .8 milligrams iron and 45 calories.

Ordering Stock

A raspberry planting traditionally lasts for 10 years, but with organic culture, many have been known to continue to produce heavily for far longer. Usually a 100-foot row will produce all the berries that can be used by a small family —from 75 to 175 quarts, depending on variety and growing conditions.

Where they are available, certified disease-free bushes should be purchased. It is better to buy new stock for starting a new patch than to run the risk of planting diseases from a neighbor's garden by using his bushes, unless you know that they are disease-free. One-year-old stock (field-grown for a year in the nursery) will produce the best roots. Number one tip layers of black varieties or number one sucker plants of red raspberries should be purchased.

Planting

Prepare the soil deeply for raspberries in the same way that it might be prepared for deeply rooted vegetables. If the bushes are set on a site that has been cultivated for 2 or 3 years there will be less trouble with deep-rooted perennial weeds. Before planting a 100 foot row of bushes, one to two tons of well-rotted manure plus a heavy application of rock phosphate and granite dust should be turned under to a depth of 10 to 12 inches on a strip 5 feet wide and 100 feet long.

In sections with mild winters, red raspberries may be planted either in fall or early spring. Young black raspberry bushes are more easily heaved during their first winter, due to their shallower roots, and are better planted in spring in all parts of the country. Planting should be done as early as the soil can be prepared.

Raspberries may be planted in the hill system, with plants spaced 5 feet apart in each direction, or in a hedgerow pattern, spaced 2½ to 3 feet apart in rows 7 to 8 feet apart.

Dig holes 12 to 18 inches in diameter and deep enough to contain the roots without crowding. If it is available, fill the lower half of the hole with rotted compost or well-rotted leaf mold. Set the plants one or two inches deeper than they grew in the nursery, but do not cover the crowns with soil. Give each bush a quart or two of water, and shake it to settle the soil around its roots before the hole is quite filled.

Care should be exercised to keep raspberry roots damp at all times. If they seem dry when the plants are received, soak them in water for 2 to 3 hours before planting. Do not allow them to lie exposed to the sun while working with them. Keep the roots covered with wet burlap or peat moss until the holes are ready to receive them. If planting must be delayed after they arrive from the nursery, store them, covered, in a cool cellar, preferably 35 to 40 degrees, until they can be planted.

As soon as they are planted, cut back any canes to 4-inch stubs. When new shoots arise, remove the stubs of the old canes completely, as a sanitary measure. If they contain any diseased material, it is better to remove and destroy them immediately.

Maintenance

Raspberries need large quantities of humus which is best maintained by a permanent mulch around the roots. A thick layer of straw, 10 to 12 inches deep when applied, or the same quantity of partly rotted leaves, will not only preserve soil moisture and supply humus, but will also replenish some of the soil nutrients. Leaves alone, in large enough quantities, can supply all the food necessary. Straw or hay should be supplemented with strawy manure or other nitrogenous fertilizer at the rate of one pound of pure nitrogen per year on a 100-foot row. This amount is contained in 10 pounds of fish meal, 12 pounds of tankage, 8 pounds of dried blood or 14 pounds of cottonseed meal.

Sawdust, wood shavings or crushed corncobs also make fine mulch for raspberries, and need not be spread more than 3 or 4 inches deep to perform the same function as 2 or 3 times that depth of straw. But unless these materials are well-aged before they are used, it is best to mix them with a nitrogenous fertilizer when they are applied. Bacteria which break down these materials need nitrogen to feed upon. Unless they are given an extra supply, they will actually rob the plants of nitrogen before they begin to supply it to them. So before fresh sawdust, shavings or corncobs are spread as a mulch, they should be mixed with one of the concentrated nitrogenous fertilizers listed above, at the rate of one ounce of pure nitrogen per bushel.

If they are well mulched, raspberry bushes do not need more water than is supplied by the rainfall in most sections of the country. In the arid sections they must be watered well between blossom time and harvest. In other sections, if there has been a drought water them well 7 to 10 days before harvest. If the bushes go into the harvest period dry, the berries may be small and seedy.

Pruning and Training

Raspberry canes are biennial, that is, they grow up in one season, make laterals which bear fruit the second season, and then die. New canes are constantly springing from the roots to replace the old canes which have fruited.

Best fruit is produced on the lowest laterals, except the first one or two, on each cane. Buds get progressively weaker as they get farther and farther from the roots, until the buds at the tips of the canes are quite feeble, and are usually winterkilled. Pruning and training raspberry bushes has the object of developing the fruiting buds on the centers of the canes, removing unnecessary growth and controlling the planting to prevent its becoming a briar patch. Because of their different habits of growth, red and black raspberries must be pruned differently.

Red raspberries are pruned twice each year, first in early spring after the end of the coldest weather and before the appearance of the first leaves; and they are pruned again after harvest. The spring pruning removes all weak and damaged canes, leaving no more than 8 to 12 strong shoots for each mature plant. Head back these shoots to 4 or 5 feet each and tie them to their supports. The second pruning comes immediately after harvest, when all of the old canes that have fruited are cut off at ground level. Do not leave stubs 3 or 4 inches tall when removing canes. These accumulate through the years to make pruning difficult, and they harbor insects and diseases in their rotting ends. No further pruning of red raspberries is needed, except the removal of any suckers which arise in the aisles between rows. If new plants are wanted, these suckers should be promptly dug and moved, before they get out of hand. Otherwise, they should be treated like weeds and pulled up.

Black raspberries need 3 prunings each year. The spring pruning, done at the same time as that of red raspberries, consists of removal of weak canes, leaving 4 to 6 strong ones to each plant. Diameter of the canes which are permitted to remain should be no less than one-half inch at the base. Laterals should also be cut back in early spring to a length of 6 to 8 inches each.

The second pruning of black raspberries must be done weekly from late spring through the early part of the summer, while new canes are making their fastest growth. As each cane reaches a height of two feet, it should be pinched back at the tip to stop its upward growth. This job should be done frequently enough so that the tips can be pinched out by hand as they reach the required height. Fewer satisfactory fruiting laterals will be formed if the canes are allowed to

Organic fruit grower Frank Shaw of California uses telephone or number 12 wire for supporting raspberry vines. He spaces raspberry rows 9 feet apart.

make too much growth before they are cut back.

The third black raspberry pruning is the same as the red raspberry post-harvest pruning. Cut out all old canes to the ground as soon as the berries have been picked. This removes a possible source of infection, in the old and dying canes, and permits the plant to throw all its strength into the development of canes for next year's crop.

To facilitate harvest and to prevent the raspberry patch from spreading out of control, canes are tied to supports in spring before growth starts. Bushes planted on the hill system may be tied to strong stakes 3 to 4 feet high, which are placed next to each bush during the second spring. Varieties which have numerous strong canes can be tied without stakes, tepee fashion. Canes are gathered together 6 inches to a foot below their tips after pruning, and they are loosely tied together, each one helping to support the others.

The usual support supplied to raspberries planted in a hedgerow is a wire trellis which runs the length of the row. This may be a simple one-wire trellis, strung 3 feet above ground from posts 30 feet apart for red raspberries, or 20 inches above ground for black varieties. Canes from each bush may be tied in one bundle to the wire, or they may be divided into several bundles and spread out to cover more space along the wire.

A more elaborate trellis system is sometimes used where frequent attention may be given to training the canes during the growing season. This consists of 2 or 4 wires strung along opposite sides of the fence-posts to make a box, through which the canes are trained to grow. If more space is needed within the box, the wires may be strung from crossarms 12 to 15 inches long. Canes are not tied to the wires, but are allowed to stand free within the framework. Where this system is used, posts are usually spaced no more than 15 to 20 feet apart, and end posts must be well-braced.

Propagation

Additional black raspberry bushes may be raised to increase the home garden planting by permitting the bushes to develop a few unpruned canes. In late summer, when the tips show signs of thickening, anchor them to the ground and cover them lightly with soil. In spring, before growth begins, sever the tips from the parent plants. Allow the tip layers to grow for one season before transplanting, or transplant at once to the new position.

As has already been indicated, red raspberry suckers may be dug and planted to form new bushes. Or either species may be increased by root cuttings. These cuttings are made in fall. Dig a few sturdy roots and cut them into 3-inch lengths. Store the pieces in damp (not wet) sawdust or peat moss in a cool cellar until spring. As soon as the soil can be prepared, plant the pieces 3 inches deep in a nursery row. By the following spring new plants should be ready to transfer to their permanent rows.

Enemies

Well-mulched and well-nourished, organically grown raspberries are usually free of disease and pests of all kinds, except perhaps an occasional stinkbug. However, we list the diseases to which the bushes may be subject if conditions are not right for them.

Among the fungus diseases are anthracnose, spur blight and rust. These should be dealt with promptly by cutting out and destroying any diseased parts. Anthracnose may be distinguished by rough grey to black cracked blotches on the canes. Where the blotches girdle the canes, they will die. Remove affected canes by cutting several inches below the diseased area. Spur blight should be similarly treated when it is discovered. It can be seen in reddish to grey spots below the buds on woody portions of the cane, or purplish spots on the green portions. Both of these diseases should be borne in mind at the spring pruning, and canes bearing them should be entirely removed at that time.

Mosaic is a virus disease which is carried throughout the plant by the bush's own juices. When this is discovered, the only cure is to dig out and destroy the whole plant, to prevent aphids from spreading it to healthy bushes. Mosaic affects the leaves, causing them to curl and show red and yellow mottling. Leaf curl, which is not so serious, also causes the leaves to curl, but does not discolor them.

Fruit worms are sometimes present in the berry cups at harvest time. If they are very bad, the bushes may be dusted with rotenone dust the following year,

The raspberries above have been mulched with wood chips. Well-mulched and organically-fed raspberries are seldom bothered by disease or insects.

3 times between blossom and harvest. Rotenone is not poisonous to humans.

A certain amount of winterkill must be expected in the tips of canes which go into the cold season with green, immature wood. Since the laterals or canes of all types of raspberries continue to grow until frost, a few inches of dead wood in spring is inevitable. But this should be limited to no more than 6 inches, if the bushes have been selected properly for their climate, and have been handled well. Where winter injury is a problem, all fertilizer should be applied in late winter or very early spring, so that growth will slow and harden before fall.

Experiments carried out at the South Dakota Experiment Station seem to show that the best winter protection for raspberries in cold areas is to bend the canes to the ground and cover at least one-third of their length with soil before the ground freezes. Bent over in this way, the canes form a snow-trap, and may be covered with snow drifts from late fall to spring. Snow not only protects the canes from the fluctuations of the winter temperatures, but also provides a blanket to protect them from drying winter winds. After the snow melts in spring, canes are raised and tied to their supports and any winterkilled portions removed with the spring pruning.

Harvest

Raspberries become soft when they are fully ripe, and must be picked daily, or at the very least every other day. They should be picked into shallow containers to prevent crushing. Rain at harvest time causes the berries to become moldy. As many as possible of the ripe berries should be picked immediately after a rain, and processed at once before they mold. If the picking is prompt, they are not damaged too badly to use for jam or to freeze. If they cannot be picked within a few hours after the rain, all moldy berries should be removed and composted at the next picking. If they are left on the plant, mold will spread to the green berries and part of the crop will be lost.

ROSE HIPS

Rose Hips. One of the most valuable sources of vitamin C is the fruit of an old-fashioned species rose, the *Rosa rugosa*. Rose hips are not much cultivated in this country nor widely esteemed as a fruit, yet, in some of the fruit-poor countries, such as Sweden, they are gathered yearly and preserved in various ways as an essential part of the winter diet. Rose hips of any species or variety of rose are rich in vitamin C, but the hips of the *rugosa* are larger and more succulent than most, and contain a high percentage of natural ascorbic acid. Depending upon variety and growing conditions, this may amount to as much as

Rose hips develop highest vitamin C content when they have reached full color in late fall. When fully ripe, flavor is mellow and somewhat nut-like.

1,200 to 1,800 milligrams per half-cup, or about 25 times the vitamin C content of the more popular form of vitamin C, the citrus juices.

Rosa rugosa in either the species or in fancy hybrid varieties makes an attractive hedge on the home grounds. The stiff, upright bushes grow from 6 to 15 feet tall, with thorny stems, rough, wrinkled foliage, fragrant and sometimes irregular flowers, followed in fall by orange to red fruit which may grow to two inches in diameter. The *rugosas* are very hardy and endure cold, heat, dryness, even poor soil and salt sea air. The flower of the species is a single pink bloom; hybrids may be obtained in white, red, yellow and pink in singles and doubles.

Culture

Although rose hips may be grown in any soil, the largest fruits are obtained when they are well-nourished and cared for. They prefer a slightly acid soil, with a *p*H of 5. to 6., and must have good drainage. Full sun and free circulation of air are also important to their health. To be at their best, they should receive 8 hours' sun a day, but they can subsist on 6 hours, provided they are morning hours.

Best plants to buy are two-year-old field-grown stock. They may be planted in fall or in early spring. Where winters are mild, fall planting is advisable. Where temperatures are low in winter it is better to plant early in the spring. If they are to serve as a dense hedge, they should be planted two feet apart. As specimen plants they may be allowed to occupy a 5-foot circle.

For a really superior rose-hip planting, prepare the soil deeply and thoroughly. Soil should be broken up and enriched to a depth of two feet in a trench as long as the hedge, and 18 inches wide. To do this, remove the top layer of soil from the trench with a spade and pile it beside the trench. Next remove another spade's depth of soil from the bottom of the trench and haul this layer away. If the soil below (which should be about 18 inches below the surface) is hardpan, break it up with a mattock and mix it with some coarse sand or gravel before returning any soil to the trench. The trench should have good drainage to provide drainage for the roses. If it is in a heavy clay soil, tiles leading away from the bottom of the trench may improve the drainage of the bed.

Now mix the topsoil which was reserved at the side of the trench with one-half rotted manure and one-half humus —compost, leaf mold, peat moss or any available well-rotted material. Add an application of rock phosphate and granite dust or greensand, and the trench is ready for planting.

Make a cone-shaped mound of the enriched topsoil in the bottom of the trench for each plant. When it is seated on the cone, the plant should be high enough so that the bud graft, if any, is at soil level. If the bush is grown on its own roots, it should be placed slightly deeper than it grew in the nursery. Damaged roots or unduly long, straggling ones may be cut back. Throw soil in loosely around the roots until the trench is half filled, then pour a pail of water around each plant to settle the soil. When the water has drained away, completely fill the trench and tamp the soil snugly around the roots. Cut back all canes to a length of 6 or 8 inches. Spread a mulch of hay, straw, crushed corncobs or peat moss over the soil for at least the first year after the hedge is planted. When the bushes become thick and vigorous, they will shade the soil and protect it from drying, and the mulch is not too important. If the hedge is started with a rich mixture of soil around its roots, an annual feeding such as is administered to the flower garden will be sufficient to maintain it in good condition.

Harvest and Use

Vitamin C content is highest in rose hips when they have attained their full color in late fall. When they are somewhat green, their flavor is mildly astringent. When fully ripe, it is mellow and somewhat nut-like. An extract of rose-hip juice, which can be canned for winter use, can be made according to a recipe developed by Adelle Davis. Miss Davis recommends the addition of one tablespoonful of extract to fruit juices served to each member of the family daily in the winter months. Her recipe is as follows:

Wash hips quickly and cut off blossom ends and stems. Plunge them into boiling water, using 1½ cups of water for each cup of rose hips. Cover the pan and simmer 15 minutes. Mash the softened hips with a fork or stainless steel or wooden potato masher. Pour the mash into a covered bowl or glass jar and allow to stand for 24 hours.

Next day strain off the extract. Unless a stainless steel strainer is available, use a jelly bag, because a badly tinned strainer may destroy the vitamin C. Bring the extract to a rolling boil, add two tablespoonfuls of lemon juice to each pint, pour into sterilized jars and seal. Store in a cool, dark cellar.

To make rose hip purée, Miss Davis advises to take two pounds of rose hips and two pints of water. Remove the stalk and remnants of the rose from the end of the berries and stew them in a saucepan until tender. This will take about 20 minutes and the lid should be kept on. Then press the mixture through a sieve and the result will be a brownish-tinted purée of about the same consistency and thickness as jam.

SALAL

SALAL. The salal, *Gaultheria shallon,* is a small blueberry-like fruit grown in the redwood section of the Pacific coast area. It is usually not cultivated as a fruit, but may be planted as a ground cover in some places, when fruit is gathered for jelly or jam as a secondary product.

The salal is a close relative of wintergreen, and the small plants are very similar to wintergreen vines in growth habit. They are members of the heath family, demanding very acid soil, a large amount of moisture-retentive humus, and partial shade or a very mild summer climate. Small pink or white bell-shaped blossoms in spring are followed by small red-purple fruits in late summer. When it finds congenial conditions, the plant spreads quickly and covers the ground with an evergreen mat of rather coarse, tough leaves. Because they last a long time after cutting, the greens are sometimes used for Christmas decoration in the West.

Salal is difficult to transplant from the wild, having very fine roots with no secondary feeding roots. It is better to start new plants from cuttings, or to buy a few potted specimens from a florist to plant in the home garden. They may also be started from seed, but the seed is fine and difficult to germinate.

SALMONBERRY

SALMONBERRY. This is the common name of *Rubus spectabilis,* a member of the raspberry genus, and closely resembling the raspberry in fruit and bush, but not in flavor. Salmonberries are native to the northern portion of the Pacific coast, where they grow wild but are not much cultivated. They are vigorous and healthy in the wild, and may be useful for hybridizing raspberries though they have no particular value in themselves. Two forms of salmonberry are found in Oregon and Washington — a yellow, rather insipid variety with large, subacid berries, and a salmon-colored or red berry which is more acid. Many more flavorful bramble berries are hardy in the sections where salmonberry grows, so there is not much purpose in giving these space in the home garden.

SATSUMA

SATSUMA. *See* ORANGE, MANDARIN.

SERVICEBERRY

SERVICEBERRY. *See* JUNEBERRY.

SHADDOCK

SHADDOCK. *Citrus maxima* or *C. grandis,* the largest of the citrus fruits, is called pompelmous by the Dutch, which has been shortened to pummelo. In this country it has always been called shaddock, after the sea captain who brought the seed to the West Indies, Captain Shaddock. The fruit was known in Spain as early as the twelfth century. It is used here chiefly by the Chinese, who traditionally decorate their houses with it on their New Year. It is grown very little in the United States except in southern Florida.

Shaddock trees are small, growing from 15 to 25 feet tall, and are usually not thorny. Their enormous fruits weighing up to 15 or 20 pounds are borne separately, distinguishing the pummelos from pomelos (grapefruit), which grow in clusters. The fruits are pear-shaped with thick rinds, leathery septa between sections, pulp not generally very juicy, but crisp and firm. Some are eaten like grapefruit, others are eaten out of hand. All the varieties mature their fruits in winter.

Strawberries need rich, light soil which is not overly exposed to late spring frosts. The strawberry patch should be in full sun; a mulch is always advised.

Three main varieties are grown in Florida: Mammoth, Pink and Tresca. Mammoth is lemon-yellow, 5 by 6 inches, with tough white flesh whose flavor is sweetish-bitter. Pink is a little larger, with lemon-yellow skin and pink flesh that is fairly juicy, its flavor bitter-acid. Tresca, whose season is latest, is small, 4 by 5 inches, with dark lemon-yellow skin, coarse pinkish flesh in which sweetness and acidity are well blended. It is possible that Tresca is a cross between grapefruit and shaddock, but its origin is not known.

Shaddocks are grown like Oranges, which see.

STRAWBERRY

Strawberry. The origin of the name of our most popular small fruit is a matter of some uncertainty among the philologists. Some say the name comes from the Old English for strew—"streawberry"—because of the plant's habit of spreading over the ground. For the same reason, others claim that the name was originally "stray-berry." As organic gardeners, we prefer the more practical explanation of the name—straw-berry—so-called because a straw mulch has traditionally been spread beneath the plants to keep the fruit clean and healthy. So far as we know, this straw mulch was one of the first generally accepted mulches used in every good vegetable garden.

The Latin name for strawberry, *Fragaria,* has an uncontested meaning. It signifies a pleasant odor, one of the strawberry's most cherished attributes. There are about 20 species of the genus *Fragaria,* but only about 4 are important to the development of our modern berries. From Europe comes *F. vesca,* the wood or Alpine berry, and *F. elatior,* or *F. moschata,* the flavorful Hautbois. *F. virginiana* is the eastern American wild strawberry and *F. chiloensis* is the Pacific coast native which has been hybridized to produce our many modern varieties.

Strawberry cultivation and development has all taken place within the last

130 years. Although the berries were known to the Romans, they gathered them from the wild. The Greeks, whose knowledge was more confined to the hot climates, make no mention of them. Strawberries were considered a delicacy through the ages, but always as products of the woods. At one time in England, they brought as much as 10 pounds for a pound of berries, which were gathered under hedgerows and brought in by horseback to London, at great risk from armed highwaymen.

It was not until the development of the Hovey seedlings near Boston in about 1830, that strawberries began to take a place in the garden. These were followed by other seedlings and hybrids in rapid succession. Hundreds of varieties are now available, some suited to every state in the Union, including Hawaii and Alaska.

Strawberries grow well in all temperate zones, the colder parts of which produce fruits of the highest quality. The plants are classed as herbaceous perennials. They are propagated by seed only as a means of developing new varieties. To propagate the characteristics of any variety, new plants are made from the runners put forth by most plants. In the few runnerless varieties, the plant characteristics must be propagated by division of the crowns.

Most varieties are stemless except for their runners. Rosettes of 3-part leaves arise directly from thick tufts of roots at ground level. The berries are produced on stems which, in the hardiest types, grow below the leaves. The fruit which we call a berry is actually a specialized swollen pistil receptacle. The actual fruit, botanically speaking, is the tiny "seed" or achene which is embedded in our berry. After fruiting, the plants send out long wiry runners which make tufts of leaves at their tips. If these tufts find a good rooting area, they send down their own roots and make plants which bear fruit the following year.

Strawberries contain a small amount of vitamins A and B and are a fairly good source of vitamin C. Little vitamin loss takes place when they are frozen. Some varieties have been found to contain more vitamin C than others, but this may be either the characteristics of the variety or of the soil and growing conditions of the samples tested. Where we have any information about these tests, we have given it in our variety chart. In the average serving of 100 grams of strawberries, may be found 100 units of vitamin A, .025 milligrams vitamin B_1, .05 milligrams vitamin B_2, .2 milligrams niacin, 50 milligrams vitamin C, 34 milligrams calcium, 28 milligrams phosphorus, .6 milligrams iron and 30 calories, all from sugar.

Ordering Stock

A 100-foot row of spring-bearing strawberries evenly divided between early, midseason, and late should produce 3 quarts of berries a day over a period of 3 weeks. For planting a 100-foot row, 50 to 75 plants are needed, depending upon their vigor and the number of runners the variety may be expected to produce. Fruit from a 100-foot row is usually sufficient for a family of 4.

In some areas everbearing strawberries are grown successfully. The so-called everbearing varieties produce light crops in spring and fall, and a few berries of inferior quality during the summer. In order to yield satisfactorily in fall, they must be supplied with plenty of water throughout the growing season, and blossoms must be removed from the plants in spring. Also, they must have time after the weather turns cool to mature a crop. It takes about 30 days to mature a berry from a pollinated blossom. It follows, then, that everbearings can produce well in the fall only in areas where a period of cool weather may be expected to last a month or more between the end of summer's heat and the first fall frosts. It is difficult to find everbearings which are certified virus-free. New varieties are constantly being developed, however, and it is possible that those with improved tolerance may come on the market at any time.

In ordering any stock, give preference to plants which are inspected and certified to be disease and virus-free. If plants are obtained from an established garden in the neighborhood, it is well to examine it carefully for signs of infection before purchasing plants. Usually plants for expanding a home strawberry patch are transplanted from runners formed by the established planting. If there has been any disease in this patch, it is better to buy new plants than to transplant possibly infected plants.

Earlier in the century, strawberry plants with perfect flowers were scarce, the first developed varieties having been dioecious, that is, with male or female but not hermaphroditic flowers. Now, all varieties offered by nurserymen are perfect-flowered, and it is not necessary to plant more than one kind for cross-pollination. In some varieties a few of the first blossoms which open may be imperfect, and they will not set fruit. But these are followed by enough perfect flowers to set a good crop.

In most sections early spring planting is best. In order to get a choice of good plants, place orders in early winter for shipment at planting time. It is best to order plants from nurseries which are located reasonably nearby. Plants take hold much better if little time elapses between digging from the nursery and planting in the garden. Also, strawberry plants seem to be set in their habits. They resent being moved from an area where they will receive a different amount of daylight than the amount with which they grew up. Plants shipped from the North seldom succeed in the South, and vice versa. And although there is no light differential, shipments between the East and West also seem to disturb the plants. Best vigor and production may be expected from plants which have not traveled far.

New plants should be one-year stock. They should have healthy tops with a few green leaves, moderately strong crowns, and vigorous tassels of straw-colored or white roots arising directly below the crowns. Blackened crowns or water-soaked, wiry roots indicate two-year-old plants or some kind of injury.

Freshly dug plants take hold more quickly than those which have been dug and stored for some time. An exception to this rule seems to be the polyethylene-wrapped plants which are being held in cold storage by some of the big nurseries for June planting. The plants are lifted while they are entirely dormant, wrapped in plastic to prevent evaporation, and are held for planting at a time when spring garden work slacks off. Experiments carried out by the Department of Agriculture show that these plants may be held 8 to 10 months without much deterioration, but that if they are planted in June or early July they do better than if they are planted later in the season. Plants set this late in spring usually require some irrigation during the summer.

Planting Site

Strawberries need rich, moisture-retentive but light soil in a warm position that is not too exposed to late spring frosts. The bed must be well-drained or the plants will be unnecessarily subjected to the risk of red stele disease—a fungus disease which infects roots on wet ground. Site of the strawberry patch should be in full sun, and removed from tall trees whose roots might penetrate the bed.

A strawberry barrel can produce large quantities of berries. Plants are inserted into the holes as barrel is filled with enriched soil. Four rows of one-inch diameter holes can be spaced about 6 inches apart around the barrel, providing space for about 50 plants.

The best position for a bed, from the standpoint of both water and air drainage, is on a gently sloping hillside where the grade is no steeper than 2 feet in 100 feet. If this spot is rather high on the slope and faces south, it is almost ideal. Being high, the cold air of late spring frosts should drain away to the valley; facing south, it should warm up early to help produce an early spring crop. Strawberries may also be planted on steeper slopes, but if they are, the rows should follow the contour to prevent the shallow roots washing out with heavy winter rains.

A new strawberry bed is easiest to handle if it is planted on soil that has been cultivated for one or two years, to remove all deeply rooted perennial weeds. But when it is planted on an old vegetable garden site, it is best to use a portion that has not been planted to strawberries during the past 5 years.

Any soil except dry sand or wet muck can be made to produce a crop of strawberries. The best soil is that which produces good potatoes—slightly acid, fertile, loose and open. Strawberries will tolerate any *p*H between 5. and 7., but prefer one between 5.5 and 6.5. Lime should not be applied unless the *p*H is 5. or lower.

Roots of strawberry plants do not penetrate deeply into the soil, but the plants must be supplied with large quantities of water, especially while the berries are growing. In order to keep this moisture within reach, the top foot of soil in the bed must have a large percentage of humus incorporated into it before the plants are set out. This is best dug in the fall before the bed is planted.

Barnyard manure is the best source of humus for a strawberry bed because it supplies nutrients to the plants as well as organic matter. On soil that has little organic content, manure may be spread at the rate of one-half ton over a 3-foot strip 100 feet long, or one bushel per 50 square feet. If the humus content of the soil is already high, half that amount is sufficient. If manure is not available, other organic residues may be dug in if they are adequately reinforced with nitrogenous fertilizer to help them break down quickly. Sawdust, wood shavings, crushed corncobs or leaves are excellent. The first 3 should be applied at the rate of 10 pounds (dry weight) per square yard, with one ounce of pure nitrogen mixed into each bushel. This may be obtained from 10 ounces of fish meal, 12 ounces of tankage, 8 ounces of dried blood, or 14 ounces of cottonseed meal. Leaves, if they are shredded and composted before being applied, do not need additional nitrogen. They may be dug in at the rate of 5 to 6 bushels per 100 square feet. An application of rock phosphate and granite dust or greensand should be mixed with the soil along with the organic matter.

If the soil is prepared in fall before the strawberry bed is planted, organic matter breaks down during the winter, and makes a good light loam which warms up early in spring. This is helpful in the spring preparation of the bed. Strawberries should be planted while they are fully dormant, in very early spring in almost all parts of the country. But the soil should not be prepared until it is dry enough to be very friable.

Planting

In the far South, strawberry plants may be set out either in fall or early spring. Where there is much freezing and thawing of the soil during the winter, newly set plants are almost certain to be heaved when they are fall-planted. Another consideration is the small yield the first spring after fall plants are set out. If 100 plants are put out in the fall, the harvest next spring will be from 100 plants. If 100 plants are set out in spring, each plant will send out two to a dozen runners, and as many as 1,000 plants may supply the harvest next spring.

Strawberries may be planted in hills or in rows. Several variations of the row system are possible. Usually the everbearing varieties which do not form runners are planted in hills; plants which form runners are planted in so-called matted rows—the runners are permitted to root freely—or in hedgerows, where the number of runners permitted to root is controlled.

Hill-planted strawberries may be set 12 to 18 inches apart in rows that are 18 to 36 inches apart, depending upon whether they will make any runners at all. If they do, they need more space; if not, 18 inches is sufficient.

Planted in matted rows, varieties which send out numerous runners may be planted 2 to 3 feet apart in rows that are 42 inches apart. Less vigorous plants may be set 18 inches apart in rows 3 feet apart. Runners from these plants are allowed to spread out in all directions. By the first fall, they should have filled in the rows with a solid mat 24 to 30 inches wide.

The gardener who wishes to devote a large amount of time to training strawberries may plant in one or another of the "hedgerow" patterns, and may rigidly control the number of runners which he permits his plants to make. In addition, he must place each runner and anchor it where it will carry out his

TABLE 84: Strawberry Varieties

Variety	Section for which recommended	Season	Description	Disease and cold tolerance	Suitability for freezing
Albritton	Southeast	Late	Large, firm, bright red to center. Subacid. Excellent dessert quality.	Hardy south of Md. Susceptible to leaf scorch.	Good
Armore	Corn Belt Northern plains	Late midseason	Large, medium red, mild, good flavor.	Leaves subject to disease.	Fair
Blakemore	Southern plains Southeast Delta States Corn Belt Appalachians Northern plains Pacific coast	Early	Medium size, light red, acid.	Resistant to verticillium wilt. Tolerant to viruses. Resists leaf spot, leaf scorch.	Fair
Brilliant	New England Great Lakes	Everbearing	Moderately vigorous plants with many runners. Medium size, light red. Best flavor in cool weather.	Cold hardy.	Good
Catskill	Alaska Pacific coast Southern plains New England Appalachians	Midseason	Large, medium quality, medium firm, very high in vitamin C. Needs much water.	Susceptible to viruses, leaf spot.	Fair
Dixieland	Southeast Appalachians	Early	Large, light red, good flavor.	Susceptible to leaf scorch.	Good
Dunlap	Alaska Northern plains New England Southern plains	Early midseason	Medium size, dark red, soft. Good flavor.	Withstands drought and cold temperatures. Tolerant to viruses. Susceptible to leaf spot, scorch.	Good
Earlidawn	Great Lakes Corn Belt	Very early	Large, tart, light red, fair dessert quality. Few runners.	Blossoms frost hardy. Susceptible to viruses, resists leaf diseases.	Good

TABLE 84: Strawberry Varieties (Continued)

Variety	Section for which recommended	Season	Description	Disease and cold tolerance	Suitability for freezing
Empire	Appalachians New England	Midseason	Large, light red, tart, firm, medium quality. Productive. High in vitamin C.	Susceptible to leaf spot.	Fair to good
Fairfax	Corn Belt Great Lakes Appalachians New England Alaska	Early midseason	Large, deep red, high quality, sweet, good flavor. Not very productive. High in vitamin C.	Susceptible to viruses, resistant to leaf diseases.	Good
Fairland	Appalachians	Early	Medium size, rich red, good dessert quality.	Resistant to red stele, virus diseases.	Fair
Gem	Northern plains New England Pacific coast Alaska	Everbearing, small spring, heavy fall crops	Large, light red, acid flavor, many runners. Fair amount of vitamin C.	Cold hardy. Susceptible to leaf spot, resists leaf scorch.	Good
Klondyke	Pacific coast Southern plains	Midseason	Dark red, small, medium firm.	Tolerant of virus, susceptible to leaf scorch, leaf spot.	Good
Klonmore	Delta States Southeast	Early	Small, medium firm, fair dessert quality.	Very resistant to leaf spot, very susceptible to leaf scorch. Tolerant to viruses.	Good
Lassen	Pacific coast	Midseason	Very large, soft, fair dessert quality.	Resistant to virus, red stele.	Fair
Marshall (Oregon Plum)	Pacific coast	Early midseason	Large, soft, dark crimson, excellent flavor.	Very susceptible to viruses, leaf spot.	Good
Mastodon	Pacific coast Southern plains	Everbearing, large spring crop	Large, medium red, fair flavor. High in vitamin C.		

TABLE 84: Strawberry Varieties (Continued)

Variety	Section for which recommended	Season	Description	Disease and cold tolerance	Suitability for freezing
Missionary	Southeast Pacific coast	Early midseason	Small, medium firm, fair dessert quality.	Resistant to leaf spot, tolerant to virus.	Good
Ogallala	Northern plains	Everbearing	Vigorous, medium to large berry, smaller in summer. Dark red. Good flavor.	Very hardy, and frost-resistant to 20°. Drought-tolerant.	Good
Plentiful	Northern plains Corn Belt	Late midseason	Large, light red, mild flavor. Productive.	Resistant to red stele.	Poor
Pocahontas	Corn Belt Southern plains Southeast Northern plains New England	Early midseason	Large, medium red, tart, good flavor, productive.	Resistant to leaf diseases.	Very good
Premier (Howard)	New England Northern plains Corn Belt Appalachians Southern plains Pacific coast	Early	Medium size, medium dark red, good flavor, medium quality. Fair amount of vitamin C.	Resistant to light spring frosts. Also to virus diseases.	Poor
Red Rich	Corn Belt Alaska Northern plains New England	Everbearing	Large, dark red, excellent flavor, vigorous, but few runners.	Susceptible to virus diseases.	
Robinson	Northern plains Alaska New England	Late midseason	Large, light red, good flavor, soft, fair quality. Fair to good in vitamin C.	Susceptible to leaf blight, stem end rot.	Fair
Shasta	Pacific coast	Midseason	Very large, medium firm, good dessert quality.	Susceptible to leaf spot, tolerant to viruses.	Good

TABLE 84: Strawberry Varieties (Continued)

Variety	Section for which recommended	Season	Description	Disease and cold tolerance	Suitability for freezing
Sitka Hybrid	Alaska		Medium to small, sweet, soft.	Most hardy variety.	Good
Sparkle	Corn Belt Appalachians New England	Late midseason	Medium size, medium dark red, fair to good flavor, firm. High in vitamin C.	Resistant to red stele, leaf spot, leaf scorch.	Very good
Superfection	Appalachians Southern plains Northern plains New England	Everbearing, large spring crop	Medium size, light red, good flavor. Many runners.	Cold hardy.	Good
Surecrop	Corn Belt Appalachians	Midseason	Large, medium quality.	Red stele resistant.	Good
Temple	New England Appalachians	Midseason	Vigorous, productive, large, dark red, good quality.	Resistant to red stele, viruses, leaf diseases.	Poor
Tennessee Beauty	Southeast Southern plains Appalachians Corn Belt	Late midseason	Medium size, firm, good dessert quality. High in vitamin C.	Resistant to virus and leaf diseases.	Very good
Twentieth Century	Pacific coast Southern plains Great Lakes	Everbearing	Vigorous, productive, but few runners. Large, light red, firm. Good quality and flavor.		
Vermilion	Corn Belt Northern plains	Early midseason	Medium size, bright red, good flavor, soft.	Resistant to red stele, leaf spot, leaf scorch.	Fair

pattern. The basic pattern for a hedgerow is to set the plants 24 inches apart in single rows spaced 30 to 42 inches apart, depending upon ultimate design. As the runners develop, all but two are removed from each plant. These two are carried along the row, one in each direction from each parent plant. When the daughter plants are all rooted, the planting consists of a single row of plants 8 inches apart.

One variation of this plan consists in spreading the original row from a single line of plants to 3 lines. This is done by permitting each plant to make 4 runners, two trained to fill in the original row as above, and two more stretched out to the sides to start two more lines of plants, one on each side of the original row. This plan may be made into a permanent bed if on alternate years the original plants, then the secondary ones, are dug out and replaced by runners, each time making the new runners take the place of their grandparents, as it were.

A much more simple plan for forming a permanent bed, and one which takes less work on the part of the gardener, is to form a matted row, allow the runners to develop freely, and to remove all of the old plants each year after the harvest. In addition, all of the wandering plants which fill in the aisles between rows may be removed or transplanted to extend the planting. Or the runners may be left to form new matted rows along the aisles, while the original rows of plants are turned under to form new aisles. Any number of permanent plans may be carried out, but all are contingent upon the bed's good health. If fungus or virus diseases get a foothold, they will make the planting "run out," no matter how careful the training. When this happens, the best plan is to choose a new site and start a new bed, using only newly purchased, clean stock.

Strawberry plants should be protected from the sun and wind while they are being planted in the garden. The work is best done on a cloudy or rainy day. If the sun is bright, cover the roots of the plants with a piece of wet burlap to protect them from drying. Make each hole large enough to spread the roots well. This can be better accomplished if a cone of soil is mounded up in the center of each hole, and the roots spread in a circle around the mound.

Set the plants exactly as deep as they grew in the nursery. All of the roots, right up to the shoulder, should be covered, but the crown of leaves should be above the soil level. If the soil is dry, give each plant a quart of water after it is set. Immediately after planting, spread a mulch around the plants. Not only will this help preserve the moisture which the young plants badly need, but it will also eliminate most of the heavy weeding chore during the coming summer. After runners have formed in midsummer, if the mulch has not become compact enough to encourage the runners to root through it, separate the mulch under the runner plants and allow them to make contact with the soil. All blossoms should be pinched off the young plants during their first summer. To do this, go over the bed once each week until July. This will permit them to make strong roots before they bear their first crop, and will channel all of their vigor into establishing a healthy bed.

Gardeners whose space is limited may be able to grow a few berries in an old-fashioned strawberry barrel, or its modern counterpart. The strawberry barrel is a vertical garden, useful especially for a small family for whom a few berries from everbearing plants may be made to supply the table in the bearing months. The barrel is ornamental and a conversation piece on a terrace, as well as being useful. The following instructions are given by the Federal Cooperative Extension Service at Corvallis, Oregon, for making a strawberry barrel:

Use a barrel made of oak or any other solid material. Bore 2 or 3 1½ to 2-inch holes in the bottom for drainage. With the same auger, bore a series of holes spaced 10 to 12 inches apart in a circle 6 to 8 inches above the bottom of the barrel. Six to 8 inches above these, make another series of holes staggered at least 3 inches to one side of the first holes. Repeat this process up the side of the barrel, making the last series 6 to 8 inches below the top. The barrel will last longer if its inside and freshly cut surfaces are painted with a wood preservative.

In order to have uniform growth and fruit production, it is best to install some mechanism by which the barrel may be easily revolved. It may then be turned frequently so that all plants get an equal amount of sunshine. This may be accomplished if the barrel is mounted on a large set of casters or rollers, a turntable from an old clothes reel, or an old wheel mounted on an axle or pipe which in turn is embedded in a heavy plank or concrete base.

At the bottom of the barrel, place a layer of gravel, broken crockery, or brick to facilitate drainage. Fill the bottom layer, to the level of the first auger holes, with a rich mixture of loam and compost or rotted manure. Tamp the soil to firm it before inserting the first plants. Place a young strawberry plant on the soil at each hole, so that the crown protrudes from the hole. Spread the roots and press them into the soil inside the barrel. Follow the same procedure as the soil is built up to each series of holes. Finally, set 4 or 5 plants in a circle in the soil at the top of the barrel. But before the barrel is entirely filled, its watering system should be installed.

After the first third of the barrel is filled with layers of soil and plants, place a pipe, tile, or wooden tube up the center. This pipe or tube should be 3 or 4 inches in diameter, and perforated with a series of small holes. Its top should be flush with the top of the barrel. These small holes will admit air and water to the soil within the barrel, and will carry fertilizer to the plants.

When the barrel is filled, thoroughly moisten the soil by running water from a hose into the central pipe. Soil should be thoroughly wet, but not waterlogged. After the first moistening, fill the center of the pipe with a loose and open mixture of straw and manure. The water will wash through the manure each time the barrel is irrigated, and some of the nutrients will wash in to the roots of the plants.

If no barrel is at hand, the same type of arrangement may be built from the ground up in pyramid form, arranged not to turn, but in low enough steps so that the sun can reach the far side over the top of the pyramid. Walls of the pyramid may be square, if built of brick, wood or stone, or they may be round if built of metal. Plants are placed on the steps of the pyramid. This form is easier to care for than the barrel, because compost may be sprinkled over the soil on the steps of the pyramid, and no turning to catch the sun is necessary.

Best varieties of plants for a barrel or pyramid are the runnerless everbearing type. They should be selected from the varieties recommended for the area in which they are to grow. If they are planted in early spring, a light crop of berries may be produced the first fall.

Winter protection is easily given to the portable strawberry barrel, which may be wheeled into a sheltered shed for the winter. Soil should not be allowed to become too dry while the barrel is under a roof. If the concentrated planting is in the form of a stationary pyramid, it should be covered for winter with a wrapping of straw or evergreen boughs, to prevent alternate thawing and freezing.

Maintenance

Under ordinary conditions, commercial growers expect their strawberry beds to bear very well one year, about half as well the following year, and then to be dug and replaced the third year. Good care, plenty of organic fertilizer and permanent mulch can make a bed last much longer than that, provided the plants are thinned and old ones removed each year. If the bed becomes choked with vegetation, the size and health of the berries produced will suffer. Plants should never be closer together than 6 inches when they go into their bearing season. Early in the spring, before they bloom, they should be thinned to stand at least this far apart. Or the thinning may be done in connection with fall pruning.

The last runners formed at the end of the runner stems make small plants which do not root deeply enough to bear many berries. In fall, before the bed is mulched for the winter, these small runner plants should be removed. At the same time excess plants may be thinned, to leave the bed in good condition for its spring growth.

Strawberries planted in sections where the snow comes early in fall and stays late in spring do not need any winter mulch except the snow, which is their best protection. In all other areas where the temperatures drop below 20 degrees, plants must be protected with a thick coating of hay, straw or pine boughs during the winter months. This protection should be applied after the first frosts have made the plants entirely dormant and have coated the soil with a frozen crust, but before the temperatures drop below 20 degrees. Fruit buds for the coming spring are already set on the plants before they go into the winter, and they may be badly damaged by being exposed to cold at 15 degrees or less if they are not adequately protected. A mulch over the bed keeps the temperatures low on unseasonably warm winter days, and prevents it from falling too low when the hardest freezes come.

If the mulch is allowed to remain fairly late in the spring, the plants will be protected from starting into growth so soon that their blossoms may be frosted. When the weather starts to warm up, watch the plants under the mulch. They will show definite signs of wanting to grow, and the leaves will begin to yellow when they need the sun. This is the time when the mulch should be removed. Pile it in the aisles between the rows, leaving enough straw around the plants to cover the bare soil. Leaves will grow up through this light mulch, which will help tc smother the weeds and to keep the berries clean. If a late frost threatens after blossoms have begun to develop, draw the winter mulch back over the plants for the night, and remove it in the morning.

After the harvest of the first complete crop from a strawberry bed, the planting is ready for its first renovation. In the small home garden, preparation may be made for the renovation as the berries are picked. The best plants for renewal of the bed are the largest, healthiest specimens which produce the best fruit. Each time that you go into the garden to pick, carry with you a pocketful of markers. Place a marker next to each plant which is yielding exceptionally well.

When all the berries have been harvested, plants should be thinned from the bed to leave one plant to each square foot of row, in a row no more than 18 inches wide. These plants should be selected from the best of those marked for survival during the harvest. All other plants should be dug and removed.

A side-dressing of fertilizer should be liberally applied at this time. This can be in the form of rich compost, rotted manure, or of a mixture of organic fertilizers which provides one ounce of pure nitrogen per 100 square feet. A good all-round fertilizer may be made by mixing 2½ pounds of bloodmeal or 5 pounds of cottonseed meal, 4 pounds of bone meal and 3 pounds of granite dust o1 greensand. Apply this mixture at the rate of 3 pounds per 100 square feet of row, or a liberal handful around each plant.

After renovation, the mulch around the strawberry plants should be renewed. The best materials for mulching a strawberry bed are wheat straw, cotton hulls, crushed corncobs, peat moss, wood shavings, pine needles or spoiled hay. If peat moss is used, it should be soaked in water before it is applied, because it is impervious to moisture when thoroughly dry, and may prevent subsequent rain from reaching the roots of the plants. Straw or hay which contains weed seeds should be spread on the ground 3 weeks before it is applied to the bed. It should be thoroughly moistened and allowed to germinate all the weed seeds. When spreading it between the plants, turn each forkful of mulch upside down and germinating seeds will die. The other materials mentioned above for mulch should either be composted for a year before they are used, or they should be mixed with a nitrogenous fertilizer when applied. The following amounts of materials are recommended for mixing with each bushel of mulch: 10 ounces of fish meal, 12 ounces of tankage, 8 ounces of dried blood or 14 ounces of cottonseed meal.

The new polyethylene mulch has been used with success in some of the commercial strawberry beds in the more arid sections, where the berries are watered by irrigation rather than rainfall. This can be used to advantage only on beds of runnerless plants, since runners have no way of rooting through the plastic. It is used for berries planted in the hill system, or those planted in spaced double or triple rows. After the plants are set, the plastic is spread over the entire row. The gardener then goes along the row with a pair of scissors with which he slits the cover over each plant and draws the leaves up through the slit. This widens the hole enough so that subsequent shoots can find their way up through it.

The advantage of polyethylene as a mulch is that the sun can penetrate it to warm up the soil for an early crop. Of course, it also serves the purposes of any mulch in preserving moisture and keeping down weeds. Its disadvantage is that it adds no organic matter to the soil, as an organic mulch would do. Where the material is available, the organic mulch is much to be preferred for its soil-building properties.

Water is needed by the developing berries during the fruiting season, and it is usually present in soil soaked with spring rains if the mulch has been well-maintained. But after the harvest, the need for water in the strawberry bed continues through the summer. Healthy new runners and fruit buds for the following year's crop are growing, and they cannot flourish without a sufficient amount of water The average loam requires one inch of rain per week throughout the summer to maintain sufficient moisture for a strawberry bed. If the loam is sandy, more is needed. If it is clayey, it may need less.

The condition of the surface soil is not always an indication of the need of moisture. Soil may be completely dry on top (though under a good mulch it seldom is) and may have plenty of water 6 inches down. If in doubt, dig a hole 8 to 10 inches deep. Squeeze a handful of soil from the bottom of the hole. If it forms a ball, it has enough moisture. If it crumbles, irrigation is needed. Except in areas where irrigation trenches are maintained year-round, the best way to supply water in summer to a strawberry bed is from an overhead sprinkler. The bed should be sprinkled early in the morning on a day that will be sunny, so that leaves may be thoroughly dried before night.

Enemies

Few insects of any importance attack strawberries. Their troubles lie principally with fungus or virus diseases. The only insects that cause much damage are strawberry crown borers and slugs.

Strawberry crown borers are white grubs that feed on the sides of the crowns and weaken or kill the plants. They are harbored by wild strawberry plants and cinquefoil in the neighborhood. Dig up and destroy any neighboring wild strawberries or cinquefoil before starting a strawberry bed. If the infestation in the bed becomes bad, the only remedy is to plow the plants under after harvest and set a new bed on another site. To be completely sure that the infestation will not follow the strawberry patch, the new bed should be as far removed from the original as possible—preferably 300 yards away. Use only newly purchased plants for the new bed.

Garden slugs which infest mulch may be a problem in the strawberry bed when the berries are ripening. Slugs, unpleasant though they may seem, have a definite place in breaking down organic mulching material to make it fine enough to mix as humus with the soil. They feed almost entirely on dead vegetation, but they can easily be persuaded to make an exception in favor of ripe strawberries. Slugs do most of their damage on the bottoms of the berries that grow close to the soil or damp mulch. They dislike dry surfaces and will seldom venture on them. If the slug damage at harvest time is bad, place fairly large pieces of wooden shingles beneath the fruit clusters. Any slugs which attack the berries overnight will crawl below the shingle when the sun comes out. If, each morning, you turn over the shingles you can kill the slugs adhering to them by sprinkling table salt on their backs.

Two fungus diseases attack leaves of strawberry plants, especially those in beds of long standing. They are much more damaging in wet years than in dry. Leaf spot is characterized by small, irregularly distributed reddish-brown or purple spots whose centers later become grey-white, bordered with red or purple. Leaf scorch is similar, but lacking the white centers. Plants on heavy or wet soils are most subject to the diseases. In a bed that is expected to remain in its position for many years, the best treatment is to mow all the leaves off the plants after the harvest, rake them and destroy them. Do not use runners from this planting to make a new bed, even if the leaves in late summer and fall seem clean. Instead, if a new bed is to be started, buy plants of varieties resistant to the fungi.

A fungus which attacks the fruit in wet seasons is called fruit rot. If it rains while the berries are ripening, some of them are almost certain to rot. Go over the planting as soon as it is dry after the rain, and remove all rotting fruit. If it is not promptly taken away, the spores from the rotting berries will spread to sound, green ones and cause much more damage.

Red stele is a root-fungus disease which can destroy a whole planting if it gains a foothold. It is most prevalent in poorly drained soils. Evidences of red stele can be found only from early spring through the harvest period, after which they disappear until the following spring. Affected plants are stunted, the fruits are small and sour, and the roots show a reddish core when they are split in two. Dig any dubious-looking plants carefully, and examine the roots, which give the only certain diagnosis of the disease. Remove all affected plants and improve the drainage of the bed, if possible. If this cannot be accomplished without digging out the entire planting, install a good drainage system for the new bed before it is planted, and put into it only newly purchased red stele resistant plants.

Several virus diseases may infect strawberry plants. These are systemic diseases, and there is no cure except to dig and burn infected plants. Most prevalent viruses are the yellows and the stunt virus. Plants infected with them have leaves on abnormally short stems, and the leaf edges may be puckered or cupped. A thin stand of plants and very low yields are characteristic of virus-infected beds. The best treatment is to plow under the entire bed, and to plant another at some distance from the first with virus-tolerant stock.

Mice sometimes take up their winter quarters in a mulched strawberry bed, and may do much damage by rooting around among the stems. In order to avoid this, thin out the plants in early fall before the mice begin to make their winter nests, and do not spread the winter mulch too early. By the time the top of the ground has a frosty crust, mice will have made their winter homes elsewhere and the winter protection may be safely spread on the strawberries.

A few principles of strawberry culture, if rigidly adhered to, will help to prevent many ailments. First, when making a new bed, plant it on soil that has not had strawberries on it for many years. Make the new bed as far as possible from old, infected beds. Prevent crowding of runners at any time. If necessary, thin them several times through the summer. Make sure that they go into the winter thinned and pruned and ready for spring.

Avoid applying high-nitrogen fertilizer in early spring to bearing beds. These increase the incidence of fruit rot and leaf spot by making the leaves and fruit soft in the wet months, and by spreading a rotting layer of material directly below the plants in the wet weather. If applied after the harvest, the fertilizer will help develop next year's crop during the dry months when rot on top of the soil cannot do so much damage.

Harvest

The best time to pick strawberries is in early morning while the fruit is still cool. Pick the berries by twisting them from the vine, holding on to the stem. Do not pull them off their stems. If the stems are removed before they are washed, a much higher loss of vitamins and nutrients takes place than if the stems are removed after washing. Pick the berries into small baskets or containers, and place the filled receptacles in the shade as soon as possible.

If the berries are to be frozen, they should be washed briefly, drained thoroughly, stemmed and frozen in as short a time as possible after picking. If possible, chill the fruit quickly before working on it. If they are for dessert, place the picked berries in the refrigerator as soon as they come from the garden. Wash and stem them just before serving.

TAMARIND

TAMARIND. *Tamarindus indica* is a large, spreading tree belonging to the family of legumes. It is planted in tropical countries for its fruit and as a dense, ornamental shade tree. It is thought to be a native of tropical Africa, but has been cultivated most widely in India for many centuries. In this country, it can be grown only in the southern part of Florida and in Hawaii, where the fruits may be bought in the markets. It grows well on limestone soils, especially if well-fertilized and irrigated in the dry season.

Tamarind trees grow slowly, but attain a fair size. In Florida, they usually are medium-sized trees. Elsewhere they grow taller. When they are young, they are extremely tender, but after they have attained maturity they can withstand a few degrees of frost for brief periods. They are useful in southern Florida because their finely divided leaves are not wind-resistant, so they are little damaged by tropical storms.

Leaves of the tamarind tree are compound, 4 to 6 inches long with 20 to 30 leaflets. Attractive clusters of yellow flowers in summer develop into fruits that resemble lima bean pods in shape. They are slightly curved brown pods 3 to 7 inches long, an inch or less wide, with a brittle outer shell and 1 to 12 shiny seeds embedded in edible flesh. The fruit matures in late spring or early summer, and will hang on the trees for months. However, the pulp is subject to attack by beetles, so the fruit is usually picked as soon as it ripens.

Pulp of the tamarind is made pleasantly acid by a combination of tartaric and citric acids. Its flavor is refreshing, and it is sometimes used as a substitute for lemon in flavoring. The fruit is used in drinks, in preserves and chutneys and, in India, it is one of the ingredients of curry. It is very high in calcium content and has a good amount of B vitamins, but lacks vitamins A and C. In spite of acidity, its sugar content is as much as 20 per cent and a cup of pulp contains almost 500 calories.

Propagation is usually by seeds, which germinate quickly and easily. Desirable fruiting trees may be side-veneer grafted. Shield budding is possible, but more difficult.

TANGELO

TANGELO. An effort to produce a grapefruit which can be eaten out of hand has resulted in many hybrids of grapefruit and tangerine, grouped together under the name of tangelo. Most of the fruits look like grapefruit, but are smaller and taste a little like each. Some of these hybrids are of excellent quality, but will not find their way into the markets for many years. They should be planted in home gardens, where their culture is very similar to ORANGE, which see. Following are some of the most successful varieties:

TANGERINE

TANGERINE. *See* ORANGE, MANDARIN.

TEMPLE

TEMPLE. *See* ORANGE, MANDARIN.

WATERMELON

WATERMELON. *Citrullus vulgaris,* the watermelon, is a member of the cucumber family closely related to the citron melon which is native to this country. Watermelons are native to tropical Africa, where they are eaten not only by natives but by the wild jungle animals as well. Their culture was unknown in Europe until the sixteenth century, but was introduced to this country by the early settlers.

At least 4 months of frost-free weather are needed to mature the larger varieties of watermelons. Where the summer days and nights are very hot, as in the South and in the interior California valleys, this time may be somewhat reduced. Where the temperatures are very moderate, the melons may not ripen even in 4 months. Most success attends their culture in the Southeast, the Delta States, the Southern plains, the southern Pacific States and along the Atlantic seaboard up through Delaware.

A limiting factor in watermelon culture in the home garden is the amount of space that the plants require. A watermelon vine needs at least 36 square feet in which to grow, and some varieties need as much as 100 square feet. From one vine no more than two melons may be expected to mature. This makes it impossible to grow them except in farm

TABLE 85: Tangelo Varieties

Variety	Season	Fruit characteristics	Flavor
Mineola	Feb. to Mar.	Orange-red, medium to large. 4 to 12 seeds.	Tangy, pleasant.
Orlando	Dec. to Jan.	Red-orange. Medium size.	Acid and sweet combined.
Nocatee	Late	Medium to large. Lemon-yellow. Good keeper.	Mingled tangerine and grapefruit.
Sampson		Small to medium, chrome yellow, loose skin. 10 to 15 seeds.	Excellent quality. Tangy subacid flavor. Sweeter than grapefruit but slightly bitter.
Thornton	Midseason	Large. Orange. Loose skin. 5 to 12 seeds.	Distinctive, unlike any other citrus fruit. Subacid, no bitterness.

TABLE 86: Watermelon Varieties

Variety	Days to maturity	Resistance to disease	Description of fruit
Black Diamond	90		Large, round, dark green skin, flesh sweet and red, seeds stippled black.
Blue Ribbon	80	Wilt	Medium size, oblong, striped green, excellent quality.
Congo	95	Anthracnose	Medium size, medium green, dark stripe, oblong, red, firm, sweet flesh, light tan seeds.
Dixie Queen	85	Wilt	Medium size, oval, light green, dark stripe, red, crisp, very sweet flesh, small white seeds.
Garrison	85		Large, long, light green with dark stripe, red flesh with good flavor, white seeds.
Hawkesbury	85	Wilt	Large, long, grayish green, pink flesh of good quality, dark brown seeds.
Irish Gray	95		Large, long, gray-green, red firm flesh, white seeds.
Kleckley's No. 6	85	Wilt	Cylindrical, bluish green, red, very sweet flesh, creamy white seeds.
Klondike R-7	85	Wilt	Oblong, solid green, medium size, very good quality.
New Hampshire Midget	70		Small, oval, medium green, red flesh, small black seeds.
Stone Mountain (Dixie Belle)	90		Large, round, medium green, scarlet, very sweet flesh, white seeds edged with black.
Sugar Baby	75		Midget, round, dark green, red, good flavor, small dark tan seeds.
Tom Watson	95		Large, dark green, red coarse flesh, brown seeds.

gardens where space is not at a premium.

Melons can be grown successfully only on very light loam or sandy soils. They do not do well on heavy soils and their quality is inferior on clay. A large amount of moisture is needed for the large fruits, but roots penetrate as much as 6 feet into the soil, so they seldom suffer from drought if the soil contains sufficient humus.

Watermelons are composed of 90 per cent or more of water and 7 to 8 per cent of sugar. They are a fair source of vitamin A, a medium-sized serving, weighing 300 grams, containing about 450 units of vitamin A, .180 milligrams vitamin B_1, .084 milligrams vitamin B_2, 22 milligrams vitamin C, 33 milligrams calcium and about 90 calories.

Ordering Seed

A packet of watermelon seed will plant 8 hills; an ounce, 25 to 30 hills. Varieties should be selected on the basis of the length and temperature of the growing season. They should also be chosen for their resistance to disease. Melons subject to anthracnose or wilt should not be planted on the same piece of ground more often than once in 10 or 12 years. Even those resistant to these diseases should not be planted in the same place more often than once in 4 or 5 years. If the soil has not previously had a crop of watermelon, the less resistant strains may be planted, but if they have ever grown on it, it is advisable to plant more resistant strains.

Planting

Soil, in addition to being light for good watermelons, should be slightly acid —*p*H 5.5 to 6.5—fertile, deep and well-drained. Preparation of the soil should begin the fall before planting. At that time manure should be turned under to a depth of 6 to 8 inches. If there is a plentiful supply of manure on hand, dig in an inch layer of it all over the watermelon bed. If the supply is limited, a few forkfuls may be dug into the hills, and left to decompose during the winter. This gives the nutrients time to leach down into the soil to a depth where the vine's deepest roots will find it. A handful of rock phosphate and one of greensand or granite dust may be incorporated in the hills at the same time. Lime should not be used unless the *p*H is below 5.0, because melons prefer the acidity normal to organic soils.

In cool areas, or where the growing season is short, seeds may be started indoors in peat- or manure-pots and moved to the garden when all danger of frost is past. A greenhouse or hotbed makes it possible to start the seeds 8 weeks before field-planting time. If they must be grown in a sunny window in the house, a 6-week start is as much as the vines can usually take without becoming overgrown and leggy. Plant 3 or 4 seeds in each pot and thin to one vine. When planting out, place 3 pots in each hill. After the vines have made a foot or two of growth, thin each hill to one or two vines.

If seeds are to be started directly in the garden, space hills 6 to 12 feet apart, according to variety planted and the fertility of the soil. On rich soil with high summer temperatures, the plants will make much more rampant growth than they will on leaner soils or with milder summers. In the South, practically all melons are started in the open. Seeds are planted 10 to 14 days before the last expected frost, so that the seedlings will come up as soon as possible after the frost. If there is any doubt that frost may overtake the seedlings, make two plantings in each hill a week apart, putting in half the seeds each time. A total of 8 to 10 seeds should be planted in each hill in a circle 8 to 12 inches in diameter. Cover the seeds with an inch of soil. After the first true leaves appear on the young plants, reduce the number to 4 or 5 plants per hill. Gradually thin them as they grow larger, until only one or two strong vines are left.

Watermelon vines may be mulched to keep down the weeds, but the mulch should not be applied until the soil is thoroughly warm. Weeds should be kept down by shallow hoeing until the mulch is spread. Straw, hay or chopped leaves are the best mulching materials to use. As soon as the soil is warm enough, spread a 6-inch mulch over the entire watermelon patch and draw it up to the base of the vines. This should be done before fruits begin to form, because the small fruits may be damaged by handling. The best time to apply the mulch is right after a rain, when the soil is thoroughly damp.

Commercial growers often thin fruits on the vines in order to produce larger and more uniform melons. This practice also speeds ripening, if no more than two melons are left on each plant. In the home garden, where melon size is not so important, the vines may be permitted to set more fruit if the gardener wishes, but late-set fruit should be removed. After the date when too few hot days and warm nights are left for maturing fruit, all blossoms should be removed from the plants before they begin to develop. The sooner these are removed, the more plant energy will be diverted to the development of the early-set fruit.

Enemies

Fusarium wilt is one of the most serious diseases of the watermelon. It is caused by a fungus which lives in the soil and penetrates the roots. Symptoms are a brown discoloration of the stems, followed by wilting of branches and the death of the plant. Planting wilt-resistant seed and long-time rotation with other vegetables are the best preventives. There is no known cure for the disease. This is also true of anthracnose, which is common to cucumbers and muskmelons as well. Leaves, fruit and stems of anthracnose-infected plants show water-soaked, sunken areas, and the plants die. Crop rotation, sanitation and planting of anthracnose-resistant seed is recommended.

Cucumber beetles attack watermelon vines, especially when they are young. As with cantaloupes, much of the damage from beetles may be prevented if the young vines are covered with plastic hot-caps or with netting until they have developed half a dozen leaves.

Harvest

Watermelons must ripen on the vines, because they do not develop more sugar or better color, after they have been taken from the vines while green. It is not easy to know exactly when they are ripe, but experience and possibly a plugged melon or two can help. Most melons are ripe when the tendril accompanying the fruit dies, but this is not true of all varieties. The color of the light patch beneath the melon is some indication—when the melon is ripe it turns from white to creamy yellow and becomes slightly rough. Best indicator is the sound that the melon makes when knocked with the knuckles. A ripe melon has a hollow sound, a green one a metallic ring. According to Mark Twain, a ripe melon says "punk" when thumped; a green one says "pink" or "pank."

WONDERBERRY

WONDERBERRY. *Solanum nigrum,* the deadly nightshade of the tomato family was developed by Luther Burbank into a fruit something like a blueberry which had a brief vogue in the early 1900's, and has now almost disappeared from gardens. Perhaps the reason for its lack of popularity is that the berries of the bush from which it was developed were thought to be poisonous unless thoroughly ripe, and the foliage also was thought to have toxic qualities. Fruit was used principally for making pies, but real blueberries, which are as easy to grow make better desserts, so the wonderberry has never been very successful.

YOUNGBERRY

YOUNGBERRY. *See* BLACKBERRY.

Section 5.

33—Organic Nut Culture: A Complete Listing

ALMOND

ALMOND. *Amygdalus communis,* the almond, is a member of the rose family and very closely related to peach. Although peaches and almonds have many characteristics in common, they have been separately cultivated as far back as our records show. Almonds originally came from North Africa and the Orient, and were known to all of the early Mediterranean cultures. Ten different varieties were grown and listed by the Greeks. The almond tree was depicted in early Greek and Egyptian paintings, and figured in Greek mythology.

Two types of almonds are classified roughly as bitter and sweet almonds, though some bitter may be found on sweet almond trees, and an occasional sweet almond is found on a bitter almond tree. Those grown for food are the sweet type. Bitter almonds are perpetuated chiefly for their rootstocks, on which sweet almond tops may be grafted or budded.

Almonds, like most nuts, are a concentrated form of food rich in protein, fat and the B vitamins. One cup of shelled nuts yields 850 calories. In that amount, there are 26 grams of protein, 77 grams of fat, 28 grams of carbohydrate, 332 milligrams calcium, 6.7 milligrams iron, .34 milligrams vitamin B_1, 1.31 milligrams vitamin B_2 and 5 milligrams niacin.

Almond trees are just as hardy as peach trees and, for decorative purposes, they may be grown in any climate suitable to peaches. They have the same winter chilling requirements as peaches. But almonds bloom 2 to 4 weeks earlier than peaches and the blossoms may be killed, not only by actual frost, but even by chilly weather at any time after buds form. As a result, the only part of the country where the trees may be depended upon to produce a crop year after year, is in the warm sections of the Pacific coast where peaches grow. Even in California, where almonds occupy the third largest acreage of all fruit and nut crops, few growers attempt their culture without making provision for heating their almond orchards in spring. For the home gardener living in the warm sections, almond trees should certainly occupy a place in the planting scheme, but they should not be depended upon to yield except in favorable seasons.

Varieties of almonds grown in California are chiefly Nonpareil, the best and earliest; IXL, Jordanolo, Peerless, Ne Plus Ultra, Drake and Texas, or Mission. All of these are self-sterile, so two or more varieties must be planted together to insure a set of nuts. In addition, Nonpareil and IXL are intersterile, and will not cross-pollinate each other. These varieties are all soft-shelled, and will produce in few places in the country outside of the Pacific coast.

Experiments have been made from time to time by growers in other parts of the country to develop hard-shelled almonds which might be induced to yield crops wherever peaches are grown. These attempts have met with some success, but the quality of nuts and consistence of the crops has not been sufficient to warrant any extensive plantings. However, a few of these varieties said to yield as far north as Michigan are Hall, Pioneer or Ridenhower. The Hall variety in particular, is convincing more northern gardeners to grow the almond tree. A dwarf variety has also been developed.

Culture

Almond trees need much the same type of soil and culture as peaches. Best situation for a tree is in deep sandy loam containing plenty of humus. They will not grow in clay, and even a heavy loam is likely to prove fatal. Where there is any doubt about whether the soil is light enough, it is better to plant trees budded on peach stock. Those budded on bitter almond will respond only to sand or sandy loam.

Trees should be given a space 20 to 30 feet in diameter when they mature. In California young trees are set out between December and the first part of March. In other parts of the country they should be planted as early in spring as possible. The soil should be damp but not wet when they are set out.

When planting almonds, prune back only the injured roots. Dig the holes large enough to contain all roots that are in good condition, and to contain them without crowding or bending the roots. As soon as they are planted, cut the tops back to a height of 24 to 30 inches. If the side branches are strong, select about 3, well-spaced, for the main scaffold branches of the tree, and cut off all others, leaving a stub, about one-fourth inch long, of each branch.

The second winter after planting, if the framework was started at planting time, branches need only be thinned out. Leave no more than two laterals for each scaffold branch after the second pruning. If the tree was pruned back to a whip when it was planted, the whole program is delayed by a year. Trees should start to bear the third or fourth year after being planted.

In arid areas, almond trees need irrigation as much as any fruit trees. Where the rainfall is sufficient to support shade trees, almonds do not have to be watered. Crops may be increased if manure is spread under the trees occasionally, but the trees do not respond as dramatically to a regular fertilizer program as many fruits do.

The outer hulls of the nuts split open in fall when almonds are ripe. When most of the hulls in the center of the tree have split, nuts are shaken or knocked out of the branches and gathered from the ground. Heavy rubber mallets may be used for jarring the nuts loose without injury to the bark. After harvest, almonds should be shelled and dried, to prevent mildew. Kernels are spread for drying in a partially shaded place, and are left until the meat is crisp. They are then stored in airtight containers in a cool place until used.

BEECHNUT

Beechnut. Nuts of the American beech, *Fagus grandifolia,* and of the European beech, *F. sylvatica,* were both at one time considered great delicacies but the trees are seldom planted for nuts today. Both species are beautiful large ornamentals, which may attain a height of 100 feet and live to a great age.

The beech is very hardy and will grow in almost any soil. It should be transplanted to its permanent place when it is very young, because it grows with a quickly-developing long taproot which must be preserved. When grown in nurseries, the professional growers transplant the young trees frequently to prune the roots, and to make possible their transplanting to the garden.

Beech plantings in the northern part of the country have been affected by a bark disease which has killed off some of the fine specimens. The disease is a fungus which is carried by the beech bark louse. It may be controlled by a dormant oil emulsion spray.

BRAZIL NUT

Brazil Nut. *Bertholletia excelsa.* Brazil nuts, also sometimes called Amazon nuts, can be grown under the near-jungle conditions of Brazil, Venezuela and Guiana, but in almost no part of the United States except possibly the very tip of southern Florida. The trees require great heat, moisture and rich soil. They reach a majestic height, and have two-foot leathery leaves, clusters of showy flowers with no petals but colored sepals and hard, woody, brown fruits 4 inches in diameter which are packed with 18 to 24 of the triangular nuts.

Brazil nuts are very rich in oil, which was, at one time, expressed and used in caulking ships. One cup of broken nut meats contains 905 calories, of which 92 grams are fat, 20 protein and 15 starch. They are quite rich in calcium.

BUTTERNUT

Butternut. *See* Walnut.

CHESTNUT

Chestnut. Within the last half century, the American chestnut, *Castanea dentata,* has practically been destroyed on this continent by a blight which swept through our stately chestnut forests in a very few years, leaving dead and dying trees in its wake. The American chestnut, now only known by a few persistent suckers which die almost as fast as they spring from the old roots, was once an important timber and nut tree through the eastern part of the country. It was one of the best of the chestnuts, for size and beauty. It is to be hoped that some nurseryman or botanist may find a strain that is blight-resistant, and that the species may still be perpetuated.

Meanwhile, a number of chestnuts native to Europe and the Orient have been tried in this country, with more or less success. None have the stately forest quality of our American chestnut tree, but some have been found to produce better nuts. The Spanish chestnut, *C. sativa,* is a 90-foot tree producing large nuts which can be grown in the southern parts of the country, roughly in the section where the inferior southern native chestnut, the chinquapin (*C. pumila*) grows. But a hardier and more successful import is the Chinese chestnut, *C. mollissima,* which can be grown in any part of the country that is right for growing peaches.

A close-up of the burrs and nuts of the Chinese chestnut showing splitting of burrs, and detail of foliage and nuts.

Chinese chestnuts are larger than the American nuts, and are equal to them in flavor and food value. Chestnuts contain about 6 per cent protein, 5 per cent fat and 42 per cent starch. The trees have a more spreading habit than their American relatives, growing no more than 50 to 60 feet tall, and with an equal spread. Flowers are borne in catkins, male and female in the same cluster. Nuts are enclosed in a prickly bur which opens in fall to allow 2 or 3 nuts to drop to the ground.

Culture

Chinese chestnuts should be planted, when possible, on a northern or northeastern slope, to delay blossoming in spring. Like peaches, their blossoms are likely to open before the weather is warm enough for good pollination and early growth.

Deep, fertile, sandy soil containing a large percentage of humus is desirable. When they are young, chestnut trees are shallow-rooted. While they will not tolerate moisture constantly standing about their roots, they are also sensitive to drought. When the top layer of the soil dries out, the young chestnut trees may quite possibly be injured or killed. Moisture-retentive soil, then, is a must. Also helpful is a heavy straw mulch over the roots of the young trees. As the trees mature, their roots go deeper and they are less susceptible to injury from drought.

Although the trees contain both male and female flowers, chestnuts need to be planted in 2's or 3's for cross-pollination. Two different varieties should be planted in the same area or no nuts will be produced.

Chinese chestnuts may begin to bear within 3 years after they are planted, if conditions are very favorable. They may be depended upon to bear, at most, 6 years after planting. Fully mature trees in the warmer sections of the country yield as much as 100 to 150 pounds of nuts per tree.

Thus far, Chinese chestnuts are propagated in this country principally by seeds. Grafting is possible, and even desirable, in cases where exceptionally good trees are available for scions. But little grafted nursery stock is available, the demand being greater than the supply of trees thus far.

Some Chinese chestnut trees may show a tendency to sucker in the garden. If the suckers are not pruned out, a shrubby plant will result. In this form, the tree may be attractive, but it is harder to care for than if the suckers are removed to leave a single trunk. Little other care is needed by the trees, after they have become established, beyond an occasional watering in time of drought.

As the nuts drop in fall, they should be picked up before they become moldy or wormy on the ground. After gathering them, put the nuts in water. Any rotten or defective ones will float and should be removed. Perfect nuts should be stored in containers which are not quite air-tight, for preservation of best flavor. Unlike most other nuts, they are best stored in a slightly damp atmosphere.

FILBERT

FILBERT. Filberts and hazelnuts stand in the same relation to each other as Chinese and American chestnuts—the filberts are imported members of the same genus as hazelnuts, and produce larger and, in some cases, better nuts than the native American species.

American hazels constitute the species *Corylus americana* and *C. cornuta,* sometimes called cobnuts, or beaked hazelnuts. Both yield thick-shelled nuts growing on shrubby trees or bushes not much more than 10 feet high, and hardy through most of the country. Three species imported from Europe and Asia produce better nuts, but they are more tender than the native American hazels. *C. Avellana,* European hazel, makes a shrubby plant up to 15 feet high with good-sized nuts. *C. colurna,* the Turkish hazel, grows in tree-form up to 60 feet, but it is used principally as rootstock for grafting filberts. *C. maxima,* native to western Asia and southeastern Europe, is the filbert, a tree or shrubby plant 10 to 30 feet tall, with large, well-flavored nuts. Commercial orchards in the states of Washington and Oregon grow principally the varieties of filbert and their hybrids.

True filbert trees, in themselves hardy, can be made to yield nuts in the East or in the northern states in any sections where peaches grow well, but they must not be depended upon to yield a money crop. As a planting in the home garden, they are attractive, useful and, if planted in favorable positions, will yield well in most years. The trees may be grown as spreading, dense-topped shade trees, as tree-clusters, like birch, on several trunks in a graceful arrangement, or as hedge shrubs, useful for screening on the home grounds.

Filberts bloom early in spring, their catkins developing on the first warm spring days. If the bloom is followed by a really cold night or two, the flowers may be killed and no nuts will set. In sections where heavy freezes are rare late in spring, filberts should bear well most years, providing winter temperatures do not fall below 15 degrees below zero. Dormant flower buds on the filberts can be killed at that temperature, even in midwinter.

Most satisfactory area of the country for filbert culture is the northern Pacific coastal area, where winters are mild, springs are fairly warm and late freezes are rare. In this section the choice varieties of filberts may be grown. The chief plantings there are predominantly of Barcelona, a variety which makes a large, spreading, productive tree with large nuts which fall easily from their husks as they ripen. Another popular variety is Brixnut, though the trees are somewhat less vigorous than Barcelona. All filberts must be interplanted with a second variety for cross-pollination, being self-sterile. Recommended varieties for cross-pollinating Barcelona are Daviana, Du Chilly, White Aveline, Du Provence, Monticello and Nottingham. White Aveline is also a good pollinator for Brixnut.

In colder sections of the country, American hazels are more reliably productive than filberts, but many of them produce small nuts which are not worth much as food. When a seedling is found in the wild which produces nuts of a good size, suckers should be dug and removed to the garden for cultivation. A few named hybrids have been developed and may be available from dealers. These are mostly offspring of the hazel varieties Rush and Winkler. Both of these produce fair size nuts, but on small bushes. When they are crossed with European filbert varieties, they may yield hybrids which are fairly hardy, yield large nuts, and develop into orchard-size trees about as large as plum trees. Some of these hybrids are Bixby, Buchanan, Reed and Potomac. Two or more of these varieties should be planted together for cross-pollination.

Culture

Where the climate is at all doubtful for filberts, trees should be planted in protected spots with northern exposures, in order to delay blooming as late as possible in spring. If the trees can be planted on a slope with good air drainage, they may be further protected from frost damage, as well as winter injury.

Filberts need a very deep, fertile and well-drained soil for long, heavy production. Roots of the mature trees go down as deep as 11 to 12 feet. If the soil is too shallow for deep penetration, growth and development of the tree will halt when the roots have reached the limit of their growth. And since the nuts are produced on new wood of the previous season's growth, a halt in tree growth means cessation of nut production. An indication of the suitability of the soil for filberts is the vigor of native hazel growth. If hazels in the area make only small, shrubby growth, the soil is not deep enough for filberts. If the local hazels are vigorous and tall, the soil is fine and filberts should thrive.

Trees should be planted in early winter in the warm areas or in very early spring in the cold sections. The distance between trees depends upon variety, as well as on the purpose of the planting. In Oregon and Washington orchards, filbert trees are spaced 25 to 30 feet apart each way. In the home grounds, if they are to be grown as trees, they should be given about 25 feet. In hedges, the plants may be set as closely as 6 feet apart. Or if they are to provide a small thicket which will serve as a screen, the plants may be set 10 to 12 feet apart and allowed to sucker freely.

Before planting a tree, prune away as much of the original layered wood as possible. Most filberts and hazels are propagated by tip layering. If the remnants of the tips or branches are cut off and the plants set so that roots are given a downward, rather than lateral, direction, fewer suckers will need to be cut away. Sift soil carefully around the roots, and pack it firmly to avoid air pockets.

All filbert trees should be protected 2 or 3 years after planting. Immediately after they are set, roofing paper or thick layers of newspaper should be wrapped around the trunks and tied loosely, as a protection from sunscald. The same precaution is needed in winter, when reflected sun from snow may raise the bark temperature in the daytime enough to cause heavy damage in the freezing night which follows.

Trees are usually cut back to 24 to 30 inches when they are planted, to balance roots and top. Any pruning and shaping, with the exception of suckering, should be completed during the first year after planting. Filberts have delicate bark which does not heal quickly over wounds, so the training should be early and severe, while only small scars are left after branches are cut back. Three to 5 scaffold branches are usually permitted to grow. Crotch breakage in filbert trees is rare, so the branches need not be as far apart as they are on fruit trees. After the first pruning, little should be necessary except the removal of root suckers, if a tree-form is desired, Suckers must be removed from the point of origin on the roots, and the job should be done several times each season during the first 3 or 4 years. Soil should be carefully pulled back around the suckers, and they should be cut back to the root. Any buds showing on adjacent roots may be rubbed off at the same time. Suckers should always be removed before the wood begins to harden.

If additional trees are desired, a few of the suckers may be permitted to grow after the tree has made enough growth to support them. When the sucker has grown for one year, it is bent down in spring and pegged to the ground a few inches from its tip. A cut or break in the bark where it touches the ground often helps to start roots from that point. Throw a few handfuls of soil over the point at which the sucker is pegged, permitting the tip to remain above ground. By the end of the season, a cluster of roots 2 to 4 inches in diameter should have formed, and the tip should have made some growth. Early in the second spring, sever the sucker from the layer. The new plant may be allowed to remain for another year, or it may be removed immediately to its new position.

Young trees should be well mulched with straw, strawy manure, alfalfa hay or residues of leguminous crops. If the mulch is grain straw, the addition of some concentrated nitrogenous material, such as cottonseed or bloodmeal, is desirable.

Filberts and hazelnuts are picked up after they have fallen to the ground in fall. If the gathering is done regularly, nutshells will be lighter in color and will be less likely to mold. After they are gathered, the nuts should be immersed in water and floaters removed. The balance of the crop is then dried by spreading it in a dry room, no more than 2 or 3 nuts deep, and stirring occasionally. When all excess moisture is dried out, the nuts should be stored in rodent-proof containers at a temperature of 55 to 65 degrees.

HAZELNUT

HAZELNUT. *See* FILBERT.

HEARTNUT

HEARTNUT. *See* WALNUT.

HICKORY NUT

HICKORY NUT. The hickories are valuable timber and nut trees of the walnut family. Their genus, *Carya,* includes pecans, as well as half a dozen nuts known as hickories, mockernuts, bitternuts, pignuts, and king nuts. All of these are native to the eastern part of the United States, and all cross-pollinate with ease. As a result there is almost no such thing as a pure hickory, pecan or pignut, except where special trees have been selected by horticulturists for asexual propagation. Some fine specimens of shellbark hickory, *C. laciniosa,* shagbark hickory, *C. ovata,* or pecan, *C. pecan,* might be found in the wild, yet, if nuts from any of them were planted, the chance of getting trees like the parents would be very small.

Best of the hickories are the shagbarks, shellbarks and pecans (*which see*). The other species may produce fine timber trees, but their nuts are too hard-shelled or have kernels too small or bitter for use as food. Shagbarks are the most hardy, with a native habitat extending up into Canada. Shellbarks are hardy through the Corn Belt, southern New England and southward. The range of the pecans is usually confined to the Appalachians, the southern plains and southward, though hybrids known as hicans can be grown north of that region, and some of the so-called northern pecans are hardy northward, though they seldom yield nuts except in very favorable locations.

Hickory trees commonly attain a height of more than 120 feet, but they are slow-growing. Male and female flowers are borne in the same tree. The nuts are the seeds of a 4-part fleshy drupe which becomes woody in maturity. Best flavored nuts are produced by the shagbark hickory, but the trees are hard to propagate and transplant so nursery stock is very limited. All hickories have a strong taproot which goes deep into the soil. When they are transplanted, this taproot must be transferred intact and must be kept straight, or the young tree will die. This is so difficult to do that it is easier to plant hickory nuts and to graft the seedlings in their places in the field, though the grafting should be done by an expert, because it is not easy to do.

When a purchased tree is planted, the hole must be dug deep enough to accommodate all the roots without bending. Special care should be taken to keep the roots moist while the work is being done. Fill the hole with rich topsoil, but do not use manure near the roots. The young tree should be tied to a stake so that the wind will not sway it during its first two years. It should be kept well-watered for several years after planting, and responds well to a deep mulch over the roots. If possible, use strawy manure in the mulch, but keep it pulled well away from the trunk, to prevent rodent damage. Pruning the young tree is confined to cutting back one-third of the top when it is planted. With no further help, it will form a shapely top which starts high above the ground in the mature tree.

Many hickory varieties have been developed, most of them hybrids. Among the hardiest for northern gardens are Wilcox and Davis, both of which yield nuts of excellent quality. Also recommended are Glover, Kentucky, Abscoda, Romig, Kirtland, Vest and Weiper. It is best to avoid hybrids with bitternut in their parentage, because some, like Beaver, yield inferior nuts. Where grafted trees are planted, bearing may start in 5 years.

MACADAMIA NUT

MACADAMIA NUT. *Macadamia ternifolia,* formerly known as the Queensland nut, is now being marketed from Hawaii under the common name of macadamia nut, by which it is becoming increasingly familiar in this country. It is native to Australia, where the handsome evergreen trees grow to a height of 40 to 50 feet in deep alluvial soil. In Florida, where their culture is being tried experimentally, the trees seldom attain much height, perhaps because the soil is not deep enough for them. In the shallow limestone soil they are sometimes uprooted by heavy winds. They retain their balance better in sandy, acid soils.

Macadamia trees are desirable ornamentals in the home grounds, in areas where frosts are few and light. Their leaves resemble dark green spiny holly leaves, and their panicles of white blossoms are sometimes a foot long. Only a few of the hundreds of tiny blossoms in the panicles ever set fruit, but these produce very hard-shelled seeds an inch or more in diameter, enclosed in a husk. With 6 to 7 months of warm weather, the seeds mature, the husk opens and the so-called nuts fall to the ground. These are gathered once or twice a week over a period of a month or more, and are thoroughly dried before storage. The seed shells are extremely hard, and cannot be cracked by ordinary home nutcrackers. A pair of vise-grip pliers is required for the job. The kernels taste like filberts or brazil nuts and have about the same nutritional value.

Macadamia nuts, with their hard shells, are slow to germinate under ordinary circumstances. They are more easily propagated if the nuts are stratified through the winter, then planted in individual pots and given bottom heat in the greenhouse in spring. Care should be taken not to disturb the roots when the seedlings are planted out in the garden. Seedling trees show a variance in form as well as in seeds, some of them being tall and slender, some low and spreading. If available, they may be grafted by side- or veneer-grafts with scions from some of the more desirable Hawaiian varieties. They thrive in deep, rich soil with plenty of moisture.

PECAN

PECAN. *Carya Pecan* is, botanically, one of the hickories, a genus of the walnut family. Its native habitat is the lower Mississippi Valley, but pecans are now commercially cultivated through almost all of the states on the southern border of the United States, and in gardens and even in the wild they can be found up the Mississippi Valley as far as Illinois, Iowa and Indiana.

A long and hot growing season is needed to mature good crops of pecans. In the southern states a season of 220 to 270 days, not only frost-free, but quite warm are considered necessary to a successful orchard. In these states—the cotton states—southern varieties may be grown. In some of the hot, sunny states, like New Mexico, the total may be cut down to 205 to 215 days. Northern pecan varieties require 180 to 200 days' growing season.

Pecan trees are relatively hardy, and blossoms open late enough so that they are seldom injured by late spring frosts. Also, the wood becomes dormant in early fall. But greatest danger lies in early freezes, with fall temperatures dropping very low. A tree which could normally be expected to stand many degrees of frost when the thermometer drops slowly may be killed outright if the fall freezes are too severe.

Twenty-year-old pecan trees often have branches spreading 40 feet, and a root structure spreading even further.

Fruit buds of the southern pecan varieties are more tender to winter injury than those of northern varieties. Some of the southern varieties are Stuart, Pabst, Moneymaker, Success, Schley, Burkett and Texas. The following are some of the recommended northern varieties: Major, Niblack, Indiana, Busseron, Green River and Posey. The hardy varieties have somewhat thicker shells and the nuts are smaller than those of the southern varieties, but the above northern varieties compare favorably with the more tender nuts.

Native pecan groves are found in deep alluvial soils which, though they may occasionally be flooded, drain quickly and thoroughly. Under cultivation pecans may be grown on a variety of soils, which meet the following requirements: soil must be deep to allow deep root penetration; it must be capable of holding moisture; it must be well-drained.

Pecan trees, like the hickories and walnuts, have long taproots which develop rapidly in the young tree. When they are very young, the taproots are almost as thick and long as the trunk of the tree, with only a few short feeder roots. After the taproot has made a deep penetration, side roots begin to grow, some only 6 inches to a foot below the surface. By the time the tree is 20 years old, its branches may have spread 40 feet, and its roots will have spread even further. Planting plans should be made to allow the mature tree a spread of 70 feet for branches and roots.

Pecan trees bear both male and female flowers, but they bloom at slightly different times so provision must be made for cross-pollination, if a crop of nuts is expected. In some trees the male flowers bloom early and the female late. In other trees the order is just reversed. It is important when choosing two varieties to be sure that their blooming seasons complement each other. A reliable nursery can supply information about the relative blooming periods of the trees it sells. If the trees are purchased from a local nursery, they should be suitable to the climate in the area in which they will be grown.

Culture

Young nursery trees should be dug with at least 3 feet of the taproot. Larger trees should have at least 5 feet of root. Lateral roots are usually trimmed to 8 to 10 inches. Holes should be dug deep enough to accommodate the taproot without bending, and to plant the tree one or two inches deeper than it grew in the nursery. Fill the hole with well-pulverized topsoil, tamped and watered in to eliminate air pockets. Leave a basin two inches deep around the trunk for frequent waterings during the first summer.

Trees with about half of their tops removed when they are transplanted make better growth the first season. The amount which should be cut depends upon the amount of root lost in the digging. After they are set, burlap or building paper should be tied loosely around the trunk to prevent sunscald. Also, a mulch of straw, hay or leaves around the base helps to keep the soil moist between irrigations.

Pruning of pecan trees should not begin until the top has made enough growth to shade the trunk. When a thick top has developed, limbs which are less than 6 feet above the ground may be cut off, one each year. Until they are removed, these lower branches may be kept headed back to keep them subordinate in growth to the upper limbs.

Pecan trees respond to fertilizing with manure by yielding good crops which increase every year until the trees reach a ripe old age. Trees not fed liberally are likely to produce well for a while and then decline when they have reached the limit of the resources of the soil. After the first year, a heavy application of strawy manure may be made to the root area to build the soil. When the trees are young, fertilizer should be spread in a zone 2 to 6 feet from the trunk. As the tree matures, the circle should gradually be widened, until it is about double the radius of the branches.

Harvest

When pecans mature the nuts separate easily from the shucks, which contain several nuts. Early-ripening nuts, in the South, complete this process early while temperatures are still high enough to shrink and dry the shells, permitting them to separate and drop to the ground.

When the season is cool, or if the nuts do not mature as early, sometimes they do not separate from the shucks by themselves, and must be knocked from the branches with light poles. Usually a tarpaulin is spread over the ground first, to catch the nuts as they fall. After they are separated from their shucks, the nuts may be stored in their shells or may be shelled, and the kernels stored in glass containers. Since unshelled nuts keep better than shelled, it is better to store them whole in a cool cellar until a week or two before they are to be used. After shelling, they keep best under refrigeration.

PISTACHIO

PISTACHIO. *Pistacia vera,* a species of the sumac family, is grown around the Mediterranean and in southern Asiatic countries. The pistachio tree attains a height of about 20 feet in Mexico and the interior California valleys, but grows somewhat larger than that in Europe. It is dioecious, having male and female flowers on separate trees, and grows well only in fairly arid areas. In some places it bears good crops only every other year, the alternate crops being blighted with fruits which lack the edible kernel for which the trees are grown. The fruits resemble wrinkled olives, are red and are about an inch long.

In California, the trees are fairly easy to grow, and do best in the sections which do not have the mildest winters. Varieties considered best for their nut-bearing qualities are Bronte and Aleppo, with Kaz as the pollinator. Unless a pollinating tree is planted with the female tree, no fruit can be set. In the home garden, if space is limited, a branch from a male tree may be grafted on the female tree, and only one tree will be necessary. Trees should be spaced 20 to 25 feet apart. They are usually grafted upon roots of Chinese pistachio, a much more vigorous tree.

QUEENSLAND NUT

QUEENSLAND NUT. *See* MACADAMIA NUT.

WALNUT

WALNUT. Half a dozen of our best known and relished nuts, including English or Persian walnuts, black walnuts, butternuts and heartnuts, are all closely related members of the *Juglandaceae* family, and of the genus *Juglans.* Most common in this country, especially in the northeast, is black walnut, *J. nigra,* a handsome hardy and stately tree native to all the states east of the Rockies. Black walnut trees produce a reliable harvest of strongly flavored, oily nuts year after year in all sections of the country, except the far south, where blossoming is sometimes too early. The trees are ornamental, the nuts good, though hard-shelled, and the timber is valuable.

English or Persian walnuts, *J. regia,* are grown mainly on the Pacific coast, except the Carpathian variety, which is more hardy. Most of the English walnut varieties are hardy only where temperatures do not go below 10 degrees below zero, even when the trees are entirely dormant. The Carpathians, however, have been known to withstand temperatures down to 30 below. The *regia* species is characterized by thinner shells which are easier to crack, milder flavor, lower and more spreading trees and greater tenderness in the fruit buds.

The butternuts, sometimes called white walnuts, are the species *J. cinerea.* These are also native American trees, growing in about the same region as black walnut. The trees have been greatly reduced in number in recent years because of a fungus infestation, which killed off many of the old trees. Butternuts are also hard-shelled and difficult to crack, but their flavor is more mild than that of black walnut.

The Japanese or Siebold walnut, *J. ailantifolia,* is similar to butternut both in flavor and in difficulty of cracking. Some Japanese walnuts are quite hardy, but some may consistently suffer frostbite in the colder areas. There is a great variation in the size and roughness of the nuts.

The heartnut is a variety of Japanese walnut, *J. ailantifolia cordiformis,* which is more consistently tender, but also yields nuts with smoother shells which are easier to crack. Heartnuts are able to withstand winters which are not too cold for peaches, so they may be grown in more protected northern areas and throughout the Atlantic coastal states, as well as in the Pacific northwest.

BLACK WALNUTS. Black walnut trees may be found growing in every country lane through the eastern half of the United States. The best of these trees are on deep, fertile soil, and form a good index to the layers of subsoil beneath the surface. Black walnuts make good shade trees for the home grounds and their branches start high enough so that early and late sun each day can shine on plantings beneath them. The trees grow rapidly in good soil, an increase of two feet per year not being unusual during the first 20 years.

Black walnuts mature in any area where the growing season is at least 150 days, with an average summer temperature of 62 degrees. Best soil is deep loam, either clay or sand with a clay subsoil, or deep alluvial soil that is well-drained. They do not succeed on infertile upland soil, or on soils with poor drainage.

About 100 or more varieties are available for planting, the best of them with comparatively thin shells and kernels that are in large sections. One of the earliest varieties developed is still one of the best—the Thomas walnuts. Also

A handsome, hardy and stately tree native to states east of the Rockies, the black walnut yields reliable harvests of richly-flavored nuts.

recommended for their nuts are Stabler, Ohio and Miller. No variety has yet been developed which is good in every respect, but these are the best of those offered at present.

Seedling black walnut trees which are to be found through the countryside may be top-worked to one of the named varieties by cleft grafting, if the trees are not too big. Trees as much as 20 years old have been worked in this way and, within a few years, were back in full production, yielding good crops of the new varieties.

When the nuts mature and fall from the tree, they are enclosed in a tough green husk which must be removed before they can be cracked. This husk stains the hands deep brown, unless they are protected with gloves. Nuts with the harder shells may be husked by spreading them on the ground, and driving a car back and forth over them. The husks should be removed before they turn black, for best-flavored meat.

After they have been husked, the nuts should be thoroughly dried to make it easier to remove the meat from the shells. After they have been air-dried for about a week, a day in an oven gently heated to below 90 degrees will complete the process. They should then be shelled, and stored in a cool place in covered glass containers.

English Walnuts, or Persian walnuts, have, until recently, been grown only in the warmest sections of the Pacific coast and in the East, south of a line drawn through Maryland and southern Pennsylvania. The best varieties are Placentia, Eureka, Chase or Ehrhardt in southern California; Concord, Payne, Franquette and Blackmer in northern California; Mayette and Franquette in Oregon and in the East. Although all varieties of the tender walnuts require some winter chilling, they will not stand temperatures below 10 degrees below zero.

Within the last 20 years, new varieties of *J. regia* have been developed, some of which may be grown as far north as Canada, and in most of the northern United States. These are the Crath Carpathians, which thrive in many of the types of soil found through the northeast. The best of the Carpathians thus far developed are Metcalfe, McKinster, Colby, Weng, Orth, Morris and Deming. In quality, size of nuts and ease of shelling, these walnuts compare favorably with the English walnuts grown in California.

Where grafted trees are planted, pistillate flowers are borne by the fifth season and, if older trees are present, nuts may be gathered from the 5-year-old trees. But the staminate flowers are not produced until the seventh year, so where the trees are self-pollinating, they will not bear a crop until that time. Carpathian walnuts have been known to produce after being exposed to a temperature of 30 below, and they are quite free of insect enemies and diseases.

Persian walnuts thrive best on rich, well-drained soil which is slightly acid, but they are likely to start into growth too early in spring, and both foliage and blossoms may suffer frost damage. Where hardiness is a problem, trees should not be encouraged to make too much vegetative growth by excessive fertilization.

BUTTERNUT. Butternuts are the hardiest of any of the northern nut trees and, if it were not for the fungus disease to which the trees are subject, would be one of the longest-lived. Butternut trees may achieve a height of 90 feet, with well-shaped, rounded heads. They grow moderately well on poor upland soil where black walnut will not grow, but are at their best on slightly acid to neutral soil that is fertile and well-drained.

The Carpathian walnut tree is able to withstand temperatures to 30 degrees below zero, produces thin-shelled, mild-flavored nuts.

Butternut trees grafted to black walnut rootstocks are said to be more vigorous and disease-resistant than those growing on their own roots. Varieties recommended for superior shelling qualities are Kenworthy, Kinneyglen, Buckley, Helmick, Craxezy, Herrick, Johnson, Sherwood, Thill and Van der Poppen. Not all of these may be available through nurserymen.

Butternuts respond well to feedings of organic fertilizers and to leaf mulches through their early years. Newly set trees should be heavily mulched as far out as the drip line, but a circle, two feet in diameter around the trunk, should be left free of mulch to discourage rodents. Soil should be prepared before planting by incorporating rock phosphate into it as deeply as possible, and a yearly application of a nitrogenous fertilizer in early spring will make the young trees grow better. Every 3 or 4 years additional rock phosphate should be worked into the top layer of soil under the tree.

JAPANESE WALNUT. The Japanese walnut makes a good shade tree, even in poor soil, but different specimens show a wide variation in their hardihood. They are not reliably hardy enough to bear yearly crops in areas as far north as New York state. They also show a great difference in their nuts, which are quite similar to butternuts, but are sometimes very rough and tough in shell. They are hardier than Persian walnuts, and hybrids between the two species may yield Persians with superior frost-resistant traits.

HEARTNUT. The heartnut is a Japanese walnut of a special variety, *cordiformis,* which closely resembles the species in foliage and growth habit, but yields nuts which are smoother and easier to shell. Some of the varieties are not hardy in the northern sections of the country, but are very satisfactory through the central, more moderate areas. Varieties include Bates, Faust, Fodermaier, Gellatly, Ritchie, Stranger, Wright and Walters. Of these, Walters and Bates are hardiest, most productive and are recommended for shelling qualities.

The trees thrive in both clay or sandy soil and make a rapid growth. They do not grow as tall as black walnut or butternut, and are of a more spreading habit. The heart-shaped nuts have smooth shells and a flavor similar to butternut. They are borne in clusters, sometimes as many as 10 on one twig, but the yields are not as heavy as black walnut. Thus far the trees do not have a record for long life. Whether organic culture will prolong their span beyond the present 30-year average, we cannot guarantee, but it is worth a try, because the nuts are superior.

Section 6.

34—Herb Gardening

By botanical definition, an herb is any plant which dies back to the root each year. By horticultural or culinary definition, an herb is an edible plant which is dried and used for its perfume, its healing properties or its flavor. As food, an herb is seldom eaten alone, but is usually taken in conjunction with other foods. (A description of each individual herb is given in the following chapter.)

Almost every plant which is not too poisonous has been used at some time as a medicine—a "simple" or "physick." And many flowers can be used for scent—more than we will list here. We are primarily interested in the herbs which are used as foods. A few are very nutritious in themselves, some others are injurious when taken in large doses. But most herbs are used in minute quantities for a subtle touch of flavor. In these tiny amounts, the vitamins which they doubtless contain will not help much in meeting the minimum daily requirement.

Herbs and spices get their strong, distinctive flavors from volatile aromatic oils, which may occur in leaves, seeds, flowers or roots, sometimes in all of them in the same plant, sometimes in only one or two. When herbs are dried properly, the oils are concentrated, so that much less of the desiccated product than of the fresh herb is needed for flavor.

Sweet spices and herbs are useful in fresh fruit desserts and drinks, where they often will take the place of sugar. In herb-flavored meat or vegetable dishes, salt is often unnecessary. For special diets, herbs and spices are much more healthful than saccharine or salt substitutes. Infusions of various herbs, which were first used by the Colonies to circumvent the hated tea tax, are still valuable in diets which prohibit coffee or tea. Recently there has been increasing interest in this use of herbs for tea.

Herb Families

The majority of herbs belong to 4 large plant families: *Labiatae,* the mint family; *Compositae,* the daisy family; *Umbelliferae,* the carrot family; and *Cruciferae,* the mustard family.

The family names are all descriptive of the plants. *Labiatae,* the mints, have blossoms which are lipped, like wide-open mouths. Their flowers are often tiny, as in the mints themselves and in lavender and thyme, but are somewhat larger in some of the sages. Stems of most of the members of the family are square, and flower parts are in 2's and 4's. Basil, rosemary and horehound belong to this group.

Compositae have blossoms which are actually a composite of two types of flowers—the disk flowers, packed tight together in the center, and the ray flowers, which are the "petals." Among the herbs in this family are camomile, costmary, tansy and pot marigold, or calendula.

The carrot family, *Umbelliferae,* is distinguished by its flat clusters, or umbels, of tiny flowers, many on hollow stems. This is by far the largest group of herbs, including in its number caraway, chervil, coriander, dill, fennel, cumin, lovage, angelica, parsley and many others.

Flowers of the mustard family, *Cruciferae,* are shaped like a Maltese cross. Beside mustard itself, the cresses are part of this family.

Planning the Herb Garden

Herb gardening employs more of the techniques of flower gardening than of vegetable growing. Many of the herbs are biennials or perennials, and there is a long list of annuals, as well as tender perennials grown as annuals in cool climates, or taken in to the window sill in the winter. Because many of the herbs are so attractive, they are often grown as part of a flower border, or are combined with shrubbery in the garden picture. A few herbs demand exactly the conditions they find in the rock garden, and many do well tucked into a dry wall, along with alpines.

Where a home gardener has enough time and space, a garden planted entirely with herbs near the kitchen door will be found very convenient, especially for the tiny quantities of fresh leaves which are added at the last minute to spark up a dish ready for the table. Such a garden may be a small corner where a dozen herbs are planted next to a sunny wall; it may be a rustic-type informal border around the service yard; or it may be as formal as an eighteenth century knot-garden.

The more formal the planting pattern of the herb garden, the more time must be devoted to its care. The dozen plants against a sunny wall may be seeded in early spring, weeded a couple of times, and allowed to grow on top of each other. But if the plants are to look tidy, and especially if they must follow a pattern, they must be carefully thinned and pruned or the pattern will never emerge, except in the eyes of the gardener. From the point of view of the cook, the tidy garden is best, because short, bushy plants which are well grown produce more essential oils than tall leggy ones crowded together on too-rich soil. Most herbs are inclined to sprawl. It is a temptation to plant them close together to make them support each other. If they occupy a few rows in the vegetable garden, no one will care how much they mix together, nor how messy they look. These considerations, and the time which the gardener can spare, should be taken into account when an herb garden is planned.

Propagation

Most plants for a formal planting should be started indoors, in the cold frame or in the outdoor seedbed and moved to the garden when they are a few inches tall. This applies to annuals, as well as perennials. Seed may be broadcast directly in the border space allotted it, but to carry out a definite pattern the exact number of plants needed should be carefully placed.

The following herbs can be started from seed with very little trouble and, if planted outside, should be seeded after the ground is warm in spring: anise, basil, borage, calendula, caraway, coriander, dill, fennel, nasturtium, summer savory. Some of these will seed themselves and come up year after year, with no further work on the part of the gardener. Seeds will not always fall where they are wanted, but they will be somewhere in the border. A few biennials are also planted every spring, especially those which, like parsley, are grown for their leaves. Burnet and clary sage are among these. Angelica, which is biennial, has seed which retains its viability for such a short time that it should be planted as soon as the seed ripens in fall. Parsley seed germinates very slowly, and should be put in the ground very early in spring, or started under cover.

Some members of the *Umbelliferae,* like parsley, have long sensitive taproots which transplant badly. These should be started in small pots in which they may grow until planted in the garden, when the whole root ball is moved with soil around it, and the roots are undisturbed.

Perennials which grow easily from seed are: wormwood, catnip, sweet cicely, horehound, hyssop, lovage, pot marjoram, sweet marjoram, oregano, clove pink, rue, sage, winter savory and common thyme. These may be spring-planted except seed of sweet cicely, which must be exposed to freezing temperatures before it will germinate. It is planted in fall outside, or stratified. To stratify them, place seeds in a container of moist sand or peat moss and bury the container in the garden during the winter. In early spring, it is dug up and the seeds are planted in the usual way, along with spring-planted seeds. This is done when seeds arrive too late for fall planting, or a new bed is to be dug.

Many of the perennial herbs may be propagated most easily by division of the roots, which is best accomplished in early spring before growth begins. Some perennials, like tarragon, whose hardiness is doubtful in the northern sections, should never be dug and divided after their shoots are more than two inches tall. By transplanting them early in spring, the roots are given every chance to spread and anchor themselves well before the next winter.

Herbs which may be divided are: lemon balm, costmary, chives, horehound, the marjorams, mints, thymes, tansy, tarragon, wormwood and sweet woodruff. Propagation by root cuttings, a somewhat different process, is also recommended for tarragon and sweet woodruff. To make these cuttings, dig a portion of the root in midsummer and store it in moist sand in a cool dark place. During the following winter, cut the root into 3-inch pieces, bring it into a warm room, planted in sand, and keep it moist. Sprouts may be set in the garden in early spring.

Stem cuttings taken from new growth may be rooted in sand to propagate many perennial herbs. This method is especially recommended for plants which will be wintered in the kitchen window sill to supply fresh herbs all winter. Cuttings for this purpose may be started in pots in a shady corner of the garden until they have rooted, then their pots may be buried in the bed beside the parent plant, until time to take them indoors. Winter pot plants grown from such cuttings will be much more satisfactory than whole parent plants lifted and potted.

The following perennials root easily from cuttings: clove pink, rose geranium, horehound, lavender, lemon verbena, sweet marjoram, sage, winter savory and common thyme.

Potted Herbs

Many annual herbs will do well on the kitchen window sill, or in any window where they will receive at least 6 hours sun each day. To have stocky plants to bring inside before the first frost, seed should be started in pots sunk in the garden some time before midsummer. Pots may be seeded and placed in a shady spot until the seed germinates, and then they may be sunk in a sunny bed until fall. A light but rich soil mixture should be used for potted herbs. Two parts loam, one part sand and one part compost is best for most. Use pots of generous size—not less than 5 inches for seeds or cuttings.

The transfer from the moist surroundings of the garden bed to the dry atmosphere of a heated house is hard on any plant, but especially hard on one or two small pots which are brought in to winter by themselves. A deep window sill entirely filled with plants, or a planter or window box with a dozen or more plants, will have healthier specimens than one or two isolated pots on even the sunniest sill. When many potted plants are grown close together, they create a humid atmosphere for each other, and each plant is happier and healthier. The atmospheric moisture in their neighborhood may be increased even more if the pots stand on 3-inch deep trays full of gravel, which are kept flooded with water. Bottoms of the pots should not touch the top of the water, unless they have an inch layer of gravel inside.

Herbs which can be grown in a sunny window sill year-round or through the winter, include the following: chives, basil, thyme, marjoram, anise, balm, borage, caraway, chervil, coriander, dill, parsley, rosemary and savory. Best temperature for robust plants is 60 to 65 degrees in the daytime, and 55 to 60 degrees at night. Once each week, the plants should be taken from the sill to the sink or bathtub for a shower. Use a fine spray on both sides of the leaves, with water that is at room temperature.

Pots of spearmint may join the other herbs on the window sill if small shoots are potted in late summer and left outside until the pots freeze solid. Then they are brought indoors and gradually thawed, in an artificial springtime. When they are ready for the sill temperature, they must have full sun, and they will send up tall shoots. Pinch them back early and they will make stocky plants.

Leave room on the sill for a few pots of seeds, which will be started during the late winter for next year's outdoor herb garden. Early in January is the best time to start seeds of the various sages. Thyme can be planted in a pot about January 15. A month later is about right for marjoram, followed soon by parsley. Annual herbs can start late in March or early in April, to complete the cycle.

Drying and Storage

Leaves, roots, seeds and flowers of many plants can be dried successfully in the time-honored way, to provide flavoring or scents through the winter. Stocks of dried herbs on cupboard shelves should be renewed from the garden each year, and old supplies thrown on the compost heap as newly dried leaves are ready to be stored.

Leaves are cut when the plant's stock of essential oils is highest. This usually occurs just before blossoming time. Tender young leaves will dry most quickly, and speed in drying is essential to the preservation of color and flavor.

Cutting should be done on the morning of a day that promises to be hot and dry. As soon as the dew is off the plants, snip off the top young growth—perhaps 6 inches of stem below the flower buds. Wash the leaves briefly if they are dusty, and hang the herbs, tied in small bunches, in the sun just until the water evaporates from them. Then immediately, before the sun starts to broil them, take them in and hang them in a hot, dry attic, away from strong light, but where the air circulates freely. If the leaves are not too small, as they are in thyme or savory, they may be removed from the stems and spread in a single layer to dry on a screen. Savory is best dried by pulling the entire plant and hanging it upside down.

Leaves are most flavorful if they can be dried in 3 or 4 days. But conditions are not always right for such quick drying. If they are not entirely dry in two weeks, they should be placed in an oven heated to 100 degrees and kept there until they will crumble into dust when rubbed between the palms of the hands.

When the drying is complete, remove the leaves from the stems and store them in tightly sealed bottles in a warm place for a week. At the end of that time, examine the bottles. If there is moisture on the inside of the glass, remove the contents and spread it out for further drying. Do not store in paper or cardboard containers, because they absorb the oils and leave the dried herbs tasteless.

Seed is dried in the same way as leaves. Heads or pods are picked when they have changed color, but before they begin to shatter. They are spread out on a screen and dried as quickly as possible in a warm, dry place. When they seem thoroughly dry, rub the heads or pods between the palms of the hands, and the seeds should fall out easily. Blow gently to remove the chaff, and store the seeds in bottles. Examine after one week and, if beads of moisture appear inside the bottle cap, spread out again for further drying.

Flowers to be dried for winter bouquets are usually picked just as they begin to open. They may be hung upside down to dry, and then covered with paper bags to keep out dust, until they are needed for winter arrangements.

Flower petals which will be used in cookery are treated in the same way as leaves. The flowers are cut on the first day they are opened. If petals alone are used, they are removed from the calyx and spread in a tray. Rose petals should have their claws—the narrow white portion at the base of each petal—removed. Flower heads used for tea, like camomile, are dried whole.

Flowers or petals for potpourris, sachets or pomanders are dried in the same way. After drying, they are mixed with spices, scented dried leaves or powdered roots, with a fixative added, and are stored for several weeks in a tightly closed container until used. The fixative, which may be salt, orris root, gum of benzoin, is mixed with the dried leaves and petals to prolong the life of the scent.

Colonial housewives scented their linen closets and drawers with sachets of dried herbs, which not only gave the linen a pleasant perfume, but was thought to chase moths. One recipe for a linen sachet contained sweet marjoram, rosemary, lemon verbena, rose geranium and lavender.

Pomander balls were also used to scent linen, or were worn on the person in lieu of perfumes. They ranged from simple concoctions of apples or oranges stuck with cloves and dried, to perforated crystal, ivory, gold or silver containers for dried petals and spices.

Herb Cookery

An herb cook should have a mortar and pestle in her kitchen for grinding some of the tougher stems and seeds, especially hard husks which are unpleasant to come upon in a soft omelet or a soup. Seed like angelica and coriander do not give their best flavor until they have been bruised thoroughly, or ground.

Herbs should be introduced into dishes with great discretion. A very slight herb flavor which is unidentifiable is an indication of the light hand of a good cook. Dried herbs are about 4 times as strong in flavoring capacity as fresh ones, and amounts should be adjusted accordingly. To get the best flavor from them, add dried herbs no more than one-half hour before the dish is ready for the table, or fresh herbs about 10 minutes before serving.

Some foods absorb the essential oils more quickly than others. Butter, eggs, milk, cheese and soup stock are especially absorbent. To get the most from the flavorings, add the herbs to these ingredients, when they are present in a dish, and allow them to stand a few minutes before introducing them to the pot.

When herbs are used to flavor cold drinks, they will give the strongest flavors if an infusion is made by pouring boiling water on them, and allowing it to stand a few minutes. Strain and ice, and then add a sprig of the same leaves, fresh, for color.

Fennel, dill, tarragon, garlic and basil are sometimes used to flavor vinegar which will be used in salad dressings and sauces during the winter. To make an herb vinegar, fill a bottle with leaves of the herb and pour in wine vinegar to the brim. Shake the bottle every day for 3 weeks, then strain off the vinegar and bottle it. Garlic vinegar is made in the same way, with as much or as little garlic as you like in your salads.

Combinations of herbs, referred to in recipes as *fines herbes* or *bouquets garnis,* usually contain parsley leaves or roots, chives and any other herb needed for the particular dish in question. A *bouquet garni* also usually includes bay leaf and thyme. The sprigs of herbs, if fresh, are tied together with a string which is trailed over the side of the pot, so that the herbs may be fished out before they go to the table. If they are dried, the herbs are tied loosely in a cheesecloth bag and a similar fishing contraption contrived, something like a tea bag.

A recipe for a *bouquet garni* used to flavor soup is as follows: one sprig each of sweet marjoram, rosemary and thyme; 2 sprigs of parsley, 2 green onions, 3 whole cloves, 1 or 2 blades of mace, 2 peppers (capsicum), 1 or 2 peppercorns, 1 stalk of celery and one-fourth of a lemon.

A "faggot of herbs" is a dried mixture which is used by the spoonful, whenever needed. It may be made by mixing one ounce each of basil and lemon peel, two ounces each of dried parsley, thyme and marjoram.

Herb butter, used on broiled or roasted meats, spread on sandwiches, or served with hot biscuits, is usually a combination of one tablespoonful of chopped parsley, chives and another herb, such as dill or marjoram, creamed with one-fourth pound of butter and a few drops of lemon juice. Any that is left over after the meal will keep in the refrigerator for several days.

Freezing Herbs

Two different methods for freezing herbs are offered here, one which will probably work well with some herbs, the other one with different herbs. We have not tested them all, so we recommend that you try both methods with your favorites, and select the one that seems best to you.

Mint and tarragon are very easy to freeze and work very well by the following method:

Gather leaves early in the morning while the dew is still on them. Wash, sort and whirl dry in a salad basket. Wrap the leaves in small squares of freezer paper, as many as will be used in one dish in each packet. Seal with freezer tape, and place packets in a box labeled on the outside, one herb to a box. Freeze as quickly as possible.

Or, after washing and sorting the herbs, tie them in small bunches and dip for one minute in boiling water, then two minutes in ice water. Drain, wrap and freeze.

These packets are so small that one freezer box is large enough to hold a winter's supply of most herbs. The flavor is very nearly that of fresh herbs, though the color is quickly lost after they are thawed. The following herbs may be frozen by one or both of the above methods: sweet basil, burnet, chervil, chives, dill, fennel, lovage, sweet marjoram, spearmint, parsley, sage, tarragon and thyme. Also, *bouquets garnis* may be tied in their bags, and frozen by the hot water method, ready to thaw and use during the winter.

The following list of herbs, any or all of which may be planted in an herb border, will attract the bees. If you are raising bees for their honey, you will find it more flavorful if the hives are in the middle of the herb garden. For the bees, plant lemon balm, basil, bergamot, borage, camomile, catnip, fennel, hyssop, lavender, pot marigold, marjoram, oregano, rosemary, sage, winter savory and thyme.

35—Organic Herb Culture: A Complete Listing

ALLSPICE

ALLSPICE. *Pimenta officinalis.* A 40-foot tree native to the West Indies belonging to the myrtle family, *Myrtaceae.* It is sometimes called Jamaica pepper, or Jamaica pimento, which serves further to confuse it with pimiento, the fruit of the genus *Capsicum,* which belongs to an entirely different family, the *Solanaceae.*

Allspice, given its name because it tastes rather like a combination of several other spices, is the dried green fruit of the tropical evergreen, *P. officinalis.* Berries are one-fourth inch in diameter when they are picked, but after they have dried for 6 to 10 days in the sun they shrink to a third that size, and turn dark brown.

Allspice is not grown in the United States, though a small amount is cultivated in Central America. The East Indies have become an important source.

Trees are cultivated on allspice plantations for 7 years before they begin to bear, but after reaching maturity they will continue to yield for as long as 50 years. After the berries have reached their full size, native boys climb the trees and toss down twigs bearing the fruit to women and children on the ground, who strip the berries and spread them out to dry.

The dried berries, which have a sweetish flavor, are used in pickles, pastry seasoning, curry powder, sachets and scents.

ANGELICA

ANGELICA. *Angelica archangelica,* a biennial herb belonging to the prolific *Umbelliferae* family. Native to Europe and North America.

Angelica is one of the few herbs not mentioned by the ancients. It was known during the Middle Ages, but was apparently not used until the seventeenth century in France and Germany.

A large handsome plant, angelica sometimes grows 6 feet tall, and is useful as a background planting in the herb garden or flower border. Its 3-part leaves grow in a heavy clump at the base, and thin out as they approach the rounded cluster of greenish flowers at the top of the stalks. It sometimes takes two years to produce flowers, after which the plant dies out and must be replaced.

Moist soil and a cool climate are the most important of angelica's requirements. Seed is viable for only a very short period after it ripens, so it should be replanted immediately, to winter outside, or it should be stratified.

Young stems are sometimes candied and used as a garnish on pastries and desserts. For this purpose they should be picked while they are still tender, and only the top few inches of each stem are used. Roots and leaves are collected the second year for drying. Seed ripens in late summer and is harvested before the heads shatter, if they are wanted for flavoring.

Fresh leaves may be used in salad, fresh fruit compotes, jams and jellies for flavor. Stems are sometimes blanched and used as a vegetable. Oil distilled from the root is used to flavor liqueurs.

ANISE

ANISE. *Pimpinella Anisum.* An annual herb of the *Umbelliferae* family native to Greece and Egypt, but not found growing wild there. Anise was known to the ancient Greeks, who used its seeds to flavor their bread and their wine. Since the seventeenth century, anise leaves have been used in Europe for salad.

Anise is one of the members of the carrot family which will mature its seeds the first year, if it has full sun and a growing season of 120 days. It grows a flower stalk that is 18 to 24 inches high. Moderately rich soil, which is fairly dry but with uniform rainfall, is best for it.

Seed should be sown one-half inch deep, 8 to 10 seeds to the inch, which are later thinned to about 8 inches apart. A row 6 to 8 feet long will grow enough seeds for almost any family. Plants will be less likely to lose their seeds to the wind if they are hilled during the flowering period. But enough seed usually escapes, no matter how careful the gardener, to self-sow the crop for the following year.

When seeds have turned grey-brown, head should be cut off and spread in a warm, dry place to finish drying. When they seem thoroughly dry, rub them between the hands to loosen the seeds and blow the chaff away.

Seeds are used to season many pastries and cookies in Italy, and may be used in any dish which needs a slight licorice flavor. In Holland, seeds are steeped in hot milk which is drunk at bedtime to induce sleep.

BASIL

BASIL. *Ocimum basilicum.* One of the *Labiatae* family, also known as sweet basil, the herb is native to western and tropical Asia. The Greeks knew it, and condemned it as an "enemy of sight and robber of the wits." The Romans cursed it as they sowed the seeds, in order to get a good crop. On the other hand, the Brahmins regard the plant as holy, and their women pray to it every day. Tradition has it in Italy that, if a girl leaves a pot of basil on her window sill, it is an invitation to her lover to visit her. Basil is now widely cultivated for its leaves, which are used for flavoring, and its seeds, which are eaten in parts of the Orient.

Basil grows from 18 to 24 inches tall, with a thick crop of succulent leaves on bushy stems, if the main stalks are pinched back early. Seed germinates in 5 to 7 days. Plant them one-half inch deep and thin the plants to stand 6 to 10 inches apart in the rows.

About 85 days after planting, just before the small purple flowers open, cut the stems 6 to 8 inches above the ground. Tie the branches in bunches and hang in a dry attic. If enough of the base of the plant is left growing, it will send up new branches and leaves, and a second cutting may be made later.

Leaves may be used fresh or dried to flavor many tomato dishes, eggs, cheese or rice dishes or spaghetti sauces. The Italians use basil so universally that they

Widely cultivated for its leaves and seeds, basil grows to about two feet tall. Pinch main stalks back early to have bushy stems.

put a fresh basil leaf in every jar of canned tomatoes. Leaves may also be used in the water in which shellfish is cooked, to give the fish a slightly spicy flavor.

BAY

BAY. *Laurus nobilis.* Also called sweet bay or bay leaf. The evergreen tree, which grows to 40 feet in warm climates, is European in origin. The Greeks knew it as the daphne tree, and wove its leaves, along with those of other fragrant herbs, into their wreaths of honor—hence the phrase, "to win one's laurels." Roman custom used a bed of laurel leaves under a wedding cake, which gave rise to a Latin proverb about finding a bay leaf in the wedding cake. This eventually became translated to our phrase, "a needle in a haystack."

A member of the *Laurinaceae* family, bay is a tropical tree which is indigenous to the Mediterranean, but is now grown in Central America and in our southern states. Further north it is sometimes grown in a container in the greenhouse in winter, and on a terrace or in a formal planting outside in summer. It will grow 4 to 10 feet high in a pot and may be trimmed to formal shape, which makes it an excellent accent plant for a patio.

Sweet bay is propagated from cuttings, and grown outdoors only in frost-free climates. It needs rich, moist soil. Its mature leaves may be picked and dried at any time.

Bay leaves are a meat, rather than a sweet spice. They are used in soups, gravies, meat or fish dishes and in cooked vegetables.

BEE BALM

BEE BALM. *Monarda didyma.* Also called Oswego tea, bergamot. Bee balm is one of the *Labiatae* family, and is native to North America, from the Alleghenies through New England and northward, westward to Wisconsin and into Canada. The Oswego Indians introduced the use of bee-balm tea to the colonists, who used it during their boycott of the British imported product. The French, who took the flower back from America, found a use for its volatile oils in perfumes and pomades. The leaves have a slight lemon flavor and are used to flavor fruit dishes as well as drinks.

Bee balm is a handsome 4-foot plant with large red flowers. Wild bergamot has similar flower heads, but the blossoms are lavender. Plants may be propagated by division of old clumps, when only the young outer roots should be replanted and the center woody ones should be discarded. Or they may be grown from seed planted outside any time from midsummer to November. Plants will grow to flowering size the following year, but for largest flowers, they should not be permitted to bloom until their second year.

For drying, the young tender leaves from the tips of the stems are picked in early morning, just before flowering. They should be dried quickly and stored in airtight containers, because their flavor is very delicate.

BORAGE

Borage. *Borago officinalis,* the leading member of the *Boragineae* family. Borage is an herb that was widely used by the Greeks and Romans, and continued to be popular through the Middle Ages. In the Elizabethan period, it was one of the most commonly grown salad herbs in England, and is still grown extensively there, though it has never been so popular in the United States. It has a flavor something like cucumber.

Borage is an annual growing 18 inches tall, with white, pink or blue flowers. In full sun and a moderately rich soil, it will do well in the vegetable or flower garden. It matures in about 6 weeks, so if it is wanted for continuous bloom or as a fresh herb, it should be sown several times in the spring and early summer. Seed may be broadcast in a corner of the flower garden to make a pink and blue clump.

Borage leaves have been used in almost any way that any herb can be used. The Greeks and Romans used it for coolness in their drinks. At one time, it was used as flavoring in a cordial which was thought to banish sorrow. Sprigs are sometimes used to garnish foods or drinks, like mint or parsley. The leaves may be cooked alone as a potherb, or they may be combined with other vegetables.

Leaves are used before the flowers bloom, for best flavor. If they are to be dried, the tender ends of the sprigs are cut and dried on cheesecloth and spread in a warm, dry place.

BURNET

Burnet. *Sanguisorba minor,* or *Poterium sanguisorba.* Also called pimpernel. A member of the rose family, native to north temperate climates. Burnet has been known from ancient times, and used medicinally, but was seldom cultivated until after the Middle Ages.

Burnet is a perennial which sends up 12 to 18 inch graceful stems with compound leaves which are topped by oblong white-rosy flower heads.

Seed may be sown in early spring or in fall in any good garden soil. Plants are thinned to stand 12 inches apart. Full sun and well-limed soil are essential. The plant is almost evergreen, so if leaves are required for salad, new seed should be sown yearly. If the leaves are to be dried, the plants may be propagated by division. The plant is hardy in most climates.

Young and tender leaves may be used in salads to impart a cucumber flavor, or may be used in drinks. Dried or fresh leaves may be used to make burnet vinegar. Dried leaves are used to make an herb tea.

CALENDULA

Calendula. *Calendula officinalis.* Also called pot marigold. A member of the *Compositae* family, but not to be confused with the genus *Tagetes,* which includes French and African marigolds. Calendula has scentless leaves, while the *Tagetes* species have foliage with a strong, rather unpleasant odor. Calendula is native to southern Europe, like many of the herbs of the Greeks and the Romans whose use has come down through the ages. Calendula petals have been used for a long time in Holland and England to flavor soups, and to color butter. The petals are often substituted for the much rarer and more expensive saffron.

Calendula is a hardy annual which will self-sow after its first year, and come back every year. Well-grown plants are one to two feet tall, but the stem is more often sprawling than erect. Yellow single or double daisy-like flowers are borne on fleshy stalks from spring until the hard freezes in the late fall or winter. Although it blooms through the summer, calendula flowers most profusely during cooler weather.

Any good garden soil with full sun will grow calendula, but it appreciates rich loam. Seed may be sown outside in late fall or in spring as soon as the soil can be worked. Thin plants to stand one foot apart. To have plants for the winter window sill, plant seeds in pots in late summer, and give them a good start by plunging the pots to their rims in the flower border, until after the first frosts. Plants started in this way are more satisfactory than garden specimens which are lifted and potted.

To dry flowers, cut them when they are fully open. Separate the yellow ray flowers, and spread them in a warm place to dry. Dry and store away from strong light to preserve the color. The petals have a mildly spicy flavor, and are used fresh in salads, dried or fresh in bouillabaisse, fish chowders, with game or rice.

CAMOMILE

Camomile. Two different members of the family *Compositae* are known as camomile. *Anthemis nobilis,* a creeping mossy ground cover, is the English or Roman camomile, used to make herb tea. The German or Hungarian camomile, *Matricaria Chamomilla,* grows to 15 inches tall, and is an annual which is sometimes used in the flower border, but is grown chiefly for medicinal purposes.

English camomile is a hardy perennial, native to Europe, but naturalized along the central Atlantic seaboard. It is an attractive plant for edging a flower border, where its daisy-like flowers with yellow disks and white rays bloom from midsummer to frost.

Seed should be planted in early spring outside, and plants thinned to stand 9 inches apart. They spread rapidly and will need to be divided every 2 or 3 years. Any good garden soil in full sun will grow them admirably.

For camomile tea, gather the flower heads when they are in full bloom and dry them. The yellow disk flowers are used for herb tea.

CARAWAY

Caraway. *Carum Carvi.* Also called kümmel. Use of this member of the family *Umbelliferae* predates recorded history. Seed was found in the ruins of lake-dweller homes in Switzerland, and were also used by Greeks and Romans. Caraway is indigenous to Europe, the Orient and northern Asia. It is now most grown and used in northern European countries and in the United States. It grows well in the northern states.

Two varieties of caraway are available for planting, an annual and a biennial. The annual variety is sown in early spring, and matures its seeds in about 70 days. The biennial may be sown from spring to midsummer to mature its seeds the following summer. Germination and early growth of both varieties is very slow.

Full sun, a fertile clay loam and a position protected from strong winds are best. Caraway will grow well where cool temperate climates prevail, and is very winter-hardy. Flower stalks, which grow to 24 to 30 inches, should be protected from the wind while seed is ripening, because the heads shatter easily.

When seed is just turning brown, cut off flower heads and permit them to dry in a warm dry attic, spread either on fine screening or on clean paper on the floor. Seed is easily loosened, when dry, by rubbing the heads between the hands, and chaff may be blown away.

Seed is used in bread, cookies, cooked with sauerkraut and with goulash and may be ground and sprinkled over liver. Leaves are sometimes added to salad for flavor, or are mixed with cooked vegetables, mashed potatoes, soups or cream cheese. Roots may be boiled as a vegetable and served in the same ways as parsnips.

CARDAMON

Cardamon. *Eletteria cardamomum maton,* a member of the tropical ginger family, *Zingiberaceae,* has been cultivated in the East Indies from time immemorial. The Arabs and East Indians, who like to eat the seeds as a sweetmeat, have been growing them since at least 800 B.C., and they are now also grown on Crete.

Cardamon seeds, actually fruits, are grown on a woody perennial or shrub 8 to 12 feet tall, which bears its fruit after the third year. After a harvest, the whole plant is dug up and transplanted. The old stem dies back, and a new stem is sent up from the roots. The next crop is ripened the second year after transplanting.

Whole cardamon seed pods are used in pickling, and are chewed by the natives where they are grown. Ground, the spice is sometimes used in pastry, and is one of the ingredients of curry powder.

CASSIA

Cassia. *Cinnamomum cassia blume* is a member of the *Laurineae* family which has been grown and used since before 2700 B.C. on tropical Oriental islands and mainland. Bark of the cassia tree, and of several allied species, is often sold as cinnamon, its flavor being similar and stronger than real cinnamon.

Culture and harvesting of cassia is the same as that of cinnamon, which belongs to the same botanical genus. In addition to its bark, cassia is grown for "cassia buds," actually the unripe fruit which is dried and used in pickling.

CATNIP

CATNIP. *Nepeta Cataria.* Also called catmint. This member of the family *Labiatae* is found growing wild in Europe and Asia, especially in the area near the Himalayas. Our name for it is thought to have come from the village of Nept in Tuscany, where it has always been profuse. It has been used since the fifteenth century as a sort of spring tonic herb, which is eaten in new green salads in Europe. Tea made from its dried leaves was brought to America in Colonial days, and the Indians who drank it liked it so well that they cultivated the plant, so that now it may be found growing wild in many parts of the country.

Catnip may be grown in almost any garden soil in sun or part shade, in a temperate climate. If it is grown in sandy soil, the flavor is stronger than when it is grown in clayey loam. The plant, which grows to a 3 to 4 foot perennial, will mature from seed the first year. Seed may be sown in early spring or late fall, but fall sowings will be more robust. Catnip may also be propagated by division, the plants set 12 to 16 inches apart. Foliage is attractive, the heart-shaped leaves being green above and grayish below, and the flowers blue-violet.

Tops should be cut for drying when blossoms are fully open. Drying and storage should be done in darkened or semidarkened places, in order to preserve the color of the leaves for the tea. If the dry leaves are stored in a cool dry cellar, they will retain their flavor over a longer period.

CHERVIL

CHERVIL. *Anthriscus cerefolium.* This member of the carrot family, *Umbelliferae,* is native to the whole Eurasian continent, and has been used for many centuries in much the same way as its near-cousin, parsley. Its flavor resembles that of tarragon, and in appearance it is a lighter green parsley, with finely cut, fern-like leaves. Like parsley, some varieties grow tuberous roots which may be cooked and eaten as a vegetable.

Chervil is an annual which grows to two feet. Seed should be planted early in spring and again in late summer for a fall crop. Like others in its family, chervil has weak seeds which germinate poorly and are a long time in the process. They should be sown rather thickly where the plants will stand, because it does not transplant well. Plants should be thinned to stand 6 inches apart. Chervil likes the shade, so it may be planted in corners near the shrubbery, where the soil is rich and moist.

The delicate flavor of chervil is better preserved by freezing than by drying. Tender young leaves for freezing may be cut when the seedlings have been up 6 to 8 weeks. For drying, they may be allowed to become somewhat more mature. The leaves should be dried rapidly or their flavor will be entirely lost. Spread them in very thin layers on screening in a warm room.

The flavor of chervil is especially good with fish, and it may be substituted in fish recipes that call for parsley. It can also be used to flavor fat which is brushed or basted over broiled or roasted chicken. Fresh leaves are added to salads, and frozen ones may be used with cooked potherbs. Tuberous roots of the fleshy-rooted varieties are cooked like carrots, and have a unique flavor.

CHIVES

CHIVES. *Allium Schoenoprasum,* one of the mildest and most delicately flavored herbs of the onion clan, *Liliaceae* family. Chives, like all the others of its family, was grown and used by the Chinese in very early times. It is mentioned in their literature as early as 3000 B.C. It was also used in other parts of Asia, and in Europe, and the plants are pictured, but they were not described there as a variety of perennial onion separate from other types until after the eighteenth century.

Chives are hardy perennials which may be grown from seed or from division of the clumps. The tubular leaves grow 8 to 12 inches tall, with round heads of lavender flowers rising a few inches taller.

Chive seed should be planted when very fresh, because it does not remain viable long. In the garden, seed should be planted in full sun in sandy loam if possible. In pots, it should be planted thickly in soil that contains a large percentage of sharp sand or even fine gravel, mixed about 50-50 with screened compost. Bulbs started in pots indoors may be later planted outside in the vegetable garden, flower border or rock garden. Bulbs should be spaced 5 inches apart when set outside, because they will quickly multiply and make small clumps wherever they are planted. Clumps may be taken up and separated every 2 or 3 years.

Pots of chives may be kept growing on the kitchen window sill all winter if clumps are lifted and potted in fall. The pots are best stored in the cold frame until a few weeks before they will be wanted in the kitchen. Several pots, brought in at intervals through the winter, will insure a constant supply of fresh chives, so drying or freezing of the tops is unnecessary.

CICELY, SWEET

CICELY, SWEET. *Myrrhis odorata.* Also known as myrrh. Sweet cicely is one of the *Umbelliferae.* Its genus name, *Myrrhis,* comes from the Greek word for perfume. Although it is native to Asia Minor, it is not the myrrh mentioned in the Bible, which was an exudation of a small tree, *Balsamodendron Myrrha.* However, the use of sweet cicely does go back as far as the Romans, who substituted it for and sometimes called it by their names for anise and chervil. It is botanically and horticulturally closely related to both of the latter herbs.

Sweet cicely is a perennial which grows 2 to 3 feet tall, with light, airy foliage, finely divided, and spidery white flowers. It needs an acid, moist soil and at least part shade. Seed should be sown in the fall as soon as it ripens, or should be stratified and planted in early spring. Or the plants may be propagated by division. Left to itself, sweet cicely will self-sow and multiply to form a fine clump when conditions are favorable.

Seeds are picked green and used to flavor liqueurs or are mixed with other herbs. Dried seeds can be used for a slight anise flavor in cooked cabbage or other dishes. Leaves are sometimes added to salads.

CINNAMON

CINNAMON. *Cinnamomum zeylanicum Nees.* This is the true cinnamon, and closely related to the many species of cassia which are sometimes substituted for it. It, too, belongs to the laurel family.

Cinnamon is one of the oldest spices known. Its history dates back at least to 5000 B.C. It has been used in religious rituals throughout the centuries. The Egyptians and Hebrews used cinnamon-scented perfumes in their temples, the Romans dedicated it to Mercury and burnt cinnamon incense in his temple, and the Catholic church today still burns incense, one of whose ingredients is cinnamon.

The short, bushy trees which yield the precious cinnamon bark grow only in Ceylon and India. Early Arab traders, whose caravans brought it to Europe, shrouded its source in mystery for many years, in order to protect their cinnamon monopoly.

Cinnamon trees, or bushes, grow to a height of about 5 or 6 feet in their first 6 years. After they have attained this size, 2 or 3 shoots may be selected for peeling. After the tree has begun to yield, 5 or 6 shoots may be cut every other year. The leaves of the bush are red when they first open, and then turn to green. When they are all green, peeling may begin. Shoots are cut from the tree, preferably after a rain, when the bark separates most easily, and the bark is first carefully scraped. Then it is peeled in long strips and dried. As it dries, the bark curls into long quills or pipes. Smaller quills are inserted into the long ones, they are sorted for thickness and quality and tied into bundles for export. Best quality cinnamon is in long quills of 3 to 4 foot length. Pieces down to 8 inches long form the next grade; chips the third; and the lowest grade consists of scrapings and trimmings.

Whole and ground cinnamon are used in cooking. The whole pieces are sold as "stick cinnamon," the broken pieces as "cracked cinnamon."

CLARY SAGE

CLARY SAGE. *Salvia sclarea,* one of the larger members of the *Labiatae* family which is useful in the flower garden as well as in the herbiary. It is native to the Mediterranean area, and is also found growing in parts of the Orient. Clary was used by the Romans, but was not known in Britain until the sixteenth century.

Clary is a biennial but, to have flowers each year, seeds must be planted annually. Flower stalks grow to a height of 3 or 4 feet from a strong stem that bears pebbled leaves, large at the base, becoming smaller toward the top of the stalk. Flowers are white with pink and blue bracts, and resemble the scarlet sage or salvia plants much used for bedding. They have a strong aromatic odor.

Ordinary garden soil and full sun are needed for clary. Seed may be started directly in the border, or may be planted in a cold frame and moved to the border in the spring. Plants should be spaced at least a foot apart when they are set out.

Clary leaves are used to impart their peculiar flavor to muscatel wine. They are used fresh in drinks or omelets; dried they may be used in sachets or to flavor soups, wines or beer. Flowers are dried and used to make herb tea.

CLOVES

CLOVES. *Caryophyllus aromaticus.* Cloves are the dried flower buds of evergreen trees of the *Myrtaceae* family which are native to the Dutch East Indies. Cloves have been known from ancient times as one of the precious spices of the East. They were used by the Chinese, Persians, Greeks and Romans, all of whom held them in high regard. In the East Indies, prosperity has always been closely allied to the spice trade, until the clove tree has become a symbol to the natives. In the Moluccas, a family plants a clove tree when a child is born. They believe that as the tree prospers, so will the child's life.

The clove tree is tall and majestic, and bears its flowers for more than a hundred years on rather delicate, frail branches. First blossoms appear when the tree is 7 years old. They grow in clusters of 10 to 50 flowers on a stem, and are picked for drying just as the calyx begins to turn pink, but before the petals open. Two harvests a year are gathered from the lower branches of the trees, and from as high as the tree will bear a man's weight. The clusters are dried in the sun in 4 or 5 days. As they dry, the buds and their stems are removed from the flower stalks.

Cloves are used whole in most pomanders, sachets and potpourris. Their oil has been in use for centuries in perfumes and cosmetics. Beside their familiar use in sweet dishes and confections, cloves are also added sparingly to meat dishes, soups and stews for just a touch of spice.

CLOVE PINK

CLOVE PINK. *Dianthus caryophyllus,* also called carnation. One of the leading members of the pink family, *Caryophyllaceae.* Pinks are natives of Europe and Asia and, though not much used in foods at present, were at one time greatly valued when cloves were the exotic and expensive spices of the mysterious East.

Clove pinks are perennials, hardy in most of the United States with winter protection. Their small carnation-like pink or white flowers grow on jointed blue-green stems from a cushion of grassy blue-green leaves. They are the most fragrant of all the *dianthus,* having a penetrating clove odor. They may be propagated from cuttings, seeds, or by layering their recumbent leaf stems.

Flowers were formerly used to make a spicy conserve and to impart a clove-like flavor to wines and vinegars.

CORIANDER

CORIANDER. *Coriandrum sativum,* one of the earliest cultivated members of the *Umbelliferae.* The Chinese used the root and seeds before 5000 B.C.; seed was found in an Egyptian tomb of the twenty-first dynasty; it is mentioned repeatedy in the Bible as having been used by the ancient Hebrews; the Greeks used it before the Golden Age of Athens; Caesar's legions carried it into northern Europe; and it was in use in England before the Norman conquest.

Coriander is an annual that grows one to two feet tall, with fine-cut foliage, umbels of delicate pale pink flowers, and a fine, tenuous root system. Seed can be matured in 3 months from planting date. Because of its very delicate root system, it does not transplant well and seed should be started in the garden where the plants are to stand.

A well-drained, sunny site with fairly rich soil is best for coriander. Protection from the wind will help to conserve the seed heads, which shatter easily when ripe. Seed germinates easily, and should be planted an inch deep. Plant after the last frost date in the North, in fall in the South. When the plants are 3 inches tall, they should be thinned to stand 10 inches apart. Thin them by pinching off at ground level, rather than by pulling up, which might disturb the roots of the seedlings which are allowed to remain.

When seed begins to turn brown, cut off plants at the base and hang them in a dry attic, suspended over a container which will catch any dropping seeds. Allow the seed to dry thoroughly, then thresh out by beating the plant heads on the inside of a deep pail. Blow away the chaff and store.

Seed must be thoroughly dry and aged for a while before it develops its best flavor. If it is eaten green, the flavor will be found unpleasant.

Whole seeds are sometimes sugar-coated and eaten as candy. They are also used to flavor liqueurs, gin and vermouth. In cooking, the seed should be crushed before using it to flavor meats, sausages, fruit dishes or desserts. Leaves are sometimes used in soups, stews and sauces.

COSTMARY

COSTMARY. *Chrysanthemum balsamita.* Also called bible leaf. A large rather coarse member of the family *Compositae* native to the western shores of the Mediterranean. Costmary was known from ancient times, and grown and used on continental Europe since the time of Charlemagne; in England, since the sixteenth century, it has been used to flavor ale and negus. Its second name, bible leaf, is thought to have arisen from the custom of the early American colonists of marking their places in their Bibles with its long, fragrant leaves.

Costmary forms large perennial clumps of long oval leaves which send up 4- to 6-foot stalks of buttonlike yellow flowers. It is a hardy plant which may be grown in almost any temperate climate. Its leaves have a strong bitter anise-like flavor and a minty odor.

Fertile but rather dry soil is best to its liking. If the plants are wanted for decorative value, they should be planted in full sun. In the shade, the leaves develop more flavor, but the plants will seldom bloom. Propagation by division of underground runners is most satisfactory, but costmary may also be grown from seed started early in spring indoors. Seedlings should be spaced two feet apart in the border when they are planted out after the last frost, and the older plants may be left undisturbed for several years.

Leaves are pulled early in spring for drying, being pulled in much the same way as rhubarb. They are hung to dry or are spread on screens away from the light to preserve their color. After they are thoroughly dry, the tender part of the leaf is stripped from the midrib and stored.

Dried or fresh leaves may be used sparingly to flavor meat or game. Dried leaves are also used to make an herb tea.

CUMIN

CUMIN. *Cuminum cyminum.* This member of the family *Umbelliferae* is native to the eastern Mediterranean, especially the upper reaches of the Nile. It follows that it was known early to the Persians, Egyptians and Hebrews, and its use followed civilization around the Mediterranean and into France and England, though it has never become so popular in the northern countries as it has always been in the Near East. At one time, when pepper was rare and expensive, the Romans substituted cumin for it. In the time of Christ, the value of the cumin had increased until it had become negotiable in payment of taxes.

Cumin needs 4 warm months to mature its seed, so it is better grown in the South than in the northern states, although a crop may be matured in the North if plants are started early indoors. Seedlings should be planted outside in warm sandy loam when the spring has grown really warm. Plants are low and sprawling, because their weak stems will not support the weight of the large heads of flowers and seed. When they are planted outside, the seed is sown at the rate of 16 to 20 seeds to the foot, and the plants are not thinned. A thick growth helps support the heavy heads, and keep them off the ground. Like most of the carrot family, cumin plants have finely divided foliage and lavender-white flowers.

Seeds ripen late in the fall. When the heads begin to turn brown, they should be cut and dried indoors. When thoroughly dried, they may be separated from the heads by rubbing between the hands. Seeds are small and may be separated from most of the chaff and stems by being strained through a sieve.

Seeds are used to flavor cheese, bread, cookies, sausage, meats, vegetables, fish and game. When ground, they are sometimes among the ingredients of curry powder and chili powder.

DAMASK ROSE

DAMASK ROSE. *Rosa Damascena.* The old-fashioned damask rose is not the only member of the family *Rosaceae* used as a flavoring herb, but it was the one traditionally sought in the old-time gardens. Fruit of many of the species roses have been used for food. Most notable at the present time is the fruit of *Rosa rugosa,* for which, see the entry in the section on fruits. Other roses whose hips have been used for fruit include *R. canina,* the brier rose; *R. cinnamomea,* the cinnamon rose; *R. macrophylla; R. nutkana; R. rubiginosa,* sweetbriar; *R. spinosissima,* the burnet rose; and *R. villosa.* Tender shoots of the ash-leaved rose, *R. fraxinellaefolia,* were eaten by the Indians in early spring. The Chinese make a ragout of the whole flower of *R. semperflorens,* the monthly or red China rose. They also use the petals of *R. centifolia,* the cabbage rose, to scent their tea.

The damask rose grows in bush form, often as high as 8 feet, with pale pink to red, double, fragrant flowers. It is one of the forebears of the modern hybrid perpetual roses. Plants flourish over a life span of about 5 years, and then usually need to be replaced.

Damask roses were grown in old-time herbiaries for their petals, which were used in sachets and potpourris, and were also used as a subtle sweet flavoring in honey, jelly and dessert-type foods. Petals were also sometimes used in hot tea or in cold drinks for flavoring.

To dry the petals, pick the flowers in early morning of the first day that they are open, as soon as the dew is dried. Pull the petals from the calyx and remove the white claw at the base of each. Spread to dry in the shade on a screen.

Fresh petals are used in France to make a delicately flavored vinegar; in England, to make preserves and they are candied; the Persians and the English both made rose-petal wine at one time; and in India they are placed between layers of rock candy in jars that are permitted to stand in the hot sun until they become a sort of jam called *gulkanda.*

DILL

DILL. *Anethum graveolens.* One of the *Umbelliferae* family found growing wild in many parts of Europe and Asia. The dill plant was grown by the Greeks, who used it in many dishes as well, as for a decorative herb which was woven into garlands and wreaths. The Romans, too, used the flowers to decorate and scent their banquet halls. Its use did not become common in England until some time in the sixteenth century.

Dill is an easily grown annual which reaches a height of 2 to 3 feet. It can be grown in the full sun in any good garden soil. Seed may be sown early in spring, or in fall to germinate the following spring. After the first year, the plants will self-sow. Also, if some of the first early-planted seed heads are allowed to shatter in the garden, small new plants will spring up in the row to take the place of the first ones which die off after forming seed.

An easy-to-grow annual, dill seed should be sown in early spring; plants will self-sow after the first year.

Seed is sown rather thickly and covered with one-fourth inch of fine soil. If 15 to 20 seeds are planted to the inch, they may not need thinning, because they germinate weakly. When necessary, thin the plants to stand about 3 inches apart.

Dried dill foliage loses its flavor. If it is needed in the winter, the best means of carrying it over is to grow pots on the kitchen window sill or to freeze it. Some dill flavor may be extracted from dried seeds, which are used in gravies, stews and meat dishes. Flower umbels may also be picked just before blossoming and dried. Picked at the same time, the umbels are used fresh to make dill pickles. Both blossoms and foliage may be used to make dill vinegar, which retains its flavor through the winter.

FENNEL

FENNEL. *Foeniculum vulgare dulce.* Fennel, or sweet fennel, which is native to Europe, is another of the *Umbelliferae* grown for its fresh foliage and seeds. Greeks and Romans both used fennel with dill for scent and for decorating their heroes. The Romans became so fond of its flavor that they used it in most of their dishes. In England, where it was introduced in about the fourteenth century, it was used as a potherb. Seeds of fennel were not used as food or for flavoring until about the fourteenth century.

In mild climates, fennel is a perennial. In areas where winters are cold, it is grown as an annual. The common fennel grows, where it is perennial, to a height of about 5 feet with heavy heads which need support when they develop seeds. Sweet fennel, the variety *dulce,* grows only about two feet tall and has broader stems which are sometimes eaten as celery. Finochio, or Florence fennel, is still smaller, and has a bulbous base used as a vegetable. See FENNEL in the vegetable section.

Perennial fennel is planted out in the garden with plants spaced about 18 inches apart. Or seed may be sown about 12 to the foot, seedlings thinned to 4 inches apart and then transplanted, when they begin to crowd each other, to the wider spacing. Full sun and good soil will grow the best fennel. The perennials will not bear flowers the first year, but after their second year they will bear every year for several years. Then the plants become exhausted and should be replanted.

Fish, salads and soups may be seasoned with fresh fennel leaves. Seed is picked to dry when it begins to turn color. The dried seed is used whole in flavoring liqueurs, cakes, cookies and apple dishes. Ground, it may be added in small quantities to cheese, fish and vegetable dishes.

GINGER

GINGER. *Zingiber officinale.* A member of the ginger family, *Zingiberaceae,* native to southern Asia but now grown in Africa, the West Indies, especially Puerto Rico and Jamaica, and even in southern Florida.

The ginger plant, which is one of the most beautiful tropical plants, was used in the Orient for centuries, but was not known to Europe until Marco Polo brought it back from his journeys. The Spanish explorers brought it to America in the sixteenth century.

Ginger resembles, in its manner of growth, the cannas which we plant in showy beds. Stout cane-like stems rise 3 or 4 feet high from fleshy rhizomes, which are the source of the spice. Leaves which sheath the base are 8 to 12 inches long, and oblongish in shape. Flowers are yellow-green with yellow-spotted purplish lips on bracted spikes. In this country, except on the southern tip of Florida, ginger can be grown only in a warm greenhouse.

Ginger plants require 10 months of hot, humid weather to mature. They must have very fertile soil, good drainage but not sandy or gravelly soil, heavy rains and high temperatures. The climate in which they are grown need not be entirely frost-free, because the rhizomes are dug at the end of the 10-month period and eyes are replanted two months later. New plants arise from the rhizomes, which are irregular in shape, wherever a tip approaches the surface of the soil.

In Florida experiments, pieces of the rhizomes are set 3 inches deep, 16 inches apart, in rows two feet apart. They are planted in February and harvested in December. Withered tops are removed and pieces of the root are washed in scalding water, then dried in the sun. This fully mature root is ground, when dried, to make our ground ginger, used in cooking and baking. For crystallized ginger, the roots are dug earlier, before they develop their full pungency, and are candied.

Ginger, with pepper, was once used to disguise the off-flavor of meats when they had been kept too long. It was also thought to help preserve meats.

GINGER, WILD

GINGER, WILD. *Asarum canadense.* This plant, which is a member of the *Aristolochiaceae* family, most of which is native to North America, was used by the American Indians before the first explorers came to this continent. The roots were used in much the same way as ginger was used in Europe, to disguise tainted meat. It was also used as a flavoring in hominy.

Wild ginger once grew profusely through North American woods, but is now becoming quite scarce. It is a low, spreading plant with heart-shaped leaves which covers the ground in moist, rich woodland. In early spring it bears purple-brown flowers, that resemble dark campanula bells and lie on the ground. It is a better subject for the wild garden than for the herbiary, but may be grown in dappled shade beneath shrubbery at the back of the herb border. Propagation is by root division.

HOREHOUND

HOREHOUND. *Marrubium vulgare.* This member of the mint family, *Labiatae,* is native to Europe, Asia and North Africa, but was brought to the North American continent by the early settlers who planted it for its medicinal properties, and has become naturalized here. It was used from early times as a throat medicine, but it was not until the nineteenth century that it was made into a candy, which was intended to soothe sore throats and coughs.

Horehound is a hardy perennial which grows into a spreading, bushy plant about 14 inches high. It has gray-green, crinkly, downy leaves with whorls of white flowers in the upper leaf axils. It will grow in all climates, but it may winterkill unless it is well-drained and fairly dry through the winter. It needs full sun and does well in light, poor soil. Propagation is by root division or seeds. Plant the seed one inch deep in early spring, and thin the plants to stand 8 to 10 inches apart.

Leaves have a bitter, strong flavor, but may be used sparingly in cooking. The fresh leaves and their juices are used to flavor cough drops. The plant is used in England to make beer. Dried leaves placed in a pot of honey give it a bittersweet flavor, and the leaves make a strong herb tea. To dry them, cut the smaller stems near the ground, or the tops of the larger stems, just before flowers open, and dry quickly in a warm, dark place.

HYSSOP

HYSSOP. *Hyssopus officinalis.* This is a member of *Labiatae* family which is often planted in the flower garden for its decorative effect. It is native to Europe and temperate Asia. Hyssop was mentioned many times in the Bible as a useful herb, and has been used ever since, principally in medicine. In early ritual, it was one of the chief herbs used in ceremonies of repentance and purification.

Hyssop is a hardy perennial growing about one to two feet high, with small pointed leaves, woody stems and small purple, pink or white flowers. If it is cut back early, it may be kept blooming all summer. The dark green leaves are almost evergreen.

Propagation is from seeds, cuttings or division of the plants. Hyssop grows easily from seed, in fact, it often self-sows. Seedlings should be set in the border 10 inches apart in well-limed soil, in full sun or part shade. Tops should be cut back often to keep tender young leaves growing.

Tender young flowering stems are cut and dried for winter use in flavoring soups, stews and herb teas. The leaves have a very strong flavor, and should be used with a light hand. They are used to flavor the liqueur chartreuse, and may be added to vegetable or cranberry juices. A very few fresh leaves may be chopped fine and added to salad.

LAVENDER

LAVENDER. *Lavandula spica, L. officinalis, L. vera.* Several lavenders, all belonging to the same genus of the *Labiatae* family, have been grown through the centuries and treasured for their fine, clean fragrance. Phoenicians, Greeks and Romans used lavender in their rituals. Although it is native to the Mediterranean area, lavender was scarce in French Royalist days, and the kings paid high prices to perfume themselves with it. Gradually its culture became widespread until every housewife in the nineteenth century tucked a lavender sachet among her linens. It was not until fairly recent times that tips of the lavender flowers were used to flavor desserts and drinks, though the Romans used a little of it in salad. "Oil of spike," distilled from *L. spica,* is an ingredient of some varnishes.

The various species of lavender are more or less hardy perennials, the most tender being *L. vera.* None of the species is completely hardy where winters may be very severe. They should all be given winter protection where the ground freezes and, where the cold is extreme, plants should be wintered in a cold frame or indoors.

Plants become woody in their second year, and sprawl to cover an area about one to two feet in diameter when mature. Flower spikes, which rise above the branches of small grey-green leaves in most species, may be one to two feet tall.

Best soil for lavender is dry and sunny, well-drained and somewhat sandy. A rock garden with gravelled mulch or the top of a dry wall are excellent spots for it. Propagation may be from cuttings rooted in winter and set out in spring, or by seed planted late in fall or in early spring. Flowers should not be cut the first year and, in fact, the plants will blossom best the second year and thereafter if the buds are pinched off and not allowed to open the first year. Life span of lavender plants is 4 to 7 years, after which they die out and must be replaced.

Flowers to be used in dry arrangements should be cut before fully open for best color. To dry flowers for sachets or flavoring, cut when first opened buds are fully open, and the others on the spike are opening.

LEMON BALM

LEMON BALM. *Melissa officinalis.* A rather weedy member of the *Labiatae* family grown by the Greeks and Romans for their bees, and for the lemon flavor which it imparted to soothing herb teas. It was introduced to English gardens in about the sixteenth century, and became very popular there, both in horticulture and in literature.

Balm may be grown as a perennial outdoors in mild climates, but is best wintered over in cuttings which are potted and kept in the window sill, where winters are very cold. The plants become rank and profuse, spreading rapidly outdoors unless they are confined. It grows one to two feet tall, and has inconspicuous white flowers.

Commercially, the leaves of lemon balm are used in perfumes and toilet waters, and in flavoring benedictine and chartreuse. Fresh or dried leaves may be used to flavor fish, lamb or beef dishes. Fresh leaves are used in salads, and dried ones to make herb tea.

Propagation is by seed or cuttings. Seed may be sown indoors in the late winter, and seedlings set in the border 18 inches apart when the soil is warm. During the first year, growth will be slow, but in the second year it will make a strong clump. Part shade and a moist soil are best. Cuttings are made, in late summer or early fall, from new growth, or from the upper portions of the older stems.

In cold areas, plants may be set in the garden in large pots which are lifted in fall and wintered in a hotbed or greenhouse.

LEMON VERBENA

LEMON VERBENA. *Lippia citriodora.* A member of the *Verbenaceae* family which is native to tropical America, lemon verbena is a woody shrub growing to 10 feet in warm climates, a 3-foot shrub in more temperate climates and a 10-inch houseplant in cool climates.

This verbena's lemon-scented leaves are long and narrow, and grow in whorls on woody stems. Its blue to pinky-lavender flowers rise in slender spikes.

Lemon verbena will grow from cuttings or from seed in any good garden soil or moderately rich potting soil. In gardens, where the climate is mild, the plants may be set out two feet apart. In the tropics, they must be spaced further apart.

Where the winters are cold, take up the plants at the first sign of cool weather, cut them back and pot them. They may be stored in a cool greenhouse or in a cellar where the temperature does not go below 55 degrees. Or the plants may be kept in pots or tubs the year-round, and repotted each spring, when they are set outside for the summer. Weak shoots should be trimmed off in spring, when the plants are potted for the garden, so that summer growth will all go into the sturdy branches.

Fresh leaves may be used to give a lemon flavor to drinks or fruit desserts. Dried leaves are used for herb tea.

LOVAGE

LOVAGE. *Levisticum officinale.* A large member of the *Umbelliferae,* flavored and used much like celery. Lovage is native to southern Europe, and a wild variety also grows on the Isle of Skye, off Scotland. The same wild lovage, called sea lovage or Scottish lovage, may be found growing on the North Atlantic coast in the country, probably naturalized from plantings in the Colonists' gardens. Lovage was known by the Greeks and Romans, who used the plant but may not have cultivated it. It was used medicinally through the Middle Ages, and as a culinary herb for many centuries in Europe.

A tall, hardy perennial, sometimes growing as high as 7 feet, lovage has dark green leaves shaped like celery leaves, and sends up umbels of yellow-green flowers in spring or early summer. It is easily propagated from seed, which may be sown in fall in the garden, or planted indoors for an early spring start. Seedlings 3 to 6 inches tall are planted outside in rich, fairly moist soil, when frost is past, spaced one foot apart. Three or 4 plants are sufficient to supply most families.

Leaves, seeds and roots are all used fresh or dried for their celery flavor in soups, stews and sauces. In addition, the hollow stems are sometimes blanched and eaten like celery.

Leaves and seeds are dried like burnet and caraway. Roots should be dug in October of the plant's second or third year. Strong offsets are trimmed and set back in the border to renew the stock. To dry the root, wash and cut the large main root into half-inch slices. Lay them on a screen or cheesecloth in a warm, dry attic where there is plenty of ventilation. Turn the pieces at frequent intervals until thoroughly dry. Store in airtight containers.

MACE

MACE. *Myristica fragrans,* Houtt. The aril or shell which covers the kernel of fruit of the nutmeg tree, that is, the nutmeg (which see). The trees are large evergreens native to the East Indies, but now widely cultivated in the West Indies. Their fruit is about the size and shape of a small peach, and delicious. When ripe, the flesh is colored yellow to orange, and the coating of the kernel is bright red. This coating is carefully removed from the nutmeg kernel and spread in the sun to dry. When dried, the pieces are referred to as "blades of mace."

Mace has a flavor much like nutmeg, but more earthy. It is used in blades or ground to flavor salty, rather than sweet, dishes. Wherever a meat recipe calls for nutmeg, mace can usually be substituted to improve the flavor. It is often an ingredient of ready-mixed blends sold as pie spices, pickling spices and poultry seasonings. It is also used in fish sauces and sausages, as well as in dishes more frequently flavored with nutmeg, such as various desserts and pastries.

MARIGOLD, POT

POT MARIGOLD. *See* CALENDULA.

MARJORAM

MARJORAM. *Origanum marjorana.* One of the 3 species of the genus *Origanum, Labiatae* family, which are grown as flavoring herbs. This one is also known as sweet marjoram. *O. onites,* pot marjoram, is a perennial which is not grown as widely, because its flavor is inferior to the flavor of sweet marjoram. *O. vulgare,* usually known as oregano and also called wild marjoram, has a somewhat different flavor, and is listed below under OREGANO.

Sweet marjoram is native to the Mediterranean region, and has been cultivated as a flavoring herb since the time of the Greeks. Pliny and Albertus Magnus both praised its fragrance. The Greeks and the Romans used it as a fertility symbol and, in India, it is ranked with basil by the superstitious.

Sweet marjoram is a perennial in the South but is usually treated as an annual in the North, because it is very tender. Plants or cuttings may be potted in fall in colder areas, and wintered in a hotbed or cold frame, or on the kitchen window sill; or seed for new plants may be started very early indoors, to be ready when the first annuals go into the border. Seed germinates slowly and should be given plenty of time to make 2- to 3-inch seedlings to set out in April.

Sweet marjoram is a low, sprawling gray-green herb with small oval leaves and tiny white flowers. It grows no more than a foot high, but should be allowed a square foot of garden space per plant because it spreads laterally. Soil need not be rich to accommodate it, but should be well-limed. Pot marjoram grows somewhat higher than sweet marjoram, and has purple flowers. Cuttings of either species may be made by pulling off branches when they become slightly woody, with the crown or heel attached. Plunge the heel into moist sand for about 5 weeks, until well-rooted. Then pot or place in the cold frame for the winter. Plants may also be wintered in a cool cellar, where they are kept dormant by excluding light and allowing the pots to become almost, but not quite, dry.

Marjoram is used principally in meat dishes. Stews, meat loafs, omelets, gravies of all kinds, poultry stuffing and sausages are improved by just a small touch of marjoram. Fresh leaves rubbed on meats before they are roasted give them a slight flavor. Occasionally a few fresh leaves may be cut into a green salad or its dressing. Game and fish dishes also profit from a dash of sweet marjoram.

Fresh leaves may be cut for kitchen use when the plants are 4 to 5 inches tall. When the plants bloom, the top 4 inches should be cut back and these tips may be dried. Or drying may be done in the fall when whole plants are cut off and hung in the attic. As soon as the leaves are thoroughly dried, they should be removed from the stems, pulverized and stored.

MINT

MINT. The genus *Mentha* of the *Labiatae* family has many useful species, all well-known herbs with more or less minty flavors. Most popular for cooking and for cooling drinks is *M. spicata,* spearmint. Mint sauce, jelly, fresh fruit desserts, iced tea, all provide vehicles for its fresh flavor.

M. piperita, peppermint, is less used in cooking than in confectionery and in medicine. Oil of peppermint has soothed many a baby's stomach through the ages.

M. Pulegium, pennyroyal, is a creeping mint with a flavor similar to that of peppermint. It is used to make teas for cough or cold remedies, but may be poisonous to people who are sensitive to it, and should be used with caution.

M. rotundifolia, woolly mint or apple mint, is a variety much more popular for cooking in Italy than in the United States, though it grows wild here in profusion.

The mints are native to the Mediterranean region, both in Europe and North Africa and are also found growing in many parts of Asia. They are widely naturalized in America.

The name of the genus comes from Greek mythology. Proserpine, the spring goddess beloved by Pluto, became jealous of a beautiful wood nymph called Menthe, on whom her lover cast languishing glances. Proserpine's solution to her conjugal problem was difficult—if she killed her rival, the next stop on her afterlife was Pluto's own domain. So she transformed Menthe into one of the lowliest of plants, which would ever afterwards be trampled under foot. Proserpine must always have been haunted by the fragrance of Menthe's spirit, which arises from mint whenever it is stepped on.

The mints are all hardy perennials 12 to 24 inches tall, with lavender flowers in terminal spikes or in wreaths around the stems at the leaf axils. Propagation is by offsets of the plants, which spread by underground runners.

To plant, set strongly rooted runners about two inches deep in fairly moist soil, in full sun or part shade. Slightly acid soil is best, from *p*H 5.2 to 6.7. New plants will arise from root nodes, and the planting will spread rapidly if the soil is kept moist. If a limit is to be set on the bed, it will best be confined by setting boards or metal strips 6 inches deep into the soil. Keep part of the bed cut back all summer long, to provide fresh tender tips for the kitchen.

Leaves may be dried away from the light, and stored for winter use. But they will yield more flavor in winter dishes if they are frozen. Wrap a few tips, as many as are used at a time, in freezer paper, seal, and store in a labeled box in the freezer. Mint keeps well through the winter without a steam or hot water bath before freezing.

MUSTARD

MUSTARD. *Brassica sinapis alba,* white mustard, or *B. sinapis nigra,* black mustard. Both kinds of mustard have been used as condiment for many centuries. Use of the seed is thought to have dated back to the time of the Greeks, who also used the plant's green leaves as a potherb and salad.

Black mustard is a hardy annual which

grows to a height of 4 feet, even in poor soil. But when grown for its seed, black mustard must be kept fairly moist and well-nourished.

White mustard is more often grown in the garden for its seed. It is a plant that grows to a height of 18 inches, and its leaves are useful for salad. It may become a pest, because it self-sows freely, so it is inadvisable to plant it in the regular vegetable garden or flower border.

When grown for the seed, which is used in pickling and as a peppery condiment, pods must be picked before they are entirely ripe, or they will shatter and the fine seed will be lost. Pods are spread on muslin to dry, and are crushed to remove seeds and hulls. The seed must be further crushed and milled to remove the fine hulls before they are used. Commercially, mustard seed is ground to a flour to provide dry mustard. This process, however, is beyond the scope of home gardeners and is not recommended.

NASTURTIUM

NASTURTIUM. *Tropaeolum minus* and *T. majus*. The only important members of the *Tropaeolaceae* family are the dwarf, *T. minus,* and the climbing, *T. majus,* nasturtiums. Both have been grown as decorative plants in the flower border, as well as for their peppery leaves, flowers and seeds, since they were introduced to Europe from Peru in the sixteenth century.

Succulent nasturtium leaves, stems and bright yellow, orange or red flowers are added to salads, canapes and appetizers for an exotic touch. In the Orient, petals of the flowers are sometimes added to tea, as are jasmine flowers. Nasturtium seeds can be picked green and pickled for use as a substitute for capers.

Nasturtiums are very tender annuals which may be planted only after the weather has become warm and settled in late spring. They need full sun, but will produce few flowers in soil that is rich. In rather dry, poor soil they bloom more freely, but the round or elliptical leaves are paler green and not so profuse as when the soil is enriched. Grown for their decorative value or their seeds, they should be starved; for salads more rich in vitamins, they should be fed.

Leaves and flowers are used fresh. Seeds and seed pods are pickled, but not dried for culinary use.

NUTMEG

NUTMEG. *Myristica fragrans,* Houtt. The dried seed in the kernel of the fruit of the nutmeg tree. (*See* MACE.) Nutmeg trees are tall evergreens native to the East Indies. Although the Romans knew about them, they did not use them as food, but only as one of the ingredients of incense. It was about the twelfth century, when they became known as a fine spice, that they became extremely valuable. The Dutch made a good business out of exporting them, as one of the main items of trade for the Dutch East India Company, in the 1600's.

Nutmeg trees bear male or female flowers, but not both on the same tree. When the trees are 9 years old, they begin to bear fruit. After a plantation starts to fruit, it may bear for 50 years if it is properly taken care of.

Fruits of the trees, which resemble peaches, are picked by means of a long pole, with prongs at the end and a basket suspended beneath the prongs. After picking, the flesh is cut away, and the aril, which is dried to become mace, is carefully stripped. This leaves the kernel, a shell containing the nut-like nutmeg. The kernels are spread to dry in the sun for several weeks, until the nuts inside have shrunk enough to loosen and rattle. Then the shells are removed, the nutmegs are dried further, and are ready for use.

Nutmeg has a delicate flavor which blends well with either sweet or salty flavors. It is used mostly to flavor dessert-type foods, but is also included in many sausage, stuffing and pickling blends. Nutmegs may be purchased whole or grated into a granular form, ready for use. Flavor is more piquant from nutmeg grated fresh as it is added to the dish.

OREGANO

OREGANO. *Origanum vulgare*. Oregano is also known as wild marjoram, although its leaves have a distinctive flavor of their own, stronger and more peppery than that of marjoram.

Oregano, like the marjorams, is a tender perennial which is usually grown as an annual in the North. In the southern states and on the West Coast, it may be grown outdoors the year round, and makes a handsome clump as high as 3 feet, with pinky lavender flowers through the 3 hottest months.

Propagation is from seed or from division of the crown. Seed is very small, so the seedlings are slow to develop and should be started very early indoors. When they are 2 to 3 inches tall and danger of frost is past, they may be given about a square foot each in full sun in the garden. Soil should be rich and moist, but well-drained.

Leaves and flowering tips should be cut for drying just as the blossoms begin to open. They should be dried quickly in a warm, well-ventilated place, and the stems should be stripped and removed before the leaves are stored.

Oregano is the herb used most in Italian dishes which have become popular in the United States. It flavors spaghetti sauces, pizzas and a variety of other Italian dishes. It may also be used in soups, stews and in many meat dishes.

PARSLEY

PARSLEY. *Petroselinum crispum,* curly-leaf parsley; *Petroselinum hortense,* plain leaf turnip-rooted parsley. Parsley is a member of the *Umbelliferae,* native to the Mediterranean, probably originating on the island of Sardinia. Both Greeks and Romans held it in great esteem, both as a food and as a decoration. Banquet halls were decked with garlands of parsley; parsley was fashioned into wreaths which were presented for heroic or athletic prowess, or were set on the brows of loved ones.

Use of parsley seems to have been continuous through the ages; after the ancients, the monks of the Middle Ages grew it in their monastery gardens for its medicinal qualities, as well as its food value. One herbal of the time recommends it because "it multiplieth greatly a man's blood . . . It is good for the side and the dropsy. It comforteth the heart and the stomach." Proof of the accuracy of these statements had to wait until the twentieth century, when nutritionists discovered that parsley's deep green leaves were excellent sources of vitamins A, B complex, C and E, in addition to calcium and iron.

Parsley is no longer considered a decorative garnish to be left on the plate along with the lamb chop's panties. One tablespoon of minced parsley, eaten at each meal, will provide the minimum daily requirement of vitamins A and C. Since it is a mild-flavored herb, parsley can be combined with many foods, several in the same meal. It may be added to meat dishes, soups, salads, omelets, vegetables, fish or poultry. No herb bouquet is complete without a generous sprig of parsley. The most flavor and nutrition is derived from it when fresh parsley is added to the dish when it is just about to be removed from the stove, or when it is being arranged for the table.

Parsley can be easily grown indoors in winter by potting a few garden plants; keep pot on a sunny window sill all winter.

Parsley may also be dried for winter use, though it loses much of its nutritional value in the process and most of its flavor. A better way to provide winter parsley is to pot a few husky plants from the garden, prune off the outer leaves for immediate use, and keep the pot on a sunny window sill all winter. Plain-leaf parsley, which is hardier than the curly species, may be planted in a window box when a hard freeze threatens the garden, and kept on an unheated porch for the winter. It will withstand frost on the porch, but not a prolonged frigid period.

Parsley seed is very slow to germinate, but is best planted directly in the garden where it will grow. It may be planted in fall and allowed to soften with the first spring thaws, or it may be soaked overnight in warm water before it is planted outside in early spring. Cover seed with one-fourth inch of soil, and thin plants to stand 3 to 6 inches apart, depending on variety. Parsley will grow in any soil and any climate, in full sun or part shade, but does best in rich moist soil in full sun. It is a biennial, but dies back in winter where the ground freezes hard.

Turnip-rooted parsley is grown for both its plain, uncurled leaves, which have a slightly stronger flavor than the curly varieties, and for its creamy-white root. The roots may be dug and stored in fall in a root pit, if desired, and cooked like parsnips. They have a flavor something like a fragrant celery. Or they may be used for flavoring soups, or cooked with other vegetables. The Italians sauté a small amount of minced parsley root and add it to vegetables like green peas, while they cook. It may also be substituted for carrot in recipes for gravies and sauces.

PENNYROYAL

PENNYROYAL. *See* MINT.

POPPY

POPPY. *Papaver somniferum,* the opium poppy, member of the family *Papaveraceae.* Opium poppies are native to Greece and the Orient. They were grown by the Greeks, Romans and early Chinese primarily for their seed, which was used in cooking and as a source of oil. A variety of poppy with very flavorful seed has been developed by the Dutch from a combination of *P. somniferum* and *P. rhoeas,* the corn poppy. This poppy is widely grown today in France, Greece and Poland, for its seed. Opium poppies are grown in India, Turkey and Persia, both for seed and for opium. The seed does not retain any of the narcotic properties, which are present only in the green seed pod.

Opium poppy is an annual which grows about one foot high. Its 4-petaled white flowers are tinged with blue and grow to 4 inches in diameter. It requires a long growing season, and cannot be matured in the North unless plants are started indoors. This is a precarious undertaking, because the roots are very sensitive and resent transplanting. Seed may be sown in pots, and the whole root ball is transferred to the garden when the weather has become warm. Six inches to a foot should be allowed for each plant, depending on variety. Soil should be rich and moist, but above all, sunny. Poppies bloom very indifferently in the shade.

Pods should be allowed to dry on the plant, but must be caught before they turn their saltcellar-like openings to the ground to spill their seed. Drying can be finished on muslin indoors. When the pods are crisp, seed may be removed by rubbing the pods between the palms. The tiny seed can be separated from the chaff by putting it in a coarse strainer and shaking.

Poppy seed is used principally for baked products of all kinds. It is also one of the main ingredients of commercial birdseed mixtures. Oil from crushed seed is used as a substitute for olive oil.

ROSE

ROSE. *See* DAMASK ROSE.

ROSE GERANIUM

ROSE GERANIUM. *Pelargonium graveolens,* one of the scented-leaf members of the genus *Pelargonium* of the *Geraniaceae* family. Others widely grown for scent are apple or nutmeg geranium, *P. odoratissumum; P. tomentosum,* or peppermint geranium; and variety *Clorinda,* a fish geranium grown mostly for its brilliant flowers. All are tender perennials from the tropics of South Africa. They were introduced to Europe during the seventeenth century, and were grown there for their scented leaves, which were used in many perfumes and cosmetic preparations.

The *Pelargoniums* grow to fair-sized shrubs as much as 4 feet tall, where they can be left outside through the winter. But in the North, they must be potted and wintered indoors, or carried over from cuttings, and seldom achieve a height of more than one foot. Leaves are lobed and cut, and rough to the touch. Rose geranium leaves smell of roses with overtones of spice.

Any good garden soil will grow husky geranium plants. For best bloom, they must have full sun. Propagation is by cuttings. Plants which must be wintered inside on the window sill will be more satisfactory if they are grown from cuttings taken in midsummer than if the parent plants are potted.

Leaves may be taken from the mature plant at any time for drying. Fresh or dried leaves are used to flavor jams and jellies. Baked apples or pears may be flavored by standing them on a fresh leaf when baking. Dried, the leaves are used in sachets, potpourris and pomanders.

ROSEMARY

ROSEMARY. *Rosmarinus officinalis,* a member of the *Labiatae* family native to the western Mediterranean. Rosemary was grown by the Greeks and the Arabs for its fresh fragrance, and was used by the Anglo-Saxons as early as the eleventh century.

Much of the sentimental lore surrounding rosemary probably arises from a misunderstanding of the meaning of Shakespeare's, "Rosemary, that's for remembrance." Shakespeare probably thought he was quoting a strictly scientific fact of his day. Alchemists of the sixteenth century claimed that the herb stimulated the mind, particularly the memory.

Medieval herbals give to rosemary many gay and endearing properties, but not sentimental ones. The boar's head at Christmas time was garlanded with rosemary. And they advised, "take the flowers thereof . . . and bind it to thy right arm . . . and it shall make thee light and merry. Also . . . put the leaves under thy bed and thou shalt be delivered of all evil dreams . . . Also make thee a box of the wood and smell to it and it shall preserve thy youth."

Rosemary is a tender perennial which must be wintered indoors in the North. Where the ground does not freeze hard, it may be left outside, where it will grow into a 3 or 4 foot shrub. It is useful in the South as a hedge plant, because it will stand much pruning. It has glossy evergreen foliage, and produces tiny light blue flowers after its second year. It rarely produces seed, so it is propagated from cuttings or crown divisions, or by layering. It makes a good specimen plant to grow in a pot or tub, which may be wintered in the North in a cool cellar or greenhouse. Outdoors, it grows best in sunny locations, in well-drained but moist soil, which is well-limed but need not be rich.

Cuttings may be taken in midsummer and potted for the winter window sill. Break off a sprig with heel attached, and root in moist sand for 4 to 6 weeks, then pot. New growth should be pruned back twice each season, and leaves pruned off may be dried for winter seasoning.

Rosemary is used in many meat dishes and dressings, and an oil extracted from it is used in medicine.

RUE

RUE. *Ruta graveolens.* A bitter herb of the *Rutaceae* family formerly used only in rituals or medicine, but more recently included among the culinary herbs. The Greeks and Romans used rue as an antiseptic, spreading branches over the floors of their temples to ward off plague. It became the symbol of repentance when it was used in the Catholic church to sprinkle holy water over sinners. Though it is no longer used medicinally, rue is known to contain rutin, a vitamin which regulates the coagulation of the blood. However, the leaves should be used with caution, because they can cause blisters like those caused by poison ivy when they are handled by people sensitive to them, particularly in hot weather.

Rue is a hardy perennial with small club-shaped blue-green leaves and yellow-green flower clusters which bloom in summer after the second year. Chalky or well-limed soil is essential to its growth, but it winterkills less easily in poor soil than in rich, and it will grow equally well in sun or shade. Plants seed themselves and spread easily. They should be allowed about a square foot of space per plant when set in the border. Their chief use is to fill corners in the border where nothing else will grow. Rue will fill such niches with almost evergreen, lacy foliage. Propagation is very easy from seed, but cuttings or root division may also be used.

Leaves may be dried for winter use at any time in the early summer, when they are most tender. The bitter aromatic flavor is sometimes added sparingly to vegetable juices, stews and ragouts.

SAFFRON CROCUS

SAFFRON CROCUS. *Crocus sativus,* a small autumn-flowering crocus of the family *Iridaceae,* which must not be confused with colchicum, called the autumn crocus. Colchicum is poisonous when eaten.

The saffron crocus is native to Greece and Asia Minor, and was introduced to France and England from the Crusades. It was mentioned in the Bible and by Homer, and the Romans spread its blossoms on the ground under the feet of their heroes. It is grown now in Austria, France, Greece, Spain, Sicily and the Far East for saffron.

Saffron, a delicate-flavored spice used as much for its yellow coloring as its flavor, comes from the 3 bright-orange stigmas at the heart of the *Crocus sativus,* which is a lavender-colored flower. The stigmas must be picked by hand, a back-breaking and tedious task. The stigmas of 25,000 flowers are needed to make a pound of saffron, which sells for justly high prices.

Best quality saffron comes from bulbs planted in soil of medium fertility. The corms may be planted in August, and blossoms will appear in October, though most of the foliage grows the following spring to renew the corm. Every 2 or 3 years, the corms should be lifted and divided. Bulbs should be planted 4 inches apart and 6 inches deep.

Flowers of the safflower are sometimes used as a substitute for saffron. (For culture *see* SAFFLOWER in the vegetable section.) To prepare dried blossoms, the individual safflower florets must be separated from their thistle-like heads and dried. Safflower may be planted in the herb or flower border, where its flowers provide a brilliant spot. They should be given a warm, sunny position, and plants should be spaced 6 inches apart. They grow one to two feet high. They are easy to grow from seed, which should be planted when the soil has become warm.

SAGE

SAGE. *Salvia officinalis,* one of the strong-flavored *Labiatae* family, native in the Mediterranean area and grown from antiquity, as much for medical as culinary use. In some places sage tea is still brewed as a spring tonic, though the present-day use of sage is primarily for flavoring meats, especially pork.

There are innumerable varieties and species of sage, all with more or less the same flavor in varying strengths, and with varying overtones. Common sage, *S. officinalis,* is a hardy and rather woody perennial, almost evergreen, with showy purple flowers in spring. Its pebbly, grayish-green leaves are about the size of dried bay leaves.

Seed of sage germinates easily. The perennial bed may be started from a few seeds indoors, which are later transplanted about two feet apart in early spring. Any good garden soil, which is well-drained and in full sun, will grow robust sage plants. After the new plants start to bloom, they will self-sow. If the planting becomes too thick, it is best to remove and discard the oldest plants, which may have become too woody. Plants should all be cut back to a height of about 8 inches in fall, to encourage young tender growth in early spring. Sprigs cut off in fall may be used as cuttings, rooted in sand in a cold frame over the winter, to be ready to set out in the border in spring. Old sage plants last no more than 4 or 5 years, so at least one young plant should be allowed to develop each year.

Best time to cut and dry sage leaves for winter use is just before flowering. Tender young leaves are cut from the tips of the branches. Fresh leaves may be picked and used at any time in the growing season. New plants should not be cut until they have reached a height of at least 8 inches.

Sage is used in many meat dishes, in stuffings, sausages, fish dishes, and is mixed with some cheeses to give them their greenish color. Sage cheeses are imported from England, Switzerland, Holland and are made in the United States. Sage tea is thought to be soothing to sore throats.

SASSAFRAS

SASSAFRAS. *Sassafras albidum.* An American member of the family *Laurineae,* native to the East from Maine to Florida and westward as far as the Mississippi Valley. Sassafras is used mainly in the United States, though a tea known as saloop is made from it in England. The Spanish explorers thought that its bark had medicinal qualities.

Sassafras trees are usually small, though a few large specimens have been known. Leaves are plain and three-lobed or two-lobed, in the shape of mittens. Racemes of fragrant yellow flowers bloom in the spring before the leaves unfold, and develop into deep blue fruits with a bloom, on fleshy red stalks. The foliage turns brilliant scarlet in fall.

Sassafras will grow in any light soil, sandy or woods mold, in a sunny position. It prefers a *p*H between 6. and 7. It may be propagated by seed sown as soon as it is ripe, by root cuttings or by suckers. The small trees are difficult to transplant because of their long taproots.

Every portion of the tree has been used as food in some form. The mature green leaves are dried, powdered and used as a seasoning known to the Creoles as gumbo filé. This is used in soups to give it an aromatic flavor, and also to thicken it, a property of the leaves due to a mucilaginous substance which they contain. Flowers are dried and used to make sassafras tea. Young shoots are cut in spring and used to make sassafras beer. Roots and bark are both used as flavoring, and to chew. The Pennsylvania Dutch cook a piece of root with their apple sauce and apple butter, both for flavor and to ward off winter colds.

SAVORY

SAVORY. *Satureia hortensis,* annual summer savory, and *S. montana,* perennial winter savory, both members of the *Labiatae.* Savory is native in southern Europe and the Caucasus, and has been used since about the third century. Both summer and winter varieties are aromatic herbs used for flavoring, the perennial species somewhat stronger than the annual.

Summer savory grows to a height of about 18 inches, with bronzy small leaves on bushy semiwoody stems. The plant has a habit of falling away from the windward side in the garden, and looks like an arrested tumbleweed. Small pinky lavender flowers bloom in early summer. Seed is sown in early spring, one-half inch deep in good garden loam in full sun. Leaves are ready for cutting and drying in 60 days from seed, and plants to be dried should be pulled before the flowers open. If fresh summer savory is needed for the kitchen, successive sowings may be made through the spring, or leaves from older plants may be used until fall, though the flavor deteriorates through the summer.

Winter savory is not quite as tall as summer, growing to about a foot in height. It has dark green, glossy, small leaves on woody stems. Its flavor is not quite as sweet as that of summer savory, and is stronger. Seed may be started indoors and seedlings planted out 18 to 24 inches apart in rather poor but sunny soil. Planted in rich, damp soil, winter savory is more likely to winterkill. Or it may be propagated from cuttings of new growth rooted in early summer. The top 6 inches of new shoots may be cut while they are tender for drying. Savory should be cut back once or twice during the season, and deadwood should be removed.

When leaves of either variety are dry, they should be stripped from the stems and stored in airtight bottles. They are very popular for flavoring almost any kind of meat dishes, beans, peas, cabbage or sauerkraut.

SESAME

SESAME. *Sesamum orientale,* a member of the *Pedaliaceae* family, native to Africa and the warmer parts of Asia. Earliest records show that sesame seed was a staple food and oil source in China, Japan and India, and was also used in ancient Greece. The Egyptians and Persians of Biblical times ground it into a kind of flour, from which they made bread. The Romans crushed the seed and used it like butter for a spread on bread. Today it is still used as a bread spread, the oil being used in the manufacture of some oleomargarines.

Sesame is a strong, slender annual growing to 18 inches, with slender darkgreen leaves and inch-long flowers, resembling foxglove, which lie along the square stem. It is a warmth-loving plant which will only mature its seeds in our southern states and Hawaii, though it may be started early in pots in the North and planted in the garden for decorative purposes.

Seed is started early enough in spring to allow it a growing season of 90 to 120 days of hot weather. Plants are thinned to stand 6 inches apart. Full sun and moderately rich soil are required. Planted as close as this, the plants grow thick and bushy and shade the soil under them.

Seed pods are picked in fall just before the first frost causes them to shatter. They are spread to complete their drying on trays, then are shaken out of the pods, hulled and winnowed.

If used in large enough quantities, sesame seeds are a very good source of vitamins C and E, of calcium and of unsaturated fatty acids. The seeds are used in many baked products, needing to be baked or roasted to bring out their nut-like flavor. In the Orient, they are often mixed with honey and dried fruits to make a sweetmeat.

TANSY

TANSY. *Tanacetum vulgare.* One of the *Compositae* family native to Europe and Asia, now naturalized in the northeastern states. Tansy was widely used in Europe during the Middle Ages as a medicinal and culinary herb. Puddings, cakes and tea were made with it in England in early spring, as a spring tonic—a "pick-me-up" really needed by people who had lived through an entire winter with no fresh greens. The herb is now grown principally as a flower border subject, and is not much used in cooking. Its flavor no longer appeals to many people, and there is some doubt about whether it contains a toxic substance which may have a poisonous effect.

Tansy is a strong-growing hardy perennial 30 to 50 inches tall, with finely divided, fernlike dark green foliage and flat clusters of tiny yellow flowers, like miniature pompon chrysanthemums. Blooming period is from July through September. It will grow in almost any soil, but prefers one that is moist. It is best planted where its spread, which may be rapid, is confined by walks or walls. For decorative value, the variety *crispum* is best.

Propagation may be by seed, sown in early fall or in spring, or by root division. Seed planted too late in the fall will not germinate until the following spring. Plants should be spaced 18 inches apart.

Tansy blossoms may be dried for winter bouquets. The buttonlike flowers are cut when they are freshly opened, and hung in loose bunches in a warm, shady place to dry. Leaves for winter use may be cut at any time for drying. Tea made from fresh or dried leaves is said to have a calming effect. Fresh or dried leaves may also be used sparingly in omelets, fish dishes or meat pie. Tansy is used commercially in the preparation of cosmetics and toilet water, and is one of the flavorings of chartreuse.

TARRAGON

TARRAGON. *Artemisia Dracunculus.* One of the *Compositae,* native to eastern Europe, the Orient and especially the region of the Himalayas. It was cultivated and used by the Greeks and Egyptians, but was not much used through the rest of Europe until after the twelfth century. It is now greatly esteemed in Italy and France, and throughout the Orient. In Persia, leaves are eaten before a meal, as an appetizer.

Tarragon is grown in this country as a perennial, and does not often make seeds. In France, where it makes seeds freely, it is grown as an annual. It is a fairly hardy perennial as grown here, but needs some winter protection in cold areas, and should be planted where it is sheltered from strong winds. It will grow in sun or part shade in soil that is moderately rich.

Propagation of the perennial tarragon is by cuttings rooted in early spring and set out before August, or by division of the roots in spring before the tops have grown more than two inches tall. After midsummer, tarragon can be transplanted in the colder areas only when whole plants are lifted with a large root ball, and the plants must be given plenty of winter protection the first year after such treatment. Roots need a long period in which to anchor the plants to the soil before winter freezes begin. Every 3 years, the parent plants should be lifted and divided in early spring, the divisions made with a minimum of two shoots each, and set 18 inches apart.

Young tarragon leaves may be picked in spring and dried for winter use, or the leaves may be frozen and will yield more flavor than when dried. Leaves give a delicate flavor to steaks, roasts and fowl, when they are rubbed into the surface before broiling or roasting. Tartar sauce, lemon butter sauce and salad dressings are all improved by tarragon. The flavor may also be introduced to salads by using tarragon vinegar in the dressing. To make the vinegar, fill a jar with leaves and pour in wine vinegar to fill to the rim. Shake every day for 3 weeks, then strain.

THYME

THYME. Many species and varieties of thyme may be grown for culinary use, perfume and sachets or for their decorative horticultural value. The thymes are especially useful in conjunction with natural rock plantings and are used, not only in rock gardens, but among flagstones, as ground cover, in dry walls and between natural rock steps, where they trail gracefully.

Most commonly grown in the herb garden are *Thymus vulgaris,* the common thyme, which grows a foot high, and some of the trailing varieties of *T. Serpyllum,* the wild thyme. Most popular variety is probably *citriodorus,* lemon thyme, with pink blossoms. Variety *coccineus* has bright red blossoms; variety *alba,* white; and variety *lanuginosus,* woolly thyme, is silvery with lavender blossoms.

Thyme is one of the *Labiatae* which is native in the Mediterranean and has been used since ancient times, more for decorative than for culinary purposes. In ancient Greece thyme was planted on the graves of heroes, and sprigs of it were burned as incense in the temples. Thyme honey originated in Sicily and the eastern coast of Italy, where wild thyme grows thick. Sheep put to graze in that region's pastures, rich in thyme, yield meat which has a distinctive flavor, actually being seasoned on-the-hoof.

Thyme is propagated most often from seeds which may be sown indoors, or outside in protected seed beds. The seed is very tiny, and must be given protection from drying until the seedlings are large enough to be easily visible. When they are 2 to 3 inches high, the seedlings are set in the garden bed from 6 to 18 inches apart, in full sun, in well-drained, rather dry and sandy soil.

Propagation may also be from cuttings, taken in summer and potted, by layering the trailing stems by pulling soil over them until they root, or by division of the crown. Thyme may be too tender for some northern gardens, and it is best always to winter a few plants in a cold frame, in case the main planting is winterkilled. For best plants for the kitchen, new plants should be started every 2 to 3 years.

To dry leaves for winter use, cut off the tips of the stems as far back as the stem is tender, just before the blossoms open or while they are in bloom. Hang the stems, tied loosely in small bunches, in a well-ventilated warm room to dry. When dry, strip the leaves from the stems and store in airtight containers.

Thyme provides the principal flavor in ready-mixed stuffing seasonings, and is good in almost any salty-type dish. It is indispensable in clam chowder, and is used in many other soups and meat dishes.

Tiny flowers are sometimes picked separately and dried on muslin in the shade for use in sachets, tea and potpourris.

TURMERIC

TURMERIC. *Curcuma longa,* a member of the *Zingiberaceae* family, is very similar in appearance and growth to ginger. Turmeric is native in tropical Asia, but is cultivated now in the West Indies, especially Haiti. It is closely related to arrowroot.

Like ginger, turmeric is grown from rhizomes which are the source of the spice. Pieces of root are planted in the fields two feet apart, and from them grow strong stems with large leaves and terminal flower spikes, which, in the Orient, are green, but, when grown in the West Indies, are brilliant orange. After the plant has matured and died back, the roots are dug, washed and dried in the sun. In good weather they are thoroughly dry and ready for grinding in a week or 10 days.

Turmeric powder is the chief ingredient of curry powder, and is used as a tangy and colorful spice in many Oriental dishes. It is sometimes added to commercially prepared mustard for flavor and color. It is also sometimes substituted for saffron.

WINTERGREEN

WINTERGREEN. *Gaultheria procumbens.* A trailing member of the *Ericaceae* family which is native to northeastern America. The red wintergreen berries were at one time sold as a sort of confection in Boston markets, but they are seldom if ever found there now. Leaves of wintergreen were used by the Maine Indians to make a tea. Principal use now is for flavoring candy and chewing gum, which is a pity, because the bright red berries are sweet, tender and flavorful.

Like most of the heath family, wintergreen needs an acid soil which it finds most often in pine woods. The acidity should be about the same as for trailing arbutus, between *p*H 4.5 and 5.5. It needs shade and a well-drained, preferably sandy soil. Because of its strict requirements, it makes a better wild garden subject than one for the herbiary.

It is very difficult to establish plants

taken from the woods. Nursery-grown stock is more satisfactory, having been accustomed to artificial conditions. It should be planted on a shady slope and mulched with 2 to 4 inches of pine needles.

Berries and leaves are both used for flavoring, but are not dried. Berries in brandy make a drink like bitters. Oil distilled from the leaves is used for flavoring and for perfume.

WOODRUFF

WOODRUFF. *Asperula odorata.* A member of the *Rubiaceae* family native to Europe and Asia, especially along the Rhine and in the forests of southern Germany. Sweet woodruff has been used, since about the thirteenth century, to flavor wines and liqueurs. It is still used to flavor May wine.

In the garden, sweet woodruff makes a delicate and beautiful ground cover in shady spots. It is semirecumbent, the stems sometimes about 12 inches long, with fine leaves in whorls around the stems and white flowers. It is a good rock garden plant, and grows well in light sandy soil which is rich and moist. Its dried leaves, when crushed, smell like a combination of new-mown hay and vanilla.

Sweet woodruff is grown as a perennial, but should be wintered in a cold frame in the coldest areas. Seed germinates very slowly, so it is most often propagated by division, or by cuttings taken from the mature plants and rooted in sand.

Foliage is cut and dried in spring when the fragrance is strongest. It is not odorous when cut, but develops its full flavor after being dried for a time.

To make May wine, pour a bottle of light wine over a bunch of dried sweet woodruff leaves and allow it to stand for a few days.

WORMWOOD

WORMWOOD. *Artemisia Absinthium.* One of the *Compositae,* grown in England and in Europe for its bitter flavor, which is used almost solely in absinthe. In this country, it is grown mostly as a gray accent in the garden.

Wormwood is a rank-growing perennial with woolly gray foliage and yellow-green, inconspicuous flowers. It will grow in any soil, in sun, part shade, or almost full shade. Propagation may be by seed, by rooted runners or by cuttings. Plants have a tendency to sprawl in late summer, and should be cut back or tied.

In addition to its use in absinthe, wormwood has a limited use in medicine.

Section
7.

36—Exotic Tropical Fruits: A Complete Listing

Fruits in this section may be grown only in the tropical and subtropical areas of Florida, California, Texas and Arizona or in Hawaii. Some are greenhouse-grown in the North, but mainly as curios rather than for their crops. Very few are grown commercially in this country so they are little known. The home gardener in the tropics has an opportunity to widen his food horizons by planting as many exotics as his garden will hold. Wherever possible, we have indicated cultural directions. These are necessarily sketchy for some fruits which have been little grown here.

AKEE

AKEE. *Blighia sapida*. The akee is a native of the Guinea coast of West Africa, and is greatly esteemed in the West Indies. The fruit takes its Latin name from Captain Bligh, who brought seeds here. It is restricted to the warm sections seldom touched by frost.

The akee fruit grows on a medium-size tree, which attains a height of 20 to 35 feet, and is of spreading habit. Its leaves are large and compound, with small greenish-white blossoms in the axils from early spring to June. The fruit, which ripens from late summer to early winter, is yellow to red in color, 3 to 4 inches long and longitudinally ribbed into 3 sections which split apart. Inside each section is a shiny black seed attached to the edible white aril, which is eaten fried in butter.

There seems to be some basis for the belief that overripe fruit and the seeds of the fruit may be toxic. Unripe fruit is indigestible, and the portions between the arils should not be eaten.

Akee trees are propagated by seed, which should be planted as soon as taken from the fruit, or by shield-budding.

AMBARELLA

AMBARELLA. *Spondias cytherea* or *S. lutea*. Also called Otaheite apple. This tree, whose fruits vary widely from one specimen to another, can be grown only in the warmer sections. It is native to Tahiti, and is grown on other South Pacific islands.

The tree grows to 60 feet, has feathery foliage and pannicles of whitish flowers near the ends of stout, brittle branches. These are succeeded by clusters of yellow to gold plum-like fruits ripening from August to February. The fruits have a single seed from which woody fibers radiate out through the flesh. Some specimens produce fruit that is strongly turpentine flavored. Others are described as aromatic, like quince, with a fresh, slightly acid flavor. The fruit may be eaten fresh or used to make sauces or preserves. Flower buds may also be eaten as sweets.

Propagation may be by seeds, but it is better to take cuttings from a tree known to have good fruit. Cuttings root very easily. In fact, it is said that, in the South Seas, a branch with green fruit, if plunged into the soil to make a new tree, will take root and ripen the fruit.

ATEMOYA

ATEMOYA. This is a hybrid of the cherimoyer and the sugar apple, with some of the best characteristics of each. The plant is larger than sugar apple, but resembles it closely. It forms a larger tree but is similar to sugar apple in foliage and flower. The fruit is like cherimoyer, and is eaten fresh. It has juicy white pulp, of an agreeable consistency and flavor, which separates easily from the half-inch brown seeds. The Page variety is recommended for Florida, where the cherimoyer does not do very well. Page sets fruit without hand-pollination, and is less susceptible to rust, but fruits which mature during the rainy season in Florida may split.

Propagation is by budding or grafting, using seedlings of custard apple, sugar apple, or pond apple as rootstocks.

BIGNAY

Bignay. *Antidesma Bunius.* The bignays, as well as several other species of the *Antidesma,* are not much grown in this country, but they are adapted to culture in southern Florida and the warmer sections of California. Bignay is native to southeastern Asia, Malaya and western Australia. It is an attractive evergreen tree with glossy green leaves, with numerous small flowers which emit a rather unpleasant odor. Male and female flowers are borne on separate trees. Fruits, which turn to red, then almost to black when ripe, are small, one-half inch in diameter, and are borne in clusters of 20 to 40. They are juicy, subacid and flavorful. Seeds, which are not always viable, are similar to cherry pits. With pectin added, the green and ripe fruit, together, is made into a very good jelly.

Propagation may be by seeds, if viable seeds are available. The fruit sets on female trees without pollination, but seeds from such fruit will not germinate. Propagation may also be done with air layers, cuttings, or grafting. Chief pests are scale insects and mealybugs.

Another *Antidesma* with a smaller, bush-like tree and larger fruit is the *A. dallachyanum,* a native of Queensland. Fruits from this species are borne in smaller clusters, but are individually larger. They are deep red at maturity, and are quite acid, but make good jelly.

A Hawaiian species, *A. platyphyllum,* may also be grown in Florida. This plant is slightly more upright, but its fruit is essentially the same as the bignay.

BREADFRUIT

Breadfruit. *Artocarpus incisa.* This is a tropical fruit, and can probably not be grown in any part of the continental United States except the Florida Keys. It has been grown in Hawaii since before the white men arrived there. It was probably brought from the South Pacific islands. Breadfruit is the staple of the Polynesian daily diet.

Breadfruit trees grow to 50 feet in height and have large, leathery leaves and fruits which are 4 to 8 inches long, brownish green to yellow when ripe. The rind is peeled to reveal a fibrous pulp which surrounds a tough central core. The fruits are usually seedless, though one variety is grown for its seeds, which are cooked and eaten.

In the South Seas, breadfruit is eaten in the milk stage, when still green and starchy. In Hawaii, where it is not used as much, it is generally allowed to ripen. Green or ripe, it is always cooked. When ripe, the pulp does not contain starch, which has all changed to sugar. It is a fair source of vitamins A and C and, when eaten in the quantities that provide most of the day's calorie intake, as it is in the Pacific Islands, it also provides a large quantity of B vitamins and calcium.

The amount of fruit produced by one breadfruit tree in the course of a year is prodigious. An acre of trees, containing perhaps 28 to 30 mature specimens, can entirely support 10 to 14 people during the 8 months when the fruit is ripening.

Breadfruit trees are usually grown from cuttings, since the fruits seldom contain seeds. The woody cuttings are placed in the ground where the tree will make its permanent growth. The trees are very difficult to transplant, even when very young. Somewhat hardier, though still subject to injury at the slightest frost, is the Jack Fruit Tree, which *see.* This is the same genus as breadfruit, but another species.

CANISTEL

Canistel. *Pouteria campechiana,* or *Lucuma nervosa.* The canistel is a native of Central America which has strayed to the southern part of Florida. The tree is small, seldom growing higher than 20 feet, but it may be upright or spreading in different specimens. Also varying from tree to tree is the size of the leaves and of the fruit, and the shape and quality of the fruit. Canistel is quite tender, and should be grown in protected locations. It is wind-resistant, however, and may be grown on sandy or on limestone soils.

Shining, bright green leaves make the tree an attractive ornamental. Leaves vary from 4 to 12 inches long. Small flowers in clusters of 2 to 5 are borne on new wood. The fruit, pointed at the bottom, is 2 to 6 inches long and orange-yellow at maturity. The flesh, which is mealy and sometimes dry, has a sweet musky flavor relished by some but not by everyone. Textures of the fruits also vary from tree to tree, some being less dry than others. Also, if the fruit is picked a few days before it is fully ripe, it is less dry when it softens. Fruit ripens from November to February, though some trees produce their fruit at other times.

To propagate a good specimen, start seedlings from the newly ripened fruit, and make a side-graft with a scion from the desired tree. Seeds are slow to germinate, sometimes requiring as much as 5 months, but are more dependable than cuttings, which are slow to root. Rust and scale are the two principal pests of the canistel in Florida.

CAPULIN

CAPULIN. *Muntingia calabura.* The capulin, a native of tropical America and the West Indies, requires a warm, frostless climate. It is a small tree, growing to about 30 feet, which reaches bearing age within two years from seed. Flowers, which are white with numerous stamens, are borne in leaf axils throughout the year, but more abundantly in summer. The fruit, a yellow or red berry, has sweet juicy pulp with many tiny seeds. It is eaten fresh, like cherries, or used for pie or jam.

The capulin is most healthy on sandy soils. On limestone it is subject to leaf spot, and its new wood dies back. Propagation is by seed.

CARAMBOLA

CARAMBOLA. *Averrhoa carambola.* The carambola is an attractive tree with pleasant fruit which can be grown only in the southernmost areas of Florida and in Hawaii. The leaves, which are compound, are sensitive to light and to touch, and fold up when they are shaded or tapped. Pink to purplish flowers are borne at 3 times during the year, early spring, midsummer and fall. The fruit is golden yellow when ripe, translucent, ribbed and star-shaped in cross-section, and may be either one of two varieties, one sweet, the other sour. Average length of the fruit is 4 to 5 inches. It is pleasantly juicy when eaten fresh, and the juice may be extracted to make a drink rich in vitamin C.

Propagation is by seed and by grafting desired scions to seedlings. Young trees are very frost-sensitive, and should be protected even in the warmest parts of Florida. When the trees are mature they are able to withstand temperatures down to 29 degrees for short periods.

A closely related species, *A. bilimbi,* is also sometimes grown in southern Florida, but it is even more sensitive to cold. The bilimbi is sometimes called the cucumber tree, due to the shape of its fruit, which is less deeply ribbed than carambola, is greenish-yellow when ripe and resembles a cucumber pickle. Its flesh is sweet and acid and is something like a plum.

CARISSA

CARISSA. *Carissa grandiflora.* Also called Natal plum. The carissa is a handsome spiny shrub useful in central and southern Florida, California and Hawaii as foundation planting or hedge material. Occasionally it attains a height of 15 feet, and is very dense and practically impenetrable. The foliage is dark green and leathery with two-pronged thorns. Fragrant two-inch white flowers are showy for several months each year. The fruits are bright crimson, streaked with dark red, one to two inches long, with firm reddish pulp. In the center are about 12 small, flattish seeds. When bruised or broken, the fruit and branches exude a sticky white latex which is harmless. Fruit is eaten either raw or cooked, and has a flavor reminiscent of raspberry. It makes a sauce which many people prefer to cranberry. Carissa is an excellent source of vitamin C. Five medium-sized fruits can provide as much vitamin C as one medium-sized orange.

Seeds of the carissa germinate quickly and easily, but seedlings are slow to make a growth. More satisfactory as a means of propagation is layering by pegging a notched branch to the ground. Cuttings are not easily rooted when taken from the bushes.

CASHEW

CASHEW. *Anacardium occidentale.* The cashew is best known in the United States for its nuts. Only in the tropical areas, where the little cashew trees are grown, are cashew apples known as the fruits of the same tree. A native of the West Indies, the cashew tree can be grown only in the warmest areas of continental United States.

The cashew is a small evergreen with leathery leaves 3 to 8 inches long. Its tiny pink blossoms are borne in loose clusters at the ends of the young branches.

The cashew apple, which matures in late spring or early summer, is 2 to 4 inches long, bright red or yellow, fragrant, juicy and somewhat astringent. Protruding from the bottom of each apple is a second fruit, shaped like a roundish kidney bean. This is the true fruit, the cashew nut. It is enclosed in a shell which has an irritating oil similar to the irritant in poison ivy. Varieties of the tree grown in the Asiatic tropics are said to be free of this oil. None of its irritating properties are released until the shell is broken. Consequently, the nuts are roasted, destroying the oil, before they are shelled.

The cashew apple may be eaten fresh, but it is usually cooked. Its juice makes an excellent fresh drink, and it may be fermented to make a very good wine. It is also used for making jams, jellies and pies.

Cashews may be grown on poor soil, but they do better when fertilized and irrigated. The young trees are very cold-sensitive and must be protected unless they are planted in very warm areas. First fruits are borne when the trees are 4 to 5 years old. Propagation is by seeds, which germinate in 3 or 4 weeks. They may also be grafted.

CERIMAN

CERIMAN. *Monstera deliciosa.* The ceriman is a rampant tropical vine closely related to the well-known house plants, the philodendrons. It is an evergreen native to Mexico, and can be grown only in completely frost-free sections of this country. It is most often grown in greenhouses, where occasionally it is persuaded to put forth its calla lily-like flowers and strange fruits.

Leaves of the ceriman are 2 to 3 feet long, almost as broad, and deeply and angularly cut at the edges. The flower, which appears between June and September, is composed of a huge waxy white spathe enclosing a green spadix. Fruits develop in 14 months after bloom. The spadix develops into a succulent fruit about the size and shape of an ear of corn, entirely enclosed in yellowish scales. The flavor of the pulp, when completely ripe, is said to be like both pineapple and banana, and sweet. Crystals of calcium oxalate, discernible to the tongue as needle-like spicules, cause a burning sensation to some people when the fruit is eaten slightly green. It ripens from the base toward the tip, not all at one time. When cut, the stems should be kept in a glass of water until the scales loosen, at which time it is ripe.

Vines are easily propagated from stem cuttings. Make the cutting with two or more segments, and plant it where it is wanted. Or the cuttings may be rooted in pots or tubs, and moved to their permanent position after growth has started.

CHERIMOYER

CHERIMOYER. *Annona cherimola.* The cherimoyer is native to the uplands of Peru and Ecuador, and is said to be one of the world's choice fruits. Unfortunately, there are few sections in the United States where it can be grown, even indifferently well, and none where it really thrives. Trees can be grown in the southern part of Florida and in California from Santa Barbara south, but they do not bear very well. The fruit is quite perishable, so it cannot be shipped to American markets from the South American areas where it grows well. It is eaten only fresh.

The trees in this country never attain a large size, though in South America they are commonly at least 25 feet tall. The deciduous leaves are light green and velvety. Fragrant yellowish flowers an inch long are borne singly or in small clusters. Pollen is usually shed late, after the pistils have ceased to be receptive, so the blossoms must be hand-pollinated to set fruit. Insects do not seem to pollinate it very well.

Fruits are 3 to 8 inches long, looking something like a tightly closed artichoke, but when cut open the pulp is creamy-white and of a custard-like texture, richly aromatic, sweet and slightly acid. It contains many dark brown to black seeds about one-half inch long. Quality and flavor of fruits from different trees varies greatly.

In order to propagate a good variety, seedlings of cherimoyer, sugar apple, or custard apple are grafted with scions of the desired tree. The trees demand well-drained soil, and are very tender both to cold and to too much heat. Also, strong winds and low humidity are uncongenial. They are susceptible to a rust fungus and are sometimes attacked and killed, when young, by an ambrosia beetle.

CHERRY OF THE RIO GRANDE

CHERRY OF THE RIO GRANDE. *Eugenia aggregata.* The genus *Eugenia* of the myrtle family includes a number of unfamiliar South American fruits, among which are cherry of the Rio Grande, pitanga, pitomba and grumichama.

Cherry of the Rio Grande is an attractive evergreen which attains a compact, upright growth to 15 feet. Pairs of white flowers blossom opposite each other along the branches, and quickly develop into reddish-purple cherry-like berries. The fruit matures in 3 weeks after the flowers open, so from April to June the tree, if well fertilized and watered at this time, yields a steady crop of cherries. The fruit is juicy, subacid, well-flavored, and contains one or two seeds, or sometimes none. It is good dessert fruit, usually eaten fresh.

Trees may be propagated from seeds, or they may be grafted.

COCONUT

COCONUT. *Cocos nucifera.* The coconut palm is a tender tropical tree which requires frost-free, humid atmosphere. It can be grown in the Florida Keys, where it has become naturalized. Elsewhere in Florida, it is primarily an ornamental. The story goes that a vessel transporting coconuts was wrecked off the Florida coast about 1840, and that the trees now growing there are offspring of the nuts which floated ashore. In Hawaii a small variety, called locally the Hawaiian coconut, is thought to have been brought to the islands by the first Polynesians who settled there. Larger and more prolific varieties are now cultivated, although Hawaii is on the northern border of the coconut zone, and the trees do not bear as abundantly as they do farther south.

Coconut is most important in the Far East, where it is their most widely grown and planted fruit. In India and other tropical Asiatic countries it is a staple of the daily diet. The fruit is used in several stages of development, starting with the flower spathes, from which a wine called toddy is made. After being fertilized, the liquid within the fibrous husk of the seed (the whole coconut is one seed) gradually forms a coating inside the husk. At first the coating is jelly-like, and is known as spoon coconut. In this stage it is considered a delicacy, and is fed to infants, old people and to those who are ill. Later the jelly becomes the hard and crisp coconut meat, and in this state the nuts are sold fresh on the market. Little change takes place in the sugar, vitamin or mineral content as the fruit develops, but the full oil content is not present until it is mature. At that time the meat consists of about 30 per cent oil and 4 per cent sugar. It has little vitamin C and no vitamin A, but contains a fair quantity of niacin and, in the early stages, a good supply of calcium.

Propagation is by planting the whole coconuts, usually where the trees are to grow. The nuts should be partly covered, lying on their sides, and the soil should be kept moist. Germination takes 4 to 5 months. If they are to be transplanted, the job should be done while the shoot is less than a foot high.

Coconuts respond well to organic fertilizers and mulch. They need large amounts of food, especially after they begin to bear at the age of 5 to 7 years, when some trees bear as many as 200 nuts in a year. Flowers are produced at any time in the year, so coconuts are constantly reaching maturity on a single tree.

CUSTARD APPLE

CUSTARD APPLE. *Annona reticulata.* The custard apple is a close relative of cherimoyer, which it resembles in tree and fruit, but not in flavor. In Cuba and some of the Central American countries, no distinction is made between the two, but in Peru and Brazil the cherimoyer is much preferred.

Custard apple trees are deciduous, and make a rather straggling growth to about 20 feet. They are frost-tender, and may be grown only where the temperatures seldom or never dip to freezing. Leaves are shed in late winter, at about the time when the fruits turn buff to reddish-brown, and mature. Pulp of the fruit is creamy in color and texture, but rather insipidly sweet unless thoroughly chilled.

Trees are easily propagated by seed, and the chief value of the species is the rootstocks which it provides for cherimoyer, sweetsop and soursop.

DOWNY MYRTLE

Downy Myrtle. *Rhodomyrtus tomentosa.* This attractive evergreen shrub gets its Latin name from its rose-pink myrtle-like flowers which appear in small clusters in spring and early summer. Fruits, which are about one-half inch in diameter, greenish-purple, sweet and agreeably flavored, are said to make very good jams and pies. They are cultivated for their fruits in Australia, Asia and the East Indies.

Downy myrtle needs an acid soil, preferably deep and sandy. It will stand a little frost, but does best in frost-free areas. Propagation is by seeds.

GRUMICHAMA

Grumichama. *Eugenia Dombeyi.* The grumichama is a tidy little evergreen tree from Brazil which produces squarish, scarlet to purple-black fruits in a few weeks after blossoming. The fruits are one-half to one inch in diameter, being larger on some trees, and still further increased in size by good organic culture. The fruit is sweet, with an agreeable flavor, thin-skinned, with delicate, soft and melting flesh, and one or two rounded seeds.

Grumichamas will stand some frost, but need fairly warm growing conditions. They do not like limestone soil, doing best on sandy, acid soil which is rich in organic matter. Propagation is by seeds which germinate in about a month.

ILAMA

Ilama. *Annona diversifolia.* A native of Mexico and Central America, the ilama is quite similar to its close relative, the cherimoyer. The trees grow a little bigger than cherimoyer, and the blossoms range in color from greenish tinged with red to solid maroon. The fruit is greenish-yellow, sometimes tinged with red or pink, and sometimes covered with a whitish bloom. The pink or white pulp is good, but large tan seeds occupy a large part of the space which, in cherimoyer, is filled with pulp. In frost-free sections, where ilamas can be grown, the fruit ripens from August to December.

Trees may be raised from seed, though, to be sure of germination, it is best to obtain imported seed. Seed from fruit grown here is slow to germinate, and its viability is uncertain. Imported seed germinates in about 30 days. Trees may also be propagated by grafting desirable scions to roots of sugar apple, pond apple and custard apple. The pond apple stock produces somewhat dwarfed trees.

IMBE

Imbe. *Garcinia livingstonei.* The imbe is primarily an ornamental, shrubby tree, which can be grown in sections where temperatures never go below 20 degrees. It bears edible fruit—an orange-colored plum with firm, acidulous, pleasantly-flavored flesh—but the layer of flesh covering the stone is so thin, that total yield from a well-grown tree is very small.

The imbe is native to Portuguese East Africa. The tree grows to a height of 15 to 20 feet and resembles, in growth habit, a much enlarged gooseberry bush. It usually has more than one trunk, and its branches, which are short and stiff, grow out at right angles to the trunk from the ground to the tip. Stiff, leathery leaves are oblong, about 1 by 4 to 2 by 6 inches, with light colored veins. Greenish-yellow flowers grow out of the leaf axils from February to April, sometimes before the leaves are out. In good years, the trees may bloom and produce a second crop later in the year. Propagation has always been from seed, though if a seedling producing fleshier fruit is discovered, it should be reproduced by grafting.

JABOTICABA

Jaboticaba. *Myrciaria cauliflora* or *Eugenia cauliflora.* Jaboticabas are popular fruits in their native Brazil, where they are as commonly grown as scuppernongs in southern United States. The fruits, in flavor, size and shape, resemble scuppernongs or other muscadine grapes. They are dark maroon to almost black, with tough, thick skin, juicy pulp and 1 to 4 seeds. Their flavor, too, is somewhat vinous, sprightly and subacid, and they are used in Brazil for making wine, fresh juice or jelly, as well as for eating out of hand. The ripe fruit freezes well.

Jaboticaba trees grow to a height of 35 feet in Brazil, but in this country they are seldom over 15 feet.

They have rather fine evergreen foliage on branches that start so low on the trunk as to give the tree a shrubby aspect. The trees grow very slowly and do not start bearing until about their tenth year. Fruit and small white blossoms are all borne inside the foliage, on the older wood. If the trees are given sufficient water they may bear half a dozen crops throughout the year. The blossoms develop into fruit in two months and, after one crop has ripened, blossoms again begin to open.

The trees grow equally well on limestone or sandy soil, but the seeds need special acid peat media in which to start. Unless they are given the proper treatment while young, the seedlings grow slowly and their leaves become yellow. Peat enriched with organic nutrients rich in iron are needed for the plant's early growth. Best source of an organic form of iron is deep-rooted weeds, which may be dug and composted before use on jaboticaba seedlings. If the trees are then mulched with these weeds, either composted or chopped, they may attain maturity in shorter time.

Jaboticaba will grow where temperatures infrequently dip below freezing for short periods. They dislike strongly alkaline soils.

JAK FRUIT or JACK FRUIT

JAK FRUIT OR JACK FRUIT. *Artocarpus integrifolia,* or *A. integra.* Jak fruit is the first cousin to breadfruit, which feeds most of the Polynesians most of the year. It is native to southern India and Malaya, and is subject to frost injury, but can be grown in the warmest sections of this country, where breadfruit cannot.

In its native habitat, the jak fruit tree achieves a height of 60 feet, but it does not usually grow so tall here. However, in southern Florida, there are a few large specimens. It has large, leathery leaves, milky sap and bears male and female flowers in separate clusters on the same tree. The fruits, which weigh from 10 to 40 pounds each, can be borne only on the larger limbs or directly from the trunk.

The fruit is as long as two feet, when grown in the tropics, but in Florida it is seldom more than one foot. It is oblong or melon-shaped with small sharp projections from its skin, like the points on a grater. Its yellowish, soft, juicy pulp is quite strong, with a distinctive flavor, which is either greatly relished or just as greatly disliked. The pulp may be eaten fresh, dried or preserved. Seeds are roasted and eaten in some of the Oriental countries.

Jak trees may be grown as ornamentals, even if the fruit is not eaten. They make attractive shade trees in the warm sections of the country. Propagation is by seedlings or cuttings. Seed should be planted in boxes or pots or directly where the tree is to stand. They cannot be dug and transplanted with a great percentage of success.

JAMBOLAN

JAMBOLAN. *Syzygium cumini,* or *Eugenia cumini.* Also called the jambolan plum or the Java plum, the jambolan is a compact evergreen tree which may be grown in the warmest portions of the United States. Two varieties are widely grown in Hawaii. The trees grow to a height of 40 to 50 feet. They are able to withstand the wind well, and may be used in a windbreak. But their large crops of olive-like purplish fruits fall to the ground and stain everything they touch, so they should not be used in a much-frequented portion of the garden, even though the trees themselves are attractive and provide good shade.

Jambolan flowers are produced in February and March in large clusters on old wood. The fruit ripens about 3 months later. Skin of the plums, which are about the size and shape of olives, is purple, and flesh may be either purple or whitish. Fruits with purple flesh are strongly flavored and make very good jelly if combined with another fruit, such as guava, which is rich in pectin. White-fleshed jambolans contain pectin of their own, but their plum-like flavor is less strong. Both varieties of jambolans are too astringent to be eaten out of hand, though the fruit of some trees is sweeter than that on others.

Both varieties provide a good supply of vitamin C and of riboflavin. They have very little vitamin A.

Jambolan trees are easily propagated from seed, but the resulting fruit may not be true to type. If a good variety is available, it may be reproduced by grafting.

KARANDA

KARANDA. *Carissa carandas.* The karanda is closely related to and very similar to the carissa, or Natal plum, but it is a native of India. It forms a large, rigid, thorny shrub which is suitable for a protective hedge. When well fertilized the bushes grow thick and rank, and become impenetrable. Fragrant pink-blushed, white flowers are borne in terminal clusters, and are followed by black fruits which resemble cherries in size and shape. The pulp of the fruit is pale red and quite acid. When cut or broken, the fruit exudes droplets of a gummy juice. It is too acid for eating fresh, but may be used for making cold drinks or jelly.

Karandas may be propagated by seeds, air-layering or cuttings. The two latter methods are preferable, because seedlings make a slow growth. In warm sections, the bushes thrive on most types of soil, and need little attention after they are established.

KEI APPLE

KEI APPLE. *Dovyalis caffra.* This is a South African fruit, also called the umkokolo there, which can be grown in the citrus sections of this country. Plants can stand temperatures down to about 20 degrees without much injury and, if frozen back, they recover quickly and may produce a crop of fruit the following season.

Kei apple is a vigorous-growing plant which may be pruned to hedge-shape if desired. As a tree, it attains a height of 15 to 20 feet. Its leaves are shiny, growing 1 to 4 in a cluster on spine-armed stems. Male and female flowers are usually borne on separate plants from February to June. The fruits are ripened in summer or autumn. They are little apples with yellowish-green skins, juicy pulp reminiscent of apricot and highly acid, with 5 or more small pointed seeds. They are borne on short stems in leaf axils of the younger wood.

The fruit is used mainly for sauces, preserves or jelly.

Kei apples may be spaced 10 to 12 feet apart, unless grown as a hedge, when they should be set closer. Propagation is by seeds, layering or grafting.

KETEMBILLA or KITEMBILLA

KETEMBILLA OR KITEMBILLA. *Dovyalis hebecarpa.* Ketembilla or Ceylon gooseberry, is very closely related to Kei apple, but is not quite so hardy. It is native to Ceylon, but grows well in Florida and in parts of California, and is found in many Hawaiian dooryards.

Like Kei apple or warb, ketembilla may be used to form a spiny hedge, though the branches are often weighted down with fruit. Leaves and fruit are both velvety, the foliage light green and the fruit, which sets and matures year-round, maroon-purple. The fruit is about one inch in diameter, with soft, purplish, acid, juicy flesh which is rich in vitamin C and contains some vitamin A. A 100-gram portion of the fruit, fresh, contains about 66 milligrams of vitamin C and 230 units of vitamin A, as well as some of the B vitamins. The fruit is usually used for jams, jellies, drinks or juice, but it is quite strongly flavored and is best when diluted with other juices or fruits.

Propagation may be from seeds, which germinate easily in about two weeks, or the plants may be multiplied from cuttings, budding, or grafting. When seeds are planted, a number of bushes should be started fairly close together in order to be sure of having both male and female flowers. Buds or scions for grafting should be taken from bushes known to produce fruit from perfect flowers. Older plants need some thinning at the top to prevent overcrowding of prolific branches.

KWAI MUK

KWAI MUK. *Artocarpus hypargyrea.* This small-fruited species of the breadfruit and jak fruit genus is native to southern China, and seems to be just slightly more hardy than its close relatives. Young trees grown in southern Florida have been known to survive a degree or two of frost.

Fruit of the kwai muk is about an inch in diameter and more or less round. It is quite soft when ripe, with a very tender and slightly fuzzy skin. Several small seeds are embedded in the deep orange, soft, pleasantly flavored flesh. When it is unripe, the fruit exudes a white sticky milk if it is punctured.

The kwai muk tree is a decorative little evergreen growing to about 20 feet in height. Its leaves are dark green and smooth, 2 to 5 inches long, and it bears spikes of male and female flowers on the same tree. Fruit ripens in the autumn.

Propagation of kwai muk is by seed,

which should be started in a pot or where the tree is to stand. Like others of its family, the kwai muk dislikes having its roots disturbed. It seems to do well on limestone soil.

LINGARO

LINGARO. *Elaeagnus philippensis.* The lingaro, which is native to the Philippines, is an attractive evergreen climber of the oleaster family. It is usually grown without support, when it attains a height of about 10 feet and a diameter of 20 feet or more. From January to March, it bears clusters of little whitish flowers which are yellow inside, on the ends of the new wood. The fruits which follow are technically nuts, but they are enclosed in a pink or red fleshy casing which is sweet and cherry-flavored. The size and shape of the fruit is about that of an olive, but the pit is very slender. It may be eaten as a dessert fruit, and it makes a highly colored jelly.

Lingaro is easily grown from seeds, which germinate in 2 or 3 weeks, or it may be grown from cuttings. The plants are not particular as to soil, doing well on sandy or calcareous loam.

LONGAN

LONGAN. *Euphoria longana* or *Nephelium longana.* The longan is closely related and similar to the lychee nut, and is native to southern China and nearby Asiatic countries. But the handsome evergreen longan tree is more hardy than lychee, and less exacting in its requirements.

A longan tree may grow to 35 feet in height, with a heavy mass of dark green leaves. Its inconspicuous flowers are in upright clusters at the tips of the branches and in axils of the compound leaves. These are followed by large clusters of small fruits one-half to one inch in diameter, with yellow-brown, brittle skin and a rather large brown seed, enclosed in an edible receptacle. The size and flavor of the fruits vary in different trees, and with different cultural practice. Like lychees, longans are eaten fresh, dried or preserved.

Trees may be propagated from seed or from air-layers. They do best on a good loam which contains plenty of humus. If they are planted on light soil, they need a heavy mulch and large amounts of water while the fruit is ripening. Fruit is improved by feeding the tree with frequent applications of manure or compost, and its size may be increased if half of the fruit clusters are thinned out immediately after flowering.

LOQUAT

LOQUAT. *Eriobotrya japonica.* The loquat is also known as Japanese medlar, or Japanese plum, although its native habitat is China, rather than Japan. It is grown as an ornamental in many of the southern and Gulf states, as well as on the West Coast, but the tree does not ripen its fruit unless it has a frost-free winter. The loquat tree itself is entirely hardy, and is not injured by several degrees of frost.

Loquat trees are attractive evergreens growing to about 25 feet tall, bearing fragrant white flowers in terminal clusters from October through February. In very warm climates, the blossoming period persists through the summer. The yellowish or reddish apple- or pear-shaped fruits are sweet, and have a distinctive flavor. Half a dozen named varieties are available for planting. Tanaka is recommended through the southeastern and Delta states, Oliver for southern Florida and Champagne for California. A seedling is also available from the Florida Subtropical Experiment Station, SES-4, which has large, pale-yellow, tart fruits that are good for cooking. Fruits of the other varieties may be used for cooking, jelly-making, preserving or eating fresh.

Loquat trees may be propagated from seeds, but are usually increased by grafting, budding or air-layering. Seedlings are usually budded or grafted. Budding is usually done in late fall, using a dormant bud, and growth forced by cutting back the stock in spring. Side-veneer, cleft or whip grafts are also used. The trees thrive on any good garden soil, so long as drainage is good.

The size of fruit can be increased if the trees are given plenty of fertilizer before fruiting, and are watered well while the fruit is developing. Occasionally the trees may be attacked by fire blight which kills the blossoms and some of the branches in spring. The best control is to prune out all affected growth, cutting well back into healthy wood. Remove and burn all diseased material.

LYCHEE

Lychee. *Litchi chinensis.* The lychee is a fruit tree native to China. It is known in this country primarily as litchi nut, the dried fruit, which bears about the same relation to fresh lychee fruit as does the raisin to the grape.

The lychee tree is an evergreen growing to 40 feet in height. It bears male, female and bisexual flowers, all on the

same tree. Observations made in South Africa seem to indicate that sometimes one, sometimes the other sex predominates on the tree, a factor which makes for heavier bearing in some years than in others, since only the female and the bisexual flowers bear fruit. Fruiting panicles bear from several to several dozen fruits.

The mature lychee fruit is about 1½ inches in diameter, with a bright-red, thin, leathery shell and one seed whose size varies in different varieties. Between the rough shell and the seed is a translucent pearl-white pulp something like grape pulp in texture.

Fresh lychee is a good source of vitamin C and a fair source of niacin. The amount of edible pulp in a fruit varies according to the size of the seed as well as that of the fruits. Seeded and shelled fruits vary from 20 to 35 to the cupful. They are eaten out of the shell, or are added to fruit cocktail or fruit salad.

Lychee is suited to subtropical climates. It does not thrive in areas of heavy frosts, though it will withstand temperatures down to 28 degrees when not in active growth. Tender young growth, and also young trees, may be killed at a temperature just below 32 degrees. But the trees need temperatures between 32 and 40 degrees for their winter rests. This limits culture in the United States to the southernmost areas in Florida, and to Hawaii, where it is grown and sold in the markets. In Florida the mature lychee is considered not quite as hardy as sweet orange, but hardier than mango or avocado. The variety most popular in Florida is Groff; in Hawaii, the Brewster, Hak Ip and Kwai Mi are all considered good.

Lychee needs heavy, slightly acid soil that is well-drained, and about 50 inches of annual rainfall, or its equivalent in irrigation.

The trees respond well to organic fertilizers and mulches. If a young tree is given 50 pounds of barnyard manure a year, and a mature tree as much as 500 pounds, best results may be expected. This amount should be spread over the entire root area. Also, a heavy straw mulch should be maintained all year, to provide decaying organic material to the soil, and to preserve soil moisture.

Trees are propagated principally by means of air-layering. Propagation by cuttings is possible, but difficult. Seeds may be germinated easily, but the quality of the fruit obtained from seedlings is uncertain. Also, seedling trees do not produce fruit for 10 to 15 years. Trees in Florida are spaced at least 25 feet apart. In the Orient they need up to 45 feet of space.

MABOLO

Mabolo. *Diospyros discolor.* The mabolo, sometimes called mangosteen, is one of the tropical relatives of the persimmon, and a member of the ebony family. It can be grown only in frost-free portions of the Florida peninsula, thriving best where the soil is sandy.

The trees are medium size, with leathery leaves. Both male and female flowers are borne separately on some trees, but some trees bear only male flowers and no fruit. The mabolo fruit is 2½ to 3½ inches in diameter, and somewhat resembles a quince. Its flesh is yellowish and aromatic with a distinctive flavor, which is more pleasant. Some trees produce dryish fruit with rather large seeds; on other trees the fruit is seedless and more moist and sweet.

Propagation may be by seeds, but a more desirable variety may be assured from grafting. Most successful are side or veneer grafts, of mature wood from near the tips of girdled branches, on seedling rootstocks.

MAMEY

MAMEY. *Mammea americana.* The mamey or mammee apple is sometimes known as the mamey of Santo Domingo, to distinguish it from the mamey sapota. Actually, the two fruits belong to separate families, the *Guttiferae* and the *Sapotaceae,* families with fairly similar characteristics and separated only by some botanical technicalities.

The mamey is a native of the West Indies and tropical South America. It can be grown only in the very warmest parts of Florida, where it forms a compact, cylindrical evergreen that may attain a height of 50 to 60 feet. Its leaves are glossy, deep green and 4 to 8 inches long. On some trees, the large waxy white flowers, which are very fragrant, are almost entirely male, and little or no fruit is produced. On other trees, the flowers may be predominantly female, or polygamous, and these trees bear good crops.

Fruits of the mamey vary considerably from tree to tree. On some trees, the flavor is sweet and rather like apricot. The fruit is large, 4 to 8 inches in diameter, with a thick, rough and sometimes russet skin, yellow to reddish flesh and 1 to 4 large seeds. Fruit of the best trees may be eaten fresh, or it may be cooked, preserved or used for jam. Trees are usually propagated by seed, which germinates in two months.

MAMONCILLO

MAMONCILLO. *Melicocca bijuga.* Fruit of the mamoncillo tree, which is grown in a limited way through the southern part of Florida, is produced only on trees in the warmest sections. A satisfactory crop can be obtained in the Florida Keys, but little is borne on trees in districts north of that.

The trees are large, but they grow slowly. Fragrant green-white flowers are borne at the tips of the branches, the flowers being unisexual. Female clusters develop during the summer and early fall into bunches of round green fruits, each about one inch in diameter, with a large round seed, and juicy, acid, whitish pulp with an agreeable flavor.

Trees are propagated by seed, or sometimes by air-layering, but they do not respond well to asexual reproduction methods.

MOMBIN

MOMBIN. *Spondias Mombin* and *S. purpurea,* respectively the yellow and the red mombin are also known as hog plums. They are tropical American fruits, but can be grown in the warmest parts of Florida. Two other species of *Spondias* are also grown for their fruit in South America, but they do not grow well in this country. They are the imbu, *S. tuberosa,* and the amra, *S. mangifera.*

Mombin trees are deciduous, 20 to 40 feet tall in the red and yellow species. They are closely related and similar to the ambarella, or Otaheite apple. Fruits vary greatly in size and quality from tree to tree, but generally the red mombins are of superior quality.

Fruits of the mombins are plum-like in form, though sometimes they are somewhat knobby and irregular in shape. The ripe fruit has a red or yellow, rather tough rind, scanty but juicy and sometimes flavorful flesh, surrounding a large seed. The seeds as well as the flesh are eaten in the tropics. Trees of either species may be propagated by seed or by cuttings of the mature wood.

MOUNTAIN APPLE

MOUNTAIN APPLE. *Eugenia malaccensis.* The mountain apple, called *ohia ai* by the Hawaiians, is a native of the Malay Peninsula and is grown in warm Oriental countries. It was planted in the Hawaiian Islands by the early Polynesian settlers, and is now naturalized throughout the islands. Fruit is brought down from the mountains and sold along Hawaiian roadsides as it ripens, from June to December.

The apples grow on small evergreen trees. They are delicate in skin, which is crimson, with pleasantly juicy, crisp, white flesh of very mild flavor, described by various people as rose-like or quince-like. One or two large brown seeds, which are freestone, are embedded in the flesh. The fruit bruises easily, and stains deep purple everything it touches.

We have no specific directions for growing mountain apples. However, others of the same genus may be propagated by seeds, and prefer acid to alkaline soil, especially when it is well-enriched with organic matter.

OTAHEITE GOOSEBERRY

OTAHEITE GOOSEBERRY. *Phyllanthus acidus.* This very tidy ornamental tree is native to Madagascar and India, but is sometimes found growing wild in the southern part of Florida. It grows to a height of 15 to 20 feet, and has leaves that appear to be compound, similar to ailanthus, though they are actually single and arranged opposite on green twigs which drop off when the leaves are shed.

Otaheite gooseberries are similar to northern gooseberries in acidity, flavor and the way they grow along the branches. Each berry is divided into 3 deeply cleft lobes, and contains a single stone in which are 6 seeds. The fruit is green or greenish-yellow when ripe. The season is April to June.

Trees may be propagated by seeds, greenwood cuttings, or they may be budded on seedling rootstocks.

PITANGA

PITANGA. *Eugenia uniflora.* The pitanga is also known as the Surinam cherry, but in its native habitat, Brazil, it is called pitanga. It is widely grown in Hawaii, and can stand the climate through most of Florida. Where it is irrigated, it may be grown in the Southwest, but it needs plenty of water while fruit is forming. In fact, in Brazil, it grows wild, mainly along the banks of streams.

Pitanga is a broad, compact evergreen shrub growing to about 10 feet. It makes an admirable hedge. The leaves are rich wine color when young, turning dark green and glossy when mature. Its flowers are fragrant, cream-white and about one-half inch in diameter. They are followed in a few weeks by berries which may be crimson to black on different plants, an inch in diameter and more or less prominently 8-ribbed. The thin skin and soft juicy pulp, both of the same color, surround a single stone, or sometimes 2 to 3 seeds pressed close together. Flavor of the fruit varies, sometimes bordering on an unpleasant turpentine flavor, sometimes very good, but always very acid. Even when they have a large amount of sugar—the carbohydrate content may be as high as 22 per cent—pitangas taste sour because they contain so much acid. They are a good source of vitamin A and a fair source of vitamin C. They are usually served cooked and sweetened, but occasionally, a few are cut into a mixed fruit compote or salad for color. They should be picked just before using, because they turn bitter on standing. In Florida, the main crop is borne through the spring, with sometimes a secondary crop harvested in fall. In Hawaii, when the rainfall is sufficient, pitangas bear fruit year-round.

Propagation is by seeds, though the color of the fruit may not come true to the parent. It is possible to graft them, but the rootstocks sucker badly and in a few years the grafted variety is lost in the seed variety. Fruit size and quality are both improved by feeding with organic fertilizers and providing plenty of water at the fruiting period.

PITAYA

PITAYA. *Hylocereus undatus.* Also called strawberry pear and, more commonly, night-blooming cereus. The pitaya is a climbing Mexican cactus used in some of the warmest sections of the United States, primarily as an ornamental to cover low walls. It clings to any support offered it by aerial roots which arise from the underside of its jointed, 3-winged stems. It has large, showy flowers which open at night, followed by oval red fruits 3 inches in diameter, which are edible but not particularly good. Propagation is from cuttings, which are easily rooted especially if they already bear aerial roots.

PITOMBA

PITOMBA. *Eugenia luschnathiana.* The pitomba, native to Brazil, is an attractive 30-foot evergreen tree with dense, leathery foliage and one-inch orange-yellow fruits. It may be grown in frost-free districts, and prefers acid to limestone soil.

The fruit, which matures from May to July, has an agreeably aromatic flavor, somewhat acid. The pulp is soft, juicy and orange-colored, and encloses a seed cavity in which are one to several seeds. Fruits are especially good for making jams and jellies and are also used in sherbets.

Propagation is by seeds, which flourish in soil rich in organic matter.

RAMONTCHI

RAMONTCHI. *Flacourtia indica.* A native of Madagascar and the warm parts of Asia, the ramontchi will grow in the United States, in sections where there is little frost, and does well even in arid sections, if it is irrigated. In the Orient it is a shrubby tree up to 25 feet tall, but here it is usually not more than 12 feet, and may be used as a hedge. It has a dense rounded growth habit; in fact, it may often need thinning. Branches of some specimens are armed with inch-long spines; others are spineless.

Fruit of the ramontchi is a berry an inch or less in diameter. It varies considerably on different plants, being red to almost black in color, sweet and pleasantly flavored to acid. Unless thoroughly ripe, all varieties are astringent. Flesh of the berries is fairly soft and contains a few soft, thin seeds. Flowers are either male or female on different plants, so it is necessary to plant at least one staminate plant for fruit setting.

Propagation may be by seeds, cuttings of mature wood or grafting.

ROSE APPLE

ROSE APPLE. *Syzygium jambos.* The rose apple tree, which is related neither to roses nor apples, is a native of the East Indies and is grown in southern Florida, more as an ornamental than for its somewhat flavorless fruit.

The tree is a handsome evergreen which grows to 30 feet, with an equal spread. Its leaves are thick and shining, and new growth is wine-colored. Flowers $2\frac{1}{2}$ inches in diameter and resembling white pompons are borne from February to April. The apples which follow are a little smaller than the flowers, pinkish-white to pale yellow, with crisp white sweetish flesh with a faint rose flavor. The fruit matures sometime between April and June. It may be eaten fresh, but is better preserved or in a jelly.

The trees can only be grown in frost-free climate, but are quite drought-resistant, and thrive with a minimum of care. They are propagated by the large brown seeds, 1 to 3 being found in the seed cavity of each apple.

ROSELLE

ROSELLE. *Hibiscus sabdariffa.* Roselle is an annual, and one of those universal plants, all of whose parts are useful for something. In various parts of the world, where the hot growing season lasts for 7 or more months, it is grown for its flower calyx and bracts, which are used like fruit in jelly making; its leaves, which are eaten like spinach; its seeds, which are eaten roasted; and the fibres of its stems, which are used for cordage.

Roselle grows 4 to 7 feet tall in this country, somewhat taller in Hawaii. Its most common use is as a fruit. Thick, fleshy sepals form a united calyx, red in color and very acid, which may be used, with or without the seeds enclosed in it, for jelly making. To make a sauce comparable to cranberry sauce, the seeds are usually removed and the calyx is thoroughly cooked.

Roselle is grown very much like eggplant. Seed is started indoors in late winter, and plants are set out in the garden in May. Fruit, which ripens in November and December, should be harvested before it becomes hard and woody, when it loses its flavor.

SAPODILLA or SAPOTA

SAPODILLA OR SAPOTA. A number of members of the family *Sapotaceae* are known under the common name of sapota in South and Central America. *Sapota achras,* the naseberry, is known as sapodilla, and in Florida is the most widely cultivated of all of them. But also popular in the tropical climates, are the mamey sapote, *Calocarpum sapota,* also a member of the family *Sapotaceae;* black sapote, *Diospyros ebenaster,* one of the *Ebenaceae* clan; and white sapote, *Casimiroa edulis,* one of the *Rutaceae.* Several varieties of the sapodilla and white sapota have been developed for planting in Florida and California. The fruits are not much known to the rest of the country, because they all have delicate, tender skins and cannot easily be transported.

Sapotas and sapodillas all are tropical or subtropical. They need a warm, almost frost-free climate, especially when the trees are young. White sapotas like acid soil; sapodillas thrive on sand or limestone, and are impervious to sea spray. Fruit is round, yellow-green to brownish and scurfy, with flesh which may be cream colored to brownish in the different species and, within each species, on different trees. Some fruits are very high in quality and flavor, and are being developed by the Florida Agricultural Experimental Stations. These include Prolific and Russell sapodillas, both superior varieties. Other named varieties in Florida are Dade, Lenz and Golden white sapotas, and in California, the Pike, Whatley and Nancy Maltby varieties of the same fruit. The black sapota, from Mexico, has not been so widely cultivated in this country.

All of the sapotas and sapodillas may be grown from seeds, but in order to propagate particularly desirable varieties, they must be grafted. White sapotas and sapodillas are readily side-grafted. Seed of the white sapota and mamey sapota should be planted immediately after removal from the fruit.

SOURSOP

Soursop. *Annona muricata.* The small tropical American evergreen, the soursop, can be grown only in the most frost-free portions of continental United States. It is grown in Cuba and the West Indies, and is common in Hawaii.

Fruit of the soursop, not borne prolifically even in the warmest parts of Florida, are large, often weighing 4 to 5 pounds. They have been described as having flesh resembling cotton soaked in an aromatic liquid. In spite of this description, they are relished as a breakfast fruit and for salads, or as a basis for iced drinks, sherbets or gelatine dishes. They contain a good supply of niacin and riboflavin, and some thiamin and vitamin C.

The mountain soursop, *A. montana,* is similar, but has somewhat smaller and inferior fruits. Propagation of either species is by seeds or grafting. Trees grafted on the rootstocks of pond apples are successful, and produce slightly dwarfed specimens.

STAR APPLE

Star Apple. *Chrysophyllum cainito.* Star apple is a relative of the sapotas, and is also native to tropical America. The trees are sometimes grown in southern Florida as ornamentals, but they can be grown only in the warmest sections. As little as one degree of frost can kill them when they are young, and 3 degrees can cause considerable damage even when they are older.

Star apples grow 2 to 4 inches in diameter, sometimes green and sometimes purple. The flesh has a good texture and flavor, mildly sweet, when thoroughly ripe. The name is descriptive of the star shape of carpels and seeds, when the fruit is cut in half across the middle. The apples ripen in April and May.

Propagation is easy from seeds, and may also be done by air-layering or cleft grafting, though the latter methods are not always successful.

SUGAR APPLE

Sugar Apple. *Annona squamosa.* The sugar apple is another close relative of one of the best fruits of the tropics, the cherimoyer. It can be grown only in comparatively frost-free sections of the South. The tree is deciduous, reaching a height of about 10 feet.

Sugar apple resembles the other annona fruits, but its segments are more strongly marked than they are in some species, each one being almost separate from its base on the receptacle, like the sections of a pine cone. At maturity, it is yellow-green, sometimes with a white or bluish bloom. The pulp is white to cream colored, custard-like, sweet and pleasantly flavored. Seeds embedded in the flesh are about one-half inch long.

Trees may be propagated by seeds, or they may be budded or grafted on self-roots or on rootstocks of custard apples.

WARB

Warb. *Dovyalis abyssinica.* The warb, a native of Ethiopia, is practically unknown in this country, though it may be grown in the frost-free sections of the South. It is closely related to kei apple and ketembilla, but is said to be no harder to grow and to produce fruit superior to either of them.

The warb grows on a 10-foot shrub, which may be armed with slender spines, or may be thornless. Male and female flowers are borne on separate bushes. One staminate plant is sufficient to fertilize flowers on several pistillate plants. A hybrid, distributed by the USDA, has both male and female flowers on one plant.

The fruit is similar in flavor and color to an apricot, but is about one inch in diameter, and contains a few small, flat seeds. We have no information on the vitamin content, but considering its color, it is possible that it, like apricot, is a superior source of vitamin A.

Plants may be propagated either by seeds or by grafting. The hybrid, which is crossed with *D. hebecarpa,* is propagated by layering or by grafting to roots of *hebecarpa.*

INDEX